1st ITB Centennial and 4th PlanoCosmo International Conference 2018

Transforming Beyond Borders, Starting the New Urban Agenda

IOP Conference Series: Earth and Environmental Science
Volume 158

Bandung, Indonesia
3 - 5 April 2018

Editors:

Fikri Zul Fahmi
Euginie Birch
Tim Bunnell
Delik Hudalah
Haryo Winarso

ISBN: 978-1-5108-6398-9
ISSN: 1755-1307

Printed from e-media with permission by:

Curran Associates, Inc.
57 Morehouse Lane
Red Hook, NY 12571

CURRAN ASSOCIATES INC. proceedings.com

Some format issues inherent in the e-media version may also appear in this print version.

Printed by Curran Associates, Inc. (2018)

For permission requests, please contact the Institute of Physics
at the address below.

Institute of Physics
Dirac House, Temple Back
Bristol BS1 6BE UK

Phone: 44 1 17 929 7481
Fax: 44 1 17 920 0979

techtracking@iop.org

Additional copies of this publication are available from:

Curran Associates, Inc.
57 Morehouse Lane
Red Hook, NY 12571 USA
Phone: 845-758-0400
Fax: 845-758-2633
Email: curran@proceedings.com
Web: www.proceedings.com

Table of contents

Volume 158

The 1st ITB Centennial and 4th PlanoCosmo International Conference

3–5 April 2018, Bandung, Indonesia

Accepted papers received: 13 April 2018
Published online: 1 June 2018

Preface

Papers

Informal Urbanism and Inclusive Development

Infrastructure and Transportation System

Peri-Urbanization and Regional Policy

Regional Innovation and Smart City System

Rural Transformation and Small Town Planning

Preface

Economic, social and cultural activities, as well as environmental and human impacts, are increasingly concentrated in and around cities. In the era of planetary urbanization and technological revolution, cities have transformed beyond their proper borders into hinterlands, city-regions, urban corridors, and mega-regions. As a consequence, development challenges have become far more complex than were anticipated in the 'traditional' conception of the 'city'. Transforming beyond their natural and human resources and institutional capacity, cities currently face significant sustainability challenges in terms of governance; land and housing; basic services; food, water and energy security; and disaster risks.

In 2016 the United Nations Habitat Agenda set out the New Urban Agenda (NUA), which seeks to create a mutually reinforcing relationship between this trend of expanding urbanization and development. In this context, urbanization will be seen to become a parallel vehicle for achieving sustainable development. The NUA attempts to address the development trends while also recognizing that cities and metropolitan areas are the major drivers of regional, national, and global economies. The idea of the NUA is to offer guidelines on a range of 'enablers' that can further strengthen the relationship between urbanization and sustainable development. In this context, infrastructure in its widest sense will be a key vehicle for achieving sustainable urban development.

Currently, the NUA has not been properly recognized and integrated in urban, regional, and infrastructure development strategies. In fact, there has been a limited recognition of the NUA in the current academic research and policy-making processes. Considering the urgency of the NUA in the planning, development, and governance of contemporary cities, we should start incorporating the agenda into academic research and policy making, as well as communicating its problems and potential solutions in a scholarly and practically relevant discourse.

Therefore, we held the First ITB Centennial and Fourth Planocosmo International Conference in the main campus of Institut Teknologi Bandung, Bandung, Indonesia, on 2-4 April 2018. The conference theme was "Transforming beyond Borders, Starting the New Urban Agenda". The aim of the conference was to discuss infrastructure, transportation, and planning issues in the context of New Urban Agenda (NUA), which are centered on: *connectivity, inclusiveness, resilience, security,* and *governance*. The conference was a special opening event to commemorate the Centenary of Indonesian Higher Education in Engineering and the 60th Anniversary of Urban and Regional Planning Education in Indonesia. The event was organized by the Urban and Regional Planning Study Program under the School of Architecture, Planning and Policy Development, Institut Teknologi Bandung and the Research Center for Infrastructure and Regional Development, Institut Teknologi Bandung.

While several papers were invited for publication in Journal of Regional and City Planning, these proceedings were prepared as the main publication outlet for most papers presented at the conference. The proceedings consist of 56 selected papers, which are distributed to several interrelated parts resembling the tracks organized during the conference.

The first part, Informal Urbanism and Inclusive Development, consists of five articles focusing on the issues of informality-formality in designing and planning urban spaces in various contexts across the globe. The second part, Urban Land and Housing System, presents six papers in the topics of housing for everyone, housing development and finance, gentrification, and urban heritage. The part of Infrastructure and Transportation System includes 14 papers dealing with infrastructure provision and management, inclusive and sustainable transportation system, technology for Intelligent Transportation System (ITS), Transit Oriented Development (TOD) and compact cities, and megaproject planning and management.

Meanwhile, nine papers in Environmental Planning and Management concentrate the discussion on the topics related to energy security and planning, coastal zone management and marine planning, climate change adaptation, socio-ecological resilience, and green cities and ecotourism. As another part of the proceedings, Disaster Risk Reduction brings together five accepted papers that pay attention to pre- and/or post-disaster reconstruction and rehabilitation planning and management and also community preparedness towards disaster.

Peri-urbanization and Regional Policy is the next part of these proceedings containing five papers, which concerns with infrastructure of peri-urban and suburbanization. Regional Innovation and Smart City System becomes the part of the proceedings comprising nine papers focused on smart city and big data, infrastructure 2.0, planning in a digital era, creativity and innovation, and sharing economy and sharing cities. The last special part is Rural Transformation and Small-Town Planning, which is dedicated to those interested in reading rural-urban linkages with three papers.

These proceedings were not possible to be completed without the restless effort of our technical team. Therefore, our sincere thanks go to our colleagues Fikri Zul Fahmi, Lisna Rahayu, Dika Fajri Fiisabiilillah, Aryani N. Chandramidi, Nabilla Dina Adharina, Yustina Octifanny, Gina Puspitasari Rochman, Sosrowinarsito and Ivanie Destila Sari. Other colleagues in the conference's organizing committee should also be mentioned here, including Adiwan Fahlan Aritenang, Tessa Talitha, Seruni Fauzia Lestari, Farida Khuril Maula, Niken Prilandita, Shanty Yulianty Rahmat, Bea Regina Marcendy, Puspita Dirgahayani, Hansen Sutanto, Saut A.H. Sagala, Nurrohman Wijaya, Sri Utami Purwaningati, Azis Hakim, Bagas Dwipantara Putra, Tri Rahayu Wulansari, Alhilal Furqan, Aliyah Alfianda Dwicahyani, Yunie Nurhayati Rahmat, Rama Arianto, Achmad Fauzan Iscahyono, Tizar Bijaksana, Tika Savitri Hasan, Alvaryan Maulana, Qurrota Aini, Alvin Noviansyah, Dewi Widaningsih, Susi Agustini, Jaja Suhendar. We are also grateful to our colleagues in the Scientific Committee, whose inputs and ideas are remarkable: Tommy Firman, Euginie Birch, Richard Haigh, Eric Sheppard, Deden Rukmana, Haryo Winarso, Pradono, Teti Armiati Argo, Tim Bunnell, Paul Jones, Shabbir Cheema, Terry van Dijk, Sri Maryati, Hastu Prabatmodjo, Harkunti P. Rahayu, Denny Zulkaidi, and Wilmar Salim. We are indebted to our Steering Committee for their kind supports and encouragement: Miming Miharja, Benedictus Kombaitan, Robert Cervero, Richard Legates, Johan Woltjer, Roos Akbar, Arief Rosyidie, Tubagus F. Sofhani, Ibnu Syabri, and Ridwan Sutriadi.

IOP Conf. Series: Earth and Environmental Science **158** (2018) 011001 doi:10.1088/1755-1315/158/1/011001

Finally, we are thankful and lucky to have dedicated partners from College of Architecture and Urban Planning, Tongji University, China, including Peng Zhenwei, Zhao Min, Li Zhang, and Chen Chen.

Dr. Delik Hudalah

Chair of the Organizing Committee
The 1st ITB Centennial and 4th PlanoCosmo International Conference

IOP Conf. Series: Earth and Environmental Science **158** (2018) 011001 doi:10.1088/1755-1315/158/1/011001

Editorial team
Proceedings of the 1st ITB Centennial and 4th PlanoCosmo International Conference on Infrastructure Development: "Transforming beyond Borders, Starting the New Urban Agenda"

Editors

Fikri Zul Fahmi (Institut Teknologi Bandung, Indonesia)
Euginie Birch (University of Pennsylvania, United States)
Tim Bunnell (National University of Singapore, Singapore)
Delik Hudalah (Institut Teknologi Bandung, Indonesia)
Haryo Winarso (Institut Teknologi Bandung, Indonesia)

Production team

Lisna Rahayu
Dika Fajri Fiisabiilillah
Aryani Nurnisa Chandramidi
Gina Puspitasari Rochman
Nabilla Dina Adharina
Yustina Octifanny
Sosrowinarsito
Ivanie Destilla Sari

English copyeditors

Martin Drenth
Sybrand Zijlstra
Zavitri Citra Wardhani
Anita Utami
Luthfi Abdurrahman
Kirana Kania
Aryani Nurnisa Chandramidi
Mohammad Hafizh Adinugraha
Mohammad Hakim Adiprasetya

Welcoming remarks

Chair of the ITB Centennial Commemoration Committee

Dear Colleagues,

Welcome to the very first opening of the series of events leading up to the Centennial Commemoration of Higher Education in Engineering in Indonesia!

Institut Teknologi Bandung (ITB) was formerly established as Sekolah Tinggi Teknik (Technische Hoogeschool, TH) in Bandung, 1920, and became the first higher education in engineering in Indonesia. For almost 100 years ITB has been leading up to the establishment of many higher education institutions in engineering, strategic industries, and research centers in Indonesia and have contributed to the creation and innovation of science and technology that aims to further develop Indonesia.

This centennial commemoration of ITB as Higher Education Engineering in Indonesia is a remarkable moment to uphold the spirit to be united by one sole mission of advancing Indonesia's science and technology and spur innovation for the betterment of our nation. Therefore, by acknowledging the success of previous three series of PlanoCosmo, without any doubt we are pleased to collaborate with the 4th PlanoCosmo to pioneer other events towards the centennial commemoration and provide inspiration for scholars, practitioners, and general public to become useful and contributive individuals towards urban and infrastructure development in Indonesia in the future.

The 1st ITB Centennial and the 4th PlanoCosmo focuses on the role of infrastructure development in implementing the goals of the New Urban Agenda. Though this event, we hope that scholars, policy makers, and innovators can communicate to cultivate new ideas, technologies, and policies concerning infrastructure development for more sustainable cities. And aside from the main conference agenda, we proudly present a magnificent exhibition from education institution, cities to national government agencies, private sectors, and association of related groups, to exhibit their innovation in development toward the implementation of the New Urban Agenda.

Prof. B. Kombaitan
Chair of the ITB Centennial Commemoration Committee

IOP Conf. Series: Earth and Environmental Science **158** (2018) 011001 doi:10.1088/1755-1315/158/1/011001

Welcoming remarks

Dean of the School of Architecture, Planning and Policy Development
Institut Teknologi Bandung

Dear Colleagues,

I am very delighted to welcome you all to the 1st ITB Centennial and 4th Planocosmo International Conference on Infrastructure Development in the stunning city of Bandung.

The series of Planocosmo International Conference has been a pride of SAPPD ITB to trigger a significant contribution to creating and disseminating scientific knowledge in planning, design and policy development of sustainable environment. The conference also greatly contributes to actualizing the vision of SAPPD ITB to be a leading, reputable and respectable educational and research institution in Asia and recognizable worldwide, in the fields of planning, design and policy development of human settlements and sustainable environment. With the main theme related to the New Urban Agenda, we mark this conference as a historic moment for SAPPD in pioneering global discussion platform towards the centennial commemoration of ITB.

I am honoured to say that in this year's conference agenda we have delegates from 15 countries around the world. We will have over 150 presentations scheduled for this year conference in Bandung, as from the 2nd-4th of April 2018.

I would like to thank the Organizing Committee for their dedication and passionate hard work for the whole year to prepare and host this promising conference. Not to be forgotten, I also would like to thanks Prof. Benedictus Kombaitan, the chair of P100, for giving an honourable yet a great opportunity to be featured as the very first opening of the series of events leading up to the Centennial Commemoration of Higher Education in Engineering in Indonesia.

Prof. Widjaja Martokusumo
The Dean of SAPPD ITB

IOP Conf. Series: Earth and Environmental Science **158** (2018) 011001 doi:10.1088/1755-1315/158/1/011001

Committees

Steering Committee

1. Prof. Benedict Kombaitan (Institut Teknologi Bandung, Indonesia)
2. Prof. Christopher Silver (University of Florida, United States)
3. Prof. Robert Cervero (University of California Berkeley, United States)
4. Prof. Richard LeGates (San Francisco State University, United States)
5. Prof. Johan Woltjer (University of Westminster, United Kingdom)
6. Prof. Roos Akbar (Institut Teknologi Bandung, Indonesia)
7. Prof. Arief Rosyidie (Institut Teknologi Bandung, Indonesia)
8. Prof. Peng Zhenwei (Tongji University, China)
9. Dr. Miming Miharja (Institut Teknologi Bandung, Indonesia)
10. Dr. Tubagus F. Sofhani (Institut Teknologi Bandung, Indonesia)
11. Dr. Ibnu Syabri (Institut Teknologi Bandung, Indonesia)
12. Dr. Ridwan Sutriadi (Institut Teknologi Bandung, Indonesia)

Scientific Committee

1. Prof. Tommy Firman (Institut Teknologi Bandung, Indonesia)
2. Prof. Eugenie Birch (University of Pennsylvania, United States)
3. Prof. Richard Haigh (University of Huddersfield, United Kingdom)
4. Pro. Eric Sheppard (University of California Los Angeles, United States)
5. Prof. Deden Rukmana (Savannah State University, United States)
6. Prof. Haryo Winarso (Institut Teknologi Bandung, Indonesia)
7. Prof. Pradono (Institut Teknologi Bandung, Indonesia)
8. Prof. Zhao Min (Tongji University)
9. Dr. Teti Armiati Argo (Institut Teknologi Bandung, Indonesia)
10. Dr. Tim Bunnell (National University of Singapore, Singapore)
11. Dr. Paul Jones (The University of Sydney, Australia)
12. Dr. Shabbir Cheema (East-West Center, United States)
13. Dr. Terry van Dijk (University of Groningen, The Netherlands)
14. Dr. Sri Maryati (Institut Teknologi Bandung, Indonesia)
15. Dr. Hastu Prabatmodjo (Institut Teknologi Bandung, Indonesia)
16. Dr. Harkunti P. Rahayu (Institut Teknologi Bandung, Indonesia)
17. Dr. Denny Zulkaidi (Institut Teknologi Bandung, Indonesia)
18. Dr. Wilmar Salim (Institut Teknologi Bandung, Indonesia)

Organizing Committee

Chair	Dr. Delik Hudalah
Co-Chair	Dr. Adiwan Fahlan Aritenang
Secretary of Academic Affairs	Dr. Fikri Zul Fahmi
	Ivanie Destila Sari, ST.
Secretary of Events	Tessa Talitha, MA
	Seruni Fauzia Lestari, ST.

IOP Conf. Series: Earth and Environmental Science **158** (2018) 011001 doi:10.1088/1755-1315/158/1/011001

Treasury	Farida Khuril Maula, MSc Dr. Niken Prilandita
Pre-Conference & Ceremonial Events	Dr. Shanty Yulianty Rahmat Bea Regina Marcendy, ST
Keynote & Plenary Sessions	Dr. Puspita Dirgahayani Hansen Sutanto, ST.
Parallel Sessions & Book Launch	Dr. Saut A.H. Sagala Nurrohman Wijaya, MSc Sri Utami P., ST. Azis Hakim, ST.
Exhibition & Business Events	Dr. Bagas Dwipantara Putra Tri Rahayu Wulansari, MT Alvin Noviansyah, ST. Eka Darma Kusuma, MT.
Tour & Post-Conference Events	Dr. Alhilal Furqan Dr. Chen Chen (Tongji University) Aliyah Alfianda Dwicahyani, ST.
Logistics, Accommodation, and Transportation	Yunie Nurhayati Rahmat, MT Achmad Fauzan Iscahyono, ST
Publications and Proceedings	Lisna Rahayu, M.Sc. Dika Fajri Fiisabiilillah, M.Sc. Aryani N. Chandramidi, M.Sc. Gina Puspitasari Rochman, MT. Nabilla Dina Adharina, ST. Yustina Octifanny, ST. Sosrowinarsito, B.Sc.
Liaison Officer	Tizar Bijaksana, MT. Tika Savitri Hasan, MA.
Documentation	Alvaryan Maulana, ST. Qurrota Aini, ST.

IOP Conf. Series: Earth and Environmental Science **158** (2018) 011001 doi:10.1088/1755-1315/158/1/011001

Overview of Conference Agenda

MONDAY
APRIL 2nd, 2018

08.00 **Registration**

09.00 **Opening**

10.00

10.15 **Keynote Lecture 1:**
The Role of (Local) Government in Establishing Sound (Local) Fiscal System to Enable Equal Redistribution of Urban Value

11.00 **Keynote Speech 1:**
Integration of Strategic Regional Development Strategies with the New Urban Agenda Targets

11.30 **Keynote Speech 2:**
The Role of International Financial Institution in Delivering Smart Urban Infrastructure Financing

12.00

12.10

13.10
- **Plenary 1.1**
  Conventional and Alternative Infrastructure Financing

- **Meet 1**
  JRCP Editorial Board Meeting

14.40

15.10 **Plenary 1.2**
Access to Financial Support for Infrastructure Projects

16.40

17.00 **Dinner Preparation**

18.30 **Welcoming and Dinner Reception**

IOP Conf. Series: Earth and Environmental Science **158** (2018) 011001 doi:10.1088/1755-1315/158/1/011001

TUESDAY
APRIL 3RD, 2018

09.00 **Plenary 2**
Sustainable Regional Development

10.30

10.45

11.00 **Plenary 3**
JRCP Plenary on Inclusive Urbanism

12.30

12.45

13.45
- **Parallel 1**
- **Book Launch 1:** Jakarta, Claiming Spaces & Rights in the City
- **Workshop 1:** Open Publication

15.15

15.45
- **Parallel 2**
- **Book Launch 2:** Teknopolis & Kota Komunikatif
- **Workshop 2:** Investing Disaster Risk Reduction for Coastal Resilience (City)

18.30 **Gala Dinner**

IOP Conf. Series: Earth and Environmental Science **158** (2018) 011001 doi:10.1088/1755-1315/158/1/011001

WEDNESDAY
APRIL 4TH, 2018

09.00 **Opening**

09.30 **Keynote Speech 3**
National Urban Policy: Cities as Engine of Growth

10.00

10.15

10.30
- **Keynote Lecture 2:**
  New Urban Agenda: Problems and Their Potential Solutions

- **Meet 2:**
  Faculty Board Meeting of SAPPD ITB and CAUP Tongji University

11.30

11.45

12.45
- **Parallel 3**
- **Workshop 3:** Presentation of Studio's Results by ITB and SU Students
- **Workshop 4**: Financing to Implement the New Urban Agenda

14.15

14.45
- **Parallel 4**
- **Poster Presentation**
- **Workshop 5:** Smart City Innovation
- **Workshop 6:** Financing to Implement the New Urban Agenda

16.15 **JRCP Best Paper Award**

16.45 **Wrap-up and Closing Ceremony**

IOP Conf. Series: Earth and Environmental Science **158** (2018) 011001 doi:10.1088/1755-1315/158/1/011001

The 1st ITB Centenntial and 4th PlanoCosmo

International Conference on Infrastructure Development:

"Transforming beyond Borders: Starting the New Urban Agenda"

Organized by:

Urban and Regional Planning Study Program
School of Architecture, Planning and Policy Development
and
Research Center for Infrastructure and Regional Development
Institut Teknologi Bandung

Co-organized by:

Collage of Architecture and Urban Planning
Tongji University, China

Secretariat of the Committee:

Ganeca Campus ITB, Sugijanto Soegijoko Building (Labtek IX A)
Laboratory of Regional Planning (KK PWD), 5th Floor
Jalan Ganeca 10 Bandung 40132 INDONESIA
Phone: +62-22-2509171 (ext. 3613)
Email: planocosmo@pl.itb.ac.id

Peer review statement

All papers published in this volume of *IOP Conference Series: Earth and Environmental Science* have been peer reviewed through processes administered by the proceedings Editors. Reviews were conducted by expert referees to the professional and scientific standards expected of a proceedings journal published by IOP Publishing.

IOP Conf. Series: Earth and Environmental Science **158** (2018) 012001 doi:10.1088/1755-1315/158/1/012001

Designing urban rules from emergent patterns: co-evolving paths of informal and formal urban systems - the case of Portugal

Paulo Silva[1]
[1]University of Aveiro, Department of Social, Political and Territorial Sciences, ID + - Research Institute of Design, Media and Culture, GOVCOPP – Research Unit of Governance, Competitiveness and Public Policies, Aveiro - Portugal

Corresponding author: paulosilva@ua.pt

Abstract. In many societies, informality has been a relevant part of the construction of the urban fabric. This is valid along a city's history and in recent urbanization processes. In the past, informality was in the origin of many of urban planning. Very soon urban planning adopted, as one of their main missions malfunctions in cities. Therefore, the need of formalization became one of the main reasons on the emergence, the control of informal processes. As an answer to informal individual solutions, urban planning responded with standardized rules and the urge of creating spaces fitting into pre-established rules instead of rules fitting into spaces. Urban planning as a discipline has gradually changed its path. The contrast between urbanization promoted under formal urban planning and informal urbanization is only one sign of the mismatch between urban planning actions and informal urbanization dynamics.

Considering this tension between formal and informal dynamics, in some cases, planning rules and planning processes continue ignoring informal dynamics; in other cases, planning rules are designed to integrate informality "without losing its face" through "planning games" [1]; and a third and less explored way in which planning systems interact with informality and from that interaction learn how to improve (we consider it a process of enrichment) planning rules while they promote an upgrade of informal interventions [2]. This latter win-win situation in which both informal and formal systems benefit from their interaction is still rare: most of the time either only one side benefits or none benefit from the interaction. Nevertheless, there are signs that from this interaction co-dependent adaptation might occur with positive outcomes for the urban system – in which co-evolutionary dynamics can be traced.

We propose to look at the way building rules have been designed in Europe in a context considered successful in the sense of dealing of informality – the one of Portugal. The country experienced a wave of informality associated with illegal urbanization since the 1960's in the main urban areas. The process of interaction between informal and formal urban systems proved to be a success in statistic terms. Slum clearance reduced the existence of informal occupations to almost zero. Informal settlements involving land tenure have been dealt with in the last two decades with considerable positive impact in the urban fabric.

Based on this, with this paper we will evaluate how informal and formal systems are impacting each other and changing along the time the shape of building and of planning rules. For this we will look at the planning tools created to formalize informal settlements in the Lisbon Metropolitan Area from the last forty years to see how urban and building rules were adapted to respond to the specific needs of informal settlements; how this adaptation moved from temporary and exceptional to permanent rules; finally, how were these new rules able to "contaminate" the general planning and building codes. We aim that these findings would help us to contribute to a "healthier" relation between formal and informal urban systems, not ignoring each other, not controlling each other but instead learning with each other. By achieving this, planning systems become more responsive; on the other hand, informal occupations can be upgraded without being destroyed with the contribution of the planning systems.

1. Introduction

Informality has been at the root of a relevant part of the construction of the urban fabric. All over the world this is valid for a city's history and recent urbanization processes, particularly in the global south [4]. Informality was the origin of many malfunctions in cities, but also the "normal" way of building them. The need of formalization became one of the main reasons for the emergence of urban planning as a discipline, to deal with the effect of rapid industrialization and population growth. Very quickly urban planning adopted, as one of its main missions, the control of informal urban development. As an answer to individual informal solutions, urban planning responded with standardized rules. The creation of spaces that fit into pre-established rules replaced rules fitting into pre-existent spaces. The contrast between urbanization promoted under formal urban planning and informal urbanization is only one sign of shift. Another is the interdependence between urban planning actions and informal urbanization dynamics.

We proposed to look at the way building rules were designed in a European context that was considered successful in the sense of dealing with informality – the one of Portugal. The country experienced a wave of informality associated with illegal urbanization since the 1960's in its main urban areas. The process of interaction between informal and formal urban systems proved to be a success in statistic terms: informal settlements involving land tenure have been dealt with in the last two decades, with a considerable positive impact in the urban fabric.

The Lisbon Metropolitan Area will be the focus of this paper for two main reasons. First, the Portuguese urban informality had its largest expression in this territory. In the 1980's, over ten percent of the population of this metropolitan area resided in informal settlements. This dynamic coincided with another one – the conceptual shift from the Greater Lisbon to the Lisbon Metropolitan Area. In 1986, Portugal joined the European Union and benefited from EU funds through the ERDF (European Regional Development Funds) program, dedicated to the construction of infrastructure. During the second half of the 1980's and during the 1990's, the face of the metropolitan area radically changed. The dependency from a single pole (the municipality of Lisbon) to the development of a formal multipolar structure put some of the former peripheral informal settlements into the centre of the LMA.

The interaction between formal and informal is therefore useful to understand the contribution of informality to the construction of the metropolitan fabric. Most approaches look at formal and informal systems separately. Comparative studies on this field, in regards to the interaction of formal planning institutions with informal settlements, are scarce [5]. It is easy to find research on formal planning systems, also on informal ones, but much less easier to find comparative ones between the two. This dichotomy between formal and informal that is based on a regulatory vision of planning narrows the interpretation of people's personal needs [6] and impeaches the improvement of planning tools.

Informality implies the existence of a formal statutory system that regulates urban transformation and expansion. Some experiences in (mainly southern) European countries are said to be relevant as inspiration for new ways of dealing with informal settlements [7]. In the case of Portugal, recent researches add evident signs of co-evolution between informal settlements involving land tenure and planning institutions [2], and the fact that the possession of land might play a relevant role on how spatial policy and planning instruments have been designed in recent decades [3]. As opposed to other situations in which "one fits all" solutions were applied independently of land tenure [9], in the Portuguese case, land tenure seems to play a crucial role in relations between formal and informal, or in practice in the relation between individuals and planning institutions.

This has an impact on urban design, by way of how territorial development is shaped and how spatial strategies are informed towards a more sustainable territorial setting, by combining the individual needs with metropolitan visions are. There is a fine grain of connection between one and the other that often escapes the planner and urban designer's scope. In fact, it is not always easy nor evident how to combine in a coherent discourse the glamour around new centralities and the much less glamorous background of informal settlements.

Based on this, we pretend to make an evaluation of how informal and formal systems are impacting each other and changing the shape of building and of planning rules. Therefore, we will look at planning tools created to formalize informal settlements in the Lisbon Metropolitan Area from the last forty years to see how urban and building rules adapted to respond to the specific needs of informal settlements; how did this adaptation change from temporary and exceptional to permanent rules; finally, how were these new rules able to "contaminate" the general planning and building codes. Our aim is that these findings would help us contribute to a "healthier" relation between formal and informal urban systems, neither ignoring, nor controlling/being controlled processes, but instead engaging on learning. By achieving this, planning systems can become more responsive; on the other hand, informal occupations can be upgraded without being destroyed by the contribution of the planning systems.

This paper is based on the planning data collected from the website of *Direção Geral do Território – Sistema Nacional de Informação Territorial* (Territorial Information National System), the Portuguese database of planning instruments, from the national to local level. We analyzed all the detailed plans approved for the municipalities that compose Lisbon Metropolitan Area. Another element was crucial to this research, the analysis of the AUGI Act. This legal document was approved in 1995 by the national parliament, culminating three decades of attempts to formalize informal settlements. To discuss it, in section 2 we will establish the formal and informal settings that constitute the ground to the debate, based on a bibliography review. In section 3 we will present the way formal and informal settlement had evolved in an interdependent way until the "moment" in which they started to co-evolve. In section 4 we will present the data regarding the use of planning tools to deal with informal settlements. In section 5 we will discuss those findings before moving on section 6, the conclusion and final remarks on advantages and limitations of the co-evolution between informal and formal systems.

2. Formal versus informal

In some cases, planning rules and planning processes continue ignoring informal dynamics. The first generation of urban regeneration policies **Error! Reference source not found.** was launched to deal with informal settlements (slums in particular) in countries that experienced the effects of the industrial revolution earlier. Slum clearance was a major movement mainly in western Europe, first in-between wars in England, later in the post second world war in countries such as France. If this worked as a significant and effective way of reducing informal settlements in many western European countries, it solved mainly physical issues. Decades later, these countries are still dealing with the negative (mainly social) effects of the radical measures and are still seeking alternatives. The global south seems to be discovering this "old" way of dealing with informality and negative social impacts can be expected from these processes [9]. In fact, in many cases, early informal settlements that were formerly peripheral became surrounded by formal urbanisation. New accessibilities and infrastructure contributed to increase the attractiveness of these areas.

In other cases, planning rules are designed to integrate informality "without losing its face" through "planning games" [1]; all over Europe, law-amendments, amnesties, and temporary windows of legalization have been opened. In many cases they are materialized in self-declaration procedures, payment of penalties, temporary legalizations and formalization with some restrictions. They concentrate more on bureaucratic procedures involving financial compensation. Most of the time, compensation is neither enough to finance the infrastructure nor represents a real penalty for landowners. In a context in which informality evolved from fulfilling basic housing needs to become the way to build secondary homes in privileged areas, in many cases this compensates. This formalization through the design of exceptional rules has worked mostly to statistically reduce the number of informal settlements than as an effective way of upgrading them.

Less frequently, planning systems have been interacting with informality. From that interaction they learned how to improve planning rules and at the same time promote an upgrade of informal occupations [3]. The research on this process added some differences in comparison with the two previous cases. This learning process was the result of several decades of interaction between informal settlements and planning institutions. It reflected the diversity of informal settlements and gave more sustainable answers to informality. Given this, which seemed to be effective in the Portuguese case, some qualities of the process emerged: the long-term process of trying and failing was also the opportunity to experiment; the length of several processes developed at the national level allowed the establishment of long-term relations between actors as a fundamental ground for trust; and finally, all this context permitted the establishment of short-chains. These undermined processes of self-organisation and co-evolution, which together with other examples from southern Europe (Italy, Spain, Greece), are considered to be relevant and innovative by international institutions such as the United Nations.

It is important to note that these are not rigid patterns. In fact, it is possible to find within the same planning system the three different approaches. In Portugal, the "old fashion" slum clearance is still implemented, while at the same time administrative mechanisms to formalize the informal is also applied (recently an exceptional regime was launched in order to allow the formalization of economic activities installed in informal buildings). At the same time, more innovative and intertwined solutions have been implemented. Several factors might explain this. The lack of land tenure plays a crucial role when it comes to adopt more radical urban design changes. When informal settlers have no property rights, planning institutions feel free to define the design settings according to other criteria than the needs and aims of future residents.

The latter win-win situation in which both informal and formal systems benefit from their interaction is still rare; mostly, either one side benefits or none benefit from the interaction. Nevertheless, there are signs that from this interaction, co-dependent adaptation may occur generating positive outcomes to the urban system – in which co-evolutionary dynamics between planning institutions and informal settlements can be traced. Since this was possible to identify at the law-making level, In the following sections we will describe and discuss the main challenges at the urban design level.

3. Formal and informal settlements in Portugal

3.1 The formal setting

The Portuguese planning system evolved from being mainly concerned with urban areas (during the 1930's), moved on to the regional scale influenced by economic planning, and later covered the entire territory at the municipal level. In the 1930's, the first planning system in Portugal was created, with the establishment of urban plans for the largest cities in the country. In the meantime, the first officially considered informal settlement in Portugal was built in the outskirts of the city of Lisbon.

The 1960's – the *Plano Diretor da Região de Lisboa* and the "informal" explosion of the city suburbia emerged. In the 1960's, the first regional plan of Lisbon was produced. That plan still followed the concept of a greater Lisbon, which concentrated all the iconic functions of a national capital, such as government structures and public agencies combined with historical features, and, in a much lower level, combined them with a mixture of incipient satellite cities combined with dormitory areas. In this conceptual frame all the urban areas depended on the city of Lisbon. The plan also showed some environmental concerns that were quite inspired by the British planning system, by introducing the idea of a green belt. Since the city of Lisbon was corresponding to its administrative limits, the green belt corresponded predominantly to the suburban municipalities. The south bank of the river was considered mainly as a protected area that was to be preserved. Unfortunately, nobody knew how to do it, and with what means (Figure 1).

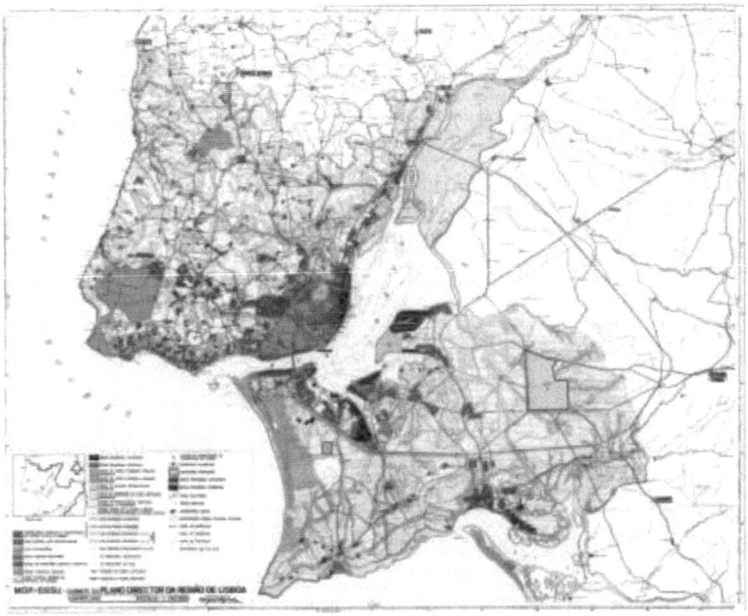

Figure 1. The Regional Plan of Lisbon (*Plano Diretor da Região de Lisboa*) from 1964: most of the areas from the green belt on the south bank actually became informal settlements.

The municipal level, which in content corresponds in general to a zoning definition, became mandatory in the 1990's and later in the following decade gained more strategic content [10]. By the 1990's an entire planning system was consolidated in theory with the publication of the decree-law 380/99 and in practice with the publication of the first version of the national spatial planning framework, known in Portuguese as *Programa Nacional de Políticas de Ordenamento do Território* (*PNPOT*). In the 2000's, a part of the national territory had regional plans approved. The recent trends on spatial planning system emerged in the present decade with the concentration of binding rules applicable directly to private (individual or collective) entities on the municipal level, while the regional and national became or remained exclusively as orientations for municipal plans.

The next regional scheme was launched in the 1990's and corresponded to the shift from the greater Lisbon to a metropolitan concept in two steps. The first and so far only regional plan that had a development scheme very similar to a mega land use plan; by analysing it one can see that it corresponds very much to the sum of the zonings of the different municipalities composing LMA. A second version of this plan was produced in the 2000's but was never published (although it reached the phase of public discussion) and the process is already interrupted in the present decade due to the financial and economic crisis that affected Portugal and postponed many public investments in the metropolitan area (namely a third bridge connecting the two sides of the Tagus river and a new international airport in the south bank). While the first regional plan had an inclusive tone, this unfinished version had a selective approach in terms of the design of a regional development model.

3.2 The informal setting

The first illegal settlements in the Lisbon metropolitan area (LMA) were associated with dense industrial areas and emerged in the 1930's. This was considered the first illegal settlement because for the first time an entrepreneur promoted a settlement to answer peoples' needs of housing, predominantly of the working-class families.

In the 1960's, informal settlements had spread all over LMA and diversified in size (from the minimum of 0,5 ha to several hundred hectares), typology (from multi-story apartments to individual detached houses), density, resistance of construction (since new structural elements were without previous calculations), uses (from primary to secondary houses), visibility (from "real buildings" to permanently settled vans and even caves without any above the ground construction), location (from closer to remote areas), tenure (from rented dwellings to self-owned houses), and environmental contexts (from river basins to slopes). In compensation for this diversity, the LMA informal settlements had a few things in common, among them the fact that they were built on rural land where construction, according to the incipient planning tools, was not allowed and that it was generally in a context of weak law enforcement.

With time passing, a specialization of areas emerged. In the 1990's, this was already clear. The single house plot of land build under a regular geometric street pattern pre-established by the seller became the primary residence of many families living in LMA when settlements were relatively close to elements which could allow metropolitan integration – accessibility, concentrations of jobs, ... In the cases where there were other values, such as in environmentally privileged areas and not so close to territories where jobs were concentrated, they became mainly areas for secondary residences. And third, the areas that were not interesting enough to attract enough numbers of either primary residents nor secondary residents and for that matter remained in limbo. This period of around four decades provides clear signs of evolution, as we will see in section 5. But before that, in the next section we will see how the planning system influenced the way informal settlements were formalized.

4. Formalization of informal settlements in the last 20 years – the case of Lisbon

In this section, we will focus on data regarding the detailed plans designed to formalize informal settlements. While in previous decades informal settlements were (1) first ignored and (2) then dealt with as if they were regular occupations, two decades ago a special regime was created to formalize this kind of settlement and which has been applied since then. With the approval in 1995 of the AUGI (acronym for *'Áreas Urbanas de Génese Ilegal*, which can be translated into urban areas generated by illegal processes) Act, a completely new setting was established. Since then a single model of formalizing was created in addition to a diversity of tools, processes and decision-making procedures.

First, the formalization of areas had to be included in municipal zoning plans as areas with vocation to become urban spaces, to be declared AUGI. When the law was published (1995), many municipalities were working on their land use plans, either concluding, discussing or approving the proposals. This was an important moment to launch the discussion between the municipalities and regional authorities about which informal settlements should be formalized and those that should not be. The general tendency was towards formalization with some exceptions. Especially on the north bank of LMA, informal settlements in land slide risk areas were excluded from this process. On the south bank the exceptions to formalization happened with the informal settlements located in coastal areas. This refusal to formalize was effective in planning terms, although it didn't lead to demolitions and clearance of those areas in the short term:

- On the north bank, up to this day not many things have changed. Constructions remain in this risk area; without a chance to be legalized, access to infrastructure is precarious and the few improvements in public spaces are predominantly done to allow circulation. It is important to say that these areas, although very visible because they are built on slopes, represent the backyard of many municipalities, including Lisbon, where some of these informal settlements are located behind the airport.
- On the south bank, the situation is significantly different, with a special program being launched for the regeneration of the ocean front where many of these informal settlements are located. The Polis Program was a nationwide instrument created to regenerate urban waterfronts, launched after and inspired by the success of the regeneration of the industrial site where the EXPO'98 took place in the east side of Lisbon. A coordinated set of planning actions was carried out and a significant part of these occupations were relocated. It is important to mention that Portugal's coastal areas are severely endangered by erosion and this plays a significant role in public debates and in public opinion in general, with a combination of general acceptance of the need of set backing with nimby local behaviours.

Once the it had been decided to formalize informal settlements, the model of management was defined by having in mind the diversity of situations referred to in the previous section. The management process could be either lead by the landowners, the municipality or by both the landowners and municipality. Some informal settlements had landowners with skills to deal with bureaucratic and technical procedures, usually because they had medium or high educational backgrounds. In other cases where those skills were missing, they could express their interest at the municipality's process leadership, creating grounds to deal with the variety of financial, technical, and juridical issues which the formalization process implied.

One of these technical issues referred to the way of formalizing the urban design layout. This could be either made (1) through the submission of a parcelling out proposal by the landowners to the municipality (initially this corresponded to the cases in which the process was coordinated exclusively by the landowners) or (2) through a detailed plan to be produced by the municipality (initially this option was chosen when the municipality was assuming the coordination of the process). The parcelling out process was, most of the time, a follow up of the work already made by landowners before the AUGI Act was published. It was easy to design and to apply in less conflicting contexts. Detailed plans were usually adopted in the cases in which some land adjustments had to be made to the binding conditions defined by the municipal land use plan. In other cases, detailed plans helped to redistribute costs and benefits among landowners.

Either in the case of parcelling out or in the case of detailed planning, the urban design proposal was usually confined to the informal settlement official limits. Moreover, the binding rules were concentrated on combining the application of spatial planning rules with landowners' recognized rights. It must be said that the application of spatial planning rules - from land use master plans to national laws defining street width or per capita collective facilities' needs – were subjected to exceptional regimes. Sometimes informal settlements were already occupied in a way that made it difficult or impossible to achieve the ratio of soil for collective needs. In other cases, different street widths were adopted because setbacks were impossible to apply (informal settlers currently use the tactic of making constructions on the edges of the informally created plot of land to secure as much as possible what they consider "theirs"). The adoption of a detailed plan was the solution found by planning institutions to validate these exceptions. It is important to also mention that planning procedures became very open since the beginning (1990's) and were subjected to a scrutiny from different levels of public administration (Figure 2 to 4). Also in the case of more binding rules, exceptions were created. Sometimes pre-existent buildings had a different height or number of floors from the plan's rules. If they proved to provide healthy conditions in terms of ventilation and insolation, then they could be accepted as exceptions. The same could occur with the amount of built square meters, of functions, and of number of units.

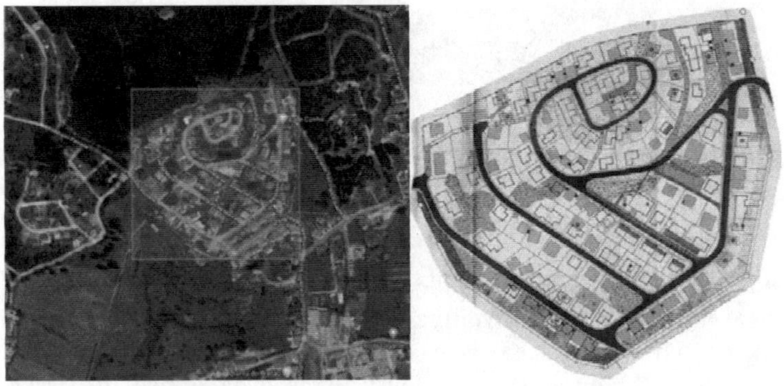

Figure 2. Detailed plan from 1992 (municipality of Odivelas, LMA, north bank): one of several examples of formalized ex-informal settlements exclusively residential.

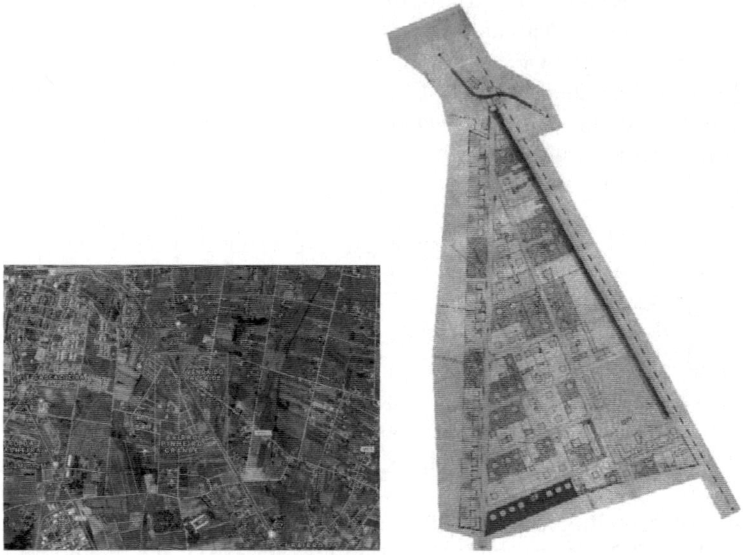

Figure 3. Detailed plan from 1988 (municipality of Palmela, LMA, south bank) – flexibility on parameters: landowners were allowed several options from new constructions with full demolition to the possibility of keeping informal constructions or adapting informal constructions and enlarging them.

Figure 4. Detailed plan from 1988 (municipality of Seixal, LMA, south bank) – flexibility on parameters: landowners were allowed to decide which building typologies to adopt, although it must respect the ratio of construction.

In many informal settlements it is possible construction-wise to identify a "before" and "after" formalization. Also, at the public space level they represent an upgrade of the urban network, compared with formal settlements. The fact that taxes paid by landowners were applied in the formalization of informal settlements had a strong impact on these areas. The impact of assuming the payment of considerable sums of money by the landowners was generally well accepted because they would see money being applied on their streets and their neighbourhoods. But leaving the ex-informal settlement reality was totally different. These areas became connected with the rest of the urban realm by poor infrastructures, with insufficient number of collective facilities and suffered from an informal stigma.

This already long process of two decades of formalization of informal settlements changed the face of the metropolitan area of Lisbon and became an effective contributor to the metropolitan urban fabric. With the consolidation of the metropolitan network they experienced new and different contexts. While some joined and suburban areas, becoming one more dormitory of Lisbon, others became new urban centres with relevance at the metropolitan level. And finally, some remained not formalized, either because they didn't follow environmental criteria or because landowners didn't mobilise themselves to formalise their settlements. In the next section, the interactions of informal settlements with planning institutions will be addressed.

5. Discussion

The process of formalization initiated more than two decades ago represented an enormous effort from both planning institutions and informal settlements to cooperate. This interaction had impacted each side of the table and affected the way they evolved. To formalize, informal settlements had to follow a strict number of rules. This inspired the binding rules created to the new formal settlement. It is striking in this respect the contrast between the diversity of informal settlements that emerged from decades of urban sprawl and the unifying character of binding rules that were designed to formalize them. In some cases, the rules were useful to control the density of some informal areas, but in some cases they also contributed to their densification.

Formalization was not achieved without mutual permanent change. Planning institutions created exceptional rules to deal with informal settlements. In some cases, informal settlements were composed by landowners who had never seen each other before. This made it difficult to design arrangements and to do simple things, such as registering land to conclude formalization. To overcome these obstacles, land registration became possible as an individual act (instead as a collective act as was before the AUGI Act). The system proved to simplify so much bureaucratic procedures that this simplification ended up being adopted also by general legislation.

The contribution for a metropolitan network revealed a process of adaptation undermined by another process of selection [11]. While time passed another process of adaptation emerged. Formal allows land right, enforces land tenures and eases transactions of land between new and old landowners. The change of the metropolitan setting put some new-formal settlements in the spot. The 1990's when formalization started was also the decade of large public investments in infrastructure. Peripheral areas, such as the ones where informal settlements were located, became opportunity areas to develop those infrastructures. They include new metropolitan highways, a new bridge over the Tagus River and relevant investments on public transportation such as train and tram lines. The scale of informal settlements was also relevant for the change. An initial wave of occupants created a functional structure, emergent from the sum of hundreds of individual actions **Error! Reference source not found.**. The combination of large scale informal settlements and good accessibility allowed the arrival of new residents to these areas. Of course, this cannot be generalized as many areas remained as dormitories, attracting low-income segments of the society, many of them immigrants.

6. Conclusions

With this paper we tried to highlight the potential for improvement of planning rules through the interaction with unplanned settlements, which are quite often confused with spaces with no rules. We chose LMA, an area where a process of informality took over in recent decades. Still, to this day, informal areas are being integrated in the metropolitan setting. We can say that those areas that were not successful in terms of formalization are those that, for one reason or another, were out of the metropolitan system, either because the accessibility network didn't reach those areas or because, despite being in the inner parts of LMA, they are enclaves where natural risks are evident. And so far, this has been a red line for formalization, although not so much has been done in alternative to non-formalization.

As opposed to coastal areas where planned retreat has started to be discussed, this did not happen in terms of inner informal settlements in risk areas. However, the setting could not be better in terms of resettling. There were good reasons for removal since the State could not ensure the safety of people and goods in land slide risk areas and therefore they could not formalize these settlements. And even with the economic and financial recovery of the last couple of years, which has attracted new investors for non-consolidated urban areas, the offer of vacant soil is still much higher than the search for it. Therefore, a displacement from risk areas to empty spaces in urban areas could be an option to be explored.

Evolution relates to adaptation in a permanent way to surrounding conditions. Co-evolution is to proceed to that change in an intertwined way with another entity, body, or system. When we look to the interaction between planning institutions and informal settlements this is clear, and the effect caused on each of them. At the legal level, planning institutions learned how to create more flexible possibilities governance-wise. Institutions shared with informal settlements the leading role of formalization, instead of concentrating, as it was happening before 1995, when the AUGI Act was approved and implemented.

If at the governance level the achievements are evident we can see that at the planning rules the tendency was to integrate the previously established codes without major changes. The reasons for that are there to be explored in further research. Possibly one reason was to ask the least of informal settlements in terms of urban rules (number of floors, ratio of construction, functions, setbacks) to simplify communication. By doing that, it was easier to communicate with landowners. On the other hand, the time to explore and test and sediment new rules was short. Landowners expected that during decades of formalization and to engage in a new experimentation process at the local level probably became more difficult to be accepted. However, it is interesting enough because this deconstructs the recently traditional perception; that the more we reach the top of the hierarchy the less effective decision-making processes are. In fact, in this case planning institutions' adaptation seemed to be easier to formalize at the national level, with the parliament approving the AUGI Act, rather than at the lowest level of planning institutions – the municipalities. One reason for this may be that the decisions taken at the national level have it easier to introduce shifts in their path because they congregate aims from multiple stakeholders, whereas at the municipal level the landowners' aims are perceived more as individuals and for that less legitimate to cater to.

When we go into detail to the urban scale, we see how different patterns emerge and how planning institutions deal with them. First, some very particular features characterize these informal settlements. Once they were formalized, they brought into light urban spaces consolidated in terms of urban grid – with streets being completed at the ground and at the infrastructure levels, while the tissue remained quite unfinished. This is a challenge for urban management. Another feature is that as opposed to planned settings in which plans define centralities, in the cases of larger informal settlements, the emergence of centrality patterns occurred in a quite different way depending not of a collective debate but instead of a sum of hundreds or thousands of individual initiatives that later defined centralities' patterns. Some shifts contradict pre-assumed ideas on informal settlements. For example, in the case of density there are many references of how high densities are in informal settlements as opposed to formal ones. Well, in the case of LMA there are many cases that the shift was inverse from expected. Instead of formalization decreasing density, in many cases the opposite occurred as part of the process of metropolitan integration. Aims and wishes from informal settlements' landowners were mostly inspired by a rural background in which the detached house surrounded by some vacant soil is the representation of their dream house. Shifting from an informal remote settlement to a formal networked settlement they attracted new residents, mainly young couples moving from more central and densified suburbia to more distant yet better connected territories where their aims of a house could be fulfilled. The match between the formalization, the metropolitan integration and new residents' aims made a new pattern of occupation emerge, the one of semi-detached houses, which doubled in practice the previous densities generated by informal settlements. Among others, informal settlements' formalization remains a fascinating field of research, opening opportunities of reinterpretation of the urban fabric.

References

[1] Calor I 2017 *A ilegalidade urbanística e o sistema de planeamento. Perspetiva comparada sobre políticas e práticas de controlo urbanístico de obras particulares* PhD thesis (Lisbon: New University of Lisbon)

IOP Conf. Series: Earth and Environmental Science **158** (2018) 012001 doi:10.1088/1755-1315/158/1/012001

[2] Silva P and Farral H 2016 Lessons from informal settlements: a 'peripheral' problem with self-organising solutions *Town Planning Review* **87** 297-319

[3] Silva P and Farral H 2016 From informal to formal: what can be learned from reviewing 50 years of portuguese models, policies and politics in Attia S, Shabka S, et al. (eds) *Dynamics and resilience of informal areas – international perspectives* (Cham: Springer Nature) pp 25-42

[4] Alzamil W 2017 The urban features of informal settlements in Jakarta, Indonesia *Data in Brief* **15** 993-999

[5] Ascensão E 2016 Interfaces of informality *City* **20** 563-580

[6] Davy B 2014 Polyrational property: rules for the many uses of land *International Journal of the Commons* **8** 472–492

[7] UNECE (United Nations Economic Commission for Europe) 2009 *Self-made cities: In search of sustainable solutions for informal settlements in the United Nations Economic Commission for Europe Region* (New York and Geneva: United Nations)

[8] Carmon N 1999 Three generations of urban renewal policies: Analysis and policy implications *Goeforum* **30** 145-158

[9] Jones P 2017a Formalizing the informal: understanding the position of informal settlements and slums in sustainable urbanization policies and strategies in Bandung, Indonesia *Sustainability* **9** 1436

[10] Campos V and Ferrão J 2015 O Ordenamento do território: uma perspetiva genealógica *ICS Working Papers* **1** (Lisboa: Instituto de Ciências Sociais da Universidade de Lisboa) [Online]. Available from: http://www.ics.ul.pt/flipping/wp2015_1/index.html

[11] Silva P 2014 Infrastructures: discontinuities, design and planning in Bartolomei L and Bortolloti A (eds.) *Ricercha e progetti per il territorio, la città e l'architettura*, (Bologna: Departamento de Arquitetura, Univ. Bologna)

[12] Silva P 1996 Tendências recentes de ocupação do território: o caso da Quinta do Conde (unpublished MSc. thesis) (Lisbon: University of Lisbon)

[13] Totry-Fakhoury M and Alfasi N 2016 From abstract principles to specific urban order: Applying complexity theory for analyzing Arab-Palestinian towns in Israel *Cities* **62** 28-40 [Online]. Available from: DOI 10.1016/j.cities.2016.12.001

[14] Calor I and Altermann R 2017 *When enforcement fails: Comparative analysis of the legal and planning responses to non-compliant development in two advanced-economy countries*

[15] Jones P 2017b *Unpacking informal urbanism: Urban planning and design education in practice* (Bandung: Penerbit ITB)

IOP Conf. Series: Earth and Environmental Science **158** (2018) 012002 doi:10.1088/1755-1315/158/1/012002

An observation on the quality of interfaces in order to understand the complexity and coherence of informal settlement: A study on Tamansari Kampung in Bandung

S Sawira[1] and T Rahman[1]

[1]The University of Sidney, Australia

Corresponding author: tanvia.arc@gmail.com

Abstract. Self-organized settlements are formed within the limited capacity of the inhabitants with or without the Government's interventions. This pattern is mostly found in the informal settlements, where occupants are the planners who are guided by their needs, limited resources and vernacular knowledge about place making. Understanding the process of its development and transformation could be a way of unfolding the complexity it offers to a formal urban setting. To identify the patterns of adaptation process, a study of morphological elements (i.e. house form, streets) could be a possible way. A case study of an informal settlement (Kampung of Tamansari, Bandung in Indonesia) has been taken to dissect these elements. Two of important components of the study area: house forms and streets created the first layer of urban fabric. High population density demanded layers of needs and activities which eventually guided the multifunctional characteristics of streets and house forms. Thus, streets create dialogue with the complex built forms-often known as interface is the key element to understand the underneath order of Tamansari. Here interface can be divided into two categories depending on their scale – small and large. Small scale interfaces are comprised of small elements such as, extended platform, fence, steps, low height wall, blank wall and elements to set above, set forth, set over in house forms. These components help to create and define semipublic spaces in the settlement. These spaces could be visually and physically interactive or no interactive which result into active or inactive spaces respectively. Small scale interfaces are common features of the settlement, whereas large scale interfaces are placed at strategic locations and act as active spaces. Connecting bridges, open spaces and contours often create special dialogue within and beyond the study area. Interfaces cater diversity in the settlement by creating hierarchy of spaces. Sense of belonging and scope of personalization of the inhabitants are integral parts of alleyways and thus they create a complex yet coherent urban fabric. Apart from the physical elements, the settlement embodies some intangible assets like social bonding, trust, kinship, empathy and sense of belonging that add value to the spatial quality which is a distinctive character of Tamansari kampung. Informal settlements are certainly complex in nature, as it is an outcome of multiple people working to accommodate multidimensional needs. Whereas in a formal system, approach to cater for need is guided by a set of rules developed by a set of professionals end up in creating prototypes irrespective of necessity, affordability and cultural diversity. Cities throughout the world, are experiencing rapid urbanization creating different urban issues. Therefore, it is highly necessary to address different need and affordability of users and come up with suitable urban solutions. Understanding Tamansari Kampung as an informal settlement will enrich the knowledge and expertise to work in complex urban settings.

1. Introduction: Informal settlement

With half of the global population living in urban areas, cities are growing and expanding beyond capacities. Concentration of economic and social activities in urban areas catalyze large scale rural-urban migration and increasing demand of job, housing and infrastructure push cities towards major transformation. In response to this rapid urban development trend, urban areas of emerging economies host informality to some extent. Due to the limitation of resources and incompetence to accommodate and provide basic civic facilities to the rising population, cities with limited resources often opt for alternative solutions in form of informal urbanism.

In general, informal urbanism refers to the outcome of autonomous urbanization from planned urban framework. Traditionally formal planning practices are fashioned on a set of centralized rules and principles, whereas socio-economic factors and bottom up approaches shape informality. Many scholars have emphasized on the process and output of informal urbanism and explored potential of informality. With the idea of more democratic and flexible manner of space utilization, informal urbanization pattern has been presented as an alternative way of shaping cities in various literatures.

The urban realities of formal and informal approaches are much more complex and dynamic. Dovey [1] and Bayat [2] overruled the general idea of informality associating with poverty and illegal urban growth. Simeone [3] defined the theory as a way forward to explore human capacities and opportunities past legislative boundaries. According to Perlman [4], informal growth has significantly contributed towards economic advancement of developing countries and informal urbanism has opened up a prolific trend of urbanization over last few decades. Roy [5] noted a niche idea about informal settlement considering it as a complete package of live, work and recreation. In purview of her, informal urbanism is a mechanism established on human needs, struggles and achievements.

In addition to that, Kamalipour [6] identified morphological study as an efficient tool to unpack informal urbanity. Micro-scale analysis of informality informs us about the interrelations among density, grain size, and mixed use of built environment in response to sociocultural context. Nevertheless, Jones [7] explored resilience of informal urbanism against wider urbanization processes. He identified persistence of resilience in informal settlements via housing forms and adaptation process. Social bonding and unspoken planning agreement among the inhabitants are the driving forces in achieving resilience in the process of obtaining maximum benefit of limited resources. Jones [7] advocated understanding informality as coping mechanism of the locals (residents of the case study, Tamansari, Indonesia) by physical and spatial elements.

Urban informality has been observed, analyzed and addressed in various literatures from different aspects, i.e. spatial growth, physical form and their relationships, assemblage and coherence and so on. The tightly knitted informal settlements have been studied to understand the complex yet coherent nature. Salingaros [8] explored coupling between built forms and connectivity at small and large scale to unfold the complexity it offers to a formal urban setting. Following analysis of urbanist Jane Jacobs, Salingaros [8] identified geometric complexity of small scale built forms cater to a coherent urban fabric and "people-people and people-object" interaction is the core component of complexity. Often, these diverse mode of interactions develop a unique dialogue with multifunctional grains of informal settlements-known as interfaces. In same spirit of Salingaros [8], Roy [5] and Kamalipour [6], this paper will analyse interfaces to understand the underneath order of informal urbanity.

2. Case Study of Tamansari Kampung, Bandung

Bandung- capital city of West Java is the third largest city of Indonesia in terms of population size [9][10]. During the Dutch colonial period, the city was developed as a resort city for Dutch tea plantation owners and was often referred as "*Parijs van Java* or Paris of Java" [9]. However, after independence the city developed rapidly and transformed from a tranquil town to a bustling mega-urban district. Despite of having a planned urban structure, a significant part of Bandung developed out of formal planning system to accommodate influx of migrants in post-World War II period. Under new governance, large localities of the city were divided into sub-districts and each sub-district consisted of large and small village groups known as " Rukun Warga (RW)" and "Rukun Teangga (RTs)" respectively. Since then, like many other Asian cities, these under serviced urban areas have been the source of refuge and income for relatively low income population in Bandung. The case study has been conducted in one of these settlements (Kampungs) of the city.

IOP Conf. Series: Earth and Environmental Science **158** (2018) 012002 doi:10.1088/1755-1315/158/1/012002

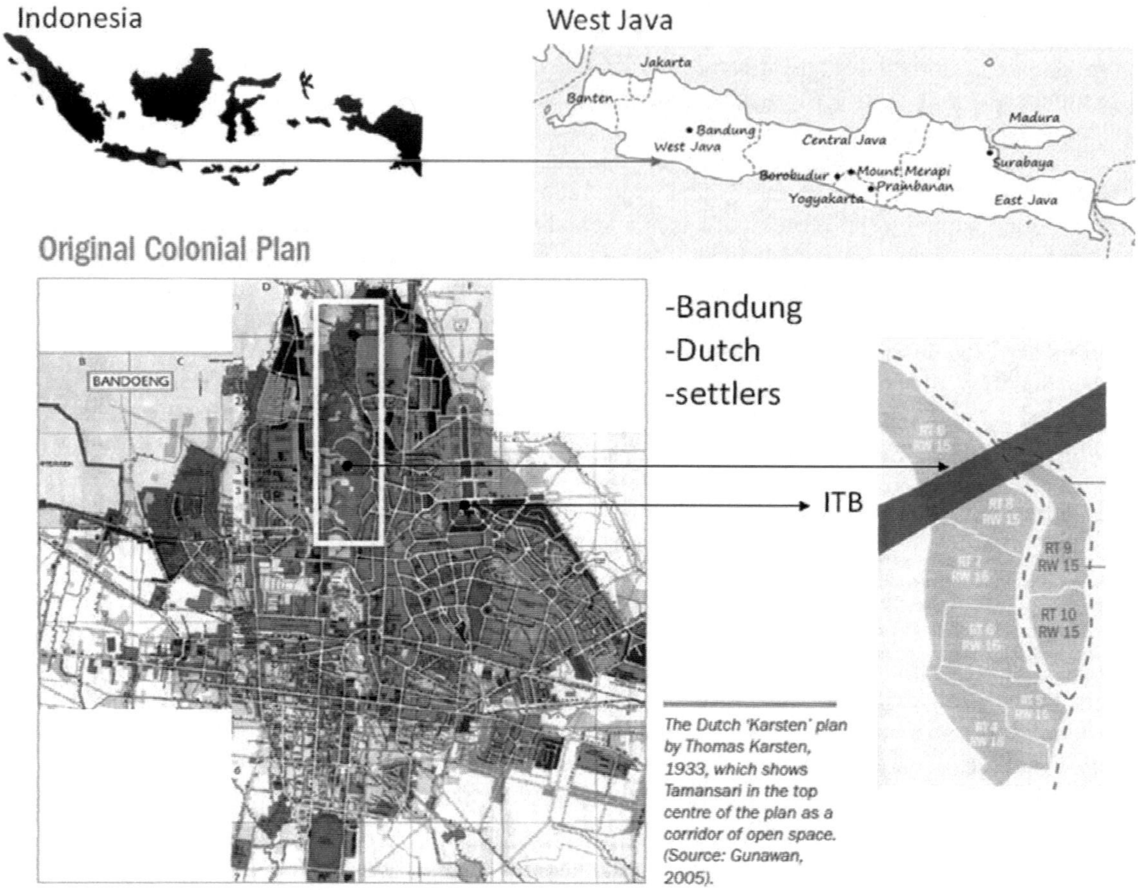

Figure 1. Research context of Tamanssari [9]

Among the kampungs of Bandung, Tamansari is positioned at the fertile valley of Cikapundung River. The precinct was formerly known as 'Lebak Gede (great valley)' and had been planned to develop as a green belt in colonial period. However, with the change in master plan to develop 'Lebak Siliwangi' as 'multi-functional precinct' instead of 'urban lungs' triggered urban growth [9]. Close proximity to city center, accessibility, increasing demand of land and housing promoted densification of Kampung Tamansari. Cikapundung River and Jalan Kebun Binatang define physical boundary of the study area. The area is wedged amongst Cihampelas Walk, ITB, few luxurious hotels and modern high-rise apartment buildings.

Both natural and manmade forces are responsible for formation and transformation of Tamansari. The Cikapundung River and irrigation channel were the initial forces behind the settlement establishment. Expansion and transformation of the area is shaped up by topography. Jalan Kebun Binatang as an important infrastructure is another driver of change.

IOP Conf. Series: Earth and Environmental Science **158** (2018) 012002 doi:10.1088/1755-1315/158/1/012002

3. An observation on morphological elements of Tamansari Kampung

At a first glance, Kampung Tamansari seems to embody the much-known outer shell of any informal settlement. It appears as a highly complex and dense area hindering the development of Bandung city (Figure 2). Low rise horizontal sprawl of Tamansari is a sharp contrast with the adjacent high rise residential blocks with large foot prints, hotels and shopping centers built under formal planning system. The development pattern of Tamansari is organic, based on the vernacular knowledge of the inhabitants. The creation and modulation of spaces are purely intuitive. But this organic character has an underneath order driven by needs of the people living here, which makes it rational.

Figure 2. An aerial view of Tamansari kampung

In order to understand the spatial logic of this settlement, an analysis on interfaces had been undertaken. Variety of interfaces could be identified in Tamansari kampung. The whole settlement is comprised of two basic elements- house forms and streets which are weaved through interfaces. Primarily, the house forms used to serve the basic need of the occupants and later, transformed into complex features in order to accommodate growing population. Likewise, the streets were used as thoroughfare for pedestrians and slow-paced vehicles. High population density demanded layers of needs and activities which eventually guided the multifunctional characteristics of streets. The streets serve the inhabitants according to their different activities throughout the day which result into an organic layout. Interfaces play a vital role in this regard. In accordance with that, this settlement has been formed to a complex yet coherent system. Therefore, it is important to have a clear idea of interfaces in order to understand the informality of Tamansari.

Here interface can be divided into two categories depending on their scale – small and large. Small scale interfaces are comprised of small elements such as, extended platform, fence, steps, low height wall, and blank wall. These small elements are used to achieve a variety of house forms and street types, according to the need of users. These elements help to create and define semipublic spaces, which act as transitional spaces between the private zone (house form) and the public zone (alleyways as streets). Here, optimisation of space usage is important as the resource is limited. Interfaces offer the flexibility to adapt the various needs throughout a day. In this way, multi-functionality being a prime character of these spaces, makes it a pattern of an informal settlement of this kind. Layers of uses give these spaces a character of complexity.

Small scale interfaces are evident in-house forms making them diverse and complex. House forms reflect the transformation process of the settlement. Previously, detached houses were scattered throughout area on large blocks (figure 3b). However, condensation started with the influx of migrants. Blocks were subdivided to accommodate rising growth and large scale urban infill took place. Built forms went through further adaptation process for optimum utilisation of space. The process is done incrementally where interfaces play a vital role having elements to set the forms forth, above and back (figure c,d,e). Here, housing growth occurred in various stages in different time phases. As the family size grows, according to the need required spaces are occupied and built. However, it falls under a formal system having a definite size of site. The same process happens in Tamansari, without any technical knowledge but using vernacular sense of space.

Figure 3. Diversity in house forms

A typical alleyway of Tamansari is similar in terms of physical form, but it is complex due to its variation in functionality. The spatial pattern of an alleyway is organic which follows needs of the users defined by small scale interfaces (figure 4 and 5). Different types of small scale interfaces support for mixed uses, in terms of the width of alley ways and dwellings, appropriate for live-work condition.

Figure 4. Diversity in small scale interfaces influencing the space quality in different parts of one typical alleyway

Figure 5. Diversity in small scale interfaces influencing the space quality in different parts of one typical alleyway

In Tamansari kampung, small scale interfaces help to create a unique dialogue between the public and private uses (figure 6 and7). Here, the flow of space is gradual to ensure a balance between privacy and socio-cultural relationship prevalent within the community. It is a best outcome of the adaptive nature of the spaces according to diverse needs. The spatial flow depicts the coherence of the entire settlement which is a dominant character of its own. In this system, if any small element of interface is being changed or withdrawn, the flow will be lost, so as the coherence.

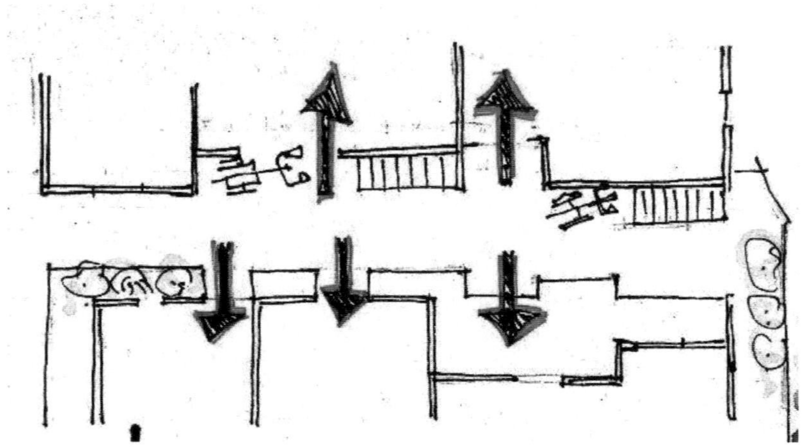

Figure 6. A plan of a typical alleyway showing flexibility and organic layout

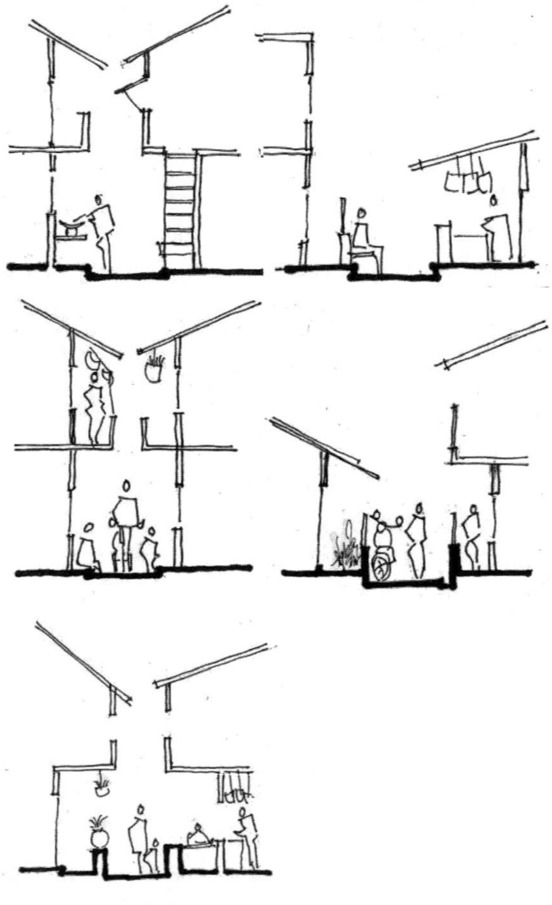

Figure 7. Different types of small scale interfaces add uniqueness to the alleyways, as well as the scope of creating diversity, multi functionality, adaptability, flexibility, fluidity and optimisation of space use. Sections through different parts of alleyways and house forms

Small scale interface is a typical feature of the settlement, whereas large scale interfaces are placed at strategic locations and act as active spaces. Connecting bridge, open space and contour often create special dialogue within and beyond the study area. Interfaces cater for diversity in the settlement by creating hierarchy of spaces. Mimicking the small-scale interfaces, the same gradual flow of space has been maintained in a larger scale.

In some parts, the public realm (surrounding areas of Tamansari) is connected to the private realm (the Kampung) with a transitional space as a bridge over the irrigation channel. Here, the bridge acts as a circulation space as well as a breathing space for the inhabitants (figure 8). In rare cases, moderate size open spaces act as interface. It acts as a transitional space between the main street and small cluster of residences (figure 9). It works as a multipurpose space such as playground, congregation space during religious and cultural festivals etc. As in some parts of Tamansari, the contour shapes up the settlement pattern, interface type is different here (figure 10). The contour acts as an interface in different levels as a physical and visual connection rather being only a physical one.

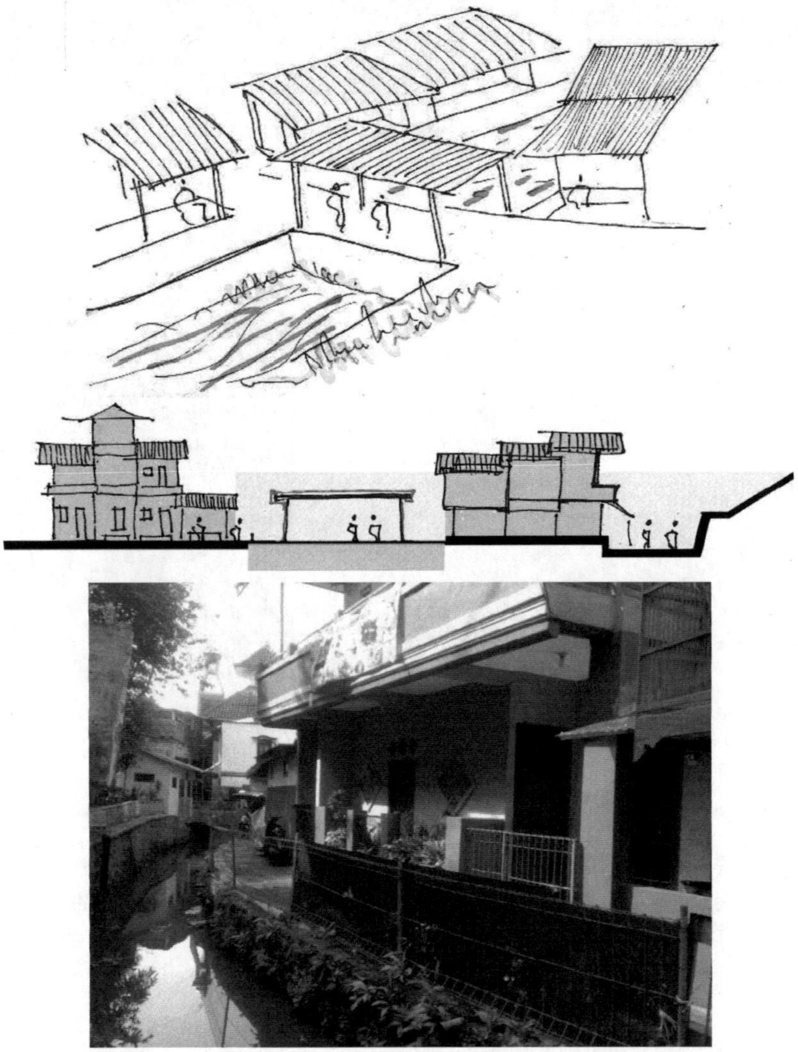

Figure 8. An example of a large scale interface as a bridge over the irrigation channel

Figure 9. An example of a large scale interface as an open space

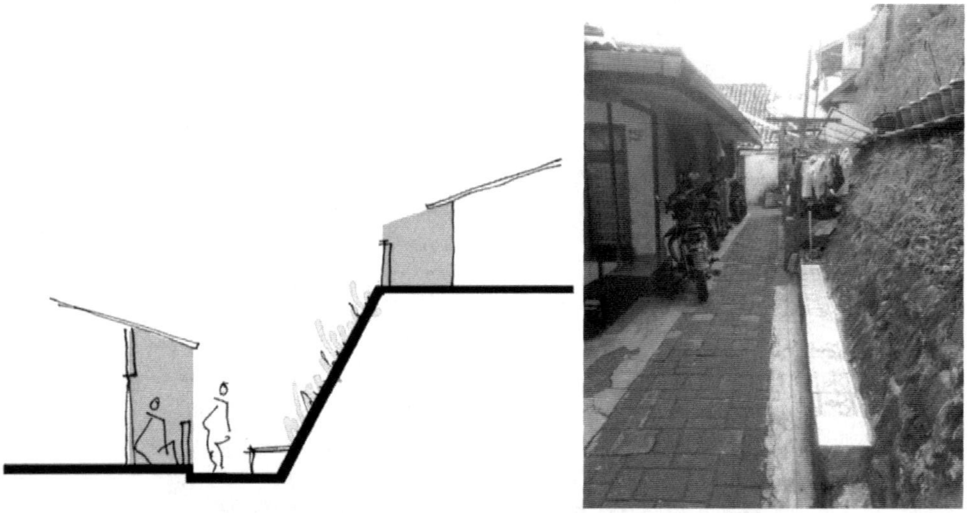

Figure 10: An example of a large-scale interface as a contour showing public and private realm in different levels

The above-mentioned characteristics of large scale interfaces offer multi-functionality of the spaces which results into a complex spatial pattern. However, the layers of uses are lesser than small scale interfaces. The large scale interfaces serve for both semipublic and public purposes of the users, where as the small scale interfaces cater only for semiprivate uses. In this way, large scale interfaces add complexity to the settlement while maintaining the coherence in a same manner as the small-scale interfaces do (figure 11).

Figure 11. An image of complexity and coherence through a mix of small and large scale interfaces in Tamansari kampung

Tamansari offers a good source of income generation where the inhabitants are able to sustain their own system of live-work condition. Interfaces play with a great deal of importance in this process. On the main street of Tamansari, the interfaces vary from set forth, set above and set forth, appropriate to house large commercial activities in one storied or double storied structures. They have steps and extended verandah over it to allow people to sit and have a small talk to let them passively contribute in the commercial activities (Figure 12a). In some cases, an extended platform acts as an interface with a floor set above. It gives a sense of enclosure which is appropriate to induce home business within the private realm of residential zones of the kampung (figure 12c). Food carts are placed fixed against blanks walls whereas (figure 12d), temporary moveable carts are drawn in pick hours along the alleyways (figure 12e). A rich variety in commercial opportunity adds different layers and complexity in Tamansari and offers a palette of choice to the users.

Diversity in spatial quality through interfaces has been achieved by negotiating with neighboring sites. Here, social relationship (for example kinship, trust, social bonding and cooperation) influences the underneath order of the settlement. Unspoken agreement prevalent within the community plays a vital role to give shape of the physical settings. Thus, analysis of Tamansari kampung's interfaces allows us to understand the socio-cultural context of this informal settlement.

The system of Tamansari kampung works on the basis of trust and social bonding within the community. Religious beliefs and cultural practices also add value to that. The occupants get the scope of personalization, which add to the diversity of this place. They get the freedom of using their own spaces according to their choices. The priority and preferences are determined by the users according to their needs. The occupants get the freedom of choice and scope of personalisation, which add value to the diversity of this place. Birdcages, tree planters, an aquarium on the extended platforms and verandas depict the different tastes of the users. They get the freedom of using their own spaces according to their choices (figure 13). This character adds richness to the pedestrian experience while walking down the alleyways.

(a) (b) (c)

(d) (e)

Figure 12. Interfaces offer variety in space quality to allow scope of small business

Many of the inhabitants are capable of moving to other parts of the city equipped with better infrastructure and services. But they choose to stay here for several years only because of the quality of social bonding prevailing here. It makes them feel safe, closely knitted and to rely on each other during their hard times. Where the formal system focuses on the physical elements first, Tamansari has been formulated through the exact opposite process. Here, the process is important than the product. Here, a set of people get engaged in the process of places making rather than someone else dictating the spatial pattern who is not a part of the community. While in formal planning system is resulting into social isolation, disrupting mental and physical wellbeing of the inhabitants, Tamansari is a great example to follow.

IOP Conf. Series: Earth and Environmental Science **158** (2018) 012002 doi:10.1088/1755-1315/158/1/012002

Figure 13. Kinship, trust and cooperation are the strengths of Tamansari kampung

4. Conclusion

Informal settlements are formed by the inhabitants within their limited resource and capacity. People who are unable to afford government services, make their habitats based on their need and necessity. Hence, diverse people with their various needs add complexity to the urban informality. It is this diversity and complexity that make an informal settlement unique and an alternative solution of complex urban issues parallel to traditional planning system. On the contrary, the formal system prioritise the rules and disciplines over people's need and individuality. The physical forms lose their very own identity which can only be achieved if importance is given to the dwellers' needs, aspirations and freedom.

Tamansari kampung as an example of informal urbanity, depicts it's own unique characteristics, as oppose to the traditional planning system. In this case, the rules are fluid and flexible which can be understood through the organic spatial pattern (figure 14). On the contrary, the formal systems have fixed rules with rigid spatial layouts. Here, the role of interface is to determine the legal boundaries of properties, where as in informal settlements interface is the place of negotiation [5]. Interface links house forms with streets. It creates the spatial flow of public-private domain. Where in formal systems, this flow of spaces is repetitive and straight forward, avoiding any kind of diversity. However, in Tamansari kampung, this flow is spontaneous and gradual, allowing scopes to create multiple functionality. This particular character makes this place diverse and a complex layout, difficult to understand at a first glance or from an aerial view.

Figure 14. Product of informal urbanism on the foreground, in contrast with the product of formal planning system in Tamansari creating the backdrop

On the other hand, the juxtaposition of uses, forms and spaces make the settlement homogeneous. Every house form depicts an identity of its dweller and it applies for every house in a typical alleyway. Thus, this variation in appearance and functionality gives a unique character to this settlement. It becomes an identity of its own where a balance is maintained without any intentional effort. In this way Tamansari kampung is working in a very harmonious way and forms an identity than the rest of the city of Bandung.

In this settlement, public- private realm is not clearly defined as it is fluid in its nature. The planning and design principles are very much place specific. It is more about social interaction and community living than only being a place of living and working in a fixed urban environment. While the formal system is resulting into social isolation, here sharing is caring is the basis of living. Overlapping of a lot of spatial components makes this place diverse and complex in its nature.

Tamansari Kampung has showed an example of a valuable source of affordable housing and income generation for different socio-cultural context and interfaces play a vital role in the process. This research takes an effort of differing with the popular belief of coupling informal urbanism with chaos and irrationality. It is rather complex in its character than being chaotic which makes the system to work as a cohesive whole.

References

[1] Dovey K 2012 Informal urbanism and complex adaptive assemblage International *Development Planning Review* **34** xx-xx
[2] Bayat A 1997 *Street politics: poor people's movements in Iran* (New York City: Columbia University Press)
[3] Simone A 2009 *City life from Jakarta to Dakar: movements at the crossroads* (xx:xx)

[4] Perlman J E 1976 *The myth of marginality: Urban poverty and politics in Rio de Janeiro* (Berkley, CA: University of California Press)
[5] Roy A 2011 Slumdog cities: rethinking subaltern urbanism *Int. J. Urban and Regional Research* **35** 223–238
[6] Kamalipour H 2016 Urban morphologies in informal settlements: A case study *Contour J.* **1** 60-75
[7] Jones P 2015 Housing resilience and the informal city in *Joint International Conference of the 3rd PlanoCosmo & the 10th SSMS*
[8] Salingaros N A 2000 Complexity and urban coherence *J. Urban Design* **5** 291–316
[9] Jones P 2016 *Unpacking informal urbanism* (Bandung: Penerbit ITB)
[10] Badan Pusat Statistik Kota Bandung 2017 *Statistics of Bandung City* (Bandung: Badan Pusat Statistik Kota Bandung)

IOP Conf. Series: Earth and Environmental Science **158** (2018) 012003 doi:10.1088/1755-1315/158/1/012003

The Hidden Structure of Organic Informal-like Settlements in Jogjakarta City: An Investigation of Socio-Spatial Relationship in an Urban Kampung

I A W Hutama

Department of Architecture and Planning, Faculty of Engineering, Universitas Gadjah Mada

Corresponding author: irsyad.adhi@mail.ugm.ac.id

Abstract. Some urban kampungs are registered as informal settlements in Indonesian cities. The existence of a kampung amidst the cities is crucial in providing an affordable access to housing for most of the city's residents. As an integral part of the city, urban kampungs face many problems such as social segregation, low and deteriorated physical quality, and exclusion from spatial plans which may worsen the residents' quality of life. Apart from this condition, an urban kampung is believed to have a strong social cohesion and resilient community which are able to persist and survive socially and economically. Towards the sustainability of the city, kampungs as part of the city structure need to be addressed spatially through proper plan and design strategy. In doing so, planners and architects must understand the unique characteristics of urban kampungs to appreciate their positive contribution to the cities. This paper is presented to understand the underlying information on the organic spatial layout of the informal kampung and its impact on social composition within the framework of lively, vibrant, and safe neighborhoods. We utilized Space Syntax (SSX) combined with activity snapshot to understand the socio-spatial relationship and then triangulated the results with a questionnaire on the residents' perception. Kampung Code, located in the inner city of Yogyakarta, was selected as the case study area as it represents the informal-like settlements. This study revealed that the complex interplays of the organic spatial layout does not significantly affect the frequency of outdoor activities. However, the presence of eight popular spaces (streets, house's terraces, local shops/taverns, small open spaces, riverbanks, guard posts, community buildings, and small mosques) distributed within Kampung Code does influence it, and it is the driving factor for the residents to engage with others in the form of social interactions. Popular places are able to accommodate residents' need for various activities from informal business, community interaction, and hobbies. Eventually, the continuous process of interaction and encounter in the same shared public places shapes the strong social connection and conducive living situation, ensuring the kampung to be a lively, vibrant, and safe neighborhood. Reflecting from this study, where organic spatial layout has little effect on social composition and encounters, planners and architects should be aware that proper spatial layout design (e.g., high integration street networks in global and local scale) could generate constant flow of both locals and passersby – enhancing the vibrancy and vitality of the public realm in the kampungs

1. Introduction

Urban kampung or kampung is a common name to address residential neighborhoods in Indonesian cities. The name "kampung" represents its unique characteristic as a traditional settlement which was built and evolved from pre-colonialization until now [1]. At the same time, rapid urbanization in the past decades led 924 million people to live in the slums or squatter settlements in the city seeking for a better life-changing opportunity [2]. This trend has a clear impact on how people sustain in the city, especially and most critically, how they reconcile the cost of living and their income at a balance. This traditional issue makes people look for both lower land value and rent housing value, yet they still expect proximity to workplaces. Hence, urban kampungs emerge as a spatial unit of human habitat which is arguably affordable for the majority of the city's residents to access the housing market. However, the continuous growth of an urban kampung without a clear plan and direction, and the fact that they are mostly erected in unclear or disputed land tenures [3] register urban kampungs as informal settlements.

Urban kampungs has their own characteristics. Commonly, it is located whether near riverbanks, railways, traditional markets, or strategically near to city's central business district. The built-up environment is worse, with low building material quality, inadequate open space due to overcrowding, sanitation issues, sporadic and irregular street layout, and secure tenure. These conditions apparently meet the five criteria for slums and informality set by UN-Habitat [4]. In some parts of a big city, urban kampungs also suffer from spatial and social segregation from older policies and spatial plans, leading to the deterioration and deprivation of its physical quality. However, apart from its poor physical quality, urban kampungs are believed to have a vibrant public life and strong social connection [3,5,6]. This is the key asset for people living in the kampung to sustain and satisfy their life.

The important role of an urban kampung in shaping a prosperous and livable city cannot be neglected since the urban kampung itself is an integral part of the city as a residential neighborhood. United Nation Development Group [7] stated that the post-2015 development agenda focused on the complex issue of informal settlements in urban areas which needs to be addressed through multi-stakeholder participation, building social inclusion, empowerment, and engaging in local partnerships. Urban kampungs now need to be retouched by urban planners and architects to improve their quality of life. They need to learn comprehensively from the Kampung Improvement Program (KIP) which has been established to improve and upgrade kampungs for about 36 years, but deemed as failed project [8]. Many experts indicate that after the project completed, people in the urban kampungs were not capable to maintain and conduct self-improvement in the long-term, resulting on the gradual deterioration that worsened the quality of life. Regarding this issue, livability, vibrancy, and safe neighborhood hence must be set as a goal for a new renewal and upgrading program.

Understanding the unique characteristics of kampungs can start from analyzing the underlying information from spatial layouts and the social composition to formulate the most proper design strategy for kampung improvement programs. To achieve this, typo-morphological analysis tools such as Space Syntax have been used in the various projects to investigate the physical and social qualities of urban space [9–12]. In a specific context like informal settlements, space Syntax has also been utilized to describe and evaluate the existing spatial layout and design proposal for adaptive regeneration scenarios [13,14]. From all these studies, the findings suggest that the design of the spatial layout was crucial in shaping a vibrant public life. However, what if the already saturated urban kampung with its organic spatial layout want to achieve a lively and vibrant neighborhood? Finally, this study was conducted to reveal how the relationship between activities and organic spatial layout of informal settlements like urban kampungs work in shaping the condition that is believed to be a vibrant and lively kampung.

2. Literature review

2.1. *A common space forms the social character of the kampung*

Typical urban kampungs in Indonesia possess a unique spatial character as well as the community living in it. Shared spaces are an inherent spatial feature of the kampung where formal-informal interactions occur, creating a smaller degree of private space and creating more semi-public to public spaces [8,15]. For example, private spaces such as houses limit access and are strict to visitors, while public spaces such as street space, parks/fields, and community buildings give the residents equal opportunity to access and use the place. Further, semi-public spaces are generally formed by rules and the consensus of the people in perceiving the common spaces [16]. In the kampung, such places can be the taverns, terraces, yards, and local shops. This transition zone between private and public is essential to promote long duration activities and social interactions [17:p.103].

Kampungs in the heart of the city have a similar characteristic. The most common one is situated along the riverbanks where the status of the land is unclear. Kampungs also have higher building density (>85% are built up), higher population density (100,000 per km^2), low housing quality with traditional and concrete structure [5]. Malek et al. [18] added that the too-high building density left a serious shortage of open space which functions as a media of social interactions. The pressure of density and the needs of interaction among neighbors create an adaptation of space into acceptable public spaces. Empirical study of kampung Ratmakan by Rahmi et al. [15] showed that communal places such as alleys/street spaces, local shops (*warung/kios*), community buildings (*balai warga*), public baths (*MCK komunal*), and mosques (*langgar*) are typical public spaces that can be found in any urban kampung. These common public places (see table 1) enable kampungs to retain continuous social activities, which later strengthen the social cohesion [16] and attachment to the place [19].

Table 1. Common Social Spaces in the Urban Kampung [15,20–22]

Common places in the kampung	Feature and Character	Description
Alleys and Pathways	Organic and spontaneous, street widths vary from wide to narrow	Pathways and alleys are directly connected with house terraces. As the majority of houses have no fence or gate, pathways and alleys form the transition zone from public to semi-public space which is common for dwellers' spontaneous social interaction.
Local shops and taverns (*kios/warung*)	Informal local economic activity, Home Base Enterprise, semi-permanent construction	Shops and taverns not only supply the basic need of dwellers but also serve as a social hub (to chat and discuss). These particular places are recognized to convey longer social interactions and building the symbolic identity of the neighborhood. [16].
Public bath and toilet	Communal use, mostly spotted in un-exposed space	Sanitary facilities such public bath and toilets are unique properties of high-density urban kampungs, as such they stimulate social relations at particular times of the day. Such a place is shared by 10 to 15 households where, for example, women use this place for washing clothes.

Social facilities (community building, hall, guard post, and mosque)	Communal use, symbolic representation of the community	Social facilities in urban kampungs obviously play an important role in accommodating social interaction as well as preserving the social system that binds the community [20]. Community buildings and halls are used for meetings among the community and other purposes. Meanwhile, guard posts are used for keeping the kampung safe.
Field and open space	Limited in size, part of residual space, and heterogeneous form	Open spaces and fields are limited in the kampung with different forms and typologies, yet they still function as a prominent social medium. Open spaces are used by dwellers for diverse activities such as cooking, selling food/breakfast, drying clothes, and especially social interaction.
Riverbank	Linear open space	The riverbank is the only natural feature for some kampung. This place is able to accommodate social and optional interaction ensuing to signify the place attachment of the kampung [19]

2.2. *Space Syntax as a tool to analyze underlying spatial layouts and their implication on social phenomena*

Space syntax (SSX) is method developed from a robust theory about the topology of space networks (spatial structure) and its implication towards social composition. The key aspects of SSX are the understanding of reciprocal interaction between the configuration of space in the system and multiple variables associated with social phenomena, including pedestrian movement. [9,23]. SSX adopts a graph-theoretic method to mathematically model how urban spaces are integrated or segregated using axial maps. Continuous open spaces are represented by individual axial lines, which are the longest visible lines cutting across urban void (e.g., street networks and open spaces) [24,25]. Based on this analysis, syntactic measurement can produce connectivity, integration or depth, angular/choice, and others. Through SSX, various social data can be added to perform further correlation analysis revealing the potential use of space and its implication.

Several SSX studies revealed that there is a high correlation between spatial layout and number of movement [9,26], land use density [27,28], crime and anti-social behavior [29,30], co-presence and encounter in street space [31]. Previous SSX studies then become a postulate for this paper such as a relatively high integration in both local and global scale generates a continuous flow of people and encounters for both locals and a mixture of visitors. Secondly, borrowing from the outcome of the first postulate, a significant number of people on the street creates a natural surveillance, reducing the potential for robbery, crime, and anti-social behavior. Moreover, highly integrated and connected street space compels businesses to run as the movements are rich and continuous.

When it comes to an organic spatial structure like urban kampung, one may think that the causal correlation is slightly biased, and thus it cannot robustly interpret the social phenomenon occurring in a disorganized spatial network. A research by Hutama [19] suggests the low correlation resulting from SSX measurement between integration and outdoor activities, implying that the spatial layout of the kampung is not always significant in promoting social activities. Socio-spatial interplays thus need to be studied comprehensively involving several factors like perception and place attachment.

2.3. *Spatial perspectives of livable, vibrant, and safe neighborhood*

To start with, livability, vibrancy, and safety is a condition wherein its establishment is partly associated with the quality and character of a space. Space with different settings produces different responses and behaviors to adapt with it. According to Gehl [17], for example, spaces installed with lovely street furniture such as benches and fountains attract people to sit or talk. Meanwhile, disorganized or empty spaces only left people to just walk away or misappropriate activities. Van Nes & Rueb [11] said "human behavior takes place in space, and the spatial layout shapes the opportunity for people to interact and utilize the space. This perspective can argue for the establishment of those three conditions.

Reciprocal interaction of man (activities) and space prevails in different urban scales [32]. On the micro scale level, design strategies such as organizing the building facing to the road creates control over street space as an active public space [17,33]. Moreover, planning a different building use in an accessible walking distance and reshaping the ground floor as a shopping street type building generate continuous activities nearby. On the macro scale level, designing a good spatial layout promotes the constant flow of people in the space. In this regard, Gehl [17] and Jacobs [34] signified that well integrated and well connected spaces are a key element in generating and encouraging encounters. This explains to some extent that a well-connected and integrated street space was naturally more favorable for spontaneous social interactions [31]. This combination between certain design of the spatial layout and mix-use evidently shapes the livability, vibrancy, and safety of the residential neighborhood [11,13,31].

Safe neighborhoods are associated with the feeling of security from disturbances, anti-social behaviors, and loss of belongings due to thievery. In a residential neighborhood, this feeling is formed from the conducive social connection among residents which acts as social control and warns against inappropriate behaviors [35]. Recent studies also reveal that safe neighborhoods are associated with the design of the built-up environment. Dense building blocks facing the streets and active ground floors for local business are always eyes on the street [34,36]. Everyone can do activities in outdoor spaces by being watched by the others. This induces natural surveillance on street spaces, thus forming a sense of safety in the neighborhood [29,33]. As such, the degree of safety can be measured from different perspectives of social cohesion, urban layout, and constant flow of activities on the street space.

3. Identification of case study

Kampung Code is an informal settlement which administratively consists of four neighborhoods of Tukangan, Gemblakan, Cokrodirjan, and Juminahan. The kampung code is registered as a slum based on Mayor of Yogyakarta Decree 393/2014. In terms of location, Kampung Code is situated strategically in proximity to the city center and sits along Code River (see Figure 1). As an informal settlement, Kampung Code has both high population and high building density wherein 10,059 inhabitants (2596 households) are housed within 25,69 hectares. In terms of occupation, 33% of dwellers were self-employed/entrepreneurs, 32% were employed by others, 12% were retired, and others were working seasonally. In addition, dwellers in the kampung were poor with a monthly income of under IDR 2 million and only 7% made a living of above 5 million. Based on the survey, majority of the residents are working in the informal economic sector nearby the kampung, taking advantage of its strategic location near the city center of Yogyakarta.

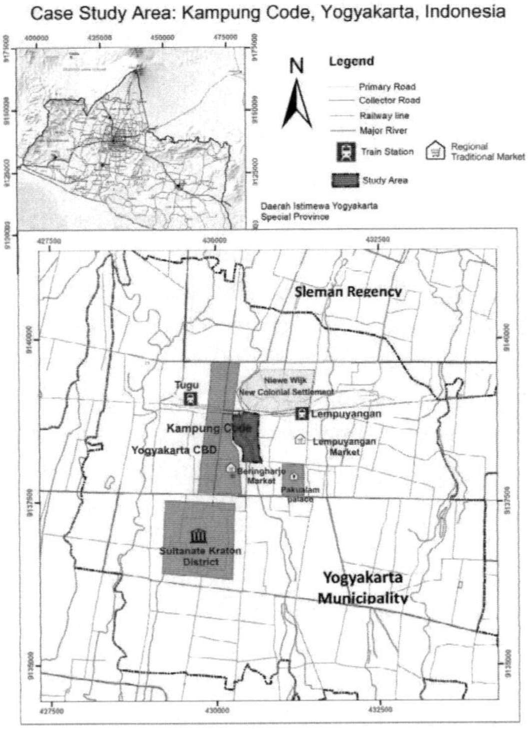

Figure 1. Case study area: Kampung Code (red) situated strategically in the city center. It is close to several locations such as the Kraton (royal palace) and Malioboro

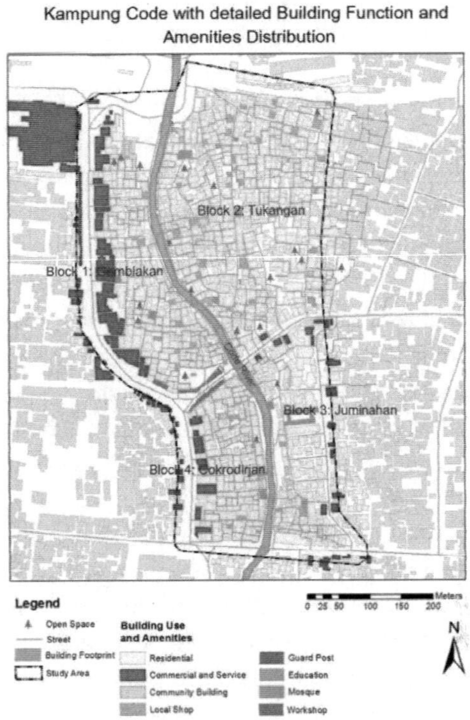

Figure 2. Map of building function shows different uses of buildings, with residential being the dominant use

Spatial properties of kampung Code are composed by organic spatial layout, irregular street width ranging from wide in some segments and narrow in others, and inadequate open space due to a too-high building density. Irregular building orientation and size. Some buildings' front doors do not face the road. Kampung Code is not a solely mono-function neighborhood, but it is diverse in terms of function. Although the majority of buildings were residential, we still can find commercial and service buildings (at the left edge of the kampung), local shops/taverns, schools, craft workshops, and mosques that can be reached within walking distance.

4. Data collection and method

This paper attempted to justify the hypothesis mentioned in the introduction by understanding the socio-spatial interplays in the informal-like settlement of urban kampung. How the organization of organic space layout in kampung affects the social interaction that frame a lively, vibrant, and safe community. In order to comprehend that complex relationship, the study area (Kampung Code) should be analyzed by considering the micro (activity and use of place) and macro (configuration of space) relationship [29]. The case study was selected as a method to understand this complex relationship through empirical observation in the field [37]. We also utilize a mixed-method approach to elaborate the varied information crossing over from qualitative measurement of spatial configuration to the social life of the kampung [38]. In this study, the axial base map model of kampung code was produced to enable the identification of the configurational value of spaces. Several SSX measures are then selected to perform this analysis. Parallel with SSX, activities were observed and registered on the map. Subsequently, the values of SSX measurements of each street space in the kampung were correlated with the sum value of activities registered on each street space with Spearman correlation. The result is then elaborated with the interview to withdraw further findings.

4.1. *Space syntax measures for analyzing the spatial layout*

SSX is a method and tool which enables the analysis and interpretation of the configuration of space and social variable. The aim of SSX is to understand how the continuous urban void that are connected to each other within the system can be calculated to interpret various forms of social phenomena [31]. Syntactic measures are the core analysis tools in SSX. In this research, we use SSX variables such as connectivity, local integration, and global integration. Connectivity measures the number of street spaces that are directly connected to other street spaces within one step. The principle is, the higher the connectivity value, the higher connectivity of street space it has to access another space in the immediate locality [39]. Integration, both local and global, measure the average depth of each street space to all other street spaces in the whole network. The lower number of street spaces represented by axial lines and fewer turn changes in the network system, the more integrated (accessible) the network system becomes [9,40,41]. To measure local integration, R3 value was chosen as the street space can be assessed in three turn changes from other space.

To perform SSX analysis, an Axial map is drawn manually in GIS environment using the AXWOMAN plug-in [42] by stretching the longest visibility lines from an arbitrary start-point to end-point in street space. This can be done with the assistance of a high-resolution image superimposed in the background of the worksheet. Once the axial map was finished, AXWOMAN is then used to execute SSX analysis. The syntactic results were automatically saved in the attribute table for each street space (axial lines). Three syntax parameter values (connectivity, global, and local integration) were visualized in gradient colors (reddish to blue). The SSX results for each variable are: Connectivity (Min: 1, Mean: 3.45, Max: 23); Local integration/R3 (Min: 0.05, Mean: 2.06, Max: 5.38); Local integration/R (Min: 0.61, Mean: 0.94, Max: 1.60) (see Figure 3 for results in maps).

4.2. *Activity mapping and questionnaires*

Observation and activity snapshot were employed to record the activities of both individuals and groups in the kampung. Referring to Can & Heath [31], this approach is effective to document all activities through photographs in the public realm where some activities could prevail longer and others spontaneously happen in short duration. For example, people working and serving food in the tavern, children playing football/kite on the streets, and group of people mingling might prevail longer and it was easy to identify. However, activities such as men encountered on the street or open space and women chatting while doing something may only happen in a short duration in random spaces. All activities and their specific locations are then registered on the A3 map during fieldwork. Afterwards, the paper maps were digitized in ArcGIS to produce an activity mapping which is spatially informative to study the behavior of residents in the various setting of space [43].

To effectively understand the socio-spatial behavior of residents in kampung, observation and activity mapping with snapshots were conducted at three different times a day which is in the morning (07.30-08.30), midday (12.30 – 13.30), and afternoon (16.30-18.00). These activities were also conducted on one weekday (Tuesday) and one Sunday to fairly create a whole picture of the daily routine in the kampung. Due to the complexity of the built-up environment in the kampung, the researcher was assisted by several surveyors and set them in the certain places within a radius of 50 m. This observation has been conducted in two steps by paying attention to some limitations. First, observations were carried out in the determined location for about 10-20 minutes to be able to take snapshots of stationary and moving activities. The next step was, walking to nearby areas to take snapshots of the activities occurring in the remaining spaces on each zone. Second, the surveyor was directed to pay attention to peculiar

linear open spaces such as the riverbank and main streets/pathways where people did several activities individually or in a group. The limitation of this method was the possibility of an un-exposed activity that was not recorded due to the small-time gap when the surveyor walked to a nearby location [22]. Activities and SSX results were correlated to give a profound explanation on how organic spatial layout influenced outdoor activities in the kampung. Prior to doing so, activity and SSX results must be recorded in the same attributes table. "Generate near table tools" were used to associate the activities point data with the nearest axial line (new data attribute). The sum operation was used to aggregate the number of people to the nearest axial line where activities take place.
Questionnaires with closed and open of question types were addressed to the heads of households (HH) and some family members to inquire the individual's perception towards the kampung in terms of lively, vibrant, and safe community. In the study, the researcher employed quota sampling (probability method) based on the division of four neighborhoods with different population densities. We took about 160 (about 7%) samples of HH with 2-4 family members. The questionnaire was divided into five chapters obtaining different types of information. The first and second chapter gathered information about the socio-economic characteristic of dwellers, wherein more than 70% are poor and mostly working in informal economic sectors. The third chapter explains the use of dwellers' habit/preference to use eight popular places in the kampung. The fourth chapter depicts the livability of kampung, the social cohesion, and followed by the perception of safety and cleanliness in the kampung.

In this study, using a mixed-method approach was useful in terms of inferring the result from different perspectives of data. The correlation results of activities and SSX triangulated with residents' perception could justify whether the spatial configuration of kampung Code impacted the lively, vibrant, and safe community or there was another reason that residents could reveal through the interview.

5. Result and discussion

5.1. *Types of activities and use of informal space in urban kampung*

The daily life of the urban kampung is shown in various activities. Through snapshots and observation, we found 45 varieties of activities which can be classified into necessary (38%), social (40%), optional (18%), and religious activities (4%). Necessary activities in the kampung are mainly related to make subsistence (e.g., selling daily needs in local shops, selling food/drinks in the taverns, and making handicrafts in a workshop). Spatially, work-based activities mostly happened on the streets/alleys and on the edge of study areas where passers-by travel frequently. Social activities (chatting, playing, doing sport) were diverse in terms of age and gender. With regards to Gehl [17], social activities were found in most outdoor spaces of the kampung and it can spontaneously occur in streets/alleys, semi-private spaces like terraces, taverns, riverbank, and irregular shapes of open spaces. However, there were formal-informal activities in kampung. Residents who were doing an activity related to work are usually accompanied or encountered by neighbors and they spontaneously have a conversation on the premises. Hence, this can be overlapped as social activity, explaining the nearly same frequency of social and necessary activities. Further, self-oriented or optional activities (e.g., fishing, gardening, reading, and contemplating) had a random spatial pattern as they tended to occur based on the personal sentiment on the place. Religion-related activities were counted very few and they occurred at specific places such as a small mosque (*langgar*).

Most of the spaces in the urban kampung are composed of intertwined irregular solid and void structure. We observed eight places in the kampung which attract activities. Those places are house terraces, streets/alleys, irregular open spaces (e.g., field), riverbank, community buildings, guard posts, and tavern or local shops, and mosque. Riverbank, house's terraces, streets/alleys, and community buildings are the places which have a high frequency of social activities among neighbors [22].

Particularly for Kampung Code, the riverbank is a place which allows dwellers to linger with neighbors and do various activities (chatting, fishing, sitting, or sightseeing). As Hickman [16] highlighted, this type of place can be considered as a "third place" where the formal open spaces were not well designed in most kampungs. House terraces are common semi-private places in the kampung which plays the important role of contributing to good social ties among neighbors. Residents may use the terrace to interact with other neighbors even without the presence of the owner. Streets/alleys and taverns are important social features in the kampung that not only connect the places but also dwellers. Similar to Rahmi et al. [15], those places have been identified as the community's tangible assets that serve as a social medium in the urban kampung.

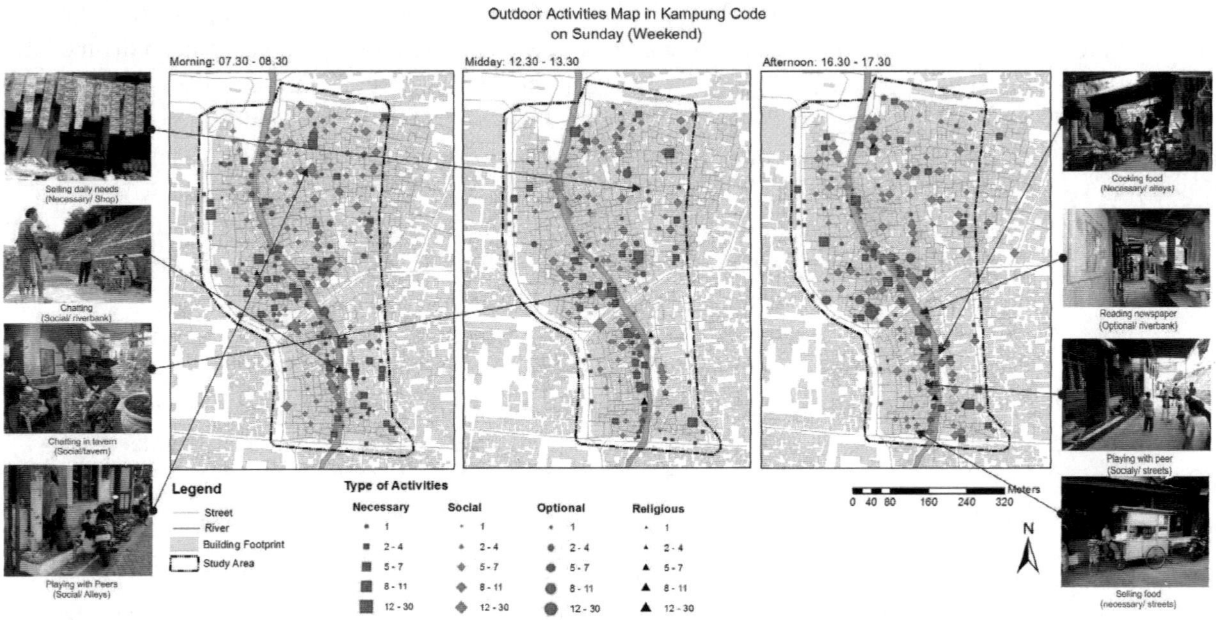

Figure 3. Map shows different activities and use of space through the day on the weekend. During weekend, outdoor activities appeared most compared to weekdays in eight popular places in the kampung.

5.2. *Correlation between activity and urban kampung's spatial layout*

The spatial configuration of the kampung can be described as organic and chaotic, without a clear pattern like a grid, curve, or linear. Spatial layout of the kampung was formed spontaneously through a gradual process of residing as dwellers come and erect buildings in the remaining land. As a result, there were residual spaces in between buildings with irregular sizes and forms. The continuous residual open spaces to some extent serves as shortcuts or channels that connect one place to other. In SSX measures, connectivity value scores 3.45 (avrg) and 23 (max) meaning that residents can move from one place to another with at least three different alternatives of streets/alleys. Only a few street segments (within the kampung) have a higher connectivity value, allowing to absorb potential of the social composition into those street segments. However, as it turns out, the activities are not fully attached to the higher street connectivity, instead clustered into several popular spaces in the kampung. In this case, only necessary activities took place in the higher connectivity street segments as a lot of passersby spotted.

Spatial analysis of integration shows a relatively high value at the edges of the kampung. This makes sense since the street at the edges like Jl. Mataram, Jl. Juminahan, and Jl. Mas Suharto in the middle has the higher street hierarchy in the city level, connecting the kampung directly to collector street network in the Central Business District (CBD) of Yogyakarta. Similar with Majumdat, et al, Majumdar, Sliuzas,

IOP Conf. Series: Earth and Environmental Science **158** (2018) 012003 doi:10.1088/1755-1315/158/1/012003

Munshi, & Brussel (2009), streets at the edges appeared to have more commercials and services. At the local scale (R3), highly integrated street spaces are located within the kampung showing the internal main structure of the kampung where the residents composed a mutual interaction.

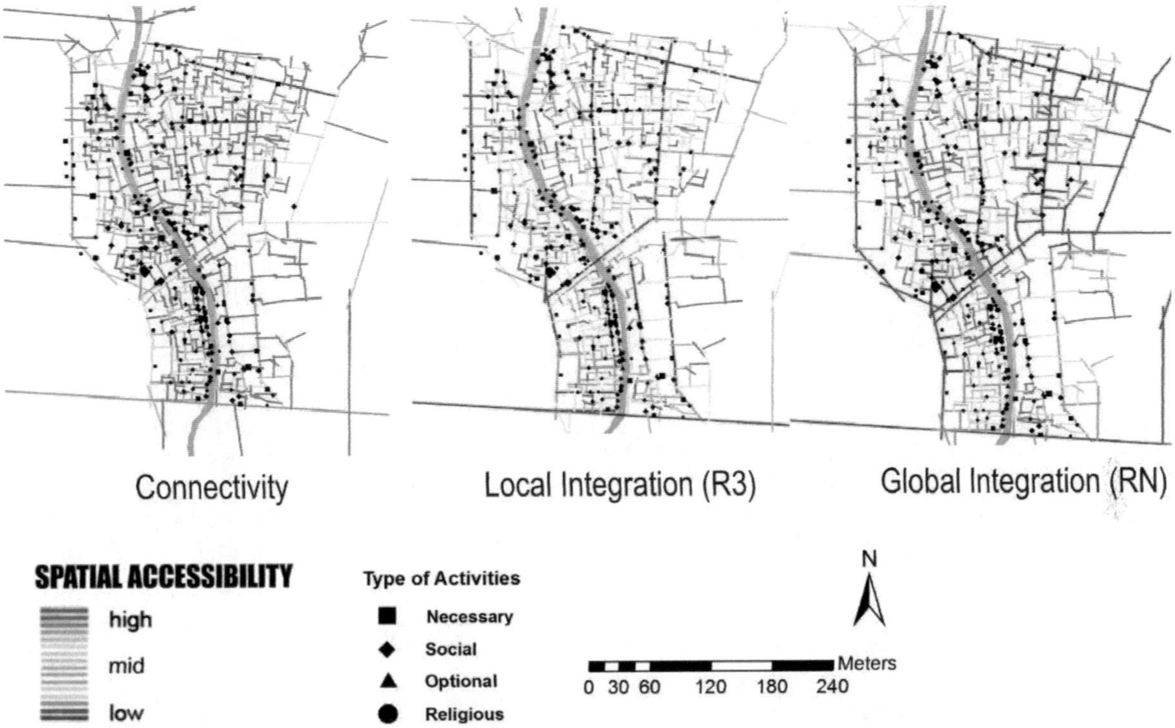

Figure 4. Map showing various spatial measurement (SS) of Kampung Code combined with activities.

Table 2. Correlation of Activity Snapshot and Space Syntax Parameters

Activity snapshots and SSX - R^2 Linear	Connectivity	L.Integ	G.Integ
Necessary (walking and working)			
Tuesday	0.211	0.121	0.006
Sunday	0.193	0.142	0.013
Social (interaction and playing)			
Tuesday	0.221	0.208	0.045
Sunday	0.257	0.238	0.066
Optional (relaxing and sitting/standing)			
Tuesday	0.175	0.102	0.006
Sunday	0.188	0.114	0.027

Spatial layout of the kampung governs the number of outdoor activities, particularly on the composition of social activities. In the correlation result, social activities are more associated with local integration (R3) than other types of activities (see table 2). It means, the higher the accessibility of the street network at the local scale, the more there are social activities. Despite the lower correlation results, the survey showed the spatial layout affects the number of people encounters on the street, initiating social activities. In this study, the relationship between organic spatial layout and activities in the kampung agreed with [31,44,45].

5.3. *Resident's perception on lively, vibrant, and safety in the kampung*

Activities in the kampung are diverse and continuous throughout a day (see **Figure 5**). On a weekday, necessary activities gradually went down from the morning to afternoon. Conversely, social activities reached its height in the afternoon as the residents returned home from work then socialized with their neighbors. During midday, outdoor activities are slightly decreased particularly for social activities due to uncomfortable heat. Dwellers tend to seek shaded space like house terraces, taverns, and guard posts (*cakruk*) to have social interactions. On the weekday, particularly in the morning and midday, necessary activities were spotted more frequently than that on the weekend; meanwhile, in general, social activities were spotted more frequently on the weekend (morning and afternoon). The dynamic and continuity of activities from morning to afternoon induces the sense of a vibrant place. It is similar with the study about people, space, and activity by Gehl [17], Jacobs [34], and Mehta [46] who stated that the constant flow and presence of residents from different genders and ages in public spaces throughout the day indicates a vibrant and livable neighborhood. This condition was signified by the fact that the mixture of building use like local shop/taverns, community buildings, guard posts, and mosque generate different types of activities [6,47]. The high density of buildings in the kampung and inadequate formal open spaces pushed the residents to adapt and use semi-formal and informal space, promoting a cluster of several activities near them.

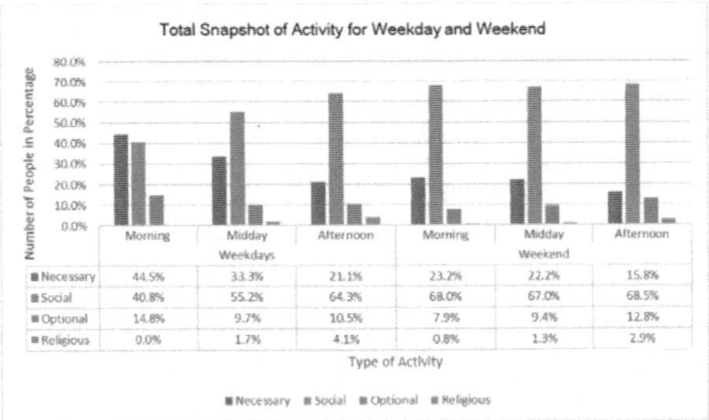

	Morning	Midday	Afternoon	Morning	Midday	Afternoon
		Weekdays			Weekend	
Necessary	44.5%	33.3%	21.1%	23.2%	22.2%	15.8%
Social	40.8%	55.2%	64.3%	68.0%	67.0%	68.5%
Optional	14.8%	9.7%	10.5%	7.9%	9.4%	12.8%
Religious	0.0%	1.7%	4.1%	0.8%	1.3%	2.9%

Figure 5. Comparison of total activity snapshots for Weekday and Weekend.

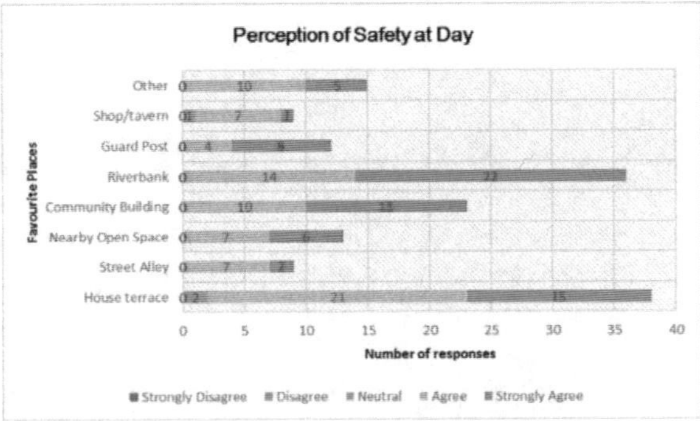

Figure 6. Perception of Safety in different popular places in the kampung.

Regardless of the condition of economic and physical poverty in the kampung, residents perceived that the kampung is safe in terms of criminal activity (theft/robbery) and anti-social behavior. Quoted from one of the residents, "*people just park their motorcycle in the alley all day and it's never gone – only a small thing happened such as thievery of pet birds*". This situation is apparently signified by the strong social cohesion which the community later established a social system such as patrol at every night (*Ronda*) and massive cleaning (*kerja bakti*). Residents spent at least one hour or more and have visited each other in popular spaces to have conversations. This routine social interaction and visit among neighbors play a pivotal role in promoting and strengthening social capital as an asset among other spatial entities in a kampung. As studied by Rahmi et al. [15] in locally urban kampung and Gilroy [48] in typical settlement of western society, social cohesion is apparently an image of the residential neighborhood where people can feel the existence of various activities in public realm. In this study, the kampung is empirically proven as a lively and vibrant human settlement.

5.4. *Socio-spatial interplays to promote lively, vibrant, and safe community in the kampung.*

We have critically described the activities and use of space in the kampung, the spatial layout, and the residents' point of view on livability, vibrant, and safe communities. This study shows the spatial layout is not statistically significant in affecting the frequency of activities taking place in all outdoor spaces in the kampung. Particularly for social activities, the number of co-presence of residents are less counted in relatively high locally integrated spaces and become lesser in the lower locally integrated street segment. It is explained by the organic spatial layout creating some convex spaces in which one needs more turns to access this space. The more organic the spatial layout, the higher possibility of the depth of space from whole system. Hence, theoretically, based on the evidence from space SSX studies [9,11,12,49] the outcome of the spatial layout (high integration on both local and global scale) which induces the relatively high flow of movement is not really relevant with the condition where a spatial layout was organically developed like in Kampung Code. This finding was signified by Van Nes et al [33] where the pre-war neighborhood was registered to be less lively and vibrant due to the less number of residents appearing in public spaces. In that case, pre-war neighborhoods shared the same character as urban kampung with organic spatial layout resulting in relatively low integration value and high depth of space.

The eight popular spaces and mixed building use for housing and local business are the factors that promotes the livable and vibrant kampung. These places are able to accommodate dynamic of interactions through the day. Particularly for the riverbank, this place is compelling for residents to stay outside, interact with each other, and satisfy their emotion towards the daily stressor of living in an overly dense building. Socio-spatial interaction then can be seen through the clear pattern of outdoor activities which clustered nearby the popular spaces and the amenities within the kampung (see **Figure 7**).

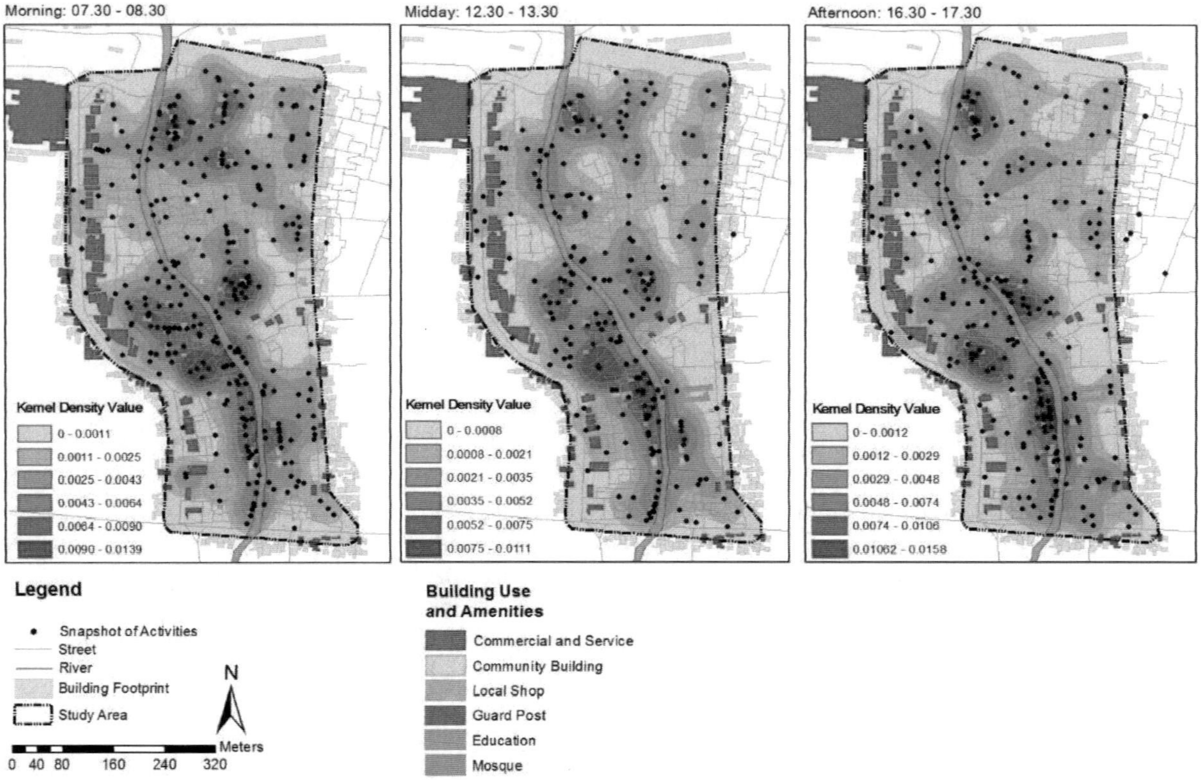

Figure 7. Kernel density map of outdoor activities creates a clear pattern of socio-spatial interaction where outdoor activities tend to be spotted nearby the eight popular spaces and amenities within the kampung.

Socio-spatial interplays in terms of safety of the community in the kampung reasonably contrast with Hillier & Sahbaz [30] and Van Ness & Lopez [29] who stated that the lower degree of integration, both local and global, reduces the number of people engaging in the street that might serve as natural surveillance. The street space with low natural surveillance at the local scale tends to be affected by thievery and anti-social behavior. Empirical evidence shows the correlation between types of activities and local integration in the kampung were low (see Table 1) meaning that the poor organic spatial layout of the urban kampung generated small chances of co-presence on the street space, which can worsen the safety issue. However, the kampung is perceived as safe by the residents themselves. This apparently departed from the social trust among neighbors as they have been residing in the kampung for 43 years on average. This resulted in the knowledge and understanding between neighbors such as their occupations, spatial behavior in the kampung, and their family profile. The safety in urban kampung hence was the product of strong social cohesion and the daily routine of social interaction.

6. Conclusion

This research suggests that kampungs in the inner city of Yogyakarta are livable, vibrant, and safe. This condition is perceived by residents as the outcome of good social cohesion, continuous outdoor activities, and conducive living neighborhood from thievery and anti-social behavior from youngsters. The organic spatial layout of the kampung is not strongly associated with the frequency of various outdoor activities, but the presence of eight popular places distributed within the kampung does encourage the residents to have the opportunity to use and share it for various activities. The streets, local shops, house terraces, guard posts, community buildings, riverbanks, small open spaces, and mosques thus are the popular places in the kampung that act as the key role in binding and accommodating social interaction. In the end, we believed that the flexibility to walk through the disorganized multi-connected street network within the kampung, the presence of humane popular places, and active ground floor for local business activities (*kios* or *warung*) has contributed to both the active use of public spaces and the strong social connection between residents.

There are key outcomes that enrich the analysis and recommendations from this research. Space Syntax (SSX) embedded with social data from activity snapshots can be used both for analysis and consultancy tools for urban planners, architects, and practitioners who are responsible for making positive changes towards informal settlements like the urban kampung. This approach is robust to understand the underlying feature of spatial layout and its impact on social phenomena like the frequency of outdoor activities. Instead of focusing on repairing the organic spatial layout of the kampung, which indeed needs much effort and cost, we can design the space and place to elevate the values and the community in the kampung. Designing some of the popular spaces especially for riverbanks, community buildings, and small open spaces with proper street furniture and little ornamental touches like wall paintings will not only attract residents but also retain them longer in the public space. These efforts are to appreciate the presence of the public realm as the common ground and lifeblood of the kampung. For further and long-term upgrading or development, let one not romanticize the condition that emerged from the scarcity of proper open public spaces and well-developed sense of community in the kampung. But, the importance of spatial layout (high integration both on the local and global scale) for generating the continuous flow of both locals and passers-by must be considered as this significantly contributes to the lively, vibrant, and safe neighborhood. Finally, the design process in every scale of the upgrading program or residential development must engage the local community to consult on what they need and what it ought to be to create positive change in their lovely kampung or neighborhood.

Reference

[1] Funo S, Yamamoto N and Silas J 2002 Typology of Kampung Houses and Their Transformation Process a A Study on Urban Tissues of Indonesian City. *J Asian Archit Build Eng* **1** 2 p 193–200.
[2] UN-Habitat 2003 The Challenge of Slums
[3] Raharjo W 2010 Speculative Settlements: Built Form/Tenure Ambiguity in Kampung Development (University of Melbourne)
[4] UN-Habitat. State of the World's Cities in 2006 [Internet]. 2006. 1-108 p. Available from: http://mirror.unhabitat.org/pmss/listItemDetails.aspx?publicationID=2101
[5] Tunas D 2008 *The Spatial Economy in the Urban Informal Settlement. International Forum on Urbanism* (IFoU); 2008. 112 p.
[6] Doherty G and Lino e Silva M 2011 Formally informal: Daily life and the shock of order in a Brazilian favela. *Built Environ* **37** 1 p 30–41.
[7] United Nations Development Group 2014 DELIVERING THE POST-2015 DEVELOPMENT AGENDA: Opportunities at the National and Local Levels. 2014;
[8] Sastrosasmito S 2009 Compact Kampungs: Formal and Informal Integration in The Context of Urban Settlements of Yogyakarta, Indonesia. *J Habitat Eng* **1** 1 p 119–34
[9] Hillier B and Hanson J 1984 The Social Logic of Space. First Edit 1-280 p (Cambridge: Cambridge University Press)
[10] Majumdar A, Sliuzas R, Munshi T and Brussel M 2009 *Exploring the Relationship Between Urban Roads, Urban Form and Economic Activity in Ahmedabad Using Space Syntax* International Workshop on GIS for Transportation (Wuhan, China)
[11] van Nes A and Rueb L 2009 Spatial Behaviour in Dutch Dwelling Areas: How Housing Layouts Affects the Behaviour of its Users. *7th Int. Sp Syntax Symp* p 1–16.
[12] Legeby A 2009 From housing segregation to integration in public space. 7th Int Sp Syntax Symp. 1–12.
[13] Karimi K and Parham E 2012 An Evidence Informed Approach to Developing an Adaptable Regeneration Programme for Declining Informal Settlements *Proc. of the Eighth Int. Space Syntax Symposium* (Saniago de Chile) p 1–27.
[14] Vasku M 2013 Generative Improvement of Street Networks Based Space Syntax Applied in a case study on an informal settlement in Jeddah. S*pat Perform Sp Syntax*.**1** p 367–74.

[15] Rahmi D H, Wibisono B H and Setiawan B 2001 Rukun and Gotong Royong : Managing Public Places in an Indonesian Kampung. In: Miao P, editor. Public Places in Asia Pacific Cities: Current Issues and Strategies p 119–34 (Kluwer Academic Publishers)

[16] Hickman P 2013 "Third places" and social interaction in deprived neighbourhoods in Great Britain. J *Hous Built Environ* **28** 2 p 221–36.

[17] Gehl J 2006 Life Between Building: Using Public Space Sixth Edit 200 p (The Danish Architectural Press)

[18] Malek N A, Mariapan M and Rahman NIAA 2015 Community Participation in Quality Assessment for Green Open Spaces in Malaysia. Procedia - Soc Behav Sci [Internet]. Elsevier B.V 168:219–28. Available from: http://linkinghub.elsevier.com/retrieve/pii/S1877042814056857

[19] Hutama I A 2016 Combining the Activity-Space Pattern with Dwellers Perception to Explore the Sense of Place of an Urban Kampung. *Int Conf Indones Archit Plan* **3** p 66–79.

[20] Yatmo Y A and Atmodiwirjo P 2013 Spatial strategies for domestic service activities in urban kampung houses. *Int J Technol* **4** 1 p 24–33.

[21] Setiawan B, Oblo D and Djati R 2010 Kampung Kota dan Kota Kampung Potret: Tujuh Kampung Kota di Jogja Urban Kampung and Kampung Urban: Portrait of Seven Urban Kampungs in Jogja 235 p (Yogyakarta: Centre of Environmental Studies UGM)

[22] Hutama I A 2016 Exploring the Sense of Place of an Urban Through the Daily Activities, Configuration of Space and Dweller's Perception: Case Study of Kampung Code, Yogyakarta [Internet]. University of Twente 2016. Available from: As every place including human habitat has its sense of place, this research is set out to investigate the attributes that are forming the sense of place of an urban kampung (informal settlement). The sense of place is formed when individuals perceive the

[23] Hillier B 2007 Space is the Machine. electronic 314 p (United Kingdom: Press Syndicate)

[24] Wang M 2012 Extending geographic information systems to urban morphological analysis with a space syntax approach

[25] Jiang B and Liu X 2010 Automatic Generation of The Axial Lines of Urban Environment to Capture What We Perceive **24** 4 p 545–558.

[26] Hiller B and Lida S 1987 Network effect and physichological effect: a theory of urban movement 490 p.

[27] Ye Y and Nes A 2014 Quantitative tools in urban morphology: Combining space syntax , space matrix and mixed-use index in a GIS framework *Urban Morphol* **18** 2 p 97–118.

[28] Kim H-K and Sohn DW. An analysis of the relationship between land use density of office buildings and urban street configuration. Cities [Internet]. 2002 Dec [cited 2016 Feb 9];19(6):409–18. Available from: http://www.sciencedirect.com/science/article/pii/S0264275102000719

[29] Van-Nes A and López M J J 2010 Macro and Micro Scale Spatial Variables and the Distribution of Residential Burglaries and Theft from Cars An investigation of space and crime in the Dutch cities of Alkmaar and Gouda. *J. Sp. Syntax* **1** 2 p 296–314

[30] Hillier B and Sahbaz O 2008 An evidence based approach to crime and urban design Or, can we have vitality, sustainability and security all at once? Des Sustain Cities Decis Tools Resour Des Oxford Wiley Blackwell p 163–86

[31] Can I and Heath T. In-between spaces and social interaction: a morphological analysis of Izmir using space syntax. J Hous Built Environ [Internet]. Springer Netherlands 2015 Mar 5 [cited 2015 May 22];1–19. Available from: http://link.springer.com/article/10.1007/s10901-015-9442-9/fulltext.html

[32] Carmona M, Tiesdell S, Carmona M and Heath T 2010 Public Places Urban Spaces: The Dimension of Urban Design 389 p (New York, United States: Routledge)

[33] Van Nes A, Lopez M, Verhagen D and Waayer S 2016 Spatial tools for diagnosing the degree of safety and liveability , and to regenerate urban areas in the Netherlands *RIUS Geo-Design* **4** p 139–56

[34] Jacobs J 1961The Death and Life of Great American Cities 462 p (Penguin Books)

[35] Stedman RC 2002 Toward a Social Psychology of Place: Predicting Behavior from Place-Based Cognitions, Attitude, and Identity *Environ Behav* **34** 5 p 561–81

[36] NextCity. Embracing the Paradox of Planning for Informality [Internet]. 2018 [cited 2018 Jan 4]. Available from: https://nextcity.org/features/view/embracing-the-paradox-of-planning-for-informality
[37] Yin KR 2003 Case Study Research Design and Methods Third Edit 181 p (London: SAGE Publication)
[38] Bryman A 2012 Social Research Methods. Fourth Edi 750 p (New York: Oxford University Press)
[39] Klarqvist B. A space syntax glossary. Nord Arkit [Internet]. 1993;2:11–2. Available from: https://fenix.tecnico.ulisboa.pt/downloadFile/3779573909551/glossarySS.pdf
[40] Vaughan L. The spatial syntax of urban segregation. Prog Plann [Internet]. 2007 Apr [cited 2015 May 21];67(3):205–94. Available from: http://www.sciencedirect.com/science/article/pii/S030590060700013X
[41] Jiang B and Claramunt C Integration of Space Syntax into GIS: New Perspectives for Urban Morphology. Trans GIS 2002 [Internet]. **6** 3 295–309. Available from: http://doi.wiley.com/10.1111/1467-9671.00112
[42] Jiang B. Axwoman 6.3: An ArcGIS Extension for Urban Morphological Analysis [Internet]. 2015. Available from: http://fromto.hig.se/~bjg/Axwoman/
[43] Gehl J and Svarre B 2013 How to Study Public Life 179 p (Washington: Island Press)
[44] Parham E. THE SEGREGATED CLASSES : spatial and social relationships in slums. In: Eighth International Space Syntax Symposium Santiago de Chile: PUC, 2012 [Internet]. Santiago; 2012. p. 1–19. Available from: http://www.sss8.cl/media/upload/paginas/seccion/8150_1.pdf
[45] Brown F E, Darjosanjoto E T. The Use of Street:Configuration, Culture, and Space-use in the Coastal Settlements of Eastern Jave. In: Space Syntax Second International Symposium. Brasilia; 1999. p. 14
[46] Mehta V 2013 The Street 229 p (Routledge)
[47] Montgomery J 1998 Making a city: Urbanity, vitality and urban design. *J Urban Des* **3** 1 p 93–116
[48] Gilroy R 2010 Planning and Diversity in the City: Redistribution, Recognition and Encounter *Plan Theory Pract* **11** 1 p132–3
[49] Penn A 2001 Space Syntax and Spatial Cognition

IOP Conf. Series: Earth and Environmental Science **158** (2018) 012004 doi:10.1088/1755-1315/158/1/012004

Exploring the Notion of the Family Friendly City

Riela Provi Drianda

Faculty of Social Sciences Waseda University
1-6-1, Nishi-waseda, Shinjuku-ku, Tokyo, 169-8050, JAPAN

drianda@aoni.waseda.jp

Abstract. There is a common perception that downtown areas will never attract families and big cities are not the best place to raise children. Particularly the downtown areas of cities been depicted as the place where criminals, prostitutes, drug-sellers, and other dangerous strangers live. People with children are more likely to look for the suburbs to find bigger housing with more affordable prices, cleaner air, richer nature, a slower lifestyle, and safer environment. However, living in the modern suburb is not always easy and cheap, especially for those who need to commute to the central city. Dealing with the long commutes can be stressful and it affects the health, happiness, and well-being of family members. As the number of modern families with both parents in the workforce is rising, the demand to live closer to the workplace is getting stronger and growing. In some parts of the world, more families increasingly want to live in the cities. This trend can be seen in the United States, Japan, Korea, and Canada. Being family-friendly has become increasingly important for modern cities as more millennial generation show the tendency to raise their families in the urban area. Moreover, it is predicted that two-thirds of world's population will live in cities by 2030. To accommodate the growing population, in particular, those with children, modern cities should be developed to suit urban families. But what criteria and qualities make one city more family-friendly than another? What would a family-friendly city look like? To date, the number studies exploring the notion of the family-friendly city has been very limited. Most studies have been focusing on the notion of family-friendly dwellings, family-friendly workplaces, or child-friendly cities. This paper brings together and examines the dominant and recurring ideas about the family-friendly city represented in the relevant literature and current urban practices. This paper also questions whether the concept of the child-friendly city is adequate to create a better environment to raise families in the city.

1. Introduction

By 2030, it is predicted that two third of the world's population will live in cities. This prediction signifies the potential increase in urban upbringing in the coming years and it highlights the necessity for modern cities to provide more conducive environments for raising families. Since 1925, cities, particularly the downtown areas, have been perceived as places where hobos, criminals, and poor people gather; thus, people believe that downtown will never attract families. The hustle and bustle of cities has been viewed as an inappropriate environment to raise a child. Once people start a family and have children, they are most likely to move to the suburbs area so they can get bigger housing and a safer environment for their child. Those, who are left in the cities, are mostly families that have no choice about where to live and are dependent on government subsidy for their well-being. In other parts, downtown neighbourhoods are mostly occupied by singles, gays, and empty-nesters. Many scholars believe that this influences the demographic balances, socio-economic conditions, and the sustainability of the city itself.

IOP Conf. Series: Earth and Environmental Science **158** (2018) 012004 doi:10.1088/1755-1315/158/1/012004

However, recently there has been a surge of young people flocking back into the city. Millennials might not always stay in the urban core but increasingly young parents choose to raise their families in the urban settings. Compared to the earlier generation, these young urban professional parents (YUPPs) are less interested in having big houses and living far away from the city centre. They prefer to live closer to the urban amenities and adjust their lifestyle by living in a smaller space and using the whole city as their backyard. The middle-class families have started to reclaim the city by constructing their own spaces and are actively engaged in shaping the public domain such as by creating informal playgrounds on the sidewalks or on a larger scale by their consumption pattern and lifestyle choices [1]. Some cities such as Vancouver, Tokyo, Amsterdam, and Toronto have recognized these growing demands and started to transform urban spaces to meet the needs of families such as by providing family apartment units, building public playgrounds and plazas, and providing more day-care, after-school hours and various family-related activities in the central area. The astonishing rise of millennial urban families signifies the necessity for urban planners to design a city that is more family-friendly. But what makes a city good for raising a millennial family? How could we define a family-friendly city?

This paper brings together and examines the dominant and recurring ideas about the family-friendly city represented in the relevant literature and current urban practices. The paper begins with a review of the ills of urban living and the reasons why many families chose to settle in suburb areas. The next section discusses the suburban decline and the tendency of young professional families to reside closer to the urban core. Following this, the paper examines the term family-friendly used in the society and questions how we should define a city that suits family life. The last part of this paper summarizes the definition of the family-friendly city and the challenges that need to be addressed to create a better city for families.

2. Finding the best place to raise a family: the city vs suburban areas

2.1 *The ills of urban living*

Since the industrial revolution, cities have been portrayed as places that are unsuitable for raising a family. The revolution had transformed the urban landscape into a harsher environment for many people to live in. During this time, cities became the centre of industrial growth and attracted many migrants to work and live in their core areas. The expansion happened very quickly and gave very little chance for the cities to accommodate the growing population. The newly arrived migrants often settled in the poorest urban neighbourhoods and occupied the slums areas. Families were huddled in small, dirty and dilapidate tenement rooms. There were no proper waste disposal systems and the newly arrived residents often did not know how to manage their environment. Cities were very dirty, polluted, crowded and people no longer knew their neighbours well. The growth of cities led to unexpected living conditions where epidemics frequently broke out and crimes rose sharply. In addition to this, the urban growth divided the cities into several types of settlements and land-use. Central cities were mostly occupied by the newly arrived immigrants, the poor working class, and ghettos residents. The immigrants settled in these areas because housing values were relatively low and the areas were closer to the factories. The urban core was portrayed by Burgess, a famous Chicago School Theorist, as the zones of slums and 'bad lands' with their poverty, degradation, disease, and underworlds of crime. Those who accumulated more economic resourced moved to the zone of better-residence. This settlement offered more spacious areas for middle-class families. During this time, the affluent families were more likely to reside in the urban outskirts. The influx of immigrants and farmers to the central cities had driven the richer communities to escape the city and seek refuge in the suburban area [2].

This practice continues in numerous countries. Even though some urban cores have become much better in terms of physical design, hygiene, and environment, they still end up in being designed as places where the residents tend to give up on having children and raising a family. The ultra-dense cities of eastern Asia and some American big cities are still childless. Big cities are still seen as inappropriate places for raising a family due to the shifting sexual demographics, and economic and moral disasters [3]. This signifies how core cities could only attract the young and fail to keep the residents as they age and start families [4]. It further signifies how cities, even with their impressive achievements and outstanding architectures, could not accommodate the whole lifecycle of many urbanites.

2.2 The suburban dream

The damage of the urban core had generated a massive exodus to the suburbs. People who want to escape from the crowds, noise, pollution, and fear of immigrants moved to the hinterlands to seek a better quality of life. Feller [5] described one suburb housing-commercial brochure published during the industrial revolution as a tempting ad. The suburbs were illustrated as areas that offer freedom from dust and smoke and one could have desirable neighbours. During this time, suburban areas were portrayed as the urbanist paradise that offers a greener, cleaner, and safer place to raise a family. Suburbs in America were labelled as a growing-up place where the communities consisted of only the affluent, single-family homeowners and people of one race [6]. Meanwhile, in other countries with homogenous races such as Indonesia, Japan, and China, suburban area not only become an escaping destination for the richer families, but also for the middle-low income families who are looking for more affordable housing in a healthier environment.

The suburb appears as a settlement that can guarantee the family's happiness and can meet their expectations whether in neighbourhood qualities or good school systems [7]. Compared to the city core, the urban hinterland offers lower density, which is often perceived as a place that is safer, more satisfying, and has a lower risk of depression [6]. Safety is one of the features offered by the suburban housing developers hence many housing complexes were developed with excessive attention to the security. Security fortresses have become a common feature of suburban housing complexes. All buildings and transport systems are inter-connected and all points of entry are protected by private security forces [8].

While suburbs seem to offer an ideal environment for raising children, some suburban residents expressed their concerns about the time and energy they lost during the commuting time. Christian [9] described how for one hour increase in daily commuting, the male respondents in his research decreased their time with their spouse with 21 to 21.8 minutes and 18.6 minutes with their children [9]. As for the female respondents, he described how their one hour increase in daily commuting is equal to a 11.9 minutes decrease in time with friends. However, Christian's [9] study showed how females must reallocate time from other usages at greater rates. For example, they need to reduce exercise or skip preparing a healthy meal.

The long commutes reduce the opportunity for one of the parents to engage in full-time work especially when they lack child-rearing support from the day-care or family members. This situation forces one of the parents to work from home or to sacrifice their career for the sake of their children. In addition to this, the cost living in suburban areas is not as cheap as expected. The commuting cost including the highway tickets and gasoline could be significant, particularly if the living area is not connected by integrated public transportation. Brown and Roberts [10] argued that long commutes also sacrifice other activities that are beneficial to well-being and participation in family life. Parents' work-stress will influence their children and it can also decrease parents' productivity in the workplace [11]. A study by Li [12] underlined how fathers' commuting activity influence children's social relationships. The farther the father's commuting distance, the less likely children will have good peer relationships. The long commutes prevent fathers to participate in child-rearing and they often report physical fatigue and strains when they come home. Li [12] argues that parents' mental and physical health is a very important resource to promote healthy child development.

Longer commuting trips are also associated with fewer socially oriented trips and less access to social capital [13]. Delmelle [14] also highlighted how commutes of 30 minutes or more have a negative effect on social satisfaction. In addition to this, suburban living also poses a risk of social isolation [15]. Miller argued that the social and spatial structure of suburbia can promote familial isolation through a lack of public space and an emphasis on home maintenance and home-centred entertainment.

2.3 Living closer to the core

In recent years, there have been some changes in the behaviour of young professional parents in some countries such as The Netherlands, the US, and Canada [16]. Compared to the previous generations, today's parents show more interest in staying closer to the central area, particularly the young urban and professional parents. Urbanized suburbs and downtown areas have become some of the favourite places where these young parents prefer to live. The urbanization of edge cities has transformed the traditional residential community into a more complex mix of domiciliary and economic functions. These shifting suburban areas are often landlocked, therefore, they could not expand their territory outward. The only choice left is to expand the territory upward and with these vertical urbanism characteristics, the distinction between the cities and the suburbs has become blurrier, particularly in large contemporary metro areas [17].

Some young urbanites started to look more comfortable in raising their kids in high-rise buildings. In many core cities, the downtown areas are no longer seen as dangerous and inappropriate areas for raising children. As for these young urbanites, the whole city is becoming their living room and hinterland areas have become destinations for their weekend getaways. Even though they probably live in smaller places than before, they have more access and freedom to explore the surrounding urban parks, museums, and community centres. Living with kids in a smaller space is stressful for them but the millennial parents make some efforts to stay closer to the city by adjusting their lifestyle such as recycling and reducing the number of furniture in their house [18].

Today's families, or also known as millennial families, are debunking the stereotype of the traditional family. The lifestyle and values they hold are quite different from the previous generation. Futurecast has divided millennial families into a number of categories, starting from those with a more stable income to the families on a tighter budget. As for the families with more stable income and higher education level, children come first and they were least likely to sacrifice time with family [19]. Even though they have enough money to spend they shop wisely and teach their kids to value money. Parents in this category are also health conscious, child-oriented, and they actively seek parenting related information from online resources [19]. In line with this, McCulloch, and McCann [20] also argued that many millennial parents greatly value parenting, about a 10 percent increase over the previous generation. This may explain why the young urban professional parents (YUPPs) tend to avoid long commutes and live closer to the core so that they can have more time with their children.

Vancouver and Toronto have started to respond to the rising demands of these YUPPs. These cities have started to accommodate the need for bigger rooms for the downtown condominiums after discovering that the residents have difficulties to store children's stuff in the small apartment rooms. From the very beginning, the condominiums were not built and designed with the family needs in mind, thus, there is not enough space to put the baby stroller and/or children's bicycles. This situation leads many families with small children to use the bathtub to store children's stroller and bicycles. To accommodate the growing needs of YUPP families, the government of Toronto established guidelines for vertical neighbourhoods and buildings fit for families with children. Vancouver also echoes this movement as the number of Vancouverite who move into multi-family units such as apartments and condominiums is rising significantly.

The tendency of young families to live closer to the core area is also shown in other big cities such as Amsterdam and Tokyo. The rise of YUPPs who live in the downtown and urbanized city's edge has transformed the urban landscape and generated the birth of family-oriented consumption spaces [16]. This situation can be seen also in Tokyo, which faces a significant increase of families living near urban core. The ultra-convenience of the Greater Tokyo Region has increasingly attracted young families to leave the countryside and reside nearer to the capital area. To respond to their needs, numerous housing units and condominium complexes for families have been built intensively within the Greater Tokyo Region.

Even though the four cities above have started to identify the necessity to create a family-friendly environment for their young professional inhabitants, their initiatives are quite rare. Other core cities have not completely considered the needs of families or transformed their physical and social environment to suit family life. The ultra-dense cities of Eastern Asia such as Hong Kong, Singapore, and Seoul exhibit the lowest fertility rates on the planet. Kotkin [21] further argues that to flourish, these core cities need to change and be responsive to the different human needs - from birth to the end of life. Otherwise, cities will be childless and only filled with singles who will leave the area or emigrate when they grow old and start families.

3. The notion of family-friendly and the family-friendly city

The term of family-friendly was introduced mainly in the field of personnel management [22, 23, and 24]. This term is used to define the opportunity for women in particular, to juggle their career and family life. The term family-friendly is also used to describe the type of community-based service given to children and families [25]. However, there is very limited academic research investigating the term family-friendly from the perspective of urban physical and social planning. Recent studies have focused on the development of the child-friendly city instead of the family-friendly city. Compare to the term 'family-friendly city', a 'child-friendly city' has a more lucid definition and the term has been officially acknowledged by many governments and institutions as the platform to ensure the rights of every child in the city. The child-friendly city initiatives promoted by UNICEF have guided many cities in the inclusion of children's rights in their goals, programmes, components, and structures.

Table 1. Child's rights in the city. [26]

No.	Rights to
1	Influence decisions about their city
2	Express their opinion on the city they want
3	Participate in family, community, and social life
4	Receive basic services such as healthcare and education
5	Drink safe water and have access to proper sanitation
6	Be protected from exploitation, violence, and abuse
7	Walk safely in the streets on their own
8	Meet friends and play
9	Have green spaces for plants and animals
10	Live in an unpolluted environment
11	Participate in cultural and social events
12	Be an equal citizen of their city with access to every service regardless of ethnic origin, religion, income, gender, or disability

There have been magnificent and outstanding movements to make today's cities friendlier for children such as providing more spaces where children can play in their neighbourhood or even developing official children's forums in the city. However, this paper argues that the concept of the child-friendly city (CFC) itself could not solely support young urbanites to raise a family in the city. The concept does not completely represent the needs or considers the characteristics of other family members who act as the children's guardians. In addition to this, the CFC concept does not discuss family's access to economic capital. The costs of raising a child in the city from birth to maturity are quite expensive. Apart from food budget, the young parents need to rent or buy property that suits family activities and is also close enough to their workplace, groceries, and childcare facilities. The provision of housing that suits families and of spaces and places for children in a family-friendly planning idea should be followed by the effort to increase affordable and decent housing for the middle-class family and even increase the service of public transportation or other mechanisms that would support parents' trip chains.

Supporting parents' trip chains is very important to ensure that parents can balance their work and family life. In households with children, women create substantially more complex trip chains than women in households without children. This situation led women with children to travel in private vehicles to juggle multiple activities within a limited time. As women have always taken a bigger responsibility in maintaining the household, the lack of workplaces that offer family-friendly policies and benefits can also impact their choices to join the workforce or even to have a child. The attitudes toward women with children are still discriminative in some parts of the world. It is not always easy to find or change job for women with children. Inflexible working hours and long commutes often prevent women to take on full-time employment. If cities continue to be designed or end up being designed as places where women cannot balance work and family life, then women with higher education might prefer the childless lifestyle as what happened in many countries nowadays.

As important as mothers are, today's fathers also play very important roles in raising their families [27]. Unlike the previous generation, the millennial fathers are more likely to participate in child-raising. The change in fathers' attitudes signifies the necessity to improve the child-rearing facilities within the cities. The increased roles of fathers in family-raising should also be considered in the design of the family-friendly city for example by installing diaper-changing stations in men's toilets or introducing more incentives for fathers who participate in child-raising activities. The millennial fathers are also showing the tendency to participate in local movements and civic participation and they know that their voices can influence how the city works.

Based on the phenomena discussed above, this paper argues that several matters need to be considered in the development of the family-friendly city. Both physical and non-physical factors contributing to the family's well-being in the city should be addressed carefully, such as family's access to economic capital, family-friendly transportation, considering current trends such as the shifting roles of modern parents and different characteristics of millennial families.

Table 2. Key features of family-friendly planning. [28]

No.	Key feature
1	Housing
2	Spaces and places for children
3	Inclusive, high quality of public space
4	Greening the city
5	Knowledge-based urban planning
6	Children as stakeholders

4. Cities that suit families

The above discussion showed how the use of the term family-friendly is not yet uniform and similarly, there is still very limited discussion about what a 'family-friendly city' is. There are very few studies that have investigated the notion of the family-friendly city and the things that should be developed to support family's well-being and sustainability in the modern cities. Even so, for a long time, people have believed that families could not strive well in the big city particularly the core areas. To some extent, the industrial revolution had influenced people's perception of the ideal place to raise a family. The following section discusses insights from some cases of the American ideal 'family-friendly city'. Smaller towns, suburb cities, and less dense cities are considered by most American families as a good place to start a family. The latest surveys by Forbes, Niche, and Kiernan revealed the criteria used to define a family-friendly city from the perspective of American families (see Table 3).

Table 3 indicates how American families define some criteria of the family-friendly city in the same way. Some common criteria included the availability of affordable housing, urban safety, and education quality. Despite some similarities, there are also some subtle but important differences to help understand and enrich our perspectives. For instance, Forbes mentioned commuting as one of its family-friendly city criteria. As described in the previous section, millennial parents are less likely to sacrifice their time with family, thus, commuting time has become one of their considerations in choosing a home location.

However, commuting may not be the only factor contributing to the friendliness of a city for family life. A survey by Niche revealed different factors such as the importance of outdoor activities, the percentage of children under 17 in the area, and the diversity to make a city friendlier for families. These criteria are aligned with the latest research by Futurecast [19] which described how millennial families value life-experience and changed the way people look at diversity. For these families, it is important to ensure that their kids have access to great outdoor experiences and encounter different people. The survey showed that millennials may not be very rich but they would spend their money on facilitating their children to experience more in life [19].

Table 3. Criteria of Family-Friendly City (American version). [30, 31, and 32]

Criteria	Forbes	Niche	Kiernan
1	Median income	Public schools grade	Family-fun
2	Cost of living	Crime and safety grade	Health and safety
3	Housing affordability	Higher education rate	Education and Childcare
4	Commuting	Cost of living grade	Affordability
5	Owning homes	Family amenities grade	Socioeconomics
6	Crime	Housing grade	
7	Education	Outdoor activities grade	
8		Percentage of residents between the ages of 0 and 17	
		Diversity grade	

Similar but a little bit different, Kiernan also mentioned a criterion related to family's opportunity to engage in fun activities in the city. However, unlike the other surveys, Kiernan clearly specified about childcare facilities as one of a family-friendly city's criteria. As there has been a rise in dual-career couples in Western countries, there are big challenges for millennial families to balance the demands of two working partners with children at home. For these young professional parents, a family-friendly working environment and access to affordable and good child care services are very important [24].

According to the above surveys, the best cities in the US to raise a family are Overland Park, Madison, Plano, Grand Rapids, Idaho, Provo, Naperville, The Woodlands, and Columbia. Nevertheless, none of the cities mentioned above appear in the ten best cities in the world to raise a family, according to Homeday's best cities for families index 2017 (see Table 4).

Table 4. Best cities to raise a family. [33]

Rank	Name	Rank	Name
1	Copenhagen, Denmark	6	Vancouver, Canada
2	Oslo, Norway	7	Basel, Switzerland
3	Zurich, Switzerland	8	Toronto, Canada
4	Stockholm, Sweden	9	Stuttgart, Germany
5	Hamburg, Germany	10	Munich, Germany

Interestingly, cities that rank higher in the Homeday Index are mostly big cities with a large population. This finding simply refuses the idea that large cities are not appropriate for families. The criteria set by Homeday to measure the level of family-friendliness in the city are described in Table 5.

IOP Conf. Series: Earth and Environmental Science **158** (2018) 012004 doi:10.1088/1755-1315/158/1/012004

Table 5. Criteria of Family-friendly city (Homeday version). [33]

No.	Criteria
1	Housing
2	Education system
3	Safety
4	Unemployment
5	Pollution
6	Transportation
7	Maternity and paternity law
8	Healthcare
9	Happiness
10	Kids friendly airport
11	Activities for kids
12	Green spaces
13	Parents' perceptions
14	Professionals' perceptions

Compared to the surveys carried out by the American counterparts, Homeday's survey covers different but interesting matters such as transportation, maternity and paternity law, happiness, and kids' friendly travel and activities. The difference in the family-friendly criteria is probably influenced by the European culture that values the family-work balance and independent mobility. Parents' right to take a leave after childbirth is carefully considered in many European countries and fathers are also expected to take part in the child-rearing activities. Thus, it explains why paternity law is also considered as the criteria of a family-friendly city. Transportation is also an important factor because European families tend to use public transport and/or bicycles in their trip chains. Cities without proper and integrated public transportation would pose more obstacles for parents to juggle their career and family life.

The keyword of happiness is also used in Homeday's criteria of the best city to raise a family. The ability of the city to promote residents' happiness has been considered one of the important features in future urban design and planning [34]. Leyden et al. [35] underlined how cities that can provide easy access to convenient public transportation and to cultural and leisure amenities would promote happiness. In line with this, the research indicated that cities that are affordable and support urban upbringing would affect the residents' happiness. Surprisingly, a kids-friendly airport has also been considered as one of Homeday's criteria of a family-friendly city. This support another finding that millennials would rather travel than buy a home [36]. The availability of activities for kids in the city during vacation, weekends, and bank holidays appears to be very important for parents. It is obvious that for the millennial parents, travel is not seen as a luxury but more of a necessity.

5. Conclusion

This paper attempts to provide a review of millions of people's efforts to find the best place to raise a family. The industrial revolution made many families think twice about raising a child in the city, particularly in the downtown area. Families who could not endure the city life would escape to the suburbs once they had enough more money to buy or rent a bigger house. The suburbs are filled with families who are trying to find a better quality of life for their families. However, suburbs also have several problems that force some families to flock back into the city. To sum up, people seek to find a better place for their families because they are looking for the following matters, (i) a healthier environment; (ii) a safer environment; (iii) a less dense settlement; (iv) a better social life; (v) better access to jobs and other economic opportunities; (vi) longer time to spend with their family; (vii) a better education system and quality; and (viii) work-family balance. The qualities that people seek in the new place signify the ideal urban characteristics for modern families. As more people will live in the urbanized areas, it is very important to understand the kind of cities that can support the lifestyle of modern urban families. Unfortunately, there is neither a single definition of what a family-friendly city is nor some global consensus on the definition and establishment of the 'family-friendly city' yet. However, the term 'family-friendly' itself is used frequently in personnel management, community-based services, parenting, and real estate discussions.

Compared to the child-friendly city, the concept and movement to build a friendlier city for families still lacks official recognition. A shared vision of what kind of city should be developed to accommodate the growing urban population is needed. The trends show that today's families prefer to live closer to urban cores and urbanized suburbs. The young families have gradually left the traditional suburbs and moved closer to the cities to provide better experiences for their children and spend more quality time with their family members. These families would put in some efforts to adjust to the vertical living and its different lifestyles, including dealing with the constant needs of recycling and space management. Therefore, this paper strongly argues the necessity for cities to carefully address and pay attention to the growing and changing demands of families that live in high-rise settlements.

The paper underlines the necessity to develop a more solid definition and define a clearer scope of 'family-friendly city' development. The current concept of CFC has not covered a number of matters related to parents' ability and the things that parents need to raise their children in urban areas. The CFC concept is inadequate in ensuring family's well-being and sustainability. The well-being and sustainability of a family involve more complex issues ranging from the stability of parents' jobs, incomes and even their work and family balancing skills. Transportation to support parents', particularly mothers' trip chains, should also be listed in the features of the family-friendly city. However, the current family-friendly urban planning's concept has not touched upon this issue and to some extent is still much focused on fulfilling children's needs. With the women's issues in work-family balance and the shifting roles of fathers in modern society, this paper suggests the necessity to establish a new vision of a 'family-friendly city' that will help the urban planners and society to plan a friendlier city for the millennial families.

The surveys carried out by private institutions and real estate companies have given us some clues about the characteristics of cities that are suitable for raising a family. A city that is affordable, safe, and offers good education quality for children can be defined as the simplest version of what a family-friendly city should be - regardless its size and density. Even though there are some differences in families' perception about cities that are friendly to a family, there are some similar matters that need to be provided by cities to support urban upbringing such as housing that suits families and environments that are conducive for families to carry out their daily activities. The author of this paper believes that the more the city could provide families with various amenities and accessibilities, the friendlier they are to raise a family. Therefore, all criteria mentioned in the above surveys should be deeply considered. Finally, this paper suggests the necessity to research more communities to understand the growing demands of millennial families and the kind of cities that should be provided to support the young urbanites to raise a family. Further and bigger-scale research needs to be carried out to investigate more literature and current practices of family-friendly city development in various countries and continents around the world.

References

[1] Karsten L and Verstate L 2016 Play and Recreation, Health and Well-being Ed Evans E, Horton J, Skelton T (Singapore: Springer)
[2] Karakiewicz J, Yue A, Paladino A 2015 Promoting Sustainable Living- Sustainability as an Object of Desire (New York: Earthscan from Routledge)
[3] Clayton H 2013 *J. of Urban History* **39** 933-55
[4] Kotkin J 2016 The Human City-Urbanism for the Rest of Us (New York: Agate Publishing)
[5] Feller W V 2003 *Suburban Sprawl: Culture, Theory and Politics* ed MJ Lindstorm and H Bartling (United States of America: Rowman & Littlefield Publishers) p 49-63
[6] Larco N 2010 Journal of Architectural & Planning Research **27** 6987
[7] Rhodes A and Warkentien S 2017 *American Educational Research J.* **54** 168S189S

[8] Byers M 2003 Suburban Sprawl: Culture, Theory and Politics ed MJ Lindstorm and H Bartling (United States of America: Rowman & Littlefield Publishers) 23-44
[9] Christian TJ 2012 *Preventive Medicine* **55** 215-18
[10] Brown H and Roberts J 2011 Exercising choice: The economic determinants of physical activity behaviour of an employed population. *Social Science & Medicine* **73** 383-390.
[11] Hare M 2014 The relationship of parents' work-stress and child functioning in the context of spillover effects, marital and parenting stress, and parents' perceptions (United States of America: University of Central Florida)
[12] Li J and Pollmam SM 2016 *J. of Family and Economic Issues* **37** 488501
[13] Besser LM, Marcus M, Frumkin H 2008 *Am J Prev Med* **34** 207-11
[14] Delmelle EC, Haslauer E, Prinz T 2013 *J. of Transport Geography* **30** 110-16
[15] Miller LJ 1995 *Sociological Forum* **10** 393418
[16] Karsten L 2014 Journal of Economic and Social Geographic **105** 175-88
[17] Downey DJ and Smith DA 2011 *J. of Urban Affairs* **33** 2144
[18] Nethercole M and Home R 2016 Environment and Planning A: Economy and Space **48** 1581-98
[19] Futurecast. Millennials as new parents: The rise of a new American pragmatism (Internet). 2013 (Cited 28 December 2017). Available from: http://www.millennialmarketing.com/wp-content/uploads/2016/05/FutureCast_Millennials-as-New-Parents-Report-1.pdf
[20] McCulloch J and McCann A 2017 *J. of Human Lactation* **33** 570-72
[21] Kotkin, J 2016 The Human City: Urbanism for the Rest of Us. Agate Publishing.
[22] Roberts G E 2003 Review of Public Personnel Administration **23** 241254
[23] Roberts GE, Gianakis JA, McCue C, Wang X 2004 *Public Personnel Management* **33** 307-30
[24] Feeney MK and Stritch JM 2017 Review of Public Personnel Administration 127
[25] Singh NN, Curtis WJ, Wechsler HA, Ellis CR, Cohen R 1997 *J. of Emotional and Behavioral Disorders* **5** 8292
[26] Childfriendlycity homepage (Internet). (Cited 30 December 2017). Available from: http://childfriendlycities.org
[27] Yogman M and Garfield CF 2016 *Pediatrics* **138** Pe20161128
[28] Karsten L 2017 Six Features of Family Friendly City Available from The European Network of Child Friendly City
[29] Forbes. Best cities for raising a family. (Internet). (Cited 28 December 2017). Available from https://www.forbes.com/pictures/eddf45gihi/best-cities-for-raising-a-family/#524f22e22bb3
[30] Niche. Best places to raise a family methodology (Internet). NICHE. 2017 (Cited 28 December 2017). Available from: https://about.niche.com/methodology/best-places-for-families/
[31] Kiernan, J.S. 2017's best & worst places to raise a family. Wallethub. 2017. Available from: https://wallethub.com/edu/best-cities-for-families/4435/
[32] Homeday. Best cities for families index 2017 (Internet). Homeday. 2017 (Cited 28 December 2017) Available from: https://www.homeday.de/en/family-index-international/
[33] Montgomerry, C 2013 Transforming Our Lives Through Urban Design (United States of America: Farrar, Strauss and Giroux)
[34] Leyden KM, Goldberg A, Michelbach P 2011 *J. of Urban Affairs* **47** 861-88
[35] Boston Consulting Group. Millennials fuelling the experience economy. Eventbrite Poll. 2014. (Cited 28 December 2017). Available from: eventbrite-s3.s3.amazonaws.com

From Informal to Formal: Status and Challenges of Informal Water Infrastructures in Indonesia

S Maryati[1], A N S Humaira[1] and D M Kipuw[1]

[1]Regional and City Infrastructure System Research Cluster: School of Architecture, Planning and Policy Development, Institut Teknologi Bandung, Ganeca 10 Bandung 40132, Indonesia

E-mail: smaryati@pl.itb.ac.id

Abstract. Informal water infrastructures in Indonesia have emerged due to the government's inability or incapacity to guarantee the service of water provision to all communities. Communities have their own mechanisms to meet their water needs and arrange it as a self-supplying or self-governed form of water infrastructure provision. In general, infrastructure provisions in Indonesia are held in the form of public systems (centralized systems) that cover most of the urban communities; communal systems that serve some groups of households limited only to a particular small-scale area; and individual systems. The communal and individual systems are systems that are provided by the communities themselves, sometimes with some intervention by the government. This kind of system is usually built according to lower standards compared to the system built by the government. Informal systems in this study are not defined in terms of their legal aspect, but more in technical terms. The aim of this study was to examine the existing status and challenges in transforming informal water infrastructures to formal infrastructures. Formalizing informal infrastructures is now becoming an issue because of the limitations the government faces in building new formal infrastructures. On the other hand, global and national targets state 100% access to water supplies for the whole population in the near future. Formalizing informal infrastructures seems more realistic than building new infrastructures. The scope of this study were the technical aspects thereof. Making descriptive and comparative analyses was the methodology used. Generally, most of the informal systems do not apply progressive tariffs, do not have storage/reservoirs, do not have water treatment plants, and rarely conduct treatment in accordance with standards and procedures as formal systems do, which leads to dubious access to safe water, especially considering the quality aspect.

1. Introduction

It has been estimated that around 1,1 billion people in the world at the beginning of 2000 lacked access to safe water. This has resulted in the challenge of providing access to safe water to all, particularly in developing countries [1]. The government alone in low- and middle-income countries seems unable to meet this international target as a consequence of engineering factors regarding high operating costs and under-utilized investments, unplanned or disintegrated access to basic services between legal areas in cities and illegal or unregulated settlements [2]. Moreover, multiple other factors, such as neglect by and poor capacity of central and local government authorities and no guarantee of cost recovery for high upfront financial and infrastructural investments in providing water to both rural and poor urban areas caused by insecure and uncertain land tenures as well as the terrain and dispersed poor communities complicate the condition of water provision [3].

This basic uncertainty of the water infrastructure that is supposed to be served and provided by the authorities has led to self-managed informal infrastructure provision or, adapting Allen's terminology, 'needs-driven practices', which are based on some form of collective action (within or between communities and the authorities) and/or those that are undertaken individually and may rely on solidarity among households and neighborhoods or on market-based mechanisms [4].

Since recently, informal water infrastructures as unofficial service provision is seen as a 'mode of urban governance' for the underserved population, which is largely allowed and even encouraged by governments because it is increasingly viewed as a necessary and acceptable mode of urbanism [5]. This practice of providing access to informal services in order to meet basic needs and filling the gap left by the government is widely implemented, especially in developing countries. It includes cases such as household water resale activities of alternative service providers and small-scale independent providers (SSIPs) in peri-urban areas of greater Maputo, Mozambique [6]; small-scale private bottled water supply in Dhaka-Bangladesh; commercially operated public toilets in Kano-Nigeria; licensed standpipe operators in Ouagadougou-Burkina Faso; pushcart water vendors in Delhi, India and Manila, Philippines; female water vendors in Honduras [7]; domestic private sector water service delivery in Tanzania [8]; *nyelang* (those without a piped water connection buy bulk water from their neighbours) and utilization of shallow and deep groundwater managed by individual households using bore wells in Jakarta, Indonesia [9]; a local initiative in community water supply in Ashanti Region-Ghana [1]; and community-public partnerships or other community-based water governance approaches for water supply in developing countries [3].

Informal water infrastructures in Indonesia have emerged due to the government's inability or incapacity to guarantee the service of water provision to all communities. To fill this gap, communities have their own mechanisms to meet their needs and arrange it as self-supply or self-governing water infrastructure provision. In general, infrastructure provisions in Indonesia are held in the form of: public as well as centralized systems that cover most urban communities; communal systems that serves some group of households, limited to a particular area (one or more neighborhoods); and individual systems. These communal and individual systems are provided by the communities themselves and can be classified as informal in the terms of the governance framework as well as in a legal context. This self-supply water infrastructure by communities is widely varied in the context of technical, management, and institutional aspects.

In accordance with the global target established in the sustainable development goals (SDGs), the government of Indonesia has set the main goal in the water sector, which is translated in the *National Medium-Term Development Plan* (RPJMN), as a target of 100% of all households having adequate access to safe drinking water. Adequate access in this context is defined as access to a water supply with a piping system according to regulations, standards and procedures, and normative principles in providing water infrastructure networks. In connection with the target set in RPJMN, the existing informal water infrastructure should be incorporated and transformed into a formal infrastructure. Referring to the water service provision in West Java Province (one of the leading provinces in Indonesia), in 2015 only 20% of households were served by the public provider *Perusahaan Daerah Air Minum* (PDAM). Meanwhile, 47% of households acquire water from a communal piping system, either delivered by the government or independently served, while the remaining 33% of households obtain their water from unprotected non-piping infrastructure networks.

Formalizing the informal is mostly done in land development. In this field, formalizing is supported by the theory of Soto in the form of granting land rights [10]. The absence of land rights limits the poor in increasing the value of their property. Formalizing informal land ownership is still being debated in relation to its effectiveness. In some cases the granting of property rights does not guarantee security of ownership. Infrastructure services have to be added to land rights to ensure a positive impact on landowners [11]. Formalizing the informal sector in infrastructure is mostly done in the waste management sector. The main reason for formalization is that the informal sector contributes greatly to waste management, but its presence is unknown in the formal system [12]. Especially for waste management, informal sector workers generally work in poor conditions.

The aim of this paper is to examine the existing status and challenges in transforming informal water infrastructures to become formal infrastructures with the main focus on technical aspects. This paper does not contribute to the debate on formalization in the context of land development but focuses on the next step after formalizing land status, which is formalizing infrastructure, in this case water supply infrastructure. Formalizing the informal has become an issue because of the limitations of the government to build new formal infrastructures. On the other hand, global and national targets state access to safe water for 100% of the population in the near future. Formalizing the informal seems more realistic than building new infrastructures. In this study, the water infrastructure system managed by PDAM as public water supply in the form of a centralized service is categorized as formal infrastructure. This 'formal' category refers to the governance framework. PDAM is the local water enterprise that has the authority to provide water through a public system. PDAM is owned by local governments. Water provision, in terms of providing a basic service, is an obligation of the government. Therefore, informal water infrastructure in this paper refers to water infrastructure systems that are not managed by PDAM. They can be managed by a community for a communal system and/or individually by each household, or they may be served by other parties or a partnership between stakeholders.

2. Methods

This study was focused only on the technical aspect of transforming informal infrastructures into formal infrastructures. Data were collected through observation of informal infrastructures in Indonesia and were documented based on related studies. Descriptive and comparative analysis were the methodologies used in this study. Descriptive-qualitative analysis was used to describe the status of the informal infrastructures, while comparative analysis was used to identify some challenges in transforming informal infrastructures into formal infrastructures. Comparative analysis was basically used to compare the current condition of informal infrastructures to an ideal condition, which is defined as formal infrastructure. Formal infrastructure in this context was not defined in legal terms, but more in relation to the fulfillment of the requirements for a formal infrastructure, which was defined more according to the physical aspects. The criteria for formal infrastructure are listed in Table 1.

This study first examined the typology of existing informal water infrastructures in Indonesia characterized by technical criteria in water infrastructure provision, consisting of the source, production, distribution/transmission, and water delivery service. This typology was extracted from a previous study. The first step in building the typology is identifying the core of the typology; in this research the core of the typology are the technical or physical criteria. The identified typology then was used as a tool to investigate the challenges in formalizing informal infrastructures. These challenges were formulated by comparing the existing status and the norms/standard in terms of the technical aspects in providing a formal water infrastructure, as defined in Table 1, with the current condition of each typology. The findings demonstrate a number of informal water infrastructure typologies in Indonesia and the status of water infrastructure in terms of a formal infrastructure.

3. Policy and technical guidance on DWSS

3.1. National policy of drinking water service in Indonesia

Drinking water is a basic human need that must be fulfilled. The government is obliged to provide drinking water, including its infrastructure. Therefore, sufficient access to safe water for 100% of the population by 2030 is included in Goal 6 of the sustainable development goals established at the United Nations Sustainable Development Summit. In the context of Indonesia, this target is translated into various levels of policy documents, such as the *National Medium-Term Development Plan* (RPJMN) 2015-2019 that is derived from the detailing in the *Strategic Plan of the Directorate General of Human Settlements*, Ministry of Public Works and Housing. In this strategic plan, the target of 100% of the population having access to safe water by 2019 has been set as the main objective for the water supply sector [13].

This target is to be achieved by combining approaches on the supply side (optimization and new development) and on the demand side (increasing efficiency of drinking water provision and creating a conducive environment) [13]. The first approach is done by facilitating drinking water supply systems (DWSS) managed either by PDAM or non-PDAM, developing new DWSS in some special areas, specifically urban slum areas (661,600 house connections), fishing areas (66,200 HC), and water prone areas (1,705,920 HC), and developing regional DWSS, DWSS in urban areas, and community-based DWSS.

The demand side approach is conducted by implementing the principles of water saving and water conservation, which are applied in all areas in Indonesia. This last approach was effectuated by formulating the *Master Plan of Drinking Water Supply System* (long term), the *Strategic Plan of Regional Drinking Water* (medium term), and the *Annual Plan of Drinking Water* (short term) for each region in Indonesia as a guidance in DWSS development, improving the management and organization of data and information regarding drinking water as a baseline in planning and budgeting, and facilitating the regulatory development in all regions that ensures the provision of drinking water services throughout the country.

3.2. Technical guidance for DWSS

To ensure access to safe water, a series of technical aspects needs to be identified and then implemented to provide a guaranteed service of DWSS. From the literature and theoretical review (regulations, standards and procedures, and normative principles), a number of technical aspects resulted that must be addressed in developing DWSS, as illustrated and explained in Figure 1 and Table 1.

Table 1. Guidance of technical aspects for developing DWSS.

No	Aspect	Explanation
1	Source	Sources of water utilized for drinking water can be divided into 3 categories, i.e. surface water, groundwater, and rainwater. Choosing the most suitable source depends on the raw water availability and sustainability in any season (dry and rainy) from the source closest to the service area [14].
1.1	Surface water	Water that is open to the atmosphere and results from overland flow (i.e. runoff that has not yet reached a definite stream channel). In other words, surface water is the result of surface runoff. For the most part, surface water (as used in this context) refers to water flowing in streams, rivers, lakes, fresh ponds, and oceans [15]. Different types of intake (depending on the physical conditions) are used to collect the raw water from the surface water source to be utilized as drinking water [14][16].
1.2	Groundwater	The precipitation that falls on land infiltrates the land surface, percolates downward through the soil under the force of gravity and becomes groundwater. Groundwater collected in saturated layers under the earth's surface is called an aquifer. Three types of aquifers exist: unconfined, confined, and springs [15]. Source catchment (spring), shallow wells (dug-well and tube-well), deep wells (bore-wells), etc. are the structures to catch the raw water from groundwater [14][16].
1.3	Rainwater	Rainwater can be used and harvested as a second option or alternative for the areas that are classified as water prone areas with fluctuating availability of raw water. One of the requirements utilizing rainwater is a minimum precipitation of 1,300 mm/year [16]. Rainwater catchment basins are needed to harvest rainwater [14][16].
2	Production	Production units are facilities and infrastructure that can be used to process raw water into drinking water through physical, chemical or biological treatment, including processing and equipment buildings, operational equipment,

No	Aspect	Explanation
		measurement tools and monitoring equipment, and drinking water storage buildings [16].
2.1	Treatment	The purpose of a water treatment system is to bring raw water up to drinking water quality. The treatment process will vary between sources depending on the quality of the source water [17].
2.1.1	Complete treatment	Complete treatment for surface water [17]: - Screening Screening of the source is done to remove relatively large floating and suspended debris. - Coagulation Chemical treatment that adds a coagulant to the water in a mixing tank to encourage suspended solids to adhere to each other and form larger particles that settle more easily. - Flocculation The process of gently mixing the water and coagulant that allows the formation of large floc particles. - Sedimentation After flocculation, the water flows through a sedimentation basin or clarifier that is designed to hold the water for a long enough time so that gravity allows most of the suspended solids (floc) to settle down. - Sludge processing Where the mixture of solids and liquids collected from the settling tank are dewatered and disposed of. - Filtration The effluent from the tank is then filtered in a rapid-sand filter (better known as filtration unit), which consists of a number of layers. Other processes, such as adsorption, continued flocculation and sedimentation in the pore space, are also important removal mechanisms. - Disinfection The final step is disinfection of the liquid effluent to ensure that the water is free of harmful pathogens (kill any remaining pathogenic organisms because only a small amount of the bacteria are removed in the previous processes). - Hardness removal This process can be added to the generalized process if needed, yet surface water seldom has high hardness level, so it is not usually part of the water treatment process. A commonly used method in this final stage is chlorination. Chlorine is completely effective against bacteria and microorganisms. In some larger systems, secondary disinfection may be required by maintaining chlorine residuals within the distribution system [18]. Besides, another approach should be implemented to process seawater into drinking water using seawater reverse osmosis (SWRO) [16]. Complete treatment for ground water [17]: Ground water usually has high water hardness and alkalinity caused most notably by the presence of calcium and magnesium ions. Therefore, softening is a part of water treatment processes prior to disinfection (if needed). There are two common approaches to water softening: 1. Lime-soda process Either quick lime or hydrated lime is added to the water, which will increase the pH to about 10.3 and convert soluble bicarbonate ions to insoluble carbonate. The carbonate will then be precipitated. Most of the precipitates formed are removed in a sedimentation basin.

No	Aspect	Explanation
		2. Ion exchange process Hard water is forced through a column containing solid resin beads made of naturally occurring clays called zeolites or synthetic resins. In an ion exchange unit, the resin removes Ca2+ and Mg2+ ions from the water and replaces them with sodium ions, which form soluble salts. These salts do not caused hardness.
2.1.2	Simple treatment	For the case of small-scale (communal and individual) DWSS, raw water treatment is still required, only using simpler methods such as filtering and disinfection, slow sand filtration, ultraviolet light, membrane filtration, peat water purification, infiltration galleries, distillation, and standard practices for household use [16].
2.2	Storage	Raw water that has been processed into drinking water will then be stored in a reservoir or a basin before distribution.
2.2.1	Operational storage	Operational storage is used to provide water during normal operation when other sources of supply are not delivering water [18].
2.2.2	Standby (or emergency) storage (optional)	Standby storage is used to provide water to a system during unusual or emergency conditions when a source of supply may be temporarily unavailable [18].
2.2.3	Equalizing storage (optional)	Equalizing storage is used to supply the difference between peak-hour demands due to normal diurnal flow variations and the capacity of the source of supply [18].
2.2.4	Fire flow storage (optional)	Fire flow storage is used to provide the required volume of water to meet fire-fighting needs within the water system's service area [18].
3	Transmission and distribution	Transmission and distribution process depend on the terrain between the water source and the service area. A gravitation system, or pumping system, and/or combination of both gravitation and pumping can be utilized. On the other hand, the distribution system consists of a piping system (house connections) and also public hydrants/communal water taps/communal toilets. To choose between a house connection or a communal service depends on the demographic, economic, social, and physical characteristic of the service area.
3.1	Piping system (house connection)	The water transmission system consists of pump stations and larger diameter piping that conveys water from a source of supply to the point of entry into the distribution system. This point of entry may be a reservoir, pump station or the first service connection to the system. The distribution system consists of a network of pipes, booster pump stations and pressure-reducing valves that convey flow from the system point of entry to the users [18]. Some densely populated areas that have a predominantly flat terrain will be suitable for this type of water distribution system.
3.2	Public hydrant/communal water tap/ communal toilet	A communal drinking water service facility in the form of a reservoir of water placed above the ground or on a foundation [14]. Sparsely populated areas, low-income settlements, and countered areas are more suitable to implement public hydrants/communal water taps/communal toilets to distribute the water.
4	Water delivery service	Water delivery services must guarantee access to safe water with 3 main indicators, i.e. quality, quantity, and continuity. Besides, water is a basic right for human beings and therefore should be affordable yet the user/consumer also has the obligation to contribute in terms of operational and maintenance costs.
4.1	Tariff/billings	- All water customers/connections are metered [19]. - Written procedures for billing and collection [19]. - The water rate is structured among users (progressive tariff for residential, commercial and industrial users) [19]. - For domestic use, consider affordability for users according to the global standard of maximum 4% of their average monthly income [20].
4.2	Safe access (quality, quantity,	- Quality (can meet the water-quality standard in accordance of required health standards) [21].

No	Aspect	Explanation
	and continuity)	- Quantity (can meet the needs of daily water consumption) [21]. - Continuity (water is available for 24 hours, 7 days a week) [21].

Based on the literature review, the technical aspects of DWSS can be divided according to: source, production, transmission, and distribution. Major sources of water used in Indonesia are surface water, groundwater, and rainwater. The production system consists of treatment and storage; transmission and distribution system (piping system or non-piping system); water delivery service. The aspects of the latter that will be further explored in this study are water tariff and safe access.

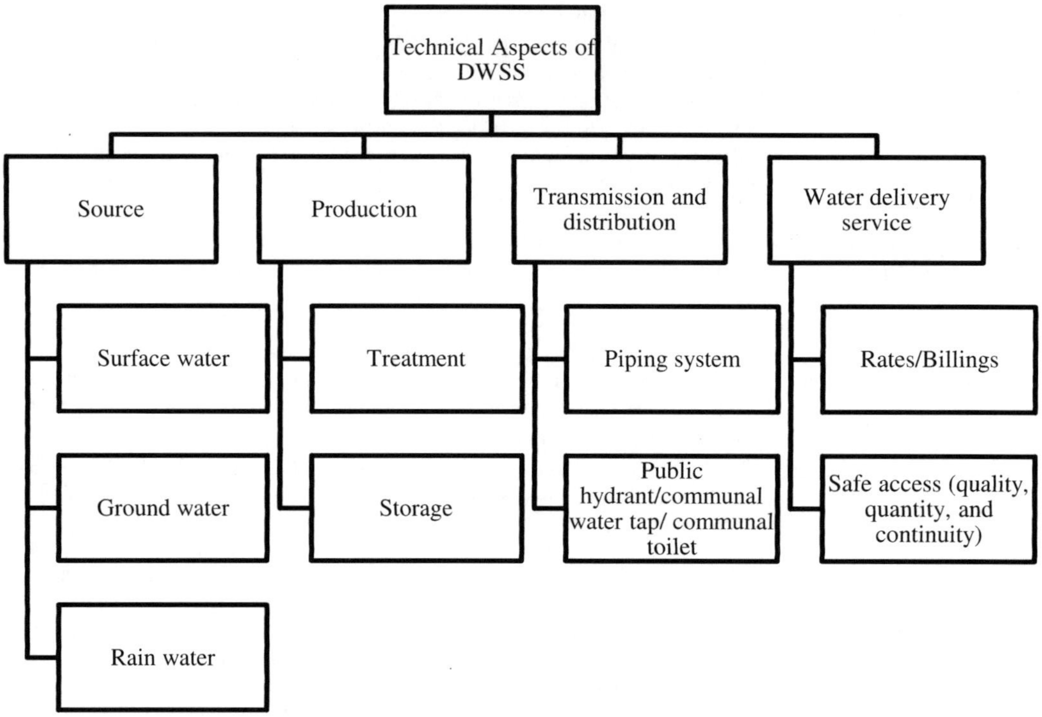

Figure 1. Technical aspects for developing DWSS

4. Discussion

4.1. Status of informal water infrastructure in Indonesia

In 2015, only 18% of households were served by the public water supply system (PDAM), where most of the served consumers were living in urban areas. This condition is caused by the infrastructure not being optimally utilized, resulting in a large amount of idle capacity (37,900 liter/second) [13]. This is due to several factors, among others the decrease of raw water capacity, unoptimized operational capacity of production units, and unallocated budget of distribution pipes and house connection installment by local governments.

In terms of funding, current drinking water investments rely on government funding rather than alternative funding sources such as PPP (public private partnerships) or B-to-B (business-to-business) schemes. PDAM does not function optimally due to the ineffectiveness of the institutional system and the low drinking water tariff, which is below the production cost. In 2014, there were 182 PDAMs with good or 'healthy' performance, 103 PDAMs with 'less healthy' performance, and 74 PDAMs were categorized as having 'bad' performance. The Directorate General of Human Settlements has encouraged the development planning of drinking water infrastructure through the formulation of the *Master Plan of Drinking Water Supply System* (RISPAM), which thus far has been formulated in 468 regions in Indonesia.

On the other hand, DWSS that are not managed by PDAM (let us say 'informal DWSS') have shown a vast development (82%), but nevertheless their growth still requires assistance from the government and other parties. These informal DWSS emerge as a response as well as consequence of the inability of the government to provide water service to 100% of the population. Some of the informal water infrastructures have been appropriately established but most of the rests are not in accordance with the standards and procedures of DWSS from a technical point of view. Table 2 presents a brief description of the condition of informal water infrastructure in Indonesia seen from its technical aspects and divided according to provider.

Table 2. Typology of water informal infrastructure practice in Indonesia based on technical aspect

Provider	Typology and location	Technical aspect					
		Source	Production		Transmission and distribution	Water delivery service	
			Treatment	Storage		Tariff/ billings	Safe access
Individual	**Typology A:** Household in Lebak Siliwangi, Bandung City, West Java	Shallow groundwater	No	No	Pumping system	No	No lab test
	Typology B: Household in Lebak Siliwangi, Bandung City, West Java	Artesian well (deep groundwater)	No	No	Pumping system	No	No lab test
	Typology C [9]: *Nyelang* (buy from the neighborhood or person to person agreement) in Penjaringan, North Jakarta	Artesian well (deep groundwater)	No	No	With hose and water bucket	No fixed rate, rough estimation (Rp13,000 to Rp89,000 per m^3)	No lab test
	Typology D [9]: *Nyelang* (buy from the neighborhood) in Penjaringan, North Jakarta	From PDAM	Yes	No	With hose and water bucket	No fixed rate, rough estimation (Rp13,000 to Rp89,000 per m^3)	Yes (with lab test)
	Typology E: Solidarity-based (water as a gift in special cases) in Lebak Siliwangi, Bandung City, West Java	Groundwater (shallow and deep)	No	No	With hose and water bucket	No (free)	No lab test
	Typology F: Solidarity-based (water as a gift in special cases) in Lebak Siliwangi, Bandung City, West	From PDAM	Yes	No	With hose and water bucket	No (free)	Yes (with lab test)

Provider	Typology and location	Technical aspect					
		Source	Production		Transmission and distribution	Water delivery service	
			Treatment	Storage		Tariff/ billings	Safe access
	Java						
	Typology G* [22]: Individual in Margahayu Raya Bandung City, West Java	Artesian well (deep groundwater)	Slow sand filter	Yes	Piping and pumping systems	No	Yes (with lab test)
Communal	**Typology H** [23]: Community-based in Ciburial Village, Bandung Regency, West Java	Spring water	No	Yes	Piping and gravitational systems	Fixed rate (no water meter)	No lab test
	Typology I [23]: Community-based in Griya Bukit Mas Residential, Bandung Regency, West Java	Artesian well (deep groundwater)	No	Yes	Piping and pumping systems	Progressive block tariff systems with water meter	No lab test
	Typology J: Community-based in Lebak Siliwangi, Bandung City, West Java	Artesian well (deep groundwater) and shallow well	No	Yes	Pumping systems, communal toilet, public hydrant and water tap	Fixed rate (daily, weekly, monthly)	No lab test
	Typology K: Community-based in Lebak Siliwangi, Bandung City, West Java	From PDAM	Yes	Yes	Communal toilet, public hydrant and water tap	Fixed rate (daily, weekly, monthly)	Yes (with lab test)
	Typology L* [22]: Community-based in Pesantren Oemardyan, Indrapuri, Aceh Besar Regency, Aceh	Stream intake (surface water)	Yes, slow sand filter	Yes	Pumping and piping system	Fixed rate (no water meter)	Yes (with lab test)
	Typology M* [22]: Community-based in Wisma Sanita,Pejompongan, Jakarta Pusat	Shallow and deep groundwater	Yes, slow sand filter	Yes	Pumping and piping system	No	Yes (with lab test)
	Typology N* [22]: Community-based in Giri Cahyo Village, Gunung Kidul Regency, Yogyakarta	Underground river	Yes, PV pumping system	Yes	Pumping and piping system	Fixed rate (no water meter)	Yes (with lab test)
	Typology O* [22]: Community-based in Oheitel Tual Village, Maluku Province	Groundwater	Yes, infiltration galleries and PV	Yes	Pumping and piping system	Fixed rate (no water meter)	Yes (with lab test)

Provider	Typology and location	Technical aspect: Source	Production: Treatment	Production: Storage	Transmission and distribution	Water delivery service: Tariff/ billings	Water delivery service: Safe access
			pumping system				
	Typology P* [22]: Community-based in SP Transmigrasi Jejangkit Cs, Barito Kuala, South Kalimantan	Peat water (surface water)	Yes, peat water filtration and PV pumping system	Yes	Pumping and piping system	Fixed rate (no water meter)	Yes (with lab test)
Water vendors	**Typology Q:** Refill water in Sukajadi, Bandung City, West Java	Refill water from groundwater	Yes (RO, UV, UF, ozone, etc)	Yes	Gallon package	Rp4,000- per gallon	No lab test
	Typology R: Branded bottled water in Sukajadi, Bandung City, West Java	Spring water	Yes (filtration, UV, and UF)	Yes	Gallon package	Rp15,000- per gallon	Yes (with lab test)
	Typology S: Water push cart seller in Kota Tua, Semarang City, Central Java	Shallow groundwater	No	No	Jerry can and push cart	Fixed rate per jerry can	No

Note:
*Categorized as best practice along with 20 other best practice cases of informal DWSS in Indonesia, which are not discussed in this study. These cases seem infrequently implemented, so that they can be referred to for other informal DWSS in other locations with similar characteristics.

Based on Table 2, the informal systems can be classified as systems with and without treatment, systems with and without tariff, and systems with and without storage. Most systems are systems with treatment and tariff. The systems in this category are: bottled water, community-based water supply with river as water source, and water from vendor. Most systems do not have a reservoir. Systems that apply a tariff system usually use a fixed rate; only one system applies a progressive tariff. Figure 2 shows the classification of the systems mentioned in Table 1.

Treatment, No Tariff Typology F, G, M	**Treatment, Tariff** Typology D*, L, N, O, P, Q, R
No Treatment, No Tariff Typology A*, B*, C*, E*	**No Treatment, Tariff** Typology H, I, J, K, S*

Note: *system without storage

Figure 2. Classification of Informal System Typology

4.2. Challenges of formalizing the informal water infrastructure in Indonesia
As stated above, in this study, the formal water infrastructure system refers to the technical practices conducted by PDAM in providing drinking water to its customers. Thus far, most PDAMs conduct procedures in accordance with the standard established in many regulations in Indonesia as well as the normative guidance as listed in Table 1. Based on Table 2, there are various typologies seen from a technical point of view that can at least portray informal water infrastructure practice in Indonesia conducted by individual households, community, and water vendors. Among these, production (no treatment and no storage) and water delivery service (tariff/billings: no water meter and rough estimation; safe access: no laboratory tests even though most of the consumers reveal that they are quite satisfied with the water consumed in terms of quantity, quality, and continuity) are technical factors that seem to be absent from informal DWSS. Yet, among the informal practices there are several best-practice cases that meet the requirements of the technical aspects of developing DWSS such as implemented in a formal system by PDAM among others.

In order to implement the formalization of informal systems in the context of their technical aspects, the current informal practices face some challenges in duplicating the formal infrastructure system. First of all, it is necessary for the government to have a simultaneous monitoring and control of the aquifers exploited by communities for drinking water and relay that information to society in order to raise awareness of water saving, water reuse, and water conservation. Most of the informal practices (70%) rely on groundwater in fulfilling daily consumption, yet society often ignores that the availability of raw water has been decreasing year by year.

Secondly, many simple methods and technologies of water treatment can be applied in small-scale DWSS, both at household and at communal level. Technical assistance and guidance should become routine homework for the government and transfer of information and knowledge on simple and basic water treatment should be widely spread to society. Thirdly, developing and implementing regulations for applying water meters in any informal installation, both in piping systems (house connections) and public hydrants/water taps or communal toilets. That way, water theft can be reduced and rough estimations can be avoided with a more fair calculation system. Besides, with a proper billings method, affordability, fairness, cost recovery, water efficient consumption, protection of raw water, and transparency as well as accountability can be provided [20]. Lastly, regulations are also needed for informal practices to oblige all water consumers to test their water in a laboratory to ensure its quality, which must meet the health standard established by the Minister of Health to guarantee that the water consumed is safe enough.

5. Conclusion
Seen by the comprehensiveness with which they meet the formal criteria, several informal cases in different locations were found to be practically the same as the formal system, with some interesting notes to be explored. These cases tend to be best practices and can be a good model to be duplicated in other locations with similar characteristics. Generally, as far as the technical aspects are concerned, most of the informal systems do not apply a progressive tariff, do not have a storage/reservoir, have no water treatment plant and rarely conduct treatment in accordance with the standards and procedures that the formal system uses. It was also found that informal systems mostly provide dubious access to safe water, especially according to the health quality standard, since most of the informal systems do not test the water utilized for drinking water with laboratories tests. Government as an enabler, provider, regulator, and facilitator supported by and cooperating with other stakeholders (donors, NGOs, and private sector parties) has a central role to play in guiding society to attain access to safe water, especially in the production and water service delivery aspect. The most important improvement is transforming informal systems into formal systems in terms of water quality standard and water tariff.

6. Acknowledgments
The research, on which this article is based, is funded by Institut Teknologi Bandung under the scheme of 'Program Penelitian, Pengabdian kepada Masyarakat dan Inovasi (P3MI) ITB 2017'. The title of the research is 'Can Informal Infrastructure Overcome Infrastructure Provision? Case: Provision of

Infrastructure Informal Lebak Siliwangi'. The authors would like to thank all the stakeholders involved in this program. However, the authors alone are responsible for any mistakes and shortcomings.

References

[1] Nyarko K B, Awuah E, and Ofori D 2009 Local initiative in community water supply: Case study in Ashanti Region, Ghana *J. Desalination* **248** 650-65
[2] Allen A, Davila J D, and Hofmann P 2006 The peri-urban water poor: citizens or consumers? *J. Env. & Urb.* **18** 333-351
[3] Adams E A and Zulu L C 2015 Participants or customers in water governance? Community-public partnerships for peri-urban water supply *J. Geoforum* **65** 112-124
[4] Allen A, Hofmann P, Mukherjee J, and Walnycki A 2016 Water trajectories through non-networked infrastructure: insights from peri-urban Dar es Salaam, Cochabamba and Kolkata *J. Urb. Research & Practice* **10** 22-42
[5] Wutich A, Beresford M, and Carjaval C 2016 Can informal water vendors deliver on the promise of a human right to water? Results from Cochabamba, Bolivia *J. World. Develop.* **79** 14-24
[6] Matsinhe N P, Juizo D, Macheve B, and Santos C D 2008 Regulation of formal and informal water service providers in peri-urban areas of Maputo, Mozambique *J. Phys. And Chem. of the Earth* **33** 841-849
[7] Kjellen M and McGranahan G 2006 *Informal water vendors and the urban poor* (London: International Institute for Environment and Development-IIED)
[8] United Nations Development Programme 2011 *Services and supply chains: The role of the domestic private sector in water service delivery in Tanzania* (New York: United Nations Development Programme)
[9] Kooy M, Walter C T, and Prabaharyaka I 2016 Inclusive development of urban water services in Jakarta: The role of groundwater *J. Hab. Int* 1-10
[10] De Soto H 2000 *The mystery of capital: Why capitalism triumphs in the west and fails everywhere else* (New York: Basic Books)
[11] Choplin A and Dessie E 2017 Titling the desert: Land formalization and tenure (in)security in Nouakchott (Mauritania) *Habitat International* **64** 49-58
[12] Aparcana S 2017 Approaches to formalization of the informal waste sector into municipal solid waste management systems in low- and middle-income countries: Review of barriers and success factors *Waste Management* **61** 593–607
[13] Directorate General of Human Settlements, the Ministry of Public Works and Housing 2016 *Strategic plan of the Directorate General of Human Settlements 2015-2019* (Jakarta: Directorate General of Human Settlements, the Ministry of Public Works and Housing)
[14] Ministry of Public Works and Housing, Indonesia 2016 *Evaluation standard document. Attachment IV Regulation of Public Works and Housing Minister Number 27/PRT/M/2016 regarding development of drinking water provision* (Jakarta: Ministry of Public Works and Housing)
[15] Spellman F R 2003 *Handbook of Water and Wastewater Treatment Plant Operations* (Boca Raton: CRC Press LLC)
[16] Directorate General of Human Settlements, the Ministry of Public Works and Housing 2007 *Development of Simple Drinking Water Supply System* (Jakarta: Directorate General of Human Settlements, the Ministry of Public Works and Housing)
[17] Masters G M 1991 *Introduction to Environmental Engineering and Science (*Englewood Cliffs: Prentice-Hall, Inc)
[18] HDR Engineering Inc 2001 *Handbook of Public Water Systems*: *Second ed* (New York, Chichester, Weinheim, Brisbane, Singapore, Toronto: John Wiley & Sons)
[19] US EPA 2002 *Capacity Checklist for Public Water Systems* (Washington DC: US EPA Archive

Document)

[20] Ministry of Home Affairs, Indonesia 2016 *Regulation of Home Affairs Minister Number 71 2016 regarding Calculation and Stipulation of Drinking Water Tariff* (Jakarta: Ministry of Home Affairs)

[21] Ministry of Public Works and Housing, Indonesia 2016 *Evaluation Standard Document. Attachment IX Regulation of Public Works and Housing Minister Number 27/PRT/M/2016 regarding Development of Drinking Water Provision* (Jakarta: Ministry of Public Works and Housing)

[22] Directorate General of Human Settlements, the Ministry of Public Works and Housing 2014 *Buletin Ciptakarya: Karya Cipta Infrastruktur Permukiman Ed.11/Tahun XII/November 2014* (Jakarta: Directorate General of Human Settlements, the Ministry of Public Works and Housing) pp 26-33

[23] Maryati S and Humaira A S 2015 Co-benefits of community based water supply management *Proc. Int. Conf. on The 5th Environmental Technology and Management Conference "Green Technology towards Sustainable Environment" (Bandung)* OP/NR/018-2-7

IOP Conf. Series: Earth and Environmental Science **158** (2018) 012006 doi:10.1088/1755-1315/158/1/012006

Medan City: Informality and the Historical Global City

N Sudarmadji, B Tyaghita, P T Astuti, D Etleen

Tunas Nusa Foundation, Bandung, Indonesia

Corresponding author: novinoviantari@gmail.com

Abstract. As projected by UN that two-thirds of Indonesia's population will live in urban areas by 2050, rapid urbanization is happening in Indonesian cities. Initial research on eight Indonesian Cities (which includes Medan, Jatinegara, Bandung, Surakarta, Yogyakarta, Surabaya, Balikpapan, and Manado) by Tunas Nusa Foundation since 2012 shows that urbanization of each city has happened throughout history creating cultural, economic, and environmental networks that are distinct from one city to another. While the networks remain until today and continuously shapes the urban agglomeration pattern, not all parts of the city could undergo subsequent development that confirms the existing pattern, leading to the creation informality. Nor could it make future planning that comprehends the nature of its integrated urban dynamic beyond its current administrative authority.

In this paper, we would like to share our study for Medan, North Sumatra as it shows a portrait of a city with a long relationship to a global network since the Maritime trade era. Medan has become home to many ethnic groups which have sailed and migrated as part of a global economic agenda creating a strong economic network between port cities along the Malacca Strait. The city has kept its role in the global economic network until today, to name a few, becoming the frontier for the Indonesia-Malaysia-Thailand Growth Triangle. While we celebrate Medan's potential to become a global city with major infrastructure development as well as cultural assets as its advantage in the future, we argue that microscale cohesion supported by government policy in agreed planning documents are fundamental for the city to thrive amidst the challenges it is facing. Yet, these cultural assets, as well as micro scale cohesion in Medan City today, are still undermined. Thus, informality in Medan exists as result of ignorance and marginalization of certain socio-cultural groups, abandoning places and identity, as well as the marginalization of small economic and transactions.

The hypothesis is that the informality within an urban network and the agglomeration pattern in Medan today are evidence of external stimulation in combination with limited micro scale cohesion built throughout history. Using historical timeline analysis, we would like to identify the informality in an urban network and its relation to history. The aim is to understand the fabric of Medan city made by the trans-cultural-economical and environmental network. This study is expected to give the foundation of Medan's character as the basis of her peoples' livelihood and for a better planning/development in the future.

IOP Conf. Series: Earth and Environmental Science **158** (2018) 012006 doi:10.1088/1755-1315/158/1/012006

1. Introduction

Medan is the provincial capital of North Sumatera, Indonesia. It is the third biggest city after Jakarta and Surabaya, the largest in Sumatera Island with an area of 265,2 km2 and a population of more than two million people. Medan is also known as a metropolitan city with a long history of cosmopolite community since the era of maritime trade at the beginning of the century, which made Medan hold the potential to be a global city. Even before colonization in the 17th Century, Medan has highly diverse residents which consist of Malayan, Batak, Tamil, Chinese, Arabs, Indian, Turks, Persian, Javanese, and European.

The name of Medan, however, was mentioned in around 16th century when Islam started to grow within the Malay Archipelago. Kampong Medan Baru built by Guru Patimpus at the meeting point of Deli and Babura River dated July 1st, 1590 started the formal identity of Medan. However, Deli Port in the estuarine of Deli River, or today Labuan, built by Chinese merchants to relate with their settlement of Kota Cina had predated the name of Medan in the 12th century; while a settlement of the native Batak people had already appeared in the interior of Sumatra near Toba Lake in the 3rd century. Other than that, the global trade network had brought many communities around the world to make a colony along the East Coast and West Coast of Sumatra since the first century, which in turn initiated a robust cultural atmosphere, in this focus, of Medan City today.

Medan gained its status as a city only in 1886 after it became an administrative centre for the Dutch Indies' production system of plantations in East Sumatra within the late colonization era. The rich resources and the wealth it had gained from Medan and its surrounding hinterland subsequently initiated the formal city planning for the importance of Dutch-Indies' representative in Medan City at the beginning of the 20th century.

This paper starts with the premise that Medan was a global city that has been connected to the global economic network since the ancient maritime era. Until today, it holds the legacies of a multicultural community and transnational economic network which spans at least to the extent of Malacca Strait. Yet, we argue that these legacies are potential drivers that are not, yet, well acknowledged and comprehended in the recent development scheme of Medan City. Thus, might raise the opportunity for informality and social issues.

Therefore, the objective of this study is to develop a better understanding of how the global economic network has molded and given foundation to the making of Medan City. The hypotheses are: (1) that the evidence of its connectivity to the global economic network lies in the urban network and agglomeration pattern of Medan City, which represents not just a physical entity, but also social, cultural, and economic cohesion in the making of Medan City; (2) that informality has happened throughout Medan City's formation which also is an evidence of how external factors has been a stimulant to the process. It is expected that this understanding will be a contribution towards better planning and development of Medan City.

2. Understanding the Context of Global City and Informality

2.1. Global City

The concept of a global city has been studied since the 1960s and gained more attention in the 1980s. The concept associates a city with globalization and suggests its significance in a global economy. In short, a global city is a significant production point of specialized financial and producer services that make the globalized economy run [1]. Deriving of what might entail the global economy, [2] points out that through the process of making the world economy, there have long been **cross-border economic processes** - to include within are flows of capital, labour, goods, raw materials, and tourists - in which a global city is considered as the top hierarchy of urban entity in population, networks, and sophistication of services that lead to important implication for economic development.

The criteria of "what makes a global city" is continuously evolving as the world progresses [3]. This suggests that the concept might be applied distinctively for cities in the context of ancient times, cities today, as well as cities in the future. A literature discussing cities in the ancient world suggests global cities as cities ***"where globalization takes place in much more pronounced ways than anywhere else:*** *as economic exchange, migration, communication, technological development and political conflict, as cultures clashing and amalgamating, and also as a violent process"* [4]. Meanwhile, The Global Cities Index poses five critical dimensions that is used to examine the performance and ranks of cities in relation to the prominence of the cities in the global economy, those are business activity, human capital, information exchange, cultural experience, and political engagement [5].

Sassen has introduced her hypotheses through which she organized the theorization of a global city model. In the hypotheses, she introduces the impact of the growing fortune of the city on the making of new urban territory and social and cultural dynamic. As a city becomes a significant specialized service firm in an economic network, it results in the strengthening of cross-border city-to-city transactions and networks which might be the beginning of **the formation of transnational urban systems [2]**.

In relation to the nation-state, the independent connectivity of the city to the global economy, or at the least within the territory of the transnational urban system, might strengthen to the extent of increasingly disconnecting the city from their broader hinterlands or even their national economies. This, however, might happen in the other way around. Sassen suggests that both national and global are mutually exclusive space, yet since the **independent connectivity with the global economy and nation-state, one might begin where the other ends [2].** Another study suggests that a global city is, however, able to meet the criteria of having far-reaching significance for only certain periods of time which can be short or long [4].

Sassen also suggests that the growing number of high profit-making and high-level professionals contributes to the raise degree of **spatial and socio-economic inequality** evident in the city [2]. **Informalization** is regarded to find an effective demand to survive under conditions in which their profit rates do not allow them to compete with the high-profit making firms at the top of the system. At this point, the growing inequalities between highly provisioned and profoundly disadvantaged sectors and spaces of the city will raise questions of power and inequality.

Sassen concludes that a focus on networked cross-border dynamics among global cities sets a more readily **growing intensity of transactions in other domains, to name it out: political, cultural, social, and criminal [2]**. She adds, *"Global cities around the world are the terrain where a multiplicity of globalization processes assume concrete, localized forms. These localized forms are, in good part, what globalization is about. Recovering place means recovering the multiplicity of presences in this landscape. The large city of today has emerged as a strategic site for a whole range of new types of operations—political, economic, "cultural," subjective. It is one of the nexi where the formation of new claims, by both the powerful and the disadvantaged, materializes and assumes concrete forms."*

2.2. Informality

A closer look at informality in urban space is mostly associated with the phenomenon of the process of marginality [6]. The term was firmly established in the early 1960s as an alternative to the functionalist urbanism proposed by the Congrès Internationaux d'Architecture Moderne (CIAM). It came up to address population growth and the rapid spread of spatial contestation along with the loss of certainties regarding the paradigms of urbanism and modern architecture [6]. **Informality is defined as groups of people, economic activity, spatial entity and urban growth process which are considered outside the formal frameworks by those who have power whether in the economy, social and political sector** [6][7].

Other literature state that informality has emerged as a concept in the 1970s and refers to the movement of labour to the city during 1950-1960 in which informality within urban space was often associated with immigrants and poverty which were denied access to wage employment [7]. Informality which refers to certain practices done by groups of people consists of petty traders, street vendors, coolies and porters, small artisans, messengers, barbers, shoe-shine boys and personal servants at alleyways or street sides [7][8]. The constant incoming of immigrants which participate in the trade-sector within urban space contributes to the growth of spatial entity of informality which may consist of informal trade areas that are more dynamic and slum settlements or informal housing areas which were more permanent. Informal settlements often found occupying what is considered 'terrain vague' within urban space, which may consist of interstitial land, marginal use, unmapped land, and/or the land available for change of use [7][8]. The local factor contributing to the growth of informality may consist of economic factor such as in favela, Rio; political factor in most of the Middle East, privatization of lands both in the Middle East and South East Asia and communication revolution in Karachi. However, what is similar in every case of informality in Latin America, the Middle East, and Asia is how globalization is affecting and contributing to the growth and characteristic of informality which is locally based [7].

Lutzoni names out three phases that sets the perspective towards informality. The first phase is between the 70s and 80s with dualist school predominance in which informality is seen as a group of marginal activities excluded from the formal economy. The dualist approach considers informal and formal as two different sets of activities and/or sectors, this stemmed from ILO research in 1972 regarding informality [6][7]. This approach argues that informal sectors growth as the residue of formal activity [6]. De Soto in 1989 argues that informal sectors grow because of excessive state regulation, the foundation of the legalist approach. Both approaches develop another approach regarding the relation between formal and informal. The second phase is during the 80s to 90s with legalist approach and structuralist school predominance in which; respectively; informality is seen as a set of positive forces in a formal context linked with power strategies and considering informality as an integral part of a single system yet referred solely to the economic field. The structuralist, developing its approach also from ILO research, however, while admitting there are formal and informal sectors, argue that those two are part of one system. Thus, their approach proposed that the role of the state was to equalize the difference [7]. The latest perspective is in the debate of the twenty-first century which conceived the relation of informality with globalization processes that are changing the economic, social, and political geography of the world. The relational approach that prefers to put the dichotomous view of informality into the background and explore the relationship between formal and informality [6].

As we list out the perspective towards informality, in the case of this paper, the informality in Medan city might have occurred due to diverse reasons as all paradigms suggested. The issues of informality were not only how to stop it, clean it or even legalize it. **Informality has grown as a way of survival for those which were outside the 'formal frame' and has shown its contribution to the local economy [8].** Or, using the understanding by Wirth (1938) in his essay of Urbanism as Way of Life sets a city as a "place of urbanism", an urban way of life that is no longer a mere physical entity. Urbanism is considered as the outcome of a wider system of relations deriving from variables that determine the urban condition as they interact with each other [6]. This understanding might give the foundation to an alternative point of view towards informality suggesting that the discussion should begin with an alternative way of framing **informality as a concept of hybridity, simultaneousness, and coexistence [6]. Both formal and informal order may both be considered legitimate, simultaneous ways of "making the city"** [6].

Identifying and understanding informality within urban space that grows over time through urban agglomeration is necessary since informality has the organizing logic which has become pervasive as to comprise a way of life in many cities [7]. Borrowing Al Sayyad's questions, the two important questions to understand informality are 'what made it happen and what is the form of vulnerability that is being endangered by these processes of informality.

3. Historical Analysis of Medan City

The discussion in this paper will be delivered as a historical analysis framing Medan in four periods of contexts which are: (1) the time of maritime trade and early settlements; (2) the time of the establishment of Islam, colonization, and the city making; (3) the time towards sovereignty to the centralized authority in the context of Indonesia as the new nation; and (4) the time of reformation to the recent era in which cities are taking back its significance in the developing economically, physically, socially, and cultural (from nation-state).

We will start the discussion with the geographical context of Medan in the Malay Archipelago, Malacca Strait, and Sumatra Island as well as the natural context with the richness that drives trade. Deriving from the literature of global city and informality as well as towards the objective of this paper, we will then identify from each period: (1) the context that suggests the driver or stimulant of the situations; (2) actors to suggest subject of cohesion and power contest; (3) urban network and/or agglomeration in the form of settlement of traders, labourers and others, office or administration cluster of firms, plantation and/or production source, ports, waterways and/or roads, railway and stations - to suggest physical evidence of the interactions between actors; (4) network of economy to suggest evidence of economic scale; as well as (5) movement, organization, and conflicts as social evidence of the interactions between actors.

3.1. The Maritime Trade Era and Early Settlements

The Maritime Trade or Spice Route was a trade route known mostly to transfer spices from China to India, Arabia, Africa, and vice versa. It occurred benefitting from the monsoon winds of the Indian Ocean. The route was initially established since 3000 BC and initiated the global trade network until today.

Connecting the Indian Ocean and the South China Sea, the Malay Archipelago was the entry point of China in the East of Asia; and India, Arabia, and Africa in the West. It was also the stop-over point where goods were exchanged in the array of ports. The archipelago itself was well known as the source of clove, mace, paper, nutmeg, cinnamon, baros, camphor, benzoin, and gold.

The gate into the Malay Archipelago was the choke point of Sea of Melayu, today Malacca Strait. A funnel-shaped waterway separates- or connects- the mainland of the Malay Peninsula and Sumatra Island. The Malay Peninsula and Sumatra Island act as a natural barrier separating the Indian Ocean and the South China Sea, creating transit havens for vessels waiting for the change of the monsoon current or to connect with the overland passage trade route through the Malay Peninsula [11]. Sumatra Island, also known as Swarnadwipa which means the Island of Gold, was the place of many traders to settle and establish colonies. Along its East Coast are rivers and lowland estuaries that connect Malacca Strait to the hinterland of Sumatra. Spanning from North East to South East in the almost centre line of the island is the mountain ridge - the source of gold and other forest virtues that gives notion to the name of Swarnadwipa. An enormous volcanic lake named Toba is located in the North portion of the mountain ridge with the drained part of its valley and ancient floodplain bequested to support life. Separated by the mountain ridge, is the West Coast of Sumatra that has been known as the source of camphor, the ingredient mostly used in incense in ancient times. Although part of the East coast was a lowland swamp, its proximity to the shallow waterway of Malacca Strait made it preferable for settlement rather than the West part of Sumatra that faces directly to the Indian Ocean.

When the maritime route was established around the 1st to the 3rd century [10], the strategic situation drove early ports and kingdoms to flourish along the Malacca Strait region. To include the region within Central Southeast Asia[1], [12] stated Central Southeast Asia as a zone where traders brought spices and forest products in exchange for textiles and other manufacture from China and India.

Until the 15th century, the maritime trade network was self-regulated in nature where the term of trade was set by the merchants and the demands of the market [13]. However, at least within Malacca Strait, the trade was beyond the influence of a kingdom's or empire's policy until around 4th century. With the rise of the Gupta Dynasty in India, the concept of 'Negara' was introduced along with the spread of Hinduism, and subsequently Buddhism [17].

Amongst the earliest settlements was Kota Cina, located at the estuary of Deli river in the West part of Labuan Deli today. It was a settlement of Chinese traders who invited colonies and built a port settlement. At the West Coast, a port named Baros was well known since the early century for its commodity of camphor. Traders from Tamil, Arabs, Persian, and Javanese were involved in trading on the West coast of Sumatra. Later, it was mentioned as the place where Tamil traders established their first trader confederation [18]. Meanwhile, the native of North Sumatra, Batak Karo, was settled at the hinterland around Lake Toba and built their first settlement in 200 AD. Not much influence from foreign traders infiltrated the interior of Sumatra. Only the Tamil people had ventured into the interior.

In the time when Hinduism started to spread, the early Hindu Kingdom in Malacca Strait was probably Langkasuka. Established in 300 ADS in Kedah, it was then followed by Jiecha in 500 AD. The heyday of Hindu - Buddhist kingdoms, brought from China and India, was within the era of Sriwijaya kingdom around the 7th century.

The Buddhist kingdom of Sriwijaya managed to control Malacca and Sunda Strait with the trade activities along the waterways. Sriwijaya actively participated in the growing world economy by positioning itself as the 'Master of the Straits of Malacca' [11]. The kingdom of Sriwijaya was involved in commerce, compelling passing vessels to visit Sriwijayan ports and levying port dues and taxes upon them; while establishing strong economic ties with its giant neighbours, China and India [11].

The kingdom was not just the centre of commerce, but it was also the centre for Hindu and Buddhist learning, with its centre recorded varying between Palembang and Jambi today. A 7th-century Chinese monk I Qing noted the city as the place where more than a thousand Buddhist priests in its monasteries advised pilgrims from China to spend time to master Sanskrit and Pali. Since there were Indian and Chinese monks maintaining these language abilities, there must have been Indian and Chinese commercial communities maintaining the monks. Chinese trading communities are also likely to have helped manage the tributary trade between Srivijaya and Tang China, which was so important for the commerce of the whole region [12].

In 1000 AD, the Pane kingdom was established at the bank of Pane and Baruman Rivers which generated the East-West trade corridor. Another major kingdom in Sumatra was Aru (Haru) kingdom. It was a Hindu Kingdom established as a Batak Karo polity that drew the hinterland of the Northern part of Sumatra into the network of inter-regional commerce in 1100 AD. It put Kota Cina and today's Deli Serdang Regency under its sovereignty with Kota Rentang which was suggested to be the capital city of the kingdom. In 1300 AD, the port that had been built along with the settlement of Kota Cina became the well-known Deli Port, one of biggest ports which dominated international trade at Malacca Strait. Pane and Aru, together with the kingdom of Aceh, believed to give influence to the current Batak culture. Deli Port in the other part arguably has underpinned economic connectivity that made an initial foundation for the making of Medan City.

[1] ".. it is a place in quest of a more adequate name. It is the Bangkok-Jakarta central axis of Southeast Asia formed by the world's longest peninsula, nearly blocking the shipping route between East Asia and the rest of Eurasia and Africa, the two Straits through which it obliges that shipping to pass, and the adjacent littoral.It is thus a natural place of entrepots and meeting places, .."

After the fall of Sriwijaya in 1025 AD, Majapahit was the next political power that managed to rule several states in Sumatra, Malay Peninsula, Borneo, Celebes, Moluccas, to the extent of some part of the Philippine archipelago [11]. Ruling from the tip East of Java Island, by the mid of the 14th century, Majapahit happened to take command of the Malacca and Sunda Straits [30] and gained its wealth from agricultural produce, particularly rice; as well as economic and political network benefitted from maritime trade [11].

By the end of 16th century, Sumatra was the location of many ports with city-states and kingdoms which included; the kingdoms of Pasai, Pedir, and Deli along the East Coast; Tiku, Priaman, Padang and Selebar along the West Coast; while Karo Simalungun and Toba Batak remain in the interior close to Toba Lake.

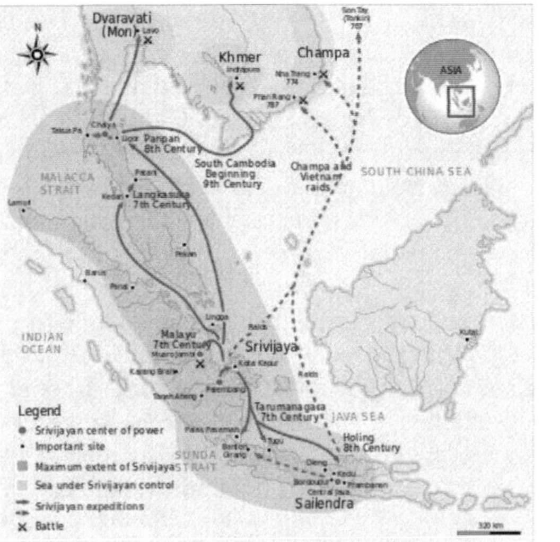

Medan and its regions under Srivijaya

Medan and its regions under Majapahit

Figure 1. Initial of Medan vicinity during Srivijaya and Majapahit time.
Source: https://upload.wikimedia.org

Reid mentioned that it (Central Southeast Asia) remains today what it has been throughout recorded history, one of the most urban-dominated zones of the world [12]. A plural community was shaped along the choke point of maritime trade, in this case, states in Sumatra and along Malacca and Sunda Straits. Cosmopolis was therefore built after successful entrepot in the zone which was governed by leaders themselves, cosmopolitan in culture and able to mediate between groups [18]. Specifically, along the East Coast of Sumatra, the trade ports flourished and, with the trader colonies, its generated cultural innovation through time to become an influence to the extent of South East Asia [18].

Reid suggested that while religion and language separate nations, the spirit of commerce unites them so that the task of cosmopolis is to mediate these two contrary impulses. He added, "the problem for cosmopolis in this part of the world was to find a form of government that would protect commercial communities rather than preying upon them; it so happened that the formula had almost nothing to do with the nation, but much with the supernatural charisma of kingship." A system displayed by Sriwijaya as it gained its golden time from the charisma of its king who was mentioned as kind and courteous, and cared for all foreigners [18]. Arguably, even when the concept of 'Negara' was indicated to be the foundation of Hindu kingdom, Majapahit reached its control over the Malay

IOP Conf. Series: Earth and Environmental Science **158** (2018) 012006 doi:10.1088/1755-1315/158/1/012006

Archipelago by Mitreka Satata strategy which means, international relations with friendly countries, to include within were Cambodia, Champa, and, presumably, India.

The interactions associated with trade has underpinned the political system and notion of a cosmopolis society along the maritime road. Represented in Siam, Sumatra, Malay, to kingdoms along Java within what Paul Wheatley called the Isthmian Age, between about 550 and 1400 AD [18]. The name of the period he suggested is to be associated with the importance of little port-states along Central Southeast Asia. Within which, the economic system was run by many traders with the most powerful state as the political power to manage and secure the trade.

3.2. The Establishment of Islam, Colonization, and the City Making

Coinciding with the dawn of Majapahit was the rise of Islam in the Malay Archipelago. The new belief introduced by Gujarati traders began to flourish in the 11th century and subsequently became a new foundation for many kingdoms in the Archipelago. The first kingdom that initiated the formal spreading of Islam was Samudra Pasai. It was located at the tip of North Sumatra with the first king, Marah Silu. He converted to Islam and built a kingdom that lasted from 13th to 16th century. Following and replacing the influence of Samudra Pasai, was the Malacca Sultanate in 14th century located in the Malay Peninsula and Aceh Sultanate in the 15th century. Afterwards, there were the kingdoms of Johor, Padang, Minangkabau, Indragiri, and Jambi. However, none of these kingdoms had immense sovereignty to master the control over maritime trade in the Archipelago as Sriwijaya nor Majapahit.

At the beginning of 16th century, the maritime trade network started to become involved in the dynamic of the world's political and/or economic interest. With the advent of the political turmoil in 15th century Europe brought by the capture of Constantinople by the Ottoman Turks, the Age of Exploration brought the Portuguese to conquer Malacca Sultanate in 1511. This event marked the start of the colonization era in the Malay Archipelago. Following the Portuguese were the Spanish, Dutch, and British that anchored sails and attempted to gain monopoly control over the trade along the Malay Archipelago.

Similar to the competing atmosphere between Europeans who came to the Archipelago, the situation between kingdoms was a power contest amongst each other. Using the circumstances, during the first century of its occupation, each European state tried to make alliances with quarrelling local powers to defeat other European states and gain control over land and/or resources. Until eventually those kingdoms who were befriended by the Dutch and British gained the most control in the Archipelago.

The Dutch wrested power in Southeast Asia after establishing a trading port in Bantam, today Banten, in 1600 Sunda Strait [11]. In 1602 the government of the United Province of Dutch provinces (there was no nation of Netherland yet at the time) created a single entity to conduct trade in Asia on the principle that this would be stronger and more powerful than the sum of the provinces and named it the Verenigde Oost-Indische Compagnie, the Dutch East Indies Company or VOC [19]. To include within the strategy were to expel Portuguese from Spice Island (Banda) and take over Malacca (that was accomplished by 1606). By maintaining good ties with the Johor kingdom, the Dutch succeeded to gain control over both Sunda and Malacca Strait [11]. While Britain maintained alliances with the Aceh Sultanate in the North West of Malay Archipelago; the VOC built the Citadel of Batavia in the North Coast of Java Island, today Jakarta, as the headquarter by which the Dutch managed to grow its power Eastward in the Archipelago.

Nevertheless, none of the European states or enterprises established control over Sumatra until 19th century. Most Europeans came as adventurers and small traders who ran businesses under the administration of the Aceh Sultanate. Monopoly attempted by the Dutch was attempted in Jambi and Palembang but resulted in resentment from the natives. Moreover, the challenging nature of the mountain ridge, dense swamps along its East Coast, and proximity to open water in the West Coast kept Sumatra from pervasive European invasion in the earliest two centuries [18].

IOP Conf. Series: Earth and Environmental Science **158** (2018) 012006 doi:10.1088/1755-1315/158/1/012006

Within the period, Deli port and Kota Cina in Labuan was under Aceh Sultanate power. Sultan Iskandar Muda sent his commander, Gocah Pahlawan, to extend the power of the sultanate in the remnant of Aru. Starting from Medan Putri kampong at the delta of Deli and Babura River, Westward from the previous city centre in Kota Cina, Gocah Pahlawan was able to expand the new power to the extent of today's Percut Sei Tuan and Medan Deli districts. He established kampongs under his control such as Gunung Klarus, Sampali, Kota Bangun, Pulau Brayan, Kota Jawa, Kota Rengas Percut, and Sigara-gara with the agreement that the vassal states were to be freed from taxes through marriage with the leader's daughter. The Deli Sultanate finally commenced its sovereignty in 1669 with a capital city named Labuhan, located 20 km East-ward from today's Medan City.

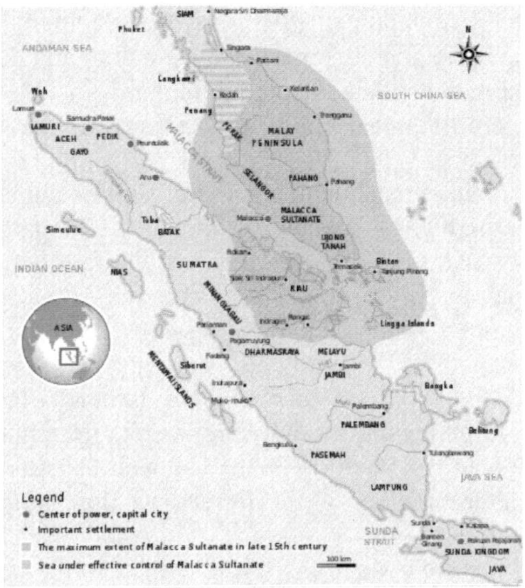

Medan and its regions under Malacca Sultanate

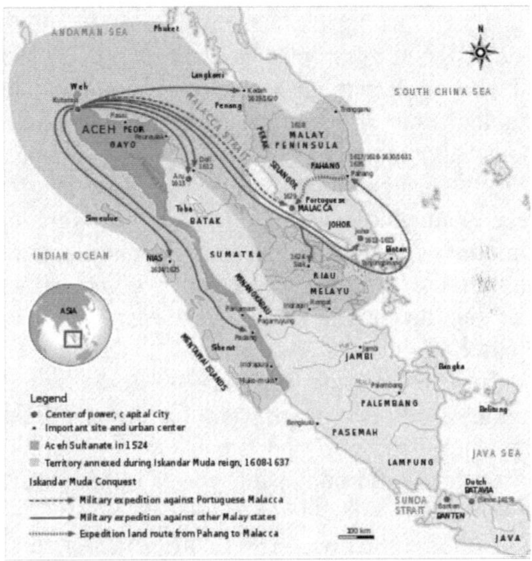

Medan and its regions under Aceh Sultanate

Figure 2. Initial of Medan vicinity during Malacca and Aceh Sultanate time.
Source: https://upload.wikimedia.org

Prior to the invasion of the interior of Sumatra, the Anglo-Dutch Treaty of London was agreed in 1824 which settled the Malay Peninsula under British authority and Sumatra Island under Dutch, which divided the Malacca Strait into Britain's at Malay side and Dutch at Sumatra side. Following the agreement, an internal conflict of local kingdoms which spurred the request of support from the Europeans was eventually seen as an opportunity for the Dutch to settle agreements with local powers to safely gain territory. Other than offensive attempts of Padri War in 1830 with Minangkabau, on 1 February 1858, a political contract was signed between the Netherlands Indies government and the Sultanate of Siak Sri Indrapura which brought Siak to come under the rule of the Dutch colonial government. Subsequently in 1862, to include within Siak dependencies, Deli, Langkat, and Serdang were claimed under Dutch territory [20]. With the instalment of Sultan Sjarif Kasim Abdoeldapalit Saifoedin as the new ruler of Siak by Resident Elisa Netscher in 1864, it began the first economic relations between the native rulers at the north-east coast of Sumatra and the Western agriculture industries [20].

Far in Europe, the new wave of the age of reason generated a new motivation for invention and exploration. New commodities were explored, transferred, and introduced between colonies, from the New World of America to the Malay Archipelago, India, Africa and/or vice versa. The invention of the steam engine, and coal fuel marked the beginning of the Industrial Revolution at the beginning of

IOP Conf. Series: Earth and Environmental Science **158** (2018) 012006 doi:10.1088/1755-1315/158/1/012006

the 19th century in Europe and soon were utilized in the colonies. The published research by William Marsden about Sumatra at the beginning of the 19th century, armed the exploration towards the interior of Sumatra to find sources of commodities as well as exploring the opportunity for the newly introduced commodity from across the seas by established plantations.

The second half of the century was marked by the opening of the Chinese market, the advent of ocean-going steamships, and the dramatic reduction in the time and cost of shipping goods to Europe after the opening of the Suez Canal in 1869. The operation of the Suez Canal further boosted the growth of plantations and once again made Malacca and Singapore Straits to become the fastest route to connect Europe and the Far East [17].

This period was also marked by the establishment of the biggest wealth that generated Medan as a city later. The Deli Maatschappij established in Labuhan by Jannsen, P.W. Clemen, Cremer, and Nienhuys had tobacco as its main commodity, other than coffee and tea, amplified by rubber and palm oil. The modern town Medan owes its existence to the decision of the Deli Maatschappij in 1869 to establish their administrative centre at a location at the confluence of the Barbura and Deli Rivers, about 10 km south of Labuan-Deli, known as Medan Putri [20].

In the last quarter of the 19th century, the development and expansion of the North East of Sumatra were on a quick rise. By 1874, there were twenty-two plantation companies in the North part of Sumatra with Deli Maatschappij remaining the wealthiest. More and more land was conferred from the Sultan of Deli to the Western companies, with headquarters in Europe or the US to the extent that, later on, the urban layout of Medan was mainly developed on the land concessions of the Deli Maatschappij [20].

The latest decade of the 19th century was closed with the invention of petroleum as new fuel for engines. The discovery of petroleum in the East Sumatra Regency was one that initiated the merger of The Shell Transport and Royal Dutch Petroleum Co. Ltd, today Shell company, as well as the new reason for world contesting power to set control over the archipelago in the immediate future of the 20th century.

Around 1880, eleven years after the establishment of the Deli Maatschappij, the embryonic urban grid was recognizable [20]. As much as other town components; such as the Esplanade, barracks of the military garrison, guest house, hospital, church, public auditorium, and colonial club; the urban connectivity of Medan was initiated by the Deli Maatschappij. Chosen for its administrative centre at Medan Putri, it pioneered the centrality of the area agglomerated with the Kampong Kesawan at nearby South in the West bank of Deli river and the main settlement of the Batak region, Sukapiring. Later on, with the centrality of the Chinese community role pioneered by Tjong A Fie, Kampong Kesawan was transformed into a commercial district characterized by the shophouse.

The Deli Maatschappij founded a railway company the 'Deli Spoorweg Maatschappij' in 1883 and in 1885 the railway between Medan and Labuan-Deli was inaugurated with the railway station located on the western side of the Esplanade, Lapangan Merdeka today [20]. In 1886 another railway was established, connecting Medan and the new port of Belawan as well as with the sources of production in the hinterland Southward towards the ridge, Northwest to Langkat, and Southeast to Siak. Passages towards the interior of Sumatra were also established to become the base for road development. This development impacted the shift of water-based transportation to land-based.

The urgency for efficiency and to control migrant labourers, also called coolie, and traders from China, Java, Tamil, Kling, Padang, Arab, and many regions in Asia, the settlers were organized in settlements based on ethnic group. Meanwhile, the centrality of Medan in the growing economy led to the transfer of the administration of Deli Assistance Residence from Labuhan to Medan in 1879, before finally in 1887 the capital of East Sumatra Regency was moved from Bengkalis to Medan.

In May 1891, Deli Sultanate palace was also moved from Kampung Bahari in Labuhan to Maimoon Palace in Medan. Built by the Dutch, as it was presumably a political strategy to maintain control over the local ruler. With the whole administrative and distribution centres operating in Medan, Labuhan was eventually left. Around 1900, the agglomeration pattern of Medan City shows an organization based on groups of ethnic and social status. The Europeans lived at the northern and western sides of the Esplanade; the Chinese population mainly lived and worked in the commercial district Kesawan and in the adjacent area on the eastern side of the railway; the Indonesian population, separated into the indigenous Malay residents and immigrants mostly of Javanese origin, lived mainly in the so called 'Sultans kampungs' in the South-east area and in kampungs on the outskirts of the town; while the small group of Arabs and other Asiatics, mainly from British India, usually lived and worked on the western side of Kesawan [20].

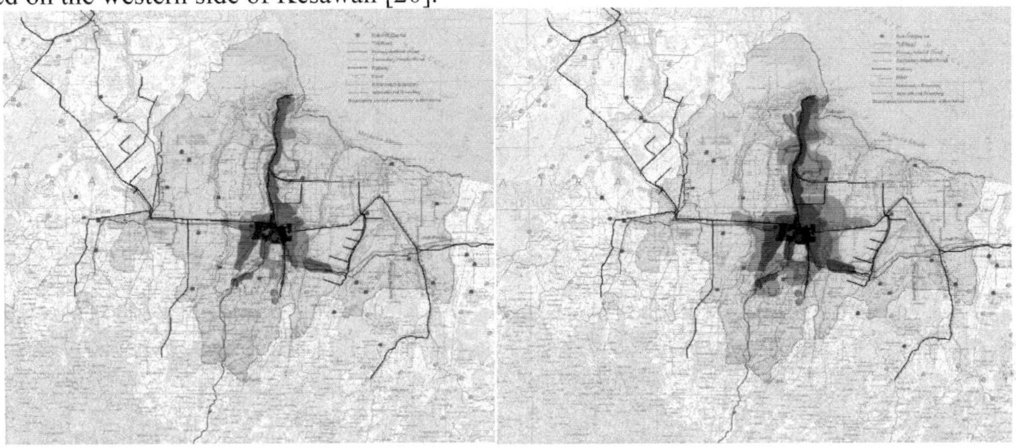

Figure 2. Comparison of Built Area in 1989 (left) and 2001 (right).

In the time when plantations were open, it had contributed to the development of settlement in Medan and the region surrounding it. The backbone of the town layout was formed in the eighties of the 19th century, and it can still be recognized very clearly nowadays. Planters and Chinese entrepreneurs gave birth to the modern city of Medan, in a turbulent process of interaction, in which the colonial administration played a subordinate role [20].

Amidst the development, most of the plantation workers and coolies had no access to land, fair compensation, knowledge, and health services. Prior to Kampong Improvement Program held by the Central Dutch Indies Government, the municipality took steps towards playing an active role in public housing and hygiene starting in 1918 [20]. Yet, with the fast-growing economy and expansion of the kampongs that was difficult to control, the municipal contribution did not make much sense with the limited financial resources and small-scale initiatives. A building model project was developed and expected to drive the kampong people to improve their own housing conditions - an idealistic point of view that was beyond the financial capability of the kampong people [20].

The time when plantations were opened, it was also the beginning of long-lasting impact to future socio-cultural tension. As Sultan of Deli started the land concessions, resistance rose from the inhabitants of the area, mainly Karo. These actions lead to the Sunggal-war from 1872 until 1897 which with the help of the Dutch, the Karo were beaten. On the other hand, although there was resistance among the Batak community in the interior of Lake Toba towards the Sultanate for Dutch authority, the missionary mission was successful resulting in more Batak communities to convert to Christianity as well as become more European educated. As they gained to be more resourceful, many of the Christian Batak communities established residences in Medan City within the first decade of the 20th century (Bruner) giving more diversity at the new urban centre of Medan City after the dominant Malayan, Chinese, Javanese, Tamil, and Arabic. Yet, transmigration policy which started during Dutch period from Java to Sumatra might be another factor that resulted to the increased tension of the socio-culture dynamic.

IOP Conf. Series: Earth and Environmental Science **158** (2018) 012006 doi:10.1088/1755-1315/158/1/012006

3.3. In the Time towards Sovereignty to the Centralized Authority of the New Nation

Despite the ambitious exploration during the 19th century, the beginning of the 20th century was endowed with ethical ideas of colonization and nationality. Entering the second decade of the 20th century, ideas of nationality and governance lead by common people flourished around the world, concern towards human rights was on the rise, and so was the competition to become the world's superpower. Prepared with a more educated youth within the Malay Archipelago, the occupancy of Japan in 1942 to 1945 finally provided an opportunity for the declaration of independence in August 1945.

However, the long and arduous journey of the Indonesian independence revolution and wars which contributed to Indonesia's independence from the Dutch and Japan caused a considerable amount of collateral damage. The main urban areas of the administrative centre of Dutch colonies were mostly burned; which reduced the available space to live. There were also wounded veterans, previous coolies, and workers which had lack of knowledge, economic capital, and proper health.

Further complicating the situation, the Batak which previously lived in the highland mostly went down to live in the lowland, changing the demographic dynamics in the lowland. As tensions were still high, complicated with the grudge which the mid-low people had with the sultanate family, resulted in the East Sumatra Social Revolution in 1946. During those times many sultanates were overthrown with mass killings of members of the aristocratic families. An armed group performed the purge to eliminate Dutch allies and the feudal social structure. The revolution also aimed to seize the wealth of the sultanate families.

In the early time after independence, Medan and its region were largely ignored. While local government entities tried to educate their people through access to knowledge and art by providing libraries, organized movie screenings and discussions, without support from the main Government at Java Island, the residents were having difficulties in competing with traders and scholars in Java.

During the 1950s to 1960s Medan was known as the capital of pulp fiction with its literature and art sectors flourishing. The influence was found across the Malacca Strait. However, without proper support from the central government, the development halted. Most of its dominant actors either moved to Singapore or to Jakarta, in which they established the textile and film industry. During another flood of LEKRA's, the artisans' organization initiated by a growing socialist-communist party, refugees from Java; the cultural scene in Medan became increasingly political. The army which was already considered important as they supported the literature and art movement, increased its financial support to LEKRA's adversaries to counteract LEKRA's political propaganda.

The Old Order period ended in 1966 with hard to say peaceful terms. Political turbulence that ended the regime was associated with the rivalry between the socialist-communist party, religious extremist, and national military. Not to mention groups from regions that disagreed with centralized political power in Java. After nationalization of assets, to include within were plantation, mine sites, government and administrative offices, the New Order regime began to rule the new nation of Indonesia after a military attempt to turn down the disputes. However, as the new nation was in its beginning to establish unity, it was made into the common idea that communism was to be blamed for all the instability. At this point, the Chinese community was associated with the idea of communism and any attributes that relate the community with their Chinese origin including names and celebrations were repressed.

Medan was one of the urban centres that needed to set development accordingly to the importance of central power. The beginning of the New Order regime marked the shift of the development orientation and centralized the decision-making process to the central power in Jakarta. This suggested the downturn of the potential economic network that local Medan had. Yet, economic and regional development was the core strategy during the 1970s to 1990s. With more exploration towards the forest hinterland, land concessions were continuously handed out. This time, supported by the central government towards economic development with main commodities of palm oil, rubber, and oil. Until today, this has led to the loss of forest area.

As the world economy slowly headed into crisis at the end of the first decade of the 1990s, the mandated leader, Suharto, had enabled his family to build businesses across Indonesia. One of it was the paper industry located near Lake Toba. These industries, along with the plantations which were inherited from the Dutch, contributed to the environmental damage across North Sumatra and further. In addition, the economic policies by Suharto contributed to the unhealthy finance and economy of the nation.

As the crises worsened in 1997, the price of oil and basic needs rose. Students protested in all of Indonesia's major cities; urging the impeachment of Suharto. When this turmoil happened, ethnic sentiment issue was on the rise again (or made for distraction from the impeachment). Once again, the Chinese community was made to be the subject of the sentiment. During that time, Medan was a city with the most Chinese population ratio thus becoming the first city where the riot took place, which then continued in Solo and Jakarta.

The 1998 riots in Medan showed that towards sovereignty and during heavily centralized authority of the new nation, Medan and its region was still heavily affected by global economy changes. The economy still became the rule along with the new authority in the form of the army and local government. Those which had less economic capital, knowledge, and authority, became informal. Most residents which were educated or already having a trade skill, would be able to survive within different rulers and/or condition whether through adaptation or by migrating to Java or Singapore. Those who could not would then continue to suffer. At this time, the term of informal was used to record at least 113 traders and 138 building that were not supposed to exist and were thus 'cleared' by the local government raid in 1998. However, the number raised into 258 informal traders and 503 informal buildings in 2000.

The terms 'illegal occupation' and 'houses without a legal permit' (wilde occupatie) were used by Dutch in Indonesia. Thus, the act of 'clearing' 'illegal' vendor and buildings were actually inherited from the Dutch. The period after independence was a period when Medan was trying to configure its identity without Dutch oppression. Limitation in knowledge, economic capital, and government support forced the populace to utilise informality as a way of survival. In Medan, the first phase of informality had started by occupying airfield peripherals and/or behind provincial buildings [9]. After it was fenced, the efforts moved to more private land to avoid eviction by the municipality. There were also cases in where informal settlements were the result of illegal subletting, such as along the railway plots in Medan. The continued efforts found through statistic records of Medan from 1998 until 2016 of eviction, showed that informality was still a means of survival.

3.4. *In the Time of Reformation to the Era of Cities*

As the big industry plantation industry of mainly rubber and palm oil kept on growing in Medan, the term of informality is meant to address the slum settlements which are associated with micro trading. Based on 2015 data regarding slum settlements across Medan, there are 42 sub-districts in which the government has identified and verified some of its settlements as slum settlements. The government indicator for slums consists of physical and non-physical indicators. The physical indicator consists of building regularity, building density, the physical feasibility of buildings, accessibility, drainage system, clean water, waste and grey water system and fire safety. The non-physical feasibility consists of legality, people density, resident's job, electrical usage, health services and education. Those which lacked in most of the indicators were then considered as a 'very bad' slum. From 42 settlements identified as slum, 18 of them were identified as the worst level of slum which consist of Tegal Sari I, Tegal Rejo, Polonia, Sunggal, Tegal Sari Mandala II, Gedung Johor, Pekan Labuhan, Hamdan, Aur, Labuhan Deli, Gaharu, Pulo Brayan Darat I, Bandar Selamat, Bantan, Belawan Pulau Sicanang, Bagan Deli, Belawan I, and Belawan II.

Informal settlements in Medan today are a mix of informality inherited from the pre-colonial period to the colonial period and the period after independence. Labuhan Deli, Sunggal, and Bagan Deli are

IOP Conf. Series: Earth and Environmental Science **158** (2018) 012006 doi:10.1088/1755-1315/158/1/012006

suspected as informal settlements which can be traced back into pre-colonial Medan. Tegal Sari I, Sunggal, Tegal Sari Mandala II, Hamdan, Aur, Gaharu and Bagan Deli are suspected as settlements that were established during the colonization period for plantation activity. Tegal Rejo, Polonia, and Pulo Brayan are informal settlements which is suspected to have been established during the void of formal planning after independence. While the rest of it was established even after there was a formal city planning in 1973.

Those which were established during the pre-colonial and colonial period era are located within the inner city as part of the old and/or main city centre or other parts of the city which has historical significant such as ports and previous plantation area. Most of the settlements located in the inner city will choose to locate near railway or toll/artery road. Those which were established during the void of formal planning after Indonesia's Independence are located near railway/toll road and the periphery of military bases. They are still located within the city, but not close to the city centre. Those which were established after the first formal city planning are mostly located at the periphery of the City, in between urban and rural context, which in the future might be developed into peri-urban.

Most of the slum settlements livelihood are trade, allegedly micro or even informal trade. Physically most of the settlements lack in building regularity, building density, the physical feasibility of buildings, accessibility, drainage system, clean water, waste and grey water system and fire safety. The accessibility has become a problem especially to those which are located at the periphery or port, as they will have difficulty to access basic services which are mostly available at city centres. There are also issues in lack of education, which still needs to be dealt with, as most of the efforts were physical.

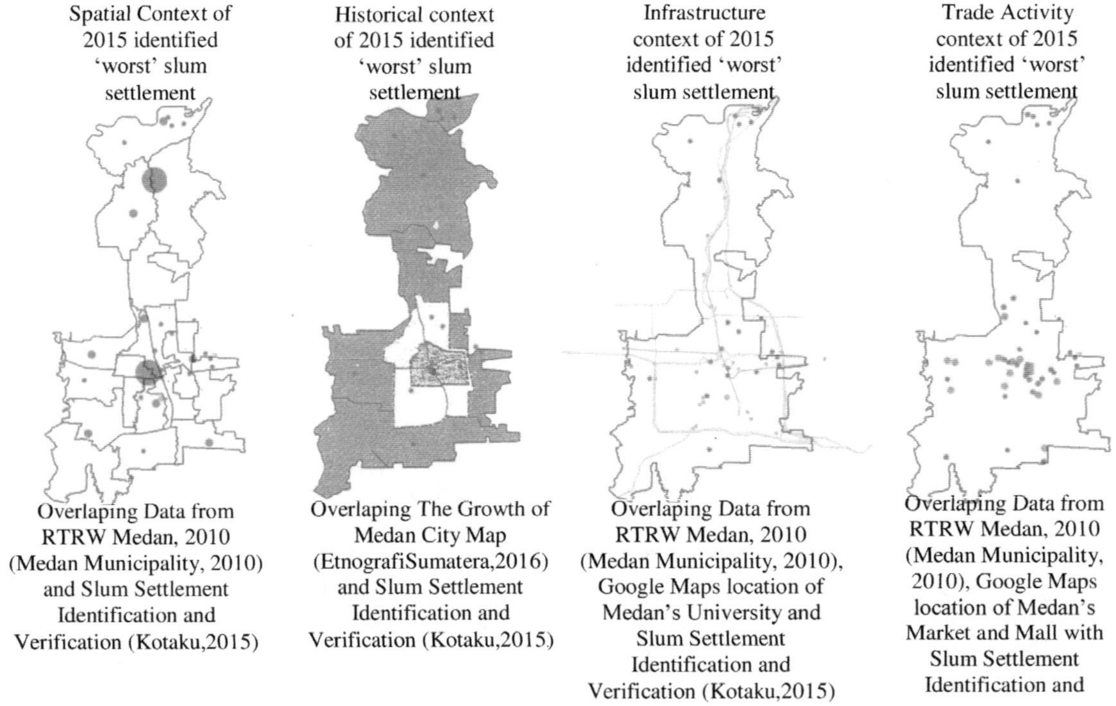

Figure 3. Spatial, Historical, Infrastructure and Trade Activity Context of Identified & Verified Slum Settlements.

While the information regarding settlements is already available, the information regarding informal trade is still lacking. Based on the information from BPS, after 1998 financial crisis and political turmoil, the number of informal trade evicted rose from 115 in 1998 up to 480 in 2005. After another change to the master plan of Medan City took place in 2005, to accommodate the changes of the governance system in Indonesia related to the implementation of the decentralisation policy, the number further rose up to 710. The increasing number of evicted informal trading might also be related to the increase of oil and gas price as the result of the policy to decrease the oil and gas subsidy. At the second half of the period the number kept on fluctuating between 42 up to 383, while the number of market stalls kept on increasing; suggesting informal trading still flourishing, even after several high-end malls were constructed and operating after 2013.

The result of USAID's quick field survey report in 2000 suggested that informality in the form of trade might be one of the factors which made Medan easily bounced back from the 1998 financial crisis. While the 1998 financial crisis contributed to the high level of unemployment in Medan, it also created new groups of educated unemployment. This group had more knowledge and social capital which benefitted them to get another job or to open their own business which started small and might be informal. The formal jobs they have consist of working in Indonesian Bank Restructuring Agency, insurance agent, electrical industry, automotive industry, developer, contractor, service such as wartel and vehicle rent, trade and renting. The rest of them worked in the informal sector until they managed to either build their own business or were accepted on another formal job.

4. Conclusion and Recommendation

Informality in Medan City evolved along with its long history and development with the economic aspect as the main driving force that was shown consistently throughout history. In terms of its early connectivity with the global economy in the pre-colonial or maritime trade era, there was no such dichotomy as informality as something that happened beyond the formality. As the economic activities were performed within the independent city-state, the activities of trade and subsequent result of the growing economic network, such as the emergence of settlements, were taken as a simultaneous way towards growing prosperity. During this time, the Cosmo polis characterized by a plural community was to be celebrated. In some early kingdoms; to name a few Siam, Majapahit, Sriwijaya, and Malacca; the plurality even adhered into the social and political organization.

In the burgeoning economy during the colonization era, the land concession made by the local ruler and aggressive investment by foreign firms driven unjust sharing of prosperity between ruler, planters or traders, and labours led to social conflict. Even when formal city planning effort was held in Medan City, it remained as an insufficient support relative to the economic development that has driven the urbanization. However, similar to other South East Asian countries, Medan's current informality was the by-product of Dutch C olonialization and Dutch legacy in urban planning. Tracing back to the pre-colonial history of Medan, informality found now might be a reestablishment of the way of life before colonialization. An irregularity which might have had some regularity and rule within it. While opportunities and life difficulties during the void period after independence and before formal planning was established, new informal settlements and trades emerged to survive. Similar with informality emerging after the 1998 financial crisis which at the end has proven to be a safety net and a way for economic survival.

As [2] stated that the potential of a global city lies on its transnational network, that a city might form its own urban system, Medan shows the similar character as Sassen's statement. Its connectivity with, mainly, Malacca straits economic network built since the ancient maritime time remains thriving until today. Yet, as it also said that its connectivity to the independent urban system might dim its dependence to the nation-state; or vice versa; seems to be occurring to Medan as well. Moreover, the ignorance of this potential might become a threat to the relationship of Medan and Indonesia as its nation-state.

However, informality might give an opportunity for small or abandoned sectors to survive. With the diverse community of Medan that also has a long historical potential, informality might be a potential glue for the society. Last, as the recommendation, we argue that within the new form of globalization and the rise of importance of the city, emerging informality might be used as the indicator for urban or city planning to succeed.

References

[1] Renn A What is A Global City? New Geography [Internet]. New Gography. 2018. Available from: NewGeography.com

[2] Sassen S 2005 The Global City: Introducing a Concept. Winter/Spring 2005. Volume XI, Issue 2. *The Brown. J. World Affairs*. Brown University.

[3] Sadler S What Makes a Global City? [Internet]. Capa Org. 2014 (Cited March 12 2014). Available from: https://capaworld.capa.org/2014/03/12/makes global-city

[4] Truschnegg B 2013 The Phenomenon of Global Cities in the Ancient World. Innsbruck University Press. *Globalization and the Cities*. p. 75-102. Available from: http://www.openedition.org/6540

[5] Kearney A.T About Global Cities [Internet]. Atkarney. 2017. Available from: https://www.atkearney.com/global-cities/article/?/a/about-global-cities-2017-article. 2017

[6] Lutzoni L In-formalised urban space design. Rethinking the relationship between formal and informal. springeropen.com. *City, Territory, and Architecture*. DOI 10.1186/s40410-016-0046-9. 2016.

[7] Roy A and AlSayyad N 2004 Urban informality: transnational perspectives from the Middle East, Latin America, and South Asia (Lexington: Lanham, Md)

[8] Dovey, K. Urban Design Thinking: A Conceptual Toolkit. New York: Bloomsbury Academics. 2016.

[9] Colombijn F, Barwegen M. Under construction. Leiden: Brill; 2014.

[10] Daggett, Adriane. The Indian Ocean: A Maritime Trade Network History Nearly Forgot. Discover Magazine [Internet]. Discover Magazine. 2016 (Cited March 20 2016). Available from:. http://discovermagazine.com/2016/nov/trading-places. October 20, 2016.

[11] Mohd R and Mohd H I 2012 Straits of Malacca and Singapore: Pride of the Malay Archipelago, Priceless Maritime Heritage of the World. Jabatan Kemajuan Islam Malaysia dan Institut Islam Hadhari, Universitas Kebangsaan Malaysia.

[12] Reid A. 2004 Cosmopolis and Nation in Central Southeast Asia. Asia Research Institute Working Paper Series No. 22. Asia Research Institute, National University of Singapore, aridir@nus.edu.sg.

[13] Green J Int'l Commerce, Snorkeling Camels, and The Indian Ocean Trade: Crash Course World History [Internet]. Youtube. 2012. Available from:.https://www.youtube.com/watch?v=a6XtBLDmPA0&index=18&list=PLBDA2E52FB1EF80C9

[14] Badan Pusat Statistik Kota Medan dan Badan Perencanaan Pembangunan Daerah Kota Medan. Medan dalam Angka 2002. Medan: Bandan Pusat Statistik Kota Medan; 2002 p. 27-35,40-45,191-193.

[15] Barus P The Historical City in Sumatra. Available from: https://www.indonesia.biz.id/barus-port/Bin Mohd Rusli, Mohd Hazmi. 2012. Straits of Malacca and Singapore: Pride of The Malay Archipelago, Priceless Maritim Heritage of The World. Jurnal Harian Special Edition (2012) 109-127. Department of Islamic Development Malaysia and Institute of Islam Hadhari. Malaysia.

[16] Bruner E M Urbanization and Ethnic Identity in North Sumatra.

[17] Wiryomartono B 2011 Urbanism in the Malay World: The Origin and Evolution of the Idea, Function, and Institution of Pre-colonial Bandar Melaka. The 5th International Conference of the International Forum on Urbanism (IFoU). National University of Singapore, Department of Architecture. Global Visions: Risks and Opportunities for the Urban Planet. Singapore.

[18] Reid A 2010 Sumatera Tempo Doeloe: dari Marco Polo sampai Tan Malaka. Komunitas Bambu. Jakarta, Indonesia.

[19] Frankopan D 2016 The Silk Roads. A New History of the World (London: Bloomsbury Paperbacks)

[20] Passchier C MEDAN: Urban development by planters and entrepreneurs (1870-1940) *Issues in Urban development* CNWS Leiden University. 1995 with some revision in 2008.

[21] Chandler D and Steinberg D In search of Southeast Asia. Sydney [u.a.]: Allen & Unwin; 1989.

[22] Dovey K Informal urbanism and complex adaptive Assemblage, International Development Planning Review, 34 (4) 2012 pp 349 - 367. doi:10.3828/idpr.2012.23. 2010.

[23] Evers H The End of Urban Involution and the Cultural Construction of Urbanism in Indonesia. Internationales Asienforum [Internet]. 2007. [cited 17 December 2017];38(1-2):51-65. Available from: http://crossasia-journals.ub.uni-heidelberg.de/index.php/iaf/article/viewFile/253/247

[24] From Distant Tales: Archaeology and Ethnohistory in the Highlands of Sumatra. 1st ed. Newcastle: Cambridge Scholars Publishing; 2009.

[25] Kratoska P South East Asia. London [etc.]: Routledge. 2001.

[26] Memory Of The World Register: Nāgarakrĕtāgama or Description of the Country (1365 AD). 2012-51. Available from:http://www.unesco.org/fileadmin/MULTIMEDIA/HQ/CI/CI/pdf/mow/nomination_forms/indonesia_nagarakretagama.pdf

[27] Nesbitt S Lack of Healthcare from Poverty in Indonesia, Remnants of Dutch Colonialism. Washington State University. History.libraries.wsu.edu. Spring 2016 (cited 28 December 2017). Available from: https://history.libraries.wsu.edu/spring2016/2016/01/19/lack-of-healthcare-from-poverty-in-indonesia-remnants-of-dutch-colonialism/#_edn2

[28] Pemerintah Kota Medan. Rencana Tata Ruang Wilayah (RTRW) Kota Medan Tahun 2010-2030. Medan: Pemerintah Kota Medan; 2010 p. IV-1 - IV-76.

[29] Pemerintah Kota Medan. Peraturan Daerah Kota Medan Nomor 2 Tahun 2015 Tentang Rencana Detail Tata Ruang Dan Peraturan Zonasi Kota Medan Tahun 2015-2035. Medan: Pemerintah kota Medan; 2015 p. all.

[30] Rothenberg A et al Rethinking Indonesia's Informal Sector *World Development.* 2016;80:96-113.

[31] Statistics of Medan Municipality. Medan Municipality in Figures 2017. Medan: Statistics of Medan Municipality; 2017 p. iii-v,46-57,321-324.

[32] Statistic of Medan City with Regional Development Planning Board of Medan City. Medan in Figures 2008. Medan: Statistic of Medan City; 2008 p. 30-33,51-100.

[33] Tarigan A et al Medan City: Development and governance under the decentralisation era. Cities. 2017;71:135-146.

[34] Truschnegg B. The Phenomenon of Global Cities in the Ancient World. Innsbruck University Press. *Globalization and the Cities.* p. 75-102. Available from: http://www.openedition.org/6540. 2013.

[35] USAID. Hasil Pengamatan Lapangan Cepat Tim Smeru: Krisis Ekonomi Dan Tenaga Kerja

IOP Conf. Series: Earth and Environmental Science **158** (2018) 012006 doi:10.1088/1755-1315/158/1/012006

Ter-PHK Lulusan Universitas/Akademi: Adaptasi Terhadap Realitas Pasar Kerja Baru [Internet]. Usaid; 2000 p. all. Available from: http://www.smeru.or.id/id/content/hasil-pengamatan-lapangan-cepat-tim-smeru-krisis-ekonomi-dan-tenaga-kerja-ter-phk-lulusan

IOP Conf. Series: Earth and Environmental Science **158** (2018) 012007 doi:10.1088/1755-1315/158/1/012007

Comparison of place attachment influence on the level of happiness of people living near residential parks

I Nurhakim, E B Kurniawan, and D K Wardhani

Department of Urban and Regional Planning Universitas Brawijaya, Malang, Indonesia

Corresponding author: ilhamnrhkm@gmail.com

Abstract. Several studies have shown that living close to a park is associated with high levels of happiness. However, there is a possible difference in the level of happiness between living close to a park (0-400 meters radius) compared to living far from the park (400-1.000 meters radius). Therefore, this study aims to compare the influence of Place Attachment on the level of happiness for people living near a park, especially residential parks in some housing areas in Malang City. Place Attachment in this study defined as a bond between an individual and a particular setting. The study demonstrates a positive influence of Place Attachment on the level of happiness of people living near the park. Respondents who live close to a park (R = 0.441; Happiness Level = 97.8%; Average Happiness = 6.833) tend to have a higher level of happiness compared to respondents who live far from a park (R = 0.326; Happiness Level = 69.9%; average Happiness = 4.148). This result shows the urgency to reconsider the provision standard of residential parks in Indonesia which only suggests one residential park for every 1,000 meters radius.

1. Introduction

Cities continue to grow every year in terms of demographics. One of the contributing factors is the trend of urbanization or the population movement to the city [1]. By 2030 it is predicted that 60 percent of the world's population will choose to live in cities [2]. The choice to live in the city has consequences, one of them relates to psychological aspects such as mental health and happiness. A study conducted by White et al. [3] showed that urban dwellers tend to have high levels of stress, poor mental health, and low levels of happiness.

On the other hand, some studies demonstrated that the provision of green space in urban areas can be a solution to problems of mental health and happiness. Several studies have also shown that living near green space has a positive influence on mental health [4,5] and happiness [6-8]. Studies conducted by Beyer et al. [5] show that by living in a neighborhood with more green space is associated with better mental health levels than living in neighborhoods with less green space. The green space also has an impact on happiness as showed by Larson et al. [8] who indicated that green space has a positive influence on the happiness of the city dwellers who live nearby (800 meters radius).

Gordon Jack's article in the book of Wellbeing and Place [9] entitled Role of Place Attachments in Well-Being revealed the relationship between place and well-being. Jack [9] mentioned that some studies have shown how places and well-being relate to one another. It reinforces several studies showing the relationship of green space (place) and happiness (well-being). The place factor can be

IOP Conf. Series: Earth and Environmental Science **158** (2018) 012007 doi:10.1088/1755-1315/158/1/012007

judged by how the city dwellers living near green space feel attached to the green space itself. The bond between an individual and a place can be assessed using Place Attachment theory [10]. The happiness of the city dwellers living near the green space itself can be assessed using the theory of Subjective Well-Being [11].

Although some studies have shown that living near green space affects happiness, there may be differences in the level of happiness between people living close to green space compared to people living far from it. It is as revealed by White et al. [3] living close to green space is associated with high levels of happiness.

Therefore, this study aims to compare the level of happiness that is influenced by the attachment of the people with the green space (Place Attachment) for people living near green space, especially residential parks. The level of happiness in this study is a dimension of happiness measured through Subjective Well-Being theory with variables of Life Satisfaction, Positive Affect, and Negative Affect. The distance in this study which is divided into close to the park and far from the park is determined using the standards of walking distance and residential park service radius [12,13]. The happiness in this study is influenced by the bond of the people living near the park with the park itself (Place Attachment). The luxury housing is chosen as a research location because of the ability of these housing developers to provide the residential parks in accordance with the standards.

2. Research methods

2.1. Data collection methods

Data collection methods in this research were divided into primary and secondary surveys. The primary surveys were conducted through field observations and questionnaires. The field observations were used to collect data on the neighborhood green open space and residential characteristics, while the questionnaires were used as a tool to measure Place Attachment and Subjective Well-Being. The secondary survey in this research was conducted to collect data on the classification of luxury housing from Real Estate Indonesia (REI) Malang. The figure below shows the distribution of luxury housing in Malang City according to REI Malang City 2016.

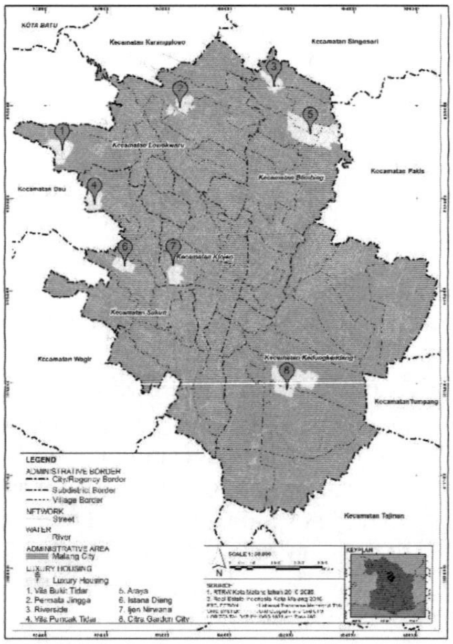

Figure 1. Distribution of Luxury Housing in Malang City.

2.2. Population and sample

The population in this study consisted of luxury housing in Malang City and people living within a radius of 1,000 meters from the chosen park. The radius was subdivided into close distance to the park (0-400 meters) and far distance from the park (400-1,000 meters). The close distance to the park was a could be reached by the park visitors with a walking distance up to 400 meters based on SNI 03/1733/2004 [12]. Meanwhile, the far distance from the park was the distance outside the walking distance (>400 meters). The researcher assumed the barrier for the farthest distance from the park based on SNI 03/1733/2004 [12] and Regulation of Public Works Minister Number 05/PRT/M/2008 [13]. Both documents stated that the service scale of the residential park with the minimum area of 1,250 square meters is a 1,000 meters radius. Based on that, the researchers assumed that the far distance from the park was the distance outside the walking distance up to the radius of the park service scale, i.e., a 400-1,000 meters radius. The housing sample was selected based on criteria covering the availability of residential parks, the residential parks with a minimum area of 1,250 square meters and the availability of houses in a radius of more than 400 meters. The table below shows the results of determining the study location.

Table 1. Study location determination.

Housing Name	Criteria			Total Score
	Park Availability	Park Area	House Availability	
Araya	1	1	1	**3**
Vila Bukit Tidar	1	0	1	**2**
Permata Jingga	1	1	1	**3**
Istana Dieng	0	0	0	**0**
Ijen Nirwana	1	0	1	**2**
Citra Garden City	0	0	0	**0**
Riverside	1	0	1	**2**
Vila Puncak Tidar	1	1	0	**2**

Based on the predetermined criteria, Araya Housing and Permata Jingga Housing were selected as study locations. The total score from Araya Housing and Permata Jingga Housing was three, which indicated that the housing fulfilled all the criteria determined by the researcher. Meanwhile, the sample of people living within a radius of 1,000 meters from the residential park was selected using the Slovin formula, which was used to determine the number of samples from known populations [14]. The Slovin formula could be used because the number of houses within a radius of 1,000 meters from the park could be known in number. In this study, one house would represent one research respondent. The table below shows the research population.

Table 2. Research population.

Housing Name	Total Population	
	Living Close to the Park	Living Far from the Park
Araya	664	1,075
Permata Jingga	613	361
TOTAL	1,277	1,436

From the population above, a research sample was determined for each distance to the park. The sample in this study was selected by a simple random sampling technique. Here the Slovin formula was used to determine the number of samples at each distance to the park.

$$n = \frac{N}{1+Ne^2} \quad (1)$$

Description:
n = Number of Sample
N = Total Population
e = Error tolerance

Table 3. Respondents sample determination.

Living Distance to the Park	Population	Sample Determination	Result
Close to the Park	1,277	$n = \frac{1,277}{1+(1,277)(0.1)^2}$	**93**
Far from the Park	1,436	$n = \frac{1.436}{1+(1,436)(0.1)^2}$	**93**
TOTAL			**186**

Based on the calculation results, the sample size for each distance to the park was 93 respondents. Subsequently, a proportional sample would be determined for each distance to the park. **Table 4** shows the proportional sample calculation for respondents who live close to the park.

Table 4. Proportional sample for close to the park respondents.

Housing Name	Population	Percentage	Proportional Sample
Araya	664	52%	48
Permata Jingga	613	48%	45
TOTAL	**1,277**	**100%**	**93**

Based on the calculations, the proportional samples for respondents who live close to the park at the Araya Housing was 48 respondents and at the Permata Jingga Housing the proportional sample was 45 respondents. The following table shows the result of the proportional sample calculation for respondents who live far from the park.

Table 5. Proportional sample for far from to the park respondents.

Housing Name	Population	Percentage	Proportional Sample
Araya	1.075	75%	70
Permata Jingga	361	25%	23
TOTAL	**1.436**	**100%**	**93**

Based on the calculations, the proportional sample for respondents who live far from the park at the Araya Housing was 70 respondents and at the Permata Jingga Housing was 23 respondents.

2.3. Stage of data analysis

The data analysis in this study was divided into six stages, i.e., identification of park and housing environment characteristics, research instrument test, analysis of Place Attachment, analysis of Subjective Well-Being, analysis of Place Attachment influence on Subjective Well-Being and Comparative Analysis of Place Attachment influence on the level of happiness based on the distance to park. The following is the elaboration of the six stages of data analysis.

2.3.1. Identification of park and housing environment characteristics. At this stage of data analysis, the characteristics of park and housing environment were identified to determine the location of the study. The location of the study was chosen based on certain criteria using a scoring method with the Gutmann scale. Based on the identification of the characteristics of the park and the luxury housing environment in Malang City in **Table 1**, Araya Housing and Permata Jingga Housing were elected as the research location.

2.3.2. Research instrument test. At this stage of data analysis, the research instrument, the questionnaire, was tested before being used. The research instrument test was conducted to 30 respondents, 15 respondents from Araya Housing and 15 respondents from Permata Jingga Housing. The research instrument test was divided into the validity test and reliability test. Validity test was used to determine whether the questionnaire was valid or not. The questionnaire was considered valid if it was able to reveal something measurable [15]. Meanwhile, the reliability test was used to measure an indicator of a variable in the questionnaire. A questionnaire was considered reliable if the respondent answered the questionnaire consistently over time [15].

2.3.3. Analysis of place attachment. At this stage of data analysis, people attachment to the park was measured using the Place Attachment method. Place Attachment is the positive bond between an individual and a particular place [10,16]. The predictors of Place Attachment consisted of Place Dependence and Place Identity. Place Dependence is the functional bond between an individual and a particular place [17], while Place Identity is an emotional bond between an individual and a particular place [18]. Each of these predictors was measured through question items in a questionnaire adapted from Williams and Vaske [16] questionnaire. Place Attachment was the result of the average Place Dependence score plus the average Place Identity score.

2.3.4. Analysis of subjective well-being. At this stage of data analysis, the happiness of people who live near the park was measured by the Subjective Well-Being method. Subjective Well-Being is the cognitive and affective evaluation of a person that represents happiness [11]. The predictors of Subjective Well-Being consisted of Life Satisfaction, Positive Affect, and Negative Affect. However, in this study, the Life Satisfaction predictors were replaced by Housing Satisfaction because Life Satisfaction was considered too broad for assessing happiness. Housing Satisfaction is a predictor of a person's degree of content in his or her housing situation [19], whereas Positive Affect and Negative Affect were affections that reflect pleasant and unpleasant feelings and emotions [20-22]. Each Subjective Well-Being predictor was measured through question items in a questionnaire adapted from the questionnaires of Diener et al. [23] and Watson et al. [20]. Subjective Well-Being itself was the result of the average Housing Satisfaction score plus the average Positive Affect score and minus the average Negative Affect score.

2.3.5. Analysis of place attachment influence on subjective well-being. At this stage of data analysis, simple linear regression analysis was done to find the influence of place attachment on subjective well-being. Simple linear regression analysis was also used to determine regression modelling of the influence of place attachment on subjective well-being. In this study, the regression modelling consisted of two forms of regression modelling. The first model was the regression model of place attachment influence on the Subjective Well-Being at a radius of 0-400 meters from the park (close to the park). Meanwhile, the second model was the regression model of place attachment influence on subjective well-being at a radius of 400-1,000 meters from the park (far from the park).

2.3.6. Comparative analysis of place attachment influence on the level of happiness. Comparative analysis of the influence of place attachment on the level of happiness based on the distance to the park was done when the results of the regression analysis indicated a positive influence between place attachment to subjective well-being. The analysis was done by comparing the value of the place attachment variable coefficients, the classification of happiness level and the average happiness value at each distance (close to the park and far from the park). The results of the analysis would answer the research question of how place attachment influences the level of happiness in people living near the park.

3. Results and discussion

3.1. Zoning of the housing
This study included distance to the park as a component in the study to group respondents into two groups. The first group was the group living close to the park, while the second group was the group living far from the park. The basis of the zone division was the living distance from the park.

(a) **(b)**

Figure 2. Housing research location.
Description:
(a) Araya Housing
(b) Permata Jingga Housing

The study location at the Araya Housing was a 2,100 square meters park located at Blimbing Indah Seltan X Street. The facilities in this park included tennis courts, park benches, and gazebo. In Permata Jingga Housing, the study location was a 3,200 square meters park located at Permata Jingga III Street. The facilities in this park included tennis courts, park benches, and children's playgrounds. The two parks were in a quite well-maintained condition at the time of the study.

(a) **(b)**

Figure 3. Residential park research location.
Description:
(a) Araya Residential Park
(b) Permata Jingga Residential Park

The research sample in each zone was chosen by using simple random sampling. Each house in each zone was given an identity number, so the sampling technique could be run by randomly selecting the identity numbers. The random selection of house identity was done using Research Randomizer. The random identification number generated from the Research Randomizer was then mapped to see the spread of the research sample houses. The number of houses sampled in each zone had been determined with the proportional sample that could be seen in **Table 4** and **Table 5**. The figure below showed the zoning and distribution of the research respondents in Araya Housing and Permata Jingga Housing.

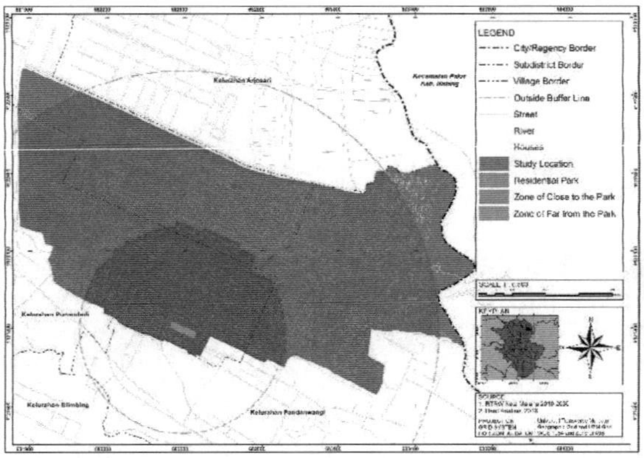

Figure 4. Zoning in Araya Housing.

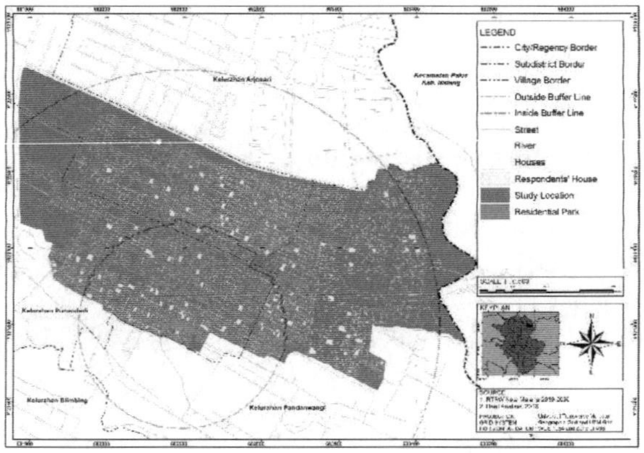

Figure 5. Araya respondents distribution.

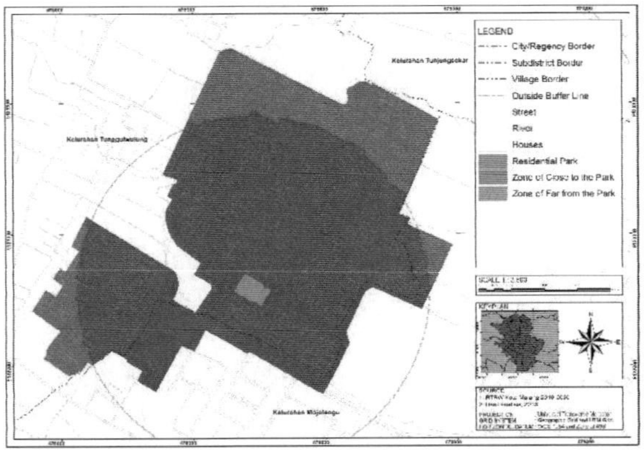

Figure 6. Zoning in Permata Jingga Housing.

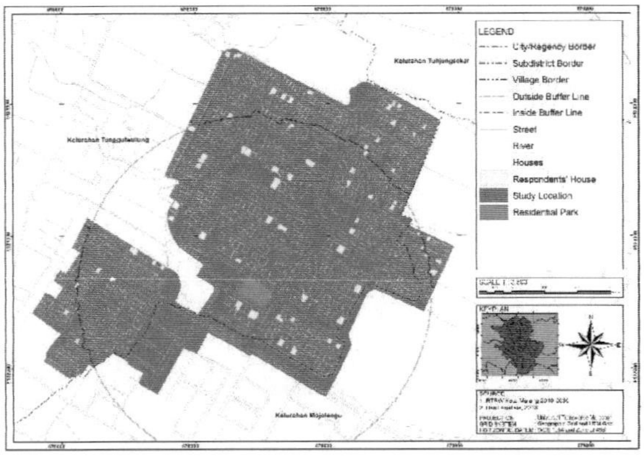

Figure 7. Permata Jingga respondents distribution.

3.2. Characteristics of the research respondents

The characteristics of the respondents were identified at each of the living distance to the park (close to the park and far from the park) by gender and age. Based on gender, at distance close to the park, 59 percent of total respondents or 55 respondents were male and 41 percent of total respondents or 38 respondents were female. At far from the park distance, 55 percent of the total respondents or 51 respondents were male and 45 percent of the total respondents or 42 respondents were female. Based on age, at close to the park distance, 28 percent of the total population or 26 respondents were aged 31-40 years, 25 percent or as many as 23 respondents were aged 41-50 years, 19 percent or as many as 18 respondents were aged 20-30 years, 18 percent or as many as 17 respondents were aged 51-60 years, and 10 percent or as many as 9 respondents were aged over 60 years. At far from the park distance, 32 percent of the total population or 30 respondents were aged 20-30 years, 28 percent or as many as 26 respondents were aged 41-50 years, 27 percent or as many as 25 respondents were aged 31-40 years, 10 percent or as many as 9 respondents were aged 51-60 years, and 3 percent or as many as 3 respondents were aged over 60 years.

3.3. Research instrument test

The research instrument test aimed to determine the validity and reliability of the research instruments before being used to collect data. The research instrument test was conducted to 30 respondents; 15 respondents from Araya Housing and 15 respondents from Permata Jingga Housing. The research instrument test was conducted through validity and reliability tests.

The validity test aimed to determine the validity of the research instruments where the score of each item was correlated with the total score. The calculation of the validity test was done using the Pearson correlation technique. The question item would be valid if the correlation coefficient of r-calculate was greater than the r-table value of 0.361. Based on the calculation of the validity test, the question items were entirely valid. This could be seen from all r-calculate values being greater than the r-table value. Thus, the question items in the research instrument were valid and could be used as data collecting tool in this research.

The reliability test aimed to determine the consistency of the research instrument as a measuring tool so that a measurement could be trusted. The reliability test calculation was done by looking at the value of Cronbach's alpha. The instrument was considered reliable when its Cronbach's alpha coefficient was greater than 0.6 [15]. Based on the calculation of reliability test, each tested variable was entirely reliable. This could be seen from all Cronbach's alpha values that were greater than 0.6. Thus, each variable of the research instrument was considered feasible to be used as a data collecting tool in this study.

3.4. Place attachment comparison based on the distance to the park

Place attachment comparison aimed to see the difference between the Place Attachment analysis on each distance to the park. The comparison of Place Attachment analysis was divided based on each variable, i.e., Place Dependence and Place Identity. The following section is a discussion of the results of the Place Attachment comparison based on the distance to the park.

3.4.1. Place dependence comparison based on the distance to the park. The comparison of the place dependence analysis was done by comparing the analysis results of respondents who live close to the park and respondent who live far from the park. The comparison was done to see the difference of the analysis results based on the distance of respondents to the park. The following is the comparison result of the Place Dependence analysis based on the distance to the park presented in the form of a table and graphs.

Table 6. Place dependence comparison based on the distance to the park.

Questions	Code	Average Answer Value	
		Close to the Park	Far from the Park
This park is the best place for what I like to do.	**PD1**	3.62	3.22
No other place can compare to this park.	**PD2**	3.55	2.75
I get more satisfaction out of visiting this park than any other.	**PD3**	3.58	2.92
Doing what I do at this park is more important to me than doing it in any other place.	**PD4**	3.66	2.90
I would not substitute any other area for doing the types of things I do at this park.	**PD5**	3.81	2.82
The things I do at this park I would enjoy doing just as much at a similar site.	**PD6**	3.77	2.91

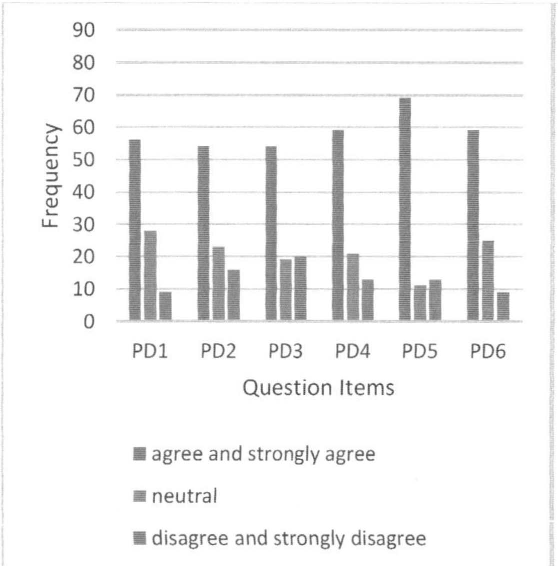

(a)

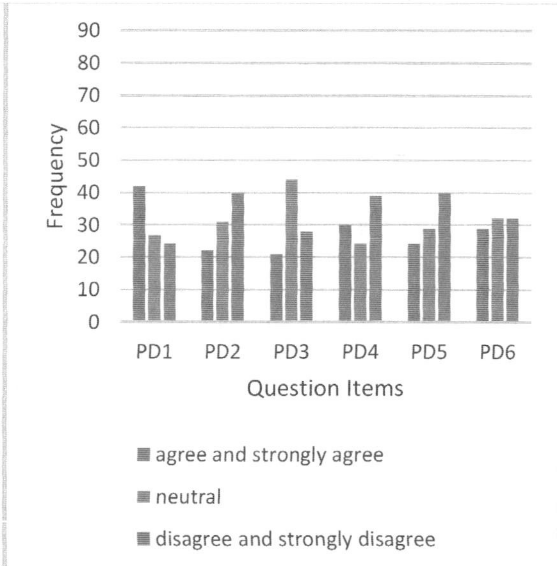

(b)

Figure 8. Place dependence comparison based on the distance to the park.

Description:
(a) Place dependence of respondents who live close to the park
(b) Place dependence of respondents who live far from the park

In the comparison table, the average value of respondents who live close to the park was greater than the average value of respondents who live far from the park for all question items. In the comparison graph, the responses 'agree' and 'strongly agree' were more frequently answered by respondents who live close to the park compared to respondents who live far from the park for all question items. Based on these points, it can be concluded that the respondents who live close to the park have more functional bonds to the park compared to the respondents who live far from the park.

3.4.2. Place identity comparison based on the distance to the park. The comparison of the place identity analysis results was done by comparing the analysis results of respondents who live close to the park and respondent who live far from the park. The comparison was done to see the difference of the analysis result based on the distance of respondents to the park. The following is a comparison of the results of the Place Identity analysis abased on the distance to the park presented in the form of table and graphs.

Table 7. Place identity comparison based on the distance to the park.

Questions	Code	Average Answer Value	
		Close to the Park	Far from the Park
I feel this park is a part of me.	PI1	3.49	2.66
This park is very special to me.	PI2	3.37	2.57
I identify strongly with this park.	PI3	3.69	2.67
I am very attached to this park.	PI4	3.62	2.51
Visiting this park says a lot about who I am.	PI5	3.62	2.38
This park means a lot to me.	PI6	3.75	2.67

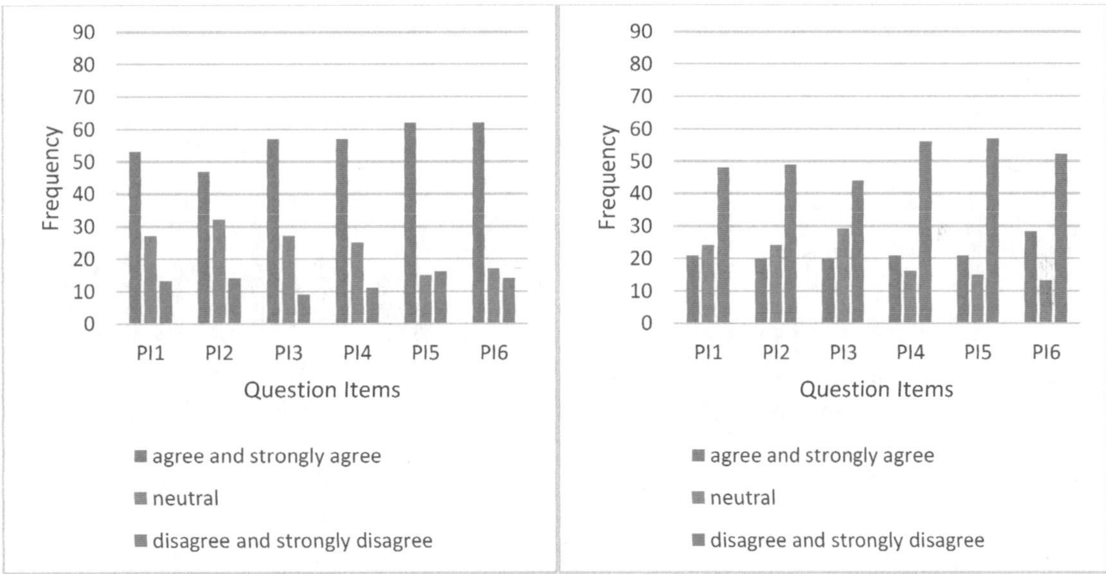

(a) **(b)**

Figure 9. Place identity comparison based on the distance to the park.

Description:
(a) Place identity of respondents who live close to the park
(b) Place identity of respondents who live far from the park

In the comparison table, the average value of respondents who live close to the park was greater than the average value of respondents who live far from the park for all question items. In the comparison graph, the responses 'agree' and 'strongly agree' on all question items was more frequently answered by respondents who live close to the park compared to respondents who live far from the park. Based on these points, it could be concluded that the respondents who live close to the park have more emotional bonds to the park compared to the respondents who live far from the park.

3.5. Subjective well-being comparison based on the distance to the park

The subjective well-being comparison aimed to see the difference between the results of subjective well-being analysis on each distance to the park. The comparison of subjective well-being analysis as divided based on each variable, i.e., Housing Satisfaction, Positive Affect, and Negative Affect. The following is a discussion of the result of the Subjective Well-Being comparison based on the distance to the park.

3.5.1. Housing satisfaction comparison based on the distance to the park. The comparison of housing satisfaction analysis was done by comparing the results of the analysis of respondents who live close to the park and respondents who live far from the park. The comparison was done to see the difference of the analysis result based on the distance of respondents to the park. The following was a comparison of housing satisfaction analysis results based on the distance to the park presented in the form of a table and graphs.

Table 8. Housing satisfaction comparison based on the distance to the park.

Question Items	Code	Average Answer Value	
		Close to the Park	Far from the Park
In most ways, my housing situation is close to my ideal.	**HS1**	4.06	3.35
The conditions of my housing situation are excellent.	**HS2**	4.12	3.39
I am satisfied with my housing situation.	**HS3**	4.28	3.48
So far, I have gotten the important things I want in my housing situation.	**HS4**	4.15	3.41
If I could choose another housing situation, I would choose my current housing situation.	**HS5**	4.20	3.31

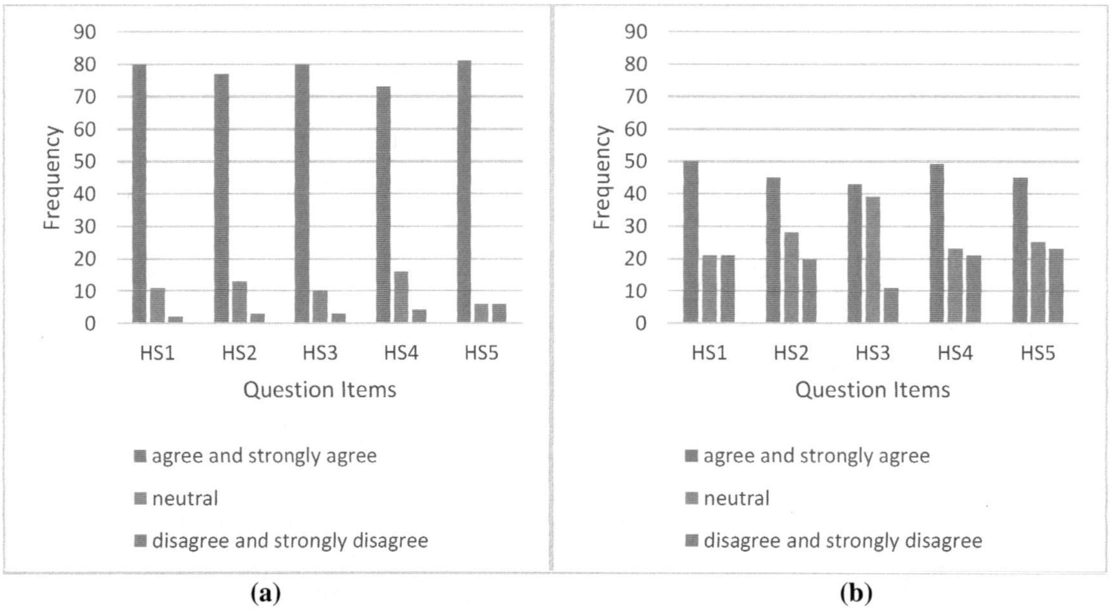

Figure 10. Housing satisfaction comparison based on the distance to the park.

Description:
(a) Housing satisfaction of respondents who live close to the park
(b) Housing satisfaction of respondents who live far from the park

In the comparison table, the average value of respondents who live close to the park was greater than the average value of respondents who live far from the park for all question items. In the comparison graph, the responses 'agree' and 'strongly agree' on all question items was more frequently answered by respondents who live close to the park compared to respondents who live far from the park. Based on these points, it could be concluded that the respondents who live close to the park were more content with their housing situation compared to the respondents who live far from the park.

3.5.2. Positive affect comparison based on the distance to the park. The comparison of positive affect analysis was done by comparing the results of analysis of respondents who live close to the park and respondents who live far from the park. The comparison was done to see the difference of the analysis result based on the distance of respondents to the park. The following is a comparison of the positive affect analysis based on the distance to the park presented in the form of a table and graphs.

Table 9. Positive affect comparison based on the distance to the park.

Question Items	Code	Average Answer Value	
		Close to the Park	Far from the Park
Interested	PA1	3.81	3.35
Excited	PA2	3.98	3.37
Strong	PA3	3.98	3.25
Enthusiastic	PA4	4.12	3.53
Proud	PA5	4.05	3.43
Alert	PA6	3.87	3.57
Inspired	PA7	3.99	3.37
Determined	PA8	4.08	3.42
Attentive	PA9	4.13	3.56
Active	PA10	4.16	3.51

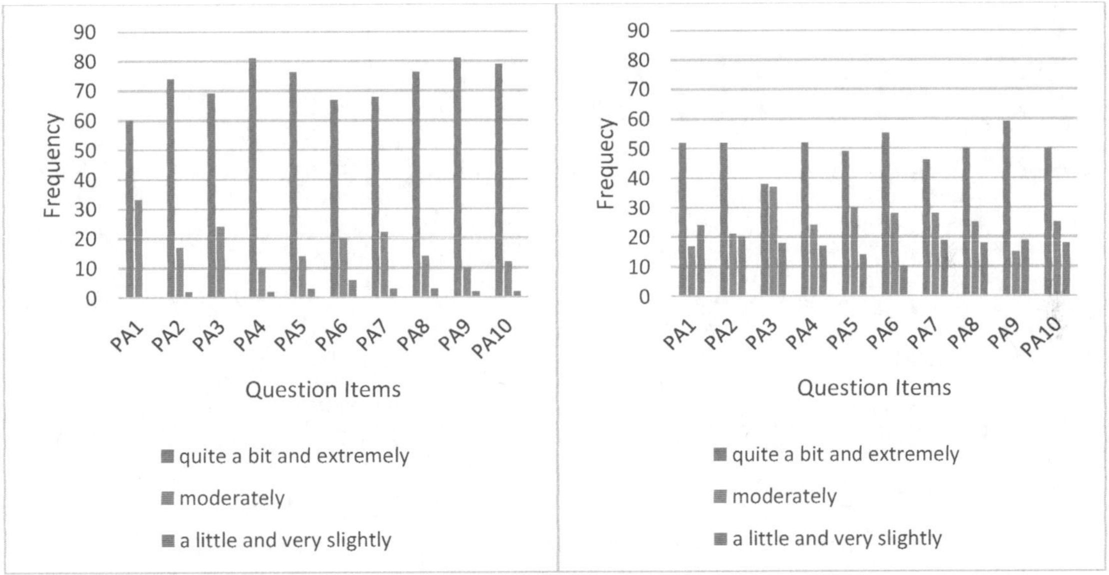

(a) **(b)**

Figure 11. Positive affect comparison based on the distance to the park.

Description:
(a) Positive affect of respondents who live close to the park
(b) Positive affect of respondents who live far from the park

In the comparison table, the average value of respondents who live close to the park was greater than the average value of respondents who live far from the park for all question items. In the comparison graph, the responses 'quite a bit' and 'extremely' were more frequently answered by respondents who live close to the park compared to respondents who live far from the park for all question items. Based on these points, it could be concluded that the respondents who live close to the park were more dominant reflecting the pleasant feelings and emotions compared to the respondents who live far from the park.

3.5.3. Negative affect comparison based on the distance to the park. The comparison of the negative affect analysis was done by comparing the results of analysis of respondents who live close to the park and the respondents who live far from the park. The comparison was done to see the difference of the analysis results based on the distance of respondents to the park. The following is a comparison of the negative affect analysis based on the distance to the park presented in the form of a table and graphs.

Table 10. Negative affect comparison based on the distance to the park

Question Items	Code	Average Answer Value	
		Close to the Park	Far from the Park
Distressed	**NA1**	1.73	2.33
Upset	**NA2**	1.58	2.33
Guilty	**NA3**	1.45	2.34
Scared	**NA4**	1.45	2.43
Hostile	**NA5**	1.48	2.37
Irritable	**NA6**	1.40	2.34
Ashamed	**NA7**	1.46	2.42
Nervous	**NA8**	1.39	2.39
Jittery	**NA9**	1.38	2.33
Afraid	**NA10**	1.48	2.39

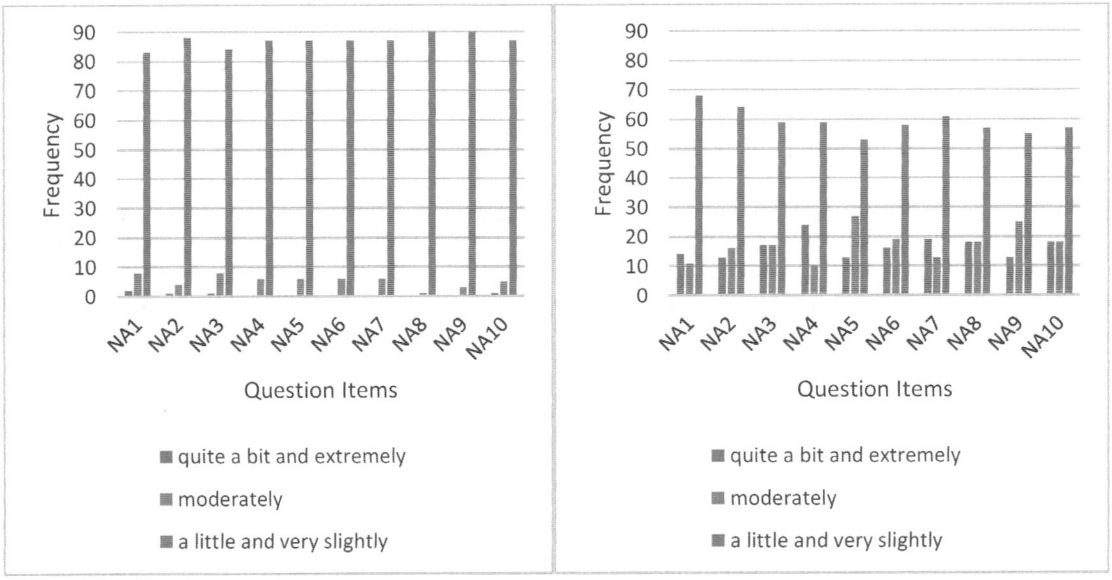

(a) **(b)**

Figure 12. Negative affect comparison based on the distance to the park

Description:
(a) Negative affect of respondents who live close to the park
(b) Negative affect of respondents who live far from the park

In the comparison table, the average value of respondents who live close to the park for all question items was not greater than the average value of respondents who live far from the park. In the comparison graph, the responses 'a little' and 'very slightly' on all question items were more frequently answered by respondents who live close to the park compared to respondents who live far from the park. Based on these points, it could be concluded that the respondents who live close to the park did not have more unpleasant feelings and emotions compared to respondents who live far from the park.

3.6. Simple Linear Regression Analysis
A simple linear regression analysis showed the influence of place attachment on subjective well-being in people living close to the park and far from the park. Simple linear regression analysis in this study

was done using SPSS software. The following section is a summary of the simple linear regression analysis based on the distance to the park.

3.6.1. Simple Linear Regression Analysis of respondents who live close to the park. The following is an equation from a simple linear regression analysis on the data of respondents who live close to the park.

$$Y = 3.861 + 0.441 X_1 \quad (2)$$

Description:
Y = Subjective Well-Being
X_1= Place Attachment

The regression equation from the simple linear regression analysis above showed the following.

- The constant of 3.861 indicated that if the Place Attachment variable was constant (unchanged), the magnitude of the Subjective Well-Being change was 3.861.
- Place Attachment Coefficient of 0.441 showed that Place Attachment positively affected Subjective Well-Being. This meant that the better the Place Attachment then the Subjective Well-Being would tend to improve.

3.6.2. Simple Linear Regression Analysis of respondents who live far from the park. The following is an equation from a simple linear regression analysis on the data of respondents who live far from the park.

$$Y = 2.378 + 0.326 X_1 \quad (3)$$

Description:
Y = Subjective Well-Being
X_1= Place Attachment

The regression equation from the simple linear regression analysis above showed the following.

- The constant of 2.378 indicated that if the Place Attachment variable was constant (unchanged), the magnitude of the Subjective Well-Being change was 2.378.
- Place Attachment Coefficient of 0.326 showed that Place Attachment positively affected Subjective Well-Being. This meant that the better the Place Attachment then the Subjective Well-Being would tend to improve.

3.7. Comparison of place attachment influence on the level of happiness based on the distance to the park

The simple linear regression analysis of the data of the respondents either living close to the park and far from the park indicated a positive influence. Based on this, the researcher could compare the influence of Place Attachment on the level of happiness based on the distance to the park. This was done to answer the research question of how the Place Attachment influences the level of happiness in people living near the residential park. The comparison involved comparing the values of the Place Attachment variable coefficients, the classification of the happiness level and the average happiness value at each distance to the park. Based on the results of simple linear regression analysis, the following was the value of place attachment variable coefficients at each distance to the park.

Table 11. Coefficient value of place attachment

Coefficient Value of Place Attachment	
Close to the Park	**Far from the Park**
0.441	0.326

The coefficient value of the Place Attachment variable on the results of respondents who live close to the park was greater compared to the respondents who live far from the park. The value of place attachment was directly proportional to the value of happiness. It showed that the influence of the Place Attachment variable on the results of respondents who live close to the park had a greater influence on the value of happiness compared to the respondents who live far from the park.

The classification of the happiness level of each respondent based on the distance to the park could be determined by utilizing the mean and standard deviations from the Subjective Well-Being analysis. The value was used to classify the level of happiness into three classifications of high, medium, and low. The following **Table 12** shows the mean and standard deviation value of the Subjective Well-Being analysis of respondents' data who either live close to or far from the park.

Table 12. Mean and standard deviation values

Minimal Score	Maximal Score	Mean (M)	Standard Deviation (SD)
1.2	8.6	5.58	1.69

The mean and standard deviation values were used to derive the range of values used as the basis for determining the classification of happiness levels. The classification of happiness levels was used to compare the respondents' happiness level at each distance to the park. The following **Table 13** shows the calculation of the happiness level classification at each distance to the park.

Table 13. Classification of happiness level based on the distance to the park

Level of Happiness (Y)		Living Distance to the Park			
		Close to the Park		Far from the Park	
Classification	**Range of Value**	**Frequency**	**Percentage**	**Frequency**	**Percentage**
High	$Y > M + 1SD$	36	38.7%	2	2.2%
Medium	$M - 1SD \leq Y \leq M + 1SD$	55	59.1%	63	67.7%
Low	$Y < M - 1SD$	2	2.2%	28	30.1%
Total		**93**	**100%**	**93**	**100%**

The cumulative frequency of high and medium happiness levels in the respondents who live close to the park were 36 respondents and 55 respondents or 38.7% and 59.1% respectively. The cumulative frequency of respondents who live close to the park was greater than the respondents who live far from the park which was as much as 2 respondents and 63 respondents or 2.2% and 67.7%. Conversely, the frequency of low happiness level on the respondents who live close to the park was only 2 respondents or 2.2%. The frequency was smaller than the respondents who live far from the park, which were 28 respondents or 30.1%. This shows that respondents who live close to the park tend to be happier compared to the respondents who live far from the park.

The argument could be reinforced by looking at the average happiness value at each distance to the park. The average happiness value was obtained by running a simple linear regression equation model on each distance to the park. The following **Table 14** shows the average happiness value at each distance to the park.

Table 14. Average happiness value

Statistics Analysis Results	Living Distance to the Park	
	Close to the Park	**Far from the Park**
Average Place Attachment Value	6.74	5.43
Regression Equation Model	$Y = 3.861 + 0.441\, X_1$	$Y = 2.378 + 0.326\, X_1$
Average Happiness Value	**6.833**	**4.148**

The average happiness value of respondents was obtained by entering the average Place Attachment value into the regression equation model. Based on the calculation, the average happiness value of the respondents who live close to the park was greater than the respondents who live far from the park. It showed that respondents who live close to the park tend to be happier compared to the respondents who live far from the park.

4. Conclusion

The discussion of simple linear regression showed a positive influence of place attachment on happiness (subjective well-being) of the respondents who live close and who live far from the park. This was the basis for further study, which compared the influence of place attachment on the level of happiness based on the distance to the residential park. The comparison involved the value of place attachment variable coefficients, the classification of happiness level, and the average happiness value at each distance to the park. Based on the comparison result, it could be concluded that the respondents who live close to the park tend to have a higher level of happiness compared to the respondents who live far from the park. This was because the influence of Place Attachment variables on the analysis result of respondents who live close to the park had a greater influence on the level of happiness compared to respondents who live far from the park. The study shows that the people living in a radius of more than 400 meters from the park are less happy. Moreover, the provision standard of residential parks in Indonesia only suggests one park for every 1,000 meters radius, which will result in a disparity of happiness level influenced by the park. Therefore, this study suggests to reconsider and review the residential park provision standards in Indonesia.

References

[1] United Nations 2015 *World Urbanization Prospect* (New York: United Nations Department of Economic and Social Affair)
[2] United Nations 2016 *Urbanization and Development: Emerging Future* (Nairobi: United Nations Human Settlements Programme)
[3] White M P, Alcock I, Wheeler B W and Depledge M H 2013 Would you be happier living in a greener urban area? Fixed-effects analysis of panel data *J. Psychological Sci.* **24** 920-28
[4] Alcock I, White M P, Wheeler B W, Fleming L E and Depledge M H 2013 Longitudinal effects on mental health of moving to greener and less green urban areas *Environmental Sci. and Tech.* **46** 1247-55
[5] Beyer K M M, Kaltenbach A, Szabo A, Bogar S, Nieto F J and Malecki K M 2014 Exposure to neighborhood green space and mental health: evidence from the survey of the health of Wisconsin *Int. J. Environ. Res. Public Health* **11** 3453-72
[6] Jorgensen A and Gobster P H 2010 Shades of green: Measuring the ecology of urban green space in the context of human health and well-being *Nat. and Culture* **5** 338-63
[7] Irvine K N, Warber S L, Wright P D and Gaston K J 2013 Understanding urban green space as a health resource: a qualitative comparison of visit motivation and derived effects among parks users in Sheffield UK *Int. J. Environ. Res. Public Health* **10** 417-42
[8] Larson L R, Jennings V and Clouther S A 2016 Public parks and wellbeing in urban areas of the united states *PLos ONE* 11(4): e0153211
[9] Jack G The role of place attachments in well-being. In: *Wellbeing and Place* (Farnham (UK): Ashgate Publishing) p 89-104
[10] Altman I and Low S M 1992 *Place Attachment* (New York: Plenum Press)
[11] Diener E 1984 Subjective well-being *Psych. Bull.* **95** 542-75
[12] National Standardization Agency *Procedure for Urban Housing Environmental Planning SNI 03-1733-2004* 2003
[13] Minister of Public Works *Regulation of the Minister of Public Works No 05/PRT/M/2008 concerning on Green Open Space Provisions and Utilization in Urban Area* Jakarta: Department of Public Works 2008.
[14] Sugiyono 2006 *Research Methods of Quantitative, Qualitative and R&D* (Bandung: Alfabeta)
[15] Ghozali I 2013 *Application of Multivariate Analysis with IBM SPSS 21 Program* (Semarang: Universitas Diponegoro publishing agency)
[16] Williams D R and Vaske J J 2003 The measurement of place attachment: validity and generalizability of a psychometric approach *For. Sci.* **49** 830-40
[17] Stokols D and Shumaker S A 1982 The psychological context of residential mobility and well-being *J. of Soc. Issues* **38** 149-71
[18] Prohansky H M, Fabian A K and Kaminoff R 1983 Place-identity: physical world socialization

of the self *J. of Environmental Psychology* **3** 57-83
[19] Toscano V and Amestoy V A 2007 The relevance of social interaction on housing satisfaction *Soc. Indicators Res. J.* DOI 10.1007/s11205-007-9107-5
[20] Watson D, Clark L A and Tellegen A 1988 Development and validation of brief measures of positive and negative affect: PANAS *J. of Personality and Soc. Psychology* **54** 1063-70
[21] Diener E, Scollon C N and Lucas R E 2004 The evolving concept of subjective well-being: the multifaceted nature of happiness *Advances in Cell Aging and Gerontol.* **15** 187-219
[22] Diener E 2006 Guidelines for national indicators of subjective well-being *Appl. Res. in Quality of Life* **1** 151-7
[23] Diener E, Emmons R A, Larsen R J and Griffin S 1985 The satisfaction with life scale *J. of Personality Assess.* **49** 71-5

IOP Conf. Series: Earth and Environmental Science **158** (2018) 012008 doi:10.1088/1755-1315/158/1/012008

Sustainable dimension adaptation measure in green township assessment criteria

R Yaman[1,3], S Thadaniti[1], N Ahmad[2], F M Halil[3], N M Nasir[3],
[1] Environment, Development, and Sustainability, Graduate School, Chulalongkorn University, Bangkok, THAILAND.
[2] Kulliyah of Architecture & Environmental Design, IIUM, MALAYSIA.
[3] Faculty of Architecture Planning & Surveying, UiTM, MALAYSIA.

Corresponding author: rostamyaman1@yahoo.co.uk

Abstract. Urbanized areas are typically the most significant sources of environmental degradation, thus, an urban assessment criteria tools aiming at equally adapted sustainability dimensions need to be firmly embedded in benchmarking planning and design framework and upon occupancy. The need for integral systematic rating is recognized in order to evaluate the performance of sustainable neighborhood and to promote sustainable urban development. In this study, Green Building Index Township Assessment Criteria (GBI-TAC) will be measure on holistic sustainable dimension pillar (SDP) adaptation in order to assess and redefine the current sustainability assessment criteria for future sustainable neighborhood development (SND). The objective of the research is to find-out whether the current GBI-TAC and its variables fulfilled the holistic SDP adaptations towards sustainable neighborhood development in Malaysia. The stakeholder-inclusion approached is used in this research in order to gather professional's stakeholders' opinions regarding the SDP adaptations for sustainable neighborhood development. The data were analysed using IBM SPSS AMOS22 Structural Equation Modelling. The findings suggested an adaptation gap of SDP in current GBI-TAC even though all core-criteria supported SDP adaptation, hence lead to further review and refinement for future Neighborhood Assessment Criteria in Malaysia.

1. Introduction

The magnitude of sustainable development in the built environment sector, particularly in urban vicinities was established for quite a long time [1]. Urban development plays an important function in pursuing sustainability via socio-economic growth and technological innovation [2]. Sustainable development at urban level suggest a well-balanced and broader specific requirement of its inhabitant which include earnings equity, job opportunities, accommodation, basic amenities, public infrastructure, accessibility connectivity and also protection to the environment [3]. Sustainable environment can be achieved at different steps and level in the urban development from the inside out of the interiors and the buildings, neighborhood/township and cities. Neighborhood are the areas and systems of cities, made up by their own built environment, social context and financial dynamics [4]. Neighborhood are perimeter to a specifically-defined boundary, and a communal acquaintance exists between their occupants [5]. To enhance neighborhood/township sustainability, an understanding of its buildings, communal spaces, public infrastructure [6], the promotion of a managing principles [7], and collaboration amid its components is essential.

Sustainable evaluation criteria comprise of indicators acquired from literature review. Several analyses on evaluation indicators suggest the intent designations and characteristic [8][9], development means of a recent indicator categorizes [10][11][12][13][14], indicator significances threshold [13] and indicators framework [15]. Presently, due to the availability of many rating method approaches, it is rather challenging for stakeholders to actually assess which one has the most general application. Diverse sets of requirement, significances threshold, physical factors and locale guidelines renders distinction. Currently, sustainable city development growth are mainly encouraged by governmental institution through relevant agendas, strategies, outlines, policies, plans, programs and incentives [16]. On the other hand, collaboration by third party establishment such as building industry related professional body and institutions efforts, the development of urban sustainability evaluation systems which are driven by market approach has achieved significance and has been progressively implemented by commissioners and developers of development projects.

The constraint for complete systematic evaluation systems is established in order to assess the implementation of sustainable conurbation and to encourage the overall sustainable urban growth. However, the current available sustainability assessment criteria systems is normally based on low carbon city framework (LCCF), the future of sustainable township development should be beyond the current LCCF; not only heavily weighted on reducing carbon related issues but towards a well-balanced sustainable pillars which addressed thoroughly economic pillar, environment pillar and social pillar. In this study the GBI-TAC is measured and then to address the missing gap of SDP adaptation in the current sustainability assessment criteria systems, it is a way forward and towards future sustainable urban development in Malaysia. Thus, the need for integral measure in SDP adaptation is important in measuring the certified sustainable neighborhood for continuing sustainable agenda aiming at professional stakeholders which act as sustainable input implementers in sustainable development. The Green Township Assessment Criteria targeted at developers may serve as sustainable label in obtaining green development status, but the real measure on SDP adaptations lies on professional stakeholders, key actors in implementing this guideline..

2. Research background

2.1 Sustainable Neighborhood Assessment Criteria

Sustainable township/neighborhood assessment criteria is an instrument that assesses the sustainable functions of designated neighborhood versus sets of benchmarking criteria [17]. There are two categories of urban sustainability evaluation approaches;

- The embedded decision making instrument into township scale development such as EcoDistricts, EcoDistricts Performance and Assessment Toolkit, Ecocity, HQE2R, One Planet Living, SCR, SPeAR, Cascadia Scorecard; and,
- The present third-party developed sustainable evaluation approaches and systems such as BREEAM Communities, LEED-ND, CASBEE, ECC, Qatar Sustainability Assessment System, (QSAS) Neighborhoods, Green Star Communities, Green Neighborhood Index (GNI), Green Mark for Districts, and GBI Township.

These sustainable township/neighborhood assessment criteria are typically developed by a building's related professional institute (normally not a government agency) or an independent third party institution. The common aims of third-party sustainable township/neighborhood assessment criteria are driven by: (1) an objective and environmentally-aware system in cooperation with market participants, without limiting the freedom of built environment design, (2) the mitigation of the impacts of development on the environment, (3) creation of a sustainable, (4) responsible local community etc. [18][19][20].

2.2. GBI Township Assessment Criteria In Malaysia

GBI Township Assessment Criteria were established by Malaysia Institute of Architects (PAM) and Association of Consulting Engineers Malaysia (ACEM) intended for Malaysia's primary sustainable urban rating instrument which progressively will be updated and reviewed towards sustainable future. This joint initiative between Malaysia Institute of Architects (PAM) and Association of Consulting Engineers Malaysia (ACEM) in green building indexing aims to support the construction development industry with regard to its sustainable development progression. GBI Township Assessment Criteria environmental rating approach is established to:

- Distinguish sustainable township criteria by instituting a generic nomenclatures and standard evaluations;
- Promote integration, holistic design; acknowledge and remunerate environmental leadership effort;
- regenerate built environment by reducing the environmental vulnerability of the development; and
- Ensure that the development of new township continues to play a key role in the near future while rejuvenating existing township development.

Sustainable township rating criteria are perceived as a guidelines in assisting urban planners, architects, engineers, builders, building owners, government agencies, housing developers and finally the end users in understanding the effect of each design project option and problem-solution with regard to being a more sustainable- responsive. The Malaysian indexing in green building was established to deliver a generic and verifiable means in sustainable benchmarking within the building industry of local context. GBI-TAC and its designed framework sets sustainability to additional level and lay out a vision for coherent approach in built environment paradigm. Its provide direction that will support all involved stakeholders in delivering sustainable urban developments.

2.3. Study Statements

Over the years, the Malaysian construction and development sector has emerged and thrived in the direction of a more progressive sustainable urban agenda. The subject of sustainable neighborhood/township benchmarking approach in the field of urban development in Malaysia is relatively new. Although GBI-TAC has been developed and implemented but there is no measure being conducted in assessing the SDP adaptations in its variables? There has been lack of study done to measure the greenness of the tools. Even though there are studies conducted on the dimension of environment, few are about what the social and economic values of environment, and/or its application in the sustainable urban development context. Do sustainable neighborhood or township evaluation criteria and frameworks in common signify sustainable holistically? [21]. Proficiency and comprehension on urban sustainability benchmarking criteria among the Malaysian sustainable urban development actors are vastly still low. Studies have indicated one of the main obstacles that hinder sustainable township/neighborhood development in ASEAN nations is the deficiency of knowledge in sustainable concerns subjects in relation to the involved building profession [22].

Past studies conducted on sustainable building rating system potential in Malaysia also show that key stakeholders in Malaysia's construction and development sector have insufficient understanding on sustainability development evaluation, benchmarking & indexing approach [23]. Due to this, many green certified urban development project in Malaysia claim sustainability merely for label advertisement, marketing tools and higher premiums instead of fully addressing the sustainable pillars. Certified GBI township projects normally a high-end urban development projects, study has indicated that sustainable certification do improve leasing and selling rate of developed properties, but this outcome is further substantial for end-users who are more innately concerned with sustainability, or pushing their 'green' appearance [21]. Even though the noble foundation for sustainable township/neighborhood is to promote and applied sustainable development, but there is no study was undertaken to address and gauge this issues. Thus, embark the problem of the research: whether the existing GBI Township Assessment Criteria fulfilled the SDP adaptation level according to holistic dimensions Pillars of Sustainability.

3. Research methodology

3.1. Stakeholders-Inclusion Approach

This research applies the Stakeholders-Inclusion Approach to deliver opinions concerning SDP dimensions of existing green township assessment. Therefore, Stakeholders-Inclusion Approach is appropriate and adopted for this research. For Stakeholders-Inclusion Approach opinion on SDP Adaptation, it constitutes a team of policy-makers and urban development professionals in accordance of their intellectual merit, rank in the order for decision-making and their competency or experience in sustainable urban development field. The conventional surveys methods experience richness in issues, meanwhile in Stakeholders-Inclusion Approach expert's surveys will basically supply richer feedback data, that is due to numerous iterations and responses, where the experts participating in Stakeholders-Inclusion Approach are positive towards additional enquiries for further elaborations [24]. The term 'stakeholders' in this study are basically an individuals or representatives from sustainable building industry related institutions or professionals such as urban planners, architects, engineers, designers, builders, manufacturers and academicians. These stakeholders are key players in implementing GBI Township Assessment Criteria whether directly or indirectly involved in sustainable urban neighborhood development in Malaysia. The respondents of Stakeholders Surveys is listed mainly from list of participants of sustainable workshops and continuing professional development programs conducted by Malaysian Green Building Confederation and other professional building related profession in Malaysia.

The Stakeholders Surveys was conducted via online questionnaires surveys using Google Documents software package. The online surveys was emailed to the list respondents for feedback. A total 100 online Stakeholder Surveys was disseminated to the listed respondents with a target of 50 returned response. The returned response in this phase is analyzed using IBM-SPSS AMOS 22 statistical software package.

3.2. Research Process

The research is about conducting comprehensive studies on current GBI township assessment criteria versus sustainability pillar dimension (SPD) aspects within urban development project. This study will look on adaptation of a well-balanced sustainability pillar dimension (SPD) which addresses economic, environment, and social impacts/aspect and loading factors. The research process includes content analysis, professional's stakeholder survey, SDP adaptation measure using Structural Equation Modeling analysis and finally derived to findings and discussions of the study (Figure 1).

4. Results & discussions

4.1. Full Structural Model of SEM

A full structural model is the element of the general path model that commends the relationship between a proposed set of unobserved latent variables [25][26]. In distinction with the measurement model (CFA), the full structural model (SEM) concerned of how the specific constructs (SDP Adapt factors) affect the latent construct (SDP Adapt) grounded on the underlying study theory.

The analyses of SDP adaptations in pre-occupancy sustainable township/neighborhood assessment criteria (GBI-TAC) are based on professional stakeholder's opinion in the Klang Valley/Greater KL. Since sustainable township/neighborhood is relatively new in Malaysia, most of the development and pool of professionals who implemented this sustainable guidelines are concentrated in this center region. Furthermore, the only available case studies of certified sustainable neighborhood development which was certified using GBI-TAC and occupied for more than one year is in the Klang Valley/Greater KL. Given there are six constructs to be analyzed, the analysis of structural equation models are particularly suited for this intention as it is to compare and distinct the outcomes in capturing the actual endogenous variables that affect the exogenous dependent variable. The SEM path model is applied to Stakeholders-Inclusion Approach – Professional Stakeholder Surveys set data. The fit indexes results reported in this SEM analysis are *P*-value, RMSEA, GFI, AGFI, TLI, NFI and ChiSq/df (Table 3).

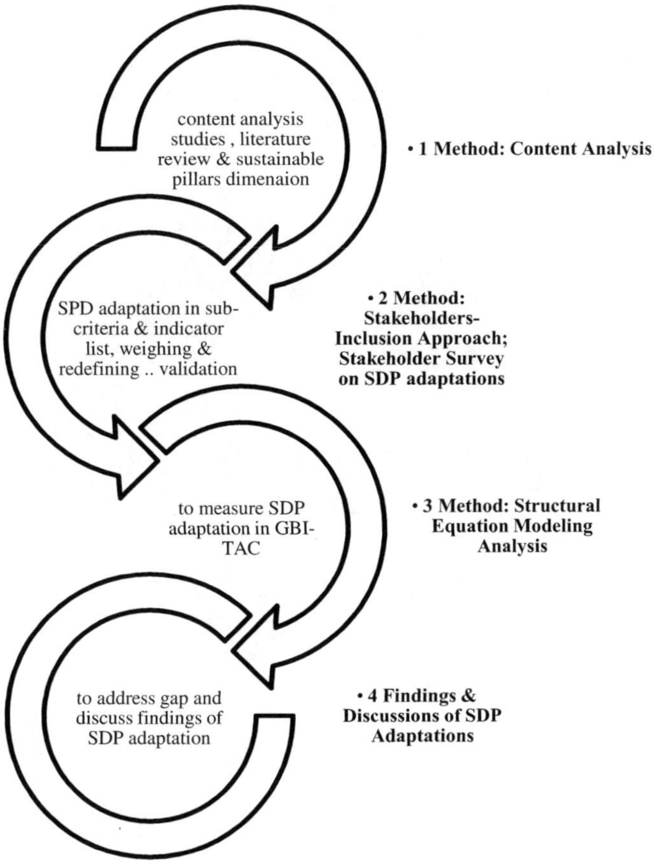

Figure 1. Final deliverables of mix method research process

The analysis emphasizes on measuring the different dimension of SDP Adaptations and variation across the assessment core-criteria of GBI-TAC. In this context, for each sustainable dimension, there is a necessity for researcher to assess the complex relationships within the SDP Adaptations and to compare the value of constructs across the assessment core-criteria in terms of correlations. Figure 3 represents the full Structural Equation Model (SEM) for SDP Adaptation in pre-occupancy assessment criteria of GBI-TAC.

4.2 SEM Analysis Result of SDP Adaptation

The generated outcome result of full structured model yielded the *P*-value = .000, RMSEA = 0.042, TLI = .977 and ChiSq/df = 1.148 which highly fulfilled the threshold outscores. The other reported indexes values is GFI = .882, AGFI = .740 and NFI = .845 are tolerable for an acceptable path model even though the index values are lower than constraint level of 0.9 (GFI, AGFI and NFI) respectively and ≤ 5.0 (ChiSq/df). The full structured model describes that the data are appropriate where all the given paths co-efficient (*P*-value) of the underlying structure were statistically highly significant at 0.001 (***) value and statistically significant at ≤ 0.05 value. Therefore, the parameter estimates of the hypothesized full structure model were free of any offending value. However, for Q4.3ENP <-- TRC that indicated only significant correlation between construct. The positive coefficient sign with P-value (0.042) in Q4.3ENP <-- TRC path is statistically significantly related. Therefore, all hypothesized paths of the full structured model show significant correlation implying that all hypotheses are supported. The positive co-efficient sign of *P*-value at (***) or ≤ 0.05 is due to the sufficient sample size of Stakeholders-Inclusion Approach – Professional Stakeholders Surveys which covers the main center region of Klang Valley/Greater KL as a study area. A structured model profiling such as sample size affects the parameter accuracy of estimates and model fit indexes. The structure model with larger sample size can generate lower values of RMSEA index. Table 1 renders the parameter estimates of the full structural model of SDP Adaptations.

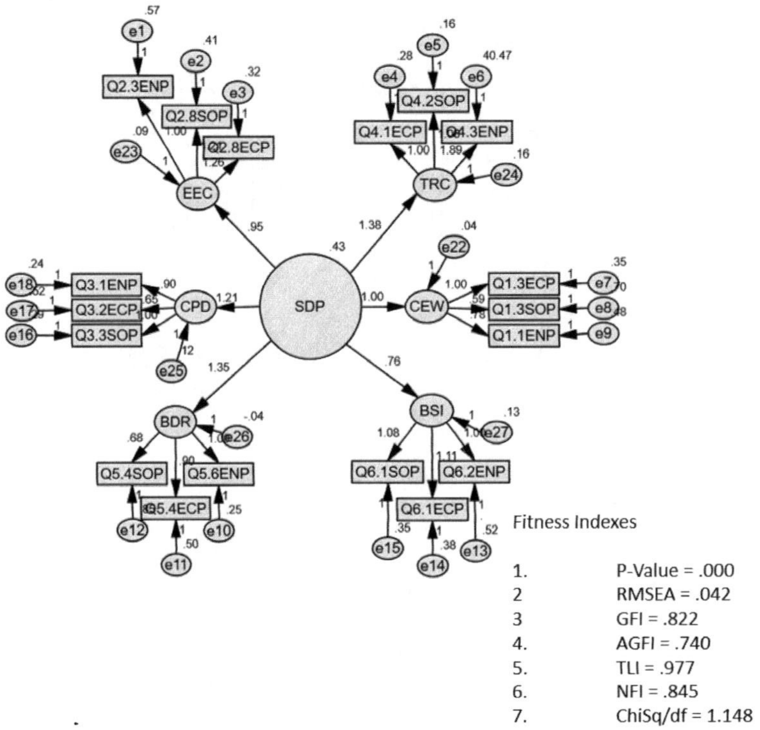

Figure 2. The standardized coefficients of Full Structural Model for SDP Adaptations

Table 1. Parameter Estimates of the Full Structural Model of SDP Adaptations in GBI-TAC

Construct &Variables			(a)Unstd	(b)Std	S.E.	C.R.	*P*
TRC	<---	SDP	1.458	.931	.239	6.109	***
EEC	<---	SDP	1.003	.938	.209	4.809	***
CEW	<---	SDP	1.000	.863			
CPD	<---	SDP	1.225	.905	.213	5.759	***
BDR	<---	SDP	1.365	1.019	.214	6.376	***
BSI	<---	SDP	.798	.860	.185	4.317	***
Q2.3ENP	<---	EEC	1.000	.664			
Q2.8SOP	<---	EEC	1.271	.810	.225	5.648	***
Q2.8ECP	<---	EEC	1.328	.874	.215	6.176	***
Q4.1ECP	<---	TRC	1.000	.886			
Q4.2SOP	<---	TRC	1.054	.932	.095	11.044	***
Q4.3ENP	<---	TRC	1.756	.263	.864	2.031	.042
Q1.3ECP	<---	CEW	1.000	.799			
Q1.3SOP	<---	CEW	.560	.441	.172	3.258	.001
Q1.1ENP	<---	CEW	.745	.624	.161	4.627	***
Q5.6ENP	<---	BDR	1.000	.851			

Construct &Variables			(a)Unstd	(b)Std	S.E.	C.R.	*P*
Q5.4ECP	<---	BDR	.942	.755	.130	7.244	***
Q5.4SOP	<---	BDR	.720	.558	.148	4.874	***
Q6.2ENP	<---	BSI	1.000	.637			
Q6.1ECP	<---	BSI	1.149	.739	.238	4.832	***
Q6.1SOP	<---	BSI	1.099	.726	.233	4.710	***
Q3.3SOP	<---	CPD	1.000	.845			
Q3.2ECP	<---	CPD	.657	.618	.123	5.358	***
Q3.1ENP	<---	CPD	.866	.822	.112	7.758	***

Source: This Research
Note:
(a) Estimated unstandardized regression coefficients
(b) Standard error of estimated unstandardized coefficient.
(c) Probability of a t value equal to or greater than actual t value in a two-tailed test for significance of coefficient under the null hypothesis that the true value is zero. The symbol *** indicates that the null hypothesis is rejected at the .001 level of significance.

4.3 Climate, Energy & Water (CEW)

The standardized (b) loading estimates of the endogenous construct (observed) yielded for Climate, Energy & Water (CEW) are as presented in Table 1. Economic Dimension (ECP) yields the highest significant value at .799 indicating that SDP Adaptations in this dimension is of importance or highly adapted. This is followed by Environment Dimension (ENP) with significant value at .624 and Social Dimension (SOP) with significant value at .441. CEW pre-occupancy assessment sub-criteria of GBI-TAC concerns heat island design principles, efficient streetscape & greenspace lighting, on site energy generation, renewable energy, reduction in waste water and reduced water use. Even though the CEW assessment sub-criteria seems to incline to Environment Dimension, however, the loading estimates from Stakeholders-Inclusion Approach – Professional Stakeholder Surveys full structured SEM modeling indicating Economic Dimension is highly adapted or most significant in SDP Adaptations.

4.4 Environment & Ecology (CEE)

Overall, the outcomes imply that EEC is within the loading values level that suggest considerable SDP Adaptations in pre-occupancy GBI-TAC. Two out of three observable endogenous constructs yielded above significant loading value of regression weight. The full structured model standardized loading estimates for Economic Dimension (Q2.8ECP) is highly significant at .874 value, followed by Social Dimension (Q2.8SOP) which is also highly significant at .810 value and lastly the Environment Dimension (Q2.3ENP) which is less significant at .664 value. The SDP Adaptations in EEC sub-criteria of GBI-TAC concerns with the preservation of biodiversity reserve, land reuse strategy, ecology availability, flood management & avoidance, wetland & water body conservation, agricultural land preserve, hill slope development principles, sustainable storm water design & management, proximity to existing infrastructure services, services infrastructure provision and light pollution control. The less significant outcome in Environment Dimension is due to lacking implementation of sustainable management and provision efforts of Environment & Ecology assessment sub-criteria in most of township/neighborhood development projects in Klang Valley/Greater KL.

All specific path co-efficient of the causal CEE full structure model are statistically significant at 0.001 (***) values, the parameter estimates of the hypothesized full structured model are also exempted from offending statistical values including the SDP Adapt ---> CEE path. Therefore, this hypotheses is supported. The less significant values yielded reveal that the Environment Dimension for pre-occupancy EEC assessment core-criteria of GBI-TAC sustainable projects implementation need improvement to further support the sustainable township/neighborhood development growth parallel with the nation's sustainable development agendas.

4.5. Community Planning & Design (CPD)

For CPD Construct, two out of three standardized loading estimates is above 0.7 significant level as suggested by Hair (2010) [27]. The Social Dimension (Q3.3SOP) obtained high significant value at .845 and Environment Dimension (Q3.1ENP) also yielded high significant value at .822. Only Economic Dimension (Q3.2ECP) generated less significant value but closed to acceptance level at 0.7 where the standardized loading estimates values is at .657. This outcomes suggest high SDP Adaptations in Social Dimension and Environment Dimension in pre-occupancy assessment criteria of GBI-TAC while least adaptation and consideration for Economic Dimension. Averagely, CPD Construct yielded at .774 value for standardized loading estimates which signifies significant.

The regression weight values vary among the designated sustainable dimension indicating that SDP adaptation differ in relative advantage as for the case of CPD core-criteria, Social Dimension and Environment Dimension appears to be the significant factor that brings the CPD assessment core-criteria towards an enhanced sustainable neighborhood development. The less significant regression weight value yielded for Economic Dimension shows that this dimension need improvement in SDP Adaptations to support sustainable growth and minimizing the CPD assessment core-criteria sustainable adaptation gap. This is due to lack of consideration in SDP Adaptations and implementation of Economic Dimension related factors in CPD pre-occupancy assessment sub-criteria which concerns greenspaces, compact development, amenities for communities, provision for universal accessibility, secure design, health in design, recycling facilities, community diversity, affordable housing, community thrust and governance.

4.6. Transportation & Connectivity (TRC)

The standardized average outcomes indicated that TRC has the slightly lower loading factor and satisfactory level of acceptance within the overall pre-occupancy assessment evaluation core-criteria of GBI-TAC. The standardized loading estimates for Environment Dimension is at .263 value, hence, insignificant or has the lowest consideration in SDP Adaptation. However, the standardized loading estimates yielded for Social Dimension and Economic Dimension is highly significant where the estimates value at .932 (Q4.2SOP) and .889 (Q4.1ECP) respectively. On average, TRC construct standardized loading estimates yielded .693 value which indicates just significant or slightly below 0.7 value suggested for acceptance level by Hair (2010)[27].

Pre-occupancy assessment sub-criteria of TRC construct concerns assessment sub-criteria that includes green transport masterplan, availability and frequency of public transport, facilities for public transportation, pedestrian networks, cycling networks and alternative transport options. The low standardized loading estimates for Environment Dimension is due to poor implementation of sustainable transportation masterplan and public transportation management and facilities. Another reason is due to minimum application of non-vehicular network systems and alternative options in the certified neighborhood development. Hence, the insignificant values yielded reveal that Environment Dimension is highly less adapted in SDP Adaptations of TRC construct. Its needs improvement in this dimension to better facilitate and support the sustainable growth of township/neighborhood development and its pre-occupancy assessment criteria and certification.

4.7. Building & Resources (BDR)

For BDR Construct standardized loading estimates, two sustainable dimension standardized loading estimates are above 0.7 significant level as suggested by Hair (2010). The Environment Dimension (Q5.6ENP) obtained high significant value at .851 and Economic Dimension (Q5.4ECP) also yielded high significant value at .755. Economic Dimension (Q5.4SOP) generated less significant value but closed to acceptance level at 0.7 where the standardized loading estimates values is at .558. This results imply high SDP Adaptations in Environment Dimension and Economic Dimension in pre-occupancy assessment criteria of GBI-TAC while least adaptation and consideration for Social Dimension. Averagely, BDR Construct yielded at .721 value for standardized loading estimates which denotes significant.

The variation of regression weight values within the defined sustainable dimension representing SDP adaptation priority concerns for BDR pre-occupancy assessment core-criteria. Environment Dimension and Economic Dimension shows the priority rank significant factor of the BDR assessment core-criteria in sustainable neighborhood development. The less significant regression weight value yielded for Social Dimension shows that this dimension least considered in SDP Adaptations in BDR pre-occupancy assessment criteria. This is due to less involvement or public participation through Social Dimension related factors in BDR pre-occupancy assessment sub-criteria which concerns low impact material (building & infrastructure), regional material, quality in construction, construction waste management, site sedimentation and pollution control, sustainable construction practice and GBI certified building. The low standardized loading estimates for Social Dimension is due to indirect relationship of social participation in the implementation of sustainable construction management and practice. Another reason is due to minimum application of low impact material and regional material during the development stage. Thus, the less significant values yielded reveal that Social Dimension is less adapted in SDP Adaptations of BDR construct.

4.8. Business & Innovation (BSI)

Lastly, is the Business & Innovation (BSI) generated outcomes. Economic Dimension (ECP) yields the highest significant value at .739 (Q6.1ECP) indicating that SDP Adaptations in this dimension is of importance or highly adapted. This is followed by Social Dimension (SOP) with significant value at .624 (Q6.1SOP) and Environment Dimension (ENP) with significant value at .637 (Q6.2ENP). BSI has the least pre-occupancy assessment sub-criteria of GBI-TAC. The assessment sub-criteria of BSI concerns on business, innovation and GBI Facilitator. BSI assessment sub-criteria inclined to Economic Dimension and Social Dimension, hence, the loading estimates from Stakeholders-Inclusion Approach – Professional Stakeholder Surveys full structured SEM modeling indicating these two Dimension is highly adapted or most significant in SDP Adaptations while less significant in Environment Dimension.

The less significant regression weight value yielded for Environment Dimension shows that this dimension need improvement in SDP Adaptations to support sustainable development growth and minimizing the BSI assessment core-criteria Environment Dimension adaptation gap. The lack of consideration in SDP Adaptations and implementation of Environment Dimension related factors in BSI pre-occupancy assessment sub-criteria should be improved in particular the innovation sub-criteria where there is an avenue for this dimension further enhancement.

4.9. Summary of Full Structural Equation Model

Following the full structural model discussed above, and the hypotheses formulated as the base of the model, the following table summarized the hypothesized model in terms of its P-value. All path coefficients of the causal structure were statistically significant when the P-value indicated less than 0.001 (***), showing that the hypotheses are all supported. Table 2 presents the hypotheses results formulated for Phase 2 full structural model.

Table 2. Summary of Hypotheses

Hypothesis	Hypothesis Path	P-value	Results
Hypothesis 1:	SDP Adapt --> CEW	***	Supported
Hypothesis 2:	SDP Adapt --> EEC	***	Supported
Hypothesis 3:	SDP Adapt --> CPD	***	Supported
Hypothesis 4:	SDP Adapt --> TRC	***	Supported
Hypothesis 5:	SDP Adapt --> BDR	***	Supported
Hypothesis 6:	SDP Adapt --> BSI	***	Supported

5. Conclusion

The Stakeholders-Inclusion Approach view on SDP Adaptations data were analyzed using IBM SPSS Amos 22. The generated outcome result of full structured model yielded value is as shown in Table 3 below. The full structured model of SEM analyses describes the data are appropriate where all the given paths co-efficient (*P*-value) of the underlying structure were statistically highly significant at 0.001 (***) value and statistically significant at ≤ 0.05 value. Thus, hypothesized paths of the full structured model for SDP Adaptations study show significant correlation implying that all hypotheses are supported.

Table 3. The SDP Adaptations model fit and their level of acceptance

Study Area	Name of index	Yielded Value	Level of acceptance
SDP-Adapt	RMSEA	0.042	RMSEA < 0.08
	GFI	0.882	GFI > 0.90
	AGFI	0.740	AGFI > 0.90
	CFI	0.982	CFI > 0.90
	TLI	0.977	TLI > 0.90
	NFI	0.845	NFI > 0.90
	ChiSq/df	1.148	Chi-Square/ df < 3.0

For Stakeholders-Inclusion Approach – Professional Stakeholder Surveys full structured SEM modelling, six constructs are tested. The standardized (b) loading estimates for CEW suggested that Economic Dimension (ECP) yields the highest significant value, indicating that SDP Adaptations in this dimension is of importance or highly adapted. The loading estimates from Phase 2 indicated that Economic Dimension is highly adapted or most significant in SDP Adaptations. The overall outcomes for EEC construct imply that it is within the loading values level that suggest considerable SDP Adaptations in pre-occupancy GBI-TAC. All specific path co-efficient of the causal CEE full structure model are statistically significant at 0.001 (***) values, the parameter estimates of the hypothesized full structured model are also exempted from offending statistical values including the SDP Adapt ---> CEE path, therefore this hypotheses is supported. For CPD Construct, **t**he regression weight values vary among the designated sustainable dimension indicating that SDP adaptation differ in relative advantage as for the case of CPD core-criteria, Social Dimension and Environment Dimension appears to be the significant factor that brings the CPD assessment core-criteria towards an enhanced sustainable neighborhood development. The standardized average outcomes indicated that TRC has the slightly lower loading factor and satisfactory level of acceptance within the overall pre-occupancy assessment evaluation core-criteria of GBI-TAC. For BDR Construct standardized loading estimates, the results imply that high SDP Adaptations in Environment Dimension and Economic Dimension in pre-occupancy assessment criteria of GBI-TAC while least adaptation and consideration for Social Dimension. Lastly, is the Business & Innovation (BSI) generated outcomes. The standardized loading estimates yielded for BSI from Phase 2 full structured SEM modelling indicating Economic Dimension and Social Dimension is highly adapted or most significant in SDP Adaptations while less significant in Environment Dimension.

Table 4. Summary of Hypotheses for SDP-Adapt. Study

Study Area	Hypothesis	Hypothesis Path	P-value	Results
SDP-Adapt	Hypothesis 1:	SDP Adapt --> CEW	***	Supported
	Hypothesis 2:	SDP Adapt --> EEC	***	Supported
	Hypothesis 3:	SDP Adapt --> CPD	***	Supported
	Hypothesis 4:	SDP Adapt --> TRC	***	Supported
	Hypothesis 5:	SDP Adapt --> BDR	***	Supported
	Hypothesis 6:	SDP Adapt --> BSI	***	Supported

Following the full structural model of SDP Adaptations study and as discussed in previous sub-sections, and the hypotheses formulated as the base of the model, the following table summarized the hypothesized models in terms of its P-value. Table 4 presents all path coefficients of the causal structure were statistically significant when the P-value indicated less than 0.001 (***), showing that all the hypotheses are all supported. However, to conclude, there should be an improvement in environmental dimension for CEW, CEE, TRC core criteria; economic dimension for CPD core criteria; and social dimension in BDR core criteria for further review and refinement of GBI-TAC in the near future.

Acknowledgement
The author would like to express gratitude to EDS Chulalongkorn University, FSPU UiTM and KAED IIUM for providing platform and grant in completing this research.

References

[1] Leipzig charter on sustainable European Cities. Final draft, document prepared by the German Presidency of the European Union, Leipzig. Gov. Regist. 2007
[2] Wu J 2014. Urban ecology and sustainability: The state-of-the-science and future directions *Landscape and Urban Planning* **125** 209-221
[3] Hiremath R B, Balachandra, P, Kumar B, Bansode S S, and Murali J 2013 Indicator-based urban sustainability—A review *Energy for Sustainable Development* **17** 555-563
[4] Schnur O 2005 Exploring social capital as an urban neighbourhood resource: empirical findings and strategic conclusions of a case study In Berlin-Moabit. *Tijdschrift voor economische en sociale geografie* **96** 488-505
[5] Berg P G and Nycander G 1997 Sustainable neighbourhoods—a qualitative model for resource management in communities *Landscape and Urban Planning* **39** 117-135
[6] Mattarozzi S and Antonini E 2011 Sustainability assessment: An integrated approach with inhabitant participation *Proc. Engineering* **21** 560-569
[7] Luederitz C, Lang D J, and von Wehrden H 2013 A systematic review of guiding principles for sustainable urban neighborhood development *Landscape and Urban Planning* **118** 40-52
[8] Valentin A and Spangenberg J H 2000 A guide to community sustainability indicators *Environmental Impact Assessment Review* **20** 381-392
[9] Harger J R E and Meyer F M 1996 Definition of indicators for environmentally sustainable development *Chemosphere* **33** 1749-1775
[10] Mori K and Christodoulou A 2012 Review of sustainability indices and indicators: Towards a new City Sustainability Index (CSI) *Environmental Impact Assessment Review* **32** 94-106
[11] Dahl A L 2012 Achievements and gaps in indicators for sustainability *Ecological Indicators* **17** 14-19
[12] Alkan O J, Bockstaller C, Stapleton L M, Ewert F, Knapen R, Therond O, Bezlepkina I 2009. A goal oriented indicator framework to support integrated assessment of new policies for agri-environmental systems *Environmental Science & Policy* **12** 562-572
[13] Niemeijer D 2002 Developing indicators for environmental policy: data-driven and theory-driven approaches examined by example *Environmental Science & Policy* **5** 91-103
[14] Repetti A and Desthieux G 2006 A relational indicatorset model for urban land-use planning and management: Methodological approach and application in two case studies *Landscape and Urban Planning* **77** 196-215
[15] United Nations 2007 *Indicators of Sustainable Development: Guidelines and Methodologies* (New York: United Nations)
[16] Shen L Y, Ochoa J, Shah M N, and Zhang X 2011 The application of urban sustainability indicators – A comparison between various practices *Habitat International* **35** 17-29
[17] Sharifi A and Murayama A 2013 A critical review of seven selected neighborhood sustainability assessment tools *Environmental Impact Assessment Review* **38** 73-87

[18] Council, U. S. G. B. 2006 *LEED for Neighborhood Developments* (Washington, DC: US Green Building Council)
[19] Garde A 2009 Sustainable by Design?: Insights From U.S. LEED-ND Pilot Projects. *Journal of the American Planning Association* **75** 424-440
[20] Hamedani A Z and Huber F 2012 A comparative study of DGNB, LEED and BREEAM certificate systems in urban sustainability *The Sustainable City VII: Urban Regeneration and Sustainability* ed(s) Pacceti M, Passerini G et al pp 121-132.
[21] Sullivan L J, Rydin Y, and Buchanan C 2014 *Neighbourhood sustainability frameworks-a literature review* (London: University College London).
[22] Yahya W K and Hashim N H 2011 The role of public awareness and government regulations in stimulating sustainable consumption of Malaysian consumers. *Conf. Business, Engineering and Industrial Applications (ICBEIA).* E-ISBN: 978-1-4577-1280-7
[23] Shari Z, Jaafar Z, Fakri M, Salleh E, and Haw L C 2008 The potential of sustainable building rating System in the Malaysian building industry *Int. J. Sustainable Tropical Design Research & Practice* **3** 3-14
[24] Okoli C and Pawlowski S D 2004 The delphi method as a research tool: an example, design considerations and applications *Information & Management* **42**15-29
[25] Bollen K A 1989 *Structural equation with latent variables* (New York: Wiley)
[26] Byrne B M 1998 *Structural equation modelling with LISREL, PRELIS and SIMPLIS: basic concepts, applications and programming.* (New Jersey: Lawrence Erlbaum Associates)
[27] Hair J F 2010 *Multivariate data analysis* (New Jersey: Prentice Hall)

IOP Conf. Series: Earth and Environmental Science **158** (2018) 012009 doi:10.1088/1755-1315/158/1/012009

Gentrification and the Vulnerability of Betawi Community

Santy Paulla Dewi[1]

[1]Department of Urban and Regional Planning, Faculty of Engineering, Diponegoro University, Semarang, Indonesia

Corresponding author: santy_paulla@yahoo.com

Abstract. Betawi community is the local community in Jakarta, residing in the city since the 17th century. Yet, since 1970s their number has declined significantly, and now it only remains 20% of the Jakarta total population. They have been displaced by several government urban development projects, and have moved to peripheral kampongs. Moreover, the displacement of this local community was triggered by gentrification due to the presence of middle-income class (gentrifiers), who come to Jakarta as professional workers, and further demand housing. Gentrifiers prefer to stay in the inner-city kampongs to save transportation cost even though the kampongs are not equipped proper infrastructures. Then, the gentrifiers renovate the house and shape the physical appearance of the neighbourhood differently. This has made the land price surge, and the competition to occupy house increase. Also, the standard of living has gone up. Many local communities could not adapt with the situation and consequently sold the house to get money; most of them consist of uneducated and unskilled informal workers, who can be included as poor-class. Living in kampongs, which are considered as informal settlements, the community do not get any government assistance to improve the neighbourhood quality; hence, the neighbourhood improvement only relis on the gentrifiers. As a result, the displacement continues to occur. Consequently, the Betawi community becomes the most vulnerable in the gentrification process, and are very sensitive to the neighbourhood transformation. Using a case-study research, his paper tries to reveal the relation between the gentrification process in the inner-city kampong and the Betawi community as the vulnerable community. This research focused on Kampong Kramat Asem, Matraman district as densely a populated district in DKI Jakarta Province. Furthermore, field observation and interview were conducted to the local community and gentrifiers to find out the community sensitivity factors and to picture the physical neighbourhood transformation. Moreover, an analysis of population trends was required to figure out displacement process.

1. Introduction

Betawi community has existed in Jakarta since 1673 or the late of 17th century. They resided in many inner-city kampongs in Jakarta. Since 1970, their number has declined significantly because of some urban development projects, such as highway, road, and buildings developments. They were evicted to other parts of Jakarta and their previous places were changed into skyscraper offices, parks and other public facilities. Likewise, they have also been displaced along with the influx of newcomers who bought their houses.

The number of newcomers in Jakarta increases every year.It can be seen from the population growth in 2015; it increased by 2 % and the total population was 12.7 million at noon and 9.9 million at night [1]. Mostly, newcomers are professional worker; bachelor or master graduated from well-known university and skilled workers. They prefer to live in inner-city because it has high accessibility to their workplace, whereby they can save their transportation cost. Moreover, they bought houses in the inner-city kampongs where the land price is cheaper than the price of new apartment units provided by private developers. Hereafter, they renovated the houses based on their styles and accordance with the number of the family. This renovation has shaped the physical face of the neighbourhood, increased the neighbourhood quality, and risen the land price. However, this situation influenced the existence of the Betawi community. The risen standard of living has given great pressures to the Betawi community, whom can be included as the poor or the low-income class in the society. As a result, they sold the house and displaced. The transformations of the physical condition, the pressures on the economic aspects, and the displacement of Betawi community indicate that the gentrification have occurred in the kampongs. In this case, the gentrification is caused by the increasing demand of housing and the competition to occupy houses.

Gentrification is considered as a transformation of the neighbourhood process which is faced by almost all of big cities [2]. It has two sides; the positive side is in the term of physical neighbourhood improvement, whereas the negative side is in the context of displacement of the local community [3]. However, the improvement of the neighbourhood cannot be perceived by all of the community; it is only experienced by the newcomer (gentrifier), while the local community is displaced.

Gentrification continues to occur, signed by the displacement of the local community; however, the government has done nothing to respond this along with the neighbourhood upgrading level. Who gets benefits from all of these conditions? In the gentrification process, the local community is often neglected and have no bargaining power. Living in the kampongs since they were born with all of their extended family is their desire. Yet, the economic pressure has caused them sell the house; working in informal sectors with salary of less than 2 million rupiah has made them unable to adapt with the increasing living standard. This community has become the most vulnerable actors in the gentrification process.

2. The definition of vulnerability

There are many definitions of vulnerability from various perspectives; namely, social, economic, politic, and technology. Vulnerability can be described as a condition where a person or a community is powerless, insecure and exposed to risk, shocks and stress [4]. The vulnerability level is difficult to assess because it is not in mathematic or statistic form; it is more complex and related to social relations and economic dynamics [5].

Vulnerability is related to risks and hazards that are addressed differently among communities; it is seen from the community's readiness in facing the hazards, ability to adapt, and capacity to rise from the problems. Generally, in urban development, poor people tend to be vulnerable because they do not have affordability to bargain with the changes of situation. Yet, it does not mean that vulnerable community always have to be poor.

In this paper, the vulnerability assessment adopted the vulnerability and capacity analysis ([6] that consists of three principles. First, material or physical vulnerability can be seen from the vulnerability of finance, food, land, jobs and access to loan. Second, the social institutional vulnerability is observed from social networking, family bonding and family welfare institutional. Third, the attitude or the motivational vulnerability consist of self-confidence, authority and affordability. Besides, it is important to identify the community adaptation related to the transformation of the neighbourhood as one of the vulnerability parameters [7].

3. Gentrification Definition and Implications

Most metropolitan cities are experiencing gentrification process, which is a part of a city's life cycle [2]. Having dissimilar entities and characteristics makes the gentrification process and implication different in each area [8]. Gentrification is a displacement process and neighbourhood transformation caused by an intervention of the government, private developers, or individual renovator [9]. Government intervention can be seen by giving incentives (low tax development activities, housing loan, and low property tax), while private developers activities which triggers gentrification include developing new residential areas and redeveloping business districts of inner-city. Individual renovators' role in the gentrification process can be seen in the form of the housing demolition and redevelopment, individual's preference, and life style.

Giving incentives in the form of lower tax for development activities, housing loan, and low property tax are some incentives from the government related to urban development. It can be inferred from the previous statement that it is not only government who can be a pioneer but also private developers who build the neighbourhood. The pioneer of gentrification can also be many actors, resulting from the elaboration of the private developers' attitude, government' policies and also housing market. Furthermore, this elaboration leads to the economic growth of the neighbourhood, which is followed by the social transformation.

Gentrification occurs in inner-cities inhabited by low-income class who had low economic (financial) capability to improve their house and the neighbourhood. Moreover, the lack of facilities and infrastructure worsens the neighbourhood quality. Then the government with its urban development projects also influences the values of the neighbourhood. This situation attracts gentrifiers to stay; consequently, their number increases every year and they displace the local community.

The displacement of the local community (low-income class) and the transformation of economic, social, and physical condition of the neighbourhood are the signs of gentrification. Some gentrifying neighbourhoods can also be distinguished by the range of the location from the CBD (two miles); this includes inner-city, land tenure, land use, and community's characteristics such as income, educational background, and job [10].

The perceived implication of the gentrification can be considered as the two sides of a coin; there are positive implications and negative implications. On one side, the improved neighbourhood quality, new job variations, and economic growth are considered as positive implications, whereas on the other side, the displacement of the local community, the uncontrolled surge of land price and land rent, the land speculators, and the extinction of the local culture are the negative implications. These implications affect the community differently with respect to gender, educational backgrounds, races, ages, incomes, and community relationship [11]. The low-income class and women are the most vulnerable actors in the gentrified neighbourhood, especially in the displacement process.

4. Research Method

A single case study strategy was used in this research, taking Kampong Kramat Asem as the research location. This research emphasized on how the Betawi people as the local community got displaced and became the most vulnerable actor in the gentrified neighbourhood. The vulnerability of Betawi community was seen from their displacement, the community's sensitivity factors and their adaptation, as well as the dimensions of vulnerability. The displacement analysis concerned with the justification and the reasons why the Betawi community decided to sell their houses or to stay in the kampong; it was revealed from the financial affordability (income), the community relationship, and job. Meanwhile, the community's sensitivity was focused on the main aspect that made the Betawi people susceptible in the gentrification process, the community's adaptation concerning with how the Betawi people who stayed in the kampong adapted with the new neighbours, new lifestyle, and new physical appearance of the kampong. Then, the dimension of vulnerability was seen as the stages of gentrification process.

A depth-interview was conducted to collect data related to the community's (i.e. the Betawi people's and gentrifiers') perception and the local government's perspective on the gentrification process. Some interviewees were chosen, including both Betawi people who stayed in kampong and Betawi people who moved from the kampong, gentrifiers, the head of the kampong, and the head of the village. The key person of this depth-interview was the head of the kampong, which then snowballed to others. Moreover, a field observation was also conducted to clarify the physical transformation of the kampong, the community's relationship, and the community's (cultural)

IOP Conf. Series: Earth and Environmental Science **158** (2018) 012009 doi:10.1088/1755-1315/158/1/012009

activities. Besides, secondary data were also collected from some government institutions such as East Jakarta Central Bureau of Statistics and Utan Kayu Selatan district office.

The descriptive qualitative method was used to reveal the community's adaptation and perception on the gentrification. Moreover, statistical data were also used to enrich and to support the analysis.

5. Gentrification in Kampong Kramat Asem: an overview

Kampong Kramat Asem is one of the urban kampongs in Jakarta which experienced gentrification. This kampong is constituted a part of Utan Kayu Selatan village, Matraman sub-district, East Jakarta municipality. The Utan Kayu Selatan village is 1.12 $km^{2,}$ wide, whereas kampong Kramat Asem is 143.165 m^2 wide [5]. This village consists of 174 RT (*Rukun Tetangga*) and 14 RW (*Rukun Warga).* Having a good connection and accessibility to both Jakarta's downtown (7.9 km) and Jakarta's peripheral areas is the main pulling factor that attracts gentrifiers to reside in these kampongs. It is not only passed by the public transport (bus rapid transit – TransJakarta) but also by the highway access.

Kampong Kramat Asem has already existed before Indonesian Independence Day (1945). Based on the history, in 1920 there was an *Asem* Tree (a Tamarind Tree), which was considered sacred by the Betawi community. Hence, this kampong was named Kramat Asem. Yet, in 1970s there was a land acquisition by the central government for road and government buildings development. Some parcels of land of the community, open space, the *Asem* tree and the sacred grave were included as redevelopment areas. As a consequence, there was no identity of the kampong and the community lost their belief related to tree and the grave. Only few of Betawi old generation keep their belief by putting some flowers and *sesajen* (the ritual offering) in previous location of the grave. At this moment, the *Asem* tree and grave location has changed into a mosque, an Islamic boarding school, and tenements.

Concerning the number of the population, this kampong has the highest share of population in Matraman sub-district (24%) with 43,396 people in 2011. This number consists of Betawi people (25 %) and gentrifiers. In detail, the gentrifiers are the professional workers, the Betawi people worked as employees, labours, and small-scale entrepreneurs. These small-scale entrepreneurs use the front part of their houses as stalls for selling traditional foods, snacks, and small grocery stores. The number of Betawi people in the kampong continues to decline along with the presence of gentrifiers.

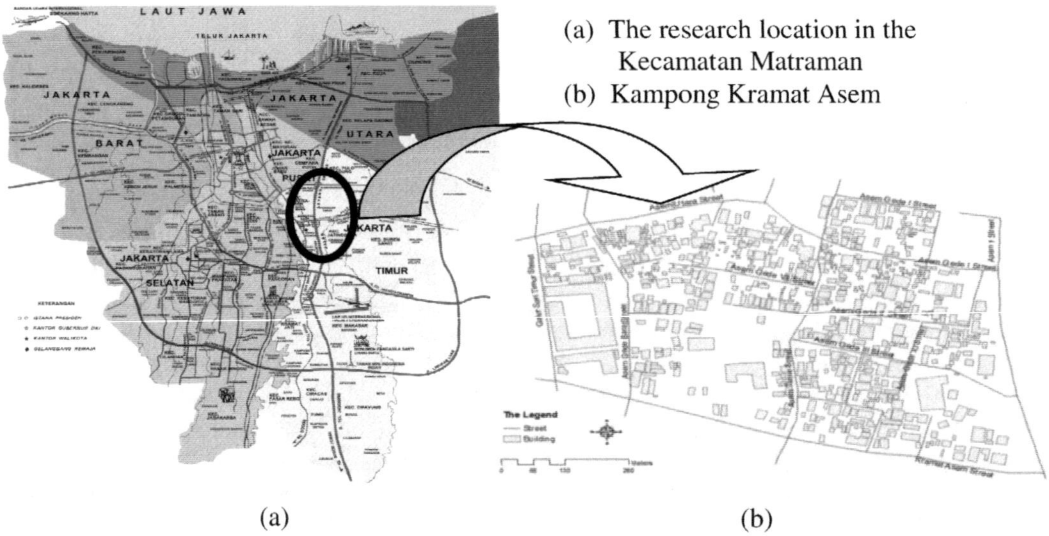

Figure 1. Research Location; Kampong Kramat Asem [12]

The land use pattern is dominated by settlements; informal settlements with an organic pattern. Since 2005, many houses have changed from settlements into offices, particularly on the main kampong street. The facilities provided are limited, such as a mosque and a community hall that is also used as a kindergarten. Likewise, the infrastructures are also limited to street and drainage, while water and waste infrastructure were provided by the community itself. Furthermore, the open space in this kampong is in the form of kampong street, whereas the green open space is very limited (less than 10%). Some fields exist in the kampong, yet they are used as parking lots for the gentrifiers who did not have carport in their houses. Hence, the children do not have playground and use roadside as the playground.

This kampong was chosen to study because of some reasons. First, the displacement of Betawi people as the local community from their hometown where they were born, grew up, married, and had their own family. They were replaced by the gentrifiers from various cities in Indonesia who had dissimilar educational backgrounds, jobs, and lifestyle. Second, the kampong resides in the strategic location with high accessibility.Further, the demand of the land in the area has also increased. Third, this kampong is considered as one of the dense kampongs in the dense sub-district.

6. The Vulnerability of Betawi Community

6.1 The decreasing Number of Betawi Population; The Consequences of Displacement

Displacement is a direct implication of gentrification, which transforms the social, economic, and physical structures of a neighbourhood. The Betawi community could not avoid the displacement process, which as though a "natural" occurence. On one hand, there was no pressure and no compulsion for the Betawi people to sell their houses; they even felt that they had been helped by the gentrifiers who were willing to buy their houses. It was not easy to sell the house without legal land certificate. In addition, houses in kampongs were not equipped with proper water and waste infrastructures, suffering flood annually, and located in narrow alleys. On the other hand, gentrifiers preferred to stay in accessible areas, which does not only mean less traffic jams, but also better access to good public transport connections. Therefore, they were willing to accept the inconvenience to stay in the kampong [13]. As a consequence of displacement, the number of Betawi people in Kampong Kramat Asem continues to decrease, along with the increasing number of gentrifiers who move there. The Betawi population was replaced by gentrifiers coming from Sumatra, Java, and Borneo islands.

However, there are differences related with the displacement influence to the Betawi people. Generally, Betawi people are respectively classified into *orang gedong* and *orang kampong*. While *orang gedong* are considered as rich Betawi people who are landlords that occupy broad land and big houses (more than 100 m^2), *orang kampong* are categorized as the poor Betawi people who stay in small houses. The dissimilarities of gentrification influence between *Betawi gedong* and Betawi *kampong* can be seen from the job variations and inhabitation preference.

The presence of the gentrifiers in the kampong who renovate the house, and improve the neighbourhood physically has significantly increased the land price and housing rent started due to the high demand. This situation is perceived by the Betawi people as an opportunity to get profit as much as possible. They start to sell their houses to the gentrifiers; they move to other kampongs in Bekasi, Bogor, or Tangerang, which are included as Jakarta's peripheral area. This displacement can be included as a voluntary displacement [14] since there is not any pressure from others to sell their houses and they really want to gain profit. Most of *orang gedong* sell their house and buy broader land in their new places, and then build new houses with some tenements. Renting tenements becomes their "new" job, which can increase their incomes. This situation is not found among *orang kampong* because they only have small houses- the one and only property they have. Also, they must share this property with all of the family members that usually consists of five siblings. They cannot afford to buy new houses in the new places. Moreover, most of *orang kampong's* houses are located in the densed and narrow alleys (in Asem Gede 1 Street), so it was more difficult for them to sell the houses because less people want to buy them. This is different with *orang gedong's* houses that are situated in the kampong main street (Asem Gede 2 Street), and hence have higher values.

IOP Conf. Series: Earth and Environmental Science **158** (2018) 012009 doi:10.1088/1755-1315/158/1/012009

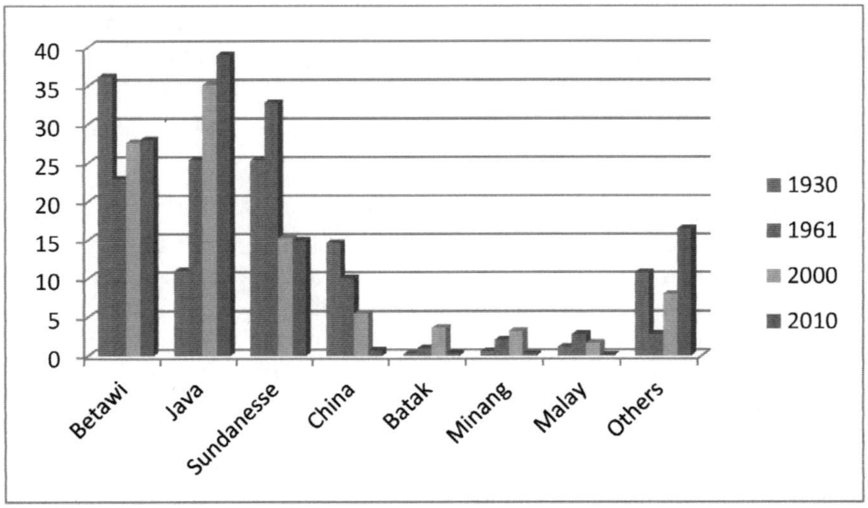

Figure 2. Total Population by ethnic in Kramat Asem
(in terms of percentage) [12]

The *orang gedong's* houses are offered to the gentrifiers individually. This means the houses are not sold in groups, such as simultaneously putting one block houses on sale. Instead, each year, the houses are sold one by one. The housing sale is also influenced by the relocation the work place. DKI Jakarta government's regulation allocates the sites for industries and companies to the peripheral areas because they attempt to reduce the density and congestion in the inner-city. It can also be triggered by the development of peripheral areas by considering the development of these industries and companies as the new growth centre. Living close to the work place is preferable for the *orang gedong* so that they choose to sell their houses in the inner-city kampongs.

Different situation is found among *orang kampongs* who sell their houses. Selling the houses is their last option when they do not have any options. Some of them may need money to pay the hospital fee when they are sick, or to fulfil their daily needs as their income as labourers is not enough, and some of them may be stuck in debt with high interest either to private or individual moneylenders. Given the choice to relocate, some of them move to other peripheral kampongs and try to find a new job while others choose to stay in rented houses in kampong Kramat Asem because they do not want to lose their job; they realize that it is difficult for them to find new jobs. It is really ironic that they must stay in rented houses in their previous places (which is their home town). Besides, some of *orang kampongs* who sell the house choose to buy affordable new houses in other inner-city kampongs (located adjacent with kampong Kramat Asem) even when the new houses have no clear status (they do not have legal land certificates). Most of these people are food street vendors that run their business in kampong Kramat Asem.

Most of Betawi people are uneducated and unskilled because of their low awareness to the education. Moreover, the women are marginalized, which makes the women position in the community very susceptible. They are considered as the families' guardians who must do all of house chores, such as cooking, washing all of family clothes, and cleaning the house. Going to school is the last option for them, which is not prioritized because it will be useless; they will not go work and will only become housewives. Thus, their educational background is unimportant. This situation is faced by both the *orang gedong* women and *orang kampong* women. While educations for the Betawi's men are slightly similar with the women; they go to school until they only graduate from the elementary or junior high school. Then, they must work; feed all of the family and fulfil all of their big family daily needs. Less education and less skill make the job occupancy very limited for them. Whereas *orang gedongs* choose to have their own small business by using their family financial support, the *orang kampongs* work as public transportation drivers who may not have any driving license, illegal parking attendants, and motorcycle taxi drivers. Yet, after the presence of the gentrifiers, the new jobs offered changes, such as the gentrifiers' drivers, gardeners, and housemaids (for women). Therefore, the replacement of the community shapes the job variation, which also increases the number of Betawi women workers. Previously, most of *orang kampong* women were housewives who spent their whole day at home. Currently, however,, many of them work as gentrifiers' housemaids. The reason is not only because their "work places" are very close to home, but also because they can contribute to increase the family income.

6.2 Sensitivity Factors of Betawi's Community

The rise of middle-income class in Indonesia since 1998 has influenced various new job vacancies offered in inner-cities, including Jakarta. Besides, the shift from manufactured industry to service-based industry has also demanded new workers suitable for the jobs. Consequently, the Betawi people who are mostly less educated and less skilled could not be absorbed in this type of jobs. Therefore, many young professionals from other cities come to Jakarta to pursue this jobs. Residing in inner-city houses close to the workplaces is preferable for them. Hence, they choose to stay in inner-city kampongs where the land price and the rent are still affordable.

The increased land price and housing rent make the properties in the kampongs high in demand. Realizing this opportunity, many Betawi people sell their houses to the gentrifiers to get profit, while the gentrifiers capture this situation to invest as much as possible; they buy their second or even their third house from Betawi people, and then renovate it as boarding houses or tenements. Then, the gentrifiers become land speculator who try to buy the Betawi people's houses . The rented houses or tenements in the kampong Kramat Asem are high in demand. This can be seen from the current rented houses and tenements that are always fully occupied. The rent price started from one million rupiahs to two million rupiahs per month per room; the gentrifiers can have more than ten rented rooms. Renting rooms or tenements is the gentrifiers' side income because they still work in formal sectors.

Theoretically, gentrification can be caused by the rent gap [2]; the significant range between existing rent price (land price) and the potential rent price that could be got. The rent gap in Kampong Kramat Asem is captured by the gentrifiers as the opportunity to gain additional income and as an investment. The land price in kampong Kramat Asem is lower than the land in other inner-city kampongs. It is not only caused by the fact that kampong is not equipped with proper infrastructures, but also by the fact that most of the Betawi community who do not have any legal land certificates. Then, it makes the land value decline; the buyers take a high risk if they buy land with no legal land certificate because they are vulnerable to face land disputes. Yet, the gentrifiers are willing to accept this risk and prioritize more on the profit than can be gained. Previously, the gentrifiers bought the Betawi's people house for 150,000 rupiahs per m^2 in the 1980s, which became up to 1 million rupiahs in 1990s [15]. In 2016, the land price jumped into 12 million - 16 million rupiahs per m^2.

Many Betawi *kampong* people cannot not stand in their kampong and sell the house because they are poor. Most of them work in informal sectors as a handyman, freelance drivers, labourers, and housemaids. Gaining income of less than 2 million rupiah/month is really difficult to fulfil all of their daily needs. As the standard of living in the kampong rises along with the improvement of the physical neighbourhood by the gentrifiers,it becomes more difficult for the Betawi kampong people to survive. The poverty makes them very vulnerable to get displaced in the gentrification process. Although not all of the vulnerable community is a poor community, the poor community is more vulnerable than other income class. Vulnerability is categorized as the central element of poverty [16]. Poverty, low education, and no legal land certificate make their bargaining position in the neighbourhood low. Similarly, they have less bargaining position when they sell the houses to gentrifiers.

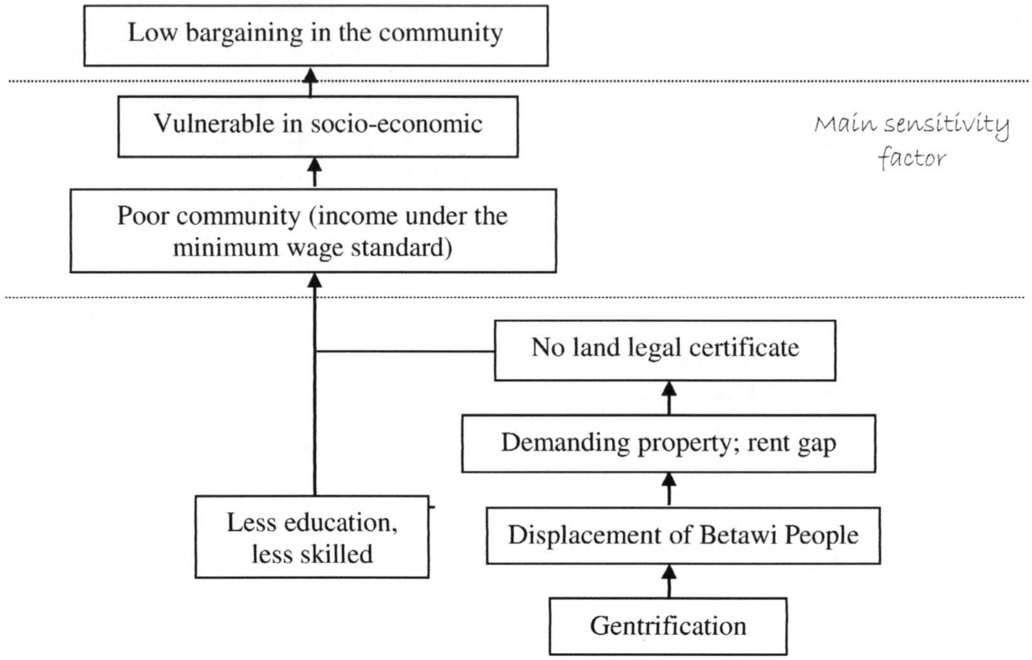

Figure 3. Sensitivity Factor of the Vulnerable Betawi Community

However, the most vulnerable community in gentrification process is the Betawi kampong people who reside in the kampongs. They live with all of their extended family; i.e. grandparents, parents, siblings, uncles, and aunts, in an inherited house; there are more than 7 people in a house, which is less than 70 m^2 wide in a very dense and narrow alley. They have no choice other than staying in the kampong because if they sell the house they must divide the money to all of the siblings and it will not be enough to buy a new house. Most of their time is used to work, even not only in one job but also second or third job. For instance, they work as handymen in the morning until afternoon, and then as drivers (taxi bike) from afternoon until midnight. Most of their kids are elementary school dropout and choose to make money as motorcycle and car washers, labourers, and illegal parking attendants. Selling the house to fulfil their daily needs is the last option for them.

Likewise, Betawi kampong people who stay are also vulnerable in social aspects. Working almost all day makes them have less time to have social relationship with their family and their neighbours; they participate less in community activities. Previously, Betawi people were well-known as a community that had strong relationship and preserved their culture; they have their own language, food, custom home, custom clothes, performances (*ondel-ondel* and *rebana)* and traditional dances. Now, preserving the culture is considered as a waste of money, so that they prefer to skip the cultural ceremonies in some community activities. Similarly, the Betawi *gedong* people have also started to leave the cultural ceremonies because most of them are married to people from other tribes who have different cultures. Therefore, they prefer to choose modern style for their wedding ceremonies, invite modern music performances (pop or *dangdut* band*)* in the community activities rather than *ondel-ondel* or *rebana* performances.

The low Betawi's cultural preservation and the weak community relationship are not only influenced by the money that they must spend and the limited time, but also by the gentrifiers' lifestyle. As a middle-income class, the gentrifiers preferred to have modern and individualistic lifestyle. They prefer to have interaction with other members of middle-income class. In addition, the television programs, internet, and other social media have also blown up the news about celebrities with their modern and up to date styles. Hence, the Betawi people are affected by this modern lifestyle which is seen as update preferences.

6.3 Dimensions of Vulnerability

In this research, the dimensions of vulnerability are seen from the gentrification implication to the Betawi community. Betawi people's houses are gadually sold was occurred; one house at a time. When a house is offered to sell today, next couple of months or next couple of years other houses are offered. The house are not simultaneously sold in blocks; instead a unit at a time is offered. The sold houses are then renovated by the gentrifiers and consequently change the physical appearance of the neighbourhood. The new rent houses and tenements replace the previous houses. This has not only changed the housing functions, but has also generated new sources of income (economically).

The improvement of the physical appearance leads to the increased price of land and housing rent. Subsequently, the standard of living has also risen. Moreover, the Betawi people are influenced by the modern life style, which affects their orientation of job, education, and health. Besides, the DKI Jakarta's government are more concerned to improve the community's quality by giving Jakarta's Smart Card (KJP) and Jakarta's Health Card (KJS) for the low-income class, including the Betawi kampong people. The Betawi people start to realize the necessity of having a health insurance, because they do not want to face the same problem as the previous generations of Betawi *kampong* people who sold the houses to pay the hospital fee.

Furthermore, the DKI Jakarta's government is concerned to improve the community's education by giving the Jakarta's Smart Card (KJP), which is similar to a debit card with some balance, for students from poor families. They can use the money to buy shoes, uniforms, bags, other school supplies, and even nutritious food. Yet, dropped out kids from the second grade of elementary school are still found in kampong Kramat Asem. Unfortunately, the Betawi *kampong* people are still less concerned with education. Besides, the marriage of young couples (youth aged less than 20 years) are also still found among Betawi *kampong* people. School dropouts and early-age marriage force them to work in a narrow job choice. Hence, they are still included as poor families, and unfortunately this pattern also applies to their next generation.

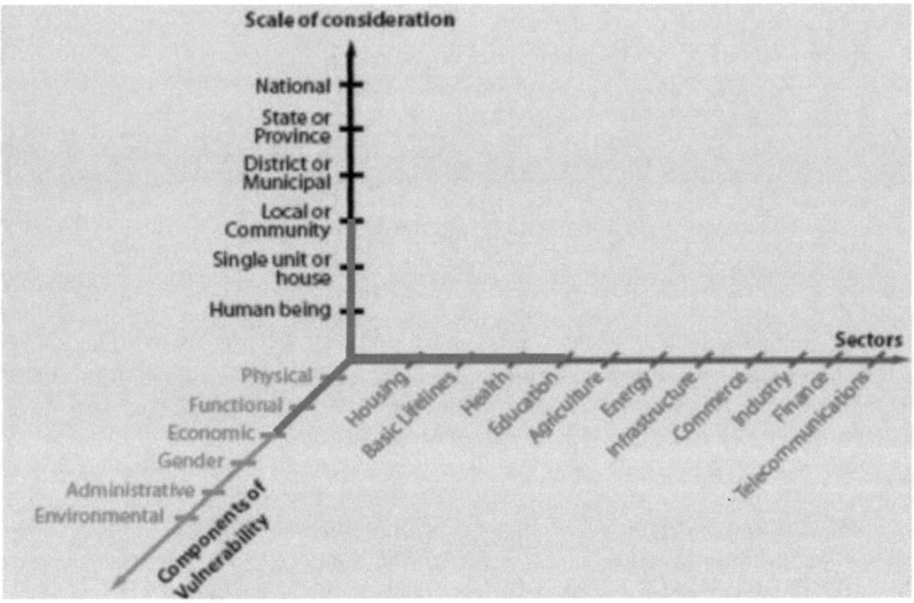

Figure 4. The Dimensions of Vulnerability [17]

6.4 The Betawi Community's Adaptation

Living with new neighbours coming from other cities, other islands, and other cultures makes Betawi kampong people try to adapt. Moreover, being shifted from the majority population to becoming the minority population has also influenced their way of life. Previously, there was a strong social bonding among these Betawi people; they used to consider all of Betawi people as a part of extended family in that they usually shared food, drinks, and money. They used to have outdoor activities with other Betawi neighbours for chatting, playing, and gossiping at the weekends. Nowadays, most of their gentrifiers neighbours have high fenced houses and keep their door closed almost all day long. Somehow, some of them do not know each other; the name, the job, and the family. Yet, this situation does not cause the social gap or social conflict because the Betawi people have the flexibility to adapt with the changes; they try to understand why their gentrifiers neighbours cannot attend the community meetings. This understanding happens because the gentrifiers have also tried to contribute to the community by giving some money to support the community activities; something that is difficult to do for the Betawi people.

Considering the continuous decline of their number in Jakarta, the Betawi people try to maintain their existence. Jakarta is their home town, so that they prefer to live there, and even become leaders for the newcomers. Therefore, many Betawi community organizations have established media for the Betawi people to discuss and cooperate. Through these organizations, they struggle to become the householders (hosts) in their own (previous) land even though their number has decreased and they become the minority in their own homeland. Yet, these organizations are somehow misused by the political powers. As a result, there have been some conflicts between the Betawi people in these organizations with other communities. However, declaring and getting involved in some Betawi community institutions show that they adapt with the changes in their city.

Moreover, the presence of gentrifiers in their neighbourhood has changed the job orientation of the Betawi people. Previously, among the Betawi people, mostly only men worked, whereas the women stayed at home and did household chores. Yet, along with the rising standard of living, these women have tried to work to help fulfil their family financial needs. Almost all of gentrifiers women are workers so they need help from others to do the household chores. Then, many Betawi kampong women work as housemaids for the gentrifier families from 8.00 in the morning until 14.00 in the afternoon. The gentrifiers often give extra money to their housemaids; they also support the housemaids' children school by giving money to buy the new uniform, new shoes, new bags, and nutritional foods for the kids. The variation of jobs have also been experienced by the men of Betawi people; they ask to be gentrifiers' drivers, gardeners, or painters of the houses.

The adaptation of the Betawi people have also been found in lifestyle; they are tempted by the consumerism lifestyle. It is because the gentrifiers' influencethat mostly are university graduates and hold modern lifestyle; they prefer using the latest mobile phone models and cars, wearing branded clothes, bags, and shoes, shopping to supermarkets to traditional markets, going to restaurants to cooking at home, and going picnic to other cities and even to overseas. Likewise, most of the television programs and social media have also showed the modern lifestyle, which tend to promote consumerism. The Betawi people also prefer adapting to modern style to preserving cultural patterns, such as in the wedding ceremonies and other community activities. In addition, they preferred shopping at supermarkets to shopping at traditional markets. All in all, they show little effort to preserve their culture.

7. Conclusion

The gentrification in Kampong Kramat Asem occurs gradually. It causes the displacement of the Betawi people. The housing investment has caused the rent gap, which has then triggered the gentrification. The Betawi people, as the Jakarta's local community who resided in the kampong, bocome the most vulnerable actor in the gentrification process. This vulnerability shapes not only the physical aspects but also the social and behaviour aspects. The physical vulnerability can be seen from the Betawi people's lost residence; they can no longer live in their previous places where they were born, grew up, got married and bore children. They cannot not afford to live in their previous places.

The social vulnerability occurs after the presence of gentrifiers, which has increased significantly and the decline number of the Betawi population. The social bonding among the Betawi people has decreased along with the fact that many Betawi people get married with people from other cities or other islands. Moreover, Betawi people who stay in the kampong spend most of their time to work so that they only have limited time for having social relationship with their neighbours. As a result, the Betawi's culture start to extinct and get replaced by the modern lifestyle.

Another sign of vulnerability can be observed from the Betawi people's attitude such as their self-confidence, authority and affordability. Realizing their status as the low-income and less educated community has made the Betawi kampong people choose the gentrifiers as the community leader (head of RT and head of mosque committee). The gentrifiers believe that they can contribute to the improvement of the community and neighbourhood. Considering the kampong development, the gentrifiers always take the Betawi people as their partners to discuss together. Therefore, there has not been any social conflict in the community, which improves the Betawi people's trust to the gentrifiers.

Acknowledgement
This research was part of Gentrification Process in DKI Jakarta Province research supported by Ministry of Research and Technology Higher Education of Republic Indonesia (BLN 2013). Moreover, the participation of the Kampong Kramat Asem community and the government had significantly contributed to this research. Finally, I do appreciate the full support of the Department of Urban and Regional Planning, Faculty of Engineering, Diponegoro University, Indonesia.

References

[1] Provinsi DKI Jakarta 2015 *Profil* http://www.kemendagri.go.id/pages/profil-daerah/provinsi/detail/31/dki-jakarta

[2] Smith N an William P 1988 *Gentrification of the City* (USA: Allen and Unwin Inc)

[3] Chaskin R J and Mark LJ 2013 "Positive"gentrification, social control and the "right to the city" in mixed-income communities: Uses and expectations of space and place *International Journal of Urban and Regional Research* **37(2)** 480-502.

[4] Pugh C 2000 *Sustainable Cities in Developing Countries* (UK:Earthscan)

[5] Adger W N and Kelly P M 1999 *Mitigation and Adaptation Strategies for Global Change* (Netherland: Kluwer Academic Publishers) chapter 4 pp 253-266

[6] Cannon T 2000 *Vulnerability Analysis and Disaters Chapter in Parker D.J* (ed) Floods (London: Routledge)

[7] Adger W N 2006 Vulnerability *Global Environment Change* **16** 268-281

[8] Zukin S 1982 *Loft living: culture and capital in urban change* (Baltimore: Johns Hopkins University Press)

[9] Atkinson R 2003 Introduction: Misunderstood saviour or vengeful wrecker? The many meanings and problems of gentrification *Urban Studies* **40(12)** 2343-2350

[10] Hamnett C and Randolph B 1988 *Tenurial transformation and the flat break-up market in London: the British condo experience* (New Jersey: Blackwell Publishing)

[11] Skaburskis A 2012 Gentrification and Toronto's changing household characteristics and income distribution *Journal of Planning Education and Research* **32(2)** 191-203

[12] Pemerintah Provinsi DKI Jakarta 2013 *Provinsi Dalam Angka* http://jaktimkota.bps.go.id

[13] Schlichtman J J and Patch J 2014 Gentrifier? Who, Me? Interrogating the gentrifier in the mirror *International Journal of Urban and Regional Research* **38(4)** 1491-1508

[14] Brown-Saracino J 2010 *The gentrification debates* (New York:Routledge)

[15] Kompas 2014 *Ini Perbandingan Harga Lahan di Kawasan Favorit* http://properti.kompas.com/read/2014/11/27/155450721/Ini.Perbandingan.Harga.Lahan.di.Kawasan.Favorit
[16] Béné C 2009 Are Fishers Poor or Vulnerable? Assessing Economic Vulnerability in Small-Scale Fishing Communities *Journal of Development Studies* **45(6)** 911 – 933
[17] De Leon J C V 2006 *Vulnerability: A Conceptual and Methodological Review* (Bonn: Institute for Environment and Human Security)

IOP Conf. Series: Earth and Environmental Science **158** (2018) 012010 doi:10.1088/1755-1315/158/1/012010

Whose Sense of Place? Re-thinking Place Concept and Urban Heritage Conservation in Social Media Era

Christin Dameria, Roos Akbar, Petrus Natalivan Indradjati

Department of Urban and Regional Planning.
School of Architecture, Planning, and Policy Development.
Bandung Institute of Technology.

Corresponding Author: christindameria@gmail.com

Abstract. A change in a conservation approach that is now more focused on the conservation of a place than on a single object has placed an understanding of a place as an important concept that must be understood in urban heritage planning. However, the urban place perspective has been shifted, as a result of the change of attitude of people living in the urban area due to the rising of social media. This paper argues the concept of place in the heritage conservation planning, especially in the area designed to be tourism objects, which needs to be revisited. The very dynamic urban people as a result of technology and information must be considered. In order to understand the man-place relationship, the sense of place concept is the most common concept used and in the current context of heritage conservation, the review of place concept could be traced by understanding who is the owner of a sense of place in the social media era. In the current academic literature, the common discourse says the local community as the owner of the sense of place because visitors have limited chance to own the sense of place. This paper also argues that the shift of place perspective due to social media could be the catalyst in creating a strong sense of place for visitors. To support the argument, this paper uses the study of Concept of Mediatization and Concept of Parochial that have successfully explained how social media provides indirect experience based on people-place interaction and a sense of familiarity of an unknown or strange place. Therefore, this paper states that in the sense of place context: (1) the experience factor, as one of the factors in the creation of a sense of place, does not need to be physically experienced but it can be built based on other person's personal reflection. It also gives an opportunity to make a self-interpretation about the value of the heritage place; (2) sense of familiarity, which is believed to be owned by the locals only due to the period of staying in the area, can be owned by persons who have never visited the place. That is why when the tourists come to visit the heritage place with their own interpretation of experience built previously they are very potential to have a strong sense of place when doing activities in such short time of visit. In the context of heritage planning, this change will certainly have an impact on (1) the change of the stakeholder mapping of urban heritage area conservation, and (2) the change of place-making concept in an effort to conserve the public space of heritage areas, where heritage planning needs to make room for digital dimension contribution as one form of advancement in information technology that has become the urban lifestyle of today.

1. Introduction

The heritage conservation approach has now undergone a change of view in considering the objects to be preserved. The conservation concept that was originally static (in the form of archaeological objects) is now dynamic with a wider scope covering the environment, areas, and even cities [1]. The understanding of the issue of historical object conservation has changed, from only the protection of structures to the protection of places, of which the scope is wider and more significant for the community. Coleman [2] stated that conservation at the present is not only related to physical changes but also acts as an effort to create a new relationship between tangible and intangible aspects where the quality and success of this relationship are important for the sustainability of conservation.

As a place, the existence of heritage areas contributes to creating a sense of place which is one important factor to form a city's identity. The historic environment has a resonance in creating a sense of place and cultural identity which contributes to the quality of the environment in which people live and work [3]. Cohen [4] also said that by questioning the sense of place of a heritage area is also one way to measure the quality of heritage conservation in the urban area. In order to make a successful heritage conservation program, Orbasli [5] stressed the importance of public involvement in the program, including the local people, property owner, business and other people who benefit from the area. Therefore, in the conservation effort of heritage areas, the focus on urban community as the users and their interrelation with places that have become the locations of their dwellings or activities is crucial to do because humans are one of the decisive factors in the success of a conservation activity. This paper argues that the importance of understanding the existence of sense of place in urban heritage areas can begin from the knowledge of who the actual owner of that certain sense of place is, although Jiven and Larkham [6] explained that there is still a lot of debate about who actually the owner of sense of place is in the context of heritage areas as a place.

On the other hand, the existence of social media, as a result of the development and progress of communication and information technology, has influenced the behavior of urban community in the public space, including how they perceive a place. New media and information and communication technology are influencing the way that we develop social networks; understand places and location [7]. By combining the two issues, namely the issue of owner's sense of place in urban heritage areas and the issue of urban community's activity changes due to the influence of social media, this paper seeks to review the concept of place in urban heritage areas. This paper also seeks to see its effect on the ownership of sense of place and the conservation effort of urban heritage areas in social media era.

This paper is the results of a descriptive analysis library research using a systematic review methodology to synthesize the secondary data used.

2. Literature Review

According to Seamon [8], the meaning of place is considered to play a role in shaping human experience while Bott et al. [9] wrote: "The relationship to a place is a fundamental feature of human existence". Therefore, place-based knowledge will be an important foundation for institutionalizing adaptive management and human awareness of place as well as a reference to comprehend the changes that will be posed to the environment and their impacts on humans. For that, this literature review discusses the position of a place in planning and the perspective shifting of place due to the influence of social media:

2.1. The Complexity of Place in Planning

According to Trancik [10], a place is formed if it has a contextual meaning which is derived from the local or regional content, whereas Tuan [11] believed that the structure of a place without humans is only a geographic location, so the concept of place is only significant when there is human existence. To complete Trancik and Tuan, Canter in [12] referred to Relph's theory stated that place is a function of 'activities' plus 'physical attributes' plus 'conceptions'.

Even though the concept of place is difficult to be limited in one discipline, Castello [13] wrote that "place is a well-accepted theoretical construct of the areas of spatial studies". Therefore, an analysis of place should be able to help understanding the spatiality of a city through the life in the urban space. Puren et al. [14] explained that in the end, spatial discipline must be involved in applying the concept of place through place-making activity, which is an effort to create the quality of a place so that it can be integrated with the planning of the built environment. In that effort, intangible aspects like meanings and values should be included in the formulation of spatial guidelines [14].

Understanding the concept of place in spatial planning means understanding the culture and characteristics and hallmark of urban space so that it can be considered in planning, for example in an effort to provide an intimate space to its inhabitants. However, Graham and Healey [15] questioned the concept of place in the context of spatial planning, especially in urban areas, with the question: "In a world of tumultuous economic, social, cultural, technological and physical change, how can we best conceptualize the dynamics of places and the role of planning action in shaping them? With globalization apparently 'stretching' and deepening the relations between places, tying them into multiple webs of capital, technology, data and services, human interaction, and ways of thinking, at proliferating spatial scales, how might spatial planners translate new understandings of socio-spatial relations into their practices?" Criticizing the classical approach, Graham and Healey [15] argued that the approach to the concept of place in planning practice should be done with a new approach, based on relational constructs which are dynamic, rather than the Euclidean approach, deterministic, which has only one dimension, the legacy of thought from a scientific approach in the 1960s and early 1970s. Therefore, as long as urban space is still thought in terms of a city-centred view of space as an object and is still often conceptualized as the 'container' to limit the activities occurring there, Graham and Healey [15] argued that ultimately planning practices have difficulty in capturing the dynamics of complex contemporary urban changes.

In the context of heritage planning, Fredholm [16] stated that complexity of heritage values in the city and regional development have not been adopted by the planners who tend to think that technical problems can be solved using rational decision. Forester [17] said, in reality, planning that only considers physical and standard rules fails to address the problems within the society and is considered as a traditional type of planning that does not fit with the dynamic of the people. Critics on urban heritage planning are actually based on the idea that it should be focused on the value of the place itself related to the culture.

The question is how to understand the place itself. Jorgensen and Stedman [18] state that despite a number of theories on the relationship between spatial setting and human, the Sense of Place Concept is the most common one used to explain the existence of place. In conservation, the attempt to explain the sense of place with historical character is seen to be more important than detail restoration of the urban physical element in urban conservation [19]. The traditional approach in conservation that only focuses on buildings preservation on conservation list will fail to save urban heritage and urban sense of place [see [20][21]). Martokusumo and Zulkaidi [22] address the needs conservation focus that relates to the application of a sense of place concept in the main urban area, to create strong conservation program that is sustainable as well. However, Jorgensen and Stedman [18] admit that the relationship between place construction theories is not easy to understand. Despite some of the relationships that have been studied, still, it is difficult to reach universal agreement on place concept.

2.2. *The Perspective of Place in the Social Media Era*

The challenge of information and communication technology that i still evolving as of today has made its impact on the perspective of place and the effort to translate that understanding into the planning practice continues to be a debate. Wilken [23] argued for the need to examine "the way that mobile media influence and shape places and place experience, and the way that mobile phone use is integrated into the flows of everyday life". Abdel-Aziz et al. [24] stated that there is an opinion that the advances in information and telecommunication technology have led to the activities in the public space beginning to decrease, due to information being easily accessible from home using the internet.

In their paper, Abdel-Aziz et al. [24] divided the Information and communication technology (ICT) into four categories: Wi-Fi Network, Social Media, Interactive Public Display and Smartphone Application in Public Space, but this paper only focuses on social media role, which has recently dominantly influenced the behavior of community in urban areas. Social Media can be defined as "Service which enables to share various news, information, and opinion in ways of multimedia" [25].

The content is often a media content like photographs or videos which by commenting it enables discussion [24]. For South East Asia region, for example, wearesocial.com in 2017 [26] notes “Mobile social media continues to grow with the number of monthly active users accessing social platforms up 34% year-on-year”. Brenner and Duggan [27] state that the number of social media users in an urban area is significant. To explain the effect of social media on the concept of place, this paper refers to two pieces of paper that have discussed this topic, namely the papers by Tussyadiah and Fesenmaier [28] and Humphreys [29].

2.2.1. Online Video Shared and the Concept of Mediatization. Currently, images, videos, and movies available in the virtual world can provide a variety of messages, as well as serve as a mediator in the experience of traveling. Tussyadiah and Fesenmaier [28] studied the role of online shared videos applications that are used as the mediator of travel experiences, by using the analysis method on the videos published online by tourists visiting a tourism site. The data collected in this study followed a purposive sampling process to select all videos containing tourism activities in New York City, a major tourist destination in the United Stated.

In their analysis, Tussyadiah and Fesenmaier [28] explained that the video shared application offers the possibility for viewers to have a travel experience as described by the tourists in the videos. When a video is available for viewing, there is a communication and an interaction between the sharing party and the viewers through the threaded comments. Viewers enjoy how they interact with the person or object in the setting that is actually an unfamiliar place to the viewers. Other fun viewers get from watching the video is the viewers’ imagination as they put themselves as the actors and conduct tourism activities guided by the illusion of spatiotemporal narrative illusion offered in the video. In the narrative, the video maker shares various information related to a trip rich in personal opinion or perception of a place.

The perception from imaginative access provided by social media could increase interaction between users and urban setting as a place. That is why Tussyadiah and Fesenmaier [28] found that after watching an online video, most viewers are planning to visit the city of New York. This condition is driven by a process called by Escalas [30] as "narrative transportation", which is a concept of virtual mobility in which viewers feel taken mentally to a place. In other words, the experience of tourists in a technologically mediated place has succeeded in creating the mental pleasure of the living fantasy, the self-interpreted result of virtual mobility itself.

According to Tussyadiah and Fesenmaier [28], the experience in the built environment is often mediated by information technology that reshapes and changes the interpretation of a place to allow for mediatization. The concept of mediatization, according to Jansson [31], occurs when the globalized media culture allows people to travel mentally and emotionally (one of them is experiencing tourism activity) without moving in physical geography

2.2.2. Mobile Social Network and the Concept of Parochial. Humphreys [29] discussed the case study on the use of mobile social network called Dodgeball as well as its relation to spatial practices in urban areas. Spatial practices refer to the everyday lived experience of and movement through social and physical space [32]. Dodgeball is a service, owned by Google and based in New York City, which allows users to let personal networks of friends know where they are via their mobile phones [29]. This service facilitates the exchange of location information among its users to encourage face-to-face interaction. In his research, Humphreys [29] believed that the increasing use of mobile social networks will affect and even change the way people gather and interact in the public space (refers to non-domestic spaces such as dance clubs, parks, restaurants, bars, cafes, the street, and so on).

Humphreys [29] explained that: “Mobile social networks can help to turn public realms into parochial realms through parochialization”. According to him, parochialization can be defined as the process of creating, sharing and exchanging information, social and locational, to contribute to a sense of commonality among a group of people in public space. In urban social space, parochial is the space between the private and public spaces. Parochial spaces are territories characterized by “a sense of commonality among acquaintances and neighbors who are involved in interpersonal networks that are located within communities" [33]. The use of Dodgeball has, in fact, turned public space into parochial space, foreign but familiar (a sense of familiarity) because of the information shared about that public space encouraging users to conduct activities in the public space.

In conclusion, Humphreys [29] stated that social networks allow new types of information to flow into public spaces, changing the way people gather and interact in public spaces, resulting in a rearrangement of social and spatial practices. The advancement of new communication technology may have changed the spatial and temporal boundaries, but the location and socio-spatial factors still actually form communication even through the media.

3. Discussion: Sense of Place, Urban Heritage Area, and Social Media

Litvin [34] stated that the historic area is a combination of places that have heritage values. Smith [35] also believed that the 'effect' of place helps us to understand the meaning of heritage site and wrote: "Heritage as a place, or heritage places, may not only be conceived as representational of past human experiences but also of creating an effect on current experiences and perceptions of the world. Thus, a heritage place may represent or stand in for a sense of identity and belonging to particular individuals or groups". Similar to that, Davis et al [36] said that whether we refer to such places as heritage sites; there are undoubtedly historic, contemporary, natural and cultural features in the landscape that hold special meaning that contributes to the creation of a sense of place.

Heritage areas indeed have a close relationship with the presence of a sense of place because these areas contribute greatly in creating a sense of place in urban areas, which is one of the important factors for shaping urban identity. In addition, to create an identity, sense of place is also one of the quality indicators of cities with historical values. Montgomery [37] stated that one of the lists of urban qualities that community value for a historic city is a sense of place. In support of this, Orbasli [5] stated that: "Historic areas remain familiar in a changing environment and provide a sense of place".

What is a sense of place? The sense of place is a concept that explains the relationship between man and place. The sense of place begins with a sensation received by a person from a physical environment where man interacts, an impression that later would be perceived and to create a subjective meaning of the place. The sense of place is a thorough concept where a man could sense a place, create a perception and connect to the meaningful place [38]. Although Robertson [39] believed the sense of place has the role to make people stay longer, there is no exact measurement on a sense of place concept because itis highly dependent on cognitive aspect and subjective perception. Therefore Stokowksi [40] argued that the sense of place concept would remain in change, adaptable and unique for each place.

In the conservation context, the understanding of the sense of place could be started from the idea of who owns the sense of place. As Canter [41] mentioned, if we are to understand peoples' responses to places and their actions within them, it is necessary to understand what (and how) they think. Therefore, the question to discuss is: Who is the true owner of a sense of place in urban heritage areas within the existing academic studies? Does the change of behavior of urban community in looking at places, due to the influence of social media previously described through the Concept of Mediatization and the Concept of Parochial, influence the ownership of the sense of place in urban heritage areas? Then, what are the impacts on the conservation of urban heritage areas?

3.1. Whose Sense of Place?

Based on the debates on the owner of a sense of place in the academic literature, this paper classifies two groups that potentially have a sense of place: the local community and visitors or tourists. The local community is seen to be the group that has the highest potential to own a sense of place because there is awareness that the sense of place comes after a sense of belonging (see [42]) or social insideness due to familiarity with both the community and the landscape (see [43]). That explains why a sense of place is often related to duration of staying. Ching et al. [44] also found that the local residents (owners and building tenants of heritage shophouses) in Armenian Street George Town Penang Malaysia are willing to spend money for renovation because they have a sense of belonging and social insideness due to the familiarity with the environment where they live. In line with that, Davies et al. [45] believed that adults who live in areas of higher concentrations of the historic environment and young people who are interested in the historic environment are more likely to have a stronger Sense of Place.

It is interesting when an emerging opinion stating that sense of place is not always owned by the local community. In their study in Taiwan, Davis et al. [36] found that local community even has the potential to have no sense of place. The attention given by the local community to an area is minimal because the environment is considered as something regular and routinely seen. They wrote: "Although they are local people, frankly speaking, their understanding of Hsin-hua is limited to the town centre. They have never paid much attention to other parts, even when they walk past; they have never paid any attention". In that case, is there any other party who has more sense of place than the locals? The question has brought us to the second group, the tourists.

Currently, heritage areas have become one of heritage tourism destinations in urban areas due to the changing trend in the world of tourism today, which according to Cahyadi and Gunawijaya [46], occurs because of the desire of tourists to see different cultures, view local attractions, buy locally produced goods, meet with local communities, and attend local events. Therefore, Jakle in [6] believed that the most appropriate party to have a sense of place is the tourists and not those living in these places as tourists do have the intention to deliberately come to a place and seek the experience offered by that place.

Mediating it, Kianicka et al. [47] stated based on their research that sense of place may actually be owned by locals and tourists, but the form of their sense of place is different because the meaning and significance of those places are captured differently by both parties. They found that the sense of place owned by the local population related to their everyday life such as work, property, and social ties, is also linked to memories of childhood and adolescence. Different from the locals, the sense of place owned by tourists is formed from the aesthetics and characteristics factor of a place when they get the experience of a recreational activity carried out at the place. Tourists have an "outsider's" view and therefore may perceive and value landscapes differently (Stremlow in [47]).

Based on the above explanation, this paper believes that sense of place does not only belong to the local community but also belong to tourist. Tourists have the possibility to have a sense of place and to partake in heritage conservation program. However, the role of tourists that involve in the heritage area activities has not been well studied. The limited study on the tourists' sense of study is probably due to the idea that tourists, before coming to the place, have only idea of what the place is, so the opportunity to create a sense of place is very limited. The chance that they develop a sense of place is limited and the chance that this affects their identity is even smaller [48]. In many instances, tourists are seen as the enemies of heritage conservation – their presence will not only enhance wear and tear of the sites, but also change the aura of the heritage site [49].

3.2. What is the Effect of Social Media on Sense of Place?

In postmodern thought, a place has the potential to have a shift in meaning due to the advances in information and communication technology. When the meaning of the place has shifted, then this paper believes that it should influence the creation of a sense of place. Although there have been many new concepts discussing places as the result of the effect of information and communication technology advances, there are only a few that review the influence of social media on the existence of a sense of place, when the heritage areas become heritage tourism destination.

According to Relph [43], sense of place is the result of man's interaction with (1) Physical setting; (2) Activity; and (3) Meaning and the ability to give identity to the place. Montgomery in [13] constructs the three components of sense of place in the context of urban design by explaining (1) physical setting as Form (such as: scale, intensity, permeability, landmarks, and space to building ratio, etc); (2) Activity (such as: street life, café culture, events and local traditions, etc); and (3) Meaning as Image (such as: symbolism and memory, imageable and legability, sensory experience and associations, and knowledgeability, etc) as result of cognition, perception, and information. In regard to the sense of place component, Carmona et al [12] state that conception of each place will have variation from the Relph's components. It is possible that in a certain place, the physical setting proportion is bigger than other components and in other place, activity and meaning have bigger scores in creating the sense of place.

The experience is one of the determining factors of Meaning, a component to create a sense of place. When a person interacts with a place and receives a sensory experience, the person has the potential to get the meaning of a place (see [50] and [12]), but because of the concept of mediatization in the social media era, that experience could be accessible without direct interaction. Tussyadiah and Fesenmaier [29] stated that this social media has succeeded in engaging viewers at a certain level, hence the viewers manage to produce an enjoyable experience of a place imaginatively. Currently, social media can function as a means of 'transportation' to a place and provide imaginative access, based on the viewers' perceptions of the place.

The experience factor has now become something interesting to review in the era of social media when it turns out that experience does not have to be obtained directly through physical sensation of a place and is built on someone else's personal reflexive consideration. This needs to be revisited because the current paradigm believes experience can only be gained physical sensation from direct interaction with a place. Tussyadiah and Fesenmaier [29] wrote: "Sharing digital reviews broadens the access to experiences of 'touristhood' so that tourists do not only see their travel experiences in contrast to their own day to day activities, but also in many different contexts based on other travelers' experiences".

The trends show that one person's shared experience becomes another person's inspiration for dreaming and the cycle begins all over again [51]. Even so, the indirect experience gained is not necessarily accepted by the viewers, it first goes through a process of observational learning. To generate the perception of an experience, and individual does not automatically mimic (mechanically) after observation but use the consideration based on their previous experience, morality, perspective or thought (Bandura in [52])

Graham [53] stated that the existence of physical setting in a place is manifesting through their presence in both material spaces and electronic spaces. Cresswell [50] also explained how individuals can gather a sense of place from filmic representations of a place. Therefore, experience, though indirect, which tends to encourage viewers to interact directly in a place ultimately potentially creates a strong sense of place as they become tourists in the place.

Another concept, the Concept of Parochial, has given a new perspective in seeing how humans interact in the public space due to the influence of social media. The development and proliferation of mobile social networks have the potential to transform ways that people come together and interact in public space [29]. The use of social media has successfully transformed public space into a space that is *Parochial,* foreign but familiar, so eventually, it is able to encourage the users to have activities in public spaces. Information and communication are two essential factors of interest and attraction specific to urban environments [24]. Digital technologies have changed the way we inform, entertain and educate ourselves, as well as our ways of working, carrying out research, doing business, and keeping in contact with each other [54].

So, what is the impact of the sense of familiarity to a sense of place? The familiarity and insideness can form Meaning, which is one condition to create a sense of place. Many people believe that the familiarity sense can only be owned by the local as result of the duration of staying (see [43]). As a matter of fact, when parochial rooms are created, the outsider also has the opportunity to experience the insideness factor. When the shifting process of the perspective of place due to the influence of social media is experienced by the tourists before visiting urban heritage areas, this paper argues that the tourists have the potential to have a strong sense of place when interacting directly with the heritage areas. At some level, a sense of place has the potential to no longer be acquired by just spending time at a place, such as by staying in it, because currently, people can live in one place but do most of their social activities in another place. They can be physically 'here' while the environment they are building up the familiarity with is out 'there' or brought in here through the digital screen (Toft in [29]).

Other than providing an indirect sense of familiarity, social media also provides an arena for visitors to do interpretation in the heritage area. Interpretation is an important factor in meaning that creates a sense of place. Supporting this, Ponting and McDonald [55] stated that media-driven repertoires of images and narratives not only form the basis of tourist attractions but also implicitly influence the interpretation of sites and cultures and the ways in which they are experienced, recorded and communicated by tourists. Social media has made heritage area as an area that not only has historical value, but also can be interpreted with own understanding, by both the local community and the visitors.

In the end, this paper argues that social media has the potential to (1) create experience factor (in the Activity component) without the necessity to do direct interaction; (2) produce familiary factors (in the Meaning component) without having long duration in a place; and (3) help to make an interpretation (which produces the Meaning component) without having directly interaction with the place. So, when social media is successfully helping to create a sense of place components, social media has actually the potential to be the catalyst of a strong sense of place when the visitors visit the area.

3.3. What Is the Impact on Urban Heritage Area Conservation?

Urban heritage conservation is significantly different from building conservation because area conservation tends to be more complex, involving various aspects such as socio-economic aspect and the impacts on the local community. Orbasli [5] states that one different factor between the two is urban heritage conservation needs to have a social dimension and involves all users. An inhabitant, property owner, business and other people involved in the area are part of the conservation process [5]. Therefore, the description of the sense of place of the stakeholders, particularly the presence of tourists who have a strong relationship with social media, is to be reckoned with in the conservation planning. The existence of this sense of place will provide an opportunity to explore the social and cultural processes that affect the environment and landscape assessments, including the opinions and judgments of the community in landscape planning policy widely [56].

The role as visitors, as an element that has strong potential to own a strong sense of place, must also be considered as a stakeholder in the urban heritage conservation program. This is against the thinking of tourist as a cause of negative externalities in the urban heritage area. Although tourists often cause heritage sites to become commodities, reviews show that they are also concerned with conservation; tourists are thus not "enemies" of heritage [49].

Other than bringing up the role of visitors, social media has also raised the issue of heritage to the cyberspace. A virtual community evolves as different people recall and interpret their heritage experiences and the construction of heritage has now extended into cyberspace [49]. This has helped the establishment of many communities in the cyberspace that share interest and concern on heritage area. The cyberspace-based community is also important to be considered as a stakeholder in the conservation effort.

In the context of public space, due to the importance of the use of digital networks in everyday life, Abdel-Aziz et al.[24] believed that the addition of a digital layer to an existing urban landscape is needed, enabling the experience of an urban environment accessible through direct contact or through the mediation of digital technology. There is a redefinition of public space that has a physical layer and a digital layer in it. This condition is called as the new social public place.

Can this "new social public place" be accommodated in the heritage areas? This paper believes that place-making efforts for the public space of urban heritage areas that have the concept of "the new social public place" are interesting ideas and can be one of the programs in conservation efforts. This idea is contrary to the opinion stating that communication and information technology will encourage individualistic characteristics. Abdel-Aziz et al. [24] actually also believed that the new social public place will make a new public space that can invite individuals to interact. The existence of the digital layer along with the physical layer will further encourage the existence of heritage areas in cyberspace, especially in the flow of information and communication in social media.

4. Conclusion

The spaces will turn into places at the point where the setting's physical and cultural characteristics meld with the individual's affective perceptions and functional needs [9]. Therefore, Malpas [57] believed that regardless of time, history, geographical situation, technology, even social condition, humans will always need a place because owning and identifying places are integrated with human identity. The place is a space that is given meaning by the users, so the concept of place is physical as well as psychological. The concept of place, with all its complexity and debates within it, is a well-received theoretical construction in the spatial field of study, thus a sense of place is the most commonly used concept in understanding a place because it can describe the relationship between spatial setting and humans that conduct activities in it.

Escobar [58] emphasizes the dichotomy of place between place as a conceptualization of identity, our mental image or 'category of thought' about a locality; and place as a physical entity, 'a constructed reality'. When a place becomes a physical reality constructed through social media, then social media can reduce or even increase the construction. The shift of place perspective due to social media is explained through the Concept of Mediatization (see [28]) and the Concept of Parochial (see [29]).

The Concept of Mediatization explains how someone's direct experience in a place can be mediated by social media, re-shaped until it can change the interpretation of viewers of that place. The factor of experience, which was originally a direct result of physical sensation, can now be owned based on indirect experience, namely in the form of cognitive knowledge on the direct experience of the place shared in social media. The Concept of Parochial also describes the shift of construction of place when social media has succeeded in causing a person to have a sense of familiarity towards a place that is still foreign. Other than that, social media has also given chances for the person to do own interpretation with own interpretations of heritage objects.

This paper argues that in urban heritage context, social media has the potential to be a catalyst in creating a strong "sense of place" at the time a person visits the heritage area. Social media is playing an increasing role in shaping travelers' expectations, and thus in shaping their heritage experiences [49]. Therefore, heritage conservation planning must give a space to the digital dimension, which has been part of modern lifestyle. In the context of urban heritage conservation, the sense of place factor is an important knowledge because it would change the stakeholder mapping. Tourist and heritage communities in the cyberspace play important role in the conservation thus they can be considered as a stakeholder.

As social media has impacted people's way of viewing a place, this paper believes the meaning of place in heritage area needs to be revisited. As a matter of fact, in an urban heritage conservation context, place concept must also include the dynamics of the people. As explained by Rapoport [59], the meaning of built environment is strongly influenced by the popular culture. The objective of this review is not only to neglect but also to enrich the understanding of urban heritage conservation for tourism. Therefore, the argument that the postmodern community, through technological and cultural corrections, can now ignore place, must be questioned existentially. From this point of view, the argument that the postmodern society, through technological and cultural correctives, can now ignore place is questionable existentially and potentially devastating practically, whether in terms of policy, design, or in terms of popular understanding [60].

References

[1] Sudikno A 2004 Pendekatan Sejarah dan Konservasi Perkotaan Sebagai Dasar Penataan Kota *J. PlanNIT* Jurusan Planologi Fakultas Teknik Sipil dan Perencanaan Institut Teknologi Nasional vol **2** no 2 December p. 98-112

[2] Coleman V 2004 *Heritage and Sustainability: A Discussion Paper* (Parramatta: The Heritage Council of NSW and the NSW Heritage Office) Available online: www.heritage.nsw.gov.au.

[3] Russell H, Smith A, and Leverton P 2010 *Sustaining Cultural Identity and A Sense of place – New Wine in Old Bottles or Old Wine in New bottles?* CEM Occasional Paper Series (United Kingdom: The College of Estate Management)

[4] Cohen N 1999 *Urban Conservation* (MIT Press)

[5] Orbasli A 2000 *Tourists in Historic Towns Urban Conservation and Heritage Manage* (London: E & FN Spon)

[6] Jiven G and Larkham P 2003 Sense of Place, Authenticity and Character: A Commentary *J. of Urban Design* vol **8** no 1 (Carfax Publishing) p 67–81
[7] Houghton K 2010 Augmenting Public Urban Spaces : The Impact of The Digital Future on The Design Of Public Urban Spaces *Queensland Planner* vol. **50** 4 (PIA Australia Queensland Division) p 19-23
[8] Seamon D and Sowers J 2008 Place and Placelessness Edward Relph *Human Geography* ed Hubbard P et al (London: Sage) p 43-51
[9] Bott S, Cantrill G J and Myers E O Jr 2003 Place and the Promise of Conservation Psychology *Human Ecology Review* vol **10** 2 p 100-12
[10] Trancik R 1986 *Finding Lost Space* (New York: Van Nostrand)
[11] Tuan Y F 1977 *Space and Place The Perspective of Experience* (Minenesota: University of Minnesota Press)
[12] Carmona M, Heath T, Oc T and Tiesdell S *2003 Public Places Urban Spaces, The Dimensions of Urban Design* (Elsevier: Architectural Press)
[13] Castello L 2006 City & Time And Places: Bridging The Concept Of Place To Urban Conservation Planning *City & Time* vol **2** (Centre for Advanced Studies in Integrated Conservation) p 59-69
[14] Puren K, Drewes E and Roos V 2008 An Exploration of Sense of Place As Informative for Spatial Planning Guidelines: A Case Study of The Vrederfort Dome World Heritage Site South Africa *International J. of Social Science* vol 1 p. 188-95
[15] Graham S and Healey P 1999 Relational Concepts of Space and Place: Issues for Planning Theory and Practice *European Planning Studies* vol. **7** 5 p. 623-46
[16] Fredholm Sussane 2017 Assets in the Age of Tourism: the Development of Heritage Planning in Ghanaian Policy *J. of Contemporary African Studies* vol. **34**
[17] Forester J 1989 *Planning in the Face of Power* (University of California Press)
[18] Jorgensen B and Stedman R 2001 Sense Of Place as An Attitude: Lakeshore Owners Attitudes Toward Their Properties *J of Environmental Psychology* vol 21 p. 233-48
[19] Martokusumo W 2014 *Kota (Pusaka) Sebagai Living Museum* Temu Pusaka Indonesia Galeri Cemara (7-9 September Jakarta)
[20] Idid S Z A 2005 *Urban Conservation Approach for A Multi Cultural Historic City: The Urban Planning and Design Perspective Case Study on the Conservation Guidelines for the Historic City Of Melaka Malaysia* PhD Thesis (University of Tokyo: Department of Urban Engineering)
[21] Shamsuddin S and Sulaiman A B 2002 *The Importance of Conserving the Old Town Centre in Achieving a Sustainable Built Environment of the Future* Seminar Nasional in Built Environment: Sustainability through Management and Technology (5-6 Agustus Kuala Lumpur)
[22] Martokusumo W and Zulkaidi D 2015 Heritage List: Some Notes on Area-Based Conservation. Lesson Learned from Bandung. International Journal of Built Environment and Sustainability *IJBES* Vol **2** (2) (Universiti Teknologi Malaysia Faculty of Built Environment) p 125-130
[23] Wilken R 2008 Mobilizing Place: Mobile Media, Peripatetics, and the Renegotiation of Urban Places *J. of Urban Technology* vol **15** 3
[24] Abdel-Aziz A A, Abdel-Salam H and El-Sayad Z 2015 The Role Of Icts in Creating the New Social Public Place of The Digital Era *Alexandria Engineering J.* vol **55** p 487–93
[25] Hosio S, Kukka, H, Jurmu M, Ojala T and Riekki J 2010 Enhancing Interactive Public Displays with Social Networking Services *Proc. of 9th International Conf. On Mobile and Ubiquitous Multimedia MUM 10* (ACM) p. 23:1-23: 9
[26] Kemp S 2017 Digital in SouthEast Asia in 2017 Special Reports wearesocial.com https://wearesocial.com/special-reports/digital-southeast-asia-2017
[27] Duggan M and Brenner J 2013 *The demographics of Social Media Users-2012* Pew Research Center's Internet & American Life Project p. 1–14 (Washington DC)

[28] Tussyadiah I and Fesenmaier D R 2009 Mediating Tourist Experiences: Access to Places via Shared Videos *Annuals of Tourism Research* vol **36** 1 p. 24–40
[29] Humphreys L 2010 Mobile Social Networks and Urban Public Space *New Media & Society* vol **12** 5 p. 763-78
[30] Escalas J 2004 Imagine Yourself in the Product: Mental Simulation, Narrative Transportation, and Persuasion *J. of Advertising* vol **33** 2 p. 37–48
[31] Jansson A 2002 Spatial Phantasmagoria: The Mediatization of Tourism Experience European *J. of Communication* vol **17** 4 p. 429-43
[32] Certeau M de 1984 *The Practice of Everyday Life* translated by S Rendall (Berkeley: University of California Press)
[33] Lofland L H 1998 *The Public Realm: Exploring the City's Quintessential Social Territory* (New York: Aldine de Gruyter)
[34] Litvin S 2005 Streetscape Improvements in an Historic Tourist City A Second Visit to King Street Charleston South Carolina *J. of Tourism Management* vol **26** p 421–29
[35] Smith L 2006 *Uses of Heritage* (Oxon: Routledge)
[36] Davis P, Huang H, and Liu W 2010 Heritage, Local Communities and the Safeguarding of 'Spirit Of Place' in Taiwan *Museum and Society* vol **8** 2 p 80-9
[37] Montgomery J 1998 Making A City: Urbanity Vitality and Urban Design *J. of Urban Design* vol **3** 1 p 93-116
[38] Hashemnezhad H, Yasdanfar S A, Heidari A A and Behdadfar N 2013 Comparison the Concepts of Sense of Place and Attachment to Place in Architectural Studies *GEOGRAFIA OnlineTM Malaysia J of Society and Space* vol 9 1 p. 107-17
[39] Robertson K 1999 Enhancing Downtown's Sense of Place *Main Street News* September 1-4 p 12-13
[40] Stokowski P 2002 Languages of Place and Discourses of Power: Constructing New Senses of Place *J. of Leisure Research* vol **34** 4 p 368-82
[41] Canter D 1991 Understanding, Assessing and Acting in Places: Is An Integrative Framework Possible? *Environmental Cognition and Action: An Integrated Approach* ed T Garling and G Evans (NY: Oxford University Press)
[42] Hawke S K 2011 *Sense of Place, Engagement with Heritage and Ecomuseum Potential in the North Pennines AONB* Thesis International Centre For Cultural And Heritage Studies School Of Arts And Cultures (Newcastle University)
[43] Relph E 1976 *Place and Placelessness* (London: Pion)
[44] Ching S, Ochiai C, and Kobayashi H 2013 *A pilot Program of Participatory Historic Urban Conservation Case from Armenian Street of George Town Penang* Reports of the City Planning Institute of Japan No 11 February
[45] Davies J, Whimster R and Clayton L 2009 *Heritage count 2009* (London: English Heritage)
[46] Cahyadi R and Gunawijaya J 2009 *Pariwisata Pusaka Masa Depan Bagi Kita, Alam Dan Warisan Budaya Bersama* United Nations Educational Scientific and Cultural Organization (UNESCO) and Program Vokasi Pariwisata Universitas Indonesia
[47] Kianicka S, Buchecker M, Hunziker M and Müller-Böker U 2006 Locals' and Tourists' Sense of Place A Case Study of a Swiss Alpine Village *Mountain Research and Development* vol **26** 1 Feb p 55–63
[48] Vanneste D 2015 About *A Share Sense of Place, Connectedness and Heritage Heritage*, Tourism and Hospitality International Conference 2015
[49] Munar A M and Ooi C 2012 *What Social Media tell Us About the Heritage Experiece* CLCS Working Paper Series Centerr for Leisure & Culture Services Department of International Economics and Management (Copenhagen Business School)
[50] Cresswell T 2004 Place: *A Short Introduction* (Blackwell Publishers)
[51] World Tourism Organization (UNWTO) Affiliate Members 2011 *AM Reports Technology in Tourism* vol **1**

[52] Suroso 2014 Teori Belajar Observasi Menuju Belajar Mempertajam Rasa *Buletin Psikologi* Tahun XII No 1 Juni 2014 (Universitas Gadjah Mada) p 16-32

[53] Graham S 1998 The End of Geography or the Explosion Of Place? Conceptualizing Space, Place and Information Technology Centre for Urban Technology Department of Town and Country Planning *Human Geography* vol **22** 2 (University of Newcastle) p 165-185

[54] Maldonado A M F 2005 *ICT Infrastructures as A New Challenge for the Urban Planning Profession* ICT Infrastructures 41st ISoCaRP Congress

[55] Ponting J and McDonald M G 2013 Performance, Agency, and Change in Surfing Tourist Space *Annals of Tourism Research* vol 43 (Elsevier Ltd) p 415-34

[56] Soini K, Vaaralab H and Poutaa E 2012 Residents' Sense Of Place and Landscape Perceptions at The Rural–Urban Interface *Landscape and Urban Planning* vol **104** p 124–34.

[57] Malpas J 1999 *Place and Experience: A Philosophical Topography* (Cambridge: University Press)

[58] Escobar A 2001 Culture Sits in Places: Reflections on Globalization and Subaltern Strategies in Localization *Political Geography* vol **20** p 139-74

[59] Rapoport A 1982 *The Meaning of The Built Environment: A Nonverbal Communication Approach* (Sage Publications)

[60] Relph E 1993 Modernity and the Reclamation of Place *Dwelling, Seeing, and Designing: Toward a Phenomenological Ecology* ed Seamon D (Albany NY: Sunny Press) p 25-40

IOP Conf. Series: Earth and Environmental Science **158** (2018) 012011 doi:10.1088/1755-1315/158/1/012011

Land-Price Dynamics Surrounding Large-Scale Land Development of Technopolis Gedebage, Bandung, Indonesia

A Hasanawi [1], H Winarso [1]

[1]Urban Planning and Design Research Group, School of Architecture, Planning and Policy Development, Institut Teknologi Bandung, Indonesia;

Corresponding author: filah@students.itb.ac.

Abstract. In spite of its potential value to governments, detailed information on how land prices vary spatially in a city is very lacking. Land price in the city, especially around the development activity, is not known. There are some considerable studies showing that investment in land development increases the land market price; however, only a few are found. One of them is about the impact of large-scale investment by Sumarecon in Gedebage Bandung, which is planning to develop "Technopolis", as the second center of Bandung Municipality.This paper discusses the land-price dynamics around the Technopolis Gedebage Bandung, using information obtained from many sources including an interview with experienced brokers. Appraised prices were given for different types of residential plot distinguished by tenure, distance from the main road, and infrastructural provision. This research aims to explain the dynamics of the land price surrounding the large-scale land development. The dynamics of the land price are described by the median land price market growth using the Surfer DEM software. The data analysis in Technopolis Gedebage Bandung shows the relative importance of land location, infrastructural provision and tenure (land title) for dynamics of the land price. The examination of data makes it possible to test whether and where there has been a spiraling of land prices. This paper argues that the increasing recent price has been consistently greater in suburban plots than that in the inner city as a result of the massive demand of the large-scale land development project. The increasing price of land cannot be controlled; the market price is rising very quickly among other things due to the fact that Gedebage will become the technopolis area. This, however, can indirectly burden the lower-middle-class groups, such as they are displaced from their previous owned-land, and implicate on ever-decreasing income as the livelihood resources (such as farming and agriculture) are lost.

1. Introduction

Bandung's land price market, like any other large, fast-growing cities in the developing, is terribly disorganized–no one knows the shape of land prices from the city center to suburbs. No one also finds out how infrastructure availability or having a clear title affects the values of residential plots. With up-to-date and accurate land price information, government can estimate how much money can be proposed by property tax system. With land price information, the government can estimate the costs of acquiring rights-of-way for transit lines. If land price information is tabulated according to the availability of infrastructure and type of title, the estimates of the potential benefits (as measured by higher land prices) of infrastructure investment schemes and land titling programs can be made. In addition to that, the basis for recovering the cost of such programs can be provided. The private sector can benefit as well; land price information system would be a boon to real estate appraisers, banks, and investors by providing information for rapidly assessing property values.

The land is a strategic property for the development of area or city. The urbanization rate and the increasingly unrestrained population growth of the city have caused the urban sprawl phenomenon in Bandung. Land demand increases while land availability always remains. Therefore, there is a competition in urban land use where the highest-bid land user can get the desired place. Expansion of Bandung through the utilization and use of land causes densification of settlements that are getting bigger and supported by the increasing population in the city of Bandung. The population of Bandung in the period 2005-2013 experienced a growth rate of 2.04 percent, while the growth rate of West Java Province was only 1.49 percent [1]. However, in the previous period, i.e. 1990-2000, there was a drastic decrease in population growth rate of Bandung from 2.4 percent to 1.3 percent.

The development of large-scale land not only provides benefits for the development of the land itself, but also leads to an increase in land value. The development of the land will cause the value of land use to increase due to the increased intensity of activities in the location. Indirectly, the development of large-scale land in a region has resulted in the region and the surrounding area to be a strategic area and having more value because the access to various activities becomes easier. As the value of land, land prices are the reflection of value of land increased as well.

Despite the obvious benefits of having good land price information, no public- or private-sector agency in Bandung is systematically tabulating land price market. There is no central recording of property transactions, and tax records are incomplete and inaccurate. Trying to fill this gap, we applied an appraisal survey technique used previously in Jakarta [2]. The method involves asking experiences brokers in neighborhoods to appraise the probable selling price of the typical residential plot. A separate appraisal is made for plots with varying levels of infrastructure and security of title. The appraisals were completed in 2017 and retrospectively from 2000 until 2016. This report demonstrates the application of low-cost research method for generating interest in applying the technique in other countries among economists and planners.

This paper argues that the increasing land price in Technopolis Gedebage Bandung cannot be controlled by the government. The market price is affected only by the knowledge that there will be a large investment by private sector in Gedebage. This,in turn , could indirectly burden lower-middle group in the area. This paper is presented in four parts. After this introduction, the method of this survey is presented, followed by a discussion on the survey result. In the end, a conclusion and policy implications are presented.

2. Survey Methods

The purpose of the survey was to obtain price appraisal of hypothetical plots of residential land within the boundaries of large-scale development, Summarecon Technopolis Gedebage, Bandung. The survey was carried out by using *RW* (Rukun Warga, Community Association) as the unit of analysis. The survey covered 120 *RW* within *Kecamatan Gedebage* and was carried out over a 4-week period (13 February to 13 March 2017) by five Indonesian surveyors.

The technique to get average per-*RW* land values was by interviewing land brokers (market intermediaries who match buyers and sellers) and asking them to appraise the hypothetical plots of land with specific characteristics. As the occupation of land brokers, contracts between surveyors and brokers were made through a variety of informal channels (i.e. asking at coffee shops or through recommendations of friends) and, in some cases, by making inquiries through *RW*, *Kelurahan,* and *Kecamatan* government offices.

The consistent standard for qualifying our informants, we used information from the brokers for each RW. We relaxed this minimum in a few instances (such as where a broker of long standing had been hired by *kelurahan* office to work on land development issues) where it was apparent that the broker was truly knowledgeable of local land prices. Once three brokers were interviewed in an *RW*, the middle value of the three responses was chosen to be the representatives land prices for that *RW*.[1] By choosing

[1] In cases where we only had two values for a particular type of plot, a mean was calculated between themand the difference between the two was no more than 20 per cent.

the middle value, rather than calculating the mean of the three values, we reduced the effect of 'outliers', or cases where one person's response was unreasonably high or low compared to the other two.

In the interviews, brokers were asked to appraise typical plots of residential land which has different distance to the main road, infrastructure level, and tenure. Each of these categories consists of three different types (three different plot sizes, three level of infrastructure availability, three distance to the main road, and three grades of tenure claim). In addition, brokers were queried about land prices in 2016 as well as current prices, obtaining a total of 140 possible values recorded in whole *Kecamatan*.

In making their appraisals, brokers were asked to consider four standard conditions for the hypothetical plots:

(1) Plots are on residential streets;
(2) Plots are located in mid-block;
(3) Plots are on small streets;
(4) The unit price of land price is Rupiah per m2. However, there are some areas where the land price is still calculated per tumbak. Therefore, the price of land per tumbak is converted to m^2 (1 tumbak = 14 m^2)

This type of method is quantitative interviews of the land market price in the study area through a land broker and basic tax data from *Badan Pengelola Pendapatan Daerah (BPPD)* (Management Board of Regional Revenue) in Bandung. The research method is descriptive research of the land-price dynamics surrounding large-scale land development of Technopolis Gedebage, Bandung, Indonesia. The research method used consists of data collection methods and data analysis methods.

3. Discussion

The data collected from the survey are used in the analysis below for three basic purposes: to examine land price changes over the last 16 years with certain characteristics, to analyze the effect of large-scale land development to land price, and to estimate the contribution of land value (market price) to these characteristics.

Two different analytical methods were employed in this study to assess systematically how land prices vary. Firstly, the average land prices were calculated according to three types within each category (for example, a subset was made of each tenure type), so that price differences by category could be examined. The other method to estimate values for particular characteristics of land was multiple regression analysis which can be used to estimate values for the dependent variable (current land price) in terms of proportional contribution of each series of independent variables (the characteristics of land). This same technique was used for constructing a land price gradient for the city, which relates to current prices and for the past 16 years it is far from the center of the city. Each of this methods,comparing the means of subset categories and regression analysis, is discussed in greater detail in the following sections.

These analytical methods allow us to examine the interrelationships between location, urban infrastructure, land tenure and land prices. Such assessment can be useful to policymakers for understanding the potential economic benefits of land title registration and infrastructure improvements. The methods of data colletion are as follows (see Table 1):

1. Literature readings to obtain literature review related to large-scale land development area Summarecon Gedebage Bandung, the value and price of land.
2. Secondary data collection was conducted to obtain information on land prices in the study area and developments occurred in the study area within the period of 2000-2016.
3. Primary data survey was carried out by interviewing land brokers and local communities to obtain information on land price developments in the study area.
4. Observation was done to see the condition of existing land use in the study area.

Table 1. Target, data requirement, data Source, data collection method

No.	Target	Data Requirement	Data Source	Data Collection Method
1	Identified changes in land prices around large-scale land development areas in time series over the last 16 years.	Map of Sub Region City (Sub Wilayah Kota (SWK)) Gedebage Bandung City Spatial Planning Office Secondary data survey	Spatial Planning Department (*Dinas Penataan Ruang*) of Bandung	Secondary data survey
		The development of large-scale land development includes: - Use Land - Location	• National Land Agency (Baand Pertanahan Nasional (BPN)/ATR) • *Kecamatan Gedebage*, *Kelurahan* Rancabolang, *Kelurahan* Cisaranten Kidul, *Kelurahan* Rancanumpang, and *kelurahan* Cimincrang • PT. Summarecon Agung Tbk.	Secondary data survey
			• Community/ Land Brokers • Government (Camat, Lurah)	Primary data survey (semi-structured interview)
			Observation	Observation
2	Analyzed the effect of large-scale land development on the price of adjacent land	Land price is based on the market year 2000-2016	• *Kelurahan* Rancabolang, *Kelurahan* Cisaranten Kidul, *Kelurahan* Rancanumpang, and *Kelurahan* Cimincrang • Community	Primary data survey (semi-structured interview)
		Basic tax of land 2000-2016	Regional Revenue Processing Agency (Badan Pengolahan Pendapatan Daerah (BPPD)) Bandung	Secondary data survey
3	Interpreted the analysis of the effect of large-scale land development on the price of adjacent land	-	-	- -

As the case of large cities in market economies, the land values of Technopolis Gedebage, Bandung are extremely high because it is near the main road and infrastructures; the prices will decline with distance from the main road. In this section, prices of residential plots in both formal- and informal-sector residential areas are presented. Formal-sector plots are defined as plots with high levels of both infrastructure and registered title. Informal-sector plots are those with low levels of infrastructure and weak title claims. Subsequent sections will discuss the role of location, infrastructure, and tenure in shaping land prices in more detail. Figures 1 illustrates the land price zoning of geographical area (based on the distance to the main road, infrastructure, and type of tenure). The zoning below does not reflect commercial or market land values located along main roads.

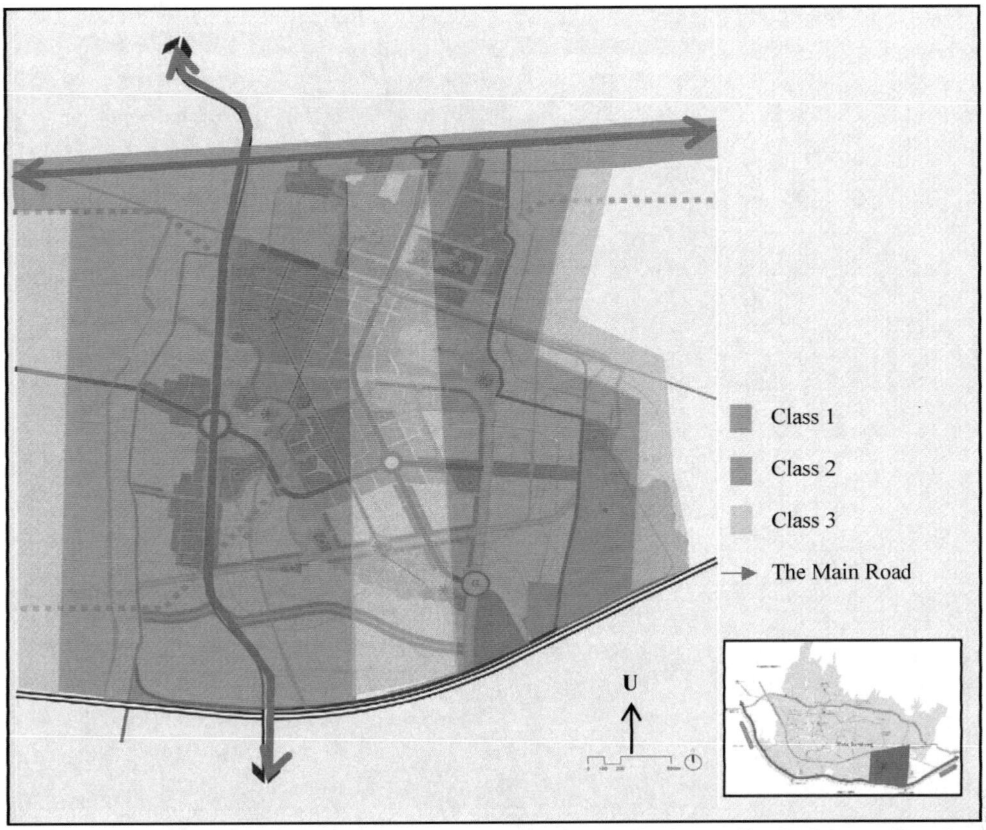

Figure 1. The geographical area zoning of land price in Technopolis Gedebage, Bandung

Overall (formal and informal) residential land prices in Technopolis Gedebage, Bandung are presented in Figure 2 [6]. Worldwide researchers have found that land values decline with distance. It illustrates that land users are willing to pay more for a square meter of land if it is to main road and infrastructure. In 2000-2005, land prices tended to be homogeneous in every *RW* in the sense that there was no big difference, except in *RW* 1 and *RW* 11, whose land prices were lower than other *RW*s. However, in 2010 the price of land in *RW* 1 increased rapidly even it became the highest land price of all *RW*. In addition, in each *RW* there is no significant difference for regions 1, 2, and 3 so it can be concluded that the differences in accessibility with Summarecon's main roads have little effect on the development of land prices in the region. However, in 2016, the price of land in *RW* 11 which previously had lower land price than other *RW* experiencing a drastic increase in area 1; while in area 3, the price of land was still the lowest land price. The same is true in *RW* 1 and *RW* 4 so that land prices within the *RW* vary between high and low land prices. This situation indicates that last year the location factor began to affect the development of land prices. To compare the development of land prices in each *RW*, the following graph shows the average price of land applicable in *Kecamatan Gedebage* (see Figure 2).

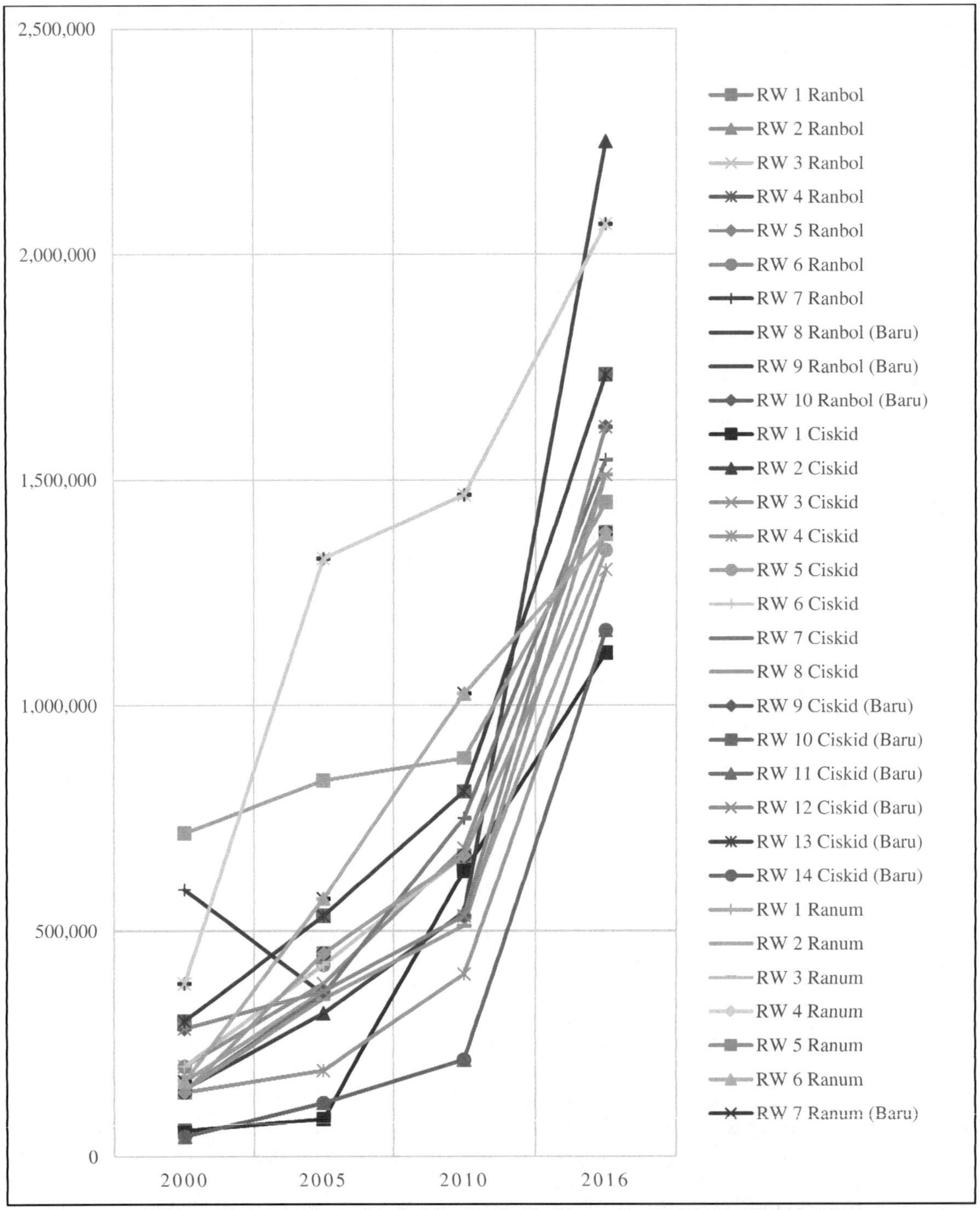

Figure 2. Average market land price in Technopolis Gedebage, Bandung

The pattern of land-price dynamics occurs due to several things. *RW* 2 and *RW* 1 are the closest *RW* to main roads and Summarecon, so that both *RW*s have higher land price and the price increased significantly in 2010 (after the construction of Summarecon began to take place) until 2016. In *RW* 11, located far from Summarecon, most of the land is still agricultural; it has worse infrastructure and the land price is the lowest among other *RW*s. Meanwhile, the land price of *RW* 1 increased significantly after 2010. This could be due to the fact that Summarecon bought potential land (before 2010) in *RW* 1.

The process of purchasing land price by Summarecon still continues until now [5]. The increasing of the land prices occurred in *RW* 3, 4, 5, 6, 7, 8, and 9 were almost the same and there was no significant increase in 2000-2005. In 2010-2016 the increase was higher than the previous year caused by Summarecon influence indirectly and the issue of Technopolis Gedebage as the second primary center of Bandung.

From the land price information, we can visualize isovalue of land price in *Kecamatan Gedebage*, Bandung. In the isovalue map, the land price of the area is represented as a point and as input to create hills on the map. The purpose of the isovalue map is to identify the difference of land prices at each location, whereas the hilltops are not only the highest point but also the highest price in each location.

From Figure 3 and Figure 4, it could be seen that in 2000 the land price hill was found only in the northern region, i.e. the area near Soekarno-Hatta Road (main road/National Road). The next five years, land hills began to rise in the southern region, although the highest land prices remained in the northern region of *RW* 2 of *kelurahan* Rancabolang and *RW* 02 of *Kelurahan* Cisaranten Kidul. Besides, there was no significant change in land prices because most of *Kecamatan Gedebage* land was agricultural land. There were only two residents built since the monetary crisis (in 1998), they are Bumi Adipura housing residents (in south of Technopolis Gedebage Road, *kelurahan* Rancabolang) and Riung Bandung housing residents (close to Soekarno-Hatta Road, northern area of *Kelurahan* Cisaranten Kidul).

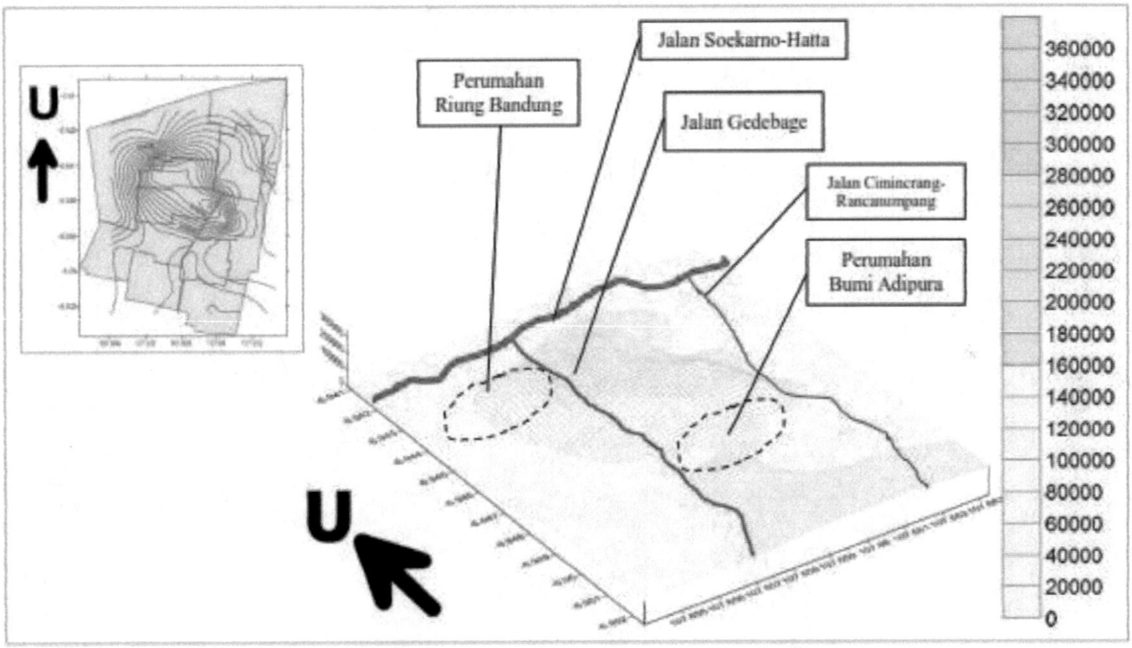

Figure 3. Isovalue land prices 2000

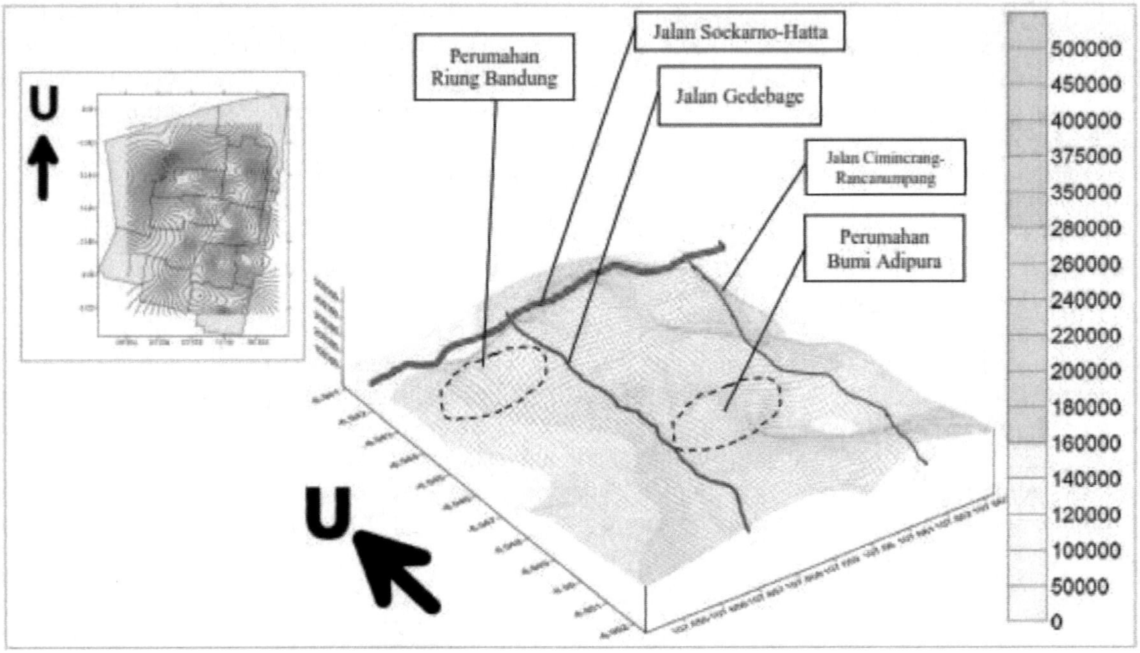

Figure 4. Isovalue land prices 2005

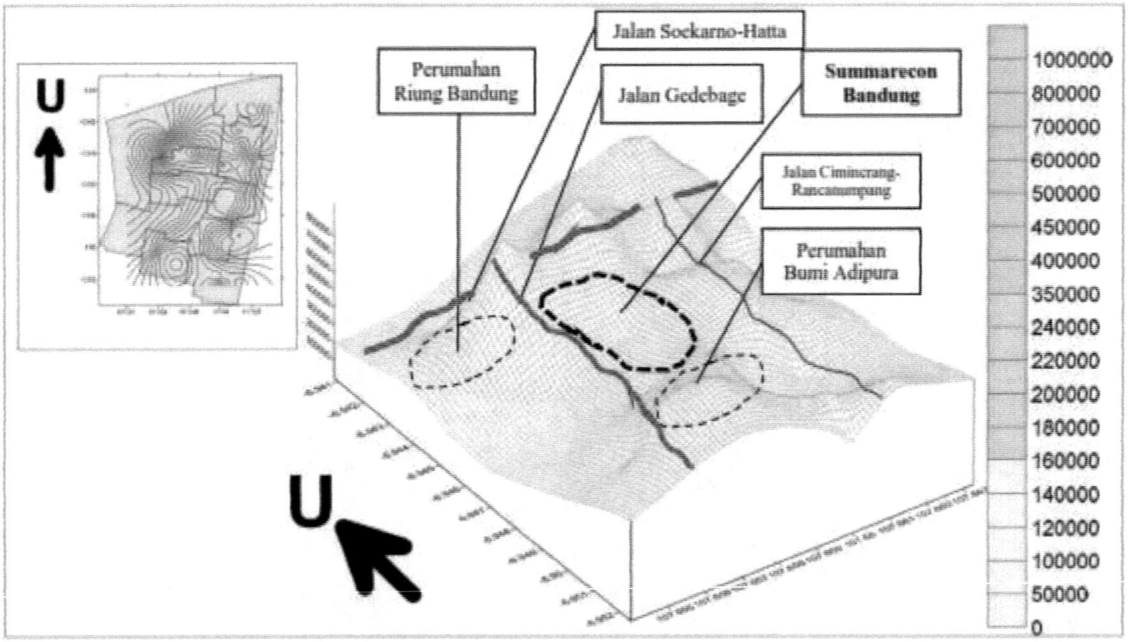

Figure 5. Isovalue land prices 2010

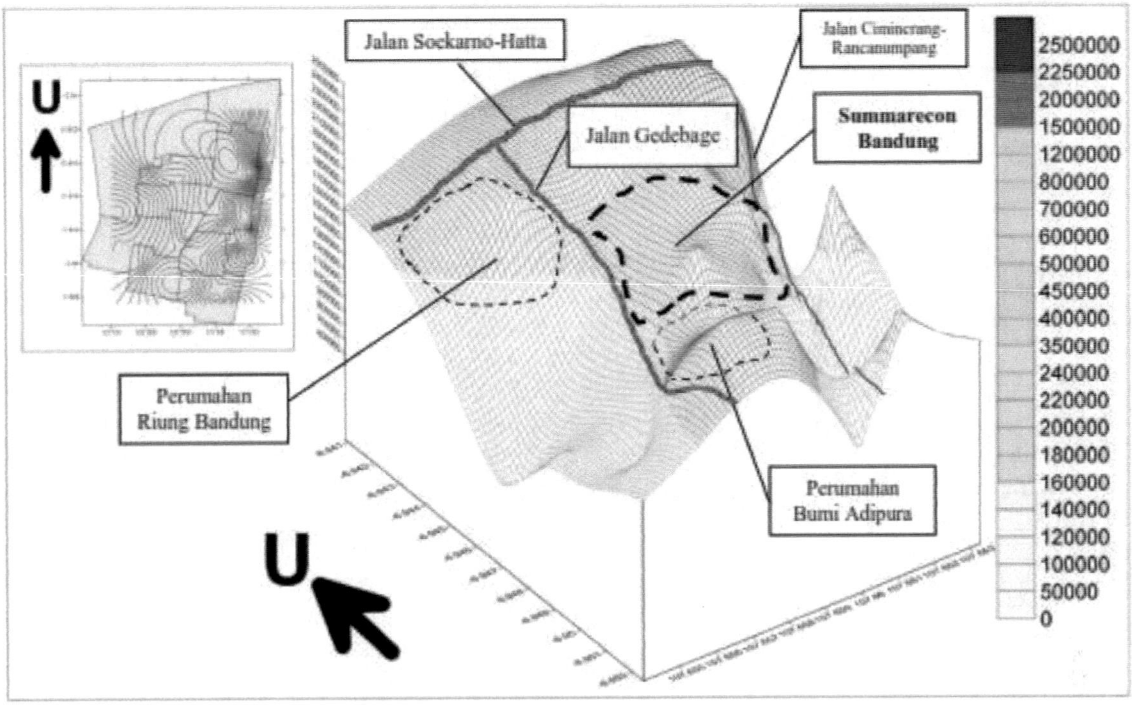

Figure 6. Isovalue land prices 2016

Summarecon has bought land in *Kecamatan* Gedebage, partly in *Kelurahan* Rancabolang (*RW* 1 and 2) and *Kelurahan* Cisaranten Kidul (*RW* 1, 2 and 4) since 2010. So, there was a significant increase in land prices from 2010 to 2016 (see Figure 5 and 6) in those *kelurahan*. Summarecon Bandung construction project has started since 2010 (still in progress of land maturation).

In general, almost all regions have larger prevailing land price than basic tax (*Nilai Jual Objek Pajak* (NJOP)). This shows that land pricing is more dynamic than NJOP as it is influenced by various factors. The most commonly used method for determining the value of land is by using market price approach or comparison as described above. The difference calculation of price which applies with NJOP is to see different pattern that occurred. The difference in the average market price of land with NJOP is presented in the graph below.

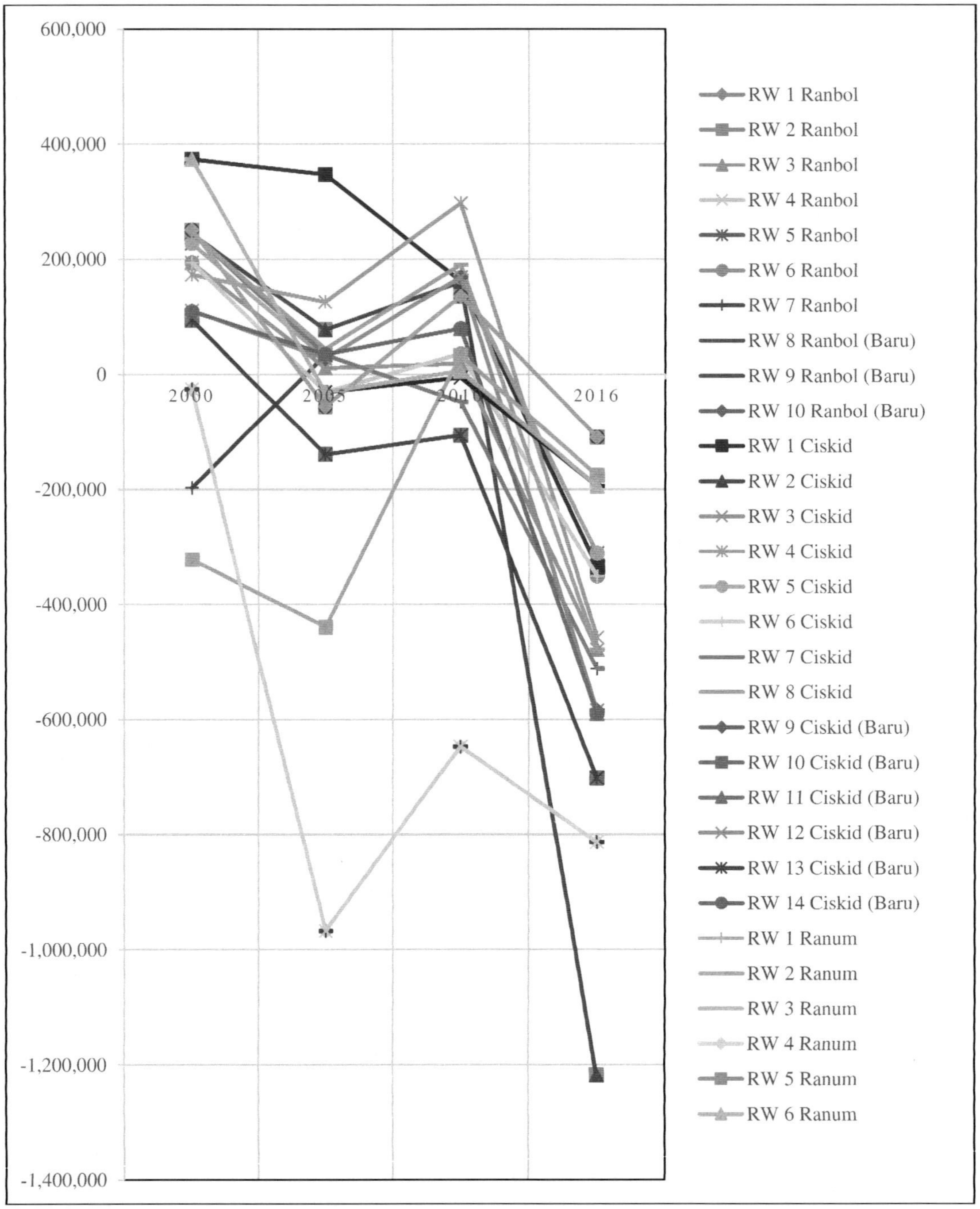

Figure 7. Difference in average market price of land with basic tax (NJOP) in Technopolis Gedebage, Bandung

The comparison of the values suggests that residential land prices in Technopolis Gedebage are lower than those in other location in Bandung, i.e. Dago, Alun-Alun (CBD), and Kosambi, where the land market is extremely constrained but higher than in Bangkok, where the land market is very responsive to housing demand. If adjustments are made for differences in income, the Karachi case is even more severe. Because of gross national product (GNP) per capita of $350 (1986) (US dollars), it would require approximately 13 percent of GNP per capita to purchase 1 m2 of residential land IO km from the CBD. In contrast, a similar situated plot in Bangkok would require 4 percent (GNP per capita of

$810 in 1986). In Bandung, it would require 9 percent (GNP per capita of $490 in 1986). Worldwide researchers have found that land values decline with distance, illustrating that land users are willing to pay more for a square meter of land which is closer to the center of the city.

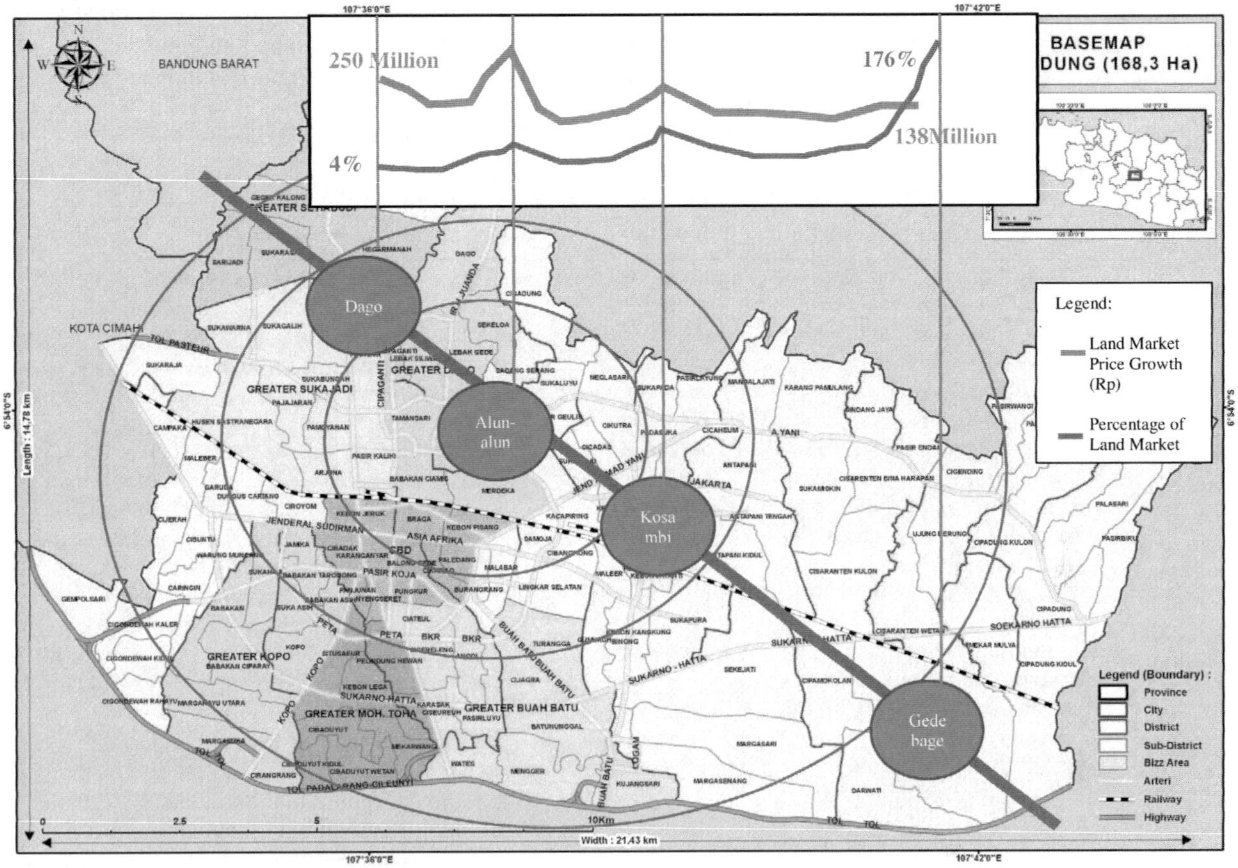

Figure 8. Land price map of several areas in Bandung

In comparing the market price of land in the study area with other areas, there are several criteria that are made equal in each region i.e. the price on the same soil quality with the building of the same type 45/72. The selection of the same quality and type aims to minimize the bias estimating the price of market land in that location. The land market price in Gedabage is compared with other areas in Bandung (with linear lines, Figure 8), those areas are:

- Dago area, North Bandung
- Alun-Alun Bandung (CBD)
- Kosambi area
- Technopolis Gedebage

The price of land in each location is as follows:

Table 2. Land price market of several locations in Bandung, 2010-2017 [7]

No.	Location	Land Price Market (Rp/m²)		
		2010	2013	2017
1	Dago, Northern Bandung	240.000.000	245.000.000	250.000.000
2	Alun-Alun Bandung (CBD)	250.000.000	275.000.000	300.000.000
3	Kosambi	150.000.000	180.000.000	200.000.000
4	Technopolis Gedebage	50.000.000	96.000.000	138.000.000

Source: http://rumahdijual.com, 2017

4. Conclusions and Policy Implications

This paper has reported the application of technique for generating residential land price data around large-scale development area. The field-work and data coding took approximately 1 months with 5 Indonesian surveyors. The results of the project clearly illustrate the applicability of the technique, and how its results can be used to support policy and investment decision making. The analysis of the land price data shows that infrastructure and titling programs confer considerable benefits on property owners. The findings, conclusions, and recommendations of this study are summarized in Table 3.

Table 3. Finding, conclusion, and recommendation

Finding	Conclusion	Recommendation
1. The dynamics of land price market around the main road and Summarecon Bandung continued to increase rapidly after 2010 (the land was purchased by PT Summarecon Agung Tbk.) 2. The highest land market price is in *RW* 2 Kel. Rancabolang and *RW* 2 Kel. Cisaranten Kidul while the lowest market price is in *RW* 11 Kel. Cisaranten Kidul. 3. Land prices are affected by the type of housing (formal housing and *kampung*), proximity to the main road, proximity to Summarecon, and proximity to large-scale land development activities.	(+) Developers develop settlements approaching the main road (-) Developers only buy and/or benefit *RW-RW*s close to the main roads	• Recommendations to the Government: Before making a plan it should first conduct data collection such as land price (in time series) and in-depth analysis of the area to be developed, so that the development of these areas can benefit the community, not just developers, such as building public infrastructure (such as the KM 149 toll gate plan, TOD terminal, etc.) which is actually more exploited by the developer. To minimize the externalities of the private sector, a mechanism such as zoning bonus should be made around the Semanggi ring toll road, Jakarta. • Recommendation to the Private: It should not only build private facilities, but also build in-situ phases (such as the Zoning Bonus mechanism) and pay attention to neighboring areas, especially during land ripening. • Recommendations to the community: The Technopolis Gedebage community needs to contribute in determining the direction of regional development because the indigenous people are excluded from their territory by being forced to sell the land (with speculation of selling at high prices) so they can not directly feel the impact of future Technopolis Gedebage development.
4. Basic tax is much cheaper than the market price	Basic tax does not follow the market	• Recommendation to the Government: It should be adjusted to NJOP , not too far away from the market price for example correcting the inflation happened. If the NJOP increases very rapidly, then the area is likely to become a commercial area, while the designation for Technopolis Gedebage is the settlements and offices. • Recommendations to brokers/people involved in the sale and purchase of land: Market price speculation should be avoided because it could harm the price of land in the region in the future.

The final conclusion of the analysis is that prices of plots, which are close to maid road, large-scale development project, infrastructure and has the security of tenure, increase faster than those plots unserved by the formal sector in fully serviced subdivisions. It is likely that the cause of the higher

IOP Conf. Series: Earth and Environmental Science **158** (2018) 012011 doi:10.1088/1755-1315/158/1/012011

inflation rate is the strong demand (for either investment or use) for low-priced plots provided by the informal sector. In order to moderate price increased and improve the accessibility of the poor to low-cost land for housing, the government should explore how the supply of informal-sector plots can be increased. Guided land development is one tested method that is well-known for Bandung's planners.

References

[1] Central Statistics Agency of West Java Province (2000-2015). Jawa Barat Dalam Angka.

[2] Dowall D E and Michael Leaf 1991 The Price of Land Housing in Jakarta *Urban Studies* **28** 5 p 707-722

[3] Directorate of Urban Directorate General of Spatial Planning Ministry of Public Works. 2013. Technical Material Spatial Detail Plan (RDTR) Bandung. Jakarta: Urban Directorate of Directorate General of Spatial Planning Ministry of Public Works

[4] Dowall D E 1989 Bangkok: Profile of An Efficiently Performing Housing Market *Urban Studies* **26** p 327-339

[5] PT. Summarecon Tbk. 2013. Annual Report of Summarecon. Jakarta: PT. Summarecon Tbk.

[6] The Government of Bandung 2015 Government Regulation of Bandung Number 10 the Year 2015 about Spatial Detail Plan and Zoning Regulation of Bandung 2015-2035 (Bandung: Bandung Government)

[7] Admin RumahDijual. Jual beli rumah jadi mudah dan cepat [Internet]. Rumahdijual. [cited 2 May 2017]. Available from: https://rumahdijual.com/carirumah.php?transaksi=BELI&jenis=RUMAH&kota=Bandung&minprice=&maxprice=&ltmin=&ktmin=&q=SEMUA&sort=0

Becoming Heritage: A Place-Making Study of Old Neighbourhood Marketplace in Bandung

A S Ekomadyo, A Nurfadillah, A Kartamihardja, A J Cungwin

Institut Teknologi Bandung, Indonesia

Corresponding author: ekomadyo@itb.ac.id

Abstract. The urban architectural heritages of Bandung represents the city's historical layers. A unique urban architectural element found in Bandung is the heritage neighborhood marketplace. In the style of European post-renaissance urban design, a heritage neighborhood marketplace is initially designed as housing clusters with open spaces as neighborhood centre to create community privacy. But along the time, the neighborhood centre open their enclosed spaces and landmark buildings, transforming itself into small marketplaces with intensively daily activities. Through everyday place-making methods, this article investigates the history and meaning of neighborhood marketplace as everyday urban artefacts through observation, historical study, and interview. The study reveals are three main forces assembling the current form of neighborhood marketplaces in Bandung: Dutch-Indies colonialization, Chinese society trading network, and local traders' tradition. Dutch-Indies colonialization in Bandung brought European architectural knowledge into neighborhood morphological design that contributed well-defined urban spaces. The Chinese trading network has been at place since the diaspora era and is still continued until now, bridging past and present life through economic activities. Local traders, who came after Dutch-Indies colonialization, fill the empty spaces of neighborhood centres with intensive formal and informal economic activities. Bandung heritage neighborhood marketplaces represent the city's urban constellation, where global and local political-economic and socio-cultural forces meet in an urban process. Here, the urban heritage is redefined as living artefacts experienced in everyday urban life.
Keywords: urban heritage, place making, neighborhood marketplace, Bandung, urban assemblage, living artefacts.

1. Introduction

In architecture, studies about space and place aims to define and describe the relationship between human beings and its surrounding physical elements. Architectural forms define space and create images to conceive of a place. The study of place has expanded from merely looking at how forms, activities, and images intersect into, by looking at psychological aspects of human beings, understanding deeper the phenomenology, psychometric, or social-constructivism approach. In urban culture, the architectural perspective of space and place contributes to uncover how power works in everyday urban life [1].

This paper analyzes the everyday life of heritage neighborhood marketplaces in Bandung. These marketplaces are unique to Bandung. While there are many another colonial cities with neighborhood clusters, but only in Bandung the clusters are constructed with enclave configuration where the

IOP Conf. Series: Earth and Environmental Science **158** (2018) 012012 doi:10.1088/1755-1315/158/1/012012

marketplaces become the clusters' center. A study found that these clusters shown three cultural layers: Dutch colonial, Chinese, and local traders [2].

By analyzing space, place, and cultural assemblage of Bandung, this paper will read urban spaces are formed and creating meaning to represent the urban cultural assemblage. As working paper, this article tries to open discussion about mutual relation among human and physical elements that construct urban places. It is one step to build thought about urban learning, which is concerning to the learning process on urban life by relation among human and non-human agents [3]. The concept of urban learning would hopefully unravel the deeper connection between human beings and its physical environment and conceive improved ideas of sustainability and better future of the city.

2. Method

This paper is based on short research about urban architecture phenomena. This short research uses descriptive method, collecting its data through morphological map, field observation, interviews, and urban historical study. Analysis of space formation is conducted by space/form framework from [4], and place-making by intersecting the forms, activities, and image [5]. At the end of discussion, findings will discuss urban cultural assemblage by associating it with *city shaped* [6] and *urban assemblage* [3].

3. Literature review: space, place, and urban cultural assemblage

In architecture, the concept about space is introduced as a mono-dualism with form, a unity between two opposing things, like the concept of Yin and Yang. Space will be defined by form, and vice versa, form can be seen and enjoyed from the space created [4]. Architects have the duty to compose form, here space is defined when architectural forms can be seen and enjoyed.

The relation between space/form is be broken down systematically into three parts: space forming, space organization, and form/space hierarchy. Schematically, space includes horizontal elements (base plane, elevated base plane, depressed base plane, and overhead plane) and vertical elements (vertical linear elements, single vertical plane, L-shaped plane, parallel planes, U-shaped plane, and closure). Space composition defines patterns in a certain organization, such as centralized form, linear form, radial form, clustered form, and grid form The relation of these form/space can also create hierarchy, where a form can be more dominating than other forms, through size, shape, or space of the form [4].

In architectural design, the concept of place is used to differ space and place, where place is space with meaning. Place is created because the composition of form in that space, user activities, and the images of the usage of space [5]. Place-making themed architectural design is simple way is composing form/space by imagining how that place will be livelier with the presence of activities and gives off good vibes by the users.

Studies of place-making develop broader to uncover the relation among humans and built environment in real practice. Relating with psychological school thought, studies of place can be mapped into three main approaches: phenomenology that concern with humanistic aspects of built environment, psychometric that measuring the relation between human and physical environments, and social-construction that see mutual relation among construction of spatiality and sociality [7]. Assemblage thinking of place studies can be classified in social-construction or –in more neutral term-socio-spatial approach. Assemblage thinking generally refers to two French philosophers, Deleuze and Latour, although they have different roots of thought [8]. Assemblage thinking sees different elements come together, and in place studies it concerns on how human dwell as socio-material activities assembled in everyday practice [3]. Here, place is investigate how it experienced, structured, and constructed through flows of desires and power in everyday activities [9].

By assemblage thinking, we can investigate the detail of urban process in everyday life. The term of urban process is used by Kostof to remind the urban planner and designer who tend to create ideal city that never been built [10]. Through historical studies, Kostof urges to understand urban patterns and elements that are assembled by political, economical, social, and cultural through history [10]. Here, the urban physical forms can not be avoided from the social contents as an organized "matrix" in everyday activities.

4. Case: old neighborhood marketplaces in Bandung

Bandung, with the nickname and brand "Parisj van Java", is a colonial city in Indonesia. Bandung was designed as "Indisch" concept in central area and "garden city" in neighborhood clusters. These clusters represented the ideas of "tropical Europe" concept. Several ones were designed with closed enclave formation and open spaces and landmarks to signify the clusters centers and gates [11]

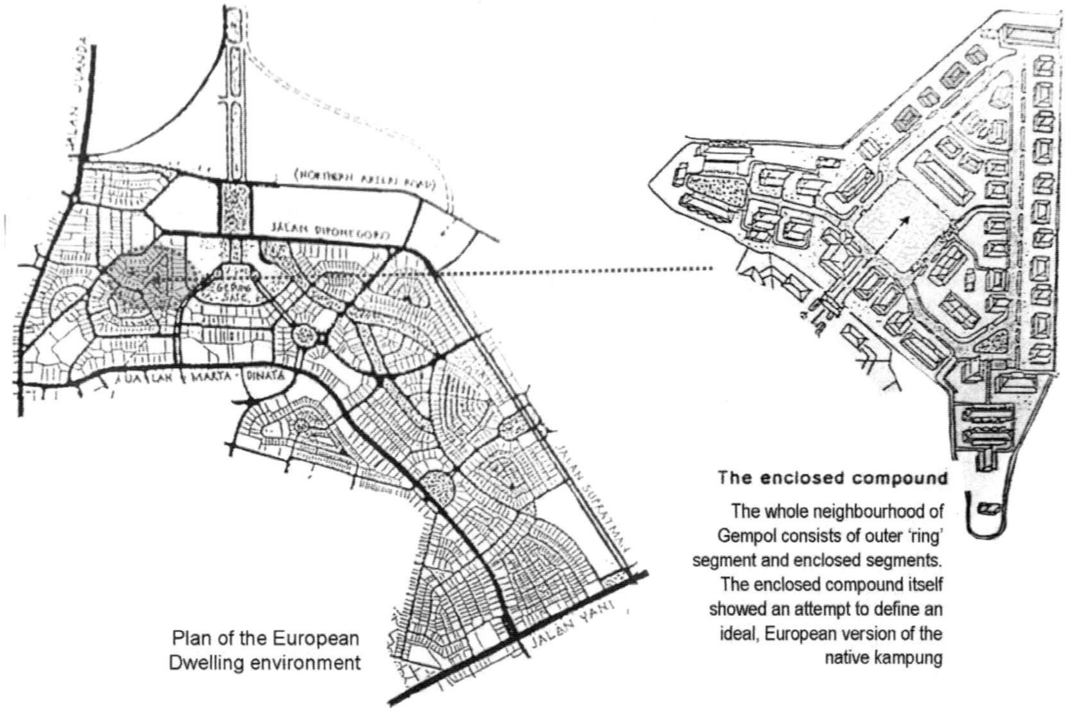

Figure 1. Map of Bandung as colonial city and a neighborhood cluster [12]

Some neighborhood clusters have special character with Chinese old shops as landmarks in the clusters centers. Those neighborhood clusters are Cihapit, Gempol, Cilaki-Rasamala, Tikukur, and Mangga. These shops are owned by descendants of Chinese traders and named as Toko Tjihapit, Toko Cahaya, Toko Widawati, Toko Engseng, and Toko Cairo. Their architectural appearances are quite diverse: Toko Tjihapit and Widawati have Chinese architecture, Toko Cairo has colonial architecture, while Toko Cahaya dan Toko Engseng. Their position at neighborhood center attracts passers-by and crowd, making them lively places (figure 2).

The old shops owners organize their business as family inheritance. Toko Tjihapit, owned by a fourth generation of owners, has stood for almost a hundred years. The owner organizes this old shop because as an inheritance passed over from generation to generation and because of the shop's considerable number of customers. The owner and customers of Toko Tjihapit have good social relationship with the surrounding community, which is an important factor to the shop's survival despite the increasing competition from more modern competitors.

Toko Cahaya in Gempol Cluster has been open for 80 years. This shop was previously named Toko Enghwa, but the aftermath of political turmoil in 1966 required the shop to have itself renamed with an Indonesian name. The current owner runs the shop as legacy from her father as well as to provide livelihood for her family and pay her children's education. Toko Cahaya is known to offer cheaper prices than supermarkets. Customers of Toko Cahaya are dwellers of the Gempol neighborhood public market adjacent to this shop.

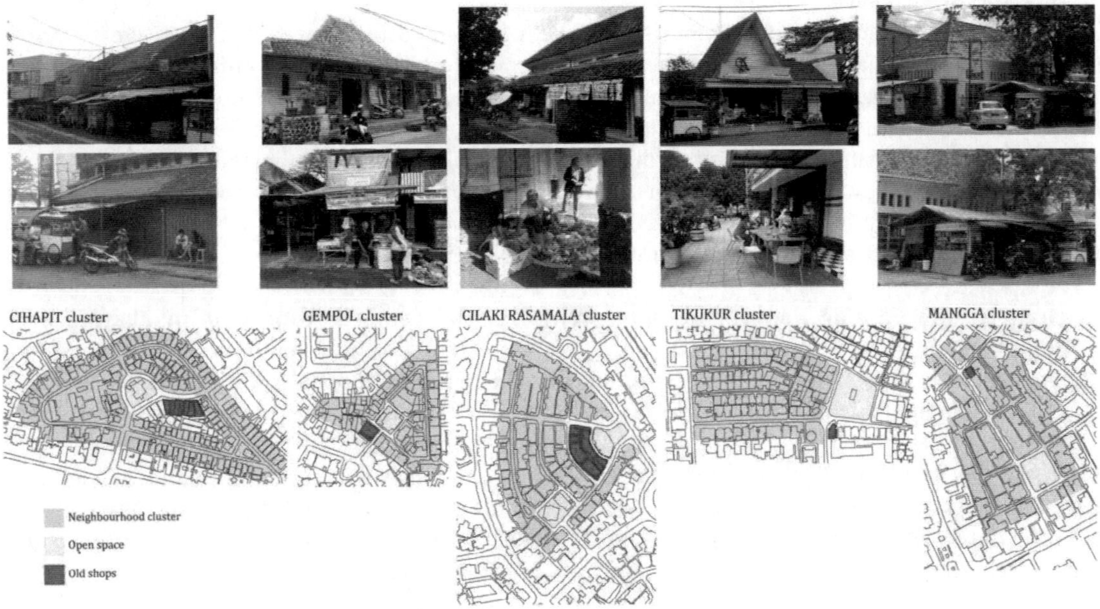

Figure 2. Colonial neighborhood clusters with old shops in Bandung

Although there is a row of shops at the Cilaki-Rasamala Cluster, only a small number are still in operation. One of them is owned by a lady named Widawati, a second-generation owner who inherited the shop from her parents-in-law. They owned the shop since the colonization of Indonesia ended in 1945. She continued to keep this store in order to put the building into good use and to keep her parents legacy. Although not as busy as it once was, it is still a major shopping place for neighbors and street vendors around it. Several buyers said that they have been their customer for decades and would find it inconvenient if the shop is closed. Widawati is not convinced of the continuity of this shop since none of her children are interested in continuing running the shop.

Toko Engseng, located in Tikukur Cluster, has been open since 1945. It is now owned by its second generation of owners, who run the shop since 1981. The owner said that she feels bad if she did not keep the shop. Toko Engseng is the legacy from her parents with many loyal customers. She hopes that her children will continue the shop to maintain their family's legacy.

Toko Cairo in Mangga Cluster is now organized by second generation of the shop founder. There is family relations between this shop and Toko Cahaya in Gempol cluster. She run this family business since 1984. The owner said that it is a customary to continue a family business. Despite the shop being the main income for her family, she is not expecting to his children to continue the shop as her children are set with their respective career.

The old shops owners are part of business network of Chinese-Indonesian community in Bandung. Historically, the immigration of Chinese community into Bandung was different from other port cities in Southeast Asia [13]. Chinese community was prohibited to enter Bandung area as the Dutch consider them as a thread for their trading monopoly. It was in the first half of the 19th Century, the Dutch government started to allow Chinese community to enter Bandung with a strict rules under surveillance and let them settled in appointed area known as Chineeschewijk or Chinese Quarter. But this racial-dividing quarter politic soon be abandoned as the Dutch government found out that this system was harmful to the settlement area and later replaced with a more proper settlement policy. At the same time, the Dutch government was building some major infrastructure to triggered economic activities and economic growth in Bandung such as the Great Post road, railway station, and new market (Pasar Baru). These infrastructures were main factors that influence the spreading of Chinese community settlement area, marked by the sharp increase of Chinese community population in the turning of the 20th Century and the growth of Chinese community settlements around the Great Post road, railway station, and marketplaces [14].

Nowadays, the old shops and neighborhood centers become urban places that intensively used by everyday informal trading activities. There are economic activities through selling and buying, also communal activities through social relation generated by these activities. Trading activities in Toko Tjihapit in Cihapit Cluster and Toko Cahaya in Gempol Cluster have been integrated with neighborhood public markets adjacent to the old shops. The owners organize the business mainly through the motive of respecting family inheritance instead of the ambition of taking the most amount of profit. As a result of that cultural and relational motive, their business practice maintains the communal relations with social and non-commercial activities around their shops.

5. Discussion: space, place, and cultural assemblage of the neighborhood marketplaces in Bandung

These clusters have similar form/ space configuration: the centers are open spaces, defined by vertical planes by buildings surround them. All clusters show similarities patterns of form/space formation. Each cluster center is an open space, defined by vertical planes from the sides of the building mass surrounding the area. This enclave configuration creates an open space in the middle of area. Radial form organization is marked by roads along the settlements, oriented to the open space. The existence of the shop in the center becomes more noticeable because it has a different shape (hierarchy by shape) and having a ranged space in the middle (hierarchy by space, figure 3).

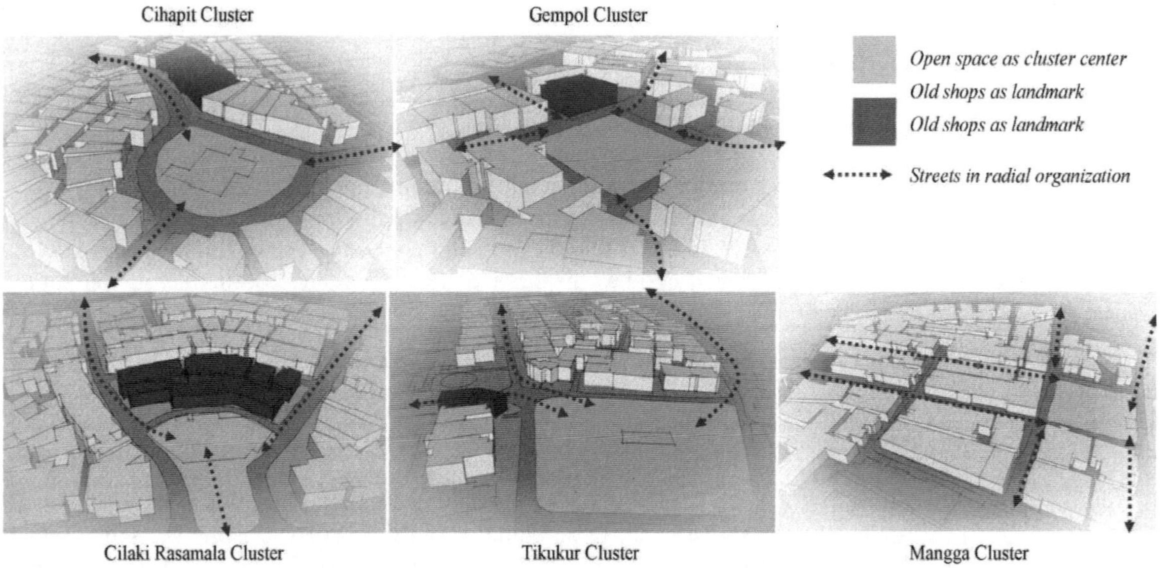

Figure 3. Form/space formation in each neighborhood clusters

This spatial configuration becomes an important formal element in creating a place at a settlement cluster's center. Psychologically, people will often gather in a space characterized by a center as its orientation and strengthened by a dominating form as its orientation. As an urban place, space configuration becomes an activity place with an important meaning to the space's user.

The spatial configuration in Mangga Cluster has a distinctive character, because the old shop (Toko Cairo) is not located at the open space as neighborhood center. This exceptions is reasonable in an urban process, considering city construction in form/space did not happen all at once. There has been an evolution and adaptation process in past, so the old shop were built on a street intersection and not in the neighborhood center. It also answers why Toko Cairo is less crowded compared to other old shops.

Informal socio-economic activities signify the everyday life of these clusters centers. This informal activities are supported by commercial function of the old shops that also create communality by social relation of informal economic activities. Such as common marketplaces, the socio-economic activities in clusters centers generated by goods, services, information, and values exchanges among traders and buyers.

These neighborhood marketplaces show different place intensity, related to other surrounding activities. Toko Tjihapit in Cihapit cluster and Toko Cahaya in Gempol cluster are most intensive places, because both are parts of the neighborhood public markets activities adjacent to them. Toko Widawati and Toko Engseng have medium intensity because there are buyers from school adjacent to them, other than dwellers of the neighborhood. Toko Cairo have the lowest intensity as its customers are strictly limited to surrounding dwellers.

The user has varying perception of the heritage of the old shops. The meaning ascribed by the buyers to these old shops are mostly as daily trading places to fulfill their needs. Only few of these buyers understand the special form of the old buildings. The owners understand that their old buildings are family inheritance as well as a source of livelihood that needs to be maintained. However, they do not understand the historical value of their building until informed by several architects who interview them for the purpose of research or exploring creative ideas.

Studying the heritage of colonial clusters of neighborhood marketplaces in Bandung allows us to understand three main cultures assembled here: European knowledge of form/space design, Chinese community trading network, and local-people informal socio-economic customs. Three cultures come together and interrelate between each other in an everyday practice of place-making. There are no Chinese old shops if no Colonial clusters designed in the past. These places intensity are determined by local people informal socio-economic activities. And these activities are taken on places that is well-defined by Colonial architecture.

A city is shaped and assembled by economical, political, social, and cultural forces through history. Kostof urged that urban physical elements can not be avoid be avoid with social content as a organized matrix in everyday life [6]. Through small architectural artifacts of old neighborhood marketplaces in Bandung, we can draw the lesson of cultural assemblage from the past until now.

Different cultures are determined by different groups. They create mutual relationship through human and non-human agents to stabilize the socio-technical configuration in a process of urban learning. Further research will uncover deeper how three cultures assembled through investigating actors and artifacts relation in everyday life, including how stabilized relation occurs. This actors-artifacts relations are constructed by a network connected only with others outside the places, to unveil what forces, mostly generated by political and economic motives, are at play to construct everyday urban places.

This short research also aims to spark discourse on how to interpret urban heritage. It is found that various actors have different understandings of the meaning of urban architectural artifacts, of whether or not they are considered as heritage. Concept of urban heritage is not a static definition; instead, a dynamic understanding on actors' attachment to the artifacts. The urban process does not consist only on the construction of physical elements, but also the collection of meanings ascribed them. The process of actors understanding the old buildings as heritage is important in urban learning perspective. Here, the everyday life of colonial clusters old shops in Bandung is not fixed or finished urban heritage, but instead in the process of becoming urban heritage.

6. Conclusion

This paper offers the method to read urban space and place formation combined with cultural assemblage thinking. With case of old neighborhood marketplace in Bandung, this short research can explore stories from inside well defined spaces. There are assembly of the "tropical Europe" urban design ideas from the past, the inclusive spirit of Chinese community in Bandung through their trading network, recent everyday life of old buildings through socio-economics informal activities by local inhabitants.

Cultural assemblage concept of urban place tries to recall the urban process thought in urban architecture studies. Studying old shops everyday activities in colonial clusters in Bandung give empirical evidences of the thought. Further thinking of urban assemblage is hoped to unravel how political, economical, social, and cultural forces work in everyday practice in urban places.

Assemblage thinking is also hoped to support architecture discipline to more contribute in real practice of urban place-making. Rather than technical discipline, assemblage thinking put architecture into socio-technical perspective, to see mutual relation among human and physical environments. By urban learning concept, the production of architectural artifacts can be put as learning devices for community how to dwell together in urban life.

By assemblage perspective, architectural heritage is not a fixed meaning: it is a process of becoming. This research found that there are many different meaning of heritage from architects, the old building owners, and building users. To be heritage, a building is seen as architectural: a form of architecture in relations with the others. Assemblage thinking focus on relation among architecture and its social context, including how heritage meaning is constructed. It is the key how the urban learning can be investigated and understood clearly.

References

[1] Low S 2017 Spatializing Culture: The Ethnography of Space and Place (New York: Routledge)
[2] Ekomadyo A S. 2018 *Membaca Ruang Kota Sebagai Tempat Rajutan Budaya Kasus: Keseharian Pertokoan Tua Pada Kluster Permukiman Kolonial Di Bandung*. Seminar Kota Layak Huni Jakarta (22 February 2018 Trisakti University)
[3] McFarlane C 2011 Learning the City Knowledge and Translocal Assemblage (West Sussex, UK: Blackwell Publishing)
[4] Ching F D K 2007 Architecture: Form, Space, & Order. (Third Edition) (Hoboken, New Jersey John Wiley & Sons, Inc.)
[5] Montgomery J 1998 Making a city: Urbanity, vitality and urban design. *J. of Urban Design* vol **3** 1 p.93-116
[6] Kostof S 1992 The City Assembled: The Elements of Urban Form through History (London: Thames and Hudson)
[7] Morgan P 2009 Towards A Developmental Theory of Place Attachment, *J. of Environmental Psychology* vol **30** 1 p. 1-22
[8] Riyadi A, Ekomadyo A S, Rusli S, Nuzzela S and Yuliar S 2017 Local Wisdom for Social Learning: Case of Rumah Sahabat Salman The Indonesian *J. of Science and Technology* (IJoST) (forthcoming).
[9] Dovey K 2010 Becoming Places: Urbanism/ Architecture/ Identity/ Power (London: Routledge)
[10] Kostof S 1991 The City Shaped: Urban Patterns and Meanings through History (London: Thames and Hudson)
[11] Siregar, S A 1990 *Bandung: The Architecture of a City in Development: Urban Analysis of a Regional Capital as a Contribution to the Present Debate on Indonesian Urbanity and Architectural Identity*. Doctoral Theses. Faculty of Applied Science Katholieke Universiteit Leuven
[12] Siregar S A 1998 *Continuity and Discontinuity in Urban Transformation*. Case City of Bandung International Symposium and Workshop on Historic Cities in Islamic Societies. Aga Khan Award for Architecture (21-23 April 1998 Yogyakarta)
[13] Kartamihardja A I 2017 *Kajian Bentuk Arsitektur Shophouse Sebagai Hasil Adaptasi Budaya Imigran Tionghoa di Kota Bandung*. Seminar Nasional Kearifan Lokal dalam Perspektif Global Medan (25-26 January 2017 University of North Sumatera)
[14] Tunas D 2007 *The Chinese Settlement of Bandung At the Turn of the 20th Century*. Master Thesis. Department of History National University of Singapore

Acknowledgments

Authors acknowledge to Institut Teknology Bandung for research funding, for Deddy Wahjudi (lecturer, Industrial Design Department, ITB) for interesting discussions about heritage bazaar in Bandung and Lanna, Tiffany, Keshia, and Marcellina (bachelor students, Architecture Department, ITB) for their assistance in interviewing owners and buyers of old shops.

IOP Conf. Series: Earth and Environmental Science **158** (2018) 012013 doi:10.1088/1755-1315/158/1/012013

Community Participation on an Urban Sanitation Program: a Comparative Study

Ismu Rini Dwi Ari[1], I Nyoman Suluh Wijaya[1], Aditya Dewanto[2]

[1]Regional and Urban Planning Department, Universitas Brawijaya, Malang, Indonesia.
[2]Engineering Faculty, Universitas Brawijaya, Malang, Indonesia.

Corrsponding author : dwiari@ub.ac.id

Abstract. Sanitation development becomes one aspect, particularly the sixth pillar, of Sustainable Development Goals that aims at ensuring the access to water and sanitation for all. The success of the sanitation program will bring about significant changes in the human life. However, it is necessary to consider community involvement in the development program in all of the steps holistically – planning, implementation, operation and maintenance. This research scrutinizes the key factors of successful urban sanitation development programs in Talangagung sub-district as compared to Ardirejo sub-district in Kepanjen district in Malang regency, Indonesia. Through field observation in the two-research areas, and questionnaire survey distributed to to104 heads of households in Talangagung sub-district and 84 heads of households in Ardirejo sub-district, an evaluation on the suitability of the program as well as the social capital measurement was conducted. The result of estimation illustrates that Talangagung sub-district has higher score than Adirejo sub-district, which respectively 470 and 340 for the whole steps of the development. Then, the study continued with the social capital measurement using three indices of social network analysis – covering rate of participation, density and centrality. In general, the rate of participation and density of the residents involved in the urban sanitation program development in Talangagung sub-district was higher value than in Ardirejo sub-district. However, the residents' participation in the development process tends to decrease from the planning – implementation – operation/maintenance. It is predicted that the low participation in the third step – operation/maintenance, is due to the lack of understanding of the whole participants, since the information is exclusively understood by the organizing committee. In other words, there is asymmetrical information occurring between the organizing committee and the residents. For the centrality, this research utilized three indices of the centrality covering degree – closeness – betweenness centrality. The estimation result at planning step draws that there are 21 central actors in Talangagung, wherein 8 of them are the committee members and 13 of them are regular residents. Meanwhile, in Ardirejo sub-disctrict, there are 5 central actors, but all of them are committee members. In the implementation step, there are 10 central actors, consisting of 7 committee members and 3 residents in Talangagung sub-district. Meanwhile, in Ardirejo sub-district, there are 7 central actors and all of them are the committee members. Finally, at the operation/maintenance step, there is no central actor for both sub-districts.

1. Introduction

Sanitation development, in principal, is not only dealing with the development and improvement of sanitation facilities, such as toilets – as a physical approach, but also with the hygienic behaviour as one basic prerequisites on how people behaviour-changing from the open-defecation or other improper defecation activities into toilet-use as indicated by the availability of sufficient sanitation infrastructure for every single house. Therefore, the success of the development has basic requirements, wherein people participation together with infrastructure development will reach the target of forming a hygienic environment [1].

One big obstacle for the housing settlement, particularly in urban areas, for having a proper sanitation infrastructure is their parcel land limitation – wherein, generally, urban housing settlements might be characterized as dense areas with respect to both housing and population. As a result, it is difficult to fulfil each house with internal sanitation infrastructure – both toilets and septic tanks. Therefore, one common solution is to establish communal sanitation in the available vacant land nearby the houses. This means that the participation of the residents as a result of the community involvement is one of the influential factor towards the success and sustainability of the sanitation development through the community empowerment [2].

In the research areas, generally every house has their individual toilet. However, they do not have sufficient space inside the house to build a proper septic tank. The SPBM Program – a national Program of Urban Sanitation Based on Community that has been implemented in the research area as one among the chosen area of National Program for Community Empowerment in Urban areas (*PNPM Mandiri Perkotaan*) and has been granted a minimum three-time cycle of The Direct Community Assistance Funds (*BLM*). The aims of the SPBM Program are i) to support the community empowerment collectively, ii) to strengthen the social network, and iii) to create open communication between the community and the government institution. [3]

Therefore, the first aim of this research is to measure the level of suitability towards the implementation of the program between the two villages, namely Ardirejo and Talangagung, located in Kepanjen District, Malang Regency. The two areas are chosen since the second location seems able to show better development result than the first one. Hence, it is based on the division into four-steps of the program according to SPBM Implementation Guideline (*Pedoman Pelaksanaan SPBM*) published by Ministry of Public Works (Direktur Jenderal Pekerjaan Umum, 2011), and best lesson learnt from the SPBM Program that might be useful for future programs. The second aim of the research is measuring the social capital of the SPBM Program through three indices of social structure, which consists of the rate of participation, density and centrality as proposed [4][5] using Social Network Analysis (SNA) approach, wherein the equation formula [6]. Finding answer of the second aim might bring a better understanding of whether the functioning of social capital might give significant influence to the success and the sustainability of the SPBM Program, whereby the better community involvement illustrates stronger social ties that make social capital work [7][8].

2. Research method

2.1. Research area

Ardirejo Village (210.74 Ha) and Talangagung Village (354.91 Ha) are administratively located in Kepanjen district, Malang regency. Based on an internal evaluation of the SPBM program, the program that has been conducted in Ardirejo village could not be done maximally due to (i) less community contribution occurrence, (ii) dissemination information at the socialization, and (iii) poor maintenance of the result. Meanwhile, the program in Talangagung village could be done in better level among the entire six villages in Kepanjen district. Therefore, in order to have a better understanding of the program, this research tries to compare the program between the two villages.

2.2. Sample

In order to have sufficient answers for the main research aims, the gathered data have been compiled in two ways – covering field observation and interview to the targeted population using questionnaire survey. For the first research question, the chosen respondents are the ones who have sufficient knowledge about the whole program. Therefore, two respondents from each village have been chosen using purposive sampling method. The chosen respondents are the heads of the Community Self-supporting Group (*KSM*) and the facilitators of the program, who know very well about the whole process of the program in the area. Then, the respondents chosen to answer the second research question are the heads of household of the participants of the SPBM program – as the program beneficiaries, and the whole committee members of the KSM. The number of the beneficiaries of the program in Ardirejo village and Talangagung village are 84 and 104 households, respectively. On one side, the number of the committee members of KSM in Ardirejo village are 7 people, covering the tasks as chairman, vice-chairman, secretary, treasury, task force, as well as logistic and public relation. On the other side, the composition of the committee members of KSM in Talangagung village are mostly similar with two more additional committee members that function as the security and budgeting officers; hence it consists of 9 people. The following Table 1 indicates the number of households as the beneficiaries program at the sub-hamlet/hamlet areas.

Table 1. Number of SPBM program beneficiaries

Ardirejo Village		**Talangagung Village**	
Sub-hamlet (RT)/ hamlet (RW)	**Number (HH)**	**Sub-hamlet (RT)/ hamlet (RW)**	**Number (HH)**
RT 1/ RW 2	38	RT 19/ RW 3	20
RT 2/ RW	46	RT 20/ RW 3	84
Total	84	Total	104

Considering the first research question [9], whereby a list of questions in the questionnaire survey were divided into 4 categories – classified based on the four-steps of the SPBM program covering Planning (input) – Implementation (process) – Post Implementation (output) – Impact Post the Program after the completion of the whole activities (impact). In the planning step, four variables were considered consisting of i) the program socialization – the number of socialization, ii) the involved actors – the number of involved actors, iii) the funding – the funding disbursement criteria, and iv) the human resources – the number of KSM committee. In the implementation step, there were five variables covering i) the planning of the work, ii) the work force arrangement, iii) the procurement of goods and services, iv) the monitoring and supervision of the construction implementation, and the activity report. In the post implementation step, the variables consist of the utilization of the constructed sanitation, ii) the improvement of the access to basic sanitation, and iii) the institution. In the step of the post impact of the program, the variables were i) changes of the community's knowledge, ii) changes of the community's attitude and behaviour, and iii) the level of public health. Then, simple scoring analysis is applied as the approach for suitability evaluation of the program, and it is continued by weighting method through classification rank.

Then, the second research question was addressed using Social Network Analysis approach that required affiliation data, wherein the whole respondents were asked about their participation in the program – which then classified into three steps: Planning – Implementation – Operational and Monitoring. The type of the affiliation data here might be classified as informal network since the social network that had been formed was due to their relation on the events for each step of the program. The procedure of analysis might be described as follows. First, the affiliation data were collected in the shape of rectangular matrix labelled as incidence matrix of respondent – event (nxm) that illustrated the participation or involvement of each respondent at every step activity of the program. Second, the incidence matrix was transformed into a square matrix called the adjacency matrix of respondent-by-respondent (nxn). Third, the affiliation data in the shape of adjacency matrix as input to measure the three indices using software called UCINET 6 version 6.365. For the second research aim, the number of respondents became set of actors labelled as n and the number of activities at each step became set of events labelled as m.

3. Estimation result and discussion

3.1. Suitability analysis

The following Table 2 illustrates the results of the scoring analysis of the two villages classified into 5 categories of suitability – very low (100-180), low (181-260), enough/pretty (261-340), suitable (341-420) and high (421-500). In general, based on the expert's evaluation – the heads of KSM and facilitators, Talangagung village has higher score than Ardirejo village for the whole four-steps.

Table 2. The evaluation of the SPBM program

Step	Ardirejo village			Talangagung Village		
	Score	**Classification**	**Suitability**	**Score**	**Classification**	**Suitability**
Planning Process (input)	330	3	Enough/Pretty	430	5	High
Implementation Process (process)	340	3	Enough/Pretty	453,33	5	High
Post-implementation Process (output)	333,3	3	Enough/Pretty	366,67	4	Suitable
Impact Post the Program (outcome)	383,3	4	Suitable	433,33	5	High

Table 3 indicates the total score for each village at the whole four-steps. The total score for Ardirejo Village is within the range of 261 – 340, meaning the SBPM Program has been conducted in enough/pretty suitability. Meanwhile, for Talangagung Village, the total score is between 421-500. It means that the SPBM Program in Talangagung village had been done in high suitability. Thus, we might assume that the SPBM Program in Talangagung village has been done successfully with higher suitability level than the one conducted in Ardirejo village.

Table 3. The scoring calculation for the whole steps

Step	Ardirejo village			Ardirejo village		
	Weight	**Classification**	**Scores**	**Weight**	**Classification**	**Scores**
Planning Program (input)	12,5	3	37,5	12,5	5	62,5
Implementation Program (process)	17,5	3	52,5	17,5	5	87,5
Post-implementation Program (output)	30	3	90	30	4	120
Impact Post the Program (outcome)	40	4	160	40	5	200
Total			340			470

3.2. Social network analysis

3.2.1 The rate of participation. The purpose of the analysis is to obtain descriptions of participation level of the whole community in the research area through the respondents' affiliation data at each step of the program in each village. The measurement of the participation rate in Ardirejo village was based on 9 activities covering 3-steps: Planning – Implementation – Operational and Maintenance, which consisted of 6 – 2 – 1 composition, respectively. Meanwhile, similar measurement in Talangagung village consisted of 10 activities, which were divided into 6 – 3 – 1 composition. The following Picture 1 describes the number of participation of the respondents in each village as the whole beneficiaries group.

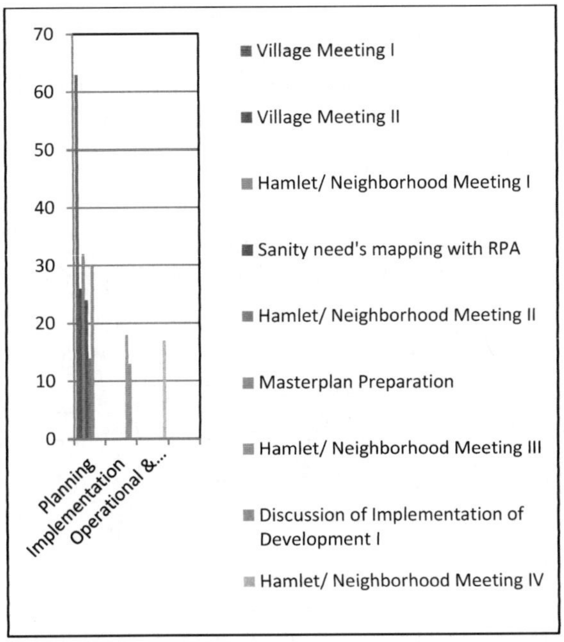

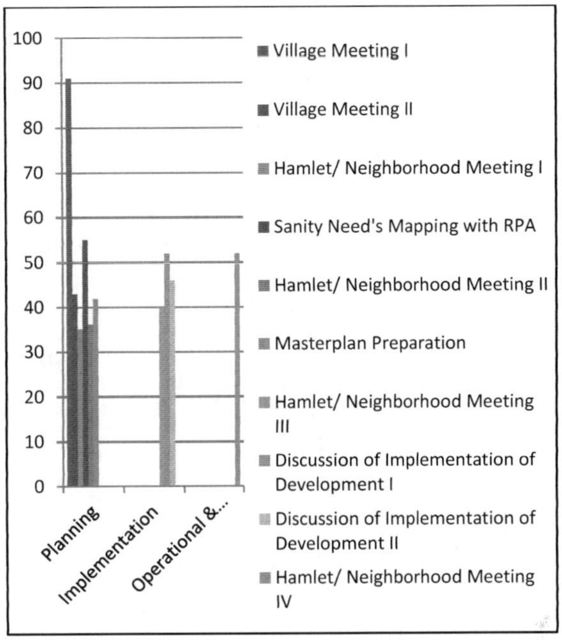

Figure 1. The number of community's participation - (left) Ardirejo village & (right) Talangagung village

Table 4 illustrates the calculation of the participation rate in the research areas. The classification of t participation rate is divided into three categories – low, middle, high in different total number of activities in each village as mentioned above.

Table 4. The rate of participation

	Ardirejo village				Talangagung village			
Step	**No. Of Respondent**	**No of Activity**	**RoP**	**Classification**	**No. Of Respondent**	**No of Activity**	**RoP**	**Classification**
Planning	84	6	2,27	Middle	104	6	2,90	Middle
Implementation	84	2	0,40	Low	104	3	1,32	Middle
Operational & Maintenance	84	1	0,20	Low	104	1	0,50	Middle

In general, the participation rate of the community in Talangagung village is higher than the in Ardirejo village for the whole three-steps. At the planning step, the average number of community's participation is 2 – 3 amongst 6 activities for both villages, with slightly higher rate of participation in Talangagung village. At the implementation step, some communities in Ardirejo village participated in one activity, but some of others did not join in any activity. Meanwhile, for the community in Talangagung village, generally they joined in 1 – 2 amongst 3 activities. At the operational and maintenance step, there was only one activity available. However, the number of participation in both research areas were not many, wherein about 80% of the communities in Ardirejo village did not participate in the activity, meanwhile, about 50% of the communities in Talangagung village participated on it. Hence, in the communities in Talangagung village, one in every two community members participated in the activity. For the whole steps, the communities in Ardirejo village joined 3 – 4 amongst 9 activities. Meanwhile, the communities in Talangagung village participated in 5 – 6 amongst 10 activities. Therefore, we may infer that within the whole steps, the communities in Talangagung village can be classified as having middle rate of participation, while the communities in Ardirejo village can be classified as middle rate only in one out of the three steps.

The declining participation in the activity could be explained with two reasons. First, it was caused by community reluctance to participate in each step due to clashes between the working time and the communal work activity schedules. Second, some communities also thought that representativeness of the committee members of KSM was more than enough to run the program, particularly in the operational and maintenance steps.

3.2.2 Density. The value of density is between 0 – 1, which, in this research, is classified into three level covering low – middle – high. The purpose is to attain a clear illustration on how dense communities' informal network at each step of the program within the village. The following Table 5 describes the density of the SPBM program in the research areas.

Table 5. Density

Village	Classification	Step	Density	Category
Ardirejo	0-0,33 (Low)	Planning	0,675	High
	0,34-0,66 (Middle)	Implementation	0,062	Low
	0,67-1,00 (High)	Operational & Maintenance	0,039	Low
Talangagung		Planning	0,918	High
		Implementation	0,477	Middle
		Operational & Maintenance	0,248	Low

The density at the planning step illustrates high level of participation for both villages. This means that the majority of the community were involved in the 6-activities at the planning step. The highest attendance was present at the village discussion I, wherein the information dissemination was firstly conducted. At the planning step, only less than 10% of communities in Talangagung village did not involve in the activity, and it had much higher density as compared to Ardirejo village. As a result, the flow of information as well as resources might be disseminated more effectively among the communities in Talangagung village than in Ardirejo village.

At the implementation step, the density of the communities in Talangagung village can be classified as middle level (0.477).This level is very much higher than density in the Ardirejo village, which shows low density level (0.062). This result might represent that the majority of the communities in Ardirejo village (93.8%) did not participate in the second step of the program. Meanwhile, only about 47.7% of the communities in Talangagung village participated in the program. At the last step, the value of density is even lower, wherein communities in both villages indicated low level of density. However, the level of density of the community in Talangagung village still show higher level than the one in in Ardirejo village. It is about 24.8% : 3.9% communities involvement between the two villages, respectively. The majority of community who wwere involved in the operational and maintenance step came from the committee members of KSM. It might represent that the flow of information as well resources is dominated by the committee members. As a consequence, in one hand, the sustainability of the program might fully depended upon the committee members of KSM, and on the other hand, the common community might only act as pure users of the result of the program.

3.2.3 Centrality. In this research, there are three different types of centrality – degree, closeness, betweenness that were measured in order to find a prominent actor for each step of the program. Similar with the density, the index values of centrality is between 0 – 1, and they are classified into three levels too. The following Table 6 illustrates the results of the three-centrality measurement for both villages at the three-steps.

The degree of centrality measures the number of networks of an actor to the rest actors within the networks, in order to identify the most popular actors in the network. Among 84 respondents in Ardirejo village, there were 63 respondents with high level of centrality degree at the planning step. In addition, the whole respondents had high levels of centrality degree. It might be referred that most of them knew each other, with almost similar influence to share information as well as any resources to others. The closeness centrality exemplifies actors that have the closest distance to the rest of the actors in the network that referred to how fast the central actors disseminate the information and resources to the rest of actors in the network. In this measurement, an actor who did not participate in any activity has to be omitted from the calculation, since this actor could contribute nothing to the network. Since the number of community members participated lessened from the first step to the second and the third step, the closeness levels also decreased. This means that the number of prominent actors with regard to their closest 'geodesic' distance in the network was also getting lower from first step to the rest of steps. The last measurement of centrality is betweenness aimed at scrutinizing the prominent actors that are indicated through their position as a bridging of the information flow as well as resources to the rest of actors in the network. The more mediating position played by an actor, the higher the level of betweenness they have. However, since most of the respondents have direct relation to each other, which is shown by themajority high degree centrality, the degree of betweenness is in the low level. It implies that in the network among the communities in Ardirejo village, there is no single actor who played the significant role as mediators.

Table 6. Centrality

	Ardirejo village			Talangagung village		
Step	***Degree***	**Closeness**	**Betweenness**	***Degree***	**Closeness**	**Betweenness**
Planning	• Low: 17,8% • Middle: 7,14% • High: 75%	• Middle: 22,61% • High: 77,38%	• Low: 100%	• Middle: 5,76% • High: 94,23%	• Middle: 0,96% • High: 99,03%	• Low: 100%
Implementation	• Low: 100%	• Low: 69,04% • High: 30,95%	• Low: 100%	• Low: 17,30% • Middle: 63,46% • High: 19,23%	• Low: 0,17% • Middle: 7,69% • High: 75%	• Low: 100%
Operational & Maintenance	• Low: 100%	• Low: 79,76% • High: 20,23%	• Low: 100%	• Low: 50% • Middle: 50%	• Low: 50% • High: 50%	• Low: 100%

The similar pattern is also found for the centrality measurement in Talangagung village, particularly for betweenness centrality. It means that there is no prominent actor functioning as the mediator in the network. For the degree and closeness of centrality, there are some prominent actors (high level) that could be found for both implementation as well as operational and maintenance steps. This means that the flow of information and resources might able to disseminate through the prominent actors for more effective result.

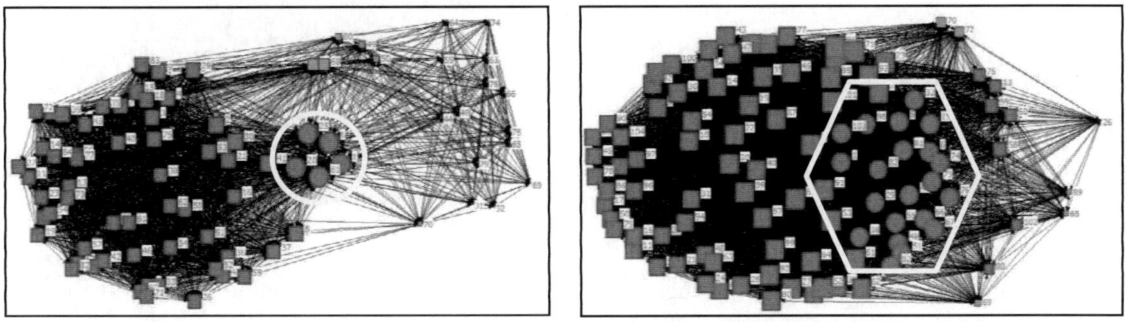

Figure 2. Netdraw Centrality at the Planning Step - (left) Ardirejo village & (right) Talangagung village

Graph 2 describes the net draw centrality at the planning step for both villages. There are five prominent actors at the planning step in Ardirejo village. All of them are the committee members of KSM covering chairman, vice-chairman, secretary, treasury and logistic section. Thus, none of them is common community member in the area. In case of Talangagung village, twenty one prominent actors are found, wherein seven of them are the committee members of KSM in the village.

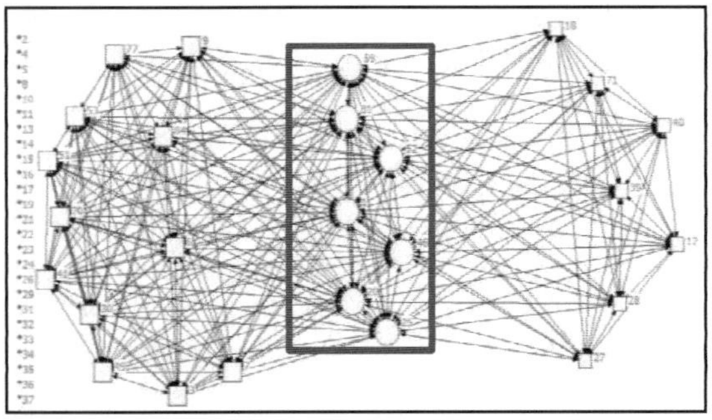
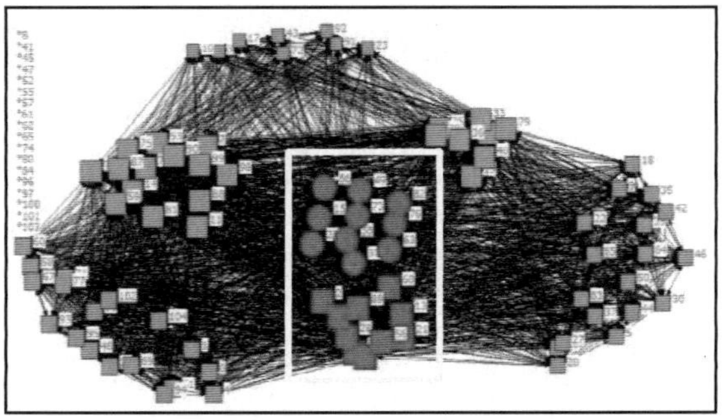

Figure 3. Netdraw Centrality at the Implementation Step - (left) Ardirejo village & (right) Talangagung village

From the graph 3 we can assume that at the implementation step, there are seven prominent actors in Ardirejo village – covering the whole committee members of KSM. So, once again, none of the prominent actors comes from common community. In Talangagung village, there are 10 prominent actors that consist of 7 of the 9 committee members of KSM plus 3 prominent actors from common communities. Thus, we may infer that in general, the committee members of KSM play important roles at the implementation step of the program.

Graph 4 illustrates that at the operational and maintenance step, we could not find any central actors in the two villages. In other words, it means there is no central actors in the research areas, even though 17 respondents in Ardirejo village and 52 respondents in Talangagung village participated at the step. Nevertheless, we are still able to find from the calculation that in Talangagung village thatthe respondents have middle level of degree of centrality and high level of closeness centrality, but it still could not place them as central actors.

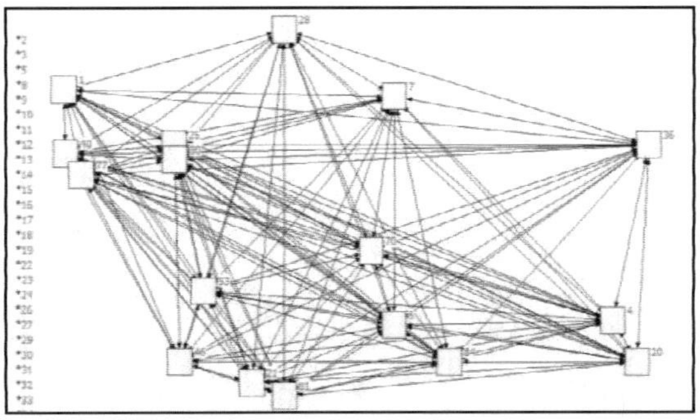
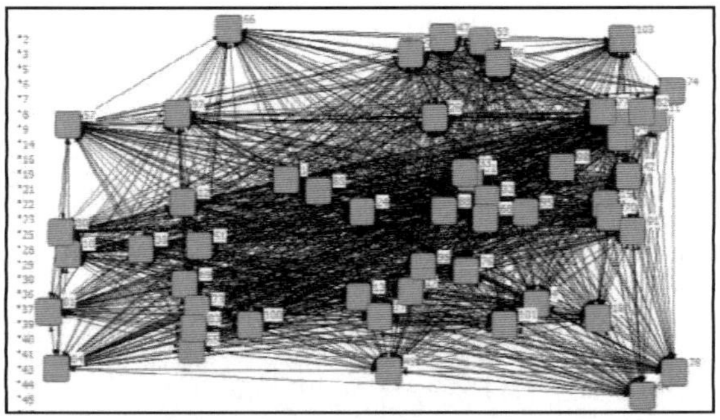

Figure 4. Netdraw Centrality at the Operational & Maintenance Step - (left) Ardirejo village & (right) Talangagung village

4. Conclusion

Research found that community in Talangagung Village had done their community participation on public sanitation program in the higher level of sustainability than the community in Ardirejo Village [5]. Social Network Analysis with three indices – rate of participation, density and centrality illustrated that community in Talangagung Village had higher social tie through their informal network towards the whole steps of the program than the community in Ardirejo Village. Hence, we might conclude that the strong social tie of the community in Talangagung Village is one of the important factors that assures the success of the program.

As a recommendation, in order to guarantee the success and sustainability of the SPBM program, as well as other participation development programs that required community involvement, this research might suggests three important notes as follow. First, inclusiveness involvement of the community at the whole steps of a kind of empowerment development program is necessary. The role of prominent actors are also important in the sense that the prominent actors might play as a strategic node to foster the flow of information and resources from the government to community – the vertical network as well as among communities – the horizontal network. Second, the symmetry of information for all is also a necessity. One big obstacle for a kind of empowerment and development program is the availability of one group that is very dominant to reach access to the information as well as resources. As a result, it will bring burden for the whole process since these kind of development programs requires solid support from the whole aspects in the community. Third, knowing prominent actors as the booster to the more effective progress of the development programs are necessary too. Government might utilizes them as important channels in the network, in order to foster social structures in the network into stronger community movements.

References

[1] Sofyan I, Soewondo P, Kunaefi T D, and Handajani M 2014 *Faktor-faktor penting pengelolaan sanitasi oleh masyarakat di kawasan kumuh perkotaan Bandung Raya.* (Bandung: Institut Teknologi Bandung)

[2] Baker J L 2013 Indonesia: Evaluasi program pembangunan berbasis masyarakat (CDD) perkotaan. (Jakarta: PNPM Support Facility)

[3] Ari, I R D, Ogi K, Matsushima K, and Kobayashi, K. 2013 Community participation on water management, Case Study: Singosari District, Malang Regency, Indonesia. Procedia Environmental Sciences **17** 805-813

[4] Ari, I R D, Jeong H, Matsushima K, and Kobayashi K 2014 Network analysis of community based water management. community based water management and social capital. ed(s) Kobayashi K, Syabri I, et al. (London: IWA Publishing) pp 197-216

[5] Wasserman, S and Faust K 1994 *Social network analysis: methods and applications* (New York: Cambridge University)

[6] UNICEF 2017 WASH: Water, Sanitation and Hygiene [Online]. Available from: www.unicef.org

[7] Coleman J S 1988 Social capital in the creation of human capital American J. Sociology **94** S95-S120

[8] Ostrom E and Ahn T K. 2003 A social science perspective on social capital: social capital and collective action *Revista Mexicana De Sociologia* **65** 155-233

[9] Direktur Jenderal Pekerjaan Umum 2011 Pedoman pelaksanaan sanitasi perkotaan berbasis masyarakat (Jakarta: Ministry of Public Works)

IOP Conf. Series: Earth and Environmental Science **158** (2018) 012014 doi:10.1088/1755-1315/158/1/012014

Spatial Equity in Trans Jogja Performance in the Yogyakarta Urbanized Area (YUA)

D P Ramadhani, Y Herwangi

Urban and Regional Planning Program, Department of Architecture and Planning, Universitas Gadjah Mada, Yogyakarta, Indonesia

Corresponding Author: dinda.rmdhn16@gmail.com

Abstract. The availability of facilities and supporting infrastructure in a region is an absolutely necessity. Public transportation is important in accommodating the movement of low-income groups or captive users who have limited mobility options. However, most of the low-income people in the Yogyakarta Urbanized Area (YUA) currently prefer to use motorcycles rather than Trans Jogja. This is expected to be caused by the poor quality of public transport and the lack of equity in Trans Jogja services. This research focuses on transport equity by overlaying the effectiveness of Trans Jogja as the public transportation and the distribution of low-income communities per sub-district in the YUA. This study found that the performance of Trans Jogja based on the indicators of affordability, availability, accessibility, and acceptability is already effective. However, in the agglomeration of Sleman and Bantul Regency, the affordability aspect is still less effective. Meanwhile, the appraisal of the public transport equity found that there are some areas that are fair, with a large low-income population and effective performance of Trans Jogja. Some other areas are considered unfair because the low-income population is high but the performance of Trans Jogja is less effective.

1. Introduction

The Yogyakarta Urbanized Area (YUA) emerged due to the expansion of the city of Yogyakarta towards the surrounding periphery. The area is currently quite crowded and growing. Moreover, the region is designated as a National Activities Center in the Spatial Plan of the Special Region of Yogyakarta. Therefore, the presence of facilities and supporting infrastructure is an absolute necessity, including the availability of public transportation services. Currently, to accommodate the movement in the YUA, Trans Jogja is already available, which applies the Bus Rapid Transit (BRT) concept.

The public transport in YUA is also needed to accommodate the movement of low-income groups. According to [1] , most of the captive users in public transport are low-income groups. Captive users are a class of people who are forced to use public transportation because of the limitations of private vehicles. This gives low-income groups a low mobility because of their limited ability. Therefore, one of the main functions of public transportation is to narrow the gap between the mobility of high-income and low-income communities.

However, in many big cities, motorcycles are currently the main choice for the most of the middle-low income groups to meet their needs for movement [2]. Motorcycles are considered to have the advantages of a low price and high flexibility. However, the use of motorcycles as the main mode of transportation in low-income communities puts a financial burden on them [2]. This condition is caused because the low-income groups still have to bear the cost of private vehicles ownership. In other words, low-income people are forced to use motorcycles and bear the costs due to the limited choice of modes for their movements.

The current dependence of low-income people on motorcycles indicates that the accessibility of public transport is lower than the access to motorcycle ownership. This condition is suspected to be an implication of the low effectiveness of Trans Jogja, which is a form of urban public transportation in the YUA. The low effectiveness of Trans Jogja is also coupled with the allegation that, currently, there's inequity in Trans Jogja services in the YUA, especially for low-income people. Therefore, this study will discuss the effectiveness of Trans Jogja's performance and assess whether the existing Trans Jogja services are already implemented fairly.

2. Transport Equity

Equity in transportation refers to the distribution and impacts of the transportation facilities and infrastructure in a region. To analyze transport equity, the indicators that are commonly used are accessibility and mobility. According to [3], transport equity is divided into three categories: horizontal equity, vertical equity with regard to income and social class, and vertical equity with regard to mobility needs and ability.

Horizontal equity considers that every group in the society in a region has equal ability and needs in terms of transportation. According to this definition, equal individuals and groups should be treated the same in the distribution of resources, benefits, and cost burdens. Vertical equity, or so-called social justice, is the distribution of the impact between different individuals and groups in terms of abilities and needs of transportation. Individuals or groups who have the same abilities and needs will get the same impact. Meanwhile, individuals or groups with different abilities and needs will feel different effects. Vertical equity itself can be divided into two groups, namely vertical equity with regard to income and social class and vertical equity with regard to mobility needs and abilities.

According to [4], the analysis of transport equity uses four variables namely affordability, availability, accessibility, and acceptability. Affordability refers to people's financial ability to reach public transport services. It also refers to the efforts made by the people to obtain the public transport services. Availability relates to the existence of routes, the frequency of public transport services and all things related to time, e.g., waiting time or headway. Accessibility describes the ease of a community in a region to obtain public transport services either in terms of distance or availability of information about the public transport services. Acceptability relates to user convenience and comes from the perception of the public transport users.

The concept of fairness in transportation is a concern both in developed and in developing countries. Inclusive transportation planning was introduced in transport planning in Europe and America [2]. Transport planning no longer focuses only on economic and environmental aspects but also on social aspects. It aims to anticipate the emergence of marginalized groups of transport services especially for women, children, the elderly, the disabled, and low-income communities.

3. Methods

3.1 *Study Area and Design of the Research*

The study area in this research is the Yogyakarta Urbanized Area (YUA). YUA is a functional area in the Special Region of Yogyakarta, which comprises the City of Yogyakarta and some sub-districts in the regencies of Bantul and Sleman. YUA emerged from the expansion of Yogyakarta City to the surrounding periphery. As a result of the development process, the suburbs of Yogyakarta city underwent a spatial transformation in the form of the built-up area around the city of Yogyakarta where the city of Yogyakarta, as the core, still affects the activities in the surrounding area. In the regional spatial structure plan, as stated in DIY Regional Regulation Number 2 of 2010, YUA will be developed into a National Activity Center. YUA's population was 1,439,898 in 2015 with an area of 324.91 km^2.

In this research, YUA is divided into seven zones based on the similarity of land use in each zone. The division is used to assess the effectiveness of Trans Jogja's performance.

Table 1. The division of the study area.

Zone	Location	Main activities
Central Area Zone (BWK 1)	Kraton and Malioboro Area	Commercial, offices, and social services
West Area Zone (BWK 2)	Tegalrejo Sub-district and the surrounding area	Residential, commercial, and offices
East Area Zone (BWK 3)	Gondokusuman Sub-district and the surrounding area	Residential, commercial, offices, and social services
Southeast Area Zone (BWK 4)	Umbulharjo Sub-district and the surrounding area	Residential, commercial, industry, and social services
Southwest Area Zone (BWK 5)	Mantirejon Sub-district and the surrounding area	Commercial and social services
Sleman Regency Zone	Depok, Gamping, Mlati, Ngaglik, and Ngemplak sub-districts	Educational, residence, and commercial
Bantul Regency Zone	Sewon, Banguntapan, and Kasihan Sub-districts	Educational and residence

In this research consists of the following steps to analyze spatial equity in Trans Jogja performance:

- Conducting data analysis from the field survey and questionnaire to assess the effectiveness of Trans Jogja performance. The assessment uses the Likert scale to score the Trans Jogja performance.
- Overlaying the results from the appraisal of Trans Jogja performance effectiveness with the distribution of low-income groups in YUA.
- Making interpretations and conclusions on from the analysis and appraisal of the spatial equity of Trans Jogja performance in the form of narration and maps.

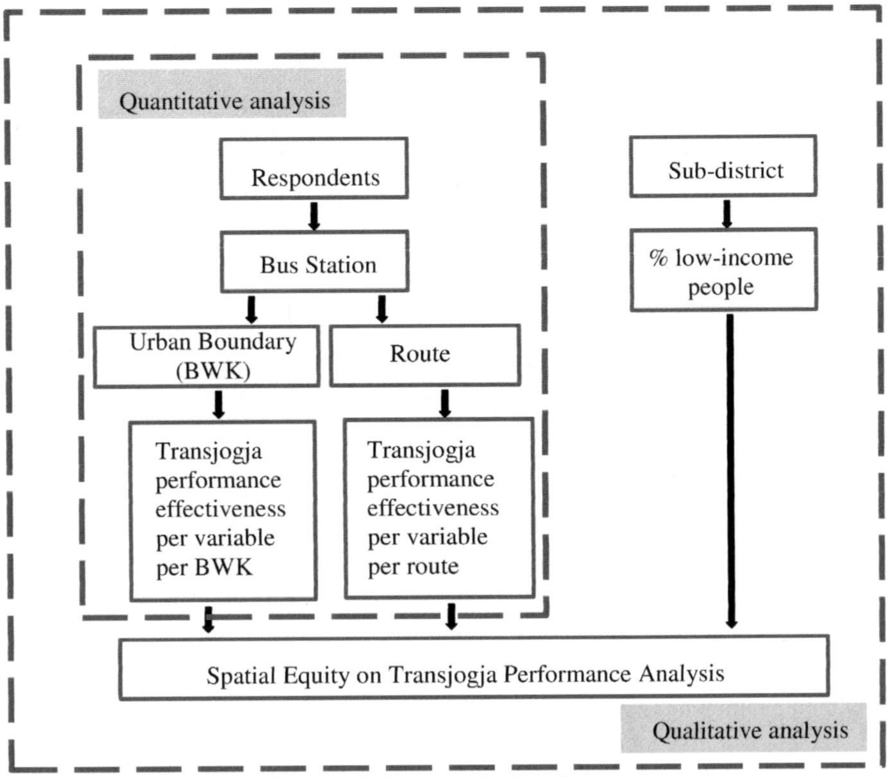

Figure 1. Data analysis scheme.

3.2 Trans Jogja performance effectiveness

The variables to analyze the effectiveness of Trans Jogja's performance are based on the literature review of theories concerning the criteria of public transport performance. It is based on several expert theories and standards of institutions in several countries, i.e., World Bank standards, MoUD India guidelines, the Decision of the Directorate General of Land Transportation Number SK.687/AJ.206/DRJD/2002, Government Regulation Number 10/2012 concerning Minimum Service Standards for Road-Based Mass Transit [5][6][8][9]. The selected variables and indicators are then used for collecting data through observations and the questionnaires.

The observations are conducted at bus stations to assess the Trans Jogja performance in the form of the areas covered, the frequency of service, headway, the number of vehicles in operation, passenger load factor, and bus station load factor. Meanwhile, questionnaires are used to gather data in the form of waiting time, fare, and distance to or from the nearest stop.

The population in this study includes Trans Jogja and also Trans Jogja users. To determine the sample at Trans Jogja stops and buses, the sampling method used is purposive sampling. Purposive sampling is used to determine which stops are used as samples by considering the distance between the locations of the stops to each other. The selected stops are not located close to each other, have diverse land uses, and are traversed by different routes. There are 112 Trans Jogja stations in the study area is. By using the reliability level of 0.1 then the sample size for Trans Jogja stops is 54 stations. Thus, in each zone, the number of stop samples was 7 to 8.

Meanwhile, to determine the sample of Trans Jogja users, the convenience sampling method was used. Three respondents are taken at each sample stop in the previous step. So, the total sample of Trans Jogja users is 162 respondents.

The Likert scale is used to process the data that has been obtained from the observation and questionnaires. With the Likert scale, the variables to be measured are divided into indicators and parameters. Each parameter is given a score to be able to calculate the total score to measure the effectiveness of Trans Jogja performance on each variable. Table 2 shows the indicators for assessing the effectiveness of Trans Jogja performance along with the parameters and scores on each indicator.

Table 2. Variables, indicators, and criteria for Trans Jogja Performance Effectiveness.

Variable	Indicator	Criteria and *scoring*
Affordability	Fare(*)	• < 10% of total income = 3 • 10-25% of total income = 2 • >25% of total income = 1
	Area covered(**)	• >75% area covered = 3 • 50-75% area covered = 2 • <50% area covered = 1
Availability	Waiting time(*)	• <20 minutes = 3 • 20-30 minutes = 2 • >30 minutes = 1
	Frequency of service(**)	• >6 bus/hour = 3 • 4-6 bus/hour = 2 • <4 bus/hour = 1
	Headway(**)	• <15 minutes = 3 • 15-20 minutes = 2 • >20 minutes = 1
	Number of vehicles in operation(**)	• >82% vehicles in operation = 3 • 65-82% vehicles in operation = 2 • <65% vehicles in operation = 1
Accessibility	Walking distance to and from the bus station or terminal(*)	• <400 meters = 3 • 400-600 meters = 2 • >600 meters = 3
Acceptability	Passenger load factor(**)	• < 1,25 = 3 • 1,25 – 1,5 = 2 • >1,5 = 3
	Bus station load factor(**)	• <4 people/m^2 = 3 • 4-6 people/m^2 = 2 • >6 people/m^2 =1

(*) The data that have obtained from questionnaire
(**) The data that have obtained from observation

3.3. Spatial Equity of Trans Jogja Performance

The analysis of the spatial equity of Trans Jogja uses the data of the distribution of low-income people in each sub-district and the effectiveness of Trans Jogja performance per variable for each BWK (urban boundary). The data of the distribution of low-income people is obtained as a secondary data and the data of Trans Jogja performance effectiveness per variable for each BWK is obtained from the calculations in the previous step. The next step is to overlay both data using ArcGIS. The results of this process can be used to determine whether the existing Trans Jogja services have been implemented equally or not.

Table 3. Indicators of spatial equity in Trans Jogja performance in YUA.

		Low-income people		
		High	Moderate	Low
Effectivity Performance Trans Jogja	Effective			
	Moderately effective			
	Less effective			

The columns with the lightest colors illustrate equal conditions for regions with many low-income people and effective Trans Jogja performance. Meanwhile, the dark-colored columns illustrate the unequal condition, where a region has a high amount of low-income people but the Trans Jogja performance is classified as moderately effective and less effective.

4. Result

4.1 Characteristics of the Respondents

From the data of the questionnaires, most of the YUA community uses Trans Jogja to go to work or educational activities. As much as 46 percent of the respondents are students who use Trans Jogja to go to school and college, while the use of Trans Jogja for working is 27 percent. Beside these two activities, Trans Jogja also used by the community to go to social activities, health facilities, and as a connector mode of transportation to other cities or districts. By looking at the high number of trips for educational and work activities, the fluctuations of users of both types of activity is stable because both are routine and scheduled activities.

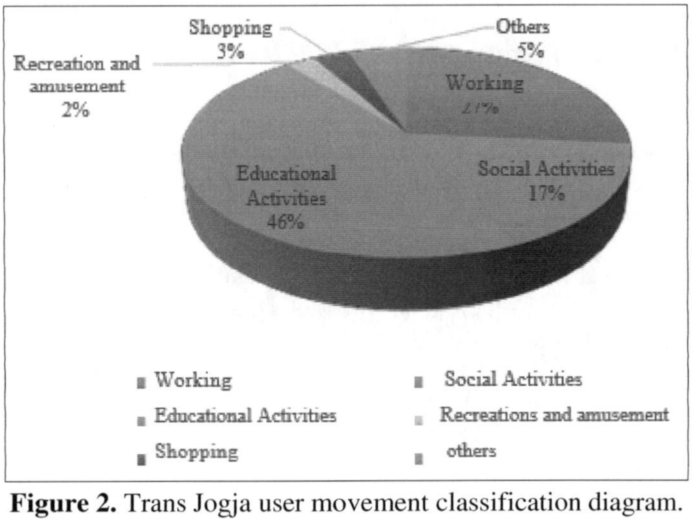

Figure 2. Trans Jogja user movement classification diagram.

Moreover, 65 percent of respondents use Trans Jogja every day. Most of them are students and people who use Trans Jogja to work. Meanwhile, 22 percent of the respondents use Trans Jogja 1-4 times per week. Most of the users who use Trans Jogja with a frequency of 1-4 times per week has the intention of traveling in the form of shopping, tutoring, or social activities such as visiting a friend's house. The rest are respondents who use Trans Jogja with less than once per week or not necessarily.

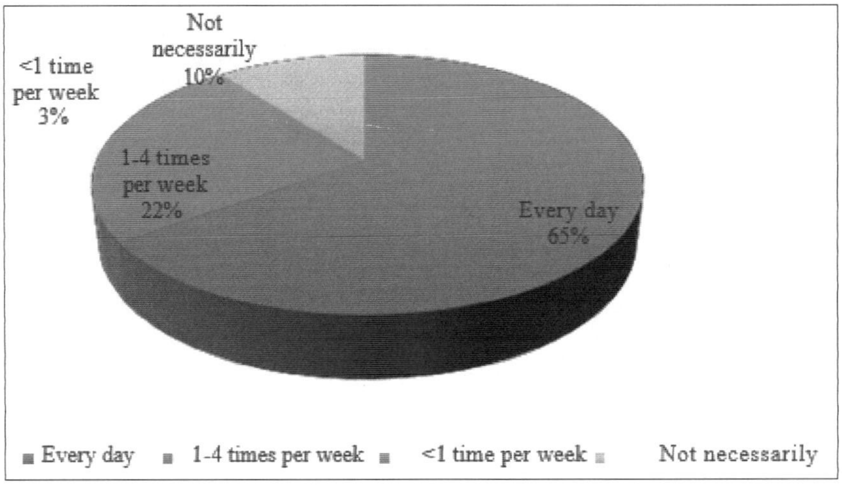

Figure 3. Percentage of travel frequency chart using Trans Jogja.

4.2 Trans Jogja performance effectiveness analysis

Trans Jogja's performance effectiveness is analyzed from the four criteria of equity in public transportation services: affordability, availability, accessibility, and acceptability. The four criteria will be discussed for each zone (BWK) and each existing Trans Jogja route. The Likert scale is used to determine the effectiveness of these criteria.

4.2.1 Trans Jogja performance effectiveness per variable per BWK

4.2.1.1 Affordability

The affordability of Trans Jogja is assessed from the fare adjustment indicator and the area covered by Trans Jogja routes. Fare adjustment is used as one indicator seen from the aspect of economic ability in reaching Trans Jogja. Meanwhile, the area covered is an indicator of the affordability of Trans Jogja in terms of radius coverage.

In terms of fare adjustment, it was found that in all BWK most of the respondents said that the cost for Trans Jogja is less than 10% of their total income. This makes Trans Jogja service effective in terms of fare suitability at all BWKs in YUA.

IOP Conf. Series: Earth and Environmental Science **158** (2018) 012014 doi:10.1088/1755-1315/158/1/012014

Table 4. Fare Effectivity Appraisal.

	<10% of total income (scoring)	10-25% of total income (scoring)	>25% of total income (scoring)	Maximum score	Total score (Effectivity level)
BWK 1	63	6	0	72	69 (Effective)
BWK 2	51	8	0	63	59 (Effective)
BWK 3	60	2	0	63	62 (Effective)
BWK 4	93	14	0	114	107 (Effective)
BWK 5	24	2	0	27	26 (Effective)
Sleman Regency Zone	96	8	3	117	107 (Effective)
Bantul Regency Zone	24	2	0	27	26 (Effective)

To assess Trans Jogja in terms of service coverage, a buffer method is used with GIS. The first step is digitizing all bus stations located in the YUA. After that, each point was given a 600 meters buffer. This analysis shows a difference in the area covered in each BWK. Areas that are effective in terms of the area covered by Trans Jogja's are covered by more than 75 percent. The moderately effective and less effective areas are covered by 50 – 75 percent and less than 75 percent. BWK 1 has effective area coverage. Meanwhile, BWK 2, BWK 3, and BWK 4 are classified as moderately effective. The worst condition occurred in BWK 5 and the agglomerations of Sleman Regency and Bantul Regency. The total area covered by Trans Jogja in these three BWKs is only 45.45 percent, 20.4 percent, and 4.2 percent respectively, which is less effective coverage. Figure 4 shows the areas covered by Trans Jogja in each BWK.

IOP Conf. Series: Earth and Environmental Science **158** (2018) 012014 doi:10.1088/1755-1315/158/1/012014

Figure 4. The area covered by Trans Jogja in each BWK

4.2.1.2 Availability

The variables used to assess availability are waiting times, headway, the frequency of service, and the number of vehicles operating. Waiting time is the time spent by Trans Jogja users from the moment of arrival at the bus stop until entering the Trans Jogja bus. This indicator of waiting time is one of the things that greatly affect people's interest to use Trans Jogja services. The results of the questionnaires showed that the average waiting time for all Trans Jogja routes ranged between 10-20 minutes at non-busy times and 15-30 minutes at busy times. Effective waiting times occurred at BWK 1, BWK 2, BWK 5, and the agglomeration of Bantul Regency. This can be caused by several factors such as the number of operating vehicles, travel speed, and traffic conditions. Meanwhile, a moderately effective waiting time occurred at BWK 3, BWK 4, and the agglomeration of Sleman Regency. Thus, the waiting time of Trans Jogja in YUA is moderately effective.

Table 5. Waiting time Effectivity Appraisal.

	<20 minutes (scoring)	20-30 minutes (scoring)	>30 minutes (scoring)	Max. score	Total score (effectivity level)
BWK 1	45	18	0	72	63 (Effective)
BWK 2	27	16	4	63	47 (Moderately Effective)
BWK 3	24	18	4	63	46 (Moderately Effective)
BWK 4	39	34	9	117	82 (Moderately Effective)
BWK 5	27	0	0	27	27 (Effective)
Sleman Regency Zone	75	16	6	117	97 (Moderately Effective)
Bantul Regency Zone	27	0	0	27	27 (Effective)

Other indicators used to assess availability are headway and service frequency. Headway is the time difference between the arrival of one bus and the next bus, while frequency is the number of Trans Jogja buses per hour. From the calculation showed that in all BWKs, Trans Jogja bus service headway is the moderately effective category. This shows that the average of Trans Jogja bus headways in all YUA areas ranges from 15-20 minutes. Meanwhile, the frequency of Trans Jogja services to all BWK is also classified as moderately effective. However, different conditions occur in BWK 3 where at 3 BWK Trans Jogja bus service frequency classified as less effective.

The last indicator to assess availability is the number of vehicles operating. The number of vehicles operating on each route can affect the waiting times, headways, and frequency of Trans Jogja busses. A higher number of vehicles operating can shorten the waiting time and headway. Thus, the bus service frequency can also be increased. The effectiveness of the number of vehicles that operate is calculated by comparing the number of vehicles that operate with the number of vehicles that should operate on each route according to the planning document. The planning document used is the Regulation of the Governor of Yogyakarta Special Region No. 22 of 2014 concerning Trans Jogja Urban Route Network. At each BWK the number of vehicles passing on each route is almost 100 percent of the planned number. However, on some routes, the number of vehicles operating is under the standard at 82 percent. The 3B route has the lowest percentage of vehicles operating. As this condition occurs thoroughly all BWKs, the overall classification for the effectiveness of the number of vehicles operating at each BWK is classified as moderately effective.

4.2.1.3 Accessibility

Accessibility describes the ease of the society in a region to obtain public transport services. Ease of obtaining public transportation services is assessed from the time the user goes to the station location from where they are doing the activity. Accessibility is also assessed from how far the location of public transport services is from the places of community activity and how to get to the transit location. The accessibility of Trans Jogja was analyzed by looking at the distance indicator to or the nearest Trans Jogja stop.

Table 6. Accessibility Effectivity Appraisal

	<400 meters (scoring)	400-600 meters (scoring)	>600 meters (scoring)	Max. score	Total scoring (effectivity level)
BWK 1	48	16	0	72	64 (Moderately effective)
BWK 2	42	8	3	63	53 (Moderately effective)
BWK 3	48	8	1	63	57 (Effective)
BWK 4	75	18	4	114	97 (Moderately effective)
BWK 5	12	4	3	27	19 (Moderately effective)
Sleman Regency Zone	6	14	12	117	86 (Moderately effective)
Bantul Regency Zone	12	4	3	27	19 (Moderately effective)

The analysis shows that in almost all BWKs the classification of Trans Jogja accessibility is moderately effective. The ideal distance that is used as a parameter for the measurement of accessibility to Trans Jogja is a maximum of 600 meters, which is an ideal distance to walk. At BWK 1 and BWK 3, the accessibility is effective. This indicates that the Trans Jogja stations at BWK 1 and BWK 3 are located close to where the community is moving. Meanwhile, the accessibility at other BWKs is moderately effective.

4.2.1.4 Acceptability

The indicators for measuring acceptability are the bus load factor and the stop load factor. The bus load factor is the ratio between the number of passengers on the bus and the capacity of the bus. Each Trans Jogja bus has a capacity of 40 passengers, consisting of 20 seated and 20 standing passengers. The calculation of bus load factor is done at three different times, i.e., morning, noon, and afternoon due to the difference in load factor at different times.

The bus load factor is below 100 percent on all Trans Jogja routes. This indicates that the load factor level of Trans Jogja busses on each route is still quite good. It also shows that the convenience of Trans Jogja buses is good because there is no occurrence of passenger accumulation. The 1 A route has the largest load factor of 40-60 percent on average. Meanwhile, the highest load factor is during the afternoon rush hour when the buses reach more than 50 on average. This is due to an increase in the number of users in the afternoon and coincides with the time of returning home. BWKs 1, 2, and 3, and the agglomeration of Sleman Regency have the largest load factor. There are many types of activities in those zones, especially activities that have a pull factor to travel in the morning and the push factor for travel in the afternoon. Therefore, Trans Jogja bus users in that BWK are quite many, resulting in a high load factor.

Meanwhile, the load factor of the bus stop is the stop fullness which is obtained by comparing the number of users in the bus stop with the size of the bus stop. The above load factor measurements are taken during peak hours which represents the fullest condition. The stop fullness is measured at this time because if the load factor value in the peak hour is good, then the load factor at non-peak hours will also be good.

The calculations showed that the average load factor of stations in the YUA is 1-1.5 people/m^2. This is still below the station load factor comfort standard of 4 people/m^2. The low level of this load factor indicates that the stations in the YUA can accommodate the users of Trans Jogja quite comfortably. The BWKs with the highest station load factor are BWK 3, BWK 4, and the agglomeration of Sleman Regency. This is because there are several stops at that BWK that are used as transit locations such as hospital stops Bethesda, Condong Catur, and Yos Sudarso. With the conditions as described above, it can be said that the load factor stop effectiveness is classified as effective.

The following table shows Trans Jogja performance effectiveness per BWK.

Table 7. Compilation of Trans Jogja Performance Effectiveness per BWK.

Zone	Affordability		Availability			Accessibility	Acceptability	
	Fare	Area covered	Waiting time	Headway	Frequency	Walking distance	Load factor bus	Load factor station
BWK 1	●●●	●●●	●●●	●●	●●	●●●	●●●	●●●
BWK 2	●●●	●●	●●●	●●	●●	●●	●●●	●●●
BWK 3	●●●	●●	●●	●●	●	●●	●●●	●●●
BWK 4	●●●	●●	●●	●●	●●	●●	●●●	●●●
BWK 5	●●●	●	●●●	●●	●●	●●	●●●	●●●
Sleman Regency Zone	●●●	●	●●	●●	●●	●●	●●●	●●●
Bantul Regency Zone	●●●	●	●●●	●●	●●	●●	●●●	●●●

●●● = Effective
●● = Moderate Effective
● = Less Effective

4.2.2 Trans Jogja performance effectiveness per variable per route

Table 8. Trans Jogja Performance per Route Appraisal.

Route	Waiting time	Headway	Frequency	*Load factor* bus
1A	Effective	Effective	Effective	Effective
1B	Moderately	Effective	Effective	Effective
2A	Effective	Effective	Effective	Effective
2B	Moderately effective	Moderately effective	Moderately effective	Effective
3A	Effective	Effective	Effective	Effective
3B	Moderately effective	Less effective	Less effective	Effective
4A	Effective	Less effective	Less effective	Effective
4B	Moderately effective	Less effective	Less effective	Effective

The table above is the result of Trans Jogja performance calculation per route using the Likert method, which was also used to calculate the effectiveness of Trans Jogja performance per BWK. The indicators are waiting time, headway, frequency, and bus load factor. The calculation of the waiting time shows that the waiting time on almost all of Trans Jogja routes is effective. Routes 1B and 2B are the two routes that are still classified as moderately effective. This indicates that the gap between one bus and the other bus on both routes ranges from 20-30 minutes. The longer waiting time for the 1B and 2B routes is thought to be due to the length of both routes and the number of buses operating so that one travel is relatively long compared to other routes. The is worse during the peak hour of the afternoon when the waiting time for each route can increase by about 5-10 minutes due to the more crowded traffic conditions at that hour when people return home from work and school.

Meanwhile, the calculation of headway and frequency of bus service Trans Jogja shows that routes 1A, 1B, 2A, and 2B are is effective in terms of headway and frequency. Meanwhile, the less effective routes are 3B, 4A, and 4B. The headway is considered effective if it is less than 15 minutes and bad if it is more than 20 minutes. Meanwhile, the frequency of service is good if there are more than six buses per hour, and less effective the number of buses is less than three per hour. The length of the headway from one bus to another and the low frequency on the 3B route due to the length of the route makes travel time becomes longer. Meanwhile, on the 4A and 4B routes, the duration of headway and low frequency is due to the limited number of buses on both routes, i.e., only four buses per route. The condition of headway and frequency of Trans Jogja buses affects the waiting time that passengers spend.

Table 9. Load factor Effectivity Appraisal.

No	Route	Load Factor			Average
		Morning	Afternoon	Evening	
1	1A	27.3	45.9	65.0	46.0
2	1B	17.6	32.1	34.1	27.9
3	2A	24.0	38.8	53.3	38.7
4	2B	21.0	32.6	41.9	31.8
5	3A	24.5	42.6	59.2	42.1
6	3B	34.5	32.1	34.8	33.8
7	4A	15.6	20.3	26.4	20.8
8	4B	7.1	21.8	19.2	16.0

The last indicator used is the bus load factor, which is relatively low on each route. This confirms that currently, Trans Jogja is not the main choice of transportation in YUA. The lowest load factor average is on route 4B, which is suspected because it is influenced by the long headway between each bus that causes a long waiting time. In addition, on the routes with a low load factor value, the route is likely not long. It makes the route offered less varied and less able to reach a wider area. Meanwhile, the high load factor of the bus is on the 1A and 3A routes. This is because both routes have a wide range.

4.3 Spatial Equity in Trans Jogja Performance

Spatial equity means the distribution of impacts in accordance with socioeconomic conditions of the society in a region. In other words, equity is the vertical justice where there are certain groups that should gain more benefits than other community groups due to their limited economic ability. The spatial equity in Trans Jogja service in YUA is analyzed by overlaying Trans Jogja performance effectiveness with the distribution of low-income groups in YUA. Figure 5 shows the distribution of low-income people in each BWK in YUA.

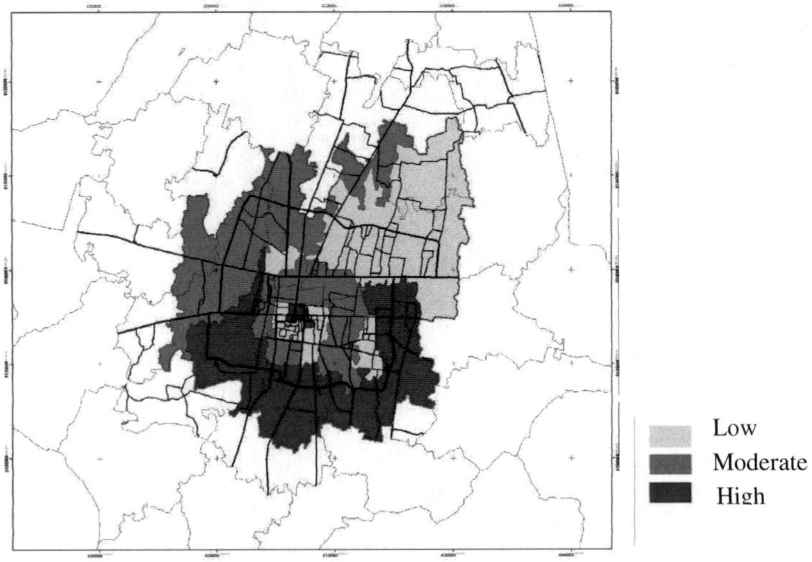

Figure 5. The distribution of low-income people in YUA

The map above shows that the concentration of the low-income people in YUA mostly occurs in the agglomeration of Bantul Regency, namely in Sewon District, Kasihan Sub-district, and Banguntapan District. Meanwhile, the lowest percentage of low-income people is in BWK 1, BWK 2, BWK 4, and the agglomeration of Sleman Regency. The other areas have a moderate classification. To conduct further analysis on spatial equity in Trans Jogja service in YUA, the data of low-income people is overlaid with Trans Jogja performance effectiveness. The aspect of spatial equity is assessed from each indicator on Trans Jogja's performance itself. Each indicator is discussed below.

IOP Conf. Series: Earth and Environmental Science **158** (2018) 012014 doi:10.1088/1755-1315/158/1/012014

4.3.1 Affordability

Fare adjustment is effective

Figure 6. Trans Jogja fair adjustment effectiveness in YUA.

High amount of low-income people and fare is

Moderate amount of low-income people and fare is effective; Low low income people and fare is effective

Figure 7. Spatial equity in term of fair adjustment in YUA.

From the result of the overlay between Trans Jogja effectiveness in terms of fare adjustment with the distribution of the low-income people, it is found that there is an inter-regional equity in terms of the fare adjustment. This is because in the areas with a high low-income population, the cost incurred to obtain Trans Jogja services is still moderately effective. In other words, the overwhelming majority of the public spends less than 10 percent of their total income on Trans Jogja. This can be seen from the areas that have a lighter color on the map above. In addition, this condition also implies that the current fare for obtaining Trans Jogja services are still relatively light and does not burden the public. Low-income people who are categorized as captive users in transportation can still reach Trans Jogja in terms of costs.

On the other hand, the overlay between the effectiveness of the area covered by Trans Jogja services with the distribution of low-income people is equal yet. This can be seen on the map below where the only area that is equal in terms of the area covered by Trans Jogja services (indicated by light colors) is the Gondomanan Sub-district in BWK 1. In addition, the darker color shows the areas that have a large number of low-income people but the areas covered by Trans Jogja services are less effective. This will increase inequality, especially when it is compared with the condition at the agglomeration of Bantul Regency where the number of low-income people is high and the area covered by Trans Jogja services is relatively small. Thus, in the agglomeration of Bantul Regency, the coverage of Trans Jogja must be further enhanced considering the large number of low-income people who are very dependent on the existence of such public transportation.

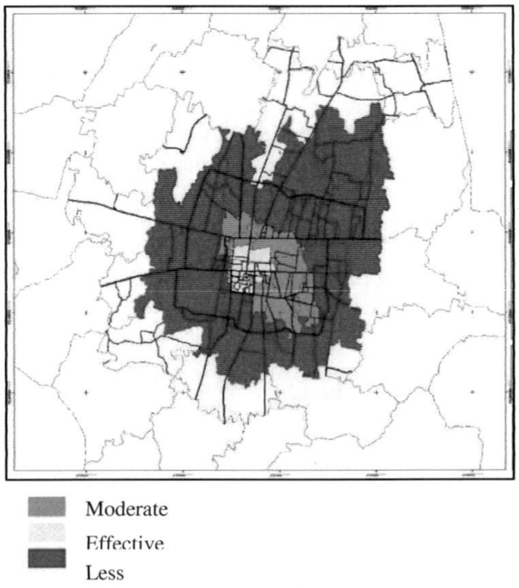

Figure 8. The area covered of Trans Jogja effectiveness in YUA.

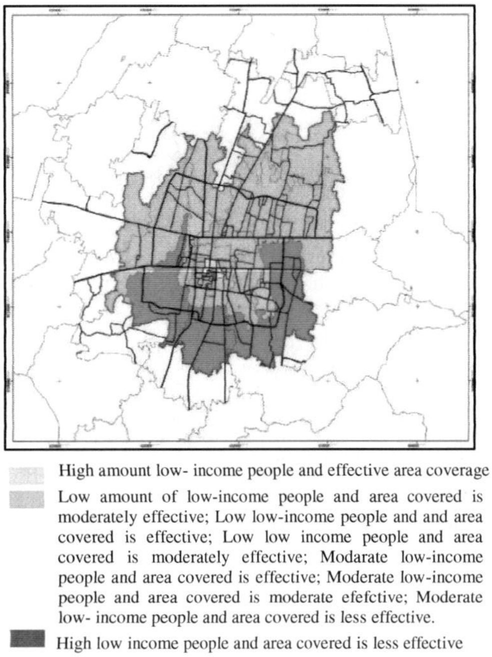

Figure 9. Spatial equity in term of the area covered in YUA.

4.3.2 Availability

The first indicator to measure the availability of Trans Jogja performance is the waiting time. The overlay of the effectiveness of the Trans Jogja bus waiting time and the distribution of the low-income people shows that inter-regional equity has been achieved in terms of the effectiveness of Trans Jogja service waiting time. This can be seen on the map which shows that in areas where the number of low-income people is high the waiting time for Trans Jogja bus service is fast, i.e., less than 20 minutes. These conditions occur in the Gondomanan District in BWK 1 and in the agglomeration of Bantul Regency. Overall, the YUA has the same condition in terms of waiting time equity. There is no inter-regional inequality in terms of waiting time; each region gets a waiting time service in accordance with the conditions of low-income people in the region. However, the condition still needs to be improved considering the waiting time for Trans Jogja buses in almost all YUA areas is still classified as moderately effective.

Figure 10. Waiting time of Transjogja effectiveness in YUA.

Figure 11. Spatial equity in term of waiting time in YUA.

The frequency of Trans Jogja buses is classified as moderately effective. After overlaying it with the low-income people, it can be seen that there is a gap between regions. This can be seen from the map where the agglomeration of Bantul Regency has a dark color which means that in that region the number of low-income people is high but the frequency of Trans Jogja buses is less effective. This negatively affects people's mobility in that region. Meanwhile, in other BWKs, the conditions are good enough but not yet good because the frequency of Trans Jogja buses is not effective yet.

The same condition happened for the headway of Trans Jogja buses. The map shows inequity between regions in terms of effectiveness of Trans Jogja bus headway. In the agglomeration of Bantul Regency, the number of low-income people is high but is not be supported by short Trans Jogja bus headway. The headway times should be improved in the areas where have the number of low-income people is high enough so that people do not have to wait long for Trans Jogja buses. Meanwhile, in other BWKs the conditions are better than in the agglomeration Bantul Regency. This is because in other BWKs the number of low-income people is moderate or low and the effectiveness of headway is also quite effective.

Moderate
Less

Figure 12. Frequency of Transjogja effectiveness in YUA.

Low low income people and requency is moderately effective; Modarate low income people and requency moderately effective; Moderate low income people and

High low income people and requency is moderately

Figure 13. Spatial equity in terms of frequency of Transjogja buses in YUA.

Moderate

Figure 14. Haedway of Transjogja effectiveness in YUA.

Low low income people and headway is moderately effective; Modarate low income people and headway moderately

High low income people and headway is moderately effective;

Figure 15. Spatial equity in terms of headway of Transjogja buses in YUA.

4.3.3 Accessibility

In terms of accessibility, the maps below show that there is inequity in terms of accessibility from and to the nearest Trans Jogja stops. The areas that are fair in terms of accessibility so far only reside in Gondomanan District in BWK 1. Although the low-income population is high in the area, this is supported by the proximity of the distance from and to the location of Trans Jogja shelter. Meanwhile, the opposite occurs in the agglomeration of Bantul Regency. In the maps below, the areas with the darkest color indicate that the number of low-income people is high but accessibility is classified as less effective. There should be an effort to make Trans Jogja more accessible especially in the areas where the amount of low-income people is high so that Trans Jogja could support their mobility in those areas.

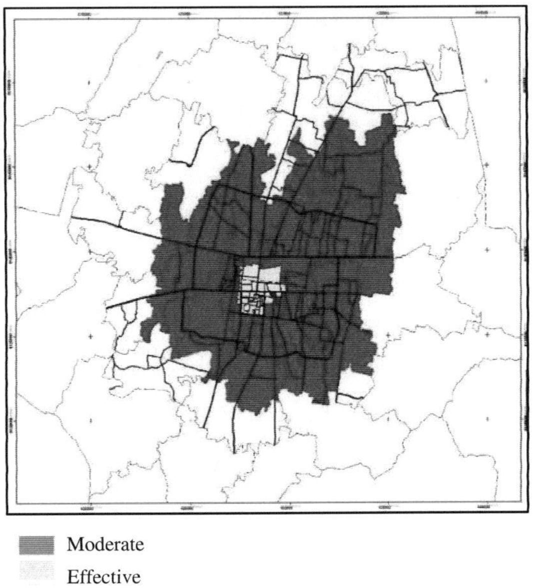

Figure 16. Accessibility of Transjogja services effectiveness in YUA.

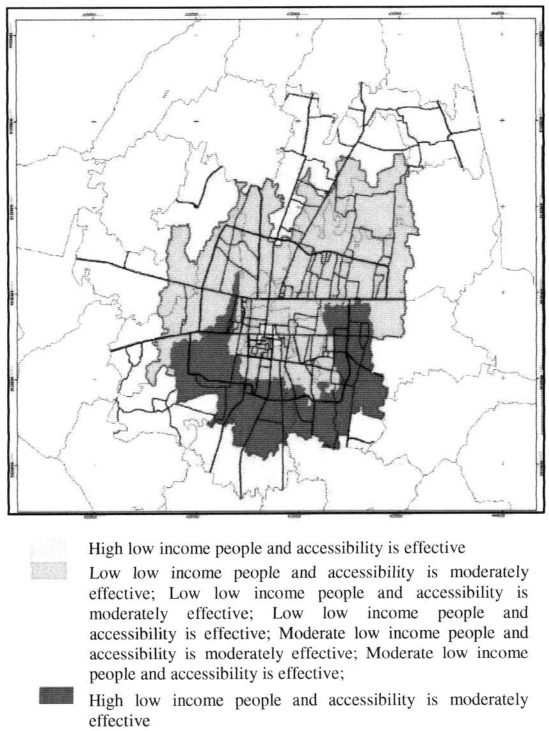

Figure 17. Spatial equity in terms of accessibility.

4.3.4 Acceptability
The first indicator used to assess equity in terms of acceptability is the bus load factor. The map below shows an inequity in all YUA areas in terms of Trans Jogja bus load factor. In the agglomeration of Bantul Regency, the existing condition is fair as the bus load factor is effective. This is because there has not been a buildup of passengers on buses that pass through the area, thus providing opportunities for low-income people to access the services. However, the condition also shows that Trans Jogja is not yet used massively by the people in YUA. This is reflected from the absence of buses on each route that exceeds the limits of bus load factor, especially during peak hours.

Effective

Figure 18. Bus load factor of Trans Jogja effectiveness in YUA.

High low income people and bus load factor is effective;

Low low income people and bus load factor is effective; Modarate low income people and effective bus load factor is effective

Figure 19. Spatial equity in terms of bus load factor.

The same condition occurs in the indicator of the load factor of the bus stop. After overlaying the map of the distribution of low-income people and the effectiveness of the load factor of the stop, shown that there is an inequity in the load factor of the bus stop. In most BWKs conditions are classified as moderately effective, but in BWK 1 the condition is effective. This indicates that at BWK 1, the stops can already accommodate the movement of Trans Jogja users. Meanwhile, just like the value of the bus load factor, the value of the load factor of Trans Jogja bus stop in YUA is still low. This reinforces the indication that Trans Jogja is not currently the primary choice for the community.

5. Discussion
This research shows a gap between the northern and southern regions from the aspect of equity in Trans Jogja. The southern region especially in the agglomeration of Bantul Regency has a relatively high number of low-income people but is not balanced with effective Trans Jogja services. The opposite occurs in Jogja City and the northern areas (the agglomeration of Sleman Regency) where the concentration of low-income people is low and Trans Jogja service is effective and moderately effective. It indicates that the spatial equity of Trans Jogja in YUA area has not been achieved yet.

The assessment of the equity of Trans Jogja shows that on the covered areas, the frequency of bus services, bus headway, and accessibility is not equal. In the southern region (the agglomeration of Bantul Regency) the four variables of Trans Jogja are classified as less effective whereas the concentration of poor people in the region is high. This is a proof that the purpose or function of public transportation to accommodate the movement of captive user groups in society is still not well implemented in the YUA.

In addition to the fact that Trans Jogja's performance has not been effective in the south, urban public transport policy in Yogyakarta Province is focused more on accommodating the movement of society for education and tourism activities. Both activities are currently concentrated in the central and northern areas of Yogyakarta. This causes the southern region to lag behind in terms of availability of urban public transport services considering that there are not many educational and tourism activities because the area is more dominated by settlements.

The inequity of Trans Jogja's existing services has resulted in the high dependence of the people on motor vehicles. This is especially true in low-income communities with limited mobility options. It is also influenced by the difference between the costs incurred when using Trans Jogja with private vehicles, especially motorcycles. According to [10], the difference in costs incurred for work ranges from IDR 141.796 - 203,522 per month, while for educational activities the costs ranged from IDR 33,602 - 178,192 per month. This condition makes Trans Jogja service less competitive than motorcycles. The bigger cost that must be spent is because the Trans Jogja has not reached the area around the settlements yet so that additional costs are required to access Trans Jogja by other public transportation services such as city buses.

The dependence on private vehicles also presents a financial burden for low-income communities. However, due to the limited available options, these low-income communities still feel unencumbered with the financial burden they have to incur. In other words, dependence on private vehicles for low-income groups is a condition that forces them to use private vehicles. However, if the area around where the low-income people live has been traversed by an effective public transportation service, then the dependence on private vehicles will decrease and will indirectly ease the financial burden for low-income people.

The discussion above highlights some aspects that are still less effective. This research offers the following recommendations:

- Add routes and stops at BWK 5, the agglomeration of Sleman Regency, and the agglomeration of Bantul Regency. The Regulation of the Governor of Yogyakarta Special Region Number 22 of 2014 concerning the Trans Jogja Urban Route Network already includes plans for expanding the network to those areas. However, progress is constrained by funding problems in the procurement of buses and other supporting facilities.
- To overcome the condition of the point above temporarily, it is better to use feeder buses especially for the communities in the agglomerations of Sleman Regency and Bantul regency. It is also necessary to optimize the integration with other urban transport.
- Optimize the availability of park and ride facilities and infrastructure and increase the number of shelters with park and ride facilities aimed at improving the user's coverage.

6. Conclusion

From the discussion about spatial equity in Trans Jogja service in YUA, there are some things that can be concluded. Overall, Trans Jogja's performance in the YUA has different effectiveness scores. However, in BWK 5 and the agglomerations of Sleman Regency and Bantul Regency, Trans Jogja is still ineffective in covering all areas. This is due to the lack of stops in these three areas and the lack of routes that pass through BWK 5.

Meanwhile, the overlay of Trans Jogja service performance and the number of low-income people per BWK indicates that overall there is still inequity in Trans Jogja service. In some aspects of equity, the areas that are high in the number of poor people have not yet received an optimal service. Some areas have not even been covered by Trans Jogja. This makes many low-income people unable and not wanting to use Trans Jogja. The agglomeration of Bantul Regency is classified as having inequity because the region has a relatively high number of low-income people but for some indicators, Trans Jogja is still less effective.

References

[1] Welch T F 2013 Equity *n Transport: The distribution of transit access and connectivity* Atlanta Georgia Institute of Technology

[2] Herwangi Y, Pradono P, and Kustiawan I 2015 *Making the connection between transport disadvantage and motorcycle usage of low income people in Yogyakarta Urbanized Area Indonesia.* Journal of the Eastern Asia Society for Transportation Studies **11**

[3] Litman, T 2017 *Evaluating transportation equity: Guidance for incorporating distributional impacts in transport planning* **volume 8 no 2**: 50-65 Victoria Transport Policy Institute

[4] Ciommo F D and Shiftan Y 2017 *Transport equity analysis* Transport review **37**:139-151

[5] World Bank Institute 2002 *Dasar-dasar Analisis Kemiskinan. Edisi Terjemahan* Jakarta Badan Pusat Statistik

[6] Decision of the Directorate General of Land Transportation Number SK.687/AJ.206/DRJD/2002 *Pedoman Teknis Penyelenggaraan Penumpang Umum di Wilayah Perkotaan dalam Taryek Tetap dan Teratur* Jakarta

[7] Ministry of Transportation Republik Indonesia 2002 *Government regulations Number 10/2012 about Minimum Service Standard road-based mass transit* Jakarta Kementrian perhubungan

[8] Hamzah M 2011 *Evaluasi Efektvitas Pelayanan Moda Transportasi Umum Trans Jogja Pada Pusat-Pusat Kegiatan* Universitas Gadjah Mada

[9] Maryeni Y 2016 *Aktivitas Pelayanan Halte Trans Jogja Pada Pusat- Pusat Permukiman Di Kota Yogyakarta* Universitas Gadjah Mada

[10] Herwangi Y, Pradono P, and Kustiawan I 2014 *Peran dan pola penggunaan sepeda motor pada masyarakat berpendapatan rendah di kawasan perkotaan Yogyakarta* Seminar Cities

Importance-Performance Matrix Analysis (IPMA) Of Transport Disadvantage Variables on Social Exclusion in a Rural Context

[1]Ophilia Larasati, [2]Dr.Eng. Puspita Dirgahayani

[1]Transportation Master Program, School of Architecture, Planning and Policy Development, Institut Teknologi Bandung, Indonesia
[2]School of Architecture, Planning and Policy Development, Institut Teknologi Bandung, Indonesia

Corresponding author: pdirgahayani@sappk.itb.ac.id

Abstract Transport services are essential to support daily life. A lack of transport supply leads to the existence of transport disadvantaged (TDA) groups who are vulnerable to social exclusion, which happens when a particular group or individual is having difficulties to access certain activities that are considered normal in society. To tackle this phenomenon, the understanding of the influence of TDA variables on social exclusion is needed. The aim of this study is to analyze the influences of TDA variables on social exclusion in a rural context, with Cibeureum Village (Bandung Barat Regency) and Bunikasih Village (Subang Regency) as the study case. Both case studies provide different characteristics of accessibility. Partial Least Squares (PLS) Structural Equation Modeling (SEM) is chosen as the method to analyze the influences of TDA variables on social exclusion. The PLS-SEM model is developed according to the social exclusion variable and four TDA variables, i.e., accessibility, individual characteristics, private vehicle existence, and travel behavior. IPMA is done after the PLS-SEM model is evaluated. The study reveals that among four of the TDA variables, accessibility has the most influence on social exclusion, hence interventions related to improving accessibility are needed to tackle social exclusion. More specifically, the provision of alternative modes is needed in both study areas, while in Bunikasih Village the cost of travel is also an important variable to consider.

1. Introduction

The essential aim of transportation is to provide access for people to goods and activities needed to support their daily lives [1]. Thus, transportation services need to be accessible to everyone. However, in some locations, the demand for transportation cannot be fulfilled by the available transport services. The lack of transport supply causes transport disadvantaged (TDA) groups and makes these groups vulnerable to social exclusion, which is called transport-related social exclusion (TRSE) [2]. People are considered socially excluded when they cannot participate in the activities that are considered normal in society [3].

To prevent someone from being socially excluded, it is important to learn how TDA aspects relate to social exclusion. Studies about the relationship between transportation and social exclusion are rarely conducted in developing countries [2]. Until now, published research regarding TRSE, especially in Indonesia, has only been conducted in an urban context. In reality, research in a rural context is very important considering the differences in characteristics between urban and rural people, so that social exclusion is more likely to happen to rural people. TDA aspects may influence social exclusion differently in an urban or rural context. Therefore, the study of TDA aspects and their relationship with social exclusion is very important, especially in a rural context. The aim of this study is to identify the relationship between TDA aspects and social exclusion in a rural context. As a result, the effort to reduce the vulnerability to social exclusion of people who lives in rural areas will be effective.

2. TDA and the Concept of Social Exclusion

TDA groups can be defined as having transportation needs that are not fulfilled by the available transportation services [4]. TDA is a multidimensional condition reflected by characteristics such as location, mobility, or individual characteristics [5]. The measurement of TDA based on location usually considers distance or travel time to reach the center of activities (e.g., workplace, healthcare, education, or shopping center). Meanwhile, the measurement of TDA based on mobility considers the level of people's difficulties to access transport services to support their daily lives (e.g., ownership of private vehicles, availability of public transportation). The other measurement is based on individual or group characteristics (e.g. disability conditions, age, family structure). Studies usually include two or three aspects to achieve a more comprehensive understanding of TDA [5]. Furthermore, TDA is formed by factors influencing immobility, such as the lack of access to private vehicles, demographics, and public transportation availability [6].

There are several reasons why TDA relates to social exclusion [7], e.g., (1) the concept of social exclusion can describe the social consequences of the deprivation of access that leads to a lack of exposure to transport facilities which are important to improve people's quality of life. (2) This concept can be used to distinguish individual characteristics and the characteristics of transportation services so that the formulated policy will be effective, and (3) study findings on social exclusions can be a valuable input for transport planners and decision-makers in describing social exclusion as a multidimensional concept, deeply related to transportation and is dynamics.

Social exclusion can be defined as an accumulation of processes containing various aspects that make an individual feel inferior to their surroundings [8]. In other words, social exclusion happens when an individual or group forcefully cannot participate in normal activities in society [6] [9]. Those activities consist of [3] (1) consumption activities, which is the ability to purchase goods and services at the minimum level that is considered normal in society; (2) savings activities, for example savings, pension rights, and properties; (3) production activities, which is an involvement in social and economic activities, such as paid work, education, training, having a family; (4) political activities, which is an involvement in order to improve or protect the social or physical environment; and (5) social activities, which is an involvement in social interactions on the household or society, or community level.

3. Study Areas

In order to analyze the relationship between social exclusion and transportation, in this case, transport disadvantage, it is needed to compare two contrasting study areas in terms of required accessibility (i.e., one study area must be a TDA and the other study area must be a non-TDA). In this study, study areas were chosen mainly based on their locations. The main considerations for choosing the study areas are its easiness to reach basic and transportation facilities.

Based on those considerations, two study areas were chosen. First, Bunikasih Village which is located in Subang Regency, as a TDA, and second, Cibeureum Village, which is located in West Bandung Regency, as a non-TDA. The difference between the two study areas is especially seen from their accessibility to/from the nearest major city. The access to Cibeureum Village mostly in moderate condition (56.13%), several segments are in slightly bad condition (43,87%), and no segments are in poor condition. Whereas the road condition to Bunikasih Village are mostly in poor condition covered with slippery dirt or stones (41.53%), slightly bad condition (32.82%), while only a few segments are in moderate condition (25.65%). The comparison of accessibility characteristics can be seen in Table1.

For both areas, public transportations are no option since they are unreliable. In Bunikasih Village, only informal modes exist, usually used to go to the market. However, the residents rarely use them because the fare is extremely expensive. One trip to the market costs them IDR 75.000, with travel time around 2 hours. Considering their job only earns them IDR 30.000 to IDR 50.000 per day, that fare is way beyond their ability to pay. Meanwhile, in Cibeureum Village, public transportation does exist, but the operational hours are short, only from around 4 AM to 10 AM. Hence, they cannot fulfill the respondents' demand.

Table 1. Accessibility Characteristics of the Study Areas

Characteristics	**Bunikasih Village**	**Cibeureum Village**
Public transportation availability	None	Limited
Road condition	Mostly in poor condition	Mostly in moderate condition, several segments are slightly bad
Basic facility distribution	Located outside of walking distance range	Spread out within walking distance range

Although both study areas have different accessibility characteristics, it turns out that the characteristics of the residents are similar. Most of the movements are internal, meaning that the respondents perform daily activities inside the village. This makes sense considering the majority of respondents are casual workers (63.3%) and most of them are farm workers. For both areas, walking remains the main mode of transportation (71.2%) to perform the daily activities. Due to this habit, unusual perceptions exist, especially in Bunikasih Village. The residents of Bunikasih Village are used to walking long distances to reach basic facilities which are located in neighboring villages (approximately 4 km), so walking for such distance is not a problem in their perception. Despite motorcycle ownership, it turns out that the motorcycles are not used regularly because many destinations are impossible to be reached by motorcycle.

Regarding the basic activities, the residents in Cibeureum Village stated that it is not very difficult to fulfill their basic needs due to the availability of many local shops providing basic needs for a reasonable price. Conversely, the residents of Cibeureum Village stated that it is hard to access the bank in the district capital. A bank is one of the most important places for them because they have to access it regularly (once every three months) to collect grants from the government. The existing public transportation cannot fulfill this demand due to its short operational hours. Specifically in Bunikasih Village, the residents claim to have much difficulty in doing shopping for their basic needs. They rarely go to the central market (only once or twice per year) and rely hardly on the only local small stall owned by one of the residents to buy groceries.

4. Research Methodology

This study is explorative with a quantitative approach. Partial Least Squares Structural Equation Modeling (PLS-SEM) was chosen as a method of analysis. SEM is one of the multivariate statistical methods that combines factor analysis and path analysis to analyze the relationship between several dependent and independent variables simultaneously [10]. PLS-SEM is effective in explorative studies with undeveloped basic theory [11]. In this study, PLS-SEM modeling was done by using SmartPLS software version 3.

The sampling method was based on a literature review on the minimum sample required for PLS-SEM. The rule of thumb regarding the minimum sample required in PLS-SEM is the ten times rule which means that the minimum sample required is ten times the largest number of formative variables in particular of the latent variable [11]. According to the model developed in this study, the minimum sample required is 5 times 10, which equals to 50 samples. The survey that was done in both study areas resulted in 139 data set, 57 dataset were taken in Bunikasih Village and the rest in Cibeureum Village. Due to its small populations (34 families in total), the data in Bunikasih Village is based on population, not sample. Meanwhile, the data from Cibeureum Village is based on a sample using the random sampling method.

The data were acquired in February 2017 by doing a primary survey in the form of questionnaires. The questionnaires were filled in by the head of the family and other family members who are actively mobile, so that the questionnaires were filled in by a minimum of two family members in one house, usually the husband and wife. Table 2 lists the indicators that were used in this study. The questions were formulated based on the indicators used to describe TDA and social exclusion and were answered by a Likert scale ranging from 1 to 4 where 1 means strongly disagree and 4 means strongly agree. Accordingly, all variables included in the model were in ordinal scales.

Table 2. Variables and Indicators Used in PLS-SEM Modeling

Variable	**Code**	**Indicators**	**Statements**
Accessibility	Aks_Jalan	Road condition	The roads to/from where I live are in good condition
	Aks_PT	Public transportation availability	My area is served well by public transportation
	Aks_AltModa	Alternative modes availability	There are options/alternatives of transportation modes to support my daily activities
	Aks_Biaya	Travel cost to basic facilities	The costs to reach basic facilities is low
	Aks_Jarak	Distance to basic facilities	The distance to basic facilities is short
	Aks_Waktu	Travel time to basic facilities	The travel time to basic facilities is short
Travel Behavior	Per_Biaya	Average travel cost for daily movement	I only spend a little amount of money in my daily movements
	Per_Ekst	Frequency of external movement	I rarely do external movements
	Per_Jarak	Average distance for daily movement	I only travel short distances in my daily movements
	Per_Waktu	Average travel time for daily movement	I only spend short travel time in my daily movements
Individual Characteristics	Ind_Wajib	Role in family	I do not have any obligation related to my family that inhibit my daily movement
	Ind_Sulit	Difficulty of movement due to physical condition	I do not have any physical difficulties that trouble me to walk or use public transport
	Ind_Penyakit	Illness	I do not suffer specific illness that hinders me from doing movement
	Ind_Kel	Family structure	I have big families
	Ind_Difab	Disabilities	I do not have disabilities and/or mental illness that inhibit me from doing movement
Motorcycle Availability	Motor_Mampu	Ability to use motorcycle	I have a license to ride a motorcycle
	Motor_Milik	Motorcycle	I own a motorcycle that can be used in my

Variable	Code	Indicators	Statements
		ownership	daily movements
	Motor_Frek	Frequency of motorcycle usage	I often use a motorcycle (either self-owned or borrowed) for my daily activities
	Motor_Guna	Usage of motorcycle	I use a motorcycle daily
Social Exclusion	Eks_Prod	Production activity	I am able to do production activities easily
	Eks_Kons	Consumption activity	I am able to reach shopping facilities
	Eks_Invest	Investment activity	I am able to do saving activities
	Eks_Politik	Political activity	I am able to participate in political activities
	Eks_Sosial	Social activity	I am able to visit my relatives whenever I need or want to

Each variable has its own functions and is formed by indicators used to describe intangible variables. The main aim of the model is to identify the influence of each TDA variable on social exclusion. Hence, the TDA variables act as independent variables while social exclusion is the dependent variable. One thing to be noted is the motorcycle availability variable, which can be seen as a mediator variable. In other words, the variable of motorcycle availability can act as an independent or dependent variable. This is based on theoretical studies that explain that motorcycle availability is affected by other TDA variables but also affects the variable of social exclusion.

At the early stage of model formulation in the software, several indicators were deleted before the evaluation, such as disability (Ind_Difab) due to zero variance. In addition, the variance of every indicator in both study areas needs to be checked before IPMA is conducted. There were two indicators with zero variance in latent accessibility variables in both study areas, i.e., road condition (Aks_Jalan), and public transport availability (Aks_PT). The indicators were deleted from the beginning so that the evaluation can be done without making any changes before the IPMA step. According to the previous explanation, Figure 1 shows the model created in SmartPLS software.

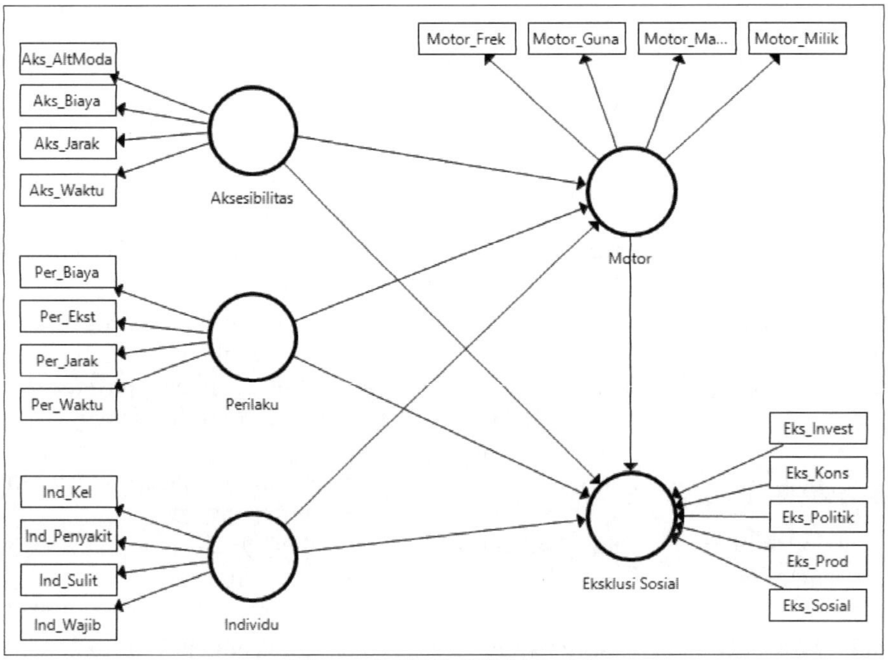

Figure 1. PLS-SEM Model

5. PLS-SEM Model Evaluation

In PLS-SEM, the model is evaluated step-by-step starting from the outer model for each latent variable either reflective or formative. Then the inner model is evaluated to identify the relationship between the latent variables. A reflective indicator is a tangible variable that can be seen as an effect of a latent variable, whereas a formative indicator is a tangible variable that can be the cause of a latent variable. Model evaluation in PLS-SEM is done systematically with a two-step approach of measurement model evaluation and structural model evaluation [11].

5.1. Measurement Model Evaluation

Reflective measurement model evaluation assesses its reliability and validity. Composite reliability (CR) is a common parameter used to explain them in PLS-SEM whereas convergent validity can be seen from each indicator loading and AVE (average variance extracted). Additionally, discriminant validity, which can be seen from each indicator cross-loading, also needs to be assessed. The result of the reflective measurement model evaluation is presented in Table 3.

The deleted indicators are the number of family members (Ind_Kel) and individual role in the family (Ind_Wajib) due to the low loading in every latent variable. In other words, those two indicators did not belong in any latent variable. The deletion improved the validity and reliability.

Table 3. Result of Reflective Measurement Model Evaluation

Latent Variable	Indicators	Accessibility	Individual Characteristics	Motorcycle Availability	Travel Behavior
Accessibility	Aks_AltModa	**0.861**	0.457	0.456	-0.234
	Aks_Biaya	**0.906**	0.237	0.312	-0.197
	Aks_Jarak	**0.969**	0.245	0.245	-0.144
	Aks_Waktu	**0.962**	0.284	0.258	-0.152
Individual Characteristics	Ind_Penyakit	0.187	**0.886**	0.310	-0.206
	Ind_Sulit	0.384	**0.941**	0.401	-0.218
Motorcycle Availability	Motor_Frek	0.313	0.340	**0.948**	-0.512
	Motor_Guna	0.361	0.348	**0.948**	-0.496
	Motor_Mampu	0.308	0.400	**0.888**	-0.584
	Motor_Milik	0.221	0.333	**0.818**	-0.318
Travel Behavior	Per_Biaya	-0.131	-0.095	-0.300	**0.716**
	Per_Ekst	-0.238	-0.373	-0.663	**0.913**
	Per_Jarak	-0.087	-0.057	-0.269	**0.823**
	Per_Waktu	-0.021	0.156	-0.166	**0.730**
Convergent Validity	CR	0.960	0.910	0.946	0.875
	AVE	0.857	0.853	0.814	0.639
Discriminant Validity?		Yes	Yes	Yes	Yes

Formative measurement model evaluation is similar to reflective measurement model evaluation, which is to assess its validity and reliability. Firstly, convergent validity was assessed by correlating the formative and reflective indicators from the same latent variable. This analysis is called redundancy analysis [11]. In this study, the formative latent variable is social exclusion. Thus, redundancy analysis was performed by correlating the formative indicators (as the independent variable) and the reflective indicators (as the dependent variable) from social exclusion.

The formative indicators of social exclusion consist of five normal activities as explained previously. Meanwhile, the reflective indicators of social exclusion are several factors influencing people's social exclusion state that are easy to be measured (e.g. income, education, housing condition). Specifically for the income level indicator, the data is obtained by collecting perceptions of respondents about the adequacy of income levels to meet their household needs just in case they are unable to mention the exact amount of their income. Other indicators such as education and housing condition are made into a categorical-ordinal type so that there are ranks or levels in each answer as well as any other statements related to perception. Figure 2 shows the result of the redundancy analysis. Although the value is below the expected value, it can still be accepted since the value is above 0.5.

The next step is to perform collinearity assessment. The assessment of the formative indicator, which can be done by evaluating the VIF value and tolerance, is necessary since it is the cause of the formative latent variable. The relations between a formative indicator and its latent variable is similar to a multiple regression equation that requires all independent variables to not be highly correlated.

Table 4 shows that the VIF value and tolerance for all formative indicators are at the expected value, which is VIF < 5 and tolerance > 0.20 [11]. Therefore, each formative indicator is not highly correlated. Next, the outer weights need to be evaluated to find out the significance and relevance of the indicators. The weight can be obtained from a bootstrapping process. The T statistic value from the bootstrapping process can be used to assess the significance of an indicator in forming its latent variable. Table 5 shows the result of the bootstrapping process. According to [11], the Eks_Prod indicator did not have enough empirical evidence to be included in the analysis. Therefore, it was deleted.

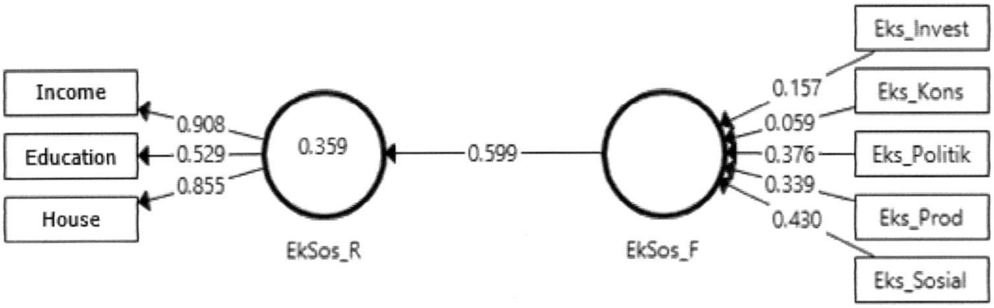

Figure 2. Results of Redundancy Analysis

Table 4. Collinearity Evaluation

Formative Latent Variable	Indicators	Tolerance	VIF
Social Exclusion	Eks_Prod	.747	1.338
	Eks_Kons	.560	1.786
	Eks_Invest	.548	1.826
	Eks_Politik	.511	1.955
	Eks_Sosial	.683	1.464

Table 5. Result of Formative Measurement Model Evaluation

Latent Variable	Indicators	Outer Weights	Outer Loadings	t Statistics	Sig. Level	p Value	Confidence Interval
Social Exclusion	Eks_Invest	0.215	0.781	3.105	***	0.002	0.080, 0.350
	Eks_Kons	0.294	0.778	4.992	***	0.000	0.178, 0.410
	Eks_Politik	0.515	0.904	7.596	***	0.000	0.382, 0.648
	Eks_Prod	0.067	0.487	1.227	NS	0.220	-0.039, 0.173
	Eks_Sosial	0.163	0.644	2.673	***	0.008	0.043, 0.283

*p < 0.1; **p < 0.05; ***p < 0.01; NS: non-significant
Bootstrap confidence interval for error probability of 0.05 ($\alpha = 5\%$)

5.2. Measurement Model Evaluation

The structural model evaluation was done to find the relationship among latent variables in the model [11]. This step consists of several assessments, including significance and relevance of the inner model, determination coefficient value (R^2), f^2 effect value, Q^2 value, and q^2 effect value. To measure relevance from a significant relationship, it was necessary to see the path coefficient value and compare it to the relationship among other latent variables. A path can be significant but the path coefficient is so low that it is considered irrelevant [11]. R^2 is a parameter used to measure model accuracy and describe the influence of all independent variables on the dependent variables. The R^2 value ranges from 0 to 1 where a higher value means higher accuracy [11].

The f^2 value is obtained from R^2 and is used to measure the level of importance of an independent variable in explaining a dependent variable. The Q^2 value is used to predict the existing indicator of reflective dependent latent variable and dependent variable with a single indicator. The expected Q^2 value is above zero, which indicates that the independent variable can reflect every indicator in a dependent variable in the model [11]. Table 6 and 7 show the result of the structural model evaluation.

Table 6. Value of Path Coefficient, f^2 Effect, and q^2 Effect

	Motorcycle Availability			**Social Exclusion**		
	Path Coefficient	**f^2 Effect**	**q^2 Effect**	**Path Coefficient**	**f^2 Effect**	**q^2 Effect**
Accessibility	0.174	0.053	0.033	0.901	3.383	0.784
Individual Characteristics	0.233	0.075	0.050	0.073	0.020	0.002
Motorcycle Availability	-	-	-	-0.102	0.015	-0.004
Travel Behavior	-0.456	0.323	0.201	-0.009	0.000	-0.002

Table 7. R^2 and Q^2 Value

Latent Variable	**R^2 Adjusted**	**Q^2**
Social Exclusion	0.798	0.462
Motorcycle Availability	0.386	0.295

The f^2 and q^2 values can be compared with the other dependent and independent latent variables. In the motorcycle dependent variable, the accessibility variable and individual characteristic have a low effect, whereas the behavior variable has a quite high effect. This means that the use or ownership of a motorcycle is most affected by the travel behavior. Among all independent latent variables, accessibility has the highest f^2 and q^2 value to predict social exclusion. According to this study, generally in both study area, accessibility is the most important point to consider in affecting social exclusion.

6. Importance-Performance Matrix Analysis (IPMA)

In order to obtain a more comprehensive result, IPMA is needed. IPMA is an additional analysis to capture the level of independent variable importance to a particular dependent variable. In addition, IPMA can be used to see the performance of variables in the model. In IPMA, the data of each latent variable is averaged and rescaled to find the performance value of each variable on the target variable. This value, combined with the importance value obtained from previous analysis, can be used to formulate intervention priority to improve the performance of the targeted variable, which in this case, is social exclusion. The IPMA result for both study areas of the determined indicator is shown in Figure 3 and 4.

In both study areas, it can be seen that accessibility variable plays an important role in the social exclusion variable. The accessibility variable is expected to be the most noticed variable to intervene social exclusion. Furthermore, the performance value is still below average. This means that an intervention in accessibility is going to highly affect social exclusion.

From Figure 3, it can be concluded that the indicator that needs to be prioritized is accessibility in relation to the availability of alternative modes. Cibeureum Village respondents stated that there are some difficulties in traveling due to the lack of alternative modes. Respondents believed that the available public transport cannot fulfill their demand so it is difficult for them to travel outside at particular times. Alternative modes do not only mean the provision of public transport because it is sometimes considered inefficient to provide public transport in rural areas [12].

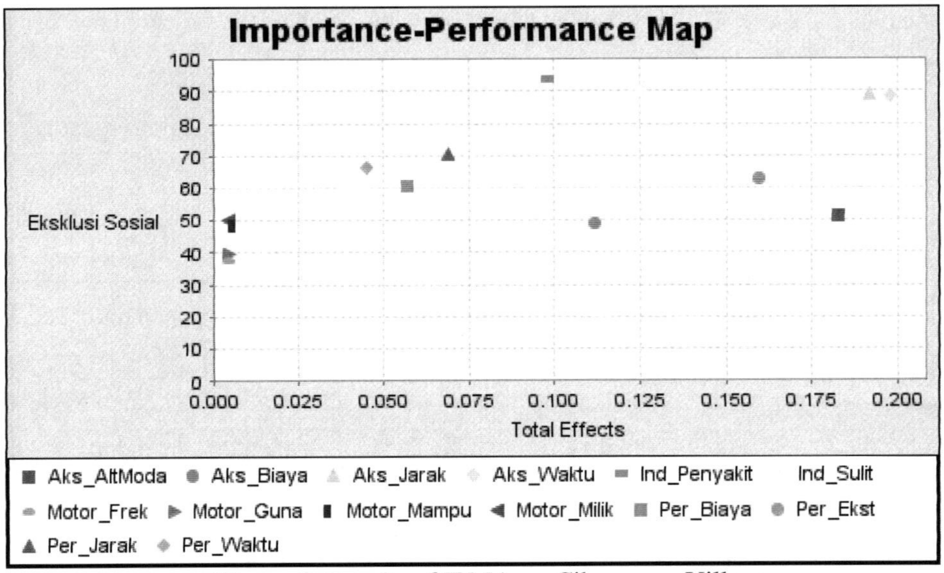

Figure 3. Results of IPMA on Cibeureum Village

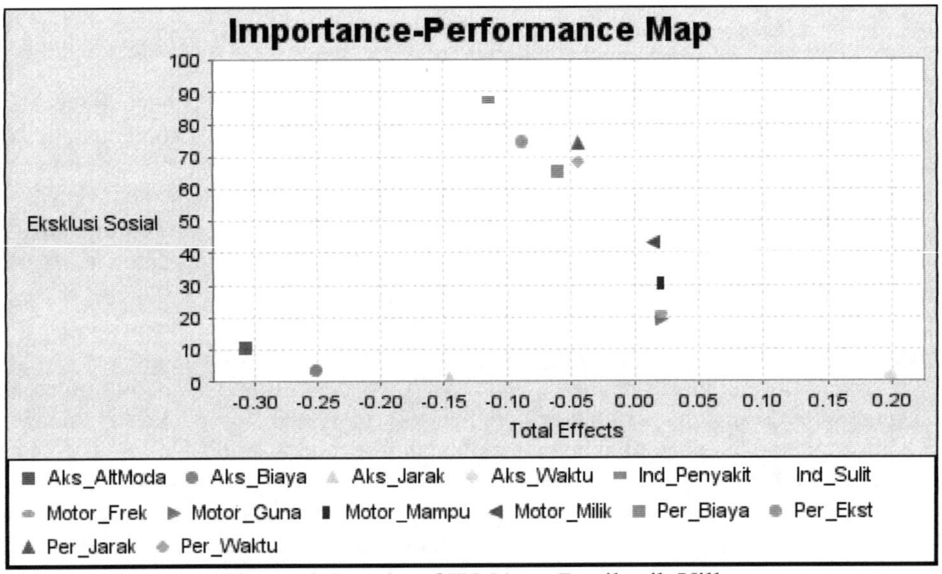

Figure 4. Results of IPMA on Bunikasih Village

A similar result is obtained from IPMA in Bunikasih Village. It is known that almost all accessibility indicators in this area have low performance values. From the IPMA chart, the most important indicator is accessibility in terms of alternative modes availability and travel cost. In Bunikasih Village, there is no public transportation or any other alternative modes to support people's movement. The existing transportation in Bunikasih Village is informal such as rental motorcycle (ojek) and car with very high costs. The lack of alternative transport modes and high travel cost makes villagers reluctant to do external movements. Though in reality, the chance of being socially included becomes higher when people do more external movement where people can access normal activities. For example, in the case of consumption activity, the price of goods in the central market will be more affordable so that people do not need to spend much money.

7. Conclusion

Based on the study findings, it can be seen that every TDA variable has a different influence on social exclusion. In general, three TDA variables, i.e., individual characteristics, motorcycle availability, and travel behavior have a low influence on social exclusion. Hence, intervention in these variables will not have much impact on reducing people's vulnerability to social exclusion. The developed theory regarding TDA and social exclusion often involves the four variables that are included in this study. However, the study found that accessibility variable has a very strong influence on social exclusion in a rural context, more than any other variables that were analyzed. Furthermore, a low value of performance makes this variable even more important to be prioritized so that villager's vulnerability to social exclusion can be reduced. According to this study, an improvement of accessibility in both areas in terms of the provision of alternative modes and travel cost can affect people's ability to access normal activities in society.

Interestingly, in both areas, it turned out that motorcycles have only have little influence on the relationship between TDA variables and social exclusion. The availability of a motorcycle does not increase the respondent's mobility or accessibility in both study areas, especially for daily activities such as working and shopping. The respondents in both study areas do have motorcycles but they rarely use it. This is due to the condition of destinations that can be reached by foot or because it is impossible to reach the destinations using a motorcycle. Another reason relates the driver and the motorcycle itself, such as the absence of license and registrations of the motorcycle.

Specifically, in Cibeureum Village, the important accessibility variables that influence social exclusion are distance and travel time to basic facilities. However, those two variables already perform well so it is better to look at other important variables with bad performance. In this village, the provision of alternative modes that can fulfill villagers' demand is the most improvement to assist their movement to the central market and the bank which are located quite far from where they live. This makes sense because the existing public transportation is very limited, both in the terms of fleet number and service hours. Meanwhile, in Bunikasih Village, the accessibility variables that are considered important are the provision of alternative modes and the reduction of travel costs. Since those two variables are in bad performance, they have to be prioritized. In other words, in Bunikasih Village, the provision of alternative modes is not enough. They also require affordable transport modes so that their vulnerability to social exclusion can be reduced. As previously explained, specifically in this village, the main activity with high urgency to be intervened is shopping activity.

Conclusively, this study has confirmed that improving accessibility through providing alternative affordable modes in a rural context is essential. However, this may be challenging and it may require demand-based alternative modes, rather than route-based ones to fulfill their specific travel pattern which can vary in space and time.

Acknowledgement

This study is based on SAPPD ITB master program thesis with the title "Pengaruh Variabel Transport Disadvantage Terhadap Eksklusi Sosial Pada Konteks Perdesaan (Studi Kasus: Dusun Cibeureum, Kabupaten Bandung Barat, dan Dusun Bunikasih, Kabupaten Subang)".

References

[1] Fitzgerald G 2012 *The Social Impacts of Poor Access to Transport in Rural New Zealand* (Wellington: New Zealand Transport Agency)

[2] Herwangi Y, Pradono, Syabri I and Kustiwan I 2013 *Transport-Related Social Exclusion and Motorcycle Ownership in Low-Income People: A Conceptual Framework* (Malang: Plannink)

[3] Burchardt T, Grand J L and Piachaud D 1999 Social Exclusion in Britain 1991-1995 *Social Policy & Administration* **33(3)** 227-244

[4] Rashid K, Yigitcanlar T and Bunker J M 2010 *Minimising Transport Disadvantage to Support Knowledge City Formation: Applying the Capability Approach to Select Indicators* (Melbourne: QUT Digital Repository)

[5] Delbosc A and Currie G 2011 Transport Problems That Matter - Social and Psychological Links to Transport Disadvantage *Journal of Transport Geography* **19** 170-178

[6] Kenyon S 2003 *Understanding Social Exclusion and Social Inclusion* (Bristol: the Institution of Civil Engineers)

[7] Lucas K 2011 Making the Connections Between Transport Disadvantage and the Social Exclusion of Low Income Populations in the Tshwane Region of South Africa *Journal of Transport Geography* **19** 1320-1334

[8] Estivill J 2003 *Concepts and Strategies for Combating Social Exclusion: An Overview* (Geneva: International Labour Office)

[9] van Wee B and Geurs K 2011 Discussing Equity and Social Exclusion in Accessibility Evaluations *European Journal of Transport and Infrastructure Research* **11(4)** 350-367

[10] Latan H 2013 *Model Persamaan Struktural: Teori dan Implementasi AMOS 21.0* (Bandung: Penerbit Alfabeta)

[11] Joseph F, Hair J, Hult G T M, Ringle C M and Sarstedt M 2014 *A Primer on Partial Least Squares Structural Equation Modeling (PLS-SEM)* (California: Sage Publications Inc)

[12] Ellis S D 1997 *Key Issues in Rural Transport in Developing Countries* (Berkshire: Transport Research Laboratory)

Train users' perceptions of walking distance to train station and attributes of paratransit service: understanding their association with decision using paratransit or not towards the train station

R. Syafriharti [1,4], B. Kombaitan[2], I. P. Kusumantoro[3], and I. Syabri[2]

[1]PhD Student of Transportation, SAPPK, Institut Teknologi Bandung, Indonesia
[2]Urban and Regional Planning, SAPPK, Institut Teknologi Bandung, Indonesia
[3]Transportation, SAPPK, Institut Teknologi Bandung, Bandung, Indonesia
[4]Urban and Regional Planning, FTIK, Unikom, Bandung, Indonesia

Corresponding author: romeiza.syafriharti@email.unikom.ac.id

Abstract. Access mode is an important factor in public transport systems. Most of the train users from Cicalengka to Padalarang via Bandung use paratransit as access mode. Access modes under this study are only paratransit and walking. This study aims to explore the relationship between access mode choice to the station and the perception about walking distance to station, perception about attributes of paratransit service quality which consist of accessibility, cheapness, comfortable, swiftness, safety, security and easiness. Of all the variables tested, walking distance to the station is the only variable relating to the mode access choice. So, a person will tend to use paratransit when his/her perception of walking distance to station is relatively far away. While perceptions about the quality of paratransit service can not determine whether a person will choose paratransit or not.

1. Introduction

Access mode affects the use of public transport [1-2], so that the integration of public transport as the main mode with access mode is an important factor that determines the success of the public transportation procurement objectives. Accessibility to public transport system is an important aspect that takes effect to the system's effectiveness and attractiveness [3].

Access mode to public transport is generally distinguishable for short distances and longer distances. Accessibility for short distances is fulfilled by making it easier for people to walk in walking distance range around public transport facilities. While for a longer distance accessibility to public transport can use private vehicles or provided some feeders. Therefore, walking distance is an important factor in the provision of public transport services [4-6].

In Bandung Metropolitan Area (West Java, Indonesia), train is a popular public transport used for commuting from Cicalengka to Padalarang and vice versa. For train users commuting from Cicalengka to Padalarang via Bandung, there are several access modes that can be used, i.e. walking, *ojek* (motorcycle taxi), *angkot* or urban transportation, and private vehicle (usually a motorcycle). *Ojek* and *angkot* are two modes of transportation which belongs to paratransit group. Among the two paratransits, *angkot* is the dominant access mode chosen by train users boarding from Cicalengka station [7], and the paratransit which refers to in this study is *angkot*. *Angkot* classified as paratransit is land-based mode owned and operated by private companies or individuals [8]. *Angkot* in Indonesia is like a *jeepney* in Philippines or *tuk-tuk* and *songtaew* in Thailand.

For a more macro purpose, ie sustainability, train users who live in the walking distance range around the train station are not expected to use the vehicle to the railway station, even if it is a paratransit. Choosing walking as access mode will bring many benefits, not only for the walking individual but also for the environment. Access mode share with paratransit to Cicalengka train station reaches 48.3%, while with walking is as much as 11.3%; the rest uses motorcycle taxi or private vehicle. At glance, the proportion of access mode share shows that the rail users to Cicalengka Station are only a few who live around the station because built environment around the station is not walkable. Although it is not yet clear, there is a relationship between walking and built environment [10-12].

If it is only based on the walking distance range, the rail users outside walking distance range will use paratransit or other vehicles to Cicalengka train station. In conditions outside the walking distance range, the attributes of paratransit services may influence the decision to use paratransit or not. Furthermore, the satisfaction of public transport users is influenced by the quality of public transport services [13,14], especially the user perceptions of service quality [15,16], in addition to the influence of socio-economic characteristics. Simply stated that the perception of something will affect people to use or do something. Therefore, the variables to be analyzed in relation to the selection of paratransit as an access mode are in the form of the perception of traveler.

However, there is a study that cannot be ignored, the finding of scientists at Stanford University in the US revealing that Indonesians are the laziest for walking among the 46 countries [17]. If indeed Indonesians are lazy to walk, despite the perception of walkability especially about the distance of a good category included, there will be inconsistent perceptions of walking distance with the mode of choice. Simply stated, although the perception of walking distance is good, they still do not want to walk.

The data used in this research is the railway user data that depart from Cicalengka train station using paratransit and walking as access mode. The purpose of this study is to find consistency/inconsistency of rail users who ride from Cicalengka train station in selecting paratransit as access mode with their perception of walking distance to station and their perceptions about the quality of paratransit services. If there is a relationship between the selection of paratransit as an access mode and the perception of walking distance to station, then there is a consistency. In other words, the rail users who ride from Cicalengka train station are not among Indonesian people who are lazy for walking. They will only choose paratransit as an access mode if they think the distance to the station is far or far enough. If there is no relationship, there is inconsistency which means reinforcing the results of research scientists at Stanford University; the Indonesian people are lazy to walk and prefer the paratransit although the close range and the services of paratransit are considered not good. Then, if there is a relationship between the selection of paratransit as an access mode to a station and the quality of paratransit service, there is a consistency. Thus, it can be found out what factors are related to the decision to choose a paratransit as an access mode, whether the perception of walking distance to train station or perception of the quality of paratransit services. The ideal condition is access mode related to walking distance with every attribute of paratransit services. Consistency will be achieved when people will ride paratransit and the distance cannot be reached by walking and good paratransit service.

This paper was begun with the formulation of research background, including the importance of access mode in the public transport system, the description of popular paratransit, previous research which becomes the basis of this research, as well as the research objectives. The earlier research related to this study is outlined to formulate the variables used in the analysis. Methodology and results are described in the next section. The last is outlined in the discussion and conclusions of this study.

2. Literature Review

The reviewed previous studies are divided into two sections. The first is the walking distance in public transport service. The second is the attributes associated with public transport service.

2.1. Walking Access in Public Transport System

It is important to understand the influence of walking distance to public transport as it is a key element of establishing equitable access to public transport [4]. This type of mode is the most important determinant of walking distance. In a research conducted in Sydney, Australia, it was found that more people walk to the station than to the bus stop. This shows that the distributions of walking distances are different for each mode and the most important feature of walking distance is the availability of stops and stations. In research on travel behavior, distance to transit is one of the D of 5 D's used to analyze the influence of built environment on travel behavior, especially walking and cycling [12].

The behavior to decide to walk or not to a public transport facility may vary. It is hard to find the same pattern. As well as research conducted by Walton and Sunseri [18], walking distance did not affect the decision to walk or not to public transport facilities. Factors that affect precisely is the weather and comfort for the car park. This finding is reasonable because the respondents are people who usually drive, although residing maximum 1 km from public transport facilities.

Accessibility to public transport is represented not only by walking distance, but also by other factors that overall reflect walking accessibility [19]. In a study conducted by Wibowo and Olszewski, walking accessibility represents not only the actual walking distance but also generalized walking effort. The assumption used is for every distance to the terminal so there is a probability of walking to access. Type of walking routes to access public transport are distinguished by walkways, sideways, road crossings, etc., that influence the effort of walking.

Studies on the effect of walking distance on the use of public transport for older adults in California and New York have been conducted [20]. There are several findings, namely, the drivers are more sensitive to walking distance than nondrivers. Walking time affects the use of public transport. In addition, older adult men, not white, and low income, more often use public transport.

In addition to walking distance and walking efforts, accessibility for walking to public transport can also be represented by the time required to walk to the station or bus stop or other public transport facilities. There are different ranges for walking distance and walking time for male and female [21]. Another study conducted in Brisbane, Australia, found that walking time is the most sensitive among the part-time workers, high-income earners, and elderly travelers. While among postsecondary students who are studying and working at the same time, walking time is least sensitive [22].

2.2. Quality of Public Transport Service

Previous research on the quality of public transport service varies greatly, both on variables and methods. However, the quality of public transport service is mostly discussed based on the perception perceived or the experience of public transport user. The attributes used to analyze the quality of public transport service are bus stop availability, route characteristics, frequency, reliability, bus stop furniture, overcrowding, cleanliness, cost, information, promotion, safety on board, personal security, personnel, complaints, environmental protection, bus stop maintenance [13]. The method used is the structural equation modeling.

The importance of integrated multimodal in public transport [23] is basically related to access and egress mode. Through user perspective, authorities can develop attractive transfer routes with comfortable transfers. The theory of planned behavior (TPB) was adopted to investigate the role of perceived behavioral control (PBC) in travelers' intentions to use public transport (PT) routes with transfers.

In the transportation domain, the guarantee for safety is absolutely essential. Bad experience about safety can affect the frequency of use of public transport [24].

There are times when individual perceptions can be the same as or different from reality (objective measurement of the quality of public transport services). In a study conducted in Tallahassee, Florida, it was found that the perception of changes in transit services is greater than reality [15].

In Indonesia, the perceptions of paratransit users are examined with six variables [16]. These variables are availability, accessibility, information, customer service, fare, and negative experience with crew attitude. Path analysis is used to indicate the association between variables, including testing the loyalty of paratransit users. Another public transport service attribute is used to analyze the user perceptions in Indonesia, involving users from Bandung, Jakarta and Yogyakarta. The attributes are cleanliness in vehicle, cleanliness in the station/stop, condition in the vehicle, comfort in the vehicle, comfort in the station/stop, route coverage, ease of payment, safety and security in the vehicle, security in the station/stop, access to the vehicle, driver's skill, punctuality, attitude of the staff, and ease of getting the mode [25].

The government of Indonesia issues minimum service standards for public transportation. Minimum service standards for urban public transport are security, safety, convenience, affordability, equality, and regularity [26]. Each of these service components is detailed to facilitate use in performance evaluation of services. The safety and comfort components was revised in 2015, with more detail on both components of the service. [27].

3. Methods

The respondents were the train users departing from Cicalengka train station to Bandung and Padalarang, departing at 06.15, 07.25, 09.00, and 09.45. The survey was conducted for 3 days, Monday - Wednesday in November 2016. The survey involved 20 surveyors. Questionnaires were prepared in advance to be filled by respondents. Most of the questionnaires were filled over trains during the trip by respondents with the help of surveyors. Only a few questionnaires were filled at the station because most of them came to the station before the train departures.

In accordance with the purpose of the study, the respondents are limited to rail users who use paratransit or walking as an access mode to the station. Of the nearly 203 filled questionnaires, only 116 questionnaires could be used. Access mode was then analyzed in relation to other variables using crosstabulation analysis. Significance level (α) used is 5%. SPSS is used for data processing and analysis. Thus, Ho will be rejected if the probability value of Asymp .Sig. (2-sided) is < 0.05, which means there is a relationship between the variables tested. Furthermore, Spearman Correlation was used to determine the level and mark correlation. Although crosstabs and Spearman Correlation are simple, they are sufficient to achieve research objectives due to the fact that what will be explored is still limited to the relationship between access mode and other variables. However, to get a more complete result, it needs to use another method. For example, if we want to know the effect of walking distance and attributes of paratransit service on paratransit selection as access mode and by involving latent variables, structural equation modeling can be used.

Access mode is set as an ordinal variable, with value 1 representing paratransit as access mode and value 0 for walking as access mode. The determination of value 1 for paratransit choice is due to other variables that will beanalyzed including the quality of paratransit services and the focus of research on selecting paratransit as an access mode instead of walking.

Based on the previous research, the minimum service standards for public transport in Indonesia, and the conditions in the field, the tested variables which are related to access mode to determine what factors determine the decision to use paratransit as access mode are:

3.1.1 Perception of walking distance to the station. In the previous research, walking distance is the dominant factor in the public transport system [4-6], i.e. people will walk to the station if the distance to the station is close for walking. It means that people will not walk to the station if they assume the distance to the station is not close enough. Paratransit can be an option if the distance to the station is not close enough to walk. However, other studies suggest Indonesians are lazy to walk [17]. If the perception of respondents about walking distance is not close enough and they still ride paratransit, it means they are lazy to walk. Therefore, the perception of walking distance to station is one of the factors tested to access mode selection.

The respondents were required to provide an assessment of the requested agreement on 5-point Likert scale from "strongly agree to strongly disagree". If the respondents feel they can walk to station, they will answer strongly agree, and the point given for this answer is 5. Point 1 is given for strongly disagree answers from the respondents. With the answer of the agreement, this variable also includes ordinal variables.

The assumption for this variable is if the respondents disagree that distance to station is close enough for walking, there will be a bigger chance of election of paratransit as access mode. This assumption is based on previous research which states that walking distance to station is a factor that determines whether people will walk or not to station [4, 6, 18]. If in a short distance people tend to walk, in the longer distance, they will prefer doing the paratransit to walking to the station. Thus, the sign of correlation between the access mode relationship and the perception of walking distance is negative.

3.1.2 Perceptions about attributes of paratransit service quality. The variables tested in the study were determined based on previous studies, applicable regulations, and characteristics of paratransit services. The assumption is that the better the user's perception of the attribute of paratransit services, the satisfaction level is higher and the greater the chance that paratransit is chosen as access mode [13]. The relationship between access mode and every paratransit service attribute is positive. These variables are: accessibility [14, 15, 25], cheapness [13, 14, 26, 28], comfortable [13, 14, 15, 25, 26, 28], swiftness [14], safety [13, 14, 15, 25, 26, 28], security [13, 14, 25, 26, 28], and easiness [15, 25].

- Accessibility is all about the convenience of reaching the paratransit, not only the walking distance, but also the other attributes of accessibility, including comfort, safety, and security reaching the paratransit.
- Cheapness associated with pratransit rates.
- Feeling comfortable on the vehicle, so it does not include comfortable headed or waiting for paratransit.
- Swiftness concerns about the quality of travel time of the paratransit to the station.
- Safety is the safety of traffic during a trip with a paratransit or during on a vehicle.
- Security is the security of crime during travel with paratransit.
- Easiness is a practical measure of using paratransit, including the need for special shelters along the route of service.

The respondents were also required to provide an assessment of the requested agreement on 5-point Likert scale from "strongly agree to strongly disagree" for all variables including attributes of paratransit service quality.

As for walking distance to station, attributes of paratransit service quality were also measured through user perceptions. User perceptions represent what they feel. Perceived and actual distance can be different and transport mode choice is correlated with perception [29]. The perception of paratransit service also influences attitude and travel behavior more than 'reality' [15].

Table 1. Variables

Access Mode to Train Station
Paratransit = 1
Walking = 0
Perceptions about walking distance to station
5 point scale from strongly agree to strongly disagree
• Strongly agree if the distance to station is very close by walking = 5
• Agree if close distance to station is reached by walking = 4
• Ordinary if distance to station is not far or not close to go by = 3

- Disagree if distance to station is far by walking = 2
- Strongly disagree if distance to station is very far by walking = 1

Perceptions about attributes of paratransit service quality
Accessibility, Cheapness, Comfortable, Swiftness, Safety, Security, Easiness
Each attribute of paratransit service quality is given a 5-point scale from strongly agree to strongly disagree

- Strongly agree = 5; Agree = 4; Ordinary = 3; Disagree = 2; Strongly disagree = 1

All variables tested include ordinal variables, so the correlation test used is Spearman Correlation. Furthermore, to analyze the relationship of choice mode access and other variables, the relationship hypothesis is determined and the correlation mark. There are eight hypotheses formulated, including the sign of correlation and assumptions used.

Table 2. Hypothesis

No.	H_0	Sign of Correlation	Assumption for Correlation
1.	There is no relationship between Access Mode Choice and Walking Distance to Station.	-	The further walking distance to station, the greater the chance of selecting paratransit as access mode.
2.	There is no relationship between Access Mode Choice and Accessibility to Paratransit	+	The better the perception of accessibility to paratransit, the greater the chances of selecting paratransit as access mode.
3.	There is no relationship between Access Mode Choice and Cheapness of Paratransit	+	The better the perception of Cheapness of Paratransit, the greater the chance of selecting paratransit as an access mode.
4.	There is no relationship between Access Mode Choice and Comfortable of Paratransit	+	The better the perception of Comfortable of Paratransit, the greater the chances of paratransit selection as access mode.
5.	There is no relationship between Access Mode Choice and Swiftness of Paratransit	+	The better the perception of Swiftness of Paratransit, the greater the chance of selecting paratransit as access mode.
6.	There is no relationship between Access Mode Choice and Safety of Paratransit	+	The better the perception of Safety of Paratransit, the greater the chance of paratransit selection as access mode.
7.	There is no relationship between Access Mode Choice and Security of Paratransit	+	The better the perception of Security of Paratransit, the greater the chance of paratransit selection as access mode.
8.	There is no relationship between Access Mode Choice and Easiness of Paratransit	+	The better the perception of Easiness of Paratransit, the greater the chance of selecting paratransit as an access mode.

Note: H_1 = There is a relationship between pair of variables tested

4. Data and Results

Most respondents chose the paratransit as access mode to station, almost four times the size of walking to station. The perception of walking distance to station is almost the same as strongly agree, agree, ordinary, and disagree. Almost all perceptual-related variables are few or none answered strongly disagree. Many respondents answered disagree for the attributes of paratransit quality service. In general, it can be said that paratransit services based on the perception of rail users are good.

Table 3. Description of Variables

Variables	Description
Access Mode	Paratransit = 96 (82.76%); Walking = 20 (17.24%)
Perception of Walking Distance to Station	5 = 21; 4 = 33; 3 = 27; 2 = 31; 1 = 4
Perception of Accessibility to Paratransit	5 = 19; 4 = 66; 3 = 29; 2 = 2; 1 = 0
Perception of Cheapness of Paratransit	5 = 13; 4 = 55; 3 = 42; 2 = 6; 1 = 0
Perception of Comfortable of Paratransit	5 = 14; 4 = 23; 3 = 64; 2 = 15; 1 = 0
Perception of Swiftness of Paratransit	5 = 12; 4 = 26; 3 = 60; 2 = 17; 1 = 1
Perception of Safety to Paratransit	5 = 9; 4 = 30; 3 = 58; 2 = 18; 1 = 1
Perception of Security of Paratransit	5 = 10; 4 = 29; 3 = 55; 2 = 20; 1 = 2
Perception of Easiness of Paratransit	5 = 11; 4 = 52; 3 = 45; 2 = 8; 1 = 0

Of the seven attributes of paratransit service quality, accessibility to paratransit received the best rating from respondents, because those who answered agree and strongly agreed reached more than 70%. Security, safety, swiftness, and comfortable are good enough according to the perception of respondents. However, most respondents' answers are ordinary and disagree.

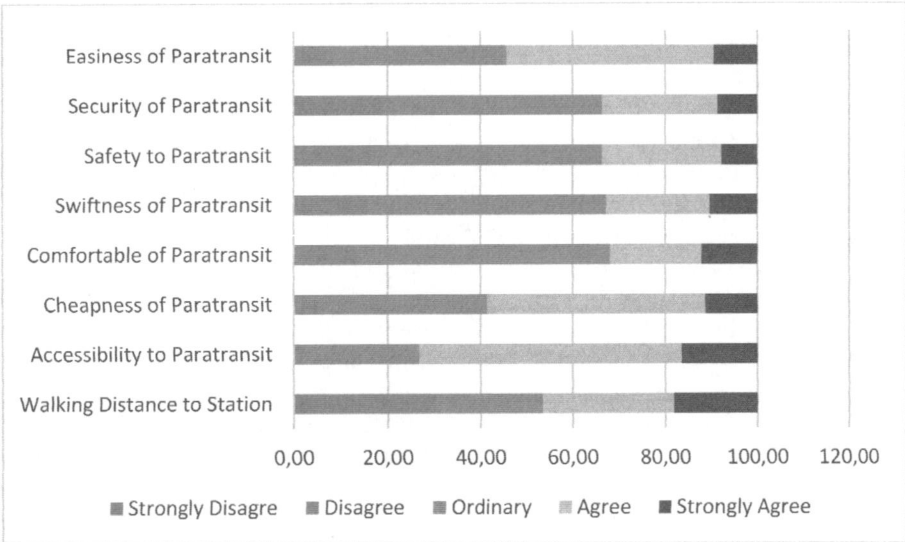

Figure 1. Perceptions Percentage of Walking Distance to Station and Paratransit Service Quality Attributes

From the characteristics of socio-economics respondents, the number of male respondents is more than female respondents. The respondent are mostly 25-50 years old. Most of the respondents have a vehicle. It shows that many respondents choose to walk or use paratransit to the railway station though they have vehicle.

Based on the analysis of crosstabs, Asymp values. Sig. (2-sided) is > 0.05 which means there is no relationship between access mode choice and walking distance to station since $H_{0,}$ is rejected. With correlation test result of -0.375 and the minus sign, it can be explained that the better perception of walking distance to station, the more people tend to choose walking than paratransit as access mode to station.

The correlation between access mode and security is very small (0.099). Thus, further analysis is needed to obtain better results. Meanwhile, the female is more likely to choose a paratransit than a male. The correlation value is sufficient to state that there is a relationship between gender and the selection of paratransit as an access mode.

Table 4. Result

Variables	Asymp. Sig. (2-sided) - Pearson Chi-Square	Related or not	Spearman Correlation
Access Mode vs Walking Distance to Station	**0.001 (< 0.05)**	Related	-0.375
Access Mode vs Accessibility to Paratransit	0.414 (> 0.05)	Not related	
Access Mode vs Cheapness of Paratransit	0.336 (> 0.05)	Not related	
Access Mode vs Comfortable of Paratransit	0.072 (> 0.05)	Not related	
Access Mode vs Swiftness of Paratransit	0.132 (> 0.05)	Not related	
Access Mode vs Safety of Paratransit	0.161 (> 0.05)	Not related	
Access Mode vs Security of Paratransit	**0.017 (< 0.05)**	Related	0.099
Access Mode vs Easiness of Paratransit	0.211 (> 0.05)	Not related	

5. Discussion and Conclusion

The results of this study is different than of the earlier study which presents that Indonesians are lazy for walking [17]. The Indonesians are not lazy to walk. It isproven by the railway users case departing from Cicalengka train station because there is a connection between access mode choice and walking distance to station. The respondents chose paratransit if the perception about distance to station is far enough to walk. Instead, they tend to walk if perceptions about the distance to the station are relatively close. However, because in this study walking distance is a perceived distance, there is a gap possibility between the perceived distance and the factual distance. It is possible that the factual distance is still within walking distance range. They do not like to walk then their answer about the distance to the station is far enough if they travel by walking. Therefore, for further research it is necessary to compare between the perceived distance and the factual distance so that the actual walking distance can be captured to find the relationship and its effect on the access mode choice. More complex methods are needed to obtain a true description of Indonesians' walking habit, particularly towards and from the public transport facility. In addition to walking distance, it also needs to consider the other built environment factors for walking, such as comfort, safety, security, cleanliness, and even the beauty of the environment.

From the results obtained, it can be said that there is no relationship between the quality of paratransit services and the access mode choice. It means that the respondents choose a paratransit not because the quality of paratransit service is good, or they choose to walk not because the quality of paratransit service is not good. Thus, the quality of the paratransit services is not the determinants of the access mode choice. Good or bad perception, respondents do not necessarily choose or do not choose paratransit. Although the access mode choice is related to the security of paratransit, the correlation value is very small. To obtain better result about the relationship between access mode and security, it requires further statistical analysis. For service quality, it is also necessary to compare between perceived and reality. In case of access mode to Cicalengka train station, the paratransit is not selected because of the quality of service. The user is reluctant to walk, including due to distance factors. Therefore, regulators, the central, provincial and district/city government should still pay attention on the quality of paratransit services, including formulating a more aplicable service standard.

In conclusion, respondents are consistent when expressing their perception of walking distance to station by selecting access mode to station. Conversely, there is an inconsistency between an assessment of the quality of paratransit services and access mode.

References

[1] Givoni M and Rietveld P 2007 The access journey to the railway station and its role in passengers' satisfaction with rail travel *Transport Policy* **14(5)** 357-365
[2] Krygsmana S, Dijsta M and Arentzeb T Multimodal public transport: an analysis of travel time elements and the interconnectivity ratio *Transport Policy* **11(3)** 265-275
[3] Tuan V A and Son D T 2015 Accessibility to Public Transport Systems in Developing Countries – An Empirical Study in Ho Chi Minh City Vietnam *Journal of the EASTS* **11** 1240 – 1258
[4] Daniels R and Mulley C 2013 Explaining walking distance to public transport: the dominance of public transport supply *JTLU*, **6(2)** 5-20
[5] Ratanawaraha A, Chalermpong S and Chullabodhi C 2015 Walking Distance of Commuters after Modal Shift to Rail Transit in Bangkok. *Journal of the EASTS* **11** 1467-1478
[6] Chia J, Lee J and Kamruzzaman M 2016 Walking to public transit: Exploring variations by socioeconomic status *Int J of Sust Transp* **10(9)** 805–814.
[7] Syafriharti R, Kombaitan B, Kusumantoro, I P, and Syabri I 2017 *Relationship between train users' perceptions of walkability with access and egress mode choice* Proceedings *The Third Int Conf on SIBE* Bandung
[8] Joewono T B and Kubota H 2007a User Satisfaction with paratransit in competition with motorization in Indonesia: anticipation of future implications *Transportation* **34** 337-354
[9] Althoff T, Sosič R, Hicks J L, King A C, Delp S L and Leskovec J 2017 Large-scale physical activity data reveal worldwide activity inequality *Nature* **547** 336-341
[10] Saelens B E and Handy S L 2008 Built Environment Correlates of Walking: A Review *Med Sci Sports Exerc* **40 (7)** S550–S566
[11] Sun G, Oreskovic N M and Lin H 2014 How do changes to the built environment influence walking behaviors? a longitudinal study within a university campus in Hong Kong *Int J of Health Geographics* **13** 28
[12] Cervero R, Sarmiento O L, Jacoby E, Gomez L F and Neiman A 2009 Influences of built environments on walking and cycling: lessons from Bogotá *Int J of Sust Trans* **3** 203-226
[13] Eboli L and Mazzulla G 2007 Service Quality Attributes Affecting Customer Satisfaction for Bus Transit *JPT* **10(3)** 21-34
[14] Javid M A, Okamura T and Nakamura F 2015 Public Satisfaction with Service Quality of Daewoo Urban Bus Service in Lahore *J of the EASTS* **11** 1097-1108
[15] Bhattacharya T, Brown J, Jaroszynski M and Batuhan T 2015 The Effects of Perception vs. "Reality" on Travel Behavior after a Major Transit Service Change: The Case of Tallahassee, Florida *JPT* **17(2)** 1-26
[16] Joewono T B and Kubota H 2007b User Perceptions of Private Paratransit Operation in Indonesia *JPT* **10(4)** 99-118
[17] Althoff T, Sosič R, Hicks J L, King, A C, Delp S L and Leskovec J 2017 Large-scale physical activity data reveal worldwide activity inequality *Nature* **547** 336-339
[18] Walton D and Sunseri S 2010 Factors Influencing the Decision to Drive or Walk Short Distances to Public Transport Fasilities *Int J of Sust Trans* **4** 212-226
[19] Wibowo S S and Olszewski P 2005 Modeling Walking Accessibility to Public Transport Terminals: Case Study of Singapore Mass Rapid Transit *J of the EASTS* **6** 147 – 156
[20] Hess D B 2009 Access to Public Transit and Its Infuence on Ridership for Older Adults in Two U.S. Cities *JTLU* **2(1)** 3-27
[21] Yang R, Yan H, Xiong W and Liu T 2013 The Study of Pedestrian Accessibility to Rail Transit Stations Based on KLP Model *Procedia - Social and Behavioral Sciences* **96** 714 – 722

[22] Chia J, Lee J and Kamruzzaman M 2016 Walking to public transit: Exploring variations by socioeconomic status *Int J of Sust Trans* **10(9)**805–814
[23] Chowdhury S, and Ceder A 2013 A Psychological Investigation on Public-transport Users' Intention to Use Routes with Transfers *Int J of Transp* **1(1)** 1-20
[24] Marteache N, Bichler G and Enriquez J 2015 Mind the Gap: Perceptions of Passenger Aggression and Train Car Supervision in a Commuter Rail System *JPT* **18(2)** 61-73
[25] Joewono T B and Santoso Dj S 2015 Service Quality Attributes for Public Transportation in Indonesian Cities *J of the EASTS* **6** 1064 – 1081
[26] Standar Pelayanan Minimal Angkutan Orang dengan Kendaraan Bermotor Umum Dalam Trayek. Final rules. Peraturan Menteri Perhubungan Republik Indonesia. 2013
[27] Perubahan atas Peraturan Menteri Perhubungan Republik Indonesia Nomor PM 98 Tahun 2013 tentang Standar Pelayanan Minimal Angkutan Orang dengan Kendaraan Bermotor Umum Dalam Trayek. Final rules. Peraturan Menteri Perhubungan Republik Indonesia. 2015
[28] Grujičić D, Ivanović I, Jović and Đorić V 2014 Customer Perception of Service Quality in Public Transport *TRANSPORT* **29(3)** 285-295
[29] Hernández D and Witter R 2015 Perceived vs. Actual Distance to Transit in Santiago Chile *JPT* **18(4)** 16-30

IOP Conf. Series: Earth and Environmental Science **158** (2018) 012017 doi:10.1088/1755-1315/158/1/012017

Discrete-Choice Modeling Of Non-Working Women's Trip-Chaining Activity Based

Amelia Hayati[1], Pradono[2], Heru Purboyo[2] and Sri Maryati[2]

[1]PhD Candidate, Institute of Technology Bandung, Bandung, Indonesia
[2]Institut Teknologi Bandung, Bandung, Indonesia

Corresponding author: amelia.hayati@fe.unpad.ac.id

Abstract. Start The urban developments of technology and economics are now changing the lifestyles of the urban societies. It is also changing their travel demand to meet their movement needs. Nowadays, urban women, especially in Bandung, West Java, have a high demand for their daily travel and tend to increase. They have the ease of accessibility to personal modes of transportation and freedom to go anywhere to meet their personal and family needs. This also happens to non-working women or as housewives in the city of Bandung. More than 50% of women's mobility is outside the home, in the term of trip-chaining, from leaving to returning home in one day. It is based on their complex activities in order to meet the needs of family and home care. While less than 60% of male's mobility is outdoors, it is a simple trip-chaining or only has a single trip. The trip-chaining has significant differences between non-working women and working-men. This illustrates the pattern of Mom and Dad's mobility in a family with an activity-based approach for the same purpose, i.e. family welfare. This study explains how complex the trip-chaining of non-working urban women and as housewives, with an activity-based approach done outdoors in a week. Socio-economic and household demographic variables serve as the basis for measuring the independent variables affecting family welfare, as well as the variables of type, time and duration of activities performed by unemployed housewives. This study aims to examine the interrelationships between activity variables, especially the time of activity and travel, and socio-economic of household variables that can generate the complexity of women's daily travel. Discrete Choice Modeling developed by Ben-Akiva, Chandra Bhat, etc., is used in this study to illustrate the relationship between activity and socio-economic demographic variables based on primary survey data in Bandung, West Java for 466 unemployed housewives. The results of the regression, by Seemingly Unrelated Regression approach methods, showed the interrelationship between all variables, including the complexity of trip chaining of housewives based on their daily activities. The type of mandatory and discretionary activities, and the duration of activities performed during the dismissal in the series of trip chains conducted are intended for the fulfillment of the welfare of all family member.

1. Introduction

Historically, travel is a derived demand from the need for consumption and usually uses the main transport. Along with the changing pattern of consumption to be complex life needs, the changes in travel patterns are increasingly complex as well. Based on physical activity in meeting needs and consumption, activity-based models and the study of trip-chain patterns become one of the focuses on transportation models development.

To achieve a policy statement that is in the process of friendly citizens, the authors believe it is important to understand the travel behavior activity-based of two groups of women from Bandung City. Although there are many studies on non-worker travel behavior, but have not received deep attention, especially women as housewives. Group of non-working women is an important part of Bandung's population. In comparison, [1] focused on the behavior of individual travel-activities of the Jakarta Metropolitan Region in Indonesia. Create a lower dependency on personalized mode, reduce slightly more travel, and allocate less time to travel. Then [2] in his analysis of long-term activities of the Osaka Metropolitan area of Japan, found non-workers and their activities. The selection found differences in daily patterns of non-workers between the Indonesian city and the U.S. [3].

The paper examines a typology of trip chains based on the spatial-temporal structure of trips and activity type at the destination and contains as follows. First, some introduction notions regarding the study of travel behaviors and trip chaining are provided. Then, the methodology is defined. An entire section is devoted to the presentation of a trip chain typology along with a summary analysis of the occurrence of these various types of chains for the population under study. The relations between activity-based and socio-demographic variables are then examined. A conclusion follows [4].

2. Background TheoryAnother section of your paper

2.1 Activity models

Typical transportation models rely on the notion of single trips. The widely used four-step model predicts the number of trips that will occur in a study area and relies on this simple concept through the four steps in order to estimate the level of usage of various transportation infrastructures. Faced with these limitations, the increasing complexity of travel behaviors and the increasing capacities of technology, new modeling approaches are being developed to enhance the relevance of models and outputted results. Activity-based models are one of the directions taken by the research community in order to enhance modeling capacities. These models rely on the notion that trips are a derived demand for the necessity, or desire, of individuals to perform out-of-home activities in various locations. One of the strong points of this model is its disaggregate nature since it performs the analysis at the individual level (set of trips made by an individual).

[5] identify three important and positive features of this new generation of models:

- tour-based travel where the tour is the basic unit for modeling travel instead of the single trip.
- rely on an activity-based platform implying that travel is derived demand within a general framework of the daily activities undertaken by households' member, especially housewife.
- micro-simulation techniques are applied at the disaggregate level of a person based on household's needs.

Behaviors of individuals of households are examined simultaneously to account for combined trips and the negotiation between individuals to gain access to mobility tools. With this underlying framework, it is easier to address more complex behaviors. It is also easier to make comparisons between each of households and not just between sectors. Figure 1 presents the typical structure of an activity-based framework.

Almost all travel demand analysis is rooted in applied and suspended econometric techniques on a discrete choice model. It has had an understanding of the process. Econometrics, more than microeconomic theory, is inspired by the study of the actual situation in which data are available and which are of policy interest.

Even so, this paper micro-economically discusses the development of travel demand activity-based, most of which are travel for shopping, recreation, socializing, eating out and other activities. Current transport analysis has experienced problems with travel patterns and complex activity scheduling from conceptual and empirical perspectives. This has spawned a variety of applied studies on travel patterns that bring their cues directly with econometric techniques [6].

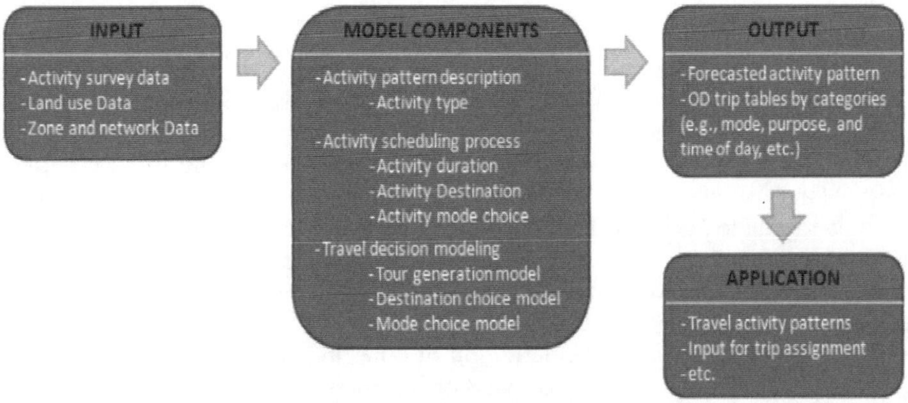

Figure 1. Input, model components, output and application of activity based approaches [4]

2.2 Tour or trip chain

In [4], activity-based models support analysis at the tour or trip chain level. Several authors have conducted studies on daily behaviors for typical days and propose various definitions of a trip chain. For instance, [7] suggested that a trip chain is a series of trips between two anchor points, home and work. Others suggested that a trip chain includes all trips between leaving home and returning to it. It is the case for [8]. [9] is mostly in agreement with this previous definition but adding that a trip chain is not necessarily only “home – activity – home”, but can also be “home - activity 1 - activity 2 - activity N – home”.

In addition, several authors have examined statistical relations between trip chains and variables such as gender, age, present of children in the household, mode of travel, residence location and others. First, gender greatly influences trip chains. According to [10], women have a higher likelihood of performing multiple activities during their trip chains than men. In addition, women do more trip chains per day than men. [8] focus their analysis on the trip purposes of men and women. The results show that women perform a greater number of simple trip chains for shopping, drop-off and pick-up than men. Household composition also greatly influences trip chains [4].

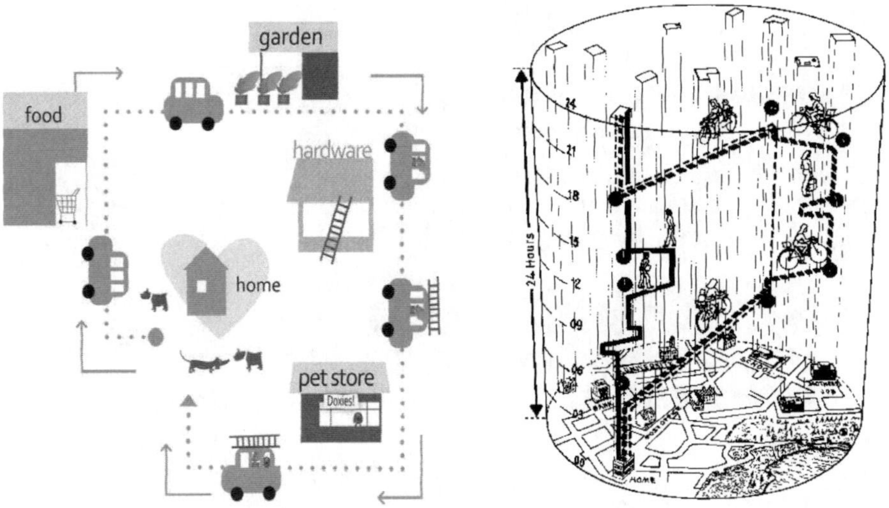

(a) (b)

Figure 2. Women’s trip chaining activity-based approaches (a) Natural approach (b) Space-Time approach [11-12]

3. Methodology

3.1. Concepts

First, trip chains are based on anchor points. These anchor points are important and fixed spatial locations. For the purpose of this study, three locations can be set as anchor points: home location, workplace and school. The home location is set as the anchor point at the beginning and end of every trip chain. Work and school activities are often considered as mandatory. Many people in the examined population segments will, during a typical weekday, travel for at least one of these purposes if not for both.

Workplace and school are therefore important locations in the daily travel behaviors of individuals; thus, the proposed typology will need to take these anchor points into consideration. Moreover, intermediate trips will often be linked to mandatory trips leading to these anchor points. This leads to the definition of other important concepts: mandatory and discretion activities. The mandatory activity of a trip chain is the one with the longest duration. However, if work or study is in the trip chain, it becomes the mandatory activity even if it is not the longest activity. Typically, mandatory activities are located at anchor points. However, other trip purposes may result in mandatory activities, especially when there are no work or study trips in the chain. In these cases, the activity with the longest duration will be set as the mandatory activity. The discretion activities are all the other activities conducted between home location and the mandatory activity of a trip chain. Thus, the trip chain will include a single mandatory activity, and one or many discretion activities [4].

Second, the method used in this paper is to examine the travel patterns and their complexities travel spatially with the Space-Time Prism Method to illustrate the trip-chaining model in question. To explain the variables affecting women's activity by household background and the complexity of travel patterns, a regression approach with Seemingly Unrelated Regression is used.

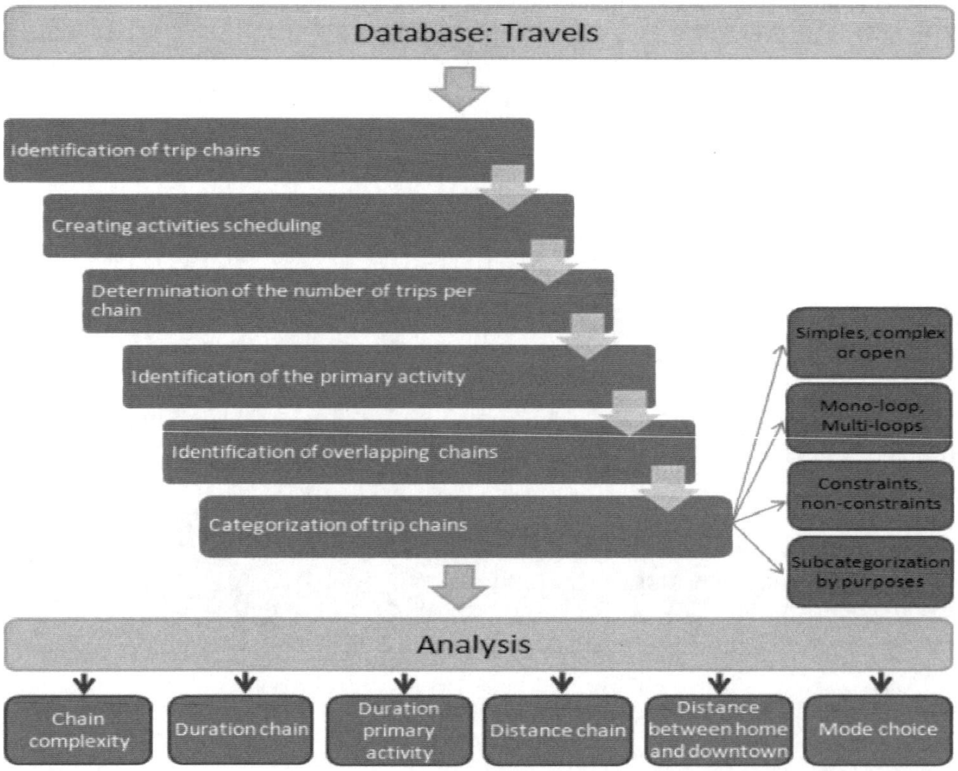

Figure 3. Various steps involved in the enumeration and classification of individual trip chains [4]

3.2. Space-Time Prism Method to capture Trip Chaining Model

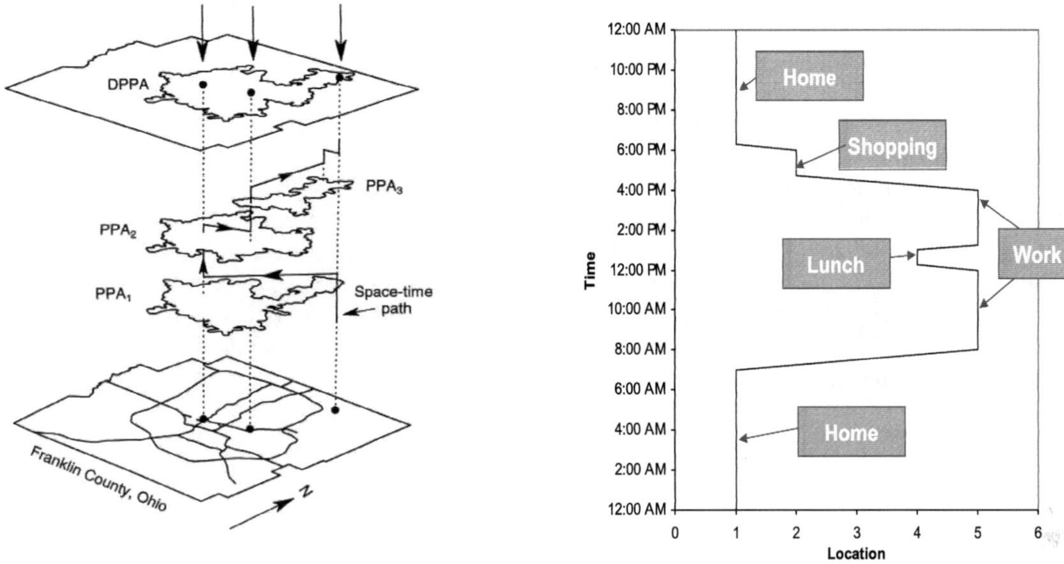

(a) (b)

Figure 4. Trip chaining map (a) Space-Time Path with Map approach (b) Space-Time without Map approach [13-14]

Space-time prism method is to describe the pattern of trip chaining performed by women as housewives with a spatial approach based on distance and time. This method can explain the complexity of daily trips by looking at the number of stop for activity-based discharges.

3.3. Seemingly Unrelated Regression Model to capture Trip Chaining Figure
Random Utility Maximum Model is a kind of Economic model to capture the women travel behavior in mathematical and statistical approaches. Random utility maximum by time and task optimization can describe a hectic mobility of urban women to meet their needs in term of making household members' satisfaction.

Random Utility Maximum Model can be explained by econometric technically as Seemingly Unrelated Regression (SURe). SURe model is very useful for this case because it is considered capable of accommodating ordinal and nominal variables according to data taken in the questionnaire adopted, such as VISTA (Victorian Integrated Survey of Travel Activity) and simplified. SURe is used because it is considered to be the most accommodating research needs for multi data both input (Independent Variable), as well as output (Dependent Variable). This Quantitative Data Processing uses E-VIEWS software.

Regression Model:

$$Y = \alpha_0 + \beta_i X_i + \varepsilon \quad (1)$$

Estimation Model

$$\hat{Y} = \alpha_0 + \beta_i X_i \quad (2)$$

Regression Model by Seemingly Unrelated Regression Approach :

$$\widehat{TMAct} = \alpha_0 + \beta_1 Age + \beta_2 Educ + \beta_3 Child + \beta_4 Fam + \beta_5 House + \beta_6 Eltc + \beta_7 Motor + \beta_8 Bns + \beta_{10} TDAct + \beta_{11} ToJ + \beta_{12} TOD + \beta_{13} Inc \quad (3)$$

$$\widehat{TDAct} = \alpha_0 + \beta_1 Age + \beta_2 Educ + \beta_3 Child + \beta_4 Fam + \beta_5 House + \beta_6 Eltc \beta_7 Motor + \beta_8 Bns + \beta_9 TMAct + \beta_{11} ToJ + \beta_{12} TOD + \beta_{13} Inc \quad (4)$$

$$\widehat{ToJ} = \alpha_0 + \beta_1 Age + \beta_2 Educ + \beta_3 Child + \beta_4 Fam + \beta_5 House + \beta_6 Eltc + \beta_7 Motor + \beta_8 Bns + \beta_9 TMAct + \beta_{10} TDAct + \beta_{12} TOD + \beta_{13} Inc \quad (5)$$

4. Area of study
The study area, Bandung city, is the capital city of West Java Province in Indonesia. The population of Bandung City is nearly 2.5 million. As the same time, information from the Bandung Central Bureau of Statistics. Bandung City covers an area of approximately 167.67 km^2 with density is about 14.768/km^2. This city is the biggest metropolitan area in West Java Province. The largest ethnic is Sundanese for more than 60% of population with most religion is Islam (88.72%), Protestant (7.32%) and Catholic (3.01%).

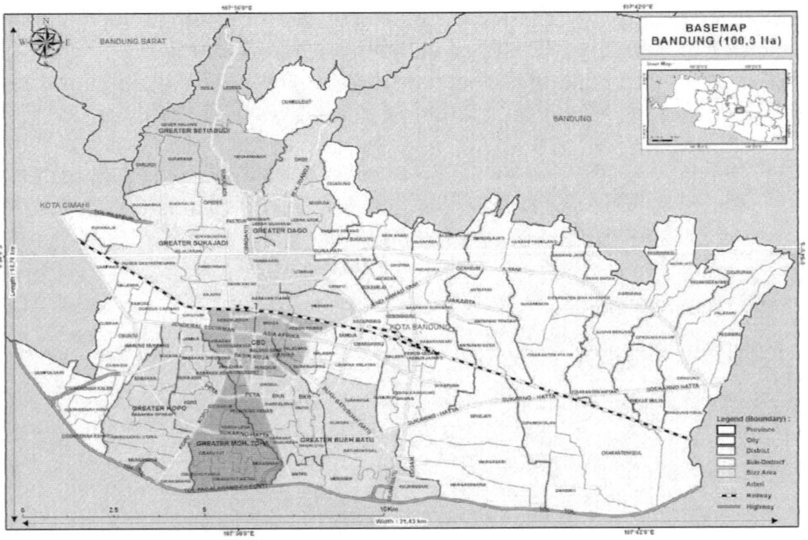

Figure 5. Bandung Base-Map [15]

4.1. The Data

This primary data were surveyed by the authors and survey team in 2 clusters of Bandung City with 'Face-to-Face' interview method for 89% response rate. Interviews were conducted with random respondents from 30 sub-districts that include socio-demographic trends in 2017 WJPMDM region of 466 women as housewives from 466 households. The applied random sampling method was used to collect demographic data and household data.

4.2. Characteristics of the sample

Out of 500 households, the selected number of 466 households was valid, 466 non-working women for household purposes were available for analysis. Summary of household features is presented in Table 1. Most households have at least one child. An important part of monthly household income data ranges from 1,000,000 to over 10,000,000 rupiahs. Most non-workers live with husbands and families.

The share of motorcycles has a very high household in the cities of Indonesia, including Bandung, as seen from the table. However, more than 80% of households have one two-wheeled motorcycle available for use.

5. Analysis

5.1. Analysis of activity–travel behavior

A preliminary investigation into travel behavior activity based on non-working women as housewives in a week (Monday, Thursday, and Sunday) of the survey is the focus of this section. The travel behavior activity-based indicators are considered in the analysis by different activities (travel behavior purposes), time allocation for various activities, mode choice, daily patterns, and time options for travel or trip departure.

A summary can see significant differences between clusters of household and private socio-demographic factors in the context of downtown and suburbs. Such factors among many others that influence them are the determinants of individual activity-based travel behavior.

5.1.1. Data of Socio-economic. The most age characteristic of travellers is lower than 40 years, a very productive age. For educational level, it is a bit difference between the first cluster and second as shown in the table. It means the first cluster are more educated persons. Educational characteristic is followed by income in which the first cluster has higher income. There is 4.17 % of women in the first cluster who are not married, but none in the second cluster. The family planning program in Bandung seems successful in reducing the bigger population of Bandung's household.

Table 1. Bandung Urban Non-working Women Socio-economic Characteristics
Percentage (%) of Socioeconomic Household Characteristics
For Non-Working Urban Women (N=466) *)

Household Characteristics		Cluster 1 (Radius 0 – 10 Km)	Cluster 2 (Radius 10 – 20 Km) in the area of Bandung City
Age of women	< 30	18.52	32.40
	31 – 40	47.69	37.20
	41 – 50	26.39	23.20
	51 – 60	6.94	6.00
	> 60	0.46	1.20
Education	Elementary School	6.48	18.00
	Junior High School	12.04	40.00
	Senior High School	62.50	27.60
	University	18.98	14.40
Household Income (Millions Rupiahs/MR)	< 2.5	21.76	38.40
	2.5 – 5	62.96	43.60
	5 – 10	14.35	12.80
	>10	0.93	5.20
Marriage Status	Unmarried	4.17	0.00
	Married	95.83	100
	No Child	4.63	2.80
Number of Children	1 Child	29.63	28.40
	2 Children	38.43	38.00
	>2 Children	27.31	30.80
Number of Family Member	2 – 4 persons	68.98	74.40
	>4 persons	31.62	25.60
Vehicle Ownership	No Vehicle	7.32	14.40
	2 Wheels Vehicle	47.73	67.60
	4 Wheels Vehicle	25.44	1.60
	Mixed Vehicles	19.51	16.40

*) Result Processing

5.1.2.Characterizing on different types of activitie. From the data in table 2, every woman spends more time to do her activities for the whole day as it is shown by time duration for more than 2 hours. Both of those clusters, more than 45% of the individuals participate in discretional activities in Bandung City; whereas 61.30 % of the individuals participate in a discretional activity in the first cluster. All of the women population in Bandung do their travel on weekday to cover their daily needs of transportation with two-wheels motorized vehicle.

Table 2. Bandung Urban Non-working Women of Trip-Chaining Activity-Based Percentage (%) of Non-Working Urban Women Trip Chaining Activity-Based*)

Characteristic of Variables		Cluster 1 (Radius 0 – 10 Km)	Cluster 2 (Radius 10 – 20 Km)
Type of Trip Chain	Simple	12.50	60.40
	Complex	86.57	39.20
	Open	0.93	0.40
Time Duration of Activity	< 1 hour	2.78	4.68
	1– 2 hours	11.57	40.24
	>2 hours	85.65	55.08
Type of Activity	Mandatory	38.70	53.75
	Discretion	61.30	46.25
Type of Day Trip	Weekday	72.63	78.24
	Weekend	24.37	21.76
Mode Choice	Non-Motorize	1.85	5.62
	Public Transport	10.56	11.80
	2 Wheels	57.34	65.32
	4 Wheels	30.25	17.26

*) Result Processing

5.1.3. Trip Chaining for Daily Pattern

The analysis is also presented at the daily pattern level. Consistent with observations in samples, three typologies of trip-chaining are used for analysis. They are Simple Single Chains where individual one-stop visits for today, Simple Multiple Chains/Complex where individual visits for more than one stop per tour, and an open travel chain that cannot be monitored within the study area. All tours come from home. The samples were sampled under each of the categories in Table 2. More than 86% of women did Trip chains in the first group but only 39.20% in the second group.

As summary, exploratory analysis generates behavior that does not work in the activity behavior of each cluster in the city of Bandung. In cluster 1, the weekday travel chain is longer than with a relatively large number of stops. On the weekend, it is also not too much different even though there is a longer tendency. While in cluster 2, the daily business day-trip chain is shorter with fewer stops, but over the weekend.

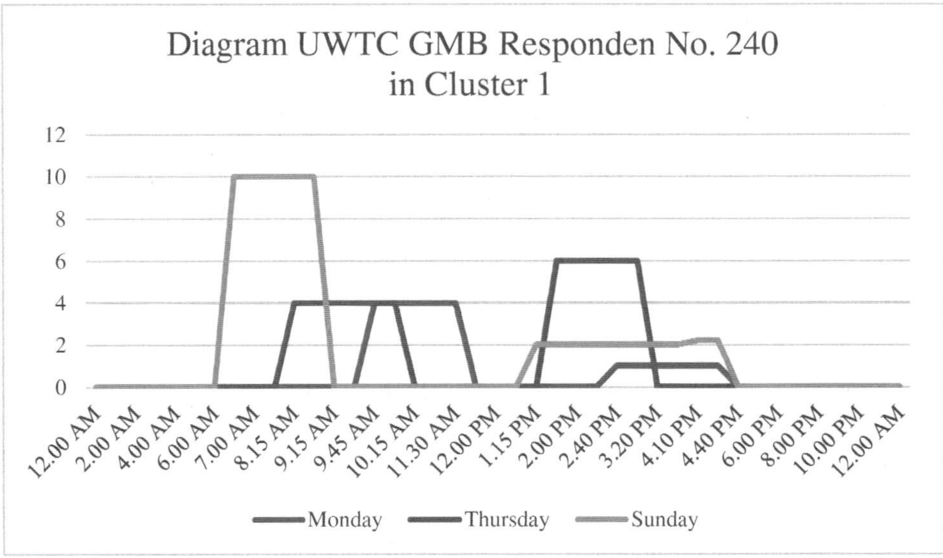

(a)

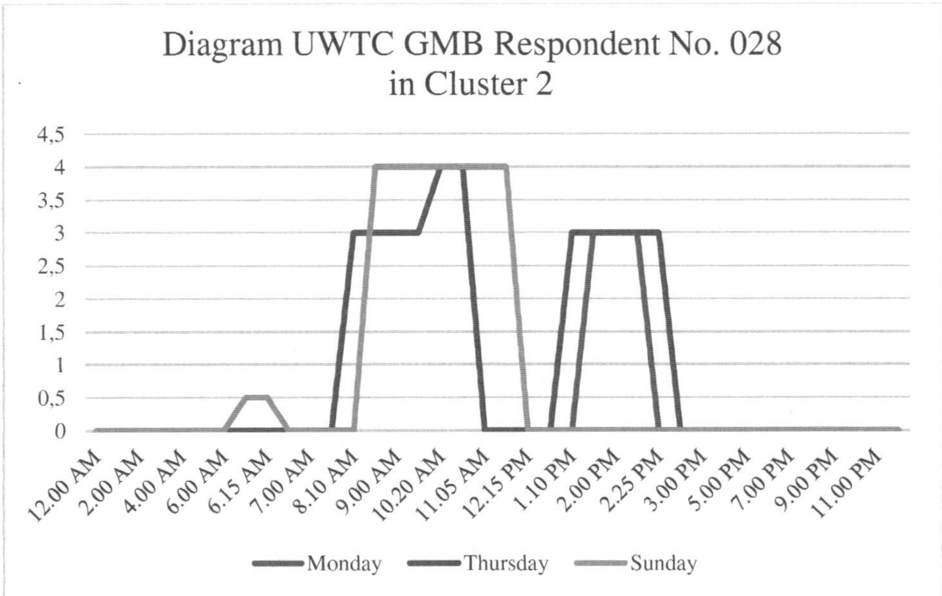

(b)

Figure 6. Samples of Individual Trip chaining map (a) Diagram of the respondent no.240 in Cluster 1 (b) Diagram of the respondent no.028 in Cluster 2

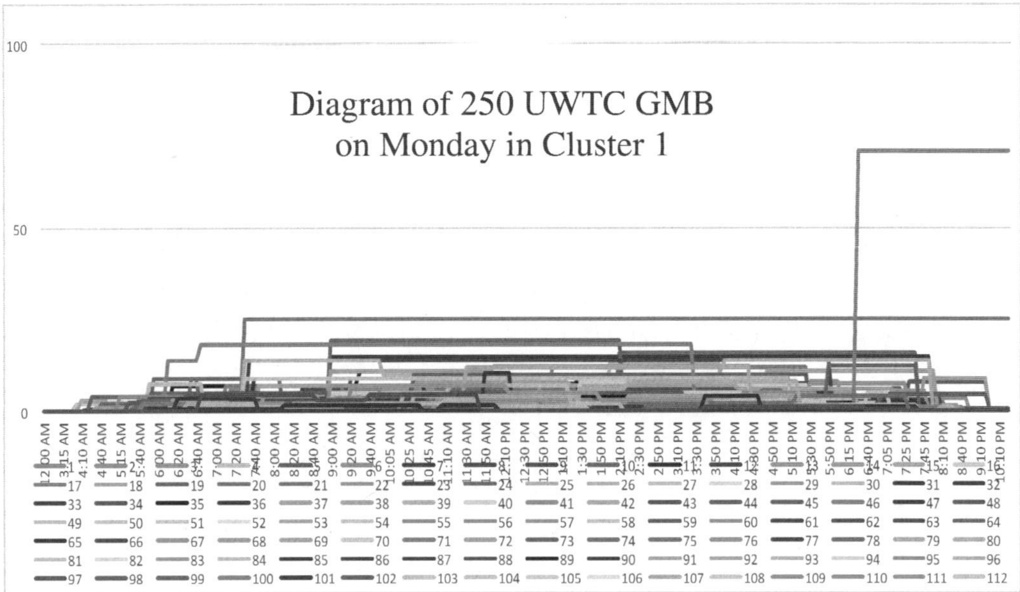

(a)

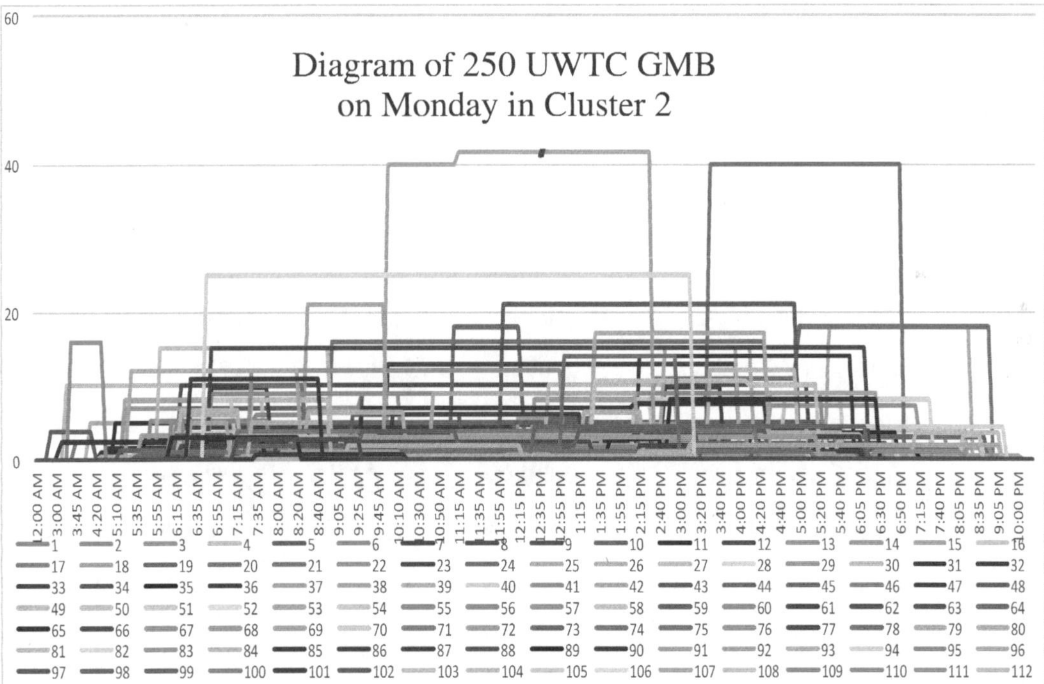

(b)

Figure 7. Samples of Aggregate Trip chaining map (a) Diagram of 250 UWTC on Monday in Cluster 1 (b) Diagram of 250 UWTC on Monday in Cluster 2

5.2. Analysis of activity participation behavior

The objective of this section is to analyze the influence of household, individual, and land use on the travel behavior activity based on non-working women as housewives. Child dropping, shopping, and personal activities are the focus of this section. The regression approach with the Seemingly Unrelated

IOP Conf. Series: Earth and Environmental Science **158** (2018) 012017 doi:10.1088/1755-1315/158/1/012017

Regression model is used to find out the relationship between the variable attributes of the daily travel behavior activity based on household utility. It can generate 3 regression models that can accommodate the interrelations between these variables.

5.3. A simultaneous model of activity–travel behavior

It can be seen from Table 3, for cluster 1 models of SURe, that the Model 1 is the best model of all that belongs to "Time of M(Mandatory) Act" as the dependent variable. In that model, many variables are significant, these are Mode-choice(motor), business (buss), Time of D Act, and Time of journey, with having t-significantly by 1% (***), 5% (**), and 10% (*).

For model 2 in table 3 where Time of Journey is the predictor variable, this proves that many independents variable such as education, business, time of mandatory activity, and time of discretion activity can significantly affect the time of journey. It means several socio-economic of households are significantly relevant to the time of journey to meet their activities in Bandung City, even on weekday and weekend.

Table 3. Result of SURe Approach for Non-working Urban Women Daily Activities in Cluster 1

	Model 1	Model 2	Model 3
R^2	0,314	0,229	0,203
F_{stat}	8,868***	5,754***	4,930***
(Constant)	601,441***	479,259***	14,642
Age (Yr)	-1,812	-0,227	0,274
Education (Yr)	7,730	-10,597*	9,063*
No. of Children (Pr)	-9,272	18,103	5,093
No. of Family (Pr)	0,341	6,723	-8,407
Housing (Y/N)	57,946	-57,826*	38,767
Electricity (Watt)	-0,054	-0,008	-0,022
Motorcycle (Y/N)	-158,292**	19,005	1,698
Business (Y/N)	232,096***	-0,963	-49,930**
Time of M Act (Mnt)	DV	-0,246***	0,177***
Time of D Act (Mnt)	-0,473***	DV	0,247***
Time of Journey (Mnt)	-0,584***	0,424***	DV
Time of Outdoor (mnt)	---	---	---
Income (M Rp)	-4,005	-3,905	7,192

Compared with the SURe Model of Cluster 1, we found different model of Cluster 2 in table 4 describing outside area around the Cluster 1. For model 1 where "Time of M(Mandatory) Activities" as the dependent variable, we found that "motorcycle", "business", "Time of D (Discretion) Activities" and "Time of Journey" variables significantly affect to "Time of M (Mandatory)

Activities". It means that mode-choice, added activities to business, discretion activities and time of journey for daily travel significantly affect Time spent on Mandatory activities. Those variables support Mandatory activities i.e. Child dropping and or primary shopping.

However, the resulting model in Cluster 2 (see table 4) is significantly different. The best model is Model 3 where the "Time of Journey" is the dependent variable. Variable "Business" and "Time of D(Discretion) Activities become the significant variables to support the dependent variable. It means discretion activities and business support the dependent variable for the non-working women in Cluster 2. When we see the distance to the downtown in radius, it can be logically to do.

Table 4. Result of SURe Approach for Non-working Urban Women Daily Activities in Cluster 2

	Model 1	**Model 2**	**Model 3**
R^2	0,045	0,162	0,139
F_{stat}	0,892	3,660***	3,071***
(Constant)	609,071*	313,875**	-34,279
Age (Yr)	6,982	-1,949	1,430
Educ (Yr)	26,029	5,262	4,542
No. of Children (Pr)	49,117	-18,776	-1,340
No. of Family (Pr)	-47,041	4,641	5,627
Housing (Y/N)	-11,432	36,479	31,714
Electricity (Watt)	-201	-0,012	-0,017
Motorcycle (Y/N)	21,658	67,626	34,185
Business (Y/N)	111,265	-105,425***	52,205**
Time of M Act (Mnt)	DV	0,007	-0,004
Time of D Act (Mnt)	0,051	DV	0,157***
Time of Journey (Mnt)	-0,80	0,393***	DV
Time of Outdoor (mnt)	---	---	---
Income (M Rp)	4,096	2,660	6,019*

5.4 Planning and policy implications

The model of trip chaining of non-working urban women can be predicted by socioeconomic data and regression model with the SURe approach. The analysis of the findings has important implications for transportation planning and policy for Bandung city. Table 1 and 2 report the aggregate effects of various exogenous variables. The findings are timely relevant as Bandung City has been experiencing rapid socio-demographic transition.

From the SURe approach, we found that several certain variables significantly affect to "Time of Mandatory and Discretion Activities", these are business, mode-choice, and time of journey. It explains that the raising of socio-economic of household whose non-working urban women will increase the travel time of journey on weekday or weekend. Motorcycle modes that they have may cause rapid travel behavior with increasing the number of stop place.

The suggested policy is the facility of parking area in the stopping place for women, in the example: around the school area, the mall or mini market area, and the coffee shop, café and recreation area to support the stop area in the journey of non-working women daily trip-chaining. If they have bigger facilities for parking, so they can easily come-in or come out in terms of their stop as the part of their daily trip-chaining. The government has to support this facility to increase of ease women's mobility, especially non-working women as housewives.

6. Summary and Conclusion

This paper presents exploratory and statistical analyses of the activity–travel behavior of non-working urban women in Bandung City using the primary survey to almost 500 respondents. Only 466 questionnaires are valid by Victorian Integrated Survey for Travel and Activities (VISTA). The exploratory analysis presents a summary of socio-economic and time travel and activities indicator, such as age, education, household income, marriage status, number of children, number of family, and vehicle owner/accessibility, type of trip-chain, time duration of activity, type of activity, type of day-trip, and mode choice.

The study also compares the summary measures with case studies from different areas of Bandung City that is divided into Cluster 1(less than 10 km radius from Zero point of city center), and Cluster 2 (10 – 20 km radius from Zero point of the City center inside the border of Bandung City area) focusing on the activity–travel behavior of non-working urban women who live within the area of Bandung City.

The analysis shows that, by comparing the case studies from Cluster 1 and Cluster 2 areas, non-working urban women in Bandung city were observed to have 5 activity participation types (in terms of , type of trip-chain, time duration of activity, type of activity, type of day-trip), and a distinct time-of-day preference for departing to activity locations (depend on clustering area). Whereas the summary measures show similarities (by socio-economic and activity-based) and differences (time allocation and mode choice) with the case studies from different clusters.

After the summary of activity–travel attributes, statistical models are developed for analyzing the effects of individual and household socio-demographics, and land use attributes on the activity–travel behavior of non-working urban women. Important findings from the statistical analysis are summarized below:

1. Household socio-demographics and individual's role place have influences on different activities participation decisions of non-working urban women; whereas individual's decision and household socio-demographics determine personal/household business activity participation decision, in case in Cluster 2.
2. Individual socio-demographics and Cluster characteristics influence the selection of 'simple in multiple tours' daily patterns; on the other hand, household socio-demographics, activity participation characteristics, travel contexts and location of the house in the different cluster influence the selection of 'complex' daily pattern.
3. Household income and vehicle ownership govern the selection of private modes for travelling, and apart from the relevant socio-demographics land use attributes appear to influence the decision to go traveling.

Overall, the study contributes to the literature by providing a brief exposition of the activity–travel behavior of non-working urban women in Bandung City, and drawing policy implications of the study in the present urban and transportation systems contexts. Apart from divulging the differences in activity–travel behavior of non-working urban women, the study also sheds light on the influence of in-home activities and the ownership of two-wheels vehicle.

The findings of the two models demonstrate the need for integrating different attributes of activity–travel behavior in a single framework as the observed relationship between many variables (in independent models) can be manifestations of complex relationships through many variables. Further, if non-workers do not have a specific role in residential self-selection, the analysis shows that land use attributes affect the activity participation behavior of non-workers.

Another extension of this study can be investigating the role of non-working group in residential self-selection, and revisiting the empirical models presented here with its influence. The investigation of neighborhood-level attributes on the activity participation behavior of non-workers is another fruitful area. Furthermore, the impacts of transportation systems on time allocation to in-home and out-home activities with activities defined at disaggregate level should be investigated since the recent studies show that multiple activity categories yield interesting insights into the activity scheduling choices (e.g., [16]).

References

[1] Dharmowijoyo DBE, Susilo YO, and Karlström A 2014 Day-to-day interpersonal and intrapersonal variability of individuals' activity spaces in a developing country *Environment and Planning* **41(6)**

[2] Susilo YO and Cast O 2017 Exploring key determinant of travel satisfaction for multimodal trips by different traveler group *Journal of Transportation Research Part A*

[3] Manoj M and Verma A 2015 Activity–travel behavior of non-worker from Bangalore City in India *Transportation Research Part A* **78** 400–424

[4] Valiquette F and Morenchy C 2010 Trip chaining and its impact on travel behavior *12th WCTRS Proceeding* (Lisbon: WCTR Society)

[5] Davidson W, Donnelly R, Vovsha P, Freedman J, Ruegg S, Hicks J, Castiglione J, and Picado R 2007 Synthesis of first practices and operational research approaches in activity-based travel demand modeling *Transportation Research Part A* **41** 464-488

[6] Anas A 2007 A unified theory of consumption, travel and trip chaining *Journal of Urban Economics* **62** 162–186

[7] McGuckin N, and Murakami E 1999 Examining trip-chaining behavior: a comparison of travel by men and women *Transportation Research Record* **16**

[8] Primerano F, Taylor M A P, Pitaksringkarn L, and Tisato P 2007 Defining and understanding trip chaining behavior *Transportation* **35(1)** 55-72

[9] Holzapfel H 1986 *Trip relationships in urban areas* (Brookfield: Gower Publishing)

[10] Kumar A and Levinson D M 1995 Chained Trips in Montgomery County, Maryland *ITE Journal* **65(5)** 27-32

[11] Christine R 2014 http://www.christinerainsdesign.com/infographics/

[12] *Women's trip chaining activity-based approaches* geog_user.org

[13] Kwan M P 2000 Gender differences in space time constrains *Area* **32** 145-156

[14] Auld J and Mohammadian A 2009 Framework for the development of the Agent-based Dynamic Activity Planning and Travel Scheduling (ADAPTS) model The International Journal of Transportation Research **1** 245-255

[15] Bappelitbang Bandung 2017 *Bandung Base Map*

[16] Whitehead C and Kockelman K 2010 American's time use: A focus on women and children rearing Via SEM *Journal of Transportation Research Record* **2163** 32-44

IOP Conf. Series: Earth and Environmental Science **158** (2018) 012018 doi:10.1088/1755-1315/158/1/012018

Optimation of Operation System Integration between Main and Feeder Public Transport (Case Study: Trans Jakarta-Kopaja Bus Services)

M. Miharja[1], **Y N Priadi**[2]

[1] Regional and City Infrastructure System Research Group, School of Architecture, Planning and Policy Development, Institut Teknologi Bandung, Bandung, Indonesia
[2] Undergraduate Program in Regional and City Planning, School of Architecture, Planning and Policy Development, Institut Teknologi Bandung, Bandung, Indonesia

Corresponding author: mimingm@pusat.itb.ac.id

Abstract. Promoting a better public transport is a key strategy to cope with urban transport problems which are mostly caused by a huge private vehicle usage. A better public transport service quality not only focuses on one type of public transport mode, but also concerns on inter modes service integration. Fragmented inter mode public transport service leads to a longer trip chain as well as average travel time which would result in its failure to compete with a private vehicle. This paper examines the optimation process of operation system integration between Trans Jakarta Bus as the main public transport mode and Kopaja Bus as feeder public transport service in Jakarta. Using scoring-interview method combined with standard parameters in operation system integration, this paper identifies the key factors that determine the success of the two public transport operation system integrations. The study found that some key integration parameters, such as the cancellation of "system setoran", passenger get in-get out at official stop points, and systematic payment, positively contribute to a better service integration. However, some parameters such as fine system, time and changing point reliability, and information system reliability are among those which need improvement. These findings are very useful for the authority to set the right strategy to improve operation system integration between Trans Jakarta and Kopaja Bus services.

1. Introduction

The topic of "transport integration" has been discussed in different contexts of transport both theoretical and practical. Among others, [1-8] point out that transport integration can be seen from various perspective such as technical, economic, organizational, policy-based and so on. However, the basic purpose of every transport integration analysis is to assure the connection among different transportation modes operating in a certain transportation. Transport integration system is intended to facilitate passengers/goods transfer between the modes and assure the safe, smooth and efficient flow of passengers/goods from their origins to their destinations [3,9]. In particular topic of public transport integration, the EU projects [10-11] define urban public transportation as an organizational process by which elements of the passenger public transportation system (network and infrastructure, fares and ticketing systems, information and marketing components) and a variety of carriers which serve different transportation modes to interact more closely and efficiently to generate an overall improvement in service quality level and enhance the performance of the combined public and individual transportation. According to [12], the implementation of different transport integration solutions may result in the following benefits: reduction of travel times, transportation costs, traffic congestion and environmental pollution. Transport integrating solutions may improve the accessibility of urban public transportation system and overall competitiveness, and assure better utilization of different transportation means and infrastructure.

The topic on the optimization of urban public transport integration is very important in Indonesian urban transport plan and development. This urgency is due to the fact that today's urban transport problem is basically triggered by predominant use of private car in this area transport service system. Thus, serious effort to increase public transport service quality, including inter public transport mode integration, is crucial to attract more private car user to use public transport. As an effort to improve public transport service quality in Jakarta Capital Region, since June 2015 the Government of Jakarta Province has introduced an operational integration program between PT Transjakarta and Kopaja. Through this program, it is expected that an integrated service system between the feeder and the main transport will improve the quality of public transport service as part of the grand strategy to attract private car users. Feeder transport is a type of bus service which is designed to transport passengers in the local area and delivers passengers to the transfer point where they continue their journey in the main corridor [13]. In its implementation, the operational integration between PT Transjakarta with Kopaja has been proceeded through the reordering of bus route integration, tariff adjustment, e-ticket development and new payment system by replacing kilometer based payment to "setoran" (the daily-based fee the driver must pay) payment system.

This study focuses on the optimation analysis of operation system integration of main-feeder transport in the case of operation integration between PT Transjakarta and Kopaja. The perspective of operation integration parameter is based on the existing optimation theory of public transport operation integration. More specifically, the objectives of this paper are: (1) to know the fulfilment level of the operation system integration by Transjakarta-Kopaja operator based on the parameters of operational integration, and (2) to set the optimation strategy of Transjakarta-Kopaja main bus-feeder operation system.

2. Method

2.1. Data Collection Method

In this research, primary and secondary data are both examined. Primary data were obtained from a questionnaire filled by an operational officer in the field and in-depth interview with the shareholders using purposive sampling method; whereas, secondary data were obtained from Transportation Agency of DKI Jakarta, PT Transjakarta and Kopaja. Secondary data were used to analyze empirical facts in their suitability with the stated target in the operational integration plan that has been stipulated.

2.2. Data Analysis Method

Questionnaire data analysis was performed through descriptive qualitative analysis method, whereas the analysis of interviews data was performed using content analysis method. Secondary data analysis were developed to analyze system performance by comparing the empirical evidence with the operation system parameters. Meanwhile, the content analysis would produce parameter priority that is still to be optimized based on company policy consideration. Shareholders were also asked to give their score on the degree of operational integration with a scale of 1-5 (1 = very weak; 5 = very strong). This step is done as part of triangulation method through cross-check of analysis results of two different analytical methods. Finally, a component in the operational integration parameter was analyzed thoroughly to formulate recommendation in the form of proposal of handling priority as an effort to optimize main-feeder transport operation system integration in the future.

3. Literature Review on the Integration of Public Transport System

Public transportation planning or other public services should be composed of several aspects such as efficiency, effectiveness, and satisfaction [14]. According to [15] the concept of smart feeder is a system with easy access to fast transportation by studying the market demand for short-haul feeders or short-range feeders. This feed system will provide services that operate reliably and quickly as part of a door-to-door passenger transport by means of a good and synchronized transfer. The integration of public transport system occurs in three important levels, i.e. institutional, operational and physical. In line with the focus of this research, integration that will be discussed is in operational level which involves the application of management technique to optimize the allocation of public transport resources and services coordination.

There are nine criteria of system arrangement in the process of public transport operational integration in [16], i.e:

1) *Rationalization of Redundant Services,* i.e. an effort to reduce duplication in public transport operation to increase competitiveness and efficiency of resources usage through reduction of headways at the existing route and expand services to new areas
2) *Matching Modes to Service Requirements*, i.e. an effort to arrange the service pattern where large capacity supply system is operated in the main area with high density and corridor with high level of demand, while park-and-ride facility is provided in low density area
3) *Unification of The Fare Structure*, i.e. the stipulation of tariff system that enables customers to buy one ticket at the beginning of the trip and to transfer freely among all modes and services network in a system.
4) *Fare Discount*, i.e. an effort to reduce tariff to achieve three objectives, i.e. (a) tariff subsidy to a certain user group, such as elderly, physically handicapped, weak economy; (b) divert some of the users to off-peak time; (c) simplification of weekly or monthly payment procedure.
5) *Honor Payment Fare Systems*, i.e. developing an honorarium system to save operational cost by eliminating ticket checking process by the conductor to be more efficient by increasing the number of entrance gate rather than only one entrance gate as the main control point.
6) *Coordinated Public Information Systems*, i.e. development of integrated information system on route, schedule, tariff, and point of transfer by publishing route map, vehicle schedule, tariff schedule and promotion information, road signage and vehicle identification, stop point, transfer point, station location, etc.
7) *Reserved Bus Lanes and Streets*, i.e. limited use of special lane and street for bus with the purpose of promoting the use of public transport and limiting the use of private vehicle from entering Central Business District (CBD) area.
8) *Parking Controls Transit Ridership,* i.e. support for the public transport user through controlling the number of parking space and fee for private vehicle. Arrangement through parking disincentive scheme is generally more effective than increasing gasoline tax or vehicle ownership cost.
9) *Changes of Work Schedules,* i.e. development of strategy through the distribution of user peak demand so that the amount of demand can be adjusted with the capacity of public transport. Included in this scheme is work schedule arrangement. Scheduling is one of the strategic policy stages in achieving user satisfaction of public transport [17] scheduling system with maximum synchronization level aims to provide bus services on the main route network in accordance with the demand at that time [18].

4. Optimation Analysis on the Integration of Transjakarta-Kopaja Bus Operation System

4.1. Rationalization of Unnecessary Service

This parameter is divided into two, i.e. passenger loading-unloading system and tariff payment system. Currently, there are still many passenger loading-unloading systems outside the official bus stop related to bus stop infrastructure insufficiency, many of the stopping points are only equipped with signage, such as in Karet, Casablanca City, and Pantai Indah Kapuk bus stops, without adequate facilities. Second, tariff payment is still a dual system with 'stick and go' payment method using electronic card or e-money and cash payment in the vehicle using EDC.

Table 1. Analysis of Rationalization of Unnecessary Service

Parameter 1	Rationalization of Unnecessary Service
Changes in service system of Kopaja, then vs transjakarta feeder bus, now Passenger loading-unloading system; Tariff payment system	
Description of factor that has already been fulfilled	Description of factor that has not been fulfilled
• Tariff payment system is still divided into two; payment at the ticket box and in the bus • Passenger transport system: at the bus stop and appropriate signage	• Passenger unloading system: unloading is still conducted freely near the last bus stop • Information system for bus users: 4 out of 9 bus stops do not install information on Transjakarta Feeder Bus route; 2 out of 9 bus stops do not install information on feeder bus ticket at the ticket box 7 out of 9 bus stops do not have real-time information screen on transjakarta feeder bus

4.2. The Appropriateness of Feeder Transport to Support Main Transport

Based on the survey, out of nine main transjakarta feeder bus stops, only seven have access to facilitate a transfer to another mode within the radius of 400-800 meters with the percentage of public transport availability of 31.08%.

Table 2. Matrix of Public Transport Availability

Public Transport Radius 400-800 Meter	KOPAJA	Metromini	Angkot	Mikrolet	KRL station
National Monument	X	X	X	X	X
Pantai Indah Kapuk	X	X	X	X	X
Ragunan	1. P-19 2. S-68 3. S-602 4. 605A 5. S-612	S-77	KWK S-15A	M-42	X

Public Transport Radius 400-800 Meter	KOPAJA	Metromini	Angkot	Mikrolet	KRL station
Dukuh Atas	X	X	X	X	Dukuh Atas Train Station
Lebak Bulus	1. P-20 2. B-86 3. S-615	X	X	X	X
Senen (through Cikini station)	1. P-20 2. U-27 3. S-615	X	X	X	Dukuh Atas Train Station
PGC	1. T-57 2. S-68 3. T-505 4. T-503 5. T-501	1. 58 2. 60 3. S-64	Cililitan Bus Stop 1. T-02 2. T-17	Cililitan Bus Stop 1. M-19 2. M-29 3. M-04	X
Harmoni	X	X	X	1. M-12 2. M-08	X
Tugas	X	1. T-47 2. 077	X	X	X

Table 3. Analysis of Appropriate Transportation Mode as Main Transportation Support

Parameter 2	Transportation Mode Appropriate as Main Transportation Support
Supporting transportation around the KOPAJA feeder bus stop is to connect the trip of Transjakarta bus users (example: regular KOPAJA, metro mini, angkot, mikrolet, KRL.)	
Description of factor that has already been fulfilled	Description of factor that has not been fulfilled
There are seven transjakarta feeder bus stops within the radius of 400-800 meters that enable users to continue the trip	There are two bus stops of the feeder bus in National Monument and Pantai Indah Kapuk that do not have supporting transportation mode.

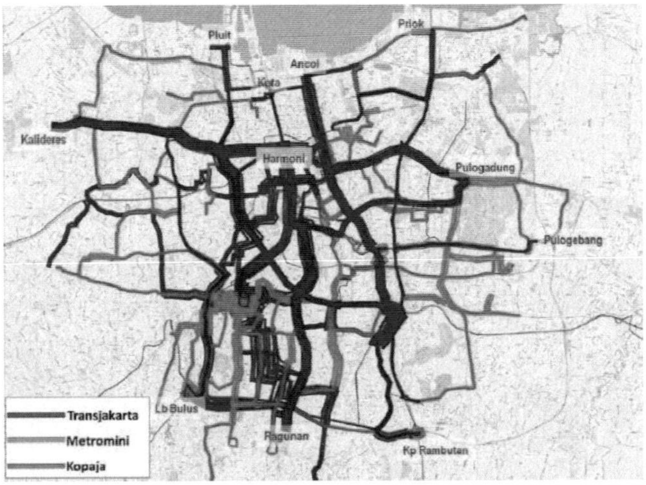

Figure 1. Service coverage of Transjakarta, Kopaja, Regular and Metromini Transport

4.3. Unification of Transport Tarif Structure

Kopaja that becomes the operator of Transjakarta bus feeder experiences tariff adjustment integrated with the tariff structure owned by PT Transjakarta. The tariff structure of regular transjakarta bus (not feeder) has two-time divisions, i.e. 05.00 – 07.00 is Rp. 2,000, whereas 07.00 – 23.00 is Rp. 3,500. However, Kopaja feeder bus does not have time-based tariff division. During the operation from 05.00 – 23.00, Kopaja bus has similar tariff to regular transjakarta bus, i.e. Rp. 3,500. The entire ticket payment systems for Kopaja bus are done by officials from PT Transjakarta and given to on-board officials of PT Transjakarta and the data are stored using EDC. With this system, PT Transjakarta has full responsibility for managing ticket money of the bus users and tariff that has been stipulated.

Table 4. Analysis of the Unification of Transportation Tariff Structure.

Parameter 3	The Unification of Transport Tariff Structure
The making of one integrated payment system for the bus users; Comparison between the KOPAJA tariff bus, then and transjakarta feeder bus tariff, now	
Description of factor that has already been fulfilled	Description of factor that has not been fulfilled
Transport tariff structure has been fulfilled by paying the same tariff as regular transjakarta bus with a tariff of Rp. 3,500	-

4.4. The Enactment of Tariff Discount

According to Chien & Yang (1999), the criterion for optimizing feeder routes on non-regular networks (not dedicated lines) is to lower total costs, including supply and passenger costs. Tariff discount is enacted to facilitate minority users, such as elderly, physically handicapped using wheel chair, and toddlers. However, both on regular transjakarta bus and feeder bus there is no discount for those minority users, except for children under two years old are exempt from paying duty; whereas, tariff discount for off-peak hour is only enacted for regular transjakarta bus service. Since the number of transjakarta feeder bus route is still too few to reach profit level, discount for user group above is not enacted.

Table 5. Analysis of the Enactment of Tariff Discount

Parameter 4	The enactment of Tariff Discount
- Discount for elderly, physically handicapped, and under age; - Tariff discount during the off-peak hour. - simplification of feeder bus ticket payment system (weekly or monthly system)	
Description of factor that has already been fulfilled	Description of factor that has not been fulfilled
In general, there is no tariff discount for transjakarta feeder bus, or for off-peak hour, except on DKI Jakarta birthday on June 22 enacting free of payment for the entire transjakarta transport.	-

4.5. Fine Payment System for the Users

The fine payment system of the users is enacted to the Kopaja bus users who violate the payment obligations. However, since transjakarta bus corridor is relatively small and directly close to ticket gateway, there is not much violation. Likewise, inside the feeder bus during the trip, on-board officials make sure that passengers have payment receipt from the EDC. If there is a violation, the users are given sanction by having to pay directly to the officials of PT Transjakarta. This system prevents the possibility of a free rider.

Table 6. Analysis of Fine Payment System for Bus Users

Parameter 5	Fine Payment System for Bus Users
Monitoring system of the feeder user and fine for violation by the users	
Description of factor that has already been fulfilled	Description of factor that has not been fulfilled
Not many violations from the users at the transjakarta bus stop.	There is no official sanction yet for fine payment of bus users.

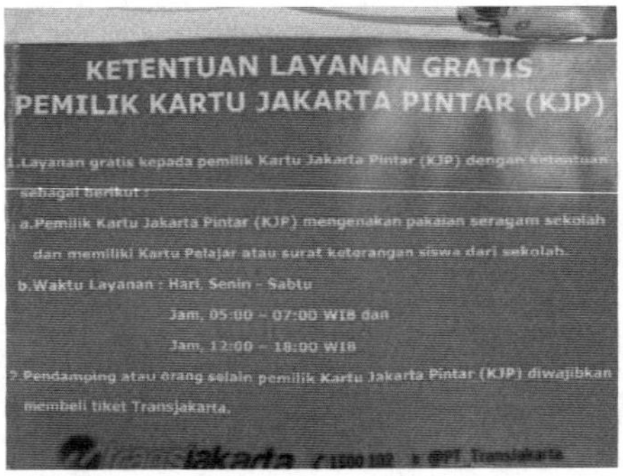

Figure 2. Static Information for the user of Smart Jakarta Card

Figure 3. Static Information of the Transjakarta Passenger code of conduct

4.6. Information System of Coordinated Public Transportation

Bus passenger satisfaction is closely related to the reliability of information system informing the time of departure and arrival from and at every bus stop. With the support of digital technology, it is easier to develop a real-time information system for users in the form of electronic panel, digital information on LED panel and other dynamic information that can be accessed with celular pone or smartphone. In addition to dynamic information, static information for users is also important such as signage leading to the bus stop, sticker of route number on the bus, bus network map installed at the bus stop, bus route map, local area map, emergency indication, and other information.

Analytical result of the performance of operational integration on this aspect shows that the first and dynamic information currently installed at the Transjakarta bus stop, particularly information for Kopaja feeder bus, is still minimal. Digital information screen notifying the estimation of bus arrival with bus fleet and delay information is only installed at the main corridor. Official information provided by PT Transjakarta regarding the operation Schedule of Transjakarta bus and its feeder uses official account of PT Transjakarta twitter at @pt_transjakarta that provides an update transjakarta bus activities and all changes from the account. However, this account system is still categorized as manual and has a very low control system.

The second option in obtaining information on integrated feeder transport is by using GO-BUSWAY which is an application of GO-JEK. GO-BUSWAY has better information platform. Itis easy to use through smartphone and integrated with the use of GO-JEK. The current GO-BUSWAY application is a real-time application capable of looking at transport trip through GPS tracker. However, the disadvantage of this application is that it cannot show real-time changes due to information management factor which is indirectly done by PT Transjakarta.

Table 7. Analysis of Information System of Coordinated Public Transportation.

Parameter 6	Information System of Coordinated Public Transportation
Are the information on route, schedule, tariff and transfer point for Transjakarta feeder bus adequate or not?	
Description of factor that has already been fulfilled	Description of factor that has not been fulfilled

Parameter 6	Information System of Coordinated Public Transportation
Go-Busway application is available There is an update of transjakarta bus trip on twitter account; There is a digital information board on the main route of transjakarta bus.	Official application directly held by PT Transjakarta provides real-time information of bus trip.

Figure 4. Real-time Information Screen of Transjakarta Bus.

Figure 5. Application of Go-Busway from Go-Jek Application

4.7. Specific Street and Lane for Bus

According to [20], in a more operational sequence, the transport carrier operating system may use a "Demand Adaptive System" consisting of a service line defined by trip scheduling serving the predefined core corridor. Basically, feeder transport is intended to carry feeder outside the main corridor, therefore, PT Transjakarta has fleet with wide service coverage in the urban area. Currently, feeder transport operation is still in the main corridor route until undetermined time. The presence of feeder transport in the same corridor as regular transjakarta bus will add headway time and the corridor will be congested.

Specific lane can be segregated and applied in different ways according to permeability (example: delineator, lane with different color, etc. In a certain design, the bus stop can have a role as a barrier. Several systems using permeability can be suggested since sometimes the broken bus can blockade busway corridor and should leave the corridor.

Table 8. Analysis of Information System of Coordinated Public Transportation

Parameter 7	Specific Street and Lane for Bus
Specific lane designed for bus in certain areas, reliability, and speed of public transportation to fix the quality of transjakarta bus service	
Description of factor that has already been fulfilled	Description of factor that has not been fulfilled
-	Clear route for feeder transport

4.8. Parking Control for Public Transportation Users

The percentage of private users in Jakarta reached 45.2%, a combination of twowheeler and fourwheeler. This fact underlies the importance of parking facility development for the convenience of the transit of private vehicle user to public transport, including transjakarta bus.

Out of nine bus stops that were observed, there are three bus stops owning parking location that can be accessed by bus users. The criteria for the calculated parking location are the location that is close to the bus stop; normal parking tariff is Rp. 4,000 – Rp. 5,000 per hour and it does not need ID card to park. Dukuh Atas bus stop is located next to Car Insurance Office with no parking area for two-wheeler vehicle that is commonly used by 72% of the transjakarta bus users. Meanwhile, the National Monument bus stop is located close to government offices that require specific ID to park. Finally, the TU Gas bus stop is new route so that the location makes it difficult to get access to a good parking space.

Tabel 9. Parking control for Public Transport Users

Parameter 8	Parking control for Public Transportation Users
The presence of parking location which is accessible to transjakarta bus user around transjakarta feeder bus stop.	
Description of factor that has already been fulfilled	Description of factor that has not been fulfilled
Pantai Indah Kapuk Ragunan Lebak Bulus Senen (Cikini Station) PGC Harmoni	Dukuh Atas National Monument TU Gas

5. Analysis on the Selection of Transport System Optimation of Jakarta Feeder Bus

The optimation assessment of Transjakarta Feeder Bus is a process of assessing the result of observation on feeder bus conducted in the field. The form of optimation system selected is an optimation system with the most critical score to be developed for transjakarta feeder bus. This shows that the selected form is an optimation form of transjakarta feeder bus that is the most potential one to increase feeder transport service aggregately and to attract more new users. The range of component point is 1 – 5 (minimum point 1 and maximum point 5) with the definition as follows:

1 = Very unnecessary to be optimized

2 = Unnecessary to be optimized
3 = Maybe optimized or not
4 = Necessary to be optimized
5 = Very necessary to be optimized

The assessment of each criterion is also considered by the shareholder from PT Transjakarta and Kopaja. There are some assumptions used to conduct an assessment of every parameter. Fisrt, rationalization parameter of unnecessary service means that to optimize Transjakarta feeder transport; the emphasis should be on the rationalization of unnecessary service because of critical factor to be fixed and should have an impact significantly on the conveniency of transjakarta bus users.
Second, parameter of transportation mode that is appropriate as feeder of main transportation explains that the most important factor to be optimized for transjakarta bus transportation is the one that facilitates users to transfer between bus stops by using other transportation.

Third, the unification of the most appropriate transport tariff structure as parameter of optimation system of transjakarta feeder bus is meant to use transjakarta with similar tariff to regular transjakarta bus, and cheaper than regular transport such as Kopaja, Metromini, angkot, and Microlet.

Table 10. Optimation Assessment of Transjakarta Feeder Bus System by PT Transjakarta

NO.	PARAMETER	ASSESSMENT
1	Rationalization of unnecessary service	5
2	Appropriate mode of transportation as main transportation support	5
3	Unification of transport tariff structure	1
4	The enactment of tariff discount	1
5	System of user fine payment	3
6	Information system of coordinated public transportation	5
7	Specific street and lane for bus	2
8	Parking control for public transportation users	1

6. Conclusion

The analysis of optimation system of Transjakarta feeder bus transport shows that the integration of operation system implemented by PT Trans Jakarta for Kopaja freeder transport has not yet run at maximum level. The analysis result shows that feeder transport still operates at the main corridor of Transjakarta bus and has not fulfilled its function as feeder system that serves the main Transjakarta service corridor.

The number of feeders targeted to operate is 320 Kopaja buses, but only 200 buses are in operation. It shows the failure of Kopaja management to fulfil the operational requirement that has been stipulated by PT Transjakarta.

Based on the assessment of the second parameter, transjakarta feeder bus has increased its quality service, such as passenger loading and unloading in bus stop with orderly tariff payment system at the bus stop ticket box only, and uses electronic card to reduce cash money by the officials. However, feeder transport operation still needs integration with another mode of transportation from regular Kopaja, Metromini, angkot and mikrolet whose fulfilment is not maximal due to the overlapping inter mode route.

Third parameter fulfilment shows that, in general, the tariff structure of the feeder is the same as transjakarta bus, i.e. Rp. 3,500. Thus, it is quite successful. The fourth parameter has not really been achieved, special tariff for elderly and children under five years old has not been equated to the system of transjakarta bus. The fifth parameter has not really been achieved as well , the system of fine payment for those who violate rules at the bus stop or inside the bus is not clear. PT Transjakarta has only installed a poster not to violate the rules. It should show a clearly stated fine and is not limited to verbal reprimand for those who violate.

Sixth parameter achievement, information on feeder bus at the transjakarta bus stop is still minimal. This is measured by the presence of poster, digital information screen and smartphone for Transjakarta bus information. The manual poster that is normally used for tariff notification and route information is not installed equally at feeder bus stops. It causes users' ignorance regarding the availability of the feeder service, which results in the decline in feeder passenger. Currently, information screen is only available in the main corridor due to high maintenance cost. This information screen provides departure and arrival time of buses passing certain bus stops which is important for the bus service reliability. Currently, the crucial aspect of Transjakarta bus information system is the unavailability of official application from PT. Transjakarta for regular Transjakarta and its feeder system. Considering the smartphone is now the most practical information media, it is predicted that, if all required information regarding the Transjakarta service system, such as schedule, seat capacity, etc. is accessible through smartphone, the number of passengers using transjakarta will most probably increase.

The seventh parameter is a special lane of transjakarta bus that explains whether the presence of feeder transport is fully within the special lane of transjakarta bus corridor or it is a mixed-traffic way with other vehicles. The analysis shows that the majority of feeder transport should operate in a mixed-traffic way because it will reach locations that do not have transjakarta bus corridor in the feeder area.

Finally, parking control parameter is useful for bus users to support park-and-ride scheme. It supports private car user accessibility to transfer to public transportation. However, at the moment, there is no parking control officially provided by PT Transjakarta for the users and officials at the bus stop.

In general, it can be concluded that optimation system of feeder bus transport operation which involves PT Transjakarta as an institution overseeing Kopaja shows that the parameter operational integration is not been fully applied, so that the feeder operation system is not optimum.

Meanwhile, based on the scoring done by the management of PT Transjakarta, there are critical parameters that require optimation programs for a better operation system integration, i.e.: 1) rationalization of unnecessary service; 2) appropriate mode of feeder transportation as the support of main transportation; and 3) a better information of integrated public transport services system.

References

[1] Hine J 2000 Integration planning for sustainable and integrated transport systems in the new millennium *Transport Policy Elsevier* **7(3)** 175-177

[2] Janic M and Reggiani A 2001 *Integrated transport systems in European Union: An overview of some recent developments* Transport Reviews **21(4)** 469-497

[3] Ibrahim M 2003 Improvements and integration of a public transport system: the case of Singapore *Cities* **20(3)** 205-216

[4] Stead D 2003 Transport and land use planning policy in Britain really joined up? *International Social Science Journal* **55(2)** 333-347

[5] Potter S and Skinner M 2000 On transport integration: a contribution to better understanding *Futures* **32** 275-287

[6] Hull A 2005 Integrated transport planning in the UK: from concept to reality *Journal of Transport Geography* **13** 318-328

[7] Dydkowski G 2005 *Integration of the local collective transportation Proceedings of the*

National Conference: Transportation systems Theory and practice (Gliwice: Silesian University of Technology) 107-116

[8] Givoni M and Banister D 2010 *Integrated transport: from policy to practice* (Abingdon: Routledge)

[9] GUIDE 1999 *Urban interchanges - A good practice guide project funded by the European Commission under the Transport RTD Programme of the 4th Framework Programme Final Report*

[10] Hilferink P, Roest Crollius A and Van Elburg J C 2003 *Integration and regulatory structures in public transport Final Report Study commissioned by European Commission DG TREN Rijswijk*

[11] Żak J, Fierek S and Kruszyński M 2014 Evaluation of different transportation solutions with the application of macro simulation tools and multiple criteria *Procedia Social and Behavioral Sciences* **111** 340 –349

[12] PROSPECTS 2003 *European Union research project: Procedures for recommending optimal sustainable planning of European city transport systems Final Report Leeds.*

[13] Verma A and Ramanayya T V 2015 *Public Transport Planning and Management in Developing Countries* (Florida : CRC Press).

[14] Savas E S 1978 On Equity in Providing Public Services *Management Science* **24(8)** 800-808

[15] Yim Y B and Ceder A 2006 Smart Feeder Feeder/Shuttle Bus Service: Consumer Research and Design *Journal of Public Transportation* **9(1)**

[16] Vuchic V R 2007 *Urban Transit Systems and Technology* (Canada : John Wiley & Sons, Inc)

[17] Ceder A and Wilson N H M 1986 Bus Network Desain *Transportation Research Part B: Methodological* **20(4)** 331-344

[18] Ceder A 2001 Operational Objective Functions in Designing Public Transport Routes *Journal of Advanced Transportation* **35(2)** 125-144

[19] Chien S and Yang 1999 Optimal Feeder Bus Routes on Irregular Street Networks *Journal of Advanced Transportation* **34(2)** 213-248.

[20] Malucelli F, Nonato M and Pallottino S 1999 Demand Adaptive Systems: Some Proposals On Flexible Transit *Operations research in industry* 157-182

Development of Methodology to Evaluate TOD Feasibility in Built-up Environment (Case Study: Jakarta and Bandung, Indonesia)

P Dirgahayani[1] and D N Choerunnisa[1]

School of Architecture, Planning and Policy Development, Institute Technology of Bandung, Jl. Ganesha 10, Bandung, 40132, Indonesia

Corresponding authors: pdirgahayani@sappk.itb.ac.id

Abstract. The major Indonesian cities, among which Jakarta and Bandung, are in the process of constructing and planning urban railway systems, including developing transit-oriented development (TOD) concepts for their stations. For ensuring the targeted ridership and financial sustainability, these TOD plans should be integrated with the mass transit development plan. However, to adopt this is quite challenging since the planned railway lines run through dense built-up environments with limited vacant land available. This study aimed to develop a methodology to evaluate the feasibility of designating built-up environments as TOD areas. The methodology used consists of two elements: (i) a method for determining the suitable TOD typology; and (ii) a method to assess the feasibility of TOD in potential areas based on the requirements of its TOD typology. For the first this study adopted a mixed-method approach comprising of two steps: (i) formulating a set of criteria and indicators based on basic TOD characteristics and the 5 Ds (Calthorpe, 1993; Ewing & Cervero, 2010) as well as related national and local regulations; and (ii) conducting a gap analysis of the transit location characteristics (existing and as part of the spatial plan) based on these criteria and indicators. The methodology was developed based on case studies of potential TOD areas in Jakarta and Bandung. It was found that there are differences between the development requirements based on area characteristics and TOD typologies.

1. Introduction

Transit oriented development (TOD) or transit-based development planning often emerge in discussions on increasing mass transportation system development in the major cities of Indonesia. Basically, TOD is a developmental concept that presents mixed land use integrated with public transportation networks [1].

Developing an area based on TOD aims to minimize the movement of vehicles by ensuring that the population living in the TOD area can work or do their other activities within the area by walking. Meanwhile, if they need to do activities outside the area, they can access public transportation easily and comfortably as the development is designed to make connectivity to and from transportation nodes easier, whether by walking, cycling, or using public transportation systems.

TOD can be implemented in three types of locations [1], i.e. re-developable sites, infill sites, and new growth areas. Re-developable sites are developed areas that can be revitalized. Infill sites are empty spaces surrounded by developed urban areas. New growth areas are undeveloped spaces that are going to be developed, commonly found in newly developed cities or regions. There are main five principles for implementing TOD, known as the 5 Ds [2], i.e. *density*, *diversity*, *design*, *destination accessibility* and *distance to transit*. The 5 Ds should be implemented in areas that use TOD in such a way that the development can be in line with expectations.

Generally, most TOD in Indonesian cities happens in built-up environments, either re-developable sites or infill sites. Transforming a built-up environment into a TOD area along with its attributes is challenging. From a regulation perspective, in Indonesia there are two legal foundations to guide such transformation: (i) Regulation of the Ministry of Public Works No. 6/2007 on Building and Environmental Design Plan General Guidelines (RTBL); and Regulation of the Ministry of Agrarian and Spatial Planning No. 16/2017. However, these policies do not explain how to develop and apply the TOD provisions according to the area's characteristics.

Accordingly, this study was aimed at developing a methodology for evaluating TOD feasibility specifically in built-up environments, departing from the hypothesis that planning challenges are different in different areas, especially in built-up environments. In some cities where TOD was implemented, mistakes were made, rendering TOD less effective. Examples are Coral Gables, Miami, where pedestrian needs were insufficiently considered, and the development feasibility as reported in *City Lab* [3].

This paper consists of two parts. First a suitable TOD typology is determined and the composition of diversity and floor-area ratio (FAR) of the planned TOD locations are calculated to meet the ridership requirements. A set of development criteria and indicators is formulated based on the 5 Ds and the basic interests of TOD. These criteria and indicators are then used to evaluate the existing characteristics of the planned TOD locations and associated with their spatial plans. The second part contains a gap analysis of the areas (according to existing conditions and spatial plans) based on the criteria and indicators. The output of this paper is the recommendation of a methodology to evaluate TOD feasibility in built-up environments based on the criteria and indicators of the TOD principles and interests, existing characteristics, and spatial plans.

2. Case Studies

Jakarta as the capital city of Indonesia and Bandung as the capital city of West Java Province will develop their railway stations using TOD as part of their spatial plan. In Bandung, TOD will be used for its LRT stations. Bandung City has inserted TOD in their 2011-2031 general spatial plan and detailed spatial plan. TOD was used to support LRT development in the greater Bandung metropolitan area as written in Mayor's Decree Nr. 1175 2015 on Transportation Master Plan of Bandung City. Inside this master plan, nine LRT corridors are planned, two of which are prioritized. In accordance with the LRT development within the transportation master plan, the development is also supported by a presidential draft decree on the acceleration of building an integrated LRT system in the greater Bandung area [4]. Cimindi, Bandara Husein, Pajajaran, Martadinata, Antapani, Gedebage, and Tegalluar are 7 of the 23 planned stations.

In Jakarta, TOD will be implemented for MRT and LRT stations. Jakarta's TOD plan is also included in the macro spatial plan (RTRW) and detailed spatial plan (RDTR). TOD in the MRT line connecting Lebak Bulus and Kampung Bandan will be implemented in two phases, with Lebak Bulus, Fatmawati, and Dukuh Atas Station as well as the areas of Cipete and Blok M in phase I and Depo Kampung Bandan-Ancol in phase II. Furthermore, TOD along the LRT line connecting Kelapa Gading and Velodrome will also be done for 6 stations, i.e. Depo Station in Pegangsaan Dua Kelapa Gading Street, Mall Kelapa Gading Station, Kelapa Gading Boulevard Station, Pulomas Station, Pacuan Kuda Station and Velodrome Station.

3. Methodology

This study used a mixed-method approach to collect and analyze the data. Firstly, a secondary survey among relevant agencies was conducted to identify the prevailing spatial plans and transportation plans. Secondly, a literature study was conducted to compose criteria and indicators based on the 5 Ds and

TOD related regulations. Thirdly, field observation was done to understand the existing conditions in the TOD areas.

Generally, formulating the methodology to evaluate TOD feasibility in built-up environments was done in three main steps, as shown below.

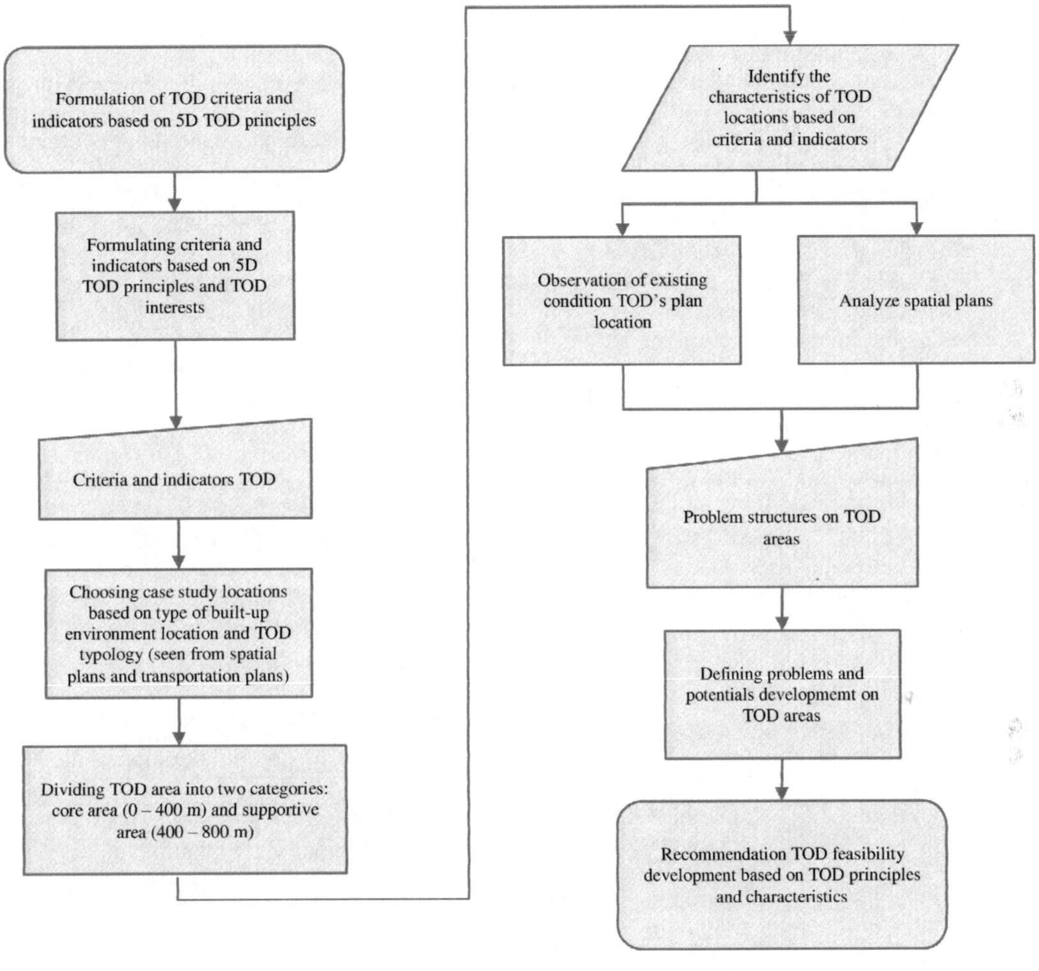

Figure 1. Methodology flow chart.

Basically, the different conditions of the core and its supporting areas will follow several conventions:

- The farther the area is from the station, the more the density and height of the buildings will decrease;
- The nearer the area is from the station, the more diverse and attractive land use will be;
- The denser the area, the wider the pedestrian ways will be;
- There will be limited parking space, especially in the core area. Park-and-ride will be placed in the supporting areas adjacent to the core area;

- The core area will mostly be served by public transportation;
- There will be several feeder public transportation nodes in the supporting areas for the core area.

The next process in composing these guidelines is to compose criteria and indicators based on the 5 Ds. Ewing and Cervero [2] describe these 5 principles as follows:

a. Density: the density of the area surrounding the radius of the transit service area that supports the main mass transportation is such that this system works adequately.
b. Diversity: diversity of land use and building use per area width, floor width, or employment field.
c. Design: integrated design of the spaces between one area and another depends on pedestrian movement and includes traffic patterns, pedestrian facility availability, and bicycle facility availability.
d. Destination accessibility: easy access to destinations both inside of the TOD area itself and between TOD destinations citywide.
e. Distance to transit: proximity between home or workplace and station, bus station or other public transportation nodes.

Based on the components of each principle, the criteria and indicators can be defined as reference for TOD. The criteria and indicators were composed through reviewing literature on TOD and TOD precedents. For example, *MARTA* [5], *Reconnecting America's Center for Transit Oriented Development or FTA* [6], *Department Planning, Property, and Development City of Winnipeg or DPPDCW* [7], *Land Use Planning and Policy Department City of Calgary or LUPPDCC* [8], and *ITDP* [9]. The criteria and indicators are also based on the TOD typologies. Even though the basic principles of TOD seem to be the same in all contexts, their specific application greatly differs in form, function and impact, calling for context-based TOD. The typology of TOD can help to map local specificities and better focus policy interventions [9]. Below are the criteria and indicators that were composed to identify the conditions at the planned TOD locations.

Table 1. TOD Criteria and Indicators. [2, 6, 7, and 12]

Criteria	Indicators				
	Regional Center	**Urban Center**	**Urban Neighborhood**	**Suburban Center**	**Special Uses**
Density					
	Minimum FAR is 5.0 or 4 – 30 stories	Minimum FAR is 2.5 or 2 – 20 stories	Minimum FAR is 1.0 or 2 – 5 stories	Minimum FAR is 4.0	Minimum FAR is 2.5
Building Density and Intensity Levels	Building Density and Height: • Core Area: High • Supporting Area: High-Mid	Building Density and Height: • Core Area: High-Mid • Supporting Area: Mid-Low	Building Density and Height: • Core Area: Mid - Low • Supporting Area: Low	Building Density and Height: • Core Area: High-Mid • Supporting Area: Mid	Building Density and Height: • Core area: mid, depend on their function • Supporting area: mid-low, depend on their function
Diversity					
Land Use Diversity to Support Various Activities	At least consists of: • Employment	At least consists of: • Residential buildings	At least consists of: • Residential buildings	At least consists of: • Residential	At least consists of:

		• Residential buildings • Supporting Facilities With commercial land use as the dominant land use	• Employment • *Mixed Use* • Supporting Facilities	• Commercial/ office buildings • *Mixed use* • Supporting facilities With residential land use as the dominant land use	• Commercial • Employment • Supporting Facilities	• 20% residential buildings • 10% retail, commercial or public facilities.
		• Target number of residential buildings: 8,000-30,000 • Target number of employment: 40,000-150,000	• Target number of residential buildings: 5,000-15,000 • Target number of employment: 5,000-30,000	• Target number of residential buildings: 2,500-10,000 • Target number of employments: -	• Target number of residential buildings: 2,500-10,000 • Target number of employment: 7,500-50,000	• Target number of residential buildings: 2,000-5,000 • Target number of employment: 7,500-50,000
Design						
Walk-ability	Encourages walkability	• Safe and comfortable pedestrian ways • Standardized pedestrian way width and wider pathways for denser areas • Continuous and distinct pedestrian ways • Connected with cross pedestrian ways				
	Parking system that discourages private transportatio n use	Provision of parking facilities applying the travel demand management principle in order to improve mode transfer to public transportation with different standards and with applicable parking standards. Off-street facilities can be used as park-and-ride.				
	Supports bicycle uses	There is a special bicycle lane and parking/rental center to safely and easily borrow bicycles				
	There is an open space or park to support place making	There is an open space or park that is accessible by walking and able to support interactions between people				
Destination Accessibility						
Easy to access public transportation		Served by all types of public transportation (interprovince, intercity, city, and country)	Served by intercity, city and country public transportation	Served by city and country public transportation	Served by intercity, city and country public transportation	Connected to modes that connect residential and activity centers to special use locations
		There are transport nodes/hubs in the TOD area				
Distance to Transit						
Accessible transit location from the activity core		Coverage of transportation services is accordance with walking distance				

IOP Conf. Series: Earth and Environmental Science **158** (2018) 012019 doi:10.1088/1755-1315/158/1/012019

4. Analysis and Discussion

4.1. Identification of TOD Typology

In identifying the normative TOD typology, there are two planning documents that should be consulted, i.e. the regional spatial plans – both macro (RTRW) and detailed (RDTR) – and the transportation system master plan. Based on the regional spatial planning the activity center plan around the planned TOD location was analyzed that will have an impact on the development of the TOD location. This refers to the spatial structure plan. Meanwhile, land use diversity refers to the spatial pattern plan. The transportation master plan provides to other public transportation plans, besides the LRT, that will affect the TOD location (potentially becoming a feeder node). The transportation master plan usually also includes other plans and policies regarding transportation, such as parking space, travel demand management, etc. that can be relevant to TOD implementation.

The following table shows the urban spatial plan and transportation plan at each TOD location.

Table 2. Development and Transportation Plan in TOD Plan Areas.

No	TOD Location	Urban Spatial Plan	Transportation Master Plan
1	Husein Sastranegara TOD Area*	Located in the area of Husein Sastranegara International Airport, which is the only airport in Bandung City. Connecting Bandung City with other regions.	Connect with LRT Corridor II Bandung and will be traversed by Bandung City's bus line plan.
2	Martadinata TOD Area*	There is no specific plan for activity center development around the Martadinata TOD area. However, the Martadinata TOD area is located in the center of Bandung City in a strategic location for commercial and service purposes.	Connect with LRT Corridor II Bandung and will be traversed by Bandung City's bus line plan.
3	Antapani TOD Area*	A residential area development is planned in the Antapani TOD area.	Connect with LRT Corridor II Bandung and will be traversed by Bandung City's bus line plan
4	Kelapa Gading TOD Area**	The Kelapa Gading area is one of the secondary activity centers in DKI Jakarta. Secondary activity centers play a role in serving provincial scale activities. The Kelapa Gading area is located on the outskirts of DKI Jakarta in East Jakarta.	Connect with LRT Jakarta but there is no specific transportation development planning that will traverse the Kelapa Gading TOD area. However, this TOD area can be connected to the TransJakarta BRT.
5	Ancol TOD Area**	The Ancol TOD area is part of the Taman Impian Jaya Ancol coastal tourist destination development planning, thus this area is focused on tourism.	Connect with LRT Jakarta but no specific transportation development is planned that will traverse the Kelapa Gading TOD area. However, this TOD

No	TOD Location	Urban Spatial Plan	Transportation Master Plan
			area can be connected to the TransJakarta BRT.

*located in Bandung
**located in Jakarta

Based on the development characteristics plans and transportation plans of both cities, as mentioned in the table above, we can classify the TOD locations into a TOD typology.

Table 3. Typology of TOD Plan Areas

No	TOD Location	TOD Typology
1	Husein Sastranegara TOD Area*	Regional Center
2	Martadinata TOD Area*	Urban Center
3	Antapani TOD Area*	Urban Neighborhood
4	Kelapa Gading TOD Area**	Sub-urban Center
5	Ancol TOD Area**	Special Uses

*located in Bandung
**located in Jakarta

4.2. Identifications Characteristics of TOD Areas

4.2.1. Characteristics of TOD Areas in Bandung

A. Characteristics of Husein Sastranegara TOD Area (Regional Center Typology)

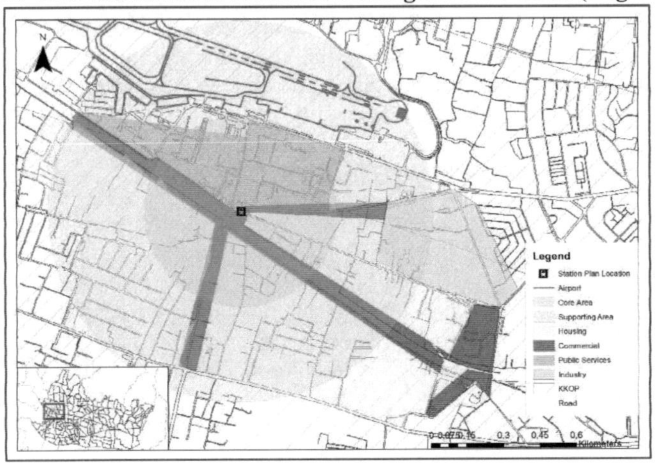

Figure 2. Husein Sastranegara's TOD Map.

Husein Sastranegara TOD area is classified as a regional center according to the urban and transportation development planning. The existing conditions in the Husein Sastranegara Airport area do not fulfill all TOD indicators, particularly those related to the typology of a regional center. Only 4 out of 15 indicators are fulfilled. The indicators density, diversity and design are unfulfilled, but the design principle can be fixed because in Bandung City's *RDTR* there are revitalization plans for the pedestrian ways, which can improve the pedestrian way quality according to TOD standards. As for density, the average building has 1-2 floors in the core area and 1-3 floors in the supporting areas with an average FAR of 1.2-2.0, which is usually smaller closer to the core area. Furthermore, the location is within a flight operation safety zone, which imposes limitations on development over a certain height. That is why buildings in this location cannot be developed as high-rise buildings, especially in the core area. The core activity has not been planned yet as one of the attractions in this TOD location. As for diversity, land use in Husein Sastranegara is not diverse. Land use is dominated by settlements, while regional centers are supposed to be dominated by commercial land use. It is necessary to adjust the land use according to the TOD principles. Several public transportation modes, such as micro buses (*angkot*) and

IOP Conf. Series: Earth and Environmental Science **158** (2018) 012019 doi:10.1088/1755-1315/158/1/012019

city buses, pass Husein Sastranegara. They could function as feeders for the main transit system, which is the LRT. According to the public transportation routes, Husein Sastranegara is well connected with the main hubs of Bandung City.

B. Characteristics of Martadinata TOD Area (Urban Center Typology)

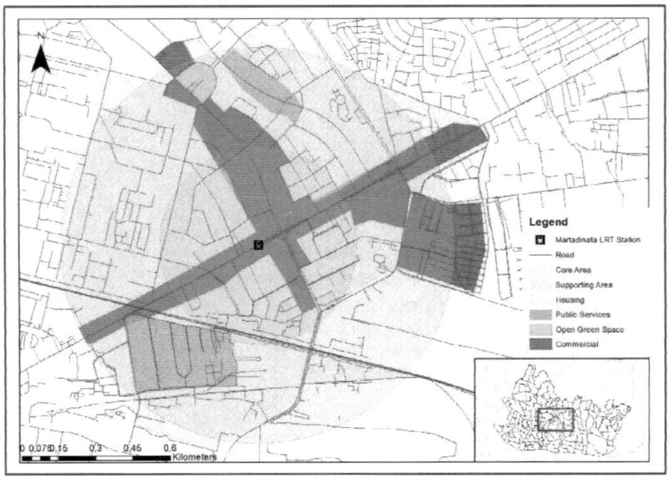

Figure 3. Martadinata's TOD Map.

The Martadinata TOD area is classified as an urban center according to the urban and transportation development planning. The existing conditions do not fulfill all criteria; only 6 out of the 15 indicators are fulfilled. Especially the density and design principles are not met. As for density, building intensity in this location is still below the minimum intensity that should be in a TOD, especially according to the typology of an urban center, and the intensity in supporting the areas is higher than in the core area. As for design, the existing conditions can be fixed because Bandung City's *RDTR* contains pedestrian way revitalization plans, which can improve the pedestrian way quality according to TOD standards; parking management plans; and bicycle lane plans. Hence, development in this location can fulfill all the TOD principles.

C. Characteristics of Antapani TOD Area (Urban Neigborhood Center Typology)

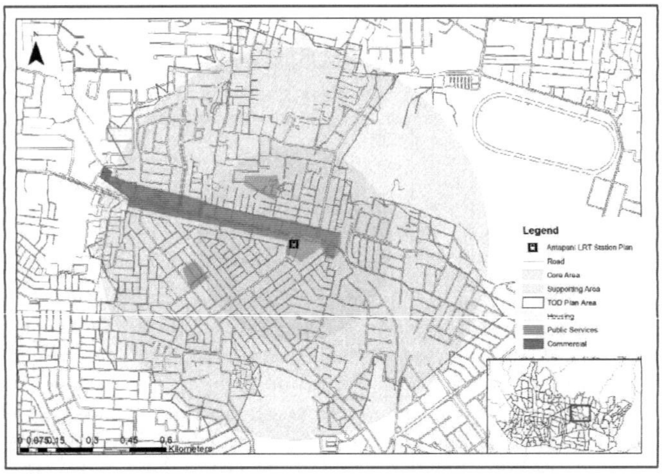

Figure 4. Antapani's TOD Map.

The Antapani TOD area is classified as an urban neighborhood center according to urban and transportation development planning. The existing conditions do not fulfill all criteria; only 5 out of the 15 indicators are fulfilled. The existing conditions mostly do not fulfill the design and distance to transit principles, either of which cannot be fixed because in Bandung City's *RDTR* there is no transportation planned that traverses this area. However, tthere are pedestrian way revitalization plans, which can improve pedestrian way quality according to TOD standards; parking management plans; and bicycle lane plans. As for distance to transit, in Bandung City's *RDTR* there is no plan for core activities or addition of *angkot* routes (as feeder).

4.2.2 Characteristics of TOD Areas in Jakarta

A. Characteristics of Kelapa Gading TOD Area (Sub-Urban Center Typology)

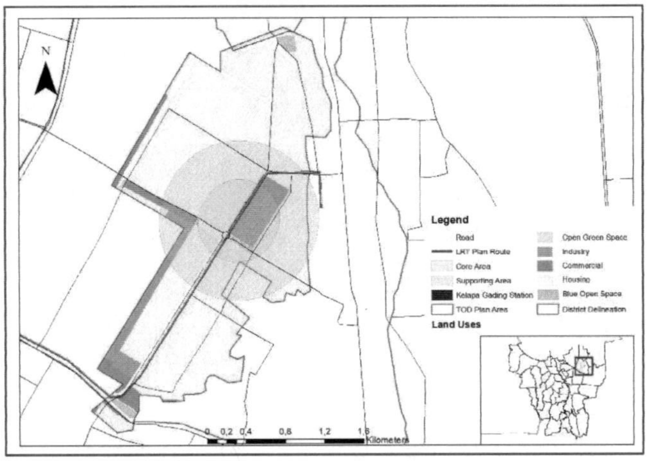

Figure 5. Kelapa Gading's TOD Map.

Kelapa Gading TOD area is classified as a sub-urban center according to the urban and transportation development planning. The existing conditions of the Kelapa Gading TOD area do not fully fulfill the criteria in terms of density, diversity and design. The existing conditions only fulfill 5 out of the 15 indicators. However, the diversity principle will be fulfilled by the development planned included in DKI Jakarta's *RDTR*. As for density, the building floor coefficient given in DKI Jakarta's *RDTR* is still below the FAR provision for sub-urban centers, therefore the development is not in accordance with the provision of this typology. As for design, the existing conditions also do not meet the required indicators but some criteria will be fulfilled because they are listed in DKI Jakarta's *RDTR*.

B. Characteristics of Ancol TOD Area (Special Uses Typology)

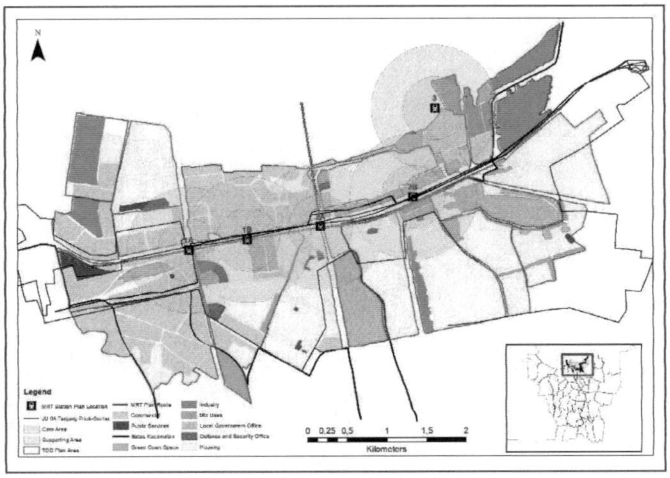

Figure 6. Ancol's TOD Map

The Ancol TOD area is classified as a special use area are according to the urban and transportation development planning. The existing conditions in Ancol do not fulfill all criteria; only 5 out of the 15 are fulfilled. The existing conditions do not meet the density, diversity and design principles. Generally, the existing conditions can also not be fulfilled by DKI Jakarta's *RDTR*, therefore the development of Ancol still has obstacles. Moreover, because of the special use typology, Ancol TOD development is also expected to support tourism development, thus the activities and transit system developed in Ancol should be able to support coastal tourism.

5. Problem Structure in TOD Areas

This section analyzes the gap between the 5 TOD principles of each typology and the existing location characteristics as described in the previous section. Based on this gap, we obtained the problem structure for each location representing each typology. The following table summarizes the result.

Table 4. Problem Structure of TOD Plan Areas.

Principles	Regional Center		Urban Center		Urban Neighborhood		Sub-Urban Center		Special Use	
	Core Area	Supporting Areas	Core Area	Supporting Areas	Core Area	Supporting Areas	Core Area	Supporting Areas	Core Area	Supporting Areas
Density	The density and height of the buildings do not correspond to the typology, neither in the existing conditions nor in the planning	The intensity of the buildings is not appropriate and the supporting area's intensity is higher than that of the core area	The density is lower than that of the supporting area	The density is higher than that of the core area	There is no problem with the density, but the intensity can still be optimized		The density and height of the buildings do not correspond to the typology, neither in the existing conditions nor in the planning	The intensity of the buildings is not appropriate	The density and height of the buildings do not correspond to the typology, neither in the existing conditions nor in the planning	
	For locations that have special provisions, make those special provisions a priority		The development is not yet vertical and compact oriented. Vertical and compact development are possible because the intensity in the zoning regulations of Bandung City's *RDTR* can be optimized				The development is not yet vertical and compact oriented. The intensity in the core area should have higher intensity than the supporting area		For locations that have a special use, the special use should be made a priority	The intensity in the supporting area should be planned to support special uses in the core area
Diversity	Land-use in the existing conditions and in the development, plans are not various enough and cannot support regional scale activities	Because there are limitations in the land use of the core area, the supporting area is more diverse than the core area	In the core area, land use is dominated by commercial and service activities. This is caused by the location in the city center, which serves as a destination. However, the commercial and service activities are not attractive enough to attract many visitors		There is no problem in diversity, but it can still be optimized and made more attractive				The diversity of land use should support the special use of the area. The majority of land use could become special use area	Land use in the supporting area is planned to support the special uses in the core area and to improve transit system ridership
Design	• The pedestrian ways that are available are not sufficient to be used safely and comfortably by pedestrians.									

Principles	**Regional Center**		**Urban Center**		**Urban Neighborhood**		**Sub-Urban Center**		**Special Use**	
	Core Area	**Supporting Areas**	**Core Area**	**Supporting Areas**	**Core Area**	**Supporting Areas**	**Core Area**	**Supporting Areas**	**Core Area**	**Supporting Areas**
	• There should be shared parking lots that are located on the border of the core area and the supporting area. Shared parking should be located at the back of building									
Destination Accessibility	The types of existing public transport serving the core area is still insufficiently diverse. But, there are plans to introduce new routes and modes through the core area. The new public transport needs to be integrated with existing public transport.	There are mass public transport developmen t plans that will traverse the supporting area. They need to be integrated with the existing public transport in order to become good feeders for the core area.	There are no problems with destination accessibility in the core area.	Existing public transport that functions as feeder is not sufficient to serve the core area.	Existing public transport is unable to support movement towards to center of activities.		There are no problems with destination accessibility in the core area.	The closed gate system in residential areas reduces connectivity and access to public transportation .	There are no problems with destination accessibility in the core area.	Residential areas are not yet sufficiently served by public transportation to support the core area.
Distance to Transit	• The distance from the TOD location to existing transportation nodes is ± 2.4 km to Bandung Station and ± 4.5 km to Leuwi Panjang Terminal. Husein Sastranegara Airport can be reached by foot. • Existing public transportation does not serve the entire area inside the TOD but only operates along the main road.		• The distance from the TOD location to existing transportation nodes is ± 3.2 km to Bandung Station and ± 5 km to Leuwi Panjang Terminal. This is quite far. • The availability of transit locations in the northern part is better than in the southern part.		• The distance from the TOD location to existing transportation nodes is ± 7 km to Bandung Station and ± 8 km to Leuwi Panjang Terminal. This is very far. • Existing public transportation does not operate along the major street and there is only one route serving this area. The availability of transit in		• The distance from the TOD location to existing transportation nodes is ± 2 km to the Pulomas TransJakarta Shelter. It is quite isolated but it is connected by a microbus service (*angkot*).		• The distance from the TOD (1st point) location to existing transportation nodes is ± 400 m to Kampung Bandan Station and from 2nd point is ± 200 m to Ancol Station and can be reached by foot. • The distance to Jakarta's primary activity center (Mangga Dua) is ± 1.2 km and could be a potential strength for TOD.	

IOP Conf. Series: Earth and Environmental Science **158** (2018) 012019 doi:10.1088/1755-1315/158/1/012019

Principles	Regional Center		Urban Center		Urban Neighborhood		Sub-Urban Center		Special Use	
	Core Area	Supporting Areas	Core Area	Supporting Areas	Core Area	Supporting Areas	Core Area	Supporting Areas	Core Area	Supporting Areas
					the northern part is inadequate.					

From the analysis summarized in the table above, every location has different problems. However, in general, none of the areas that were studied fulfill all 5 TOD principles. The plans being drawn in the micro spatial plan, both for Bandung and Jakarta, do not accommodate these principles in the future development of the case study areas.

6. Findings and Discussion

This section further discusses the problem structures encountered at each TOD location if the TOD plans are implemented. From the five TOD locations, as mentioned above, none fulfill the TOD principles. This implies that, in the Indonesian context particularly, when a TOD area is located in a built-up environment, it is necessary to take into account the existing conditions when formulating the best-suited development strategies. While managing the area's development, it is also important to integrate the city development with both the existing and planned transit systems serving the TOD area. More importantly, the fact that the spatial plan does not or limitedly accommodate any of the TOD principles in the future, TOD development will impose a significant bottleneck, among others, in obtaining construction permits.

Based on the problem structures and with the intention of solving the problems that may be encountered in other areas with similar characteristics, the following TOD development guidelines were formulated based on the location plans and typologies.

A. Regional Centers: TOD Located around an Airport

1. In the realization of the density principle in re-developable areas, land consolidation is needed. Limitations on height and density of buildings according to their specific local provisions are prioritized to accommodate the specific provisions. In locations with specific provisions, such as an airport, the density in the core area will be lower than the density in the supporting area, so that the development will be more intensive in the supporting area.
2. Development of the diversity principle can be conducted in accordance with *RDTR* zoning regulations. The diversity of land use should be able to accommodate a variety of activities, especially in the core area. Diversity can be improved both through horizontal and vertical development. However, if there are limitations in realizing high-rise buildings, then diversity can still be done horizontally despite not being a compact form of development. In locations with limited land availability, the adjacent supporting area can be used as service location for the TOD area.
3. Development of the design principle is very feasible if the width of the roads within the TOD area is adequate. The main design principle is to maintain the comfort, security and safety of pedestrians moving within the TOD area. On major roads, vehicles will be relatively faster, in which case a more spacious pedestrian environment with a barrier between road and pedestrian environment should be created.
4. For the development of destination accessibility, the availability of various public transport modes on a regional and local scale as well as the availability of interchange modes to cater to the urban areas are essential. In addition, more than one alternative access to the core area should be provided, considering the large scope of services in this area.
5. Improvement of distance to transit needs to be realized by increasing the coverage of public transport services to support the TOD activities and locations. Besides, there should be an integrated public transportation system that connects both internal and external areas to reach transit locations.

B. Urban and Sub-urban Centers: TOD Located in a City Center

Urban centers and sub-urban centers have similar problem structures and thus the guideline is the same for both cases. The points are:

1. Development of the density principle in a core area with infill site locations is relatively easy because in the core area these are usually located on vacant plots, especially transit

locations. In the supporting areas, density will tend to be adjusted to the conditions in the core area. For example, if the core area is at a location that allows high-rise building in compliance with the *RDTR* zoning regulations.

2. The diversity principle can be developed by locating infill sites. Because the selected locations are on vacant land, it would be more convenient to allow the land use activities to meet the requirements of the core area. On the other hand, if such land use cannot be accommodated, the allowed activities should be optimized in such a way to make the area more attractive.
3. The main design principles that should be prioritized are pedestrian ways and parking management. Pedestrian ways must be developed to maintain the security, safety and comfort of pedestrians, especially in the core area, to encourage walking. Parking management is done to reduce the use of private vehicles. On-street parking is provided in commercial zones for fast shopping, such as minimarkets. Off-street parking is provided in the supporting areas and parking buildings should also be used for other purposes. If possible, parking is preferred for shared use (shared parking).
4. The destination accessibility principle is developed through the provision of various modes of transportation to serve users towards regional and local services. More than one access road to the transit location in the core area should be provided in order to ease the travel to the transit location.
5. Development of the distance to transit principle can be improved by creating activity centers within the area or integrate the activity centers outside the TOD area with emphasis on the destination accessibility principle. There should be a feeder system to connect the TOD location to existing nodes.

C. Urban Neighborhood Centers: TOD in Locations Dominated by Residential Areas

1. The density principle for residential areas with infill sites can be developed by building vertical housing, such as apartments, especially in the core area. Also, these vertical housings can be combined with other activities so that the diversity principle may be enhanced and compact development is obtained.
2. The diversity principle for residential areas means that the major land use is housing, but other activities should also exist to support the main environmental activity. Mixed land use is preferred, for example vertically for commercial and offices, so that the existing space is optimized for residential areas and public facilities. Other functions that may support the TOD area should be placed in the core area.
3. The design principle development should be focused on providing open spaces/parks that can be used for public activities and as interaction space. For parking, there can be secondary streets with on-street parallel parking in the core area so that the number of parking spaces can be controlled in the context of travel demand management. Park-and-ride would also help, since residential areas are mostly found in the countryside.
4. The destination accessibility principle development can be done by adding more options of transportation routes to support people mobility.
5. Because the distance between existing nodes and the TOD location is far, transit locations and feeders (planned and existing) should be integrated in the plan to connect the TOD area with existing nodes.

D. Special Use Centers: TOD Located in Locations with Special Use Characteristics

TOD in special use areas should be managed to support the specialty or main attraction of the area. The rest will be relatively similar to the aforementioned types of centers in accordance with the scope of service of the area, either regional, urban, or local.

7. Conclusion and Recommendations

TOD is suitable for urban areas that are characterized by high density, compact and mixed land use, possess efficient and high quality public mass transportation and an ideal pedestrian environment. The purpose of TOD is to centralize employment, housing, services and increase comfort levels around the major public transport facilities. Since TOD is about integrating transit (T) and development (D), a wholesale TOD policy and planning for a region must addresses two issues [13]: firstly, identifying areas where urban development has high transit orientation but poor access to high-quality transit, and secondly, identifying areas that surround high-quality transit but where the transit orientation needs improvement. To achieve these goals there are principles that must be fulfilled. This study adopted the 5D TOD principles (diversity, density, design, destination accessibility and distance to transit) as proposed by Ewing and Cervero [2].

TOD is an emerging concept in urban development in Indonesia, because some of the major Indonesian cities, including Jakarta and Bandung, are in the process of constructing urban railway systems. This study has developed a methodology to evaluate the feasibility of adopting TOD in built-up environments because there is a lack of guidance in this matter. Exercising this method is expected to ensure the effectiveness of TOD in the long run by considering the principles of TOD, existing conditions, spatial plans, and transportation plans.

This study specifically attempted to formulate a method that can be used for evaluating the conditions of a TOD plan area so that its development can be integrated and fitting. There are three main steps: (i) identifying the location characteristics based on criteria and indicators; (ii) formulating the problem structure at each location based on a gap analysis; and (iii) formulating recommendation guidelines for TOD. In identifying the location characteristics, there are some things to be highlighted. Firstly, the TOD typology needs to be determined as a basis for the development scale. The typology can be determined based on the local spatial planning policy as well as the local transportation development plan. Secondly, criteria and indicators are derived from the 5 TOD principles, although other concepts may be available.

Learning from the case studies divided into five typologies, different problem structures were found. The characteristics of each case are influenced by location, spatial policy, zoning regulations, and existing conditions (whether the location still has to be developed or not). In particular, areas in locations with special restrictions, such as flight operation safety regulations for an airport or other functions, the development of density and diversity may be restrained. In this context, the remaining 3 Ds (design, destination accessibility, and distance to transit) should be the focus of development and optimization.

If a planned TOD area is located in an urban or hub area, land use is dominated by commercial and service activities. At these locations, all 5 Ds can be developed optimally. The composition of diversity should be designed as attractive as possible, while the destination accessibility principle is also important because the TOD area at these locations must be easy to access from several locations both inside and outside the area.

For TOD locations dominated by residential areas, the 5 D principles can also be developed optimally. However, the diversity principle should be further observed; the diversity of land use in this these TOD areas must be protected and developed such that it fulfills the needs of the local residents. Furthermore, residential areas must be connected well with transit hubs.

In conclusion, a methodology to assess the feasibility of TOD application in potential areas was developed by synthesizing a theoretical framework and empirical evidence. This method is expected to be adopted especially in the context of built-up environments, to assist the planning and development of TOD areas for which some principles should be followed without setting aside the characteristics of the existing area or spatial plans. The results can be applied for formulating more detailed guidelines on how to implement TOD, particularly in Indonesian cities. This study can also provide insight into the complexity of TOD implementation in built-up environments from the perspective of integrating them in the spatial planning process, because it is problematic when TOD is not part of the paradigm of creating a sustainable urban form.

References

[1] Calthorpe, Peter. (1993). The Next American Metropolis: Ecology, Community and the American Dream. New York: Pricenton Architectural Press.
[2] Ewing, Reid., Cervero, Robert. (2010). Travel and the Built-up environment: A Meta Analysis. *J. of the American Planning Association*, First published on: 11 May 2010
[3] Benfield, Kaid. (2013). An Epic Transit-Oriented Development Fail. Accessed from https://www.citylab.com/design/2013/03/epic-transit-oriented-development-fail/4932/ on February 2018
[4] Metropolitan Atlanta Rapid Transit Authority (Marta). (2010). Transit-Oriented Development Guidelines.--.
[5] Reconnecting America's Center for Transit Oriented Development. (2008). Station Area Planning: How to Make Great Transit Oriented Development Places (Laporan No. FTA CA-26-1007). California.
[6] Department Planning, Property, and Development City of Winnipeg. (2011). Transit-Oriented Development Handbook (Laporan). Winnipeg: PB's Place Making Group.
[7] Land Use Planning and Policy Department City of Calgary. (2005). Transit Oriented Development Policy Guidelines (Laporan). The City of Calgary
[8] Department of Transportation Bandung City. Pre-Feasibility Study Bandung Monorail Development Corridor 1 and 2. Bandung: 2014
[9] Lyu, G., Bertolini, L., & Pfeffer, K. (2016). Developing a TOD Typology for Beijing Metro Station Areas. *Elsevier, J. of Transport Geography 55* (2016) 40-50.
[10] Florida Department of Transportation. (2012). Florida TOD Guidebook. Department of Transportation, State of Florida.
[11] Department of Infrastructure and Planning Queensland Government. (2010). Transit Oriented Development Guide (Laporan). Australia
[12] Singh, Y. J., Lukman, A., Flacke, J., Zuidgeest, M., & van Maarsseveen, M. (2017). Measuring TOD Around Transit Nodes – Towards TOD Policy. *Elsevier*, *Transport Policy 56* (2017) 96-111.
[13] Presidential Inactment Draft About The Acceleration of Building An Integrated LRT System in Bandung Raya Area. 2016

IOP Conf. Series: Earth and Environmental Science **158** (2018) 012020 doi:10.1088/1755-1315/158/1/012020

Determining the potential for Transit Oriented Development along the MRT Jakarta corridor

W Budiati[1], A B Grigolon[2], M J G Brussel [2], S Y Rachmat[3]
[1]Magister of Regional and City Planning, School of Architecture, Planning and Policy Development (SAPPK), Institut Teknologi Bandung, and Urban Planning and Management, Faculty of Geo-Information Science and Earth Observation (ITC), University of Twente
[2]Department of Urban and Regional Planning and Geoinformation Management, Faculty of Geo-Information Science and Earth Observation (ITC), University of Twente
[3]Regional and City Planning, School of Architecture, Planning and Policy Development (SAPPK), Institut Teknologi Bandung

Corresponding author: wenybudiati@gmail.com

Abstract. Rapid population growth has increased the mobility of people, which leads to traffic congestion. Transit-oriented development (TOD) is a planning approach that is being adopted in many cities as it favours to reduce private vehicle usage by developing mass transit stations around high-density mixed land use and walkable (and cycle-able) environments. However, TOD development needs good strategies learned from historical and current conditions. In this research, we measured the current conditions of proposed TOD areas in thirteen MRT stations in Jakarta to determine the potential TOD development. These thirteen stations were divided into three types based on environmental characteristics of the station location such as CBD, urban, and neighbourhood characteristics. The 5D of built environment variables were used as indicators to examine the potential of TOD areas. Then, stakeholders were involved to fit and weigh the indicators. The weights of the indicators were applied to the Geographic Information System (GIS) platform for quantifying all spatial indicators. TOD indexes for each station were examined to identify potential improvement strategies. Each station is classified as having a high, medium or low TOD index. The results showed that three out of thirteen stations have high levels of TOD; six are classified as having a low TOD level. The low index proposed TOD stations should improve their built-up area, for example by intensifying the floor area ratio (FAR), adding green space, and enhancing the transit and pedestrian-oriented design. By understanding the specific strategies for different TOD indexes, this paper is intended to support those development areas to improve local urban development towards higher levels of TOD.

1. Introduction

Jakarta as the capital city of Indonesia and the centre of governmental, economic, and political activities in Indonesia experiences mobility problems, among which congestion is prominent. Congestion in Jakarta is predicted to get worse as 1,000 to 2,000 new motorcycles and 500 new cars appear on the roads every day [1]. If there is no improvement in transportation planning and/or infrastructure development, vehicular gridlock will truly happen in Jakarta. Consequently, Jakarta aims to reduce the road capacity problem by promoting mass public transport such as BRT, MRT and LRT, and encouraging transit-oriented development (TOD) around public transport stations. TOD can be an opportunity to reduce the number of motorized trips, increase the share for non-motorized trips and reduce travel distance [2]. The construction of a new MRT in Jakarta is an opportunity to plan for TOD and to gain maximum benefit of urban development from TOD planning. The planning of TOD should be examined carefully since it will not automatically follow when mass public transportation is built.

The indicators to examine whether TOD has been developed to support the function of mass transit and to accommodate the liveability of an area are determined by the variables of the built environment. The built environment is defined as "the humanitarian-made space in which people live, work, and recreate on a day-to-day basis". The "built environment encompasses places and spaces created or modified by people, including buildings, parks, and transportation systems" [3]. The dimensions of the built environment that promote successful TOD are density, diversity, design, destination accessibility, and distance to transit [2][3]. For example, the D for design refers to the presence of amenities such as lighting, benches, and trees on the street which creates areas that are suited for non-motorized trips [2]. Likewise, building commercial areas or shopping centres typically increases land use diversity which may lead to a higher proportion of shorter trips for which non-motorized modes are suitable. These trips may substitute trips outside the local area [5]. Similarly, creating a concentration of business activity in the compact commercial core in the centre will lead to a high density and short trip distances which encourages people to walk [6].

In this research, we investigate the potential for TOD planning along the Jakarta MRT corridor. We want to measure the current condition of the proposed TOD areas along the corridor and, based on the results, develop strategies for TOD development to encourage people to use non-motorised modes of transportation. In order to determine the future potential of TOD development, we first establish a baseline of current TOD conditions. This will reveal weaknesses and strengths of those development areas, which will provide us with inputs into concept strategies to improve local urban development towards higher levels of TOD.

2. Case study description

A tremendous number of people commute every day from surrounding cities Bogor, Depok, Tangerang, and Bekasi (also referred to as BODETABEK) to the city centre of DKI Jakarta (Figure 1). The number of commuters to Jakarta has increased tenfold between 1985 and 2002 [7], and 1.5-fold from 743 thousand trips in 2000 to 1.105 million trips in 2010 [8] and has further grown to 1.382 million trips in 2014 [9]. Most of the trips are made by personal modes of transportation such as car and motorcycle; a relatively low number of trips are made by public transport. Despite congestion, the popularity of personal vehicles keeps increasing; the JUTPI Commuter Survey shows that car and motorcycle ownership has increased from 2002 to 2010, at the expense of public transport. This condition calls for the improvement of transportation planning and the strengthening of public and non-motorised modes of transportation.

The development of MRT and LRT in Jakarta is meant to facilitate a shift from private to public transport but also allows local area based planning around the rail-based station areas to promote TOD. The operator of MRT Jakarta wants to promote TOD development in every station but the Spatial Planning Policy 2030 of Jakarta mentions that stations in which TOD will be promoted shall be designated (1) in the intersection of mass transportation corridors (two or more): (2) in the areas with predicted high economic value; and in the areas that are planned or defined as activities centres [10]. TOD is planned in every station of MRT which is legitimated in the Urban Guide Line of MRT Jakarta (Figure 2a). This study focuses on MRT Corridor 1 (Figure 2b).

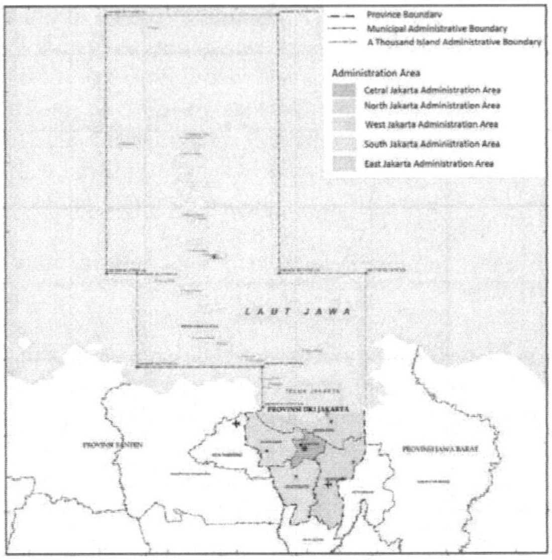

Figure 1. The city centre of DKI Jakarta

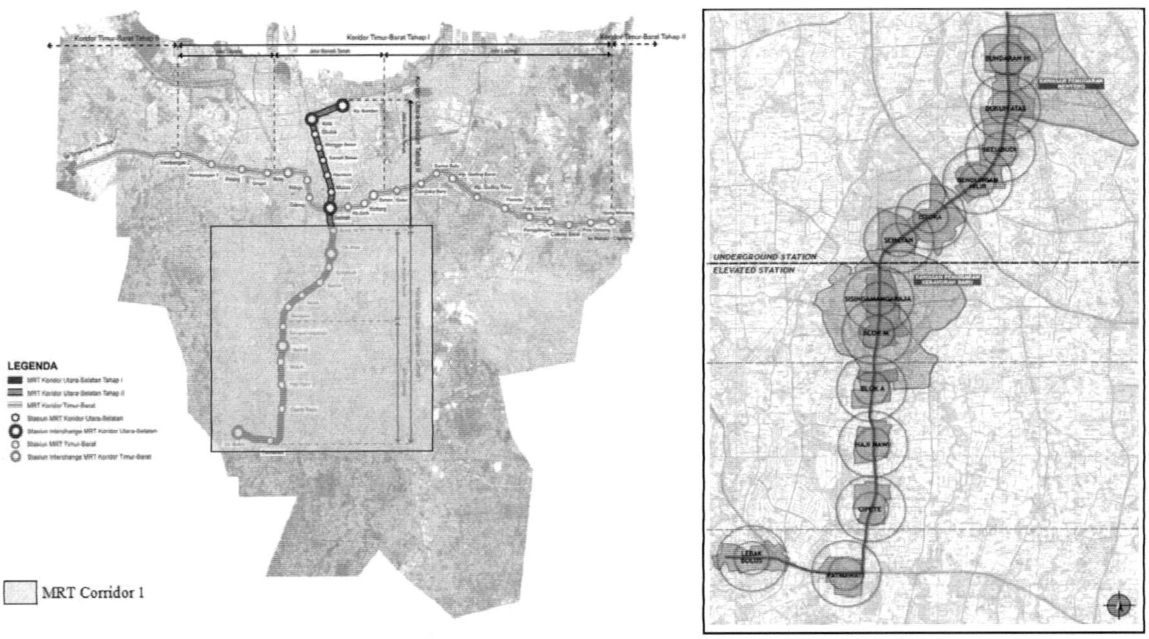

Figure 2. (a) MRT corridors (b) The first phase of MRT Jakarta corridor 1 [11]

In this study, a typology of stations is proposed based on three sections of the MRT Corridor because the station locations have different prominent characteristics. They are clustered in three sections, i.e., Fatmawati street to Panglima Polim Street, Sisingamangaraja Street and Jenderal Sudirman Road to M.H. Thamrin Road. Typical stations for each of these sections are Cipete Raya station, Sisingamangaraja station and Bundaran HI station which are shown in figure 2. Lebak Bulus and Fatmawati station, which are located at Kartini Road, are categorized similar to the stations located at Jenderal Sudirman and M.H. Thamrin Road because they have similar environmental characteristics. Table 1 shows the station's type and their location.

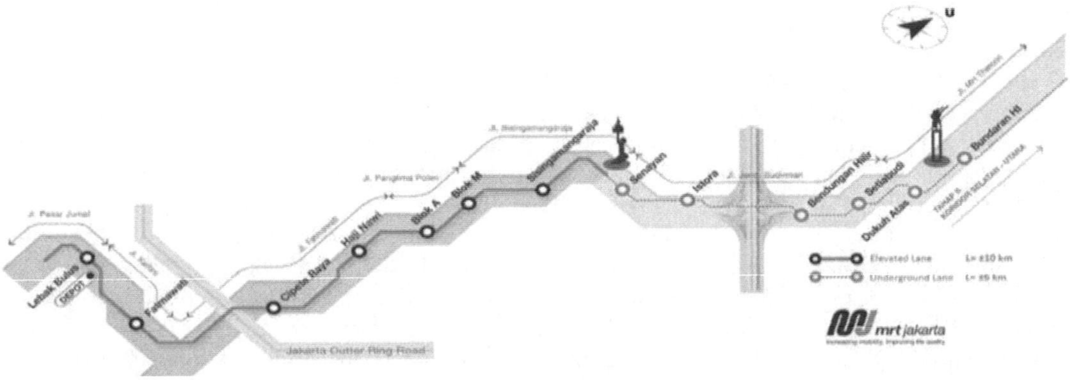

Figure 3. MRT Jakarta corridor and stations [12]

Table 1. Station types

Stations	Station type	Location
Lebak Bulus	3	Kartini Road
Fatmawati	3	
Cipete Raya	1	Fatmawati and Panglima Polim Street
Haji Nawi	1	
Blok A	1	
Blok M	2	Sisingamangaraja Street
Sisingamangaraja	2	
Senayan	3	J. Sudirman and M.H. Thamrin Road
Istora	3	
Bendungan Hilir	3	
Setiabudi	3	
Dukuh Atas	3	
Bundaran HI	3	

Type 1 Stations

Type 1 Stations are represented by Cipete Raya station. In this area, many interior decoration shops; a traditional market; clinic and hospital; house equipment and tool shops; sanitary shops; and a small mall can be found. The area in the vicinity of the street has one to two land uses per building with building characteristics as low-rise residential buildings and shopping buildings. The housing types are mid-rise, low-rise, town houses and small-lot single family houses. There are two types of open space: park and small park. This street is served by public transportation such as metromini, kopaja and small minibus.

Type 2 Stations

Type 2 stations are represented by Sisingamangaraja station. It is a conservation area and development is heavily restricted. This area has one to two land uses per building. There are offices, parks, schools, and a big mosque which attract people. The building characteristics of this area are mid-rise buildings and low-rise buildings. The housing types are mid-rise and low-rise townhouses. The open spaces are medium to large size parks. Various public transport services are provided in this street such as TransJakarta, city bus, metromini, and kopaja.

Type 3 Stations

Type 3 stations are represented by Bundaran HI. These stations located at Jenderal Sudirman and M.H. Thamrin road within the Jakarta CBD. The buildings in the vicinity of this road have more than two land uses per building with functions such as offices, big malls, hotels, embassies, university, sports fields, sports halls, stadium, concert space, exhibition hall, and car-free day location. High-rise and mid-rise apartments and condos and low-rise buildings can be found in this area. This area is served by public transport systems like TransJakarta, city bus, metromini, and kopaja. Open spaces in this area function as small plazas, sports fields or parking space. In this road, Dukuh Atas station is located. This is a local transport hub where many public transportation systems connect. It is expected to become the largest TOD location in the city because it will combine the MRT, LRT, a commuter line, TransJakarta, and the speed train to Soekarno-Hatta Airport. Lebak Bulus is a residential area, but it tends to develop into offices and commercials area although the development is limited due to the area's function as a water catchment area in the Detailed Spatial Plan of Jakarta.

3. Research methodology

This study measures the potential level of TOD implementation in the MRT Jakarta corridor based on built environment characteristics – the five dimensions mentioned by Ewing and Cervero (2010). We determined the measurement by reviewing indicators from researches, such as measuring TOD by SMCA approach in city region Arnhem and Nijmegen, in The Netherlands and some variables of the built environment [4]. By combining these views and other previous research, the potential of future TOD for the MRT Jakarta corridor can be determined.

3.1 Measuring TOD

The indicators selected for the present study were derived from the 5Ds of the built environment indicators: density, diversity, design, destination accessibility, and distance to transit [4]. These TOD indicators were spatially quantified within a spatial scale of a 700-meter buffer around each transit station. The size of the buffer corresponds to a 10 minutes' walk to/from the stations. Spatial Multi-Criteria Analysis (SMCA) was used to compute a TOD index for each station. All calculations were done using the software ArcGIS.

First, **density** was expressed by residential density, commercial density, and employment density. These indicators reflect the intensity of residential and employment development, which are measured for the potential of commuting in a TOD area. The original data for these density indicators are the number of jobs and land use data in Jakarta, measured using a data apportion for non-conterminous polygon method. This operation is used for disaggregating data from regional data into the area of interest for residential and commercial density, while the employment data was converted from small zone data into the area of interest. A higher density value contributes to a higher TOD level.

Second, the **diversity** of land use influences travels demand, because people can reach the facilities and the services within short distances. For instance, locating residential development near markets or health facilities and locating offices near shops, retail services, restaurants, banks, etc. makes the land use of an area diverse. There are different methods to calculate this indicator such as the notion of entropy, the popular Simpson index of land use diversity and the proportion of land use in one area [13][14]. To calculate the diversity of land use, this study uses the entropy method that was applied in the Dutch spatial context by Ritsema Van Eck & Koomen [15], which was also adapted by [13]. A high score indicates that the area has a higher mixed-ness of land use.

$$LU_d(i) = -\sum_i Q_{lui} X \ln(Q_{lui})(\ln(n))^{-1} \quad (1)$$

$$Q_{lui} = S_{lui}\,(S_i)^{-1} \quad (2)$$

LU_i = Land use class within the buffer area i

Q_{lui} = The share of specific land use within the buffer area i

S_{lui} = Total area of the specific land use within the buffer area i

Third, the **design** dimension for the built environment is generally related to the street network. Street network design includes the shape of streets and sights of sidewalks on highly interconnected streets toward transit [4]. The convenient and pleasant design of streets positively influences pedestrians to walk. In this paper, street design indicators refer to the sidewalk condition and the presence of trees, benches, and lamps for pedestrians on sidewalk information. The data was collected during fieldwork and supported by google street view, while the presence of green open space data was acquired from Jakarta Capital City Government. Street indicators were sampled within half the radius of the TOD area, i.e. 350-meter buffer from each station because according to UDGL MRT Jakarta, this is the core area of transit movement that pedestrians are willing to walk to reach an MRT station.

Fourth, **destination accessibility** measures potential access to destinations. It can be categorized as regional and local destination accessibility [5]. Regional destination accessibility refers to the access from residential locations to business centres or the city centre, while local destination accessibility refers to access from residential areas to the nearest shops. The interaction potential is measured using gravity measurements considering population, employment, and travel time. This is measured with a formula adapted from Higgins & Kanaroglou [14]. For station areas oriented to population, the total interaction potential is quantified as:

$$DA_{ij} = \ln \sum_{\forall j=i}^{j} (Pop_i)\,(Emp_j)(T_{ij}^{2})^{-1}$$

Pop_i= The total population in the labour force in station i

Emp_j= The total employment in station j

TT_{ij} = The travel time on transit between station i and j

The population data used for this indicator is the population in the labour force within the age range of 20 – 54 years old. We adopted another indicator for destination accessibility that seeks to measure the interchange possibilities to other bus-based systems at each MRT station. In Jakarta, these systems are the TransJakarta corridors, LRT corridors, the speed train of Indonesia-China (KCIC) to Soekarno-Hatta Airport, commuter lines (regular train) and intercity or regular bus stations. Like Singh [16], interchange to other transit system indicator was calculated by measuring the number of different mode stations and the average of walking distance to the other mode stations. If there is more than one station in the same mode or corridor within the station area, the nearest station will be adopted. A higher number of different mode station implies a higher destination accessibility. which is illustrated by a normalized value of 0 and 1, i.e., lowest to highest. In contrast, the higher the average walking distance of different mode station the lower the destination accessibility. Therefore, the normalized value from 0 to 1 represents the highest to the lowest contribution of this indicator. The total score of the interchange to other transit modes is the sum of the normalized values of both abovementioned indicators.

Fifth, **distance to transit** relates to the distance from origin to transit. This indicator shows how a station is reachable within 10 minutes walking in a circular buffer area [14]. It was calculated using the “closest facility’ function of Network Analyst in ArcGIS. The closest facility was calculated from the centre point of each building to the nearest MRT station, through the road network. The average value of the distances was calculated for every TOD area.

The **TOD index** was computed using the abovementioned indicators, within a spatial scale of a 700-meter buffer. SMCA was used to calculate the index at each station from the indicators computed using ArcGIS. In addition, we hypothesize that not every indicator contributes the same weight to the TOD index. Therefore, we have conducted a survey with 14 local stakeholders (Jakarta Capital City Government, Non-governmental Organizations (NGO), urban planners, academia/university, MRT Jakarta operators, and the private sector). The questionnaire asked respondents to rank the indicators for the three types of MRT stations. The result is a weighed TOD score for each of the 13 MRT stations.

Table 1. TOD indicators, data, format and source

Characteristics	Indicator	Data requirement	Year	Format	Data Source
Density	Residential density	Land use map	2010	Shp	Government of Jakarta
	Commercial density	Administrative boundaries	2010	Shp	Government of Jakarta
	Employment density	Number of employment	2011	Shp	ITC (Previous research)
Diversity	Land use diversity	Land use map	2010	Shp	Government of Jakarta
Design	Sidewalk condition	Sidewalk figures	2015	image	Google Street view
	Green open space	Green open space area	2016	Shp	Government of Jakarta
	Trees, benches, lamps	Sidewalk figures	2010	image	Google Street view
Destination accessibility	Interaction potential	Population	2015	Pdf	Jakarta Bureau for Statistics
		Number of employment	2011	Shp	ITC (Previous research)
		MRT lane and stations	2014	Shp	Government of Jakarta
	Interchange to other transit	Road network map	2010	Shp	Government of Jakarta
		LRT plan Map	2015	Pdf	Government of Jakarta
		BRT lane and station	2015	Shp	Open Street Map
		Speed train to airport lane	2015	Pdf	Government of Jakarta
		MRT lane and stations	2014	Shp	Government of Jakarta
		Big station of buses	2015	Shp	Open Street Map
		Commuter line	2015	Shp	Open Street Map
Distance to transit	Distance to transit	Road Network Map	2010	Shp	Government of Jakarta
		MRT lane and stations	2014	Shp	Government of Jakarta
		Building footprint	2017	Shp	Open Street Map

3.2 Determining the potential for TOD

After calculating the TOD index for each station, the next step is to define strategies for TOD improvement. Stations with higher levels of TOD were used to learn how these stations can reach a high level of TOD in terms of its characteristics. An evaluation of these strategies is required to understand whether the strategies are applicable in the field or not. The strategies are focused on the spatial planning of land use, infrastructure, and urban design. Table 3 shows the documents that were used to support the definition of the detailed strategies for TOD improvement at each station level.

4. Results

4.1 Measuring TOD

The result of the TOD index measurement is a single index for every station. For visualization purposes, the index was classified into three levels, representing low, medium and high levels of TOD. In addition to the map of Figure 3, spider diagrams (Figure 4 to 7) show the performance of each TOD indicator. Based on the calculation, Haji Nawi station has the lowest TOD index and Istora station has the highest TOD index (0.33 and 0.74, respectively).

Table 2. Documents revised for the determination of strategies

Documents	Type	No/ Year
Spatial Planning of Jakarta 2030	Regional Regulation	1/2012
Detailed Spatial Planning of Jakarta	Regional Regulation	1/2014
Urban Guide Line Design MRT Jakarta Phase I	Governor Regulation	182/2012
Guide Line of TOD Development area	Draft Guideline	2017
SITRAMP for JABODETABEK Phase I	Report of Masterplan study	2001
Regional Spatial Arrangement of Kebayoran Baru	Governor Regulation	146/2016
Preservation and utilization of environmental and cultural heritage buildings	Regional Regulation	9/1999
Granting Compensation exceeds of the building floor coefficient	Governor Regulation	210/2016

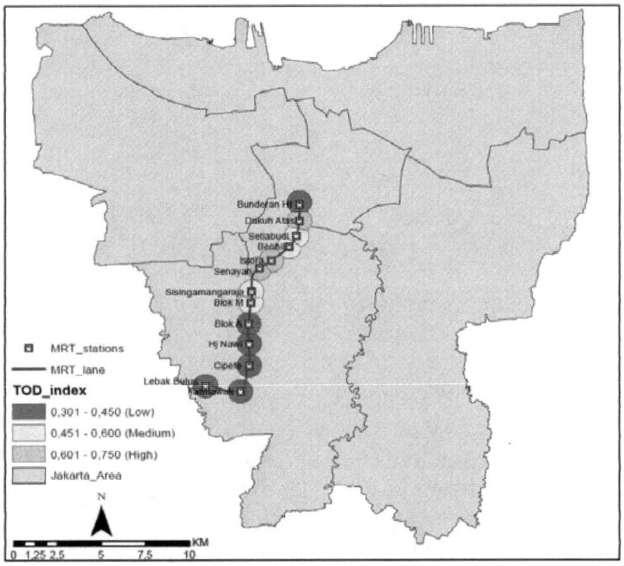

Figure 4. TOD Index [12]

The Top Three Stations

Figure 4 presents spider diagrams of the top three stations (stations with highest TOD indexes). All stations have high values for the diversity indicator, expressing the high mix of land use around each station area. For destination accessibility, Dukuh Atas station has the highest score because it contains many other transit points with adjacent walking distance from the MRT station and it has high interaction with other stations. Istora scores the highest for the design indicator, followed by Senayan. These two areas score high in this dimension because the station areas are adjacent to a sports arena, with huge green open spaces. However, the influence of the density characteristic on Istora is lower than for the two other stations because few people work or live in the sports arena and exhibition area.

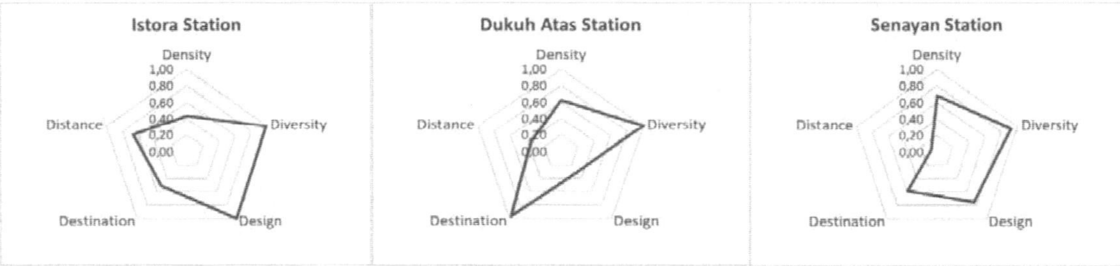

Figure 5. Spider diagram for the stations with high-level TOD

The Medium Four Stations

The stations classified as medium-level TOD are Blok M, Sisingamangaraja, Setiabudi and Bendungan Hilir station. The scores of each TOD indicator are shown in figure 5. In all cases, indicator scores for density, diversity, and design are medium (around 0.5), except for one station, Blok M, which has a high destination accessibility due to its good interaction with other transit services. However, distance to transit is low for all stations, meaning that the stations cannot be reached easily by pedestrians.

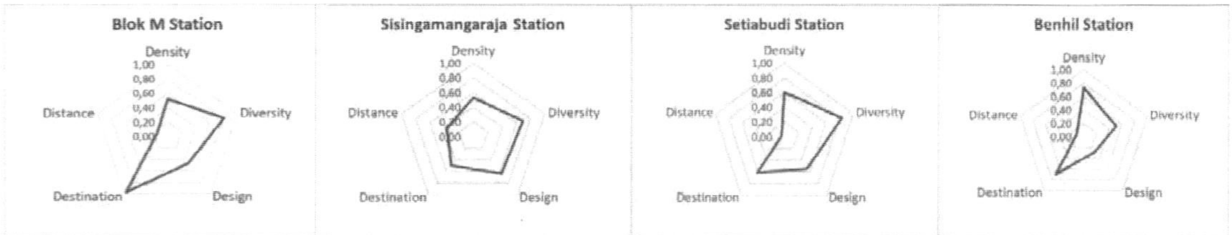

Figure 6. Spider diagram for the stations with medium-level TOD

The Bottom Six Stations

The stations classified as low-level TOD are Bundaran HI, Cipete, Lebak Bulus, Fatmawati, Blok A and Haji Nawi. The indicator scores of the stations are shown on the spider diagrams in figure 6. In general, all of these stations score low for almost all of the indicators.

Surprisingly, Bundaran HI station being in the CBD, with a huge number of activities, has an overall low-level TOD. The distance to transit indicator scores lowest, meaning pedestrians cannot easily reach the station. This could be because this area has long blocks and pedestrians must walk farther to access the station. Cipete, Blok A, and Haji Nawi station have an almost similar pattern of TOD measurement but the distance to transit of Cipete station is prominently higher than the others. Fatmawati and Lebak Bulus have a similar spider diagram pattern. Overall, all stations need much development toward TOD implementation.

IOP Conf. Series: Earth and Environmental Science **158** (2018) 012020 doi:10.1088/1755-1315/158/1/012020

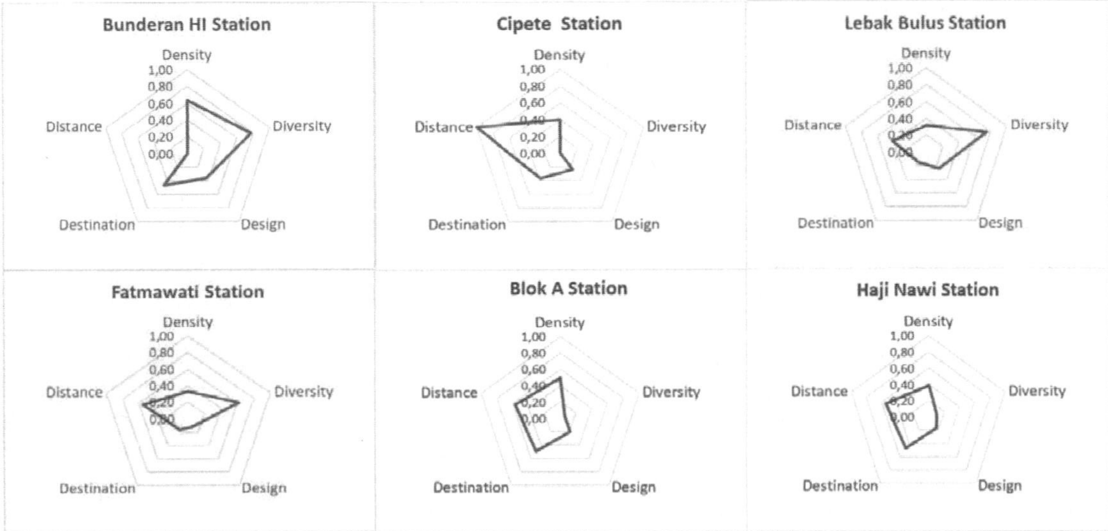

Figure 7. Spider Diagram of Stations With Low-Level TOD

4.2 Determining potential for TOD

To determine the improvement strategies, we made comparisons between the low and the high-level TOD of each station type. The idea is to make an internal benchmark, i.e., comparing the TOD index within a typology because we assume that stations of the same type may have same environmental characteristics and therefore will get the same treatment from people and government. Strategies will potentially be easier to implement when we learn from the strength of other stations with a higher index. After that, the strategies were evaluated using the applicable regulations and policies of the city. The typologies were discussed previously in section 2.

Strategies for Stations Type 1

All three stations of this type were classified as having low TOD level. Based on these results and after revising the regulations, Cipete, Haji Nawi, and Blok A should be developed as high density mixed areas with a dominant socio-cultural residential function. This can be achieved by building additional floors in one-story residential or landed houses, which are intended for multi-cultural and multi-social strata residents. This strategy is in line with stakeholder opinions, who suggest the balancing of socio-economic diversity for high-quality TOD. However, these areas still need other functions or attractions to improve the diversity characteristic. Some areas that have purely residential land use should consider mixing the land use. The timeline for land use improvement needs further studies, although some of the interviewees said that the land use changes are possible within a five-year period.

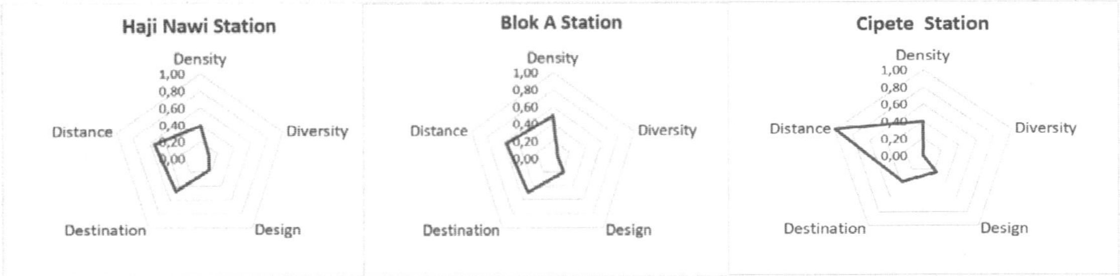

Figure 8. Spider diagram for the stations with high-level TOD

Some areas around Blok A station include conservation areas and cannot expand development, implying that the density characteristic can only be improved by maximizing the potential building floor area from special provisions on zoning regulation and utilizing the incentive of additional floors for buildings, or utilizing the active corridor of this area for commercial functions to increase density. Diversity can be increased by the development of green space areas on land acquired, especially because the vision of this conservation area is "a green and dynamic restoration area" [17].

One problem for diversity development is the land acquisition in these three stations. According to the stakeholders, the stage of land acquisition is one of the problems for construction, for example, the development of pedestrian paths toward the stations and the rearrangement of the road network within the TOD area, which will influence the distance to transit characteristic. Consequently, if the land is available, it will easier to develop all TOD characteristics. Another implication for the design dimension, according to interviewee 5 (urban planner), is that the microclimate should also be considered because it influences people to walk. Jakarta has a tropical climate; the highest temperature in October is 30,21°C and the lowest in February is 27,76°C with humidity ranging from 54 to 98 percent [18]. It is not pleasant to walk, especially in the afternoon. The strategies include mass building arrangements such as arranging the height of buildings to accommodate wind flow, planting big trees, and considering building facades, wall paint colours, glass, and pavement types.

The other aspects that influence the design dimension are the fence, frontage, and side back of the building. First, the building system in Indonesia recommends fences for safety. According to a stakeholder, these fences increase the distance between the building and the sidewalk, which gives a perceived uncomfortable feeling because pedestrians feel alone when they walk. Second, the frontage building design of, for example, shops or commercial buildings with glass design in the shop window, makes pedestrians feel more comfortable and not alone. In addition, because almost all buildings in Jakarta have air conditioners; when the shop door is open this makes people feel refreshed from the heat. Figure 8 shows people sitting in front of buildings to take a rest and feel the air conditioner from the entrance of the buildings. Also, these stations are elevated, which suggests to close off the stations and complement them with canopy for the vertical circulation and platform structure to protect passengers from rain and high temperatures [11].

Regarding the destination accessibility indicator, according to SITRAMP and network planning from the government, there is no plan for mass public transportation besides the MRT along Fatmawati Street (where the Haji Nawi, Cipete, and Blok A stations are located). However, improvements can still be achieved by improving the services and existing infrastructure in the proposed TOD area, for example, the planned LRT, TransJakarta or local bus/express bus. This suggestion for mass transit development is also mentioned in the Jakarta spatial planning 2030 document and the draft guideline of TOD.

Strategies for Stations Type 2

All three stations of this type were classified as having low TOD level. Based on these results and after revising the regulations, Cipete, Haji Nawi, and Blok A should be developed as high density mixed areas with a dominant socio-cultural residential function. This can be achieved by building additional floors in one-story residential or landed houses, which are intended for multi-cultural and multi-social strata residents. This strategy is in line with stakeholder opinions, who suggest the balancing of socio-economic diversity for high-quality TOD. However, these areas still need other functions or attractions to improve the diversity characteristic. Some areas that have purely residential land use should consider mixing the land use. The timeline for land use improvement needs further studies, although some of the interviewees said that the land use changes are possible within a five-year period.

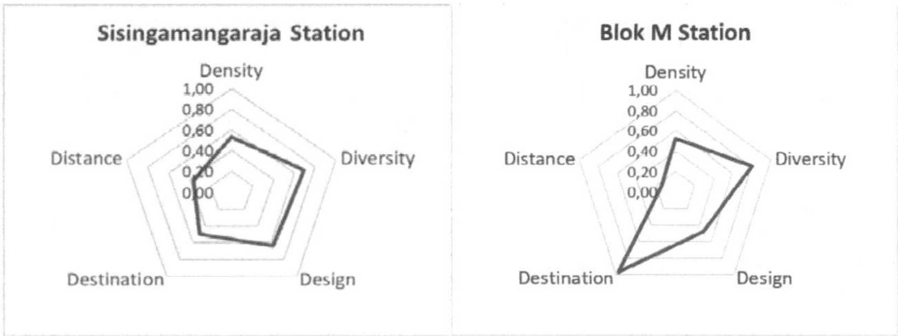

Figure 9. The lowest and the highest TOD index of type 2 stations

In addition, contributing to the design indicator, pedestrian paths should be improved through the base floor of commercial buildings, yards, or parks and new outdoor pedestrian paths to make easier for pedestrians to reach the stations. This will also benefit the distance to transit characteristic. The design dimension will also improve with increasing the quality of pedestrian paths and equipping them with trees, benches, lamps, canopy, and green spaces alongside the paths.

For density, the strategy could be to utilize the incentive of additional building floors and utilizing activities allowed in the zoning regulation within a radius of no more than 300 meters from the station. As for diversity, some areas can still be developed into a residential function. For Blok M, the existing residential land use is only 37 percent. Additionally, improving green open spaces can improve both diversity and design of this area.

Strategies for Stations Type 3

Six stations have typology 3, but to help the visualization and discussion, stations with lower-level TOD are illustrated in Figure 10 whereas high-level TOD stations are seen together in Figure 11.

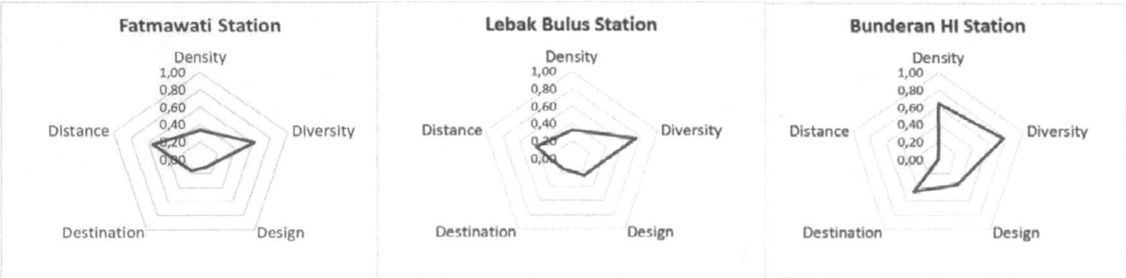

Figure 10. Type 3 stations with low-level TOD

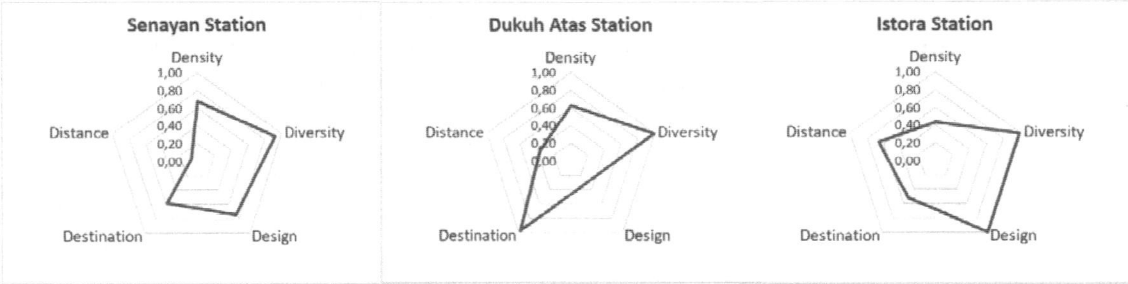

Figure 11. Type 3 stations with high-level TOD

The same strategies as mentioned before are applicable for improving the design dimension, such as constructing new green open spaces, and good pedestrian paths toward stations. Additionally, the fences of buildings and the building frontage should be more pedestrian-friendly.

"...to beautify the sidewalk, it is not only wide but has attractive frontage such as no fences and no car parks which cause people to not be interested to walk... for the outdoor, having scenery can make people distracted about distance or creating a sidewalk with a roof like a canopy can stimulate people to walk more." Interviewee - Planner

For the diversity indicator, an adjustment of land use in these proposed TOD areas is necessary to meet the requirements of the Draft TOD Guideline from the Ministry of Agrarian and Spatial Planning of Indonesia, which plans a composition of space utilization of 30 percent for residential and 70 percent for other uses. Meanwhile, the proposed TOD areas should have less existing residential land use. Some of residential land use should be changed to office and services uses, as well as social and public facilities function to place infrastructure that supports transportation. According to an interviewee, the adjustment of land uses can be done in five years from this detail spatial plan set up, together with the interventions of the density characteristic. In addition, utilizing the building floor incentive is applied along the Fatmawati road.

"The Detailed Spatial Plan of Jakarta should be revised. It will be revised in 2019." Interviewee – Government official

For the destination characteristic, Fatmawati TOD area, which is only served by MRT as mass public transportation, is located at the edge of Jakarta, where people from suburbs come into Jakarta. Consequently, it will increase the destination accessibility characteristic. For distance to transit, the strategies are to avoid constructing large blocks on the side of the MRT corridor, to corporate with building owners and to incorporate the incentive of building floors. As for the design characteristic, the objective is to provide pedestrian paths through the base floor of buildings towards the station, so that pedestrians can easily reach the station.

"Private sectors are very enthusiastic to support TOD development. Compensation exceeding KLB due to MRT construction should be used for MRT development". Interviewee – NGO

5. Conclusions and recommendations

This study was conducted to determine the potential for TOD development along the new corridor of MRT Jakarta. To measure the potential for TOD at each station, the concept of the 5D characteristics of the built environment was adopted [2][4]. This concept provides a comprehensive way to incorporate important aspects of TOD to measure the existing TOD level. This was done by developing a TOD index, which was calculated using Spatial Multi-Criteria Analysis with stakeholders' involvement to determine the importance of the criteria, which was used to calculate weighted scores for the index.

The literature review provided insight into the important characteristics and indicators related to the study area that would need to be considered to measure the TOD index. Stakeholders were involved in two parts: first, evaluating the weight of the ten indicators that were selected to measure TOD index and second, providing additional information for the feasibility of TOD development in the study area. From the prioritized calculations based on stakeholders' questionnaires, the three dimensions of density, diversity, and design are the key aspects of the pedestrian-friendly characteristic for ideal TOD development. These "3D" indicators are well established in the literature as the most prominent TOD criteria [2]. The stakeholders expressed that design and distance accessibility characteristics related to the transit system are the most important for TOD development in the study area.

The spatial analysis calculation shows that only three out of thirteen stations have high levels of TOD, while six are classified as having low TOD levels. Consequently, the proposed TOD areas need improvement. The areas that score a low TOD level need improvement in all 5D characteristics. With the use of spider diagrams, we have visualized which improvements would need to be made in the low scoring areas for these to achieve higher TOD values.

The type 1 stations should emphasize the new construction and development of the land acquisition because new development is crucial for sustainable TOD. The type 2 stations should focus on the development within a 300m buffer from the station, with for example a recreation area and the development of green space. The stations of type 3 should emphasize good connections to the stations including rearranging the fences to create pedestrian-friendly environments. For example, Fatmawati station needs a lot of improvement because it has a low level of TOD but it is located as a gateway to the city centre for people from suburban areas. The improvement should include regulations that can support the implementation of the TOD developments that have been suggested in this paper, such as the policy about FAR, minimum green space percentage in TOD area, incentives for buildings with good design and accommodating pedestrian flows. Also, the regulations should cover the cooperation and business model between the government, operators, and the private sector in the TOD area to assure the operational implementation because the cooperation and the business models of these three institutions are very important for implementing the strategies that promote TOD development successfulness in the study area.

References

[1] Zuhro L 2015 Mampet! Tahun 2020 seluruh Jakarta macet total [Online]. Available from: http://www.otosia.com/berita/mampet-tahun-2020-seluruh-jakarta-macet-total.html
[2] Cervero R and Kockelman K 1997 Travel demand and The 3Ds: density, diversity, and design. *Transportation Research Part D: Transport and Environment* **2** 199-219
[3] Ayoola P 2016 Kenneth Severn 2016. [Online]. Available from: https://www.academia.edu/22131463/Kenneth_Severn_2016_Essay_Peter_Ayoola
[4] Ewing R and Cervero R 2010 Travel and the built environment: a meta-analysis. *J. the American Planning Association.* **76** 265-294.
[5] Handy S L 1992 Regional versus local accessibility: neo-traditional development and its implications for non-work travel *Built Environment* **18** 253-267
[6] Boarnet M G, Joh K, Siembab W, Fulton, W, and Nguyen, M T 2011 Retrofitting the suburbs to increase walking: evidence from a land-use–travel study *Urban Studies* **48** 129-159
[7] JICA and BAPPENAS 2004 *The study on integrated transportation master plan for Jabodetabek (Phaseii) final report main report volume 2: Pre Feasibility Study* (Jakarta: JICA, BAPPENAS)

[8] JICA and BAPPENAS 2012 *JABODETABEK Urban transportation policy integration project in the Republic of Indonesia*, (Jakarta: JICA, BAPPENAS)
[9] Badan Pusat Statistik Provinsi DKI Jakarta 2015 *Statistik transportasi DKI Jakarta 2015* (Jakarta: BPS DKI Jakarta)
[10] DKI Jakarta spatial planning 2030. Jakarta Capital City Gov. Regist. [regulation on the Internet] 1/2012. Available from: http://www.jakarta.go.id/jakv1/application/public/download/bankdata/PERDA_NOMOR_1_TAHUN_2012_TTG_RTRW_2030.pdf
[11] Jakarta Capital City Government 2012 *Urban Guide Line Design, Development of MRT Jakarta Corridor Phase 1*. (Jakarta: Jakarta Capital City Government)
[12] MRT Jakarta 2017 MRT Jakarta [Online] Available from: http://www.jakartamrt.co.id/tentang-mrt-jakarta/id
[13] Singh Y J, Fard P, Zuidgeest M, Brussel, M, and van Maarseveen, M 2014 Measuring transit oriented development: a spatial multi criteria assessment approach for the City Region Arnhem and Nijmegen. *J. Transport Geography* **35** 130-143
[14] Higgins C D and Kanaroglou P S 2016 A latent class method for classifying and evaluating the performance of station area transit-oriented development in the Toronto region. *J. Transport Geography* **52** 61-72
[15] Ritsema J and Koomen E 2008 Characterising urban concentration and land-use diversity in simulations of future land use. *Ann. Reg. Sci.***42** 123-140.
[16] Singh, Y J 2015 *Measuring Transit-Oriented Development (TOD) At regional and local scales – a planning support tool*. (Enschede: University of Twente Faculty of Geo-Information Science and Earth Observation)
[17] Jakarta Capital City Government 2016 *Penataan ruang Kawasan Kebayoran Baru 2016* (Jakarta: Jakarta Capital City Government)
[18] Jakarta Bureau for Statistics 2016 *Jakarta in Figures* [Online]. Available from: http://jakarta.bps.go.id/backend/pdf_publikasi/Jakarta-Dalam-Angka-2016.pdf

IOP Conf. Series: Earth and Environmental Science **158** (2018) 012021 doi:10.1088/1755-1315/158/1/012021

Multimodal Freight Transport Regulations in Indonesia And Its Implementation (A Case Study of Tanjung Priok Port)

N Budisiswanto[1], M Miharja[1], B Kombaitan[1], P Pradono[1]
School of Architecture, Planning, and Policy Development, Institut Teknologi Bandung, Indonesia

Corresponding author: budi87.siswanto@gmail.com

Abstract. The purpose of this paper is to present the current situation of multimodal transport in Indonesia and the various ways in which it is implemented. This paper will also examine the existing regulations and overcome barriers in implementation for providers and users of services. The paper also aims to correlate the elements of government regulation into the logistics chain, to understand the need for the application of Multimodal Transport. Indonesia has realized the benefits that can be achieved by implementing multimodal transport, an important alternative to improve logistics performance. As the government plays an important role in offering the legal and institutional framework, this paper assesses the implementation of regulatory linkages, infrastructure, and why multimodal transport is still not implemented properly.

1. Introduction

Efficient transportation is a key factor in logistics performance. Transportation greatly affects what happens in the trading corridor. Logistics is an indicator of the difference in the performance level of a company or country (macro context) in reducing costs and increasing productivity. Indonesia is plagued with the problem of high average logistics cost. The total cost of logistics in Indonesia in 2016 is quite large: 24 percent of Indonesia's gross domestic product [1]. In addition, the high transaction cost makes Indonesia to be the archipelagic country with the biggest cost of goods inter-island distribution of goods [2]. The World Bank data in 2016 reveals that Indonesia's logistics performance is 63rd out of 160 countries. That data also reveals that while Indonesia is in the top 10 performers for lower-middle income countries, Indonesia is ranked lower than other ASEAN countries, namely Vietnam (48), Malaysia (25) and Singapore (5). Indonesia's container handling cost is highest among all ASEAN countries. Despite being an archipelagic country, the majority of Indonesia's ground transportation does not support inter-island interconnections. As a result, the cost of Indonesia's intercity and inter-island transportation is much higher than that of Singapore. Dwelling time at Tanjung Priok Port, which counts at the 3.9 days range, occurs due to the long forwarding process of goods from consignee (consignee) and consignor (freight forwarder service). This process is generally undisciplined, causing the accumulation of goods and significant container queue. Delay leads to additional incremental costs which means additional costs for consignees and consignors.

The government of Indonesia has implemented efforts to build a multimodal transportation system in support of national logistics system, among others, through the issuance of several regulations related to national logistics system (Perpres 26/2012) [3] and multimodal transportation system

(Government Regulation 8/2011 and Permenhub 2012) [4 and 5]. Unfortunately, they haven't seemed to run smoothly as reflected in the relatively low performance of the logistics system as a whole.

There are various factors contributing to the success of multimodal transport. A key constraint of multimodal transport is the lack of willingness to cooperate and coordinate between operators of different modes. Collaboration among logistic stakeholders is strongly influenced by institutional aspects especially for logistic decision-making processes (3). A number of studies discuss the importance of institutional aspects in the functioning on multimodal transport [6, 7, 8, 9, 3, 10, 11, 12, 13, and 14].

Multimodal transport and logistics are cross-sectoral and the institutions that manage them must also be cross-sectoral and cannot be managed partially, because of their nature. Multimodal transport requires an institutional approach that explains cross-cutting of multimodal transport, such as the role of the Port Authority as a regulator, indicates the formation of Multi Operators in order to create a more balanced competition between different modes of transportation, investment regulation, financing, government roles, the role of the private sector, and matters related to coordination in multimodal transport where the role of the port occupies a very strategic and important role in Multimodal Transportation. This paper investigates the structural problems with the regulatory and institutional aspects of the multi-modal development at Tanjung Priok Port.

As such, Multimodal Transport requires an approach that explains the macro and cross-cutting of multimodal, especially in terms of regulation and institutional aspects. The purpose of this paper is to link government regulations, institutional aspects, and logistical transportation chains in analyzing the implementation of Multimodal Transport in Indonesia. This study aims to identify regulatory issues and the development of multimodal transportation that impede its implementation.

2. Methodology and Data Collection

This paper employs the content analysis methodology. Qualitative Content Analysis aims to sort relevant data and clarify information obtained from different sources. This analysis includes summarizing, explicating, and structuring [15]. The technique is used to analyze secondary and primary data about legislation and related institutional issues. In regulatory analysis, this technique is used to look at existing legislation both vertically and horizontally and analyze how the interrelationship between regulations – whether there are regulations that encourage the implementation of multimodal transport, whether existing regulations are not aligned, or whether the implementation of those regulations are not conducted properly. Thus, through this analysis, we identify that one of the empirical problems of multimodal transportation in the Tanjung Priok Port. On the institutional issues of multimodal transport, this technique is used to analyze the role of the actors involved, who are the actors involved, the role of each actor, who is the leading stakeholder, and what factors are the main causes of the impediments of multimodal transport implementation. A comparative analysis is performed to provide reference and benchmark on the institutional issues surrounding multi modal implementation in other countries.

The data collection is done by survey on secondary data collection and primary data on the regulation and institutional issues of multimodal transportation in relation to the logistics system activity at Tanjung Priok Port. This research is intended to analyze regulation and institutional aspects in multimodal logistic transportation in Tanjung Priok Port. The data collection for this study is based primarily on interviews with staff of Port Authority conducted on April 2017, and secondarily by collecting data from documents such as relevant regulations.

3. Multimodal Freight Transport Concept

Multimodal transport is a transport that uses at least two modes of transport to deliver goods to places to be shipped under the Multimodal Transport Contract. Goods are handled by and under the responsibility of Multimodal Transport Operators [16]. Multimodal Transport Operators is an entity that holds a Multimodal Transport Contract, acting as the sole responsibility for the assigned tasks in the contract of carriage either alone or through other parties under its supervision.

In contrast to segmented transport where each document is required for each mode and other logistics activities such as warehousing, the use of more than one transport document describes the number of bureaucratic activities, which also explains how and to what extent the responsibilities of the actors play a role. In this segmented transport every mode of transportation, as well as other supporting logistics activities, is responsible for specific activities [17].

In SISLOGNAS document (2012) [18], by the end of 2025, we will have seen the conception of a multimodal transport system (illustrated by the diagram below) developed using the paradigm and perspective of multimodal transport which takes into account the type and characteristics of the transport system used and with consideration to the efficiency, effectiveness and ease of system operation, with the ultimate aim of delivering a highly competitive transport system.

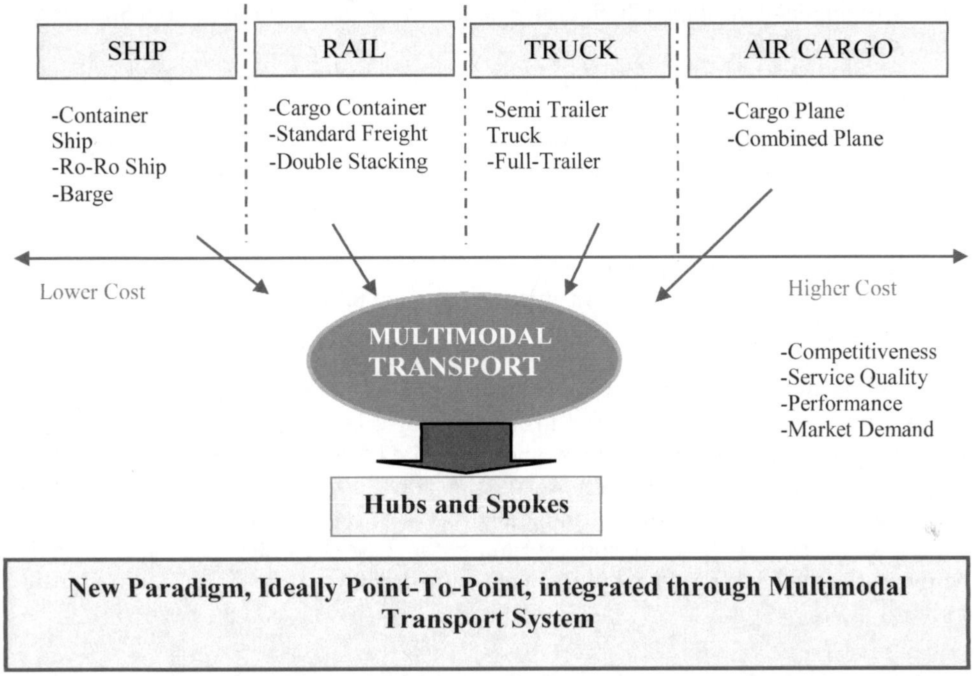

Figure 1. Multimodal Freight Transport Concept. [18]

The implementation of multimodal transport is carried by the multimodal transport operator (Multimodal Transport Operator, MTO) under the name of Multimodal Transport Provider (*Badan Usaha Antar Moda,* BUAM) as per Government Regulation No.11 of 2011. This legal entity (or its legal representations thereof) realize and perform multimodal transport contracts. Although MTO is the entity responsible for an entire chain of logistics, in its implementation, MTO can divert some or all to the transportation operators such as transport trucks, train, river and ferry, sea and air transport. Each node of transportation and logistics in such a system must build multimodal transportation facilities physically.

4. Multimodal Freight Transport Advantages

Logistics is an indicator of the difference in the performance level of a company or country (macro context) in reducing costs and increasing productivity. In that context, multimodal transport has an important role. Multimodal transport leads to the adoption of integrated and door-to-door transportation systems.

Companies that do not use multimodal systems typically opt for commercial transport schemes like conventional CIF (Cost, Insurance, Freight) scheme and FOB (Free on Board) scheme, where only a few chain logistical transport operations are not door-to-door.

The advantages of the Multimodal Transport Operator (MTO) will be more pronounced when compared with other transport alternatives. For a company operating with a fleet of self-owned logistics and equipment, MTO will assume responsibility for fleet purchases, maintenance, financing, human resources training and the provision of infrastructure for logistic activities [16].

If a logistics operator is used, whether segmented or multimodal transport, upon hiring they will negate the need for investment and logistic activities will be categorized into transport service purchase. If the company has its logistics, operating costs will be borne by the company. If using the MTO, the operational costs will be divided by the MTO another customer. The greater the volume of cargo, the greater the apportionment of direct and indirect costs.

Differences of Multimodal Transport which is implemented by the MTO is a problem that can be mentioned the following benefits:

- Better Purchase and Contract
- Better Use of transport capacity
- Energy efficiency of their blend modes
- Better use of IT
- Benefits of scale of use;
- Better use of logistic infrastructure
- Gain MTO experience and the maintenance of the trade procedures
- Reduce overhead costs

5. Comparison of Multimodal Freight Transport Regulation in Other Countries

Some countries in the world has joined forces with several regionally agreement and other have set their own laws. Multimodal Transport in Germany and the United States are good examples. Germany has not promulgated a new regulation for Multimodal Transport. Instead, Germany streamlined all of their disparate regulations into one simplified, integrated body of regulation. The United States does not make specific laws for Multimodal Transport system but instead streamline the implementation of existing regulations effectively, supported by adequate infrastructure, proper transfer terminals and good national security [20].

Table 1. Multimodal Transport Regulations in Some Country. [20]

COUNTRY	REGULATIONS	APPLICATION
Brazil	Law 9611 of February 19, 1998 on Multimodal Transport of Goods.	This regulation covers national and international multimodal transport.
German	Transport Law Reform Act, 1998	This regulation applies to the multimodal contract for all nodes of cargo transport: road, air, water and, maritime.
United States of America	The United States does not have a specific law on Multimodal Transport	The US has an effective multimodal transport supported by adequate infrastructure, transfer terminals and good national security.

6. Multimodal Freight Transport Regulations in Indonesia

Referring to the UN Convention on multimodal transport "Multimodal Transport Convention" in 1980, the Indonesian government has planned regulations on multimodal transport. The first regulation on multimodal transport, the National Transportation Systems [21] promulgated in 2005, regulated Indonesia's door-to-door transport. Indonesia also referred to the ASEAN Framework Agreement on Multimodal Transport (AFAMT) in 2000 in developing multimodal transport. Indonesia's primary laws regulating multimodal transport are as follows:

- ASEAN Framework Agreement on Multimodal Transport (AFAMT) [21];
- National Transportation Systems (Minister Regulation 49/2005) [22];
- Article 147 paragraphs (1), (2), and (3) of Law 23/2007 on Railways, governing the integration of multimodal transport [23];
- Articles 50-53 paragraphs (1) and (2), Article 54, and Article 55 of Law 17/2008 on Sea Transport, governing the integration of multimodal transport [24];
- Article 182 paragraph (11) governing multimodal cohesion responsibility and Articles 187-191 governing multimodal transport of Law 1/2009 on Flights [25];
- Article 165 of Law 22/2009 on Road Transport, governing the integration of multimodal transport [24];
- Government Regulation 8/ 2011 on Multimodal Transport
- Transportation Minister Regulation 8/2012 on Implementation and utilization of Multimodal Transport

The implementation of multimodal transport is mandated by the regulations governing traffic and transportation (road, shipping, railways, and aviation). Law 22/2009 on Road Transport mandates multimodal transport in the following articles:

- Article 2 item h, stating that the traffic and road transport was organized with unified principles;
- Article 33, paragraph 1 states that for the smooth movement of people and goods with multimodal integration in certain place, it can be built and held terminals;
- Article 93 paragraph 2 point e stating that management and traffic engineering done by integrating various modes of transport;
- Article 165 paragraph 1 and 2, among others stated that public transport on the road which is part of a multimodal transport performed by legal body/entity of multimodal transport operator and implemented based on an agreement made between the legal entities of road transport and multimodal transport or legal entity of other modes.

Law 23/2007 on Railways mentions the integration of multimodal transport in the following:
•. Article 2 point e stated that the railway is a part of the integral national transportation system was organized based on the principle of integration;
•. Article 6, paragraph 3 states the order of the public railway must be integrated with other transport modes;
• Article 147, paragraph 1, paragraph 2 and paragraph 3 states among other things that rail transport can be part of a multimodal transport performed by legal entity of multimodal transport, and implemented based on an agreement between the organizers of railways facilities with legal entities or organizing multimodal transport and other modes

In the Law 17/2008 on Sea Transport, multimodal coherence was mentioned in the following:
•. Article 2, item g, stated that the cruise organized based on the principle of integration;
• Article 9, paragraph 4 point d, stating that the route network and common sea transport established based on the integration of intra and multimodal transport;

• Article 18 paragraph 4 states that the rivers and lakes transport activities prepared and conducted in an integrated way with due regard to intra and multimodal transport
• Article 22 paragraph 2 point t stated that the determination of crossing freight traffic carried by considering the sea transport route network to gain integration of transport
• Article 50-55 states that water transport can be part of a multimodal transport
•. Article 68 item c said that the port has a role as a place of activity over the modes of transport;
•. Article 96, paragraph 2, states that the development of seaports must fulfill attention to the integration of intra and multimodal transport.

While in Law 1/2009 on Aviation regulated multimodal alignment:
• Article 123, paragraph 2, which states that the route network and flights abroad, among others, pay attention to the integration of intra and multimodal;
• Article 182, responsibility to multimodal transport;
• Article 187-191 states that air transport can be a part of a multimodal transport, and implemented based on an agreement made between the enterprises of air transport and multimodal transport entity or entities other modes;
• Article 194 states that the airport has a role as a place of activity over the modes of transport;
• Article 214 states that the construction of airports should consider the integration of transport

Internationally Indonesia has ratified ASEAN's Framework Agreement on Multimodal Transport so that the multilateral agreements, this considered as to strengthen legal basis and step the implementation of the multimodal freight transport in Indonesia.

Here is a detailed explanation of the implementation of multimodal regulation in Indonesia:

Table 2. Implementation of Multimodal Transport Regulation in Indonesia.

Regulation	Mandate	Progress in the Implementation
Government Regulation 8/2011 About Multimodal Transportation	Development of Multimodal Transport Operator (MTO or BUAM) and Preparation of STC (Standard Trading Center)	None of logistic provider has requested MTO permission yet. In the case of transport documents, intermodal transport in Indonesia also has not used a single document. Each mode of transportation in general still uses its own transport documents that are not yet integrated with other transport documents, so the responsibility of the carrier is often unclear. In addition to this, the process of completion of the document is relatively long because it is still done manually, so it tends to inhibit the smooth flow of goods.

Regulation	Mandate	Progress in the Implementation
Transportation Minister Regulation 8/2012 on the Implementation and Undertaking of Multimodal Transportation	Order documents, Facilities and Infrastructure Multimodal Transport Operator (MTO/BUAM)	The purpose of multimodal transport is to realize one stop service with single operator-tariff-document for freight transport. Logistic Actor No one has used port railway service in Tanjung Priok Port even though it has been developed, the motive and cost rationality. Lack of commitment from cargo companies to use the hinterland service under development (railway from port to hinterland), so the volume is not reached so that new channels cannot be sold beyond the minimum volume limit.
Law and Human Rights Minister Regulation 20/ 2013 on the minimum terms and conditions of service (Standard Trading Conditions-STC) in the Multimodal Transportation Field.	STC is a reference and guidance in the preparation of multimodal transport documents.	Logistic actors do not have sufficient means of information to coordinate. The most generally relevant coordination issue between rail and truck transport is the inadequate exchange of information and coordination between the Shipping Line Container, the Port Operator and the rail transportation company.
Regulation of Port Authority UK 112 On Procedures of Shipping and Handling [26] Transportation Minister Regulation 49/2005 on National Transportation System (Sistranas) Law 22/2009 on Road Transport: inter-mode road connections is arranged in the fifth section of the multimodal transport of Article 165. Law 17/2008 on Sea Transport arranged in the tenth section Law 1/2009 on Aviation that specifically regulates the multimodal pattern in paragraph 11 Law 23/2007 on Railways, especially in article 147	These government regulations are issued to support the national logistics system through the presence of multimodal transport of goods and services in operations in various modes. Multimodal transport supports business entities in carrying out multimodal transport activities that may cooperate with road, crossing, railway, shipping, or aviation business entities.	In the process of transporting responsibility still resides in each segment / section of transport activities, because the existing legislation cannot yet cover all aspects of the responsibility of the parties involved in the transport chain, or if there is often not enough clear so sometimes giving rise to different interpretations / conflicts. This is partly evident from the fact that there are two main constraints still in place, namely sole responsibility and the use of one transport document (single document).

7. Analysis

Regulations mentioned above demonstrate that there are legal frameworks in play when it comes to multimodal transport. When it comes to implementation, however, the pace is still very slow due to two aspects: quality of regulation/institution and quality of implementation.

Indonesia does not have a good implementation of multimodal freight. Despite having regulations on procedures for the transportation of multimodal goods, field implementation of these regulations and procedures has not been strong enough. At the same time, there is access problem for multimodal transportation, such as when goods are unloaded at Tanjung Priok Port, the subsequent access of goods transport is only via road. Due to limited road traffic capacity, Tanjung Priok Port congestion, and this happens despite the government regime change. Access to existing port railways is not yet effective even though rail lines are available, so there is no alternative to increase the effectiveness of the distribution of goods unless it still rests on the trucking mode. In addition, access from the Tanjung Priok Railway is still not accessible directly to the container yard from and to other stations, for example, Gede Bage Bandung, causing any goods transfer to undergo through two customs handling. In addition, there are still inadequate transit warehouse, both at port and at sea port.

Besides the difficulty of achieving equality in the regulation based on what has been established, there are regulatory barriers to making more effective multimodal transport in Indonesia.

1. Multimodal transport operator regulation
    - In Indonesia, no one has requested permission to become MTO/BUAM. There is a a problem of motivation and perception of the regulation, which is considered complicated. Besides, there is still conflict of interest motivated by economic factors, resulting in very low motivation to use railway as end container shipping. Ultimately, the use of the railway is under review by the maritime minister and find a development strategy.
    - These situations encourage inefficiencies in logistics, where the transportation is done by issuing many documents (and a lot of layers of bureaucracy).
    - Standardization and compatibility of multimodal traffic increasingly as important role. The key to the smoothness of multi-modal transportation is the compatibility between transport modes, ranging from liners, container handling equipment, to the transporter truck. This compatibility should also apply to rail and air transport, and even inter-trucking so multi-modal transportation really can be run efficiently.
    - Preparation of legislation and policies on multimodal through optimization policy making role dry port existing (Gedebage, Rambipuji, Solo Jebres, and so on) as the terminal multimodal, standard setting units and dimensions to improve the efficiency of transportation equipment and facility operational support multimodal transport and logistics, the review and preparation of guides and standardization in order to realize compatibility of conveyance and operational support facilities multimodal transport and logistics, as well as policy formulation development / construction of multimodal terminal and logistics centers.
2. Actor perception and information regarding multimodal transport logistics
    - Perceptions of multimodal transport for logistic actors have not been well established. In using the service, only few logistics actors who use railway logistics service from Tanjung.Priok Port to Cikarang Dry Port since when it was developed. The use of the train (to Cikarang Dry Port) which has been operated earlier this year but is still not effective, it is because the mode of the truck is still competing with this railway mode. The theory expressed by De Langen in 2008 [28] also elucidated a lack of commitment from the freight company to use the services hinterland are being developed (train from the port to the hinterland), so that the volume is not reached so that this new logistics channel cannot be sold because the limits the minimum volume.
    - Logistic actors do not have sufficient information tool for coordination. Ministry of Transportation has to propose a multimodal transport operator and coordinate with many actors including cargo owners, shippers and 3PL.
3. Tax incentives

    Other barrier to using multimodal transport is as a matter of tax incentives or cost. There is no clear distinction between using a multimodal transport with unimodal. In multimodal transport, cargo transported by multiple modes, before it reaches its final destination, there are a few differences in the rules between each mode in terms of tax collection, in this case more burdensome taxes train users. Train use would incur charges imposed on providers amounting to 30-40 percent of the operating cost. This charge is categorized as a state revenue outside of tax set by the Ministry of Transportation. Train use would incur net cost of approximately IDR 5.5 million per 40 feet container while truck use incurs IDR 4.4 million per 40-feet container. That train is more expensive because of this non-tax charges.

4. Dispute resolution
The formulation of the responsibility of the parties in the port of goods involved in multimodal transport are still unclear and thus potentially lead to conflict, despite the provision of the Civil Code. Claims for compensation on goods transported by land transport modes in practice difficult to solve satisfactorily, especially in trucking.

5. Infrastructure policy aspects
The existing transport infrastructure in Indonesia is an obstacle to the integration of different transport modes. Transportation of goods (logistics) in Indonesia is still dominated by road transport, these conditions lead to frequent traffic accidents and the rise in road damage. It is not based on considerations of the territorial dimension of the Indonesia as archipelagic country. Bappenas [29] reported the largest proportion of about 84.1 percent was served by a road mode, as shown railroad serves only about 7.3 percent of the national transport market, the mode of sea and air modes respectively serve only 1.8 percent and 1.5 percent. The figure 2 below, shows the transportation proportion in Indonesia.
Indonesia's transportation operation is inefficient, one of the main causes is the high economic costs, mainly driven by: (a) multimodal load irrational, resulting in excessive burden on the road network and the development of other transport modes, (b) transportation infrastructure conditions that have worsened due to the economic crisis have led to declining service performance, (c) multimodal transportation does not run perfectly, mainly because of the lack of interconnection between modes, operation management is not optimal, and support systems are inadequate, and (d) role-sharing between modes.

Figure 2. Transportation Sector Participation in Indonesia.

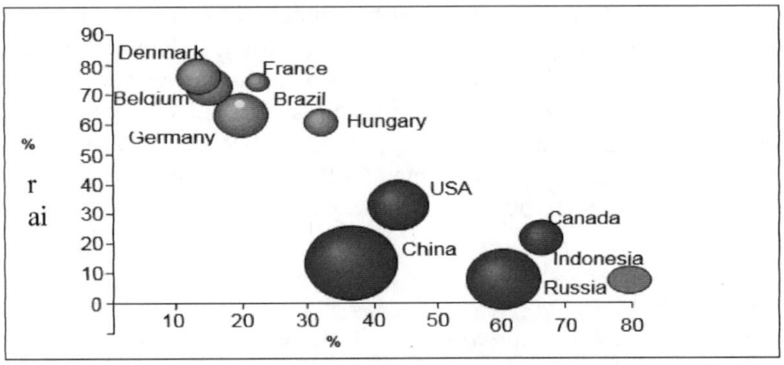

Figure 3. Mode of Transportation of Participation in Various Countries. [30]

The image above shows the portion of transport in some countries. The development of multimodal transport is to develop an effective operation of multi-modal transportation network that connects the logistics node, and the smooth transport accessibility of basic and strategic commodities as well as the leading commodity export in each economic corridor to an international hub like the Tanjung Priok Port.

8. Conclusion

This paper assesses the implementation of multimodal transport in Indonesia by examining the case of Tanjung Priok Port from the regulatory and institutional perspective. The analysis finds out that there are still some impediments and issues to implement the concept. The reasons are, among others, (1) that the existing regulations are not strong and encouraging enough and (2) there is no dedicated institution that is able to fix complicated problems with multimodal transportation, a problem that involves many actors such as port managers, port authorities, governments, logisticians, and logistics service providers. In addition to involving many actors, referring to best practices in other countries, multimodal transport actually combines aspects of the transportation system, the type of mode provided by logistics service providers, business decisions of logistics actors, regulations and coordination between the actors. Of course, the latter need to be underlined, that the coordination aspect needs to be intensified between various public and private sector actors, through various platforms and ways, so that an efficient multimodal system can be implemented.

In improving the performance of multimodal transport that has not been implemented yet, there are several actions to be taken:

- The development of multimodal transport network and logistics centers to improve the smoothness of freight traffic from the production center to the outlet-inlet import-export, through standard-setting unit and dimensions to improve the efficiency of transportation equipment and facility operational support multimodal transport and logistics, development of guidelines and standards in order to realize compatibility of transportation equipment and facility operational support multimodal transport and logistics, and the development / construction of the terminal multimodal and logistics center.
- Mapping functions and roles of the stakeholders in the development of multimodal transport and empowerment strategies and the strengthening of the respective businesses that handle or related to multimodal transport.
- Standardization of procedures and legal guidelines for the formulation of responsibility, particularly in resolving disputes.
- Formulation of the application of relevant tax exciting multimodal transport service providers and users; Build incentive fees
- Assign roles and competitive conditions of different modes in order to create fair competition;
- Improve coordination between stakeholders can be strengthened through a specialized institution that handles multimodal transport
    - Creating an attractive investment environment in the field of multimodal transport;
- Improving the management of the legal, institutional and economically more efficient flow of goods, with the continuous updating of procedures, reducing the execution time and bureaucracy in the port. Multimodality is dependent on the efficiency of the use of multiple modes and combinations of modes that ultimately aim at reducing transport and logistics costs. Overall door-to-door integration runs to reduce costs and optimize the flow of goods. Multimodal transport must be implemented in the shortest time by taking action to eliminate legal obstacles and accelerate the construction of multimodal infrastructure, especially from the investment side.

Acknowledgement

Thank you to the support and data provided by Tanjung Priok port authorities so that it can be published in this paper and become a credible evidence.

References

[1] World Bank, 2016, Logistics Performance Index, **http://lpi.worldbank.org/**

[2] Pusat Pengembangan Logistik dan Rantai Pasok ITB-World Bank, 2015, State of Logistics Indonesia 2015. Jakarta

[3] Peraturan Presiden Nomor 26 Tahun 2012 Tentang Cetak Biru Sistem Logistik Nasional

[4] Peraturan Pemerintah Nomor 8 Tahun 2011 Tentang Angkutan Multimoda
[5] Peraturan Menteri Perhubungan Nomor : PM. 8 Tahun 2012 Tentang Penyelenggaraan dan Pengusahaan Angkutan Multimoda
[6] Tsamboulas dan Kapros, 2000. The Decision making Process in Intermodal transport. *Transportation Research Record*, 1707
[7] Monios, Jason, 2014, Institutional Challenges to Intermodal Transport and Logistics Governance in Port Regionalisation and Hinterland Integration. Ashage England
[8] Hanssen, T. E. S. & Mathisen, T. A. 2011. Factors facilitating intermodal transport of perishable goods - transport purchasers viewpoint. European Transport - Trasporti Europei.
[9] Woodburn, A., Browne, M., Piotrowska, M. & Allen, J. 2007. Scope for modal shift through fiscal, regulatory and organisational change. University of Westminster and University of Leeds.
[10] Macharis, C., Caris, A., Jourquin, B. & Pekin, E. .2011. A Decision Support Framework for Intermodal Transport Policy. *European Transport Research Review*, 3.
[11] Taylor dan Jackson, 2000, Conflict, Power, And Evolution on The Intermodal Transportation Industry's Channel of Distribution. *Transportation Journal*, Spring.
[12] Hicks Associate, 2009, Institutional Arrangements for Freight Transportation Systems, Transportation Research Board, Washington DC.
[13] Ng Adolf K Y, Pallis Athanasios A. 2010. Port Governance Reforms in Diversified Institutional Frameworks: Generic Solutions, Implementation Asymmetries. Department of Logistics and Maritime Studies, The Hong Kong Polytechnic University, Hong Kong, China
[14] Budisiswanto, N et.al 2017. Institutional Analysis for Multimodal Transport to Support Logistics System in Tanjung Priok Port. *Int. J. of Supply Chain Management. Excellingtech*
[15] Flick. 2006. An Introduction to Qualitative. Research London: Sage Publications
[16] United Nations, 1980. Convention on International Multimodal Transport of Goods. United Nations, Geneva
[17] Keed, Samir. 2001. Int. Transport Logistics. Publisher Aduaneiras.
[18] Kementrian Perhubungan, Cetak Biru Sistem Logistik Nasional (SISLOGNAS). Planning Document, 2012
[19] Komarova, Anna Danielevna Hernández, 2000. Multimodal Transport of Goods: Analysis of Alternatives. Master Thesis, Instituto Militar de Engenharia - IME, Rio de Janeiro.
[20] Zwidwijk, Antonio. 2002. Multimodal Transport Perspective. International Multimodal Transport Association - IMTA, Geneva.
[21] ASEAN Framework Agreement on Multimodal Transport (AFAMT)
[22] Sistem Transportasi Nasional, Permenhub Nomor: KM. 49 Tahun 2005
[23] UU Nomor 23 Tahun 2007 Tentang Perkeretaapian
[24] UU Nomor 17 Tahun 2008 Tentang Pelayaran,
[25] UU Nomor 1 Tahun 2009 Tentang Penerbangan,
[26] UU Nomor 22 Tahun 2009 Tentang Lalu Lintas Angkutan Jalan
[27] Peraturan Kepala Kantor OP UK 112 Tentang Tata cara Pelayanan Kapal dan Bongkar Muat
[28] De Langen, P.W. 2008. Ensuring Hinterland Access: The Role of Port Authorities, Discussion Paper of OECD Int. Transport Forum, Paris.
[29] Kementrian PPN/ Bappenas 2013. Kajian Evaluasi Pembangunan Bidang Transportasi di Indonesia.
[30] UNCTAD, 2001. Implementation of Multimodal Transport Rules", United Nations, Geneva, 2001.

Revenue management of air cargo service in theory and practice

S Budiarto[1], H P Putro[2], P Pradono[2], and G Yudoko[2]
[1]Doctoral Candidate, School of Architecture, Planning, and Policy Development, Institut Teknologi Bandung, Indonesia, Nurtanio University
[2]School of Architecture, Planning, and Policy Development, Institut Teknologi Bandung, Bandung, Indonesia

Corresponding author: samsul_26@yahoo.co.id

Abstract. This study examines the air cargo service by comparing existing theories from previous research with the conditions on the ground. The object of the study is focused on the freight forwarder and the airport management. This study reviews the models and results of previous research that will be summarized and used to identify any issues related to the characteristics of air cargo operational services, as well as observing and monitoring literature with airlines, shipping companies, and airport management to explore and see the gap between prior research and implementation of its process in the air cargo service. The first phase in this study is to provide an overview of the air cargo industry. The second phase analyzes the characteristic differences between air cargo services and air passenger operating services. And the third phase is a literary bibliography study of air cargo operations, where the focus is on the studies using quantitative models from the perspective of the object of the study, which is the optimization of revenue management on air cargo services. From the results of the study, which is based on the gap between theory and practice, new research opportunities which are related to management of air cargo service revenue in the form of model development are found by adding booking timelines aspects of cargo that can affect the revenue of cargo airline companies and airports.

1. Introduction

The development of air transport activities around the world has increased the demand for airport services and the needs for a more efficient process of aircraft, passenger, and cargo services. During this decade, the demand for air cargo transportation has grown rapidly. This is reflected in the increase of air cargo fleets, the addition of cargo terminals, and the number of new cargo routes.

All airlines seek to expand their operations efficiently at airports to reduce costs and improve the customer service quality [1][2][3] explained that when carrying out their activities, airports aim to maximize aircraft mobility to improve the efficiency of operations in the competitive environment where they operate in. IATA revealed their forecast at the end of 2016 in preparation for the increasing demand in 2017, as depicted in the market distribution table below.

This could be proven through the monthly data provided by the IATA Air Cargo Market Analysis. In October 2017, the growth of Global Freight Tonne Kilometres (FTK) was at 5.9%, as shown in the graph below:

IOP Conf. Series: Earth and Environmental Science **158** (2018) 012022 doi:10.1088/1755-1315/158/1/012022

Air freight market detail - December 2016

	World share[1]	December 2016 (% year-on-year)			
		FTK	AFTK	FLF (%-pt)[2]	FLF (level)[3]
TOTAL MARKET	*100.0%*	9.8%	3.2%	2.8%	46.1%
Africa	*1.6%*	13.6%	5.3%	2.2%	29.6%
Asia Pacific	*37.5%*	9.8%	5.7%	2.1%	55.1%
Europe	*23.5%*	16.4%	5.9%	4.5%	49.7%
Latin America	*2.8%*	-1.0%	-7.9%	2.6%	36.5%
Middle East	*13.9%*	11.2%	5.9%	2.1%	44.9%
North America	*20.7%*	3.7%	-1.4%	1.8%	36.3%

Figure 1. Air Freight Market

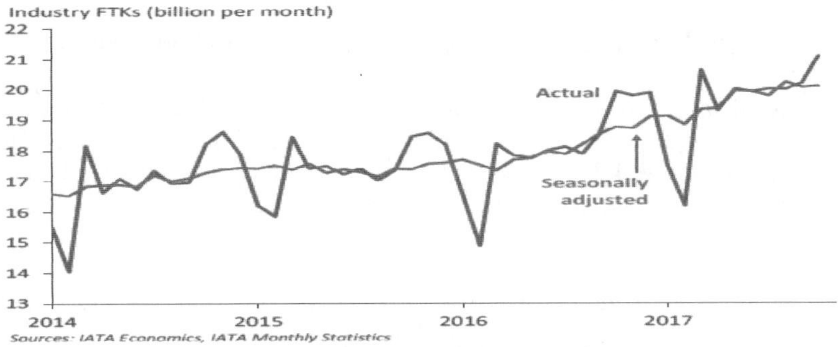

Figure 2. Growth of cargo market

The graph above shows the high growth of the cargo market in 3 years. Hence the cargo carriers must be able to keep the dynamic and keep up with the trend to face such a market pattern. This situation is consistent with research conducted [1] which reveals, with the graph below, the basis of air cargo research:

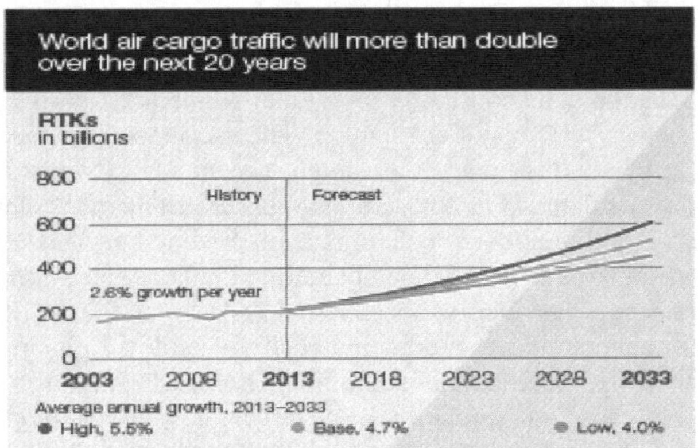

Figure 3. Growth of world air cargo

This research [1] not only shows that the gross income from Asia's air cargo is the largest, but also the flow of Asia's air cargo is the largest within the Asian continent as well.

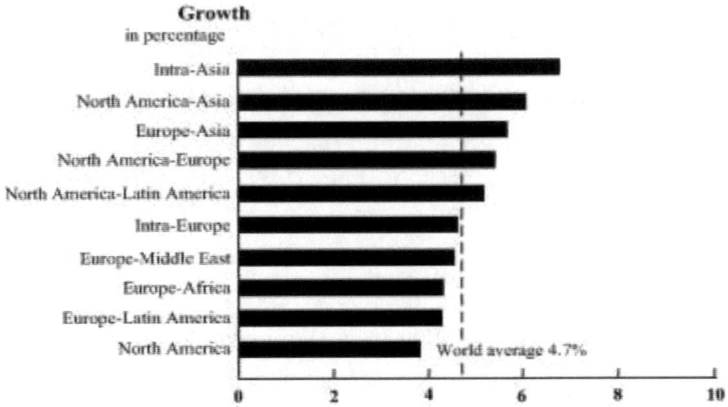

Figure 4. Asian cargo markets continue to lead industry growth

Boeing also conducted a forecasting analysis in the World Air Cargo Development 2016-2017. Boeing stated that between 2005-2015, air cargo growth reached 2%, while the estimated air cargo development from 2015 to 2035 concluded 4.2 %. Besides that, the research also explained that although the Chinese market growth rate has decreased from 7.6 % to 6.2% in a future forecast, the Chinese market still holds the highest air cargo growth in the world.

In response to that condition, this study aims to determine the challenges faced by the air cargo transportation industry by looking thoroughly over papers on aspects of air cargo operations.

The complexity of increasing revenue problems faced in carrying out the operation of air cargo encouraged a large number of research regarding air cargo. The research done by [4][5][6] [7] explained the uniqueness and complexity of operating the air cargo business in comparison to passenger transportation.*)*, [7] [8] explained the capacity of cargo in terms of the issue of overbooking. [9] did a research regarding air cargo revenue management based on weight and air cargo volume. [10] in their research considered the increased rate to maximize income. [11][12] explained the function of booking demands with cargo capacity, and [30] elaborated on the optimization of air cargo volume based on the container to be transported.

However, with the complexity of the issues that occur, most issues in this field remain unsolved in the end, especially on the optimization of income, or more commonly known as the Management Revenue. This is a culture that must be done constantly over time to increase revenue through the optimization and understanding of needs and demands of the market. The success of increased revenue on cargo shipping services for both the airline and the airport air cargo management (forwarder) cannot be separated from the operating system, as the demand factor with few optimizations in the quality, cost, and delivery must be taken into account. Additional cost(s) that arise in the income management can be converted into components of production costs that could be charged to the consumer in the form of the product selling price. The role of the cost is attached to the internal carrier itself so it depends on the airline how to handle its cargo. If managed well, the production cost will be lower, if not, the production cost and the selling price will be high with low competitive ability. Being successful in the revenue management for the airlines and airport cargo managers mean a larger size and a larger number of goods can be transported and distributed, hence more income for both. On the other hand, this also shows the success of positive contribution on general distribution activities, transportation, and development.

Therefore, the focus of this research is to develop a model of revenue optimization on air cargo operations by looking at the variables which affect it in overall air cargo operations so that it can be used in decision making, then conduct a comparison from previous theoretical research with practical experience in real life.

2. The decision problem and research questions

The common problems related to revenue management is about how the company manager makes decisions with the complexity of the air cargo problems with the uncertainty factors of arrival, quantity, and volume of goods ordered in the limited time, weight, and volume of goods to be transported on a plane. The decision whether the order of goods received can be accepted or rejected determines the level of income. Meanwhile, the research question here is: How does the problem of air cargo Management Revenue Optimization based on the coherent of existing theory and actual market activities.

3. Research method, process and procedures air cargo

3.1 Research Method

Steps to accomplish this research are: The first section of this study gives an insight on the air cargo industry, both its current condition and the problems faced. Then the second part will explain the concepts and methods of research, and explain the process and identification of air cargo operations. Section three focuses on reviewing previous research and performing air cargo literature tabulations which consists of theories and models to review the compatibility with current theories. Section four presents issues that occur in the actual practice of air cargo operations and compares it with theory to see the gap between theoretical researches and facts in the field, so that in the end we will provide new research opportunities. Finally, section 5 presents the summary.

The literature review method used in this research is to first by separate the study group based on the implementation of the concept of revenue management and the process of revenue management. The second phase of the literature review focuses on the process of revenue management that classifies the literature categories based on three main components. The next stage is to classify the existing literature based on the component group, and the following are the components and literature groups:

Preliminary studies on revenue management on air cargo based on research conducted by [14][17] [20]. In the three main components literature the revenue management process are as follows :

- Forecasting Components. This process estimates demand, available capacity, demand, and rate of price increase. [2][3][13][14]
- Spatial Capacity Component. In this process, how much space should be sold under long-term contracts and ensure which requests should be received by the airlines.[8][9][13]and [16]
- Overbooking component. In this process, accepting orders more than the available physical capacity to avoid losses due to order cancellation. Refers to research conducted by [11][12] [18] and [19]

By grouping and categorizing the existing literature articles selection, it will provide a comprehensive literature description in the field of revenue management.

3.2 Process of air cargo handling

1. First, reserve to the registration operator with the shipping instruction documents as proof that the agent/shipper has made a booking and confirm the type and quantity of goods in detail to the Airlines.
2. Then, the goods are stored in the warehouse. After it is received, there will be inspection of the goods' packaging to categorize where the products would be separated, whether into the group of Special Cargo/General Cargo.
3. After that, the labeling process would label the goods with the name, the exact address of the shipper, and the consignee. Label/cargo stickers are used to mark the packaging. All fees are charged on the freight forwarding services and must be paid directly by the sender using cash upon receipt of the cargo.
4. Last, the examination of required documents, such as the Airway Bill & Shipper Document (for Dangerous Goods), then followed by Build Up process by setting up a ULD (Unit Load Device) which will be put into the trunk of the plane. If the goods are categorized as Special

Cargo, then the sender would need to prepare a document NOTOC (Notification to Captain). The goods are then ready to be sent.

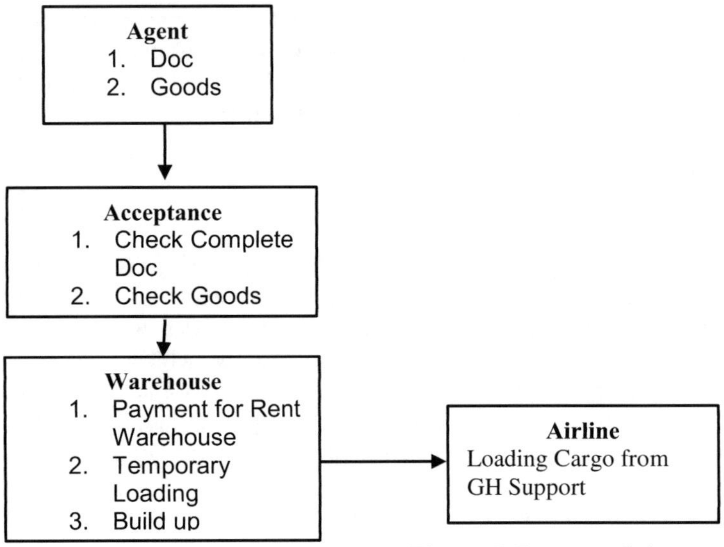

Figure 5. Process of air cargo handling

3.2 Procedure of cargo departure

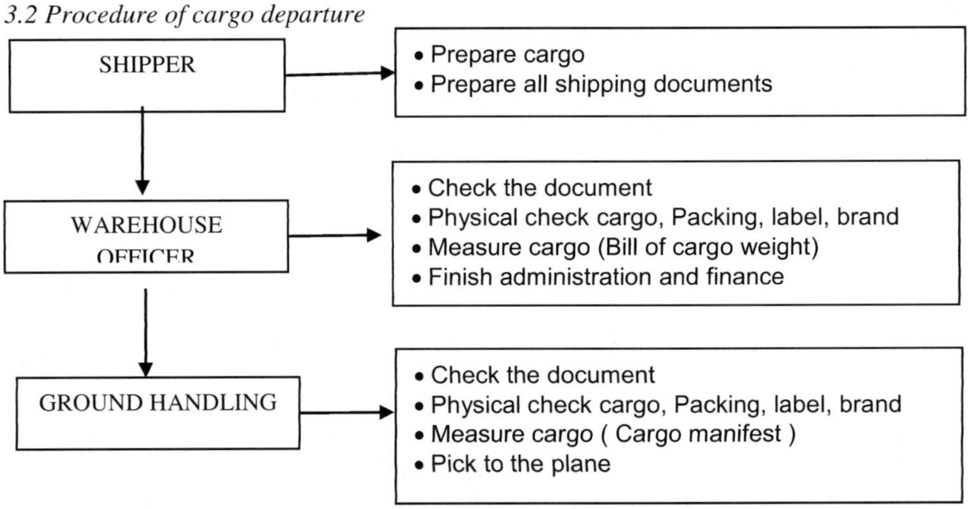

Figure 6. Procedure of cargo departure

Several research studies related to air cargo operations as undertaken by [4][5][6] [7] described the uniqueness and complexity of air cargo operations in comparison with passenger transport. Some of the things studied involve aspects of Uncertainty, Complexity, and Flexibility.

a. Uncertainty
Air transport is a combination of cargo and passenger, in which the passenger is more prioritized compared to the cargo. Thus, the level of uncertainty on air cargo is higher due to the availability of capacity that is influenced by the number and size of the airline's passengers [15]. disclosed that the shipper must ensure the use of cargo capacity long before the flight. The number of goods

to be shipped is actual and not order, and this shows the high fluctuation rate in capacity management on air cargo.

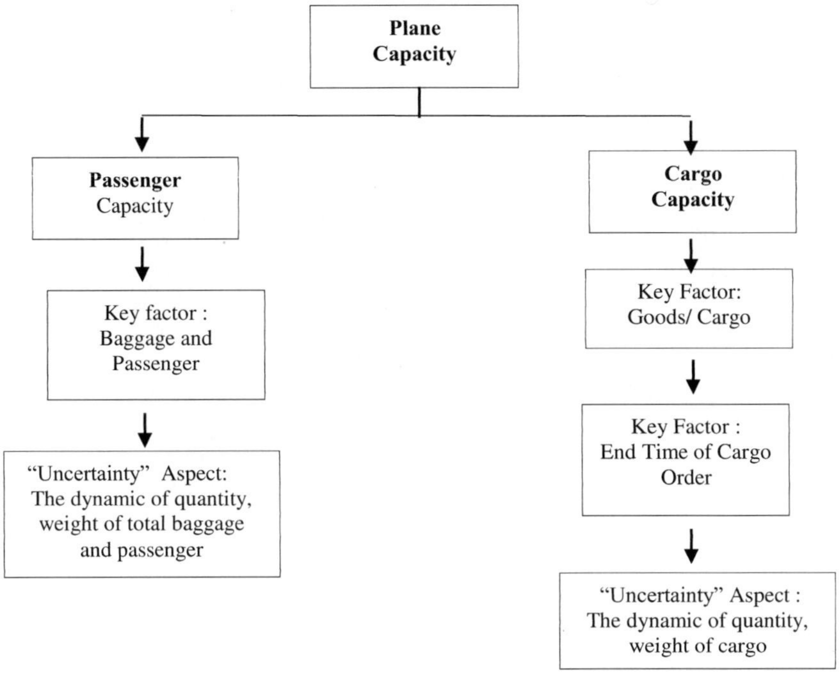

Figure 7. Uncertainty of plane capacity

b. Complexity

Determining cargo capacity through forecasting methods is significantly more complicated than estimating the capacity of a passenger aircraft. This is because before we identify the cargo capacity, which depends on the cargo space and the remaining size of the total weight of the aircraft, we must first know the number of passengers on the plane. Meanwhile, calculating the capacity through passenger aircraft can be determined from the number of seats in use, the cargo capacity depending on the type of container used, which are called unit load devices (ULDs), and which are further defined by several dimensions, such as weight, volume, type, and center of gravity[2] [21].

c. Flexibility

The delivery schedule from the origin to destination in air cargo is more flexible than passenger transport. Basically, the delivery can be done by using any kind of aircraft without having to consider the number of transit stops, as long as it arrives in the specified time; on the other hand, having several transit stops to reach a destination would be very risky for a human passenger [15] In optimizing the use of network capacity, the airline could make a delivery plan by simply determining the origin airport, the transit airport(s), and the destination airport to the forwarder.

4. Research on air cargo

Several researches on air cargo have been done with different focuses, such as the air cargo system, its operation process, and the development of the air cargo industry. The growth of the air cargo market is pushing the current research further into optimizing revenue by optimizing available capacity in the form of a quantitative decision method with a mathematical model for air cargo operations. There are several articles related to air cargo that would be discussed in the following literature.

4.1 Concept of management revenue

Revenue management is defined as the integrated management of prices and inventory to maximize profitability. According to [22] in his article, "Revenue Management Hard-Core Tactics for Market Domination," Revenue Management is the application of various tactical disciplines to maximize revenue growth. While [23] qualify Revenue Management into two categories, quantity-based and quality-based. Capacity control is to allocate capacity optimally to different classes of demand, while overbooking is when a company accepts reservations more than the available physical size in order to avoid losses due to order cancellations. The study of cargo space control was first done by [9]. In his research, he used a mathematical model with a heuristic method, HD Method, separating the two-dimensional problem of cargo into one separate dimension and later combining them together. The discussion of revenue management is always associated with passenger and freight transport. [14] stated that the revenue management is the management of passenger and seats' fare along with the levels of cargo and cargo space. In the same study, [14] described the process for air cargo revenue management into the following three steps, which are: (1) forecasting capacity available for sale, (2) ordering more forecasting capacity to compensate ordering behavior regarding cancellation and absence, and (3) allocating extra capacity to different products or markets that can maximize total revenue or profitability. This is called space allocation or capacity allocation.

Table 1. Air Cargo Revenue Management Publication.

Reference	Model	Focus	Method	Variable
Kasilingam (1996)[14]	Stochastic programming	Overbooking	Discrete or continuous probability distribution	Capacity , spoilage cost
Kasilingam (1997)[39]	Stochastic programming	Overbooking	Discrete or continuous probability distribution	Capacity , spoilage cost
Popescu et al. (2006)[8]	Nonparametric distribution estimation and forecasting	Overbooking	Discrete distribution	rate (weight or volume)
Popescu (2006)[27]	Probabilistic nonlinear dynamic programming	Accept or reject policy	Rates as a function of weight and cargo level	Capacity (weight and volume); classes
Amaruchkul et al.(2007)[9]	Stochastic dynamic programming	Accept or reject policy	Joint distribution of volume and weight	Capacity shipment type; terminal value
Gupta (2008)[25]	Stackelberg game	Capacity contract	Deterministic demand	Forwarder's effort level; freight rate
Wang and Kao (2008)[7]	Fuzzy reasoning	Overbooking	Fuzziness method	Capacity spoilage cost, over-sale cost, show uprate
Han et al. (2010)[37]	Stochastic dynamic programming	Accept-or reject Policy	Joint distribution of volume and weight	Capacity (weight and volume); profit rate;
Huang and Chang (2010)[10]	Stochastic dynamic programming	Accept-or reject Policy	Discrete distribution	Capacity, profit rate; shipment type;
Amaruchkul and Lorchirachoonkul (2011)[15]	Stochastic dynamic programming	Accept-or reject Policy	Joint distribution of volume and weight	Capacity (space)
Amaruchkul et al.(2007)[9]	Random demand	Capacity contract	Principle-agent game	lump-sum payment and refund rate
Qin et al. (2012)[26]	Dynamic programming –	Overbooking	Continuous distribution	capacity
Hellermann et	Game	Capacity	Demand is a function of	Booking level, reservation

Reference	Model	Focus	Method	Variable
al.(2013)[11]		contract	exogenous parameters and price	number
Totamane et al.(2014)[12]	Certain demand	Demand forecasting	Multiagent game	Weight, number of airlines, cargo capacity

[24] observed the development of Revenue Management from the 1960s until 1990s. The earliest usage of Revenue Management was by the American Airlines company, when they developed an online reservation system called Saber (Semi-Automated Business Research Environment) in the 1960s. In another study on American Airlines, a typical revenue management system is expected to generate an additional income of 3-5% [11]. Excess requests create substantial revenue and income management profits.

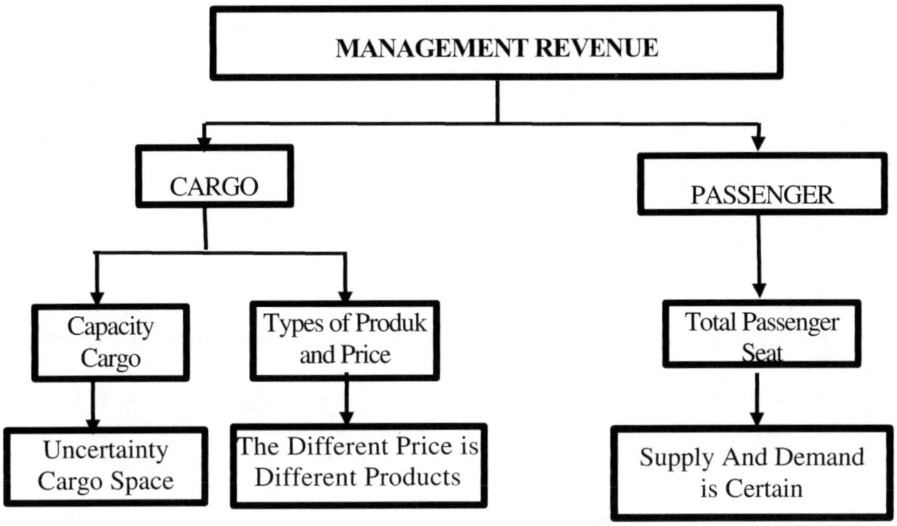

Figure 8. Management revenue

4.2 Previous studies of air cargo

The study of air cargo services from several researches, is as follows: [25] explained that the current problem with cargo capacity is that the air cargo supply chain consists of the shipper, freight forwarder, and airline. The shipper sends the order to the sending aircraft and is responsible for contacting the carrier and obtaining space to transport the cargo as the sender requires. This process often takes a long time because of the lack of coordination between the aircraft sender and carrier. This could be easily solved if an integrator or a transport plane has their own fleet.

[28] proposed the management of results with the number of passengers as an indicator of air cargo performance. An example of the case is American Airlines, which could gain $ 1.4 billion profits in three years and earn an expected profit of $ 500 million per year.

[14] correlated the level of cargo capacity with forecasting results. In his research, the capacity forecast affects the amount of capacity an airline could provide, and the demand forecast affects the number of decomposers and the quantity of freight cancellation.

[17] The cargo transport capacity is sold based on two models: (1) Contracts of assured capacity: consumer approval involving guaranteed capacity (defined in expenses and volume) on specific flights/working days. (2) Free sales: no capacity warranty, it is usually based on a specific order. The airline may receive a booking request or space booking for more profit in the future.

[25] in his doctoral dissertation, stated that cargo costs and capacity could affect the level of cargo services. The booking fee occurs if the airline has not set up space for cargo transport. A pending fee happens if the carrier can no longer haul the cargo due to limited capacity. Both costs could result in the reputation of air cargo services being questioned.

[27] Stated that air cargo capacity assessments may occur due to expiration costs and the cost of cargo carrying capacity cancellation. Expiration costs occur if the cargo cannot be delivered on time, so it will decay or expire on the road. The price of freight cancellation is charged if the cargo is already in place, yet cannot be transported to the plane. Both costs are different, although they can sometimes happen during the same time. If this happens, the airline (shipping service company) would get into trouble because consumers would feel that they are being disadvantaged, which will ultimately affect the company's reputation.

[29] Also stated that the market cannot be separated from scheduling. Scheduling is required for the airline's target market to be met. These two supports are the scheduling of vehicles and the amount of cargo. The number of vehicles/aircrafts is vital because without knowing the exact number of aircrafts in operation, the airline may not be able to arrange the scheduling properly. The amount of cargo should also be categorized by priority setting and mapping. Some loads that must be sent immediately must be transported first by the aircraft that can pass the route at the earliest. Other loads can be adjusted according to the allocation of aircraft and directions taken.

Fixed bookings or free sales may affect overall air cargo capacity performance, where network optimization is a leading indicator of air cargo performance [13].

5. Real condition revenue management in air cargo

5.1 The Operational of Air Cargo

In this research, the study is conducted in Husein Sastranegara Bandung Airport, Indonesia, and involved several airlines; Implementation of Air Traffic Management Revenue at Husein Sastranegara airport has its own uniqueness related to its actual condition and is as follows:

1. Air cargo demand in Husein S. reached an average of 25 tonnes/day while the ability to deliver is at 20 tonnes/day (source: Husein's Cargo Warehouse). The current situation shows imbalance between supply and demand; although the numbers are very profitable, but in reality, the amount of sent goods are not well distributed to all available flights, only to one or two destinations because of the limited loads.
2. The types of goods that can be sent from the Husein S. airport are all goods types for General Cargo and Special Cargo (source: Husein Cargo Sheds). This condition will provide a large room for revenue increase attempt but will probably be constrained by the cargo space arrangements and packaging problem.

Table 2. Type Of Cargo

Airline	General Type						
	General Cargo	Special Cargo					
		Valuable goods	Live animal	Human remain	Perishable goods	Strongly smelling goods	Dangerous goods
G.A	Yes	Yes	Yes	Yes	Yes	Yes	Yes
L.A	Yes	Yes	Yes	Yes	Yes	Yes	Yes
A.A	Yes	Yes	Yes	No	Yes	Yes	Yes
CL A	Yes	Yes	No	Yes	Yes	Yes	No
N.A	Garment	No	No	No	No	No	No
E.A	Yes	Yes	Yes	Yes	Yes	Yes	Yes

3. The aircraft used for cargo is a passenger aircraft type: Air Bus A320, Boeing 737 - 700 and - 800 and ATR (source: AP II Husein S.). The limited size of the aircraft and the runway reflects on the limited loading of freighter cargoes.

Table 3. The Cargo Load Limit

No	Airline	Type	Max Weight
1	G.A	B737	4 Tonnes
2	L.A	B738	2 Tonnes
3	A.A	A320	2 Tonnes
4	CL A	A320	4 Tonnes
5	N.A	B737	600 Kg
6	E.A	B737	2 Tonnes

4. The cargo storage system in the aircraft is by bulk storage (not using a Container or Pallet) (source: Husein Cargo Warehouse). This condition resulted in ineffective utilization of cargo space.
5. The absence of airlines that operate a special cargo aircraft.
6. The aircraft used is a combination transporting plane for passengers and cargo.
7. The limited size and number of the transport aircraft.

Based on the conditions stated above, it can be seen that the potential benefits are large, but it will be difficult to optimize in accordance with the theory of revenue management. In fact, the cargo business at Husein Sastranegara airport is still considered an optional business after passenger service. The problem in using a passenger and cargo combined transporting plane with an aircraft of limited size and the optional condition of having cargo as payload, is that the number of the goods loaded will be very dependent on the number of passengers and baggage loads carried by the passenger. This means that the optimization that could be done is by calculating the remaining capacity of the cargo space and the weight of goods to be loaded and by applying an overbooking strategy in an effort to anticipate losses due to the occurrence of spoilage (if the cargo goods already on the booking did not arrive, then it will cause losses due to unused space) and oversale (a condition which happens before the departure of the aircraft, as the weight and volume of cargo exceeds the capacity of the plane, then some cargo loads must be unloaded, reducing the revenue of the airline). Other facts at Husein S Airport are large market potentials not being utilized by applying pricing strategies to optimize revenue. The application of the price made static is not distinguished by the time of booking and urgency. This causes and is reflected by a not optimal income.

6. New research opportunities

After comparing the theory and practice, there are a few interesting points to be researched to obtain optimal results, such as: the various types of cargo shipments with different weight and volume in its activities, following the distribution with random variables. Aircrafts have limitations on weight capacity and volume capacity. Airline pursue a policy towards different booking and loading timeframe. Each booking request arrival is independent of the booking horizon of each carrier. Upon receiving the booking request, the airline must pay attention to the maximum load of the aircraft by looking at the number of passengers to be transported, so in this case the airline must decide whether to accept or reject the incoming booking request.

There is a condition that: (1) When a booking request is received, the airline does not know exactly the type, weight, and volume of the cargo before the departure of the plane. (2) Each received booking request could be cancelled anytime or does not arrive when the plane is departing.

This situation, of course, will cause losses for the airline if the existing capacity is not utilized correctly. Implementation of overbooking is one way to overcome this. However, excessive overbooking will cause an oversale penalty that will create more significant losses than the shortage. To overcome this problem, we need to find the optimal cargo overbooking limits obtained.[12]

Further works to do in relation to research opportunities are:

1. By observing the amount of capacity wasted associated with weight loss and transported passengers' baggage, it would provide opportunities to optimize the load capacity by allocating the remaining capacity for cargo loads.
2. Search for the optimal amount of time towards the cargo booking deadline (booking time close), considering the possibility of wasted load capacity by the total load of passenger and baggage. It is expected that further research will be able to answer these problems to enhance the airline overbooking policy over the years.

Knowing the optimal time limit of the cargo booking time would impact on: (1) Increasing revenue for airlines and airport managers. (2) Reducing queue time and the duration of airport parking lot usage. (3) Reducing the occurrence of delayed flights. (4) Improving service efficiency for airlines and airport managers. (4) Having goods flowing faster to reduce the number of loads dwelling in the cargo warehouse. (5) Increasing the capacity of cargo warehouses and airports as a whole.

7. Summary

In this paper, we reviewed literature on revenue optimization through capacity utilization in air cargo operations and compared recent theories with the practical practice.

The first thing we did was to map the air cargo industry, and then we analyzed the characteristics of air cargo and compared it with air passenger operations. Later, we conducted a literary bibliography survey on the optimization of air cargo operations capacity, where we focus on studies that used quantitative models from the perspective of airlines, freight forwarders, and air cargo. Furthermore, we identified the main decision issues in air cargo operations and discussed the gap between previous research and practical practice based on literature review and in-depth interviews with airlines and forwarders. The literature has investigated some real-world problems in the air cargo industry.

This review identified several significant differences between theory and practice in relation to capacity optimization in air cargo operations. The theories examined in this paper still provide room for them to be more refined, given the studies conducted under different situation, conditions, and characteristics. Models often differ from operational reality.

Two major points to conclude this study. First, the overbooking policy to increase airline revenue could be achieved by expanding the utilization of aircraft capacity. Second, the overbooking policy would be more optimal if we calculate the optimum booking time limit. The broader the booking time horizon, the higher the revenue opportunity would be.

Research opportunities that could be done are: developing the model of revenue optimization by considering the booking time limit on the air cargo service and the utilization of the wasted load capacity due to the remaining load of passengers and baggage transported in the aircraft (the number of passengers and baggage loads are smaller than the available size).

References

[1] Oum T H, Yu C, and Fu X 2003 A comparative analysis of productivity performance of the world's major airports: summary report of the ATRS global airport benchmarking research report—2002 *J. Air Transport Management* **9** 285-297

[2] Thelle M H and la Cour Sonne M 2018 Airport competition in Europe *J. Air Transport Management* **67** 232-240

[3] Humphreys I and Francis G 2002 Performance measurement: a review of airports *Int. J. Transport Management* **1** 79-85

[4] Bartodziej P, Derigs U, Malcherek D, and Vogel U 2009 Models and algorithms for solving combined vehicle and crew scheduling problems with rest constraints: an application to road feeder service planning in air cargo transportation *Operation Research-Spektrum* **31** 405-429

[5] Leung L C, van Hui Y, Wang Y, and Chen G, 2009 A 0–1 LP model for the integration and consolidation of air cargo shipments *Operational Research* **57** 402–412
[6] Li Y, Tao Y, and Wang F 2009. A Compromised large-scale neighbourhood search heuristic for capacitated air cargo loading planning *European J. Operational Research* **199** 533-560
[7] Wang Y J and Kao C S 2008 An application of a fuzzy knowledge system for air cargo overbooking under uncertain capacity *Computer and Mathematics with Applications* **56** 2666-2675
[8] Popescu A, Keskinocak P, Johnson E, LaDue M, and Kasilingam R 2006 Estimating air-cargo overbooking based on a discrete show-up-rate distribution *Interfaces* **36** 248–257
[9] Amaruchkul K, Cooper W L, and Gupta D 2007 Single-leg air-cargo revenue management *Transp. Sci* **41** 457–469
[10] Huang K C and Chang K C 2010 An approximate algorithm for the two-dimensional air cargo revenue management problem *Transp. Res. Part E* **46** 426–435
[11] Hellermann R, Huchzermeier A, and Spinler S 2013 Options contracts with overbooking in the air cargo industry *Decis. Sci.* **44** 297–327

[12] Totamane R, Dasgupta A, and Rao S 2014 Air cargo demand modeling and prediction *IEEE Syst. J.* **8** 52
[13] Feng Y and Xiao B 2000 A Continuous-time yield management model with multiple prices and reversible price changes *Management Science* **46** 644–657
[14] Kasilingam R G 1996 Air cargo revenue management: characteristics and complexities *Eur. J. Oper. Res.* **96** 36–44
[15] Amaruchkul K and Lorchirachoonkul V 2011 Air-cargo capacity allocation for multiple freight forwarders *Transp. Res. Part E* **47** 30–40
[16] Levin Y, Nediak M, and Topaloglu H 2012 Cargo capacity management with allotments and spot market demand *Operations Research* **60** 351-365
[17] Slager B and Kapteijns L 2004 Implementation of cargo revenue management at KLM *J. Revenue and Pricing Management* **3** 80-90
[18] Becker B and Wald A 2010 Challenges and success factors of air cargo revenue management *J. Revenue & Pricing Management* **9** 171- 184
[19] Moussawi, L., & Cakandyildirim, M. (2012). Optimal overbooking limits of a two-dimensional cargo problem: A profit maximization approach. Journal of Revenue and Pricing Management, 11(4), 453-476.
[20] Nielsen K 2004 Revenue management at virgin atlantic cargo *Presentation at the AGIFORS Cargo Study Group Meeting* (Washington DC: AGIFORS)
[21] Leung L C, van Hui Y, Wang Y, and Chen G 2009 A 0–1 LP model for the integration and consolidation of air cargo shipments *Operational Research* **57** 402–412
[22] Cross R G 1997 *Revenue management hard-core tactics for market domination* (New York: Broadway Books)
[23] Tallury K T and van Ryzin G J 2004 *The theory and practice of revenue management* (New York: Kluwer Academic Press)
[24] Donovan A W 2005 Yield management in the airline industry *J. Aviation/Aerospace Education & Research* **14** 11-18
[25] Gupta D 2008 Flexible carrier–forwarder contracts for air cargo business *J. Revenue and Pricing Management* **7** 341-356
[26] Qin C R, Luo L, You Y, and Xiao Y X An optimization model of the single-leg air cargo space control based on Markov decision process *J. Applied Mathematics* **2012** 1-7
[27] Popescu A 2006 *Air cargo revenue and capacity management* PhD Dissertation (Atlanta: Georgia Institute of Technology)
[28] Chang Y H, Yeh C H, and Wang S Y 2007 A survey and optimization-based evaluation of development strategies for the air cargo industry *Int. J. Prod. Econ.* **106** 550–562.

[29] Luo S, Cakanyildirm M, and Kasilingam R G 2009 Two-dimensional cargo overbooking models *European J. Operational Research* **197** 862-883

[30] Smith B C, Leimkhuler J F, and Darrow R M 1992 Yield management at American Airlines *Interfaces* **22** 8-31

[31] Hwang T-L, Jeng C-R, and Lee I-L 2013 Optimization of the imported air express cargo distribution problem *Int. J. Development and Sustainability* **2** 63-85

[32] Patil J T and Patil M E 2016 Review on cargo space optimization *International Conference on Global Trends in Engineering, Technology and Management (ICGTETM2016)* 188

[33] Haidar L M and Cakanyildirim M 2011 Optimal overbooking limits of a two-dimensional cargo problem: a profit maximation approach *J. Revenue and Pricing Management* **11** 453-476

[34] May A, Wu Y, Marshall A, Chipulu M, Anslow A, and Ojiako U 2014 Optimisation of key performance measures in air cargo demand management *J. Transport and Supply Chain Management* **8** 1-9

[35] Boeing Company 2014 World air cargo forecast 2014–2015 [Online] Available from: http://www.boeing.com/assets/pdf/ commercial/cargo/wacf.pdf

[36] Chan F T S, Bhagwat A, Kumar N, Tiwarib M K, and Lam P 2006 Development of a decision support system for air-cargo pallets loading problem: a case study *Expert Syst. Appl.* **31** 472–485

[37] Han D L, Tang L C, and Huang H C 2010 A Markov model for single-leg air cargo revenue management under a bid-price policy *Eur. J. Oper. Res.* **200** 800–811

[38] Huang K C and Chi W 2007 A Lagrangian relaxation based heuristic for the consolidation problem of airfreight forwarders *Transp. Res. Part C* **15** 235–245

[39] Kasilingam R G 1997 An economic model for air cargo overbooking under stochastic capacity *Comput. Ind. Eng.* **32** 221–226

[40] Mongeau M and Bes C 2003 Optimisation of aircraft container loading *IEEE Trans. Aerospace Electron. Syst.***39** 140–150

[41] Tallury K T, and van Ryzin G 2004 Revenue management under a general discrete choice model of consumer behavior *Management Science* **50** 15-33

IOP Conf. Series: Earth and Environmental Science **158** (2018) 012023 doi:10.1088/1755-1315/158/1/012023

Developing a Probabilistic Model for Constructing Seaport Hinterland Boundaries

Russ Bona Frazila[1], Febri Zukhruf[1*], Jzolanda Tsavalista Burhani[1]
[1]Department of Civil Engineering, Institut Teknologi Bandung, Jl. Ganesha No. 10, Bandung, 40132, Indonesia

Corresponding author: febri.zukhruf@ftsl.itb.ac.id

Abstract. Seaport competitiveness has been attracting a great deal of attention. It is shaped not only by a port's performance and shipping liner calls but also by its hinterland. However, planners face a number of difficulties for identifying the boundaries of the hinterland, as these are determined by several factors, such as commodity types and final destinations, hinterland connections, inland transport costs, travel costs, and sea shipping costs. Hence, hinterland analysis is often simplified by relating it to the administrative boundaries, which do not correctly represent the actual condition of seaport demand behaviour. Furthermore, the boundaries do not behave deterministically in response to changing factors, which creates a challenge when estimating hinterland boundaries. This paper presents a deterministic model for constructing seaport hinterland boundaries, which considers hinterland connections, generalised transport cost from any point related to the transportation network. Thus, the boundaries do not necessarily coincide with the administrative region. Moreover, a model is proposed for reflecting the stochastic decisions of shippers to select a seaport, which may be utilised for approximating port demand variation. A case study is also presented for exploring the applicability of the proposed model.

1. Introduction

The rising uncertainty of global trade and sustainable infrastructural investment are placing massive competitive pressure on modern ports, especially those in mature markets in well-developed regions [1]. Against this background, managing the seaport hinterland has been shown to be an influencing factor in the sustainable development of seaports. For example, Tan [2] analyses the evolution of the port city of Singapore, which has elevated the port of Singapore to one of the two leading container hub ports in the world.

The theoretical meaning of the port hinterland is a fuzzy concept, since many definitions can be found in the literature and it can be adjusted with respect to different variables as well (e.g. type of cargo). A port's hinterland is the inland area where it acquires the majority of its business [3]. Concretely, the catchment area of a specific port is the scatter of inland cargo origin/destination points generating traffic flow passing through it. In abstract terms, the traditional concept of the hinterland represents it as an area whose contour is a continuous line bounding the port's economic influence on the shore.

Recent research states that the port hinterland is of particular importance to port managers for several reasons [4]. Firstly, port managers can obtain knowledge on hinterland accessibility and hence improve their port's market share by analysing how the transportation network affects the extent and size of the hinterland. Secondly, port managers can identify potential customers (i.e. shippers) located in the hinterland and negotiate long-term contracts with them to build customer loyalty and retain more customers. Thirdly, on top of game-theoretic analysis methods, the port hinterland concept offers a new way of looking into a port's competitiveness. Port managers can understand a port's standing compared to its major rivals through hinterland analysis and identify ports they might collaborate with in order to enjoy a coalition surplus.

Several studies have been published to discuss the issue of seaport hinterland. Wan et al. discussed how hinterland accessibility can affect a port's efficiency using the case of U.S. container ports [5]. Nazemzadeh and Vanelslander applied the analytical hierarchy process method to obtain port selection criteria in descending order of importance: port cost, geographical location, quality of hinterland connections, and so on [6]. Zhuang and Yu used the breakpoint model and Huff's model (a gravity model) embedded in ArcGIS to derive the hinterlands of two ports, but neither of the models incorporates the uncertainty in transportation cost or time [7]. Kramberger et al. proposed an interesting mixed linear integer model to identify port hinterlands from a shippers' point of view [8].

This paper discusses a deterministic and a probabilistic model for delimiting the boundaries of the seaport hinterland. The models were developed based on the generalised cost incurred for sending goods to the port and the utility value for reflecting the seaport competitiveness. In the case of the deterministic model, the boundaries are constructed by connecting transport nodes according to generalised cost. Since the transport nodes are possibly denoted not only at the macro level of a zone (e.g. the administrative region) but also at the micro level (e.g. intersections of links), the analysis hence promisingly provides more precise boundaries of the seaport hinterland. In addition, as shippers do not behave deterministically, also a probabilistic model is proposed for reflecting the stochastic way of decision-making of shippers to select a seaport. The model may be used for estimating the demand range of a seaport considering randomness in shipper decision-making.

Knowledge about hinterland boundaries can have great benefits. It can help local governments to design an efficient and accessible transport network for attracting more cargo flow. In addition, port managers can identify key factors that affect the size and extent of the port's market area. Furthermore, it also allows shippers to better choose the proper port to handle their goods after gaining knowledge about seaport accessibility.

The rest of this paper is arranged as follows. In the following section, the modelling framework is described. In the third section, a case study is elaborated for investigating the applicability of the model. Finally, in the fourth section, the methodologies, results and analyses in the paper are summarized.

2. Modelling Framework

2.1. Deterministic-based Model

The deterministic-based model was formed by considering the unit transport cost, where a particular point, i, is determined as belonging to the hinterland of a port, j, only if the transportation cost is smaller compared to the transportation cost between i and other ports (elements of J). The transportation cost to be considered is the unit transportation cost of the shortest path between i and j. In the transportation network, with (N, A), set of nodes $n \in N$, arc a $(s,t) \in A$ and cost in arc c_a, the common shortest path formulation is:

$$\text{minimize} \sum_{(s,t)\in A} c_{st}x_{st}$$

subject to:

$$\sum x_{ts} - \sum x_{st} = 1$$
$$\sum x_{ti} - \sum x_{it} = 1 - n$$

(1)

where:

c_{ij} : cost in the arc, which can be in the form of distance, *d*, or time, *t*, or generalised cost, *g*
x : index for each arc

The arc represents multimodal links. The deterministic model of the hinterland may then be formulated as follows:

$$D_{ij} = \begin{cases} 1 \ if \ u_{ij} = \min\{u_{i1}, u_{i2}, \dots, u_{iJ}\} \\ 0 \ \ otherwise \end{cases} \quad (2)$$

where,

u_{ij} : shortest-path cost from node-*i* to seaport-*j*,
D_{ij} : binary variable with value 1 if node-*i* is part of the hinterland of seaport-*j* and 0 otherwise,
J : total number of available seaports.

The shortest-path cost is fixed for a certain condition of the network. Therefore, the seaport has a permanent hinterland as long as there are no changes in the network, which is formulated based on the binary variable in Eq. (2). The hinterland boundaries of seaport-*j* are then determined by connecting the nodes or interpolations between the locations that have the same transportation cost for accessing adjacent ports. A node can be defined as an administrative region (i.e. subdistrict), a freight terminal, an link intersection or any point in the land area as long as it can be related to the network. Since a node may represent a smaller area than an administrative zone, seaport hinterlands do not necessarily coincide with administrative regions, as is commonly implemented.

2.2. *Probabilistic-based Model*

Different from the deterministic-based model, the probabilistic model is based on the assumption that nodes are not fixed as hinterland members of a certain seaport but possess the possibility to be hinterland member of more than one seaport. The probability is influenced by the shortest-path cost to access the available ports. Lower cost will give a seaport a higher possibility of being chosen by a region for distributing goods. The equation below illustrates the probability-based model.

$$p_{ij} = \frac{e^{-\alpha u_{ij}}}{\sum_{j}^{J} e^{-\alpha u_{ij}}} \quad (3)$$

where,

p_{ij} : probability of node-*i* to be part of the hinterland of seaport-*j* or proportion of demand for node-*i* using port-*j*
α : coefficient

As inferred from Eq. (3), the node probability is calculated by dividing the utility value for transporting goods from node-*i* to seaport-*j* with the total utility value for sending goods from node-*i* to other seaports. By considering the other seaport utility values, this model is expected to provide a more realistic representation for capturing the decision-making of shippers and to be better at making port demand predictions.

3. Case Study

For a better illustration of the proposed model of seaport hinterland identification, it was implemented for a hinterland analysis for container transportation in Java Island (Indonesia), being the island with the highest container traffic (comprising more than 60% of total Indonesian container traffic). There are 5 provinces with three existing main container ports (Tanjung Priok Port in Jakarta, Tanjung Emas in Semarang, Central Java Province, Tanjung Perak Port, Surabaya, East Java Province) servicing both domestic and international container traffic. There is also a container port still under construction, the Patimban Port, in Subang, West Java (see the figure below).

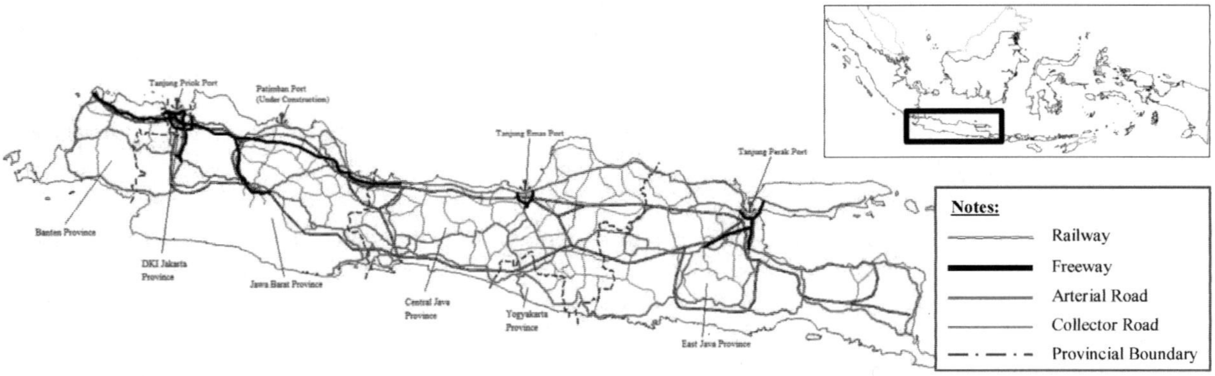

Figure 1. Transportation network and container ports in Java Island.

By the location, the three ports seem designated to service each province it is situated in and its surroundings. In other words, the hinterland for container traffic of Tanjung Priok Port includes the provinces of Jakarta, Banten and West Java. The hinterland of Tanjung Emas Port includes the provines of Central Java and Yogyakarta, while the hinterland of Tanjung Perak Port would be the province of East Java.

3.1. Deterministic Port Hinterland by Shortest-path cost

Referring to Eq. (1) shortest-path cost (from any point to access a port) may be based on distance, travel time and generalised cost (through the network). To calculate these costs, the transportation network of Java Island was modelled with 962 links and 573 nodes, as shown in Figure 2. Each link is attributed with end nodes, length and average speed of trailer (container truck), taking into account the width or capacity and existing traffic. Meanwhile, the unit cost was derived from the trucking unit fare, which is set by the Trucking Company Associations (see Figure 3) and time value for converting travel time to monetary value using the Container Storage Tariff, which is Rp. 60,000/day.

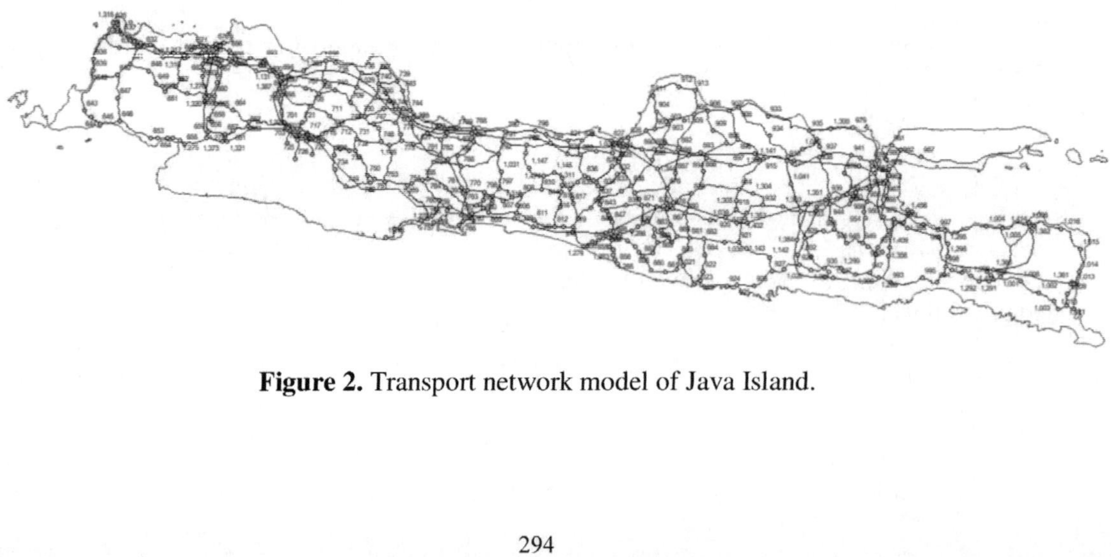

Figure 2. Transport network model of Java Island.

IOP Conf. Series: Earth and Environmental Science **158** (2018) 012023 doi:10.1088/1755-1315/158/1/012023

Distance (km)	Trucking Fare (Mill. Rp./Box)
2.5	1.4
10.0	1.8
20.0	1.9
50.0	2.1
100.0	2.3
200.0	3.5
400.0	5.8
600.0	9.0
1,000.0	15.0
1,500.0	20.0

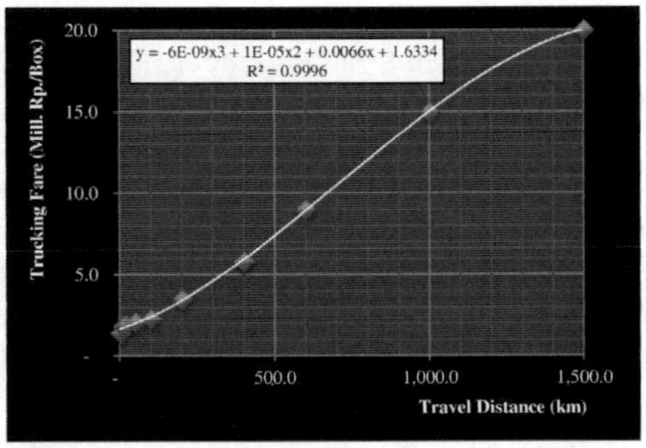

Figure 3. Unit fare by distance.

The least cost by distance, travel time and generalised cost from each node to the three available container ports in Java Island was then plotted (with interpolation between adjacent nodes) as in Figures 4, 5 and 6 with the particular hinterland boundaries.

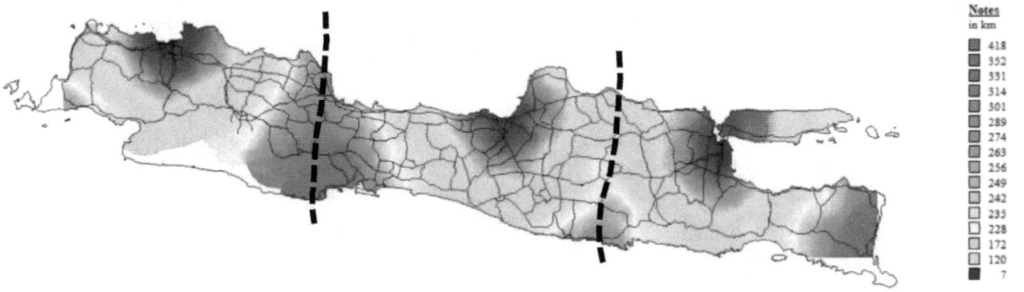

Figure 4. Distribution of least distance to container ports in Java Island.

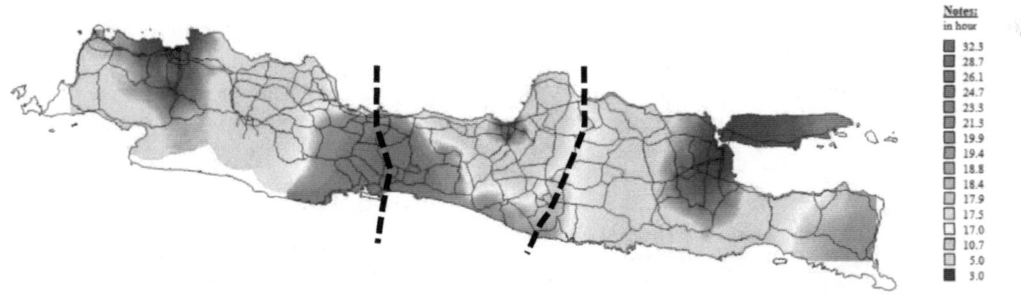

Figure 5. Distribution of least travel time to container ports in Java Island.

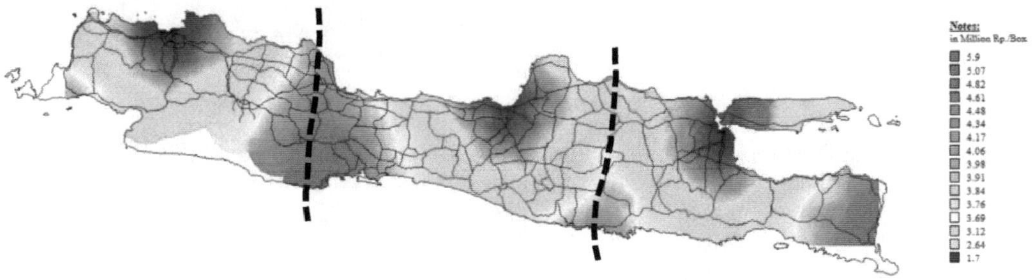

Figure 6. Distribution of least generalised cost to container ports in Java Island.

It can be seen that the hinterland boundaries by generalised cost are very close to those by distance, since the trucking fare unit is almost linear to the distance and the time cost value that was used is very low compared to the other costs. For commodities that are considered time-sensitive, the hinterland boundaries will be closer to the hinterland boundaries by travel time.

By this approach, the connectivity level of points or areas to the ports can also be identified (in this case container ports). The areas with a red colour have the worst connectivity to the main container ports. If in those areas there are some industrial estates or other activities that require accessing a main container port, then some measures need to be taken to reduce land transportation costs.

The operation of a new port will surely reduce the areas with the worst port connectivity, but for the case of Patimban Port, which is currently in the construction stage, its close location to Tanjung Priok Port makes the hinterland boundary hazier.

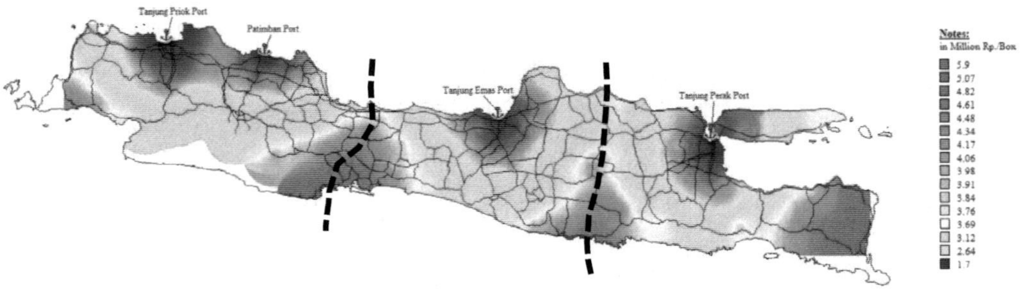

Figure 7. Distribution of least generalised cost to the container port in Java Island with Pantimban Port in operation.

3.2. Probabilistic Hinterland for Port Demand Prediction

The probabilistic approach in Eq. (3) is not meant for distinguishing the boundaries of the hinterland but is more useful for port demand prediction. The result of this approach can also be plotted to indicate the probability of areas to use a particular port. Furthermore, if the demand originating from or attracted to an area is known, the number of shippers that will use a particular port can be predicted by multiplying the demand with the probability value.

For the Java Island case, the model was applied for predicting the probability of using the Tanjung Priok Port without and with Patimban Port in operation, as can be seen in Figures 8 and 9.

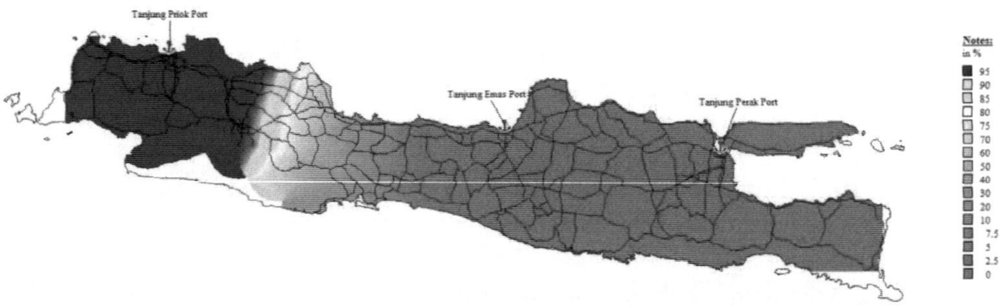

Figure 8. Probability distribution for using Tanjung Priok Container Port by area.

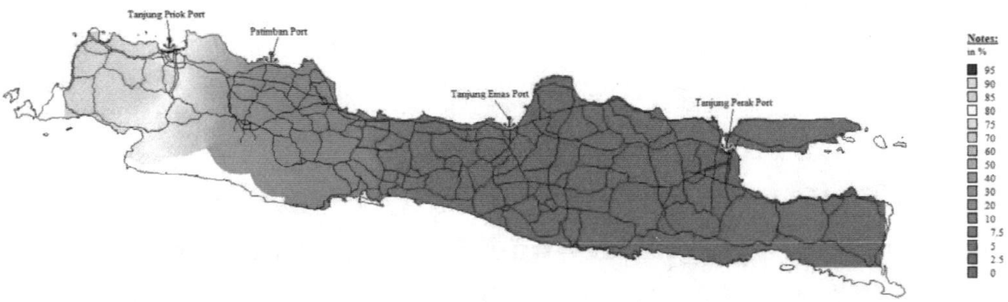

Figure 9. Probability distribution for using Tanjung Priok Container Port by area, with Patimban Port in operation.

It can be seen that with the new Patimban port in operation there will be no more 'almost-certain' area as the Tanjung Priok Port hinterland. Regardless of the capacity and port performance of Patimban Port, considering the location of the port and the industrial estates around its hinterland, its introduction may influence traffic flow in Tanjung Priok Port significantly.

4. Conclusions

The hinterland of a seaport should be determined based on a number of factors. One of the most important ones is the transportation network, which greatly influences the transportation (land access) cost, which determines the competition between ports. Depending on the commodity, the cost may be time-sensitive or not. The proposed deterministic approach for identifying the port hinterland is useful for prediction, not only regarding the seaports but also regarding port access and the surrounding transport networks. In particular, it can also be applied for measuring the connectivity level of any area to a seaport. Meanwhile, the probabilistic approach is more useful for predicting port demand, taking into consideration that the port user is not limited by certain boundaries. However, by determining the probability limits it can also be used for identification of the boundaries of the seaport hinterland.

References

[1] X. Shi dan H. Li, "Developing the port hinterland: Different perspectives and their application to X. Shi and H. Li 2016 Developing the port hinterland: Different perspectives and their application to Shenzhen Port, China *Research in Transportation Business & Management* **19** 42–50

[2] T-Y. Tan 2007 Port cities and hinterlands: A comparative study of Singapore and Calcutta *Political Geography* **26** (7) 851–865

[3] C. Ferrari, F. Parola, and E. Gattorna 2011 Measuring the quality of port hinterland accessibility: The Ligurian case *Transport Policy* **18** (2) 382–391

[4] X. Wang, Q. Meng, and L. Miao 2016 Delimiting port hinterlands based on intermodal network flows: Model and algorit*hm Transport Research Part E Logististics and Transport Review* (88) 32–51

[5] Y. Wan, A. C. lok Yuen, and A. Zhang 2014 Effects of hinterland accessibility on US container port efficiency *International Journal of Shipping and Transport Logistics* **6** (4) 422

[6] M. Nazemzadeh and T. Vanelslander 2015 The container transport system: Selection criteria and business attractiveness for North-European ports *Maritime Economics & Logistics* **17** (2) 221–245

[7] J. Zhuang and S. Yu 2014 The Hinterland Spatial Structure Evolvement of Competitive Port Based on ArcGIS Springer, Berlin, Heidelberg 1143–1153

[8] T. Kramberger, B. Rupnik, G. Štrubelj, and K. Prah 2015 Port Hinterland Modelling Based on Port Choice *PROMET - Traffic & Transportation* **27** (3) 195–203

Exploring the Role of Online 'Ojek' In Public Transport Trips: Case of Jakarta Metropolitan Area Rail Users

A F Saffan[1], M Rizki[2]

[1]Consultant, Cardno, Jakarta, Indonesia
[2]Alumnus, Civil Engineering, Parahyangan Catholic University, Bandung, Indonesia

Correponding author: farissaffan@gmail.com

Abstract. The invention of information technology shapes people's lives in the city, including the recent invention of online motorcycle taxi service. Known as 'ojek', it offers a reliable means of tactical commuting in dealing with traffic congestion in Greater Jakarta. Moreover, recent development in the online system improves the punctuality and reliability of ojek services. The emergence disrupts the transport landscape, resulting in commuters' dependency to this transport mode. However, academic sources are insufficient to answer the policy debate among planners and policy maker regarding this issue. This paper probes the role of online ojek in Jakarta Metropolitan Area Rail users' trip from their perspective, especially for the first mile and the last mile of their trip. The intensive commuter survey is currently being done at selected stations to gain information of their first mile and last mile trip diary. Discriminant analysis will be exercised to explain that to some extent online ojek is able to work to fill the gap produced due to minimum proper pedestrian facilities and a lack of public transport connection.

1. Introduction: Contextualizing Online Ojek in the Setting of Jakarta Transport

Unreliable public transport has triggered an explosion of the use of cars for daily commuting that is driven by powerful forces of inequality. Trunk public transport network does not fully cover all Greater Jakarta area (see Figure 1). Inevitably, people in Jakarta, especially the urban poor, are trapped in the consumerist use of low-cost motorcycles. A study reveals the remarkable fact that low-income people predominantly use motorcycles for daily transport [1].

The public transport share is lower than 40%, mostly used by low and medium income individuals. The new high-income people mostly tend to take their trips by car. Among the low-income, the motorcycle is the most dominant mode of transportation at 56% of all trips [1].

The circumstances of pedestrian facilities in Jakarta does not encourage people to walk for the first or last mile of their trips. People who walk in Jakarta must be ready for any risks related to traffic accidents and robbery. Good pedestrian pathways are only present in the center of the business district. Moreover, street vendors occupy many existing pathways throughout the city, leaving just a narrow space for pedestrians.

In consequence to the lack of pedestrian facilities and the lack of full public transport coverage, the market produces alternative modes of transportation such as online ojek to fill the gap. Combining IT an transport service into a smartphone application, online ojek emerged in early 2014. Technically, ojek is a motorcycle taxi service that functions as one of Jakarta's paratransit systems. It offers many benefits to commuters. It is easy to order, and easy to ride. Meanwhile, in cities in most developing countries, e.g., Jakarta, the government suffers from limited budget for an ideal public transport infrastructure, limited planning capacity, and insufficient maintenance and control [2]. The user just needs to order a trip using a smartphone application featuring transparent pricing and a clear destination map.

A transformation in transportation is inevitable, because people need more utility and effective instruments to make their life easier. Technology and users shape one another, and they somehow reconstruct human culture [3]. Fundamental application of smart applications, for instance, implies a change of way to life. The politics of innovation on public transportation triggers a renewal of constellation and learning on new procedures, which is not easier than the existing ways [3].

In terms of legal standing, ojek has a judicial bias that makes transportation policy development problematic. Naturally, ojek is a 'free-rider' because they simply pick up passengers from the street and disregard regulations [4]. On the one hand, many people still rely on this means of transportation. Also, ojek sustains the life of the urban poor [4]. By using ojek, commuters are able to cut significant transport cost instead of using a less regulated public transport, e.g., a microbus[5]. On the other hand, ojek operations are hard to monitor and are not controlled by authorities.

This paper intends to explore the relevance of ojek trips taken by commuter rail users with particular factors: income level, user attitude, quality of pedestrian facilities around the station, and quality of public transport services around the station. There are two questions that mainly construct this research: i) How does the quality of access plays a role in commuters' choice to use online ojek? and ii) How does online ojek play a role in connecting commuter rail and their competition with public transports and private vehicles? We explore these questions in the sense of individuals' decisionmaking behavior and connecting it to their travel and activity participation in the context of trip chain. The use of the paper is designed to enrich the academic discussion on commuter rail (KRL) trips and online ojek trip interaction discussion.

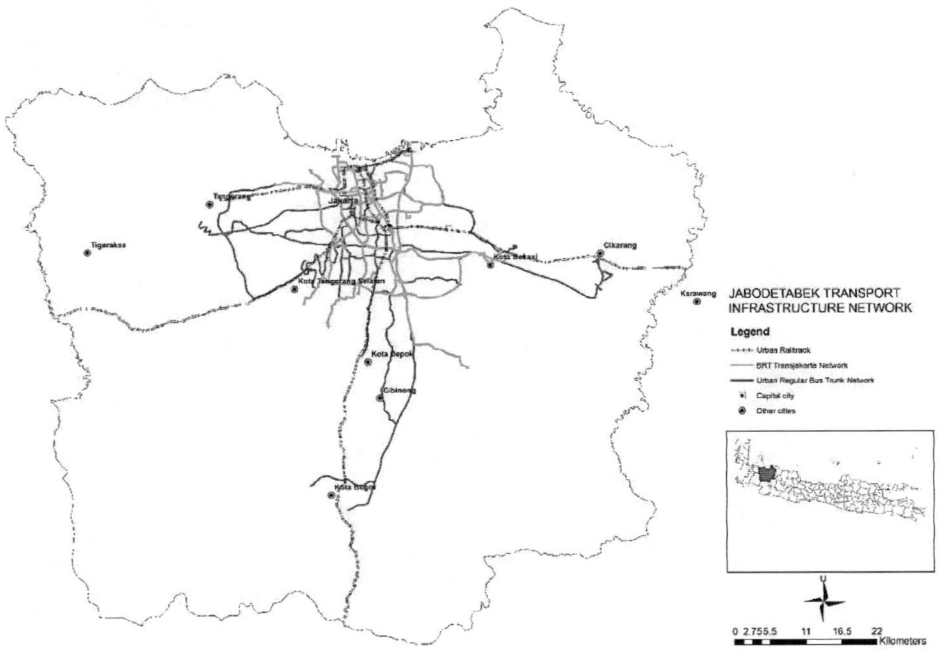

Figure 1. Jakarta public transport trunk network

2. Methodology

2.1 Data collection

Data for the analysis are obtained from questionnaire distribution. The concept of the first mile and last mile of the trip in the questionnaire is developed based on its interaction with accessibility *[6]*. The component of transport accessibility plays an important role in providing the needs of individuals' activities at their point of origin and destination, which for public transport users are in the first and last mile of their trip. The questionnaire consists of three parts: first is the individuals' social and demography characteristics. The second part of the questionnaire gathered information on the characteristic of the individuals' first mile of the trip, from their home to the rail station, and the last part is characteristics of the last mile of the trip, from the rail station to their activity location.

This research obtained more than 1015 samples from face-to-face questionnaire interviews collected at 22 selected Jakarta Metropolitan Rail stations to capture their users. The questionnaire was prepared to gather data on rail users' trip information. The survey was conducted consistently in the morning to get a typical trip chain as shown at Figure 2. Every interviewee was asked to imagine their routine travel chain in morning peak hour, and express it in the questionnaires. The trip chain approach in the questionaire was inspired by the concept of structural trip chaining[7,8].

The selection of stations considers the variance of characteristics of land near the stations: urban business district, urban mixed use, or suburban settlements. Before the analysis is performed, the data will be evaluated based on the variance and completeness [9].

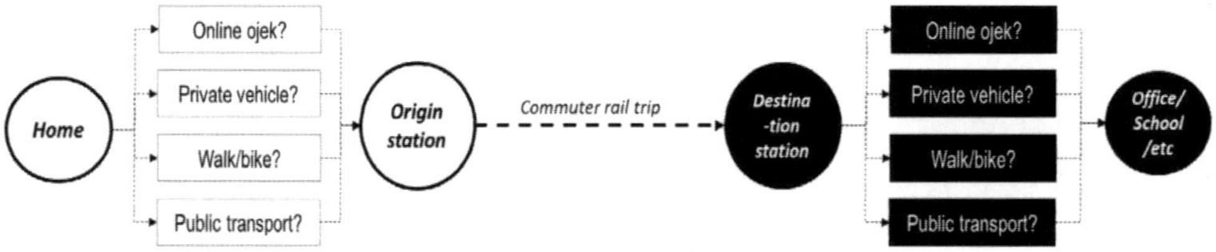

Figure 2. Typical morning trip chain

2.2. Selected indicators

The basic idea of this research is to try to analyze the behaviour of commuter rail users and their pedestrian and transit quality against online ojek trips. Presumably, every commuter who uses public transport has the largest possibility of effecting travel change, either by walking or using paratransit.

Table 1. Selected indicators of station connectivity

Indicators	Description	References
First mile and last mile cost	Transit fares (usually users will compare the cost with the service quality). The fares significantly affect affordability.	Dell' Olio, et al, 2010; Lai and Chen, 2011; Sumaedi et al., 2014
Transit travel time	Tight headway and minimum dwelling	Joewono and Kubota, 2007; Dell' Olio, et al, 2010; Tyrinopoulos and Antoniou, 2008
Transit reliabillity	How easy to get adequate transit service; close to station with minimum walking	Joewono and Kubota, 2007; Parasuraman et al., 1985; Tyrinopoulos and Antoniou, 2008; Sumaedi et al., 2014

Transit safety and security	The perception of feeling secure from crimes and accidents while travelling on board	Dell' Olio, et al, 2010; Joewono and Kubota, 2006; Zegras, et al 2015
Crossing safety	The perception of feeling safe while crossing junctions or mid-road links.	Olszeweski and Wibowo (2005)
Walking accessibility	How easy it is to walk to the station. The variable is determined by comprehensive perspective or perceived distance and ease. The indicator could be represented by the quality of walking pathways, green shades, etc.	Olszeweski and Wibowo (2005); Alshalalfah and Shalaby (2015)

This research uses classification analysis to determine the first mile and last mile mode choice with the individual perception of facilities at the station. Individuals' decisionmaking consists of a series of psychological steps which are formed by an interaction between perceptions and beliefs based on available information, and influenced by affect, attitudes, motives, and preferences that produce a choice [10]. The variables are analyses in the form of utility theory and the evaluation of the choice is done under the utility maximization concept [11–13].

The walking and transit service indicators are constructed based on intensive literature analysis. Many studies emphasize the measurement of service quality through passenger reflection to be a fundamental method, because they can be considered as the most experienced judges [14–19]. Also, some studies revealed the interaction model of service quality indicators and the effect to passengers' behavioral intentions of using public transport or Paratransit [17,20,21]. The accessibility will be asessed by a perceived scale [22].

3. Sample socioeconomic characteristics

Table 2 shows the descriptive analysis of the 1015 samples that had been collected. From the survey results, the average travel time from home to destination is 15 minutes by any mode. The average distance of the first mile is 4.2 km and the average distance of the last mile is 3.7 km. In average, the respondents spend 13-15 minutes for their last mile and first mile. Particularly, there are some respondents who spend 90 minutes for their first mile to station. This figure is typical for Greater Jakarta where many commuters live in the suburbs and work in the city of Jakarta or industrial areas such as Bekasi. In addition, not many suburban settlements are covered by commuter stations. Commuters who live in Indonesia's smaller cities might have 5 km as their full journey.

The average income of the sample is close to IDR 6.68 million per month, where the maximum monthly income is IDR 30 million or approximately EUR 2000. The data also shows that ownership of private vehicles is maximum 2 units. This figure reflects that the sample likely come from middle class families. The standard deviation of income is close to IDR 41 million. The average vehicle ownership is 1-2 units of private vehicle, and the maximum is 8. This vehicle ownership number does not differentiate by vehicle type, so it could be car or motorcycle.

Table 2. Descriptive analysis

Variables	Minimum	Maximum	Mean	Std. Deviation
Distance from home to station (km)	.10	30.00	4.2028	3.90
TT from home to station (min)	1.00	90.00	15.0020	12.07
First-mile cost (IDR)	.00	50000.00	7000.7220	6839.29
Distance from station to workplace (km)	.10	32.00	3.7626	4.48

TT from station to workplace (min)	1.00	90.00	13.9901	10.68
Last-mile cost (IDR)	.00	55000.00	5452.7113	6353.50
Family monthly income (IDR)	300000.00	30000000.00	6688267.14	4100301.5
Vehicle ownership (units)	.00	8.00	1.8150	1.21
Age	10.00	35.00	32.5764	17.24

TT= Travel time; N= 1015

The data shows the average spending for the first-mile of the trip (7000 IDR or ±0.5 EUR), and the last-mile of the trip (5000 IDR or ± 0.32 EUR). There are a few samples that do not spend any money. Typically, this data occurs when they travel on foot, or picked up by their family.

The first/last-mile cost does not represent the perceived value of satisfaction. The perceived value is more significantly affecting satisfaction instead of actual fare [21]. The typical fare for using feeder buses is 4000 IDR. If the commuters are able to use Transjakarta, transfer will not be charged again. However, if the route is surfed only by microbuses (called angkot) or regular buses, transfer will be charged. Thus, the total fare can be more expensive, instead of using online ojek. Online ojek applies distance-based fares. They also quite often offer discount. Moreover, the microbus and regular bus usually dwell at particular spots to gather passengers. The delay is unpredictable. On the other hand, online ojek offers punctuality. In peak hour, online ojek will increase the fares due to high demand. Therefore, the trade off between attitude and cost will be investigated further.

4. Perception of access

The station accessibility is assessed by exploring the quality of pedestrian access and quality of transit. The pedestrian access quality towards the station is not only affected by constant distance, but also affected by crossings, route choice, pathway quality, and conflict with other street elements [19,23]. The feeder transit quality is assesed by exploring the perception of accessibility, reliability, fare affordabillity, and safety. The transit reliabillity is the most significant service parameter that applies for all user categories [24].

Figure 3 shows users' perception of station accessibility. The parameters consist of transit safety, transit accessibility, transit reliability, transit affordability, pedestrian pathway quality, crossing environment, and pedestrian access which also means continuity of protected pathways. The pathway quality is more associated with pavement with proper betterment: flat and spacious. The doughnut diagram contains 2 circles: outer and inner. The inner circle stands for home to station trip, and the outer circle represents the trip from station to activity place.

In general, the survey data reveals poor perception of pedestrian environment connecting the commuter station to other places. In the case of pedestrian pathway quality, 75% of respondents answered that the quality at their destination station is poor, and 65% answered poor for their origin station. Both crossing environment and pedestrian access are also perceived as poor by more than 60% of the respondents. These results is relevant with the current situation in the most part of Greater Jakarta.

Typically, pathways around the central business district has a proper image and quality. The pathway is more or less 3 metres wide. However, the continuity of descending pathways and crossings is still an issue. Traffic and reckless driving behavior threaten every pedestrian who cross the road. Instead of calming the traffic, the design of the city tends to use pedestrian bridges, which are troublesome for persons with disabilities, senior citizens, and pregnant women. In areas farther away from the city center, pathways are typically encroached by street vendors. All these create contribute to conflict in pedestrian traffic and create a perception of a longer trip. Thus, those conflicts may lead to a longer perceived distance for commuters.

Figure 3 also indicates that the feeder transit service to stations mostly have adequate frequency and sufficient supply, but still lack coverage area, safety, and affordability. In the case of feeder service, 84% of the interviewees stated that transit service for the last mile of the trip to have poor affordability, and 76% claimed poor affordability for the first mile. The high transit service fares occur because the transit ticket is not integrated to the commuter train ticket. The passenger will be charged for an extra trip. Transferring to two different bus routes might cost a similar amount to or even more than the cost of online ojek. The samples also mostly perceive transit accessibillity to be poor, expressed by 86% for the last mile of the trip, and 65% for the first mile of the trip. In addition, most interviewees also think that the transit is not safe. Unlike other parameters about transit, most respondents answered that feeder reliability is adequate, as expressed by 77% for home to station trip, and 73% for station to activity place trip.

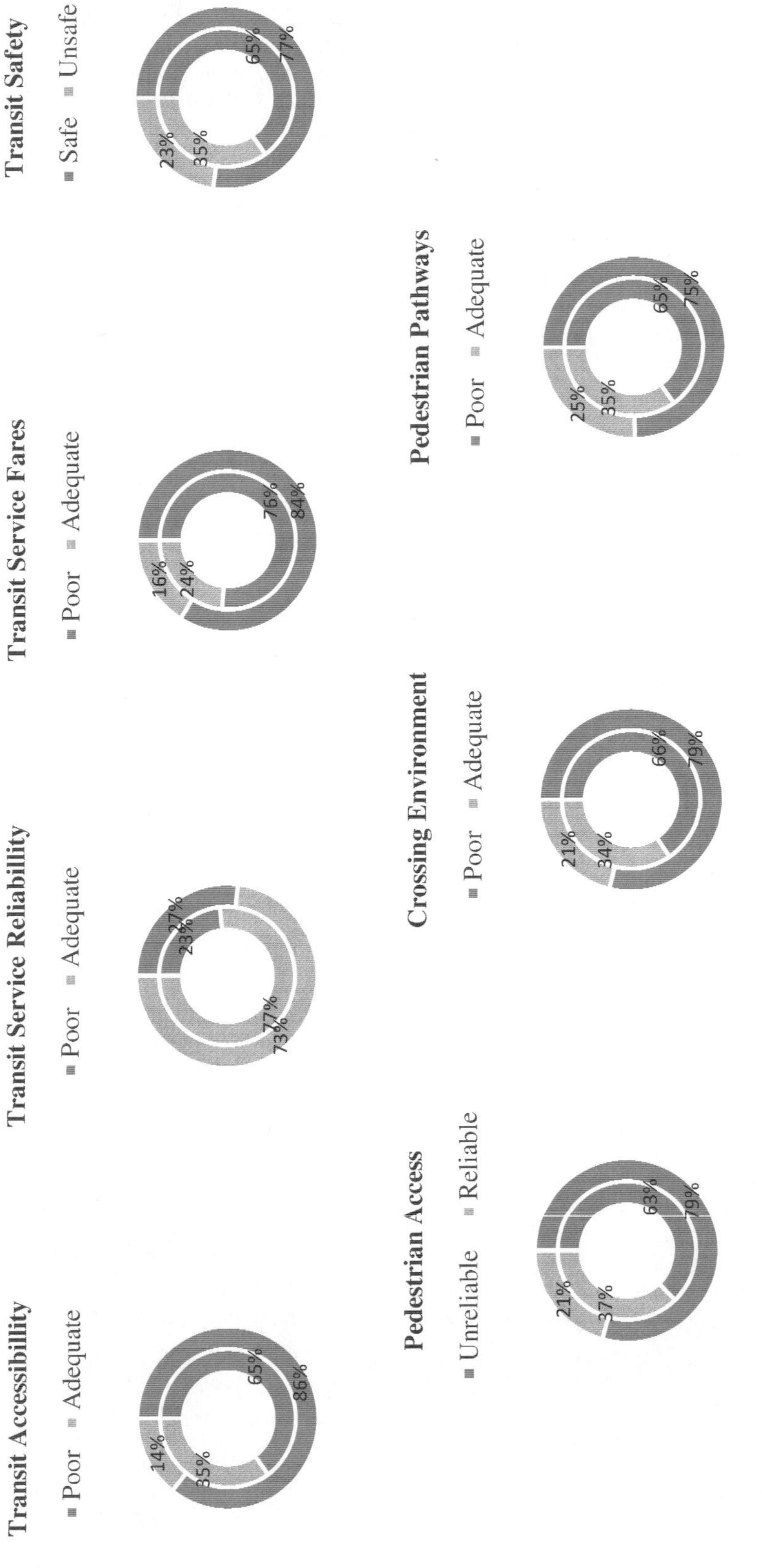

Figure 3. Station transit and perceived pedestrian quality; inner diagram (home to station trip); outer diagram (station to activity place trip)

5. First mile and last mile mode of transportation share at stations

The presence of online ojek is obviously seen at many commuter rail stations in Jakarta. Typically, online ojek flocks at spots nearby the stations to meet their customer. They encroach some spaces and claim those spaces as their informal stations. Technically, online ojek can only carry one passenger each.

Figure 5 shows the share of mode of transportation choice from house to first station, and last station to destination. The figures reveal that online ojek plays a significant role in commuter rail users' trip chain, especially for the first mile and last mile of the trip. The mode choice is categorized into online ojek, private vehicle, active transport, public transport, and others, e.g., taxi.

In the case of the first mile of the trip, private vehicles are majorly used (33%), followed by online ojek (24%), walking (16%), and public transport (12%). On the other hand, most respondents are walking (31%) and using online ojek (28%) for the last mile of their trip from station to destination. There are a few samples (8%) who use their private vehicle to travel from the station to their activity place. This figure occurs since there are some commuters who park their motorcycle in their last station chain. The shares of feeder transit users for the last mile of the trip (23%) is bigger than the first mile of the trip (12%).

The mode share of public transport is less than online ojek for all of the trip chain. This figure indicates that online ojek leads the share of the first mile and last mile of the trip chain. Besides, this figure also shows obvious interaction between commuter rail trip with online ojek trip. Most rail users spend many minutes of their travel time on board. Online ojek is able to cut off their travel journey because it is able to surf narrow alleys, and find shortcuts to avoid congestion [5].

The online ojek shares between the first mile and last mile of the trip differ. The utilization of online ojek for the first mile (home to station) seems less than the last mile (station to destination). This result indicates that most rail commuters prefer using their private vehicle for the first mile (30%). More commuter rail users 'park and ride', instead of using online ojek. It also indicates that private vehicles, e.g., driving a personal motorcylce, is more convenient than online ojek. Motorcycle one-day parking typically costs half the online ojek fare for 5 km. On the other hand, one-day parking for a car can cost double the online ojek fare for 5 km. The perceived cost of private motorcyle is cheaper than online ojek. Other factors, e.g., convenience, likely affect the preference of most commuters to use a private vehicle. Further analysis of these results will be explained in section 6.1 and 6.2. The next part explains the discriminant analysis of the mode used for the first mile and last mile, which focuses more on understanding the factors that affect the mode choice of rail commuters. In the case of the first mile of the trip, this research is not able to explain more about private vehicles versus online ojek, since there are no significant indicators that allow a discriminant anaylsis to compare online ojek and private vehicles at the first-mile case.

Figure 4. Online ojek, parked around the Kebayoran Station, wait for their passengers

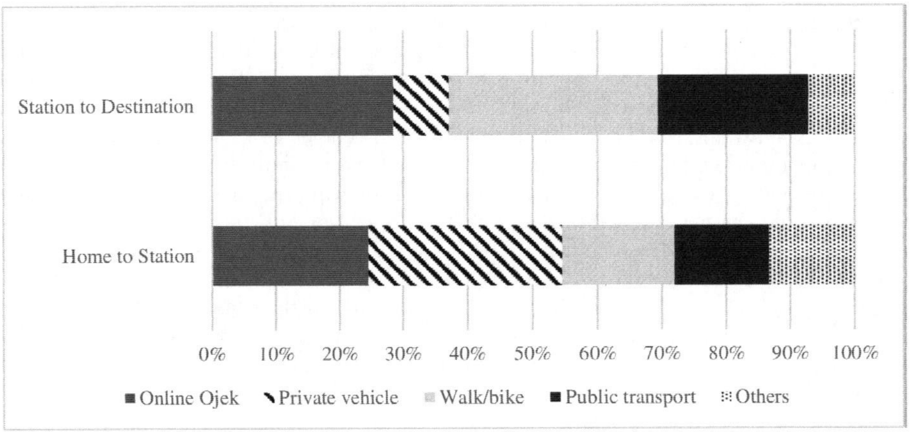

Figure 5. Mode shares at stations

6. Ojek user classification analysis

Individuals' choices could be understood through many aspects of human behavior, which attracts many researchers in the past decades. The classification analysis used in this research reveals the pattern of commuters' mode of transportation choice. This analysis is divided into two types of trip: first mile and last mile. As explained above, the first mile of the trip consists of the trip from home to the rail station, while the last mile is from the rail station to the activity location. The interpretation of the model refers to a broad literature of classification analysis in the sense of transportation [8,25,26]. Variables in every classification analysis are chosen based on the Mahalanobis value using the stepwise method. Thus, the number of variables in first mile and last mile discriminant model can be different, which depend on the optimum Mahalanobis value. Furthermore, the ratio of cases to independent variables in every model turned out to be more than 5:1 as stated in [9]. In addition, the structural matrix interpretation considers a loading value that is higher than 0.30, which is based on [27].

Regarding the exploration of the findings, the independent variable is constructed by several types of data. The first part is the attitude of commuter rail users, represented by a preference for comfort, safety, time efficiency, and price affordability. Those attitudes basically show how commuter rail users consider the criteria to choose their last mile and first mile mode of transportation. The second part is the number of private vehicle ownership, converted into a dummy variable of the total unit ownership. The third part is the distance from station to activity place, and the distance from home to station. The value of the variable is converted into a dummy variable and categorized to a distance of more than 5 km, and less than 5 km. The value of 5 km is used because it approaches the 4.3 km average distance showed in Table 2. The fourth part is the cost of the last mile and first mile of the trip. This variable can be the total of ojeks fare, transit fare, or parking charge. The next part is the perception of pedestrian environment around the station, consisting of accessibility, crossing, and pedestrian pathway reliability. The last part is the feeder transit service quality perception. The transit quality is explored through analyzing safety from crimes and accidents, reliability, accessibility, and fare.

The discriminant analysis uses the mode choice as the dependant analysis. The mode options for the function consists of private vehicle, online ojek, public transport, and non-motorized transport (walking or cycling).

A further part of this paper will explain the classification analysis result. Table 3 and Table 4 show two discriminant results based on the trip chain type. This discriminant analysis uses the structural matrix approach or also known as canonical loadings to classify the categories that correlated to the first mile and last mile mode choice.

6.1. First mile classification analysis

Table 3 shows the result of the classification analysis of mode choice for the first mile of the trip. In terms of fitting to the models, the result of Box's M test shows that the null hypothesis of the equal population covariance matrices is rejected. When using Box's M test, the non-significant M expect the variance-co-variance matrices to be equivalent. Due to the large sample collected, a significant result is not regarded as serious [27]. This discriminant analysis uses the structural matrix approach, also known as canonical loadings, to classify the categories that correlated to the first mile/last mile mode choice. The canonical correlation of the three functions are 0.538, 0.260, and 0.155 which can be interpreted as there being 28.94%, 6.7%, and 2.4% of the variance in the dependent variables can be explained by the model. Although canonical correlation indicators show weak correlation, the parameter of Wilks' lambda of this model indicates that the function is highly significant. Therefore, the model is acceptable to interpret [9].

The mode choice variable consists of online ojek (OO), private vehicle (PV), public transport (PT), and non-motorized transport (NMT). As explained above, the interpretation of the models depends on the function on the group centroid. It simply considers the most positive value and the most negative value of each function. Function 1 separates respondents who travelled with NMT (1.065) from respondents who travelled with OO (-0.943). We ignored the comparison with the other groups which are distinguished by another function. Function 2 separates respondents who travelled with PV (0.272) from respondents who travelled with NMT (-0.435). Furthermore, function 3 separates respondents who travelled with NMT (0.124) from respondents who travelled with PT (-0.380).

The evaluation of function 1 showed that the separation between online ojek and NMT is heavily influenced by the pedestrian infrastructure characteristics. It reveals that unreliable pedestrian facilities and crossing environment lead to the decline of the number of respondents who choose NMT, supported by negative loading values (-0.246 and -0.428). Regarding the travel cost, findings show respondents who have more than IDR 10,000 fare to stations are more likely to use online ojek than NMT, which does not have a significant cost to reach the stations (loading value 0.82).

In the sense of the second function, we found that the choice pattern between private vehicle and NMT is influenced by attitudinal characteristics of public transport such as fare and accessibility. Respondents who experience poor public transport services are more likely to choose private vehicle. The parameter 'high transit fare' shows an obvious loading value (0.811), indicating high transit fare leads the commuters to use a private vehicle. In addition to that finding, it is also found that respondents aged 26-40 tend to choose NMT than private vehicle. This finding is possibility related to the early respondent career stage that influences individuals' mode choice.

Furthermore, the third function found the factors which separate NMT and public transport choice. It reveals that the attitudinal variables found separate the mode choice significantly. For example, respondents who prefer comfort, safety, and time efficiency of mode are more likely to use NMT than public transport. Furthermore, it is interesting to note that even with people who suffer negative experiences with public transport (e.g., tiring and not safe), they are still less likely to use NMT than public transport. This finding could be explained by exploring the value of group means in every choice which shows that rather than public transport, respondents are more likely choose online ojeks. In terms of age, it is found that older respondents are less likely use NMT than public transport. It is also noticeable that a negative attitude to walking leads to declining NMT use by respondents.

Various findings of the first mile of the trip model emphasizes a strong association between infrastructure characteristics and individuals' attitude to the mode choice. Lacking and poor service from the pedestrian environment has discouraged commuter rail users to use NMT instead of online ojek. A strong reason could be related to the actual experiences of commuters, where the pedestrian environment affects the actual distance, creating a perceived distance. However, although pedestrian facilities are poor, individuals still choose NMT over public transport. The interpretation of public transport pattern choice shows that feeder transit quality merits commuter rail attitude. Despite perception of good quality perception, however, commuter rail users still tend to use online ojek rather than public transport.

Table 3. Discriminant analysis of online ojek and commuter rail users (home to station)

Variables	OO	PV	PT	NMT	Structural Matrix		
					F1	F2	F3
Prefer Comfort [D] [a]	.1034	.1596	.1263	.2000	.133	.031	.197*
Prefer Safety [D] [a]	.0172	.0423	.0105	.1053	.188	-.158	.516*
Prefer Time Efficiency [D] [a]	.3218	.4202	.2842	.3895	.083	.246	.456*
Prefer Affordability [D]	.0517	.0814	.1368	.0000	-.103	.110	-.240*
Female Commuters	.4655	.4886	.3474	.3895	.009	-.044*	.027
< 18 Years Old [D]	.0460	.0293	.0000	.0105	.011	.027	-.086*
18-25 Years Old [D]	.2759	.2476	.2105	.2000	-.056	.137*	.018
26-40 Years Old [D] [a]	.5057	.5081	.5053	.6737	.126	-.281*	.249
> 40 Years Old [D]	.1724	.2150	.2842	.1158	-.102	.193	-.295*
Do Not Have Private Mode	.1264	.0684	.1263	.1053	-.017	-.101*	.008
Have Private Mode	.8736	.9316	.8737	.8947	.017	.101*	-.008
< 1 km station distance from home [D]	.1782	.2182	.2105	.3789	.054*	-.002	.018
1-5 km station distance from home [D]	.5460	.5831	.5684	.4947	-.037	.015	-.048*
> 5 km station distance from home [D]	.2759	.1987	.2211	.1263	-.010	-.016	.040*
Poor crossing environment [a]	.6034	.3909	.4316	.1789	-.428*	.049	-.247
Poor Pedestrian Facilities	.5460	.4430	.4421	.2105	-.246*	.152	-.231
Unaccessible with Public Transport	.3678	.4137	.3368	.1684	-.112	.291*	-.284
Public Transport Poor Services	.7644	.7427	.6947	.7368	-.048	.078	-.142*
High-fare of Public Transport	.2644	.3876	.2947	.0842	-.098	.811*	-.190
Public Transport is Not Safe	.5287	.4300	.4316	.1474	-.213	.233	-.254*
Work/School Trip	1.2414	1.2020	1.3158	1.2000	-.176	-.202*	-.004
Number of Trip to Stations	.9540	.9805	.9789	.9895	.001	.049*	-.027
< IDR 5,000 fare to Station [D] a	.1897	.5863	.6737	.9053	.820*	-.012	-.438
> IDR 10,000 fare to Station [D]	.3563	.1270	.1158	.0000	-.360*	.011	.097
Unreliable Pedestrian Access	.6034	.4593	.4737	.2000	-.285*	.072	-.220

Fit to Parameters		Mode Choice	F1	F2	F3
Box's M [F;df1;df2;*p-value*]	409.636 [4.760; 84; 348714.517; 0.000]	Online ojek	-.943	-.198	.055
Eigen Values [Canonical Correlation]	0.408; 0.072; 0.025 [0.538; 0.260; 0.155]	Private Vehicle	.155	.272	.048
Wilks' Lambda F1 through F3; F2 through F3; F3[p-value]	0.646; 0.910; 0,976[0.000; 0.000; 0,006]	Public Transport	.162	-.083	-.380
		NMT	1.065	-.435	.124

D: 1 if yes, 0 otherwise; OO: Online Ojek, PV: Private Vehicle, PT: Public Transport; NMT: Non Motorized Transport
[a]: Variable in the analysis based on Mahalanobis D^2; *: Largest absolute correlation between variable and any discriminant function

6.2. Last mile classification analysis

Table 4 shows the result of discriminant analysis using the mode choice in the last mile of the trip. In terms of fitting the models, the result of Box's M test shows that the null hypothesis of equal population covariance matrices is rejected. When using Box's M test, the non-significant M expect the variance-co-variance matrices to be equivalent although it is not crucial since the sample is large enough.

This discriminant analysis uses the structural matrix approach, also known as canonical loadings, to classify the categories correlated to the last mile mode choice. The canonical correlation of the two functions are 0.720, 0.553, and 0.218, which can be interpreted as there being 51.84%, 30.58%, and 4.7% of the variance in the dependent variables that can be explained by the model. Although canonical correlation indicators show weak correlation, the parameter of Wilks' lambda of this model indicates that the function is highly significant. Therefore, the model is acceptable to interpret [9].

The interpretation of the model depends on the functions in the group centroid which are strongly related to the discriminant function. Function 1 separates respondents who traveled with NMT (1.434) from respondents who traveled with OO (-1.041). Function 2 separates respondents who traveled with PV (0.535) from respondents who traveled with NMT (-1.030). Furthermore, function 3 separates respondents who traveled with NMT (0.666) from respondents who traveled with PT (-0.1216).

By evaluating the first function which discriminated between the choice of NMT and online ojek, respondents whose travel cost is less than IDR 5,000 are more likely use NMT than online ojek (loading value 0.722). In contrast, respondents with a travel cost of more than IDR 10,000 are more likely to use online ojek than NMT. Short distance last mile commuters tend to choose NMT rather than online ojek, which is shown by the loading value of 0.574 for users who travel less than 1 km from the station to their destination.

Furthermore, the second function shows the separation between online ojek and public transport choice. It reveals that in a shorter travel distance, respondents tend to choose online ojek than public transport, which is shown by a loading value of 0.544. In addition, the value of group means shows that the value of OO (0.2599) is less than the mean value of NMT (0.8069). Those figures indicate that distance affects the commuter rail decision on how to choose their last mile mode. On the other hand, respondents whose travel cost is less than IDR 5,000 travel cost are more likely to choose public transport (-0.480).

In addition to the second function, several attitude characteristics also influence the mode choice. Negative experiences with public transport, i.e., 'public transport is tiring' (0.367) and 'public transport is not safe' (0.411) indicates that respondents are more likely to choose online ojeks. This finding supports the previous model in proving that the level of service will define the role of public transport in commuters' trip. Regarding to pedestrian facilities, respondents who experienced poor quality of pedestrian bridges tend to use online ojeks than public transport.

According to the evaluation of the third function's structural matrix, the discrimination of private vehicle and online ojek is associated to individuals' and infrastructures' characteristics. It is interesting to note that although commuters have private vehicles, they still tend to choose online ojek than private vehicle. The possible reason is related to the type of trip, as the last mile of the trip of rail users depends on the first mile of the trip. For instance, private vehicle users will probably park their vehicle at the first station and rely on alternative modes at the last station. Hence, it is important to note that mode integration in the last mile has a crucial influence on the whole journey.

Several important notes found in this model fill the gaps of the previous model. Pedestrian facilities, which is a crucial infrastructure in public transport trips, similarly influence commuters in the first and last mile trip. While supported by previous findings, this model also underlined the new fact of the last mile of the trip. As individuals' trip chain characteristics are related from one to another, individuals' decision of mobility in the first time of the day is influenced by their evaluation of the whole journey. Thus, the analysis emphasizes that public transport journey is related to the high level of pedestrian environment and other public transport integration in every transportation hub.

Table 4. Discriminant analysis of online ojek and commuter rail users (stations – activity location trip)

Variables	OO	PV	PT	NMT	Structural Matrix		
					F1	F2	F3
Prefer Comfort [D]	.1366	.1803	.1602	.1386	-.001	.052*	.028
Prefer Safety [D]	.0044	.0328	.0000	.0050	-.073	.089	.180*
Prefer Time Efficiency [D]	.3700	.3607	.2818	.2376	-.014	.015	.079*
Prefer Affordability [D]	.1498	.1639	.1381	.0891	-.061*	.006	.005
< 1 km station distance from activity location [D] [a]	.2599	.1311	.1657	.8069	.574*	.544	.098
> 5 km station distance from activity location [D]	.1894	.2787	.1547	.0248	-.075	-.174*	.044
< IDR 5,000 fare to Station [D] [a]	.2247	.4754	.8011	.9406	.722*	-.480	.294
> IDR 10,000 fare to Station [D]	.2379	.1148	.0939	.0149	-.287*	.213	-.112
< 18 Years Old [D] [a]	.0264	.0820	.0000	.0248	-.036	.087	.558*
18-25 Years Old [D]	.2687	.3115	.1878	.2228	.008	.037*	.032
26-40 Years Old [D]	.5595	.4262	.4807	.5743	-.020	.165*	-.057
> 40 Years Old [D] [a]	.1454	.1803	.3315	.1782	.030	-.277*	-.174
Have Private Mode [a]	.9339	.7705	.8619	.9307	.039	.180	-.515*
Female Commuters	.4493	.4754	.3591	.4802	.080	.081*	.031
Unreliable Pedestrian Access	.3040	.3770	.1657	.1485	-.162	.158	.272*
Too long to walk [a]	.7137	.8033	.7845	.5248	-.181	-.224*	.037
Unreliable Pedestrian Bridge	.4097	.4918	.0773	.1386	-.279	.307	.582*
Poor Pedestrian Facilities	.2731	.3934	.2431	.2574	-.143	.076	.281*
Unaccessible with Public Transport	.1938	.2295	.1105	.1980	.019	.111	.124*
Poor Public Transport Services	.7181	.6066	.6243	.7822	.075	.184*	-.131
High Public Transport Fare	.2819	.3443	.0884	.1782	-.097	.262*	.081
Public Transport is Not Safe	.3172	.2459	.0884	.4208	.106	.411*	.201

Fit to Parameters		Mode Choice	F1	F2	F3
Box's M [F;df1;df2;*p-value*]	488.311 [5.217; 90; 100448.317;0.000]	Online ojek	-1.041	.535	-.121
Eigen Values [Canonical Correlation]	1.079; 0.441; 0.050 [0.720; 0.553; 0.218]	Private Vehicle	-.975	-.268	.666
Wilks' Lambda F1 through F3; F2 through F3; F3[p-value]	0.318; 0.661; 0.952 [0.000; 0.000; 0.000]	Public Transport	.034	-1.030	-.119
		NMT	1.434	.403	.041

D: 1 if yes, 0 otherwise; OO: Online Ojeks, PV: Private Vehicle, PT: Public Transport; NMT: Non Motorized Transport
[a]: Variable in the analysis based on Mahalanobis D^2; *: Largest absolute correlation between variable and any discriminant function

7. Discussions and conclusions

This article reveals that online ojek somehow creates reliability for commuter rail users' first mile and last mile tactical commuting in dealing with undermined public transport service quality and inadequate pedestrian environment.

The body of the reserach reveals several findings on the role of online ojek in the trip chain of commuter rail users. First, respondents experience inadequate walking environment and feeder service quality to access stations, although transit reliabillity receives a different assessment (see Figure 3). Second, online ojek shares occur dominantly at the first mile and last mile travel of commuter rail users' trip chain (see Figure 5). Third, Table 3 and Table 4 explain the interaction of commuter rail attitude, background, and also station accessibility that affect the last and first mile mode choice. The discriminant analysis reveals the natural dependency of commuter rail users to online ojek services.

It is obvious that most the commuter rail users are dependent to online ojek service. It is able to support rail users who face the serious problem of transit service area gap in Greater Jakarta. The usage of online ojek exceeds other public transport shares, e.g., TransJakarta and microbus at majority commuter rail stations for first mile and last mile of the trip. Meanwhile, the share of walking is still surviving for the last mile of the trip where the trip occurs in urban CBD. In this situation, online ojek shows a paradox for Jakarta's transport setting. The online ojek role is competing with private vehicles in the first mile of the trip that typically start in sub-urban areas. In addition, this paper also reveals that online ojek threatens feeder transit sustainability. It seems that online ojek could lead to a natural displacement of the role of public transport feeder that connects home to station and vice versa. Some transit feeder routes connecting to stations are still less regulated with undermined service performance.

The role of online ojek tends to be more superior than walking and feeder transit. The discriminant analysis shows that poor pedestrian and feeder transit experience leads to online ojek needs. The walking distance is not only the single variable that is significant to the mode choice classification between OO and NMT (see Table 4). There are also commuters' attitude and the economic aspect that affects mode choice to station. It is also noticeable despite the nature of NMT being chosen for short and medium walking distance, when the pedestrian accessibily (e.g., pedestrian bridge, etc.) is poor, online ojeks prevail over NMT.

The preference to safety is not really significant to mode choice. Even so, motorcyles are not the safest vehicle in traffic, especially for long distance travel. Even so, some studies reveal that motorized two wheelers, as well as ojek, are the most vulnerable mode of transportation in traffic [16,28]. Quick travel time is trade off for the safety issues faced by commuters.

Inadequate crossing environment around the station is statistically signficant in discouraging commuters to walk. This result is relevant with a study in Bangkok that found not only acceptable walking distance from stations to home or workplace, the walking environment also influences the commuters to decide whether to use ojek, or walking [23]. Even so, findings also show that some commuters prefer walking up to a certain distance. People can walk longer while there are certain features on the street, e.g., shaded trees, clear pathways, and any interesting features [29].

The condition of pedestrian pathways in Greater Jakarta CBD close to the station varies. Some of them are deteriorating, and reduced because of street vendor encroachment, some of them are in proper condition. Neverthess, crossing streets in Jakarta CBD is difficult. There is heavy and reckless traffic.

Continous transit service quality is required to attract commuters to use public transport for the entirety of their trip chain. Transit passenger loyalty significantly relies on satisfaction encouraged by good safety image, adequate frequency, good stop amenities, and integrated connectivity to all mainplaces [21]. Accessibility problem at the first mile and last mile of the trip discourages them to walk and continue riding transit. In fact, Greater Jakarta still faces a 'double standard' in transit quality. One of them is Transjakarta which is subsidized, and follows a minimum service quality. The other one is a less regulated system that operates based on driver willingness.

Beyond these findings and relevance for planning, the competition among transport modes occur dependent on the urban settings. Each mode of transportation is struggling in their operation and gaining revenue. There are different settings of the market rules that are based on the ability of the government to control and finance all the services. More less-regulated local microbus owners and drivers are struggling to gain passengers to run the microbus daily operation, or at least meet their breakeven point. On the other hand, other public transport systems, e.g., Transjakarta, which is subsidized by the Government of Jakarta, also severely struggle to achieve their cost recovery. In contrast, online ojek companies are ran by professionals who are purely profit-oriented. They provide some features in the app for driver rating and deal with complaints from their customer, to evaluate the driver performance.

The government needs to ensure a win-win solution between the transit and the online ojek companies. Even the state is usually weaker, enabling regulations to control the use of online ojek, since the demand and needs of the people keep growing. The government could add more subsidy for feeder transit that connects to commuter rail stations. Nevertheless, it needs to pledge carefully, since subsidies also may have a stronger effect on reducing business viability and technical efficiency rather than generating more income [30].

In the case of land use planning, Greater Jakarta needs to reshape its city structure. Typically, the urban CBD are close to the main street and public transport line, but suburb settlements are mostly behind in the main transport network [31]. Most of the population are moving to the outskirts to find more affordable housing and more convenient environment. In fact, urbanization in Greater Jakarta is polarized, and built by different developers which do not reallly shape efficient structures [32]. Land use development and public transport development are not perfectly connected. Proper infrastructure for the public transport network shall foster rapid development to bundle those incremental development. Bus Rapid Transit can be the cheaper alternatives to reshape the land use planning, as well as Curitiba experience [33].

Urban planners might see an opportunity in this new transport market setting. The role of microbus, which has been controlled by informal regulations, is getting weaker. An integrated public transport reform can replace the informal regulation microbus market with a more reliable service feeder service. Legitimizing online ojek in the traffic law may become a strategy to control its operation by applying formal registrations and taxes. Thailand is the first country in the world that acknowledged motorcyle taxis in their regulation, and they effectively succeeded to intervene with the fare which affect the demand reduction[34]. Therefore, further research on regulating online ojek and its effects may be necessary to improve Jakarta's transport.

References

[1] JICA 2012 Project for the Study on JABODETABEK Public Transportation Policy Implementation Strategy (Jakarta: Directorate General of Land Transportation Ministry of Transportation The Republic of Indonesia)

[2] Cervero R and Golub A 2007 Informal transport: A global perspective *Transp Policy* **14** 6 p 445–57.

[3] Nahuis R 2007 The politics of innovation in public transport: issues, settings and displacements (Utrecht: Koninklijk Nederlands Aardrijkskundig Genootschap, Copernicus Institute for Sustainable Development and Innovation)

[4] Harjoko T Y, Dikun S and Adianto J 2012 Spatial contestation and involution: a case of the public transport, with particular reference to the Kampung Melayu, Jakarta *Dev. Ctry. Stud.* **2** p 94–107

[5] Syabri I, Pradono B and Soegijanto T 2013 Embracing Paratransit in Bandung metropolitan area, West Java, Indonesia Case Study *Prep. Glob. Rep. Hum. Settl.*

[6] Geurs K T and van Wee B 2004 Accessibility evaluation of land-use and transport strategies: review and research directions *J. Transp. Geogr.* **12** p 127–40

[7] Strathman, J and Dueker K J 1995 Understanding Trip Chaining, Special Reports on Trip and Vehicle Attributes: Based NPTS Data 1990 (Portland: U.S. Department of Transportation,)

[8] Joewono T B and Rizki M 2015 Classification Analysis of Students' Trip Chain in Bandung, Indonesia and Its Interpretation *J. East. Asia Soc. Transp. Stud.* **11** p 392–410

[9] Hair J F, Black W C, Babin B J, Anderson R E and Tatham R L 2006 Multivariate Data Analysis (Prentice Hall, New York: Pearson)

[10] Ben-Akiva M E and Lerman S R 1985 Discrete Choice Analysis: Theory and Application to Travel Demand (Cambridge; Massachusets: MIT Press)

[11] Koppelman F and Bhat C 2006 A Self Instructung Course Mode Choice Modelling Multinominal and Nested Logis Models
[12] Hensher D A, Rose J M and Greene W H 2005 Applied choice analysis: a primer (Cambridge ; New York: Cambridge University Press)
[13] Greene W H 2012 Econometric analysis (Boston: Prentice Hall)
[14] Dell'Olio L, Ibeas A and Cecin P 2011 The quality of service desired by public transport users *Transp. Policy* **18** p 217–27
[15] Eboli L and Mazzulla G 2011 A methodology for evaluating transit service quality based on subjective and objective measures from the passenger's point of view *Transp. Policy* **18** p 172–81
[16] Joewono T B and Kubota H 2006 Safety and security improvement in public transportation based on public perception in developing countries *IATSS Res.* **30** p 86–100
[17] Mouwen A and Rietveld P 2013 Does competitive tendering improve customer satisfaction with public transport? A case study for the Netherlands *Transp. Res. Part Policy Pract.* **51** p 29–45
[18] Tyrinopoulos Y and Antoniou C 2008 Public transit user satisfaction: Variability and policy implications *Transp. Policy* **15** p 260–72
[19] Wibowo S S and Olszewski P 2005 Modeling walking accessibility to public transport terminals: Case study of Singapore mass rapid transit *J. East. Asia Soc. Transp. Stud.* **6** p 147–156
[20] Sumaedi S, Bakti I G M Y, Astrini N J, Rakhmawati T, Widianti T and Yarmen M 2014 Public Transport Passengers' Behavioural Intentions (Singapore: Springer Singapore)
[21] Lai W-T and Chen C-F 2011 Behavioral intentions of public transit passengers—The roles of service quality, perceived value, satisfaction and involvement *Transp. Policy* **18** p 318–25
[22] Lättman K, Olsson L E and Friman M 2016 Development and test of the Perceived Accessibility Scale (PAC) in public transport *J. Transp. Geogr.* **54** p 257–63
[23] Pongprasert P and Kubota H 2017 Switching from motorcycle taxi to walking: A case study of transit station access in Bangkok, Thailand *IATSS Res.* **41** p 182–90
[24] Dell'Olio L, Ibeas A and Cecín P 2010 Modelling user perception of bus transit quality *Transp. Policy* **17** p 388–97
[25] Ortúzar S. J de D and Willumsen L G 2011 Modelling Transport (Chichester, West Sussex, United Kingdom: John Wiley & Sons)
[26] Pas E J 1985 State of the Art and Research Opptunities in Travel Demand *Transp. Res. Part Gen.* **19A** No 5/6 p 460–4
[27] Burns R B and Burns R A 2008 Business Research Methods and Statistics Using SPSS 2008 (London: Sage Publication)
[28] Dandona R, Kumar G A and Dandona L 2006 Risky behavior of drivers of motorized two wheeled vehicles in India *J. Safety Res.* **37** p 149–58
[29] Jiang Y, Christopher Zegras P and Mehndiratta S 2012 Walk the line: station context, corridor type and bus rapid transit walk access in Jinan, China *J. Transp. Geogr.* **20** p 1–14
[30] Zhang J, Niu J, Buyantuev A and Wu J 2014 A multilevel analysis of effects of land use policy on land-cover change and local land use decisions *J. Arid Environ.* **108** p 19–28
[31] Mochtar M Z and Hino Y 2006 Principal Issues to Improve the Urban Transport Problems in Jakarta *Mem Fac Eng Osaka City Univ* **47** p 31–38
[32] Firman T 2009 The continuity and change in mega-urbanization in Indonesia: A survey of Jakarta–Bandung Region (JBR) development *Habitat Int.* **33** p 327–39
[33] Cervero R and Dai D 2014 BRT TOD: Leveraging transit oriented development with bus rapid transit investments *Transp. Policy* **36** p 127–38
[34] Oshima R, Fukuda A, Fukuda T and Satiennam T 2007 Study on regulation of motorcycle taxi service in Bangkok *J. East. Asia Soc. Transp. Stud.* **7** 1828–1843

Acknowledgement
Authors would like to convey our gratitude to BPTJ (Greater Jakarta Land Transport Authority) for supporting this research. Also, we extend out appreciation to Anugrah Ilahi, Muiz Thohir from BPTJ, and Dr. Muhammad Prayudianto for giving us insights to enhance our research design.

IOP Conf. Series: Earth and Environmental Science **158** (2018) 012025 doi:10.1088/1755-1315/158/1/012025

Motorcycle dependency index at household level: case of Yogyakarta urbanized area

Y Herwangi[1], S P Putri[1], and P S Ronita[1]
[1]Urban and Regional Planning Program, Department of Architecture and Planning, Universitas Gadjah Mada, Yogyakarta, Indonesia

Corresponding author: yherwangi@ugm.ac.id

Abstract. Dependency on private vehicles has become a prevalent phenomenon in big cities experiencing urban sprawl. Related to that, there are still many unknown factors affecting the dependence on motorcycles. Various factors are suspected to influence this, ranging from spatial factors to aspatial factors. This research was conducted in Yogyakarta Urbanized Area (YUA) by taking 175 samples. Binomial Logistic Regression is used in order to find the factors that affect motorcycle dependency. The results showed that the index of dependency in YUA can be quite high. Motorcycle usage, bicycle ownership, and perception about the increase of fuel price are the factors that have a significant influence on motorcycle dependence in YUA. Even though the correlation between spatial factors and motorcycle dependency was weak, it cannot be said to have no effect. These factors are most likely to be influential if other indicators are included with more suitable proxies.

1. Introduction

The phenomenon of personal vehicle dependence is prevalent in major cities in the world experiencing urban sprawl, such as cities in developed countries like United States, Australia, and Canada [1]. Cities in Developing Nations (NSB) are also experiencing the phenomenon. Among them are countries in the Asian region such as Indonesia, Malaysia, Vietnam, Thailand which are experiencing the rapid growth of motorcycle usage. The dependence is influenced by various factors, ranging from spatial factors [2][3], socioeconomic factors [4][3] and lifestyle factors [5][6]. Nevertheless, existing studies related to private vehicle dependence only explain the phenomena occurring in developed countries but have not fully explained the phenomenon of motorcycles as a mode of transportation in developing countries in Asia.

The phenomenon of motorcycles in developing countries is allegedly due to differences in the characteristics of motorcycles and cars, which became the object of research in developed countries. The differences are, first, the motorcycles are relatively cheaper and more easily obtained through credit purchase schemes [7][8][9]**Error! Reference source not found.**. Second, the design of a small motorcycle makes it easy and agile in avoiding congestion that occurs in densely populated settlements with narrow roads **Error! Reference source not found.**. Third, the motorcycle is easier to operate than the car, so both women and men can use it [9]. In addition, there are differences in travel characteristics in developed and developing countries where in developing countries the condition and quality of public transport services is less suitable to the needs of the population. One of them relates to the cost, where the cost of using a motorcycle is cheaper than the cost of using public transportation [7]**Error! Reference source not found.**[8][9], and travel time by motorcycle is considered faster than public transportation **Error! Reference source not found.**[9].

In this regard, there is still a need for research on factors that affect the dependence on private vehicles, especially motorcycles in developing countries that have different travel characteristics to developed countries. Previous research in the urban area of Yogyakarta [11] by taking the location of Ngampilan Village, Notoprajan, Panggungharjo, and Sedangadi Villages has found that dependence on motorcycles is influenced by the availability of pedestrian paths to the shelter location, limited choice of residence location, job type, public transport, transportation costs using public transport, and public

IOP Conf. Series: Earth and Environmental Science **158** (2018) 012025 doi:10.1088/1755-1315/158/1/012025

transportation waiting time. However, there are still other factors that have not been included in the model, such as distance to the city center, the condition of pedestrian pathways, and the presence of alternative modes. Therefore, in this study the model will be tested on a wider sample, and also try to find the weight for each factor in the dependency index, as well as to know to what extent spatial factors influence dependence on motorcycle.

2. Factors affecting private vehicle dependency

Based on previous researches, the degree of dependence of private vehicles is influenced by several factors such as spatial factors, socioeconomic factors, and factors of public transport services.

2.1. Spatial factors

One of the spatial factors that affect dependence on private vehicles is the affordability of public facilities from residence. The layout of the city and the placement of public facilities are often a factor affecting the accessibility of a place on foot or by cycling or compulsion to use private vehicles (cars or motorcycles). According to [13], distance of residence to the city center and the stop is one of the things considered in the decision making of travel behavior. The farther the location of the facility, the more it will encourage someone to use private vehicles. In addition, the availability of parking area at a low cost can also encourage the use of private vehicles.

Other factors, such as road design and parking availability are also spatial factors that influence dependence on private vehicles [14]. The wide road design becomes one of the driving force for using cars. This is exacerbated by the design of roads within the city area that has no room for pedestrians, thus increasing the dependence on cars.

2.2. Socio-economic factors

The degree of dependence on private vehicles is influenced by the different socio-economic characteristics of each person. Social characteristics of society in relation to travel behavior can be seen from several indicators such as ownership and use of private vehicles in the form of cars, motorcycles, bicycles; use of public transport; and use of pedestrian paths [13][14][14]. Meanwhile, the economic characteristics can be seen from the magnitude of income [15].

2.3. Transportation factors

[17][14][14] explain the factor of transport services in terms of market system that benefit private vehicles, and the quality of alternative transport modes as a substitute for private vehicles in meeting travel needs. Alternative transportation can be described as mass public transportation, bicycle, and pedestrian. A good pedestrian is considered a support facility for the community to walk to a nearby destination, while mass transportation and bicycles can reach further destinations so as to reduce the use of private vehicles.

3. Research methods

This section will explain the unit of observation and analysis of this study, sampling frame, and data analysis methods.

3.1. Unit of observation and unit of analysis

The observation unit in this research is the Yogyakarta Urbanized Area (YUA), consisting of 23 sub-districts (14 sub-districts in Yogyakarta City, 6 sub-districts in Sleman Regency, and 3 sub-districts in Bantul District). The unit of analysis in this study includes the variables and indicators that have been determined through the deduction of theory. The variables and indicators of this study are described in Table 1.

Table 1. Variables of motorcycle dependency

Variables	Indicators
Spatial	Distance from home to public transport stops Distance from home to city centre Availability of parking area Availability of pedestrian way
Socio-economy	Vehicle usage (to school, workplace, city centre, public facilities) Private vehicle ownership Bicycle ownership and usage Income
Transportation	Travel distance The use of public transport (to school, work place, city centre, public facilities) Accessibility to public transport Level of satisfaction of public transport services Quality of alternative transport modes (pedestrian, bicycle, public transport) Market system that benefits private vehicle users Traffic condition

These variables are tested in the model as predictors for the dependent variables, which are dependent or not dependent on motorcycle. Dependent group is defined as people who stated that they cannot do at least 3 of 4 activities (working, educational activities, shopping, and recreation) without motorcycle. Non-dependant group is people who can do at least 2 of 4 activities without motorcycle.

3.2. Population and sampling method

The population of this study is the residents of Yogyakarta Urbanized Area. The number of samples is determined using confidence level of 90%, which is 175 samples. Respondents in this research are households selected with convenience sampling method. Respondents were selected based on physical characteristics of the house and income. The standard income for the low-income people referred to in this study is not more than US $ 2 / day / person, which is the standard of middle income according to the Asian Development Bank (ADB).

3.3. Data analysis method

The results of data collection were analyzed using inferential and descriptive statistical methods with the help of SPSS software. The inferential technique used is binomial logistic regression. This statistical technique is used to analyze factors affecting dependence on motorcycles in low-income households in Yogyakarta Urbanized Area, while the descriptive technique is further used as a supporting analysis to help interpret the results of hypothesis testing conducted before through inferential statistical techniques.

The assumptions used in the binomial logistic regression method are not as many as the assumptions on linear regression. The binomial logistic regression method does not assume homocedasticity, normality of errors, equality of covariance matrices, or multivariate normality such as in linear regression (Osborne, 2008). However, the assumptions for logistic regression have been fulfilled in this study, which are dependent variable is measured in dichotomous scale, independent variables are in categorical or continuous scale, and the dependent variable has mutually exclusive categories. Another assumption that must be met in binary logistic regression is linearity between the transformed criterion (i.e. log-odds) and a function of continuous independent variables in the model, which in this research has been done with Box-Tidwell test.

4. Result and discussion

Based on the results of data processing using statistical methods, the results are as follow:

IOP Conf. Series: Earth and Environmental Science **158** (2018) 012025 doi:10.1088/1755-1315/158/1/012025

4.1. Characteristics of the respondents
The sample of the research is 175 samples from all districts in Yogyakarta Urbanized Area (YUA). Based on the analysis of the samples, the level of motorcycle dependency in YUA is currently quite high. This can be seen from the proportion of people who depend on motorcycles, which is more than 50%. Both household groups that depend and do not depend on the motorcycle have the same family structure consisting of 4 people per family.

Nevertheless, there are slight differences in these two groups in terms of age and occupation. The non-dependent group was dominated by people with older age. This is because the older the age of a person, the ability to use the motorcycle is reduced so that the chance to depend on the motorcycle is smaller. In terms of work, both groups are dominated by traders. However, differences occur at the distance to their workplace locations. The motorcycle dependent group tends to engage in work activities that are far from where they live, whereas people in the non-dependent group work only in the vicinity of where they can walk. This difference can affect their level of dependence. The characteristics of the respondents can be seen in Table 2.

Table 2. Characteristics of respondents

Characteristics	Motorcycle Dependent Group	Motorcycle Non-dependent Group
Average family member	4	4
Average HH income	Rp 1,64 million (US$ 120,5)	1,75 million (US$ 128,7)
Average age of the respondents	44 years old	48 years old

4.2. Socio-economic factors
Motorcycle use among respondents can be categorized as high, which is about 79.5% of the total respondents. They claim to routinely use the motorcycle in every activity. Most of the respondents in the dependent group use the motorcycle for all their activities from school to recreation and sports. However, respondents in the non-dependent group use motorcycles more for shopping and work activities (Figure 1).

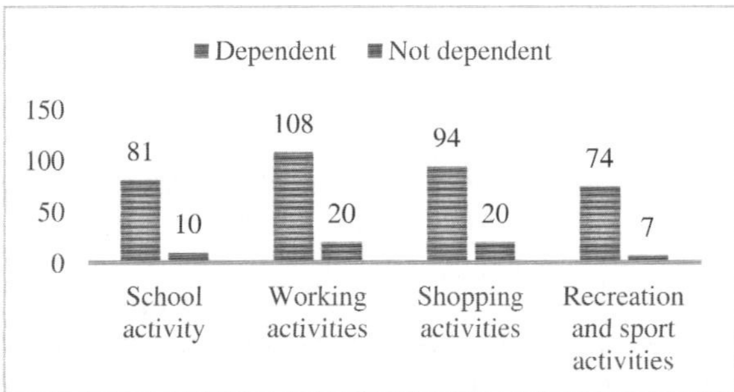

Figure 1. Motorcycle usage in various activities

Motorcycle ownership among respondents is also quite high, i.e., as many as 98.5% of respondents have at least one motorcycle. As many as 58% of them even have more than one motorcycle. This means that people who do not rely on motorcycles for their daily activities may also have more than one motorcycle. That way, the non-dependent group here are not entirely people who do not use motorcycles. A person in the motorcycle non-dependent group in YUA is person who only uses the motorcycle for activities that are considered important only and can only be done on a motorcycle.

In addition, based on travel destinations, it can be said that the pattern of travel is still local. This means that most activities are done in the local environment from school, work, shopping, to recreation. This condition is evidenced by most respondents (81%), both who are dependent and non-dependent on motorcycles, having a total travel time of less than 3 hours per day.

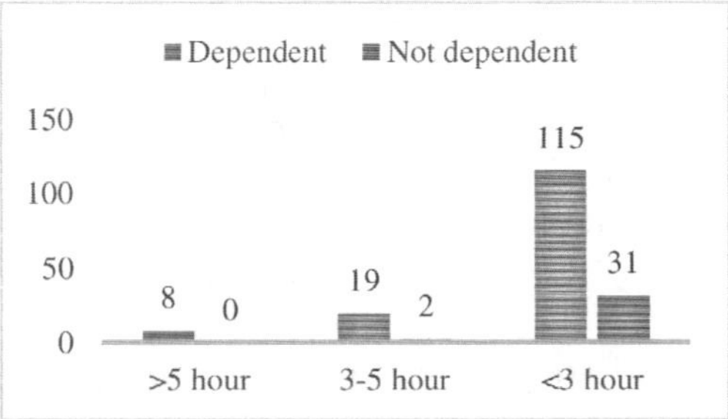

Figure 2. Duration of motorcycle usage per day

4.3. Transportation factors
In this study, the majority of respondents live in neighbourhoods with low to moderate traffic speeds, as can be seen in Figure 3.

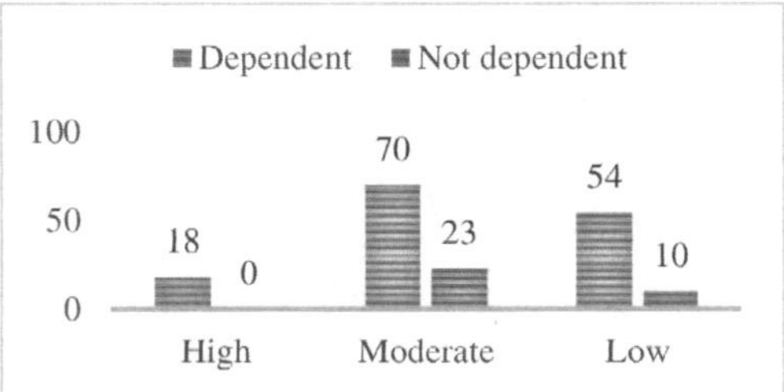

Figure 3. Traffic speed

With such conditions, the chances of people not to rely on motorcycles should be greater. With low traffic speeds, pedestrian and cyclists' activities and other non-motorized alternatives are still possible in this area. But the reality in the Yogyakarta Urbanized Area is not the case. Both in neighbourhoods with low to moderate traffic speeds, the proportion of people who depend on motorcycles is not much different. Likewise, the case is the same with people who do not depend on the motorcycle.

On the other hand, there are other alternative modes that can be used in YUA. Two of them are bicycles and buses.

1. Bicycle
Unlike the less desirable pedestrian spaces, bicycles are an alternative transportation that is widely selected and still used today in YUA. This condition applies especially to groups of people who are not dependent on motorcycles. As many as 57% of the group of people who did not rely on motorcycles were recorded to have bicycles in this study. In general, bicycles are used for school activities to recreational and sports activities. Bicycle usage for various activities can be seen in Figure 4.

Figure 4. Bicycle usage for various activities

As many as 40% of the group of people who depend on motorcycles also have bicycles. However, in general this bicycle is limited to use for school activities. Until now, bicycles are still considered suitable for use in the YUA region as an alternative transportation. That way, bicycle access can affect the condition of their dependence on a motorcycle.

2. Bus

Buses are also an alternative transportation besides bicycles that people use in YUA to perform activities from school to work. People who do not rely on motorcycles are more likely to use buses than people who do not rely on motorcycles. As many as 55% of the dependent group stated that they often used buses because of the urgency and had no option to use private transportation, such as having no motorcycle, broken motorcycle, no motorcycle parking, and so on. Therefore, the use of public buses in this study cannot be used as a reference level of one's dependence on motorcycles.

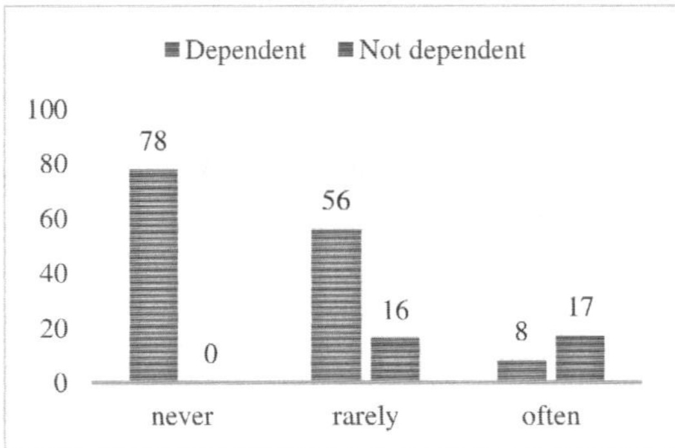

Figure 5. Public transport usage

Most respondents argued that comfort was also a determinant factor in their decision to use public bus transportation. As many as 47% of them argued that the current quality of the bus is less convenient to use. This may be due to personal circumstances (travel sickness when traveling by bus) and the efficiency of travel time (inefficient routes, long bus waiting times, and unreachable destinations by bus routes). This condition makes 57.5% of respondents feel less satisfied with public transportation buses at YUA. This factor contributed to the use of bicycles more than the use of Transjogja buses, in addition to the habits of people who are lazy to walk to the Transjogja shelter.

4.4. Spatial factors

The city center is always identified as a center of public facilities. In this study, most respondents live within a radius of more than 5 kilometers from the city center as can be seen in Figure 6.

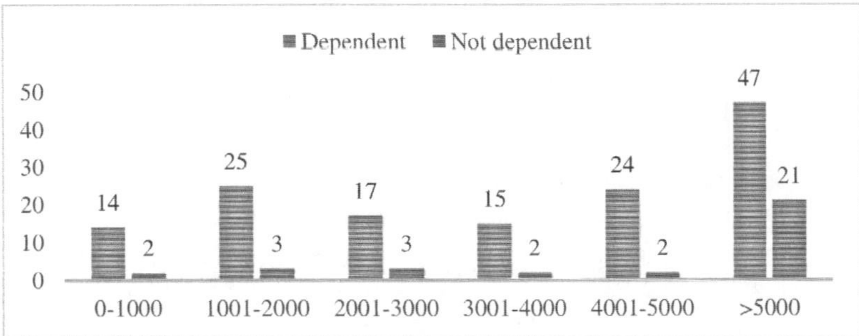

Figure 6. Distance to the city center (in meter)

This condition applies equally well to the motorcycle dependent and non-dependent groups. This indicates that motorcycle use is not affected by distance to the city center. Apparently, this happens because not all their activities can be done in the city center. With regards to pedestrian activity, as many as 54% of respondents considered that their neighbourhood has no pedestrian access. Pedestrian access is only utilized by 35% of respondents despite its poor condition. The reasons ranged from inappropriate pedestrian design, pedestrian closing by street vendors (PKL), and motorcycles that often crossed the pedestrian path. Pedestrian lane conditions that are less comfortable is what makes the pavement has not been fully utilized to reach the facilities that are within walking distance. Although they also live in a neighbourhood served by public bus routes, this condition does not make people switch to using buses.

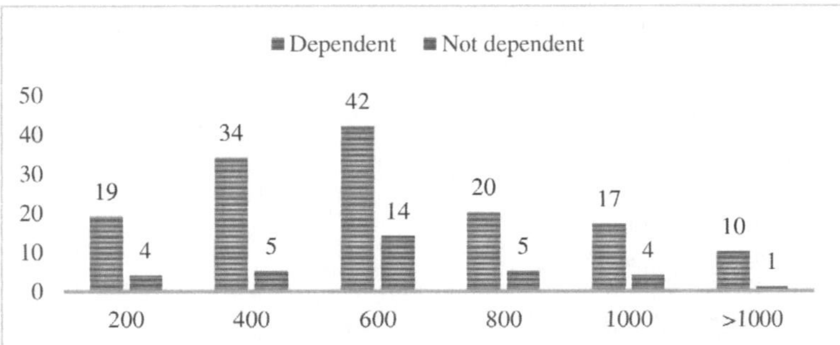

Figure 7. Distance to the nearest bus stop (in meter)

Based on Figure 7 it can be said that the majority of correspondents are within walking distance to the bus stop. However, this condition did not affect their decision to use the bus. This condition is also supported by the ease in finding parking spaces.

IOP Conf. Series: Earth and Environmental Science **158** (2018) 012025 doi:10.1088/1755-1315/158/1/012025

Figure 8. Parking space availability

Both dependent and non-dependent groups have the similar opinion that it is currently still easy to find available parking lots. This is what makes people from both groups still choose to use a motorcycle and only some who choose to use other alternative modes.

4.5. Motorcycle dependency model

In this study, there are 19 independent variables used with categorical measurement scales. Meanwhile, the dependent variable used is a dichotomous variable with the criteria of "dependent" and "not dependent" on the motorcycle. In the independent variable verification process, 11 independent variables are declared feasible. The test criteria used are: if the significance level of p_value is less than 0.25 then the variable can be considered a predictor in the model (Table 3).

Table 3. Verification results of independent variables

Variable	Chi-Square	p_value
Motorcycle usage	7,636	0,022
Motorcycle ownership	1,311	0,034
Bicycle access	11,660	0,001
Duration of motorcycle usage	3.736	0,154
Pedestrian access	9,056	0,011
Speed of traffic	6,682	0,035
Quality of bus	4.906	0,086
Satisfaction of public transport	4.623	0,099
Quality of bicycle	8.428	0,015
Economic burden	3.727	0,155
Fuel price	9.115	0,005

In this study, binomial logistic regression analysis was performed in twelve iterations to produce twelve models. This process is done to obtain the required level of significance that is less than 0.05. Twelve models are then selected to produce the best model. Selection process is done through several stages so that it produced only one model that is considered the most representative in explaining the factors that affect motorcycle dependence. In the selected model, the binomial logistic regression test output is generated as can be seen in Table 4.

Tabel 4. Logistic regression output

		B	SE	Wald	df	Sig.	Exp(B)
Step1[a]	Motorcyle_usage			6.973	2	.031	
	Motorcyle_usage(1)	2.164	.965	5.029	1	.025	8.705
	Motorcycle_usage(2)	2.306	.837	6.972	1	.008	10.031
	Fuel price(1)	.953	.447	4.545	1	.033	2.593
	Bicycle_ownership(1)	1.380	.446	9.563	1	.002	.251
	Constant	-.717	.862	.693	1	.405	.488

Based on the output, the equation is:
Y= a + β1X1 + …… + βiXi
Y= -0,717+2,164 x motorcycle usage (1) + 2,306 x motorcycle usage (2) - 1,380 x access to bicycle (1) + 0,953 x fuel price(1)

Based on the equation, it can be concluded that from 11 independent variables tested only 3 variables that are related significantly with the variable of motorcycle dependence. The three variables are based on the following hypothesis test:
H0: $\beta = 0$ (constants not significant in model)
H1: $\beta \neq 0$ (constants of significance in the model)

The testing criteria, i.e., H0, is accepted if p value > 0.05. Since all values of p value are less than 0.05 then hypothesis 0 is rejected. Thus, it can be interpreted that the three variables are considered to have a significant effect. Three variables that significantly influence motorcycle dependence are motorcycle usage, access to bicycle, and fuel price. According to this study, these three variables are the factors affecting the motorcycle dependency in YUA.

a. Motorcycle usage factor
Based on the analysis (see Table 5), the OR number of people who rarely ride motorcycles (motorcycle_user 1) than people who never ride motorcycles (motorcycle_user) is 8.705. It can be interpreted that people who rarely ride motorcycles (motorcycle_user 1) are 8.7 times more likely to rely on motorcycles than people who never ride motorcycles (motorcycle_user). On the other hand, the OR rate of people who frequently ride motorcycles (motorcycle_user2) compare to people who never ride motorcycles (motorcycle_user) is 10,031. It can also be interpreted that people that frequently ride motorcycles are 10 times more likely to depend on motorcycles than people who have never ridden a motorcycle. In this case people who have a higher intensity in the use of motorcycles will have more chance to depend on motorcycles .

Tabel 5. Interpretation of odd ratio (OR) of motorcycle usage

		B	SE	Wald	df	Sig.	Exp(B)
Step1[a]	Motorcyle_usage			6.973	2	.031	
	Motorcyle_usage(1)	2.164	.965	5.029	1	.025	8.705
	Motorcycle_usage(2)	2.306	.837	6.972	1	.008	10.031
	Fuel price(1)	.953	.447	4.545	1	.033	2.593
	Bicycle_ownership(1)	1.380	.446	9.563	1	.002	.251
	Constant	-.717	.862	.693	1	.405	.488

b. Bicycle usage factors

Based on the Table 6, the OR number of people who do not have bicycles (bicycle access1) compared to the people who do not have bicycles is 0.251. It can be interpreted that the possibility of a person without a bicycle (Bicycle Access1) to depend on a motorcycle is 0.2 times the likelihood of a person who has a bicycle. Bicycles as an alternative mode is still a choice that is widely used in addition to using a motorcycle.

Tabel 6. Interpretation of OR bicycle access

	B	SE	Wald	df	Sig.	Exp(B)
Step1[a] Motorcyle_usage			6.973	2	.031	
Motorcyle_usage(1)	2.164	.965	5.029	1	.025	8.705
Motorcycle_usage(2)	2.306	.837	6.972	1	.008	10.031
Fuel price(1)	.953	.447	4.545	1	.033	2.593
Bicycle_ownership(1)	-1.380	.446	9.563	1	.002	.251
Constant	-.717	.862	.693	1	.405	.488

c. Fuel price factors

Based on Table7, OR people who do not consider the increase in fuel price as an economic burden (Fuel_price(1)) compared to the people who consider the fuel price increase as economic burden is 2,593. It can be interpreted that the possibility of people who do not consider the increase in fuel price as an economic burden is 2.5 times the likelihood of people who consider the increase in fuel as an economic burden. In this case, the increase in fuel prices was able to influence people's decisions to use motorcycles.

Table 7. Interpretation of OR fuel price

	B	SE	Wald	df	Sig.	Exp(B)
Step1[a] Motorcycle_usage			6.973	2	.031	
Motorcycle_usage(1)	2.164	.965	5.029	1	.025	8.705
Motorcycle_usage(2)	2.306	.837	6.972	1	.008	10.031
Fuel_price(1)	.953	.447	4.545	1	.033	2.593
Bicycle_ownership(1)	-	.446	9.563	1	.002	.251
Constant	1.380	.862	.693	1	.405	.488
	-.717					

In this model, eight other variables, including spatial variables, are said to have less meaningful effects simultaneously with other variables. Nevertheless, among the four existing spatial variables there is one variable that directly affects motorcycle dependence. The variable is the distance to the city center. It is seen from the value of p value smaller than 0.05.

Although in the selected model spatial variables are said to be insignificant, it has not been able to conclude the absence of influence from these variables. Spatial variables may influence motorcycle dependency if other variables are included in the model. For example, the city center distance variable used in this study is the distance to the center of Yogyakarta City. This variable has no effect because the population activity is not always done in Yogyakarta City Centre. Therefore, other variables that can be included in the future model is the distance to the nearest facility used such as work place, schools, shops, and recreation centre.

In addition, the distance variables to the bus stop in this study also had no effect. In the case of YUA, dependence on motorcycles at the household level is very high. Their laziness to walk and wait can also be the cause of the lack of influence of these variables. The general assumption is that the closer the person to the bus service, the more likely they will use the bus. But it is not the case in YUA. This condition depends on motorcycle ownership and location of activity. Those who own a motorcycle would prefer to use it even if it is close to the bus stop. Although they did not have any motorcycles, they even opted to borrow a neighbor's motorcycle instead of using public transport

4.6. Motorcycle dependency index

In this study, the motorcycle dependency index was measured at the household scale. This dependency index is divided into two: high motorcycle dependence and low motorcycle dependence. Based on the model generated in this study it can be concluded that the dependency opportunities are based on three main aspects: the use of motorcycles, bicycle ownership, and perception of fuel price.

A household can be categorized as having high dependence on motorcycles if frequently using a motorcycle, not having a bicycle, and assuming that the fuel price increase is not an economic burden. Conversely, a household can be categorized as having low dependence on a motorcycle if it has never used a motorcycle, owning a bicycle, and presuming that the increase in fuel is an economic burden. Based on the model, the probability of motorcycle dependency can be calculated. The probability of a household to depend on motorcycle can be seen in the Table 8.

Table 8. Probability of motorcycle dependency

Influence factors		Motorcycle usage		
		Never	Rarely	Often
Increase in fuel price is a burden	Have bicycle	0,122	0,935	1,328
	Do not have bicycle	0,488	4,250	4,898
Increase in fuel price isn't a burden	Have bicycle	0,318	2,989	2,356
	Do not have bicycle	1,266	11,02	12,75

5. Conclusion

In this research, there are only three variables that significantly influence motorcycle dependency index in YUA, namely motorcycle use, bicycle ownership, and fuel price increase. These results strengthen and correct the research conducted by Litman, Gillingwater, an Kodukula [14][13][17] stating that socioeconomic, transport, and spatial services factors can affect dependency on motorized vehicles. However, not all aspects of this study show such results in the case of motorcycle.

In terms of the use of public transport, this study provides different results in household groups with high and low motorcycle dependency. Their tendency is to follow the cheapest price they can afford. Household groups with low motorcycle dependency prefer bicycles as alternative transportation compared to public transportation. This is because the use of bicycles is considered free and its use is more flexible than public transportation in terms of travel routes and travel time. However, for groups with high dependency on motorcycles, they tend to use motorcycles rather than using public transportation. In line with previous research conducted by Herwangi et al [11], the cost of using a motorcycle is even cheaper than using public transportation.

Perceptions of fuel price also significantly influences people's decisions to depend or not on motorcycle. Those who consider the fuel price increase not as an economic burden tend to have high motorcycle dependence. A previous research by Herwangi et al [18] also mentioned that people are still willing and able to bear the fuel price increase between IDR 500 to IDR 2000 per litre. This is in contrast with non-dependant groups who prefer to limit the use of motorcycles rather than pay for fuel price increases. This condition is done because of their limited income.

Based on these results, government intervention is needed to overcome dependence on motorcycles. One of them is by applying public transportation rates that can compete with the cost of using a motorcycle. In addition, supply-side policies are also required, by limiting the number and use of motorcycles. This needs to be done because based on this research, land use factors have little role in influencing the influence of motorcycle. The number and use of motorcycles can be done through increasing motorcycle prices as well as increasing motorcycle taxes.

The spatial factor in this study does not necessarily have any effect on motorcycle dependency even though the model formed states so. Allegedly this factor is less influential because there are other indicators more appropriate to measure the influence of spatial factors, rather than indicators used in this study. Therefore, it is necessary to do further research by using other indicators. In addition, in this study the motorcycle dependency index is only measured in household level. In this case, research on area level needs to be done to test whether the spatial factor does affect the dependence on motorcycles. Therefore, further research should be conducted to see the dependence index on motorcycles in YUA as a whole.

References

[1] Newman P W G and Kenworthy J R 1989 *Cities and Automobile Dependence: An International Source Book* (Aldershot: Avebury Technical Ashgate Publishing Limited)

[2] Sarkar P P and Mallikarjuna C 2013 *Effect of Land Use on Travel Behaviour: A Case Study of Agartala City* (Agra: 2nd Conference of Transportation Research Group of India)

[3] Lin J J and Yang A T 2009 Structural Analysis of How Urban Form Impacts Travel Demand: Evidence from Taipei *Urban Studies* vol 46 pp 1951-1967.

[4] Banerjee U and Hine J 2016 Interpreting the Influence of Urban Form on Household Car Travel Using Partial Least Squares Structural Equation Modelling: Some Evidence from Northern Ireland *Transportation Planning and Technology* **39 (1)** 24-44

[5] Hong J Shen Q and Zhang L 2014 How Do Built-environment Factors Affect Travel Behaviour? A Spatial Analysis at Different Geographic Scale *Transportation* vol 41 p. 419-440

[6] Schwanen T and Mokhtarian P L 2005 What If You Live in The Wrong Neighborhood? The Impact of Residential Neighborhood Type Dissonance on Distance Traveled *Transportation Research Part D* **10** 127-151

[7] Adriyana D 2011 *Studi Fenomena Pertumbuhan Penggunaan Moda Sepeda Motor di Kota Bandung (Study of Motorcycle Usage in Bandung City)* Undergraduate Thesis, Urban and Regional Planning Programme, Institut Teknologi Bandung

[8] Aritonang H O 2007 *Studi Pengaruh Pemilihan Moda Sepeda Motor terhadap Volume Penumpang Angkutan Kota di Kota Bandung (The Impact of Motorcycle Mode Choice to the Public Transport Ridership in Bandung City)* Undergraduate Thesis, Urban and Regional Planning Programme, Institut Teknologi Bandung.

[9] Erli H and Martha K D 2006 *Studi Peluang Pengalihan Penggunaan Sepeda Motor ke Angkutan Umum (Study of Modal Shift Opportunity from Motorcycle to Public Transport in Denpasar City)* Undergraduate thesis of Urban and Regional Planning Programme, Institut Teknologi Bandung

[10] Raflis R A 2006 *Karakteristik Pemanfaatan dan Pola Pertumbuhan Kepemilikan Sepeda Motor di Kawasan Berkepadatan Tinggi: Studi Kasus DKI Jakarta (Characteristics of Motorcycle Ownership in High Density Settlement: Case of Jakarta Special Region).* unpublished undergraduate thesis (Bandung: Institut Teknologi Bandung)

[11] Novitasari, F 2011 *Peluang pergeseran penggunaan sepeda motor ke angkutan kota oleh pekerja industri di Kabupaten Karawang (Opportunity of shifting from motorcycle usage to public transportation in industrial worker in Karawang Regency)* (Bandung: Institut Teknologi Bandung)

[12] Herwangi Y 2016 *Transport Disadvantage dan Kepemilikan Sepeda Motor pada Masyarakat Berpendapatan Rendah Studi kasus Kawasan Perkotaan Yogyakarta (Transport Disadvantage and Motorcycle Ownership of Low-income People in Yogyakarta Urbanized Area)* Doctoral Dissertation of School of Architecture, Planning, and Policy Development, Insitut Teknologi Bandung

[13] Gillingwater A and Darlington C 2014 *Car dependency scorecard 2014: The top English cities for sustainable transport* (London: Better Transport)

[14] Litman T 2002 *The Costs of Automobile Dependency and the Benefits of Balanced Transportation* (Melbourne: Victoria Transport Policy Institute)

[15] Victoria Transport Policy Institute 2016 TDM Encyclopedia [Online]. Available from: https://www.vtpi.org/tdm/tdm42.htm

[16] Giuliano G and Narayan D 2003 Another look at travel patterns and urban form: the US and Great Britain *Urban studies* **40** 2295-312

[17] Kodukula S 2011. *Rising Automobile Dependency: How to brek the trend?* (Germany: GIZ)

[18] Herwangi Y 2002 Faktor-faktor yang mempengaruhi pemanfaatan fasilitas pendidikan oleh penduduk di Kawasan Pinggiran Bandung (Bandung: Institut Teknologi Bandung)

[19] Clifton K and Lucas K 2004 Examining the empirical evidence of transport inequality in the US and UK Running on Empty: *Transport, Social Exclusion and Environmental Justice* (Bristol: The Policy Press)

[20] Currie G and Delbosc A 2009 Car Ownership and Low Income on the Urban Fringe - Benefit or Hindrance? *Proc. Int. Conf. of 32nd Australasian Transport Research Forum (Auckland)*

[21] Currie G dan Senbergs Z 2007 Exploring Forced Car Ownership in Metropolitan Melbourne *Australasian Transport Research Forum 2007*

[22] Department of Environment and Transport UK 1999 *Young People and Crime on Public Transport* (London: HMSO)

[23] Johnson V Currie G and Stanley J 2009 Can Measures of Disadvantage Perpetuate the Problems They Seek to Solve? *Australian Social Policy Conference 2009: An Inclusive Society? Practicalities and Possibilities*

[24] Murakami E and Young J 1997 *Daily Travel by Persons with Low-Income* (Maryland: NTPS Symposium)

[25] Osborne J W 2008 *Best Practices in Quantitative Methods* (Louisville, USA: Sage Publications, Inc)

[26] Social Exclusion Unit 2003 *Making the Connections: Final Report on Transport and Social Exclusion* (United Kingdom: Office of the Deputy Prime Minister)

[27] Shen Q 2000 Spatial and Social Dimensions of Commuting *Journal of the American Planning Association* **66 (1)** 68-82

Ecological Footprint and Ecosystem Services Models: A Comparative Analysis of Environmental Carrying Capacity Calculation Approach in Indonesia

R M Subekti[1] and D S A Suroso[2]

[1] Center for Development Control of Kalimantan Ecoregion, Ministry of Environment and Forestry, 19A Sudirman Street, Balikpapan 76111, Indonesia
[2] Regional and Rural Planning, School of Architecture, Planning and Policy Development, Institut Teknologi Bandung, 10 Ganesa Street, Bandung 40132, Indonesia

Corresponding author: murtisisme@gmail.com

Abstract. Calculation of environmental carrying capacity can be done by various approaches. The selection of an appropriate approach determines the success of determining and applying environmental carrying capacity. This study aimed to compare the ecological footprint approach and the ecosystem services approach for calculating environmental carrying capacity. It attempts to describe two relatively new models that require further explanation if they are used to calculate environmental carrying capacity. In their application, attention needs to be paid to their respective advantages and weaknesses. Conceptually, the ecological footprint model is more complete than the ecosystem services model, because it describes the supply and demand of resources, including supportive and assimilative capacity of the environment, and measurable output through a resource consumption threshold. However, this model also has weaknesses, such as not considering technological change and resources beneath the earth's surface, as well as the requirement to provide trade data between regions for calculating at provincial and district level. The ecosystem services model also has advantages, such as being in line with strategic environmental assessment (SEA) of ecosystem services, using spatial analysis based on ecoregions, and a draft regulation on calculation guidelines formulated by the government. Meanwhile, weaknesses are that it only describes the supply of resources, that the assessment of the different types of ecosystem services by experts tends to be subjective, and that the output of the calculation lacks a resource consumption threshold.

1. Introduction

Calculation of environmental carrying capacity is needed to determine the limits of a region's capacity to support population activities, such as natural resource consumption and waste discharge into nature. In Indonesia, environmental carrying capacity is normatively described in Law of the Republic Indonesia No. 26 Year 2007 on Spatial Planning, and Law of The Republic Indonesia No. 32 Year 2009 on Environmental Protection and Management.

The calculation of environmental carrying capacity can be done using various approaches. Method selection is one of the main issues in determining environmental carrying capacity. Proper selection determines the success of the establishment and implementation of environmental carrying capacity. There are two relatively new of environmental carrying capacity calculation approaches, i.e. the ecological footprint and the ecosystem services approach. The present study aimed to compare the implementation possibility of ecological footprint models and ecosystem services models in Indonesia.

2. Research Method

The method used in this study to compare the implementation possibility of ecological footprint and ecosystem services models is descriptive research. The aim of descriptive research is to describe phenomena related to the object of study, whether natural or human engineering [1]. Descriptive research can use a quantitative approach by collecting and measuring data in the form of numbers, or a qualitative approach by representing the situation as it is using words. This research tends more to qualitative description because it explains the differences between the two models in a narrative manner.

Each model is explained based on predetermined criteria, namely government policy, conceptualization and calculation, output and usage calculation, and data availability. The necessity of analyzing in terms of government policy is related to the mandate of Law of the Republic Indonesia No. 32 Year 2009 on Environmental Protection and Management, which states that procedures for determining environmental carrying capacity will be regulated by government regulations. In order to formulate these regulations, it is important to consider the selection of the environmental carrying capacity calculation approach. Each calculation approach requires different data and produces different outputs. The calculation thus depends on data availability and the method of data collection. Meanwhile, the calculation output is related to its use, for example for spatial planning or strategic environmental assessment (SEA).

In terms of data sources, this study used secondary data, i.e. data sourced from other parties and used by the authors as supporting data. The types of secondary data utilized in this study include research reports, scientific articles, and regulations. Data were collected through literature tracking sourced from scientific journals and government agencies. Data analysis was done by content analysis, i.e. pulling information from reports or written documents to make direct conclusions. Then the data were analyzed descriptively and compiled based on predetermined criteria.

3. Sustainable development and environmental carrying capacity

The environmental carrying capacity of a region has a close relationship with its sustainability. In the context of regional development, the environment is an important aspect to be considered as supporting development and as an object affected by it.

3.1. Sustainable development

Practically speaking, development is like two different sides of a coin. On the one hand, development is as an attempt to achieve an improved situation related to physical, social, cultural and economic aspects. On the other hand, development also has a paradoxical effect in the form of environmental degradation. Development trends that only pursue economic growth are a common threat to environmental conditions.

Reflecting this, the sustainable development paradigm currently has been adopted in almost all countries. The basic definition of sustainable development was first proposed by the Brundtland Commission or World Commission on Environment and Development in 1987, namely the development of life to meet the needs of the present generations without compromising the ability of future generations to meet their needs [2]. In brief, sustainable development is development that seeks to synergize economic, social, environmental goals and ensure a good quality of life for present and future generations. There are three main pillars of sustainable development: economic, environmental, and social, which are assumed to be interconnected. These three aspects should be applied in a balanced government policy to achieve sustainable development.

3.2. Environmental carrying capacity as a pillar of sustainable development

The process of development affects the environment as supporting development and an object affected by it. If the environment is well maintained, ecological functions and availability of natural resources

IOP Conf. Series: Earth and Environmental Science **158** (2018) 012026 doi:10.1088/1755-1315/158/1/012026

can support development well. Conversely, if the environment is damaged or polluted due to the impacts of development, sustainable development will be compromised.

The environmental carrying capacity needs to be known to determine how much development the environment can support. In addition, the ability of the environment to absorb the waste generated from development activities needs to be known in order to determine the limits of disposing waste resulting from human activities in nature. If the development exceeds the capacity of available resources and the capacity to absorb waste, the development is inefficient and will reduce the quality of the environment, or it can be said to be unsustainable. Therefore, the environmental carrying capacity has a close relationship with sustainable development.

4. Environmetal carrying capacity concept

As a concept, environmental carrying capacity has a theoretical basis that continues to develop. This section discusses the development of the conceptualization and modeling of environmental carrying capacity.

4.1. Environmental carrying capacity concept

The concept of environmental carrying capacity is derived from the management of livestock and wildlife [3,4]. Environmental carrying capacity shows the magnitude of the environment's ability to support animal life expressed in number of individuals per unit area. Environmental carrying capacity can also express the number of individuals supported by their habitat to be healthy and strong. According to Rees and Wackernagel [5], environmental carrying capacity is defined as the maximum population of a particular species that can be supported in a habitat without permanently damaging habitat productivity.

Then, the concept of environmental carrying capacity was applied to human populations. In that context, the environmental carrying capacity limit is the number of individuals that can be supported by a comprehensive set of resources and an environment that can provide resources while it remains in a prosperous state [4]. In this case, environmental carrying capacity has two components, i.e. the size of the human population and the resources and environment that provide welfare to the human population. According to Guwahati [6], environmental carrying capacity is an ecological concept that also includes social and economic parameters. The quality and state of the ecosystem is affected by the social and economic conditions and likewise human life is also affected by the condition of the ecosystem. Khanna et al. [7] state that environmental carrying capacity consists of the ability to support human life (supportive capacity) and the ability to accept pollution load (assimilative capacity), as depicted in Figure 1.

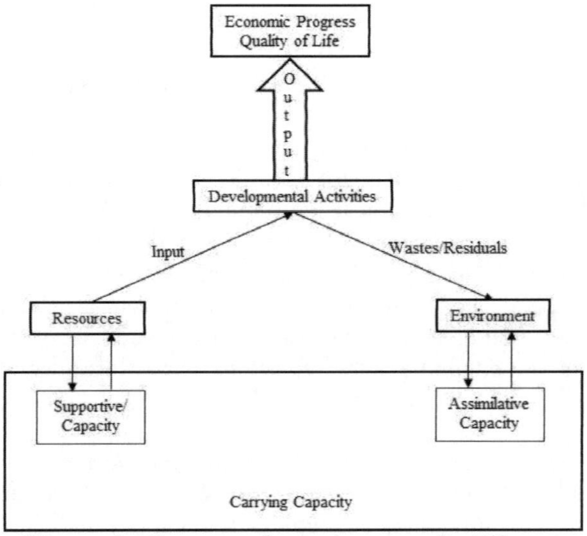

Figure 1. Elements of carrying capacity [7]

According to the above mentioned definitions of environmental carrying capacity, we can formulate the main concept of environmental carrying capacity as follows:

- environmental carrying capacity is the maximum population that can be supported in a habitat without permanently damaging the productivity of that habitat;
- environmental carrying capacity is an interaction between availability of natural resources and natural resources demand by specific populations; and
- environmental carrying capacity consists of the ability to support life (supportive capacity) and the ability to accept pollution load (assimilative capacity).

4.2. Environmetal carrying capacity policy in Indonesia

Environmental carrying capacity in Indonesia is textually stated in some rules, especially in Law of the Republic Indonesia No. 26 Year 2007 on Spatial Planning and Law of the Republic Indonesia No. 32 Year 2009 on Environmental Protection and Management. The regulations describe the use of environmental carrying capacity in spatial planning [8, 9], natural resources utilization, and implementation of strategic environmental assessment (SEA) [9].

Law of the Republic Indonesia No. 32 Year 2009 on Environmental Protection and Management divides the concept of environmental carrying capacity into two capacities, i.e. supportive capacity and assimilative capacity [9]. The supportive capacity is the ability of the environment to support humans, other living beings, and the balance between the two. Meanwhile, assimilative capacity is the ability of the environment to absorb or incorporate substances, energy, and/or other components [9].

4.3. Environmental carrying capacity model

There are various approaches or models to calculate environmental carrying capacity. Regulation of the Minister of the Environment of the Republic of Indonesia No. 17 Year 2009, containing Guidelines for Determining Environmental Carrying Capacity in Spatial Planning, stipulates land carrying capacity method, land supply and demand ratio, and water supply and demand ratio [10]. In addition there are also other environmental carrying capacity calculation approaches, such as the ecological footprint model, a graphical model, the uniconstraint model, the IPAT equation, the PSR (pressure, state, response) model [6], and the ecosystem services model [11]. Each of these has a correlation to the environmental carrying capacity concept as shown Table 1 below.

Table 1. Correlations between environmental carrying capacity models and concepts [12]

Environmental carrying capacity model	Concept[a]	Concept[b]	Concept[c]	Concept[d]
1. Land capability			v	
2. Land supply and demand ratio	v	v	v	
3. Water supply and demand ratio	v	v	v	
4. Graphical model (logistic growth)	v		v	
5. Uni constraint model	v	v	v	
6. IPAT equation			v	v
7. PSR model			v	v
8. Ecological footprint	v	v	v	v
9. Ecosystem service			v	v

[a] Maximum population supported by a sustainable environment
[b] Interaction between availability and demand resources by a certain population
[c] Ability to support life
[d] Ability to absorb pollutant load

4.3.1. Ecological footprints model. The ecological footprint concept was introduced by William Rees in 1992, which he then further developed together with Mathias Wackarnagel. Until now, research and development of the ecological footprint approach is being done by researchers who are members of the Global Footprint Network.

Rees [13] revealed that in relation to human life, the environmental carrying capacity can be seen as the maximum average consumption of resources and sustainable waste disposal in an area without damaging the functional unity and productivity of the ecosystem. The total amount of productive land needed by human activity is called the ecological footprint. No areas are seen as independent units [13]. In fact, the population of a region that has exceeded the area of environmental carrying capacity depends on trade to survive.

Calculation of the environmental carrying capacity by the ecological footprint model consists of three steps, as show in Figure 2 for a national footprint framework [14]:

- ecological footprint, which is a measurement of population demand and activity that occurs in the biosphere in a given year;
- biocapacity, which is the measured amount of biologically productive land and sea areas that is available to provide ecosystem services for human consumption as ecological budget or natural regeneration capacity; and
- ecological deficit/surplus, obtained by calculating the difference between ecological footprint and biocapacity.

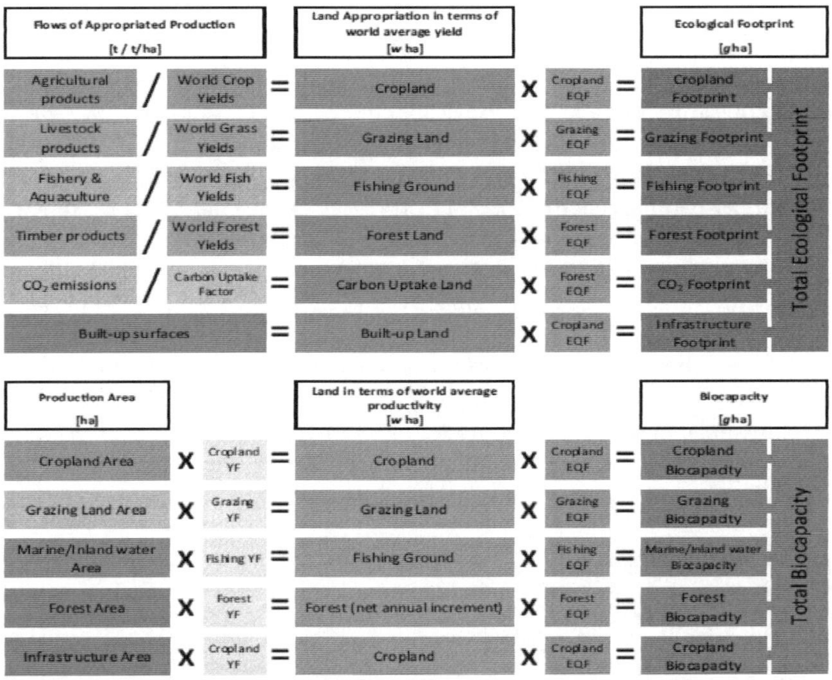

Figure 2. National Footprint Accounts (NFA) accounting framework [14]

The calculation of an ecological footprint uses two conversion factors, i.e. a yield factor and an equivalent factor. The yield factor is the ratio between the productivity within the same land category in an area and the average productivity of land in the world in the same year. The equivalent factor is a factor that converts certain local units to universal units, namely global hectares (gha). One global hectare is defined as one hectare of land (soil and water) in a given year equivalent with world average productivity, i.e. about 12 million hectares [14].

4.3.2. Ecosystem service model. The ecosystem services model was initiated by the Ministry of Environment and Forestry, Republic of Indonesia. Ecosystem services are the benefits that people obtain from various resources and natural processes, which are jointly provided by an ecosystem [11]. Ecosystem services are grouped into four types, i.e. provisioning, regulating, supporting, and cultural, which refers to the methodological framework of Millennium Ecosystem Assessment [15], as shown in Table 2 below.

Table 2. Type of ecosystem services by benefits [15]

Type of ecosystem service	Type
1. Provisioning	1. Food 2. Fresh water 3. Fiber 4. Fuel, wood, and fossil 5. Genetic resources
2. Regulating	1. Climate regulation 1. Water regulation 2. Natural hazard regulation 3. Water purification 4. Waste treatment 5. Air quality regulation 6. Pollination 7. Pest regulation
3. Cultural	1. Residential and living space 2. Recreation and ecotourism 3. Aesthetic values
4. Supporting	1. Soil formation and fertility 2. Nutrient cycle 3. Primary production 4. Biodiversity

Environmental carrying capacity in the ecosystem services model assumes that the higher the value of the ecosystem services, the higher the environmental carrying capacity. Two estimates are used for obtaining ecosystem services, i.e. landscape or ecoregion (landscape based proxy) and land cover (landcover/landused based proxy). This information is used as a basis for mapping environmental carrying capacity [11].

Calculation of an ecosystem services model produces a map of the types of ecosystem services in an area. A specific ecosystem services map looks at each ecoregion category with a specific ecosystem services index, ranging from low, moderate and high to very high. The index of the ecosystem services map uses an ordinal scale, which is used for differentiating and sorting but without indicating amount or degree.

5. Comparison of ecological footprint and ecosystem services model

As a relatively new calculation approaches of environmental carrying capacity, the ecological footprint model and the ecosystem services model need to be further analyzed in terms of 1) government policies, 2) concept and calculation approach, 3) calculation output and usage, and 4) data availability. The following comparison of these models aims to determine the possibility of its application in Indonesia.

5.1. Government policies

The ecological footprint model has not been included in any regulations of the Indonesian governement, but the government has stated that calculation of environmental carrying capacity can be established through other approaches in accordance with scientific development. According to the

Ministry of Environment and Forestry, as quoted by Subekti [12], in principle, the selection of environmental carrying capacity calculation approach should pay attention to the following:

- accuracy of information generated by needs of use and/or understanding;
- consistency of information continuation within a certain time frame for policy formulation; and
- availability of data.

In terms of policy, the ecosystem services model is in line with Law of the Republic Indonesia No. 32 Year 2009 on Environmental Protection and Management, which states that one of the strategic environmental assessment (SEA) analyses is ecosystem services [9]. Ecoregion based ecosystem services analysis is also in accordance with ecoregion zoning and environmental inventory, which are tasks of the national government and local governments. Based on this, the Ministry of Environment and Forestry has drafted regulations governing guidelines for ecosystem services based on environmental carrying capacity [12].

5.2. Concept and calculation approach

Both the ecological footprint model and the ecosystem services model have different relevance to the environmental carrying capacity concept, as can be seen in Table 1 above. The ecological footprint model conceptually shows capacity and availability of natural resources, as well as ability to absorb wastes. Capacity and availability of natural resources are shown through biocapacity, which is the productive area of land and water in the form of cropland areas, grazing land areas, marine/inland water areas, forest areas and infrastructure areas. Meanwhile, the biological capacity to absorb waste is indicated by the capacity of forests as carbon sinks. In addition to resource capacity, the ecological footprint model is also able to describe the demand for natural resources, including demands for agriculture, forestry, fisheries, animal husbandry, carbon emissions and developed land. Therefore, the model can describe supply and demand of resources in calculating environmental carrying capacity. For example, the demand for the agriculture sector is represented in the ecological footprint of agriculture related to the biological availability of land for producing agricultural products. While conceptually quite complete, the ecological footprint model also has weakness, i.e. it does not consider technological changes – one of the main factors in the utilization of resources – and does not take resources beneath the earth into account [16].

Conceptually, the ecosystem services model has common ground with environmental carrying capacity, i.e. the capacity and availability of natural resources through providing services (food and water), cultural and supporting, as well as the capacity of a region to absorb waste through regulating services (air quality, carbon sequestration, and waste absorption regulations). The ecosystem services model further represents the supply of natural resources but is not able to describe resource demand in calculating environmental carrying capacity [12].

Due to using different concepts, both models also have different calculation approaches. In terms of unit of analysis, there is not much difference between the ecological footprint and the ecosystem services model. Both models can be applied for specific areas or units of analysis, such as national (islands), provinces and districts/cities, as well as for activities or sectors. The difference is that the unit of analysis of the ecosystem services model is based on ecoregions and land cover. In terms of analysis technique used, the ecological footprint model uses mathematical formula to calculate consumption and capacity of natural resources embodied in global hectares. The ecosystem services model uses a geographic information system. This analysis produces a map of ecosystem services types within a region, which has an index or score of certain ecosystem services. The scoring method is based on expert valuation, i.e. assessment by an expert panel on the roles of land cover and ecoregions in relation to the types of ecosystem services. Furthermore, the assessment results are analyzed using pairwise comparison as part of the analytic hierarchy process (AHP) method to produce an index or weight of variables in the decision making process [11]. Assessment by experts tend to be subjective, so there is the possibility of differences in ratings between them.

5.3. Output of calculation and usage

The ecological footprint model produces values for natural resources demand (ecological footprint), natural resources availability (biocapacity), and status of environmental carrying capacity realized in global hectare units. The output of the ecological footprint model implies a threshold of resource consumption, which is not allowed to exceed the capacity of natural resources. The status of the environmental carrying capacity becomes a deficit if the ecological footprint total is greater than its biocapacity. Meanwhile, the status of the environmental carrying capacity becomes a surplus if the total ecological footprint is smaller than its biocapacity. Based on the National Footprint Account, Indonesia has had an ecological deficit since 2002. This means that per capita consumption of Indonesian society exceeds the carrying capacity of the natural environment, as shown in Figure 3.

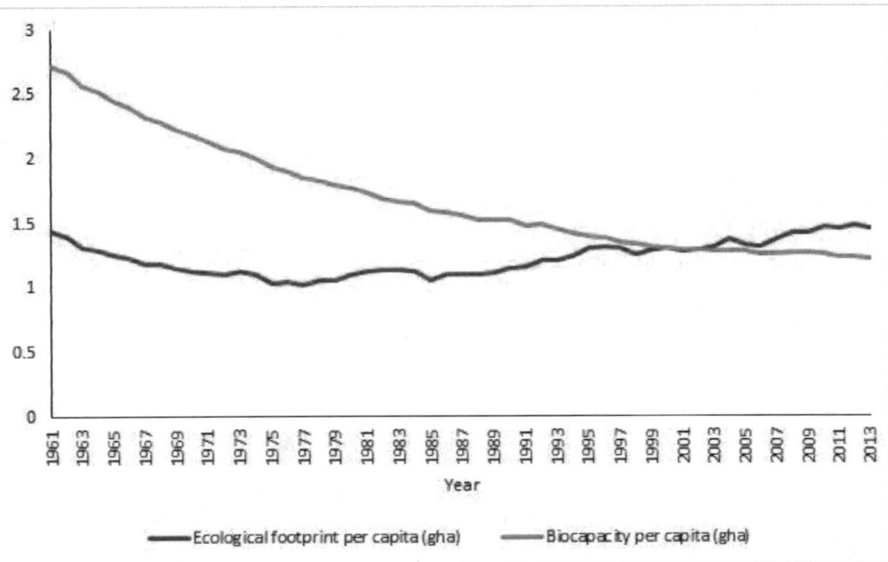

Figure 3. Graphic of Indonesia's per capita ecological footprint and biocapacity 1980-2013 [17]

In terms of information consistency, ecological footprint output can change each year, because the input data on production, exports and imports for each demand sector and the population change annually. The ecosystem services analysis produces a map of ecosystem services types. Specific ecosystem service maps show each ecoregion class with scores or an index range of particular ecosystem services, ranging from low to medium to high, as shown in Figure 4. However, the ecosystem services model is not able to set a resource consumption threshold because it uses an ordinal scale. The output of the ecosystem services model covers relatively long time periods, because the data input, including ecoregions and land cover information, change over a relatively long time period.

Based on the regulations, environmental carrying capacity calculaton is carried out in the framework of spatial planning [8,9], natural resources utilization, and implementation of strategic environmental assessment (SEA) [9]. Ecological footprint output can be utilized in such cases. For example, when there is an ecological deficit, development planning should be fixed in accordance with SEA recommendations. These recommendations can be seen in terms of supply and demand. From the supply side, allocation of space utilization must be provided and maintained, both allocation of cultivation areas and protected areas. Area allocation must be adjusted to the demand or resource consumption of the average population in the region. In the ecosystem services model, the most dominant information within a region will determine the level of management. For example, water supply as an ecosystem service in a region is dominant because of vast forests in good condition. Therefore, allocation of land uses should emphasize forest areas maintained in protected areas. Because the ecosystem service model is based on ecoregions, water supply functions must be seen in ecoregion perspective across administrative boundaries. This implies a need for coordination and

IOP Conf. Series: Earth and Environmental Science **158** (2018) 012026 doi:10.1088/1755-1315/158/1/012026

cooperation between provinces or district/cities. Between both models there is a difference in use. Firstly, the ecological footprint model can provide more scalable information than the ecosystem services model. Secondly, the ecosystem services model can see geographic distribution of ecosystem service types within a region.

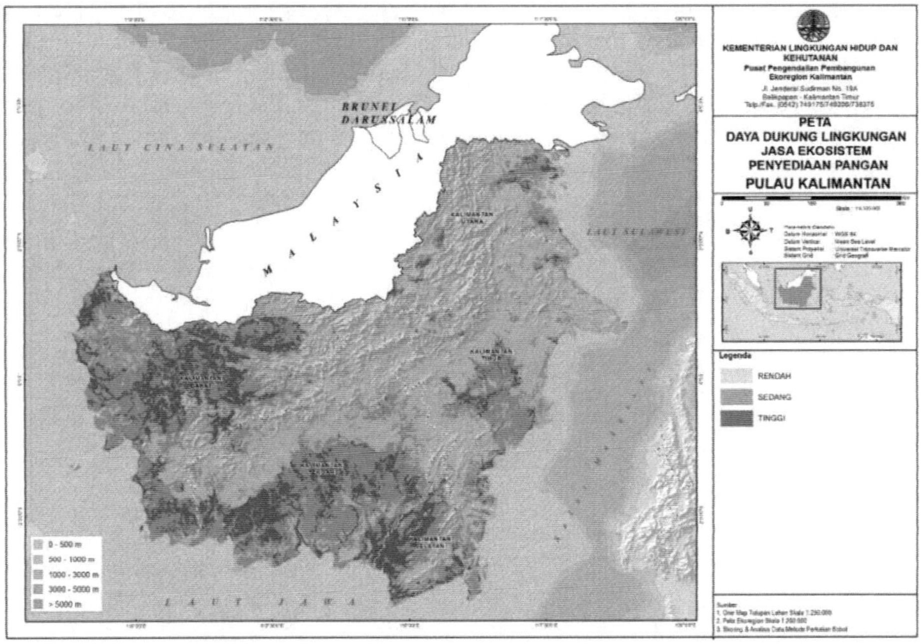

Figure 4. Map of food type services in Kalimantan [11]

5.4. Data availability

Ecological footprint calculation requires certain data, specifically statistical data in terms of production, exports and imports of commodities related to agriculture, forestry, animal husbandry and fishery. At the national level, the data can be obtained from world institutions such as FAO, IEA and UN Comtrade. But at province and district/city level these data are not necessarily available. Subekti [11] conducted a survey on the availability of data and found that data on export and import of agricultural products, fishery and forestry are generally unavailable, as well as trade data of other commodities. Referring to Borucke et al. [14], ecological footprint calculation of a particular commodity uses net consumption, i.e. the actual consumption influenced by trade (exports and imports).

The data needed for an ecosystem services model are maps indicating ecoregion, land cover, and land use. The map type must consider the scale, as it relates to the unit of analysis, such as national (island), province or district. Map availability for such units can be seen in Table 3. According to Law of the Republic Indonesia No. 32 Year 2009 on Environmental Protection and Management, each province and district has the authority to set ecoregions and environmental inventory [9]. Therefore, mapping of information on ecoregions is part of the government's tasks.

Table 3. Map availability used in ecosystems services model

Unit of analysis	Information scale	Type of map	Map availability
1. Island ecoregion (national)	Macro (1:1.000.000 to 1:500.000)	1. Ecoregion	Available, from KLHK
2. Crossprovincial ecoregion (province)	Meso (1:250.000)	1. Ecoregion (detailed)	Available, from KLHK

Unit of analysis	Information scale	Type of map	Map availability
		2. Land cover	
3. Land use/ecosystem (ditrict)	Micro (1:100.000 to 1:50.000)	1. Ecoregion (detailed) 2. Land cover 3. Land use	Differs in each district/city

6. Conclusion

The ability of a region to support human activities such as natural resource consumption and waste discharge can be determined by calculation of its environmental carrying capacity. When development activities exceed resource capacity and waste absorbtion capacity, it means that the development is not efficient and reduces the quality of the environment, i.e. the development is unsustainable. Therefore, environmental carrying capacity has a close relationship with sustainable development.

Calculation of environmentral carrying capacity can be done with different approaches. Proper approach selection will determine the success of the establishment and implementation of environmental carrying capacity. Two relatively new calculation approaches of environmental carrying capacity can be considered for use, i.e. the ecological footprint model and the ecosystem services model. The aim of comparing both models was to know their respective application possibilities in Indonesia. From this comparison, advantages and disadvantages of each model can be taken into consideration in applying either model.

The ecological footprint model has advantages, i.e. conceptually it describes demand and supply resources, including supportive and assimilative capacity, and measurable output through threshold of resource consumption. However, this model also has weaknesses, such as not considering technological change and resources beneath the earth, as well as the necessity to provide trade data between regions for calculating at the provincial and district/city level. The ecosystem services model also has advantages, such as being in line with strategic environmental assessment (SEA) in assessment of ecosystem services, using spatial analysis based on ecoregions, and a draft regulation on calculation guidelines formulated by the government. Meanwhile, the weaknesses are that it only describes supply of resources and that the outputs of the calculations are without resource consumption threshold.

References

[1] Sukmadinata N S 2012 *Hari Metode Penelitian Pendidikan* (Bandung: Remaja Rosdakarya)
[2] World Commission on Environment and Development 1998 *Hari Depan Kita Bersama* (Jakarta: Gramedia)
[3] Soemarwoto O 2004 *Ekologi, Lingkungan Hidup dan Pembangunan* (Jakarta: Djambatan)
[4] Irwan Z D 2012 *Prinsip-Prinsip Ekologi: Ekosistem, Lingkungan dan Pelestariannya* (Jakarta: Bumi Aksara)
[5] Rees W and Wackernagel M 1996 *EIA Review* **16** 223–248
[6] India Institute of Technology Guwahati 2011 *Urban Carrying Capacity: Concept and Calculation* (Guwahati: Department of Civil Engineering) 1-23
[7] Khanna P, Babu P R and George M S 1999 *Prog. Plann.* **52** 101–166
[8] The Republic of Indonesia 2007 *Law Number 26 Year 2007 on Spatial Planning* (Jakarta: Ministry of Justice and Human Rights)
[9] The Republic of Indonesia 2009 *Law Number 32 Year 2009 on Environmental Protection and Management* (Jakarta: Ministry of Justice and Human Rights)
[10] Ministry of Environment 2009 *Regulation Number 17 Year 2009 on Guidelines for Determining Environmental Carrying Capacity in Spatial Planning* (Jakarta: Ministry of Environment)
[11] Pusat Pengendalian Pembangunan Ekoregion Kalimantan 2015 *Deskripsi Peta Daya Dukung Ekoregion Kalimantan* (Balikpapan: P3E Kalimantan)

IOP Conf. Series: Earth and Environmental Science **158** (2018) 012026 doi:10.1088/1755-1315/158/1/012026

[12] Subekti R M 2017 *Exploring Environmental Carrying Capacity Approach in Indonesia, Master's Program Thesis* (Bandung: Institut Teknologi Bandung)
[13] Rees W 1992 *Environ. Urban.* **4** 121-130
[14] Borucke M., Moore D, Cranston G, Gracey K., Iha K, Larson J, Lazarus E, Morales J C, Wackernagel M and Galli A 2013 *Ecol. Indic.* **24** 518-533
[15] Millenium Ecosystem Assessment 2005 *Ecosystem and Human Well-being: Synthesis* (Washington: Island Press)
[16] Moffat I 2000 *Ecol. Econ.* **4** 359-362
[17] Global Footprint Network 2017 *National Footprint Accounts 2017 Edition* (Oakland: Global Footprint Network)

Satellite Imagery Data: Dynamic Systems Model for sustainable urban forest in area of Halim Perdana Kusuma, East Jakarta

D M Sundara[1], D M Hartono[2], E Suganda[3] and H Haeruman JS[4]

[1]Doctoral student at Environmental Science Program, University of Indonesia, Jl. Salemba Raya No. 4, Jakarta - 10430, Indonesia
[2] Faculty of Engineering, Department of Civil Engineering, University of Indonesia, Depok West Java, 16424, Indonesia
[3] Faculty of Engineering, Department of Architecture Engineering, University of Indonesia, Depok West Java, 16424, Indonesia
[4] Environmental Science Program, University of Indonesia, Jl. Salemba Raya no. 4, Jakarta-10430, Indonesia

Corresponding author: dennymsr@yahoo.com

Abstract. East Jakarta icon as a buffer and the lungs of the city is still a big dream of Jakarta. It is a classic problem that there is a struggle for land between current economic interests and sustainable environmental interests for the future. This paper discusses the development of urban forest area of Halim Perdana Kusuma, East Jakarta. The forest area according to regulations of existing city local governments is not enough to support sustainable urban development indicators. Therefore, it requires an extensive mapping of urban forest potential development accurately by utilizing satellite imaging technology. Landsat-TM satellite imagery data can provide a full picture of the potential land width for urban forest area development. The results of this satellite image will then be made into a model of urban forest as one of the indicators of sustainable urban development. This research aims to support sustainable urban development through environmental balance in the form of a green neighborhood revitalization and development of urban forests and to create socio-economic balance. This paper uses a dynamic system model to simulate the conditions of the region against the intervention performed in the potential area for development of urban forests which are derived from urban spatial analysis based on satellite image data, using GIS program as a tool. The result is a model of urban forest area which is integrated with a social and economic function to encourage the development of sustainable cities.

1. Introduction

In some countries, the urban forest has become an important component in planning sustainable urban development [13]. Developing a forest area of the city is expected to encourage the formation of an ideal city. The ideal city is often interpreted in four ways: (1) as a sustainable urban ecological system, (2) developing in a just manner, (3) the economy grows sustainably, (4) culturally able to develop a strong local identity [10]. City forest serves to mitigate urban discomfort [5]. City forest relates to vegetation function [3]. Urban forests are as carbon stocks in trees [4,7,12,15]. Urban forest and the effect of

IOP Conf. Series: Earth and Environmental Science **158** (2018) 012027 doi:10.1088/1755-1315/158/1/012027

rainwater reduce runoff [14]. In addition, city forest creates socio-economic balance [9]. The institutional framework for sustainable development involves: a) a system with functions related to its environment, b) the existence of organizational structures and procedures governing the tasks, products, communities, resources and objectives of the organization, c) preparing the organization's resilience to changes in resources resulting from economic and political relations [6]. The background of sustainability is the desire for environmental, social and economic harmony. In the context of global industry, sustainability can be defined as current development [1,2]. All economic, environmental and social components are interrelated and cannot work independently, therefore a partnership approach to all problems [6,11] should be developed. One framework strategy for achieving sustainable development is to strengthen the institution by preparing the organization's resilience to resource changes due to economic and political relations [6]. The purpose of this research is to find out the potential of urban forest area by using satellite image data analyzed with GIS tool, and then to make a model with dynamic system analysis model. The result is a model of urban forest development for sustainable urban development.

2. Methodology

The analytical methods used in this research are spatial analysis and modeling analysis, which aim to find the relationship between the broad needs of urban forest and the spatial pattern of urban forest growth, and other variables as an integral part in achieving sustainable urban development.

1. Spatial analysis

Analysis of the spatial growth is the main part of landscape ecology research, which aims to connect several complex parts between landscape patterns and processes of landscape change. It is analyzed using Geographic Information System (GIS) tools to manage geographically referenced spatial data. Data processing and data analysis were performed by rotation, scaling coordinate conversion, coordinate conversion geography, registration and spatial analysis and statistics.

The analysis in the database was done by overlaying several layers needed. To determine the condition of the existing land cover classification provided by the forest or vegetation that could potentially be a forest, land cover spatial information was obtained through satellite imagery, as shown in table 1 and table 2.

Table 1. Region with the amount of carbon dioxide produced in 2015 [16].

June 21, 1976	October 3, 1982	September 17, 2001	April 22, 2011
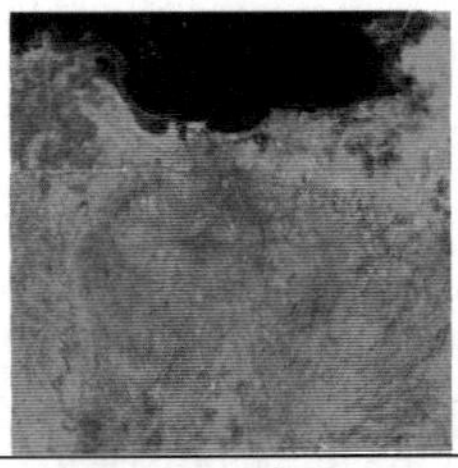	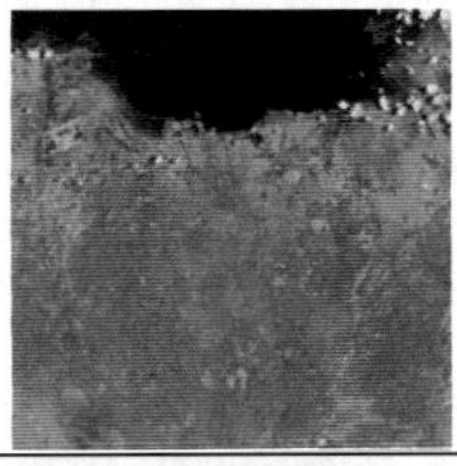	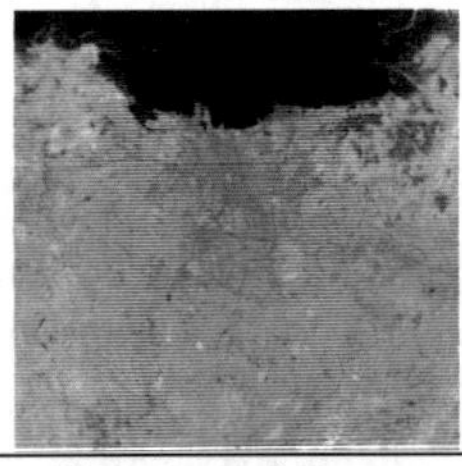	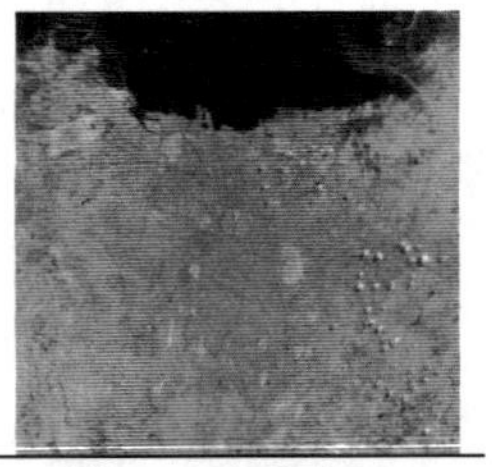
Vegetated Area Size 37.375ha (56.50%)	Vegetated Area Size 35.655ha (53.90%)	Vegetated Area Size 19.184ha (29.00%)	Vegetated Area Size 16.542ha (24.98%)

Table 2. Growth vegetated land cover size.

Region Size	**Location**	**Vegetated Land Cover Size (ha)**			
		1974	**1989**	**2005**	**2015**
2,163	Makassar	2,044	1,588	1,173	1,046

2. Modeling Analysis

The interaction between spatial land use patterns in urban areas and the pattern of urban forests is a complex system. The system dynamics method is one of the modeling techniques that can include complex and non-linear systems of multiple feedback relationships, which allow to conduct behavioral analysis and examine the long-term impact of a policy through the model [8]. The system dynamics model formulated is used for two things: (1) analyzing the environmental carrying capacity due to urban forest development, land use structure, and environmental quality as represented by ecological footprint variables. (2) examining some scenarios ahead in optimizing and controlling urban forests. The models analyzed are business as usual scenarios and compared with scenarios for applying sustainable urban development concepts through the application of eco-city concept as an alternative to the proposed scenario.

3. Result

The results of the interpretation of satellite imagery obtained by the division of the land cover categories are in the form of land vegetation (trees, bushes, meadows, fields), vacant land, bodies of water, non-vegetation (residential / buildings, roads) and no data (cloud, shadow clouds). In this study, the results of the image can be assumed as a forest of trees, while land vegetation in the form of shrubs and grass is assumed as a potential land to be reforested. The image fields and vacant land are assumed to be an area that can be converted into forest. The area of land cover based on satellite imagery in East Jakarta for each research area in the district of Makasar is presented in Table 3.

Table 3. Land cover vast area Makasar Sub-district study.

Areas of Observation	Land Cover Size (Ha)					
	Tree	Bush	Grass	Rice fields	Empty Land	Other
District Makasar	763.70	63.49	218.31	103.92	135.45	952.16
Sub-District Pinang Ranti	62.39	17.74	8.55	5.57	12.43	132.35
Sub-District Makasar	40.06	3.25	0.01	-	0.18	95.35
Sub-District Kebon Pala	59.36	4.54	0.59	4.20	2.27	149.87
Sub-District Halim PK	506.81	32.27	206.78	91.60	120.00	418.17
Sub-District Cipinang	95.08	5.69	2.38	2.55	0.57	156.42

The results of satellite image data processed with GIS are explained in Figure 1. From the results of satellite image processing, it can be described in broad outline that the area of trees with the potential to become urban forest in District Makasar is an area of 763.70 ha, and another potential area which can be developed into urban forest with a special intervention is an area of 1320 ha.

IOP Conf. Series: Earth and Environmental Science **158** (2018) 012027 doi:10.1088/1755-1315/158/1/012027

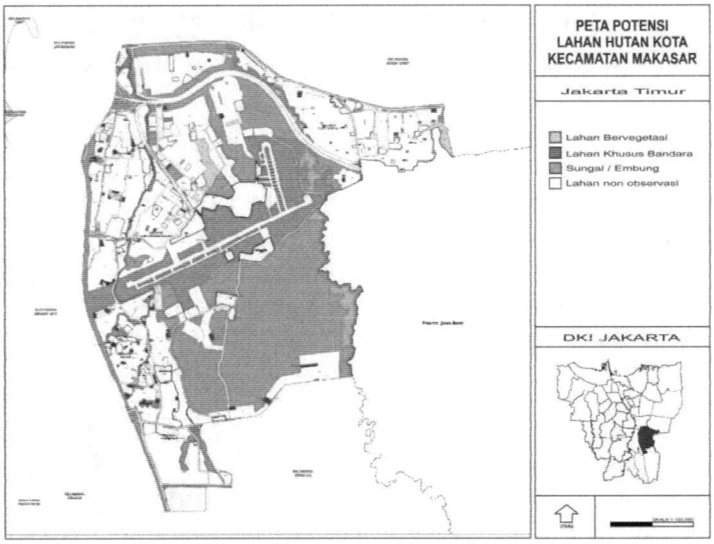

Figure 1. Map of Forest Land Potential of District Makasar

The carrying capacity of urban land is the urban limits of the city which is the object of observation. Carrying capacity of urban forest is a town's forest area limits built in accordance with the regulations that apply to the object of observation, located in East Jakarta. Population growth rate is a growth index in general. Although the intervention is done maximally, the urban forest area will still decrease because this modeling does not intervene all aspects due to other things. The intervention is assumed to increase forest cover city and get the level of GDP sustained. The intervention scenario was conducted with the Stock-Flow Diagram in modeling 'urban forests for sustainable development as depicted in Figure 2.

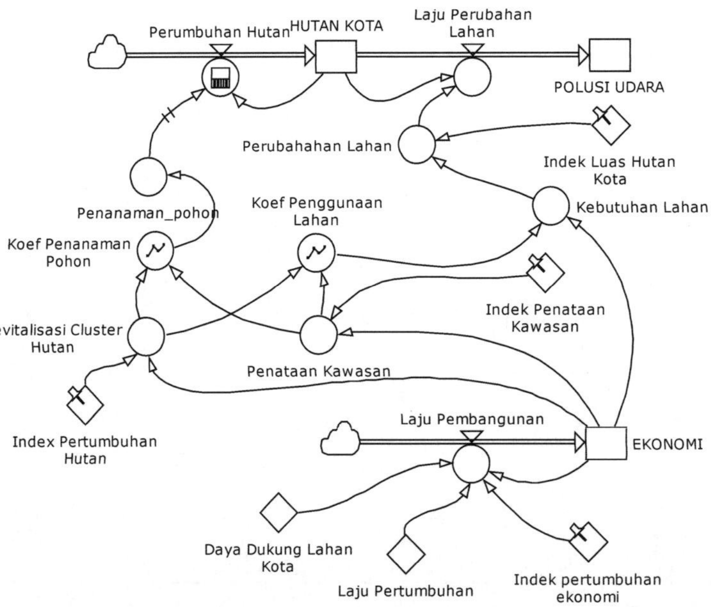

Figure 2. Stock-Flow Diagram modeling with intervention scenario.

In this intervention model, on the urban forest and GDP, the graph function coefficients are applied to the planting of trees and the coefficient of land use is a representation of the concept of regional restructuring and revitalization of the forest cluster.

Delay function is applied to the forest growth on the assumption that the growth requires 30 years (age modeling assumptions). Index of regional restructuring and revitalization of the forest cluster creates slider function to get an ideal figure and it is considered sustainable. The results of simulation models of the intervention scenario are shown in Table 4, which shows the different number of urban forests, air pollution levels and the rate of GDP growth. If the number of the urban forest is lower than the initial model simulation scenario it will keep on declining, but it is able to reduce the level of pollution air.

Table 4. Simulation models of intervention scenarios of District Makasar.

Year	Urban Forest Index	Air pollution Index	GDP Index
2015	1.19	1.88	1.05
2020	1.54	2.02	3.11
2025	1.81	2.31	4.18
2030	1.97	2.73	5.01
2035	2.03	3.26	5.73
2040	2.03	3.86	6.36
2045	1.98	3.51	6.93

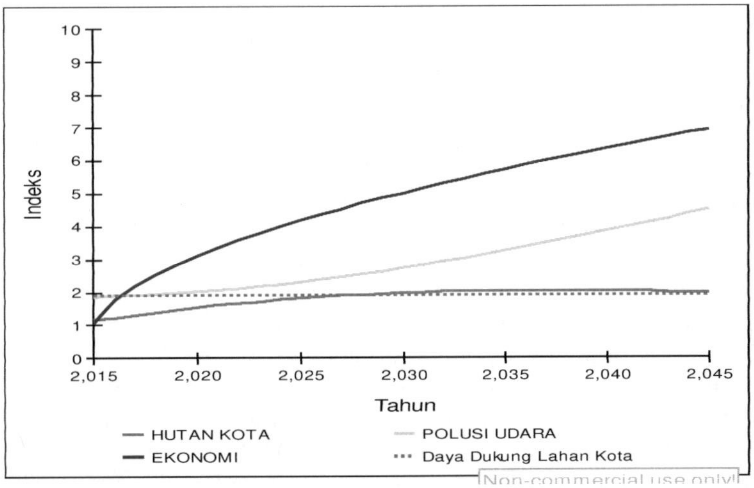

Figure 3. Graph modeling of behavior intervention scenario of District Makasar.

4. Conclusion

The result of satellite image data that wasprocessed with GIS and has the potential to be an urban forest in Makassar sub-district is 763 ha, and the potential can be developed with intervention to 1320 ha. Based on the simulation result, as illustrated in Figure 3, the effect of the intervention on changes in forest area is about the graph of modeling behavior in the intervention scenario. The graph indicates if the index of GDP increases, the air pollution index will follow to increase, but with specific interventions in the urban forest index that is in accordance with the increase of GDP index. The results of modeling with the conditions of urban forest index affect the index of GDP and air pollution. The GDP index in District Makasar increased by an average of 0:15, then followed by the air pollution index which is by an average of 0.10.The increase is equivalent to the increased levels of CO in the air which reaches an average of 3,574 tons/year.

References

[1] Coplak J and Raksanyi P 2003 *Planning Sustainable Settlements* (Bratislava: Slovak University of Technology) ISBN 80-227-1941-2

[2] Dixon T 2011 *Sustainable Urban Development to 2050:Complex Transitions in the Built Environment of Cities* Oxford Institute for Sustainable Development (Oxford: Oxford Brookes University UK) ISSN 2052-1618

[3] Irwan, Rofiqo S N, Khoisol A and Hasanbahri S, 2012 *Fungsi Vegetasi Pada Ruang Hijau Dan Hutan Kota Untuk Pengembangan Lanskap Ecopesantren* [The Function of Vegetation in Green Open Spaces and City Forest for the Development of Ecopesantren Landscape] Bagian Konservasi Sumber Daya Hutan, Fakultas Kehutanan UGM

[4] Joss S, Tomozeiu D and Cowley R 2011 *Eco-cities - A Global Survey 2011* Published by University of Westminster International Eco-Cities Initiative (Great Britain: University of Westminster)

[5] Joye Y, Willems K, Brengman M and Wolf K 2010 The effects of urban retail greenery on consumer experience: Reviewing the evidence from a restorative perspective *Journal Elsevier of Urban Forestry & Urban Greening* **9** 57–64

[6] Lejano R P and Stokols D 2013 Social ecology, sustainability, and economics Elsevier J. of Ecol. Econ. **89** 1–6

[7] Lubis S H, Arifin H S and Samsoedin I 2013 Analisis Cadangan Karbon Pohon Pada Lanskap Hutan Kota di DKI Jakarta [Analisis of Trees Carbon Reserve in the City Forest Landscape in DKI Jakarta] *Jurnal Penelitian Sosial dan Ekonomi Kehutanan* 10(1):1-20

[8] McLain R, Poe M, Hurley P T, Lecompte-Mastenbrook J and Emery M R 2012 Producing edible landscapes in Seattle's urban forest *Urban For. & Urban Greening* **11**:187-194

[9] Rapoport E and Vernay A L 2011 Defining The Eco-city ; A Discursive Approach *J. of Manage. and Innov. for a Sustain. Built Environ* Amsterdam, The Netherlands

[10] Santoso J 2009 " Kota Ideal " dalam Konteks Pengembangan Kota Berkelanjutan *Online Bulletion Tata Ruang Edisi Januari-Februari 2009*

[11] Timmer V and Seymoar N K 2006 *The Sustainable Region Initiative* International Centre for Sustainable Cities, Canada

[12] Visscher RS, Nassauer J I and Mashall L L 2016 Homeowner preferences for wooded front yards and backyards: Implications for carbon storage *Landsc. and Urban Plan.* **146** 1–10

[13] Weiland U, Kindler A, Banzhaf E, Ebert A and Reyes-Paecke S 2011 Indicators for sustainable land use management in Santiago de Chile *Ecological Indicators* **11** 1074–1083

[14] Zhang B, Xie G, Li N and Wang S 2015 Effect of urban green space changes on the role of rainwater runoff reduction in Beijing, China *Landsc. and Urban Plan.* **140** 8–16

[15] Zhou N and Williams C 2013 *An International Review of Eco-City Theory, Indicators, and Case Studies* Environmental Energy Technologies Division, Lawrence Berkeley National Laboratory Available from: http://eta-publications.lbl.gov/sites/default/files/lc_eco-cities.pdf

[16] National Institute of Aeronautics and Space 2012 *Remote Sensing Data Bank* Remote Sensing Technology and Data Laboratory

IOP Conf. Series: Earth and Environmental Science **158** (2018) 012028 doi:10.1088/1755-1315/158/1/012028

Spatial analysis related to the location characteristics of park supply. Case study: Music Park and Pendawa Park, Bandung City, Indonesia

A Malik[1], R Akbar[2], S Maryati[2], P Natalivan[2]
[1]Doctoral Student in the School of Architecture, Planning, and Policy Development Institut Teknologi Bandung, West Java, Indonesia
[2]Regional and City Planning, School of Architecture, Planning and Policy Development (SAPPK), Institut Teknologi Bandung

Corresponding author: andymalik_80@yahoo.com

Abstract. Public space plays a role in defining the character of a city and is a valuable asset for a city and one of the indicators in assessing whether a city is considered successful or not. In the context of urban sociology, high-quality public spaces with well-maintained environments can improve the quality of the heterogeneous life of urban social communities by creating economic, social, or environmental value-added. Urban societies tend to be heterogeneous, individualistic, and characterized by high competition that often causes conflicts. Another reason for conflicts is the relatively high social differentiation because of the level of religious differences, customs, languages, and sociocultural aspects brought by immigrants from various regions. In the context of space, the city is a system that does not stand alone because internally the city is a unified system of functional activities in it. Meanwhile, externally, the city is influenced by its surrounding environment. As part of the public space, park has an important role in the environmental, aesthetic, recreational, psychological, social, educational, and economic aspects of the city. Public space can be understood as open spaces in urban areas, where everyone regardless their interests and backgrounds can be intersectional and have social contact and serve as an "urban regenerator" including educational functions through innovation and technological intervention. Moreover, park can also absorb carbon dioxide emissions, produce oxygen, improve air and water quality, regulate the microclimate, reduce noise, protect soil and water, and maintain biodiversity. However, many things cause the function of parks to decrease. One reason relates to the distribution of parks related to the characteristics of their location. Research has not seen many studies on the characteristics of locations in the planning of public space. The provision of public space should consider these location characteristics. This study will use a descriptive methodology, by first explaining the policy of public space provision, the existence of the reality of public space, its function and role, the park as public space, the time of visiting the park by the community, and the various problems that occur in it. This research will look at the characteristics of neighborhood parks based on the aspect of the location of park supply and the spatial policies of Bandung City. The analysis comprised ArcGIS (Geographic Information System) and direct observations to evaluate the locational characteristics of the area around the park by using the case study of Music Park and Pendawa Park.

1. Introduction

According to Bintarto [1], the city can be defined as a system of human life network characterized by heterogeneous socio-economic strata and its materialistic features. In the context of space, the city is a system that does not stand alone because internally the city is a unified system of functional activities in it, while externally, the city is influenced by its surrounding environment. Urban societies tend to be heterogeneous, individualistic, and characterized by high competition that often causes conflicts. Another reason for conflicts is the relatively high social differentiation because of the level of religious differences, customs, languages, and sociocultural aspects brought by immigrants from various regions.

The people who are unfamiliar with each other combined with the heterogeneity of society makes the system of behavioral social supervision among members of the community difficult to control [1].

According to John Ruskin, an English pioneer of urban planning who lived in the 19th century, the greatness of every civilization and the greatness of a city can be seen from the quality of public spaces, parks, and squares. The existence of the public sphere is so important that it is the focus of the Sustainable Development Agenda by the UN Human Settlements Program. The 11th objective of target 11.7 states "(...) by 2030, provide universal access to safe, inclusive and accessible, green and public spaces, especially for women and children" [2].

1.1. The role of public space

Public space plays a role in defining the character of a city and is a valuable asset for a city. Public space is a space that can be accessed freely without a profit motive [3][4]. Public space is also one of the indicators in assessing whether a city is considered successful or not. In the context of urban sociology, high-quality public spaces with well-maintained environments can improve the heterogeneous quality of social life of urban communities by creating economic, social, or environmental value added [5]. The public spaces in question include roads, sidewalks, parks, and plazas that everyone can access [6]. According to Acasandre [7], current urban realities show the notion of "modern public space" formed from activities related to urban life and culture as a result of globalization. Public spaces are considered as living spaces, spaces of interaction and socialization for all urban communities; public spaces are social magnets. If the public space is the center of urban space then it must be "responsible" for attracting various urban activities and become the "image of the city".

To make public space "public" is very difficult. Lehtuvuori [8] describes this situation as the "unhappy marriage" between critical urban theory and the concept of space in rapidly evolving realities. Public spaces often become ambiguous in relation to the utilization of urban space. Many of the things that resulted in the functioning of the public space have switched into private spaces; conceptually and practically, this is very difficult to solve [8]. Commercialization sometimes also affects the essence and sense of the space itself. Some observations of the park as part of the public sphere indicate a transition of spatial function and activity over time. This transition often keeps the park to function continuously but the activities in it often create a certain spatial perception. Space activity is a positive activity and can be accepted by a certain circle in the community of urban communities, then this will form an image that the public space has a good function.

The urban sociologist Mark Gottdiener [9], in his book "The Social Production of Urban Space", articulates space in a social approach from the Marxist perspective by describing space as a form of translation of the capitalist ("the second circuit of capitalist"). In the end, the idea of planning as "communicative action" [10][11] states that space and place are not only understood as stable but also as events influenced by political and economic aspects as well as actors which is related to the planning itself. According to Ray Oldenburg [12], the public sphere is an informal meeting place, called the "third place" (bars, coffee shops, shops) that is essential to society and public life and to support the economic vitality of the community. While work is the "second place", the "third place" allows people to put their concerns aside and just enjoy the conversation around them. The third place "host the regular, voluntary, informal, and happily anticipated gatherings of individuals beyond the realms of home and work". This understanding can reinforce the value of daily urban life in research related to public space.

1.2. The park as public space

One of the important functions of public space (parks) is the social function. In the context of urban sociology, public space can be understood as open spaces in urban areas where everyone regardless their interests and backgrounds can be intersectional and social [13]. According to various research, parks play an important role in encouraging physical activity; parks are social bonding nodes between community groups that create a "sense of community" and improve the quality of life of urban communities [13][14][15]. The park also plays a role in reducing social problems, including urban crime [16][17].

Some studies have claimed that in successful space, in this case, public space (parks), the economic impulse factor appeals to people to live, work, recreate and invest in urban areas [14]. Since 2003, the Commission for Architecture and the Built Environment (CABE) has provided evidence showing how parks can offer economic, social, cultural and environmental benefits. Their study also confirms the relationship between high-quality green space (parks) and increased home prices; benefits in enhancing regional image and attracting investment; the contribution of open space to biodiversity; the contribution in promoting sports; the benefits to health; and the role of public space in tackling social issues such as anti-social behavior.

The role of the park has gradually changed since the mid-nineteenth century when it was first introduced as a shelter from increasingly crowded urban centers. A study of the history of green open spaces identified four types of parks based on urban characteristics, social goals, and user base [13]:

(1) "Pleasure park" (1850-1900), where environmental parks are organized to reduce the impact of the industrial cities;
(2) "Reform park" (1900-1930), which gave birth to the park concept for children of working-class families and immigrants;
(3) "Recreation facility" (1930-1965), to extend the concept of recreation to suburban and urban areas with the entry of stadiums, swimming pools, and indoor facilities; and
(4) "Open space system" (1965-1990), where public space trends such as roads, plazas, and waterfronts are used for recreational purposes.

Harnik [18] pointed out that the desire for park utilization declined after World War II because development focused more on the outskirts of the city. As a result, many urban parks have deteriorated due to lack of funds for maintenance and operations. Interest in the development of city parks was renewed in the early 1990s when urban populations began to grow. In the last two decades, social and environmental issues concerning the provision green open space are being discussed [19]. In addition to providing aesthetic value and a place for recreation, city parks play an important role in fostering economic development, improving public health, providing employment opportunities, and being self-sufficient in the use of existing resources [19][20].

1.3. Urban park issues in urban space

The park should host a variety of activities such as relaxing, playing, exercising, eating and drinking, interacting, and so forth [21]. According to Lei and Zhang [22], five main factors affect the quality of park services, namely landscape elements, sports activities, cultural quality, night activities, and traffic support facilities. Research has not seen many studies on the characteristics of locations in the planning of public space. The provision of public space should consider these location characteristics.

The strategic planning of urban green open spaces is insufficient because it is considered as a public space (common property), which is economically unfavorable. The maintenance of urban green open is inconsistent and nonroutine. Urban green open space is often regarded as a waste bin, wild hut, disease-carrying vector nest, and more of a problem than a benefit. Population growth has resulted in a rapid and uncontrolled densification of inhabitants and settlements in the city. This causes increased space requirements to accommodate its interests. The increasing demand for space, especially for settlements and waste, has an impact on the declining quality of the environment. The existing spatial plans are unable to prevent land conversion in urban areas so that the presence of green open space is increasingly threatened and the city becomes increasingly uncomfortable. The majority of green open spaces have been converted into urban infrastructure and new settlement areas. The main problem of the decrease of urban green open spaces is the too limited availability of land and inconsistently applying the spatial plan.

Although the facts state that the presence of urban green open space is far below the ideal proportions, dominant market forces change the function of land, thereby increasingly marginalizing urban green open spaces and even neglecting its functions and benefits. Spatial planning is expected to withstand the market mechanism. Thus, the legal tools governing spatial arrangements should be well implemented by decision makers and the government must be consistent in implementing spatial planning. Moreover, the provision of green open space must be adjusted to the designation specified in the spatial plan. To this end, the

Spatial Planning Law that contains sanctions can be used as a legal umbrella to meet the needs for urban green open spaces. The government should be able to provide green space for the community so as to provide comfort because of its quality environment. The identification of the availability of urban green open space needs to be done so that the government knows the availability of green space as one of the evaluation materials in determining the policy direction and protection of green open space.

1.4. Methods
This study will use a descriptive methodology, by first explaining the policy of public space provision, the existence of the reality of public space, its function and role, the park as public space, the time of visiting the park by the community, and the various problems that occur in it. This research will look at the characteristics of neighborhood parks based on the aspect of the location of park supply and the spatial policies of Bandung City. The analysis comprised ArcGIS (Geographic Information System) and direct observations to evaluate the locational characteristics of the area around the park by using the case study of Music Park and Pendawa Park.

2. Discussion
This section contains the justification of the study area, the characteristics of the region, and an analysis of the provision of the park at the study site. The initial discussion will explain the general image of Bandung, the problems and the distribution of public spaces, especially of parks. Music Park and Pendawa Park will be used as the discussion in the analysis of the provision of parks related to the use of land around the parks. The results of this analysis will demonstrate whether the spatial characteristics of the park location are an important factor in the provision of green open space urban.

2.1 Justification of study areas
Residential parks are selected as an effort to see the existence and utilization of the park as an urban facility that must be prepared based on SNI 03-1733-2004 Concerning Procedures for Urban Housing Urban Planning [23]. Based on the Regulation of the Minister of Public Works No.5/PRT/M /2008 [24], the neighborhood park is an open land with social and aesthetic function as a means of recreational activities, education or other activities at the neighborhood level. The study areas were selected by considering the social function, affordability for the users and the location of the park. The selected thematic parks are Music Park and Pendawa Park in Bandung City.

Bandung was selected as the study location based on the following considerations:
1. Based on Bandung Green Open Space Master Plan Year 2012-2031 [25] the existing park area in the city is 218.07 Ha, while parks are planned to contribute 15.92 percent (2713.9 Ha) of urban green open space requirements. This means that 2495.8 Ha of land needs to be made available for parks in the city of Bandung.
2. The data obtained from the Housing and Settlement Area, Land and Park Office of Bandung City in 2017 states that Bandung has 627 parks (including thematic parks) scattered throughout the city [26]. The provision of thematic parks in residential neighborhoods is the main driver of the new movement for public space that functions as a source of entertainment and recreation for the urban community with attractive physical design and facilities. Ideally, the provision of urban parks targets on the basis of the needs, desires, and problems faced by the urban community based on bottom-up policies [27].
3. The first selected park is a Music Park, located at Sumur Bandung Sub-district, Bandung City. This busy park one of the 20 thematic parks in the city of Bandung and is made for music lovers with a music-themed design. It was inaugurated in 2014 with an area of 2100 m2 and is equipped with Wi-Fi facilities for visitors. The park is unique because of its musical monument as a reminder of the tragedy that occurred in the AACC Building on Asia Afrika Street due to music performances that exceeded the capacity of the building.
4. The second park location is Pendawa Park, located on Pendawa street, Cicendo Sub-district, Bandung City. This park is one of the busy parks in the city of Bandung is equipped with various facilities

including children's play facilities. The park is popular among skateboarders. It is located in a residential area and has an area of 3.900 m2.

2.2 Preliminary description of the provision of parks in Bandung City

The Bandung Urban Spatial Plan 2011-2031, based on Regulation of Minister of Public Works No.5/PRT/M Year 2008 Concerning Guidance of Green Open Space Utilization and Utilization in Urban Areas [24], stipulates that Bandung needs at least 2.858 parks (with a required area of about 1,272,500 m2). This number of parks is made up of 256 Community Unit (RT/RW) parks with a required area of about 644,000 m2; 258 neighborhood parks with a required area of around 322,500 m2; 18 sub-district parks of about 162.000 m2; and 6 urban parks with a required area of about 144.000 m2.

The parks in the city of Bandung have been distributed over all its districts. The Bandung City Department of Housing and Settlement Areas, Land and Landscape (DPKP3) stated that the provision of parks in Bandung is determined by the availability of land owned by the government of Bandung. The aspect of the availability of land belonging to the city government is the basis for the preparation of the Master Plan based on the Spatial Plan and the Detailed Spatial Plans of Bandung. Table 1 below presents an overview of the distribution of parks in Bandung.

Table 1. The number and distribution of parks in the city of Bandung in 2017

Region	Sub-district	Sub-district Size (km^2)	Population in 2016	Population Density (per km^2)	Park Size (m^2)	Number of Parks
Bojonagara	Sukasari	6.27	81,908	13,063	48.356.27	24
	Sukajadi	4.30	108,375	25,203	57.759.53	29
	Cicendo	6.86	99,752	14,541	47.823.83	31
	Andir	3.71	97,553	26,295	44.368.18	15
Subtotal		21.14	387.588	15,580 (average)	198,307.81	19
Cibeunying	Cidadap	6.11	58,672	9,603	9.766.08	8
	Coblong	7.35	131,530	17,895	255.264.72	38
	Bandung Wetan	3.39	31,124	9,981	310.661.75	54
	Sumur Bandung	3.4	36,579	10,759	62.329.44	26
	Cibeunying Kidul	5.25	107,806	20,534	13.516.89	12
	Cibeunying Kaler	4.5	70,924	15,761	32.372.58	13
Subtotal		30	436.635	14,089 (average)	683,911.46	151
Tegallega	Astana Anyar	2.89	68,830	23,817	8.873.61	5
	Babakan Ciparay	7.45	147,096	19,744	3.036.00	2
	Bandung Kulon	6.46	142,411	22,045	6.709.23	3
	Bojongloa Kidul	6.26	85,668	13,685	549.20	4
	Bojongloa Kaler	3.03	120,405	39,738	10.389.83	16
Subtotal		26.09	564.410	23,806 (average)	29,557.87	30
Karees	Kiara Condong	6.12	131,972	21,564	8.866.50	17
	Batununggal	5.03	120,927	24,041	29.173.85	8
	Lengkong	5.90	71,187	12,065	60.126.90	41
	Regol	4.30	81,467	18,946	209.341.14	13
Subtotal		21.35	405.553	19,154 (average)	307,508.39	79
Ujungberung	Antapani	3.79	74,461	19,647	25.439.44	29
	Arcamanik	5.87	69,313	11,808	687.045.69	34
	Mandalajati	6.67	63,578	9,531	92.452.57	20
	Ujungberung	6.40	76,902	12,016	7.067.00	7
	Panyileukan	5.10	40,248	7,892	48.451.65	53
	Cibiru	6.32	72,016	11,395	8.409.00	6
Subtotal		34.15	396.518	12,048 (average)	868,865.35	149

Region	Sub-district	Sub-district Size (km^2)	Population in 2016	Population Density (per km^2)	Park Size (m^2)	Number of Parks
Gedebage	Buah Batu	7.93	95,108	11,993	29.491.20	39
	Rancasari	7.33	76,895	10,490	32.449.81	49
	Bandung Kidul	6.06	58,957	9,729	29.173.85	27
	Cinambo	3.68	25,231	6,856	5.745.00	5
	Gedebage	9.58	37,082	3,871	(updating)	(updating)
	Subtotal	34.58	293.363	8,588	96,859.86	120
	TOTAL	167.31	2.484.037	15,544	2,185,010.74	627

Based on this overview, the population of Bandung City is spread unevenly; Gedebage Sub-district has the lowest density with 3,871 people/km^2, while Bojongloa Kaler Sub-district has the highest density with 39,738 people/km^2. Based on data from the Bandung Central Bureau of Statistics, in 2016, Bandung had a population of 2,484,037, where Babakan Ciparay is the most populous sub-district with 147,096 people, while Cinambo Sub-district has the smallest population with 25,231 inhabitants. The region of Cibeunying has the largest parks availability with 151 parks, with Bandung Wetan is the sub-district with the largest number of parks with 54 parks. Bojonagara is the region with the least number of parks (19), while Cicendo Sub-district has the largest number of parks (31). The analysis shows that the Tegallega is the region with the smallest area of parks with 29,557.87 m^2 (although it has a number of 30 parks), compared with Bojonagara which only has 19 parks. Looking at the area in the city of Bandung shows that the area of Ujung Berung is the largest but the park availability is still below that of Cibeunying. The above analysis also shows that the parks are not evenly distributed over every sub-district in Bandung. This can be seen in Ciparay Sub-district with an area of 7.45 km^2 but only two parks, compared to Buah Batu Sub-district which has an area size of 7.93 km^2 and 39 parks).

2.3 Land use analysis of Music Park

Music Park is one of the thematic parks in Bandung. It is located on Belitung street, Merdeka Urban Village, Sumur Bandung Sub-district and has an area of 4.200 m2. Based on data from the Central Statistics Agency of Bandung in 2016, Merdeka Urban Village has an area of 1.40 km2, 9 Community Units (RW), 58 Neighborhood Units (RT) with a population of 8,697 (24.22% of the population of Sub-district Sumur Bandung), a population density of 6.12 people/km2 and a population growth rate of 0.427%. Merdeka Urban Village is dominated by settlements and has military complexes, educational facilities, and trade facilities of goods and services. The Bandung Spatial Detail Plan 2015-2025 [28] strongly supports the sustainability and existence of the Music Park in Merdeka Urban Village. These policies include the development of road access networks and light railways in anticipating congestion problems, structuring services and trading linear (informal sector activities/street vendors, pedestrian infrastructure including disability and bike lanes integrated with mass transport, providing space for small and medium enterprises are part of the development of government offices, trade, and services zone), and the policy of upgrading the electricity network (for park lighting). Figure 1 shows the land use around Music Park, while Table 2 shows the extent of land use around the park.

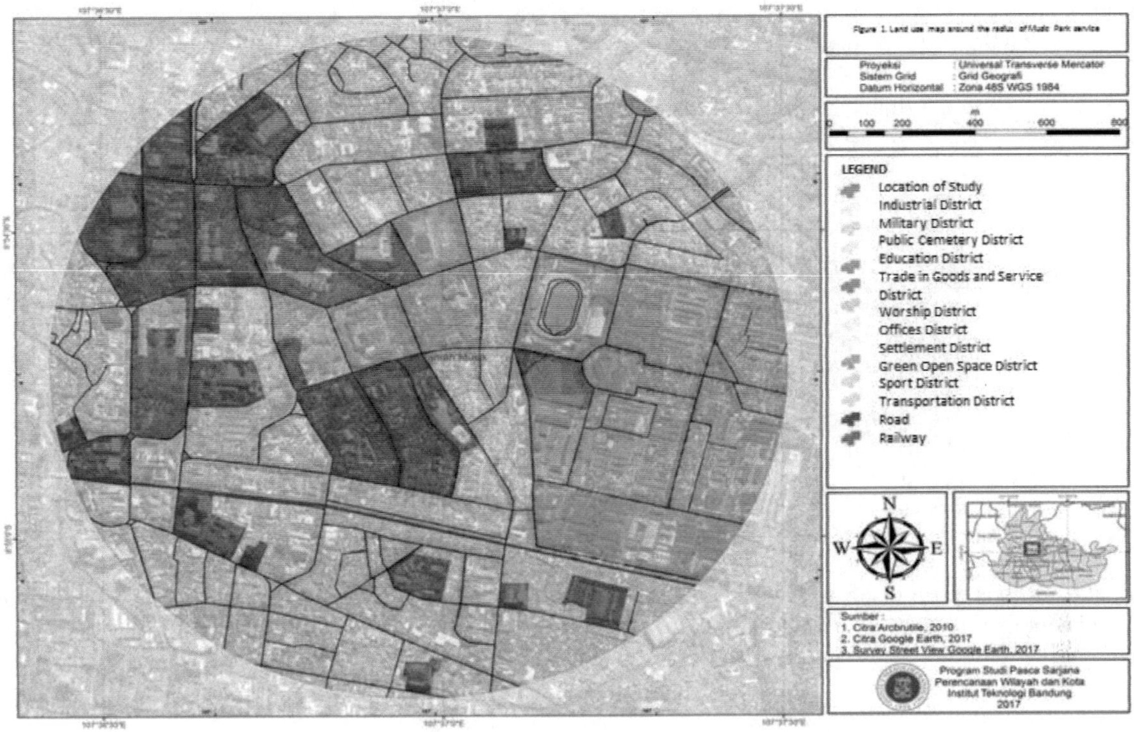

Figure 1. Land use around Music Park

Table 2. Land use around Music Park

No.	Information	Area (m²)	Area (ha)	Area Percentage (%)	Number of Buildings
1	Roads	150,278.88	15.03	4.78	-
2	Railway	15,429.56	1.54	0.49	-
3	Military Facility	554,207.53	55.42	17.62	1115
4	Education	217,124.34	21.71	6.90	133
5	Trade in Goods and Services	305,375.35	30.54	9.71	385
6	Worship	6,362.05	0.64	0.20	2
7	Offices	86,310.89	8.63	2.74	18
8	Settlements	1,553,716.51	155.37	49.40	16,787
9	Green Open Space	149,510.96	14.95	4.75	-
10	Sports	103,782.77	10.38	3.30	10
11	Study Location	3,066.11	0.31	0.10	-
	Total	3,145,164.95	314.52	100.00	18,450

The spatial analysis using ArcGIS shows that the land use around Music Park is dominated by residential area, 155.37 ha (49.40%) and military complexes, 55,42 ha (17,62%). The strategic location of the park, close to educational facilities and settlements offers great potential in presenting a beautiful environment, a place to socialize, and the location for the surrounding community to improve their economic level. Observations in September showed that the intensity of use of Music Park is highest from 14:00 to 18:00, while the lowest use is between 23:00 and 09:00, except on Saturday when the intensity of use is highest at 20.00-21.00. The pattern of intensity of use of Music Park shows that the high intensity of use of Music Park coincides with the school hours of high school students of SMA 3 and

SMA 5. Moreover, the park is used as a place to gather, study, relax, eat and play. The observation results related to the intensity of space use in Music Park can be seen in Table 3 below.

Table 3. Intensity of use of Music Park based on time of visit (per day/per hour)

Day	Time (00.00-23-00)																						
	0	1	2	3	4	5	6	7	8	9	10	11	12	13	14	15	16	17	18	19	20	21	22	23
Sunday																								
Monday																								
Tuesday																								
Wednesday																								
Thursday																								
Friday																								
Saturday																								
Estimation																								

Note:
Red: High (>50 people)
Yellow: Medium (20-50 people)
Green: Low (<20 people)

The analysis result of Music Park shows that the location of the park is in accordance with its function as a neighborhood park because most of its land use around the park consists of residential neighborhoods. However, the intensity of its usage time shows Music Park is still not utilized optimally.

2.4 Land use analysis of Pendawa Park

Pendawa Park is one of the non-thematic parks in Bandung, is located on Pendawa Road, Arjuna Urban Village, Cicendo Sub-district and has an area of 3,000 m2. Based on data from the Central Statistics Agency of Bandung in 2016, Arjuna Urban Village has an area of 63 ha and consists of 8 Neighborhood Units (RT), 68 Community Units (RW) and has 15,133 inhabitants. Arjuna Urban Village is dominated by settlements and other urban land use such as educational facilities and trade facilities of goods and services.

The Bandung Urban Spatial Plan 2015-2025 [29], related to Arjuna Urban Village strongly supports the sustainability of Pendawa Park in the present and in the future. This is evident from the development of the road network in anticipation of congestion problems around the location of the park, the arrangement of trade and linear services (informal sector activities/street vendors, pedestrian infrastructure including disability and bike paths integrated with mass transportation, and medium enterprises are part of the development of government offices, trade and services zones). Figure 2 shows the land use around Pendawa Park, while Table 4 shows the extent of land use around the park.

The spatial analysis using ArcGIS shows that the use of land around Pendawa Park is dominated by residential settlements with an area of 240,48 ha (79,46%), and trade and service 10.76 ha (3,42%). The strategic location of the park, close to educational facilities and settlements shows Pendawa Park's great potential in presenting a beautiful environment, a place to socialize and a location for the surrounding community to improve their economic level.

The observation in September showed that the intensity of the use of Pendawa Park was highest between 13.00-16.00 and lowest between 18.00-11.00. Except on Saturday and Sunday, the intensity of park visits was low. The low use of the park in the afternoon until the evening was caused by the lack of activity in the settlements of Arjuna Urban Village. One of the factors influencing the low the intensity of use at night is the absence of park lighting. This makes the Arjuna urban areas vulnerable to criminal acts such theft and robbery. The pattern of intensity of the use of Pendawa Park shows that the high use of the park coincides with the time the elementary school students of SD Kresna finish school. Moreover, the park is used to gather, relax, eat, exercise, and play. The intensity of use of space of Pendawa Park can be seen in Table 5.

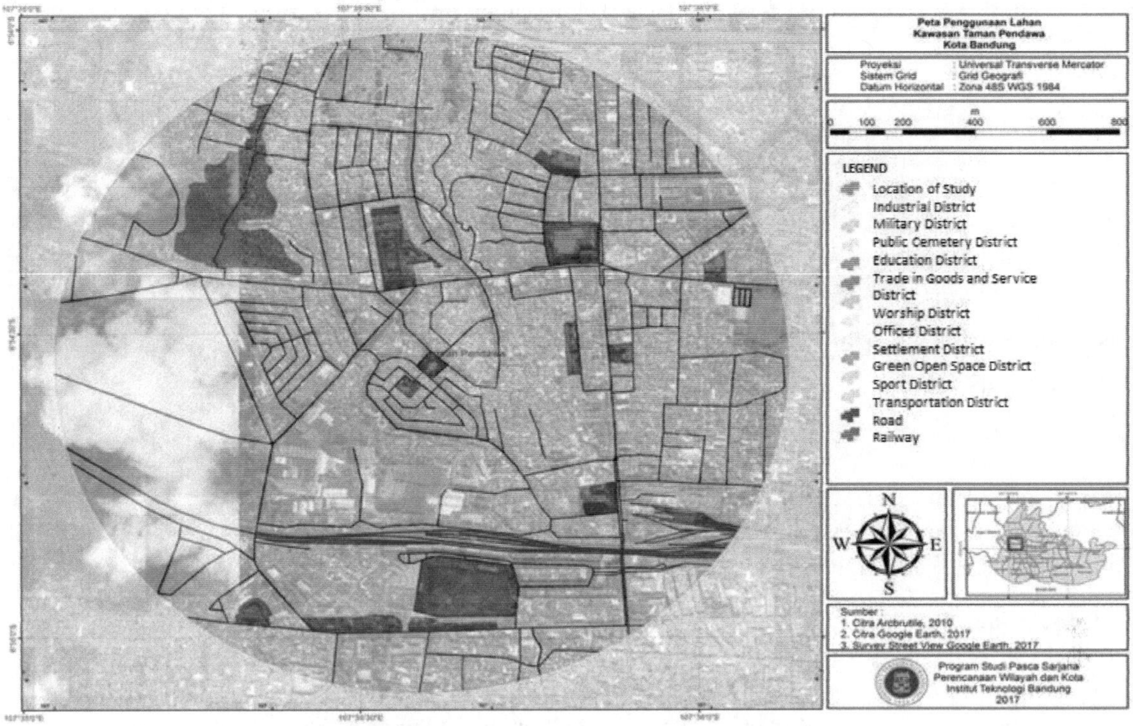

Figure 2. Land use around Pendawa Park

Table 4. Land use around Pendawa Park

No.	Information	Area (m²)	Area (ha)	Area Percentage (%)	Number of Buildings
1	Road	169,658.19	16.97	5.39	-
2	Railway	46,406.91	4.64	1.48	-
3	Industry	76,768.60	7.68	2.44	67
4	Public Cemetery	72,409.99	7.24	2.30	-
5	Education	32,532.54	3.25	1.03	30
6	Trade in Goods and Services	107,572.36	10.76	3.42	75
7	Offices	48,420.10	4.84	1.54	34
8	Settlements	2,404,772.39	240.48	76.46	28,708
9	Green Open Space	81,928.29	8.19	2.60	-
10	Sports	23,568.89	2.36	0.75	2
11	Transportation	77,941.46	7.79	2.48	28
12	Study Location	3,071.71	0.31	0.10	-
	Total	3,145,051.43	314.51	100.00	28,944

Table 5. Intensity of use of Pendawa Park based on time of visit (per day/per hour)

Day	Time (00.00-23-00)																						
	0	1	2	3	4	5	6	7	8	9	10	11	12	13	14	15	16	17	18	19	20	21	22	23
Sunday																								
Monday																								
Tuesday																								
Wednesday																								
Thursday																								
Friday																								
Saturday																								
Estimation																								

Note:
Red: High (>50 people)
Yellow: Medium (20-50 people)
Green: Low (<20 people)

Based on the analysis of Pendawa Park, the location of the park has been in accordance with its function as a neighborhood park because most of the land use around the location residential housing. However, viewed from the intensity of its usage times, Pendawa Park is not used optimally.

3. Conclusion

Based on the study, the following conclusions can be drawn:

1. Parks in the city of Bandung are distributed to all districts in the city but the analysis shows that parks are unevenly distributed over the sub-district in Bandung. This can be seen in the sub-district of Ciparay, which has an area of 7.45 km2 but the amount of park is small with two parks. In contrast, Buah Batu Sub-district has a total area of 7.93 km2 and 39 parks.
2. The Bandung Detailed Spatial Plan 2015-2025 strongly supports the sustainability and existence of the Music Park in Merdeka Urban Village. The policies in the Spatial Plan include the development of road access networks and light railways in anticipating congestion problems, structuring services and trading linear (informal sector activities/street vendors, pedestrian infrastructure including disability and bike lanes integrated with mass transport, providing space for small and medium enterprises are part of the development of government offices, trade, and services zone), and the policy of upgrading the electricity network (for park lighting).
3. The spatial analysis using ArcGIS shows that the land use around Music Park is dominated by residential area, 155.37 ha (49.40%) and military complexes, 55,42 ha (17,62%). The strategic location of the park, close to educational facilities and settlements offers great potential in presenting a beautiful environment, a place to socialize, and the location for the surrounding community to improve their economic level.
4. Observations in September showed that the intensity of use of Music Park is highest from 14:00 to 18:00, while the lowest use is between 23:00 and 09:00, except on Saturday when the intensity of use is highest at 20.00-21.00. The pattern of intensity of use of Music Park shows that the high intensity of use of Music Park coincides with the school hours of high school students of SMA 3 and SMA 5. Moreover, the park is used as a place to gather, study, relax, eat and play. The analysis result of Music Park shows that the location of the park is in accordance with its function as a neighborhood park because most of its land use around the park consists of residential neighborhoods. However, the intensity of its usage time shows Music Park is still not utilized optimally.
5. The observation in September showed that the intensity of the use of Pendawa Park was highest between 13.00-16.00 and lowest between 18.00-11.00. Except on Saturday and Sunday, the intensity of park visits was low. The low use of the park in the afternoon until the evening was caused by the lack of activity in the settlements of Arjuna Urban Village. One of the factors influencing the low the intensity of use at night is the absence of park lighting. This makes the Arjuna urban areas vulnerable

to criminal acts such theft and robbery. The pattern of intensity of the use of Pendawa Park shows that the high use of the park coincides with the time the elementary school students of SD Kresna finish school. Moreover, the park is used to gather, relax, eat, exercise, and play. The Based on the analysis of Pendawa Park, the location of the park has been in accordance with its function as a neighborhood park since most of the land use around the location residential housing. However, viewed from the intensity of its usage times, Pendawa Park is not used optimally.

6. Location characteristics are important (in addition to landscape elements, sports activities, cultural quality, night activities, and traffic support facilities) in the provision and planning of public spaces (parks). However, the study shows that the physical characteristics of the neighborhood around the park location should be supported by looking at the social characteristics of the community as users of the park. This can be seen in the use of space in Pendawa Park; although the land use is more dominated by the settlement area than Music Park, the field observations showed that its use is ineffective.

Acknowledgement
The authors are grateful for the support of the promoters Prof. Roos Akbar, M.Sc., Ph.D. Sri Maryati, ST., MIP in making this paper, and for the support from the Indonesian ministry of higher education which has assisted the author in financing the study and completion of the paper. The authors also thank the Bandung City Spatial Planning Department, and Merdeka Urban Village and Cicendo Sub-district Offices that provided assistance in the form of information and data for the analysis of this paper.

References
[1] Bintarto R 1984 *Interaksi desa-kota dan permasalahannya* (Bogor: Ghalia)
[2] UN-Habitat 2015 *Adequate open public space in cities, A human settlements indicator for monitoring the post-2015, sustainable development agenda* (New York: UN-Habitat)
[3] Garau P 2017 The biennial of public space in Rome. from the charter of public space to the post-Habitat III agenda *The J. Public Space* **2** 157-160
[4] Raipat V 2016 Re-imagining public spaces for sustainability: Ranchi, India *Cities People Places: An International Journal on Urban Environments* **1** 30-57
[5] Carmona M, Heath T, Taner O, and Tiesdell S 2003 *Public places – urban spaces, the dimension of urban design* (Oxford: Architectural Press)
[6] Lofland L 1973 *A world of strangers: order and action in urban public space* (New York: Basic Books)
[7] Acasandre, A. 2014 Central public spaces – neighborhood public spaces *The Macrotheme Review* **3** 36-51
[8] Lehtuvuori P 2010 *Experience and conflict: The dialectics of the production of public urban space in the light of new event venues in Helsinki 1993–2003* (Sweden: Helsinki University of Technology)
[9] Gottdiener M 1985 *The social production of urban space* (Texas: University of Texas Press)
[10] Healey P 1997 *Collaborative planning, shaping places in fragmented societies* (Basingstoke: Macmillan).
[11] Healey P 1993 *The argumentative turn in policy analysis and planning* ed(s) Forester J and Fischer F (Durham, London: Duke University Press) pp 233-253
[12] Oldenburg R 1991 *The great good place* (New York: Marlow)
[13] Roberts N 2014 *The modern urban park: access and programming; where have we been... where shall we go?* (San Fransisco: San Francisco University)
[14] Commission for Architecture and the Built Environment. 2005 *Start with the park: creating sustainable urban green spaces in areas of housing growth and renewal*, (London: CABE)
[15] McCormack G R 2010 Characteristics of urban parks associated with park use and physical activity: a review of qualitative research *Health and Place Journal* **16** 712-726
[16] Byrne J and Wolch J 2009 Nature, race, and parks: past research and future directions for

geographic research. *Progress in Human Geography* **33 743-765**

[17] Chitrakar R M, Baker D C, and Guaralda M 2016 Urban growth and development of contemporary neighbourhood public space in Kathmandu valley *Habitat International* **53** 30-38

[18] Harnik P 2006 *The excellent city park system: What makes it great and how to get there. The Human Metropolis: People and Nature in the 21st-Century City* ed Platt R H (Amherst, Boston, Cambridge: University of Massachusetts Press, Lincoln Institute of Land Policy)

[19] Cranz G and Boland M 2004 Defining the sustainable park: a fifth model for urban parks *Landscape Journal* **23**102-120.

[20] Walker C 2004 *The public value of urban parks* (Washington, DC: The Urban Institute)

[21] Amelie E W 2003 Future Park: Imagining Tomorrow's Urban Parks (Clayton: CSIRO Publishing)

[22] Zhang Y and Feng L 2012 Evaluation of urban park service quality based on factor analysis, *Int. J. Computer Science Issues* **9 317-321**

[23] Procedures for Urban Housing Urban Planning. Gov. Regist. SNI 03-1733-2004

[24] Guidelines for the Provision and Utilization of Green Open Space in Urban Areas. Gov. Regist. 5/PRT/M /2008

[25] Bandung Green Open Space Master Plan Year 2012-2031. Bandung Gov. Regist. 2011

[26] Bandung City Government 2017 Housing and Settlement Area, Land and Park Office of Bandung City [Online] Available from: http://diskamtam.bandung.go.id/informasi/pertamanan

[27] Muhammad A M, Zulkaidy D, Pratiwi W D 2016 Evaluasi dampak penyediaan taman-taman tematik kota bandung berdasarkan persepsi masyarakat sekitar. *Proc. Temu Ilmiah IPLBI*

[28] Bandung Spatial Detail Plan 2015-2025. Bandung Gov. Regist. [regulation on the Internet] 10/2015. Available from: http://bkprd.jabarprov.go.id/sources/hukum/daerah/d0f79-perda-rdtr_-pz.pdf

[29] Bandung Urban Spatial Plan 2011-2031. Bandung Gov. Regist. [regulation on the Internet] 18/2011. Available from: http://www.jdih.setjen.kemendagri.go.id/files/KOTA_BANDUNG_18_2011.pdf

IOP Conf. Series: Earth and Environmental Science **158** (2018) 012029 doi:10.1088/1755-1315/158/1/012029

Natural Resource Management based on Gender Perspectives and Integrating Traditional Ecological Knowledge of the Tepera in Jayapura, Papua

W A Rumbiak[1], E V Wambrauw[2]

[1]WWF Indonesia Program USAID LESTARI
[2]Universitas Cenderawasih, Papua, Indonesia

Corresponding Author: wrumbiak@wwf.or.id

Abstract. The Tepera in Jayapura Regency have a traditional ecological concept of managing their natural resources which evolved over generations. The spatial concept of their resources management is recorded visually on mental maps. The existing conditions of the landscape, forest, coastal area, and sea are considered heritage and have economic, ecological, and cultural values. The people have their own perspectives on the relationship between the resources management, cultural values, gender perspectives, and development. Thus, this research aims to identify the gender perspective in the natural resource management and environmental services; and to analyse the sustainable pattern of the land use and cultural zoning in the resources management. The methodology comprises grounded research and Participatory Action Research. This research has three findings, i.e., the tribe named the landscape; they have developed a zoning system to manage the forest traditionally; and there is a difference in perception between men and women regarding the type of forest and landscape related to food and traditional medicine sources. Therefore, it is important to incorporate the concept of managing the environment and the cultural zones of the Tepera in the programs of the local government to direct the development in sustainable way. In addition, the female participation in managing the environment should be improved, especially related to domestic aspects.

1. Introduction

Papua has an abundance of natural resources [1] and cultural diversity; it has at least 269 ethnic groups, the same amount of living local languages [2] and 1,068 clans [3]. Therefore, the perspectives of the indigenous people should play an important role in the decision-making of the development in the area.

On the provincial level, the government has developed the Mid-Term Regional Development Plan (RPJMD), a guidance that has considered the importance of culture by dividing Papua into five cultural zones based on geographical areas, i.e., Mamta, Ha-Anim, Lepago, Meepago, and Saireri. One of the tribes is called the Tepera which lives in the Mamta Area in Jayapura Regency. In Indonesian, the Tepera are also called Dewan Adat Suku Tepera (DAS) which mean tribal association. This tribe lives in the Cyclops Mountains together with other Dewan Adat Suku Tepera/DAS, the Ormu, Moy,

IOP Conf. Series: Earth and Environmental Science **158** (2018) 012029 doi:10.1088/1755-1315/158/1/012029

Sentani and Humbolt (Pt. PPMA, 2000). The map of the Cyclops Mountains Nature Reserve with the DAS can be seen in Figure 1.

The hierarchical decision-making process of the tribe follows the Ondoafian system (inherited power of leadership), although the formal institution of the tribe has not been formed yet.

Generally, the indigenous people of Papua depend on the forest [3]. Therefore, deforestation has an immeasurable impact on them, which is why the management of the forest and natural resources should involve these indigenous people to reduce the negative impact of development on them. The Tepera have their own ecological concept to manage the natural resources that considers the abiotic, biotic, and cultural aspects, which has been developed intergenerationally. Both women and men understand the knowledge based on their perspectives to name their landscape environment. The relationship between the culture of people and the landscape is called geo-cultural landscape [4]. Mostly indigenous people have a relationship with the landscape and the environment such as in Australia [5] and in New Zealand [6], the Malind–Anim in Merauke [3], and the Asmat [7]. Cognitive maps describe the knowledge of an individual of the landscape, the environment, and the area, which generally is sketched on a sheet of paper, sand, or natural material [8]. This cognitive image is developed from the notion of the imaginary map by [9]. Moreover, **Error! Reference source not found.** stated that mental maps shape the human perspective of the environment. Thus, based on the mental maps, people understand the areas they interact with. Mental maps generally describe the behaviour of people to navigate their neighbourhood, city, country, or physical space. The actions of humans and their orientation shape the geographical knowledge.

Both the Tepera women and men understand the landscape and know how to play their role in managing their ecosystem. This depends on their task and daily activities; a World Bank report stated that the rural men and women differentiate their task and responsibility to maintain the biodiversity [11]. However, due to the patriarchy system, sometimes it places men higher than women in the society, including in the decision-making that is associated with land use and environmental management. A study that sets out to analyse property right without examining gender asymmetries in rights, responsibilities, and accessibility leads to partial understanding and incorrect conclusions [12]. Thus, this research aims to understand how the traditional lore guides the local people to maintain the nature and how the traditional values shape the environment. Therefore, the perspective of women should be involved in the decision-making because women are vulnerable to the impact of the environment. Finally, the research focuses on the traditional ecological knowledge of the Tepera to manage and mention the ecosystem from a gender perspective.

2. Study Area

The Cyclops Mountains have been established as a nature reserve since 1978. It is located in two regencies (Jayapura Regency and Jayapura City) with a total area of 31,479.84 ha, and 2160 m above sea level [13]. The government also established several conservation areas through Decree of The Forestry Ministry Number 782/Menhut-II/2012 and renew the status of the Cyclops Mountains as a nature reserve. According to the Government Regulation No 20/2011 article 33, the conservation status is not only for conservation but also for other purposes such as education and research, and carbon storage. In fact, the Cyclops Mountains supply water to Jayapura and the Sentani region. At the same time, the purpose of establishing the nature reserve is to protect the environment in a sustainable way. However, before the government designated the Cyclops Mountains as a nature reserve, indigenous people have been living in the area for a long time.

The authors conducted the study of the Tepera who live inside the conservation area (see Figure 2). The participants who engaged in the research mostly have similar social and cultural characteristics.

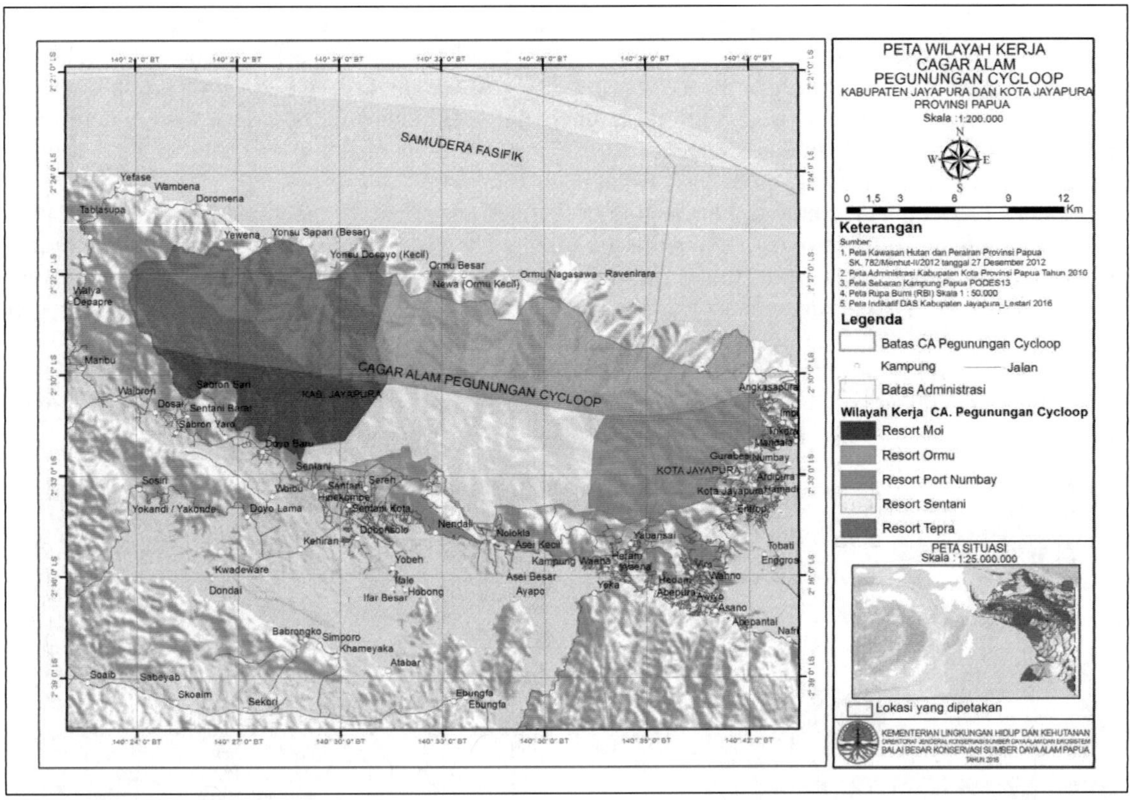

Figure 1. The Cyclops Mountains Nature Reserve (Balai Besar Konservasi Sumber Daya Alam Papua.
Source: Bureau for the Conservation of the Natural Resources of Papua Province, 2016).

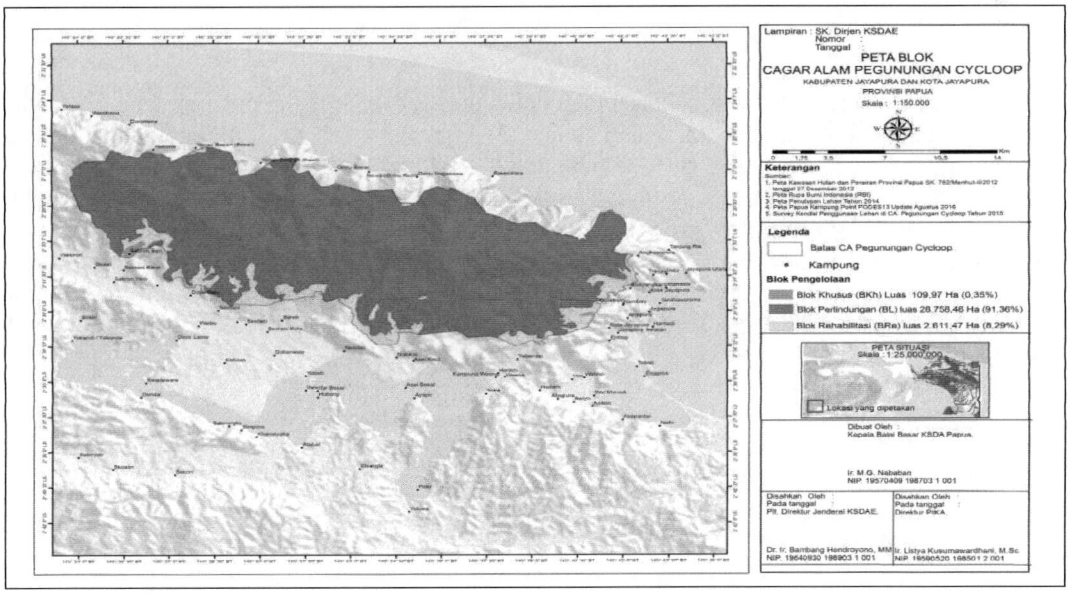

Figure 2. The Nature Reserve blocks of the Cyclops Mountains.

3. Methodology

The study is a qualitative research using the grounded approach where the researchers and the team conduct direct observations of the community. The guidelines of the research questions helped the authors to identify the concepts in the community, making it possible to compare properly the empirical approaches. In this study, the researchers collected data in a homogenous community and the respondents provided similar points of view in answering the questions. Thus, the conclusions are representative of the people a whole.

Besides the grounded method, the research comprised participatory mapping. The guidance for the mapping is produced to arrange the steps in the research such grouping based on family members, to ensure the valid representation of the family name, gender, and the customary strata. The researchers also introduced to the participants how to produce a sketch of the planned map. Thus, the participants can understand the basic knowledge in drawing their own landscape. Finally, the participants were taught the general knowledge of the Global Positioning System (GPS), so they can draw their own map of natural resources based on their perspective. Figure 3 shows the process of participatory mapping.

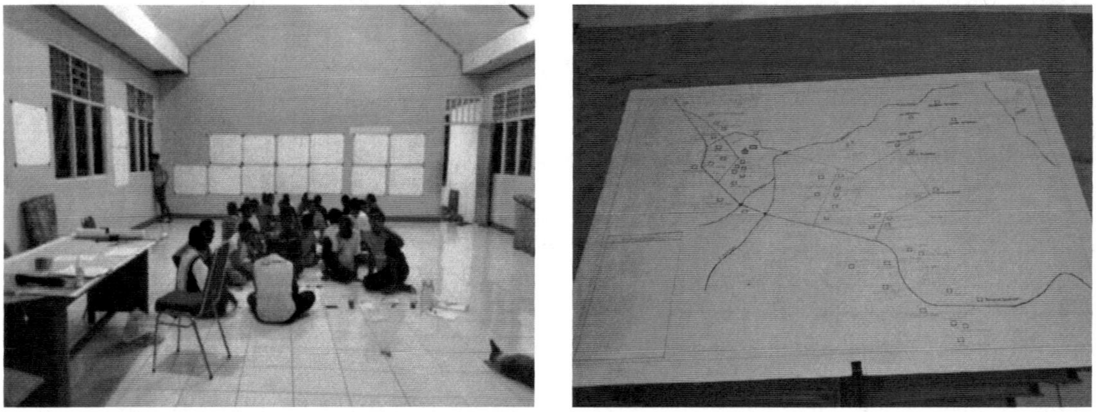

Figure 3. the process of participatory mapping.

3.1. Data

The research used primary and secondary data collection. To collect the primary data, the field study was conducted in five villages, named Yewena, Wambena, Tablanusu, Yepase, and Dormena. In Yewena, there were 51 respondents, nine people who did mapping, and two people did the drawing, while five people helped to track the areas with GPS. In Wambena, there also were 51 respondents, while ten people did mapping and two who did the drawing. There were also four informants and four people checked the location with GPS. Meanwhile, in Tablanusu, Yepase, and Dormena, there were 91, 33, and 86 respondents respectively. To map Tablanusu and Yepase, nine were people employed and ten in Dormena. The field study was conducted in November 2016. Figure 4 shows several pictures to support the information.

Figure 4. Field study during the research.

Secondary data was collected from previous studies, either published or unpublished reports. The primary data was collected by the researchers through direct observations of the physical structure, and the different perspectives and attitudes on how the Tepera manage and name the landscapes. The research also comprises Focus Group Discussion and participatory mapping. To collect information, in-depth interviews were held with key informants consisting of the elderly, women, leaders, and youth. Each interview was conducted separately to avoid interference from other community members. Next, Focus Group Discussion was done to discuss the general issues in society related to the topic of the research. Then, participatory mapping was employed to understand the patterns of how the Tepera use and manage the resources and environmental services. In this case, the key persons drew the pattern of the land tenure such as the pattern of communal land, customary land, and land permitted to use.

3.2. Analysis Data

After collecting information, the data was analysed qualitatively in four steps. First, data reduction, where the researcher selected the important and useful information and concluded into essential substance. The information is representative of the elders, the village leaders, the community leaders, and the people who understand their local wisdom. The second process was to categorize the information into similar patterns and name it as unit information. The third step was the synthesis process to connect the similarity of each information unit and label it as "theme" (topic). Lastly, all the information was used to answer the research question that was supported by the previous theory to find a novel concept.

4. Results

The Tepera believe that the earth is like a mother who feeds her children with a part of her body, such as the sea and the land. In their vernacular, this is called *Kani nekewena delrei Nameng. Nau su plre, nemeng nei kong bwo. De dalrite telrena de walri, de naute telrena de wari*

As explained before, the geo-cultural landscape shows the relation between the humans and the landscape. The people's points of view influence the names of the landscape. Geographers are keen on how people utilise their nature [14]. The study shows that the tribe names the landscape as *seko* (mountain), *meko bura* (hill), *eri bura* (slope), *paw* (valley), *meroway bu* (river), *buru rena bu baruna*

IOP Conf. Series: Earth and Environmental Science **158** (2018) 012029 doi:10.1088/1755-1315/158/1/012029

(spring), *kingkey bueaenabu* (water fall), *meraway kera-kera* (dried river), *yemo* (cape), *yepa* (beach), *nau* (sea) and *kepou* (sea stones). These terms occur everywhere in the world. However, there is a difference between how men and women name the landscape (see Table 1).

Table 1. Landscape categorisation based on perspective gender.

Landscapes	**Named by men**	**Named by women**
mountain	seko	sero
hill	meko	meko
river	meroway	meroway
lake	sensori	sensori
valley	pau	pau
village	yo	yo
old village	yo sena	yosena
water spring	butou	butou
beach	boru	
rocked beach		boru
sea		nau
cape		yemo
bay		duku pu
coral reef		sekare
estuary		buperu
sacred place		tempat keramat

Table 1 shows that women have more names for the landscape than men. It seems that women are more connected to the environment in comparison with men. The understanding of the physical environment of the people who live around the Cyclops Mountains is closely related to how they use the forest, water, land, and resources to support life. It is known that primarily women who do domestic household jobs are close to the environment. The Tepera name the types of forest as in Table 2. The people classified the type of the forest according to how they use the forest to support life.

Table 2. The types of forest from a gender perspective.

Type of Forest	Mentioned by men	Mentioned by women
primary forest	sena	sena
secondary forest	sena seke	sena seke
garden that no use anymore	seke	seke
garden	emi	emi
previous garden		emi seke
dusun (a place in the forest often used for gathering sago and natural resources)	pau	pau
village		yo

4.1. The customary law in the natural resources management based on the types of landscape

The Tepera value the mountain as a food basket, medical resource, and as a source of material for infrastructure. They believe if the forest is still in a good condition, which means having many big trees, this reflects positively on human beings. It also benefits the environment to reduce landslides and flooding and offers sustainability of human life and the ecosystem. Most indigenous people are inseparable from nature; they understand how to take care of Mother Nature by producing a customary law to protect their environment. First, this law prohibited cutting down the trees in the forest around the mountains for unnecessary purposes because it would reduce the fertility of the soil. This was followed by a prohibition to cut down the trees on the mountain due to a reduction in the number of animals living in the area. The people are also not permitted to cut down the trees on the mountain that can decrease the variety and the number of the medicinal plants. They are concerned with the disasters that are affected by unwise human behaviours, so they should not cut down the trees on the mountain slope. They understand that his action would lead to an increased risk of landslides and flooding. In addition, they try to protect the forest around the mountain to avoid draught of the river.

If a member of the community breaks the law, she or he will obtain social punishment. For example, the person who breaks the customary law will be judged by the people, which influences the psychology of the person. Even worse, the Ondoafi (the leader of the tribe) will remind the person. The tribe also has a hierarchy of resource management; besides the Ondoafi, a person responsible for the economic aspects or welfare will handle the problem and will call the person who broke the law.

Beside the regulations related to the land, the Tepera also manage the water. They call river *Meroway* and water springs as *Butou*. They are very concerned with the waterways as they understand that one of the basic need of the life is water. There are several roles in managing the water in rivers, dried rivers, waterfalls, or water springs. For example, they do not cut down the trees on riverbanks that decrease the watershed level. The tribe is also not allowed to build houses along the river, above the water spring, or above the sea because they understand the risk to either humans or to the environment, especially in the rainy season. They also recognize the importance to protect the water bodies from solid waste or garbage because it causes not only flooding in rainy season but can reduce the resource from the water.

Another important landscape of the Tepera is the village. There are two types of village, the place people live in called yo (village), and the old village (yo sena) that the Tepera believe is a place where the ancestors used to live. Usually, it has been left by the villagers and returned to the forest. The old villages have historical and sacred values. For a village, in general, the customary law states that a villager has the right to build a house in the village according to the family name, but the people are not permitted to build on the riverbank and only one family can live in the house.

Next, the Tepera are strongly connected with marine landscape ecology such as capes, beaches, the sea, and rock beaches. The Tepera understand these as places of resources. They call a cape *yemo,* a beach *yepa*, a sea *nau,* and a rock beach *boru*. They know that the places that are rich in fish are around the rocks in the sea (*boru)*. The areas have high economic values, so the Tepera mention food resources, medicinal resources, and sources of livelihood of the people. They also have a concept to maintain the sustainability of their natural marine resources, for instance, it is forbidden to use the resources for a certain period, called Tiyatiki [15]. Another role in the community is that if a member of the community catches fish, the person should share with the others; the Ondoafi (tribal leader) will help to distribute the fish to the community. Moreover, the people of the tribe are not allowed to throw garbage in the sea as their main natural resource. Next, the beach and the cape in Wambena Village have historical and sacred values related to the first missionary who came to the area. Lastly, it is not permitted to build a house above the sea, seashore, or cape.

Finally, a sacred place is called *ayenekayenanou*, a place with historical and sacred values. Only a few people can visit this place at a certain time.

4.2. The tribal roles to use the natural resources in the forest

The primary forest is called *sena* and is a place for native animals and an abundance of biodiversity. The areas of the primary forest are protected to sustain the life of the animals, plants, and, of course, the people. One of the customary regulations states that no hunting activity is allowed in the primary forest. The people can only hunt in the secondary forest, called *seke.*

The secondary forest is a good place for gardening because most of the soil is fertile and suitable for planting fruits and vegetables. Generally, the plantations are located on a slope are, to prevent damage by boars. Several roles are involved in this type of forest. First, to open a garden, the villager should get a permit from the customary government that has the responsibility for economic affairs. Then, the Tepera have a *sasi* that means it is forbidden to use the resources for a certain amount of time. Thus, to reuse the garden in this zone, people should wait for the *sasi* time, which lasts 5 to 10 years because this influences soil quality. Moreover, the people are not permitted to plant in waterways or in areas that are prone to landslides. Also, no one may hunt the Bird of Paradise except the Ondoafi. The people who hunt in this zone should share their prey. Also, they should understand the consequences of what they do, for example, the future negative impact of changing land use.

The land that is planned to be a garden is called *moko*. It is a place for food resources, so the people value it as an important place to provide their daily needs. The customary regulation is quite like the secondary forest. Both types of forest are responsible to provide resources for the people. Thus, the Tepera have developed a zoning system to manage the forest traditionally, called *sena, seke, and moko*. They developed based on the function of the forest to support life.

5. The gender perspective in the geo-cultural landscape of the Tepera.

The Tepera have an ecological concept to manage the natural resources with consideration of the abiotic, biotic, and cultural aspects, which has been developed over generations. The spatial concept of the resources management is recorded visually on mental maps. The existing condition of the landscape, forest, coastal areas, and the sea is considered heritage that has economic, ecological, and cultural values. The people have their own perspectives on the relationship between the resources management, cultural values, gender perspectives, and development. Both women and men understand the knowledge based on their perspectives.

As explained in the previous section, based on gender issues, the men understand nine types of landscape related to their daily activities, i.e., mountain (*seko*), hill (*meko*) river (*meroway*), lake (*sensori*), valley (*pau)*, village (*yo*), old village (*yo sena*), water spring (*butou*), and beach (*boru*). Meanwhile, the women understand 15 kinds of landscapes that associated to their food resources, i.e., mountain (*sero*), hill (*meko*), river (*meroway*), lake (*sensori*), valley (*pau*), village (*yo*), old village (*yosena*), water spring (*butou*), rocky beach (*boru*), sea (*nau*), cape (*yemo*), bay (*duku pu*), coral reef (*sekare*), and estuary (*bu peu*).

Based on the USAID Lestari study, the participants of the Tepera were asked to develop a weighted ranking of the most important landscape based on their personal perspectives [16]. The men valued the most important landscapes for them to be the mountain, followed by water spring, the river, the village, and the sea. Meanwhile, the most important landscapes for women are the river, village, mountain, and valley.

A priority of the people for a landscape is related to their interest; it also depends on their needs and daily activities. For example, the landscape men most prioritize is the mountains. Probably, this is the case because men go to the forest more because one of their jobs is hunting. In comparison to the women, they said the most important is the river. Women are inseparable from water when they are doing their domestic tasks.

If we compare the ranking of the forest between men and women based on the function of the nature to support life, according to the study, men chose the primary forest as the most useful landscape to provide resources but the women chose the valley or the *dusun* as a place for food resources[16].

In our understanding men are more concerned with the landscaped that is associated to the land tenure, while women are associated with domestic tasks. This is influenced by the social strata in the community. Sometimes, the poor condition occurs that men are considered superior to women, so the female perspectives are not accounted for in the decision-making. The government should involve women in the planning process. If there is a mistake in planning, the women will face a problem or issue first. For example, if the government develops the conservation area in an unsustainable way, the water levels decrease significantly. If this occurs in the dry season, the first victims are women and children. Thus, local government should involve both women and men to make a decision regarding the planned development of an area.

6. The Traditional Ecological Knowledge in Using the Natural Resources

Based on the report of USAID Lestari [16], the Tepera named their own sacred and important places, the results of this can be seen in Figure 5.

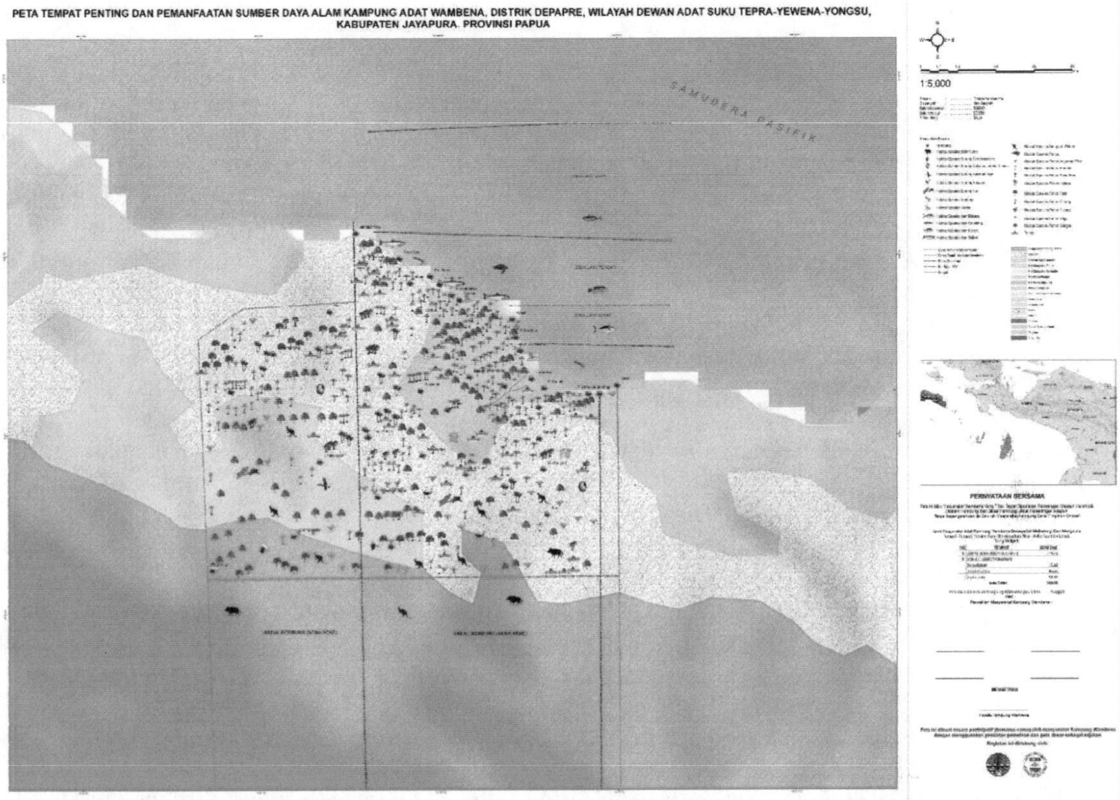

Figure 5. Important places for the Tepera in using the natural resources.

In addition to the gender perspective, when the government involves the local people, it should understand the programs that are needed in those areas. Figure 5 shows that the indigenous people understand the programmes that can be done in their areas to increase their income, while still maintaining the sustainability of their environment. This shows that the local people understand their areas and if they are involved in the decision-making process regarding their landscape, such as spatial planning policies, this leads to correct decisions for the local people.

7. Conclusion

This research has three findings. First, the tribes name the landscape as *seko* (mountain), *meko bura* (hill), *eri bura* (slope), *paw* (valley), *meroway bu* (river), *buru rena bu baruna* (spring), *kingkey bueaenabu* (water fall), *meraway kera-kera* (dried river), *yemo* (cape), *yepa* (beach), *nau* (sea) and *kepou* (sea stones). Second, the Tepera have developed a zoning system to manage the forest traditionally, called *sena, seke,* and *moko.* Third, there are differences in perception between men and women regarding the type of forest, the landscape related to food and traditional medicine sources. The zones are related to the function of the forest to support life. Therefore, it is important to incorporate the Tepera's concept of managing environmental and cultural zones into the programs of the local government to direct the development in a sustainable way. In addition, the participation of women in managing the environment should be, improved especially associated with domestic aspects.

8. Acknowledgements
The authors want to thank USAID Lestari for providing the data and funding the research about the Tepera. We also thank the USAID Lestari Team for their help in gathering data and information. Lastly, we want to thank the Tepera for sharing their amazing traditional knowledge to maintain the nature, and we thank everyone who contributed to this article.

References
[1] Tadjoeddin M Z 2007 *A Future Resource Curse In Indonesia: The Political Economy Of Natural Resources, Conflict And Development* CRISE Working Paper No. 35. (Oxford: Centre for Research on Inequality, Human Security and Ethnicity (CRISE), Oxford University)
[2] Marshall A J and Beehler B M (Eds.) *The Ecology of Papua* Part 2 (Singapore: Periplus editions (Hk) Ltd) pp 800-823
[3] Wambrauw E V 2015 *Water Resource Management in the Lowlands of Southern Papua Using a Decision Support System and Integrating Traditional Ecological Knowledge* (Auckland: Doctoral of Philosophy, The University of Auckland)
[4] Vassilopoulos A, Evelpidou N, Bender O and Krek A 2008 *Geoinformation Technologies For Geo-Cultural Landscapes European Perspectives* (London: Crc Press)
[5] Taçon P S 2010 Identifying Ancient Sacred Landscapes in Australia. *Contemporary Archaeology In Theory: The New Pragmatism*
[6] Stephenson J 2008 the Cultural Values Model: An Integrated Approach to Values in Landscapes *Int.J. Landscape And Urban Planning,* 84**,** 127-139.
[7] Wambrauw E V and Morgan T K K B 2014 Concept Alignment For Sustainability: Relevance The Mauri Model In Asmat Southern Papua *Int. J. Indigenous People* Volume 10, issue 3, pages 288-302
[8] Kitchin R 2001 Cognitive Maps Encyclopedia of Social and Behaviour Science
[9] Trowbridge 1913 *On Fundamental Methods of Orientation of Imaginary Maps Science* Vol 38 Issue 990 pp 888-897.DOI: 10.1126/science.38.990.888
[10] Goodey B 1971 *City Scene: An exploration into the image of central Birmingham as seen by area residents Research Memorandum 10 Birmingham* (UK: Centre for Urban and Regional Studies)
[11] World Bank Group 2016 Gender and Natural Resources Management [Internet]. Available from:http://siteresources.worldbank.org/INTGENAGRLIVSOUBOOK/Resources/Module10.pdf
[12] Meizen D etc 1997 Gender Property Right, and Natural Resources *World Development* Vol.25 No 8, pp 1303 -1315
[13] Petocz R G 1989 Conservation and Development in Irian Jaya ISBN 90–04–08832-6
[14] Fadiman M 2008 Natural Resource Use and Cultural Change: Nipa Hut Shingle Processing with Nypa fruticans, Arecaceae, in Palawan, Philippines *The Florida Geographer*, Vol 39
[15] Palimbunga I 2016 Partisipasi Masyarakat dalam pengembangan Parawisata di Desa Wisata Tablanusu, Kabupaten Jayapura, Papua, Master thesis Universitas Udayana Denpasar
[16] USAID Lestari 2016 Pemetaan Tempat Penting Dan Pengelolaan Sumberdaya Alam di DAS Tepera, Papua

IOP Conf. Series: Earth and Environmental Science **158** (2018) 012030 doi:10.1088/1755-1315/158/1/012030

Stakeholder Participation in Marine Spatial Plan Making Process in Lampung Province

Asirin[1,2], A M Asbi[1], V H Pakpahan[1]

[1]Urban and Regional Planning, Institut Teknologi Sumatera (ITERA), Lampung Selatan, Indonesia

[2]Resilience Development Initiative (RDI), Bandung, Indonesia

Corresponding Author: asirin@pwk.itera.ac.id

Abstract. Lampung Province has coastal areas, seas and small islands facing conflicts of interest between tourism, conservation areas for defense, environmental conservation, and the threat of unsustainable marine resource utilization. Indonesia (including Lampung Province) has committed itself to achieving the objectives of conservation and sustainable use of oceans, seas and marine resources in view of sustainable development. One of the instruments used to achieve this goal is by using marine spatial planning (MSP). The purpose of this research was to analyse the marine spatial plan making process in Lampung Province. This research also evaluated the participation process and participation level based on plan-making process criteria and the stakeholder participation ladder. This research can be useful as a recommendation in the evaluation step to improve the plan-making process in order to address conflicts of interest between various related interest groups, so that planning can be accomplished with the involvement of all relevant parties to reach consensus on how to achieve a sustainable marine environment. This research used a qualitative research method as well as a case study approach. The scope of this study was limited by the conceptual framework of marine spatial planning and the stakeholder participation ladder. The authors recommend study of the preparation of marine spatial planning in addition to a technocratic approach considering the results of the study aspects of spatial allocation and physical aspects of marine resources, while prioritizing building consensus among various interest groups related to the utilization of marine resources. Thus, it is necessary to develop technical steps to build consensus in the marine spatial plan-making process.

1. Introduction

Lampung Province has coastal areas, seas and small islands facing conflicts of interest between tourism, conservation areas for defense, environmental conservation and the threat of unsustainable marine resource utilization. The coastline of Lampung Province area is around 1,105 km. The authority of the Lampung Province management sea area extends to 12 (twelve) nautical miles, measured from the coastline at the highest tide toward the open sea and/or towards the archipelagic waters. The Lampung region contains 132 small islands. One of them, the outer islands of Betuah Island, is a Certain National Strategic Area. Indonesia (including Lampung Province) has committed itself to achieve the objectives of conservation and sustainable use of the oceans, seas and marine resources in view of sustainable development (Goal 14 in SDGs) [1]. One of the instruments used to achieve these goals is by using marine spatial planning (MSP).

The purpose of this research was to analyze the marine spatial plan making process in Lampung Province. This research also evaluated the stakeholder participation process and participation level based on plan making process criteria and the stakeholder participation ladder. The outcomes of this research can be useful as a recommendation in the evaluation step to improve the plan making process in order to address conflicts of interest between various related interest groups so that the planning can be accomplished with the involvement of all relevant parties to reach consensus on achieving a sustainable marine environment.

2. Another section of your paper

This research used a qualitative research method as well as a case study approach. The scope of this study was limited by the conceptual framework of marine spatial planning and the stakeholder participation ladder. According to Ehler and Douvere, "Marine spatial planning (MSP) is a way of improving decision making and delivering an ecosystem based approach to managing human activities in the marine environment. It is a planning process that enables integrated, forward looking, and consistent decision making on the human uses of the sea"[2].

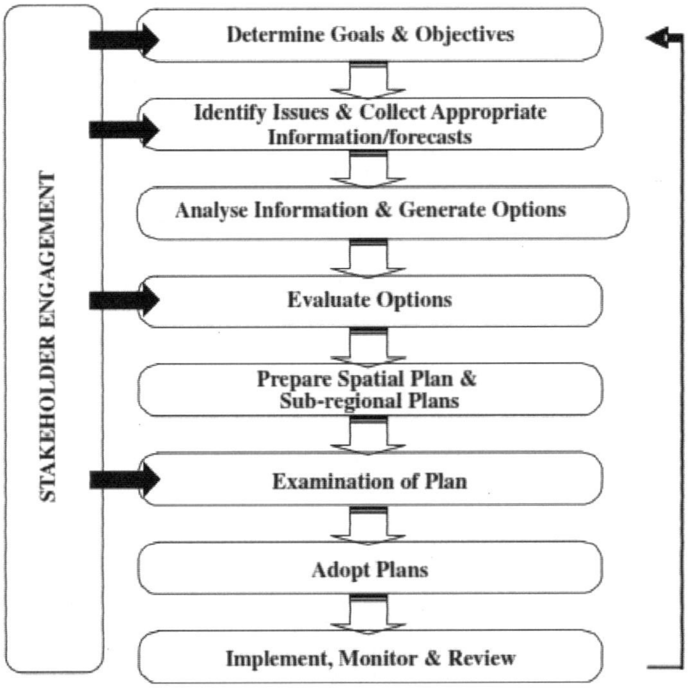

Figure 1. Normative Process of Marine Spatial Planning and Stakholder Engagement in the Marine Spatial Plan-Making Process
Source: MSSP Consortium, 2006

This study adopted the framework for the evaluation of stakeholder participation in marine spatial planning from Flannery and Cinneide in 2014 [3]. Flannery and Cinneide derived their criteria from the evaluation theory of Oels [4] and Chess [5] and from a review of collaborative planning theory developed by Chrislip and Larson [6]; Logsdon [7]; Frame, Gunton, and Day [8]; Conley and Moote [9]; Innes and Booher [10]; Innes [11]; Van de Kerkhof [12]; and Healey [13]. Flannery and Cinneide state that participation of stakeholders can be assessed in two main stages, the process of planning and the outcomes [3]. Since MSP in Indonesia happens only in the early stages of the preparation of a plan, this study only used the assessment framework at the stage of the process of marine spatial plan making. The process criteria used in this research are the process design or co-design, shared purpose, equality, interdependency, representation and participation, deliberative decision making and constructive dialogue and effective process management [3].

In the literature on participation in development, there is the concept of the participation ladder from Arnstein but the ladder is more suitable for discussing citizen participation in general in urban development. Specifically, the stakeholder participation ladder in MSP is explained by Bouamrane [14] and Pomeroy and Douvere [15]. Therefore, this study evaluated the achievement of the level of stakeholder participation using the framework of the stakeholder participation ladder by Bouamrane [14]. The levels of stakeholder participation according to Bouamrane are: communication, information, consultation, dialogue, concertation, and negotiation [14]. A detailed explanation of each of these levels are as follow.

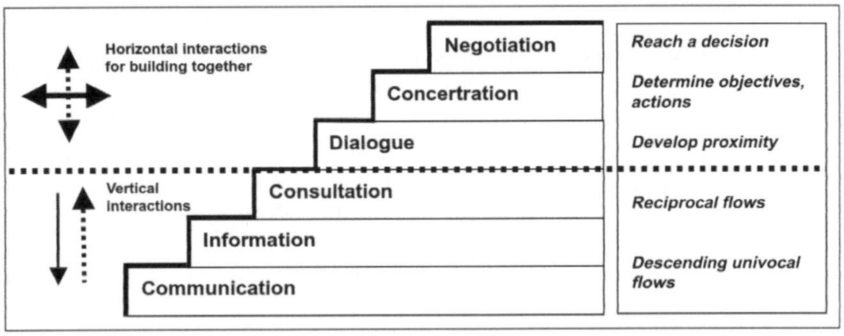

Figure 2. Stakeholder participation ladder in MSP [14][15]

Table 1. Stakeholder Participation Ladder Levels in MSP [14]

Ladder Level	Description
Communication	• The management authority wants to convey messages and get the approval of the target group on what it asserts, suggests and decides.
Information	• The management authority wants to keep the target group abreast of intentions or decisions taken, by attempting to provide a base of understanding without expecting a particular reaction from this group.
Consultation	• Exchange during which the management authority collects the opinions of the consulted stakeholders, yet there is no guarantee that the opinions expressed will be taken into account (Dziedzicki, 2001).
Dialogue	• The management authority wants to enter the field of horizontal interaction between stakeholders who are positioned as equals. • There is no precise purpose other than to better know and understand one another, but this can represent a major accomplishment.
Concertation	• Build components together with a view to better manage the biosphere reserve whether in the short or long term. • Give participants access to a true collective construction. • Let individuals play an active role in the process.

Ladder Level	Description
Negotiation	• A horizontal dialogue among participants whose goal is the collective construction of a shared focus (points of view, representations, objectives, projects) with a view to act or decide together. • 'An activity involving the interaction of several stakeholders who, faced with both divergences and interdependencies, choose to voluntarily seek a mutually acceptable solution.' (Dupont and Audebert, 1994)

According to the MSSP Consortium, the main stages of stakeholder participation in the MSP process are: (1) the planning phase; (2) the plan evaluation phase; (3) the implementation phase; (4) the post-implementation phase [16]. Because in Lampung Province has currently progressed only until the drafting of the MSP regulations, this evaluation comprises only phases (1) and (2). Here follows a description of each stage according to the MSSP Consortium [16]:

- The planning phase. Stakeholders need to be involved in and contribute to the setting of priorities, objectives and purpose of the MSP plan(s). The MSP management team can assist in setting priorities and identifying objectives through stakeholder meetings and group discussions. The idea is to identify, group and rank problems, needs and opportunities in order of priority. This can be done through criteria ranking and pairwise ranking. The output should be made available to the stakeholders and the output should be reviewed and verified with the stakeholders.
- The MSP plan evaluation phase. Stakeholders need to be engaged in the evaluation and choice of MSP plan options and the consequences of different approaches for areas of their interest. In developing the plan, a number of participatory tools and methods can be used, including focus group discussions, problem trees, and preference ranking. It is important for all stakeholders to be clear about the goals and objectives and about what can be achieved in order to focus strategies. The more participatory the process of setting goal and objectives, the greater the stakeholder acceptance and legitimacy of the MSP plan. Often the process of arriving at consensus regarding goal and objectives is an effective means of promoting an exchange of information and understanding among stakeholders. If a trained planner is not available, a plan can still be prepared based on the stakeholders' knowledge and participation.

3. Result and Discussion

3.1. Marine Spatial Plan-Making Process Practice in Lampung Province: An Overview

In Lampung Province, the marine spatial plan (MSP) is part of the Coastal Region and Small Islands Zoning Plan (RZWP-3-K). The legal form is a local regulation (perda). The current marine spatial plan making process has been done only until the local regulation draft (raperda). The process so far has been: pre-preparation of RZWP-3-K, consisting of stakeholder identification, socialization, and technical assistance; preparation of RZWP-3-K, consisting of compilation of work plan and preparation of terms of reference (ToR) and project budgeting; and preparation of the RZWP-3-K final document, consisting of data collection, field survey, data processing and analysis, description of potential and marine resource utilization, first public consultation, preparation of preliminary documents, spatial allocation, proposal determination, preparation of intermediate documents, final public consultation, final document preparation, and suggestion request; legalization of the RZWP-3-K local regulation draft (see Figure 3).

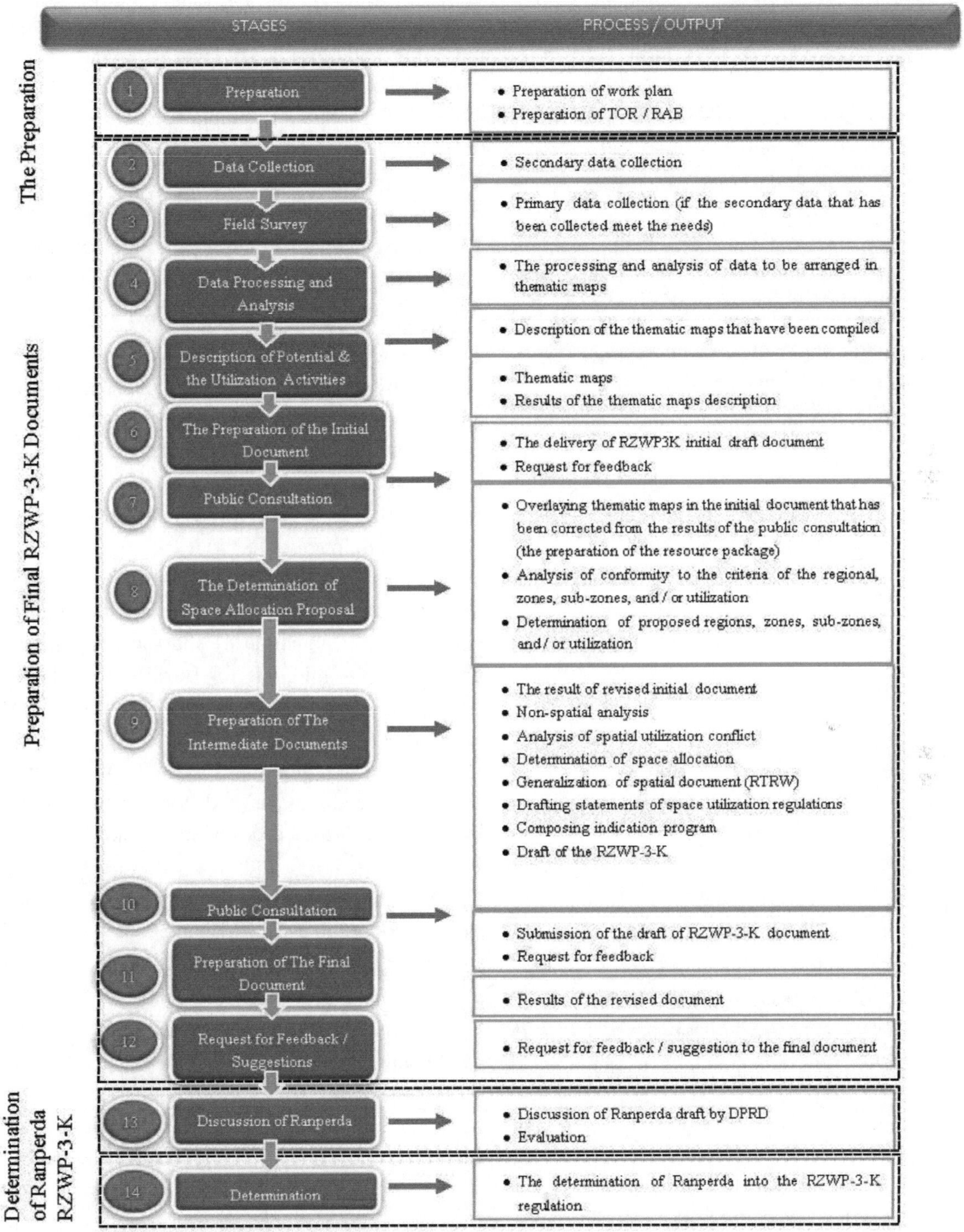

Figure 3. Marine Spatial Plan Making Process Practice in Lampung Province
Source: Fisheries and Marine Government Agency of Lampung Province, 2017

3.2. Lampung Province Marine Spatial Plan

The RZWP-3-K Raperda enters the legislation stage in DPRD, having previously received technical approval from the central government, in this case the Ministry of Marine Affairs and Fisheries and other relevant Ministries at the central level. The content of the RZWP-3-K Raperda is outlined in Table 2. The marine spatial allocation can be seen in Table 3 and Figure 4.

Table 2. Summary of RZWP-3-K Raperda of Lampung Province

Chapter	Short Description
Chapter I – General Terms	• This chapter contains descriptions of terms and definitions used in RZWP3KRZWP-3-K material
Chapter II – Scope, Principle, Duration, And FunctionPrinciples, Durations and Functions of RZWP-3-K	• This chapter contains the scope of RZWP3KRZWP-3-K rules, principles, durations and functions in maritime area development and management
Chapter III – Objectives, Policies and Strategies for Coastal Zone Plans and Small Islands Plan	• This chapter contains the objectives, the policies that need to be doneexecuted to achieve the objectives, and the strategies that need to be executedimplemented to achieve every policy that has been outlined
Chapter IV – Spatial Allocation Plan for Coastal Areas and Small Islands	• Spatial allocation plan for coastal areas and small islands, including:
Chapter V – Program Indication	• To achieve the objectiveobjectives of spatial development patternpatterns in coastal areas and small islands, indication of short and medium term programs as a guideline for Local Governmentlocal government and related agencies in utilizing spatial zones for various activities.
Chapter VI – Spatial Use Regulations	• The spatial use regulation is a condition of the spatial utilization requirements and control provisions that are prepared for each designation zone in RZWP-3-K.
Chapter VII – Small Islands	• Contains theThe terms of utilization of small islands.
Chapter VIII – Disaster Mitigation	• Provision of disaster management and disaster mitigation strategies.
Chapter IX – Supervision and Control	• Conditions of supervision and control.
Chapter X – Rights, Obligations, and Roles of Society	• Conditions of community involvement in planning, implementation, monitoring and evaluation of RZWP3KRZWP-3-K.
Chapter XI – Coordination of Implementation	• Provision of coordination among stakeholders in the planning, implementation, monitoring and evaluation of RZWP3KRZWP-3-K.
Chapter XII – Lawsuit Representatives	• Provisions on the mechanism of the lawsuitlawsuits.
Chapter XIII – Conditions of Investigation	• Conditions onof investigation.
Chapter XIV – Criminal Provisions	• Provisions on criminal sanctions for violating the

Chapter	Short Description
	RZWP3K RegulationRZWP-3-K Regulations.
Chapter XV – Transitional Provisions	• Transitional provisions.
Chapter XVI – Miscellaneous Provisions	• Other provisions.
Chapter XVII – Closing Provisions	• Closing provisions.

Source: Excerpted from Raperda RZWP-3-K Lampung Province, Fisheries and Marine Government Agency of Lampung Province.

Table 3. Marine Spatial Allocation Plan on Raperda RZWP-3-K of Lampung Province

Area	Zone	Sub ZoneSubzone
Area of public utilization	Tourism Zone (KPU-W)	• Subzone of natural tourist landscape (KPU-W-BL) • Subzone of nature underwater tourism (KPU-W-ABL) • Subzone of beach/coastal tourism and small islands (KPU-W-P3K) and • Subzone of water sport tourism (KPU-W-OR)
	Zone of settlementSettlement zone	• Sub Zone of Fisherman Settlements
	The harborHarbor zone	• Sub zoneSubzone of the Working Areaworking areas (DLKr) and Regional Environment Interestsregional environmental interests (DLKp);) and Sub zone Operational Work Areasubzone of Fishery Portoperational work area of fishery ports (WKOPP).)
	Mining Zonezone	• Sub zoneSubzone of oil and gas;
	Fishery zone	• pelagic sub zone • demersal subzone
	Cultivation fishery zone	• Sub zoneSubzone of marine cultivation.
	Industry ZoneIndustrial zone	• Sub zoneSubzone of maritime industry
Conservation area	Coastal and small islands conservation areas (KKP3K)	• core zone • limited utilization zone • other zones according to regional designation
	Water conservation area (KKP)	• core zone • sustainable fishery zone • subsequent utilization zones • other zones
	Nature reserve area	• Marine Reserve and Marine Reserve of Krakatau Islands (KSA-CA-CAL-1) • Nature Reserve of Bukit Barisan Selatan Sea (KSA-CAL-2)
Certain national strategic	Small outer islands	• Betuah Island in West Coast District (KSNT-PB)

IOP Conf. Series: Earth and Environmental Science **158** (2018) 012030 doi:10.1088/1755-1315/158/1/012030

Area	Zone	Sub ZoneSubzone
areasNational Strategic Areas National Strategic AreaAreas		
	Military training area of Teluk Lampung in Pesawaran Regency (KSN-TL-1) Sunda Strait area in South Lampung Regency (KSN-KSS-2)	
Sealine	Shipping lines	• international shipping lanes • national shipping lane • regional shipping lanes • local cruise line lanes • custom shipping lanes • Indonesian archipelagic sea lanes (ALKI)
	Submarine pipes / cables	• optical cable channel PT. Telkom, covering around Kalianda-Merak Waters (AL-APK-KT) • gas pipeline network, covering around Labuhan Maringgai Waters (AL-APK-PG) • plot of PLN / 500 Kv Electric Transmission, covering around Ketapang Waters and Sunda Strait Waters (AL-APK-KL) • PLN plot line plan, covering Legundi Island Waters (AL-APK-KL-1), Pahawang Island (AL-APK-KL-2) (Pahawang Pekon 4 Pahawang Island), Balai Sawang Balak (AL-APK-KL-3), Pekon Doh Cukuh Balak-Sawang Balak Island and Banana Island (AL-APK-KL-4) (Port of Tembakak-Pulau Pisang)
	Marine life migration	• turtle migration flow, covering around Lemong-Pulau Banana-Bengkunat-Pulau Betuah-Belimbing-Pulau Tabuhan-Kelumbayan-Punduh Pidada-Padang Cermin-Katibung-Kalianda-Sebesi-PulauMundu-Pulau Segama / Pulau Batang (AL-AMB –MP) • marine mammal migration flow, covering around Lemong-Pulau Banana Island of Betuah-Kiluan-Sebesi-Selat-Strait Sunda / Bakauheni Island-Mundu Island-Segama Island / Batang Island (AL-AMB-MM)

Source: excerpted from Raperda RZWP-3-K, Fisheries and Marine Government Agency of Lampung Province, 2017

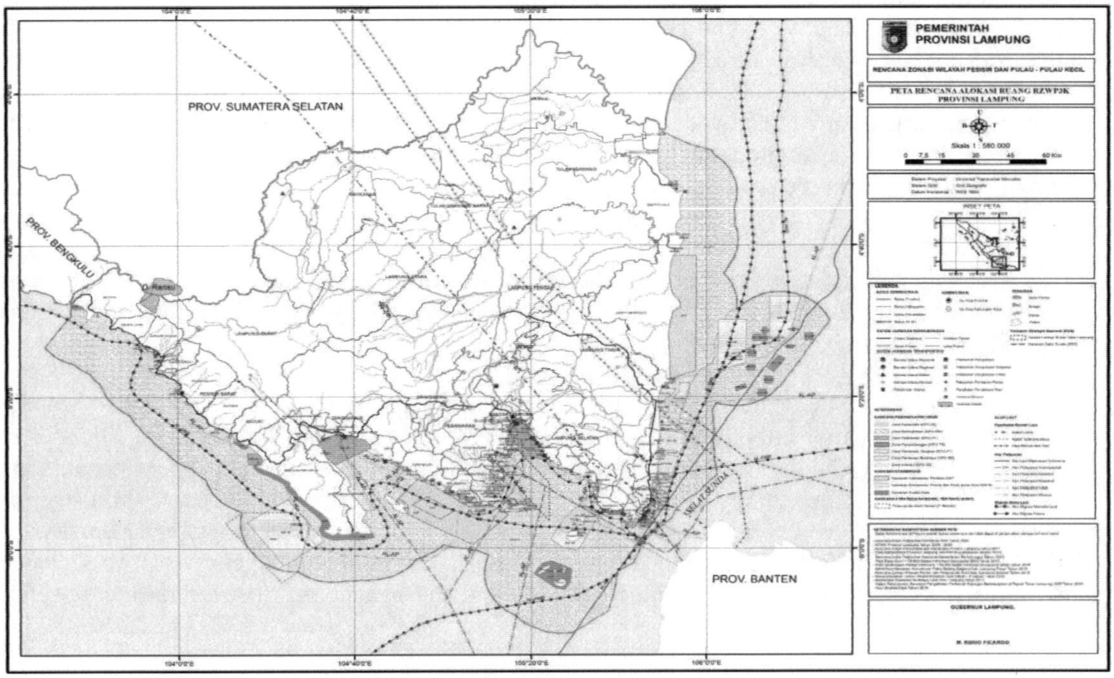

Figure 4. Map of Spatial Allocation Plan RZWP-3-K Lampung Province
Source: Raperda RZWP-3-K, Fisheries and Marine Government Agency of Lampung Province, 2017

3.3. Stakeholder Participation Evaluation in Marine Spatial Plan Making Process Practice in Lampung Province

his section details a description of stakeholder participation in the marine spatial plan making process in Lampung Province. Findings are reported in relation to each of the process criteria.

3.3.1. Process design or co-design. The process of drafting the RZWP-3-K is set out in the RZWP-3-K drafting guidelines compiled by the Ministry of Fisheries and Marine Affairs. The authority to regulate the procedure and the process of preparing the plan lies with the Ministry of Fisheries and Marine Affairs Authority. Implementing the RZWP-3-K procedure is done by the Fisheries and Marine Agency of Lampung Province. The work is supervised by a planning consultant. RZWP-3-K's initial product is a technocratic product. The next process of involving stakeholders involved in the legislation process takes place in the DPRD (legislative board). In other words, the process design is done using a top-down approach. Stakeholders only follow the involvement in accordance with established run by the Fisheries and Marine Agency of Lampung Province assisted by a third-party consultant, which in turn involves the stakeholders in discussion *meetings and public consultation meetings.*

3.3.2. Shared purpose. After the intermediate report, which is a technocratic product between the Fisheries and Marine Government Agency of Lampung Province and the consultant, the product is discussed in a public consultation. The public consultation in this stage involves various stakeholders. It is done to verify data and information and to obtain input, responses, objections or suggestions. Public consultation is a process of excavation and dialogue between local government and government, and stakeholders in the coastal areas and small islands undertaken, among others, through meetings, village deliberations and workshops. This stage is public consultation event 1 (first). The results are put into minutes meetings, supplemented with attendance lists and documentation. Public consultation at the next stage is public consultation event 2 (second), undertaken, among others, to verify the draft zoning plan, use directives and check the consistency of the RZWP-3-K draft with RTRW and other rules so that the draft spatial allocation plan can be approved by all local stakeholders. The targets to be achieved are the improvement and refinement of the draft document between and facilitate the aspirations of all relevant stakeholders as well as the determination of the allocation of spatial zones in the final document to be drawn u*p.*

3.3.3. Equality. All stakeholders present in the public consultations are entitled to the same rights and have the same obligations. All are given the opportunity to convey aspirations and suggestions regarding RZWP-3-K products.

3.3.4. Interdependency. Stakeholders and participants present at the aforementioned public consultation are representatives of any stakeholders who have been recommended in the guidelines to be consulted. Many stakeholders are involved in the drafting process, such as the Coordinating Ministry of Maritime Affairs, Ministry of Marine Affairs and Fisheries, Navy, Ministry of Transportation, Minsitry of Public Works, Ministry of Environment and Forestry, Ministry of Tourism, Minsitry of Transportation, Minsitry of Industry, Environment Agency, Local Revenue Agency, universities/academicians, community groups, and the marine and fishery business sector, journalist, etc.

Table 4. Stakeholders and Their Interests in MSP in Lampung Province

Stakeholders	Interest
Central government	
Coordinating Ministry of Maritime Affairs	Regulate and manage the various interests of both government, private parties and communitycommunities in the maritime area for the welfare of the community and the preservation of the maritime environment (sustainable development)
Ministry of Marine Affairs and Fisheries Ministry	Regulate and manage the marine areas for the welfare of fishermen and the sustainability of the maritime environment (sustainable development)
Ministry of Internal Affairs	Regulate and manage the various interests of central government agencies and local government agencies that have maritime areas for community welfare and environmental sustainability (sustainable development)
Ministry of Environment and Forestry	Regulate and manage the sustainability of the maritime environment in the maritime area
Ministry of Transportation	Regulate, use, and manage, the use of maritime areas for marine transportation and crossings
BPN (Ministry of AgrariaAgriculture)	Regulate and manage land ownership and use in maritime areas
TNI AL and POLAIRUD	Regulate, use and manage the use of maritime territories for the purpose of defense and security of maritime and state territories
BUMN (Statestate-owned enterpriseenterprises): Telkom and PLN	Uses and manages the use of maritime territory for the purposes of telecommunication and electricity networks
Local Government of Lampung Province	
Marine and Fisheries Agency	Regulate and manage the marine areas for the welfare of

Stakeholders	Interest
	fishermen and the sustainability of the maritime environment (sustainable development)
Public Works Agency	Regulate, implement and manage coastal infrastructure development in maritime areas
Forestry Agency	Regulate and manage the sustainability of coastal forest environments in the maritime region
Agriculture Agency	Regulate and manage coastal farmland in the maritime areaareas
Tourism Agency	Regulate and manage tourism activities in coastal, marine and small islands
Transportation Agency	Regulate, use, and manage, the use of maritime areas for marine transportation and crossings
Industrial Agency	Regulate, and manage, the use of maritime areas for industrial use
Environmental Agency	Regulate and manage environmental sustainability in maritime areas
Income Office Region Agency	Regulate and manages regional revenue from activities in every productive sector undertaken in the maritime region
Communication and Information Agency	Regulate and manages regional revenue from activities in every productive sector undertaken in the maritime region
Energy and Mineral Resources Agency	Use and manage the use of maritime territories for the purposes of energy networks and oil and gas exploitation
DPRD (Legislative)	Organize and manage public and private interests in the maritime area
Society and Higher Education	
Environmental NGOs	Carry out and participate in managing the environment
Organization of Fishermen Group	Carry out the use of maritime areas and participate incontribute to the welfare of fishermen
Universities	Conduct research and community service and convey ideas of innovation and technology produced for the welfare of the community and environmental sustainability in the maritime region
Organization of Marine and Fisheries Entrepreneurs Group	Carry out the utilization ofUtilize maritime territory and participate in seeking prosperity offor marine and fishery business actors
Organization of Industry and Travel Agencies	Carry out the utilization ofUtilize maritime territory and participate to seek prosperity of business actors in the field of industry and tourism

In terms of interests, stakeholders related to MSP in Lampung Province are divided into two groups: 'pro-growth' and 'anti-growth'. Pro-growth stakeholders include the private sector (business), fishermen, the industrial agency, the tourism agency, the agriculture agency, the public work agency and the mineral and energy agency, the public work agency, the transportation agency, the local revenue office, state-owned enterprises Telkom and PLN. Anti-growth stakeholders are the environmental agency, the forest agency, the Navy and environmental NGOs.

3.3.5. Deliberative decision making and constructive dialogue. Stakeholders are only entitled to submitting suggestions. Decision making is still being done by the Fisheries and Marine Government Agency of Lampung Province. In managing the process, the management of the stakeholder engagement processes in the preparation of the plan is only done by conducting public consultation events 1 and 2. Stakeholders are also given the opportunity to submit suggestions via e-mail. The Fisheries and Marine Government Agency of Lampung Province and its consultants actively solicit suggestions and input.

3.4. Level of Stakeholders Participation
With the conditions described above, it can be said that according to the stakeholder participation ladder, the stakeholder participation level is at the third level, i.e. the consultancy level. The Fisheries and Marine Agency of Lampung Province collects opinions, aspirations and suggestions from stakeholders. However, there is no guarantee that the opinions expressed will be taken into the marine spatial plan. There is no obligation to open up the debate between stakeholders in marine spatial planning.

4. Conlusion
The process of preparing the maritime spatial plan for Lampung Province involves the central government, the provincial government of Lampung, the community, and various stakeholders related to maritime resource use in Lampung Province. The main challenge in the planning process is the existence of various conflicts of interest that can be broadly grouped into 'pro-growth' and 'anti-growth'. Participation is currently at the consultation level of the stakeholder participation ladder and consensus-building efforts have not been conducted yet. These conditions make the process vulnerable to conflicts in the implementation phase when the regulation is in force. This research recommends research on the preparation of marine spatial planning in addition to a technocratic approach considering the results of the study aspects of spatial allocation and the physical aspects of marine resources, also prioritizing building consensus among various interest groups related to the utilization of marine resources. Achievement of the ladder of participation needs to be increased to the levels of dialogue, concertion and negotiation. Thus, it is necessary to develop technical steps on how to improve stakeholder participation in marine spatial planning and build consensus in the marine spatial plan making process.

Acknowledgements
This research is part of a research funded with an ITERA 2018 Research Grant. The authors would like to express their thanks to Mr. Parjito and Dinas Kelautan dan Perikanan, Lampung Province for providing support in data collection, and Mrs. Helmia Adita Fitra, B.Sc., M.Sc. and Mr. Yudha Rahman B.Sc., M.Sc. who supported the funding of participation in the conference.

References
[1] UNDP 2015 Sustainable Development Goals UNDP
[2] Ehler C and Douvere F 2007 Visions for a sea change. *Report of the first international workshop on marine spatial planning. Intergovernmental oceanographic commission and man and the biosphere programme. IOC manual and guides* no. 48 IOCAM Dossier no. 4 (Paris: UNESCO)
[3] Flannery W and Cinnéide M O 2012 Stakeholder Participation in Marine Spatial Planning: Lessons from the Channel Islands National Marine Sanctuary, Society & Natural Resources: *Int. J.* 25:8, 727–742
[4] Oels A 2006 *Evaluating stakeholder dialogues. In Stakeholder dialogues in natural resources management*, ed. S. Stoll-Kleemann and M. Welp, 117–147. Heidelberg, (Germany: Springer)
[5] Chess C 2000 Evaluating environmental public participation: Methodological questions. *J. Environ. Plan. Manage.* 43(6):769–784.
[6] Chrislip D and C Larson 1994 *Collaborative leadership: How citizens and civic leaders can make a difference* (San Francisco, CA: Jossey-Bass)

[7] Logsdon J M 1991 Interests and interdependence in the formation of social problem-solving collaborations *J. Appl. Behav. Sci* 27(1):23–37.
[8] Frame T M Gunton T and Day J C 2004 The role of collaboration in environmental management: An evaluation of land and resource planning in British Columbia *J. Environ. Plan. Manage* 47(1):59–82
[9] Conley A and Moote M A 2003 Evaluating collaborative natural resource management *Society Nat. Resources* 16:371–386.
[10] Innes J E and Booher D E 2004 Reframing public participation: strategies for the 21st century *Plan. Theory Pract* 5(4):419–436.
[11] Innes J E 1996 Planning through consensus building: A new view of the comprehensive planning ideal *J. Am. Plan. Assoc* 62(4):460–472.
[12] Van de Kerkhof M 2006 Making a difference: On the constraints of consensus building and the relevance of deliberation in stakeholder dialogues *Policy Sci* 39(3):279–299.
[13] Healey P 1992 Planning through debate: The communicative turn in planning theory *Town Plan. Rev* 63(2):143–162.
[14] Bouamrane M 2006 Biodiversity and stakeholders: concertation itineraries *Biosphere reserves, technical notes* 1 (Paris: UNESCO)
[15] Pomeroy R and Douvere F 2008 The engagement of stakeholders in the marine spatial planning process *Marine Policy* 32 (2008) 816– 822
[16] Marine Spatial Planning Pilot (MSSP) Consortium 2006 Final report [Internet]. Available from: (www.abpmer.net/mspp/docs/finals/MSPFinal_report.pdfS).

Biogas Implementation as Waste Management Effort in Lembang Sub-district, West Bandung District

H A Alberdi[1], S A H Sagala[2], Y Wulandari[1], S L Srajar[2], D Nugraha[2]

[1]Resilience Development Initiative, Jalan Imperial 2 No. 52, Dago Asri, Bandung, Indonesia
[2]Institut Teknologi Bandung, Jalan Ganesha No. 10, Bandung, Indonesia

Corresponding Author: husnul.alberdi@rdi.or.id

Abstract. As the population and economic activities increase, energy demand will increase significantly too. In the near future, Indonesia will have more limitations on fossil fuel based energy. Therefore, sources of renewable energy have to be found. On the other hand, rural areas in Indonesia suffer from lack of energy supply. Therefore, energy resilient villages need to be created. Kampung Areng in Cibodas Village, Lembang sub-district is one of the locations declared as an energy resilient village. This study focused on Kampung Areng due to previous information and studies that have shown that the farmers in this area are capable of generating renewable energy through conversion of animal waste using biogas digesters. Biogas adoption was also initiated due to the issue of Cikapundung river pollution. Some people pointed their finger at Kampung Areng as the upstream source of this pollution due to irresponsible handling of animal waste. Thus, the idea emerged that biogas digester adoption in this area could solve the waste management issue while providing an alternative energy source for the community. In this paper, the authors aimed to identify the benefit and impact of biogas adoption in Cibodas Village, particularly regarding waste management.

1. The first section in your paper

Energy is an important resource for economic development and improvement of living conditions. In the past, Indonesia has been an oil exporter with several production areas such as Balikpapan [1]. However, Indonesia has already been a net oil importer since 2004, which will likely continue in the future. From 1980 to 2010, Indonesia oil consumption has increased while its production has decreased. Since the year 2004, there has been an energy deficit and therefore Indonesia has started importing oil from other countries. If not managed carefully, this could become a threat to the energy security of Indonesia. In fact, Indonesia's Agency for Assessment and Application of Technology has predicted that Indonesia could be a net importer for fossil energy [2]. Therefore, Indonesia has to utilize other sources of energy, for instance new and renewable energy (NRE).

The utilization of renewable energy resources could be an effective and efficient solution, not only for providing energy but also for reducing pollution and emissions [3]. One of the largest environmental problems in Indonesia is waste management. As one of the countries with the largest populations in Southeast Asia, about 220 million people, which has increased in the last decade, the amount of waste it produces has also increased. There are several factors that influence waste management in Indonesia, such as regulations, capacity and dumpsite technology, low public awareness, lack of infrastructure, and low priority from local governments [4].

Indonesia has a target to achieve 23% renewable energy in the energy mix by 2025 [5], which is possible to achieve with the renewable energy potential of Indonesia. This is an ambitious target that needs involvement from various stakeholders. However, it is in contrast with the inequality of access to energy resources such as electricity, particularly for communities in rural areas. It has been shown that the electrification rate in Indonesia was 91.2% in 2016 [6] and the lower electrification rates are mostly found in eastern Indonesia. However, there are alternative resources in rural areas that can be utilized to generate renewable energy, such as biogas, biomass, and biofuel.

Biogas utilization has been done since 1895 in England using sewage treatment installations to provide street lighting [7]. Interest in biogas depends on the price and availability of fossil fuels. Interest was revived in 1973 by the mounting concern for environmental protection [8]. A number of studies [9][10] showed that Indonesia has great potential to utilize biogas in a feasible way. Subsequently, several biogas development projects were started in Indonesia. International and national organizations were involved in managing these projects with support from the government.

Indonesia has used biogas since the 1970s, and in 1981, biogas development projects have been implemented with funding from FAO in several provinces of Indonesia. Although there has been no significant effect in the last decades, the Indonesian government recently again has initiated several biogas projects. In Palembang, GoI built a biogas power plant using waste from dump sites. In other places, POME is the main material to generate biogas. In rural areas, biogas generators are managed by households or small communities. They usually use livestock manure as the resource. Several provinces use this kind of biogas technology, among which East Java, Central Java, East Nusa Tenggara and West Java.

Lembang subdistrict is one of the areas that have developed biogas utilization on a small scale for communities in rural areas. Lembang is located in West Java Province. In Cibodas village, more specifically Kampung Areng, in Lembang, the majority of the population are cattle farmers. Biogas is generated from cow manure as their source of energy due to the quite large number of small-scale cattle farms in this area. Small-scale cattle farms have a disadvantage in that it is difficult for them to handle the waste of cows wisely [11]. The development of biogas utilization in this area was also part of the problem of maintaining water quality in Cikapundung river, which passes Kampung Areng. Badly managed cow manure from the cattle farms is assumed to affect its water quality.

This study focused on Kampung Areng in Lembang subdistrict because previous information and studies have shown that the farmers in this area are capable of generating renewable energy through conversion of animal waste using biogas digesters. This is supported by the fact that most households in Kampung Areng do cow farming and have enough manure to supply and support the development of biogas digesters. Biogas adoption was initiated also due to the issue of Cikapundung river pollution; some people pointed their fingers at Kampung Areng as an upstream source of pollution due to the irresponsible handling of animal waste. Thus, the idea emerged that biogas digester utilization in this area could solve the waste management issue while providing an alternative energy source for the communities. This study aimed to identify the benefits and impacts of biogas adoption in Cibodas village.

2. Methodology

This research used a mixed qualitative and quantitative approach. The region selected for the case study in this research was Lembang subdistrict, Kab. Bandung Barat, West Bandung regency. There are currently 775 biogas installations that have been built by Hivos and Yayasan Rumah Energi (YRE) in Lembang. Tools used were questionnaires, in-depth interviews and observation. Questionnaires were spread among 81 respondents from different neighborhoods: RW 07, RW 17, RW 05, RW 04, RW 02, RW 16, RW 10, RW 14 and RW 11; in-depth interviews were conducted with seven key informants, i.e. a neighborhood head (RW 07), a women's group, an ex-leader of a farmer's group, Energi Persada, a garbage bank manager, and the current treasurer of a farmer group. Observations were done in cowsheds that have biogas reactors. Secondary data used are the list of reactor recipients from ESDM and the list of reactor recipients from the BIRU program.

For this study, data from 79 respondents were collected, from a total of 125 households that have a biodigester. Respondents had to be more than 17 years old, could understand the questions, and be the beneficiary or user of a biodigester.

The quantitative analysis used a descriptive analysis that explains the condition of the object of study for easier understanding. In addition, a content analysis was done to support the descriptive analysis to get a deeper understanding of biogas implementation in Lembang, particularly in Kampung Areng.A subsection
Some text.

3. Literature Study

3.1. Waste management in rural areas

Agriculture is usually the main livelihood in rural areas, particularly in developing countries. Waste from agriculture is often not managed well, particularly because of limited infrastructure. Agricultural activities usually produce organic waste such as animal manure, food processing waste, crop/plant waste, and other organic waste from agricultural industry. However, this waste could be a valuable resource if managed carefully.

One agricultural waste is animal manure produced daily by cattle. The large amount of animal manure has a negative impact on the environment if not managed properly. When treated poorly, animal manure can be a major source of air and water pollution [12], which often occurs in rural areas in developing countries. In developing countries, there are limitations in choosing management strategies for utilization of animal manure [13], such as:

- Public acceptance (nuisance or environmental concern)
- Acceptable integration into agriculture
- Quality control of residues being applied
- Logistics and organization
- Fulfillment of environmental regulations
- Economic viability
- Sustainability

Environmental factors are an important aspect in choosing strategies for animal waste management. Strategies for waste management that minimize water and air pollution, emission of ammonia and pathogens could provide solutions in handling animal manure [13]. Anaerobic digestion of animal manure is a method of managing this waste. This process usually produces two main products: a biogas that can be used as fuel for stoves or to generate electricity; a substrate that can be used as fertilizer.

3.2. Waste management in rural areas

Rural areas have many resources that can be developed for energy production. However, this potential relies on traditional methods, particularly in developing countries. In several areas, Indonesian communities still use firewood, which contributes to carbon emission and lack of energy efficiency. To unleash the renewable energy potential in rural areas is not an easy task. There are several problems that usually challenge the development of renewable energy, such as financial issues, coordination problems, awareness problems, and technological and knowledge problems.

The development of rural biogas utilization at household level comes with several benefits simultaneously, namely improving sanitation, enhancing ecology in rural areas, more efficient and effective energy production, decreasing greenhouse gas emission, and improving agricultural products.

Rural areas in developing countries usually have abundant biogas fermentation resources [14]. In addition, biogas is easy and simple to utilize. In the last decades, biogas utilization has grown quickly in many Asian, Latin America, and African countries, such as in China, Nepal, Thailand and Indonesia [7][14][15][16]. In China, the development of biogas for commercial use began in 1921 in Guangdong province, however its popularity increased when the Chinese government started to promote 'biogas use by every rural family' and facilitated seven million biodigesters [17].

Several types of biogas can be categorized based on its main substrate [18]:

- Natural fertilizer (cattle slurry, pig slurry, cattle manure, pig manure, chicken manure)
- Plants (maize silage, rye, grass silage)
- Products of agricultural industry (brewer grain, grain decoction, potato decoction, pomace)
- Other substrates (waste fittings, gastric contents)
- Grasses (mown grass)

In developing counties, most biodigesters in rural areas can be categorized into three major types, namely fixed dome digester, floating drum digester (telescoping digester), and plug flow digester (saused-bag or channel digester) [19][20]. Each of this type has advantages and disadvantages (Table 1). The design of biodigesters is varied based on aspects such as geographical location, availability of substrate, and climatic conditions [19]. For instance, the fixed dome model was developed in China and the floating drum model was developed in India.

Table 1. Advantages and disadvantages of three types of biodigester.

Digester Type	Advantages	Disadvantages
Fixed dome digester	• Low initial cost • Long useful lifespan • No moving or rusting parts • Compact basic design • Less land required • Low maintenance	• Requires high technical skill • Difficult to repair in case of leakage • Requires heavy construction materials • Amount of gas produced is not immediately visible
Floating drum digester	• Simple and easy to understand operation • Visible stored gas volume • Constant gas pressure • Relatively easy construction	• High material costs of extra steel drum • Short lifespan because of steel drum corrosion • High maintenance because of regular painting of drum
Plug flow digester	• Low cost • Ease of transportation • Low construction sophistication • Uncomplicated maintenance • Less subject to climatic variations than fixed dome	• Relatively short lifespan • High susceptibility to damage • Low gas pressure • Limited creation of local employment • High impact on environment

In Indonesia, several studies have shown the impact of renewable energy in rural areas. For example, biogas implementation in Boyolali district provided economic benefit for the communities as well as a clean environment and organic fertilizer for farmers [21]. In addition, biogas has an important role in providing energy to villages who have limited access [22]. Furthermore, in Sumba Island, renewable energy, i.e. biogas, provided a platform for increasing community resilience [23].

4. Study Context

Kampung Areng is located in Cibodas village, Lembang subdistrict, West Bandung district, at an altitude of 1260 meter above sea level. Cibodas village has 2.959 households and is divided into 17 small neighborhoods (rukun warga). Some small neighborhoods have a majority of residents working as dairy cattle farmers. The majority of biodigesters in Cibodas village are located in Kampung Areng (Figure 1). Most of the farmers use communal stalls, thus in one stall are cows with different owners. Below is a neighborhood level map of Cibodas village and the number of biodigesters installed in the area according to Yayasan Rumah Energi's list of beneficiaries acquired in 2015.

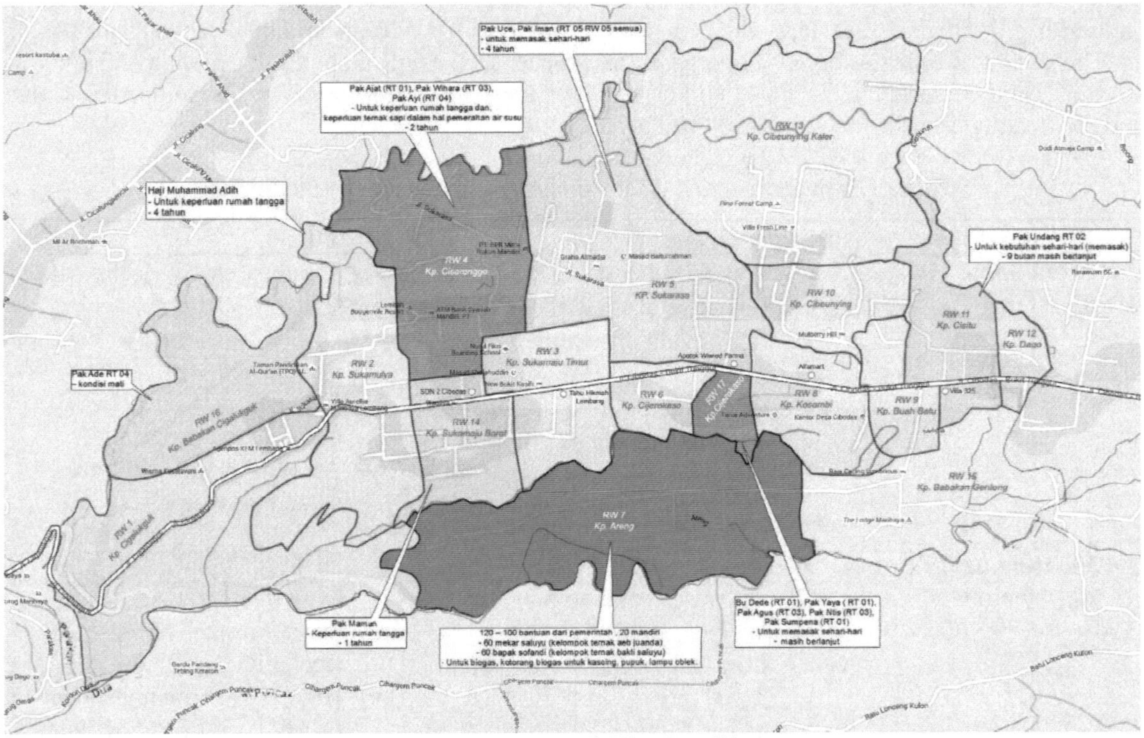

Figure 1. Map of Cibodas village
Sources: Google Maps and Yayasan Rumah Energi beneficiary list, 2015

The utilization of biodigesters in Cibodas village began with the research of a college student for Student Creativity Week. This was continued with a project from Hivos, a non-govermental organization from the Netherlands. The project is called the Indonesia Domestic Biogas Program (IDBP) and started in 2009 in West Java province. Through collaboration with the Ministry of Energy and Mineral Resources and local partner Rumah Energi Foundation, IDBP – popularly called BIRU (Biogas Rumah) – was implemented in Cibodas village.

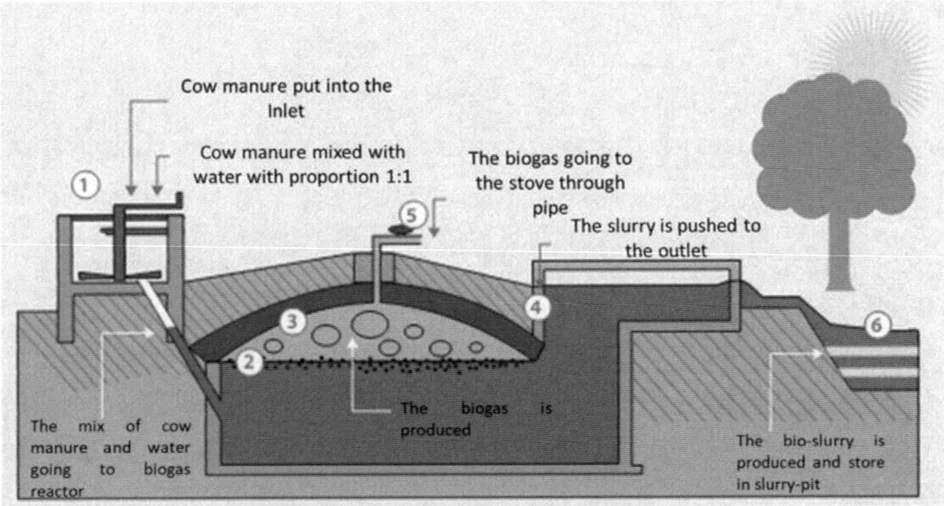

Figure 2. The process of biodigester work
Source: Yayasan Rumah Energi, 2015

Above is a diagram of the biodigester mechanism that converts cow manure to energy in a fixed dome digester [24]. Biodigesters in Cibodas village are mostly of the fixed dome type as shown above. The majority of them have a size of about 4 or 6 m2. Most biodigesters are aid from the BIRU Program or the government of West Java province. The biodigesters from the government were granted free of charge. However, beneficiaries from the BIRU program pay for construction and services, although they receive an incentive as applied under the BIRU Program, i.e. the beneficiary or potential user can receive a subsidy of about 2 million IDR. Therefore, there are some limitations and differences in the services of each program in its implementation process.

Usually, biodigesters are located near a stall to make it easier to use. The biogas is connected to a house or stall. If the distance between the house and the biodigester is far, more pressure from the biodigester is required.

According to Mekar Saluyu, ex farmer's group head, from 2004 to 2009 the adoption of biogas in Kampung Areng proceeded relatively slowly. After Yayasan Rumah Energi intervened with their BIRU program at the end of 2009, people begin to have more interest in using biogas. YRE's cooperation with KPSBU (Bandung Utara Milk Cooperation) also helped its distribution. Then, after biogas received widespread implementation, the West Java provincial government introduced the idea of Kampung Areng as an independent energy village and granted a total of 82 biogas units from 2011-2014. However, after the introduction of these programs, the interest in the community decreased and due to a lack of supervision and maintenance, many biodigesters broke down and were left unused. More recent adoption of biogas, after 2014, came from farmers who participated in the BIRU program.

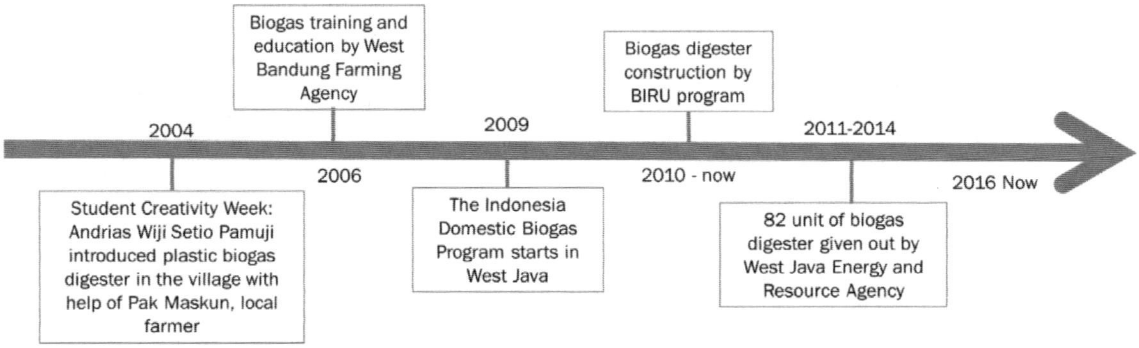

Figure 3. History of development biodigesters in Cibodas Village
Source: Interview with head of farmer group in Kampung Areng, 2016

5. Biogas as wast management

Biogas can provide environmental benefits, such as supplying a sustainable source of energy, soil enrichment, re-utilization and treatment of a variety of organic wastes, and minimization of environmental impacts of greenhouse gas emission, and even reduction of land use problems caused by organic waste disposal [15]. In Kampung Areng, the main benefit is waste management of cow manure. Due to the limited waste management infrastructure, sometimes cow manure is flushed into Cikapundung river, which is close to the area. Biodigesters can replace waste management services and even have other benefits for cattle farmers.

Implementation of biodigesters in Cibodas village is a potential solution in managing waste, specifically cow manure from dairy cattle farms. If managed correctly, cow manure can potentially be utilized for producing biogas for cooking and electricity, and even as fertilizer, as illustrated in table 1.

Table 2. Potency of cow manure utilize with biodigester

Biodigester Capacity (m^3)	Quantity (Unit)	Maximum input of digester	Total volume of manure in biodigesters
4	43	40	1.720
6	31	60	1.860
8	4	80	320
12	1	120	120
			4.020

Based on the capacity of a biodigester, effectively about 4020 kg of cow manure can be processed, about 81.21% of the total amount of cow manure produced by the cows (Table 2). However, this calculation is limited to the cows that are owned by the beneficiary of the biodigester. It is very likely that the amount of cow manure is higher than this calculation shows, because not all of the cattle farmers have a biodigester. In some cases we found that a cattle farmer who did not have a biodigester gave his cow manure to a neighbor who owned one. This calculation assumes that each cow produces about 15 kg of cow manure per day.

Table 3. Waste potentials of cows' manure.

	Number of cows	Average amount of cow manure	Potential amount of cow manure processed by biodigesters	Amount of cow manure unprocessed by biodigesters
Value	330	4.950	4.020	930
Unit		Kg	Kg	Kg
%		100	81,21	18,79

From Table 3, not all cow manure is processed due to the limited number of biodigesters. In addition, some biodigesters are not operational for various reasons. The majority of them because some parts of the biodigester are broken, such as the inset (figure 4) or the biogas pipe that connects the biodigester to a stove. Other reasons were because the cows were sold or the stove was broken, etc. Thus, the real biodigester capacity in processing cow manure is reduced; only about 73 % of total cow manure can be processed into biogas and bio-slurry.
Unprocessed cow manure can be a problem for the environment as it is a water and air pollutant. It can pollute the soil and the water of Cikapundung river near Kampung Areng that is used as water supply for Bandung city.

Figure 4. Inlet of a biodigester with a broken mixer.

Other than processing animal waste, biodigesters also have the additional benefit of slurry production. Bio-slurry can be processed further into items with added economic value such as fertilizer. In Cibodas usually bio-slurry is processed into wet fertilizer, dry fertilizer, liquid fertilizer, and medium for worm farming.

- Dry fertilizer is made using wet solid biogas dregs, which are then dried. This takes a long time because traditional methods are still used, which depend on the heat of the sun. After drying the texture of the fertilizer becomes slightly tenuous and lightweight so it is easier to transport and the price is slightly higher when sold compared to wet fertilizer. Dry organic fertilizer is sold at prices ranging from Rp. 10,000 to 15,000 per sack.
- Wet fertilizer, is non-dried and can be directly used. It is used by vegetable farmers in the study sites and yields good results. The constraint is that the wet conditions make it difficult to transport to agricultural sites that are far from the road and can be steep. Because it is in a wet condition, this type of fertilizer has a lower price at Rp. 5000 per sack.

- Liquid fertilizer is a liquid-like biogas dreg that comes out of the digester outlet and contains many nutrients that are good for plants but is still very under-utilized by the owners of digesters in Cibodas village, whereas in some other areas liquid organic fertilizer from digesters is successfully sold after screening to get rid of unnecessary by-products. This fertilizer can be applied directly to plants or diluted with water.
- Worm farm medium, i.e. solid-state biogas dregs, is used for the cultivation of the earthworm species of lumbricus rubellus, from which worms and kascing (worm-based) fertilizers can be produced. From the process of worm maintenance using biogas dregs a worm harvest is obtained that can be sold directly to collectors. These worms have a high enough selling value of Rp. 25,000 per kilogram, while good quality kascing fertilizer is valued at Rp. 25,000 per sack weighing about 30 kg. Using these worms, the resulting fertilizer dries faster than naturally dried biogas dregs and obtains a crumbly and light texture, making it easier for farmers to transport and to apply on the farm. It yields higher prices compared to dried compost. The price is set at the level of the worm farmers, where buyers buy directly from the worm cultivators. However, despite the high prices, worm farmers who also work as farmers prefer to use kascing fertilizer to meet their own fertilizer needs, especially for feeding seeds, because it produces healthy and strong seeds.

Below is a table of how the questionnaire respondents process their slurry in the area of Cibodas village.

Table 4. Bio-slurry processing in Cibodas area.

Bio-slurry uses	Yes	No	% use
Dry fertilizer	37	42	46,84
Wet fertilizer	15	64	18,99
Liquid fertilizer	3	76	3,80
	18	61	22,78

In Cibodas, dry fertilizer production is a popular way to process slurry. In the case of further processing of biogas dregs to get additional benefits that can increase the family income, 42 digester owners (53.2%) process it into dry organic fertilizer, 18 use it as a medium for kascing fertilizer, 15 use it as wet organic fertilizer and only 3 use it as liquid organic fertilizer. Meanwhile, there are only 18 (22,78%) who process their slurry as a worm farming medium.

Figure 5. Slurry processing into fertilizer and worm medium

There are some challenges in the implementation of biogas in Kampung Areng:

1. Not all of the bio-slurry is being used. Sometimes the farmer does not use the solid slurry because he does not have a farm field and the liquid slurry is thrown away into a ditch or near a neighbor's field farm. This happens because of oversupply and limited demand of bio-slurry.

2. Some of the beneficiaries of the biogas project still use LPG, particularly for biodigesters with a smaller capacity (4 or 6 m3), because their stove can only be used for 1-2 hours on biogas.
3. There is no monitoring of the implementation of biodigesters, particularly from the government, so some of the biodigesters are not operational.

One of the clear impacts of these biogas programs is empowerment of the women in Kampung Areng. Kampung Areng's Kelompok Karya Ibu has developed a business of worm farming with members consisting of mothers of farmers/ranchers, counting about 25 people. Group earnings are usually collected every two weeks from the members of the group, who collectively can produce about 100 kilograms of worms and 100 sacks (sacks) of kascing fertilizer. This is done when there is a buyer of kascing fertilizer who needs a large amount of fertilizer. Otherwise they sell directly to farmers in need or to meet their own needs. However, this case is quite rare in Kampung Areng. In addition, the women in Kampung Areng also actively manage a waste bank. They have developed an organization that manages the waste in Kampung Areng so it can be sold.

6. Discussion

To maintain a small-scale biodigester and keep it working well is not easy. It requires cooperation between the user and the provider. There are many factors that can affect the user to no longer use a biodigester, such as broken parts, cows were sold, or the family moved to another place. The BIRU project tries to manage these issues through several strategies. They provide after-sales services in the form of maintenance and monitoring the biodigesters to ensure they work well. This helps the user to keep the biodigester in good condition and to process cow manure in producing biogas and bio-slurry. However, owners of biodigesters that were provided by the government do not have this support, so some of the biodigesters are no longer operational because parts are broken and the users do not have the capability to do the repairs. This is a problem for the development of biogas as waste management solution related to cow manure in rural areas. It needs to be addressed particularly by government agencies who provide biodigesters to cattle farmers, by monitoring the implementation and maintenance of this infrastructure.

Bio-slurry as by-product of biogas can have many benefits for a community. For instance, bio-slurry is a potent organic fertilizer. It could help farmers grow their plants using organic matter and provide a viable solution to nutrient depletion of many agricultural soils. Bio-slurry can boost agriculture and horticultural production. In Kampung Areng, cattle farmers use it for their plants in the field or use it as fertilizer for growing grass as food for their cows. In some cases, they use it in the forest to maintain the growth of grass. In addition, it can reduce the costs of cattle farmers who also have a field farm.

Biodigesters could be mandatory for cattle farmers in rural areas to process their cow manure, particularly farmers who have quite a large number of cattle. It would be best if the locations of the farmers are spread so they can each distribute and sell their bio-slurry in their own area.

However, if the locations of stalls and biodigesters are agglomerated, the market system for bio-slurry needs to be developed. The excess bio-slurry needs to be distributed to other farmers, but this does not happen well. The popularity of bio-slurry as fertilizer for agriculture plants is still limited and is not insufficient to build interest of farmers to use it. Sometimes, the transport cost of bio-slurry is higher than the price of the product. If there is a market system for bio-slurry, it would give owners motivation to maintain their biodigester.

7. Conclusion

This paper shows the possibilities of biogas utilization, specifically biodigester utilization, as an infrastructure for waste management of cow manure in rural areas. Biodigesters can benefit cattle farmers by transforming cow manure to a source of energy and even fertilizer. Biogas in rural areas can become an important solution in the reduction of waste (cow manure), as found in Kampung Areng, particularly for the neighborhoods close to the upstream of Cikapundung river.

There remain some challenges in the implementation of biogas digesters. Firstly, not all of the bio-slurry is used. Secondly, some of the beneficiaries of biogas projects still have to rely on LPG, particularly in the case of biodigesters with a smaller capacity (4 or 6 m3), because their stoves can only be used for 1-2 hours on biogas. Finally, there is no monitoring of the implementation of biodigesters, particularly those granted by the government, thus some biodigesters do not operate well.

Through collaboration between the government, INGO, unions and communities, farmers in Kampung Areng could gain benefits from utilizing biodigesters. However, in some households the biodigester is not working optimally after a number of years due to several issues, such as broken parts, not having enough cow manure (cows have been sold), and more user-friendly and easier sources of energy such as LPG. Based on this study, thee is a possibility to increase the capacity of biodigesters in Kampung Areng, Cibodas village. Maintenance and monitoring of implementation of biodigesters could optimize biogas utilization and creation of a market for bio-slurry can give farmers more motivation to use a biodigester.

References

[1] Tarigan AKM et al 2017 Balikpapan: Urban planning and development in anticipation of the post-oil industry era *Cities* 60(2017):246–59.A

[2] Fitriana I et al 2017 *Indonesia Energy Outlook 2017* (Jakarta: Badan Pengkajian dan Penerapan Teknologi)

[3] Dincer I 2010 Renewable energy and sustainable development: a crucial review *Renew Sustain Energy Rev* 2000;4(2):157–75

[4] Meidiana C and Gamse T Development of Waste Management Practices in Indonesia *Eur J Sci Res* 40(2):199–210

[5] Pemerintah Republik Indonesia 2017 *Rencana Umum Energi Nasional* (Jakarta: Ministry of Energy and Mineral Resources of Indonesia)

[6] Perusahaan Listrik Negara 2017 *Rencana Usaha Penyediaan Tenaga Listrik (RUPTL) Perusahaan Listrik Negara* (Jakarta: Perusahaan Listrik Negara)

[7] Mengistu MG et al 2015 A review on biogas technology and its contributions to sustainable rural livelihood in Ethiopia *Renew Sustain Energy Rev* 48:306–16

[8] Deublein D and Steinhauser A 2008 *Biogas from Waste and Renewable Resources* (Weinheim: WILEY-VCH verlag GmbH & Co)

[9] Bajgain S 2011 *Feasibility of Biogas in Sumba* [Internet]. Hivos Org. 2011. Available from: https://hivos.org/sites/default/files/feasibility_of_biogas_in_sumba_.pdf

[10] van Ness WJ, Tumiwa F and Setyadi I 2009 Feasibility of a National Programme on Domestic Biogas in Indonesia: Final Report. *SNV Netherl Organ Netherl*

[11] Widodo TW and Hendriadi A 2005 Development of Biogas Processing for Small Scale Cattle Farm in Indonesia. *Proc. Int. Semin. Biogas Technol Poverty Reduct Sustain Dev Beijing*, 17-20 Oct 2005 (October):255–61

[12] Holm-Nielsen JB, Al Seadi T and Oleskowicz-Popiel P 2009 The future of anaerobic digestion and biogas utilization *Bioresour Technol* 100(22):5478–84

[13] Westerman PW and Bicudo JR 2005 Management considerations for organic waste use in agriculture *Bioresour Technol* 96(2):215–21

[14] Ni JQ and Nyns EJ 1996 New concept for the evaluation of rural biogas management in developing countries *Energy Convers Manag* 37(10):1525–34

[15] Aggarangsi P et al 2013 Overview of livestock biogas technology development and implementation in Thailand *Energy Sustain Dev* 17(4):371–7

[16] Cheng S et al Development and application of prefabricated biogas digesters in developing

countries. Renew Sustain Energy Rev [Internet]. 2014;34:387–400. Available from: http://dx.doi.org/10.1016/j.rser.2014.03.035

[17] He PJ 2010 Anaerobic digestion: An intriguing long history in China *Waste Manag* 30(4):549–50

[18] Kazimierowicz J 2015 The Effect of Substrate on the Amount and Composition of Biogas in Agricultural Biogas Plant *Pol Akad Nauk* (Iii):809–18

[19] Rajendran K, Aslanzadeh S and Taherzadeh MJ 2012 Household biogas digesters-A review *Energies* 5(8):2911–42

[20] Plöchl M and Heiermann M 2006 Biogas Farming in Central and Northern Europe: A Strategy for Developing Countries *Agric Eng Int* VIII(8):1–15

[21] Hnyine ZT et al 2015 Analysing the economic benefits of rural biogas adoption in Selo Sub-District, Boyolali, Indonesia (Bandung: RDI)

[22] Sagala S et al 2015 Energy Resilient Village Potential In Boyolali, Indonesia (RDI Working Paper Series; Bandung: Indonesia)

[23] Al-faruq U et al 2016 Assessment of Renewable Energy Impact to Community Resilience in Sumba Island. In: Anwar H, editor. Masyarakat Tangguh Bencana

[24] Yayasan Rumah Energi. Tentang Bio-Slurry [Internet]. Biogas Rumah. Available from: http://www.biru.or.id/index.php/bio-slurry/

Preferences mapping of household biodigester in Bandung

S Humaira[1], E Rianawati[3], S Sagala[2], M A Sasongko[3]

[1]Department of Engineering Management, Institut Teknologi Bandung, Bandung, West Java, Indonesia

[2]Department of Regional and City Planning, Institut Teknologi Bandung, Bandung, West Java, Indonesia

[3]Resilience Development Initiative, Bandung Indonesia

Corresponding author: salma.humaira71@gmail.com

Abstract. Bandung city government implemented household biodigester grants in 2015 and 2016. Unfortunately, there are some household biodigesters that still functioning well but not in use. Therefore, this study is an effort to improve the acceptance and usage rate of household biodigesters in Bandung. The purpose of this study is to know citizen's preference when it comes to household biodigester. To get the picture, we conducted survey through online questionnaire based on eight dimension of quality defined by Garvin (1987) as basis to construct factors that might be favoured by current and potential users of household biodigesters. Based on result of cluster analysis, three clusters with different preferences were interpreted and profiled through Welch's ANOVA and Games-Howell Test. This study reveals that the cluster with the largest number of members shows reliability and features as the key to determining current and potential user's preference. This study suggests the developer of household biodigester to choose cluster 1 and prioritize the aspect of reliability and feature within the development of the next household biodigester product to get higher level of public acceptance.

1. Introduction

The service capacity of garbage handling until now by PD Kebersihan (local waste management company) is still not optimal with only about 65% of the waste is served [1]. In jabarprov.go.id [2], Bandung Vice Mayor Oded M. Daniel mentioned that the city of Bandung produces waste up to 1,500 tons per day while the waste transport capacity is only about 1,100 tons.

Biodigester, a waste-to-energy power generation, generates power by turning organic waste into gases and fertilizers. In February 2015, Bandung City Government provided 100 biodigesters, each with the capacity of processing 15 kg of organic waste, for interested Bandung citizens in exchange of only a number of application documents [3].

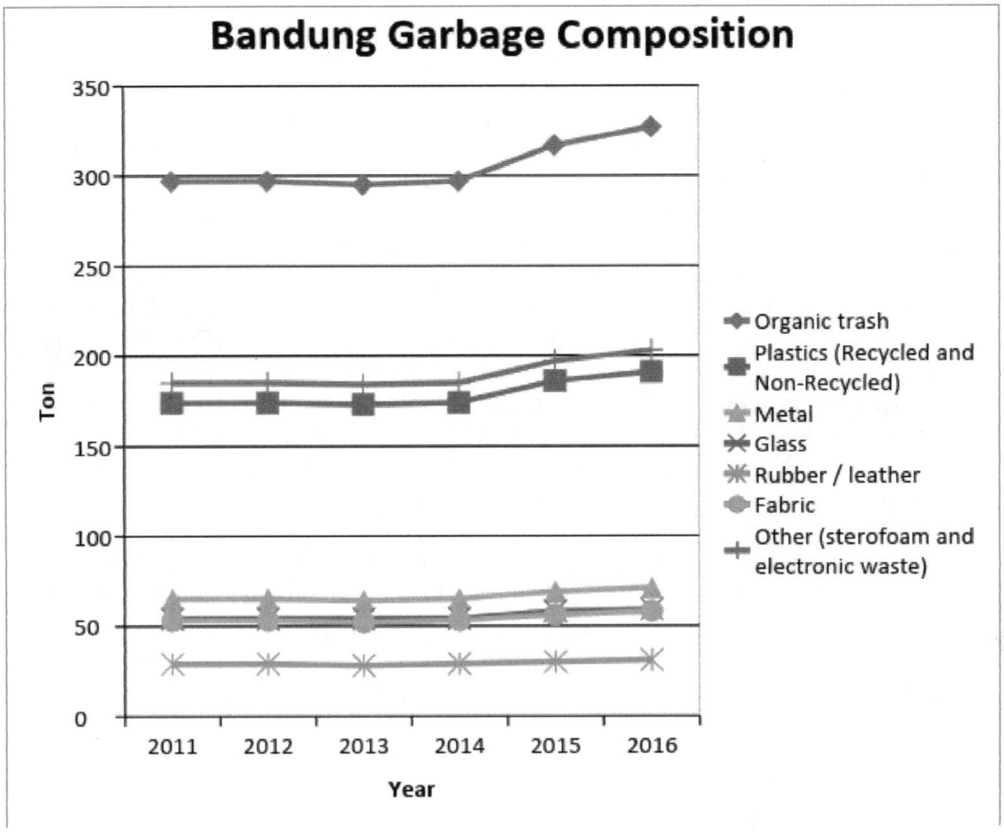

Figure 1. Bandung Garbage Composition [1]

Based on Figure 1, it is seen that the organic waste makes for the biggest percentage of total waste in Bandung. For this reason, the application of biodigester becomes highly potential even though it requires citizens' willingness to separate waste. National news channel kompas.com reported that in 2014, Bandung Mayor Ridwan Kamil made a declaration to reduce household waste. One way through which household waste could be reduced is through the use of household biodigester, which would hopefully alleviate garbage transportation cost by reducing the amount of waste that must be transported into landfills.

Prior to providing grants for household-scale biodigesters, Bandung City Government had provided grants for larger-scale biodigesters able to process 250 kg of waste. The grant program resulted in 14 250 kg biodigesters by BPLH in Bandung, but after some time, according to the survey conducted by BPLH [4] the use of 250 kg biodigester is considered to have much greater problems than household biodigesters. Those problems are:

- Community involvement
  BPLH [4] mentioned that 5 of 14 communal-scale biodigesters have no clear biodigester management responsibilities. Seven of the nine operating biodigesters are granted to an institution. Thus, the agency may assign parties within the agency to operate the biodigester. Unfortunately, in the seven institutions, the surrounding community has not been involved in donating waste to biodigesters, violating the initial goal of biodigester procurement in Bandung to reduce household waste. Therefore, the initial goal is most likely unattainable from 250 kg biodigesters.
- Land use
  A 250 kg biodigester require a minimum land size of 6.8 by 4.6 meters, therefore requiring government-owned land so as not to risk being converted. With some of the fourteen 250

kg biodigesters in Bandung still located in private lands, they face risks of conversion. For example, in Palasari Village, landowners say they want to transfer land functions in the next 5-10 years. There is also a biodigester installed in the burial ground, which makes it susceptible to land use conversion into a grave. 15 kg biodigesters occupy an area of only 1 by 1 meter and can be installed on the residents' land, for example, citizens' backyards.

- The need for organic waste
Almost all 250 kg biodigesters have not been used optimally. Some of these biodigester recipients are only able to provide 2-20 percent of the total waste needed. When compared to 15 kg biodigesters, 250 kg biodigesters grant recipients have higher difficulty in finding organic waste to meet daily waste requirements. Some 250 kg biodigesters in Bandung use biomass, mostly purchased from cattle farm waste. This is less suited to the original purpose of biodigester installation, i.e. to reduce household waste. Most people objected to transporting garbage from their homes because of the location of biodigester are quite far enough from their homes. With the neighbourhood-located 15 kg biodigesters, most residents around the biodigester's location are still willing to collect and deliver their organic waste to the within-reach location of the biodigester.
- The utilization of energy yield is not yet optimal
One biodigester installed with 250 kg of garbage input per day should be used for 5 households or more, but the energy produced by most of the 250 kg biodigesters in Bandung is only used by one party. In Rancasari Village, the biogas produced is only used for lighting in *Rukun Warga* (RW) offices. In Sumur Bandung Subdistrict, the energy generated was only utilized by managers of a landfill. Until recently, gas produced from 15 kg biodigester can only be consumed by one house, but this is worthwhile considering the much lower investment fund compared to the 250 kg biodigesters.
- Potential hazard
Unused gas produced by biodigester can fill up the gas holder. Rancasari Subdistrict suffers from potential leakage of gas holder due to the stove only being used about three times per week. This may lead to leakage as well as explosion of gas holder, releasing methane gas as contaminant into the environment.

Of the various problems that have been mentioned, 250 kg biodigester capacity is considered less feasible to reapply in the city of Bandung. Therefore, household-scale biodigester is considered to have potential that is expected to grow better in the city of Bandung because the problem posed is not as much a problem of biodigester capacity of 250 kg.

While almost all 250 kg biodigesters are not used, household biodigester has better potentials. According to research results by Cahyono, more than half of household biodigesters distributed around Bandung is still used by their respective owners. The research by Cahyono [5], seen in further details in Figure 2 below, reveals that 67 out of 103 household biodigesters are still functioning well. This fact reveals a difference between the term of "functioning biodigester" and "biodigester still in use". Some biodigesters are still functioning but not used.

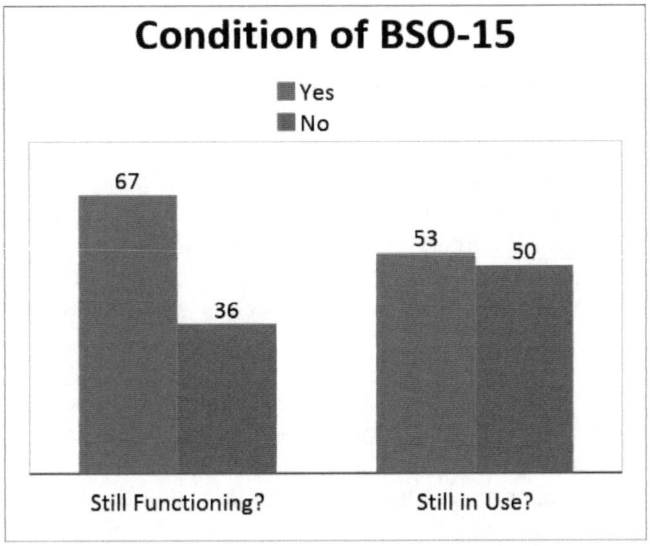

Figure 2. Condition BSO-15 in Bandung [2]

Biodigester is a technology-driven product or known as push technology product. Eppinger & Ulrich [6] implies that although push products are at the heart of a technology success, product developers need to pay attention to the preferences of users or potential users. Developing a new product carries a lot of risks that are reduced by identifying customers' needs. In household biodigesters, residents become the operators, which renders researching the suitability user preference crucial. So far, no studies have focused on capturing a clear picture on citizens' preferences for household biodigesters. Therefore, this research aims to elucidate citizens' preferences for household biodigester.

According to Gendall and Esslemont [7] market research is unable to produce explanations of consumer behaviour reliably and comprehensively, but market research can increase the probability of making the right decision. McMillan [8] states that one's interest in using a product is when the product provides the benefits they want. In a city or region, the community is the main generator of waste, rendering its role very important. The success rate of use and management of biodigesters is highly dependent on community participation. Therefore, to achieve a strong community acceptance, it is necessary to know how citizens' expectations for biodigesters. This study aims to map the preference of Bandung citizens about household biodigesters in order to improve the community acceptance towards household biodigesters.

2. Data methodology

Assumption and Scope

Biodigesters are not yet commonly known by the citizens of Bandung. As such, to increase response rate, these questionnaires are only distributed to people with in environmental preservation: environmental activists, researchers, and students who study biodigesters in their majors. This research assumes that the preference of Bandung citizen can be represented by those people. Another assumption is that preference are the same across various regions of Indonesia, because we assume that in general the people of Indonesia have the same perception about energy. To verify this assumption, available data will be tested using cross-tabulation (chi-square) which will identify any discrepancies between Bandung and non-Bandung citizens.

Design of Data Collection

The important condition for a product to succeed is that it must offer what the customer/user perceives as benefit (6). Therefore, in this research will use the quality concept of Garvin [9] which divides the quality of a product into 8 dimensions: performance, features, reliability, conformance, durability, serviceability, aesthetics, and perceived quality.

In this study, the dimensions of conformance, aesthetics, and perceived quality are not used because they are less relevant to household biodigester products. Then the operationalization of the concept conducted by describing these dimensions into 31 indicators that are considered capable of measuring these dimensions. Indicators used are obtained from various references, such as journals, articles, theories in product development, final project, and others. In addition to the literature study, the indicators were also obtained through interviews with some biodigester users in Bandung as well as the results of brainstorming (as a form of verification/content validation) with the biodigester household product developers on the technical suitability of indicators identified with household biodigester products.

2.1. Scale selection

This study uses ordinal scale which is considered as the interval scale that would be useful to process the data using parametric statistics, if the basic assumption test statistics was passed first. The type of scale selected in this study is Likert scale. To ensure the objectivity, the questionnaires were made on a balanced scale. Unbalanced scales are commonly used when obtained data has the possibility of abnormalities [10]. In order to make the respondents could choose neutral option within the scale based question answer, the odd-scale was selected because it has a middle value or a neutral value. Respondents can choose neutral because as there is a possibility that they feel neutral with these indicators. On an odd scale too, this questionnaire becomes a questionnaire that does not compel respondents because if they feel neutral about an indicator then they can choose the middle value. The scale used is 7. The scale is chosen to avoid being too narrow with 5 and to avoid respondents' being overwhelmed with too many options in the case of 9.

2.2. Questionnaire design

A good questionnaire should motivate respondents to be willing to contribute, feel involved, and provide complete, honest and accurate answers [10]. Each statement representing the indicator of this study will be marked as required. This is done to avoid missing value. Although the activated feature required this questionnaire to be impressed, respondents should not feel threatened. First, because all indicators have been identified do not require any personal data that may be too sensitive to divulge. Second, as described in the previous section, there is a neutral choice in this questionnaire. Third, data retrieval will be done using typeform.com which has a user interface that is convenient for the respondent.

The questionnaire consists of two parts. The initial part of the questionnaire contains the respondents' assessment of the indicators identified. The form of the questionnaire selected is in the form of numbers 1 through 7 with information on the highest, lowest, and middle scale to facilitate the respondents to understand each scale used in the research questionnaire.

In the second part is a request for respondents to fill out the identity. Questions about demographics are deliberately placed at the end to improve the completion rate. According to Lindemann [11], there is a common practice in making questionnaires that we must begin to avoid: do not start a questionnaire with questions about the demographics of respondents. Because it will make people lazy to fill out the questionnaire thereby lowering the response rate. Preferably in a questionnaire that was first asked not in the form of personal data.

The questionnaire for this study also uses a progress bar to show completion percentage. Progress bars are used because basically humans have a desire to achieve goals. If he does not know how far he has stepped he might stop before finishing the questionnaire so because he does not know when the questionnaire ends. According to Zhang [12] with this progress bar, when he has completed most of the questions in the questionnaire when at the end there is a sensitive question, most likely respondent will fill it because he knows he has done a lot by looking at the progress bar. So, with the design of such questionnaires, indirectly respondents are forced to provide personal data. Type of progress bar selected is the percentage because if the progress bar shows the number of questions that have been filled, then from the beginning to open the questionnaire, respondents already know the number of questions very much in the questionnaire of this study so it is feared respondents become lazy and immediately close the questionnaire without completing it first. It will be different if the progress bar is used as a percentage, then the respondent will not know the number of questions but he will have a picture of how far he has done in the questionnaire that is being filled.

2.3. Validity and Reliability Test

Quality data is obtained through a valid and correct method: the measurement is in accordance with what the researcher would like to measure. In this research, we have collected data for validity and reliability test three times with each test using about 30 respondents. The validity test used is construct validity and the results of all indicators are valid convergent. The reliability test performed is internal consistency reliability with the calculation method of Cronbach's Alpha. According to Malhotra [10], a dimension is reliable if the alpha value more than 0.6. From the calculation result is concluded that all dimensions used are reliable because all dimensions have alpha value more than 0.6.

Table 1. Reliability test.

Dimensions	*Cronbach's Alpha*	Reliable
Performance	0.6945	Yes
Features	0.7340	Yes
Reliability	0.9010	Yes
Durability	0.7063	Yes
Serviceability	0.8739	Yes

3. Collection and processing data method

After the questionnaire is proven to be valid and reliable, we collect data from 146 respondents. According to Zhao [13], the respondent required in a study should be five times the number of variables used – in this case, at least 100 respondents. After data collection, we conduct outlier test using Grubbs Test, identifying three outliers due to unusual filling that are subsequently removed. After that, we conduct the basic assumption test using the normality test, heteroscedasticity test, and linearity test. The result states that the data used pass the linearity test but cannot be assumed to be normal and heteroscedasticity. In this research, autocorrelation test is not conducted because the data used is not a time series data so it is not necessary to do autocorrelation test [14].

The tools used in this research are cluster analysis and ANOVA along with its post hoc test to know the preference of each cluster. This research also uses cross tabulation to find the demographic relation with the cluster formed. According to Malhotra & Birks [10], cross tabulation is a statistical technique that describes two or more variables simultaneously and generates a table that reflects a combination of two or more variables that have a number of different categories or values. In this research is used chi-square calculation to know the relation of demography variable with cluster formed.

Based on Santiago [15] some testing tools such as the t-test, Analysis of Variance (ANOVA), Regression, and Design of Experiment (DOE) are considered robust against normality assumptions. According to Glen [16], although ANOVA requires normality assumptions, we can still use this test if the sample size is large enough (usually more than 20). If the sample size is very small then it is better to use nonparametric testing. This study uses 143 samples (sans outliers) to justify the use of ANOVA testing.

Considering one of the assumption test results that the data obtained for this study cannot be assumed to have the same variance, this study cannot use classic ANOVA. In processing the heteroscedasticity data, Welch's ANOVA and Kruskal-Wallis (nonparametric) can be an alternative data processing. Liu (17) has conducted a study to compare what is likely to be a lower type I error and which has higher power between Welch's ANOVA and Kruskal-Wallis Test. His research concluded that Welch's ANOVA gives the best performance in the appeal of Kruskal-Wallis, and then in this study will be used Welch's ANOVA.

Welch correction is a widely used solution in testing the difference between the average of two populations when the variance is not the same. The Games-Howell test is based on Welch's correction of degrees of freedom by t-test and using studentized range statistics. This method is based on the studentized distribution range. This method uses the formula to approximate Welch's degrees of freedom (Welch's d.o.f) to obtain a confidence interval approximation and to see the difference between the two averages based on the student's t-test.

Welch correction is designed to provide valid t-test when there is an unequal population variance. This correction involves the use of a number of degrees of freedom (v) corrected to assess t-statistical significance. v is the number obtained from the following equation (1):

$$v = \frac{\left(\frac{s_i^2}{n_i} + \frac{s_j^2}{n_j}\right)^2}{\frac{\left(\frac{s_i^2}{n_i}\right)^2}{n_i - 1} + \frac{\left(\frac{s_j^2}{n_j}\right)}{n_j - 1}} \tag{1}$$

With the following information,
s_i^2 and s_j^2 represent the variance of group samples i and j
n_i and n_i are the number of samples used

And the statistical test is,

$$t_W = \frac{\overline{x_i} + \overline{x_j}}{\left(\frac{s_i^2}{n_i} + \frac{s_j^2}{n_j}\right)^{1/2}} \tag{2}$$

With the following information,
$\overline{x_i}$ and $\overline{x_j}$ is the average of the sample
s_i^2 and s_j^2 represent the variance of group samples i and j
n_i and n_j are the number of samples used

Then the critical value is as follows,

$$\frac{q_{\alpha,k,v}}{\sqrt{2}} \tag{3}$$

Where v is as mentioned in equation (1), then k is the number of treatments. The value of $q_{\alpha,k,v}$ shows the upper α point of the studentized distribution range by using the parameters k and υ. Reject H_0 if $t_W \geq \frac{q_{\alpha,k,v}}{\sqrt{2}}$ and do not reject H_0 if the opposite happens.

The post-hoc test that will be used is Games-Howell Test. According to Shingala & Rajyaguru (18), the Games-Howell (GH) procedure is the equivalent of the Tukey Kramer Test. The GH method provides the best results for pairwise comparison. Then according to Hilton and Armstrong (19), the Games-Howell test is one of the robust methods among the latest methods. With Games-Howell, pairwise comparison can be performed although there are differences in the number of samples being

IOP Conf. Series: Earth and Environmental Science **158** (2018) 012032 doi:10.1088/1755-1315/158/1/012032

compared, with the data being heterogeneously variable, and when the data does not meet the assumption of normality.

4. Processing and analysis of data

4.1. Selection of clustering variables

Cluster analysis is one type of multivariate analysis aimed to categorize objects (such as respondents, products, and others) based on the characteristics they have (20). In this study, cluster analysis is used to determine whether there are different groups in a population. This is important because by understanding the preferences of groups in society, people's preferences can be understood in more detail and in depth. Steps in cluster analysis include choosing cluster variable, determining optimal cluster number, cluster validation, outlier analysis, and profiling the cluster.

An ideal clustering variable is a variable that is not correlated with other variables. However, because in this study there are no variables that really do not correlate with other variables, the selected cluster variable is the variable with the least number of correlations. Based on the calculation of the number of correlations between variables, it is found that the selected cluster variable (to represent every dimension) can be seen in table 2 below.

Table 2. Clustering variables.

Dimensions	**Variables**	
Performance	X1	Biodigester is able to convert waste into gas that can be used for cooking
Features	X8	There is a monthly incentive for the person who operates the biodigester
Reliability	X19	The stove can still be used even when the power (electricity) is off
Durability	X20	The biodigester-generated gas has the same quality in every operation of biodigester
Serviceability	X24	There is a regular meeting with biodigester engineers for the users

4.2. Agglomeration method

The most optimal number of clusters is the number of clusters that most explain the differences between the clusters that are formed. To determine it, the tools used are hierarchical cluster using SPSS 20. The hierarchical method used is agglomerative method. This method is a grouping method that is bottom up that is done by cluster number as much of number of samples then samples are grouped one by one based on their similarity to finally form one cluster.

Table 3. Determination of optimal cluster.

Number of Clusters	Difference of Coefficient	Changes in Coefficient Difference	Difference in Changes
1	*Emptied*	*Emptied*	*Emptied*
2	152.68	108.3%	*Emptied*
3	79.02	56.4%	51.8%
4	74.78	53.8%	2.6%
5	51.75	37.5%	16.3%
6	28.64	20.9%	16.6%
7	28.13	*Emptied*	*Emptied*

The optimal number of clusters depends on the stopping rule. The cluster formation must stop when there is the biggest change in the value of heterogeneity. Table 5.12 reveals that the spike occurring in cluster 3 makes it the best cluster, as the highest heterogeneous change spike is in the difference of cluster 2 to cluster 3 (51.8 percent), followed by clustering the respondents into 3 clusters. The methods used are the same as those done when determining the optimal number of clusters using Ward's Method, distance measurement using Squared Euclidean Distance, and standardization of data using Z-score (variable). This grouping is done using Minitab 17. The proportion of members of each cluster formed can be seen in Figure 3 below.

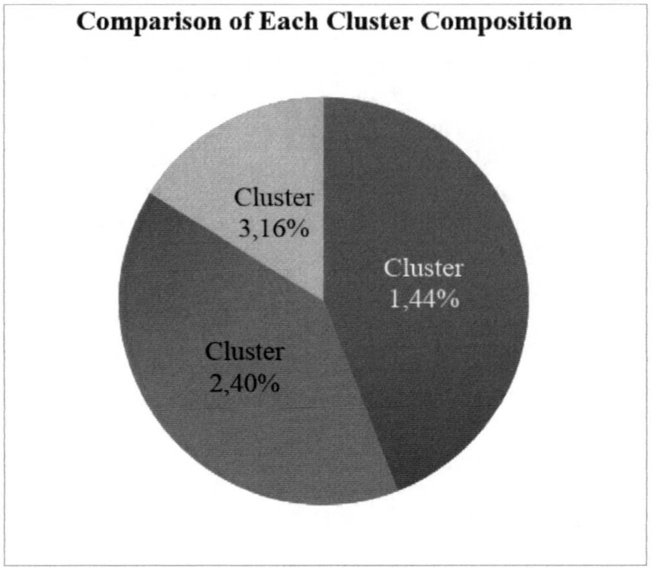

Figure 3. Comparison of each cluster composition.

4.3. Validity test
To test the validation of the clusters formed, cluster analysis with other methods (K-Means) was performed. It is known that 78.38 percent of the cluster members generated by K-Means are the same as cluster members generated through hierarchical methods. With the minimum standard validity of a cluster being 75 percent (20), it can be concluded that this study has passed the validation test.

4.4. Interpreting and profiling cluster
This study interprets and profiles cluster using Welch's ANOVA, Games-Howell Test, and cross-tabulation. The cross-tabulation test shows no difference between citizens' preference in Bandung and outside of Bandung, which means this study's assumption stands true. However, this result also makes it impossible for this research to describe member of an entire cluster. And then, to know the most important dimension of every cluster, we use the Welch's ANOVA and Games Howell Test. The result can be seen in Table 4 below.

Table 4. The most important dimensions.

Cluster	The most important dimensions
1	*Reliability*
	Features
2	*Durability*
	Reliability
3	*Performance*
	Durability
	Reliability

After that, we perform Welch's ANOVA and Games Howell Test again on the indicators of each dimension's clusters. The summary of the results can be seen in Table 5 below.

Table 5. Summaries of preferences.

Cluster	The Most Important Dimensions	The Most Important Indicator		The Most Important Indicator of Clustering Variables	
1	Features and Reliability	X14	Biodigester not easy to leak	X19	The stove can still be used even when the power (electricity) is off
		X16	Gas pipes to the stove are not easily broken	X8	There is a monthly incentive for the person who operates the biodigester
2	Durability and Reliability	X14	Biodigester not easy to leak	X20	The biodigester-generated gas has the same quality in every operation of biodigester
		X20	The biodigester-generated gas has the same quality in every operation of biodigester		
		X19	The stove can still be used even when the power (electricity) is off		
		X17	Gas distribution pipes are not easily clogged	X19	The stove can still be used even when the power (electricity) is off
		X16	Gas pipes to the stove are not easily broken		
		X15	Not easily damaged if hit by a hard object collision		
3	Performance, Durability, and Reliability	X16	Gas pipes to the stove are not easily broken	X1	Biodigester is able to convert waste into gas that can be used for cooking
		X14	Biodigester not easy to leak		
		X4	The resulting gas does not have a potent smell when used	X20	The biodigester-generated gas has the same quality in every operation of biodigester
		X17	Gas distribution pipes are not easily clogged		

IOP Conf. Series: Earth and Environmental Science **158** (2018) 012032 doi:10.1088/1755-1315/158/1/012032

Cluster	The Most Important Dimensions	The Most Important Indicator		The Most Important Indicator of Clustering Variables	
		X3	Biodigester can reduce organic waste	X19	The stove can still be used even when the power (electricity) is off

5. Processing and analysis of data

5.1. Conclusion

From the results of statistical analysis inference using Welch's ANOVA and Games-Howell Post-Hoc Test, we conclude that each cluster has different preferences. The development of future household biodigeters should consider these public preferences. Based on the cluster analysis along other tests, we conclude that there are 3 clusters of citizens' preference for household-scale biodigester. These three clusters are those who emphasize features and reliability; those who emphasize durability and reliability; and those who emphasize performance, reliability, and durability.

Cluster 1 has the biggest proportion compared to other clusters, reflecting the most common preference of the general population. Commercialization of household biodigesters should aim cluster 1 as target market due to its sizeable amount. In addition, the dimensions preferred by cluster 1 is quite easy to implement due to their simplicity, allowing household biodigester developers to focus more on delivering those preferred indicators. Cluster 1 prefers a reliable biodigester: one whose gas storage is not easy to leak and whose pipes delivering gas to the stove are not easily broken. Cluster 1 also prefers feature of stoves that can still be used even though the electricity is not on and a monthly incentive for biodigester operators. Biodigester developers should focus on developing a reliable stove with this feature for household biodigester and structure proper financial incentive for biodigester operators to maintain their willingness to operate.

5.2. Recommendation

The main stakeholder in this problem is Bandung City Government. Cross-tabulation analysis shows no demographic variables affecting user preferences. This means that the city government is able to use the result from this study without considering demographic variables.

The next question is how to structure proper management. This study reveals different preferences between different clusters when it comes to household biodigesters. Therefore, to increase community acceptance of biodigesters, Bandung City Government should design policies that match the preferences of citizens and provide household biodigesters designed in accordance with those preferences.

This study recommends cluster 1 to be the target market, emphasizing on reliability and features as the main preference. Cluster 1 mainly wants a biodigester that is not easy to leak and that has a reliable gas distributor not easily broken. By utilizing these findings, biodigesters should have a warranty against leakage and that ensures reliability of connecting pipe.

Development of household biodigester can also be done based on the most important variables among the variables used for cluster analysis. This is done because the clustering variable has a function as a differentiating variable between the clusters formed. Therefore, the use of clustering variables for the development of household biodigester is expected to make biodigesters more appropriate to the preference of the selected segment. In cluster 1, the most important of clustering variable is a stove that can still be used even though the electricity is off and there is an incentive for those who operate the biodigester. Based on this finding, Bandung City Government should provide incentives for people who are willing to operate household biodigester. Then the Bandung City Government should also ask the manufacturer (or find a capable producer) to produce household biodigesters that can be operated when the power goes out.

Then, cluster 1 can be reached by asking prospective recipients of household biodigesters to fill out the same questionnaire as used in this study. After processing the questionnaire results (using testing ANOVA and Post Hoc Test on clustering variable and on the indicators of the important dimension), the results should be compared with the cluster results formed in this study to determine whether the respondent is a member of cluster 1, 2 or 3? If the result states that the potential recipient of this biodigester belongs to cluster 1, they should be highlighted as potential beneficiary of the biodigester grant.

Information about the biodigester program should be disseminated by environmental activists because most of this study's respondents are those concerned about environmental sustainability. Therefore, it is possible that the preferences generated by this study will be more appropriate to those participating in the forums of environmental activists. However, disseminating information through local officials (heads of sub-districts, heads of neighbourhood units, etc.) should be conducted as well in order to reach the rural areas of Bandung. As some respondents of this research are ordinary citizens, preferences outlined by this study is predicted to still be in accordance with the preferences of citizens in general. For the next biodigester grant program, the recommendation of the biodigester recipient search path can be seen in Figure 4.

When disseminating information about household biodigester grant programs, warranties and additional features existing in biodigester should not be mentioned to prevent shaping the perception of potential biodigester grantees when it comes to them answering the later questionnaire. Without the mention of warranties and additional features, respondents are able to fill out the questionnaire honestly, according to their initial preferences.

Any citizen who wishes to be a biodigester grantee should be required to submit the requirement to the Bandung City Government or a designated office in person and not via courier. Upon submission, the potential grantee interested in accepting and maintaining the household biodigester should be asked to fill out the online questionnaire using a gadget provided by the government under the supervision of government officials.

By using this flow, Bandung City Government can find the right biodigester household recipients to minimize failure risks in the form of not using the provided biodigesters, because unused biodigesters are massive expenses that bleed out from the government's budget.

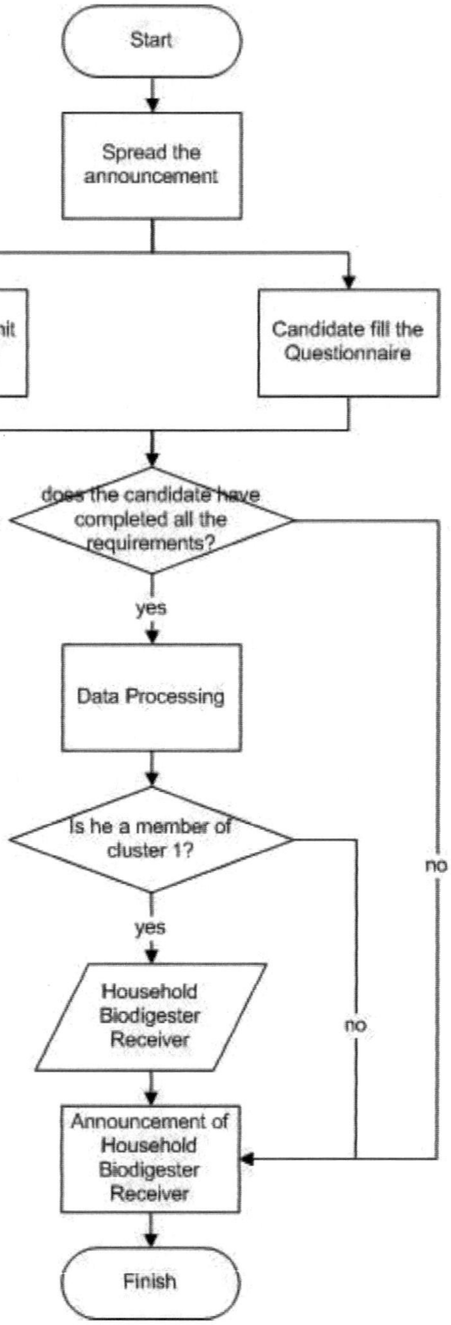

Figure 4. Recommendation of the biodigester recipient search path.

Acknowledgments

The authors would like to thank our beloved parents for their support throughout this internship program. The authors would like to thank RDI for the internship opportunity. A special expression of gratitude is in order for research supervisor Mr. Titah Yudhistira of Engineering Management ITB for his advice and direction throughout this research. Finally, the author thanks all parties involved for their help and support contributing to the completion of this paper.

References

[1] Surakusumah W 2008 Permasalahan Sampah Kota Bandung dan Alternatif Solusinya [Solid Waste Problems in Bandung City and its Alternative Solutions] *Jurnal Universitas Pendidikan Indonesia* **16**

[2] Pemerintah Provinsi Jawa Barat. [Internet] 2017 [cited 2017 June 20]. Available from: http://www.jabarprov.go.id/index.php/news/21468/2017/02/21/Bandung-Produksi-Sampah-Capai-1500-Ton-Perhari.

[3] Infobandung *Hibah Biodigester Gratis Untuk Warga Kota Bandung* [Free Biodigester Grant for Citizens of Bandung City] [Internet] 2017 Bandung [cited 2017 May 29] Available from: http://infobandung.co.id/formulir-dan-persyaratan-penerima-hibah-alat-biodigester-gratis-untuk-warga-bandung/.

[4] Badan Pengelola Lingkungan Hidup Kota Bandung 2014 *Laporan Analisis Biodigester Kapasitas 250 kg di Kota Bandung* [Analisis of Biodigester with the Capacity of 250 Kg in Bandung City Report] (Bandung: BPLHD Kota Bandung)

[5] Cahyono B. Lecture: Pengalaman Penerapan Digester Pengolah Sampah Organik di Kota Bandung [Experiences of Implementing Organic Waste Digester in Bandung City]. Lecture presented at Learning Event & Workshop Pengelolaan Sampah Organik Terintegrasi; 1st December 2016; Institut Teknologi Sepuluh November.

[6] Eppinger S and Ulrich K 2015 *Product design and development: 6th Edition* (New York: Irwin/McGraw-Hill); 2015.

[7] McMillan J. *What Do Your Customer Actually Buy?* [Internet] 2007 [cited 2017 July 2]. Available from: www.mcmillantech.co.uk.

[8] Gendall P and Esslemont D 1992 Market Research: What It Can and Can't Do *Marketing Bull.* **3**:63-66.

[9] Garvin D A 1987 Competing on the Eight Dimensions of Quality *Harvard Business Review* **65**(6) (November–December 1987)

[10] Malhotra NK, Birks DF 2007 *Marketing research: An applied orientation 5th ed* (New York: Pearson)

[11] Lindemann N *34 Ways to improve your survey response rate* [Internet] 2017 [cited 2017 August 28] Available from: https://surveianyplace.com/improve-survei-response-rate/.

[12] Zhang T *The Power of The Progress Bar* [Internet] 2013 [cited 2017 August 1] Available from: https://blog.kissmetrics.com/the-progress-bar/.

[13] Zhao N *The Minimum Sample Size* [Internet] 2009 [cited 2017 August 6] Available from: www.encorewiki.org/.

[14] Gujarati D N 2004 *Basic Econometric: 4th Edition* (New York:The McGraw–Hill)

[15] Santiago E *What Should I do if My Data is not Normal?* [Internet] 2015 [cited 2017 August 3] Available from: http://blog.minitab.com/blog/understanding-statistics-and-its-application/what-should-i-do-if-my-data-is-not-normal-v2.

[16] Glen S *Non-Normal Distribution* [Internet] [cited 2017 August 3] Available from: http://www.statisticshowto.com/probability-and-statistics/non-normal-distributions/.

[17] Liu H 2015 *Comparing Welch ANOVA, a Kruskal-Wallis test, and traditional ANOVA in case of heterogeneity of variance* Virginia Commonwealth University

[18] Shingala M C and Rajyaguru 2015 A Comparison of post hoc tests for unequal variance *Int. J. of New Technol. in Sci. and Eng.* **2**(5), 22–33

[19] Hilton A and Armstrong R A *Microbiologist* [Internet] 2006 [cited 2017 August 1] Available from: https://research.aston.ac.uk/portal/en/researchoutput/statnote-6-posthoc-anovatests(920c73af-ecb7-4213-b59f-725b6f6e4ab3).html.

[20] Hair J F, Black W C, Babin B J and Anderson RE 2010 *Multivariate data analysis: 7th Edition* (Harlow: Pearson)

IOP Conf. Series: Earth and Environmental Science **158** (2018) 012033 doi:10.1088/1755-1315/158/1/012033

The Current Observation and Challenges of Tourism Development in Batur Global Geopark Area, Bali Province, Indonesia

A. Rosyidie[1], S. Sagala[1], M. M. Syahbid[2], M. A. Sasongko[2]

[1]Department of Regional and City Planning, Institut Teknologi Bandung, Bandung, West Java, Indonesia

[2]Resilience Development Initiative, Bandung, Indonesia

Corresponding author: ariefrosyidie@yahoo.com

Abstract. Geopark is a designated physical landscape with a geological heritage value. It is one of the approaches to combine both conservation and sustainable economic development. Geopark aims to attract tourists that ultimately bring impacts to the development, especially to the local governments and communities. The first global geopark in Indonesia, which is recognized by UNESCO, is Batur Geopark. Batur Geopark is in Kintamani, Bangli District, Bali Province. The main purpose of Batur Global Geopark development is to balance between the geology, environment, social-culture, economic, and conservation. Creating geoparks as tourism destination would achieve not only scientific and natural resource preservation, but also the development of tourism. Since Geopark is considered as a new concept in Indonesia, the main question is: does geopark indeed boost tourism? If so, how significant is it compared to other tourism in a particular area, such as Bali? What are the challenges faced in the development of geopark? Having the above questions, this study selects Batur Geopark as a case study and assesses to what extent Batur Geopark contributes to the tourism development in Bangli District and Bali Province. A framework to do this assessment is set up using geopark criteria developed by UNESCO as well as trend analysis since the establishment of Batur Geopark in 2012. To get a comprehensive picture, we conducted a field survey in Batur Geopark and asked questions to the local community, local champions, businessmen, tourism actors, local government agencies and scholars in tourism development, and we also explored the potential and the challenges of the tourism development of Batur Global Geopark. The findings indicate that developing Geopark as ecotourism is a complicated issue. Geopark is seen in various and different perspectives by different actors. While the aim of geopark is to trigger tourism development, some contra-geopark and tourism activities exist, such as sand mining that makes environmental and infrastructure problems. Other activities include farming that does not fit well with the geopark aims. To address the challenges in the development of geopark, a number of activities should be considered, as follows: (i) Socialization and awareness raising on the importance of conservation in the area's geological heritage, (ii) Batur Global Geopark needs improvement, thus it creates sense of pride of local community toward their region, (iii) Creation of local enterprises, new jobs and high-quality training courses to support the economy. Meanwhile, the development of Batur Global Geopark should be coordinated with all stakeholders from local communities, local entrepreneur, local non-governmental organization and local government.

IOP Conf. Series: Earth and Environmental Science **158** (2018) 012033 doi:10.1088/1755-1315/158/1/012033

1. Introduction

The Batur UNESCO Global Geopark (BUGG) located in northeast Bali and designated in 2012 with an aim to promote sustainable development in local communities is the first UGG in Indonesia. Covering an area of 370.5 km^2, at an altitude of 920-2152 m, BUGG includes two volcanic calderas and presents a complete volcanic landscape. Between 1804 and 2000, Mt. Batur erupted at least 22 times forming a strato-volcano which is one of the 127 active volcanoes in Indonesia, and a component of the Pacific "ring of fire". With double-calderas and a crescent-shaped volcanic lake (7 km long, 1.5 km wide), Batur is known as the finest caldera in the world. This uniqueness, endemic flora and fauna, and original Balinese culture reflect an impressive combination of Earth heritages [1-2]. Batur GG, known in Bali as the Geopark Kaldera Batur (GKB), is occupied by 15 villages within the district of Kintamani, Bangli Regency. The local economy is based on agriculture and tourism. Geo-products include Kintamani sweet orange and civet coffee, wooden craft, bamboo and wood carving products. The Kintamani dog, native to the island of Bali, originates from Sukawarna Village, Batur.

The Global Geopark Network (GGN) recognized BUGG in 2012 making it Indonesia's first UGG with an aim to promote the sustainable development of local communities. This objective corresponds to the main GGN goals. The Global Geoparks Network (GGN) formed in 2004 by UNESCO aims to improve coordination of the world's geological heritage through cooperative management. According to [3] the main objectives of a geopark are (1) protection and conservation, (2) tourism-related infrastructural development and (3) socio-economic development.

The success of geopark is reliant upon local communities benefiting and participation in building harmony between people and nature. The direct involvement of local communities leads to local empowerment allowing for free expression of traditional local wisdom that reinforces local identity and cultural protection. Furthermore, geopark emphasizes inclusiveness, especially by local community involvement in conservation efforts engaging ownership of resources and locations which contribute positively to livelihoods and quality of life. UGG should holistically integrate conservation initiatives that consider sustainable resource utilization, provision of infrastructure and local socio-economic development, along with well-being (Figure 1) [4].

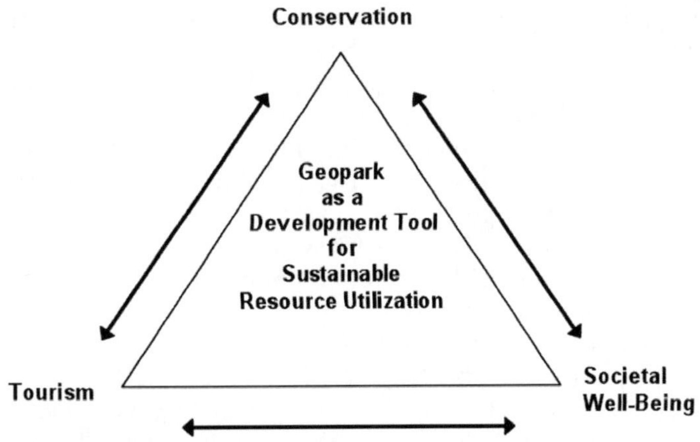

Figure 1. The Geopark and its main components [4]

Local community involvement and subsequent benefits are integral and essential to sustainable tourism development in Batur Global Geopark. The implementation of ecologically sustainable tourism safeguards ecosystems and enhances local societies and economies through ecotourism initiatives. Ecotourism encompasses the social and economic dimensions in addition to the environmental dimensions with local initiatives being influential [5-6]. Local and indigenous communities practice ecotourism as tools for biodiversity conservation, sustainable livelihoods and cultural preservation [7]. The International Ecotourism Society (TIES) refers to ecotourism as responsible travel to natural environments with actions for conservation and the sustainability of local communities' wellbeing, which are accompanied by practices of interpretation and the education of staff and visitors. Non-consumptive and non-extractive behaviors are inherent to ecotourism and foster ecological consciousness while holding eco-centric values and ethics [8]. Sustainable Tourism (ST) concerns the conservation of natural resources and social culture while distributing benefits widely among stakeholders and communities and is an important public agenda for all stakeholders in order to benefit the economy through employment, new educational opportunities and improved quality of life [5,9].

Creating geoparks as sustainable tourism destination would achieve not only scientific & natural resource preservation, but also the development of tourism. Since Geopark is considered as a new concept in Indonesia, the main question is whether geopark indeed boosts tourism? If so, how significant is it compared to the other tourism in a particular area, such as Bali? What are the challenges faced in the development of geopark? According to those issues, this study aims to explore and identify the main challenges of sustainable tourism development in Batur Global Geopark Area. This study also tries to recommend strategies for geopark as a sustainable tourism development in the specific context of Batur Global Geopark and Indonesia. Therefore, our research purpose is to identify what challenges should be recognized and what strategies should be created for geopark as a sustainable tourism development in the specific context of Batur Global Geopark and Indonesia, which is regarded as the first global geopark in Indonesia and recognized by UNESCO Global Geopark Network.

2. Methodology

The research was conducted with surveys and in-depth study in Batur Global Geopark Area, Bangli Regency (see Figure 2). The survey was conducted between July and August 2017. The surveys were conducted by interviewing forty-one key respondents, such as local communities, farmers, local vendors, local businessmen, leaders of Local Working Group (LWG), local and provincial government agencies, scholars in tourism development, and other tourism actors. In collecting data, it was divided into 5 cluster regions, which are a group of tourism destination zones grouped into five clusters based on location, geographical condition, and socio-cultural conditions of the community.

The collected data were then analyzed to know the following aspects that contribute to the sustainable development tourismreferring to sustainable tourism as encompassing the ecological, social and cultural and economic sustainability dimensions. Ecological sustainability basically focuses on development compatible with maintaining essential ecological processes, biological diversity and resources. Social and cultural sustainability ensures that development increases people's control over their lives that is compatible with their cultural values, and strengthens community identity. Economic sustainability fosters the economically efficient development and sustainable resources management to support future generations [10]. Furthermore, according to [11], sustainable tourism should ensure socio-cultural and environmental impacts which are neither permanent nor irreversible. Sustainable tourism is an approach that can be used to make all types of tourism more environmentally, socially, culturally and economically sustainable [12].

In this research we will scope the tourism sustainable development of Geoparks from three aspects, they are scientific research, local socio-economic development and environment protection.

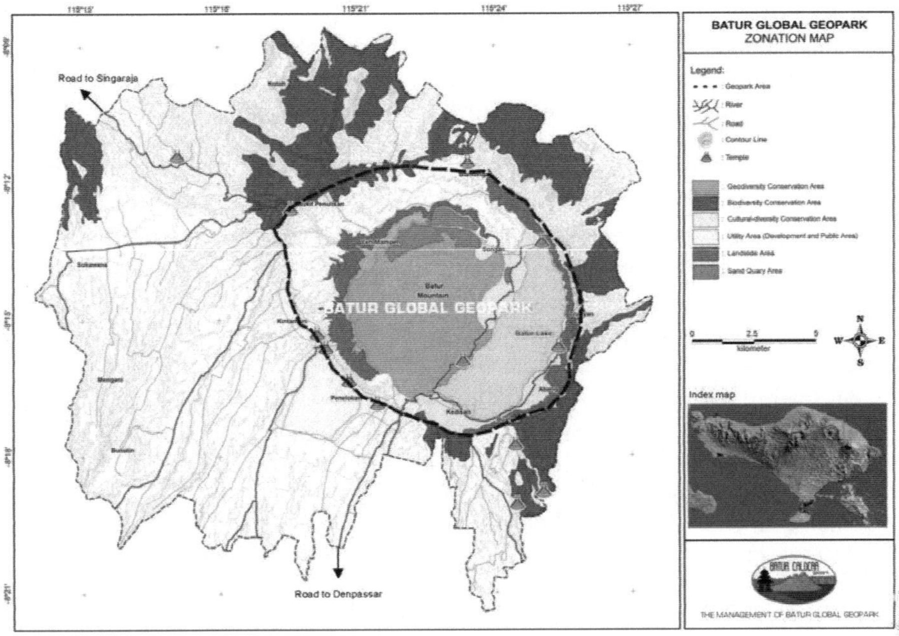

Figure 2. Map of Batur UNESCO Global Geopark [13]

3. Literature Review

3.1. Sustainable Tourism and Ecotourism Development defined

The establishment of sustainable tourism, green tourism, and ecotourism was closely followed the emergence of sustainable development and grassroots approaches to development in the early 1980s [6]. There was a consensus at World Commission of Environment and Development in 1987 regarding sustainable tourism, defined as a form of tourism which meets the needs of tourists, the tourism industry, and host communities in the present time without compromising the ability of future generations to meet their own needs [14]. [10] referred sustainable tourism as a method in which encompasses the ecological, social and cultural, and economic sustainability dimensions. Ecological sustainability basically focuses on development compatible with maintaining essential ecological processes, biological diversity and resources. Social and cultural sustainability ensures that the development process is compatible with their cultural values and aims to strengthens community identity. Economic sustainability fosters the economically efficient development and sustainable resources management to support future generations [10]. According to [11] sustainable tourism should ensure socio-cultural and environmental impacts which are neither permanent nor irreversible. In their study on ecological sustainability indicators, [12] view sustainable tourism as an approach that can be used to make all types of tourism more environmentally, socially, culturally and economically sustainable.

Local community involvement and subsequent benefits are integral and essential to sustainable tourism development. The implementation of ecologically sustainable tourism safeguards ecosystems and enhances local societies and economies through ecotourism initiatives. Ecotourism encompasses the social and economic dimensions in addition to the environmental dimensions with local initiatives being influential [5-6]. Local and indigenous communities practice ecotourism as tools for biodiversity conservation, sustainable livelihoods and cultural preservation [7]. The International Ecotourism Society (TIES) refers to ecotourism as responsible travel to natural environments with actions for conservation and the sustainability of local communities' wellbeing, which are accompanied by practices of interpretation and the education of staff and visitors. Non-consumptive and non-extractive behaviours are inherent to ecotourism and foster ecological consciousness while holding eco-centric values and ethics [8]. ST concerns the conservation of natural resources and social culture while distributing benefits widely among stakeholders and communities and is an important public agenda for all stakeholders in order to benefit the economy through employment, new educational opportunities and improved quality of life [5,9].

3.2. Geopark and geotourism as functions of Sustainable Tourism, Ecotourism and Community Based Tourism

Geotourism is a form of educational tourism with a focus on geology and landscapes including an appreciation of indigenous cultures, thus challenging the tourist to respect the preservation of local cultures and the environment [15-16]. Geotourism aims to impact on natural resources negligibly while simultaneously promoting the economic growth of indigenous communities. Geoparks are geographical areas which aim to integrate the preservation of geological heritage, biodiversity, and cultural resources through strategies for regional sustainable socio-economic and cultural development through geotourism activities [3,15,17-19].

Sustainable development of the geotourism industry requires geo parks maintained by minimal intervention with native habitats. Such undisturbed environments provide surrounding communities with vital resources for rural tourism industry development, thereby it can address rural unemployment and minimize effects from urbanization [20]. [21] found that the implementation of traditional knowledge in geotourism destination management correlates positively with the sustainability of these environmentally sensitive regions. Local and indigenous people often link their landscapes to natural phenomena, cultural artefacts, skills and customs shared through ancestral knowledge. This helps to maintain their connections with the landscape, in the context of everyday work, religious practices and common rituals. Local and indigenous peoples' involvement in geotourism destinations is integral to long-term environmental and social sustainability [21]. Recent administrative strategies encourage sustainable development of the geoparks themselves to enhance socioe-conomic development of local communities through their participation. Ecotourism in geoparks has been found to stimulate local economic growth and enhancements to community welfare by providing employment opportunities, improving local facilities and infrastructure and encouraging the production of local products [22].

GGN aims for environment conservation, geosciences awareness and sustainable growth and development of local economies with sites possessing sound management and economic development strategies enabling sustainable development especially through sustainable tourism in the form of geo and eco-tourism. Feasibility and success of geo parks will only be realised if local communities benefit and participate in building harmony between people and nature [4]. Geoparks' management plans should emphasize on inclusive elements, particularly in engaging local community involvement in conservation efforts to take ownership of resources and locations that are deemed to contribute positively to their better living [23]. Geoparks can, therefore, be pathways to achieve SDGs by promoting the holistic integration of conservation initiatives that consider sustainable resource utilisation, provision of infrastructure and local socio-economic development and well-being.

4. Result and Discussion

4.1. Current Development of Batur Global Geopark

Batur Geopark was officially accepted as a member of Global Geopark Network (GGN), supported by UNESCO on September 20, 2012 in Portugal. On November 18, 2015, it was then officially renamed to be "Batur UNESCO Global Geopark". Administratively, the Geopark is located in the district of Kintamani, Bangli Regency, Bali Province. The regency itself comprises of four sub-districts (Susut sub-district, Bangli sub-district, Tembuku sub-district, and Kintamani sub-district).

The landscape of double-calderas, active volcanic cones, lake, lava flows, cinder cones and maar is a unified volcanic morphology which is a history of the dynamics of volcanism phenomenon over tens thousands years ago to present time. Batur volcano, which is still active has a unique phenomenon, named "caldera in caldera" that is rarely found elsewhere. The Batur UNESCO Global Geopark is a tool to develop the area in a sustainable manner, mix and match among components of geology, biology and culture for the purpose of conservation, education, growth of local economic value and sustainable development. Based on *the Progress Report of Batur UNESCO Global Geopark for the year 2012-2016* , GGN has 366.4-kilometer square with population by about 192,681 people, spread over 15 traditional villages. An area of about 2,075 hectares in the Geopark region is the Natural Tourism Park. This conservation area was established by decree of the Minister of Agriculture No. 321/Kpts/Um/11/1982.

Since designated as a Global Geopark in 2012 until now, the Geopark area of 366,4 km square is unchanged. Batur Geopark become a National Geopark in 2010 and was assigned as Global Geopark in 2012. Since then, its development has been started and it has experienced significant improvement until now. The advancements can be seen from the growth of local economy value via tourism, as well as education through "Geopark to School" and "School to Geopark" programs, and conservation aspects. Regarding education in Geopark, the Regent of Bangli ordered the integration of Geopark concept within school's curriculum in Bangli (Regulation of Bangli Regent Number 18 Year 2013 on June 14, 2013 in the integration of Geopark material into the curriculum of primary schools, junior high schools, senior high schools and vocational schools;). In 2011, the Geological Agency of Ministry of Energy and Mineral Resources built "Batur Volcano Museum" at Panelokan area. This building is also functioned as Geopark Information Center. The advanced stage of development in 2013-2015 was the establishment of previous museum as "Batur Geopark Museum".

In term of tourism activities, since Batur has been designated as a Global Geopark in 2012, a number of tourists continues to increase. Significantly, the four-top tourist destinations (Batur/Panelokan, Penulisan, Terunyan Village, Batur climbing/P3GB area) have contributed to the number of Geopark tourists, either local or foreign tourists. The comparison between domestic and foreign tourists is 2:3. Foreign tourists are more interested to carry out their activities in Batur area such as climbing to the top of Batur active volcano. The data were obtained through tickets sale to four-main areas as mentioned earlier.

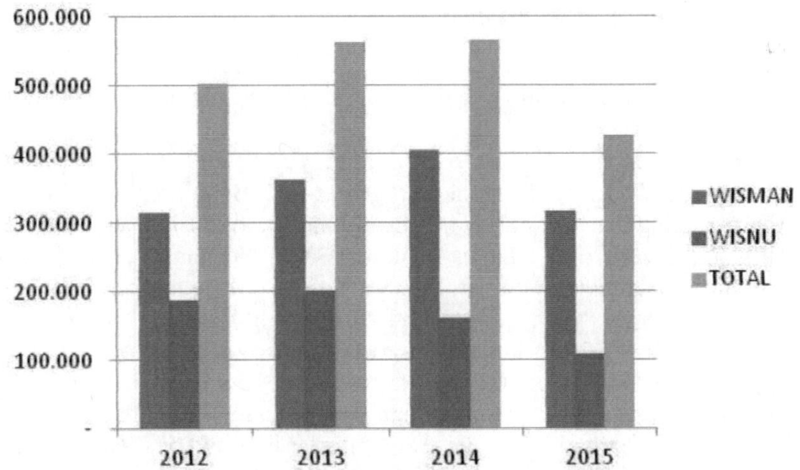

Figure 3. Number of tourists in 4 major tourist destinations [13]

Batur UNESCO Global Geopark (BUGG) is connected by geological paths or geotrails as the main attraction. Through this provided geotrails, visitors can explore the entire Geopark region. Some geotrails are managed by the local community groups, and some geotrails are guided by local community who live around Geosite and have been adequately trained. Training on guiding was conducted by Geopark Manager, in collaboration with Ministry of Tourism as well as Ministry of Energy and Mineral Resources. Sites of biology (bio-sites) and culture (cultural-sites) are dispersed in

BUGG, especially in areas of Payang Hill and Batur Lake Natural Park and in Subak Cultural Landscape which were designated by UNESCO into Cultural World Heritage site in July 2012. Twenty-one geological heritage sites (geosites) are accommodated in BUGG, both geological sites (rocks, stratigraphy) and geomorphological sites (caldera-walls, craters, lake, cinder cones, maar, volcanic-landslide). The BUGG weather is very good and clean; everything can be seen clearly from Panelokan view-point located at an elevation of outer-caldera wall. To be more effective, geosites will be grouped according to applicable terminology, after the first revalidation was completed in 2016.

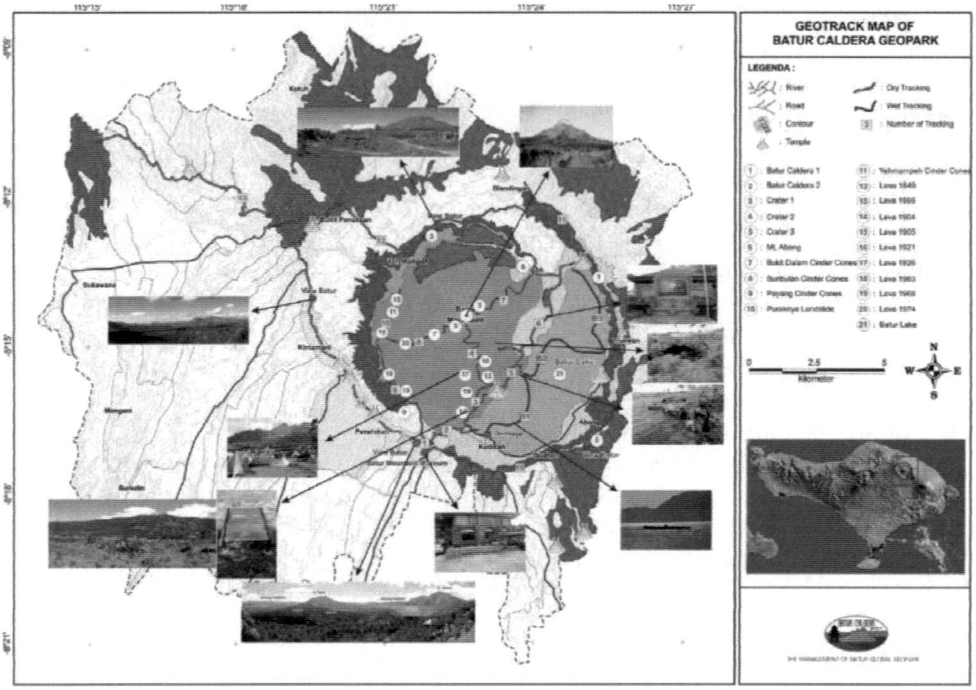

Figure 4. Geotrails within the Batur UNESCO Global Geopark [13]

The strengths and weaknesses of Batur Geopark in this region is its area and population. Therefore, the Ministry of Culture and Tourism of Indonesia together with Bangli regency government tried to re-arrange the area by involving the community in fifteen villages in one cluster Destination Management Organization (DMO). The fifteen villages include Batur Selatan Village, Central and North, Kintamani Village, Suter, Blandingan, Sukawana, Songan A Village and Songan B, Trunyan Village, Buahan, Kedisan, Abang Songan Village, Suter and Abang Batu Dinding Village.

Based on Destination Management Organization (DMO) program formed a Local Working Group (LWG) in five regions in BUGG area. Of the fifteen villages located in Bangli Regency, there is a group of tourism destination zones grouped into five clusters based on location, geographical condition, and socio-cultural conditions of the community. The five LWGs was then functionalized as janitor and guard for geosites by the regent of Bangli (Regent Decree No. 430.05/38/ 2015 dated January 2, 2015 for tourist destination development activities of the fiscal year 2015. The appointed janitors include five Local Working Group coordinators (Abang Erawang, Manuk Jambe Tarumenyan, Ulun Danu Muncar, Batur Kalang Anyar, Singa Mandawa). Each officer works in geological and non-geological heritage sites (including Geopark facilities in the form of information kiosks in five locations). Their responsibility is to maintain hygiene, safety and preservation of geological heritage sites. It is be noted that Local Working Group is a group of local people (mostly indigenous) who live around the geological/non-geological heritage sites and serve as the executor of Destination Management Organization activity or program created by the Ministry of Tourism since 2012.

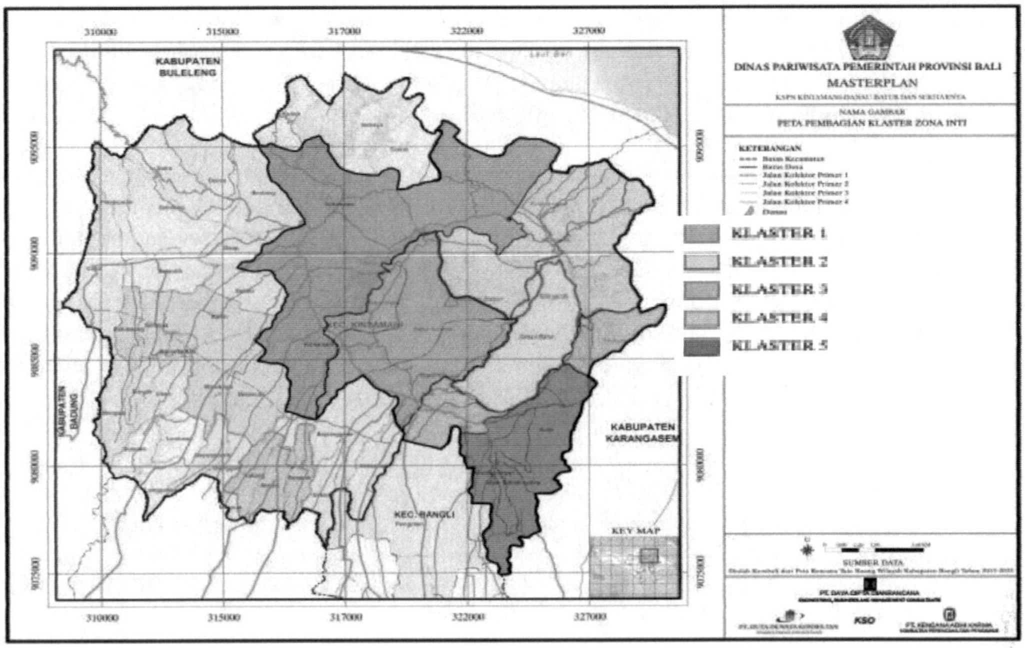

Figure 5. Map of Five Cluster of Destination Management Organization (DMO) in Batur Global Geopark (BGG) [24]

4.2. Challenges of Batur Global Geopark Development

4.2.1. The perspective of local community on social and economic impact of Geopark Development. Most of local communities in Batur Geopark area believe that Geopark is part of the massive development of tourism infrastructure. The results of interviews with local farmers in some villages of Batur Geopark area thought that the development plan of Batur geopark since 2012 has not yet provided changes in the economic improvement for the farmers. The marketing and distribution process of agricultural products have the same pattern as the time before the area of Mount Batur was still integrated as part of Global Geopark Network (GGN). Despite the lack of impact to native communities livelihoods, such as farming and mining, local communities in Batur Geopark area are generally familiar with the arrival of tourists in their area. The attitude of local people in Batur Geopark area are very welcoming to both local and foreign tourists.

> *"In general, the public has not been aware of existence of geopark. Maybe because most people are farmers, so they do not feel the impact of GGN directly. So those who understand GGN can feel it, for example, I am also a farmer, and I feel the impact, such as I can sell vegetables more easily."*

> *"The Education Agency has incorporated geopark into the curriculum, with the Geopark to School, School to Geopark program. We educate them from elementary school as well, so that the elementary school children can deliver the information to their parents."*

Local people in the Geopark area prioritize education for the next generation, which also emphasize education on local environmental and tourism knowledge. The current problem that the local people want to solve is the lack of geopark education, especially to local people who are not part of Local Working Group. This condition was clearly seen during various interview session with local people (non part of Local Working Group) and local merchants around Jalan Penelokan. A deeper knowledge of the geopark is still limited only to the Local Working Group members and local leaders.

However, there is an increasing effort on geopark education programmes in local schools around Batur Geopark area, notably in SMKN 1 Kintamani, a vocational school that has a newly established tourism major. Also, SMKN 1 Kintamani has made an agreement with the local hotels, resorts, and tours regarding further employment opportunities for its students.

4.2.2. Deteriorating environment and natural resource quality. There are ongoing waste problems faced by local people regarding the lack of garbage dump in their neighborhood area, as they still prefer to dispose of their garbage in some spots in gardens, forests, or unused lands. This garbage problem was aggravated by garbage fleet that could not accommodate the volume of existing waste produced by local communities around Batur Geopark area. So, the current waste processing is mostly done by self-burning, due to the lack of waste transportion service. It causes environmental damage, especially for water and soil. Because water in some parts of Batur is contaminated with chemicals so that it can not be used for drinking. While the land use of septicide and mining reduces soil fertility.

> *"There is still a problem of garbage transportation here. Garbage fleet only comes 3 days a week, because the capacity of the fleet is not enough. Garbage fleet currently only passes one direction, from Kintamani to Penelokan, it does not pass by other villages. The people themselves who transport the garbage, for example, I usually drive a carry car to bring garbage to the garden. There (in the garden) I sort the garbage, some is used as fertilizer and the rest is taken to the landfill."*

There is ongoing mining activity located in Songan A and B villages in Batur Geopark area. Most of the miners confess that they do not have any other option to fullfil their livelihood needs, being a miner. The government so far could not overcome the problem of mining because the activity is done on private land communities. Mining communities themselves have not been able to move to tourism since tourism gives less economic benefits than mining activities.

> *"Communities involved in excavation type-C are aware of the negative impacts of these activities, but there has been no substitute for their main livelihoods, the benefits of tourism are also not evenly distributed so that they can not replace the activities of excavation type-C. In areas where economic benefits such as Batur Village, Kedisan, and others, the community strongly opposes the activity of excavation. Because they do get more benefit from tourism."*

One of the problems faced in the Batur Geopark area is that the sand truck lanes are still same as the tourist path (despite government prohibitions and regulations), which makes it quite dangerous especially around Penelokan area where is crowded pedestrian traffic. The main route becomes a meeting place for tourist buses and sand trucks causing congestion on weekends. As an initial development of a separated pathway for tourist, villagers in the southern region of Lake Batur responded positively to the inauguration of the tourist route so that trucks carrying sand and rocks no longer passed through their area. There is also a trend of reduced sand excavation activities in the Songan A-B villages, the only problem for current excavation activitiesis that they are carried out by people on their private land.

4.2.3. Inconsistency of policy and regulation for development of Batur Global Geopark. The development of tourism in destinations throughout Batur Geopark area is partial. Rapid development takes place only in areas with great appeals, such as Penelokan Area. Meanwhile for hotel and resorts around Penelokan Area, most of the visitors only come during the lunch time, because Batur Geopark area is located around provincial street connecting southern and northern area of Bali Province; therefore most of the visitors in Batur Geopark area have limited time to stay, and prefer to continue their journeys to northern (Singaraja) or southern (Ubud, Tabanan, Denpasar, and Airport) of Bali Province.

> *"The most advanced and well-known area for those people is Penelokan area in Batur Village. People used to know Kintamani only in Penelokannya area, well then HPI members (Himpunan Pramuwisata Indonesia) asked what elsecould be developed in Kintamani. Well this is a challenge for people to develop other Geopark spots."*

The development of tourism in geopark indeed has provided economic benefits, although it is not significantly distributed equally with all parties. Tourism development from the government and tourism industry still focuses on product diversification and physical infrastructure development, while community building capacity is still not a priority. Based on the dossier of Batur Geopark that was developed together by Bangli Regency Government and UNESCO in 2012, Geopark tourism development in the short term (2012 - 2016) still focused on physical development. The physical development plan was also included in Bangli Regency Spatial Plan (2013 – 2033). Most of the critics from scholars are addressed on the physical development focusing on quantity targets; meanwhile, several media interpretation on BUGG does not function optimally. BUGG Museum as the gateway of the region has a minimal number of visitors, the interest of visits from local communities is still low, and there is nothing to sell inside the souvenir shop. The BUGG information board at Penelokan view point can not be accessed because the building was collapsed and had not been fixed for several months.

In addition to the conflict of authority, it still lacks of government coordination (between local authorities and the government above): the provincial government argues the lack of coordination from the district; on the other hand, the district government says that the provincial government gives minimal attention to the development of BUGG. In addition, government control is still weak, especially related to land use that is not in accordance with its provisions; such as Excavation Type-C mining activities and restaurants around penelokan area built on landslide-prone land (pyroclastic soil structure).

> *"For restaurant ownership, most of them are owned by the locals, only 30% are from investors, for example Cutong Restaurant whose owner is from Surabaya. Honey Sari, Gunung Sari, and Lake View are owned by the locals. In the upper area, there are still a lot of restaurants owned by local residents, while in the lower area of are owned by foreign investors."*

Regarding the regulation in the business world, business ownership such as restaurant, bike rental, homestay, hot water bathing is mostly owned by local people or managed by BUMDES. The community considers licensing affairs in Batur Regency is quite complicated, especially in the business world, resulting in the lack of business investment coming from outside Bangli regency (local and foreign). Local cultures indirectly slow down the rate of foreign investment because land and property sales policies in some places have to go through the approval of indigenous villages (Desa Adat). There is also a problem with the construction of restaurants around Jalan Panelokan mostly built on vulnerable zones/areas along the slope facing directly to Mount Batur.

4.2.4. Weak of partnership and government control to local community. When observations were made in June 2017, of five Local Working Groups (@ three villages) were established and they are still running today. Only one LWG (Cluster Abang Erawang) is considered optimal based on its performance. The government states that socialization and coaching were often directed to involve LWGs, but most of LWGs argued that the government's role was minimal (compared to the academicians involved), and the government only provided socialization without optimal coaching, especially since the focus of BUGG development from the Central Government gradually reduced.

Compared to physical development, four other LWGs need more education and mentoring from the government due to the lack of local initiatives related to tourism development.

The high conflicts of interest among pokdarwis (local tourism initiatives), especially in the management of climbing routes. Pokdarwis (local tourism initiatives) P3GB of Desa Songan at the time of observation in June 2017, still monopolized the climbing route of Mount Batur, so that other pokdarwis (such as *TCM Desa Batur*) could not open their own climbing pathway. However, the P3GB group has its own management under certain parties (although it still has the legality from BKSDA to manage the climbing routes). In consequence the conflict with other Songan A and B villagers arose due to the economic benefits that exclusively P3GB only got more. The Bangli District Government, which previously provided infrastructure assistance in the P3GB area, is still unable to resolve the conflicts of interest that occur due to the absence of a special BUGG management agency.

> *"Many of us would like to open a trekking guide business, but almost all channels are currently managed by P3GB. They have legalization as a trekking track manager from the BKSDA, which manages the conservation area of Mount Batur."*

There are at least five climbing routes to the top of Mount Batur monopolized by one of the groups that has been allowed to manage the climbing activities for 20 years so that the other climbing routes managed by other groups are considered illegal. To anticipate the reshuffle or alteration in relation to the development process of the Geopark area, the P3GB Group conducted a legalization effort to BKSDA, which is responsible for land utilization and conservation in the area of Mount Batur. This is one of the real examples of partial tourism development in Batur Geopark area. The development of partial destinations produces "little kings" in geopark area. These "little kings" try to seek the legality of ensuring the existence of their organization if there is an institutional overhaul or arrangement within the Batur Geopark area.

4.2.5. Lack of leadership role in local community groups. Most of the local champion/other parties could not assist community capacity building and tourism development of BUGG because they were not interested/in managing the affairs. While BUGG Museum, Bangli Hotel and Restaurant Association, and other LWGs as a third party with high knowledge related BUGG were not involved because of the absence of their authority to manage those affairs. They considered other stakeholders/LWGs that should be involved more, in fact, have less initiative in the development of tourism objects in their respective villages.

> *"I am here as a trigger trying to give an understanding that developing the tourist destination is not with money first, but the important things are spirit and consistency. The people initially did not believe it. There were only five people participated including me, others were from Pecalangs* (local Balinese police)*."*

4.3. Recommendation Strategies for Batur Global Geopark Development

Clearly, there is a need for additional strategies in the management of Batur Geopark area. One of the proposals currently under review is the establishment of centralized management from Batur Geopark, which has been implemented in several geoparks in Indonesia, such as the Dana Toba Authority Agency that manages Toba Geopark area. This management will assist the management of geopark areas, ranging from local revenue and funds, management of facilities, solving technical issues, setting up daily management system and strategies, regulating actors collaborations (academics, local business and entrepreneurs, local government authority, and tourism management bodies), setting up policies and regulations, regulation of local agricultural commodity prices, and other strategic programs. However, this management also has to be supported by education for the local communities and leaders regarding awareness on environmental and tourism impacts to their communities, and the importance of their own participation in the further programme from the management authority of BUGG.

To address the challenges in the development of geopark, a number of activities should be considered, as follows:

(i) Socialization and awareness raising on the importance of conservation in the area's geological heritage,
(ii) Batur Global Geopark needs improvement, thus it creates sense of pride of local community toward their region,
(iii) Creation of local enterprises, new jobs and high-quality training courses to support the economy.
(iv) The development of Batur Global Geopark should be coordinated with all stakeholders from local communities, local entrepreneur, local non-governmental organization and local government.
(v) Ensuring the implementation of educational programmes, such as Geopark to Schools is implemented evenly throughout schools in Batur Regency. Early education on environmental and tourism impacts of children's respective villages is important to raise their awareness in the future.

Figure 6. Illustration on Strategies to Achieve Sustainable Tourism Development of Batur UNESCO Global Geopark (BUGG)

To reach sustainable tourism development of Batur UNESCO Global Geopark, both top-down and grassroot approaches are required to balance between the growth of tourism demand and capacity of local communities and leaders to participate within tourism development. The capacity of local communities and leaders could be driven up by sharing lessons and experiences from local and foreign expertises, training and education of local workforces especially in tourism services, academic exchange and research collaboration of local researchers to gain more knowledge and skills on research process related to Batur Geopark and its management. From the top-down approaches, management of tourism revenue and funds, facilities, technical issues, management system, actor collaboration, policies, regulation, and strategies becomes the main steps to achieve good governance of Batur UNESCO Global Geopark, which constitutes collaboration between academics, local business entities, government bodies, and local tourism managements.

5. Conclusion

This paper has discussed the current challenges faced by Batur Geopark. We find that developing Geopark as ecotourism in Batur Global Geopark is still a complicated issue. Geopark has been seen through different perspectives by different actors. While the aim of geopark is to trigger tourism development, some contra geopark and tourism activities exist, such as sand mining that makes environmental and infrastructure problems. Other activities include farming that does not fit well with the geopark aims. The roles of local leaders are obviously important to engage the community. There is also a challenge due to negative perceptions of tourism development in the area that tourism is seen as nature exploitation. Local communities prefer to not to be involved in the tourism sector because they feel that tourism does not increase their income and quality of life. To address the challenges in the development of geopark, a number of activities should be considered, as follows: (i) Socialization and awareness raising on the importance of conservation in the area's geological heritage, (ii) batur Global Geopark needs improvement, thus it creates sense of pride of local community toward their region, (iii) creation of local enterprises, new jobs and high-quality training courses to support the economy, (iv) the development of Batur Global Geopark should be coordinated with all stakeholders from local communities, local entrepreneur, local non-governmental organization and local government, and (v) ensuring the implementation of educational programmes, such as Geopark to Schools is implemented evenly throughout schools in Batur Regency.

Acknowledgement

This research was supported by Research institutions and community service, Bandung Institute of Technology on Capacity Building Research Program. We thank our colleagues from Master Program of Tourism Planning SAPPK ITB, they are Muhammad Harits Insan Kamil and Sarita Novie Damayanti, who helped us on field survey to collect data. We also thank to Janelle Falconer who was involved in the data collection and literature review of geopark study conducted during her internship with Resilience Development Initiative.

References

[1] Batur Global Geopark 2017 https://www.baturglobalgeopark.com/
[2] UNESCO 2017c *Batur UNESCO Global Geopark, Indonesia* http://www.unesco.org/new/en/natural-sciences/environment/earth-sciences/unesco-global-geoparks/list-of-unesco-global-geoparks/indonesia/batur/
[3] Azman, Norzaini, Halim, Sharina A, Liu, Ong Puay, Komoo and Ibrahim 2011 The Langkawi Global Geopark: local community's perspectives on public education *International Journal of Heritage Studies* **17** 261- 279
[4] Halim SA, Komoo I, Salleh H and Omar M 2011b The Geopark as a potential tool for alleviating community marginality *Shima: The International Journal of Research into Island Cultures* **5(1)**
[5] Swarbrooke J 2002 *Sustainable Tourism Management* (Oxon UK: CABI Publishing)
[6] Willis K 2011 *Theories and Practices of Development 2nd* (London: Routledge)
[7] Blangy S and Mehta H 2006 Ecotourism and ecological restoration *Journal for Nature Conservation* **14(3)** 233-236
[8] The International Ecotourism Society (TIES) 2017 https://www.ecotourism.org/what-is-ecotourism
[9] ILO 2012 *Kementerian Pariwisata dan Ekonomi Kreatif Republik Indonesia bekerjasama*

dengan International Labour Organization 2012 Rencana Strategis Pariwisata Berkelanjutan dan Green Jobs untuk Indonesia (Jakarta and Geneva: International Labour Organisation)

[10] World Travel and Tourism Council *World Tourism Organization and Earth Council 1995 Agenda 21 for the Travel and Tourism Industry: Towards Environmentally Sustainable Development* (London: WTTC)Beech J and Chadwick S 2006 *The Business of Tourism Management* (Edinburgh: Pearson)

[11] Lozano-Oyola M, Blancas F, González M and Caballero R 2012 Sustainable Tourism Indicators as Planning Tools in Cultural Destinations *Ecological Indicators* **18** 659-675

[12] Bangli Tourism Agency 2017 Brundtland G 1987 *World Commission on Environment and Development: Our Common Future* http://www.un-documents.net/our-common-future.pdf

[13] Dowling R K 2011 Geotourism's global growth *Geoheritage* **3(1)** 1-13

[14] Farsani N T, Coelho C and Costa C 2011 Geotourism and geoparks as novel strategies for socio-economic development in rural areas *International Journal of Tourism Research* **13(1)** 68-81

[15] McKeever P J an Zouros N 2005 Geoparks: Celebrating Earth heritage, sustaining local communities *Episodes* **28(4)** 274

[16] Newsome D, Dowling R and Leung Y F 2012 The nature and management of geotourism: A case study of two established iconic geotourism destinations *Tourism management perspectives* **2** 19-27

[17] UNESCO 2010 *Guidelines and Criteria for National Geoparks seeking UNESCO's assistance to join the Global Geoparks Network (GGN)* http://www.unesco.org/fileadmin/MULTIMEDIA/HQ/SC/pdf/sc_geoparcs_2010guidelines.pdf

[18] Newsome D and Dowling R K 2010 *Setting an agenda for geotourism* (Oxfordshire: Goodfellow Publishers Limited)

[19] Pásková M and Dowling R K 2014 *the Usage of Local and Indigenous Knowledge in the Management of Geotourism Destinations* SGEM 2014 conference proceedings 2

[20] Khoshkam M and Jaafar M 2016 A Community-based Geotourism Entrepreneurship: A Case of Kilim Geopark Malaysia *International Journal of Social Ecology and Sustainable Development (IJSESD)* **7(4)** 47-58McKeever P J 2009. *The UNESCO global network of national geoparks: geological heritage and sustainability* (Selangor: Universiti Kebangsaan Malaysia)Bali Provincial Tourism Agency 2017

IOP Conf. Series: Earth and Environmental Science **158** (2018) 012034 doi:10.1088/1755-1315/158/1/012034

Who gets the benefits of geopark establishment? *A study of Batur Geopark Area, Bali Province, Indonesia*

S Sagala[1], A Rosyidie[1], M A Sasongko[2] and M M Syahbid[2]

[1]Department of Regional and City Planning, Institut Teknologi Bandung, Bandung, West Java, Indonesia
[2]Resilience Development Initiative, Bandung, Indonesia

Corresponding author: saut.sagala@sappk.itb.ac.id

Abstract. As an important contributor to the improvement of economic and social aspects within communities, tourism is a sector whose development must be paid careful attention to. Tourist attractions, especially for nature based tourism, have been proven to provide employment and a significant contribution to both the GDP and GNP of Indonesia. Various areas in Indonesia have natural tourist attractions. A recent popular type of natural tourism attraction in Indonesia are geoparks, which involve protection and utilization of wide geological heritage areas. One of the geoparks that have been recognized by UNESCO within its Global Geopark Network (GGN) is Batur Geopark, in the province of Bali, Indonesia. The development of a geopark should consider the local economic development within its surrounding area. In addition, the further development of a geopark area that is included within the Global Geopark Network must meet the criteria for social and economic aspects prepared by UNESCO. The establishment of these criteria is intended to ensure that the management of the geopark is also beneficiary for the development of local communities as well as contribute positively to their sustainable development. Indonesia has a number of initiatives to promote the establishment of new geoparks, such as Geopark Merangin, Geopark Ciletuh and Geopark Sewu. The most recent one to be considered is Toba Lake Area. Therefore, taking lessons learned from an existing geopark regarding its economic impacts is important. This study explored the positive and negative impacts of Batur Geopark on the local economic development of its surrounding area after its inclusion in the Global Geopark Network of UNESCO. Further recommendations are and formulated in order to enhance the mutual linkage between tourism development and economic benefits to the communities. Prior to the analyses, a literature review on best practices of geoparks was done. Qualitative analyses were conducted by interviewing 41 stakeholders that were involved in the tourism and local economic development in Bangli Regency of Bali. This study provides descriptive information on types of economic activities that have appeared after the establishment of Batur Geopark.

1. Introduction

Tourism is a fundamental sector in the economic development of many countries, as it has positive economic impacts on the balance of payment, generates employment and gross income for local communities [2]. Many developing countries, such as Indonesia, Thailand, the Philippines and Malaysia, rely on tourism as one of the main contributors to their gross domestic product. The tourism sector can generate multipliers. It plays a role in increasing regional revenue and increases the development of small and medium-sized enterprises (SMEs) that support the tourism industry. In general, tourism has three major impacts, i.e. economic, environmental, and socio-cultural, that affect local communities within the destination area [1]. Beside positive economic impacts there are also negative economic impacts of tourism. Governmental regulations are needed to reduce the negative impacts. The most important positive economic impacts of tourism are budget revenue increase, local production expansion, investment stimulation, and local population welfare improvement. Among the negative economic impacts of tourism are inflation, financial leakage, infrastructure costs, and economic dependence on tourism [8]. Identifying the positive and negative economic impacts is an important step for national and provincial governments who continually implement, monitor, update and improve policies and regulations regarding tourism, which are designed to increase national/provincial income, personal income, employment and tax revenue.

Recently, geotourism has developed quite fast. Geotourism, a relatively new phenomenon, focuses on geological resources and landscapes [4]Geoparks are a form of geotourism. Even though they are still relatively new, most influential academic and international organizations accept that geotourism and geoparks are a tool for rural development, local community participation, and poverty reduction that can be applied in developing countries with abundant geological resources. The development of a geopark area should consider local economic development aspects within its surrounding area, because the concept of geotourism and geoparks plays an important role in local economic development of rural areas where geological resources are abundant. One of the best-known examples of a geopark area that has been implemented by community based tourism (CBT) in Indonesia is Batur Geopark, Bangli Regency, Bali Province, Indonesia. This study on identifying positive and negative economic impacts selected Batur Geopark as a case study to assess to what extent Batur Geopark contributes to tourism development in Bangli District and Bali Province. To get a comprehensive picture, a field survey was conducted in Batur Geopark. The local community, local champions, businessmen, tourism actors, local government agencies and scholars in tourism development were asked questions. Also, the potentials and challenges of tourism development of Batur Global Geopark were explored. This was followed by a timeline analysis to show the sequence of economic activities and an explanation of the trends that appear. Additional information on the origin of the business actors is also discussed. The result shows that the development of Batur Geopark as a tourism destination did not affect the whole area the geopark, but only significantly benefitted selected subdistricts, such as Batur, Suter, Kedisan, Trunyan, and Songan. Unfortunately, those areas only represent a small portion of the total Batur Geopark area. Therefore, the challenge is to increase the economic benefit of the geopark to other areas in the district while also maintaining its function. This study comes up with a set of recommendations and strategies to widen the economic impact of the geopark. This set of recommendations is later to be connected with the tourism development plan of Bangli Regency and that of Bali Province.

2. Methodology

The purpose of this study is to assess the positive and negative impacts of Batur Geopark, the first global geopark in Indonesia, on the local economic development after its inclusion in the Global Geopark Network of UNESCO and formulate recommendations in order to enhance the mutual linkage between tourism development and economic benefits for the local communities. This research was conducted by conducting surveys and in-depth study on the Batur Geopark area. The survey was conducted between July and August 2017. It consisted of interviews with 41 respondents who are involved in tourism and local economic development in Bangli Regency, Bali Province, such as representatives of local communities, farmers, local vendors, local businessmen, leaders of local working groups (LWG), local and provincial government agencies, scholars in tourism development, and other tourism actors. The selection of respondents was based on the respondents' main functions, which are related to local economic and tourism development, and also done through references based on snowball sampling, where previous respondents gave suggestions for potential respondents from among their acquaintances. The list of respondents is as follows:

Table 1. List of respondents and reason for selection.

No	List of Respondents	Reason for Selection
1	Center for Natural Resources Conservation (BKSDA).	Main institution that performs conservation tasks within Batur Geopark area. Currently under Indonesian Ministry of Environment and Forestry.
2	Regional Development Planning Agency (Bali Province and Bangli Regency).	Main institution involved with the planning process of Batur Geopark area.
3	Tourism Agency (Bali Province and Bangli Regency).	Main institution that performs tourism planning for Bali Province and Bangli Regency.
4	Academician (lecturer from STP Nusa Dua).	Has experienced and completed research on geopark development.
5	Regional Secretary of Bangli Regency.	General Manager of Batur Geopark area.
6	Public Works Agency of Bangli Regency.	Main institution for the development of physical infrastructure supporting Batur Geopark area.
7	Social Agency of Bangli Regency.	Main institution involved in social affairs related to the impacts of tourism activities.
8	Manpower Agency of Bangli Regency.	Main institution involved in manpower affairs, such as regional minimum wage and labor welfare.
9	Central Bureau of Statistics (BPS).	Provided statistical data regarding local economic development.
10	Batur Geopark Museum (BMV).	One of the main attractions within Batur Geopark area.
11	Indonesian Hotel and Restaurant Association (PHRI).	Association that understands the development of tourism activities in Batur Geopark area.
12	Chairman of the Boat Driver Community.	Association that understands current tourism activities, especially around Lake Batur.
13	Toya Devasya (local business entities).	One of main attractions within Batur Geopark area.
14	3 groups of Local Tourism Initiatives (POKDARWIS).	Associations that manage several attractions in Batur Geopark area (Batur Tourism Information Centre, POKDARWIS Batur, and cycling community.
15	5 groups of Local Working Groups (LWG).	Tourism management forums in 5 cluster areas (Kintamani, Batur, Suter, Trunyan, and Songan).
16	4 village heads.	Stakeholders that understand the scope of social and economic affairs of local communities (villages).
17	Head of Kintamani Sub-district.	Stakeholders that understand the scope of social and economic affairs within Kintamani Sub-district.
18	Business actors: 2 local merchants, 2 farmers, 2 tourist spot ticket sellers, 4 local guides, and ex-sand miners.	Business actors who have had direct experience with Batur Geopark.
22	Local security (*pecalang*).	Actors involved in maintaining order and security around tourist spots.
24	Head of vocational school (SMK) Kintamani 2.	Actor involved in educational activities, especially to support tourism activities.

The collected data were then analyzed to know the economic impacts of Batur Geopark development, and to know who the stakeholders are that get the benefits from the products. The economic impact analysis basically focused on the question: *What is the contribution of geotourism activities to the economy of the region?*, and it traced tourism business activities, local product development and promotion, income trends, employment/recruitment conditions, and tax revenue/local government income that have either direct, indirect, or induced effects on local economy stakeholders (consisting of other businesses, households, and government) [11]. In this research we focused the economic impact analysis on three aspects, respectively, local business activities (investment conditions, proportion of business by sector, etc.), income trends (both of local communities and government), and employment conditions (proportion of jobs by sector, unemployment rate, skill improvement effort, etc.).

3. Literature Review

3.1. Definition and scope of geoparks

The geopark is a concept derived from sustainable development for developing a region that integrates three types of diversity: geodiversity, biodiversity, and cultural diversity. The geopark aims to build and develop the welfare of local communities based on conservation of the three aforementioned diversities [3]. The geopark is a good example of geological conservation that covers whole spatial plans and integrates all natural and cultural resources around the site, and empowers local communities and gives them the opportunity to develop cohesive partnerships to promote the area's significant geological processes, features, historical aspects linked to geology, or outstanding geological beauty [3,14]. A geopark achieves its sustainable development goals by fostering local economic development through a three-pronged approach: conservation, education, and geotourism [12].

The initial geopark program was established within Agenda 21, which was sounded at the United Nations Conference on Environment and Development (UNCED) held in Rio de Janeiro, Brazil in June 1992. The Global Geopark Network (GGN) was established in 2004 by UNESCO. It actively promotes activities around geoparks, which include geotourism and conservation, and also encourages research and development that will benefit local communities within the geopark area. The organization also seeks the membership of geoparks and has set criteria for the member selection process. Some of UNESCO's criteria are [13]: (1) the geopark must have a territory encompassing one or more sites of scientific importance, not only for geological reasons but also because of its archaeological, ecological and cultural values; (2) it has a management plan designed to support the sustainable development of socio-economic aspects within the local community; (3) it demonstrates methods that are able to conserve and enhance the sustainable use of the geological heritage and provide knowledge on geoscientific and environmental disciplines; (4) joint proposals are submitted by various stakeholders from government and private sector parties that formulate sustainable development strategies; (5) it is part of a global network that is able to share best practices regarding sustainable development strategies to other members. Evaluation based on these criteria is carried out every two years by the Geopark Committee at the International Conference on Geoparks, which was held for the first time in 2004.

3.2. Definition and scope of geoparks

The roots of geotourism can be traced back to the 1990s, when there was an initiative for the development of geoparks by UNESCO, supported by many countries, in which they agreed on a conservation plan of various geological heritages and past civilization relics. According to Farsani (2011), the establishment of geoparks can generate new jobs, economic opportunities and additional sources of income, especially in rural areas [5]. At the grassroot level, geopark development encourages small and medium-sized enterprises to produce local products and handicrafts related to geotourism. Nowadays, geotourism is seen as the best form of sustainable rural tourism marketing compared to other approaches [5-7]. One of the methods that distinguishes geotourism from other approaches is the existence of a form of cooperation between different stakeholders, ranging from the government, the private sector, academics, local businesses, local residents, outdoor companies, tour agencies, restaurants, and accommodations within geotourism development. This partnership is welcomed because it makes good economic sense and can benefit all partners [4]. There is an emergant process within geopark areas that creates new products, new jobs, and new recreational activities for local communities to foster local economic development [9].

Since geotourism and geopark development are opportunities for local economic development, they have prospects to foster employment and reducing poverty through engaging local communities in various geopark initiatives. To support the goals for local economic development, tourism and geopark authorities have to adopt regulations and policies to encourage the participation of local people.

According to Farsani (2011), various geopark authorities have adopted policies toward stimulating locals' participation for local economic prosperity and preservation of natural resources, with a positive result on local economic indicators and conservation progress within geopark areas [5]. Recent trends show an increasing interest in research and understanding on the supply side of geotourism and geopark development, but relatively less is known about the demand side for geopark products [9]. Therefore, there is a need for further research and investigation into the impacts of geotourism and geopark development, and local communities' responses towards geopark products in their area.

4. Result and Discussion

4.1. Regional overview

Batur Geopark was accepted as a member of UNESCO Global Geopark Network (GGN) on September 20, 2012 at Auroca Global Geopark, making Indonesia the 91th country in the world and the third in Southeast Asia to become a member of GGN. The official name is 'Batur Global Geopark' (BUGG). It is situated in the northeastern part of Bali Island, 70 km north of Denpasar. The caldera is bounded by latitudes 8°05'00'' and 8°40'00'' S, longitudes 115°11'00'' and 115°30'00'' E. Administratively, BUGG is located in Kintamani district, Bangli regency, Bali province. There are approximately six access routes to reach BUGG from Denpasar and one route from Singaraja.

Geologically, the Batur Geopark area has an important position within the geological history of Bali island. Past geological processes in Mount Batur and Mount Agung deposited various types of rocks in the sea and then uplifted them to become the mainland of Bali. According to Purbo-Hadiwidjojo (1971), due to these geological processes the morphology of Bali island generally has a north-south cross-section that is not symmetric, with slopes more prevalent in the south than in the north [3,10]. A very long time span of the geological processes has formed Batur Geopark area as we see can see it now. At least 25 well-documented eruptions of Mount Batur have been recorded in the years 1849, 1888, 1904, 1905, 1921, 1926, 1963, 1968, and 1974. A number of geoheritage sites within Batur Geopark are the basis of the development of the BUGG concept. One of the most well-known hallmarks of Batur Geopark is the panorama of the whole geopark that can be seen from Penelokan. Combined with the view of the lake and Mount Batur itself, this active volcanic area is rich with elements of macro- and micro-volcanic landscapes, besides volcanic rocks. In local Balinese philosophy, Mount Batur is believed to contain the four main elements of the earth: soil, water, fire, and air as the important spiritual elements of nature. Many temples have been built along the slopes as places of worship for local communities.

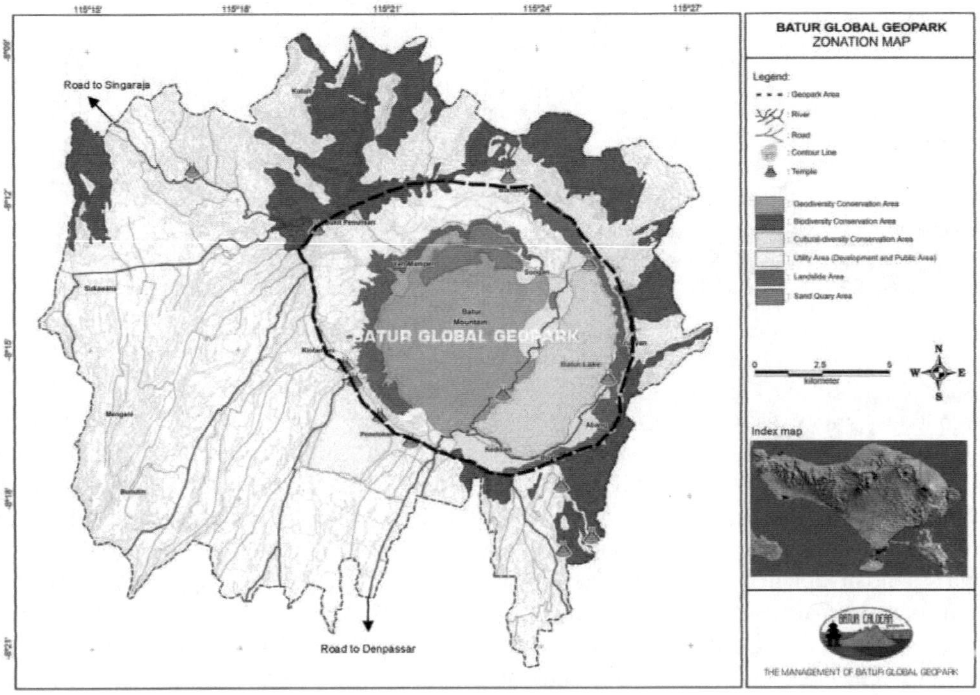

Figure 1. Map of Batur UNESCO Global Geo-park [3].

Batur Geopark area has a considerable biodiversity within its natural tourism park. Species of vegetation are mostly forest plants, including puspa (Schima noronhaea), tusam (Pinus merkusii), ampupu (Eucalyptus urophylla), mahogany (Swietenia macrophylla), sengon (Paraserienthis falcataria), sonokeling (Dalbergia latifolia), acacia (Acacia decurens), segawe (Adenanhera paronina), pine (Casuarina equsetrofolio), hibiscus (Hibiscus tilaceous), and dadap (Erytrina variegata). For the fauna, Batur Geopark area has approximately 30 species, including green jungle fowl (Gallus varius), cuckoo (Streptopelia chinensis), terocok (Gouvier ahalis), kacer (Copsycus saularis), ferret (Paradoxurus hermaproditus), porcupine (Hystrix branchura), trenggiling (Manis javanica), squirrel (Tupaia javanica), and monkey (Macaca fascicularis). Batur Geopark area is designated as a nature reserve area through the decree of the Head of Geological Agency, Ministry of Energy, Republic of Indonesia No. 37.K/73/NGL/2012.

Culturally, Batur Geopark, which covers an area of approximately 100 square kilometers and includes Mount and Lake Batur, represents the cultural diversity of processes and products related to them. The most famous and important points of interests that attract many visitors around the Batur Geopark area are Trunyan Village, Pula Hulun Danu Batur, Pura Ulun Danu Songan, Pura Dalem Balingkang, Subak, and Barong Berutuk. In terms of education on the economic value for local communities, there are some programs manifested through the interpretation of protected geological sites, establishment of a geopark museum, an information center, geotracks, a guided trip, making simple delivery media such as popular books, maps, educational materials and displays, and others. There are several community development programs, one of which trains professional guides that explain the volcanology, landscape, and human culture to the visitors of Batur Geopark area.

Figure 2. Landscape panorama from Panelokan area.

Geotourism consists of touring activities utilizing the geological phenomena in an area, mixed with its beauty, uniqueness, rarity, as well as its environment, as the main attractions [3]. The growing trend of geotourism in the Batur Geopark area is an alternative form of tourism based on the use of natural resources (geological resources) safely and sustainably. As of 2014, there are 21 geosites located in Batur Geopark area. They attract tourists for undertaking geotourism activities such as geotreking. Scenic viewpoints are regarded as the most popular locations for tourists enjoying the natural landscape of the area and are mostly located on hills, slopes, or the highest part of a region, representing the beauty of the area in a comprehensive way. In Batur Geopark area, the most representative scenic viewpoint is located at Penelokan. The Penelokan area is filled with merchandize sellers every day and on the edges of the slopes local businessmen have built restaurants and hotels. However, in the rising trend of geotourism, most local communities feel that the benefit of tourism is not distributed equally to all villages in the Batur Geopark area. Based on observations in July 2017, farming is still the major livelihood activity, where children dropping out of school to help their parents to work in the field is still prevalent within local communities. Also geotourism is not so influential and has had minimum impact in several villages that still depend on the extractive sector, such as excavation type-C in Songan A and B villages. There is also the classic reason of the difficulty of finding a job that makes many children drop out of school after graduating from junior high school to pursue work in the field. However, much potential of geotourism can still be developed in the Batur Geopark area if managed properly. All these conditions require a solution to empower the local community to manage the abundance of geological resources to increase their living standard and utilize their local economic potentials within a sustainable development framework.

4.2. The current local economic development of Batur Global Geopark

Figure 3. Restaurant and souvenir shop in Penelokan area.

Since the inauguration of Batur Global Geopark in 2012, the Batur Geopark area itself has experienced significant improvements in terms of economic and infrastructure development. Advancement can be seen from the growing number of restaurants, hotels, and other accommodations, especially around Penelokan area. Most of the already built restaurants, hotels, and other accommodations utilize the panoramic view from Penelokan as the main attraction for their guests. Most visitors, both local and international, prefer a one-day trip to visit Batur Geopark instead of staying overnight, as the area is located between the main road connecting southern and northen Bali. Points of interests for the visitor in Batur Geopark area are *adat* villages (traditional villages) and hot springs. Some attractions in the Batur Geopark area, including the hot springs, are managed daily by adat villages. Revenue from tourism management is used by the adat villages for common purposes, such as environmental management, preparation of prayer equipment, and adat ceremonies. One example of an adat village in the area of the geopark area is Panglipuran Adat Village and Batur Adat Village. Batur Adat Village currently employs more than 100 people from local villages (North Batur, Central and South Batur Village). Batur Adat Village has a form of professional operational and financial management, and has allocated its income to the construction of supporting infrastructure, operations and salaries of employees, as well as maintenance of Ulun Danu Batur Temple, which is one of the main temples in the Bangli region and Bali.

Advancement can be seen from the growing number of restaurants, hotels, and other accommodations, especially around Penelokan area. Most of the already built restaurants, hotels, and other accommodations utilize the panoramic view from Penelokan as the main attraction for their guests. Most visitors, both local and international, prefer a one-day trip to visit Batur Geopark instead of staying overnight, as the area is located between the main road connecting southern and northen Bali. Points of interests for the visitor in Batur Geopark area are *adat* villages (traditional villages) and hot springs. Some attractions in the Batur Geopark area, including the hot springs, are managed daily by adat villages. Revenue from tourism management is used by the adat villages for common purposes, such as environmental management, preparation of prayer equipment, and adat ceremonies. One example of an adat village in the area of the geopark area is Panglipuran Adat Village and Batur Adat Village. Batur Adat Village currently employs more than 100 people from local villages (North Batur, Central and South Batur Village). Batur Adat Village has a form of professional operational and financial management, and has allocated its income to the construction of supporting infrastructure, operations and salaries of employees, as well as maintenance of Ulun Danu Batur Temple, which is one of the main temples in the Bangli region and Bali.

> *"At this time, revenue from here (customary village management) is still used for the construction of supporting infrastructure for the hot springs owned by Banjar Batur in the lower regions. The details of the revenue use are as follows: 30% for employee salaries, 30% for Banjar Batur, 30% for operational costs, and 10% for maintenance of Ulun Danu Batur Temple."*
> – person in charge of Tourism Information Center, Batur Adat Village

Aside from the adat villages, there are several hotels and hot springs that are mostly owned by local businessmen. One of the most notable hotels and hot springs owned by local businessmen is Toya Devasya, located in the Toya Bungkah area, Songan B village. Toya Devasya currently employs approximately 200 employees, 99% of whom are local people from the Batur Geopark area. Toya Devasya has signed agreements with local tourism educational institutions regarding employment opportunities for graduates. Beside the local people, employees come from other cities in Bali and several come from Java and Sulawesi. Some workers from outside the island are employed because of cultural problems, such as Balinese people not wanting to work in the field of hygiene, particularly cleaning bathrooms.

> *"The workers here, 99% are local resources that we absorb from several tourism education institutions in Bangli District. The remaining are 1 from Bogor and 1 from Manado and some from Denpasar. These workers occupy all levels of staffs."* – owner of one of Toya Devasya

In terms of geotourism, since the establishment of Batur Geopark, the number of tourists continues to increase. Proportionately, the number of street merchants that flock several spots in the Batur Geopark area, such as Penelokan, Batur Lake Port, and around the main road in Batur Village (North, Central, and South Batur Village) increases as well. However, the souvenirs they sell are mostly made outside the Batur Geopark area/Bangli Regency. This trade of non-local made goods by street merchants shows that products made by local craftsmen are still less preferred. This is probably due to the absence of special and unique crafts itself from the Batur Geopark area, so there is no special demand for local crafts. Also for plantation products such as oranges, the street vendors tend to prefer to sell oranges from Java rather than local produce. They choose Javanese oranges because they are cheaper and taste sweeter (Kintamani oranges tend to have a sour taste). In addition, because of management problems related to plantation products in Kintamani it is still difficult to manage the production chain and the selling price of plantation products, especially for local orange commodities. This condition causes many local sellers to prefer selling produce imported from other regions.

> *"There are coffee, oranges, but the added value of the commodity is not there. Coffee, for example, has added value in Surabaya. So we only produce raw coffee and then we bring it out. It's outsiders who enjoy the added value. In addition, we loose competitiveness with oranges from the island of Java. Oranges from Java are cheap and sweet. We have no certainty from the side of the production chain. If there is an agency that specifically manages the production of those commodities we can start to make strategic plans, for example to set which area is best to sell our production commodities."* –youth activist from Central Batur Village

Figure 4. Street merchants in Penelokan area.

Table 2. Conditions and descriptions of local economic development.

No	Conditions of Local Economic Development	Descriptions
1	Growing number of various restaurants, hotels and other accommodations, especially around Penelokan.	Local entrepreneurs tend to open restaurants and hotels around Penelokan because of its panoramic attraction, in spite of prohibitions from the Bangli Regency government due to vulnerable conditions on the edge of a cliff. Meanwhile, many restaurant and hotel buildings on the west side of the street are abandoned by their owners.
2	Some attractions in the Batur Geopark area, including hotsprings, are managed daily by the adat village.	Revenue from tourism management is used by adat villages for common purposes, such as environmental management, preparation of prayer equipment, and some adat ceremonies.
3	Most hotels, hotsprings, and other businesses in the Batur Geopark area owned and operated by local businessmen, and employ local people.	Local businessmen already have agreements with local tourism education institutions, consisting of LPKs (course institutions) and SMKs (vocational schools), regarding employment opportunities for local human resources. Some workers from outside Bali Province also work in local hotels and restaurants.

4	Increasing number of street merchants around Penelokan and several tourism spots.	Street merchants flock several spots in the Batur Geopark area, such as Penelokan, Batur Lake Port, and around main road in Batur Villages (North, Central, and South Batur Village). They come from Bangli and other parts of Bali Province.
5	Souvenirs and plantation products (such as oranges and coffee) sold mostly come from outside of Bangli Regency.	Local craftsmen are still less preferred. This is probably due to the absence of special and unique crafts from the Batur Geopark area itself, so there is no special demand for local crafts. Also for plantation products such as oranges, street vendors tend to prefer to sell oranges from Java rather than local commodities.
6	Difficulties to manage the production chain and the selling price of plantation products.	Most farmers and plantation businessmen argue that the ones that regulate the prices of plantation products come from outside of the Batur Geopark area and even outside of Bali Province.

With these complex conditions in the Batur Geopark area, it is clear that local economic development has not been implemented evenly and in accordance with the original goals of the geopark, which emphasizes the sustainability and empowerment of all local communities. Thus, appropriate strategies are needed for the development of the tourism sector to benefit all parties involved. Table 2 provides a summary of the conditions and descriptions of local economic development in Batur Geopark. The economic impacts observed in Batur Geopark are somewhat predicted, as also observed in some other tourism developments [2,3].

4.3. The economic impacts of geotourism to local communities and stakeholders in Batur Geopark

The influence of the development of the Batur Geopark area has been felt in several areas located at the south and southwest of Mount and Lake Batur. Some villages experienced rapid development, such as North Batur Village, Central Batur, South Batur, Buahan, Trunyan, Kedisan, and Toya Bungkah area in the village of Songan B. Meanwhile, other villages, such as Kintamani Village, Pinggan, and Sukawarna are not yet considered to get significant benefits from the development of tourism in Batur Geopark. Some of these villages do not get significant benefits due to their relatively distant position from the main road passing through Batur Geopark as well as potential tourist objects that have not been commercialized.

> *"The geopark has benefited some villages, such as Toya Devasya in Songan Village, which has good development. Also in Batur there is Batur Natural Hot Spring. If this geopark is successful then who will enjoy benefits here are Buahan, Trunyan, Kedisan, Toya Bungkah. But for Kintamani, Pinggan and Sukawarna, they do not get a significant benefit. So if the question is whether the geopark has brought change (in terms of local economic development), the answer is yes. Significant or not, I would say not yet. This geopark program is like nutrition that educates people"* – youth activist from Central Batur Village

Geotourism activities in the Batur Geopark area also provide benefits for the next generation. Many young people in the Batur Geopark area are currently studying, both in primary, secondary and vocational schools, and take the time to work in part-time jobs as freelance guides or provide informant services for tourists. These activities can improve their soft skills in tourism, including English language skills. In addition, these activities also motivate students to improve the level of their education.

> *"Some of our students are either freelance guides outside school hours, or simply offering services and information to tourists for hot springs or climbing. Some of them are using free time outside school to become a daily worker in some inn or restaurant in the Songan A Village area. I think it motivates our students to raise their level of education to a higher level because they already have clarity from their field experience on how the Batur Geopark area can be utilized to generate income for themselves."* – Headmaster of SMKN 2 Kintamani (*Vocational school*)

However, some economic sectors, such as mining, have a negative impact from geotourism activities in the Batur Geopark area. Several mining activities conducted in Songan A and B villages currently begin to be restricted by local regulations, with fines as punishment. The miners think that the tourism sector in Batur Geopark area, although growing rapidly, is not yet accessible to all community groups. For example, for trekking activities in the natural conservation area of Mount Batur, the existing track has been monopolized by P3GB, a local initiative that has been legalized by BKSDA, the institution that manages the Batur Mountain conservation area.

> *"In relation to the geopark, if seen from the territorial basis of activities more home to industry activities, such as handicrafts, the geopark is like a tourism area in general, actually. But also with the existence of protected areas, it becomes less profitable for the local community as well (especially the farming and mining). Like before, there is excavation type-C mining in the Songan area. Since the establishment of BUGG it has begun to be restricted by the Bangli government. So, economically, there is also a decline in this sector as well. We see it from the whole"* – Head of Industrial Relations, Cooperatives and Labor Agency of Batur Regency

Figure 5. Excavation Type-C Mining Site in Songan B Village, and Sand Truck in Penelokan.

In terms of employment, the majority of the businesses engaged in geotourism generally employ local residents as their employees. The raise of the regional economic level can be seen from the rapid increase of the minimum wage every year. In addition, several Bangli District government offices seek to foster local small and medium sized enterprises (SMEs) through entrepreneurship training. This is one of the strategic steps to increase local economic growth, considering that the UMKM sector represents the largest proportion in terms of number of entities in Bangli Regency, especially in the Batur Geopark area.

> *"We can only nurture through entrepreneurship training, especially for SMEs. Based on the data of 2016, our minimum wage (UMK) is approximately 1,957,734, always increasing rapidly every year since 2012 (the year of BUGG establishment)."* – Head of Industrial Relations, Cooperatives and Labor Agency of Batur Regency

Various efforts have been made by the parties involved in the development of the Batur Geopark area at this time, even though there are not yet benefits for all. However, these steps are part of gradual efforts to achieve goals through the development of the Batur Geopark area, which put forward the principle of sustainable development by empowering the local communities in the area.

Table 3. Economic impacts of geotourism activities and descriptions.

No	Economic impacts of geotourism	Descriptions
1	The benefits of geopark development are not felt thoroughly by all villages in the Batur Geopark area.	Some villages are experiencing rapid development of areas such as North Batur Village, Central Batur, South Batur, Buahan, Trunyan, Kedisan, and Toya Bungkah area in the village of Songan B. Meanwhile, other villages, such as Kintamani Village, Pinggan, and Sukawarna, are considered to not yet get significant benefits from the development of tourism in the Batur Geopark area.
2	Geotourism activities in the Batur Geopark area also provide benefits for young people.	Many young people in the Batur Geopark area are currently studying, both in primary, secondary and vocational schools, and take the time to work part-time jobs as freelance guide or offer information services to tourists. These activities can improve their soft skills in tourism, including English language and communication skills.
3	Some economic sectors, such as mining and farming, experience negative impacts from geotourism activities in Batur Geopark area.	Geotourism in the Batur Geopark area, although growing rapidly, is not accessible yet to all community groups. Several geotourism activities, such as trekking within the conservation area, are monopolized by some groups.
4	The raise of the regional economic level can be seen from the rapid increase of the minimum wage every year.	Bangli District government offices seek to foster local SMEs through entrepreneurship training. This is one of the strategic steps to increase local economic growth, considering that the UMKM sector has the largest proportion in terms of number of entities in Bangli Regency, especially in the Batur Geopark area.

4.4. Recommended strategies for Batur Global Geopark's local economic development

Table 4. Observed problems and recommendations.

No	Observed Problems	Recommendations
1	- Difficulties to manage the production chain, set selling prices and marketing strategies of local craft and plantation products. - Some economic sectors, such as mining and farming, experience negative impacts from geotourism activities in the Batur Geopark area. - The increasing number of street merchants around the Penelokan area and several tourism spots.	The need to stipulate a special law that regulates the economic development of geoparks. The authorities should include within the special law: local handicraft and commodity management, including the setting of strategies for business diversification, pricing, promotion, and marketing. This law should also have the authority to manage local street merchants within the Batur Geopark area.
2	The souvenirs and plantation products (such as oranges and coffees) sold mostly come from outside of Bangli Regency.	A regulation is required that can promote economic benefits for local communities. For example by creating a guideline to train people within the geopark to be able to play a more active role. For example, besides maintaining the geopark, they can also guide tourists, make products, or develop new attractions.
3	Overlapping authorities between government institutions regarding the management of the conservation area.	Strengthen coordination so the roles between central, provincial, district agencies are well established and they can support each other. This can prevent governmental cross-border/administration issues.
4	The benefits of geopark development are not felt thoroughly by all villages in Batur Geopark area.	Establish a special geopark management body/institution. This special body/institution should also create a new benchmark for all geopark regions in Indonesia, because many of the new national geoparks are developing rapidly in Indonesia.

No	Observed Problems	Recommendations
5	The growth of various restaurants, hotels and other accommodations, especially around Penelokan, despite a prohibition from Bangli Regency government due to vulnerable conditions on the edge of a cliff.	Synchronize the Batur Geopark area planning with existing planning documents such as the *Spatial Plan* and *Detailed Spatial Plan* for the region. Synchronization with planning documents is important for establishing land use maps, which will facilitate the permit process within the Batur Geopark area.

One of the important strategies that must be implemented so that the development of geopark areas can be optimized, is the need to stipulate a special law that regulates the local economic development of geopark areas, apart from any institution and their own regulations that are involved in the development of geopark areas in particular. This law is important to regulate the management of commodities produced within geopark areas, as well as provide criteria that must be met by local businessmen, not only in Batur Geopark but also in geoparks across Indonesia, where it is necessary to ensure the participation of all local communities. The authorities within the special law should include local handicraft and commodity management, including the setting of strategies for business diversification, pricing, promotion, and marketing. This law should also have the authority to manage local street merchants. Also, a regulation is required that can promote economic benefits for local communities, for example by creating a guideline to train people within the geopark to be able to play a more active role. For example, besides maintaining the geopark, they can also guide tourists, make products, or develop new attractions.

In addition, the development of the Batur Geopark area is often obstructed by bureaucracy or different rules from different agencies that have their own special authority, territorial rights, or land tenure, so that whatever strategy is applied, as long as the agencies do not support each other, it will be in vain. For example, Batur Geopark is managed by the local government but Mount Batur itself (conservation area) is managed by BKSDA and the Ministry of Environment and Forestry using different guidelines, which overlap each other. Therefore, there is a need to strengthen coordination, so that the roles between central, provincial, district governments are well established, and they can support each other. This can overcome governmental cross-border/administration issues.

One of the management strategies that can be used to enhance the development process of the Batur Geopark area is the establishment of a special geopark management body/institution. This special body/institution will also create a new benchmark for all other geopark regions in Indonesia, because many of the new national geoparks are developing rapidly in Indonesia. Networks among geopark management bodies in Indonesia are expected to help each other in solving the problems of geopark development through solutions proposed by groups or agencies that have an interest in a geopark. If a special body of geopark managers has been established that is reinforced by the special law, then the empowerment of local communities is necessary through venues that support such specialized bodies, such as representatives/LWG, and lastly, the improvement of budgeting capabilities through training for stakeholders to increase their ability in budgeting for improvement and physical development.

Once the special geopark management body/institution is established, it could coordinate with local government agencies related to spatial planning, such as Bangli Planning Agency, to synchronize Batur Geopark area planning with existing planning documents, such as the *Spatial Plan* and *Detailed Spatial Plan*. Synchronization with planning documents is important for establishing land use maps, which will facilitate the permit process within the Batur Geopark area.

5. Conclusions

This paper discussed the benefits that the Batur Geopark area gains from the local economic development of geotourism. We found that the perceived benefits of geotourism activities so far have not been felt thoroughly by all community groups. Geotourism is seen through different perspectives by different actors. Local communities that are engaged in farming and mining have experienced that geotourism activities tend to have negative impacts on their businesses, as the Bangli Government tries to limited their activities to make room for the development of geotourism. In addition, there is the fact that the some of the current tourism sector is still monopolized by certain communities (e.g. P3GB that monopolizes trekking routes within the conservation area of Batur Geopark). Other activities include farming and plantation, which do not fit well with the geopark's aims. However, most stakeholders agree that the development of Batur Geopark is a gradual process whose benefits will be felt in the future. The roles of local leaders and communities are obviously important within the grassroot development of the Batur Geopark area. To address the challenges in the local economic development of the Batur Geopark area, a number of actions should be considered: (i) seminars and trainings, in collaboration with local universities, research institutions, and associations such as PHRI, on the importance of entrepreneurial skills for local communities, (ii) establishing a special geopark management body/institution, (iii) creating special regulations on the management of local economic development of the geopark area, which also regulate strategies of promotion and marketing of local products, (iv) creating a guideline to train people within the geopark to play a more active role, for example, besides maintaining the geopark, they can also guide tourists, make products, or develop new attractions, (v) creating participatory-planning activities that will engage local communities, (vi) establishing regulations that could strengthen coordination between central, provincial, and district agencies, (vii) synchronizing the planning of the Batur Geopark area with government planning documents such as the *Spatial Plan* and *Detailed Spatial Plan*.

Acknowledgments
This research was supported by the Institute for Research and Community Services, Bandung Institute of Technology on Capacity Building Research Program. We thank our colleagues from the Master Program of Tourism Planning SAPPK ITB, Muhammad Harits Insan Kamil and Sarita Novie Damayanti, who helped us with the field survey to collect data. We also thank Janelle Falconer who was involved in the data collection and literature review on geoparks conducted during her internship with the Resilience Development Initiative.

References
[1] Cooper C P, Fletcher J E, Gilbert D C and Wanhill, S 2008 Tourism: principles and practice (Harlow (England) : Prentice Hall Financial Times)
[2] Creaco S and Querini G The role of tourism in sustainable economic development *Proc. of the 43rd Congress of the European Regional Science Association 2003 Aug 27 – 30 Jyväskylä, Finland* Available from: https://www.econstor.eu/handle/10419/115956.
[3] Taman Bumi Global Batur, Indonesia 2014 Dinas Kebudayaan dan Pariwisata Kabupaten Bangli
[4] Dowling R. K. 2010 Geotourism's global growth *Geoheritage* **3**(1): 1-13.
[5] Farsani N T, Coelho C, Costa C 2011 Geotourism and geoparks as novel strategies for socio-economic development in rural areas. Int. J. of Tourism Res. **13**(1): 68-81.
[6] Farsani N T, Lima J, Eusébio C, Costa C and Coelho C 2013 Green Economy in Rural Areas through Geotourism and Geoparks Guy Boscarino and Derek Notte (Eds.) *Economic Developments and Emerging Markets of the 21st Century: Global Practices, Strategies, and Challenges* (USA: Nova Science Publishers) p. 211-229.
[7] Hose T A 2003 *Geotourism in England: a Two-region Case Study Analysis* (Birmingham (England): University of Birmingham)
[8] Kumar J, Hussain K and Kannan S Positive Vs Negative Economic Impacts Of Tourism Development: A Review Of Economic Impact Studies. *Proc. of the 21st Asia Pacific Tourism Association Annual Conference; 2015 May 14 – 17; Kuala Lumpur, Malaysia.* Avalaible from: https://www.researchgate.net/profile/Jeetesh_Kumar/publication/277411573_POSITIVE_VS_NEGATIVE_ECONOMIC_IMPACTS_OF_TOURISM_DEVELOPMENT_A_REVIEW_OF_ECONOMIC_IMPACT_STUDIES/links/556aa9b908aeccd7773a0110/POSITIVE-VS-NEGATIVE-ECONOMIC-IMPACTS-OF-TOURISM-DEVELOPMENT-A-REVIEW-OF-ECONOMIC-IMPACT-STUDIES.pdf

[9] Ngwira P. M. 2015 Geotourism and geoparks: Africa's current prospects for sustainable rural development and poverty alleviation In: Errami E, Brocx M, Semeniuk V, editors *From Geoheritage to Geoparks* (Switzerland: Springer International Publishing) p. 25-33.
[10] Purbo-Hadiwidjojo M. The geologic map of Bali, Indonesia, scale 1:250.000. Geological Survey of Indonesia; 1972.
[11] Stynes, D. J. 1997 *Economic impacts of tourism: a handbook for tourism professionals* (Urbana, IL: University of Illinois, Tourism Research Laboratory) p. 1-32.
[12] UNESCO 2006 *Global geoparks network* (Paris: UNESCO Division of Ecological and Earth Sciences Global Earth Observation Section Geoparks Secretariat) Available from http://unesdoc.unesco.org/images/0015/001500/150007e.pdf.
[13] UNESCO 2006 *Guidelines and Criteria for National Geoparks seeking UNESCO's assistance to join the Global Geoparks Network* (Paris: UNESCO) Internal document.
[14] UNESCO 2016 *Final Report of United Nations Educational, Scientific and Cultural Organization UNESCO Global Geoparks Council (1st Session, 24 – 25 September 2016)* (Torquay(England): UNESCO) Final Report.

IOP Conf. Series: Earth and Environmental Science **158** (2018) 012035 doi:10.1088/1755-1315/158/1/012035

The Role of Community-Based Organization in Disaster Response at Mt. Sinabung

Y Wulandari[1], S A H Sagala[2], G B Sullivan[3]
[1] Resilience Development Initiative, Indonesia
[2] Institut Teknologi Bandung, Indonesia
[3] Coventry University, United Kingdom

Corresponding author: yasmina.wulandari@rdi.or.id

Abstract. In August 2010, Mt. Sinabung in North Sumatera erupted for the very first time in 400 years; until that moment the volcano had been declared as dormant. In contrast to the communities living around Mt. Merapi in Java, the Mt. Sinabung communities had little to no preparation when faced with the sudden volcanic eruption. The Sinabung eruption caused disturbances as it ruined the farmlands and community's sources of livelihood, especially since most of the community members are farmers. Most of the time, these farmlands are located not far from their house, so displacing them means moving them away from their main source of income. This makes some of the displaced keep moving back to their old villages to farm even though they are putting themselves in danger by doing so. This condition motivated the community to initiate Beidar, an organization to monitor volcanic activities and give out early warnings to villagers nearby and within the hazard zone to evacuate. This paper will study Beidar's involvement with the community and the government, how they operate, and how they impact the community members living near the volcano. This data used in this study are primarily from interviews with Beidar members and organizers, volcanic monitoring posts, and the local disaster management. A thematic analysis of the interviews shows the history of Beidar's development as an organization proudly independent of the government; the "recruitment" of members; the important roles in the community including mediating between government and other local organizations (e.g., volcano monitoring station); specific monitoring; communication and evacuation activities (e.g., including search activities after eruptions and lahars); contacts and sharing of experience with similar groups in Indonesia; and identifying the limits of their influence in preventing local people from returning to the red zone. This research concludes that Beidar was officially acknowledged as an organization in March 2014 under the wing of the Volcanology Agency, which needed help to disseminate the volcano status information to the community. Recruiting local youth is seen as a better alternative since the communities — especially the internally displaced community members — have little trust in the government due to the crisis and deemed them 'incompetent'. Beidar, as a community-based organization, can fill the gap between the community and the government in terms of information dissemination, especially because they are local and are acknowledged as a trusted part of the community. However, this dynamic has a side effect of the villagers' feeling of security with Beidar 'watching over them'. This lowers their risk perception of the volcano, leading them to endanger themselves more by crossing over the hazard area to conducting their day-to-day activity.

1. Introduction

While Indonesia has experienced many disasters, the level of disaster preparedness varies across the country. In some places where experience with disasters is limited, the preparedness can be very low. Experience with disasters may lead to two types of responses. First, people are more aware and eventually build their capacity to deal with the disaster. Second, ignorance can increase when the impacts of disasters are considered small. Therefore, according to Lindell and Whitney, disaster experience is a good predictor to determine community preparedness for disasters [1]. The case where experience with disasters is limited can be seen from community living at Mt. Sinabung, North Sumatra.

The community characteristics, such as social capital, can also determine community preparedness to disasters. The strong ties, collaboration, and leadership help to share knowledge on awareness and eventually activities to reduce disasters. In Japan, Jishubo or Jishu Bosai is an important community activity to increase disaster preparedness [2]. Another example of a community-based disaster risk management initiative is found in India. After the Gujarat Earthquake of January 2001, a joint initiative called Patanaka New Life (PNY) Plan was commenced to ensure effective rehabilitation. It was done to train and empower local masons and communities and focuses on rehabilitation including livelihoods [3].

In some places in Indonesia, social capital also plays an important role. West Sumatra is known for the strong ties between perantauan (migrants) and their kampung (village of origin). Therefore, during the response to the 2009 West Sumatra Earthquake, many community initiatives for assistance took place [4]. In Yogyakarta, gotong royong (communal work) takes form as community preparedness and response to disasters [5], led by the leader of a village (desa) or sub-village (dusun). Examples of disaster risk management efforts initiated by the community near volcanoes are also present at Mt. Kelud and Mt. Merapi, both are located in Java Island. Both communities are started by local amateur radio clubs, i.e., Jangkar Kelud and Jalin Merapi.

In August 2010, Mt. Sinabung erupted for the very first time in 400 years; until that moment the volcano was declared as dormant. A total of 900 households or about 12.000 people were evacuated and returned home, with two causalities [6][7][8]. Later in 2013, the volcano erupted again with activity continuing up to the present. Many of the communities have been forced to relocate or stay in evacuation shelters for about three years.

The long dormancy of the Sinabung Volcano made the communities unaware of its threats and of how unprepared they were. Prior to the eruption, there has been no community-based disaster management activity; community activities are usually in form of ceremonies for weddings or births. The Karo are a unique ethnic group in Indonesia. They are part of the Batak people, an appellation created to differentiate the pagan inhabitants of the Bukit Barisan mountain range from the Malays, the Muslim lowlanders of the coastal area [9]. They have a distinct kinship and clan system, where the main patri-clan is composed of five marga: Ginting, Karo-karo, Perangin-angin, Sembiring, and Tarigan. The most important connections are the ones among Sukut (Senina or Sembuyak) agnate, a person sharing male ancestors, Kalimbubu wife-giver, and Anak-Beru wife-receiver. What the Karo society aims at, is to create and maintain the unity of Sangkep si telu, meaning the completeness of the three that hold together the entire society: kalimbubu is the head, senina the stomach, and anakberu the legs because they serve both other groups [10]. These entities are so intertwined that if one has a reason to suffer everybody does. This is why they constantly help each other in order to reinforce their unity [11]. Social capital in Karo culture is strong since they are very tightly knit rooted from the kinship culture. Decision-making in Karo culture is done by runggun, an institutionalized process of formal deliberation and decision-making by consensus within a group of people who, because of the kinship ties between them - senina, ankberu, and kalimbubu - form a unity (sangkep si telu) [12].

The Sinabung eruption caused disturbances as it ruined the farmlands and the community's source of livelihood, especially since most of the community members are farmers [13]. Most of the time, these farmlands are located not far from their house. However, the eruption displaced the communities to places far away from their main source of income. This situation created a problem since it makes some of the displaced people keep moving back to their old villages to farm even though they are putting themselves in danger by doing so [14]. The animal husbandry and fishery sectors were also affected by the disaster; many cattle are dead or went missing and there was also damage to barns and fishponds. Assistance from outside of communities would also contribute to the level of preparedness. Sagala et al. found that the level of preparedness in Mt. Merapi is determined by individual, community, and institutional variables [5]. External assistance, such as from NGOs, can play a role to increase the community's capacity through disaster simulation & education, while the government can provide immediate goods and

IOP Conf. Series: Earth and Environmental Science **158** (2018) 012035 doi:10.1088/1755-1315/158/1/012035

logistics to move people away from the disaster. However, as said before, the communities in Mt. Sinabung lack in terms of experiences as well as assistance.

Nonetheless, the length of the eruption has triggered the affected communities at Mt. Sinabung to build their own capacity by forming a group called Beidar. This name means mountain goat in the local language and it is a group formed by the people living on the flanks of Mt. Sinabung. The group initiates activities to monitor volcanic activities and give out early warnings to evacuate to villagers nearby and within the hazard zone. The emergence of this group indicates strong social capital and a need to increase community capacity to deal with the issues. How was this group formed and what has been achieved? Furthermore, to what extent has it contributed to disaster response? These are the main research questions that this paper aims to address. Therefore, this paper will study Beidar's involvement with the community and the government, how they operate, and how they impact the community members living near the volcano.

2. Mt. Sinabung eruption (2010-2017)

Figure 1. Mt. Sinabung Location

Mt. Sinabung is a volcano located in Karo District, North Sumatera Province. It was considered category "B" or 'dormant' due to its inactivity for 400 years until it erupted for the first time from August to September 2010; the volcano status rose from "Alert" to "Beware" within the same day [15][16]. The explosion was massive and authorities evacuated at least 12.000 people from high-risk areas [17]. In 2013, the volcano erupted again and has not stopped ever since.

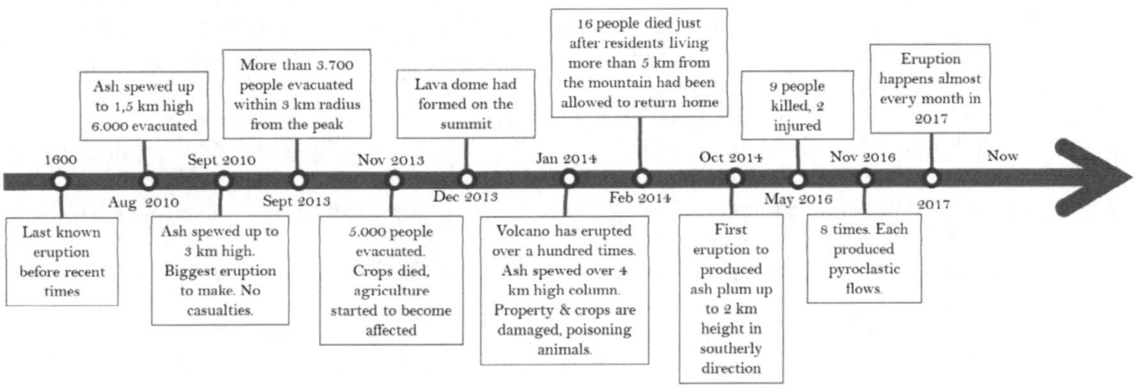

Figure 2. Mt. Sinabung Eruption Timeline

According to the Mount Sinabung Post-Disaster Action Plan 2015-2017 [18], the eruption causes damages in many sectors such as residential, infrastructure, economy, social, etc. The worst affected is the residential sector, with 9.212 damaged units consisting of permanent housing, semi-permanent housing, and non-permanent housing; causing IDR 480,21 billion (US$ 36,56 million) of damage and IDR 25,69 billion (US$ 1,96 million) of loss. Most of the damages are tin roofs that corroded by the acid in the volcanic ashes. The worst happened in Tigandreket Sub-district, Simpang Empat District, and Naman Teran District, while the most happened in Payung Sub-district with 3.719 housing units damaged.

3. Literature review

3.1. Community-based organization in disaster risk management

Community organization is a process initiated by the community, that signifies their need and they take action by developing cooperative practices. They usually focus on improving a characteristic of the community; these organizations tend to view their program in a broader community perspective [19][20]. CBOs may have different visions depending on their situation and needs at the time, one of them is disaster risk management [21].

Governments tend to respond more to political pressure than to reasoned arguments in changing their policies, which can be exerted by the people and their organizations. In disaster events, the community has more to lose by being exposed to disasters and has most to gain by reducing disaster impacts in their community. Through community-based disaster risk management (CBDRM), they have the capacity to respond to emergencies in particular [3].

CBDRM focuses on a wide variety of aspects of housing, health, livelihood, education, and so on [22]. This is anchored in the Disaster Risk Reduction framework and may cover intervention, response, assessment, projects, and programs designed by people at risk who know more about their needs and capacities [21]. CBDRM is also integral in increasing participation, obtaining resources, and changing policies and regulations [22].

3.2. Volcanic Disaster Community-Based Organizations in Indonesia

In Indonesia, there are a number of CBDRM communities with a different focus. The main focus is on risk communication and early-warning near active volcanoes. Jalin Merapi and Jangkar Kelud are two examples of CBDRM focusing on risk communication initiated by communities living near active volcanoes.

3.2.1. Jalin Merapi

Mount Merapi is an active volcano located in Central Java. During the last two centuries, this volcano has erupted explosively every 8-15 years and more violently every 26-54 years and the repose periods have not exceeded 3.5 years [23]. In 2006, the community at the Merapi underwent a period of volcanic crisis that lasted for months and in 2010 Mt. Merapi erupted again bigger than before, releasing larger SO2 emissions than any of the recorded eruptions before (from 1992-2007) [21][24].

In 2006, Jalin Merapi or Jaringan Informasi Lingkar Merapi – Information Networks of Merapi Circle was established by the local community living around Mt. Merapi. It is a social movement aimed to improve information sharing within Merapi communities, engaging three local community radio stations: K FM (Dukung, Magelang), Lintas Merapi FM (Deles, Klaten), and MMC FM (Selo, Boyolali). It has assisted the locals in developing their awareness of the Merapi hazards [25]. These three radio stations have established a network with some NGOs (Forabi, WALHI, COMBINE, Community Radio Network Yogyakarta and Central Java Community Radio) [25].

In 2010, when Mount Merapi was active again, Jalin Merapi engaged two other community radio stations, Gema Merapi FM (Cangkringan, Sleman) and Lahara FM (Salam, Magelang). Although radio is the most commonly used medium in sharing information about the Merapi in the community, it only covers one-to-one-points and is considered expensive by the community. Moreover, it only covers a particular area [26]. Jalin Merapi improvised with expanding to a website, Twitter, and Facebook. After the first eruption, they also added a volunteer form on their website. Now, they have even recruited and assigned approximately 700 volunteers to assess the needs of the internally displaced community members and help distribute relief aid [27][28]. Jalin Merapi used to only upload recent photos of the Merapi conditions but it has branched out to reach a wider audience using a web portal made using open source software [29].

Not only as an early warning system, Jalin Merapi also accurately convey important information and data to support the decision-making process. It is a network and does not interrupt the community's way of communication [24][26]. Their information is regarded as the primary information source by the community, more reliable than television and even quicker than authorized information. This high trust level increased participation and collective action, especially in the form of information sharing for relief distribution [26].

3.2.2. Jangkar Kelud

Mt. Kelud is an active volcano in East Java. It has had approximately 40 eruptions since 1000 AD. The most recent eruption happened in 2014 and saw a successful community response and an effective warning and mitigation system. A total of 166,000 people managed to evacuate after the Awas (Warning) level was issued at 21:15 local time on 13 February 2014 [30]. The communities around the volcano understand the contingency plan and risk map of the local area well [31]; supported by the community's communication this resulted in zero casualties in the 2014 eruption [32]. Similar to Jalin Merapi, Jangkar Kelud (Jangkane Kawula Redi Kelud-The Hope of the Kelud Community) is a social movement initiated by the community that is based on the communication network as disaster mitigation effort [33], that was established in August 2008 [32]. Their main tool for communication is the walkie-talky, and they also include at least 13 community radio stations in Blitar, Kediri, and Malang Districts. Nine of them use walkie-talkies; information is acquired from observation posts and disseminated in informal ways [30].

Jangkar Kelud was started by the amateur radio community and grew bigger as the members felt a sense of responsibility to be involved in disaster risk management and to raise awareness on Mt. Kelud [32]. Capacity improvement in Kelud communities was a combination of top-down support and bottom-up initiatives. Jangkar Kelud also collaborates with the Nature Lover Community (KAPPALA) and the Center for Disaster Management Study of University of National Development [30].

4. Results & discussions

This study is a part of a bigger research on the role of culture in facilitating disaster-resilient communities in the Mt. Sinabung area. The data collection is done through distributing questionnaires, observations, and in-depth interviews with community members and key informants such as the government, international and local non-governmental organizations, journalists, activists, religious figures, and village heads.

This study uses the data acquired from interviews and observations with Beidar members and organizers, the Mt. Sinabung volcanic monitoring posts (PGA), local disaster management, the Mt. Sinabung Volcanology Center and Mitigation of Geological Disasters (PVMBG), journalists, and local communities. The interview questions were a combination of exploratory and theoretically-informed questions about their work.

Data analysis was done using qualitative analysis. Data was transcribed from interviews and observations into written text. The interviews started during the disaster itself. The fallibility of memory is anticipated and remedied by cross-checking the statements with those of other interviewees, with the author's own observations, and with other documents.

5. Findings

Mount Sinabung first erupted in 2010; the sudden eruption combined with the lack of knowledge on the volcano's activity has claimed many lives and loss until present. The Volcanology Center and Mitigation of Geological Disaster Office (PVMBG) at that time decided to disperse information on volcanic activities and eruptions to communities surrounding Mt. Sinabung and send one of their partners who is a native Karo to return to his hometown and help them to communicate with locals. This person, Hasron, then founded a network to volunteer on dispersing information about Mt. Sinabung's volcanic activities and eruptions; although unnamed at the time, this network will soon become "Beidar".

Supported by the Volcanology Office, Hasron was invited to a meeting with Volcanology and UN-OCHA on October 2010, he was offered assistance to build a community focused on risk communication in Mt. Sinabung.

"Back then, our members were village youths, they are beneficial as a channel (towards the community). For example, back then, the maze of information from volcanology to community was not up. And then I was tasked to relay the information to the community, and I told them (village youths) that this is what they have to do" – NGO founder

Beidar was officially founded in March 2014 and acknowledged as an organization. Its vision is to bridge volcanic information received from the Volcanology Office to the community. Beidar is named after an endemic mountain goat species in North Sumatera. During the first eruption, hoards of mountain goats went down the mountain, as if alerting the community of the imminent danger.

Figure 3. Beidar Team Members

Beidar's main activity is to act as 'volcano's early warning agent'; its members work closely with the Volcanology Office in monitoring the lava dome and relaying information towards the community. Their main tools for communication are walkie-talkies and cellular phones, as the Mt. Sinabung communities at the time were not very familiar with social media such as Twitter. Since Beidar's members are also village youth, each member is stationed in their village on standby whenever they receive any information on volcanic activity from the Volcanology Office. Not only to the community, at times they even provide information to local newspapers on information related to Mt. Sinabung.

"I used to be a photographer, and sometimes I took photos of the lava dome's growth. I share those photos to the Volcanology Office. And then if they confirmed that the lava dome is big enough we will join them in the field and observe the development directly. We'll be on standby in case we are needed to evacuate the community" – NGO Vice Head

"The thing is, whatever the issues, we are only focused on the volcano monitoring post in Simpang Empat. So, we hear everything first, we would anticipate information on lahar or any emergency situations. We would help evacuate people and move their stuff to the car. We are never at the back of the line." – NGO Member

"Whenever there are signs of an eruption, the volcano monitoring posts would give us information on where the wind blows at the time and if the weather is good. However, if the weather is bad or foggy, they couldn't tell and we would have to check the wind direction ourselves." – NGO Member

However, information dispersal is not their only job, Beidar was also involved during evacuation and emergency response. Beidar members informed the community about the importance of using masks during ashfall and distributed masks to the communities. They are also involved in the recovery period such as reconstructing communities' houses and they donated iron sheets to be used as temporary roofing. They often assist in search and rescue, assisting the local disaster management office. Since they are located in the villages, it is easier for Beidar members to be dispatched in the process.

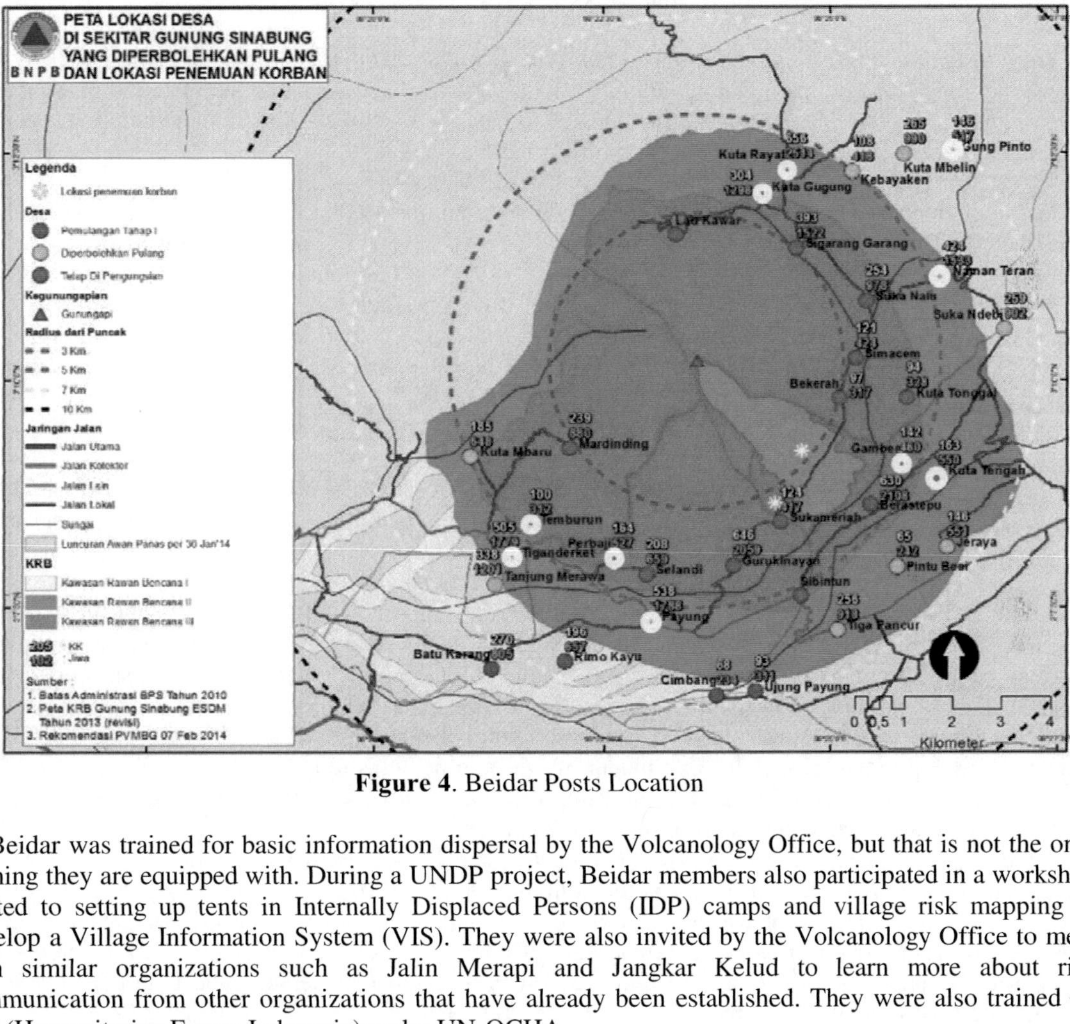

Figure 4. Beidar Posts Location

Beidar was trained for basic information dispersal by the Volcanology Office, but that is not the only training they are equipped with. During a UNDP project, Beidar members also participated in a workshop related to setting up tents in Internally Displaced Persons (IDP) camps and village risk mapping to develop a Village Information System (VIS). They were also invited by the Volcanology Office to meet with similar organizations such as Jalin Merapi and Jangkar Kelud to learn more about risk communication from other organizations that have already been established. They were also trained by HFI (Humanitarian Forum Indonesia) under UN-OCHA.

The organization maintains their independence through self-funding. They rely on members' donations for tools and uniform procurement. They refused to be funded by the government as this could make them lose their neutrality in doing their job, in the words of the founder, "… the way I see in Tanah Karo, many NGOs and community organizations have a 'shift of tone'; right now, I can freely complain and criticize the District Head with no burden … we want to give the community the same amount what we are given."

This view towards the government is apparent in Karo communities, especially in the IDP community, by being independent of government funding Beidar gained more trust from the community.

"I always tell the members, be wary of LKMD if you want to volunteer—Latih Kerih Meling Dame. Latih means tired, Kerih means you run out of money, and you will get yelled at by the mistress a lot. ... we have a joke when a member is MIA for a week, there's a 'flying magic jar' on their home" – Hasron, NGO Founder

Beidar members are predominantly men, as Karo culture has very strong patriarchal values women tend to stay at home and do domestic work. However, there are still some women in the Beidar community, mainly the wives of male Beidar members. These women are usually brought by the male Beidar members to give them an understanding of what the volunteers are working on.

"Here in Beidar we don't have income; each time we have a gathering we would bring our wife if we are married and donate some of our money to the community ... when a husband earns no income the wife usually gets angry, but by taking her to our gatherings she would understand that this is for the greater good of the community." – Firdaus, Beidar Vice Head

Hasron claimed that they have communicated the community needs to the local disaster management office but their cries have been left unanswered. This has reinforced their growing distrusts towards authority, however, their history with the Volcanology Office has left them on better terms with this office compared to the Karo District Government. Although Beidar has a close relationship with the Volcanology Office, the Head of the Volcanology post claimed that their relationship is individual, not between organizations. Even though at times Beidar helped them with monitoring, the Volcanology Office beliefs that they are better suited in information dispersal.

6. Discussions

A thematic analysis of the interviews shows the history of Beidar's development as an organization proudly independent of the government, the "recruitment" of members, and their important role in the community, especially in risk communication. Beidar is vital as a media between the government (volcano monitoring posts) and the community; their role ranges from monitoring, communication, and evacuation activities (e.g., including search activities after eruptions and lahars), contact and sharing of experiences with similar groups in Indonesia. The limits of their influence in preventing local people from returning to the hazard zone. They fit the key elements of Community-Based Disaster Risk Management (CBDRM) of community ownership; communities as ultimate beneficiaries; education and capacity building; cultural appropriateness; and participation from communities [34]. The case study of Beidar indicates how it provides a model for volunteer groups to emerge as repositories of local knowledge and wisdom, specific monitoring, community liaison, and search skills which address gaps not covered by national and local government organizations or NGOs.

Unlike their predecessors, Jalin Merapi and Jangkar Kelud, Beidar does not have background knowledge of existing hazards as much as the community. The 2010 eruption was a surprise, along with the lack of local public experience, this caused fear and chaos in the community [30]. This served as a challenge for Beidar to set up a whole new system of information dissemination. The lack of preparation for disaster situations and uncertainties reinforces a growing distrust between the Internally Displaced Persons (IDP) communities and the governments, a sentiment which is also shared by Beidar members [35]. The antipathy towards the government and authority makes the people consider Beidar as a more reliable and trusted source of information compared to the government.

The fact that Beidar members are from also part of the community helped them to earn their trusts with information and risk communication. However, the lack of synergy and trust between Beidar, the communities, and the government agencies may become an issue in the future. This situation is different to the case of Kelud, where the synergy between people and the government already existed which led to zero causalities in the 2014 evacuation. A multi-stakeholder participation and a top-down approach from the local government are vital in ensuring the sustainability and effectiveness of CBDRM such as Beidar that acts during the emergency phase [34]. A collaboration with different stakeholders could develop Beidar and utilize them for different phases in the future such as trainers for evacuation in the preparedness phase or as representatives in planning future volcano evacuation routes with the local Disaster Management Agency.

The high level of trust from the community also created a side-effect of unhealthy dependence. Many of the community members work as farmers and own a farm within the hazard zone. When they see Beidar members standing by on their posts they felt that it is safe enough for them to enter the hazard zone and visit their farmland. The government has failed time and time again to prevent people to steer away from the hazard zone; people are desperate to earn their livelihood. Beidar members are also incapable of forbidding the communities to stay within the hazard zone and can only watch and prepare to evacuate the people in case they receive volcanic activities warning from the Volcanology Office. A skewed sense of safety dawned in the community when they see Beidar member who acts as "watchful protector", making Beidar only effective in emergency response but not yet in increasing community's knowledge and capacity in recognizing the danger and risk that lie ahead.

7. Conclussions

This research concludes that Beidar was officially acknowledged as an organization in March 2014 under the wing of the Volcanology Agency. This agency needed help to disseminate information on the volcano status to the community. Hasron, the Beidar founder then recruited local youths to volunteer in passing on the information they receive from the Volcanology Agency through walkie-talkies and radios. The recruited youths are 'stationed' in their home village and are responsible for helping evacuate the villagers. Recruiting local youth is seen as a good alternative since the communities—especially the internally displaced persons communities—have little trust in the government due to the crisis and deem them 'incompetent'. Beidar, as a community-based organization, is able to fill the gap between the community and the government in terms of information dissemination, especially because they are local and are acknowledged as a trusted part of the community. However, this dynamic has a side effect of the villagers' feeling of security with Beidar 'watching over them'. This lowers their risk perception of the volcano, causing them to endanger themselves more by entering the hazard area to conduct their day-to-day activity.

Organizations like Beidar are a need-driven system initiated as a bridge between the government (Volcanology Agency) and the community. Their desire to remain independent and politically neutral is a good approach to gain the community's trust when synergy between community and authority is lacking. Although the organization was started by the Volcanology Agency, its sustainability is upheld by involving local youths that come from villages exposed to the volcanic eruptions. However, the members' shared negative sentiment of the local government could hinder the disaster risk management in the area. Learning from the evacuation in Mt. Kelud in 2014, a synergy between people and government has granted them zero causalities in the 2014 evacuation. A cooperation between Beidar and the Regional Disaster Management Agency (BPBD) could help improve the community's resilience even further. Especially with BPBD's Destana (Disaster Resilience Village) program, Beidar members could be recruited as trainers or hubs to improve preparedness towards future eruptions at the village level.

Acknowledgement

This research is funded through the PUMP-PRIME Research Grant Scheme 2015-16, Coventry University, titled "The role of culture in facilitating disaster-resilient communities: A mixed-method investigation of the community surrounding Mount Sinabung, Indonesia". The authors would like to thank Resilience Development Initiative for the academic supervision of the research. The authors would also like to acknowledge the help from Ms. Evi Syafrida and Beidar team members for providing assistance in the data collection process and their willingness to be interviewed.

References

[1] Lindell M K Whitney D J 2000 Correlates of household seismic hazard adjustment adoption *Risk. Anal.* **20** 13–25
[2] Bajek R, Matsuda Y, and Okada N 2008 Japan's Jishu-bosai-soshiki community activities: Analysis of its role in participatory community disaster risk management. *Nat Hazards* **44** 281–292.
[3] Pandey B H Okazaki K 2005 Community-based disaster management: Empowering communities to cope with disaster risks *Reg Dev Dialogue* **26** 52–57.
[4] Vanhoebrouck P Sagala S 2010 *Not so helpless victims - social capital roles in the 2009 West Sumatera earthquake recovery* (Indonesia: International Federation of Red Cross and Red Crescent (IFRC))
[5] Sagala S, Okada N, and Paton D 2009 Predictors of intention to prepare for volcanic risks in Mt Merapi, Indonesia. *J. Pacific Rim Psychol.* **3** 47–54
[6] Wahono T. 2010 Pengungsi Sinabung mencapai 12.000 orang [Online]. Available from: http://nasional.kompas.com/read/2010/08/29/08491253/pengungsi.sinabung.mencapai.12.000.orang
[7] Wahono T. 2010 Korban tewas jadi dua orang [Online]. Available from: http://regional.kompas.com/read/2010/08/29/15370269/Korban.Tewas.Jadi.Dua.Orang
[8] Suprihadi M. 2018 900 KK sekitar sempat mengungsi [Online]. Available from: http://regional.kompas.com/read/2010/08/28/19394493/900.kk.sekitar.sempat.mengungsi
[9] Steedly M M 1993 Hanging without a rope : narrative experience in Colonial and postcolonial Karoland. *Princeton studies in culture/power/history.* **16** 306
[10] Mori M 2017 *Mt Sinabung and the Karo: how cultural aspects shape resilience* (Bandung: Resilience Development Initiative)
[11] van den Steenhoven G 1973 Musjawarah in Karo-land. *Law Soc Rev.* 7 693–718
[12] Slaat P 1992 *Traditional decision-making and law: institutions and processes in an Indonesian context* (Yogyakarta: Gadjah Mada Univ Press)
[13] Sitorus H, Rianawati E, Wulandari Y, and Yohanna 2017 Final Project Evaluation: Indonesia Post-disaster Livelihood Recovery Project (Bandung: Resilience Development Initiative)
[14] Sullivan G and Sagala S 2016 Why do people still live next to an active volcano? [Online]. Available from: http://theconversation.com/why-do-people-still-live-next-to-an-active-volcano-59947
[15] Ebo A G A 2018 Status Gunung Sinabung menjadi " awas " [Online]. Available from: http://regional.kompas.com/read/2010/08/29/02252293/Status.Gunung.Sinabung.Menjadi..quot.Awas.quot.
[16] Ebo AGA. Gunung Sinabung Meletus [Internet]. Kompas. 2018 [cited 2018 Jan 21]. Available from: http://regional.kompas.com/read/2010/08/29/01452435/gunung.sinabung.meletus
[17] Reuters 2010 Thousands flee as long-sleepy Sumatra volcano erupts [Online]. Available from: https://www.reuters.com/article/us-indonesia-volcano/thousands-flee-as-long-sleepy-sumatra-volcano-erupts-idUSTRE67S06O20100829
[18] Badan Penanggulangan Bencana Daerah Kabupaten Karo 2015 Rencana Aksi Rehabilitasi dan Rekonstruksi Pascabencana Erupsi Gn. Sinabung 2015-2017. (Karo: Badan Penanggulangan Bencana Daerah)
[19] Ross M G and Lappin B W 1967 *Community organization: theory, principles, and practice* (New York: Harper and Row Publishers)
[20] Public Policy Research Institute Texas AM University 2004 Community-based organization descriptions community [Online]. Available from: https://web.archive.org/web/20090414023019/http://ccs.tamu.edu/text/cbo_classification.htm
[21] Victoria L P 2002 Community-based approaches to disaster mitigation. *Reg Work best Pract Disaster Mitig* 296–314

[22] Maskrey A 1989 Disaster mitigation: A community based approach [Online]. Available from: http://policy-practice.oxfam.org.uk/publications/disaster-mitigation-a-community-based-approach-121119
[23] Thouret J-C, Lavigne F, Kelfoun K, and Bronto S 2000 Toward a revised hazard assessment at Merapi volcano, Central Java. *J Volcanol Geotherm Res* **100** 479–502
[24] Mei E T W, Lavigne F, Picquout A, and Grancher D 2011 *Crisis management during the 2010 eruption of merapi volcano* (Santiago: Regional Geographic Conference–International Geographical Union)
[25] Surono, J P, Pallister J, Boichu M, Buongiorno M F, and Budisantoso A, et al. 2012 The 2010 explosive eruption of Java's Merapi volcano-A "100-year" event. *J Volcanol Geotherm Res.* **241-242** 121–135.
[26] Gultom D I and Joyce Z 2014 Crisis communication capacity for disaster resilience: Community participation of information providing and verifying in Indonesian volcanic eruption. *12th Bienn Aust New Zeal Third Sect Res Conf.*
[27] Motozuka T and Kanki K 2013 A study on management of the self-installation temporary stronghold after natural disaster in Indonesia: A case study of POSKO management support by Jalin Merapi. *Proc. Int Symp City Plan* 1–12
[28] Saputro K A 2016 Information volunteers' strategies in crisis communication: the case of Mt. Merapi eruption in Indonesia 2010. *Int J Disaster Resil Built Environ* **7** 102–116.
[29] Lestari P, Prabowo A, and Wibawa A 2012 Manajemen Komunikasi Bencana Merapi 2010 pada saat Tanggap Darurat *J. Ilmu Komun* **10** 173–97
[30] Andreastuti S, Paripurno E T, Gunawan H, Budianto A, Syahbana D, and Pallister J. 2016 Character of community response to volcanic crises at Sinabung and Kelud volcanoes. *J Volcanol Geotherm Res* 1-13
[31] Andreastuti S, Budianto A, and Paripurno E T 2017 Integrating social and physical perspectives of mitigation policy and practice in Indonesia. [Online]. Available from: https://link.springer.com/content/pdf/10.1007%2F11157_2016_35.pdf
[32] Lestarina F A, Soesilo T E B, and Khaerudin. 2017 Communication as an effort to disaster risk reduction for society (disaster risk reduction study with system dynamics of Kelud's eruption, East Java) *Manaj Bencana* **3** 1–21.
[33] Adila I, Dewi W W A, Tamitiadini D, and Syauki W R 2017 Disaster mitigation action plan: Digital media on improving accountability and community relationships. *IOP Conf Ser Earth Environ Sci.* **70** 1-9
[34] ECHO, UNESCAP, and ADPC. 2008 Monitoring and Reporting Progress on Community-Based Disaster Risk Management in Philippines (Manila: ECHO, UNESCAP, ADPC)
[35] Wulandari Y, Sagala S, and Sullivan G 2017 Displacement and Deadlines: Issues Found in Mt. Sinabung IDP Community, Indonesia (Bandung: Resilience Development Initiative)

IOP Conf. Series: Earth and Environmental Science **158** (2018) 012036 doi:10.1088/1755-1315/158/1/012036

Does Social Protection on Education Increase the Capacity of Communities in Facing Disasters?

D S A Suroso, S A Sagala, H A Alberdi, Y Wulandari

Institut Teknologi Bandung, Jalan Ganesha No. 10 Bandung, Indonesia

Corresponding author's e-mail address: suroso.djoko@gmail.com

Abstract. Based on the crunch model, the root causes of vulnerability to disasters include political, natural, economic, and social aspects. To target these root causes, there needs to be a focus on developing education as it plays an important role in increasing the capacity of an individual and a community. The Indonesian has prioritized the development of education and even allocated about 20 percent of its spending. In addition, the government has a social protection program that also affects the education, particularly for children namely Indonesia Smart Program ("Program Indonesia Pintar"). For this paper, Cianjur Regency was chosen as a case study for several reasons. First, Cianjur is the regency most vulnerable to disasters in West Java, particularly of floods and landslides. Second, Cianjur is one of the regencies in West Java with many social problems. The objective of this paper is to analyse the role of the Indonesia Smart Program in increasing the capacity of children to strengthen disaster resilience in Cianjur Regency. The PIP faces many challenges particularly in Cianjur Regency such as the accessibility of the beneficiaries to get the financial aid and the monitoring aspect of utilization of the money. The Indonesia Smart Program has an indirect effect in increasing the capacity of children in general. However, the effect of increasing the capacity of children in disaster risk reduction is relatively limited and indirect.

1. Introduction

Socioeconomic and demographic factors are considered the most important causes of vulnerability of communities that are exposed to disasters [1]. Various approaches have been carried out to increase community capacity and to reduce the vulnerability, particularly the social factors. Based on the crunch model [2], the root causes of vulnerability are political, natural, economic, and social aspects. To target these root causes, there is a need to focus on the development of education. In fact, various studies show the importance of education on disaster risk reduction and resilience [3].

In recent years, social protection in developing countries tends to use the strategy of cash transfers **Error! Reference source not found.Error! Reference source not found.**. Most of them use conditional cash transfer schemes in their social protection programs such as Bolsa Familia in Brazil and Familias en Accion in Columbia **Error! Reference source not found.**. Most of these social protection schemes aim to reduce the inter-generational poverty. In addition to health, education played an important role in achieving the objective. The enrollment of children in school is one of their focus areas as a requirement for a household to receive financial aid.

Education plays an important role in increasing the capacity of an individual and a community. In general, better-educated people tend to earn more, have greater wealth, and live a longer and healthier life **Error! Reference source not found.**. The Indonesian government has prioritized educating the society and even allocated about 20 percent of its spending. In addition, GoI has a social protection program that also affects education, particularly for children, namely the Indonesia Smart Program ("Program Indonesia Pintar").

The Indonesia Smart Program is a national social protection scheme for children (6 to 18 years old) by increasing their opportunities to receive a formal and informal education **Error! Reference source not found.**. The Indonesia Smart Program is a conditional cash transfer program for children. This type of program is suitable for child protection outcomes with regular monitoring and complemented with other needs of children **Error! Reference source not found.**. This program complements other social protection programs from the government for education such as School Operational Aid (Bantuan Operasional Sekolah or BOS) and Family Hope Program (Program Keluarga Harapan or PKH).

For this paper, Cianjur Regency was chosen as a case study for several reasons. First, Cianjur is the regency most vulnerable to disasters in West Java, particularly to floods and landslides **Error! Reference source not found.**. Second, Cianjur is one of the regencies in West Java that has many social problems. For instance, the child abuse in Cianjur is relatively high compared to other districts in West Java. From 2009-2015 there were 464 cases of child and women abuse, and in 2015 alone there were 74 cases. These social problems could be caused by limited knowledge of their rights and how to deal with such social problems. Therefore, education could address this issue by increasing the knowledge of their rights, values, and norm.

The objective of this paper is to analyze the role of the Indonesia Smart Program in increasing the capacity of children to strengthen disaster resilience in Cianjur Regency. The paper will present how the Indonesia Smart Program is implemented in Cianjur and what the challenges are in the process from the perspective of the beneficiaries and the government. After that, the paper will explain how the Indonesia Smart Program influences the capacity of children in facing disasters.

The main method of analysis and data collection in this study is a qualitative approach, although a small part of the analysis will be supported by simple quantitative analysis. Secondary data comprise reports, statistics, and previous studies, while primary data are collected through interviews and focus group discussion with stakeholders, such as beneficiaries (students), schools, and government agencies in Cianjur Regency.

2. Methodology

As seen in Figure 1, based on the Crunch Model, education could contribute to increasing the community's resilience towards disasters. Thus, the Indonesia Smart Program is expected to enhance the education level of communities. This paper investigates the extent to which the program contributes towards the community capacity in facing disasters. The methods of data collection consisted of in-depth interviews with the School Master, teachers, Village Secretary, staff of the Education Office and the Social Services Office, as well as focus group discussion with the beneficiaries and their parents.

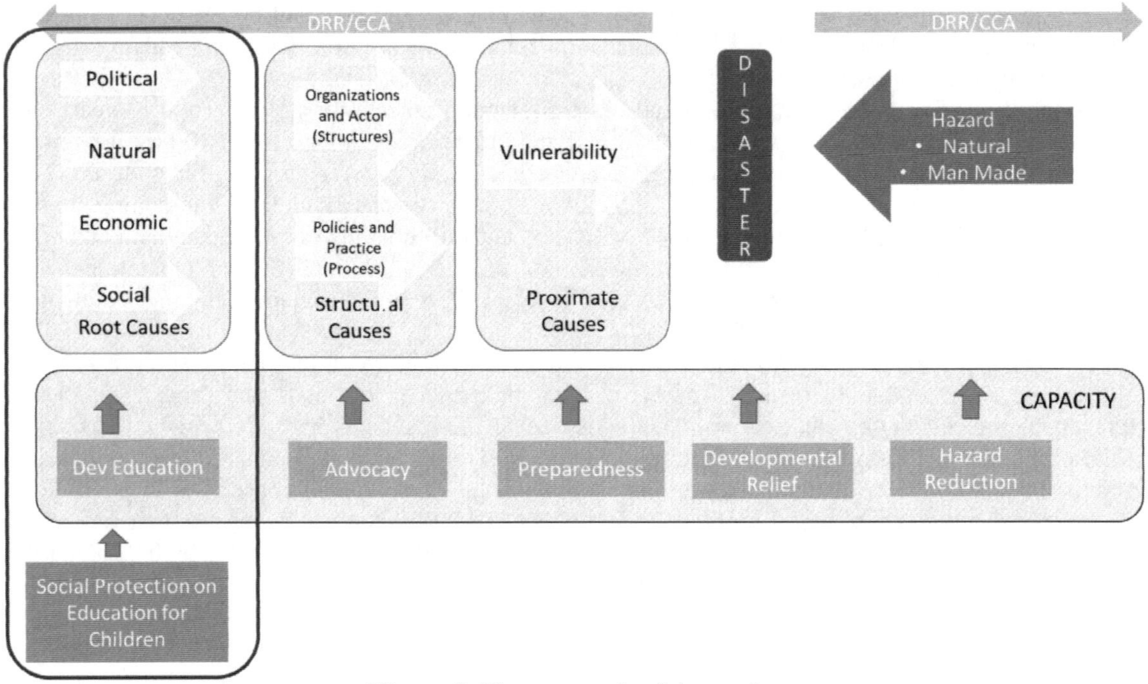

Figure 1. The approach of the study.

3. Literature Review

3.1. Social protection program on Disaster Risk Reduction

Social protection is a concept that has evolved in recent years focusing on short-term intervention to reduce the impact of unexpected shocks and is also concerned with long-term poverty alleviation **Error! Reference source not found.**. In the 1990s, social protection was in the form of safety nets for poverty, livelihoods, risk, and vulnerability **Error! Reference source not found.**. Currently, social protection is considered as an important development strategy to reduce the poverty and vulnerabilities of marginalized populations. In addition, social protection can increase the coping capabilities and decrease social exclusion in poor countries **Error! Reference source not found.**.

There are different strategies for social protection. One of the strategies that emerged in recent years, particularly in many low-income countries, is cash transfer programs **Error! Reference source not found.Error! Reference source not found.Error! Reference source not found.Error! Reference source not found.**. In general, cash transfer programs can be categorized into two types: conditional and unconditional cash transfers. Unconditional cash transfer programs have been implemented in several developing countries. For instance, the Kenyan government has designed an unconditional cash transfer program for orphans and vulnerable children that face poverty **Error! Reference source not found.Error! Reference source not found.**.

Conditional cash transfer programs usually provide cash to households that must fulfil requirements such as the enrolment of children in school, clinic attendance to monitor young children, pregnant women, and lactating mothers **Error! Reference source not found.**. This scheme arises in Latin America with several countries implementing the conditional cash transfer program for their communities such as Brazil with Bolsa Familia, Chile with Chile Solidario, or El Salvador with Red Solidaria **Error! Reference source not found.Error! Reference source not found.**. Most social protection programs invest in education as their requirement. Conditional cash transfer programs are proven relatively efficient and effective in improving educational outcomes **Error! Reference source not found.**.

Social protection has a positive impact on people's lives. Countries with social protection systems in place are better able to protect their people from the impacts of disasters **Error! Reference source not found.Error! Reference source not found.**. Social protection is aimed at assisting individuals and groups to overcome any challenges they may face by reducing vulnerability whilst also building resilience. In recent years, the national and international governments started to recognize that social protection is a means by which the impact of natural disasters can be mitigated **Error! Reference source not found.Error! Reference source not found.Error! Reference source not found.**.

3.2. Social protection program on Disaster Risk Reduction

Ghana, as a developing country in Africa, uses social protection as one of their strategies in increasing the quality of life. After their independence from colonial rule in 1957, Ghana has implemented several social protection schemes **Error! Reference source not found.**. The social protection program gives different types of aid to communities depending on their scheme such as The National Pension Scheme in 1946 which provides pensions to civil servants and army personnel; the interstate succession Law 1984 which protect the inheritance rights of widows and children; and the Education Act of 1961 that guaranteed up to six years of free primary education for all Ghanaians **Error! Reference source not found.Error! Reference source not found.Error! Reference source not found.**.

As a social protection on education scheme, the government introduced a new policy known as Free Compulsory Universal Basic Education (FCUBE) **Error! Reference source not found.**. Under this policy, pre-university education for children for was free but they still need to pay for uniforms, textbook, and PTA dues that will be a burden for poor children **Error! Reference source not found.**. Thus, the government developed new schemes under the poverty reduction strategies which gives each primary school about US$3.00 per pupil per year for their uniforms, textbooks, etc. **Error! Reference source not found.**.

Another program in Ghana is the school feeding program that began in 2005 with the objective to increase the school enrolment and retention by providing at least one daily meal to children in deprived districts **Error! Reference source not found.**. This scheme is similar to other countries' programs for feeding children. Del Rosso **Error! Reference source not found.** found that the countries which started this feeding program witnessed improved attendance, increased enrolment, and improvements in learning. This scheme also has potency in reducing hunger and malnutrition in rural communities and breaking the inter-generational cycle of poverty.

In general, these social protection schemes provided benefits to education outcomes but still face several challenges, e.g., poor record keeping, the inconsistency of money transference, and some school still charge fees for some items **Error! Reference source not found.**. The increased enrolment is one of the positive impacts of this program.

3.3. Indonesia Smart Program (Program Indonesia Pintar)

School is one of the places where children can find stability, and therefore a source of normalcy, second to home. Thus, the school can be a strategic place for a program to be installed that relates to children's resilience. From another perspective, there is an importance and urgency to support the school institutions in order to develop the children's resilience.

The Indonesia Smart Program (Program Indonesia Pintar-PIP) is one of the national social protection programs that aim to improve the enrolment in schools; increase the number of students continuing their education; and reduce discrepancies in the educational system. PIP is the improvement of the previous program named Cash Transfer for Poor Students (Bantuan Siswa Miskin or BSM) program.

The target beneficiaries of this program are students (6-21 years old) who have Indonesia Smart Card (*Kartu Indonesia Pintar* – KIP) or students from vulnerable families such as **Error! Reference source not found.**:

- Students from families that participate in the Family Hope Program (*Program Keluarga Harapan*-PKH)
- Students from families that have a Family Welfare Card (Kartu Keluarga Sejahtera – KKS)
- Students who are orphans or living in orphanages or under social care
- Students who are affected by disasters
- Students in informal education or vocational/courses institution.
- Students from vocational schools who focus on priority areas such as agriculture, forestry, and maritime affairs.

Table 1. PIP Financial Aid for Children.

Level of Education	Aid for 6 months
Elementary School (SD, MI, Paket A)	225.000 IDR
Junior High School (SMP, MTs, Paket B)	375.000 IDR
Senior High School (SMA, SMK, MAN, Paket C)	500.000 IDR

The beneficiaries are expected to use this aid to complement their study needs such as for buying books, school uniforms, and for transportation costs to go to school.

4. Study Context

Located in West Java, Indonesia, Cianjur is home to 2.5 million people. Cianjur Regency comprises 32 districts with 11 villages spread over its urban area. Cianjur is prone to disasters such as floods, landslides, and earthquakes. The regency is categorized as facing high-risk towards disasters with a score of 250 which surpasses 496 municipalities/regencies throughout Indonesia **Error! Reference source not found.**.

As one of the high-risk regencies in Indonesia, the exposure of communities to disasters is high, particularly for the most vulnerable people such as the elderly, children, people with disabilities and women, who could suffer the most from impacts of climate change and disasters. Children are some of the most vulnerable groups in society due to their dependence on adults. However, in Cianjur Regency, children face many problems that increase their vulnerability such as child neglect; street children; vagrants and beggars; human trafficking; physical and sexual violence towards children and women; poverty; malnourishment; maternal and child mortality; child marriage; and a high divorce rate (Figure. 2).

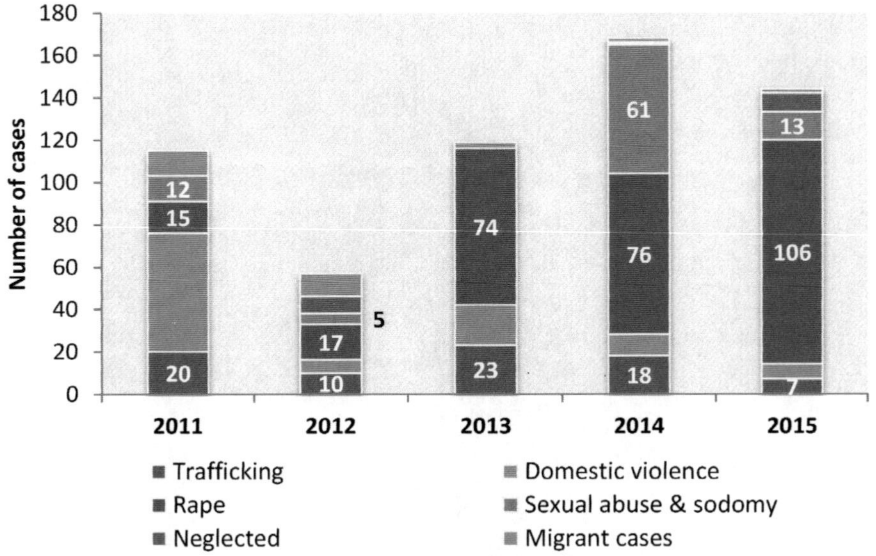

Figure 2. Social problem in Cianjur (BKKBP Cianjur Regency, 2016).

One of the ways for children to face these problems and increase their capacity while reducing their vulnerability is through education. The Indonesia Smart Program (PIP) could be a part of the solution in facing these challenges as one of the social protection programs that target children as their beneficiary to increase their participation in school. However, schools in Cianjur have many problems such as an insufficient number of schools in some areas, the number of teachers, and their educational facilities. Particularly the southern part of Cianjur has low accessibility to education compared to the northern part.

5. Indonesia Smart Program Implementation and its Challenges in Cianjur Regency

The PIP in Cianjur Regency has been implemented since 2014 and almost all districts in Cianjur are listed as the beneficiaries of PIP. Most beneficiaries are students of high schools. Figure 3 shows that the northern part of Cianjur has more beneficiaries compared to the southern part. In parallel, Figure 3 indicates that most of the northern districts are classified as high-risk in terms of disaster potential.

BRI Bank distributes the cash transfers from the PIP. The students listed as the beneficiaries are usually assisted by the schools' management to get the financial aid. As the money must be transferred directly to the student, teachers or school administrative staff accompany them to the closest BRI Bank. The Principal of State Junior High School No. 3 who used to be a teacher in the southern part of Cianjur said that "the road access to the local bank is of poor quality, so transportation costs are high and a student may spend between 10 to 30 percent of the PIP money to get the cash transfer".

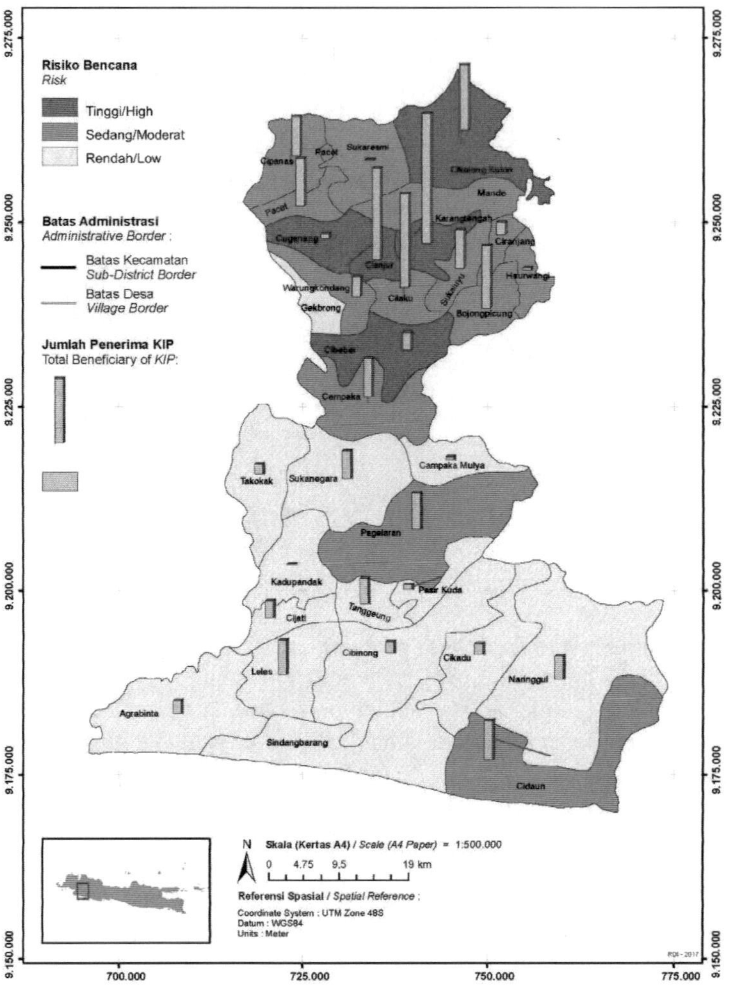

Figure 3. The mapping of the distribution of Indonesia Smart Program and Disaster Risk in Cianjur Regency 2016.

Therefore, the school usually took the initiative to help the students in preparing the administrative requirement for the payment from PIP and even coordinated a collective disbursement to be more cost efficient in terms of transportation to the local bank. However, this method to collect the cash transfer aroused public controversy as the school deducted the transportation cost from the cash transfer. One of the parents of a beneficiary during the FGD for this research actually supported such a collective way of getting the PIP aid:

> *"We actually did not mind if the school collect for transportation fees. We were really happy with the initiative of the school to facilitate the money collection from the bank...it was really helpful for us"*

FGDs with the parents and teachers of beneficiaries also reveals that there were cases when the earned money was not spent on buying books or uniforms as expected, a teacher informed:

> *"The students still used torn pants or broken shoes even after the distribution of financial aid from PIP. As a result, we established a small store which sold school*

uniforms and books to facilitate the students. However, it was seen by outsiders as too far intervening the beneficiaries. Therefore, the store was closed and we cannot really monitor the use of the cash transfer".

Another challenge is related to the question of how the government established the database of beneficiaries of the Indonesia Smart Program (PIP) **Error! Reference source not found.**. In the case of Cianjur, the communities still observed an inaccuracy of data; even though PIP has provided a feedback mechanism for beneficiaries' nomination, this mechanism did not completely work. Therefore, there were cases when students who really need the financial aid from the PIP were not listed as beneficiaries within the database. This fact has aroused jealousy among families within the communities. One of the possible factors stems from the lack of socialization of the Indonesia Smart Program so that parents are not well-informed about the program. Secondly, the required documents for PIP are sometimes not easy to fulfill by the parents, particularly for schools located in remote areas.

One of the objectives of PIP is to increase the enrollment rate of schools, which can be assessed by an indicator of the average period of school participation (*rata-rata lama sekolah*-RLS). To evaluate the extent to which PIP could contribute to the RLS index, we need to compare the data after the implementation of the previous BSM (similar to the current PIP). Figure 4 shows that although the RLS in 2016 was still 6.61 years, the RLS has continuously increased since 2004. However, the role of PIP (used to be BSM) in contributing to such a positive trend of RLS still needs further investigation as, since 2005, the government also provides Aids for School Operation (Bantuan Operasional Sekolah -BOS). BOS provides aid to help students for free enrollment to elementary school and junior high-school. Based on a national-level evaluation, BOS provides an important contribution to increasing the participation of children in schools **Error! Reference source not found.**. Therefore, the PIP may not solely contribute to the increase of RLS but complement other social protection schemes for education such as BOS.

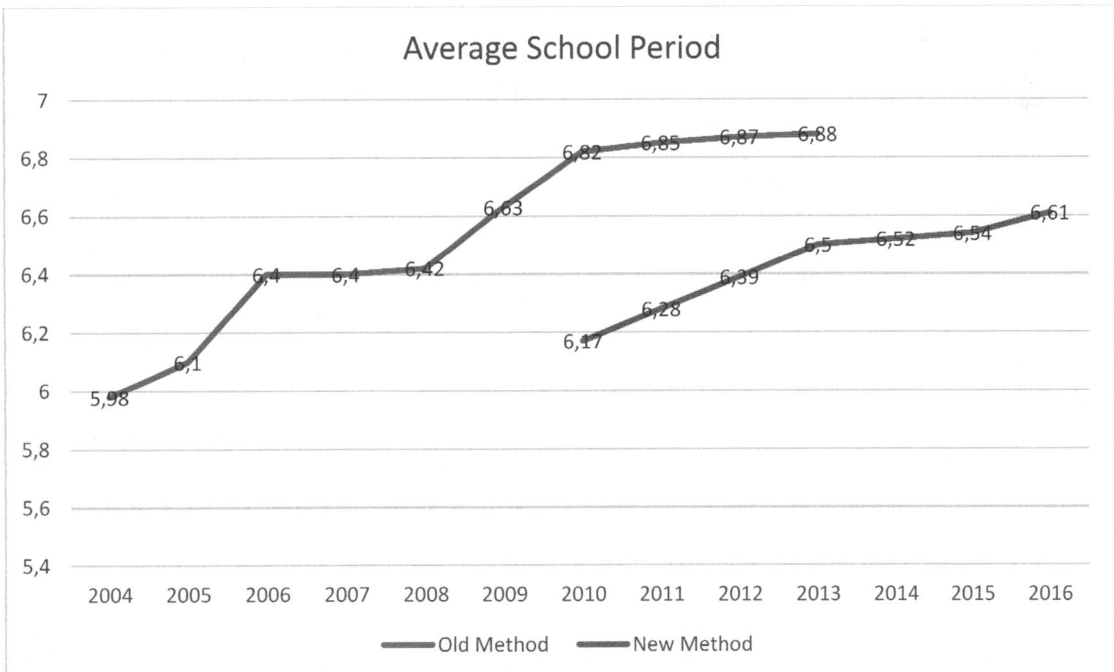

Figure 4. The average length of enrolment in school in Cianjur Regency.
Source: ipm.bps.go.id

The role of schools in the implementation of PIP is important to ensure this program works well and students get their financial aid. The teachers have the additional task to ensure the beneficiaries could access the aid. For instance, schools help students to fulfill the required documents to submit to the Education Office and to the Local Office of Bank BRI. Furthermore, school management has also acted as a facilitator to ensure the students use the financial aid in accordance with the expectation of the PIP such as for buying school uniforms, stationeries, and books.

6. Indonesia Smart Program and Community Capacity on Facing Disaster

PIP, as a social protection program that provides cash transfer for education, could increase community capacity towards disaster mitigation. In terms of disaster recovery, communities with a higher education level could recover faster than those with lower levels of education [3]. Furthermore, the resilience of communities could be enhanced through education which, in turn, could increase the capacity of communities [26][27]. In addition, other studies also showed that social protection could be used as short-term disaster relief **Error! Reference source not found.Error! Reference source not found.**.

As explained in the literature review above, one of the criteria for the PIP beneficiaries is students who were the victim of a disaster. Similarly, PKH (Family Hope Program) also targets the victims of disasters as its beneficiaries. The amount of cash transfer provided by PIP is less than the amount provided by PKH. The role of PIP in disaster mitigation cannot be immediately measured; rather, PIP has to be seen as a long-term investment to increase the capacity through elevated education levels of the communities. It means that PIP cannot immediately relieve the situation of victims of disasters in an emergency situation. In terms of emergency response, another social protection scheme like PKH may be more useful to address the needs in emergency situations. In Cianjur, the beneficiaries of PIP are usually based on the criteria of poverty, thus, the criterion of disaster victim is not commonly used.

Although the PIP provides an incentive for students to go to school, however, to build their capacity to face disasters needs complementary efforts with other programs. Through PIP, for example, students could gain general knowledge on disaster mitigation from the schools. However, for students to have more knowledge and specific skills in facing disasters needs more efforts particularly through inserting disaster mitigation topic into the school curriculum. In West Java, the provincial government has tried to insert material about disaster management in the school curriculum through local content (*muatan lokal*) since 2007 through Governor Decree 25/2007. This course could build the knowledge of students on disasters from elementary to senior high schools. Unfortunately, this local content was removed due to curriculum change in 2013. Nevertheless, our discussion with students who are the receivers of PIP shows that they have basic knowledge about disaster threats around them (e.g. landslides and earthquakes) and what they must do if a disaster happens.

7. Discussion

Social protection in developing countries faces many challenges in its implementation due to the limited capacity of the government **Error! Reference source not found.Error! Reference source not found.Error! Reference source not found.Error! Reference source not found.**. The Indonesia Smart Program, as a social protection on education, has the short-term objective to increase the enrollment rate and the long-term objective to increase the quality of life **Error! Reference source not found.**[29]. Short-term social protection usually contributes to direct coping strategies of communities through financial aid for households and in the long run has a role in building the adaptive capacity of communities **Error! Reference source not found.**. The long-term objective of PIP is also to strengthen disaster resilience. In this paper, however, it cannot be properly assessed because the program has only been implemented for about four years. PIP has a direct link to disaster mitigation by including a criterion that covers the students who experience disaster as their beneficiary. In the case of Cianjur, the number of beneficiaries of PIP is positively correlated with the level of risk in each district. The majority of PIP receivers is spatially distributed in the districts located in northern Cianjur which also have high disaster potential.

Education is always a priority in the global commitment for Disaster Risk Reduction (DRR) such as the Hyogo Framework for Action (HFA) 2005-2015 **Error! Reference source not found.** and the Sendai Framework for Disaster Risk Reduction 2015-2030 **Error! Reference source not found.**. There are substantial efforts to integrate DRR in the education sector globally **Error! Reference source not found.Error! Reference source not found.**. However, in Indonesia, there are many challenges in integrating education and DRR **Error! Reference source not found.**. In this case, PIP could be a potentially powerful tool in increasing the knowledge of students on disasters. The Safe School Program which is developed by the Education Ministry and the National Disaster Management Agency could complement PIP in strengthening the resilience of communities through the education sector.

Indonesia offers another social protection program called PKH which is similar to PIP. These two programs provide cash transfer through a bank which has offices at the village level. However, PKH has a management unit in each district which is called Implementation Unit of Family Hope Program (*Unit Pelaksana Program Keluarga Harapan*-UPPKH) [35]. The role of UPPKH in PKH is similar to the role of Schools in PIP. The Schools help the beneficiaries to understand the program guidelines so that the cash transfer can be properly received by the target groups.

8. Conclusion

The Indonesia Smart Program ("Program Indonesia Pintar" or "PIP") has a great potential to directly and indirectly increase the capacity of communities on disaster mitigation. PIP could, in the short-term, reduce the burden of students who suffer from the impacts of disasters as one of the criteria of PIP beneficiaries is disaster victims. The PIP faces many challenges particularly in Cianjur Regency such as the accessibility of the beneficiaries to get the financial aid and the monitoring aspect of utilization of the money. The Indonesia Smart Program has an indirect effect in increasing the capacity of children in general. However, the effect of increasing the capacity of children in disaster risk reduction is relatively limited and indirect. Nonetheless, it is still possible to strengthen disaster preparedness through the curriculum in schools.

References

[1] Flanagan BE, Gregory EW, Hallisey EJ, Heitgerd JL and Lewis B A Social Vulnerability Index for Disaster Management J Homel Secur Emerg Manag [Internet]. 2011;8(1). Available from: https://www.degruyter.com/view/j/jhsem.2011.8.issue-1/jhsem.2011.8.1.1792/jhsem.2011.8.1.1792.xmlA reference

[2] Wisner B, Blaikie P, Cannon T and Davis I At Risk Natural Hazards, People's Vulnerability and Disasters [Internet]. Second Edi. Routledge. London and New York; 2004. Available from: http://www.tandfebooks.com/action/showBook?doi=10.4324/9780203428764

[3] Frankenberg E, Sikoki B, Sumantri C, Suriastini W and Thomas D Education, vulnerability, and resilience after a natural disaster *Ecol Soc*. 2013;18(2)

[4] Cecchini S and Madariaga A Conditional Cash Transfer Programmes: The Recent Experience in Latin America and the Caribbean [Internet]. United Nations. Santiago; 2011. Available from: http://eprints.utas.edu.au/4774/

[5] Heltberg R Helping South Asia cope better with natural disasters: the role of social protection *Dev Policy Rev*. 2007;25(6):681–98

[6] Handa S and Davis B The Experience of Conditional Cash Transfers in Latin America and the Caribbean The Experience of Conditional Cash Transfers in Latin America and the Caribbean *Dev Policy Rev*. 2006;24(6):513–36.

[7] Valencia Lomelí E Conditional Cash Transfers as Social Policy in Latin America: An Assessment of their Contributions and Limitations Annu Rev Sociol [Internet]. 2008;34(1):475–99. Available from: http://www.annualreviews.org/doi/10.1146/annurev.soc.34.040507.134537
[8] Lutz W and Kc S *Global human capital: Integrating education and population* Science (80-). 2011;333(6042):587–92
[9] Kementerian Pendidikan dan Kebudayaan Petunjuk Pelaksanaan Program Indonesia Pintar Tahun *2017* Jakarta 2017 p 39
[10] Thompson H Cash and Child Protection: What cash transfer programming can protect children from abuse, neglect, exploitation and violence [Internet]. Cashlearning Org. 2012. Available form: http://www.cashlearning.org/downloads/discussion-paper---what-cash-transfer-programming-can-do-to-protect-children-(1)-(1).pdf
[11] BNPB Indeks Risiko Bencana Indonesia Tahun 2013 Cetakan Pe Citereup-Sentul: Badan Nasional Penanggulangan Bencana 2014
[12] Devereux S and Sabates-Wheeler R Transformative social protection *IDS Work Pap* 2004 p 36
[13] Drolet JL *Social Development and Social Work Perspectives on Social Protection* Routledge 2016.
[14] Heltberg R Helping South Asia cope better with natural disasters: The role of social protection *Dev Policy Rev* 2007 25(6):681–98.
[15] Jones N, Ahadzie W and Doh D Social protection and children: Opportunities and challenges in Ghana - Report Unicef [Internet] 2009 Available from: https://www.unicef.org/wcaro/wcaro_3798_unicef_odi_Social_Portection_Ghana-full-report.pdf
[16] Development Bank A About Social Assistance and Conditional Cash Transfers *Proc. of the Regional Workshop* 2009
[17] Ulrichs M and Slater R How can social protection build resilience? Insights from Ethiopia, Kenya and Uganda 2016
[18] A. Morley S, Coady D, Morley S and Coady D From Social Assistance to Social Development. Targeted Education Subsidies in Developing Countries. *World Development* Peterson Institute for International Economics; 2001. 1-136 p.
[19] Davies M, Béné C, Arnall A, Tanner T, Newsham A, and Coirolo C Promoting Resilient Livelihoods through Adaptive Social Protection: Lessons from 124 programmes in South Asia *Dev Policy Rev* 2013;31(1):27–58
[20] Williams A, Lamanna F and Jones N SISRI Knowledge Note No.2 Small Island States Resilience Intiative: Building Resilience through Social Protection Washington DC 2016
[21] Ulrichs M Increasing people's resilience through social protection 2016;(3). Available from: https://www.odi.org/sites/odi.org.uk/files/resource-documents/10555.pdf
[22] Osei RD, Owusu GA, Asem FE and Afutu-Kotey RL Effects of capitation grant on education outcomes in Ghana Accra Inst Soc Econ Res [Internet] 2009; Available from: http://www.researchgate.net/profile/Robert_Osei/publication/228871302_Effects_of_capitation_grant_on_education_outcomes_in_Ghana/links/0c96052287a3b36ec8000000.pdf%5Cnhttp://www.researchgate.net/profile/Robert_Osei/publication/228871302_Effects_of_Capitat
[23] Ampratwum E and Armah-Attoh D Tracking capitation grant in public primary schools in Ghana *Ghana Cent Democr Dev* 2010;10(1):1–8
[24] Miller del Rosso J School feeding programs: improving effectiveness and increasing the benefit to education-a guide for program managers 1999
[25] World Bank Memperbaiki Pendidikan melalui Program Bantuan Operasional Sekolah (BOS) di Indonesia 2015
[26] Tadele F and Bernard Manyena S Building disaster resilience through capacity building in Ethiopia *Disaster Prev Manag An Int J* [Internet] 2009;18(3):317–26. Available from: http://www.emeraldinsight.com/doi/10.1108/09653560910965664

[27] D Paton D Disaster resilience: building capacity to co-exist with natural hazards and their consequences [Internet] Disaster resilience: An integrated approach. 2006. p. 3–10. Available from: http://books.google.co.uk/books?hl=en&lr=&id=B-1VGNdM8lkC&oi=fnd&pg=PA3&dq=disaster+resilience&ots=SUsTQIHtaJ&sig=i_SSACdBMmp7SdOw1p4IbsdT8ZQ%5Cnhttp://research.arch.tamu.edu/media/cms_page_media/3494/Paton_ch1_disasterresilience.pdf%5Cnhttp://books.googl

[28] Brooks S 2009 *Social Protection and the market in Latin America* (Cambridge: Cambridge University Press)

[29] Dirjen Kementrian Agama Islam Petunjuk Teknis Program Indonesia Pintar Untuk Siswa Madrasah Tahun *2016* 2016; Available from: itjen.kemenag.go.id/sirandang/files/.../4413-6260cc99881a3e4b6cd22e311cab010

[30] UN Hyogo Framework for Action 2005-2015. Strategy [Internet]. 2005;(January):1–25. Available from: http://www.undp.org/bcpr/whats_new/rdr_english.pdf

[31] UNISDR Sendai Framework for Disaster Risk Reduction 2015 - 2030. Third World Conf Disaster Risk Reduction Sendai, Japan, 14-18 March 2015. 2015;(March):1–25.

[32] Ronan KR Advances and Continuing Challenges towards HFA2 and Post-2015: *Background Chapter Backgr Pap Prep 2015 Glob Assess Rep Disaster Risk Reduct. 2014*

[33] Ronan KR, Crellin K and Johnston D *Correlates of hazards education for youth: a replication study* Nat hazards. 2010;53(3):503–26.

[34] Amri A, Bird DK, Ronan K, Haynes K and Towers B Disaster Risk Reduction education in Indonesia: Challenges and Recommendations for Scaling up. Nat Hazards Earth Syst Sci Discuss [Internet]. 2016;(February):1–28. Available from: http://www.nat-hazards-earth-syst-sci-discuss.net/nhess-2015-344/

[35] Kharisma DD *Case Study of PKH (CCT Indonesia) in Sumba Barat and Kediri. In: Asian Social Protection in Comparative Prespective* Singapore 2008 p 1-21

Analysis of tsunami disaster resilience in Bandar Lampung Bay Coastal Zone

Alhamidi[1], V H Pakpahan[1], J E S Simanjuntak[2]

[1] Urban and Regional Planning, Institut Teknologi Sumatera
[2] Informatics Engineering, Institut Teknologi Sumatera

Corresponding authors: alhamidi013@gmail.com

Abstract. The coastal area is an area that has potential diversity of natural resources and high economic value. The coastal area is influenced by changes in land and sea so that the coastal areas are highly vulnerable to tsunami. Bandar Lampung has the potential of coastal areas of considerable potential as it is located in the bay adjacent to the Sunda Strait. Based on the study of Heru Sri Naryanto (2003), Bandar Lampung ranks third from the level of vulnerability to tsunami. Therefore, the purpose of this study to determine the readiness of the region in facing tsunami and the magnitude of the potential risks of tsunami disaster in the Gulf Coast region of Lampung in Bandar Lampung; thus, it needs to make the model or concept of tsunami disaster mitigation appropriate in terms of vulnerability and danger in creating the resilience of the Gulf Coast region of Lampung in Bandar Lampung against tsunami. The methodology used in this study was the methods of primary and secondary data collection, and the data analysis method was quantitative analysis such as spatial analysis and descriptive analysis of the data obtained from the field. The results showed that the level of preparedness in the Gulf coast region of Lampung in Bandar Lampung in facing the tsunami was still low. There are still many developed regions or houses belonging to the community either fishermen or non-fishermen located in a tsunami hazard zone. Other than that, the level of education in the Gulf coast region of Lampung in Bandar Lampung is still low where the majority of inhabitants work as fishermen. Besides, the infrastructure is old and not well-maintained so that it becomes a slum area. Therefore, the development and planning to mitigate the natural disasters tsunami using technology of IOT (Internet of Things) is an embeded system with the use of sensor seismic as a means of pre-Earthquakes vibrations, placed both on the land and in the ocean, to read the vibrations and faults in the earth's crust under the sea. With the use of seismic sensors under the sea, the vibration of the earth's crust under the sea will be detected. The sensors then will be connected to a flare marker buoys as a means to inform the disaster mitigation center. The construction of hall disaster at some point will be helpful to give first aid to those who are difficult to pass through the evacuation place since it is far away from the Gulf coast. The hall mitigation can be designed anti-earthquake and anti-tsunami. The model and concept of mitigation used is combining the Spatial Plan of Bandar Lampung and the mitigation of tsunami disaster as an integrated system of pre-disaster, during disaster and post-disaster by making the city of Bandar Lampung has the resilience to tsunamis.

Keywords: Tsunami, Resilience, and Bandar Lampung

1. Introduction

The coastal area is an area that has potential diversity of natural resources and high economic value. Based on Law No. 27 of 2007 on the Management of Coastal Areas and Small Islands, coastal region is the transition between terrestrial and marine ecosystems which are affected by changes in land and sea"[1]. The transition between terrestrial and marine ecosystems in coastal areas raises the appeal to exploit the available wealth of resources in the region. It can result in disaster if it is not accompanied by the utilization of existing policy. In addition, the coastal regions are affected by changes in land and sea so that the coastal areas are highly vulnerable to the tsunami.

Indonesia as an archipelagic country has the longest coastline in the world after Canada with a length of 81,000 km and it is accompanied by abundant natural resources [2]. In the coverage area of Indonesia, there are some areas that have a coastline with similar characteristics to Bandar Lampung. Bandar Lampung is the capital of Lampung Province which is in the southern part of Sumatra Island with an area of 192.18 km2. Bandar Lampung has high potential of coastal areas as it is located at the base of the bay which is adjacent to the waters of the Sunda Strait connecting between Java Sea and Indian Ocean. Geographically, the Gulf Coast of Lampung is located between 104°56'-105°45 East Longitude and 5°25'-5°59' South Latitude covering the land with an area of 127 902 ha and 161 178 ha of water area. Lampung Bay shoreline is along 160 kilometers of Sunda Strait.

There are 110 tsunamis in Indonesia with 100 tsunamis were caused by an earthquake, 9 tsunamis by volcanic eruptions, and 1 tsunami by landslides [2]. Tsunami in Indonesia is dominated by tsunami originating from an earthquake. The centers of earthquakes are related to the subduction (subduction) plates, so that the path of the earthquake that occurred follows the subduction zone [4]. Lampung Province, including Bandar Lampung (west) which is near the Sunda strait subduction zones, has very high levels of seismicity due to movement of the plates in the zone. Based on the study by researchers from BPPT to the tsunami disaster in the Sunda Strait region, there are five (5) vulnerabilities zones, one of them is Bandar Lampung located in the third zone (Zone III) [5]. Bandar Lampung is as one of tsunami disaster vulnerability zones because the south coast of Bandar Lampung is a bay that is prone to tsunamis and the potential for earthquakes, flooding and environmental damage on very dense settlements along the coast.

Potential tsunami disaster in the city of Bandar Lampung is derived not only from the earthquake, but also from the participation Anak Krakatau eruption in the waters of Sunda Strait so that the potential of higher tsunami in Bandar Lampung increases. Of the various potential, a handling effort in the region through a mitigation is required by paying attention to the danger of disaster, vulnerability to disasters and disaster preparedness tsunami in Bandar Lampung.

2. Basic theory

2.1 Tsunami and causes

Tsunami comes from Japanese language meaning harbor where *Tsu* and *Nami* mean ocean waves, so it can be interpreted as the ocean waves that hit the Sea Port. Tsunami is one form of disaster caused by natural activity in utilizing vibration, cracks, as well as the movement under the ocean surface that can trigger the movement of the sea waves in terms of size, height, volume, and speed of the waves. Tsunami is a series of ocean waves with the spreading capability at speeds up to 900 km per hour, mainly caused by the earthquake which occurs on the seabed [6]. Tsunamis can occur by several factors which are as follows.

1. Volcano Eruption
   Tsunami caused by volcano is the most significant cause. By the time the volcano erupts, the vibration in the earth's surface is inevitable where earthquakes are one of the causes of tsunami. In addition, land subsidence due to the volcano and landslide can cause a tsunami. Therefore, tsunami caused by the volcanic eruption can be said as the main point of tsunami.
2. Landslide
   Avalanche is the reason for tsunami. Basically, the tsunami caused by the landslide requires amount of land to drive a new wave in the ocean floor. Tsunami caused by landslide is illustrated with the decreasing soil or permafrost of the land surface to the ocean floor. In addition, encouragement and a large movement trigger the formation of a new wave at the sea surface.
3. The Earthquake
   Tsunami struck by earthquakes is the most frequent cause in some areas of the world as an echo of the earth is to balance the state of the Earth. However, the earth movement can cause tsunami waves on the sea surface. Tsunami struck by earthquakes usually occur due to the shifting plates beneath the ocean surface and depths of less than 60 km.

2.2 The tsunami disaster resilience

The resistance level of tsunami in coastal areas is assessed through the level of tsunami risk and tsunami-prone areas. The risk of a disaster has three variables, namely: 1) aspect of the threat type, 2) aspect of planning and 3) the capability to cope with [7]. Disaster Management Law Number 24 Year 2007 describes a series of disaster mitigation efforts to reduce disaster risk, either through physical development as well as awareness and increased capacity to deal with the threat of disaster [8]. Not all disasters create disaster risk. Disaster risk comes from disasters that have vulnerabilities. Therefore, the risk of disasters is the convergence between hazard and vulnerability [2].

$$\text{Risk}=\text{Hazard}\times(\text{Vulnerability}/\text{Capacity})$$

The book of Managing Country Risk Maritime Disaster in Indonesia: Disaster of Earth, Oceans and Atmospheric discusses the analysis principles of tsunami risk, the stages of work required in determining the magnitude of tsunami hazard and the vulnerability, and the vulnerability of communities and infrastructure, namely 1) assessment of tsunami hazard; 2) assessment of vulnerability to tsunami hazard; 3) analysis of tsunami risk; and 4) tsunami disaster mitigation.

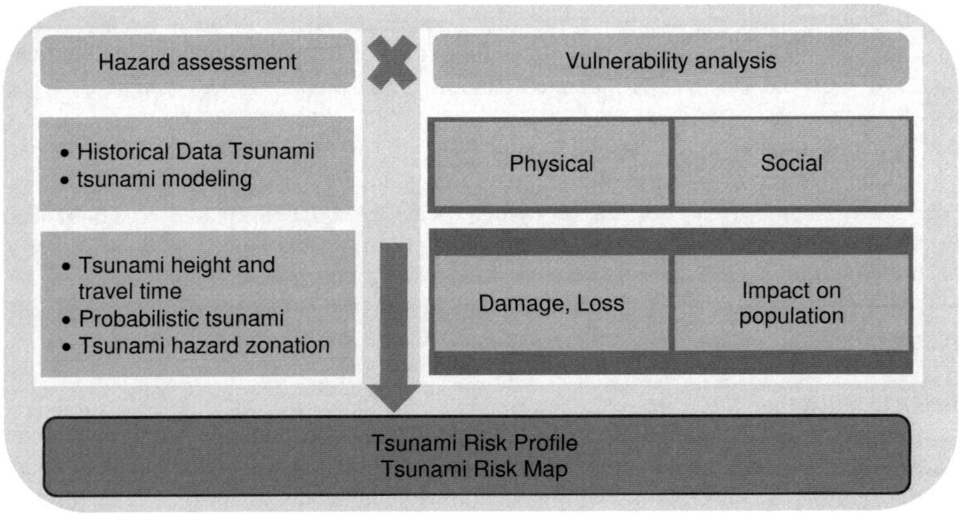

Figure 1. Framework for risk model [9]

Hazards are potential physical damage events/phenomena/human activity that may cause loss of life, property damage, social and economic disruption, or environmental degradation. Capacity is the combination of all the strengths and resources that exist within a community, society or organization that can reduce the level of risk or the impact of the disaster. Meanwhile, vulnerability is the condition and physical processes/social/economic/environment that boosts pressure and the impact of a disaster on the society or community. Then disaster risk is a potential destructive consequence that eliminates life, safety, property, livelihoods, economic activity, or the environment as the result of interaction between natural hazards and harm due to human action in conditions of vulnerability.

Coastal Community Resilience (CCR) Guide (TWCRG, 2007) is a reference guide tsunami warning center which attempt to identify, prepare, and recover tsunami and its impact [10]. Coastal Community Resilience (CCR) e is an approach done as a collaborative effort and participatory coastal communities, national government agencies and local, NGOs, private sector, and other key stakeholders to identify strengths, weaknesses, and opportunities to improve disaster resilience locally and nationally. This integral approach can be used systematically to determine the right program to increase the resilience of coastal communities in a region. The concept of coastal community resilience is applied to reduce the risk of coastal hazards and to avoid disaster and speed recovery in case of disaster. With the resistance, coastal communities can easily adapt to changes through the experience and lessons learned from the previous disasters. The analysis of the level of risk or the tsunami can be done. Then, the elements of coastal community resilience can be grouped based on assessment criteria as follows:

Figure 2. Diagram elements of coastal community resilience [10]

A. ***Governance***: Leadership, legal framework and institutions provide the conditions that allow the resilience through community involvement with the government in which the government is as a facilitator and provides space for the coastal communities to withstand disasters and come forth back from disaster quickly. The community is able to adapt to the changes so that coastal community resilience can be enhanced periodically through the intervention of government, private and civil society in the community development and the management of coastal areas by implementing disaster management therein.
B. ***Society and Economy***: People involved in a diverse and sustainable living environment are resistant to the dangers. Society and economy are an important element of disaster resilience because they have a direct relationship between economic activities and social life of the community. The activities influence any changes that occur in local and regional.
C. ***Coastal Resource Management***: Management of coastal resources, which actively supports environmental services and livelihoods and reduces the risk of coastal hazards, provides a wide range of valuable resources and sustainable for society.
D. ***Land Use and Structural Design***: Land use and effective structural design are to complement the environment. The purpose of land use and an effective structural design in reducing the risk of danger will allow people to survive from tsunami and disaster..
E. ***Risk Knowledge***: The public is aware that the hazard and risk information is useful when facing disaster. If the public knows the risk level of the dangers, the public can adapt to reduce the impact of the disaster (reducing vulnerability).
F. ***Warning and Evacuation***: People can receive notifications and alerts coastal hazards and risk warning. The warning system and evacuation are an effective response to danger by providing the opportunity for the public to take action quickly and reduce the impact of disasters.. In a warning and evacuation system, there are three parts, namely early warning systems, evacuation plans and provision of information to the public effectively and accurately.

G. ***Emergency Response***: Mechanisms and emergency response network are established and maintained to respond disasters quickly and address immediate needs in coastal communities. The level of effective emergency response makes coastal become more resilient to withstand the impact of the tsunami and other disasters. The mechanisms and emergency response planning can provide the basis for people to back up quickly from the impact of a disaster (disaster recovery).

H. ***Disaster Recovery***: Plans are made to accelerate disaster recovery, engage communities in the recovery process, and minimize the negative environment, social, and economic impacts of recovery. To build community resilience can be done with approaches covering all the elements above which are integrated into disaster management, development communities, and coastal resource management.

From the level of resiliency against tsunami hazard, one of the tools used in the tsunami disaster mitigation in today's digital era is embedded systems referring to an electronic system that is part of a site system. Embedded systems are typically found in electronic devices which are often used every day such as TV remote control, digital clock, MP3player and many more. Embedded system can be considered as part of electronic devices components, but it lacks specificity on its use. Embedded will match if it can be translated means. Embedded is different from personal computer (PC), laptop, etc. in doing an activity. It needs an adjustment and matching between devices and existing embedded systems to do a specific job. It is different from PC which can do many things during the activities and can be programmed.. Embedded system can be used and placed at various places and conditions. Prices and costs to design a series of embedded systems can be quite cheap. The provision of real-time information is also the most important factor in addressing the issues of mitigation. Embedded systems are also energy efficient because it only works on one particular job. Besides, embedded systems do not use the user interface (GUI) to do the job. It is very different from the PC display images and color which causes high power consumption. Specification of embedded system is excellent to help addressing the problems of tsunami disaster mitigation.

3. Research methodology

The method of data collection used is the primary and secondary method, which the data were obtained from the interviews and other related agencies. The method of analysis used is quantitative analysis methods such as spatial analysis (scoring) and descriptive analysis.

4. Results and discussion

4.1 Analysis of resilience and regional preparedness in facing tsunami disaster

Disasters caused by nature will have a significant impact on life on earth's surface. Every human being needs to have a concern with natural disasters in order to reduce the level of its influence on human life. Natural disasters are usually able to damage and destroy story buildings, homes, and even humans' life. Tsunami is an example of natural disaster damaging objects and creatures on the earth's surface. Total volume of tsunami waves is large enough and the high speeds can damage and sweep away buildings, homes and any other living creatures.

Primary and secondary data were obtained from the field and related agencies were processed according to the needs. The data were processed to analyze the coastal community resilience of Bandar Lampung in facing tsunami disaster and tsunami disaster mitigation. Partial secondary data are shown below, while the primary data are not shown for the confidentiality.

Table 1. Society conditions and facilities of each sub-district in Bandar Lampung in 2015 [11]

No.	Districts	Total population	Social Welfare Issues	Health facility			Land Use
				Hospital	PHC	IHC	
1	Teluk Betung Barat	29 799	3,543	0	1	26	• *Embung*
2	Teluk Betung Timur	41 645	4038	0	2	29	(retention

No.	Districts	Total population	Social Welfare Issues	Health facility			Land Use
				Hospital	PHC	IHC	
3	Teluk Betung Selatan	39 353	3,569	3	1	40	basin) • Industrial area • Protected area • General Services • Port area • General Services Region • Mining region • Trade and Services • Fishery • Government Offices • Settlement • Agriculture • RTH • Empty land
4	Bumi Waras	56 742	6,866	0	1	40	
5	Panjang	74 506	5,229	0	1	51	
6	Teluk Betung Utara	50 593	3,236	3	2	42	
7	Tanjung Karang Timur	37 108	2255	0	1	31	
8	Kedamaian	52 592	2709	0	1	30	
9	Tanjung Karang Pusat	51 126	3,640	2	2	34	
10	Enggal	28 084	1,350	3	1	23	
11	Tanjung Karang Barat	54 710	3,490	0	2	34	
12	Kemiling	65 637	3,199	1	3	42	
13	Langkapura	33 944	1,473	1	1	25	
14	Kedaton	49 055	3,085	0	1	31	
15	Rajabasa	48 027	2,645	2	1	35	
16	Tanjung Senang	45 775	1,636	0	1	25	
17	Labuhan Ratu	44 843	1,584	1	1	27	
18	Sukarame	56 921	2,033	0	3	35	
19	Sukabumi	57 334	3067	0	3	44	
20	Way Halim	61 493	2660	3	1	36	
	Bandar Lampung	**979 287**	**59 601**	**19**	**30**	**680**	

The criteria in determining the danger level of tsunami in coastal areas of Bandar Lampung are: 1) the distance to the coastal areas of tsunami sources; 2) the availability of facilities and infrastructure of protection structure; and 3) the value of coastal areas covering the economic, social, cultural, infrastructure, defense and security. Based on these criteria each variable was scored to determine the level of tsunami hazard in coastal areas of Bandar Lampung. The results of primary and secondary data were analyzed using Geospatial Information Systems (GIS), which produces tsunami hazard map of Bandar Lampung coastal regions to coastal communities (Figure 3). Tsunami hazard levels in the coastal area of Bandar Lampung are grouped into five groups, namely:

Table 2. Danger level of tsunami coastal in Bandar Lampung

No	Danger Level	Area with vulnerabilities			
		Panjang	Teluk Betung Timur	Teluk Betung Selatan	Bumi Waras
1	Very low	Srengsem, Panjang Selatan, Pidada, Pidada Utara, Way Lunik	Way Tataan, Sukamaju, Keteguhan, Kota Karang Raya, Kota Karang	Pesawahan	Bumi Waras, Sukaraja
2	Low	Srengsem, Panjang Selatan, Pidada, Pidada Utara, Way Lunik	Way Tataan, Sukamaju, Keteguhan	Pesawahan	Bumi Waras, Sukaraja, Kangkung
3	Moderate	Srengsem, Panjang Selatan, Pidada, Pidada Utara, Way Lunik	Way Tataan, Sukamaju, Keteguhan, Kota Karang Raya, Kota Karang	Pesawahan	Bumi Waras, Sukaraja, Kangkung
4	High	Srengsem, Panjang Selatan, Pidada, Pidada Utara, Way Lunik, Karang Maritim	Way Tataan, Sukamaju, Keteguhan, Kota Karang Raya, Kota Karang	Pesawahan	Bumi Waras, Sukaraja, Kangkung
5	Very high	Srengsem, Panjang Selatan, Pidada, Pidada Utara, Way Lunik, Karang Maritim	Way Tataan, Sukamaju, Keteguhan, Kota Karang Raya, Kota Karang	Pesawahan	Bumi Waras, Sukaraja, Kangkung

IOP Conf. Series: Earth and Environmental Science **158** (2018) 012037 doi:10.1088/1755-1315/158/1/012037

The regions that have potential tsunami hazard in the coastal region of Bandar Lampung are mapped to see the magnitude of potential danger of tsunami and to consider the risks and vulnerabilities to the society. Thus, from the analysis, there are four districts that have higher potential danger of tsunami than the sub-sub-districts in the city of Bandar Lampung. The indicators are slope, topography, land use, inhabitants' jobs, available infrastructure such as houses and public facilities.The higher the level of infrastructure, economic level, education communities in the region, the lower the level of the potential hazards of the tsunami. If the infrastructure level is low, the community has low education., The higher the intensity of land use the higher the potential level of tsunami hazard. Subdistricts that have potential level of tsunami are Panjang, Teluk Betung Timur, Teluk Betung Selatan, and Bumi Waras.

Table 3. Size potential of tsunami according to each group

No.	Danger level	Area (Ha)
1	Very low	36.31
2	Low	60.37
3	Moderate	101.82
4	High	23.28
5	Very high	96.89
Total Size		**318.67**

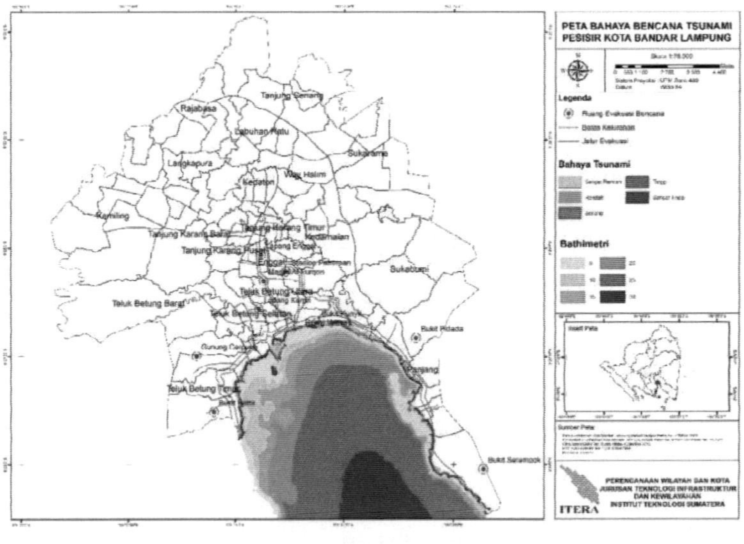

Figure 3. Map of coastal tsunami hazard of Bandar Lampung in 2017

Tsunami caused by the volcanic eruption occurred in Lampung Province in 1883; the eruption of Krakatoa in Lampung Province resulted in the Tsunami. Wave swept the beach of Lampung and Banten, where approximately 5,000 boats were destroyed and many small islands were drowned. The tsunami waves was about 40 Meter, as high as the 12-storey building, destroyed nearly 300 villages and killed more than 36,000 people. It was the biggest tsunami disaster in Lampung Province. Spatial analysis is based on the height of the tsunami.

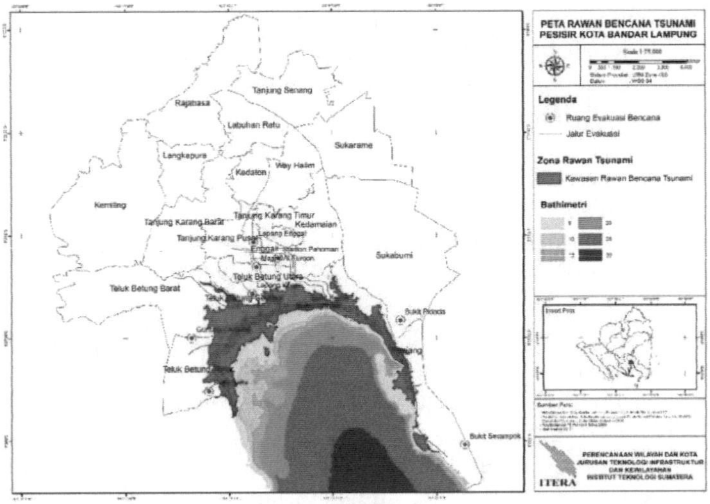

Figure 4. Map of tsunami disaster prone coastal of Bandar Lampung in 2017

Based on the tsunami hazard map of Bandar Lampung coastal areas, is the areas extremely have considerable potential because of high economic, physical, social and environmental vulnerability. There are still many developed regions or inhabitants' houses lie in the tsunami danger zone. The total population is estimated about 292.638 with a population density of about 110 person/hectare. In addition, the education level of coastal communities is low. 60% of the total population does not got to school. Infrastructure protection structure which does not meet the standard is not feasible anymore. the government policy in facing the tsunami disaster and the institution capabilities in tsunami preparedness are still low.

In social vulnerability, the education level is still low and the majority of inhabitant are fishermen. Then, in the physical vulnerability, the infrastructure is old and not well-maintained so that it becomes slum area. It indicates that the level of preparedness in facing tsunami is still low. The criteria of resistance elements are measured from governance, society and economy, coastal resource management, land use and structural, risk knowledge, warning and evacuation, emergency response and disaster recovery. These criteria of resistance elements are still low and it needs to review so that the level of community resilience in Bandar Lampung coastal areas to the tsunami is high.

From the observation and analysis, it was then applied to the concept of mitigation of the tsunami disaster in coastal area of Bandar Lampung horizontally by combining some points that can be used as a tsunami evacuation space, namely Kunyit hill, Hatta hill, Pidada hill, Serampok hill, Cerpung mountain, Al-Furqon mosque, Korpri ground, Enggal ground, and Pahoman stadium. It can be combined with the Spatial Plan of Bandar Lampung. The level of vulnerability to information and communication technologies is being developed at this time. The vulnerability of social, physical, economic, and environment can be reduced, as well as the revitalization/rejuvenation of the residential neighborhood of fishermen in coastal areas of Bandar Lampung. Then, the mitigation of hall construction needs to be done to accommodate the victims of disasters. Model and mitigation concepts used are by combining the technology and the tsunami disaster mitigation as an integrated system of pre-disaster, during disaster and post-disaster to make Bandar Lampung have resilience to tsunamis.

4.2 The tsunami disaster mitigation

Based on the history of the tsunami disaster in Lampung Province by Krakatau volcano eruption, the sea waves caused by this eruption swept the beaches of Lampung and Banten, damaged the environment, left many people dead, damage and material loss. To avoid or mitigate the tsunami disaster that can happen anytime and anywhere, it is necessary to have an attempt to reduce this so-called disaster mitigation.

Tsunami disaster mitigation measures that reducing vulnerability and increasing the capacity of a local tsunami will result in low level of risk. Some activities that can reduce the risk of tsunami are:

1. Relocation of residents living in areas prone to tsunami;
2. Disaster preparedness training for residents in the area;
3. Prepare homes or public facilities that disaster response;
4. Building anti-earthquake and anti-tsunami structure;
5. Creation and dissemination of local knowledge on disaster;

Moreover, in terms of information and communication technology can also be applied in the process of tsunami disaster mitigation. The development and planning of natural disasters mitigation, such as tsunamis, can take advantage of the IOT (Internet of Things) development, which is Embedded system. The use of seismic sensors such as ADXL335 Adruino pre-earthquakes tool vibrations placed on land and in the sea is to read the vibrations that occur on earth in the sea area. Basically, the tsunami is caused by the vibration and faults in the earth's crust under the sea, causing the flow of tsunami waves. Seismic sensors under the sea will read the vibration of the earth's crust under the sea. The sensors will be connected to a flare marker buoys as a means of information to nearest disaster mitigation center (example: lifeguard, or the Agency for Meteorological and Geophysics) using wireless sensors. The information against the danger of tsunami will spread from the center of disaster mitigation through short message transmission systems mobile devices located in the region of Lampung Bay (Coastal Zone of Bandar Lampung) and the use of sirens warning as a sign of the danger of tsunami.

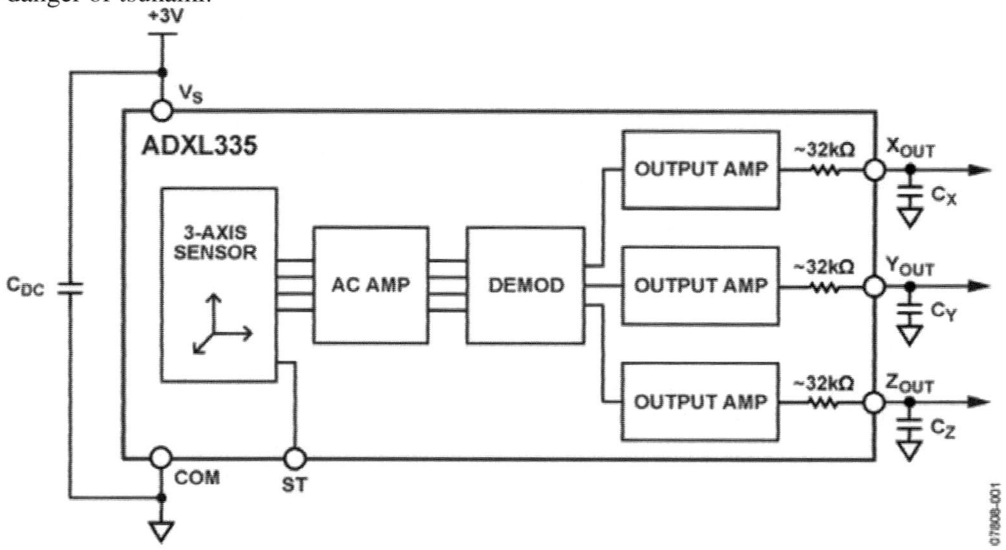

Figure 5. Accelerometer circuit of ADMXL335 [12]

The circuit uses the Arduino Uno board connected to the ADXL335 accelerometer module (connected in CON2) with its ADC input, namely A0 to the X axis, Y axis and Z axis A1 to A2. Two push-buttons via a 5V supply are connected to an Arduino Uno interrupt pins 2 and 3 and pulled to the ground through a resistor R2 and R1. This button is used for increasing and decreasing the threshold of vibration detection. LCD 16 × 2 (LCD1) is connected to a 4-wire mode with contrast control and backlight Arduino pin enabled.

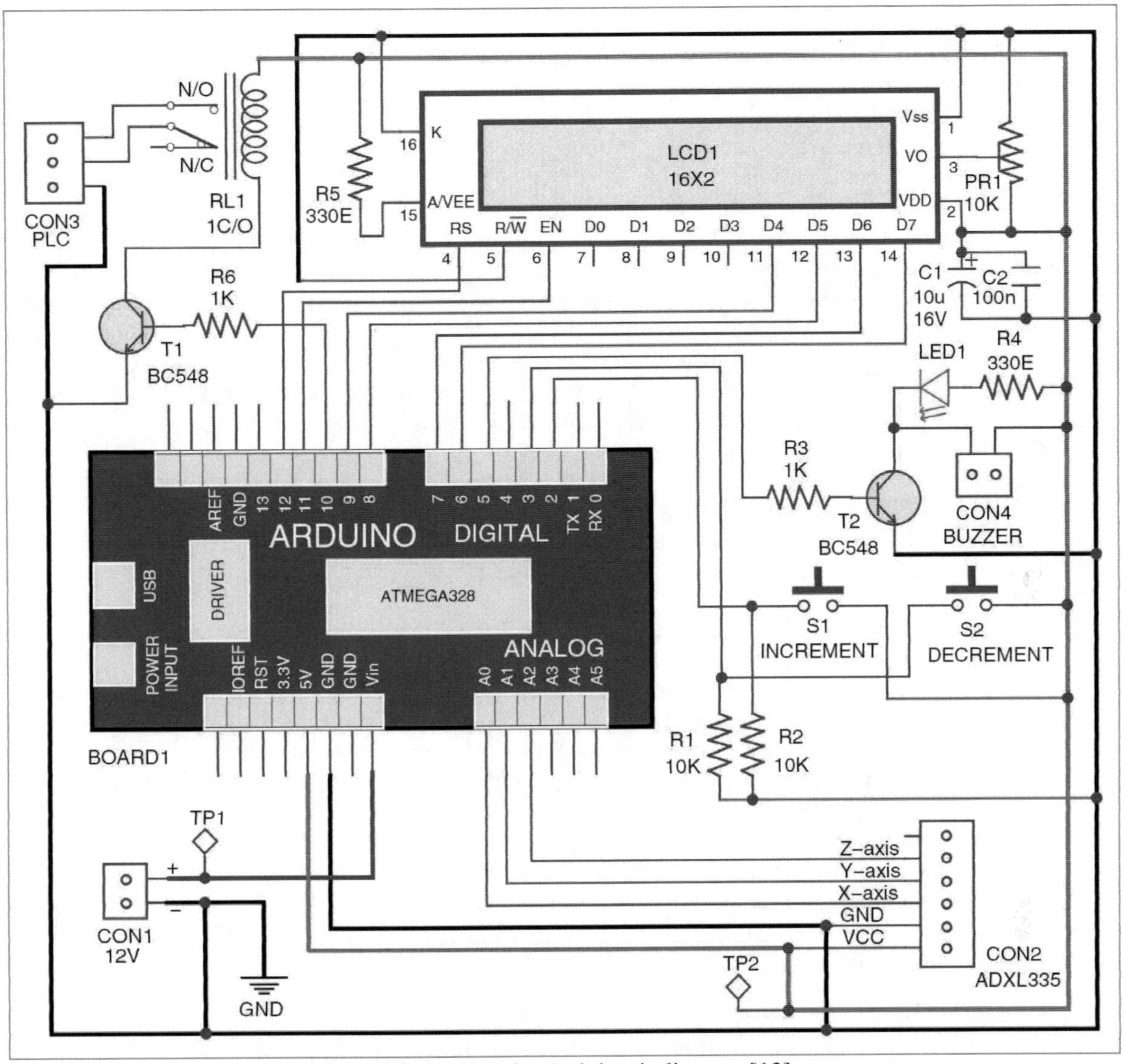

Figure 6. Earthquake Arduino indicators [13]

The wireless sensors are installed on the circuit that can deliver messages in real time with a considerable distance. The tool is typically used for volcanoes to find out when the evacuation of the vibration level of the earth. Since its function is to read the vibration, there is a possibility for the establishment of a new development to cope the tsunami by the earthquake.

5. Conclusion

The above discussion can be summed up as follows:

- Bandar Lampung is one of the vulnerable zones of the tsunami disaster because the southern coast of Bandar Lampung is a bay prone to tsunamis and has potential against earthquakes, floods, and environmental degradation in densely populated settlements along the coast. The potential of the tsunami disaster in Bandar Lampung was not only from the earthquake, but also from the volcano eruption of Anak Krakatau in the waters of the Sunda Strait, which increased the potential of high tsunami disaster.

- There are four sub-districts, namely Panjang, Teluk Betung Timur, Teluk Betung Selatan, Bumi Waras, have a higher potential tsunami hazard than other sub-districts in Bandar Lampung by considering slopes, topographic, land use, occupation dominance, and infrastructure available such as houses and public facilities.
- The height of the tsunami caused by the eruption of Mount Krakatau is a reference of spatial analysis to map tsunami-prone areas in the coast of Bandar Lampung of 1,536.02 hectares with a height of 40 meters tsunami wave covering some sub-districts,Teluk Betung Timur, Teluk Betung Barat, Teluk Betung Selatan, Teluk Betung Utara, Bumi Waras, and Panjang.
- The concept of tsunami disaster mitigation in the coastal area of Bandar Lampung City is proposed horizontally by making some points as a place for evacuation space of tsunami disaster. Spatial Plan of Bandar Lampung can be combined with the current use of information and communication technology by utilizing development of IoT (Internet of Things), that is Embedded system. The tool is designed with the addition of wireless sensor on the circuit that can deliver the actual message with a considerable distance to the vibrations generated by the earth.

6. Suggestion

To fix the lack of this research, some suggestions for further studies are given as follows.

1. Minimize the coverage area of research, to produce more representative and comprehensive results in analyzing the resistance of a region and planning the area and region to mitigate the disaster in more detail.
2. Deepen the analysis for each aspect to measure the level of resistance of a tsunami region and propagate variables/indicators to get better results and test the validity of the data.
3. The need to reduce the level of tsunami hazard risk is done by 1) the structure and infrastructure of the coastal protection building and 2) non-structural (spatial policy).
4. The government of Bandar Lampung is supposed to simplify the bureaucratic system and to revitalize the coastal areas of Bandar Lampung with attention to the inhabitants so that the coastal areas of Bandar Lampung are ready for resistance tsunami disaster in terms of the government, society, economy, and so on, especially in the area of prone.
5. In the process of disaster mitigation in the era of globalization, advanced technology should be applied to minimize casualties from the disaster. The government should pay more attention on disaster-prone areas and allocate funds in terms of preparation, planning and implementation of disaster mitigation.

Acknowledgements

The author would like to thank to all those who helped in the process of this research, and the Institut Teknologi Sumatera, relevant government institutions as well as coastal communities in Bandar Lampung which helped the process of research on the resilience of coastal communities to the tsunami disaster and mitigation of tsunami disaster in the coastal city of Bandar Lampung.

References

[1] Management of Coastal Areas and Small Islands. Gov. Regist. 27/2007

[2] Arief M D 2011 Study of coastal line change using landsat data satellite in Kendal District *J. Remote Sensing* **8** 71-80

[3] Hamzah L, Puspito N T, and Imamura F 2000 Tsunami catalog and zones in Indonesia *J. Natural Disaster* **22** 25-43

[4] Naryanto H S 2008 Analysis of potential earthquake and tsunami in West Coast Region Lampung, in relation to mitigation and regional planning *J. Science and Technology of Indonesia* **10** 71-77

[5] Naryanto H S 2003 Mitigation of the South Coast Region of Bandar Lampung against the tsunami *J. Alami: J. Water, Land, Environment, and Disaster Mitigation* **8** 24-32

[6] BNPB (Indonesian National Board of Disaster Management) 2016 *Book of disaster risk in Indonesia* (Jakarta: BNPB)
[7] Diposapto S D B 2006 *Tsunami* (Jakarta: Buku Sains Populer)
[8] Disaster Management. Gov. Regist. 24/2007
[9] Hamzah L 2009 *Tsunami risk assessment in Indonesia* (New York: World Bank)
[10] USAID 2007 *Tsunami warning center reference guide* (Bangkok: US IOTWS)
[11] BPS Bandar Lampung 2017(Central Bureau Statistics of Bandar Lampung) *Bandar Lampung in Figures 2016*, (Bandar Lampung: BPS Bandar Lampung)
[12] Tim Analog Device 2018 Analog: three-axis accelerometer [Online]. Available from: http://www.analog.com/en/products/mems/accelerometers/adxl335.html#product-overview
[13] Adhiraa R 2016 Electonicsforu: Earthquake indicator using Arduino [Online]. Available from: http://electronicsforu.com/electronics-projects/hardware-diy/arduino-earthquake-indicator

IOP Conf. Series: Earth and Environmental Science **158** (2018) 012038 doi:10.1088/1755-1315/158/1/012038

Unraveling the impact of built-environmental self-modification of the local inhabitants in their attempt to reduce the urban flood impact in Grogol, Sukoharjo*

M J Wanabakti[1,2], C R Susilo[2], M Nathania[1,2], and C Y Putri[1,2]

[1] Program Studi Sarjana Teknik Arsitektur, Universitas Atma Jaya Yogyakarta, Jl Babarsari No 44 Yogyakarta, Indonesia
[2] The Habitus Platform Initiative (Indonesia – Belgium). Principal Researcher.

Corresponding authors: maximilianwanabakti@outlook.com

Abstract. Due in part to climate change impact and the lack of environmental awareness of urban societies, urban flood disasters have increased in both frequency and impacted areas. The inevitable destructive impact is also experienced by urban inhabitants within Grogol District, Sukoharjo, Central Java. Situated within the fast-growing peri-urban area of Surakarta, the geographical condition of the district, which is located in the convergence of Bengawan Solo River and four different streamlet rivers is undoubtedly critical to flood. Since the position of the peri-urban area is along the border of Surakarta City and Sukoharjo Regency, the site is a strategic locus of the main city's urban expansion. However, its border location has made flood-reduction programs challenging to be implemented, due to unclear territorial admiration and distribution. Because of the absence of flood-reduction programs in the area, local inhabitants have been independently developing physical and built environmental modification of their settlements. By analyzing the two-months built environmental fieldwork research in Grogol Sukoharjo, this paper unravels both the success and the conflicting impact of local self-initiative modification in Grogol Sukoharjo. By combining the two-month field observation with the semi-structured interviews in three sub-districts of Grogol, this paper finds that the very pragmatic self-modification taken by the inhabitants results in short-term reductions of flood impact and increases the secure feelings of inhabitants to continue living in their neighborhoods. However, the lack of proper technical supervision from experts and the absence of inter-neighborhood coordination leads to future additional damage on the flood-prone areas of Grogol. The paper also reveals that one pragmatic physical self-modification solution taken by one neighborhood may lead to an extra flood-threat to other neighborhoods.

1. Introduction

Waterway has long had an important role in daily human activity and economics. Water is used for physical human needs, industrial, and agricultural production. Prior to and during colonialism, rivers have had important roles in the economy, transportation of goods and services, as well as human daily needs and mobility. Nevertheless, due to the escalating growth of urbanization and technology, with it bringing changes in economy, settlements, physical infrastructure and culture. With the increasing density of cities due to urbanization, more land is being covered by infrastructures. This condition has greatly reduced the natural capacity to absorb water. The loss recognition to the existence of waterways, such as rivers is one of the reasons that trigger more catastrophic natural disasters, especially the escalation of urban flood disasters.

Thanks to the increased frequency of campaigns on the impact of climate change. The existence of waterways has regained attention as an important feature necessary to be integrated into city and regional planning. Today, every agency of the city stakeholders develops concerns in responding to the increased events of urban flood. However, this paper specifically limits the result from observation on the physical responses undertaken by individuals to reduce the impact and risk of urban flood.

The National Government and many local Governments are beginning to reformulate their urban agendas in an attempt to integrate rivers, water, and flood mitigation planning into their spatial planning and urban infrastructural development. Both physical and social scenarios have been formulated to support flood reduction and flood mitigation agendas of the city. Nevertheless, the physical response remains an urgency to be executed in order to deliver immediate urban flood protection. The physical responses have been prepared and taken by the government in the macro context of the city, but there lies a need for support in mezzo-scale that are commonly handled by communities from the district level to neighborhoods and in micro-scale that is handled by individuals from neighborhood level to family.

Insofar, most Indonesian research and scenarios on physical response to floods have been frequently and generally explored in macro-scale, covering both city-scale to district planning scenarios. Ironically, the existing flood-reduction and mitigation planning and solution scenarios have frequently missed to capture the problems of urban floods at a micro-level that covers challenges of flood-reduction response in neighborhoods to the household level. In terms of physical infrastructural solutions to reduce the risk of floods, the macro-scale planning scenarios commonly propose financial support to be delivered to several selected areas of sub-district or neighborhoods based on selected priority areas, depending on the level of their vulnerability. Nevertheless, there is an absence of expert supervision to guide how the aids should be properly used to construct physical infrastructural modifications that can help neighborhoods be more protected from floods. As a result, any physical infrastructural modification in mezzo to micro level have been merely executed based on neighborhoods to individual interpretations. However, without proper supervision from experts, the existing self- modification is sometimes only able to secure the dwellings temporarily.

The provision of flood-mitigation action and scenarios in the urban context of the developing countries have to face a challenge that is generated by the dominant patterns of unplanned and organic settlements which are side by side with the other planned-and-less organic settlements. The logic that structuring the co-presence of those diverse settlements in close distance is commonly difficult to trace. Therefore, a proper scenario to provide both adequate water infrastructures and drainage networks that respond to flood-mitigation strategies are also frequently difficult to realize. As a result, the physical infrastructural scenario of flood mitigation at a neighborhood scale is commonly dependent on the decisive initiative action and scenario of each neighborhood. Therefore, the implemented flood mitigation actions tend to have a high degree of variation and disintegration among each neighborhood in close proximity. Instead of improving the capacity of urban areas to reducing the impact of floods; however, this high-degree of variation and disintegration of the self-flood-mitigation action may lead to damaging the extending environment and other nearby neighborhoods

This research highlights the urban flooding responses that have been taken place at the peri-urban area of Surakarta, particularly at Kelurahan Joyontakan of Surakarta City and at Kelurahan Grogol of Sukoharjo Regency. The fact that the uncontrolled urbanization triggers significant problems related to the increased impact of floods, and the complexity to draw flood-reduction scenarios, have been positioned in crucial attention on this paper. Joyontakan and Grogol, the focus areas of this research, are located at lowland areas where four streamlets of the Bengawan Solo River meet with the main river, and as such is at a high risk of flooding. Due to its strategic position at the busiest intercity main road connecting Surakarta City and Sukoharjo Regency, Joyontakan and Grogol have experienced rapid urbanization within the past twenty years.

Joyontakan is the last sub-district directly adjacent to Sukoharjo periphery; while Grogol is at the border of Sukoharjo, adjacent to Surakarta City. Grogol area was initially a major agricultural area of the Soloraya region. However, Grogol has grown to become the prime area of expansion for Surakarta City within the last decade. The majority of the low-land area has been occupied by real-estate oriented developments, the spread of industrial infrastructures and buildings, as well as the organic expansion of self-built urban kampong following the economic attraction generated by real-estate and industries. Since the aforementioned areas had been free of floods since 1963 to 2007, none had ever been aware on the fact that this area carried potential flood risks due to its location flanked by the Bengawan Solo River and its three streamlets. It was only after the area was suddenly hit by catastrophic flash-floods in 2007 that both government and society were made aware of the flood risk that threatens this area. Due to the wide coverage of the area that has been frequently hit by annual floods in Joyontakan and Grogol, this research tries to limit the focus of observation to within three focus areas that experience the worst damage, which covers a total of ten percent out of the entire Grogol and Joyontakan Sub-district administration areas. They are Joyontakan urban kampongs, Puri Gading and its subdivision housing clusters (including Gading Makmur and Gading Permai), and Nusupan urban kampongs.

2. Methodology

Since Indonesian urban settlements are constructed by collages of both kampongs and modern real-estate housing neighborhoods that coexist and give influence to each other; this research does not attempt to follow the previous conventional urban research which tends to observe the Indonesian urban reality in parcels of completely fragmented and separated settlements. This paper does not select one particular study focus such as only kampong, only housing or only a particular community in the research area, but instead this paper attempts to show the influence among the communities (both who live in kampongs and in housing estates) and their flood-risky living neighborhoods, which coexist and altogether deliver influences to the flood-prone areas (be it kampongs or real-estate housing neighborhoods); in order to portray the overlapping influences among diverse communities' responses to their neighborhoods and the influence of the changes in one neighborhood to the others.

The content of this paper is formulated based on the reflection of the initial result of short fieldwork research conducted in groups by student research teams of Water Urbanism Surakarta of the Department of Architecture Engineering, University of Atma Jaya Yogyakarta. To collect necessary information and data to develop validity of this paper, six research members jumped to the fieldwork locations in twenty-four days in between March 1st until May 31st of 2017, to gather primary data through necessary observation on the condition of built fabrics and infrastructures of several flood-prone neighborhoods in Joyontakan and Grogol. An interdisciplinary view, by borrowing techniques from qualitative social research was employed, which involves semi-structured interviews with local inhabitants, field observation, and documentation review of policies and plans. This qualitative method was employed to understand a rather complex social phenomenon, which cannot be identified only by visual observation. The semi-structured interview is a tool to clarify the ongoing information that has been gathered from visual information. This interview is also a tool to understand the logic and reasoning behind actions undertaken by flood-victims. Whereas, newspaper article reviews and literature studies on urban floods, especially relevant to the nature of Bengawan Solo's urban floods, has been used as secondary data to develop the analysis of this paper.

However, in collecting interviews and necessary data relevant to this research, several obstacles were also experienced, limited research time being the major constraint. The fieldwork was conducted within twenty-four days, in which the process was distributed in two-days of research per week. There was a discontinuity of observation in between weeks, resulting in a less intensive result of the interviews that had been initiated. Considering the overall research area contains parcels of very diverse characters of neighborhoods, the twenty-four days of fieldwork was still insufficient to gather as many information as is the initial plan. Despite all of that, until the end of the fieldwork time, the team members were still awaiting permission to access significant technical information from local authorities. Therefore, the research teams were unable to verify several key technical information such as, the water debit and speed of the flood stream during flood events; due to data and information which have not been made public.

The self-modifications on built environments that have been taken by flood-victim inhabitants in reducing the risk and threat of flood have been documented by the researcher in several parcels of cognitive maps. These cognitive maps are being used as a tool to build spatial understanding, informing the location and the concentration of existing flood-prone neighborhoods and the locations where physical and infrastructural self-modification exist. These spatial maps are also employed as a tool to understand the pattern of dwelling fabrics that reflect the intensity of interaction among each dwelling house in a neighborhood. The aforementioned maps also help to inform the diverse characteristic among different neighborhoods and their physical relationship, which also aided the researchers to construct a visual understanding on the distance and physical interaction between a neighborhood and the adjacent rivers and streamlets, including the distance and proximity between the neighborhood and the entry of flood. The distance and the proximity among the neighborhoods, the rivers and streamlets and the entry of flood have an apparent influence on the form of physical and infrastructural self-modification of each neighborhood.

3. A theoretical reflection on urban flooding and urbanization, a fragmented settlements and the physical intervention responding to flood

3.1. Urban flooding and urbanization

A manifold of overlapping environmental problems that decrease the permeability of land and soils is the common trigger of urban floods. The risk of floods in urban areas tends to increase simultaneously following the expansion of urbanized lands into the peri-urban and rural areas.[1,2] However, the massive transformation of rural areas into urban areas is inevitable, especially in cities of developing countries such as Indonesia. With the expansion of physical developments in rural areas, massive conversion of rural lands brings direct consequences on the duration of puddle to evaporate, on the frequency of floods, on the amount of water that flow through the water channels and on the destructive capacity of floods.[3-11] However, the impact of the aforementioned hydrological disruption varies depending on the growth level of urbanization in each spatial.[12]

The hydrological disruption commonly triggers an increasing sedimentation along water channels and rivers and the decrease in capacity of lands to support water evaporation processes due to the reduced capability of lands to absorb water during the wet season.[8,13] With increased sedimentation, the water channels suffer from lack of spaces to contain the water flow. If the capacity of the lands to absorb water has also greatly decreased, the flood hazard is also simultaneously increased.[8,13] As massive urbanization drastically consumes the spaces for greenery, rainwater loses its medium to infiltrate the earth. As a result, rainwater has become an additional source of water-flood.[14]

When the urbanization process comes rapidly in spontaneous and sporadic patterns, the conversion of lands into built areas often comes beyond the anticipation of the existing urban planning. As a result, there lacks the presence of both urban sewage and drainage infrastructure networks that service the new expansion of settlements. The spatial areas that are located on flood-risk locations such as riverbanks are commonly inherent in the aforementioned quality.[15] Therefore, the risks of flood increase according to the occupation of land along riverbanks for residential and industrial areas. [16] Since these areas usually experience a decrease in economic value, these locations attract an expansion of low-income settlements which leads to uncontrolled expansions of disorganized covered land and chaotic settlements. [15] Therefore, Serrao-Neumann et. al., affirms that effective planning policies and response can reduce the vulnerability of the disaster-victim communities to the impact of extreme weather such as the escalation of urban floods.[17]

However, the growth of urbanized land in developing countries such as Indonesia generally takes two different growth patterns; the planned urbanized land and the spontaneous and unplanned. [18] While the planned version normally covers the areas where urban planning and policies have access to regulate the settlements, the unplanned version usually takes place in areas that are occupied by squatters, slums, traditional communities and the unwanted communities. The unplanned growth of urbanized land can also take place on vacant spaces, such as at the border of two administrative regions or occupying vacant spaces along riverbanks. Hence the chaotic and uncontrolled sprawl of settlements. As for the planned urbanized areas, the pattern of its settlements normally follows a definite pattern such as the direction of rivers, roads or administrative boundary. [19]

3.2. Fragmented Settlements and Its Influence Effect to the Physical Self-Responsive to Flood

In the Indonesian urban context, kampong settlements tend to be associated with poverty, and backwardness. But the new settlements and real-estate housing clusters tend to be associated with progress, the wanted socio-economic classes, and modern societies. These two stigmas contribute to the production and construction of fragmented and segregated settlements due to differences in socio-economic classes. As a result, the aforementioned differences also reflect the socio-economical capacity of the major inhabitants of each settlement.

Differences in social and economic capabilities among the two classifications of settlements consequently build a barrier of social relationships. The predominant housing settlement of the higher socio-economic class tends to require more privacy. Hence, the tendency to limit access to other settlements and kampong neighborhoods. However, in term of economic capacity, they have better self-financial capacity to execute and construct any physical modification for flood-protection.

On the other hand, lower socio-economic class settlements tend to be more open in social relationships to other groups and settlement. As a result, community participation to handle the flood impact is more probable than the higher socio-economic class group. However, this socio-economic group has less economic capacity to conduct physical self-modifications to handle the flood due to financial limitations. As such, physical self-modifications in dwelling environments and housing units may be less in low-income settlements than higher-income settlements.

Moreover, in the course of urban flood disasters, the prevailing political regimes in Indonesian cities tend to place blame on low-income settlements at flood-prone areas as one of the main reason for urban floods along with the natural course of global climate change and increasing sea.[20] On the other hand, when land occupation involves a shift of ownership and uses correlated to business investments and large-scale land-development projects, floods are often framed as a 'must be taken' costs needed to be accepted in the name of 'a better future development for the city'. [20]

3.3. On Physical Intervention as a Response to Urban Flood

Both national and local governments usually prepare several physical interventions to reduce flood events and their impact on urban and neighborhood scales. In urban scale, the most common physical modifications that have been implemented are the construction of river embankments in hard and non-porous materials, straightening the path of the river, re-planting forests and greenery along the river path and waterways, and constructing reservoirs. However, such modifications only serve as short-term prevention from floods.[21]

On the other hand, the latest studies on flood mitigation have shown that physical modifications by reshaping and changing the morphological form of a river increase the speed of the river flow, becoming stronger than its natural condition. This condition accelerates the process of soil erosion along the river path. As a result, the river flow will deliver the eroded soil to the estuary and the lowland. The process of this erosion frequently takes place faster than the natural sedimentation and soil formation processes. [21]

Naturally, softscapes with natural soils along the river-base and riverbanks function as a medium that allows water to naturally permeate the earth. However, the construction of river embankments in hard materials restricts permeability. As a result, the river loses its ability to absorb water. In the event of an increased debit, the modified river subsequently transforms into a channel for flooding. Therefore, the construction of reservoirs helps to reduce the over-capacity of rivers, as it also functions as a container to catch the water flow from drainage channels. [22] To enhance the capacity of the reservoir in absorbing the overcapacity of flood waters, the construction of reservoirs is usually combined with softscapes, greenery, plantations, and parks.

Moreover, intervention at a neighborhood scale in Indonesia also frequently receives support from the local government. However, the local governments are usually only involved partially in supporting the physical intervention, while major actions have been handed over to the participation of the neighborhood communities. Due to the limited capacity of low-income neighborhoods, these communities usually gain priorities to receive support from the local authorities. The most common physical intervention that has been supported partially is the provision of water-gates to control water intrusion in flood-prone neighborhoods. The local government at occasion partially finances the provision of gates in selected neighborhoods based on priority. Nevertheless, this initiative is ironically and frequently handled without supervision from the government or from the experts. As a result, the water-gate often fails to secure the neighborhoods effectively and efficiently. Therefore, flood prevention actions are most frequently being executed by a self-mobilization action from the flood-victims local communities who create any alternative means of resilience to floods. [20] Such resilience initiatives are essential to raising the flood-victims' awareness on their right to the city. [20]

4. Describing and analyzing the physical self-modification of built environments in response to urban flood in the peri-urban settlements of Surakarta and Sukoharjo.

4.1. The Existing Condition of the Bengawan Solo River Flow and the Flood in Joyontakan and Grogol Areas

4.1.1. The Bengawan Solo River, Its streamlets and Flood Events. The Bengawan Solo River flows from the southern part of Java Island to the Northern part of the island. It functions as the main water channel for the provinces of Central Java and East Java. As it is recorded by the Balitek DAS Sungai Bengawan Solo, the massive conversion from open land to built-up areas occur upstream, to the areas of Wonogiri and Ngawi Regencies. This conversion plays a significant role in increasing the water level of the Bengawan Solo River when it passes the Soloraya Region. The expansion of flood on land continues to occur as it is worsened by the slow process to recede any puddle on the land. This condition is due to the massive changes of many flood-free areas following the expansion of settlements and built areas along the Bengawan Solo River, especially in Sukoharjo and Wonogiri Regency; which apparently causes the decrease in capacity of the soil to absorb water at urbanized lands and along the riverine. These regencies initially acted as water catchments and agricultural areas for Soloraya Region.

Moreover, the decreased capacity of water catchments along the Bengawan Solo River has also been triggered by the gradual deprivation of green spaces along the riverbanks. At least twelve-meter-wide green open space should be reserved on the riverbanks of the Bengawan and its streamlets; in an effort to provide spaces for natural water absorption, maintaining the natural water flow of the river in both normal and flood conditions, and maintaining an adequate amount of spaces needed for natural floodplains as well as maintaining the direct connection between the water body of the river and the floodplains when flooding occurred.

IOP Conf. Series: Earth and Environmental Science **158** (2018) 012038 doi:10.1088/1755-1315/158/1/012038

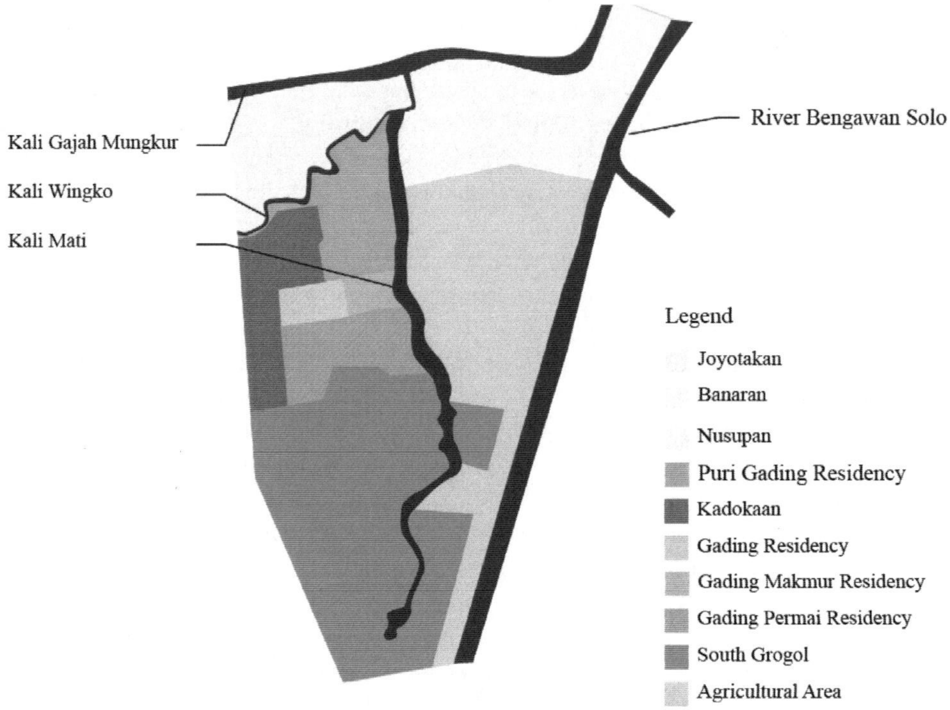

Figure 1. Research fieldwork area.

The recent development of mainstream research on urban flood has concluded that allowing natural water absorption by increasing the amount of green space along the riverbanks is considered as the most effective and sustainable way to reduce the impact of floods to settlements. This way is considered more effective than constructing hard-engineering solutions, such as concrete embankments. If this ideal condition can be achieved, direct flooding impact to settlements can be avoided. However, in practice, the construction of levee and expropriation of space along riverbanks for extended dwellings remain dominant on the riverbanks of the Bengawan and its streamlets for many years. As a result, these hard structures cover the absorption surfaces of the riverbanks, interrupts the direct river connection with its necessary floodplains during a flood event and consequently reduces natural spaces necessary to hold the overcapacity of water flow. Although the latest environmental campaigns promote the revitalization of riverbanks as green open space floodplains as the most recommended solution to tackle the increased urban flooding; however, this solution is still difficult to realize by many cities in developing countries due to high financial and social costs. Its realization in cities in the developing world has been challenged by inevitable massive settlements needed to be relocated to give spaces for the programs, which brings consequences on high expenses needed to be prepared by the local government. As a result, a hard engineering solution such as the construction of additional levees to protect settlements from water still become the favorable solution.

IOP Conf. Series: Earth and Environmental Science **158** (2018) 012038 doi:10.1088/1755-1315/158/1/012038

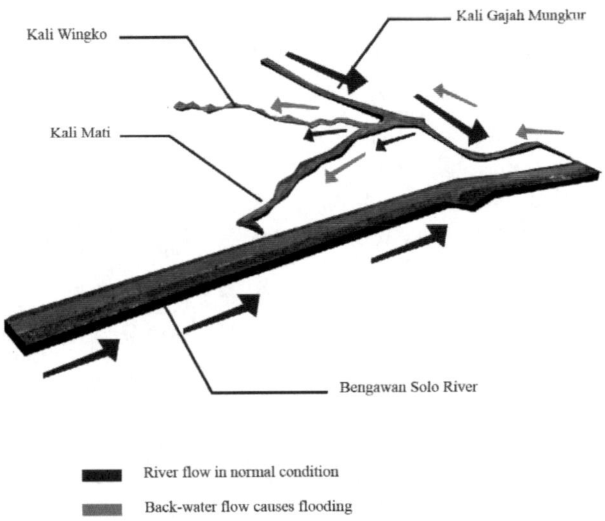

Figure 2. An extra water-flood caused by back-water flow.

Moreover, the water flow of Bengawan Solo River and its streamlets also tends to increase following the increased intensity of rainwater. As a result, the maximum water level of the Bengawan Solo River and its streamlets shows an increasing trend every year during the wet season. The possibility that the water from streamlets being unable to enter the Bengawan Solo River is evident since the main Bengawan River has reached its maximum water-level capacity. This condition triggers an additional flood threat to settlements nearby the rivers due to a potential back-water flow from the Bengawan Solo to its streamlets and then entering the smaller water channels in settlements. An extra water-flood caused by back-water is inevitable. The role of the back-water in urban floods of the Bengawan Solo has delivered an equally significant impact to neighborhoods as the true flood-water from the river.

4.1.2. The Precondition of the Flood-risky Settlements of Joyontakan and Grogol. The location of Joyontakan and Grogol is within lowland areas to its surrounding neighbors. Towards the north, the ground level is lower than the southern areas. These areas are surrounded by two streamlets of the Bengawan Solo River; Kali Gajah Mungkur and Kali Wingko. These areas are also adjacent to the former Bengawan Solo water channel, also known as Kali Mati, which is located behind the two sub-districts. Moreover, the recent channel of the Bengawan Solo River is located within a 500-meters distance from the observed area. This geographical location consequently positions Joyontakan and Grogol within critical risk conditions from the rivers. Nevertheless, almost every inch of land in and around those three streamlets have been occupied by dense [and mostly legal] settlements for many years. Therefore, any scenarios to give spaces for greenery are difficult and almost unrealistic to be executed.

The three streamlets contribute to flash-floods to the area in two different ways, which are complementing to each other in escalating the scale of the floods. Kali Gajah Mungkur and Kali Wingko are the streamlets of the Bengawan Solo River that actively function to channel the water flow from Klaten Regency to the Bengawan Solo River. Unlike the two streamlets, Kali Mati no longer acts as a water channel in normal days. In Indonesian term, Kali Mati means a dead river or a former riverway that is no longer functioning.

Kali Mati is located nearby the confluence of Kali Wingko and Bengawan Solo River. Its location is near the entry location of Kali Wingko to Bengawan Solo River. Kali Mati and Kali Wingko are smaller in dimension than Bengawan Solo. The riverbanks of both streamlets are occupied by dense settlements, both kampong settlements and real-estate settlements (including Joyontakan Kampung and Puri Gading

Housing), and [un-]official urban dumpsters. Most settlements along the two streamlets are located below the water level of the streamlets during maximum capacity. Thus, limiting space along the waterways available to accommodate the overflow of water when the streamlets are at maximum capacity.

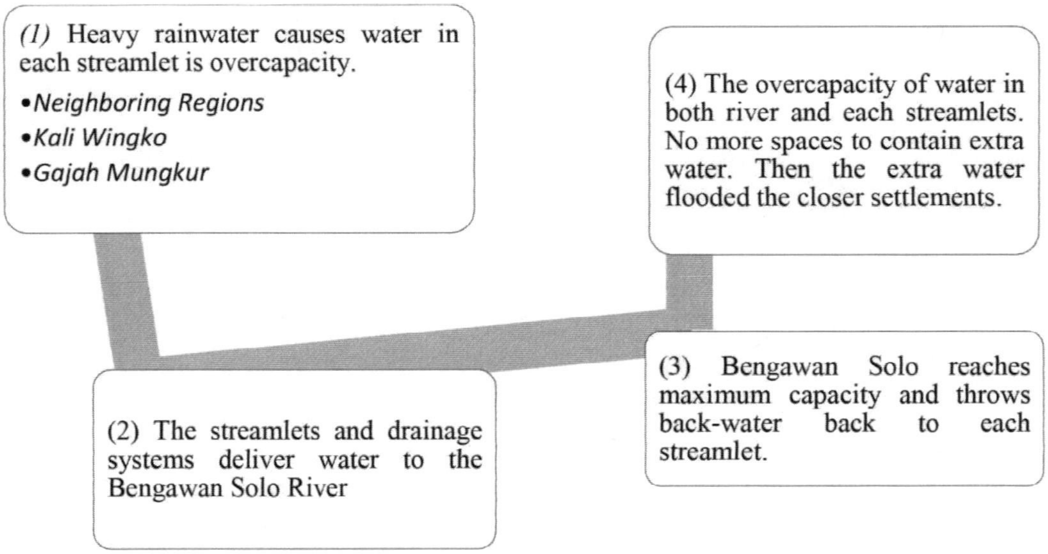

Figure 3. A diagram showing the flow of back-water reaches the settlements.

When rainwater comes in high capacity and in long hours in Soloraya Region to its neighboring regions, Kali Wingko, Kali Gajah Mungkur and other streamlets of the Bengawan Solo are at maximum capacity. The streamlets will try to deliver water entering all inlets to the Bengawan Solo River. As soon as the Bengawan Solo River reaches maximum capacity, water from the streamlets can no longer enter Bengawan Solo. The water-flow from Bengawan Solo River, which is stronger than the water-flow of the streamlets, will force the water back to the streamlets. In the case of settlements along Kali Wingko, the overcapacity of the river cannot find extra spaces to accommodate the water or to redirect the water to spread to other directions.

Therefore, when rainwater comes in high capacity and in long hours, the overflow from Kali Wingko and rainwater fills the residual space of Kali Mati and transform the space into a dead-end pond. When the debit of water in Kali Wingko, the Bengawan Solo, and Kali Mati have reached maximum capacity, this temporary dead-end pond of Kali Mati turns into a source of floods to the surrounding settlements. Therefore, Joyontakan and Puri Gading areas experience longer pools during the rainy season and frequently suffers more flood-impact damage. While the two aforementioned settlements suffer by floods from Kali Wingko, Kali Mati, and Bengawan, the Nusupan Kampong is more affected by flood from Bengawan Solo River and Gajah Mungkur River, even though the northern areas of Nusupan are also impacted by the dead-pond of Kali Mati. Nusupan and the southern areas of Grogol are less affected by floods compared to the northern settlements of Grogol. However, the lack of proper drainage in this area has delivered equal complexity on the problem of seasonal floods.

4.2. Common Physical Responses to Deal with Flood and the Influence of Fragmented Settlements

The settlements of Joyontakan and Grogol are the result of uncontrolled urbanization of Surakarta city, formed from a collage of diverse parcels of neighborhoods, in which each parcel has diverse socio-economic backgrounds. Several neighborhoods were formed spontaneously and organically in the form of peri-urban [traditional] kampong, while others were intentionally built by local developers in the form of real-estate housing neighborhoods. Each neighborhood has its own socio-economic and resilience capacities to deal with the impact of floods.

The diverse socio-economic precondition of each settlement in Joyontakan and Grogol has also brought different treatments from the local government in delivering aid and support to in terms of flood response. The kampong neighborhoods consist of Joyontakan, Nusupan, and Kadokan have been targeted as priority neighborhoods to receive government support and flood-reduction programs. On the other hand, the real-estate neighborhoods of Puri Gading, Gading Permai, Gading Makmur, and Gading Regency, although they are also impacted by floods, have been unsubsidized by the government's flood-reduction program and have been left to handle the floods by themselves.

However, to both communities, being flood-victims for years have brought a life-learning experience on how to deal with flood depending on their individual capacities. Therefore, with or without the presence of official flood-reduction programs from the local government, these communities have actively developed knowledge and actions to manage flood, either through physical infrastructural modifications or through socially oriented actions and approaches.

Table 4.2 Socio-economic precondition brought different treatments from the local government in delivering aid and support to in terms of flood response.

No	Study Area	Type of Settlements	Level of flood distruction	High to middle Priority	Low Priority
1	Joyotakan	Kampong	+++++	✓	
2	Banaran	Kampong	+++	✓	
3	Nusupan	Kampong	+++++		✓
4	Purigading Residency	Housing	+++++		✓
5	Kadokaan	Kampong	+++	✓	
6	Gading Residency	Housing	+++++		✓
7	Gading Makmur Residency		+++		✓
8	South Grogol	Kampong	++	✓	

4.2.1. The Diverse Characteristic of Neighborhoods and Its Impact to the Physical Response of the Inhabitants in Creating a Flood-Protection for the Neighborhoods Environments. The account below describes and analyses the physical self-modifications and actions on the dwelling environments that have been implemented by the inhabitants of Joyontakan and Grogol in their attempt to reduce the impact of floods in their neighborhoods. The diverse condition of the site is grouped into three groups of settlements following the dominant characteristic of each settlement. They are the Joyontakan, Puri Gading, and the Nusupan settlements.

Joyontakan Kampong Settlement

Joyontakan consists of several kampong neighborhoods that grew gradually and mainly organic for decades; some others were formed by [former] squatters and are considered as poor neighborhoods. This condition has positioned the area as a target priority for flood-reduction programs from the government.

Due to the irregular, unplanned, and chaotic formation of this settlement, the drainage systems and network of the settlement are also chaotic and disconnected into several fragments. As a result, the settlement suffers from a very poor and ineffective drainage system. Hence Joyontakan has always been the most vulnerable to floods before 2010 when the Surakarta Government delivered a flood-reduction program by reconstructing the drainage system.

The flood threat in Joyontakan is accelerated by the fact that the location of this settlement is at the lowest land of its surrounding region, acting as a basin for the surrounding sub-districts. Joyontakan's critical position to flood-threat is also accelerated by the fact that this settlement is also flanked by Kali Gajah Mungkur and Kali Wingko at its northern and southern parts, as a consequence of this condition, Joyontakan acts as natural rainwater collector for the immediate region. However, the major land of this basin has already been covered by dense settlement for years, blocking the process of natural hydrological absorption of rain and river water.

Three major flood-reduction projects have been executed by the Surakarta local government in 2010 to secure the settlement from floods. The river embankment, constructed with a hard-engineering approach along the riverbank of Kali Gajah Mungkur in an attempt to secure the flood-plain from the direct intrusion of settlements' expansion and its dwellers. The drainage system of the neighborhoods was also reconstructed; and lastly, funding to purchase, build and install neighborhood water gates and water pumps was also delivered to the kampong communities of Joyontakan, all in an attempt to control the intrusion of flood-water entering the neighborhoods during flood events.

However, there was an absence of proper supervision from the government regarding the installation process of the water gates and pumps; affecting the positioning of the water gates and the pumps to be less effective in controlling the water flow in the settlements, deeming the existing water gates and pumps insufficient to secure the settlement from flood. Moreover, due to the absence of communication and coordination between the neighborhoods of Joyontakan and Puri Gading regarding the selection of locations for water gates and pumps, the overcapacity of water that was discharged to Kali Wingko has frequently contributed as an extra source of flood to other settlements, especially to Puri Gading, which is located adjacent to Joyontakan and directly borders Kali Wingko.

Figure 4. Direct flooding flow from physical responses to the drainage system.

Figure 5. Watergate in Joyontakan and the event when waterflood began entering the neighborhood.

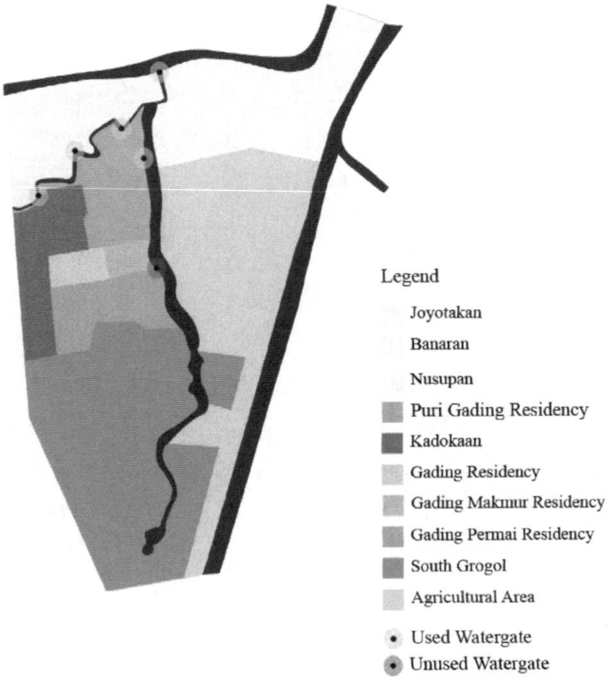

Figure 6. Distribution of used and unused water gates in Grogol.

Moreover, due to the installation of water gates and pumps in Joyontakan, the extra flood-waters intruding the settlement has been disposed directly to Kali Wingko. In flood-critical conditions, this water discharge ironically contributes to additional water capacity of the streamlet, as a result, it has also contributed in accelerating floods and increases the potential development of the flood in becoming a more catastrophic flash flood, especially during heavy rains.

Puri Gading [real-estate] Housing Settlement

Puri Gading has different characteristics from Joyontakan, experiencing a more direct flood threat due to its location that is surrounded by Kali Wingko and Kali Mati. However, this housing neighborhood has not been targeted for priority support of Government flood-reduction program. Therefore, the neighborhoods within Puri Gading have to resort to private initiatives and financial support. Among several real-estate neighborhoods in Grogol, only Puri Gading has actively tried to find a solution to reduce flood-water infiltration into their neighborhoods and protection from flood damage.

Since the flood threat has increased in frequency within the last ten years, the residents of this neighborhood have decided to protect their settlement by constructing flood-protection retaining wall along the boundary of the settlement and the rivers by themselves through financing the construction by community participation. However, since the riverbanks lose space for water infiltration due to the retaining walls, the water capacity of Kali Wingko and Kali Mati increases during the flood. The water surface of the streamlets increases over the normal capacity of the streamlet, extending the duration of the water to return to its normal state. Since both water from the streamlets and runoff from Puri Gading try to find a way to flow to the immediate rivers, the water pressure in the soil of the neighborhood increases drastically, as it tries to create new capillary waterways to the river. The increasing pressures from both groundwater and the streamlets have gradually created extra force and pressure on the immediate substructures of the buildings, the flood protection retaining wall and the soil of the existing of the neighborhood. Findings during field observation have identified a steady increase of deterioration to substructures of houses and soil landslide in the neighborhood, most commonly to houses that are located adjacent to Kali Wingko and Kali Mati, indicating an apparent new threat from the side effects of the aforementioned environmental modification in this settlement.

Figure 7. Landslide in Puri Gading.

Apart from constructing retaining walls, the residents of Puri Gading have also tried to protect their neighborhoods by installing numerous water pumps and water gates in an attempt to dispose of and to reduce flood water intrusion from streamlets entering the drainages of the neighborhood. The intrusion of flood water in Puri Gading has been disposed to Kali Mati and Kali Wingko, the flood water in both streamlets remains in overcapacity and keeps delivering pressure to enter the neighborhood.

Nusupan Kampong Settlement

Unlike Joyontakan and Puri Gading, the environmental modification for reducing the impact of floods is still absent in Nusupan, even though the settlement has also suffered from flood. Nusupan settlement is located immediately to the retaining embankment of the Bengawan Solo River. The Bengawan is located along the southern and the eastern boundaries of Nusupan. The settlement is located below the normal level of the river, hence, when the Bengawan has reached its maximum water capacity, water will intrude the neighborhood through the existing water gates and drainage. Moreover, the northern part of Nusupan also receives flood threat from the end part of Kali Mati. Therefore, both the water intrusion from the Bengawan and the Kali Mati are the main causes of flooding.

Nusupan is the newest settlement in Grogol area, recently growing in the past ten years due to the concentration of many industrial buildings in this area. The existing drainages have a lack of depth even only to flow normal water capacity. During fieldwork, fragments of disconnected drainages were found throughout the kampong. In several alleys, the drainage abruptly ends without connection to the urban drainage network. Moreover, since its location is on the immediate lowland of a major river, the already insufficient drainage system is further complicated due to clogging from mud and soil from the intrusion of river water.

Interviews that were taken from the site revealed that the northern settlements are more aware of the danger of floods. Therefore, the self-initiative of the communities to develop protection from floods are also stronger. However, the southern settlements' inhabitants such as Nusupan have a lesser sense of urgency to modify their settlements in order to create a more protected neighborhood. The decrease in urgency is manifested through their acknowledgment of what conditions are considered as being hit by floods. The psychological adaptation of each community in these flood-victim settlements plays a significant impact on the way they acknowledge a flood event. When floods are below 50cm, it is yet to be considered as a flood. They will consider the event as a flood event when the water has entered their houses once it reaches at least 30cm above the ground floor or once their houses are full of water. Therefore, the physical self-modification to secure the dwellings from floods in Nusupan generally only exists inside individual housing units.

Figure 8. Flood in Nusupan

Furthermore, the interviews in Nusupan also revealed that for the residents of Nusupan, the source of floods on their settlement does not come from the location of their settlement that is immediate to the Bengawan Solo; or due to the lacking conditions of the drainage system. According to their understanding on the condition of their settlement, the source of floods that hit their settlements have been triggered by the construction of embankments along Kali Gajah Mungkur in Joyontakan and also due to the installation of the main water gate at the intersection of Joyontakan and Kali Mati. Prior to the construction of those physical interventions, their settlement was free of floods. Their opinion shows an apparent indication that the physical intervention at the other neighborhoods may lead to a negative side effect to their neighboring settlements. Therefore, a proper comprehensive planning and study covering larger extending environments need to be completed before the next physical infrastructural modification is executed.

4.2.2. The Diverse Socio-Economic Characteristic of Neighborhoods and Its Impact to the Physical Response of the Inhabitants in Creating a Self-Modification of Housing Units to Reduce the Impact of

Flood. The diverse financial capacity of each household and neighborhood has delivered various capacities to execute physical self-modification on neighborhoods and houses. It also influences the social relationship characteristic of each neighborhood and the way this social relationship brings influence to community flood response.

The most common dwelling modification implemented is to raise the ground floor. Differences in financial capacity contribute to the scale of modification undertaken. Middle to lower-income households tend to elevate the ground floor level at least from 50 to 100 cm above street level, while upper-middle to high-income households elevate their ground floor level up to and above 300 cm above street level. Despite all of that, many households have also decided to renovate their houses into two-story houses. Additionally, the proximity of each house to the inlet spot of water from the nearby river also brings influence on the level of height of the ground floor.

Moreover, the residents of flood-victim housing clusters in real-estates housing in Grogol tend to request more privacy for their dwelling units and neighborhoods than the residents of immediate kampongs. Therefore, an increasing number of houses are protected by individual fencing. With the increasing threat of flood in this area, more fences have been rebuilt to become waterproofed fencing barriers. These fences also function to prevent flood-water from the street entering the plot. Again, in this regard, diverse financial capacities of each household influence both the quality and quantity of the modifications that have been undertaken. The wealthier a family household, the higher the level of the ground floor and protective fencing built. As of today, these two self-modifications are predominantly found in real-estate housing neighborhoods, especially in Puri Gading. However, similar modifications are also found in other housing clusters and at the nearby kampongs, depending on household financial capacity.

The aforementioned physical modifications are merely taken in an attempt to block the floodwater entering houses. By increasing the height of the ground floor and by constructing waterproofed fences, the neighborhoods still experience a decrease in the amount of natural ground with open soil surfaces. As a result, the space for water absorption and infiltration has steadily decreased. During heavy rains and during flood events, puddles on the ground and floodwater become more difficult to infiltrate the soil; which leads to the increased duration of floods.

Unlike the modifications that have been taken by many houses in real-estate housing, the majority of kampong houses in southern Grogol applies a different modification approach. Only few houses elevate their ground floor or renovate their houses to become two-story houses. But more and more households have decided to use the space of their attics as a *ranggon*; an emergency space to store their belongings. Those who do not have an attic in their house use tables as a space of emergency to avoid the flood water in damaging their belongings. The two latest approach is commonly being applied by households with lower economic capacity.

5. Conclusion

Learning from the case of Joyontakan and Grogol, any response and action to deal with flood impact need more than an individual and group's awareness and actions. Massive and uncontrolled urbanization, play a significant role in adding the threat of flood. Their existence continues to decrease the amount of land and soil that are available to absorb water. Therefore, handling the threat of flood cannot only be solved by hard-engineering solution or individual physical responses.

Table 5. Summary.

	JOYONTAKAN	PURI GADING	NUSUPAN
Urban fabric formation	Dense Irregular kampong	Less dense Regular housing	Less dense New settlements mix of regular and irregular neighborhoods
Socio economic level	Low to middle income	Lower middle to upper middle income	Low income
Location	Below normal water level	Above normal water level	Below normal water level
Support from government	Early warning system Water gates and pumps Construction of river embankment Water channel improvement	Early warning system	Early warning system Water gates and pumps Water channel improvement
Self - actions	Water gates and pumps installation Individual house modification Maintaining water channels	Water gates and pumps provision and installation Individual house modification Construction of fences Construction of perimeter wall	Maintaining water channels
Awareness to the threat of flood	++++	++++++	++

There is a need for comprehensive solutions that consider future hydrological cycles of regional-scale rivers. The solution must also cover the influence and impact of the implemented actions to neighboring settlements and regions. In the case of Joyontakan and Grogol area, the solution cannot be merely provided parcel by parcel, because one solution that was applied to one particular settlement may trigger extra flood-threats on other settlements. As is the case for Joyontakan and Grogol, a scenario to provide extra open land and green spaces to increase land capacity to absorb water needs to be prepared by comprehensive urban planning. Moreover, the engineering solution to the existing urban flood also needs additional physical infrastructural solutions on the mezzo level. If constructing river embankments remains the macro, there is a need to provide a mezzo-scale solution to secure the district scale level from floods. For Joyontakan and Grogol, preparing water catchment reservoirs may help to secure the existing settlements from overloading backwater from the Bengawan Solo and the streamlets.

Moreover, preparing communities to be aware of the cause and the actions to reduce flood threat is undoubtedly useful to secure the inhabitants from damage at micro-level. However, proper coordination between different neighborhoods and supervision from experts and governments remains critical in order to reduce the negative impact of individual actions on the other neighborhoods. Therefore, flood mitigation planning and guidelines that cover from macro to micro level needs to be prepared properly.

Acknowledgements
This research is fully funded by the Habitus Platform Initiative. The intellectual property rights of this research finding are collectively shared among the Habitus Platform and the researcher members of this research. The knowledge valorization through conference participation of this research finding is partially supported by the University of Atma Jaya Yogyakarta.

References

[1] Aklema D 2003 Flood risk assessment for EIA: an example of a motorway near Trento, Italy *Acta Geologica* **78** 147–153;

[2] Mitchell JK 1999 Natural disasters in the context of mega-cities In: Mitchell JK (ed) *Crucibles of hazard: mega-cities and disasters in transition* (Tokyo, United Nations University) pp 15–55

[3] Suriya S, Mudgal BV 2012 Impact of urbanization on flooding: the Thirusoolam sub watershed–a case study *J. Hydrol.* **412–413** 210–219

[4] Du N, Ottens H, Sliuzas R 2010 Spatial impact of urban expansion on surface water bodies – a case study of Wuhan, China *Landscape Urban Plan.* **94**(3–4):175–185

[5] Dewan AM, Yamaguchi Y 2008 Effects of land cover changes on flooding: example from Greater Dhaka of Bangladesh *Int. J. Geoinf.* **4**(1):11–20

[6] Nirupama N, Simonovic SP 2007 Increase of flood risk due to urbanization: a Canadian example *Natural Haz.* **40**(1):25–41

[7] Burns D, Vitvar T, McDonnell J, Hassett J, Duncan J and Kendall C 2005 Effects of suburban development in runoff generation in the Croton river basin, New York, USA *J. Hydrol.* **311**(1–4):266–281

[8] Weng Q 2001 Modeling urban growth effects on surface runoff with the integration of remote sensing and GIS *Environ. Manage.* **28**(6):737–748

[9] Paul MJ, Meyer JL 2001 Streams in urban landscape *Annu. Rev. Ecol. Syst.* **32** 333–365

[10] Leopold LB 1994 *A view of the river* (Cambridge (USA): Harvard University Press)

[11] Hirsch RM, Walker JF, Day JC, Kallio R 1990 The influence of man on hydrological systems. In: Wolman MG, Riggs HC (eds) *Surface water hydrology*, vol 0-1. Geological Society of America, Boulder pp 329–359

[12] Kibler DF, Froelich CD, Aron G 1981 Analyzing urbanization impacts on Pennsylvania flood peaks. *J. Am. Water Resource Assoc.* **17**(2):270–274

[13] Parker D 1999 Flood. In: Ingleton J (ed) *Natural disaster management.* (Leicester (England): Tudor Rose) pp 38–40

[14] Goudie A 1990 The human impact on the natural environment. The MIT Press, Cambridge

[15] Miguez M G, Verol A P, Sousa M M and Rezende O M 2015 Urban Floods in Low Lands - Levee Systems, Unplanned Urban Growth and River Restoration Alternative; A Case Study in Brazil. *MDPI Sustainability Journal*

[16] Hollis GE 1975 The effects of urbanization on floods of different recurrence interval *Water Resource Res.* **11**(3):431–435

[17] Serrao-Neumann S, Crick F, Harman B, Schuch G and Choy DL 2015 Maximising synergies between disaster risk reduction and climate change adaptation: Potential enablers for improved planning *Environmental Sci. & Policy* **50** 46-61

[18] Wulangsari A 2014 Tipologi Segregasi Permukiman berdasarkan Faktor dan Pola Permukiman di Solo Baru, Sukoharjo *Jurnal Pembangunan Wilayah dan Kota* **10**(4):387-399

[19] Jati, V I 2012 Kajian Perkembangan Permukiman Wilayah Peri Urban di Sebagian Wilayah Kabupaten Sukoharjo Tahun 2001 – 2007[A Study on the Development of Housing in the Peri Urban Area of a section of Kabupaten Sukaharjo] *Jurnal Bumi Indonesia* **1**(1):1-8

[20] Padawangi R and Douglass M (2015) Water, Water Everywhere: Toward Participatory Solutions to Chronic Urban Flooding in Jakarta *Pacific Affairs* **88**(3):517-550

[21] Suharjo, Anna A N, Cholil M and Rudiyanto. 2015. *Analisis Morfologi dan Morfostruktur serta Pengaruhnya Terhadap Banjir Luapan Sungai Bengawan Solo Hulu Tengah* [Morphological

and Morphostructure Analysis and its Impact on the Flooding of Upstream Bengawan Solo River] University Research Coloquim.
[22] Tucci C E 2007 *Urban Flooding Management* (Porto Alegre: Water Meteorelogical Organization)

IOP Conf. Series: Earth and Environmental Science **158** (2018) 012039 doi:10.1088/1755-1315/158/1/012039

Heritage planning and rethinking the meaning and values of designating heritage sites in a post-disaster context: The case of Aceh, Indonesia

Z D Meutia, R Akbar and D Zulkaidi

School of Architecture, Planning, and Policy Development (SAPPD), Institut Teknologi Bandung, Bandung, West Java, Indonesia

Corresponding author: dyenameutia@gmail.com

Abstract. Heritage has become a driver of development as stated in the New Urban Agenda 2016 report. A starting premise of most recent studies of the concept of heritage suggests that its nature is not as a static inheritance with fixed and enduring values. Rather, the identification of sites as heritage requires a process of identification, or heritage creation. Heritage is a fluid phenomenon rather than a static set of objects or sites with fixed meanings. This paper uses theory from Smith [1] who argued that there is no such thing as a heritage; heritage is essentially a cultural custom and social process. Today, site-based heritage planning only considers the values of old towns and lacks clarity in terms of values that create criteria for the designation of cultural heritage sites in another context. Yet, this approach is needed as a way to maintain urban assets that significantly contribute to the establishment of values and quality parts of the city. Heritage planning is also the act of communicating and remembering the past for the present and the future in the public domain. This paper aims to formulate a conceptual heritage planning of designating heritage sites that challenges the traditional notion of heritage which considers age as a key element in heritage, the privileges monumentality and grand scale, with scientific/aesthetic expert judgment as a requirement of heritage designations. The limited idea of heritage based on exclusive values as something ancient, grand-scale, historical, and with other exclusive values has excluded many places as heritage in communities emerging from disasters. Debates within the critical heritage studies movement argue that heritage is a cultural product linked to activities of remembering and is an act of communication. The dominant hypothesis is that heritage values cannot remain to exist if the physical or material aspects of sites are destroyed and this hypothesis feels flawed. This paper asks us to acknowledge the importance of understanding how local values create heritage, that heritage is not only about material and physical aspects but it is also a cultural process. The methodology used is ethnography with in-depth interviews to uncover meaning and values in this context. This paper uses the examples of how the sites in Aceh that were affected by the tsunami disaster in 2004 such as the PLTD Apung ship and the Mosque of Ulee Lheue are sites that a mutual and role of historical city relating to the potential as signs of civilization, historical values, commemorative values, religious values, and values of consensus within the community as outstanding local values, as a basis of consideration in rethinking the designations of heritage sites of less than 50 years old.

1. Introduction

Heritage is one of the key elements in the establishment of places and as a driving force of development as stated in the NUA 2016 report [2]. The newly adopted Agenda 2030 highlights several processes, each of which – whether deliberately or not - have been and will be enlightened by culture and heritage and each of which will undoubtedly have deep meaning for cultural heritage. Hereafter, while respecting and admiring the recognition of an explicit heritage target, cultural heritage, and historic urban areas support urban life in a multiplicity of ways by providing reasonable options for urban assets through reuse, restoration, and rehabilitation to improve public spaces and provide land regeneration through mixed uses ([2],[3]). Consequently, there is much value in associated cultural heritage to several of the other goals, because culture and cultural heritage are integrally connected with urbanization, health, climate change, energy issues, and poverty reduction. Whether and how the cultural heritage aspects in the New Urban Agenda 2030 are realized and implemented at all levels, from local to international, will eventually influence both the success of the Agenda itself and the prospects for preserving the earth's valuable heritage resources ([2],[3]).

It starts with a study of the conceptual heritage which states that the paradigm of heritage has shifted into something dynamic rather than only a static material object that merely contains historical value and aesthetics. This understanding is supported by expert criticism (see [4]-[9]). Now, the paradigm of heritage has shifted from mere buildings to a wider scope covering all urban spaces and its contents [10]. On the other hand, it is necessary to have a planning direction as part of the management and support from the community and various stakeholders into something to be followed up upon. The purpose of this paper is to redefine the concept of urban heritage values from a public perception for the sustainability of a city by looking at the contrasting values of the post-disaster area case studies such as age, authenticity, integrity, stakeholders' interests, etc. Traditionally, in determining urban heritage, it is necessary to identify the significance of values inherent in objects, sites, or areas that are considered heritage [11]. These values are understood from the value of old age as a key element in designating the heritage, privilege value, monumental value and, mostly, the presence of a heritage expert as the party who has the right to determine what is heritage as well as its aesthetic value [9]. This limited understanding of inheritance is considered odd in from a wider perspective. The existence of historical value and other significance values have ignored many places that emerged after a disaster as something valuable and considered as heritage by the community [12]. This dominant hypothesis states that heritage values will not be sustainable if the physical and material aspects of the site have been destructed [13]. This is considered imprecise because the community perception with its memory and commemoration is far more valuable in establishing the heritage. Therefore, this paper asks us to be aware of the importance of understanding how local values have contributed to shaping heritage and that heritage is not only about material and physical aspects but also a socio-cultural process in society and contains different values depending on context and time.

This paper is organized into five parts. The first part introduces the discourse of urban development and the current situation in JMA. The second part reviews the literature about heritage planning and re-examines the significance of heritage in a post-disaster context. The third part provides a brief explanation of the data from the case study and the methodology used in this research. The fourth part discusses Rethinking the Meaning and Values of Designating Heritage Sites in a Post-Disaster Context. The final part presents the conclusion and its implication.

2. Heritage planning and re-examining significance and values of heritage

2.1. *Heritage planning*

Heritage planning does not presume the presence of a principle and incompatible divergence between the process of conserving aspects of the past and developing and assisting new directions for the city. Heritage planning is categorized in this type of development planning, implicating the maintenance of a set of urban activities, linked to the historical characteristics of the city and serves as resources for a major urban industry [14]. Heritage planning hence is not necessarily a matter of protection from change;

it is aligned with the concept and management of aspects of change and the nature, quantity, and tempo of such changes. Heritage offers an important contribution to the establishment of urban planning goals and the election of an achievable strategy for their accomplishment through intentional interference [14]. Therefore, urban heritage planning is a management strategy for urban sites and because of that has to be defined in relation to other strategies designed to operate both on forms and functions in a particular area. Planners have an important role in organizing and maintaining the interests of cultural heritage and buildings in the community. According to O' Connell [16], if we do not participate in urban planning and management, this assumes a lack of community involvement in maintaining urban assets.

Heritage preservation is one of the aspects in spatial planning that still lack the focus of attention as argued by Thompson [17]. This is partly related to the involution of identifying which aspect of the built environment should be preserved for future generations, as well as adjusting to conflicting options on how to do this. It can also be difficult to assign definitive meanings to heritage values and significance, partly because the meanings of heritage have shifted from a narrowly aesthetic and historical understanding to a wider scope that includes social, economic and cultural components [17]. The interpretation of heritage as a collection shifted towards heritage as "makeover" and, subsequently as a "cultural representation" [18]. These changing interpretations have repositioned heritage in spatial development from a focus on (isolated) preservation to (integrated) conservation and finally, a broader notion of heritage planning ([19],[20]). Nowadays, urban heritage and its values and meaning increasingly receive attention as the New Urban Agenda (NUA) states that heritage is one of the drivers of development ([2],[3]).

2.2. The re-examined significance meaning and values of heritage

Debates on the understanding of the definition of heritage have been dominated by the western perception such as from Europe and America that emphasizes the physical material and tangible objects. Thus, it is often stated that something is heritage if it has objects, buildings, monuments, and sites of more than 50 years old that preserved its existence and maintained its authenticity ([21]-[23]). As for the Eastern perception of heritage, including Asia, Larkham [24] explained that authenticity is not only in the realm of material objects (fabric). The concept of authenticity in Asia is considered to be in terms of sustainability, how to maintain the traditions and building techniques of objects, buildings, and sites, as well as on the sustainability of functions and usages that ensure the existence of objects and sites so that their functions are used continuously [25]. Pendlebury [26] and Orbaşli [27] as experts provided a case example of the Shinto shrine area in Japan. The destruction and reconstruction of wooden group structures in the Shinto shrine is scheduled every 20 years is a tradition to preserve the values inherent in the Shinto shrine, as implied in the Nara document [28]. This is done as a way to preserve the tradition of the building which does not always refer to material things (tangible) but illustrates the life cycle that is full of meaning and values (intangible). It can be interpreted that in the eastern world, the age and value of the authenticity of buildings and sites is not an ideal indicator in determining heritage. In the understanding of the definition of heritage, it turns out that experts argue that heritage is neither a tangible or intangible thing as interpreted by western and eastern interpretation. Heritage is defined as a process of social, culture, communicative action, or actions that are associated with the past where we can negotiate with our social values, and individual identity, and collective memory (see [1], [29], [30], [4]). Through their work, they changed the understanding of heritage as something that is understood as a "verb" [30], i.e., the activity of doing heritage to produce the product (tangible) and (intangible). Experts disagree if heritage is considered as (tangible) only that forgets past memories of the present and the powerlessness of the present in maintaining physical forms by preserving its functions [1]. Criticism in the study of heritage asks us to include ideas and understanding as a social process of remembering the past by looking at the values of the heritage area so that it forms tangible and intangible heritage [31].

The cultural significance itself is a statement about why it is important to preserve heritage objects. From the discourses and empirical experiences that evolved so far, Martokusumo [32] argued that the determination of significance is done through researching the values contained in the heritage object and site. These researches generally cover a number of specific aspects and considerations and result in an explicit statement about its value and importance to be conserved, revealing past phenomena, explaining its relevance to the present, as well as its benefits for the future understanding [33]. Moreover, empirical experience indicates the assumption to not build anything in the heritage area, or just maintain it without any creative effort. This happens because the values invested are often much different from the context, as well as how the related societies interpret them. Experts argue that there is a flaw in the theory of heritage values because of the decisions are taken by the stakeholders based on an incomplete understanding of the heritage and its values (see [11],[34],[35],[36]). This is an issue in establishing urban heritage because it is not going through their implications for the collection of information, synthesis, and communication or actual requirements placed on the significant statements of heritage conservation practices.

The values of integrity and authenticity are the basis of the heritage-based approach as a determinant of whether an object is worthy as heritage and of preservation ([37],[38]). The value of integrity states that heritage conservation needs to be done with consideration of the aspect of wholeness since the physical forms of the heritage are a relic of the past contain important information of its days. The term integrity in the Venice Charter [39] is understood as a whole, defined as logic/rational or completeness, and interpreted as the expression of material aspects in heritage [40]. The integrity value is used to trace the functional and historical conditions of a site [41] or understand heritage object capability to maintain its significance over a certain period of time [42] so that it becomes one of the main values that the object or site has inherited. In this decade, differences of views on cultural heritage and the natural heritage between Western and Eastern thought as described above are a problem in heritage study [38]. In the western world, the hegemony of its values separates the natural and cultural because the cultural realm focuses more on monuments and sites whereas the natural heritage scope is limited in the realm of science. It concerns the idea of a natural environment separated from its relationship to the society. This understanding is quite different with ASEAN countries that embrace eastern understanding because of the close relationship between the local community and its environment, cultural-environment/nature relations and tangible-intangible is an important element of appreciation effort between the cultural and biodiversity aspects in the environment management and utilization [38].

In this study, it is important to assess the heritage values of community perceptions. Perception is the attempt to understand the sensations captured from reality and recognize them by linking to previously collected knowledge [43]. According to May [43], the perception of society results from the perceptions of many individuals from an area attached to their memory and generates certain meanings and values. Public perceptions of urban heritage are used to generate images of objects, sites, and historic areas and require interpretation. The interpretation utilizes direct experiences, presentation, and also an interesting hue. Getting factual information is important in the interpretation process, as well as expressing emotions, dynamic and interactive nature. Interpretations can play a role in giving, entertaining and perhaps also changing people's attitudes and behavior [43]. The facts stored in various places are not always meaningful information. It requires an appropriate interpretation to overcome them, in order to prevent a person from having a wrong view so that the identification of perception is also important. Interpretation can be used as a bridge to explain the views of certain cultural groups to other groups or interpret something from the past to the present. In a place that has historical value, spaces are created and have meaning where "existential space is a constant of production and reproduction through the movement and activity of society members" [12]. Space is not passive but moving and embodies the experience. Living space is experienced and created through life activities; it is a sacred, symbolic and mystical space that is full of social meaning that surrounds local topographical buildings, objects, and features, providing points of reference and areas of emotional orientation for attachment and human engagement [44]. Therefore, if space and memory are lost and not preserved through heritage then society will lose meaningful places and future generations will lose their identity and experiences of the past.

2.3. Heritage in a post-disaster context

In the face of catastrophe, a heritage site tends to be considered as having lost its authenticity, integrity, values, and significance. If a building has been partially or completely damaged then it can no longer be considered or defined as heritage. Moreover, through the perspective of a global conservation effort, the extent of damage caused by disasters is recognized and mapped based on the survival of, and damage to, the physical environment. Thus, the reconstruction approach – whether it is a tabula rasa or a facsimile of the original – is determined by the existence or absence of material aspects that are assumed to act as embodiments of memories, values, meanings, and identities (see, [45]-[47]). The concern of heritage that is understood to be at risk may avoid the protection of locally grown heritage embodiment that does not satisfy the established global understandings of heritage constructs. On one side, heritage meaning and values can be seen as a practice to monumentalize, refuse, or adjust to the strength of destruction. This is a contrariness that is innate to the idea of heritage, and inherent in established global, regional, and local heritage discourses and practices in the work of reconstruction. On the other side, locally emergent heritage constructs that are left to operate independently of aforesaid programs may become caretaker to memories and narratives that describe, interpret, and preserve collective memory experiences of destruction, catastrophe, and decay in the context of natural disasters (see, [45]-[47]).

According to Byrne [3] and Kalstrom [3], heritage in Southeast Asia is largely understood as intangible objects. However, little attention is given to cultural issues in the context of disaster, especially in a post-disaster context. This needs to be studied, especially on the issue of human resilience [3]. There is a shortage of social theorists and cultural roles that play a role in the act of remembrance where disaster survivors are resilient because of the influence of the post-disaster context. Therefore, the role of culture in shaping heritage is very important. Especially tangible heritage reconstruction depends on the resilience of the community to commemorate a major event [3]. Familiarity with places can help the community restore post-disaster trauma or vice versa depending on how resilient the survivors are (see [4]-[7]). Different from the exploration Samuels [3] is that the role of the non-physical environment has contributed to the process of rebuilding the physical aspects of a city. Rico [3] examined the concept of heritage at risk in the face of disasters and argued that heritage is a sustainable cultural model in shaping resiliency. This is in contrast to the opinion of Leeuwen [3] who declared that heritage with disaster characteristics has little effect on the community memory network and plays a small role in the reconstruction process as mentioned in the above definition.

The relationship between heritage and disaster is very closely related to the intangible aspect as presented by Byrne [48] and Kalstrom [49] who argued that the culture of Southeast Asia regards materiality as impermanent. In a context in which disasters are experienced more regularly, these interests are increasingly understood. The important issue of the heritage literature on disaster is determining the inclusive way to involve their heritage to improve the resilience of survivors after a disaster. Moreover, the tendency of experts to focus on the realm or urban structures and materiality alone will only reduce the well-being of survivors from disaster ([50],[51]).

3. Data and methodology

This study uses data from in-depth interviews using an "ethnographic conservation" method. Ethnography emerged to resolve problems of the social environment and involves observing a group of people in their natural environment [52]. This research prioritizes the sense of realities of researchers, the process of deep thinking, and the interpretation of facts based on the concept used, to develop it with a deep understanding and with prioritizing the values under study [53]. Therefore, to get results that are appropriate to the purpose of this study, frequently this method prioritizes the long-term mixing between researchers with the object [53]. Most ethnographic methods focus on cultural studies in a holistic sense; this research focuses on the subject view as the research object [53]. Usually, heritage planning uses ethnographic methods for assessing cultural heritage values, depending on the association of approaches to interviews and participant observation from physical and behavioral mapping. The objective of this is to disseminate cultural representations in the heritage experience [54]. Heritage ethnography does not need to be conceptualized as a record but may creatively build on existing ethnographic traditions [54].

Case study

Banda Aceh is one of the cities most affected by the 2004 tsunami, which has been one of the greatest disasters of the 21st century [55]. The disaster has changed the urban landscape in Banda Aceh and left debris, artefacts, and sites that emerged in the aftermath of this tsunami, such as the monument of the PLTD Apung ship, the mass burial area, ships on residential houses, the tsunami museum, the area of the mosque that survived from the tsunami, and many others. Normatively, according to Cultural Heritage Act No. 10 Year 2011, the legislation on heritage in Indonesia is too basic as it stipulates that heritage must be at least 50 years old, represent the shortest style of 50 years, and has special meaning for history, science, education, religion, and cultural values to strengthen the nation's personality. Then, if we capture the reality of post-disaster destruction that ruined artifacts, sites, and areas, related to age and periods of style, then is it cannot be established as a heritage? Which style periods can represent the area defined as heritage? Does society agree with the current physical forms and how are people's perceptions of seeing and treating them? Questions like these make people recall how important they are to be preserved through re-functioning with activities, used for the public benefit and social learning for the future. New knowledge that emerged from the tsunami disaster where the remnants of the relics become a reminder, commemorate and make people aware to rebuild the city, disaster reduction efforts with changes in landscape settings such as mangroves reforested by citizens with the government, knowledge of evacuation routes to save themselves if there are signs of natural disaster and the use of tsunami-affected buildings as a forum for discussion and meetings between the residents.

This study aims to understand the perception of the Banda Aceh community in seeing and using physical forms such as buildings, sites and places, both old and newly emerged ones after the disaster. This study uses data from in-depth interviews consisting of 10 questions with 20 respondents from the age range of 20 to 45 years old with diverse status. The majority of respondents are people who have lived in Banda Aceh for a long time and the sample also includes some students. This paper does not claim that the respondents are a proper representation of the community in Banda Aceh but views the perspective of the community as represented by the respondents. Based on the post-disaster data and the reconstruction of the city, Indonesian historic places got recognition from the community as a familiar place and not much changed in the city. However, many new memorial places emerged post-disaster as new tourist attractions and to commemorate the tsunami tragedy. This indicates that the historical places have a significant role to form the intimacy of the place. However, a few new places that appear can trigger negative memories in the community of the big disaster in 2004.

Figure 1. PLTD Apung ship dragged by the tsunami in densely populated residential areas [61].

Figure 2. Tsunami Museum in the city center [61].

Figure 3. Ulee Lheue Mosque, the only one building that survived the 2004 tsunami in this area [61].

Table 1. Identification of heritage values and percentage of respondent perception

No	Identification of Heritage Values	Agree	Percentage	Disagree	Percentage
1	The importance of the remnants of a tsunami relic to become heritage	V	69.7 %		27,3 %
2	The meaning of urban heritage as tangible and intangible	V	84,8 %		9,1 %
3	The age-depth of value is not the ideal criteria for designating heritage sites and buildings	V	68,3 %		31,2 %
4	Become heritage and tourism attractions	V	57,6 %		33,3 %
5	The values of memory/	V	69,7 %		21,2 %

No	Identification of Heritage Values	Agree	Percentage	Disagree	Percentage
	Commemoration				
6	The values of spiritual and religious	V	90,9 %		9,1 %
7	The values of authenticity and integrity to restore its original form		27,3 %	V	54,5 %
8	Good memory or bad memory	V	91,3 %		6,1 %
9	Conserve heritage together not only by government efforts	V	66,7 %		27,3 %
10	The values of local and global knowledge, social learning and preserving tradition	V	74,2 %		18,5 %

The data shows that the majority of respondents considers it very important to maintain the places that emerged in the aftermath of the disaster as heritage or tourist area. There are a number of values that are full of meaning like memory values which have positive and negative images because respondents think the 2004 Tsunami disaster event can be remembered as a form of socio-cultural learning in the future and to build awareness. The emergence of new knowledge triggered by this great and unexpected natural event is an aspect of consideration in changing lifestyles, rearranging life and urban spatial planning. Examples are the designation of coastal areas and river banks as green open spaces both as mangrove forests, or pedestrian and urban forest. The community has finally realized the importance of landscape arrangements to mitigate future disasters after the disaster. The respondents also agree with the current physical forms even though their authenticity is changing and the traditional theories do not include these in their definition of heritage. However, the community still considers these places as valuable and heritage that is used for various activities such as discussion forums, venues for big events and religious gatherings, learning about mitigation, tourist attraction and so on. Based on the data, the people are well aware that their places, traditions, lifestyles, habits, and norms have changed and some have disappeared through significant changes in physical and symbolic aspects after the tsunami. Similarly, for identifying the symbolic added value from many changes and reconstructions that have occurred after the tsunami. The aspect that arises when people talk about heritage sites is that historical change is considered as a part of the heritage. The most important indicator of authenticity is not material like the western understanding but rather its experience, the unforgettable memory of the society, the activity and the sense of place people talk about. Thus, in addition to physical aspects, the knowledge in the invisible values becomes a more important aspect of building awareness of disaster risk reduction and disaster response at any time, as social learning is passed on to the future and the spiritual value that makes people closer to God. From the community perceptions gained through this research, I argue that the perception of societies about the traditional values of western-dominated understandings is imprecise in this context and different from the expert view. Although the authenticity of the existing physical forms has changed, the places still function related to values in a society that run continuously for the sake of the sustainability of the city. Once a place becomes a heritage place, its physical form helps that place signify its importance to us. Heritage is a way of expressing the unique relationship between the Acehnese and their environment which comes from recent history in the process of becoming "materialized" through the commemoration of prehistorical, historical, and post-historical schemes in its urban landscape.

4. Rethinking the meaning and values of designating heritage sites in a post-disaster context

The literature and the community perceptions of the case presented above show that the value approach in shaping a sustainable urban heritage is important and is the basis for creating a high-quality, contextual, and inclusive environment [56]. The determination of the value and meaning of a site to be preserved needs to reflect the perception and interpretation of the community as the user and owner. Nowadays, the concept of society is changing; it refers not only to the residents and the community but also includes the commuters, tourists, migrants, and others [57]. Regarding the determination of the boundaries of the conserved area, it is not merely determined by physical boundaries as beyond that there are invisible values that are inscribed in the hearts and experiences of each member of society [57]. With unique local contexts, the heritage management mechanisms can respond to ambivalence and realize the continuity of urban heritage in the context of long-term development. From a critical understanding of urban heritage, the existence of a place setting or an urban area can illustrate the formation of a superiority layer which is the result of a series of long processes as a matrix between cultural values and natural elements [56]; it is nothing but the construction of the cultural heritage itself. Thus, we need a typology identification of heritage values so that the drafting can be more measurable. The heritage values approach should be reconsidered to concentrate on the relationship between the community and its region for the sustainability of the urban environment. The heritage material can be a unified expression as an ongoing work and, thus, urban heritage can be viewed as a continuity of the past for the future. In the context of sustainability, the boundaries between the past, present, and future are eliminated. The purpose of maintaining these values is to maintain the continuity of the relationship between the current society with a site or the area around it.

According to the value approach that has been criticized by many experts for its incomplete and overly western understanding that emphasizes only historical and aesthetic material aspects, the present value approach can be rethought from the starting point of maintaining and managing the relationship between the community and the environment (see [58],[59]). This view extends the meaning of the heritage values within the society to perpetually keep heritage as a living inheritance. The idea of heritage suggests to grow the values within the community that treats heritage as a process of remembering and changing lifestyles for the better and responding to changes in urban structures as an integral part of social and cultural processes ([1],[30]). This paper aims to be neutral and considers that forming the heritage cannot be separated from one another. However, physical forms as visualizations can evoke memories and perceptions of people who see it and be near the object or area. Similarly, the values embedded in these physical forms should be identified as the basis for establishing an inheritance whether the physical form is for sight or functionality. From the post-disaster perspective of the case study, this paper is aligned with the issue of sustainability and considers the most important aspect of post-disaster sustainability inheritance to be the re-establishment of habits that existed before. Opinions based on community perceptions are supported by the 'principle of sustainability'. Omer and Alon argued that value identification aims to preserve and restore functional, historical, and interpersonal sustainability at the level of individuals, families, organizations, and communities [60]. This paper agrees with this basic thinking, and feels that embracing the former social and cultural context is fundamental for the continuity of the object; the heritage site is tangible and intangible.

5. Conclusion

From the literature review, this study concludes that the uniqueness of the character of a place can be shaped by some values of memories and symbols, cultural and historical diversity, the values prevailing in society, and the quality of the public environment that can be captured and utilized by society so that it can become a heritage value. Heritage contributes to a strong uniqueness and offers experiences in the urban sphere; the wealth of these experiences further provides the quality of a place. This paper aimed to formulate a concept of heritage planning in designating heritage sites which contrasts the traditional notion of heritage that considers age as a key element in heritage, as well as the monumental privileges and grand scale, scientific/aesthetic expert judgment as a requirement of heritage designations. The narrow idea of heritage as a thing that is ancient, grand-scale, historical, and among other exclusive values has excluded many places as heritage in communities emerging from disasters and the meaning and values of designating heritage sites in a post-disaster context need to be reconsidered. This paper provides a conceptual understanding of urban heritage planning and its values of designating sites on the basis of critical thinking that rejects rigid and static values that do not consider aspects other than time and context. There are many values we can learn from post-disaster cases closely related to incorporate heritage in urban planning. These values are very useful for the practice of urban heritage preservation to understand the constellation of places that emerged post-disaster because it is full of meaning and values so that it can also build community cultural heritage such as changing traditions and a lifestyle toward a better direction without forgetting history. These sites and emerging sites are part of the everyday social life arrangement which is not only to be seen but functions with positive activities, giving people access to a better life. Many sites and objects in this context actively play an important role in the production of essential daily social functions, such as the provision of space for community consultation, conflict resolution, decision-making, and disaster mitigation learning; all of which are fundamental to the process of sustainable development. This involves not only community leadership, social networks, and hierarchies but also material conditions in which people know how to interact. Without all of the above, public relations with meaningful and comprehensive heritage sites are impossible.

Acknowledgments
This paper is part of the Ph.D. thesis of Zya Dyena Meutia, titled "Beyond Meaning and Values of Heritage Planning of Designations Heritage Sites in context post-disaster" in the School of Architecture, Planning, and Policy Development (SAPPD), Institute of Technology Bandung, Indonesia.

References
[1] Smith L 2006 *Uses of Heritage* (New York: Routledge)
[2] HABITAT III 2016 Urban Culture and Heritage *United Nations Conf. on Housing and Sustainable Urban Development 17 to 20 October Quito Ecuador*
[3] ICOMOS 2016 *The Principles for the Conservation of Heritage Sites in the World*
[4] Byrne D 2009 *A Critique of Unfeeling Heritage* In: L. Smith and N.Akagawa (Eds.) *Intangible Heritage* (London: Routledge)
[5] Cleere H 2001 Uneasy Bedfellows: Universality and Cultural Heritage In: R. Layton, P. G. Stone & J. Thomas (Eds.) *Destruction and Conservation of Cultural Property* (London: Routledge)
[6] Meskell L 2002 The Intersections of Identity and Politics *Archaeol. Annu. Rev. of Anthropology* **31** 279-301
[7] Labadi S 2007 Representations of the Nation and Cultural Diversity in Discourses on World Heritage *J. of Soc. Archaeol.* **7**(2):147-170
[8] Aikawa-Faure N 2009 From the Proclamation of Masterpieces to the Convention for the Safeguarding of Intangible Cultural Heritage In: L. Smith & N. Akagawa (Eds.) *Intangible Heritage* (New York: Routledge)
[9] Waterton E and Smith L *Heritage, Communities and Archaeology: Duckworth Debates in Archaeology* (London: Gerald Duckworth and Co)
[10] Taylor J 2015 Embodiment Unbound: Moving beyond Divisions in the Understanding and Practice of Heritage Conservation *Studies in Conservation* **60**(1): 65–77
[11] Fredheim L H and Manal K 2016 The significance of values: heritage value typologies re-examined *Int. J. of Heritage Studies*
[12] Daly P and Rahmayati Y 2012 Cultural Heritage and Community Recovery in Post-tsunami Aceh In *From the Ground up: Perspective on Post-tsunami and Post-conflict Aceh* edited by Daly P, Feener RM and Reid A (Singapore: ISEAS Publishing) p 57–78

[13] Dewi C Rethinking Architectural Heritage Conservation in Post Disaster Context *Int. J. Heritage Stud.* **23**(6):587-600
[14] Swensen G 2012 Integration of historic fabric in new urban development-A Norwegian case-study *Landsc. and Urban Plan.* 107(4):380-388
[15] Sjöholm J 2017 Authenticity and relocation of built heritage: the urban transformation of Kiruna, Sweden *J. of Cultural Heritage Manage. and Sustain. Dev.* **7**(2):110-128
[16] O' Connell, D. *"Tending the City; A Handbook of Urban Principles,"* Dublin: An Taisce, 1989.
[17] Thompson S 2012 Introduction In: S. Thompson & P. Maginn (Eds) *Planning Australia : An Overview of Urban and Regional Planning* (Sydney: Cambridge University Press) p 1-14
[18] Grijzenhout F 2007 *Erfgoed: de geschiedenis van een begrip* (Amsterdam:Amsterdam University Press)
[19] Ashworth G J 1991 *Heritage Planning: Conservation as the Management of Urban Change.* Groningen (The Netherlands: Geo Pers)
[20] Bosma K 2010 Heritage Policy in Spatial Planning In: Bloemers J H F, Kars H, van der Valk A and Wijnen (Eds) *The Cultural Landscape and Heritage Paradox: Protection and Development of the Dutch Archaeological - historical landscape and its European Dimension* (Amsterdam: Amsterdam University Press) p 641-652
[21] Morris A 1877 *Manifesto of The Society for the Protection of Ancient Buildings* (London: Society for the Protection of Ancient Buildings)
[22] Ruskin J 1899 *Seven Lamps of Architecture* (London: George Allen), 1899.
[23] Feilden B 2003 *Conservation of Historic Buildings* (Oxford: Elsevier)
[24] Larkham P J 1996 *Conservation and the City* (London: Routledge)
[25] Jokilehto J 1999 *A History of Architectural Conservation* (Oxford: Butterworth Heinemann)
[26] Pendlebury eds. 2009 *Valuing Historic Environments* (Farnham (UK): Ashgate)
[27] Orbasli, A. "*Architectural Conservation,"* Oxford: Blackwell, 2008.
[28] Nara Document. "*The Nara Document on Authenticity,"* The Getty Conservation Institute, 1994.
[29] Dicks B 2000 *Heritage, Place, and Community* (Cardiff (Wales): University of Wales Press)
[30] Harvey D 2001 Heritage Pasts and Heritage Presents: Temporality, Meaning and the Scope of Heritage Studies *Int. J. of Heritage Stud.* **7**(4):319–338
[31] Winter T 2013 Post Conflict Heritage in Asia; Shifting Geographies of Aid In: P. Basu &W. Modest (Eds.) *Museums, Heritage and International Development* (London: Routledge)
[32] Martokusumo W 2015 *Arsitektur dan Pelestarian: Menuju Pengelolaan Berkelanjutan Bangunan dan Lingkungan Cagar Budaya* [Architecture and Conservation: Towards a Sustainable Management of Buildings and Cultural Heritage Sites] Orasi Ilmiah Forum Guru Besar ITB 25 September 2015
[33] Martokusumo W 2017 *Pemaknaan Tempat dalam Pelestarian Arsitektur* [Meaning of Place in Architecture Conservation] Seminar IPLBI
[34] Rudolff B 2006 *'Intangible' and 'Tangible' Heritage: A Topology of Culture in Contexts of Faith* PhD diss., Johannes Gutenberg-University of Mainz
[35] Poulios I 2010 Moving beyond a Values-based Approach to Heritage Conservation *Conservation and Manage. of Archaeol. Sites* **12**(2):170–185
[36] Walter N 2013 From Values to Narrative: A New Foundation for the Conservation of Historic Buildings *Int. J. of Heritage Stud.* **20**(6):634–650
[37] Jokilehto J 2006a Considerations on authenticity and integrity in world heritage context *City & Time* **2**(1):1

[38] Lennon J L and Taylor K (eds) 2012 *Managing Cultural Landscapes* (London: Routledge) 2012.
[39] Venice Charter 1964 *International Charter for the Conservation of Monuments and Sites (The Venice Charter 1964)* 2nd International Congress of Architects and Technicians of Historic Monuments Venice

[40] Pereira H N 2007 Contemporary trends in conservation: culturalization, significance and sustainability *City &Time* **3**(2):2
[41] Jokilehto J 2006b World Heritage: Defining the Outstanding Universal Value *City & Time* **2**(2):1
[42] Stovel H 2007 Effective use of authenticity and integrity as World Heritage qualifying conditions *City & Time* **2**(3):3
[43] May M 2007 *Sensation and Perception* (New York: Chelsea House)
[44] Fredericks V 2011 Remembering Katyn: Mourning, Memory, and National Identity *Memory Connection* **1**(1),197-210

[45] Read P 1996 *Returning to Nothing: The Meaning of Lost Places* (Cambridge: Cambridge University Press)
[46]Samuels A 2010 Remaking Neighbourhoods in Banda Aceh: Post--tsunami Reconstruction of Everyday Life In: M. Clarke, I. Fanany & S. Kenny (Eds.) *Post-- Disaster Reconstruction: Lessons Learn from Aceh* (London: Earth scan, Ltd)
[47] Spelman E V 2008 Repair and the Scaffold of Memory In: P. Steinberg & R. Shields (Eds.) *What is a City?: Rethinking the Urban after Hurricane Katrina* (Georgia (USA): University of Georgia Press)
[48] Byrne D 2014 *Counterheritage: Critical Perspectives on Heritage Conservation in Asia* (New York: Routledge)
[49] Karlström A 2005 Spiritual Materiality: Heritage Preservation in a Buddhist world? *J. of Soc. Archaeol.* **5**(3)
[50] Barakat S 2007 Post War Reconstruction and Recovery of Cultural Heritage: Critical Lessons from the last Fifteen Years in Cultural Heritage in Post War Recovery In: N S Price (Ed.) *Cultural Heritage in Postwar Recovery* (Vol. 6) (Rome (Italy): ICCROM Conservation Studies)
[51] Ascherson N 2007 Cultural Destruction by War, and Its Impact on Group Identities In: N S Price (Ed.) *Cultural Heritage in Postwar Recovery* (Vol.6) (Rome (Italy): ICCROM Conservation Studies)
[52] Clavir M 2002 *Preserving What is Valued: Museums, Conservation, and First Nations* (Vancouver: UBC Press)
[53] Harris M and Johnson O 2006 *Cultural Anthropology : 7th Edition* (Allyn& Bacon)
[54] Ervin A W 1997 Trying the Impossible: Relatively 'rapid' methods in a city-wide needs assessment *Human Organization* **56**:379-387
[55] Kenny S 2010 Reconstruction through Participatory Practice? In: Clarke M, Fanany I and Kenny S(Eds.) *Post-Disaster Reconstruction: Lessons Learn from Aceh* (London: Earth scan, Ltd)
[56] Cane S 2009 Why Do We Conserve? Developing Understanding of Conservation as a Cultural Construct In: *Conservation: Principles, Dilemmas and Uncomfortable Truths* edited by Richmond A and Bracker A (London: Butterworth-Heinemann) p 163-176
[57] Caple C 2009 The Aims of Conservation In: *Conservation: Principles, Dilemmas and Uncomfortable Truths* edited by Richmond A and Bracker A (London: Butterworth-Heinemann) p.25-31

[58] Avrami E, Mason R and de la Torre M eds. 2000 *Values and Heritage Conservation* (Los Angeles (CA): Getty Conservation Institute)

IOP Conf. Series: Earth and Environmental Science **158** (2018) 012039 doi:10.1088/1755-1315/158/1/012039

[59] Clark K 2006 From Significance to Sustainability In: *Capturing the Public Value of Heritage: The Proceedings of the London Conference* edited by Clark K (Swindon (England): English Heritage) January 25-26 p 59-60
[60] Omer H and Alon N 1994 The Continuity Principle: A Unified Approach to Disaster and Trauma *American J. Of Community Psych.* **22** 273-874
[61] Rahman F 2017 Documentation of observation Post Disaster Tsunami Pictures, Aceh

IOP Conf. Series: Earth and Environmental Science **158** (2018) 012040 doi:10.1088/1755-1315/158/1/012040

Beyond the Early Stage of Post-suburbanization: Evidence from Urban Spatial Transformation in Jabodetabek Metropolitan Area

E Sadewo[1,2], I Syabri[1], Pradono[1]

[1]School of Architecture, Planning, and Policy Development (SAPPD), Institut Teknologi Bandung, Indonesia
[2]Badan Pusat Statistik, Indonesia

Corresponding author: erie@s.itb.ac.id

Abstract. In the theory of urban transformation, there has been growing attention to the development of metropolitan outskirts. While the debate of the post-suburbanization process has already settled, the knowledge on the development beyond is still questionable. This paper examines the urban spatial pattern transformation beyond the Jabodetabek Metropolitan Area post-suburbanization. We use the medium and large enterprise (MLE) data from the economic census (EC) and population data from village potential census (PODES) with two cross sectional times of reference. The 2005 and 2006 data are assumed to represent the post-suburbia situation, and the 2014 and 2016 data represents the situation beyond. We analyzed the extent of which post-suburban spatial pattern in JMA would develope by utilizing the Exploratory Spatial Data Analysis (ESDA) method. The result shows that the polycentric urban structure of JMA has strengthened. The low order service function which was previously clustered in suburban areas is recentralized in the urban core, leaving the manufacturing sector as the main function in the post-suburbia. It is implied that the post-suburban status has reached a steady state for a long term before its next transformation. Those transformations are followed by the shift in population dynamics, whereas workers tend to polarize based on the proximity to their job location. The findings recall further study regarding the post-suburban commuting pattern.

1. Introduction

Globalization and changes in socioeconomic conditions have shaped the appearance of the urban region. The emerging urban development and its postmodern society result in the urban spatial transformation of metropolitan areas. Regarding this issue, current attention is focused on the transformation dynamics in metropolitan outskirts. Many evidence show that what was formerly known as suburban areas have become more obsolete as the dichotomy of urban and suburban areas is increasingly blurred. Urban development has been turned inside out, whereas suburbs are becoming more attractive, and the traditional urban core is losing its importance. This phenomenon has been coined in several terms, such as 'edge city', 'edgeless city', or 'technoburb', but Wu and Phelps [1] argued that the term 'post-suburban' is the most comprehensive concept, as it generally encompasses the previous. It cannot be seen as a single definition, but rather as a signal of various processes of potential change at the metropolitan periphery [2].

Post-suburbanization can occur in any location with different scale and manner [3]. Phelps and Wu [4] argued that in their relationship with settlement typology, the urban development process can be divided into three scenarios. The first stage is termed as modern city, in which urban fringes grow and develop into suburban areas. The second stage, termed as late modern city region, describes the situation of suburbs that had resulted from the first trajectory is transformed into post-suburban. In the third

IOP Conf. Series: Earth and Environmental Science **158** (2018) 012040 doi:10.1088/1755-1315/158/1/012040

trajectory, termed as the city region of second modernity, there are five different probabilities of transformation: (1) from post-suburban developed into city; (2) developing suburbs, growing into post-suburban and then becoming new cities; (3) a surviving prosperous and stable suburb; (4) the suburbs shrink into sub suburbs; and (5) the city decentralizing into suburbs. While the first two stages describe the urban transformation into post-suburbia, the last stage is an assumption of what could happen beyond. The difference in transformation patterns resulted in the question of whether post-suburbanization would result in a series of new cities with a relatively steady polycentric pattern, or if it's just a transitory phase leading into a scattered intra-metropolitan urban form [5]. Thus, the geographical aspect of post-suburbia has become more important since its different forms or appearance in different locations may obscure valid points of comparison [2].

Urban transformation implies changes over time, so that the understanding of present development requires some reflection on the past course of a city's growth and development [6, p. 2]. But the concept of urban growth itself is far from clear. Reis et al [7] argued that there are at least three different and widely used concepts of urban growth in the urban and regional planning literature: population change, economic performance, and the spatial expansion of urban areas. Such concepts in which several characteristics can be identified as the elements of global post-suburbia (see [1], [8]–[10]). However, Bontje [11] argued that the main purpose of understanding post-suburbanization is to explore the possible contribution of new employment centers in the 'metropolitan periphery' to a more sustainable development of metropolitan regions. Without neglecting the importance of population factors and spatial expansion, this perspective led recent research on post-suburbanization to focus on its relation to economics aspects (for examples see [12]–[15]).

The Jakarta Metropolitan Area (JMA) is widely known as an Emerging Mega Region (EMR), but the existence of post-suburban elements have been increasingly visible within its development process since the early of 2000's [16]. The post-suburbanization process in JMA involves a planned deconcentration of high technology industry and multinational companies, creating a spatial implication that emerged from the change of agricultural restructurization into manufacturing sectors [12]. Such processes make an impact on mixed settlements and change its commuting pattern, which is considered to be at the early stages of post-suburbia [17]. It is further assumed that the post-suburbanization process in JMA would be followed by service sectors [12], leading to the development country post-suburban trajectory. In other words, the JMA hinterland would develop into multifunctional areas with more independence from the traditional core, resulting in a new city. Such situation is argued to be the sign of the end of a *desakota* region, which is unfavorable because it would threaten the JMA regional sustainability cantilever [18]. Although the theoretical foundation of spatial transformation towards a post-suburban form is already known, the extent to which urban transformation towards beyond is still questionable. Recent evidence insufficiently covers only a limited location which is considered to be post-suburbia (see [12], [16], [17]). Furthermore, various researches measuring urban spatial pattern in JMA are often limited to population and land use aspects (see [19], [20]). There is little known about the actual JMA urban spatial structure and its transformation in terms of economic activities.

This paper aims to make a contribution in the discourse of urban spatial transformation by exploring the spatial pattern resulted from the process beyond post-suburbanization. We explore Hudalah and Firman's [12] hypothesis of whether post-suburban areas in JMA would follow the developed country trajectory. We address three questions regarding the urban transformation of JMA. The first is to reveal the actual recent urban form of JMA and its alteration. The second is questioning whether post-suburbia in JMA would be developed into a new city or just maintain or strengthen its current form. The last question concerns population dynamics regarding the urban transformation, the extent of which could explain the next stage of post-suburbanization in the context of a developing country such as Indonesia.

We assess the urban transformation process of JMA by comparing the distribution of some socio-economic characteristics at the village level from two census periods within 2005-2016. The village level usage is very useful because it can provide analysis up to the smallest level of geographical area. We utilize the data from the 2006 and 2016 Economic Census (EC) to recognize the spatial behavior alteration of non-agricultural medium and large enterprises (MLE) in JMA, whereas the 2005 and 2014 village potential census (PODES) data are used to identify some village characteristics including its population. The comparative study is conducted using Exploratory Spatial Data Analysis (ESDA) method.

This paper is organized into five parts. The first part introduces the urban development trajectory discourse and its current situation in JMA. In the second part we review the literature about post-suburbia and how it relates to its urban form and spatial structure. In the third part we make a brief explanation about the data and methodology used in this research. In the fourth part we discuss the transformation of the JMA urban form and spatial structure in the last decade. The last part will present the conclusion and its implication.

2. Post-suburbanization and urban spatial transformation

2.1. Towards post-suburbanization

In the stage of urban transformation, many experts agree that in most metropolitan areas, the existence of "suburbia" is now increasingly difficult to find. The landscape between urban and rural areas has undergone a fundamental change, in which traditional forms of urban and rural areas no longer exist [21]. Bontje [11] argued that suburbia is gradually changing into an increasingly detached location from traditional urban centers. The low-density settlements are increasingly mingling with densely populated settlements, jobs, shopping locations and recreational facilities. In modern urban areas, the growth and development move out of the city center to the surrounding area. But in the postmodern era, the logic seems to be reversed, with the growing development of new urban centers in the metropolitan outskirts [22]. The result is a different look from what was previously recognized as suburbia but is described as neither urban nor suburban.

Kling et al [8] coined the term post-suburban to refer to a region that has several centers of activity which are mutually exclusive and different between settlements, shopping malls, or industrial estates, and provide specific attractions. Each center has a special function and is separated by a travel distance of 15-30 minutes by car, with a balanced frequency of within and intercity trips. A post-suburban territory is integrated into the global capitalist market system and transformed from an agricultural area into a society dominated by industrial and post-industrial workers in a relatively short time. The development that occurs in a post-suburban area occurs in three phases of the village network in rural areas, transformed into suburban areas, then changed into a complex and decentralized area that is a mix of urban, suburban, and rural.

One trigger for the shift from suburbia to post-suburbia in America and other developed countries is the technology revolution that began in the early 20th century. The development of cars made people able to travel from one corner of the metropolis to another corner on a regular basis every day. Advances in telecommunications and computer technology have the potential to replace face-to-face needs as well as other physical movements. This made decentralization work even more so that the American suburbs gradually began to lose attachment to the established city center [12]. Ongoing decentralization occurred in major cities in the United States, such as Portland, Oregon, Philadelphia, and Pennsylvania. But in those cities, the transition from a monocentric structure to polycentric then diffusion, did not occur linearly, but had local differences [23].

There is a significant difference between post-suburbia characteristics in North America, Europe, and Asia. This could not be separated from the influence of the culture and the political system chosen in development. The development of metropolitan suburbs in mainland Europe differs from that of North America because their urban core remains a dominant center compared to America. The revitalization of urban centers and productive zones keep urban areas in Europe attractive and capable of promoting urbanization [24]. The Netherlands experience shows how spatial planning provides the initiative of the population to live in urban suburbs [21]. From 1960 to the mid-1980's, the Dutch spatial planning policy of 'concentrated deconcentration' has supported the development of growth centers around big cities.

The growth center was basically intended to provide a suburban housing area for those who wanted to get out of the big city to prevent an urban sprawl by collecting the construction of suburban housing at a limited number of locations. The growth center that lied farthest from the urban core had the additional function to build an economic center of an independent new urban area, and to promote economic growth in marginalized suburbs. However, Bontje [11] found that the new job location actually had the following characteristics. First, the new location was relatively close to the city and the suburban limits were closest to the city limits. As a result, the new growth center was close to the big city at a location that was originally intended only for the provision of suburban settlements and necessary services. Second, the opportunity for inhabitants of the suburb to find a job near their home was only found by highly skilled workers. Third, most of the new job sites built in the last few periods were located close to the highway, so there was a tendency to increase the use of cars as a mode of transportation.

In a global perspective, Wu and Phelps [1] summarized post-suburbia traits into six characteristics. First, there is a separation, or fragmentation, of the suburbanization process, which mainly comes from the construction of settlements. Second, post-suburbia is characterized by land use as well as mixed urban morphological elements. Third, the dynamics of urban development do not seem to be seen if it happens to be bordered by the boundaries of local government. Fourth, local governments on certain aspects play an important 'entrepreneurial' role in post-suburbia development. Fifth, post-suburbia has a balance of work location and a larger residence compared to suburbia. Sixth, the post-suburban form is associated with extensive and conflicting land use between housing, manufacturing and services industries, open land, agriculture, and parks that are characterized by the expansion of metropolitan areas in Southeast Asia. Although these six characteristics can be said to be comprehensive, Phelps and Wu [2] found that this phenomenon could have varying values in different location settings.

2.2. *Urban spatial transformation*

Post-suburbia has three dimensions, which can be distinguished from other terms in urban transformation [25], since two of them are closely related with changes in spatial form. First, based on the process, post-suburbia is an indication of breaking the linkage with the pattern or process of previous suburbanization. Second, post-suburban can be distinguished based on the spatial shape of the concentrations contained in the suburbs and edge cities. Third, in terms of politics, post-suburban represents the function of the city, where the role of the business organization's interest in post-suburban politics is often greater than expected. Generally, post-suburban spatial forms lead to a multi-centered metropolitan region [8]. Therefore, the polycentric spatial structure becomes an important element in the existence of post-suburbia [1], [3], [12], [23].

The traditional urban spatial structure is assumed to be monocentric, consisting of a traditional urban core or central business district (CBD) as a production center location, and all workers living around the area and commuting to work. The monocentric model explains that population density and land rent costs will decrease monotonically as distance away from the center of the city [26]. Economic agglomeration will function well in the CBD area, and its influence will decrease with increasing distance from the CBD [27]. As a result, the CBD is dominated by mature companies. The a priori assumption of a centralized CBD form, fueling the ease and flexibility of a monocentric model use in analysis, and made it the most dominant urban economic development paradigm of the 20th century [28]. However, the assumptions used are judged to be no longer in accordance with the development of modern cities [26]. Davoudi [29] argues that contemporary urban systems show a multi-nodal structure.

Generally, Solis [30] distinguished the production of post-industrial urban regions' emerging spatial forms in four ways. The first type is generated from the generation of new employment centers or subcenters. The second is generated by the integration of large or big cities and their surrounding suburban hinterlands. The third is generated by a functional network of a series of cities and towns which are physically separated. The last is generated by the intertwining of a metropolitan urban system and a medium sized city. Urban transformation occurs when sub-urban areas experience rapid population growth so that the number exceeds the city center [31]. In its process grew two types of sub-centers i.e. coherent additional urban areas and edge city [29], which can be considered as parts of a post-suburban form.

2.3. The JMA development dynamics
The Jakarta metropolitan area (JMA) is an administrative definition of the urban area of Jakarta and its surrounding area. The region plays important role in the development of Indonesia, not only socially, but also economically. With an area of 6.392 km^2, the JMA population in 2015 is estimated at 30.97 million , making it the largest metropolitan in Indonesia and one of the largest in the world. In economy, 25 percent of the 2010 national GDP was contributed by JMA [18]. As the capital of Indonesia, and also the center of national governance, politics, business, education, and culture, Jakarta plays the main role as the traditional urban core of JMA. However, Firman and Fahmi [17] has shown that the development of JMA has a tendency of moving towards its peripheral area.

The urban spatial pattern transformations in JMA and other Asian metropolitans are different from those in Europe and America. It is not only a mega-urban functional system [18], but also a city-region dominated by Jakarta as a Mega-city [32]. As one of the largest metropolitan areas in Southeast Asia, the development of industrial land is the key to urbanization in the Greater Jakarta area [33]. In 1985, nearly 60 percent of the oil manufacturing industry was located in the region [34]. But recent evidence has shown that there has been industrial deconcentration process from the JMA core towards suburban areas, not only the large industries [12] but also for the small industries [35]. The urban transformation is driven by economic expansion in the form of new industrial complexes and satellite cities, which produce peri-urban areas with mixed land use [20]. High demand for land has encouraged private developers to do land banking, causing development to become less controlled. Decentralization of population and economic activity continue towards the peri-urban and sub-urban areas, even further into the *desakota* zone, thus forming additional metropolitan areas or mega-region cities [36].

The urban development dynamics in JMA is strongly influenced by different planning policies but are similar in the lack of government role. Various studies conducted in the JMA showed that during the New Order regime (1966-1997) urban development was marked by the absence of planning and regulation [37]. Towards a liberal economy, policies have produced the duality of urban forms: planned and unplanned [38]. By the time of post New Order era, the JMA urban development was marked by a decentralized euphoria that encouraged fragmentation and land use conflicts [17]. As of today, the JMA region governance is a combination of 10 cities and 3 regencies from three provinces. Weaknesses in terms of governance institutions make land development very much controlled by the private sector [33], [39], [40]. Consequently, development in the peri-urban region undergoes spatial segregation and occupation [40], thus encouraging the early phases of the post-suburbia phenomenon [17].

Firman [16] was among the first to show several important elements which indicated the existence of post-suburbia at JMA. First, population and urban localities grew rapidly in the outskirts since the beginning of 2000, while Jakarta itself was experiencing the opposite. Large-scale residential development conducted by the private sector in the periphery also encouraged the occurrence of this phenomenon, resulting in a satellite city that has a strong economic base. Second, the swift FDI flow in the aftermath of the economic crisis, has encouraged the growth of industrial estates to cover a very wide area on the outskirts of Jakarta, such as Bekasi in the early to mid-2000's. Third, the change in commuting patterns, show the movement of Jakarta residents to the surrounding small towns at the beginning of 2002. Although urban transformation in JMA began much earlier, it can be said that it was not until the middle of 2000 that the existence of these elements became noticeable.

3. Data and Method
Various researches aimed at measuring the spatial structure of JMA were often limited by data availability. In terms of economy, the most complete information of non-agricultural economic activity in Indonesia was gathered by the Economic Census (EC). The census was held by BPS (Indonesian Statistical Agency) every 10 years since 1986. From the location perspective, the data resulted could provide precise locations of several important establishment population characteristics. Although the data has been widely utilized for regional economic analysis, but these advantages has surprisingly a lack of use in exploring the economic activity pattern on the urban scale. This research would be among the first to utilize EC data to measure the urban transformation process in Indonesia.

To measure the transformation in urban form, we examined the change on spatial pattern of non-agricultural economic activities (establishment and its employment) density and distribution of 1501 village areas in JMA. Village level is the smallest geographical area in which EC aggregated data were available. The economic activities which referred to the International Standard of Industrial Classification (ISIC) were divided into three categories: manufacturing industry, low order services (LOS), and high order services (HOS). The industrial category consisted of manufacture and recycling activities. While we used the low order services category, which consisted of other non-agricultural activities as comparison, we were also interested on following Coffey et al. [41] to explore high order service distribution, which consisted of financial, insurance, and real estate (FIRE) activities. We compared EC data from two periods of reference: 2006 and 2016. Following the findings of Firman [16], the 2006 condition represented the post-suburbanization condition, while the latter represented the situation beyond. Furthermore, we not only explored the change in economic activity, but also identified its impact on population dynamics. Since EC does not provide information of population characteristics, we made a different approach to the data using the village potential data (PODES). PODES was collected every three years, right before the government conducted the census, to obtain information of the targeted location characteristics at the village level. We utilized PODES data from 2005 and 2014, both conducted as part of prior preparation for EC.

The 2016 EC data has captured more than 2,80 million non-agricultural economic activities in JMA. This number has increased by 7.88 percent from a decade earlier. Nearly 99.95 percent of the numbers were dominated by micro and small enterprises (MSE). But since worldwide post-suburbanization processes were driven mainly by global economic interaction (see [1], [8], [10], [12]), we decided to limit the research only on medium and large enterprises (MLE). Such an approach was used by Hudalah et al [33] to map major suburban industrial centers in the Greater Jakarta area. However, we couldn't utilize the whole 2016 EC data because the data processing completion was still in the finishing stage. Therefore, while the 2006 EC data was used completely, we used only 127,010 units, or nearly 98.31 percent of the total MLE population in 2016. The utilization of MLE data has an advantage, since it has the ability to provide a large number of employment and generate a higher value compared to MSE. Table 1 shows that the existence of 3 to 4 percent of MLE in JMA could provide up to 30 percent employment for its working population. Moreover, the overall existence of MLE is established based on long term strategic decisions, so that the change in its location decision behavior is more easily observed.

Table 1. Some characteristics comparison of JMA.

Characteristics	2006	2016
Number of village	1.493	1.501
Population	24.287.420	32.409.811
Working population	9.045.651	13.823.856
Non-Agriculture economic activity (unit)	2.519.292	2.801.893
Non-Agriculture MLE unit	80.966	129.187
Non-Agriculture MLE employment	2.713.133	4.845.641
Average per capita GDRP (000 Rp.)	35.587	133.621

Following the regulation, BPS distinguished the definition of non-agricultural businesses in EC 2006 into industry-based and non-industrial. For the industrial sector, MLE is a business that has more than 20 laborers, while in the non-industrial sector, business is classified based on annual turnover. MLE is a business that has a turnover above 1 billion. By 2016, the definition difference occurs only in non-industrialized UMB where the turnover limit used increases to 2.5 billion rupiah. This change is done as a form of adjustment to the current economic situation without losing its comparability.

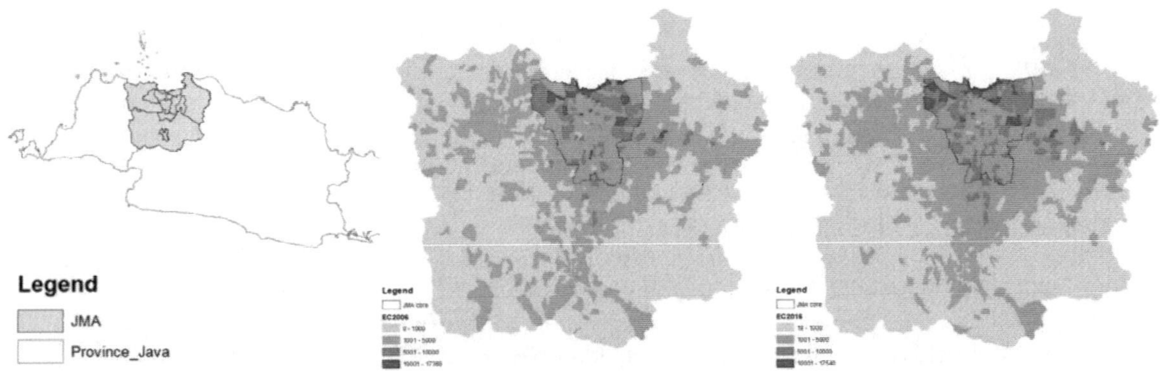

(a) Administrative map of JMA (b) Spatial distribution of non-agricultural economic activities in JMA 2006 (c) Spatial distribution of non-agricultural economic activities in JMA 2016

Figure 1. Area of Study.

There are several approaches to measuring the changes in urban form, both qualitative and quantitatively. We used a quantitative approach to measure spatial patterns because it is proven to be able to provide more objective, less ambiguous results, and contributed a better understanding of these suburban phenomena [19]. Among several quantitative methods which have been used, spatial statistics is important in resolving another study's main gap [42]. It tests whether certain types of areas (e.g. density, land use types, activities) are evenly (or randomly) distributed across the urban area. Reis et al [7] argue that spatial autocorrelation metrics are very useful to measure some important properties of urban decentralization and re-urbanization patterns and various urban sprawl studies. In the context of urban areas, positive spatial autocorrelation means that zones with high employment are clustered together, while spatial heterogeneity means that economic behaviors are not stable over space [43].

Exploratory Spatial Data Analysis (ESDA) is the most important method in this group [42], whereas the Moran's coefficient (I) and local Moran coefficient (Ii) are the two most used autocorrelation measures [7]. This method has been used in several researches of urban spatial pattern such as [42], [44]–[47]. To implement ESDA tools, spatial interdependence between observations need to be modelled using the means of a spatial weight matrix [43]. In this research, we implemented a simple contiguity-based weight matrix using the queen method. It considers an area related to each other if they share the same borders. Building from the work of Anselin [48], Moran's I is defined as:

$$I = \frac{\left(\frac{n}{S_0}\right)\sum_{i=1}^{n}\sum_{j=1}^{n} w_{ij}(x_i - \mu)(x_j - \mu)}{\sum_{i=1}^{n}(x_i - \mu)^2} \tag{1}$$

Where n is the number of observation, S_0 is the sum of spatial weight matrix w_{ij}, which is equivalent to a binary matrix with ones in position *i,j* whenever observation *i* is a neighbor of observation j, and zero otherwise.

$$S_0 = \sum_i \sum_j w_{ij} \tag{2}$$

where x_i is the observation unit i, and μ is the mean of all observation across spatial unit in which summation over *j* so that only neighbor values of *j* are included. The local version of Moran's I statistic for each observation *i* can be written as:

$$I_i = \frac{(x_i - \mu)}{m}\sum_j w_{ij}(x_j - \mu) \text{ with } m = \sum_i \frac{(x_i - \mu)^2}{n} \tag{3}$$

then,

$$I = \sum_i I_i / n \tag{4}$$

The local Moran's I value, which has a positive spatial correlation relative to mean (high-high/HH or low-low/LL) suggests the existence of spatial cluster, whereas location, which has a negative spatial correlation relative to mean (high-low/HL or low-high/LH) suggests the existence of spatial outlier [49]. We calculate the spatial association index by utilizing GeoDa [50].

4. Urban Spatial Transformation of JMA

4.1. The amplification of polycentric urban form

The growth in the number of MLE in JMA began to show a very rapid increase after the emergence of the New Order regime. The national development policies implemented through economic and trade liberalization as well as foreign direct investment have been established for MLE growth in JMA (Figure 2.a.). In 1976, the number of new MLE were less than 3,500 units, of which 74 percent of shares were concentrated in Jakarta as the core. Until the 80's, the growth in the number of MLE was still concentrated in the core area, but since the period of 1990 there were also significant development in the nearest fringe area.

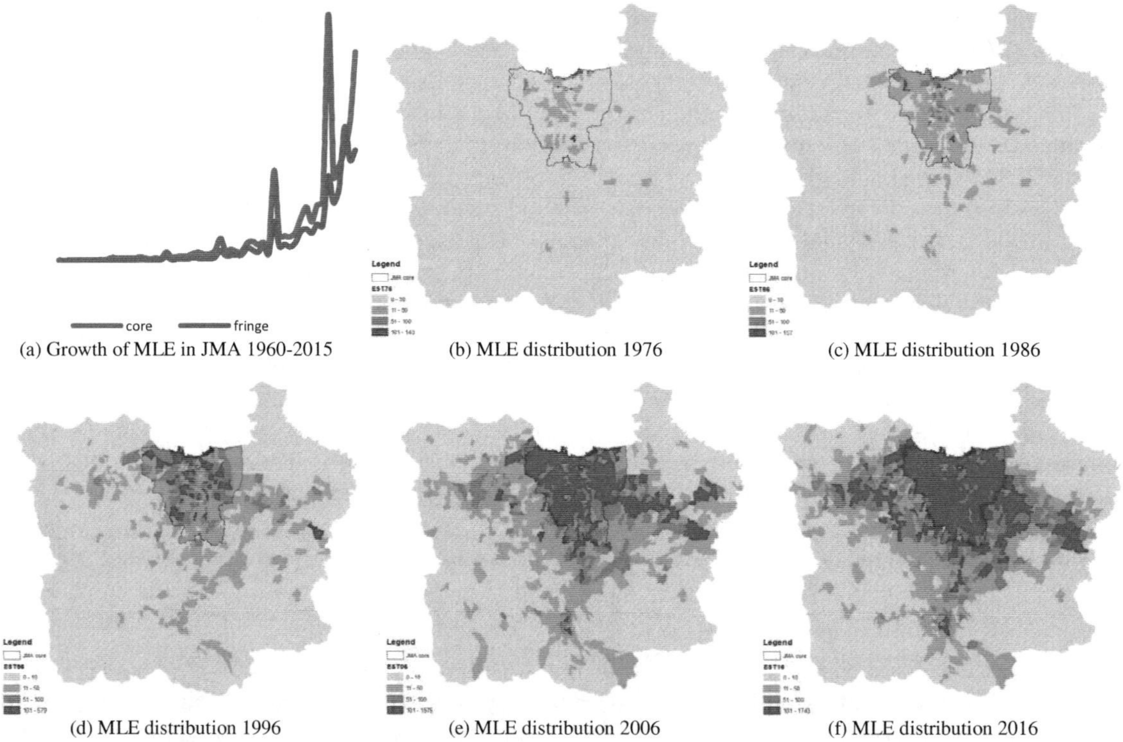

(a) Growth of MLE in JMA 1960-2015 (b) MLE distribution 1976 (c) MLE distribution 1986

(d) MLE distribution 1996 (e) MLE distribution 2006 (f) MLE distribution 2016

Figure 2. The spatial pattern of MLE development in JMA 1976-2016.

The development of the toll road to Bogor in the south, the Merak direction in the east, and the Cikampek direction in the west has helped the growth pattern of MLE expand to the nearest fringe along the main transportation route. After the year 2000, MLE growth began to occur massively in the suburbs, whereas some even developed in the edge area. Previously, agricultural areas on the JMA, edge such as Cikarang in the east and Cikupa in the west, have turned into new MLE concentrations. Nearly 40 decades later the number of MLE has reached more than 129 thousand units, or had an average increase rate of 93 percent per year. The MLE expansion occurring in JMA has a pattern similar to the pattern of land use change proposed by Rustiadi et al. [18].

In terms of polycentric urban structure existence, JMA has entered the post-suburban since 2000. Even though every administrative area within JMA has its own multifunctional economic center, but in terms of MLE, the scale is not sufficient enough to be identified as sub-center. In this research, sub-center is defined as the center of MLE employment. During that time, the urban structure was characterized by the emergence of urban polycentric structures resulting from the growth of new employment centers. The urban structure in JMA transformed into a stronger polycentric form, indicated by the increasing value of Moran's index from 0.31 to 0.49 in 2016. This finding would strengthen the earlier emerging polycentric metropolitan structure thesis of Hudalah et al [33].

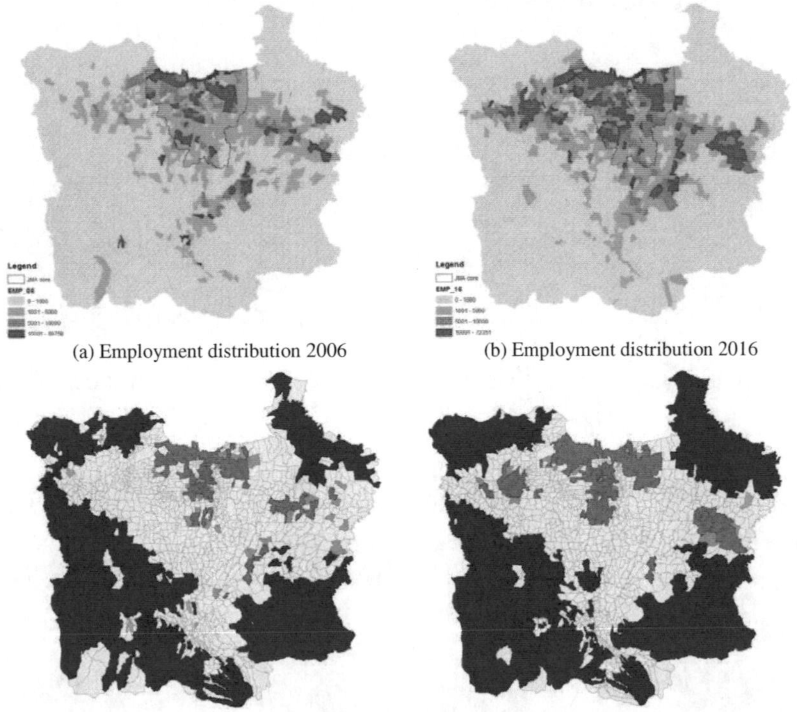

(a) Employment distribution 2006 (b) Employment distribution 2016

(c) I(i) map of employment distribution 2006 (d) I(i) map of employment distribution 2016

Figure 3. Comparative spatial pattern of employment distribution in JMA.

Although Jakarta as the core remains the largest employment concentration, but several other sub-centers have also grown in the suburbs. In 2006, there were at least 12 sub-centers of employment of various sizes. The majority were located in the eastern and southeastern parts of Jakarta. A decade later, there were processes of re-concentration, resulting to six significant employment sub-center. Several small sub-centers began to lose its significance, whereas in some sub-centers such as Cikarang the employment concentration became even larger. Another interesting point is that the emergence of the Cikupa sub-center as one of the new employment centers showed that the expansion of development during this period was shifting westward.

4.2. Towards a long term industrial post-suburban of JMA

While it has been known that there is a change of urban structure during the 2006-2016 period, the next question emerged, of whether this situation will encourage the growth of new cities, or instead perpetuate the existence of post-suburbia in JMA?. Hudalah and Firman [12] argued that the development of post-suburbanization in JMA initiated by the industrial sector will be followed by the service sector. To answer this question, it is important to know whether the sub-centers identified in 2016 are multifunctional areas or have limited functionality. In other words, will the industry agglomeration in the suburbs be followed by the service sectors? Figure 4 shows that in 2006 there were polycentric structures of every economic sector: manufacturing, LOS, and HOS. The industrial centers are mainly located in the north and western parts of Jakarta, with an equal magnitude in Gunung Putri and Citeureup. There were several other cluster of manufacturers in Cikarang and Cikupa, but both covered only limited areas. The LOS clusters are spread uniformly in Jakarta, whereas some other clusters are located in Cikarang, Tambun, and eastern parts of Bekasi City. Meanwhile, HOS is located mainly in the golden triangle of Jakarta and its adjacent areas, while a limited area of cluster is also found near Tambun in Bekasi. The coexistence of manufacture and LOS cluster in the neighboring area of eastern parts of Bekasi and Cikarang emphasized the presence of functionally-specific suburban areas. At that moment, such areas have become the only existing post-suburbia in JMA.

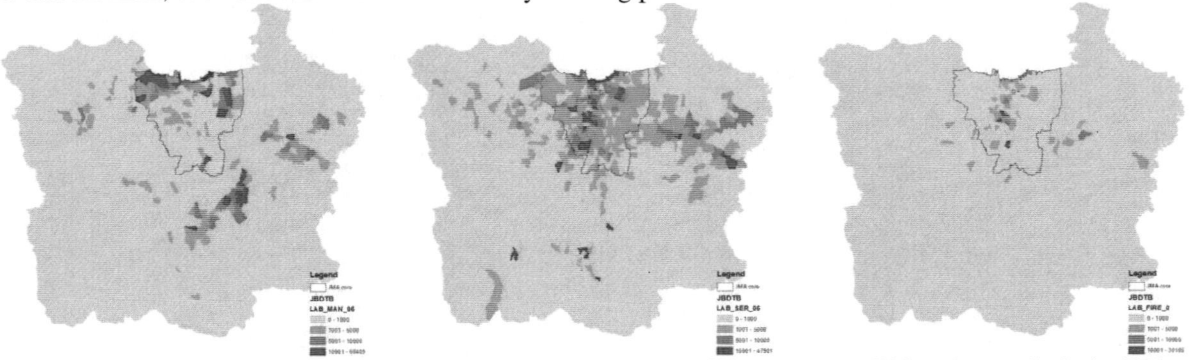

(a) Manufacturing employment distribution 2006

(b) LOS employment distribution 2006

(c) HOS employment distribution 2006

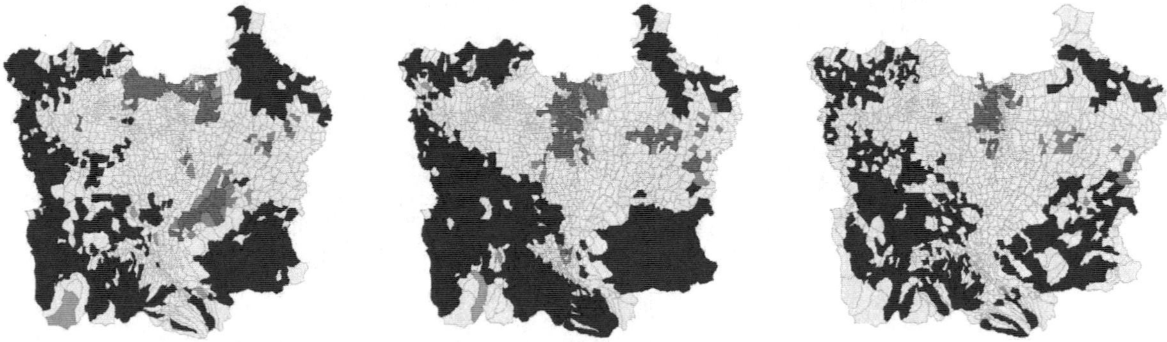

(d) I(i) map of manufacturing employment 2006

(e) I(i) map of LOS employment 2006

(f) I(i) map of HOS employment 2006

Figure 4. The spatial pattern of MLE sectoral employment in JMA 2006.

In 2016 there was a significant change of urban structure in the whole economic sector. The manufacturing sector, originally concentrated in four locations of JMA, have now grown into seven sub-centers. The decline of the Sentul cluster was followed by strengthening agglomeration in Cikarang and the rise of the Cikupa sub-center. While the economic reurbanization into the JMA core was experienced by LOS clusters, the HOS cluster haven't changed much. The LOS employment has a strengthening agglomeration in Jakarta as a JMA core, decreasing its significance in the suburbs and leaving only small contributions. The assumption of how Jakarta has reach its saturated point for further land development holds [12], but it is only true for manufacturing growth rather than for the whole economic activities. Such a situation illustrates that the transformation in the suburbs, which are encouraged by economic activities, is heterogeneous in its process.

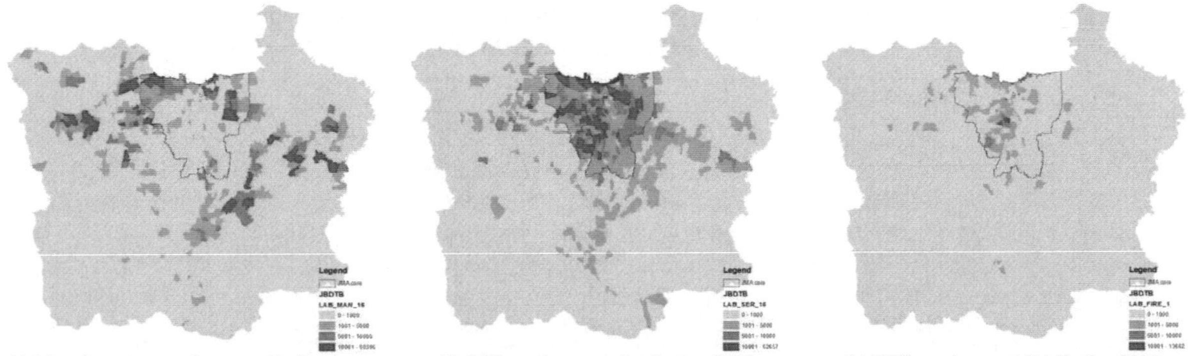

(a) Manufacturing employment distribution 2016 (b) LOS employment distribution 2016 (c) HOS employment distribution 2016

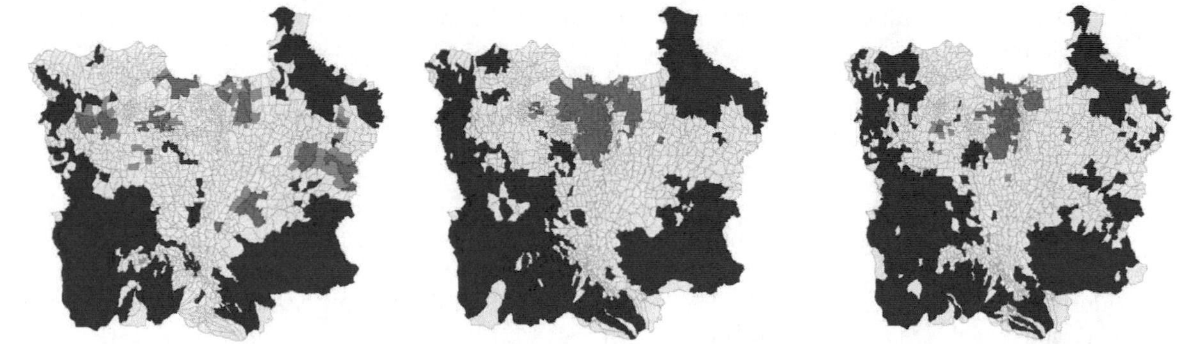

(d) I(i) map of manufacturing employment 2016 (e) I(i) map of trade employment 2016 (f) I(i) map of service employment 2016

Figure 5. The spatial pattern of the MLE sectoral employment in JMA 2016

Table 2 shows the agglomeration of the whole sectors that are getting stronger, indicated by the increase of 2016 Moran's I index compared to 2006. Figure 5, however, shows that agglomeration does not lead in the same direction. It was assumed that that the growth of LOS and HOS would occur in the suburbs where manufacturing sites did exist in the past, but that has changed in a decade. What happened was the reurbanization of LOS in the core, reflecting the need for a more suitable environment with other globally organized firms, related institution, and associated amenities. This increasingly substantiates Jakarta’s existence as a global city. In this case, the post-suburban form has transformed from a multifunctional suburb into a monofunctional one. The remaining significant subcenters which are left on the outskirts of JMA currently hold the main function as a manufacturing location only. Musil [14] argued that such phenomenon illustrates what Kunzmann called the Urban Archipelago, in which suburban areas hold the role as the functional backstage of a metropolis, which provides the whole support for the urban core. It would take a long time for the JMA post-suburban before its next spatial transformation.

Table 2. Comparative spatial autocorrelation statistics of the MLE sectoral employment distribution in JMA 2006 and 2016.

Index Value		2006			2016		
		Manufacture	LOS	HOS	Manufacture	LOS	HOS
Moran's I		0,19	0,31	0,21	0,26	0,55	0,44
I (i)	Not significant	1.009	877	1.105	1.044	841	965
	High-High (H-H)	67	118	60	70	155	91
	Low-Low (L-L)	381	465	305	341	491	425
	Low-High (L-H)	38	36	29	44	14	17
	High-Low (H-L)	6	5	2	2	0	3

4.3. Towards a polarized population

It is argued that the transformation which occurs through the decentralization of economic activities will have an important impact on population dynamics. The post-suburbanization process takes place when employment is decentralized to the suburbs and creates a certain functional landscape. This encourages the population to adapt through one of two ways: adjusting the residential location closer to their working location or changing their job to reduce the duration of work commute. Such a situation is reflected in the shift of the dominant population job in each village of JMA. During 2006-2016, the population in JMA grew rapidly at of 3.3 percent annually. This growth caused pressure on population density, which rose from 6,552 people/km^2 to 7,999 people/km^2. Figure 6 (a) and (b) shows that the population density hotspot is increasing eastward with rapid growth in area that are adjacent to the industrial cluster. This explains why the JMA population exhibit a similar spatial pattern as identified by Rustiadi et al [18] and Asri and Hidayat [51].

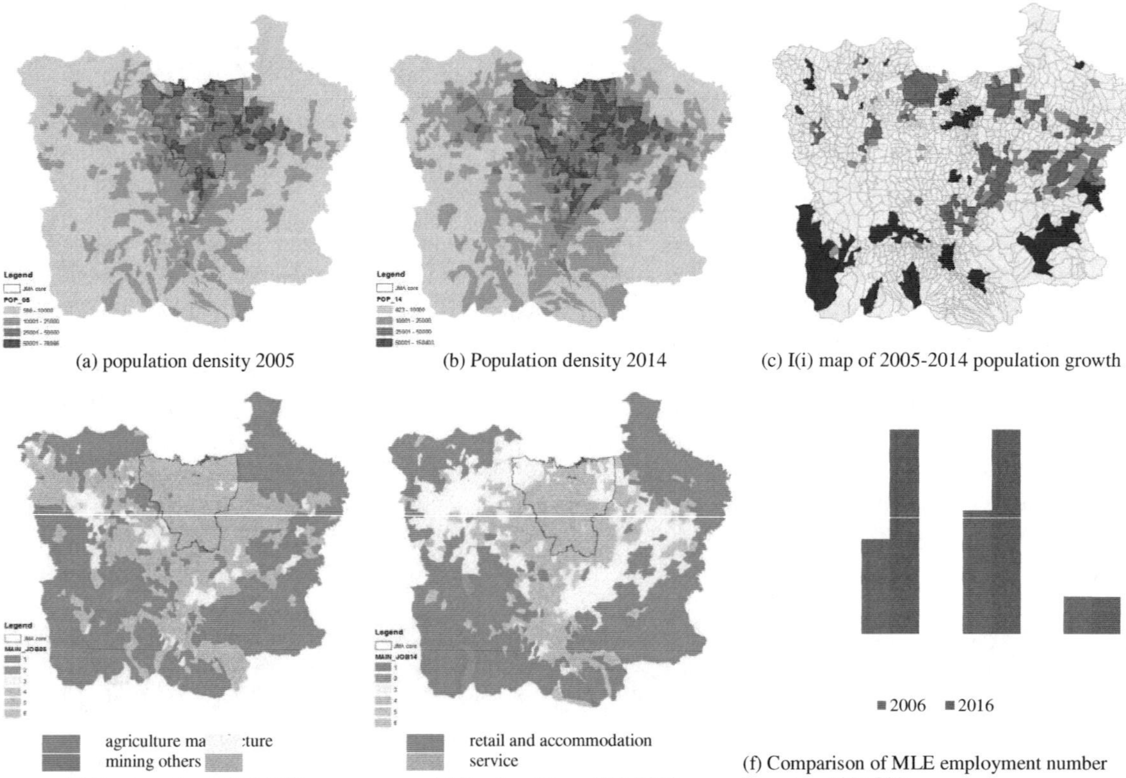

(a) population density 2005 (b) Population density 2014 (c) I(i) map of 2005-2014 population growth

(d) Dominant type of job 2006 (e) Dominant type of job 2016 (f) Comparison of MLE employment number by sectors 2006-2016

Figure 6. The spatial pattern of MLE population characteristics in JMA 2016.

The dominant occupation, which originally revolved around retail, accommodation, and service, has changed into manufacture in a decade. Populations working in the manufacturing sector are becoming increasingly dominant to occupy locations adjacent to the new existing manufacturing sub-centers, while those working in the service sector tend to get closer to the core. It is not surprising, since manufacturing employment has the most rapid growth compared to trade, whereas service sector employment remains constant. This implies that the available employment can only be absorbed by low skill workers. Urban transformation beyond post-suburbanization has created suburban population dynamics in JMA which tends to polarize based on its sub-center specialization. With regards to its commuting pattern, this phenomenon should have influenced a more efficient urban structure but has the potential to increase social segregation.

5. Conclusion

The post-suburbanization process has occurred in urban transformations worldwide. Nevertheless, there are varied understandings regarding its process across different locations. This suggests that geographic and location settings hold an important factor in the post-suburbanization process. While there has been much knowledge about post-suburbanization process, the urban transformation beyond post-suburbia is still lack of attention. Using JMA as a case study, we addressed three questions regarding urban transformation. The first is to reveal the actual recent urban form of JMA and its alteration. The second is to identify whether post-suburbia in JMA would rather develop into a new city or maintain its current form. The last question concerns population dynamics in regards to urban transformation. We make a contribution in the discourse of urban spatial transformation by providing empirical evidence of the spatial pattern beyond post-suburbanization, the extent of which could explain the next stage of post-suburbanization in the context of a developing country such as Indonesia.

Beyond the post-suburbanization era, reserach shows that JMA already consists of a polycentric urban structure since 2006. The manufacturing MLE decentralization is still ongoing outward Jakarta as the urban core. Further transformations suggest that those structures are getting stronger. There is an increase in the number of manufacturing jobs in industrial clusters in the edges of the metropolitan, resulting on a new significant manufacturing cluster in Cikarang, Bekasi and Cikupa, Tangerang. On the other hand, the spatial transformation of LOS shows opposite behavior. Though there was an increase in the number of MLE outward Jakarta, but the overall jobs provided is highly concentrated in the urban core. Such conditions make the cluster of suburban LOS formed in 2006 become insignificant a decade later. This recentralization symptom indicate that both sectors are still bound to the urban core, as it provides a more suitable environment with other globally organized firms, related institutions, and associated amenities. Jakarta is becoming a more global city and its hinterland is likely to form a steady state post-suburbia, which specializes in manufacturing sectors. Contrary to Hudalah and Firman [12], the urban transformation in JMA is characterized as a second modernity urban region, whereas a multifunctional city has decentralize into a mono functional post-suburban. Such phenomenon depicts an urban archipelago notion.

The recentralization of both sectors was also followed with workers adjusting their settlement towards the nearest suburban area from Jakarta. The population dynamics of JMA is dominated by the rapid growth of industrial workers in the fringe area, whereas commercial and service workers agglomerate in the core. Such a pattern creates a polarized population based on its sub-center specialization, which should contribute to a more efficient commuting pattern and better environment quality. But Hellbich and Leitner [9] argue that post-suburbanization spatial impact would affect not only the mix of population but also the change in commuting pattern, and the latest evidence shows that JMA is facing chronic traffic congestion that could worsen in the future [51]. These conflicting facts arise the question about the impact of urban transformation beyond post-suburbanization on job-housing balance. Several evidence suggests that there is a possibility of employment decentralization which doesn't took place in the suburban area where workers reside [52]. Thus, the spatial mismatch would increase commuting duration and create environment problems. Therefore, future research should be directed to study how beyond post-suburbanization transformation affects suburban commuting patterns.

Acknowledgment
This paper is a part of dissertation research by Erie Sadewo in The School of Architecture, Planning, and Policy Development (SAPPD), Institut Teknologi Bandung, Indonesia, and has been presented at the 4th Planocosmo, April 2018 in Bandung. The author would like to thank Roy Suerlianto for the data contribution, and Dr. Delik Hudalah for the valuable comments.

References
[1] F. Wu and N. A. Phelps, "From suburbia to post-suburbia in China? Aspects of the transformation of the Beijing and Shanghai global city regions," *Built Environment*, vol. 34, no. 4, pp. 464–481, 2008.
[2] N. A. Phelps and F. Wu, "Conclusion: Post-suburban Worlds?," in *Int. Perspectives on Suburbanization: A Post-Suburban World?*, N. A. Phelps and F. Wu, Eds. London: Palgrave Macmillan UK, 2011, pp. 245–257.
[3] E. W. Soja, *Postmetropolis: Critical Studies of Cities and Regions*. Wiley, 2000.
[4] N. A. Phelps and F. Wu, "Introduction: International Perspectives on Suburbanization: A Post-suburban World?," in *Int. Perspectives on Suburbanization: A Post-Suburban World?*, N. A. Phelps and F. Wu, Eds. London: Palgrave Macmillan UK, 2011, pp. 1–14.
[5] R. Shearmur, P. Apparicio, P. Lizion, and M. Polese, "Space, Time, and Local Employment Growth: An Application of Spatial Regression Analysis," *Growth and Change*, vol. 38, no. 4, pp. 696–722, Dec. 2007.
[6] D. C. Thorns, *The Transformation of Cities: Urban Theory and Urban Life*. New York: Palgrave Macmillan, 2002.
[7] J. P. Reis, E. A. Silva, and P. Pinho, "Spatial metrics to study urban patterns in growing and shrinking cities," *Urban Geography*, vol. 37, no. 2, pp. 246–271, Feb. 2016.
[8] R. Kling, S. C. Olin, and M. Poster, "The Emergence of Postsuburbia: An Introduction," in *Postsuburban California: The Transformation of Orange County since World War II*, Paperback edition., R. Kling, S. C. Olin, and M. Poster, Eds. University of California Press, 1995.
[9] M. Helbich and M. Leitner, "Spatial Analysis of the Urban-to-Rural Migration Determinants in the Viennese Metropolitan Area. A Transition from Suburbia to Postsuburbia?," *Applied Spatial Analysis and Policy*, vol. 2, no. 3, pp. 237–260, Oct. 2009.
[10] I. Keseru, "Post-suburban transformation in the functional urban region of Budapest in the context of changing commuting patterns," PhD Thesis, University of Szeged, Szeged, 2013.
[11] M. Bontje, "From suburbia to post-suburbia in the Netherlands: Potentials and threats for sustainable regional development," *J. of Housing and the Built Environment*, vol. 19, no. 1, pp. 25–47, 2004.
[12] D. Hudalah and T. Firman, "Beyond property: Industrial estates and post-suburban transformation in Jakarta Metropolitan Region," *Cities*, vol. 29, no. 1, pp. 40–48, Feb. 2012.
[13] A. Gospodini, "Post-industrial Trajectories of Mediterranean European Cities: The Case of Post-Olympics Athens," *Urban Studies*, vol. 46, no. 5–6, pp. 1157–1186, May 2009.
[14] R. Musil, "Globalized post-suburbia: Service firms and global enterprises in Vienna's suburban zones: a contribution to the global integration of the core?," *Belgeo*, no. 1, pp. 147–162, Jan. 2007.
[15] M. Helbich, "Beyond Postsuburbia? Multifunctional Service Agglomeration In Vienna's Urban Fringe," *Tijdschrift voor economische en sociale geografie*, vol. 103, no. 1, pp. 39–52, Feb. 2012.
[16] T. Firman, "Post-suburban Elements in an Asian Extended Metropolitan Region: The Case of Jabodetabek (Jakarta Metropolitan Area)," in *Int. Perspectives on Suburbanization: A Post-*

Suburban World?, N. A. Phelps and F. Wu, Eds. London: Palgrave Macmillan UK, 2011, pp. 195–209.

[17] T. Firman and F. Z. Fahmi, "The Privatization of Metropolitan Jakarta's (Jabodetabek) Urban Fringes: The Early Stages of 'Post-Suburbanization' in Indonesia," *J. of the American Planning Association*, vol. 83, no. 1, pp. 68–79, Jan. 2017.

[18] E. Rustiadi, D. O. Pribadi, A. E. Pravitasari, G. S. Indraprahasta, and L. S. Iman, "Jabodetabek Megacity: From City Development Toward Urban Complex Management System," in *Urban Development Challenges, Risks and Resilience in Asian Mega Cities*, R. B. Singh, Ed. Tokyo: Springer Japan, 2015, pp. 421–445.

[19] E. Rustiadi, K. Mizuno, and S. Kobayashi, "Measuring Spatial Patterns of the Suburbanization Process," *Journal Of Rural Planning Association*, vol. 18, no. 1, pp. 31–42, Jan. 1999.

[20] E. Rustiadi and D. R. Panuju, "Spatial Pattern of Suburbanization and Land Use Change Process: Case Study in Jakarta Suburb," in *Land-use Changes in Comparative Perspective*, Y. Himiyama, M. Hwang, and T. Ichinose, Eds. Enfield (NH): Science Publishers, Inc., 2002.

[21] A. Borsdorf, "On the way to post-suburbia? Changing structures in the outskirts of European cities," in *European Cities Structures: Insights on Outskirts*, A. Borsdorf and P. Zembri, Eds. 2004, pp. 7–30.

[22] M. Dear and N. Dahmann, "Urban Politics and the Los Angeles School of Urbanism," *Urban Affairs Review*, vol. 44, no. 2, pp. 266–279, Nov. 2008.

[23] B. Lee, "'Edge' or 'edgeless' cities? Urban spatial structure in US metropolitan areas, 1980 to 2000," *J. of Regional Science*, vol. 47, no. 3, pp. 479–515, 2007.

[24] J. Burdack, "New economic poles in the periphery of European metropolitan areas," in *ERSA conf. papers*, Dortmund, 2002.

[25] N. A. Phelps, A. M. Wood, and D. C. Valler, "A Postsuburban World? An Outline of a Research Agenda," *Environment and Planning A*, vol. 42, no. 2, pp. 366–383, Feb. 2010.

[26] D. P. McMillen, "Testing for Monocentricity," in *A Companion to Urban Economics*, Blackwell Publishing Ltd, 2006, pp. 128–140.

[27] R. W. Helsley and W. C. Strange, "Matching and agglomeration economies in a system of cities," *Regional Science and urban economics*, vol. 20, no. 2, pp. 189–212, 1990.

[28] D. Arribas-Bel and F. Sanz-Gracia, "The validity of the monocentric city model in a polycentric age: US metropolitan areas in 1990, 2000 and 2010," *Urban Geography*, vol. 35, no. 7, pp. 980–997, Oct. 2014.

[29] S. Davoudi, "European Briefing: Polycentricity in European spatial planning: from an analytical tool to a normative agenda," *European Planning Studies*, vol. 11, no. 8, pp. 979–999, Dec. 2003.

[30] E. Solís Trapero, I. M. Sanz, and J. M. de U. Francés, "Global Metropolitan-Regional Scale in Evolution: Metropolitan Intermediary Cities and Metropolitan Cities," *European Planning Studies*, vol. 23, no. 3, pp. 568–596, Mar. 2015.

[31] A. Romein, "The contribution of leisure and entertainment to the evolving polycentric urban network on regional scale: towards a new research agenda," presented at the the 45 th Congress of the European regional Science Association, Free University Amsterdam, 2005.

[32] A. A. Laquian, "The planning and governance of Asia's mega-urban regions," *Population Distribution, Urbanization, Internal Migration and Development: An International Perspective, United Nations Department of Economic and Social Affairs, Population Division, New York: United Nations*, pp. 302–322, 2011.

[33] D. Hudalah, D. Viantari, T. Firman, and J. Woltjer, "Industrial Land Development and Manufacturing Deconcentration in Greater Jakarta," *Urban Geography*, vol. 34, no. 7, pp. 950–971, Nov. 2013.

[34] T. G. McGee, "The Spatiality of Urbanization: The Policy Challenges of Mega-Urban and Desakota Regions of Southeast Asia," UNU-IAS, Working Paper 161, 2009.

[35] Z. Ansar and D. Hudalah, "Dekonsentrasi Industri Kecil Di Metropolitan Jabodetabek," *J. Perencanaan Wilayah dan Kota*, vol. 2, no. 3, pp. 687–700, 2013.

[36] I. M. Robinson, "Emerging spatial patterns in ASEAN mega-urban regions: Alternative strategies," in *Mega Urban Regions of Southeast Asia*, Vancouver: UBC Press, 1995, pp. 78–108.

[37] D. Rukmana, "The Change and Transformation of Indonesian Spatial Planning after Suharto's New Order Regime: The Case of the Jakarta Metropolitan Area," *Int. Planning Studies*, vol. 20, no. 4, pp. 350–370, Oct. 2015.

[38] H. Winarso, "Urban dualism in the Jakarta metropolitan area," in *Megacities*, Springer, 2011, pp. 163–191.

[39] T. Firman, "Major issues in Indonesia's urban land development," *Land Use Policy*, vol. 21, no. 4, pp. 347–355, Oct. 2004.

[40] H. Winarso, D. Hudalah, and T. Firman, "Peri-urban transformation in the Jakarta metropolitan area," *Habitat International*, vol. 49, pp. 221–229, Oct. 2015.

[41] W. J. Coffey, R. Drolet, and M. Polèse, "The intrametropolitan location of high order services: Patterns, factors and mobility in Montreal," *Papers in Regional Science*, vol. 75, no. 3, pp. 293–323, 1996.

[42] M. Alidadi and H. Dadashpoor, "Beyond monocentricity: examining the spatial distribution of employment in Tehran metropolitan region, Iran," *Int. J. of Urban Sciences*, pp. 1–21, Jun. 2017.

[43] R. Guillain, J. Le Gallo, and C. Boiteux-Orain, "Changes in spatial and sectoral patterns of employment in Ile-de-France, 1978–97," *Urban Studies*, vol. 43, no. 11, pp. 2075–2098, Oct. 2006.

[44] C. Baumont, C. Ertur, and J. Gallo, "Spatial analysis of employment and population density: the case of the agglomeration of Dijon 1999," *Geographical analysis*, vol. 36, no. 2, pp. 146–176, 2004.

[45] I. Hakim, "The spatial structure of employment and its impacts on the journey to work in the Jakarta Metropolitan Area: a Southeast Asian Extended Metropolitan Region (EMR) perspective," The University of New South Wales, 2009.

[46] J. I. Kim, C. H. Yeo, and J.-H. Kwon, "Spatial change in urban employment distribution in Seoul metropolitan city: clustering, dispersion and general dispersion," *Int. J. of Urban Sciences*, vol. 18, no. 3, pp. 355–372, Sep. 2014.

[47] L. Salvati, G. Venanzoni, P. Serra, and M. Carlucci, "Scattered or polycentric? Untangling urban growth in three southern European metropolitan regions through exploratory spatial data analysis," *The Annals of Regional Science*, vol. 57, no. 1, pp. 1–29, Jul. 2016.

[48] L. Anselin, "Local indicators of spatial association-LISA," *Geographical analysis*, vol. 27, no. 2, pp. 93–115, 1995.

[49] L. Anselin, I. Syabri, and O. Smirnov, "Visualizing multivariate spatial correlation with dynamically linked windows," *Urbana*, vol. 51, p. 61801, 2002.

[50] L. Anselin, I. Syabri, and Y. Kho, "GeoDa: an introduction to spatial data analysis," *Geographical analysis*, vol. 38, no. 1, pp. 5–22, 2006.

[51] D. U. Asri and B. Hidayat, "Current transportation issues in Jakarta and its impacts on environment," in *Proc. of the Eastern Asia Society for Transportation Studies*, 2005, vol. 5, pp. 1792–1798.

[52] A. Aguiléra, "Growth in commuting distances in French polycentric metropolitan areas: Paris, Lyon and Marseille," *Urban studies*, vol. 42, no. 9, pp. 1537–1547, 2005.

IOP Conf. Series: Earth and Environmental Science **158** (2018) 012041 doi:10.1088/1755-1315/158/1/012041

Infrastructure Provision by Residential Developers in The Peri-Urban Neighbourhoods of Metropolitan Bandung Raya

A Vitriana

Regional Agency for Research and Development , West Java Provincial Government, Bandung, Indonesia

Corresponding author: avitriana@yahoo.com

Abstract. Peri-urban areas are often regarded as a potential land for urban housing needs. Private sectors' involvement in peri-urban residential development is often viewed by the government as a solution for housing supply provision. However, their complicity is suspected to generate an unstructured and fragmented infrastructure network. This study aims to find out how housing infrastructure provision in micro-spatial areas could form an unstructured network of macro-spatial ones. The selection of case studies is based on three typologies of residential developer. Examination aspects include the availability and accessibility of residential infrastructure. The research was conducted in two villages, each located in the southern and northern part of peri-urban areas of Metropolitan Bandung Raya. The data were collected through observation and interviews with old and new inhabitants. The analytical method uses a qualitative approach. Infrastructure availability is examined by the national standard, and the accessibility is examined by the absence of obstacles. The results showed that only a few aspects of housing infrastructure components fulfilled the standard, and the accessibility of most housing infrastructures for outer residents was restricted through the gated concept. From this research, we found that even though the developer has provided the residential infrastructure properly, without an integrated planning network it would not bring any spatial improvement.

1. Introduction

Peri-urban residential development might be driven by several factors, such as population growth [1] and migration [2], better services[3,4], environmental factors[5,6], economic price [7,8], geographical factors and spatial planning [9], as well as housing policies[10,11]. Yet, many developing countries experience difficulty in handling the problems that emerge from the uncontrolled mushrooming of residential development [12] in peri-urban areas, especially in the formation of large areas such as the metropolitan region. The transformation of peri-urban land into residential areas has brought some critical issues that are commonly discussed, including high congestion[13,14,15,16], lack of infrastructure[17,18], diminished production land [19], environmental degradation[20,21], fragmentation, and segregation [22], as well as social changes[1,8,23, 24]. In metropolitan regions, the peri-urban problems are combined with the charge of several local governments due to the large coverage of development area.

With the concept of decentralization, local governments have the authority to regulate their own territory. However, as Webster's research on some countries in East Asia [13] noted, local governments often do not have sufficient capacity to deal with large-scale developments. Not to mention that local governments also sometimes pay less attention to peri-urban areas because their location is not as salient as the city center. Peri-urban areas, which originally are virgin land, oftentimes are not equipped with a strict regional concept yet, nor mature infrastructure planning. As a result, developers can develop infrastructure without restraint, sometimes unconnected to the surrounding environment, or even make connections at will. Developers also may not consider the adequacy of existing infrastructure since they are financially able to create new infrastructure subsequently. As a result, the new peri-urban infrastructure independently built by private developers is often not integrated into the existing infrastructure network, which is either already provided by the government or has been built by the former developer. As we can see, the spatial manifestation of these practices in the macro scale has created structural fragmentation and infrastructure deficiency [25].

The impact of private sector domination in the residential development of peri-urban areas is mostly not clearly realized by local governments. They are often more focused on the benefits derived from the investment of the private sector in peri-urban housing development, because their involvement can bring in local revenue sources and also provide solutions for housing supply and infrastructure development in the peri-urban area [26]. The weakness of peri-urban planning and the low capacity of local government have indirectly supported the private sector to produce peri-urban housing in a more flexible arrangement.

The failing of infrastructure planning in peri-urban housing can be seen from the inequality of residential infrastructure provision. In Indonesia, this problem may be caused by the lack of technical guidance or the inconsistent application at the local level. Within the national level, there are at least 6 central government rules related to the provision of housing infrastructure, including Law 1/2011 on Housing and Settlement, Government Regulation 14/2016 on the Implementation of Housing and Settlement, Minister of Home Affairs Regulation 9/2009 concerning Guidelines on The Procedures of Infrastructure Handover in Local Regions, Minister of Public Housing Regulation 22/2008 on Minimum Service Standard of Housing Sector in the Region, Minister of Public Housing Regulation 34/2006 on General Guidelines for Implementation of Integration of Infrastructure for Housing Neighborhood, and Indonesian National Standard (SNI) 03-1733-2004 on Procedures for Urban Housing Neighborhood Planning. The entire regulation explains the comprehensiveness of infrastructure provision in the neighborhoods which include public facilities, basic infrastructure, and utilities. Referring to these regulations, SNI 03-1733-2004 shows a more detailed technical guidance on standard housing infrastructure provision [27]. However, this housing standard guidance is rarely applied as the main reference for housing infrastructure development at the local level.

Previous studies about residential infrastructure development in peri-urban areas are generally discussed in terms of macro-spatial contexts[3,9,15,17,28,29,30,31,32,33]. Research that specifically studied the condition of the existing housing infrastructure in the micro-spatial context such as neighborhood is rarely found [34,35]. Handayani and Widyastuti identify the availability of residential facilities in the urban fringe area of Pudakpayung in Semarang city, but the research did not focus on housing infrastructure built by the developer. Meanwhile, Putri tried to identify the characteristics of basic infrastructure provision by developers in the residential neighborhood of Parongpong District in West Bandung Regency, while she only focused on the aspect of physical availability and did not discuss any aspects of its utilization for the surrounding community. Therefore, this research specifically aimed to examine the characteristics of infrastructure in the more detailed aspects of availability (by quantity indicators) and usability (by accessibility indicators) for the surrounding community in the neighborhood. This research is important as a source of information on how the development pattern of housing infrastructure in the micro-spatial generated the structural fragmentation on the macro level. The unstructured infrastructure network which is formed in the macro spatial level would be the manifestation of a bunch of irregular residential infrastructure development at the micro-spatial level.

To discover the characteristics of housing infrastructure practice in peri-urban neighborhoods, this study specifically focused on micro-empirical study. This study was obtained through the identification of housing infrastructure characteristics provided by the private sector in two case study locations in the inner peri-urban area [36] of Metropolitan Bandung Raya (MBR), with each representing a different administrative area in peri-urban MBR, including Bandung Regency and West Bandung Regency. We determined the three housing estates in each location to represent the three typologies of housing developers.

2. Infrastructure Provision in the Residential Peri-Urban Area Neighborhood in MBR: Characteristic Identification and Accessibility

This research takes place on the neighborhoods located in the fringe areas of Bandung Municipality. Bandung city is the urban center of MBR, which experienced population growth and urban expansion pass through the surrounding areas. In recent years, some regencies adjacent to Bandung city have undergone rapid growth in residential neighborhoods. MBR is known as the third largest metropolitan area in Indonesia [20]. Based on Local Regulation 12/2014, MBR covers the areas of Bandung city and Cimahi city, almost half of the areas of Bandung Regency and West Bandung Regency, and some area of Sumedang Regency. According to the Technical Report on the Preparation of Masterplan for MBR's Public Housing Provision, Bandung Municipality's developed area has reached 90.27%, while Cimahi City, Bandung Regency, West Bandung Regency, and Sumedang Regency respectively covered 86.37%, 16.39%, 15.21%, and 3.04% in 2014 [37]. These numbers indicate that the regencies located on the fringe area of Bandung city are potentially developed into a peri-urban residential area of MBR. The development of MBR indicates a growing population that drives the increase in residential needs.

This paper uses the housing typology categorization according to the study by Winarso when he tried categorize the typology of developers in Jabotabek Area (2000) [38]. Based on his study, the two categories of residential housing developers are network-extenders and foot-holders. Network extenders are those who involve themselves in the land development industry and have matured into investors and viewed the land as a long-term managed financial asset. While the foot-holder developer category referred to the type of developers that get involved after seeing the success of other developers. In his research, Winarso explained the developer typology with the aspect of investment characteristics, development characteristics, and company characteristics. In accordance to the housing development phenomenon that occurred in most of the peri-urban areas of Greater Bandung, the study refers to the terminology of foot-holder developers. Thus, the developers' typology in this research only focused on the aspect of development characteristics, specified in the area developed. Therefore, the typology of developers in this research was according to developed aspects of the wide area that consist of 1) small developers with a development area scale of less than 5 Ha, 2) medium developers with a development area of between 5 to 15 Ha and 3) large developers with a land development area of more than 15 Ha. Identification of residential infrastructure characteristics was acquired in accordance with the national standard requirement.

Through this research, it would be known whether the private sector has been able to provide appropriate housing infrastructure in accordance with the standard provision. According to SNI 03-1733-2004, we classify 13 aspects of housing infrastructure, consisting of 7 aspects of basic infrastructure and 6 aspects of housing facilities. The examination of infrastructure provision will be seen from the aspect of physical availability, while the examination of infrastructure accessibility was addressed to the external inhabitants' perception for the ease to gain the infrastructure provided by the housing developers. The research on housing infrastructure characteristics in the peri-urban neighborhood of MBR would be identified through the aspects of provision and accessibility. The use of urban housing procedures in this study can be explained as the peri-urban residential neighborhood are generally designed for urban housing needs. The availability of housing infrastructure should fulfill the equality and evenness, because the principle of the provision of basic services and public facilities, should be usable for all the community. The notion of accessibility that positively contributed to this

IOP Conf. Series: Earth and Environmental Science **158** (2018) 012041 doi:10.1088/1755-1315/158/1/012041

study refers to the lack of physical barriers (access) as well as strategic spatial placement (distribution). In this case, the housing infrastructure built by developers eventually would be handed over to the local government. The list of observations and interviews of infrastructure aspects can be seen in Table 1.

Table 1. The aspects of housing infrastructure characteristics identification in peri-urban Metropolitan Bandung Raya (Adaptation from SNI 03-1733-2004)

No	Aspects	Availability Components (Observation and Measurement Methods)	Accessibility Components (Interview)
1	Road network	The provision and dimension of 1) Pavement, 2) Roadside, 3) Pedestrian way, 4) The area of road benefit, 5) The area belonging to the road, 6) The area of road control, 7) The building borderline	There were 1) main road improvements, 2) no roads cut off, 3) no alleyway adjacent, 4) no one-way road network
2	Drainage network	1) The dimension of drainage channel, 2) The provision of body of water	There were 1) main road drainage construction and widening, 2) continuous drainage
3	Sewerage Network	The provision of 1) Septic Tank and 2) WWTP	There were 1) WWTP that can be used by surrounding communities, 2) No direct waste water disposal to drainage
4	Solid Waste Disposal and Management	The provision of 1) Trashcan 2) Garbage transport 3 times/week	There were 1) Temporary disposal sites that can be shared with the neighborhood, 2) collective transport system
5	Clean Water Network	The provision of 1) Water pipe network, 2) Public tap water (250 inhabitants), 3) Hydrant	There was clean water network that can be shared with the neighborhood
6	Power Grid	The provision of 1) Power supply, 2) Electricity network	There was 1) Street lights, substations and power poles that can be shared with the neighborhood
7	Phone Network	The provision of 1) Cable phone network, 2) Cell phone network	There was a shared telephone network
8	Religious Means	The provision and dimension of 1) Islamic Prayer Room or Musholla (250 inhabitants) or 2) Mosque (2500 inhabitants)	There were 1) Easy / no physical barrier to access the facility 2) Strategic on-site placement
9	Public Open Space	The provision and dimension of 1) Private green yard, 2) Public green space (RT/Neighborhood Association = 250 Inhabitants and RW/Citizen Association = 2500)	
10	Public Facilities	The provision and dimension of 1) Meeting hall (250 inhabitants), Security post (2500 inhabitants), Power House (2500 inhabitants), Public phone and mailboxes (2500 inhabitants), Public parking (RT = 250 inhabitants and RW = 2500 inhabitants)	
11	Educational Facilities	The provision of 1) Kindergarten (1250 inhabitants), Elementary School (1600 inhabitants), Reading Playground (2500 inhabitants)	
12	Health Facilities	The provision of Clinic (2500 inhabitants)	
13	Trade Facilities	The provision of Store (250 inhabitants)	

3. Methods

This research uses three approaches: 1) the method to determine the case study locations 2) the method of data collection and 3) the method of analysis.

3.1 Location Determination Method on Case Study Research

This paper uses a case study approach to extract lessons learned. Referring to Creswell [39], case study research is an exploration of a bonded system or case / multiple cases through in-depth data collection and involving rich sources of information in a context. The research follows a multi-level case selection

IOP Conf. Series: Earth and Environmental Science **158** (2018) 012041 doi:10.1088/1755-1315/158/1/012041

process. The stages of determining the location of the case study are the following. First, we looked for a housing list and all the supporting data from the technical departments of the Local Governments of Bandung Regency and West Bandung Regency. Second, we categorized and classified the secondary data of housing based on the typology of developers and unit settlement area (by sub-district and by village). Third, we filtered the location of case studies using the existing parameters for the three developer typologies in each village. At this step, we obtained three villages in West Bandung Regency and six villages in West Bandung which contain all three housing developer typologies.

The selected three villages in the West Bandung Regency were Cilame Village in Ngamprah Sub-district, Ciwaruga Village in Parongpong Sub-district, and Jayamekar Village in Padalarang District, while the six villages in Kab. Bandung consist of Cipagalo and Lengkong Village in Bojongsoang Subdistrict, Cinunuk Village in Cileunyi Sub-district, Sangkanhurip Village in Katapang Sub-district, Sukamenak Village in Margahayu Sub-district and Bojongloa Village in Rancaekek Sub-district.

Forth, the determination of the case study location. In this step, we determine one village per regency by considering the location that best represents the peri-urban areas of MBR. We selected Ciwaruga Village in West Bandung Regency, and Cipagalo Village in Bandung Regency as the case study locations for the following reasons: 1) the two areas contain the three typologies of housing developers, 2) both are also directly adjacent to Bandung City as the urban center of MBR, 3) those two villages have the combination of urban-rural land use as the physical characteristic of peri-urban region, and 4) their geographically locations contrast each other (northern boundary and southern boundary of Bandung city). Fifth, we assigned three formal housing neighborhoods that best represented the three typologies of developers from the two selected villages. The site selection was acquired based on the recommendations of village officials by considering the information and data retrieval.

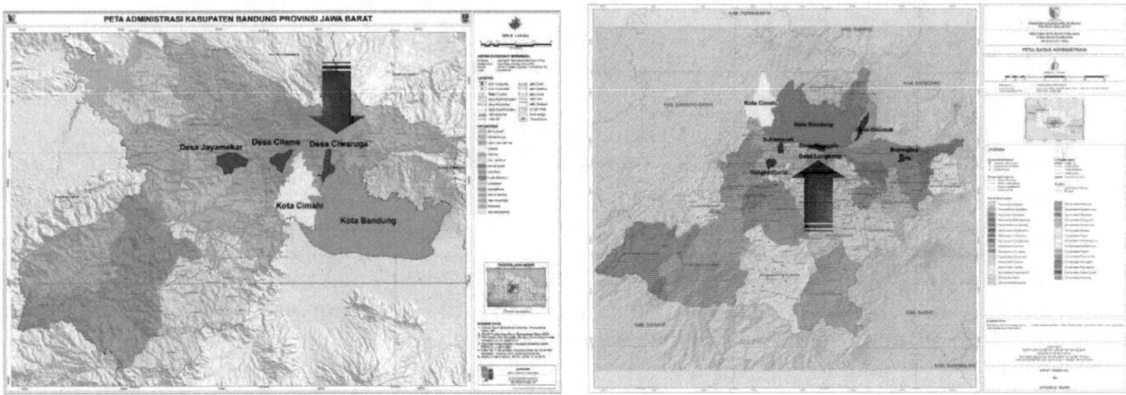

Figure 1. The village location points with the three typologies of housing developers in West Bandung Regency and Bandung Regency [40]

IOP Conf. Series: Earth and Environmental Science **158** (2018) 012041 doi:10.1088/1755-1315/158/1/012041

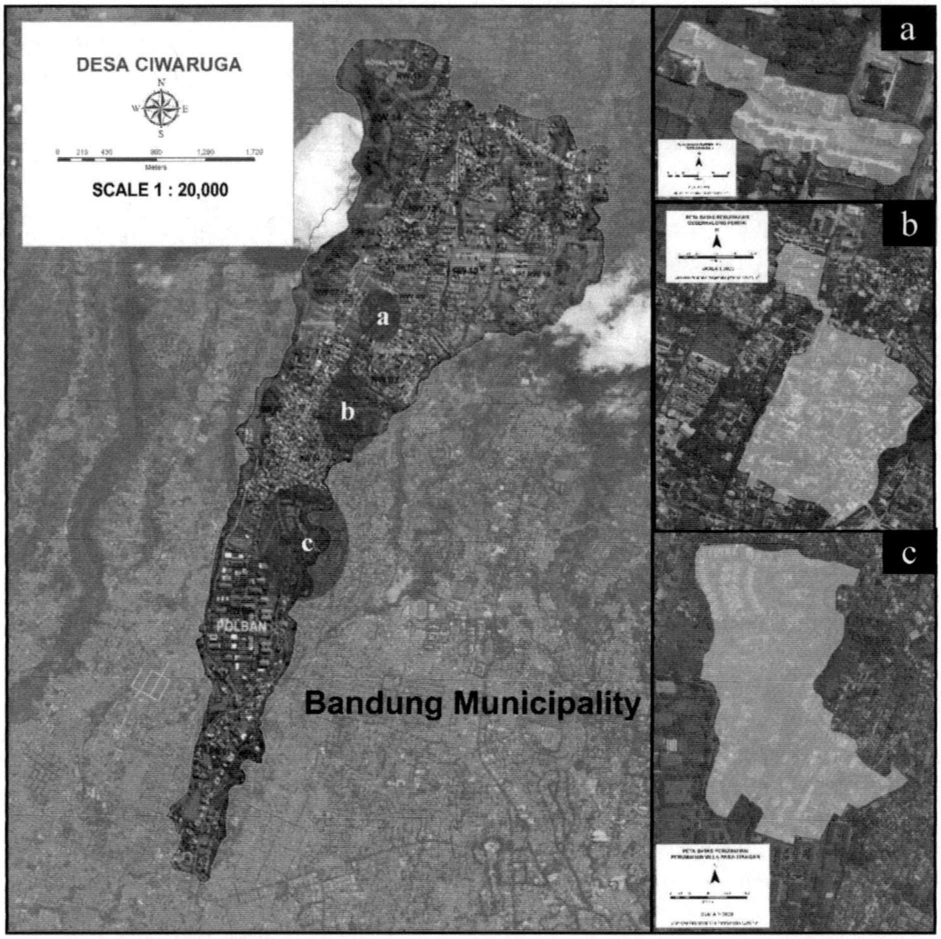

Figure 2. The location of housing estates case study in Ciwaruga Village, West Bandung Regency
Source: Result from Map Processing and Google Earth

Representation of the three typologies of developers in Ciwaruga Village included (a) Griya Amanda 3 housing estates (small-scale housing developer typology), (b) Geger Kalong Permai housing estates (medium-scale housing developer typology), and (c) Parahyangan Rumah Villa Estates (large-scale housing developer typology). Meanwhile, in Cipagalo Village the representative housing estates included (d) Surya Asri Residence housing estates (small-scale), (e) Pesona Ciganitri housing estates (medium-scale), and (f) Cherry Field Housing (large-scale). The location of housing estates in each village can be seen in Figure 2 and Figure 3.

3.2 Data Collection Method

The research uses primary and secondary sources. In the preparation phase, we collected secondary data from the Housing Department Affairs of Bandung Regency and West Bandung Regency. The data consists of housing data list, spatial plan, and regional map. In the implementation phase, we collected primary and secondary data. We collected primary data through observation and measurement of all infrastructure aspects as mentioned in SNI 03-1733-2004. Meanwhile, we interviewed old inhabitants, the management of housing estates, and the community around the housing estates. We obtained the secondary data, including housing maps and information, from the village office as well as the housing management.

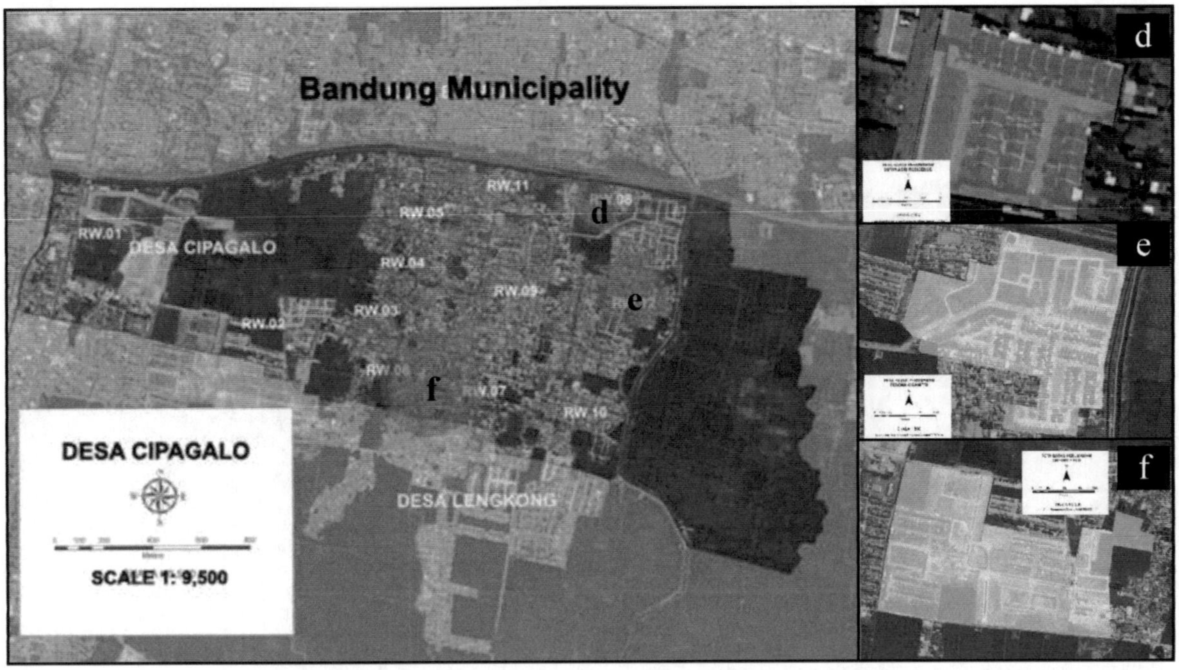

Figure 3. The location of housing estates case Study in Cipagalo Village, Bandung Regency
Source: Result from Map Processing and Google Earth

3.3 Analytical Methods
This paper relies on qualitative analysis. The data were examined using mixed qualitative methods. The analytical methods include standard qualitative techniques (coding-grouping-conclusion) and qualitative content analysis. We used standard qualitative techniques to process the physical availability data, while content analysis was used to process the interviews results.

4. Results and Discussion
Based on the data collection, we performed data processing, classification, and grouping to deliver analytical models for the housing infrastructure availability and accessibility. In the infrastructure availability model, the grouping indicators are divided into 1) available and meeting standards, 2) available but not fully meeting standards, 3) unavailable and not meeting standards, and 4) unavailable and not yet necessary. Figure 4 shows the analytical model of the housing infrastructure availability at the two case study villages of peri-urban MBR and Figure 5 shows the assessment model for infrastructure accessibility with the parameters of positive or negative contribution.

From Figure 4 and Figure 5, we can see two aspects of road components that are significantly in accordance with the standard, i.e., the dimensions of pavement and roadside vegetation. Meanwhile, the dimensions of the roadside (emergency parking) and pedestrian way was unavailable for most of the housing. Other aspects of the road network, such as drainage, road benefit area (damaja), road-owned area (damija), road-supervised area (dawasja), and GSB, indicate unfulfilled availability standards. A significant matter related to the results of road identification is related to the dimensions of the road. When we investigated the road network on a wider scale, we found that there were dimensional gaps between the housing estates' roads with the main road. While the housing estates' roads have adequate dimensions, even wider than the standard, the main road has smaller dimensions, creating congestion.

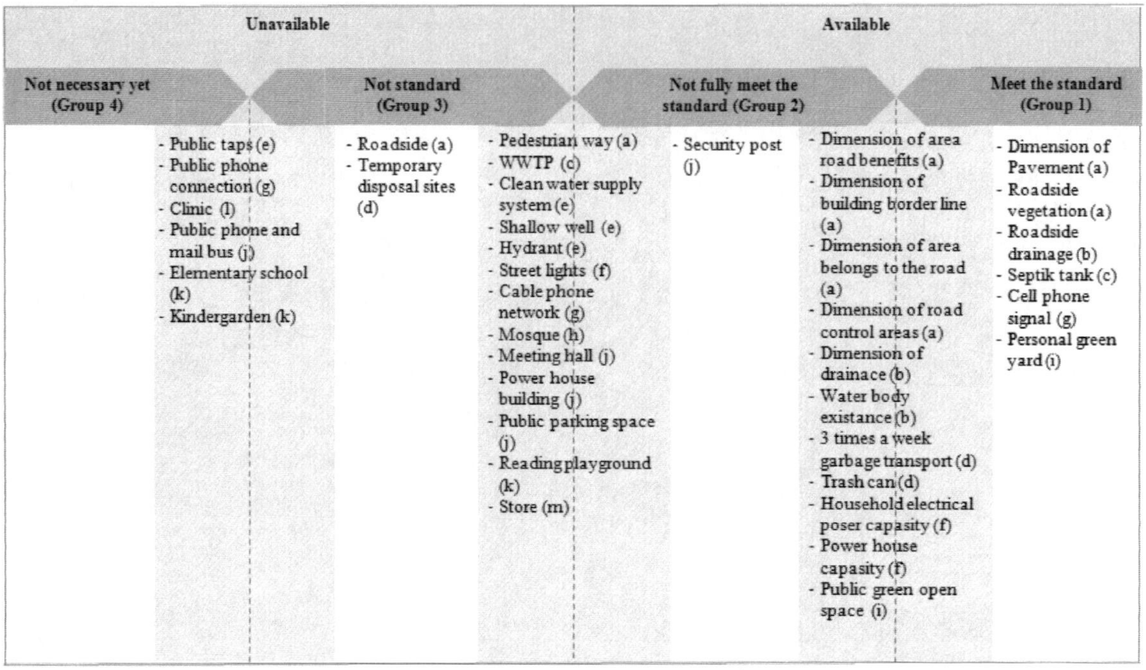

Figure 4. Analytical model of housing infrastructure availability in the MBR peri-urban neighborhoods

Notes:
List of Components
a. Road network
b. Drainage network
c. Solid waste disposal and network
d. Sewerage network
e. Water supply network
f. Electrical network
g. Phone network
h. Phone network
i. Religious facilities
j. Open space facilities
k. Public service facilities
l. Education facilities
m. Health facilities
n. Commercial facilities

There were also lots of sidetracks along the main road, which appended the traffic jam along the roads. We can predict that the number of congestion-prone points at the peri-urban transportation network will still increase in the future if such practices still happened without any solutions. It can be argued that the road improvements that have been provided by developers for the housing estates' roads has no significant effect if not supported by comprehensive infrastructure network planning.

In terms of accessibility, the involvement of housing developer, especially a large one, is generally appreciated by local inhabitants since they helped to improve the main road with paving and widening the access. However, there was a different response to the usage of inner roads of housing estates. For outsiders, the existence of inner roads in housing estate neighborhoods cannot be entirely utilized by the surrounding inhabitants, especially in gated community housings which only has one-access system and equipped with a security post. In such limited access systems, ultimately, it would lead to inefficient road network infrastructure. The presence of narrow alleys around the housing estates' barrier walls shows the presence of separate usage of road access. It indicates the inefficient supply and inequal practice of road network provision in the peri-urban neighborhood.

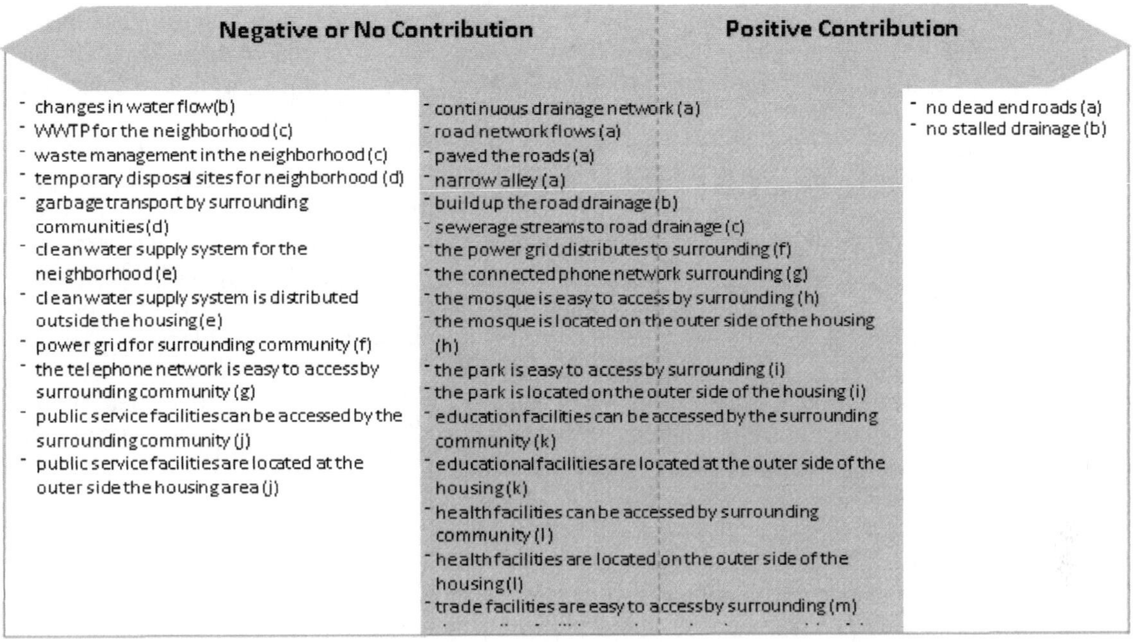

Figure 5. Analytical model of housing infrastructure accessibility in the MBR peri-urban neighborhoods

Most housing estates developers meritoriously arranged and constructed the main roadside drainage channel along with the main road improvement. But the development did not always follow sufficient dimensions or appropriate body water disembogue (such as tributary, river, or canal). The housing estates drainage channels are generally not directly connected to the existing network because developers preferred to build new channels since the land was formerly a plantation or paddy field with no drainage but irrigation channels. Switching irrigation channels into drainage channels often occurred in new housing development, even though these practices should not happen. Irrigation channels have a different water flowing system compared to drainage since irrigation divides water while drainage collect water. The practice of unplanned water channel transformation and the unstructured development of drainage channels generated from housing estates development has contributed to creating new puddles or even some flooding in the area of the neighborhood.

The infrastructure's checking tools for sewerage system consists of the availability of septic tanks and household WWTP as well as its accessibility. Every house in the residential neighborhood already had a good septic-tank provision, either personal or communal septic system. However, it was quite different with the household wastewater system. Most existing housing estates did not provide WWTP management system, either personal or communal. The housing majority drained sewage directly into drainage channels. This was certainly a bad practice since it would lead to water and air pollution and also disrupted the hygiene of the surrounding environment.

One of the main problems with the solid waste management was related to the difficulties of acquiring a temporary disposal site in residential areas. Even though disposal sites were not required yet for small and medium housing scale typologies (because the requirement for the provision of temporary disposal sites in SNI 03-1733-2004 is for 2500 inhabitants), its existence is very important as a means of collecting communal garbage. As a result, every housing estate occupant seeks for their own solution to solve disposal problems. However, not all housing estates can resolve these problems. In fact, one housing estate experienced piling up garbage that was not transported out until two to three weeks. In this case, most developers found it was difficult to locate temporary disposal sites since there was rejection from most inhabitants to stay nearby the disposal sites, since there was no guarantee for cleanliness and standard procedure for waste transport management. In terms of disposal management accessibility, it can be said that the presence of housing estates still cannot provide a positive contribution of solid waste management for the surrounding community in the neighborhood.

Most housing estates in the study case area indicated the ownership of shallow wells due to the inadequate availability of water supply networks / regional water pipelines. The water quality of shallow wells also differs between the northern and southern part of peri-urban MBR. Most of the shallow wells in the southern part of Bandung produced water to improper for consumption. It indicated there was an inadequate supply of clean water for the community. The large housing typology served with a clean water pipeline network generally stated that the water also did not regularly flow. With those problems on clean water provision, the government should pay attention more to the water problems because clean water adequacy a public good that must be fairly accessible for all the community. In terms of infrastructure accessibility, the existence of housing estates still does not have positive contribution to the provision of clean water for the surrounding neighborhood. In fact, the presence of housing estates which are mostly equipped with shallow wells raised groundwater scarcity for the surrounding neighborhood.

Most housing estates have been equipped with electrical substations. After the year 2000, most of the electrical installations for the residential neighborhood has already used Posts Transformer Substation as an alternative substitute for a powerhouse. Meanwhile, the usage of cable phone indicates the decline in recent years since the rise of cell phone usage. It is visible from the number of Base Transceiver Station (BTS) pole pops built around the neighborhoods. Most informants reveal that the cable phone network existence is no longer important since they already get used to cell phones. Even though they have a detached power grid system, the community find the advantage of such indirect benefits from the housing estates inclusion to the village to be due to the widespread use of electric network. With the presence of housing estates, local inhabitants find that getting electricity connection to their house is no longer hard.

The general characteristics of all housing estates in the case study neighborhoods have shown the unavailability of formal school facilities (kindergartens and primary schools) as well as health facilities (i.e., clinics). Some residents claimed that housing estates' managements do not allow the presence of clinics for the privacy reasons. The existence of public open spaces (parks and garden) were commonly provided by all the developers of housing estates, even though some of them are supposed not to meet the standards. For some developers, the green open space provision is only to meet the term of availability. Some of them make the 'remaining land' as a green open space which only functioned as a public garden, but not as a park. For other facilities, most of the medium-large scale housing developers have provided religious facilities (mosque). At the category of government and public facilities and services, the provision of appropriate standard-meeting facilities has been served by large-scale housing. Medium-sized housing is equipped with meeting halls, even though the existing building seems to not meet the standards. Other public service facilities that should be provided in the neighborhood are public parking. All the large housing typology shows the readiness of this kind of public facility. According to the number of occupants, this facility is still not required in the small-medium housing typology. However, if there is no public parking in the housing area, cars would be parked the roadside, hampering the flow of the road. To serve commercial and trade facilities, there is a tendency that developers generally provide them in the form of shophouses. Commercial facilities are usually placed in the public area of housing estates. Referring to the study findings, the trend of shophouses provision appears in the southern area of peri-urban MBR which is generally built after the year 2010. A public service that most housing estates had is security post. Some of them even had more than one security posts. Housing estates with clustering models have some other security posts in every sub-cluster section. However, only few security posts are already equipped with toilet facilities as those required by the SNI standard.

The accessibility concept of the housing facilities is highly correlated with the placement or the position of the facilities in the sites. If the facilities are located on the outer side or at the public open space of the housing site, it would provide a better access for outside communities. In the one-gated community concept with most facilities placed on the inner side of the housing area, the outer community hesitates to use the facility. The mosque is a means of worship with a high social function, yet most of the outer community is reluctant to use the facility if it was located on the inner side of housing estates. It can be said that the presence of security posts at the entrance of the housing estates has become 'a restriction mark' for other communities to access the housing facilities.

5. Conclusion

The research on housing infrastructure availability and accessibility of the three typologies of neighborhoods in the two representative areas of the peri-urban MBR shows that there are still many shortcomings in the implementation of standard regulations in the provision side by housing developers. It means that although developers seem to have already provided all of the housing facilities, in reality there are only a few aspects that are in accordance with SNI 03-1733-2004 as the national standard of housing infrastructure in urban neighborhoods. This might happen because not all local government uses this kind of standard as reference for housing infrastructure provision. With so many regulatory terminologies related to the housing infrastructure provision, it would eventually cause bias on the compulsory reference that should be followed by all stakeholders. This study also revealed that although some of the housing infrastructure aspects have already met the standards, without integrated planning network it would not bring any spatial improvement. In terms of accessibility, the surrounding communities generally indicate that they were not very affected by the presence of housing estates due to the detached neighborhood. For local inhabitants, the style of a gated community housing estate become a symbolic barrier that outsiders cannot simply come and enjoy the housing estate facilities. Most of the outer community has no other choice than just receiving the housing estates and adapting to their presence. Consequently, the model of gated communities housing is often called exclusive housing estates, which generated the fragmented residential infrastructure development.

6. Recommendation

This research affirms the accelerated development of peri-urban settlements should be managed with a clear infrastructure development plan. Without prejudice to environmental issues, the prospect of peri-urban development to meet the housing needs is very high. The local community also showed some appreciation for the infrastructure improvements that have been generated by housing estates development in peri-urban areas. The role of private sectors' involvement in peri-urban development as the government partner can be seen as an alternative solution for the provision of housing and infrastructure networks. However, the concept is totally different with the dominance of the private sector which then undermines the role of the government. Since there are so many obscure implementations, it seems that the government should evaluate the housing infrastructure provision system mechanism, whether due to the imperfect planning regulation or weak technical implementation. In this case, the government as an institution authorized to implement housing and settlements (Law No. 1 of 2011) has more roles than just a regulator. The government has to describe every regulation in more detail for housing infrastructure planning and applicable technical standards.

References

[1] Appiah D O, Bugri J T, Forkuor E K and Boateng P K 2014 Determinants of Peri-Urbanization and Land Use Change Patterns in Peri-Urban Ghana J. Sustain. Dev. 7 6 p 95

[2] Kleemann J, Inkoom J N, Thiel M, Shankar S, Lautenbach S, and Fürst 2017 C Peri-urban land use pattern and its relation to land use planning in Ghana, West Africa Landsc. Urban Plan 165 p 280–294

[3] Akrofi E O and Whittal J 2011 Land for Peri-urban Infrastructure in Customary Areas : A case study of Land for Peri-urban Infrastructure in Customary Areas : A case study of in FIG Working Week p 1–16.
[4] Firman T 2009 The Continuity and Change in Mega-Urbanization in Indonesia: A Survey of Jakarta – Bandung Region (JBR) Development Habitat Int 33 p. 334
[5] Hudalah D 2010 Peri-urban Planning in Indonesia (University of Groningen)
[6] Webster D, Cai J, and Muller L 2014 The New Face of Peri-Urbanization in East Asia : Modern Production Zones, Middle-Class Lifestyles, and Rising Expectations J. Urban Aff. 36 S1 p 315–333
[7] Woltjer J 2014 A Global Review on Peri-Urban Development and Planning J. Perenc. Wil. 25 1 p 1–16
[8] Zhao P 2013 Too complex to be managed ? New trends in peri-urbanisation and its planning in Beijing Cities 30 p 68–76
[9] Liu Z and Robinson G M 2016 Land Use Policy Residential development in the peri-urban fringe : The example of Land use policy 57 p 179–192
[10] Ruth N and De Duren L 2018 Why there ? Developers â€TM rationale for building social housing in the urban periphery in Latin America Cities 72 November 2017 p 411–420
[11] Sun L, Li C, Gwilliam J A, and Jones P J 2013, Sustainable peri-urban residential settlement development in china: Evaluation of three cases in Tianjin Int. J. Sustain. Dev. Plan. 8 4 p 449–463
[12] Millward H 2002 Peri-urban residential development in the Halifax region 1960-2000: Magnets, constraints, and planning policies," Can. Geogr. 46 1 p 33–47
[13] Webster D 2002 On the Edge : Shaping the Future of Peri-urban East Asia (Standford)
[14] Hendrowati A R 2013 The Infrastructure Management of The Peri-urban Interface Within the Intergovernmental Cooperation Framework J. Ilm. Adm. Publik 14 1 p 593–607
[15] Webster D and Muller L 2002 Challenges of Peri-urbanization in the Lower Yangtze Region : The Case of the Hangzhou- Ningbo Corridor, no. May (Stanford, California)
[16] Sun L, Li C, Gwilliam J, and Jones P 2011 Challenges to sustainable peri-urban settlement development in China: an analysis by empirical evidence in Tianjin WIT Trans. Ecol. Environ. 167 p 3–14
[17] Hogrewe W, Joyce S D, and Perez E A 1993 The Unique Challenges of Improving Peri-urban Sanitation (Washington D.C)
[18] Marshall F, Waldman L, MacGregor H, Mehta L, and Randhawa P 2009 On the Edge of Sustainability: Perspectives on Peri-urban Dynamics (Brighton: STEPS Center)
[19] Hudalah D, Winarso H, and Woltjer J 2007 Peri-urbanisation in East Asia : A New Challange for Planning Int. Dev. Plan. Rev., 29 10 p 503–520
[20] Firman T 2009 The Continuity and Change in Mega-Urbanization in Indonesia: A Survey of Jakarta - Bandung Region (JBR) Development Habitat Int. 33 p 327–339
[21] Vitriana A 2017 Increase in Land Value due to Spatial Transformation in the Northern Part of the Bandung – Cimahi Peri-urban Region J. Reg. City Plan. 28 1 p 73–83
[22] Winarso H, Hudalah D, and Firman T 2016 Peri-urban transformation in the Jakarta metropolitan area Habitat Int. 49 November p 221–229
[23] Abramson D B 2016 Periurbanization and the politics of development-as-city-building in China Cities 53 p 156-162
[24] Maryati S, Nisaa A, and Humaira S 2015 Increasing the Infrastructure Access of Low-Income People in Peri-Urban of Bandung Metropolitan Area Int. J. Built Environ. Sustain. 2 3 p 224
[25] Hudalah D, Winarso H, and Woltjer J 2007 Peri-urbanisation in East Asia : A New Challenge for Planning? Int. Dev. Plan. Rev. 29 10, p 508
[26] Zhang L, Ye Y, and Chen J 2016 Land Use Policy Urbanization , informality and housing inequality in indigenous villages : A case study of Guangzhou Land use policy 58 p 32–42

[27] National Standardization Agency 2004 SNI 03-1733-2004 on Procedures for Urban Housing Planning. Indonesia
[28] Owusu S E and Asamoah P K B 2005 Servicing Land for Housing Development in Peri Urban Areas of Kumasi, Ghana : Theory Versus Practice J. Sci. Technol.25 1 p 77–85
[29] Smith M D, Sohail M, and Saywell D L 1998 Basic services in peri-urban areas in AGUA'98 Int. Conf. and Seminar
[30] Sheng Y K 2014 Challenges of Peri-Urbanization in the Asia-Pacific Region Yap Kioe Sheng Int.l Policy Workshop on Rural-Urban Poverty Linkages September p 1–17.
[31] Shivendra B T and Ramaraju H K 2013 Sanitation in periurban areas of developing countries J. Acad. Ind. Res. 1 January p 476–480
[32] Toteng E N 2014 Access to Services and Infrastructure in Peri-urban Areas in Botswana : Case studies of Mogoditshane and Tlokweng Botswana Notes Rec.41 2009 p 64–75
[33] Cáceres C 2014 Exploring the periurban residential areas in Santiago de Chile: The asymmetric residential model between Corporate and State residential areas in Lampa 2000-2010
[34] Handayani A D and Yuliastuti N 2014 Identifikasi Ketersediaan dan Kualitas Sarana Prasarana Lingkungan di Urban Fringe Area Kelurahan Pudakpayung J. Wilayah. dan Lingkungan 2 3 p 197–208
[35] Putri H T 2016 Karakteristik Penyediaan Infrastruktur Dasar oleh Developer di Lingkungan Perumahan Kecamatan Parongpong, Kabupaten Bandung Barat (Institut Teknologi Bandung)
[36] Tian L, Ge B, and Li Y 2016 Impacts of state-led and bottom-up urbanization on land use change in the peri-urban areas of Shanghai : Planned growth or uncontrolled sprawl ? JCIT
[37] Directorate of Housing Provision and Planning, "Technical Report on the Preparation of Masterplan for Metropolitan Bandung Raya's Public Housing Provision," Jakarta, 2015.
[38] Winarso H 2000 Residential Land Developer's Behaviour in Jabotabek, Indonesia (University of London)
[39] Creswell J W 2014 Research Design: Qualitative, Quantitative, and Mixed Methods Approaches, 4th ed (California: Sage Publication)
[40] Provincial Regulation No. 22 Year 2010 on Spatial Plan of West Java 2009-2029

Acknowledgments
The authors gratefully acknowledge that the present research is supported by the Regional Agency for Research and Development, West Java Province of Indonesia

Housing Development on the Urban Fringe and its Challenges to Sustainable Urban Growth

RWD Pramono

Regional Planning Studies, Department of Architecture and Planning, Gadjah Mada University, Jl. Grafika Utara no 2, Sekip Yogyakarta, Indonesia

Corresponing Author: pramono.wid@ugm.ac.id

Abstract. Urban development is one of the substantial factors that shape the quality of human life and its sustainability. However, the intensive growth of housing in the urban fringe seems to threaten this. Meanwhile, the application of spatial design procedures as part of the regulating process has not been successful (Polyzos S., Minetos D., and Niavis S., 2013). Our failure to properly comprehend the phenomenon probably contributes to this failure. Rather than the macro land use approach applied all this time, an intervention in micro land use of the actors may be more effective. Supporting this strategy, this research focuses on the characteristics of housing development, its ecological aspects and its capacity to support the community's capability through field observations and questionnaires. The study indicates that the planned cluster development on the fringe of Yogyakarta City, although supporting the capability of its people, may potentially threaten the ecological balance and long-term macroeconomic situation of the

1. Introduction

Almost all big cities in Indonesia and middle-sized cities in Java experience spatial developments, sprawling rapidly on their fringes. Observations of cities such as Jakarta, Surabaya, Makassar, Medan, Semarang, and Yogyakarta show that this is a common phenomenon. This infiltration of urban elements on the urban fringe does not only transform land from agriculture into non-agriculture land use but also alters the economic, spatial, and social dimensions [1]. These big cities exhibit at least two phenomena; agglomeration and urban sprawl.

Academic discussions on agglomeration generally focus on the city size and its spatial organization. These discussions affirm that the agglomeration process in western countries is not principally driven by urban sprawl or what is commonly called leapfrog development. Instead, it takes place gradually. This process is induced by rapid population growth and horizontal land use change [2].

Meanwhile, academic discussions on urban sprawl, especially those on cases in developed countries, generally reveal that urban sprawl is promoted by lifestyle [3]. People want to move from crowded and busy urban area to a serene, green, and spacious landscape.

Urban sprawl in developed countries such as the US, Japan, and in Europe does not always form an agglomeration [4]. Thus, the focus of discussions is on transportation and energy [5], as well as the infrastructure network caused by planned clusters in agricultural areas. Therefore, despite the high transportation costs, the value of urban fringe housing is more related to the hedonic attribute of the property. People enjoy living surrounded by green farmland [6]. Only a few contradict it with the issue of loss of agricultural land.

Is the development of urban fringe housing in the US still motivated by lifestyle? Kirkman in his article "Did Americans Choose Sprawl?" [3] states that until today, sprawl is still trending due to the choices of people in the US, although he says that it is more likely to be the result of piecemeal decisions and consequences, rather than intended or unintended. Regarding the waste of transportation energy and food availability, Kirkman's later opinion is interesting to be discussed as an issue of the implementation of the sustainable development paradigm. The awareness of sustainable development is generally based on public interest and will take a long time to materialize. Therefore, often it only becomes a study in the public sector, activism, or an academic field. As a result, it has the potential to contradict the nature of sprawl, which is, as Kirkman says, driven by aggregated individual choice, piecemeal, and unintended. It only considers individual profit and most of the time it is short-termed. It tends to ignore bigger issues such as economic stability and ecological sustainability.

How about the experience of cities in Indonesia? What are the driving factors? What is the final shape? What are the non-physical impacts? This article will try to answer these questions by looking at the example of Yogyakarta.

To answer these questions, the author has conducted a research to explain 1) the spatial pattern of housing on the urban fringe; 2) the motivations and context/circumstances surrounding it; and 3) place's influence on the quality of life of new immigrants or the native villagers around the place.

Based on the research findings, this article aims to expose people's perceptual evaluation, either good or bad, regarding the sprawl of planned clusters on the urban fringe. The last part of this article will discuss the consequences for the direction of future development. It will also describe the challenges of sustainable development and what should be the attitude of public sectors as well as the actions to be taken.

Theoretically, this research attempts to discover whether the paradigm of sustainable development is consistent at the subconscious level of people in general. This can be revealed through a capability survey and public discussion involving stakeholders to bring out their awareness level. For the planning practice, this exposure provides a map for the things that are invisible on the surface, but latent and existing in people's subconscious mind.

The prospect of sustainability depends on the collaboration and participation of stakeholders. It goes along with the argument that stakeholder collaboration and participation, social sustainability, and continuous development are intertwined and together can contribute to social change [7]. Meanwhile, according to Arnstein, only genuine participation that safeguards stakeholders' needs is effective and is contextual to their problems and aspirations [8]. The problem Arnstein mentions is not fully resolvable and is still relevant nowadays as can be seen from the sustainable development goals that focus on addressing interlinked human needs. The problem is that local regulatory measures and instruments fail to transpose people's aspiration as 'half' of the concept of sustainable development [9]. That said, it is not easy, even for stakeholders, to recognize the real needs and problems, particularly the long-term ones. The reason is perhaps that stakeholders do not automatically consider the needs involving long-term communal interests, such as food security and sovereignty - that serve as of the core of sustainable development. As Kirkman says, currently, the unintended processes are relevant for sustainable development [3]. This research confirms that people's decision to sprawl brings a dilemma to sustainable development.

The consensus is that sustainable development [10] comprehensively and proportionally serves the growth of economic wellbeing in fairness to all communities and involves the harmony of social interaction, and the protection of the environment. Practically, the important core of this concept is the protection of the sustainability of well-being itself – including the sustainability of food security and the quality of amenities - and the sustainability of energy resource due to the wise use of it. Therefore, human civilization will continue to improve. To achieve this, every human action must be done in awareness. This is why it is important to understand the reason behind every phenomenon of civil development - including sprawl - from the view of people's subconscious mind. Understanding it may

IOP Conf. Series: Earth and Environmental Science **158** (2018) 012042 doi:10.1088/1755-1315/158/1/012042

produce some considerations for public policies. It can direct people's decisions including the choice for living space and finally, it will result in the development of sustainable settlements.

The development of housing on the urban fringe can be studied from the perspective of sustainable development by evaluating the spatial pattern and its effects upon people's capability and wellbeing, not only in the short but also in the long term.

2. Method

The spatial pattern on the urban fringe can be revealed based on the macro observation of the growth of newly planned clusters in the southern part of Yogyakarta, using both secondary data and direct observations (i.e., grand tour). For a more precise scale, the data are gathered from a random sampling of 50 out of 275 clusters in the surveyed area. To find out their motivations, the inhabitants of clusters and the villages around them were asked to answer some questionnaires. The objective is to find out the reason for living in that area and to understand the background and the process of land use transformation.

A capability survey [11] using closed questionnaires was applied to describe the influence of the place on people's lives. In addition, the survey can reveal whether the elements of place serve as positive or negative assets for its people or the city/region in general. The categorization of positive and negative assets can be the basis of the analysis of the advantages and disadvantages of planned cluster development in the green area of the urban fringe.

In the framework of the capability survey, the Community's Capabilities Index (Cci) scale is prepared as an instrument to indicate the quantitative level of well-being, which is basically abstract and relative in nature. The data can also be used to measure the level of a place's supporting capacity and in this context is called Place's Capability Supporting Index (Pcsi). This index measures the conformity of a place in facilitating the creation of opportunities to choose and perform a variety of functions wanted by the inhabitant (capability level). Used jointly with data of quantitative assets or transferable commodities, Pcsi can function as the coefficient of wealth *multiplier* or *deflator*. The coefficient is determined by the proportion of Pcsi in defining Cci.

Considering the dichotomy of good-bad and helpful-constraining, the capability level index can be translated as a vector with two directions. Its tendency toward 'helpful' perception is represented by a positive number and 'constraining' by a negative (see [12]) for a similar research). For this purpose, a scoring technique is applied to convert the scale of people's perceptual evaluation. A range between -2 to 2 is proposed for this need. The objective is to simplify the calculation and the interpretation of questionnaires into some calculation index. A difference of one point in each option is expected to still reflect the contrast of people's evaluation regarding the availability and accessibility of an asset. It can be done by giving a score of 2 for the answer of 'better', and -2 for 'worse' and '0' for 'not so much different' (no changes). Zero can also mean the asset is not possessed/not accessible/not perceived by the respondents. The scale of -2 to 2 is also used to measure the scale of people's perception in evaluating the level of opportunity offered by the asset. A score of 2 means the asset is very helpful, 1 means helpful, -1 means constraining, and -2 very constraining (see table 6.2). The general outline of the questionnaires is shown in table 1, whereas the whole variety of assets and the formulas of CI, Pcsi, and Place's coefficient (multiplier or deflator) can be found in table 7.

Table 1. Questionnaires Design as Capability Survey Tools.

No Assets	Components/Assets	Availability								
		Better	Not So Much Different to previous/No Improvent	Not Exist/ Not Accessible/ Unclear Inflences	Worse	Very Helpfull	Helpfull	Unclear Influences	Constraining	Very Constraining
	Score	2	1	0	-2	2	1	0	-1	-2
1	Space/land to reside									
2	Space/land for bussiness/farming									
.....										
40	Transportations cost									

The capability questionnaires were disseminated in September 2017. The typology of opinions and the evaluation of the respondents gathered from the capability survey can also give initial information regarding people's preferences. It can indicate the trends that may take place in a long run; whether people will drive or correct the ongoing development pattern. Correlated with common attitudes of the government and other stakeholders (property developers and environmental activists) obtained from the FGD results, it will provide a description on how hard or how easy the urban growth management is on this phenomenon.

3. Results

3.1. The Speed of Development toward the Urban Fringe

Yogyakarta City experiences relatively rapid development. Based on the research conducted by Bappeda Yogyakarta in 2015, for three decades, the urban built-up area has grown 1,1% from 1990 to 1995, almost reaching 5% from 2005 to 2000. The built-up area in 2000 is seven times larger than in 1995. Although the growth rate has been slowing down since 2000 at 3,3%, the additional built-up areas are still huge, increasing from about 5600 ha in 2000 to 9000 ha in 2010. Recently, the radius of agglomerated urban housing is around 12 km. If we count in leapfrog housing clusters, the radius has reached 18 km (see figure 1).

IOP Conf. Series: Earth and Environmental Science **158** (2018) 012042 doi:10.1088/1755-1315/158/1/012042

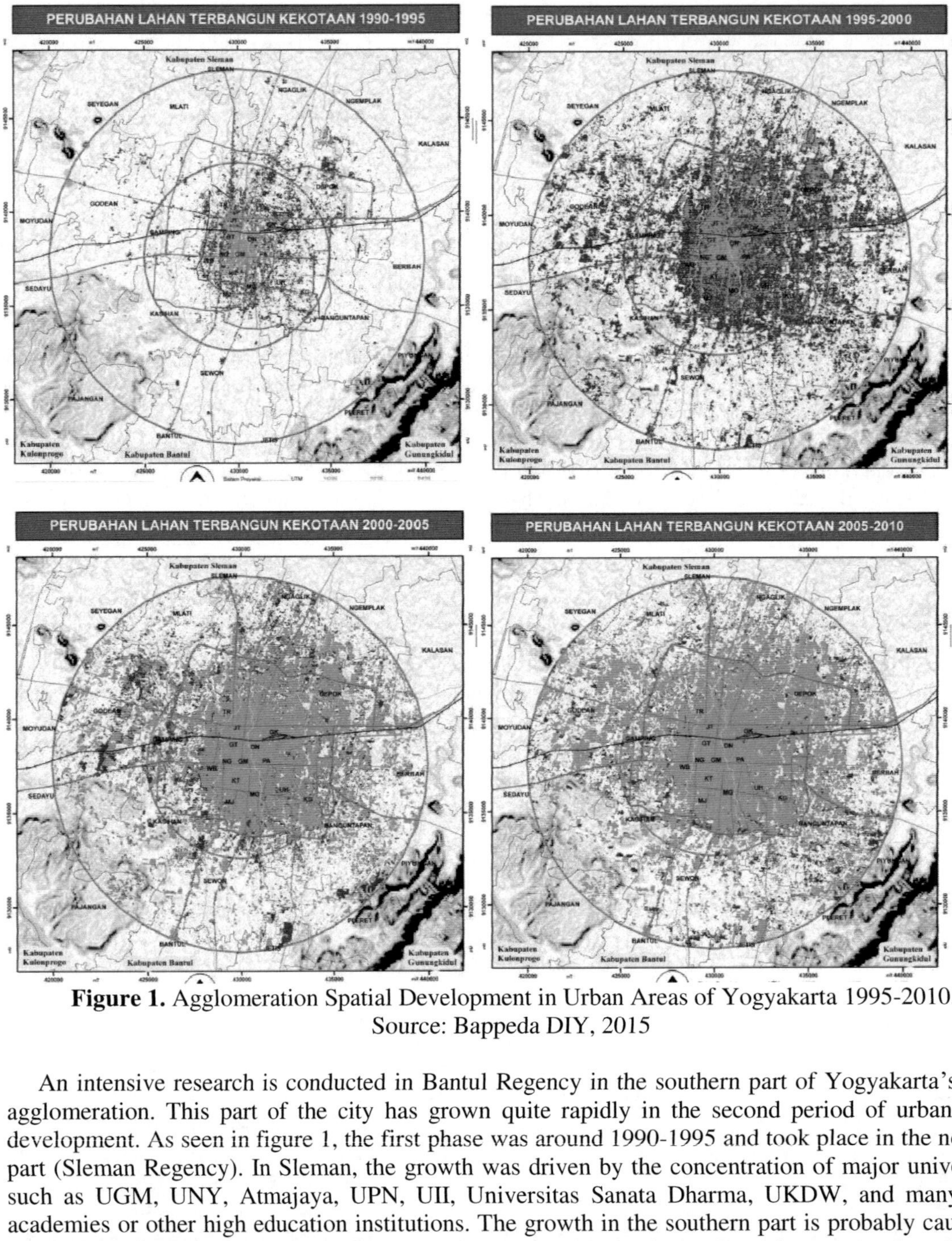

Figure 1. Agglomeration Spatial Development in Urban Areas of Yogyakarta 1995-2010.
Source: Bappeda DIY, 2015

An intensive research is conducted in Bantul Regency in the southern part of Yogyakarta's urban agglomeration. This part of the city has grown quite rapidly in the second period of urban fringe development. As seen in figure 1, the first phase was around 1990-1995 and took place in the northern part (Sleman Regency). In Sleman, the growth was driven by the concentration of major universities such as UGM, UNY, Atmajaya, UPN, UII, Universitas Sanata Dharma, UKDW, and many other academies or other high education institutions. The growth in the southern part is probably caused by the immensely high price of land in the northern part of Yogyakarta. Land large enough to develop clustered housing is more and more difficult to find in this area.

The population of Bantul is around 955,015 with 2% population growth during the last 5 years. The average population density is 1,884 people/km^2. However, the highest concentration lies in the districts adjacent to Yogyakarta such as Kasihan, Banguntapan, and Sewon. The population density in those districts may reach up to 2766 people/ha.

Figure 2. The distribution of newly planned clusters in Bantul Regency.

Based on the data from Bappeda, from about 2005 up to 2015 275 planned clusters applied for permits and were finally developed in Bantul Regency. The details are outlined in table 1 and its distribution is presented in figure 2.

Table 2. Data on the cluster development in Bantul Regency.

Year	<2010	2011	2012	2013	2014	2015
Cluster Units	170	23	15	29	18	20

Based on direct observations, it can be seen that some points around the new clusters, as outlined in the map, rapidly turn into urbanization-triggering spots. The urban atmosphere marked by the small and dense housing and the non-agricultural population is getting stronger; slowly these clusters merge with native villages and form a bigger economic agglomeration.

3.2. Reasons behind Location Choice

Based on the questionnaires distributed to 110 respondents, the profile of people in the Bantul urban fringe can be mapped out. The respondents consist for 33.6% of immigrants, 94.6% of them come from the same province and 5.7% from outside the province. Among the immigrants coming from the same province, only 11.4% comes from Bantul; the rest comes from other regencies, mainly Yogyakarta city and Sleman. Most of them (70.2%) have purchased and own the land, 13.5% lease land, and 2.7% got it as a grant, while the rest does not mention the origin and the status of their houses. The main motivations of immigrants to move in this area are:

Table 3. Reasons to migrate to urban fringe in Bantul.

Motivation		**Answer (%)**
Reasonable Price	45	40.91
Near Campuses	24	21.82
Near Working Places	15	13.64
Strategic Location	11	10.00
Comfortable Place	11	10.00
Good Accessibility	3	3.64
		100.00

3.3. Motivations for Releasing Land

As mentioned above, 70,2% of the inhabitants are immigrants who acquired their houses by buying them. Some of the respondents (45.9%) are buyers; 21.6% are sellers, while 5.4% have sold and bought as well. The interesting thing is that 24.3% of them often act as a broker.

The subsequent data reveals that 72% of the lands bought by either big developers, small developers, or by private individuals were previously farmlands, and only 11% were empty land ready for building a house on. Data from the government confirm that the conversion rate from farming land into housing is 40 ha per year.

To find out the background of the sale of land, particularly the ones that are previously farms, a further interview was conducted upon the 27% of respondents who stated that they had sold their land). The interviews were conducted using open questions, i.e., asking the respondents to mention their reasons for selling their land. They were encouraged to give more than one answer considering that there are possibly several reasons behind the decision to sell the land. Grouped based on similarity, the types and the frequency of the answers are tabulated as seen in the table below.

Table 4. Frequency of answers on the motivation for selling land on the urban fringe in Bantul.

Reasons	Percentage of being mentioned
There is an urgent need (sickness, paying tuition fee, finding a job for the children, and making a house for the children).	80.0
Highly increasing price of land	77.5
There is no one to take care it of	60.8
To be bestowed as inheritance in the form of money	45.8
Too small to be cultivated	44.2
Not so fertile, no irrigation facilities	22.5
To buy another farm in another location (farther, but larger)	11.7

The data in table 5 above affirms that the key motivation for people to sell their land is a combination between economic deprivation and lack of government incentives in the agricultural sector to push the productivity and economically beneficial scale of farming. The high prices offered for the land also add to this motive.

3.4. Physical Organization of Urban Fringe Housing

Most the respondents' houses (85.83%), owned by both immigrants and native villagers, are one-story buildings on plots of 100-200 m^2 of land. A few of them (23.33%) are bigger (more than 500 m^2) and the rest is between those two categories. A deeper study upon the answers shows that the bigger plots are owned by the natives. It can be concluded that 100 m^2 to 200 m^2 is the optimum size compromising between the standard minimum need of living space and affordability of the buyers.

As previously mentioned, most of the cluster locations were previously farmland. Therefore, it is natural that the three sides of the 81% clusters are bordered by a farm, a yard, and a village. Meanwhile, 14% of the clusters are bordered by other new planned clusters, and 5% are developed on previously occupied land. Generally (86%) the new clusters are fenced around; only 14% are not and are adjacent to a yard or merged with an old village area. One cluster is usually not correlated to other clusters as one unit of housing. For example, 3% of clusters share common streets, while 14% are adjacent but have their own street access. The most dominant (83%) are clusters located far apart. These facts affirm that the development has a sprawling and leapfrog pattern. On the other side, only a few clusters are developed by an infill method, i.e., developing a cluster in an empty land ready to be built on so it fills in the area instead of opening up a farmland.

Although most clusters are developed on farmland, the majority (78%) is developed without new access roads. Only 22% clusters build their own roads and at least 5% do not build any roads and use the existing roads around the village. Meanwhile, 21% use some trail in the farms called farming paths which are actually intended to facilitate farming sectors. Some clusters (47%) release the farm on the sides of primary collector (provincial) roads or secondary collector (district) roads and 3% release the farm on the sides of primary artery (national) roads.

Survey result shows that the government control over productive land for farming is so weak that they can barely protect it. The government is practically absent in managing physical order to form a physically integrated neighborhood. As a result, the development of the new clusters commonly puts an additional burden on the existing infrastructure in the urban fringe.

3.5. Quality of Life of Urban Fringe Residents

Based on the respondents' answers regarding their perceptual evaluation on the proposed 40 assets, the Capability Index (CI) for both communities – in the native villages and in the newly planned clusters – is quite good. They are 0.32 for the village community and 0.40 for the residents of the new clusters. According to their evaluation on their scale of happiness (between 1 to 10, 1 for very unhappy and 10 for very happy) regarding the condition of the asset in question, the rate of their happiness is also good; 7.43 and 7.63 for villagers and residents of new clusters respectively. Keeping in mind that capability is the level of freedom from the inadequacy to fulfill their needs and the freedom to choose, this CI level indicates that people in the research area is quite prosperous (1 means perfect capability, -1 means the most helpless, and 0 is the lowest limit of the capability to fulfill living standard). This condition may correlate with the expression of happiness which is a direct expression based on the experience of all respondents.

Place's components formed by Tangible Public Assets, Social Institutional Assets, and Economic Institutional Assets synchronically contribute a great deal to people's capability. This is confirmed by the Place's Capability Supporting Index (Pcsi) which is relatively high (0.35 for the village and 0.41 for the planned clusters). Even, if the place's coefficients for both communities are taken into account, the place's assets can function as a multiplier of their capability. The value is 13.01 for the villagers and 37.71 for the immigrants. It means places have multiplied the values for them.

Table 5. Indicators of happiness, capability and their factors.

Variable	Formula	Village	Planned Cluster
Happiness Score (H)	Average	7.43	7.63
Assets Availability (A)	Average	1.08	1.26
Assets Function (F)	Average	1.55	1.59
Capability Index (CI)	$(AxF^2)/8$	0.32	0.40
Average Score of PTA. SIA. EIA) Availability (Ap)	Average	1.06	1.23
Average Score of PTA. SIA. EIA) Function (Fp)	Average	1.62	1.63
Place's Capability Supporting Index (Pcsi)	$(ApxFp^2)/8$	0.35	0.41
Place' Coefficient (Multiplier/deflator)	Pcsi/ABS(Pcsi-Ci)	13.01	37.71

The data in table 6 indicates that both villagers and the new residents perceive and feel that the development of planned clusters in this area as a living place is quite profitable. However, taking a look at every group's asset forming the Place's Asset, we can see quite a conspicuous notion, particularly in the variable of assets coefficient in the groups of Economic Institutional Assets and Social Institutional Assets. Although the values are always positive in both communities (the EIA value can even serve the multiplier up to 4.22 for the villagers and 2.66 for the immigrants), it contradicts the development of public infrastructure in the group of Public Tangible Asset. The score for this is only 0.32 for villagers and 0.54 for the immigrants. The value below 1 means that infrastructure is the deflator for the assets and people's capability.

Table 6. The Roles of Place's Assets Class to Support Capability.

	Village				Pllaned Cluster				Assets Class Coefficient for All Community y
	Score of Assets Funcion	Score of Assets Function	Village Supporting Index of Assets Class	Village Assets Class Coeficient	Score Assets Availability	Score of Assets Function	Planned Cluster's Supporting Index of Assets	Planned Cluster's Supporting Index of Assets	
Tangible Public Assets (Infrastructure)	0.29	1.31	0.08	0.32	0.64	1.46	0.17	0.76	0.54
Social Institutional Assests (social capital)	1.50	2.00	0.75	1.75	2.00	1.94	0.94	1.73	1.74
Economic Institutional Assets	1.38	1.56	0.42	4.22	1.56	1.03	0.21	2.66	2.66

The asset elements considered unavailable, unfelt or even decreasing in quantity and quality can be traced back to the compilation of respondents' answers regarding all 40 assets shaping capability as tabulated in table 7 below. It can be seen that negative values apply to 1) Availability of public transportation infrastructure; 2) Quality of public transportation; 3) Availability of public library. Those three aspects are considered worse regarding their availability and functions by both communities. Meanwhile, the asset of farmland productivity/quality is perceived as worse by the villagers and the asset of transportation cost is considered worse by the immigrants. Aside those, some assets are scored 0 and 1, meaning that their availability is not too obvious or there are no changes and these assets are not considered constraining. They are 1) Government's and community's disaster management; 2) Easy access to clean water; 3) Availability of recreation facilities including lack of open public space; 4) Availability of the internet for community; 5) Easiness to get raw material for economic production; and 6) Easiness to market products.

Table 7. Average Perceptual Evaluation Score on Assets.

			Kampong				Planned Cluster			
Category	No Assets	Components/ Assets	Assets Availability	Score of Asset Class Availability	Score of Asset Class Function	Score of Asset Class	Assets Availability	Score of Asset Class Availability	Assets Function	Score of Asset Class Function
	1	Space/ Land to reside	1.13		1.77		2.00		2.00	
	2	Space/ Land for business farming	1.43		1.90		1.57		1.50	
	3	Owning transportation means (bikes, motorbikes, car)	1.07		1.13		2.00		2.00	
Tangible Individual Asset	4	Owning communication means (telephones, cell phone)	1.10	1.43	1.10	1.55	2.00	1.76	1.87	1.67
	5	Owning devices to access information (newspaper, magazines, radio, TV, and internet)	1.70		1.17		1.33		1.37	
	6	Income/ Daily needs fulfillment	1.73		1.80		1.43		1.20	
	7	Savings	1.87		1.97		2.00		1.73	
	8	Environment quality (cleanliness, water, air, public park)	1.00		1.83		1.00		1.93	
	9	Productivity/quality of farm land	-1.27		1.50		1.40		1.80	
Tangible Public Asset	10	Availability of public transportation infrastructure	-1.00	0.29	1.00	1.31	-0.67	0.64	1.00	1.46
	11	Quality of public transportation	-1.00		1.00		-0.67		1.00	

Category	No Assets	Components/ Assets	Kampong				Planned Cluster			
			Assets Availability	Score of Asset Class Availability	Score of Asset Class Function	Score of Asset Class	Assets Availability	Score of Asset Class Availability	Assets Function	Score of Asset Class Function
	12	Government/ community readiness for disaster management/ freedom from disaster	0.80		1.00		1.07		1.60	
	13	Availability of education facilities	1.73		1.90		1.87		2.00	
	14	Availability of health facilities	1.47		1.97		1.80		2.00	
	15	Availability of market facilities	1.20		1.40		1.27		1.57	
	16	Easy access to clean water	0.33		1.13		0.40		1.07	
	17	Availability of recreation facilities	0.33		0.73		0.33		0.77	
	18	Availability of public library	-0.13		1.23		-0.20		1.27	
	19	Availability of internet for community	0.07		1.00		0.13		1.47	
	20	Level of people's education	1.57		2.00		2.00		1.87	
	21	Quality of people's health	1.90		2.00		2.00		1.93	
Individual Intangible Assets	22	Capability in speaking foreign language	1.20	1.15	1.63	1.36	1.79	1.42	1.37	1.31
	23	Creativity and innovation of goods/ services	1.10		1.17		1.30		1.37	
	24	People's fighting spirit to get a better life	1.07		1.33		1.77		1.73	
	25	Kinship/ Social/ Cultural organization	1.53		2.00		2.00		1.87	
	26	Trust among society including mutual respect	1.47		2.00		2.00		2.00	
Social Intangible Assets	27	Cooperation by numbers of community regarding social, cultural, and economic matters	1.50	1,50	2.00	2.00	2.00	2.00	2.00	1.94
	28	Solidarity	1.50		2.00		2.00		1.88	
	29	The existence of discussion forum	1.50		2.00		2.00		1.97	
	30	Easy access to get bank credit, etc.	1.03		1.60		1.10		1.40	
	31	Easiness to get jobs	1.57		1.70		1.13		1.47	
	32	Opportunity to start business	1.63		1.77		1.13		1.40	
	33	Easiness to get daily consumption goods	1.97		1.93		1.47		1.70	
	34	Easiness to get raw material	0.07		0.77		0.53		0.90	
Economic Institutional Assets	35	Easiness to market products	0.80	1.38	0.90	1.56	0.53	1.03	1.03	1.49
	36	Opportunity to get government air/ subsiis for healts, ducation or business	2.00		2.00		1.93		1.97	
	37	Easiness to get health insurance	1.60		1.70		1.47		1.80	
	38	Capability to pay education fee	1.87		1.97		1.98		1.97	
	39	Capability to but/ get / or rent a house	1.67		1.83		1.67		1.73	
	40	Transportation cost	1.00		1.03		-1.57		1.00	
		Average	1.08		1.55		1.26		1.59	

The capability survey indicates that, in general, both the immigrant and the natives feel that the developments on the urban fringe benefit them. Only in some sectors people think that this development causes them setbacks and even so they believe this is overcome by infrastructure enhancement by the government. However, it is important to note that this capability survey is basically people's evaluation based on their experience of the past and present situation because principally people can only evaluate things based on their past experiences. However, this evaluation result is quite useful because it provides information regarding people's preference that can be utilized to predict their attitude toward the options of development direction offered by the government. In line with their function, the government, through the public planning board, must not merely follow people's preference, but also has to direct their choices to compromise between short-term needs and safeguard the well-being in the future. In fact, the main role of the government is to manage sustainable development.

3.6. Cost and Benefits Indications & Growth Management Challenges

The analysis of the perceptual data shows that the development of newly planned clusters in the urban fringe area is felt to provide positive changes in terms of the formation of social assets and economic institutional assets for the respondents. It is probably related to the process of magnification of housing scale that also enlarges the scale of people's economic activities. Regarding the social institutional asset, the immigrants feel a higher increase, but it is the native villagers who feel the higher benefits of it. In general, the changes in social and economic aspects have been considered to increase the people's capability and well-being. However, in general, it can be said that both the immigrants and the natives feel happy about this development. It is also indicated by the expressed happiness level in relation to the previously evaluated asset conditions.

One of the negative aspects of this that respondents see is particularly related to transportation, even for the immigrants. It is not only about the availability of the facilities, but also about the costs. Meanwhile for the natives, the negative aspect they strongly perceive is the downgrade of farmland productivity/quality. Regarding how insufficient public infrastructure (public tangible aspect) support is, people think it is more about the phase of development that is not yet to come. The government has not implemented it because it is still in the initial phase. Therefore, the unavailability of infrastructure is not regarded as a loss at this point.

The respondent's answers to the capability questionnaires only represent the individual interests. To reveal the advantages and disadvantages of this development more comprehensively, a Focus Group Discussion with stakeholder groups is necessary. One FGD meeting was held a few months before the capability survey, in January 2017 in office of the Regional Planning and Development Body (Bappeda) of Bantul and was attended by the staff of Bappeda, the Human Settlement Department and the Water Resource Department of the Office of Public Works, the Agricultural Office, the National Land Agency, the representatives of Indonesian Real Estates Yogyakarta (REI), heads of districts, and heads of villages. Based on this FGD, the potential advantages and disadvantages of this development and the stakeholders' attitude can be categorized.

The Agricultural Office, the Water Resource Department of the Office of Public Works, and local officials (heads of the districts and villages) tend to view the development of the planned clusters in their area as a disadvantage. The Agricultural Office is concerned about the loss of farmland in the Bantul area. According to their data, the rate of land transformation in Bantul is up to 40 ha per year. Meanwhile, this regency bears a duty from the Provincial Government to provide as much as 13,000 ha of sustainable farmland as the implementation of Law UU no 41/2009 regarding Protection of Sustainable Farm Land which has been followed up by Regional Regulation no 1/2011. The Agricultural Office argues that even without this duty, the loss of farmland can cause high prices of rice and other commodities in the future. Aside from that, they are concerned with the future ineffectiveness of the irrigation function. Meanwhile, the investment in irrigation and its maintenance is quite costly. The Water Resource Department of the Office of Public Works shares these concerns.

The heads of the districts and villages see the negative impacts of planned clusters development in their area in regard to the decline in supporting capacity of infrastructures, such as roads, drainage, and cemeteries. In their opinion, the growth of newly planned clusters will cause traffic congestion both in the villages and the collector roads connecting this area with downtown areas.

On the other hand, Bappeda and REI are in dilemma. Bappeda sees this growth as a generator of regional economic development, but on the other hand, they also see threats to farm sustainability. REI finds that they can only develop planned clusters in this type of area where farmland is still abundant because the land in urban areas is far too expensive. They state that in the last 5 years, it was still possible for them to find a land of IDR 500,000 to 1,000,000 per meter square in this area (around 36 USD to 73 USD per meter square), whereas it is very difficult to find such land under IDR 3 million (around 220 USD) in the agglomerated northern part of the city or in the town center. Farmland is much more affordable; in the same area, farmland costs only half or at most 2/3 of dry agricultural land. Farmland vast enough to develop an extensive cluster is also easier to acquire.

At the end of the FGD, there was a tendency by the participants to find a compromise for this problem. Their idea is to push housing investment in this area using the "infill" method, of developing clusters on dry agricultural land ready to build on and integrated it with the old villages. It should be also accompanied by investment collaboration to increase the supporting capacity of jointly utilized infrastructure. In addition, large-scale planned cluster development is only allowed on unoccupied land that, according to physical analysis, is not fertile or equipped with irrigation facilities and other agricultural infrastructure. This idea is quite interesting; however, all participants know how hard it is to implement.

4. Lessons Learnt, Policy and Theoretical Implication

Using the framework of the capability survey, this research discloses the impacts experienced by the community, from the perspective of both the immigrants and the natives. In general, people feel that the development of planned clusters is beneficial. However, the impacts are short-term. Indeed, human's subconscious mind can only assess what it has experienced and is experiencing. The long-term significance that humans think about usually concerns their descendants as reflected in the survey regarding education.

Only the experiences that people have gone through or are going through in their life can underlie most of their actions throughout their life. Regarding long-term interest – particularly the ones unrelated to the well-being of people's descendants – only the public sector can take care of it. The government staff in the public sector is concerned especially when it is their duty by law. For instance, in this case, the staff of the agricultural department is concerned with farm sustainability because it is an obligation under Law No. 41/2009 and Government Regulation No. 1/2011 regarding Establishment and Transformation of Sustainable Farm Land. The staff of the Water Resource Department of the Public Work Office feels responsible for the economic benefits of the irrigation channels they have built.

The above condition implies that a long-term interest as an essential principle of sustainable development must be institutionalized and campaigned to be the part of public awareness. The campaign should start with some issues people face directly, for instance, transportation costs and farmland fertility. Both of them are evaluated as getting worse in the capability survey. People should know the long-term consequence, for example by seeing cost projections. They should realize that their choices will not only affect them and their family but also the wider economy and ecology. They can see it through cost-and-benefit calculations that include the direct and indirect cost in a long run. The approach of monetary valuation such as the increase of food price due to the loss of farmland and the energy scarcity due to overuse will probably be easier to be comprehended by the community. This consciousness also growing in American thought. Atkinson and Ted Oleson [13] summarized that the costs of sprawl are becoming widely recognized across the social spectrum in the US. They note that in 1995, the Bank of America, the Resources Agency of California, the Greenbelt Alliance, and the Low-Income Housing Fund teamed up to produce a report titled "Beyond Sprawl: New Patterns to Fit the New California", which shows the costs of sprawl to taxpayers, businesses, suburban and central city residents, farmers, and the environment. The final report shows that continued economic development due to urban sprawl is not sustainable.

This research shows a gap between development concepts, especially the ones influencing the future, in the mind of public planner and people's "daily" decisions. It is rooted in the absence of methods and operational indicators that can bridge people's opinion dominated by their needs and present well-being and the government's responsibility to assure the sustainability of long-term improvement of the quality of life.

5. Acknowledgment

The author expresses his gratitude to Bappeda Bantul, which has collaborated in the research of planned cluster development on the fringe area of the Yogyakarta urban agglomeration at the end of 2016 and the beginning of 2017 as the part of the evaluation process upon the moratorium policy of planned cluster development investment permits in this area. Some of the data and the analysis presented in this paper are obtained through this collaboration project.

References

[1] Giyarsih S R 2001 Gejala Urban Sprawl sebagai Pemicu Proses Densifikasi Permukiman di Daerah Pinggiran Kota (Urban Fringe Area) (Journal PWK (Vol. 12, No 1/Maret 2001))

[2] Fang C and Yu D 2017 Urban agglomeration: An evolving concept of an emerging phenomenon, *Landscape and Urban Planning* (Vol 162, June 2017, pp 126-136)

[3] Kirkman L 2010 Did Americans Choose Sprawl? *Ethics & the Environment* (Volume 15, Number 1 Spring 2010) pp 123-142)

[4] Carrion C and Irwin Elena F 2004 Determinants of Residential Land-Use Conversion and Sprawl at the Rural-Urban Fringe, *American. J. of Agricultural Economic* (volume 86, Issue 4, 1 November 2004, pp 889-904)

[5] Ewing R Hamidi S Grace B J and Wei D Y 2016 *Does Urban Sprawl Hold Down Upward Mobility?, Landscape and Urban Planning.* 1 48, 2016 pp 80-88

[6] Deaton B James and Vyn. J Richard 2010 *The Effect of Strict Agricultural Zoning on Agricultural Land Values: The Case of Ontario's Greenbelt (*American Journal of Agricultural Economic) volume 92, Issue 4, 1 July 2010, pp 941-955

[7] Martinez C and Olander S 2015 *Stakeholder Participation for Sustainable Property Development In K Kähkönen (Ed.)* (Procedia Economics and Finance) Vol 21, pp 57-63 Elsevier. DOI 10.1016/S2212-5671(15)00150-1

[8] Arnstein Serry R 1969 A Ladder of Citizen Participation *J. American Planning Association* 35:4 pp 216-224

[9] Everard M and Longhurst J 2017 *Reasserting the primacy of human needs to reclaim the 'lost half' of sustainable development* (Science of the Total Environment)

[10] United Nations 2017 The Sustainable Development Goals Report

[11] Pramono R W D 2016 *Capability Approach for well-being Evaluation in Regional Development Planning: Case Study in Magelang Regency. Central Java, Indonesia (* Yogyakarta: UGM Press)

[12] Huppert F A & Whittington J E 2003 Evidence for the independence of positive and negative well-being: Implications for quality of life assessment *British. J. health psychology* 8(1) pp 107-122.

[13] Atkinson G and Oleson T 1996 Urban Sprawl as a Path Dependent Process *J. Economic Issues*, Vol. 30, No. 2,Jun., 1996, pp. 609-615, Taylor & Francis, Ltd.

IOP Conf. Series: Earth and Environmental Science **158** (2018) 012043 doi:10.1088/1755-1315/158/1/012043

Logistics system as an effort of integration in the Mandalika Special Economic Zone and its surroundings

T Suheri[1] and W G Melinda[1]
[1] Urban and Regional Planning Department, Indonesian Computer University, Dipatiukur Street 112-116, Coblong, Lebakgede, Bandung, West Java, 40132 Indonesia

Corresponding author: goldiemelindaw@gmail.com

Abstract. Indonesia as one of the maritime countries with its territory in the form of an archipelago. A SEZ with an integrated logistics system is very influential on the value of tourism, especially marine tourism. The creation of logistics activities in the Mandalika SEZ as a tourism area formed from the export and import of products as a form of tourism promotion will attract many domestic and foreign tourists. The importance of a logistics system for both people and goods can support the investment system in the area of Mandalika SEZ. The logistics system can provide agglomeration of facilities and infrastructure to support tourist activities in terms of cargo distribution both locally and globally. Mandalika SEZ requires an integrated logistics system to improve national competitiveness in the form of physical system support such as commodity availability and city/regional infrastructure as well as non-physical infrastructure such as service providers and regulation in order to be sustainable. The logistics system also plays a strategic role in synchronizing and harmonizing progress between economies and between regions in order to avoid an imbalance with the surrounding area. This study aims to find out the performance of the existing logistics systems in the Mandalika area and its surrounding areas in order to formulate policies to integrate the logistics systems within and outside the region. This research used content analysis methodology.

1. Introduction

The logistics system has a strategic role in synchronizing and harmonizing progress between economic sectors and between regions to realize inclusive economic growth and create a bastion of sovereignty and national economic authority and security [1]. Logistics is the process of anticipating customer needs and wants; acquiring the capital, materials, people, technologies, and information necessary to meet those needs; optimizing the goods or service-producing network to fulfill customer requests; and utilizing the network to fulfill customer requests in a timely way. The main actors of tourism logistics are transportation and tourism agents and suppliers. The actualization of a logistic system depends on the available infrastructure and transportation network to shift goods. The Mandalika Special Economic Zone is one of the districts of Central Lombok and is prominent in tourism spots and resorts, fisheries, and transportation surrounding the area. Thus, it requires a proper logistics to distribute people and goods effectively and efficiently.

The integrated logistics system named Special Economic Zone is very influential on the value of Indonesian investment as a maritime country. The Special Economic Zone has a function and economic criterium as tourism development area. The development of logistics activities in the Mandalika SEZ as a tourism area is formed from the activities of exporting and importing products in the form of tourism promotion which will attract many domestic and foreign tourists. The purpose of the Special Economic Zone is to accelerate regional development and trigger new economic growth.

The Mandalika Special Economic Zone is prepared to become a center of Meetings, Incentives, Conventions, and Events (MICE) on an international scale. A logistics system for people or goods can be achieved through the investment system in the Mandalika Special Economic Zone, and it can provide agglomeration of facilities and infrastructure to support tourist activities in terms of cargo distribution both locally and globally. In other words, the Mandalika Special Economic Zone must provide a cargo port to fulfill trading needs and services such for hotels, bars, restaurants, crafts (souvenirs), shopping centers, and other support facilities. This logistics system also has a strategic role in synchronizing and harmonizing the economic development within the region to prevent the disparities phenomena. This also increases the Competitiveness Index of the Mandalika Special Economic Zone. Mandalika SEZ requires an integrated logistics system to improve national competitiveness in the form of physical system support such as commodity availability and city/regional infrastructure as well as non-physical infrastructure such as service providers and regulations in order to be sustainable.

The objective of this study is to find out the performance of the existing logistics systems in the Mandalika area and its surroundings in order to formulate policies to integrate the logistics systems both within and outside the region. The performance of the logistics system is studied based on the Logistics Performance Index (LPI), and policy strategies.

2. Methodology

This research is conducted to gain in-depth knowledge about the performance of the logistics system in the Mandalika Special Economic Zone and its surroundings. The research uses a qualitative approach to content analysis to get in-depth data, i.e., data that contains meaning [2] of a social phenomenon [3]. The Logistics Performance Index (LPI) approach is used in the data analysis to find out the performance of the logistic system within and around the SEZ.

2.1 Method of Data Collection

This study uses primary and secondary data. Primary data collection is done through field observation in SEZ Mandalika. Meanwhile, the secondary data collection is done through literature study and data collection from related institutions.

2.2 Method of Data Analysis

The data analysis in this study is conducted through content analysis to understand and interpret data of the field conditions in Mandalika SEZ. The result of the content analysis and then compared with the field conditions of Mandalika SEZ. The data analysis method used is Logistics Performance Index (LPI). The information to be extracted from the Logistics Performance Index consists of six indicators that are divided into two categories. Table 1 below shows the Logistics Performance Index:

Table 1. Logistics Performance Index (LPI) [4]

No	Category	Indicator
1	Area	Efficiency of the clearance process
		Quality of trade and transport-related infrastructure
		Ease of arranging competitively priced shipments
2	Process	Competence and quality of logistics services
		Ability to track and trace consignments
		Timeliness of shipments in reaching the destination within the scheduled or expected delivery time

3. Mandalika Special Economic Zone

Mandalika SEZ consists of several zones, i.e., export processing, logistics, industry, technology development, tourism, and energy. This, the Special Economic Zone needs to build any other facilities and houses for workers. Meanwhile, micro, small, and medium scale businesses need to be provided a location to support the activity in Mandalika SEZ. The Special Economic Zone is prepared as a new geo-economic prominent and geo-strategic area to support industrial activity, export, import, high economic value activities, and international competitiveness [5]. The Mandalika Special Economic Zone is needed to accelerate the economic development in West Nusa Tenggara and support national economic development. Figure 1 shows the potential distribution area of Mandalika SEZ.

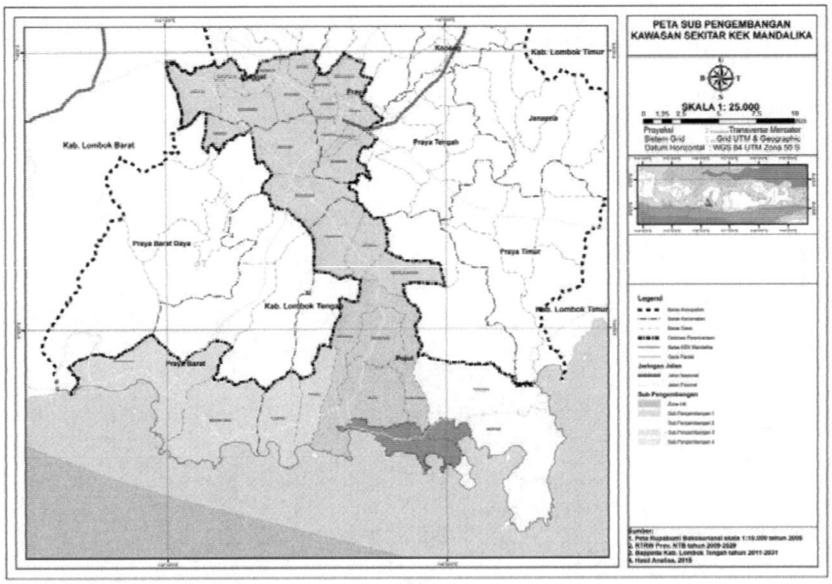

Figure 1. Potential distribution area of the special economic zone

The potential distribution area surrounding Mandalika SEZ is divided into several sub-area developments. The core area is located in Mandalika SEZ, which has many sea tourism sites like beaches, and promontory in SP-1, which is focused on supporting local economic development. The development process will involve local resources by encouraging cultivated economic productivity. The tourism development concept in SP-1 requires basic facilities by providing accommodation (hotels, resorts, guest houses), restaurants, and bars surrounding the site. SP-2 is the development area that is focused on fishery based on local resources. The development process will increase complementary facilities such as cold storage to keep fish and lobster. SP-3 focuses on local economic activity based on tourism and cultural sites with local engagement.

Accelerating the economic development (export and import) requires logistic principles in tourism management. The main principle of the logistics system is to tourist activities effective and to increase benefits on the micro scale. However, the higher tourism principle of scale is to utilize natural resources and local people surrounding Mandalika SEZ to be sustainable.

The sustainability of tourism relates to the functional logistic structure of tourism and includes traditional aspects, like logistics, and manufacturing. The purpose of the logistics system is to facilitate the product flow to the production area and tourism consumption. This requires the important step to identify the consumer flow in Mandalika SEZ and its surrounding area so that all potential in Mandalika SEZ will be integrated into the logistics system to efficiently use time and money in supporting restaurants, homestays, and hotels in Mandalika SEZ. In the next five years, homestays, restaurants, cafeteria, and retail areas are planned in the western zone which can absorb up to 5.000 local employees [3]. West Nusa Tenggara has Islamic local values and is well-known as a global halal destination for tourism. Mandalika SEZ will implement Islamic local values and become a Muslim-friendly zone. Surakarta village, Jonggat district offers cultural tourism. This village attracts many tourists domestically and from overseas as a songket village. In Mandalika SEZ, this village contributed in human resources; every woman in this village is capable of creating songkets.

In developing tourism activity, the development of the area will gradually cause a demand for hotels, guest houses, restaurants, and other facilities which requires a supply of commodity production in agriculture (paddy fields). Based on observations, there are several districts could support Mandalika Special Economic Zone especially in food security, i.e., Praya district, particularly West Praya, East

Praya, and Central Praya, which has a large horticulture area, fisheries, lobster, and seagrass. The water supply can be obtained from Rinjani Mountain, Batuja, and the Pengge Dam in West Praya.

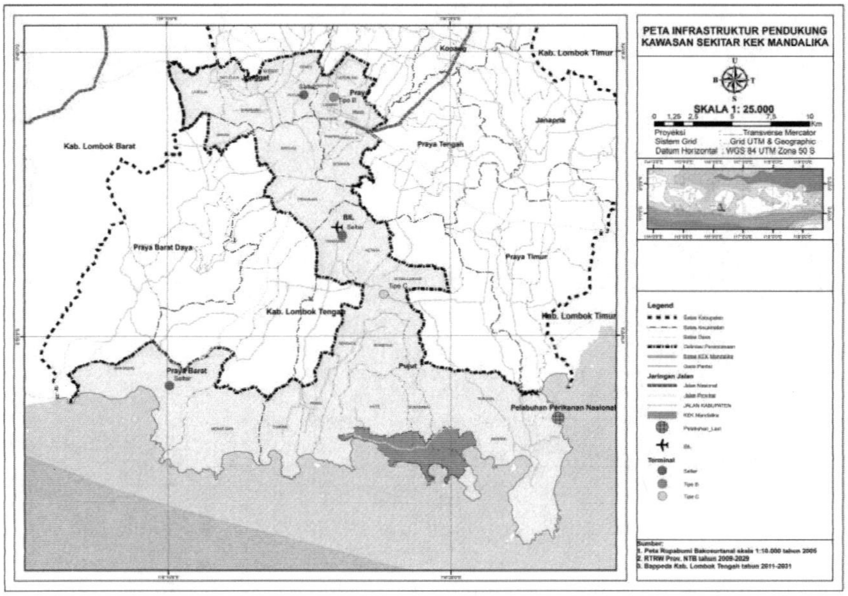

Figure 2. Distribution of supporting infrastructure around Mandalika Special Economic Zone

To serve the tourism activities in Central Lombok Regency, the International Airport Lombok (BIL) is already available. This airport already serves foreign countries and is located in Pujut Sub-district. In addition, there is a Type B Terminal in Praya District, a Type C Terminal in Pujut District, and Type C Terminals in East Praya Sub-district. Moreover, there are also supporting shelters located in Jonggat Sub-district, Pujut Sub-district (near BIL), West Praya Sub-district, Southwest Praya Sub-district, and Janapria District.

4. Logistics system

Mandalika Special Economic Zone was developed for marine tourism and religious tourism in Central Lombok Regency. Its location adjacent to the island of Bali is one of the geographic advantages that can promote the Mandalika Special Economic Zone to domestic and international tourists. Figure 3 below explains the flow of tourists in the tourism logistics system. This travel activity forms the logistics flow of tourism activities of travel agencies and tour operators.

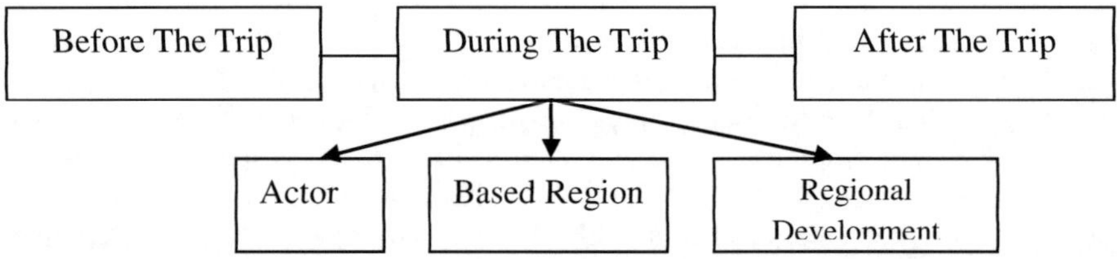

Figure 3. Tourist flows in tourism logistics

The SISLOGNAS blueprint as a system that supports supply chain management consists of six important aspects, i.e., commodities; transportation infrastructure; technology and communication; logistics HR; logistic regulation; and national logistic institution. The main discussion is about the availability of transportation infrastructure and regulatory support in the field of logistics in the Mandalika Special Economic Zone. Tourism activities cannot be separated from the provision of logistics systems; it is necessary to develop a good logistics system in order to provide supply and demand for goods to meet the needs of actors (lodging and restaurant service providers) in the Mandalika Special Economic Zone, the support of the surrounding area, and the encouragement of regional development. Thus, it takes a strategic formula how to meet the needs of these service providers by utilizing the potential around Mandalika Special Economic Zone in the form of natural and human resources to improve the economic growth of the region.

For supporting the movement, there is a provincial road network that connects between districts. Tourism area cannot be separated from the logistics system. To offer a complete holiday experience the aspect of cargo movement must be available. The support of the tourism logistics system for the Mandalika Special Economic Zone is divided into several levels. i.e., local, micro, meso, and macro. The regional structure of tourism logistics reflects its six levels of regional (geospatial) systems:

1. Local level (tour logistics);
2. Micro-level (tour operator logistics);
3. Mesoscale-level (tourist and resort zones and tourist district logistics);
4. Macro-level (national tourism industry logistics);
5. Mega-level (global macro-regions logistics);
6. Meta-level (global tourism logistics) [6].

With the system of tourism activity, a demand for lodging services and restaurants will emerge around the tourist sites, thus, this requires the supply and of distribution of main of agricultural commodities and products such as rice. Based on the observations, there are several sub-districts that can support service activities in the Mandalika Special Economic Zone in providing food resources in Praya Sub-district, partly in Praya Barat Daya, West Praya Barat, Praya Timur and Central Praya, which has land use of food crops and horticulture and support from the fisheries, lobster, and seaweed sectors. The need for clean water sources can be fulfilled using spring water from Mount Rinjani, the Batuja Dam and the Pengge Dam in Praya Barat Daya Sub-district.

To develop an area of international-level maritime tourism, a macro logistics system planning is needed, which includes the optimization of local, micro, meso and macro logistic services. In the Mandalika Special Economic Zone several restaurants, hotels, and resorts are planned to be built. To meet the needs of these services activities requires support from the area around Mandalika Special Economic Zone.

IOP Conf. Series: Earth and Environmental Science **158** (2018) 012043 doi:10.1088/1755-1315/158/1/012043

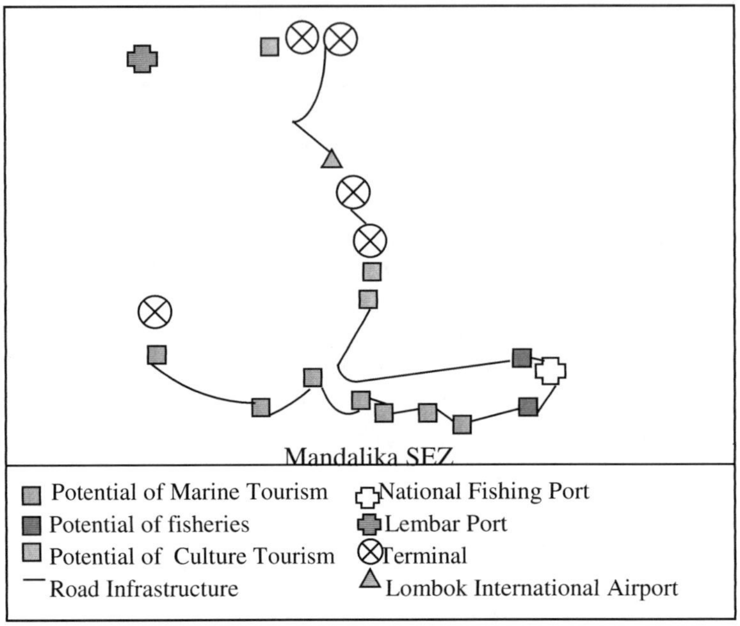

Figure 4. Infrastructure and transportation support in Mandalika SEZ

Figure 4 explains the infrastructure and transportation availability to support the distribution of logistics in tourism activities within the Mandalika Special Economic Zone. The infrastructure and transportation support that is available for tourist travel is through land, sea and air transportation. In the city of Mataram, the Port of Lembar-Gili Mas can serve tourists from and to SEZ Mandalika. In addition, land transportation services are also available; road transportation uses the provincial and district road network as a liaison between the port and the International Airport of Lombok to SEZ Mandalika.

The distribution of major commodities for the Mandalika Special Economic Zone from the surrounding areas requires a good logistics system that is related to production costs (availability of manpower and production areas); warehousing costs and strategic warehouse location; commodity quality; and the availability of various modes of transportation including cost and time. Central Lombok Regency supports the tourism logistics system in the Mandalika Special Economic Zone, in accordance with Central Lombok Regency Spatial Plan by providing the infrastructure system and transportation of the National Fishery Port located east of Mertak Village, Pujut Sub-district. The provision of cargo movement at the port is an important tool in the storage of major commodities supporting services activities in the Mandalika Special Economic Zone; it requires a flexible efficient and well-organized supply chain. A good logistics system will have an impact on the economic growth and investment of Mandalika SEZ both from within and outside the region. Economic growth cannot be separated from the quality support of the surrounding communities related to the use of human resources. The efforts in improving the quality of human resources in the Mandalika Special Economic Zone is done by building vocational high schools around the region. The improved skills in the community are needed to optimize services for local and international tourists in the Mandalika SEZ. The emergence of multiplier effects will impact various sectors including logistic services.

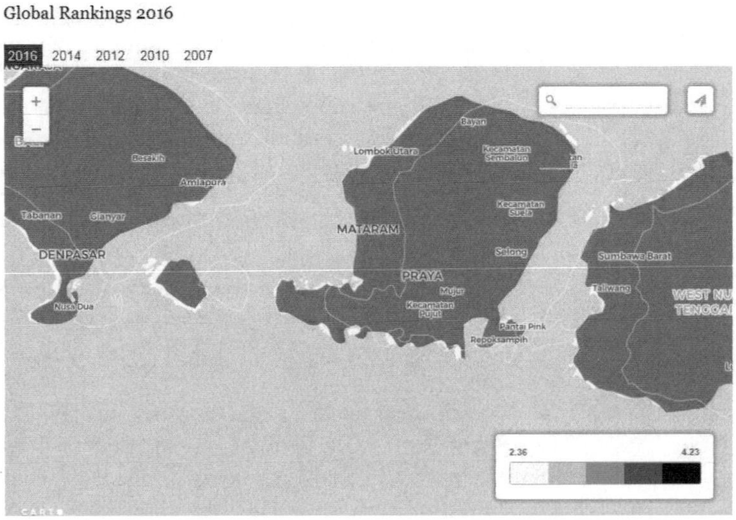

Figure 5. International Logistic Performance Index [4]

The World Bank has developed the Logistic Performance Index (LPI) measurement based on six indicators and grouped into two categories

Table 2. Logistics Performance Index on Mandalika Special Economic Zone

No	Category	Indicator	Result
1	Area	Efficiency of the clearance process	The efficiency and ease of customs provided by the Indonesia Tourism Development Corporation (ITDC) in Mandalika Special Economic Zone is also offered to investors in the Mandalika Special Economic Zone with a good service system. Law No. 39 of 2009 on Special Economic Zone, explained the ease given to investors to invest in Special Economic Zones (SEZ), namely: (a) The import of goods is granted facilities in the form of suspension of import duty, excise duty as long as the goods are raw materials or auxiliary materials production, no levy of Value Added Tax (VAT), Sales Tax on Luxury Goods and Import PPh. (b) The requirement of taxable goods from elsewhere within the Special Economic Zone may be granted a facility not subject to VAT and Sales Tax on Luxury Goods under the provisions of laws and regulations; (c) The transfer of taxable goods from the Special Economic Zone to other places within the customs area to the extent not addressed to the party receiving the facilities of VAT is subject to VAT or VAT and Sales Tax on Luxury Goods in accordance with the provisions of the law; (d) Reduction of Land and Building Tax in accordance with the provisions of the Laws and Regulations; (e) Any taxpayer conducting business in the Special Economic Zone shall be given incentives in the form of exemption or relief of local taxes and regional levies in accordance with the provisions of laws and regulations; and (f) to be granted ease and relief in the business licensing, business, industrial, trading, port and immigration sectors for foreigners of business and security facilities.

No	Category	Indicator	Result
		Quality of trade and transport-related infrastructure	The Mandalika Special Economic Zone is supported by infrastructure and transportation over land, sea, and air. Infrastructure and land transportation is supported by the condition that most of the main roads that are traversed are in excellent condition. The main roads, especially those connecting districts, are the primary arterial roads that have the status of National Road and Provincial Road. The main roads connecting these districts include the main infrastructure in Central Lombok Regency. The land surrounding Mandalika Special Economic Zone is dominated by watermelon plantations and corn and empty land along the road to Mandalika Special Economic Zone. Infrastructure and sea transportation is supported by ports. Sheet Port - Gili Mas supports sea transportation from and to Mandalika. Gili Mas port will be used as one of the tourist entrances to the Special Economic Zone (SEZ) Mandalika, in Central Lombok Regency, West Nusa Tenggara. The development of Gili Mas Port as an expansion of Lembar Port aims to facilitate overseas vessels with a large capacity that bring tourists to Lombok. In addition, there is also an air infrastructure support, namely the availability of Lombok International Airport (BIL).
		Ease of arranging competitively priced shipments	The shipping of goods to and from SEZ Mandalika is now easy because of the availability of logistics services in Mandalika area. The shipping price from and to Mandalika SEZ is still quite expensive because the distance is quite far and the quantity transported is still small, so the cost is still high for the delivery of goods or cargo.
2	Process	Competence and quality of logistics services	In the Mandalika Special Economic Zone and surrounding areas services are available such as logistic delivery service from and to Mandalika offered by PT Mandalika Gress, PT Wahana Prestasi Logistik, and Lombok Anugerah. This will help service activities in the Mandalika Special Economic Zone and surrounding areas as there will be 18 investors who will build hotels in Mandalika Special Economic Zone
		Ability to track and trace consignments	The Inaportnet online application system into the field of marine transportation facilitates tracking shippings by online logistics transporters. The Inaportnet application helps in the form of data and information about tracking and tracing of logistic goods. Lembar Port has been registered in the Ministry of Transportation's online system along with 43 other ports for easy searching of data due to centralized ship master data and easy monitoring of ship activities. However, tracking and tracing efforts have not reached the Inaportnet application stage.
		Timeliness of shipments in reaching the destination within the	The timeliness of delivery in achieving the objectives within the scheduled or expected delivery time for logistics services from and to Mandalika is good as evidenced by the existence of Standard Operating Services (SOS) in the providers of freight services available in Mandalika and its surroundings.

No	Category	Indicator	Result
		scheduled or expected delivery time	

Several indicators such as the efficiency of clearance processes and the quality of infrastructure related to transportation and have the best performance among the four other LPI indicators. The following strategies can be formulated to improve the logistics performance index in the Mandalika SEZ and surrounding areas.

Table 3. Logistics Performance Index on Mandalika Special Economic Zone

No	Category	Indicator	Strategy
1	Area	Efficiency of the clearance process	The implementation of the clearance process has been supported by the policy of the ITDC for investors, so it is expected that the utility to support the quality of facilities and infrastructure of service activities will be increased to attract investors in Mandalika SEZ.
		Quality of trade and transport-related infrastructure	The infrastructure related to trade and transportation around the Mandalika Special Economic Zone must be integrated in order to support the competitiveness of international trade between land, sea, and air infrastructure. Currently, the transportation from the Port of Lembur - Gili Mas to the SEZ Mandalika is not as well such as the connecting road network from the Port to SEZ. Thus, this requires the access roads to shorten the time and reduce the cost incurred by tourists and/or logistic delivery from and towards SEZ Mandalika.
		Ease of arranging competitively priced shipments	The ease of arranging competitively priced shipments is related to improving the quality of infrastructure. Integrated infrastructure will facilitate the delivery of goods so that the costs incurred are small. Currently, infrastructure that can support logistic delivery activities to support SEZ is by land and air transportation. Shipping by land transportation is supported by the Lombok International Airport (BIL). Improving the available road network infrastructure relates to support facilities such as lighting around the roads from and to the airport needs to be improved so that the delivery of goods can be smooth and goods can be delivered in a good condition. Improvement of this indicator can support the logistics aspect of tourism activity in the area of SEZ Mandalika and its surroundings.
2	Process	Competence and quality of logistics services	The logistics services in Mandalika are not competitive because of the limited availability of cargo transportation through sea transportation for long distances and over land for short distances. Some of the average logistics services are the delivery from and to the provincial capital of Mataram City. The strategy that can be developed is to provide a logistics delivery route from and to Mandalika SEZ.
		Ability to track and trace consignments	The tracking and tracing consignments require support of data and information on vessels and transport logistics. This data and information can be acquired easily through the online applications of the Ministry of Transportation, especially Sea

	Transportation (Hubla). This effort is the implementation of the policy of the Minister of Transportation Regulation No. PM 157 Year 2015 on the Application of Inapòrtnet For Service Ships and Goods at the Port. Inaportnet is a single electronic internet-based information service system to integrate standard port information systems in serving vessels and goods from all related institutions or stakeholders at the port (including the Port Enterprise Service system). It is expected that Lembar Port can be registered in the Inaportnet tracking application like other registered ports such as the ports of Belawan, Palembang, Teluk Bayur, Panjang, Pontianak, Banjarmasin, Balikpapan, Tanjung Priok, Tanjung Emas, Tanjung Perak, Makassar, Bitung, Ambon, Banten, Gresik, and Sorong.
Timeliness of shipments in reaching the destination within the scheduled or expected delivery time	The timeliness of shipments in reaching destinations within the scheduled or expected delivery time can be increased by improving Standard Operational Procedure (SOP) of logistic service companies respectively related to receiving points, transportation and delivery control processes and handover of goods. The process of submitting and handover of this item is related to logistic carrier information; the time (date and time) goods arrived at the warehouse; the time (date and time) goods start unloading; the time (date and time) goods are unloaded; the details of goods received (item name/item code/quantity/ volume/weight); condition of the goods received and the warehouse officer receiving the goods. This activity should be transparent and accountable so that the quality of on-time scheduled delivery can run well in Mandalika SEZ.

5. Results

The result of content analysis about the performance of the logistic system in Mandalika SEZ based on six indicators from the Logistics Performance Index (LPI) is explained in Table 4.

Table 4. Results of the Logistic Performance Index in Mandalika SEZ

No	Indicator	Logistics Performance Index
1.	Efficiency of the clearance process	Good
2.	Quality of trade and transport-related infrastructure	Good
3.	Ease of arranging competitively priced shipments	Not good
4.	Competence and quality of logistics services	Not good
5.	Ability to track and trace consignments	Not good
6.	Timeliness of shipments in reaching the destination within the scheduled or expected delivery time	Not good

This assessment is based on the conformity of indicators with existing field conditions in the SEZ Mandalika and supported by policy review.

6. Conclusion

Mandalika Special Economic Zone was developed as a tourism development economic concept. It needs a logistics system to integrate the Mandalika Special Economic Zone and its surroundings. The study found the following conclusions.

1. The performance of the existing logistics system in Mandalika and its surrounding areas in general still requires physical and non-physical improvement and policies in order to integrate the logistics systems inside and outside the region.
2. Based on the "area" indicators of the Logistics Performance Index (LPI), the efficiency of the clearance process and quality of trade and transport related infrastructure is already good in SEZ Mandalika.
3. The availability of logistics services in West Nusa Tenggara Province only serves from Lombok Island only to Mataram City, so the cost of sending logistics to SEZ Mandalika is high because the cost incurred if converted in units of distance traveled is increased.

7. Recommendations

The recommendations for further study of logistics system as an effort of integration in the Mandalika special economic zone and its surroundings are as follows:

1. In the short term, the process of cargo logistics in Mandalika through land transportation and facilities can still be controlled but if the demand increases, the logistic supply to the Mandalika Special Economic Zone will be insufficient by land route. Thus, demand must be fulfilled by the areas surrounding the Mandalika Special Economic Zone by sea and air.
2. It is necessary to improve the quality of infrastructure and sea transportation in the Mandalika Special Economic Zone.
3. Central and local government support can optimize the logistics system in Mandalika Special Economic Zone.

References

[1] Blueprint of National Logistics System Development. Pres. Regist. 26/2012
[2] Sugiyono 2008 *Metode Penelitian Kuantitatif, Kualitatif dan R&D* (Bandung: Alfabeta)
[3] Creswell J W 2013 *Research design: qualitative, quantitative, and mixed methods approaches* (California: SAGE publications)
[4] World Bank 2016 Logistics Performance Index [Online]. Available from: https://lpi.worldbank.org/
[5] Indonesia Tourism Development Corporation 2017 Mandalika [Online] Available from: https://www.itdc.co.id/the-mandalika/home-page/
[6] Kochadze T, Dangadze I, and Zaqareishvili V 2013 The role of logistics in the market for transportation and tourist services *Machenes, Technologies, Materials* **4** 1-3

IOP Conf. Series: Earth and Environmental Science **158** (2018) 012044 doi:10.1088/1755-1315/158/1/012044

Modelling Container Terminal Resilience Measurement by Considering Hinterland Losses

F Zukhruf[1], R B Frazila[1]

[1]Department of Civil Engineering, Institut Teknologi Bandung, Jl. Ganesha No. 10, Bandung, 40132, Indonesia

Corresponding author: febri.zukhruf@ftsl.itb.ac.id

Abstract. Due to its significance, specifically in archipelagic countries, seaports have been long believed to play a considerable role in the freight transport network. The ongoing expansion of world trade also promotes container transport, which offers a more efficient movement process. Therefore, over the years, numerous studies elaborated the framework concept and practical strategies to improve the efficiency of container terminal (CT). However, CT is recently challenged by reliability issues resulting from natural disasters and disruptive man-made events. The resilience level of CT is then substantially required to investigate its system ability to handle disruption events. Since CT disruption directly impacts the hinterland, this paper then proposes a model to measure CT resilience by considering the hinterland losses, which are rarely explored. This paper also incorporates the customs process, the transport link between CT and hinterland, and its quantitative measurement for calculating hinterland losses. To represent the uncertain condition of transport links, the Monte Carlo simulation structure is invoked to generate the random probability event.

1. Introduction

As gateways of domestic and international trade, seaports play a vital role in the economic effectiveness of many countries. Its performance is practically believed to greatly influence logistics costs (see [1-2]), which shape the economic effectiveness and competitiveness, and thus, the topic of port efficiency attracts a huge number of studies (e.g., [3-7]). In addition, the rapid change of trading market reflects containers' essential position as transport equipment, which then seriously restructures trade patterns and manufacturing strategies [8]. Another report also shows the considerable increase of container shipment, specifically in the developing countries [9].

Despite its essential increase, container transport faces challenges relating to the seaport disruption, which may be caused by natural phenomena and man-made disasters (e.g., labor unrest, technological accidents) as well. The risk of disruption possibly directly impacts not only port operations but also several supply chain parties, even influencing regional and global activities. For instance, after the 2011 tsunami, around 7% of total containers handled were impacted by to the seaport shutdown [10]. In addition, vehicle manufacturers are severely affected by port disruptions, especially in terms of export activities and material supplies [11-12]. This fact then conveys a conclusion about the close relation between ports and their hinterlands. Several researches have tried to capture the relation between a seaport and its hinterland (e.g., [13-14]). However, the impact of port disruption to its hinterland within the framework of resiliency still leaves room for exploration.

Literatures define the term 'resilience' in several ways. [15] describes resilience as a system's ability to reduce the probability of disruption, the impact of disruption, and its recovery time. [16] states that resilience is the ability to recover to a certain level after being affected by a disruption event. Other scholars illustrate resilience based on its key features, namely, robustness, redundancy, resourcefulness, and rapidity [17-18]. Based on the concepts of resilience, several papers on port disruption analysis were thus written. [13] investigated the quantitative measurement of port resilience from the shipper's perspective. [16] proposed a strengthened conceptual framework of seaport for dealing with a small-scale disaster. Bayesian belief network has also been used to model seaport related variables as well as to investigate resilience strategies [19]. The utilisation of real time and statistical port data has been conducted by [20] to develop an integrated system in order to forecast and to evaluate the seaport risk in extreme storms.

Although recent studies calculated the impact of port disruption to the hinterland, less attention was paid to the incorporation of CT operation. As the performance of CT depends on the operation productivity [21], which is constructed by equipment activities, disruption to certain equipment delivers different impacts to the hinterland. For instance, the disruption of QC activities may deliver a more severe impact compared to other activities. This paper thus proposes a measurement model which not only involves hinterland impact analysis but also takes into account the operation process to estimate the impact better.

The rest of the paper is arranged as follows: the following section describes the system and modelling framework. The third section elaborates the case study to investigate the impact of disruption to CT by involving CT operation and hinterland analysis. Finally, the fourth section summarizes the methodologies, results, and analyses in the paper.

2. Measurement Model

2.1. Definition of resilience and measurement model

By deliberating various definitions of resilience, this paper then delineates the term 'resilience' to build the measurement model. The resilience of CT refers to the system ability, with instant recovery response, to maintain performance level during disruptions as well as to reduce the economic losses of hinterland activities. The required time to transfer container to hinterland is thus selected as the performance indicator, which has a direct connection to the hinterland losses. Therefore, the resilience level in the u-disruption event is formulated as below:

$$R(u) = \frac{\varsigma z_0 - \Delta z(u)}{\varsigma z_0} \tag{1}$$

s.t.

$$\Delta z(u) = z_0 - z(u) \tag{2}$$

$$z(u) = \left[\frac{1}{M}\sum_{m=1}^{M} W_m^1\left(q_m^1, p\left(n_m^1, u\right)\right) + \frac{1}{M}\sum_{m=1}^{M} W_m^2\left(q_m^2, p\left(n_m^2, f_g, C_g, u\right)\right) + W^3\left(q^3, p\left(n^3, u\right)\right) + W^4(\theta, \tau) + W^5\left(f_l, C_l, \theta, \tau\right)\right] \tag{3}$$

where,

$R(u)$: resilience level against disruption event u,

z_0 : transport time to arrive at hinterland without any disruption,

$z(u)$: transport time to arrive at hinterland in u-disruption event

$\Delta z(u)$: container delay time due to u-disruption event,

u : vector set of disruption events,

ς : acceptance level of CT performance,

W_m^1 : required time to transfer container from ship to TTU by QC at dock-m,
W_m^2 : required time to transport container to the container yard (i.e., CY) by TTU at dock-m,
W^3 : required time to handle container in the CY by RTGC,
W^4 : elapsed time at the customs process,
W^5 : transport time of container from CT gate to the hinterland using transport link,
q_m^1 : number of containers served by QC at dock-m,
q_m^2 : number of containers served by TTU at dock-m,
q^3 : number of containers served by RTGC,
p : productivity of equipment for handling container,
n_m^1 : number of QC at dock-m,
n_m^2 : number of TTU at dock-m,
n^3 : number of RTGC,
τ : random value that follows distribution θ,
f_g : TTU traffic flow of link g at CT,
$C_g(u)$: capacity of link g during u disruption event,
f_l : traffic flow of transport link l,
C_l : capacity of transport link l,
M : total number of docks.

The resilience level is calculated by considering not only the delay time due to the disruption u but also the performance acceptance level (i.e.,ς). The level illustrates the persistent performance level that needs to be preserved. For instance, if the acceptance level is set equal to two, the delay time must be maintained to be less than twice of moving time in the normal condition. In the case that the delay is greater than the persistent performance level, it can be stated that the CT system has failed to handle the disruption event. The required time to move container to hinterland is simply estimated by summing the transferring time in the CT (i.e., W_m^1, W_m^2, W^3 ,W^4) and in the transport links (i.e., W^5).

As inferred in Equation (3), the disruption event possibly strikes the Quay Crane (QC) operations, links capacity, which serve truck trailer unit (TTU), as well as the rubber tyre gantry cranes (RTGC) operations.

2.2. *Definition of resilience and measurement model*

The CT operation model is then expressed in this section, which is constructed by considering the import processes in the CT. The demand is set deterministically to represent the planned order of industrial raw material, even though the vessel size may be varied. The size then determines the number of QCs assigned, where each ship is assisted by 2 or 3 QCs. The QC productivity also varies based on certain distributions, which also occurs in terms of TTU and RTGC productivity. The horizontal movement of container between ship and container yard is handled by TTU, in which its productivity is directly impacted by the travel time along the horizontal link in the CT. Since the travel time is a function of link capacity and link flow, the reduction of link capacity in the disruption events possibly toward the increasing delay, and then, it truthfully affects the bigger process of port operation. Hence, the Bureau of Public Roads (BPR) function (see Equation (4)) is included to illustrate the flow and capacity interactions, as it is described below.

$$a_g = \sum_{g}^{G}\left(a_g^0\left(1+0.5\left(\frac{f_g}{C_g(u)}\right)^{1.5}\right)\right) \tag{4}$$

where:

a_g : travel time (hour) of link-g,

a_g^0 : travel time of link g at free flow condition.

The container is continued to the CY, which is managed by RTGC. The handling time is calculated based on the queuing theory by considering the fluctuation of RTGC productivity. To illustrate the actual process of container transportation in CT, the model also takes into account the customs processes of import product, which divides containers into four line classes (i.e., red, yellow, green, and priority line) based on the risk of commodity. The red line is specified for the commodity with a high risk and importers with insufficient track records, which then requires physical and document assessments. On the other hand the green and priority lines are set for the lowest risk of commodity as well as importers with trusted profiles, and thus, it needs a shorter time to pass the customs processes. Such classes are then invoked in the model by generating a random number by following a certain distribution, which is derived from the CT actual condition.

2.3. Resilience definition and measurement model

To link the CT with hinterland, the model considers three different modes of transportation for the container, namely truck, train, and ship/vessel. The ship mode is closely related to the inland waterways mode, which obtains growing attention in Indonesia. The travel time of railway mode in this case might be considered deterministic, while for the ship mode, the speed variation during the voyage is randomly propagated based on a certain distribution:

$$a_{ship} = \frac{d_{ship}}{v_{ship}(\tau,\theta)} \tag{5}$$

where:

u_{sea} : travel time of ship mode

d_{ship} : trip distance

v_{ship} : vessel speed

Unlike the ship mode, the variation of traffic flow and road capacity should strongly affect the truck travel time. Hence, the BPR function is also used to estimate the travel time at the connecting link by reflecting the uncertainty of traffic and road capacity. As disruption of CT operation influences the hinterland activity, estimating the losses is thus beneficial. The hinterland is further simplified as the industrial cluster, which requires raw material to ensure production activities as well as to continue their supply to the costumers.

The losses probably lie on the direct and indirect loss, in which the direct loss corresponds to the commodity damage, and the indirect loss relates to the production/manufacture delay due to lack of raw material. To illustrate such conditions, the model then invokes the economic losses estimation, which is proposed by [14]. Equation (6) shows the direct loss estimation, which accounts the unit value of commodities, its volume, the transport cost, and the storage cost.

$$L_{direct} = p_{raw}V_{import}\Delta z(u) + c_{raw}V_{import}\Delta z(u) \tag{6}$$

where,

L_{direct} : direct loss due to commodities damage,

IOP Conf. Series: Earth and Environmental Science **158** (2018) 012044 doi:10.1088/1755-1315/158/1/012044

p_{raw} : unit value of commodities,

c_{raw} : unit extra cost for handling the damaged commodities,

V_{import} : volume of import commodities.

The losses which are further considered are derived from the production delay in the CT hinterland. Since the delay which occurs in the CT sequentially affects the production in the hinterland, Eq. (7) illustrates the three different types of impact. In case of delay time due to disruption below the remaining time of raw material inventory, the disruption then gives no impact to the production, which is described by the first row of Eq. (7). The second row explains the condition that the delay time is bigger than the raw material buffer time, though the product stock is still sufficient to provide the supply. In this condition, the manufacturer in the hinterland incurred the losses in term of production process. If the product stock is insufficient to maintaining the normal flow to the costumer, the manufacturer then suffers the cancelation losses.

$$L_{idrect} = \begin{cases} 0 & \text{for } \Delta z(u) < I_{raw} \\ p_{prod}V_{prod}\left(\Delta z(u) - I_{raw}\right) & \text{for } I_{raw} < \Delta z(u) < I_{raw} + I_{prod} \\ p_{prod}V_{prod}\left(\Delta z(u) - I_{raw}\right) + r_{cancel}\, p_{prod}V_{prod}\left(\Delta z(u) - I_{raw} - I_{prod}\right) & \text{for } I_{raw} + I_{prod} < \Delta z(u) \end{cases} \quad (7)$$

where,

L_{idrect} : production loss due to the delay,

p_{prod} : unit value of product,

V_{prod} : volume of production,

I_{raw} : buffer time of raw material stock to continue the production,

I_{prod} : buffer time of product stock to provide the costumer supply

r_{cancel} : cancellation rate

3. Numerical Experiment

The proposed model for measuring the resilience is then tested in a network (see Figure 1), including a CT and an industrial cluster as its hinterland. The CT operation incorporates the transferring process from the vessel to the gate facilities. The process involves the handling activities of QC, TTU, RTGC and customs. The attribute of handling equipment is derived from the actual CT data set, which are utilized as the simulation input.

In this experimental case, in the normal condition, CT is assumed to be armed by 1 dock with the simultaneous operation of 3 units of QC, 15 units of TTU, and 15 units of RTGC. In terms of customs process, the type of class and its required times are varied, where the random number follows the triangular distribution, which is generated within the framework of Monte Carlo simulation. The demand is assumed equal to 1500 containers that are carried by a single vessel.

Table 1. Simulation input characteristics

Variables	Units	Min	Max	Mode
QC Service Time	minutes	2.00	3.00	2.50
TTU Speed	km/hr.	10.00	25.00	15.00
RTGC Service Time	minutes	5.00	8.00	7.00
Customs Process	day	4..64	11.07	6.76

The connecting link to the hinterland is only operationalised by the truck mode, where the traffic flow is set based on an actual data set of connection link in Indonesia. The industrial cluster is equipped by the following information to estimate its losses due to the disruptions of CT.

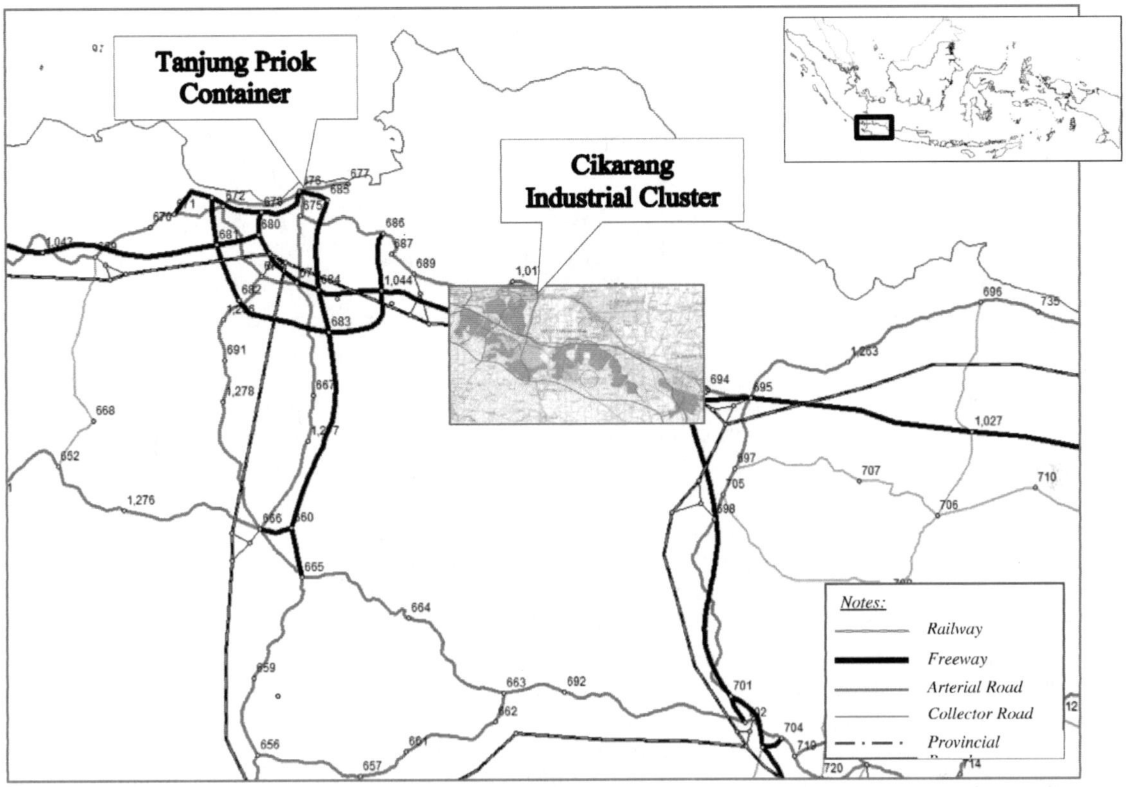

Figure 1. Test Network

Table 2. Input for calculating the hinterland losses

Parameters		Value
p_{raw}	unit value of commodities	1 bil. Rp./unit cont.
c_{raw}	unit extra cost for handling the damaged commodities	0.1 bil. Rp./unit cont.
p_{prod}	unit value of product	1 bil. Rp./unit
V_{prod}	volume of production	100 unit/day
I_{raw}	buffer time of raw material stock to continue the production	2 days
I_{prod}	buffer time of product stock to provide the supply	4 days
r_{cancel}	cancellation rate	1 %/day

Before discussing the resilience measurement more deeply, it is important to present the base condition of CT performance and its hinterland connection. In the normal condition, the container needs on average 140.11 hours (i.e., 5.84 days) to arrive in the hinterland, with the longest time being spent at the customs process (i.e., 112.21 hrs).

The resilience level of CT is then estimated in the case of QC shutdown, decrease of link capacity in the CT, and RTGC operation shutdown, in which the acceptance level is set as 1.5. The summary of scenarios is illustrated at below.

Table 3. Disruption Scenarios

Scenario	Number of Available Facilities		Decrease of Link Capacity
	QC	RTGC	
Sce. A	1	15	0
Sce. B	2	15	0
Sce. C	3	6	0
Sce. D	3	9	0
Sce. E	3	15	35%
Sce. F	3	15	70%
Sce. G	1	6	70%

As shown in Table 3, the first and second scenarios (i.e., Sce. A and B) reflect the situation of QC shutdown, in which the performance of other facilities remains unchanged. Furthermore, such kinds of scenarios are positioned in the Sce. C to Sce. F to illustrate the single disruption event of RTGC and link capacities. The RTGC disruption is modelled by setting the number of available units to be less than the normal condition. The interruption of TTU productivity is demonstrated by the decrease of link capacity, which directly influences its travel time for moving container within horizontal area of CT. To capture the simultaneous disruption events, the last scenario (i.e., Sce. G) is proposed by putting together the worst scenarios of disrupted facilities.

The defined disruption events are then tested in the network to investigate its impact to the CT performance and the resilience level. The required time to transport the container is firstly elaborated, which specifically receives a huge attention from the container owners. Since the model incorporates the CT and hinterland connection, the required time can be estimated not only in terms of the time to pass CT, which is generally conducted by other studies, but also the time to arrive at the hinterland.

As can be seen in Figure 2, the required time to transport the container to the hinterland corresponds to the type of facilities disrupted, in which the disruption of QC activities toward the highest increase of required time. The shutdown of available QC, which is illustrated by Scen. A, possibly increases the required time up to 25.58%. Moreover, the decrease of available RTGC also substantially raises the arrival time in the hinterland, where the container experiences an arrival delay of around 1.12 days. The capacity decrease of link is estimated to affect the CT operation, though its impact is less than the disruption of QC and RTGC. In addition, the simultaneous disruption of facilities seriously shapes the delay that multiply the arrival time to 29%.

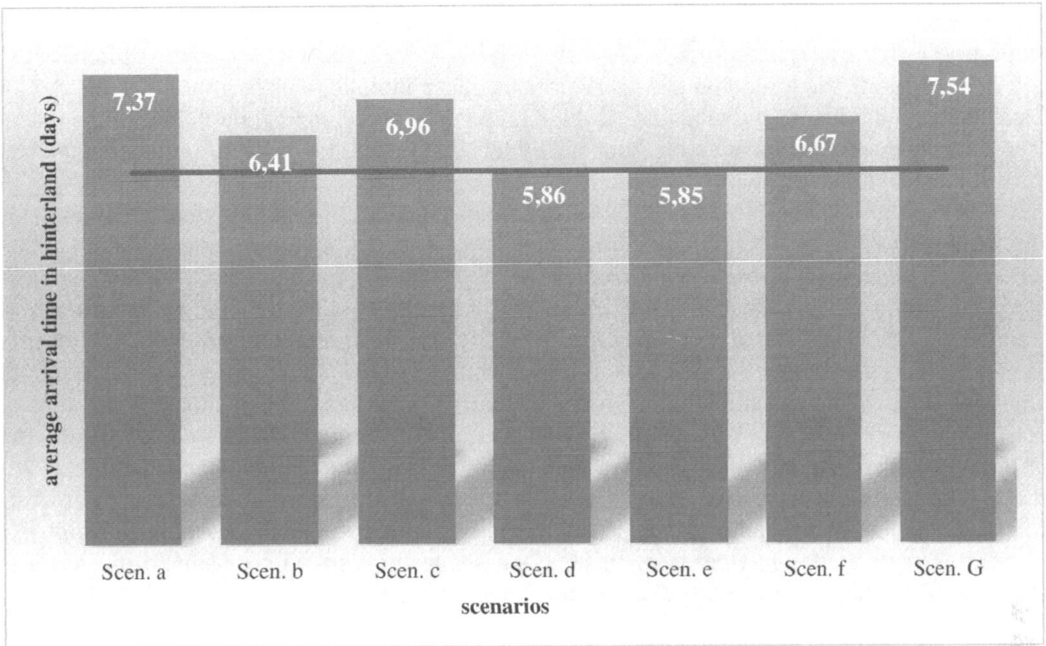

Figure 2. Average arrival time in the hinterland based on the scenario

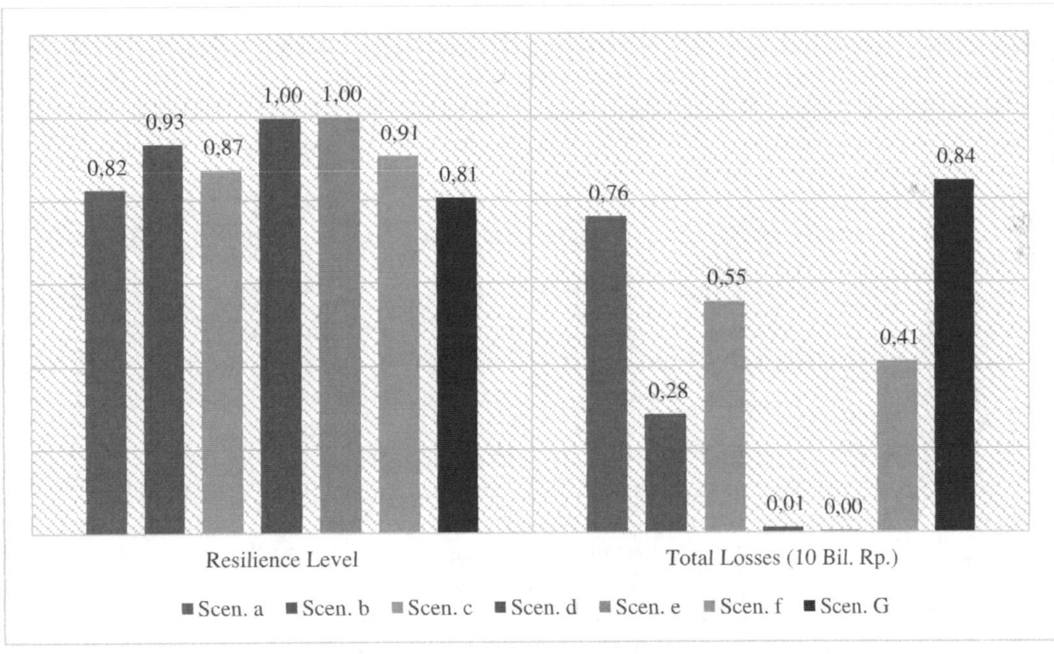

Figure 3. Resilience level and total losses incurred based on the scenario

The similar tendency is also indicated by the resilience level, in which the shutdown of QC and RTGC significantly reduces the resilience level of CT (see Figure 3). The decreasing of available QC successfully degrades the CT system ability to handle container, which is illustrated by the decrease of resilience level. As the hinterland losses calculation is included, it can be inferred that the lesser value of resilience level delivers a higher total loss incurred by hinterland. The decrease of resilience level, which subsequently escalates the arrival time, directly suffers the hinterland production and distribution performances that are implied by total losses. Hence, the protection of resilience level is strongly needed to alleviate the significant impact to the hinterland. Furthermore, the distraction of link capacity does not disturb the resilience level, which conveys a conclusion relating to the retrofit or recovery activities that need to be prioritised. Such facts can further be utilized to program the retrofit and recovery activities to deal with the disruption events.

4. Conclusions

This paper proposes a measurable model to calculate the resilience level of CT by taking into account the hinterland losses. Different to previous studies, the CT facilities productivity and its interaction are also modelled, where the operation of QC, TTU, RTGC and custom process are taken into consideration. To investigate the disruption impact, the model also incorporates the quantitative measurement to calculate the hinterland losses, which is constructed by considering the direct and indirect losses. The connection between hinterland and CT is provided by the transport links, where the uncertainty of transport links is integrated by invoking the Monte Carlo simulation-based procedure. The resilience level is then presented by capturing the potential delay due the disruption event. The numerical examination reveals that the lesser resilience level provides the higher total losses suffered by hinterland. Therefore, this model is may be applied to evaluate the retrofit and recovery activities when encountering disruption events.

References

[1] Wilmsmeier G, Hoffmann J and Sánchez R 2006 The impact of port characteristics on international maritime transport costs *Res. Transport. Econ.* **16(1)** 117–140

[2] Frazila R B and Zukhruf F 2017 A Stochastic Discrete Optimization Model For Multimodal Freight Transportation Network Design *International Journal of Operations Research* **14(3)** 107-120

[3] Kutin N, Nguyen T T and Vallée T 2017 Relative Efficiencies of ASEAN Container Ports based on Data Envelopment Analysis *Asian Journal of Shipping and Logistics* **33(2)** 67–77

[4] Suárez-Alemán A, Morales Sarriera J, Serebrisky T and Trujillo L 2016 When it comes to container port efficiency, are all developing regions equal? *Transportation Research Part A: Policy and Practice* **86** 56–77

[5] Figueiredo De Oliveira G and Cariou P 2015 The impact of competition on container port (in)efficiency *Transportation Research Part A: Policy and Practice* **78** 124–133

[6] Yuen A C L, Zhang A and Cheung W 2013 Foreign participation and competition: A way to improve the container port efficiency in China? *Transportation Research Part A: Policy and Practice* **49** 220–231

[7] Song D P, Lyons A, Li D and Sharifi H 2016 Modeling port competition from a transport chain perspective *Transportation Research Part E: Logistics and Transportation Review* **87** 75–96

[8] Rodrigue J P 2013 *The Geography of Transport Systems* (New York: Routledge)

[9] World Bank 2014 *Gross Domestic Product database* http://data.worldbank.org/

[10] Reuters 2011 *Impacts of the Great Earthquake in Japan on Energy, Bulk Commodity, and Ports* http://cn.reuters.com/article/2011/03/24/idCNCHINA-402992011 0324?pageNumber=1

[11] Ministry of Economy, Trade and Industry Japan (METI) 2011 *Economic Impact of the Great East Japan Earthquake and status of Recovery* http://www.meti.go.jp/english/earthquake/recovery/pdf/20110516_impact.pdf

[12] Forbes 2011 *Japan Quake, Tsunami Take Heavy Toll On Toyota* https://www.forbes.com/sites/greatspeculations/2011/04/08/japan-quake-tsunami-take-heavy-toll-on-toyota/#2234e4eb61b4

[13] Chen H, Cullinane K and Liu N 2017 Developing a model for measuring the resilience of a port-hinterland container transportation network *Transportation Research Part E: Logistics*

and Transportation Review **97** 282–301

[14] Zhang Y and Lam J S L 2016 Estimating economic losses of industry clusters due to port disruptions *Transportation Research Part A: Policy and Practice* **91** 17–33

[15] Labaka L, Hernantes J, Laugé A and Sarriegi J M 2013 Policies to improve resilience against major industrial accidents *Lect. Notes Comput. Sci* **6983** 187–199

[16] Gharehgozli A H, Mileski J, Adams A and von Zharen, W 2017 Evaluating a "wicked problem": A conceptual framework on seaport resiliency in the event of weather disruptions *Technological Forecasting and Social Change* **121** 65–75

[17] Bruneau M, Chang S, Eguchi R, Lee G, O'Rourke T, Reinhorn A, Shinozuka M, Tierney K, Wallace W and von Winterfelt D 2003 A framework to quantitatively assess and enhance seismic resilience of communities *Earthquake Spectra* **19** 733–752

[18] Zobel C W 2010 Representing perceived tradeoffs in defining disaster resilience *Decis. Support. Syst.* **50** 394–403

[19] John A, Yang Z, Riahi R and Wang J 2016 A risk assessment approach to improve the resilience of a seaport system using Bayesian networks *Ocean Engineering* **111** 136–147

[20] Repetto M P, Burlando M, Solari G, De Gaetano P and Pizzo M, 2017 Integrated tools for improving the resilience of seaports under extreme wind events *Sustainable Cities and Society* **32** 277–294

[21] Zukhruf F, Frazila RB and Burhani JT 2017 A stochastic discrete optimization model for designing container terminal facilities *AIP Conference Proceedings* 1903 (1) 060007

IOP Conf. Series: Earth and Environmental Science **158** (2018) 012045 doi:10.1088/1755-1315/158/1/012045

Towards Smart and Resilient City: A Conceptual Model

Y Arafah[1,2], H Winarso[1], D S A Suroso[1]

[1]Urban and Regional Planning, School of Architecture, Planning, and Policy Development, Institut Teknologi Bandung, Jl. Ganesha 10, Bandung 40132, Indonesia.
[2]Departmen of Architecture and Planning, Faculty of Engineering, Syiah Kuala University, Jl. T. Abdurrauf, No.7 Darussalam, Banda Aceh 23111, Indonesia.

Corresponding author: yunitaarafah@s.itb.ac.id

Abstract. This paper aims to compare five smart city models selected based on a number of specific criteria. Following the comparison and assessment performed, we draw conclusions and further linkages identifying the components and characters found in resilient cities. The purpose of this analysis is to produce a new approach and concept: the "smart and resilient city." Through in-depth literature study, this paper analyzes five conceptual smart city models deemed to have a background, point of view, and benchmark towards software group, as they focus on welfare, inclusion, social equality, and competitiveness. Analyzing the strategies, methods, and techniques of five smart city models, this paper concludes that there has been no inclusion of resilience concepts in the assessment, especially in the context of natural disasters. Basically, the models are also interrelated and there are some things that overlap. As a recommendation, there is a model that tries to combine the components and character of smart city and resilient city into one entity that is embedded as a whole in a conceptual picture towards the new concept, the "smart and resilient city". The concept of smart city and resilient city go hand in hand with each other and thus are interrelated. Therefore, it is imperative to study that concept deeper, in this case primarily in the context of disaster.

1. Introduction

The rapid development of the smart city concept, has led to the concept being widely used in various fields. However, this is also offset by the many criticisms that emerge about smart city from various researches. Some researchers say that the concept of smart city is still abstract, bias, unclear, and without consistent interpretation ([1]; [2]; [3]; [4]; [5]; [6]; [7]). Some mentions that there has been no mutually agreed, multi-dimensional, universal definition of smart city, and that the problems smart cities attempt to solve are not relevant existing problems ([4]; [8]).

The discussion about smart city concept has a developed greatly as of late, seen from the dramatic increase of the number of research on it. Based on the latest data from Colding and Barthel, (2017) [9], publications on the concept of smart city has increased sharply in the last decade (from 2006 to 2016) (Figure 1).

Publications on smart city has also spread on various aspects. One of the most prominent discussions is the concept of the smart city that focuses on information communication and technology (ICT) when in fact other fields are covered by the discussion, for instance social issues, environment, energy, business, and so on. The increased number of research yields an increase of resulting publications containing various conceptual models.

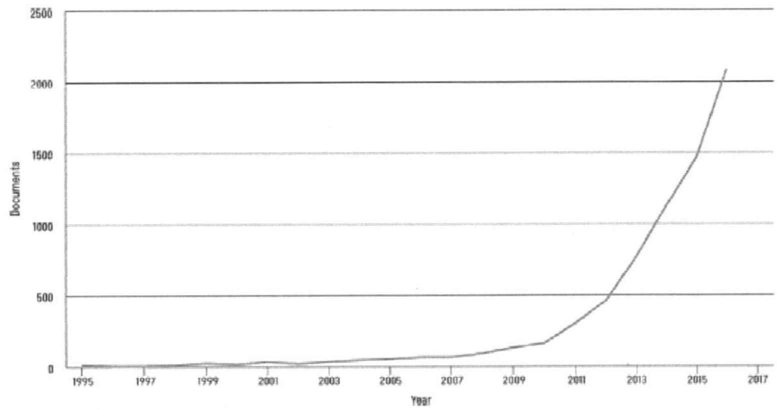

Figure 1. Number of publications on “smart cities” based on SCOPUS database [9].

Anthopoulos, et al. (2015) [10], classified these different models based on their factors and ideas, for instance, by relating them to sustainability, resilient, global city performance, local government effectiveness, urban competitiveness, and good urban governance. This comparison is made to draw a wide perspective on benchmarking method in order to measure smart city on different perspectives and clarify each different perspective in which issues are focused on and assessed in. Those perspectives are the society, government, economy, mobility, environment, and life [10].

This article will compare and analyze smart cities perspectives through in-depth literature study. We will select the models by exploring, through Google Scholar, articles about smart city concepts written in 2010-2015. From those articles, we identified five models that we subsequently analyze further and compare between each other. Unlike the model comparison done by Anthropolus et al., (2015) which uses different perspectives in comparing models, this article specifically focuses on articles providing the fundamental concept of smart city.

The five models are: 1) Europe smart city model by Giffinger and Gudrun (2010); 2) The basic component of smart city model by Nam and Pardo (2011); 3) Smart city initiatives framework by Chourabi et al., (2012); 4) the basic element of smart city model by Dameri (2014); and 5) relationship component and characteristic smart city model by Manville, et al., (2014).

The aim of this article is to compare existing models in smart city concept for further analysis, followed by linking the conclusion of the analysis with elements and characters found in resilient cities. The expected outcome of this paper is to introduce a new approach combining concepts of smart and resilient city. Resilience is an important factor as we are facing rapidly changing natural and social conditions, which require cities to be more resilient. This resilience concept in the context of a city refers to the city’s ability to absorb, adapt, and respond to any changes in urban system. Therefore, a resilient city is able to withstand the impact of shocks, hazards, and pressures through adaptability or transformation to ensure the long term sustainability, basic functions, characteristics and the structure of a city [11].

Resilience has become an important element in establishing smart city because resilience is a key factor in creating a better quality of life, sustainable urban development, and improving environmental condition. Therefore, this paper will analyze and evaluate existing smart city models in order to clearly identify elements of resilience so as to create a new, and highly-applicable, model of smart and resilient city.

The first part of this paper will introduce the concept and background of a smart and resilient city. The second part will dissect the analysis method: a literature study on the approaches, research strategy, method, and techniques used in five smart city models. The third part of this study will narrate the whole understanding on three concepts – smart city, resilient city, and the combination of the two – and assess the five selected models based on smart city and resilient city characteristics. From the assessment above it can be concluded that these models are interrelated and have some elements that overlap. These models also have not incorporated the concept and elements of resilience intentionally in the discussion and consideration when forming the model, especially in the context of natural disasters. Finally, the paper will talk about conclusion and recommendation about the importance on having resilience elements in smart city concept in order to form a new concept, the smart and resilient city concept. This new concept is realized through a model in which there are 25 characters that are integrated and embedded within the scope of smart and resilient city concept. To achieve a complete concept, the characters must not be separated and should always be attached. This model is still a recommendation, therefore more in-depth research is required for exploration and model testing.

2. Research Approach

This chapter will elaborate on analytical approaches used in determining the selected smart city model, what strategy will be used and how will the method and technique use in comparison and discussion be done. The research approach has five tiers, each one increasing in specificity: 1) approach; 2) strategy; 3) method; 4) technique; and 5) the outcome model. This is done in order to view the model comprehensively, so the discussion will be more objective and structured.

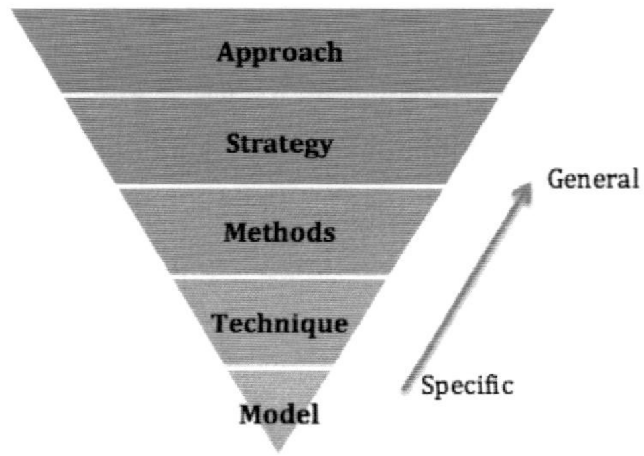

Figure 2. Research approach in analyzing model.

2.1. Approach

The approach is a very general state used to benchmark the use in forming a model. In smart city concept, there are 2 main approaches: hard infrastructure and soft infrastructure [12]. The hard infrastructure approach focuses on the use of technology and ICT managing urban infrastructure such as transportation, clean water management, sewage, electrical, waste, energy, security service, etc. The soft infrastructure approaches focus more on the community, inclusion, community social capacity development, and human resources, education quality improvement, society participation, social innovation, social equity etc.

Out of more than two thousand publications [9], most of the existing researches are in computer science and engineering, followed by researches in social sciences, mathematics, and various other fields. Looking from the data, it is obvious that researches on computer science and engineering still dominate existing publications, compared to studies, for instance, in social sciences. Hence, publications and researches in social science should be encouraged to increase the number and broaden the studies.

This paper will centralize its research on soft infrastructure, with hard infrastructure as supporting element.

Some of the existing definitions or understanding on smart city which focus on software approach, had many times stated about the importance of having such role as the utilization of information and communications technology, the awareness and smartness of society, welfare, participation, effectiveness, and competitiveness in implementing smart city, all of which has the aim to achieve a higher quality of life and improve the quality of the environment [13]. Out of five selected smart city models, all generally use approaches focusing on soft infrastructure and hard infrastructure as support. Thus, to delve more on the soft infrastructure approach, it is crucial to be familiar with its strength and weakness.

Table 1. Strength and weakness of soft infrastructure on smart city concept approach ([12]; [14]; [15]; [16]; [17]).

Strength	Weakness
- Improve human resources. - Community empowerment. - Intelligence and theory invention. - Social capital improvement. - Social and digital inclusion sustainability. - Possess an equal portion of belonging sense and responsibility to their city. - Humanity approach in representative technology. - Enhance skills and user's interest. - Respect individual along with their diversity.	- Unequal access of society participation. - Limited community involvement. - The great availability on data and information does not ensure the improvement of knowledge and integrity of society. - Cyberspace is not mainly and purely a public space.

2.2. Strategy

Strategy is the next discussion after the analysis approach. General strategy will be those related to ideas, planning, and activity performed. In this article, strategies to be undertaken in discussing each smart city conceptual model will at least refer to some strategic elements, which are [18]: 1) identification on specification and result qualification, also targets to be achieved; 2) main approaches and considerations taken in achieving the objectives; 3) steps taken from the start to the end; 4) confirmation on benchmark and criteria.

Strategy is important in conducting research, moreover when it is supported by the development in procedural and contextual analysis so that it will create a wider and more convincing development [19]. The four elements of strategy mentioned above will be prominent in discussing and analysing strategy of each smart city conceptual model.

Description:

I = Identify the specifications and qualifications of the results, as well as the targets to be achieved.
II = Main approaches and considerations taken in achieving the objectives.
III = Steps taken from start to finish.
IV = Setting benchmarks and criteria.

Table 2. Research strategies analysis on each model.

Elements of Strategy	Europe smart city model, by Giffinger and Gudrun (2010) [20].

I	- Ranking a Smart city is done through a different approach. This approach is used as an effective instrument in identifying the weaknesses and strengths of cities in improving city competitiveness through relevant strategic efforts. - A smart city is a city with good performance in six characteristics, built on a combination of intelligent support, a decision of community activity in self-determination, independence, and awareness.
II	- Smart city ranking approach not only provides city ranking data, but also the types of characteristics used as the basis of discussion in the strategic plan. Six characteristics in this conceptual model are: smart economy, smart people, smart environment, smart mobility, smart living, and smart governance.
III	- Transparency in ranking from selected groups of cities. - Elaborate and illustrate the specific characteristics and profile of each city. - Develop existing benchmarks among selected cities. - Identify strengths and weaknesses for strategic discussions and policy advice.
IV	- Using 6 characteristics that characterizes the city with smart city concept. - Using 31 relevant factors that reflect an important aspect of each characteristic, then each factor is empirically defined through 74 indicators used to operationalize and incorporate relevant factors.
Strategy elements	The basic component of smart city model by Nam and Pardo (2011) [8].
I	- The conceptual model is based on three multidimensional components, namely: technology, society, and institutions. These components become the basic factor in creating a smart city concept. - A city will be smart when there is investment in human/ social capital and information technology infrastructure that can improve the quality of life and promote sustainable growth through participatory governance. This is an embodiment of the relationship of the above three factors.
II	- Besides technology, social factors are important factor in the smart city concept, therefore it takes a socio-technical view in the context of smart city. - In creating a smart city initiative requires a thorough understanding of the complexity and the interrelationships between social factors and technical factors of service and the physical environment in a city.
III	- Identify the current conditions of cities that have already begun to apply the smart city concept successfully, discuss the meaning and definition of the smart city concept, and find a variety of concepts that are similar or have similarities.
IV	- The strategic principles are aligned with the three main dimensions (technology, people, and institutions), namely: integration of infrastructure and services through technology media, social learning to strengthen human infrastructure, and governance for institutional and citizen engagement.
Strategy elements	Smart city initiatives framework by Chourabi et al., (2012) [2].
I	- Using a conceptual framework to understand the smart city concept as a whole.

IOP Conf. Series: Earth and Environmental Science **158** (2018) 012045 doi:10.1088/1755-1315/158/1/012045

	- Identify eight very important factors of smart city: management and organization, technology, governance, policy context, people and community, economy, built infrastructure, and natural environment.
	Continued...
Strategy elements	Smart city initiatives framework by Chourabi et al., (2012) [2].
II	- Creating a conceptual framework that can be used to characterize how to create the vision and design of smart city initiatives. - This vision is realized through shared services, and through the direction given to the challenges that the city faces.
III	- Develop an integrative framework to explain relationships and influence between factors and smart city initiatives. - Each factor is spelled out through the derivation of factors, the dimensions, and the challenges it faces.
IV	- This conceptual model provides direction and agenda for the researcher in the smart city field and provides an overview of practical implications for government professionals.
Strategy elements	The basic element of smart city model by Dameri (2014) [21].
I	- The desire to create a broader vision of the existing smart city phenomenon. - Developing a conceptual model from the previous theory of Nam and Pardo (2011) and smart city dimensions by Giffingger and Gudrun (2010), became a conceptual model of smart city consisting of four core components: land, infrastructure, society, and government.
II	- Compare the studies and considerations on how to form a smart city. - How to design a smart strategy and how to measure a smart action really can shape public value for society and create a better quality of life in urban space.
III	- Searching for the growing of smart city ideas, then defines the meaning of smart city from existing literatures. - How existing initiatives can improve intelligence and can achieve the ultimate goal of the smart concept. - Measure the intelligence and creation of public value through clear and measurable objectives and processes.
IV	- In an effort to create a city with a better concept of smart city, then a city must improve the "smartness" of each component of the city. - Benchmarks used are, effectiveness, environmental balance, and innovation.
Strategy elements	Relationship component and characteristic smart city model by Manville, et al., (2014) [22].
I	- The goal is to generate strategies and design of smart city initiatives. - Smart city as a city that can solve public problems through ICT-based solutions based on multi-stakeholder cooperation and city-based partnerships. These solutions are developed and refined through smart city initiatives, either as separate projects or as a network of overlapping activities.
II	- Mapping cities with the concept of smart city in Europe. - Making connections between characteristics and components in the smart city concept.

III	- Model the relationship between smart city project, smart city initiatives, and smart city concept. - Defining smart city based on the existing literature and redefine it in accordance with the goals and objectives to be achieved. - Defining the characteristics of smart city. - Seeing the relationship between the characteristics and components of smart city in certain cases. - Identifying smart city for mapping study. - Conducting smart city mapping in Europe.
Strategy elements	Relationship component and characteristic smart city model by Manville, et al., (2014) [22].
IV	- In smart city strategies and initiatives, at least one characteristic must be included as either the objective or the mode of operation, namely: Smart Governance, Smart People, Smart Living, Smart Mobility, Smart Economy and Smart Environment. - This characteristic will be a goal for stakeholders that participating in smart city initiatives. For example: to solve environmental problems.

2.3. Methods

In each study, the method is intended as a form of concrete and practical activities undertaken to achieve the research objectives. The methods used in the existing models generally use the method of desk study and literature review method. This method is done by summarizing and analyzing the results of previous studies that present comprehensive and balanced facts and theories. The results of the study will generally be summarized through data integration and then expected to generate new theories or concepts in a deeper and more thorough understanding level [23]. The following descriptions and explanations of the methods used in each model.

Table 3. Description of the methods in each model.

No	Model	Methods
1.	Europe smart city model.	- Field research, interviews, and mostly use desk research. - Analysis using primary and secondary data. - Weight factors and indicators are generally used in their influence for the purpose of ranking. The method used is a mix qualitative and quantitative. Qualitatively in determining the characteristics, indicators, and factors, and quantitatively calculate the weighting for the urban ranking. - City rankings were made with the aim of comparing characteristics and to identify the strengths and weaknesses of a city in a comparative way.
2.	The basic component of smart city model.	- Using qualitative methods through desk study. The study was conducted using a literature study, the previous study of smart city and the projects that have been undertaken.
3.	Smart city initiatives framework.	- Using qualitative methods through the study of literature from various disciplinary fields.
4.	The basic element of smart city model.	- Using qualitative methods through in-depth analysis of the various literatures. Furthermore, using the five basic frameworks in the literature searching process by Vom Brocke et al, namely: the definition of scope;

		conceptualizing topics; literature search; literature analysis and synthesis; and research agenda.
5.	Relationship component and characteristic smart city model.	- Conventional approach with the strong emphasis on desk study method. - Using in-depth analysis.

2.4. Technique

Technique is how a researcher implements a specific method. Research techniques usually consist of how the data obtained, how the data is processed and how the procedure performed. The following explanation of research techniques used in each model.

Table 4. Description of research techniques used in each model.

No	Model	Description of the technique used.
1.	Europe smart city model.	- Selects the cities, identifies the indicators and data sources, and standardizes and aggregates data. - Using cities of medium size or second city in the scale of European cities. Town with population 100.000-500.000 population.
2.	The basic component of smart city model.	- The conceptual framework is obtained through literature studies with the technique of defining smart city through the meaning of intelligence in urban contexts, exploring new definitions, and understanding smart city as a broad concept. - Then, observe the key components of smart city and discuss the strategic principles that can create success in smart city initiatives.
3.	Smart city initiatives framework.	- Explores the literature from various fields such as e-government, information science, urban studies, and public administration. It further identifies key trends, identifies and discusses challenges, success factors, and the impact of initiatives to create smart cities.
4.	The basic element of smart city model.	- Using in-depth analysis techniques through various literatures. Redefine the research methods used for the purpose of identification and organization of the articles to be used for the literature review analysis. - Then test the collected data to answer specific questions through timing analysis, terminology, typology, and geographical analysis.
5.	Relationship component and characteristic smart city model.	- Using in-depth analysis to provide an actual, accurate, and objective picture of the patterns and trends that occur, and gain an understanding of the factors that contribute to the success of smart city initiatives as the basic of smart city.

3. Discussions and Findings

Based on analysis and discussion in the previous chapter, it can be concluded from the five models, each model is related and has some similarities with another. This is due to several studies in more recent years citing several items from previous researchers. Furthermore, the discussion in this chapter will be divided into three sub chapters, namely: the smart city concept, the resilient city concept, and combination of the two.

IOP Conf. Series: Earth and Environmental Science **158** (2018) 012045 doi:10.1088/1755-1315/158/1/012045

3.1. The Smart City Concept

As discussed earlier, the smart city concept has grown rapidly and extensively to various fields. Therefore, the understanding and meaning of smart city that emerged also expanded and varied. Based on the previous discussion on the five smart city models, it can be concluded that the smart city concept that used towards resilience is the smart city concept with soft infrastructure approach.

There are several meanings and concepts of smart city in an approach that focus on soft infrastructure (such as participation, inclusion, and citizen intelligence), among others: 1) The main purpose of smart city concept is to improve sustainability with the help of technology [24]; 2) Smart city is a concept that can run a city well through its perspective on the economy, society, government, mobility, environment, and life, it's built using an intelligent combination of support and independent community activity, and society awareness [25]; 3) The concept of smart city uses the role of information and communication technology (ICT) to achieve prosperity, effectiveness, and competitiveness. [12]; 4) Smart city is built on the political understanding of technology, it takes a process approach to manage the city, and focus on economic benefits and other public values [7]; 5) The concept of smart city actually does not match with the character of a city which is difficult to be predicted, informal, and can lead to the emerge of social inequality [26].

After concept approach, the research strategy in smart city concept also an important thing to be considered. From the analysis conducted on the model research strategy, it can be concluded that the five smart city models can be grouped into two main groups, namely: 1) group of models that describes dimension and main component in smart city concept; and 2) group of models narrating the process toward the smart city and generates measurements in the form of city rankings (table 5). Each model group is important, because these two groups are supporting the creation of models in applying smart city concept. Prior to the process of measuring the level of smart city, it is necessary to identify the dimensions, components, and important factors in shaping the concept of smart city.

It can be concluded that generally these models use a qualitative approach in research, with the method of in-depth analysis into desk research and literature review. Techniques used to obtain data, data processing, and processing procedures also vary. General data is obtained using literature documents, interviews and surveys. Data processing and procedures employs the identification of previous smart city, finding new definitions, identifying indicators and standardization, identifying smart city components, identifying trends and patterns, identifying existing challenges, identifying factors success, using literature with multiple perspectives (such as information science, government, public administration), and using various analyses (such as timing analysis, typology, terminology, geography).

Based on the meaning of each character, we assessed five smart city model from existing documents. From this assessment we will see which model represents the most smart city concept.

After concept approach, we look at the research strategy in smart city concept. We conclude that the five smart city models can be classified into two groups: 1) group of models that describes dimension and main component in smart city concept; and 2) group of models narrating the process toward the smart city and generates measurements in the form of city rankings (table 5). Each model group is important, because these two groups are supporting the creation of models in applying smart city concept. Prior to the process of measuring the level of smart city, it is necessary to identify the dimensions, components, and important factors in shaping the concept of smart city.

It can be concluded that generally these models use a qualitative approach in research with the method of desk research and literature review through in-depth analysis. Techniques used to obtain data, process data, and process procedures also vary. In obtaining general data using literature documents, interviews and surveys. In the processing data and procedures, here are some of the main techniques: identifying the previous smart city and then finding new definitions, identifying indicators and standardization, identifying smart city components, identifying trends and patterns, identifying existing challenges, identifying factors success, using literature with multiple perspectives (such as information science, government, public administration), and using various methods of analysis (such as timing analysis, typology, terminology, geography).

IOP Conf. Series: Earth and Environmental Science **158** (2018) 012045 doi:10.1088/1755-1315/158/1/012045

Based on the meaning of each character, an assessment was conducted to five smart city model based on existing documents. The Assessment is done by seeing whether from each element of the character becomes one of discussions or considerations in discussing and forming a smart city model. From this assessment we will see which model represents the most smart city concept (tabel 6).

Table 5. Grouping of models.

<table>
<tr><th>No.</th><th>Model</th><th>Image/ illustration</th><th>Category</th></tr>
<tr><td>1.</td><td>Europe smart city model.</td><td>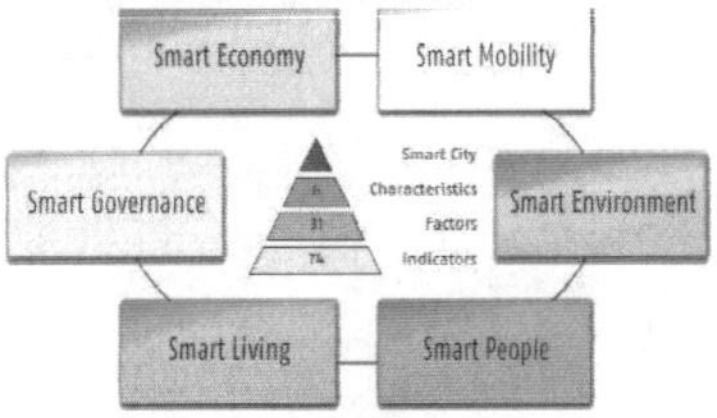
</td><td rowspan="2">The process towards smart city and city ranking measurement.</td></tr>
<tr><td>2.</td><td>Relationship component and characteristic smart city model.</td><td>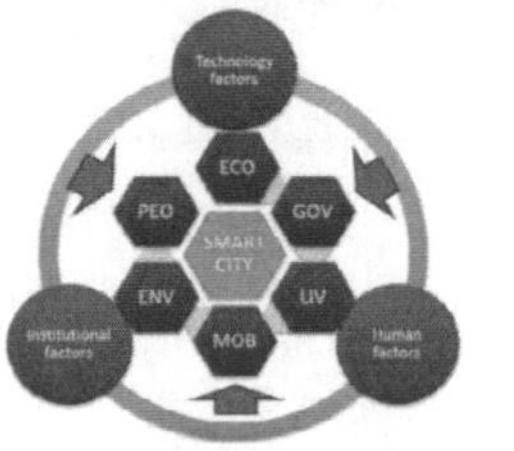
</td></tr>
<tr><td>3.</td><td>The basic component of smart city model.</td><td>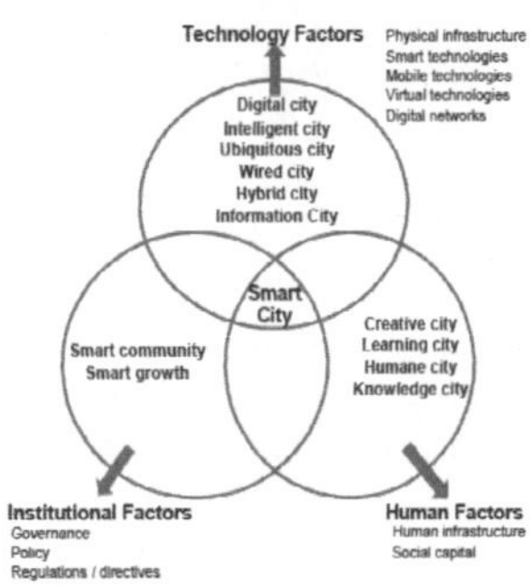
</td><td rowspan="3">Dimensions and main components in the smart city concept.</td></tr>
<tr><td>4.</td><td>Smart city initiatives framework.</td><td>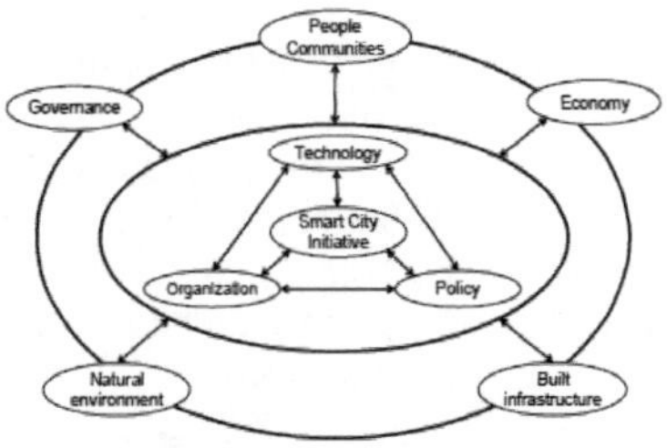
</td></tr>
<tr><td>5.</td><td>The basic element of smart city model.</td><td>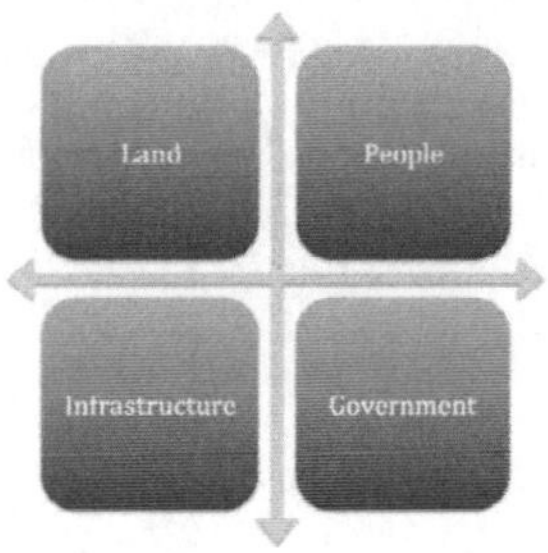
</td></tr>
</table>

Table 6. Assessment of models based on the character of smart city concept.

No.	Traits of a smart city	Europe smart city model	Basic component model	Smart city initiatives framework	Basic element model	Relation of component & characteristics
1.	Efficiency	1	1	1	1	1
2.	Flexibility	1	1	1	0	1
3.	Adaptability	1	1	0	0	1
4.	Networking capacity	1	1	1	1	1
5.	Learning capacity	1	1	1	1	1
6.	Diversity	1	1	1	0	1
7.	Innovation capacity	1	1	1	1	1
8.	Creativity	1	1	0	0	1
9.	Participation	1	1	1	1	1
10.	Awareness	1	1	1	1	1
11.	Collaboration	0	1	1	1	1
12.	Fairness and social inclusion	1	1	0	1	1
13.	Monitoring capacity	0	1	1	1	1
14.	Reliability	0	0	0	0	1
15.	Anticipation	0	0	0	0	0
		11	13	10	9	14

Based on the results of the assessment above on the five models, it can be concluded that, "the model relation of component and characteristic" is a model that strongly represents the characters in the smart city concept. The second model that has the closest character of the smart city concept is "the basic component model". Both models seem to represent their respective groups. The model by Manville represents the group "process towards smart city and city rank measurement", and the model by Nam and Pardo represents the group of "key dimensions and components in the smart city concept ".

3.2. Resilient City Concept

The explosive pressure of the population is one of the major problems in the world. The human population is growing rapidly: we are expected to reach 7.4 billion in 2015 and is expected to reach about 8.3 billion people by 2030, and 8.9 billion by 2050 [27]. Urban populations are generally a major challenge, cities in the world must be able to cope with such diverse pressures as urbanization, climate change, air pollution, water crisis, increased risk of natural disasters, and so on. To face this, the city is required to solve the problem, it can learn to adapt in facing the challenges ahead in an uncertainty era by building the concept of resilience.

The concept of resilience was originally developed by C.S. Holing in 1973 in ecology, developed in the field of disaster science ([28]; [29]), planning [30], social psychology by Walsh in 1998 [31], resilience studies in socio-ecology ([32]; [33]), and resilience concepts leading to social studies ([34]; [35]; [36]). The meaning of resilience in the city context refers to the ability to get, adapt, and respond to the changes in the urban system. A resilient city is a city that is able to detain or absorb the impact of the disasters, shocks and pressures through adaptation or transformation, in order to ensure long-term sustainability, basic functions, characteristics, and urban structures [11].

In the field of planning, the concept of resilience is developed in various sub-areas, such as: 1) The context of catastrophic disasters, terrorism, energy crisis, and climate change ([37]; [38]); 2) the context of socio-ecological resilience with communicative planning [39]; and 3) the context of collaborative planning for governance in addressing resilience [40]. In the context of disaster and other contexts that are still related, some researchers interpret the concept of resilience as follows:

- Resilient city is a system of physical networks (artificial and natural components) and human communities (social and institutional components) that can survive and continue to be sustainable under conditions with extreme and unique pressures. Physical system is a network of muscle and community network that is like the brain of a city, without any resilience of these two systems, then a city will be very vulnerable to disaster. [37].
- Resilience has three main features that must be owned, namely: persistence, adaptability, transformability [41].
- Resilient city will reflect the capacity of the city through individuals, communities, institutions, companies, and systems to survive, adapt and develop. It does not matter with any kind of heavy pressure or any strong shock they have experienced. [42].
- Resilience refers to the capacity of individuals, households, population groups or systems to anticipate, absorb, recover from the dangers and impacts of climate change, other shock effects, long-term stress and potential long-term prospects. Resilience is not fixed, but it is a dynamic set of conditions and processes [43].
- Resilient city has a vision to enhance the city's ability to respond to various pressure factors (climate, environment, energy and economy), with the ultimate goal of ensuring a higher quality of life and sustainable urban development. [44].
- Resilience is the capacity of the social or ecological system and its component parts to overcome dangerous and pressure shocks, in a timely and efficient manner, in responding, adapting, and changing the ways that restore, nurture, and even enhance important functions, structures, and identity while maintaining capacity for growth and change. [45].

The resilience of a city in the face of disasters can be built through strategies and key factors: building cross-sectoral leadership for disasters, adding the possibility of greater earthquake risks to building codes, increasing resilience through communities, and taking into account the importance of equilibrium natural resources under certain conditions. [42]. Another resilience strategy is proposed by Wilkinson (2012) [46] by creating four strategies: assumption of changes and uncertainty about the arrival of disasters, maintaining conditions of recovery and repair after a disaster, a combination of different types of knowledge is needed to create learning, and creating opportunities for self-organization.

In strengthening the analysis and discussion on the existing model, the next analysis will identify characters of resilient cities in the five models. These characters are spelled out through three main factors in resilience [47]: persistence, adaptability, and transformability. Further described and interpreted one by one from the side of resilient city by Papa, et al. (2015) [44] in detail. There are a number of different characters and similar characters between smart city and resilient city.

Generally, these five models have not deliberately used the perspective of resilience or incorporated the elements of resilience in the discussion of the formation of smart city concept models. However, since there are some elements in the concept of smart city that are also contained in the concept of resilience, on the other hand, the assessment of the resilience elements in the five smart city models will be done through the identification of the character of the concept of resilient city.

As mentioned earlier, the assessment will be done by looking at whether each character of the resilient city character is considered informing a smart city model. From this assessment, it will be found which model represents the most or use elements of the resilient city character.

Table 7. Assessment of models based on the character of resilient city concept.

No.	Traits of a resilient city	Europe smart city model	Basic component model	Smart city initiatives framework	Basic element model	Relation of component & characteristics
1.	Efficiency	1	0	0	1	1
2.	Flexibility	0	0	0	0	0
3.	Adaptability	1	0	1	0	1
4.	Networking capacity	0	1	0	1	1
5.	Connectivity	0	1	0	1	1
6.	Learning capacity	0	1	0	0	1
7.	Diversity	0	1	0	0	1
8.	Innovation capacity	1	1	1	1	1
9.	Creativity	0	0	0	0	0
10.	Participation	0	1	1	0	1
11.	Awareness	1	0	0	1	1
12.	Knowledge	0	1	0	1	1
13.	Collaboration	0	1	1	1	1
14.	Persistence	1	1	0	1	1
15.	Resistance	0	1	1	1	1
16.	Modularity	0	0	0	0	1
17.	Redundancy	0	0	0	0	1
18.	Memory	0	0	0	0	0
19.	Robustness	0	1	0	0	1
20.	Resourcefulness	0	1	1	1	1
21.	Transformability	1	1	1	1	1
		5	11	7	8	17

Based on the assessment above (table 7), it can be concluded that out of the 21 characters resilient city, "the model relation of component and characteristic" is a model that strongly represents the characters in the resilient city concept. The second model that most has character of the resilient city concept is "the basic component model". Just like the results of smart city character analysis, both models seem to represent their respective groups. The model by Manville represents the group "process towards smart city and city rank measurement", and the model by Nam and Pardo represents the group of "key dimensions and components in the smart city concept".

3.3. Smart and Resilient City Concept

The vision of the smart city concept and the resilient city concept is in line and aligned, it's aimed to create a higher quality of life, sustainable urban development, and improving the quality of the environment that is judged to be declining ([44]; [3]; [16]; [17]). Some perspective approaches to the concept of smart city also use urban sustainability and resilience factors, with the aim of indicating an effort to both directions [10]. Resilience is an important element that must enter into the complexity of a city, therefore a city must be built with its capacity toughness. The Capacity consists of multicomponent, namely process and interaction that put it above the physical boundary of the city. It's Important for a city to be resilient when it is said as a smart city [48]. Therefore, the concept of resilience is one of the important factors in smart city planning.

One of the meaning of smart city concept based on resilience are: the concept of the city that utilizes the role of information communication and technology (ICT) in raising awareness, intelligence, welfare, and citizen participation, when facing pressure, shocks and hazards, in order to survive, adapt, tough and transform, so that later can lead to quality of life and environment , which is better and sustainable in the face of an uncertain future [13].

The concept of smart city and resilient city have transformed and changed over time. This occurs in the different approaches, characters, and focus of each concept. From three main groups of smart city and resilient city concepts, the closest approach to this discussion is a mix of software and hardware-focused groups and groups that focus on adaptive capacity, transformation skills, learning and innovation.

The discussion of the concept of smart city that leads to the concept of resilience concept, is marked by the start of studies that link between the concept of smart city and resilience. The relationship between theory and concept of smart city and resilience has been done by Baron (2012) [49], which sees the importance of resilience from the perspective of business praxis, urban studies, and territoriality, through literature studies and empirical studies. The existence of theoretical relationship between the smart city concept and city resilience is expected to be the starting point for further research and reflection that concentrate on city development strategies that want to achieve intelligence and toughness.

Papa et al. (2015) [44], also looked at the synergies and inconsistencies between the smart city concept and resilience through the identification of key characteristics of the smart city and resilient city urban systems, then compiling them into a conceptual model in case of mitigation and adaptation in the context of confronting change climate. This model is expected to serve as a guide for planners and decision-makers in building smarter and more resilient cities, and to increase capacity responses from complex urban systems in facing the climate change.

In addition, Viitanen and Kingston [50] criticized smart cities using the perspective of green growth to achieve resilience, which is used by big companies through public and private partnerships. The municipal system became a digital marketplace where consumer-citizen participation increased by accident and increased hegemony of global technology companies. This makes the city less resilient in facing future social and climate risks.

Some of the studies discussed above are studies that connect the concept of smart city and resilience in the context of soft infrastructure. From these studies, no studies have specifically linked the smart city concept and resilience in the context of disaster with a soft infrastructure approach.

The closest model is the resilience study made by Desouza and Flanery (2013) and Papa et al. (2015). They set up city resilience benchmarks through a design, planning, and management approach to perfecting city resilience, including dynamic process and cultural evaluation within a city. Next, they compile the resilience framework of a city by dividing the two components of the city, namely: resources and processes (physically); and communities, institutions, and activities (socially). This model describes the city as a complex adaptive system and then divides it into components and elements of analysis, making it easier to achieve its goal of becoming a more resilient city through the design, planning and management of urban systems. Existing physical and social processes are understood as spatial and temporal interactions across networks to improve what is beneficial and reduce something harmful. There are four kinds of stresses in this model: natural stressors such as earthquake, tsunami, hurricane, etc.; economic stressors such as reduced employment opportunities, rising poverty levels, decreasing business investment, monetary crisis, etc.; technological stressors such as failure and error complex technical network system; and finally, human stressors such as terrorism, war, crime, riot, etc.

In this study, there are two pressures that become important points and affect the smart and resilient city concept, namely natural pressure and technological pressure. Natural disasters are one of the types of pressure that fall within the natural stress group, which will be related to disaster resilience. Technological pressures arise due to the increasing complexity of existing technical network systems in cities with smart systems, such as smart cities, will create interconnected network systems, so vulnerable to failures caused by errors that can spread through the system.

4. Conclusions and Recommendations

4.1. Conclusions

This study looks at the concept of smart city as a whole and towards resilience city through one part of many fields on resilience concept, that is the disaster field. One side of disaster, especially natural disaster, is expected to fill the gap in bridging the importance of resilience element in smart city concept.

Based on the discussion in the previous chapters, it can be concluded that this study more clearly directs the smart city concept to the approach of soft infrastructure and hard infrastructure as a supporter. With the concept of resilient city that leads to disaster resilience group through approach to resilience based on non-equilibrium or social-ecological resilience.

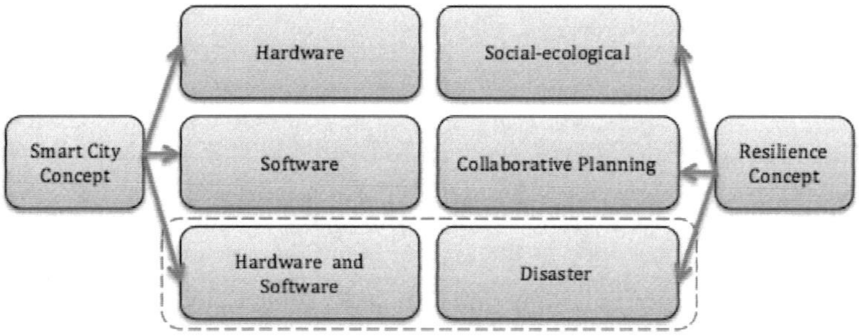

Figure 3. Approach in shaping the smart and resilient city concept.

Out of five smart city models analyzed using approaches: strategies, methods, and techniques, it can be concluded that from the whole model there has been no inclusion of resilience concepts in the discussion, especially in the context of natural disasters. Basically, the models are also interrelated and there are some things that overlap. Based on the assessment of the five models using the characteristics of smart city concept and resilience concept, the results obtained are in the following table.

Table 8. Model rating in smart city concept and resilient city concept.

No.	Model Name	Rank in smart city	Rank in resilient city
1.	Europe smart city model	3	5
2.	Basic component model	2	2
3.	Smart city initiatives framework	4	4
4.	Basic element model	5	3
5.	Relation of component & characteristics	1	1

“The model relation of component and characteristic” is a model that is ranked first and considered to highly represent smart city concept character. This model also ranked first in the assessment on the resilient city character. In the second rank of smart city concept, there is "the basic component model". This model also ranked second based on the assessment of the character of resilient city.

In this case it can be concluded that the concept of smart city and resilient city is a concept that goes hand in hand and has a close relationship and linkage. Therefore, it is important to do a more detailed and in-depth study, in this case primarily in the context of disaster.

4.2. Recommendations

To add value in existing models of smart cities, we need to add components and traits for resilient cities, particularly when it comes to resilience against natural disasters.

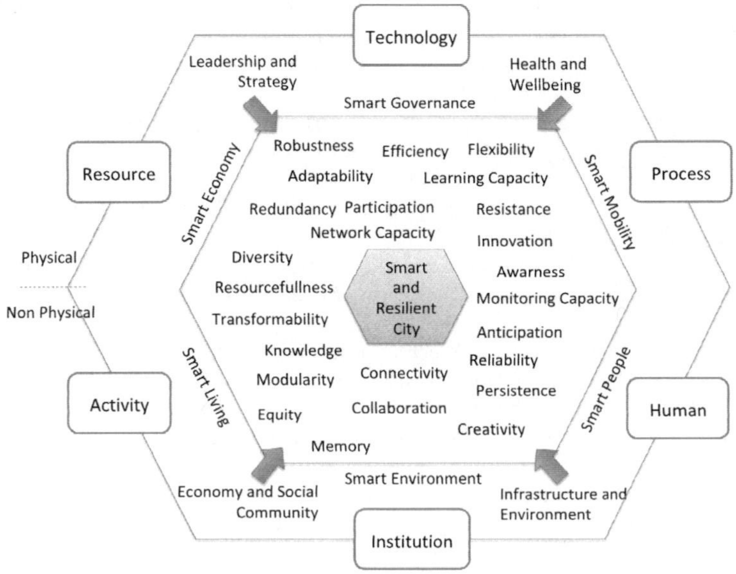

Figure 4. Smart and resilient city model.

In this smart and resilient city model, there are 25 characters that are integrated and embedded within the scope of smart and resilient city concept. To achieve a complete concept, the characters cannot be separated and will always be attached. Manville, et al., (2014) [22] mention that there is at least one characteristic of the six characteristics (smart governance, smart economy, smart environment, smart living, smart mobile and smart people) because it can lead to a multi-dimensional strategy to synergize and support each other. In the new conceptual model, four dimensions by ARUP (2013) [42] are added, and as a whole must be included in shaping the model toward the smart and resilient city concept.

Components of the city are divided into physical and non-physical components, namely: resources, processes, and technologies included into the physical group; people, institutions, and activities entered into non-physical groups. Certainly, with the note that all components of the city are not equivalent to its position. Human component that has the greatest role than other components, because humans that determine the creation, governance, and maintenance of all the components that exist [48]. Minimizing the impact on communities and their efforts to re-emerge from shock is a critical criterion evaluated when measuring the resilience of a city. This model tries to combine the components and character of smart city and resilient city into a new, integrated concept: the "smart and resilient city". This model is still a recommendation that requires more in-depth research, exploration, and model testing.

5. References

[1] M. Angelidou, “Smart cities: A conjuncture of four forces,” *Cities*, vol. 47, pp. 95–106, Sep. 2015.
[2] H. Chourabi *et al.*, “Understanding Smart Cities: An Integrative Framework,” 2012, pp. 2289–2297.
[3] A. Caragliu, C. Del Bo, and P. Nijkamp, “Smart Cities in Europe,” *J. of Urban Technology*, vol. 18, no. 2, pp. 65–82, Apr. 2011.

[4] R. G. Hollands, "Will the real smart city please stand up?: Intelligent, progressive or entrepreneurial?," *City*, vol. 12, no. 3, pp. 303–320, Dec. 2008.

[5] M.-L. Marsal-Llacuna, J. Colomer-Llinàs, and J. Meléndez-Frigola, "Lessons in urban monitoring taken from sustainable and livable cities to better address the Smart Cities initiative," *Technological Forecasting and Social Change*, vol. 90, pp. 611–622, Jan. 2015.

[6] R. S. Wall and S. Stavropoulos, "Smart cities within world city networks," *Applied Economics Letters*, vol. 23, no. 12, pp. 875–879, Aug. 2016.

[7] A. Meijer and M. P. R. Bolívar, "Governing the smart city: a review of the literature on smart urban governance," *Int. Review of Administrative Sciences*, vol. 82, no. 2, pp. 392– 408, Jun. 2016.

[8] T. Nam and T. A. Pardo, "Conceptualizing smart city with dimensions of technology, people, and institutions," 2011, p. 282.

[9] J. Colding and S. Barthel, "An urban ecology critique on the 'Smart City' model - ScienceDirect," *J. of Cleaner Production*, vol. 164, no. 5 October 2017, pp. 95–101, Jun. 2017.

[10] L. Anthopoulos, M. Janssen, and V. Weerakkody, "Comparing Smart Cities with Different Modeling Approaches," presented at the International World Wide Web Conference Committee (IW3C2), Florence, Italy., 2015.

[11] UNISDR, *How To Make Cities More Resilient A Handbook For Local Government Leaders*. Geneva: United Nations, 2012.

[12] M. Angelidou, "Smart city policies: A spatial approach," *Cities*, vol. 41, pp. S3–S11, Jul. 2014.

[13] Y. Arafah and H. Winarso, "Redefining smart city concept with resilience approach," *IOP Conf. Series: Earth and Environmental Science*, vol. 70, p. 012065, Jun. 2017.

[14] A. M. Townsend, *Smart Cities: Big Data, Civic Hackers, and the Quest for a New Utopia*, 1 edition. New York: W. W. Norton & Company, 2013.

[15] A. Cocchia, "Smart and Digital City: A Systematic Literature Review," in *Smart City*, R. P. Dameri and C. Rosenthal-Sabroux, Eds. Cham: Springer International Publishing, 2014, pp. 13–43.

[16] European Commission, "Communication from the commission. Smart cities and communities – European innovation partnership," Brussels, 2012.

[17] IEEE (the Institute of Electrical and Electronics Engineers), "IEEE Smart Cities," 2014. .

[18] Newman, Logan, and Hegarty, *Strategy: A Multi-Level, Integrated Approach*. Cengage South-Western, 1988.

[19] A. M. Pettigrew, "The character and significance of strategy process research," *Strategic Management Journal*, vol. 13, no. S2, pp. 5–16, 1992.

[20] R. Giffinger and H. Gudrun, "Smart cities ranking: an effective instrument for the positioning of the cities?," *ACE: architecture, city and environment*, vol. 0, no. 12, 2010.

[21] R. P. Dameri, "Comparing Smart and Digital City: Initiatives and Strategies in Amsterdam and Genoa. Are They Digital and/or Smart?," in *Smart City*, R. P. Dameri and C. Rosenthal-Sabroux, Eds. Cham: Springer International Publishing, 2014, pp. 45–88.

[22] C. Manville, G. Cochrane, J. Cave, J. Millard, and J. K. Pederson, *Mapping Smart Cities in the EU*. Brussels: Policy Department A: Economic and Scientific Policy European Parliament, 2014.

[23] A. Perry and N. Hammond, "Systematic Review: The Experience of a PhD Student.," *Psychology Learning and Teaching*, vol. 2 (1), 2002.

[24] H. Ahvenniemi, A. Huovila, I. Pinto-Seppä, and M. Airaksinen, "What are the differences between sustainable and smart cities?," *Cities*, vol. 60, pp. 234–245, Sep. 2016.

[25] R. Giffinger and N. Pichler-Milanović, *Smart cities: ranking of European medium-sized cities*. Centre of Regional Science, Vienna University of Technology, 2007.

[26] M. Krivý, "Towards a critique of cybernetic urbanism: The smart city and the society of control," *Planning Theory*, p. 1473095216645631, Apr. 2016.

[27] United Nations, Department of Economic and Social Affairs, "The impact of population momentum on future population growth," United Nation, Departement of Economic and Social Affairs, 2017/4, Oct. 2017.
[28] P. Timmerman, *Vulnerability, resilience and the collapse of society: a review of models and possible climatic applications*. Toronto: institute for Environmental Studies, University of Toronto, 1981.
[29] G. A. Tobin and B. E. Montz, *Natural Hazards: Explanation and Integration*, 1 edition. New York: The Guilford Press, 1997.
[30] A. Wildavsky, *Searching for Safety*. Transaction Publ., 1988.
[31] F. Walsh, *Strengthening Family Resilience, Second Edition*, 2nd edition. The Guilford Press, 2011.
[32] C. S. Holling, "Understanding the Complexity of Economic, Ecological, and Social Systems," *Ecosystems*, vol. 4, no. 5, pp. 390–405, Aug. 2001.
[33] B. Starzomski, "Berkes, F., J. Colding, and C. Folke. 2003. Navigating social-ecological systems: Building resilience for complexity and change. Cambridge University Press, Cambridge, UK.," *Ecology and Society*, vol. 9, no. 1, Jan. 2004.
[34] W. N. Adger, "Social-Ecological Resilience to Coastal Disasters," *Science*, vol. 309, no. 5737, pp. 1036–1039, Aug. 2005.
[35] B. Obrist, C. Pfeiffer, and R. Henley, "Multi-layered social resilience: a new approach in mitigation research," *Progress in Development Studies*, vol. 10, no. 4, pp. 283–293, Oct. 2010.
[36] R. K. Larsen, E. Calgaro, and F. Thomalla, "Governing resilience building in Thailand's tourism-dependent coastal communities: Conceptualising stakeholder agency in social–ecological systems," *Global Environmental Change*, vol. 21, no. 2, pp. 481–491, May 2011.
[37] D. R. Godschalk, "Urban Hazard Mitigation: Creating Resilient Cities," *Natural Hazards Review*, vol. 4, no. 3, pp. 136–143, Aug. 2003.
[38] L. J. Vale and T. J. Campanella, Eds., *The Resilient City: How Modern Cities Recover from Disaster*, 1 edition. New York: Oxford University Press, 2005.
[39] B. Goldstein, "Resilience to surprises through communicative planning," *Ecology and Society*, vol. 14, no. 2, p. 33, 2009.
[40] J. E. Innes and D. E. Booher, *Planning with Complexity: An Introduction to Collaborative Rationality for Public Policy*, 1 edition. Milton Park, Abingdon, Oxon ; New York, NY: Routledge, 2010.
[41] C. Folke, "Resilience: The emergence of a perspective for social–ecological systems analyses," *Global Environmental Change*, vol. 16, no. 3, pp. 253–267, Aug. 2006.
[42] ARUP, "Arup | Publications | City Resilience Index," 2013. [Online]. Available: http://publications.arup.com/publications/c/city_resilience_index. [Accessed: 19-Feb-2017].
[43] M. Turnbull, C. L. Sterrett, and A. Hilleboe, *Toward Resilience: A Guide to Disaster Risk Reduction and Climate Change Adaptation*. Practical Action Pub, 2013.
[44] R. Papa, A. Galderisi, M. C. Vigo Majello, and E. Saretta, "Smart and Resilient Cities. A Systemic Approach for Developing Cross-sectoral Strategies in the Face of Climate Change," *TeMA - J. of Land Use, Mobility and Environment*, vol. 8(1):19-49, 2015.
[45] ICLEI, "Resilient Cities - ICLEI: Glossary of key terms," 2016. [Online]. Available: http://resilient-cities.iclei.org/resilient-cities-hub-site/resilience-resource-point/glossary-of-key-terms/. [Accessed: 18-Feb-2017].
[46] C. Wilkinson, "Social-ecological resilience: Insights and issues for planning theory," *Planning Theory*, vol. 11, no. 2, pp. 148–169, May 2012.
[47] C. Folke, S. Carpenter, B. Walker, M. Scheffer, T. Chapin, and J. Rockström, "Resilience Thinking: Integrating Resilience, Adaptability and Transformability," *Ecology and Society*, vol. 15, no. 4, Nov. 2010.

[48] K. C. Desouza and T. H. Flanery, "Designing, planning, and managing resilient cities: A conceptual framework," *Cities*, vol. 35, pp. 89–99, Dec. 2013.
[49] M. Baron, "Do We Need Smart Cities for Resilience," *Journal of Economics & Management*, vol. 10, p. 32, Mar. 2012.
[50] J. Viitanen and R. Kingston, "Smart Cities and Green Growth: Outsourcing Democratic and Environmental Resilience to the Global Technology Sector," *Environment and Planning A*, vol. 46, no. 4, pp. 803–819, Apr. 2014.

IOP Conf. Series: Earth and Environmental Science **158** (2018) 012046 doi:10.1088/1755-1315/158/1/012046

Crowdsourced Smart Cities versus Corporate Smart Cities

Tooran Alizadeh
University of Sydney

Corresponding author: tooran.alizadeh@sydney.edu.au

Abstract. Considering the speedy growth of smart-city promises and practices, there is an urgent need to take a critical approach and offer an integrated vision for an otherwise fragmented and sectoral concept. In particular, the literature warns about a critical deficit around the theorization of the smart city because discussions of relevant smart city theories or frameworks are few and fall short of offering alternative practical resolutions to the dominant discourse. In developing a response to such a deficit, this paper takes up the challenge to broaden theoretical insights into smart cities, by offering a bottom-up understanding of the 'smart city' concept with special attention to the potential of passive crowdsourcing based on the ocean of mostly untapped and unutilized available data in the public domain. Crowdsourced smart cities are proposed as an alternative to enable public engagement in smart city debates and decision-making – especially when dealing with global digital corporations.

1. Introduction

The concept of the smart city has emerged at the intersection of debates on the future of urban places, new technologies, and infrastructures – as a solution to offer clean, livable, technologically advanced, economically robust cities [1, 2]. The popularity of the concept is based on a mix of various factors, including the availability of substantial public financial resources (such as the EU Strategic Energy Technology Plan) to fund smart city initiatives; the tendency of global corporations (such as Cisco, Google and IBM) to heavily invest in urban digitization projects [3, 4]; and, finally, a growing range of complex urban challenges that need advanced technology-enabled solutions [5-7].

Nevertheless, the literature warns about the lack of both theoretical insights and empirical evidence required to fully understand the opportunities, challenges, and implications of smart cities. Research in this field is in its infancy [1, 4], fragmented along disciplinary lines [8, 9] and based on limited city case studies [6, 10].

Given the fast growing global attention given to smart cities and significant practical implications in the form of smart city initiatives and projects, there is an urgent need to take a critical approach and offer an integrated vision for this otherwise fragmented and sectoral concept [1, 2]. Here, the danger is that urban visioning is increasingly reduced to a single technology-centric vision that is simplistic and does not account for the socio-economic and spatial complexity of cities [3, 11, 12].

The literature specifically puts an emphasis on the critical deficit around theorization of the smart city [1, 12, 13]. Discussions of relevant smart city theories or frameworks are few; analyses lag behind actual practice and then fall short of offering alternative practical resolutions to the dominant discourse.

In developing a response to this deficit, this paper takes up the challenge of broadening theoretical insights into smart cities by offering a bottom-up understanding of the concept. Firstly, it examines how the smart city is currently defined in the literature and points out different – and sometimes contradictory – approaches taken towards smart city practice around the world. Secondly, it focuses on the definition and evolution of crowdsourcing during its relatively short existence. Thirdly, it brings the two earlier parts together by proposing the 'crowdsourced smart city' as an alternative theoretical perspective that has practical means of enabling people's voices to be heard in smart city decision-making and evaluation processes. This alternative theoretical perspective also has the capacity to empower informed smart city policy development for governments of all levels and guide them in their dealings with global corporations' increasing interest in the concept of the smart city.

2. Smart cities

2.1. The issue of definition

Despite their ongoing popularity and influence, there is no universal definition for smart cities [3, 14]. Indeed a number of related terms such as 'intelligent city', 'knowledge city', 'digital city', and 'eco-city' are used interchangeably in the literature [15-17]. All of these terms seem to stem from an earlier fascination with the technological aspect of the smart city concept, promising a new kind of technology-led urban utopia, which in turn has been criticized for being overly simplistic and ambiguous [10, 18]. Nevertheless, a growing stream in the literature has attempted to address the over-simplification of the smart city concept by offering holistic, multidisciplinary perspectives around key areas of smart infrastructure, smart economy, smart mobility, smart governance, smart environment, smart living, and smart people [2, 19]. This broad approach of defining the smart city has in turn been criticized for being out of reach and unattainable [20].

Amidst ongoing efforts to reach a comprehensive shared definition for smart cities, there is also a trend of making people's voices be heard in smart city design and decision-making processes, calling for direct participation of local actors and multi-stakeholders in planning and executing social, technological and urban transformation for smart cities [12, 14]. There are, however, serious questions about how this alternative approach – of giving a voice to citizens in the smart city discourse – can be implemented in urban development processes already in place and using technological advances already at hand. In order to contribute to the ongoing dialogue around such questions, this paper undertakes a review of the existing, diverse and sometimes contrary smart city practices and approaches to shed light on how smart technologies can be used to enable citizens' voices to be heard. This could then promote an alternative vision on smart cities in which citizens' voices are taken into account – a vision that is not compromised and yet attainable.

2.2. Smart city practice

Despite the lack of a shared definition, the smart city practice has been spreading around the world. The spread is important for a number of reasons, including:

- It is not limited to specific geographies, as smart city projects and initiatives now extend across the Global North and South [6]. There is, however, a far more limited understanding of the different ways in which smart cities are being rolled out in cities of the Global South [1, 21, 22].
- It is not limited to global cities, as medium-sized cities in different parts of the world have actively engaged with and invested in smart city projects and initiatives [23, 24].
- Smart city practice covers an incredibly diverse ranges of topics, including e-governance [25, 26], smart transport [27, 28], efficient production of urban services [29], and open data in cities [30].
- Smart city practice comes in various sizes: examples of large-size city-wide plans include Singapore's iN2015 (intelligent nation) project, South Korea's Songdo, Guangzhou Knowledge City in China, Masdar City in the UAE, and Barcelona in Spain [9, 31]. Among the large-scale city-wide initiatives, Barcelona is of special interest due to its apparent desire to integrate smart city efforts with the city's urban planning and policy thinking under the four main topics of smart governance, smart economy, smart living and smart people. More importantly, Smart City Barcelona is a collaborative movement among its corporations, universities, government, and the residents of Barcelona [9]. In contrast, some of the out of scratch smart city projects in Asia have attracted wide criticism as top-down practices with little room for people's voices in the process [32, 33].
- Smart city practice is funded by quite different private and/or public sources around the world. Over the last few years, the European Union (EU), in particular, has devoted constant efforts to support smart city initiatives – investing millions of euros in research, development and pioneer projects that try to link smart city initiatives to environmental targets (such as reduction in greenhouse gas emissions) and economic development (such as innovation and

employment) [2, 3, 11]. Smart city practices around the world have opened up new horizons in the problematic relationship between the public and private sectors in the management of cities [3, 34]. A critical line of scholarship has emerged around the role of global technology corporations in developing smart cities; and how they have tried to manipulate urban governments while engaging in providing philanthropic smart city services [7, 10, 35, 36].

Finally, the literature notes that the utmost majority of smart city projects and initiatives around the world are still in the planning phases or early phases of implementation [15, 23] and no city claims to be fully-fledged smart [12]. This means further research is required to assess progress of smart city projects and their implications on the overall success or failure of cities over time.

2.3. A critique of smart cities: corporate smart cities

Despite the lack of a universal definition for the smart city, which has resulted in heterogeneous smart city practices worldwide, the critiques of the concept seem to be quite focused on what is labeled as 'corporate smart cities' [3, 37-39]. From a critical perspective, we are reminded that the term 'smart city' was coined in the US inside two global high-technology corporations, IBM and Cisco. This indicates a point of view in which a combination of the profit motive for global corporations, on the one hand, and entrepreneurial governance – as coined by urban geographer Harvey [40] – on the other hand, drives the leading narrative of the smart city discourse [11, 31]. According to this view, the smart city has crystallized into an image of a technology-led urban utopia permeated with top-down and centrally controlled technological infrastructure, with the aim to improve the urban environment in terms of efficiency, security and sustainability [14, 41].

As noted earlier in this paper, critiques of the corporate smart city [2, 42] vigorously warn against the new smart cities being constructed from scratch in Asia, the Arab world and even Europe led by giant corporations (e.g. Cisco, IBM, and Siemens) using examples such as Masdar in the United Arab Emirates (UAE), Songdo in South Korea, and PlanIT Valley in Portugal. The underlying assumption with these cities – which is strongly questioned in the literature [33, 43] – is that IT can automatically make cities more economically prosperous and equal, more efficiently governed, and less environmentally wasteful – with little room for ordinary people to participate [22, 37]. The corporate vision of smart cities has promoted a trend towards cities selling themselves, being 'open for business', and increasingly becoming a backdrop to corporate advertising and the privatization of public space [3, 35].

In fact, it might be argued that in the corporate vision of smart cities, citizens are seen as barriers in the race towards smartness and that they need to be educated as to the benefits IT can bring. This lack of concern with democratic decision-making, and real citizen involvement and participation has encouraged a new trend in the literature to search for an alternative version of smart cities and provide a counter-point to the corporate vision [31]. This alternative vision has emanated from small-scale and fledgling examples of participatory community-based types of smart initiatives [44, 45].

2.4. Alternative smart cities and their shortcomings

In response to the lack of people's voices in the corporate vision of the smart city, a new trend both in academia and practice has started to offer alternative visions. Examples can be identified mostly in the US and Europe in the last few years: Metalab is a non-profit innovation center based in Vienna offering a physical space for free exchange of information and collaboration between technology enthusiasts, hobbyists and hackers. Another initiative that can be linked to the alternative smart city paradigm is the Medialab-Prado – a collective innovation laboratory in Madrid – interested in the production, research and dissemination of cultural projects to sustain an active community of engaged citizens [14]. Similar initiatives have been developed in Amsterdam and Barcelona [2] to encourage bottom-up initiatives.

Amidst the excitement and flashy titles for the alternative visions of the smart city, the characteristics below give a united front against the corporate vision of the smart city [2, 13, 14, 46]:

- A democratic bottom-up approach: to promote participatory urban technologies, greater social inclusion and a substantial shift in power from corporations to ordinary people and their communities.
- Reliance on dynamic public-private partnership: with an emphasis on participatory governance rather than entrepreneurial governance.

- A tendency to identifying the urban problem first and only then reaching out for the relevant technological solutions: with emphasis on the capacities of each city and its distinct cultures, histories and political economies.
- Associated with the free-software and open-access movement.
- In the preliminary phase: far from being mature and mainly existing in seed form.

In other words, alternative visions of the smart city put strong emphasis on multidisciplinary smart city investments in human and social capital, traditional (transport) and modern (ICT) communication infrastructure to fuel sustainable economic development and a high quality of life with a wise management of natural resources through participatory governance [12, 14, 47]. An essential element of these alternative visions is the emphasis on citizen engagement beyond the simple delivery of services [2, 13].

The problem with the alternative smart city visions offered in the literature is the distinct lack of attention to implementation. Rather, the core mission of giving a voice to the people seems to fall short, as implemented successful examples are scattered, very small in scale and get lost against the scale and fast spread of the corporate smart city vision [31, 48]. This core problem in part stems from the difficulty involved in gathering people's voices and identifying urban issues that are indeed important to citizens and not just corporations.

3. Crowdsourcing as a solution?

Given the issues around the practicality of gathering people's voices in smart city decision-making processes and practices, this section examines the relatively new concept of crowdsourcing. Here, crowdsourcing is seen as an alternative technologically-enabled way of making people's voices be heard in decision-making. It should be highlighted that the aim is not to offer an all-inclusive account of the ever-growing concept but rather to provoke further discussion around the concept of the crowdsourced smart city.

Below we offer an account of how crowdsourcing has evolved in a short period of time. A brief review of the current landscape of crowdsourcing with a focus on urban problems then follows. Combing these two will then provide the foundation for the articulation of the crowdsourced smart city as an alternative vision later in this paper.

3.1. Crowdsourcing: Evolution of a concept

Howe [49] first coined the term crowdsourcing in a Wired Magazine article as "the act of a company or institution taking a function once performed by employees and outsourcing it to an undefined (and generally large) network of people in the form of an open call". Since then there have been many attempts to define crowdsourcing based on the diversity of its practice and different theoretical models [50]. In search for a common definition of crowdsourcing, Estellés-Arolas [51] found eight common characteristics of crowdsourcing: the task at hand; the recompense obtained; the crowdsourcer or initiator of the crowdsourcing activity; what is obtained by them following the crowdsourcing process; the type of process; the call to participate; and the medium. Based on these common characteristics he proposed the following definition [51]:

> "Crowdsourcing is a type of participative online activity in which an individual, an institution, a non-profit organization, or company proposes to a group of individuals of varying knowledge, heterogeneity, and number, via a flexible open call, the voluntary undertaking of a task."

This 'task-oriented' definition of crowdsourcing has been revised several times. Revisions were proposed in response to emergent crowdsourcing models based on the prominence of social media and support by ubiquitous mobile devices [52]. For example, Thapa et al. [53] introduced two types of crowdsourcing: 'selective' and 'integrative' (task-oriented), based on different types of citizens involved. In an integrative crowdsourcing non-professionals are able to participate due to the simplicity of the tasks. In contrast, in selective crowdsourcing the complexity of the problem reduces the target group to professionals with special expertise.

However, the most significant evolution of the crowdsourcing concept stems from a shift from a task-oriented approach to what can be described as the crowdsourcing of opinions [54]. In this second approach, crowdsourcing is no longer about getting a certain task done with the help of the public. Instead, crowdsourcing of opinions is used to gauge opinions, ideas or perceptions of the public in different forms of polling, sentiment analysis, and opinion mining. Sentiment analysis uses language processing and machine learning to identify which topics different groups talk and care about the most. Social media in general, and Twitter in particular, are rich sources of opinions; and have been used by both private and public sector in the crowdsourcing of opinions. There are numerous examples of companies using crowdsourcing of opinions – via social media – in their marketing efforts [55, 56]. More importantly, the crowdsourcing of opinions also serves governance and participatory planning and decision-making processes in different forms [57].

The crowdsourcing of opinions, in turn, is then categorized in two broad categories, namely active and passive. In terms of the difference between active and passive crowdsourcing, Loukis and Charalabidis [58] argue that the active crowdsourcing of opinions is more like mainstream private-sector crowdsourcing, which actively stimulates discussion and content generation by citizens on specific topics. Meanwhile, passive crowdsourcing is more compatible with the public sector; it passively collects information, knowledge, opinions and ideas concerning hot topics of the day and important public policies created by citizens without any initiation, stimulation or moderation from government postings [57-59]. Social media monitoring (SMM), as systematic, continuous observation and analysis of the ocean of data already available and mostly untapped, is the main source of passive crowdsourcing in the public sector [60].

In this paper, we acknowledge the evolution of crowdsourcing as a concept and yet mainly focus on passive crowdsourcing of opinions as a way to inform and evaluate smart city decision-making processes and plans. However, this requires a better understanding of the current landscape of crowdsourcing elaborated in the following section.

3.2. Crowdsourcing: Current Landscape

Several studies [56, 61, 62] have tried to make sense of the rapidly evolving crowdsourcing landscape. Different categorizations have been offered to articulate the variety of directions crowdsourcing has been taking in the public, private and non-for profit sectors. Dawson and Bynghal [56], for example, summarize the current landscape of crowdsourcing in eight crowdsourcing models that cover 22 categories. Table 1 shows a summary of the current landscape of crowdsourcing as articulated by Dawson and Bynghal [56]. The models include Crowd Service, Crowd Ventures, Media and Content, Marketplaces, Crowd Platforms, Crowd Processes, Content and Products Markets, Non-profit. Crowd Service refers to services that are delivered fully or partially by crowds and includes Managed Crowds and Labor Pools. Crowd Venture refers to ventures predominantly driven by crowds, including idea selection, development, and commercialization. Media and Content is another model, focusing on the creation of media, content, data and knowledge shared by crowds. Non-profit models include citizen engagement for governance purposes, volunteering for disaster management and crowd contribution to science. Marketplaces match buyers and sellers of services and financing through mechanisms, including bidding and competitions. The Marketplace model covers crowd-funding, microtasking, innovation markets, innovation prizes, services marketplaces and competition markets. Platforms are software and processes to run crowd works and crowd projects for use with internal or external crowds, including idea management, prediction markets and crowd platforms. Crowd processes are services that provide value-added processes or aggregation to existing crowds or marketplaces. Content and Product refers to sale of content or products that are created, developed or selected by crowds.

Table 1. Summary of the current landscape of crowdsourcing as articulated by Dawson and Bynghal [56]

Crowdsourcing Models	**Crowdsourcing Categories**	**Examples**
Crowd Services	Managed crowds	Thinkspeed, GeniusRocket
	Labor pools	BzzAgent, UTest
Crowd Ventures		Globumbus, SENSORICA
Media and Content	Data	Data.com, Rootwireless
	Content	Wikipedia, DemandMedia
	Knowledge sharing	Quora, ChaCha
Marketplaces	Equity crowd-funding	SeedUps, CrowdCube
	Crowd-funding	IndieGoGo, Fundedbyme
	Microtasking	Mechanical Turk, CloudFactory

Crowdsourcing Models	**Crowdsourcing Categories**	**Examples**
	Innovation markets	IdeaConnection, Innocentive
	Innovation prizes	Cisco i-Prize, DARPA
	Services marketplaces	Freelancer.com, Guru
	Competition markets	Zooppa, TopCoder
Crowd Platforms	Idea management	IdeaScale, BrightIdea
	Prediction markets	CrowdWorx, ConsensusPoint
	Crowd platforms	Crowdicity, CrowdEngineering
Crowd Processes		CrowdFlower, Smartsheet
Content and Product Markets	Content markets	Threadless, Shapeways
	Crowd design	Ponoko, Quirky
Non-profit	Contribution	Kiva, Crowdrise, Causes
	Science	GalaxyZoo, Phylo
	Citizen engagement	CrisisCommons, Ushahidi

While elaborating the diversity of the ever-growing crowdsourcing models in the current landscape is beyond the scope of this paper, below we offer a few examples of the different types of crowdsourcing used to tackle urban problems.

3.2.1. Crowdsourcing in emergency/disaster management. There have been several cases around the world where crowdsourcing has been used as an additional, complementing tool in emergency/disaster management [63-65]. Emergency responses in several cases – including but not limited to the recent earthquakes in Nepal and Haiti, and floods in Myanmar and India – have shown that different crowds with different skills and expertise can get involved differently in crisis management. Disaster-affected people can bring local and directly observed information and detect precise location (crowd as sensors); diasporas can offer their implicit socio-cultural knowledge in detecting needs within the affected area (crowd as a reporter); and, digital GIS volunteers contribute to processing and managing crisis data (crowd as a micro-tasker) [66, 67].

3.2.2. Active crowdsourcing of opinions in strategic planning. There is a growing number of planning departments at different levels (e.g. local and state) that use crowdsourcing to seek public opinions, ideas and feedback on their – mostly strategic – draft planning documents; to promote public participation in policy making and decision-making processes [68, 69]. In most cases, specially designed digital platforms are used to facilitate active crowdsourcing of ideas, which are often expensive to design and maintain and more importantly have to compete with the well-established social media platforms (i.e. Facebook and more aggressively Twitter). For instance, the City of Vancouver used an online platform to seek feedback as part of the participatory process involved in the development of its first urban digital strategy document [17, 70].

3.2.3. Active crowdsourcing of data in local planning/development. The most widespread form of crowdsourcing used in response to urban challenges is when data are crowdsourced from citizens to have a better understanding of the community conditions to facilitate better evidence-based decision-making [71]. For example, Street Bump is a crowdsourcing platform, produced by the Mayor's Office in Boston, which collects real-time road condition data from residents while they drive to improve their neighbourhood streets by fixing problems or planning long-term investments [72].

3.2.4. Passive crowdsourcing of opinions by government. There is a relatively new approach that calls for governments to exploit the extensive political content continuously created in numerous social-media platforms by citizens – without government stimulation – to better understand public needs, issues, opinions and arguments concerning a particular domain of government activity or public policy. [57]. This novel approach towards e-participation based on passive crowdsourcing of data is rarely used by urban governments [73, 74].

4. Crowdsourcing and smart cities

In principle, crowdsourcing has great potential in participatory planning; promotes many elements of smart cities including open government; and can be used as expansion of e-governance to we-governance by facilitating citizen-to-government support, citizen reporting, and citizen-government co-production of cities [75-77].

Nevertheless, the problem is the scale of uptake of crowdsourcing in urban governments. Previous studies have pointed out that the slow uptake of technologically-enabled solutions in e-governance at the urban level, in part, is related to the lag in technical skills at the local government [25, 78, 79]. It is argued that staff and organization structure of urban governance lack technical readiness and some of the well-established technical advances are missing at the local level [80]. In the last few years, however, we have witnessed an influx of technologically advanced smart city projects and plans around the world and cities of all sizes have shown interest to jump on the smart city wagon [6, 81]. Such a trend suggests that urban governments are catching up on technological issues somehow – or at least join forces with those (companies) who have the technical skills required [11, 31].

Moreover, financial restrictions and resourcing issues at the local government level have been pointed out as another set of reasons slowing down the rate of technologically-enabled solutions in e-government and we-government [75]. However, passive crowdsourcing requires less preparation and financial resources in comparison to many other smart solutions around that require sophisticated advanced equipment [58, 82]. So the problem must be somewhere else.

In this paper, we argue that the slow uptake of passive crowdsourcing capabilities to promote citizen's voices in urban decision-making is mainly a conceptual problem at the core of how the smart city is defined. Therefore we argue that the slow uptake of passive crowdsourcing capabilities, at least in part, is because it is seen as a side issue and not integrated with the popular smart city concept.

In the next section, we identify and respond to two reasons that have limited the role crowdsourcing has played in the smart city discourse so far.

4.1. Inclusionary concerns

Over the last few decades, a stream in the literature has consistently raised concerns about the inclusionary limitations of any online platforms, with reference to the digital divide [83, 84]. More specifically, it has been argued that there are limitations to how citizens can systematically, reliably and consistently get involved with crowdsourcing [85, 86]. One of the important challenges of crowdsourcing is finding, motivating and retaining participants. It has been noted that to prevent bias in results, participation in crowdsourcing should be inclusive and take in people from all walks of life and sectors of society affected by a problem [73, 87-89].

Having said this, we argue that inclusionary concerns over the use of crowdsourcing in urban decision-making seem to be mitigated. This, by no means, is to suggest that the digital divide between urban vs. rural at any national level, or Global North vs. Global South at the international level does not exist [90-92]. However, participatory technologies at the local level have been growing in popularity with an increasing number of cities and planning agencies using technology to engage the public in planning processes [93, 94]. This is partly due to the enormous popularity of the social-media platforms, which are no longer limited to a limited elite and educated sector of society [73, 95]. The popularity of social media basically paves the way for passive crowdsourcing by the public sector [68] as the voices captured are quite diverse [55, 87].

4.2. Corporate agenda

Socio-technological advances in the last decade or so have mitigated some of the inclusionary concerns around the role that crowdsourcing could play in the smart city discourse. We, however, argue that corporate agendas [31, 38] at the center of the dominant smart city discourse is the main barrier against crowdsourced cities. In other words, the dominant corporate smart city discourse mainly holds its power by claiming that the digital corporations have the solutions and the skills required in a complex urban world [35, 96]. In contrast, in this paper, we join the stream in the literature arguing that smart cities should build on democratic decision-making [20] and contribute to transparency [93, 97] through public participation. Participatory processes enabled by crowdsourcing help urban governments to identify and respond to wicked problems [43, 98], democratize decision-making, and increase social capital [99].

This focus on the citizen-centricity of the crowdsourced smart city [41, 100] emphasizes the potential of online participatory technologies to allow citizens to actively engage in shaping their city.

Empowered citizens, in turn, then empower local governments in their negotiations with digital corporations that have clear vested interests in running the show in the smart city discourse. As a first step in dealing with the dominant corporate smart city vision, passive crowdsourcing has the potential to evaluate some of the smart city projects already in place to assess their success based on public opinion. In other words, the crowdsourced smart city can shift the power from corporations to citizens, help local governments to avoid undervaluing their skills and expertise, make more strategic investments informed by public opinions, and ensure that smart city projects and initiatives have equitable outcomes for all citizens.

5. Conclusion: crowdsourced smart city vs. corporate smart city

This paper started by acknowledging an urgent need to take a critical approach to the concept of the smart city [1, 3]. In particular, it pointed out the deficit around theorization of the concept [12, 13]. In developing a response to such a deficit, we took up the challenge of broadening theoretical insights into smart cities by redefining the concept of the smart city with special attention to data that are available in the public domain.

In so doing, we examined both 1) the current dominant – yet mostly exclusive and technocratic – and also 2) the alternative – yet mostly limited and not comprehensive – definitions of the smart city. This paper then discussed the evolution of the crowdsourcing concept and its practices since it was introduced and built upon its latest definition to tap into the streams of underutilized urban data, as citizens are using online channels to engage on topics that they feel passionate about.

In conclusion, we propose the crowdsourced smart city as an alternative to the corporate smart city. The crowdsourced smart city provides the response so desperately sought in the literature to shift from the corporate smart city paradigm to a bottom-up, socio-technological inclusive approach that produces non-planned forms of citizen empowerment in urban governance [61]. Passive crowdsourcing has the capability to harnesses the ocean of untapped data already available in social-media platforms to generate collective intelligence [101] and improve planning, evaluation and functioning of smart cities. At this stage, we call for urban governments around the world – interested and invested in smart cities – to verify any smart projects and initiatives already in place; to use passive crowdsourcing to evaluate the impact of their earlier and ongoing smart city efforts; and to modify any future smart city projects and plans based on public needs, opinions, and priorities.

References

[1] Luque-Ayala A, Marvin S. Developing a critical understanding of smart urbanism? Urban Studies 2015;52(12):2105-16.

[2] Papa R, Gargiulo C, Galderisi A. Towards an urban planners' perspective on Smart City. *TeMA J. of Land Use, Mobility and Environment.* 2013;6(1):5-17.

[3] Vanolo A. Smartmentality: the Smart City as disciplinary strategy Urban Studies. 2014;51:883-98.

[4] Luque A. The smart grid and the interface between energy, ICT and the city. In: Dixon T, Eames M, Hunt M, and Lannon S, editors. *Urban Retrofitting for Sustainability*. London: Earthscan; 2014. p. 159-73.

[5] Alizadeh T, Irajifar L. Towards Gold Coast Smart City: A Combination of Local Planning Priorities and International Best Practices. State of Australian Cities National Conf.; 28-30 November; Adelaide, Australia 2017.

[6] Alizadeh T. An investigation of IBM's Smarter Cites Challenge: What do participating cities want? Cities. 2017;63:70-80.

[7] Paroutis S, Bennett M, Heracleous L. A strategic view on smart city technology: The case of IBM Smarter Cities during a recession. *Technological Forecasting and Social Change*. 2014;89:262-72.

[8] Alawadhi S, Aldama-Nalda A, Chourabi H, Gil-Garcia JR, Leung S, Mellouli S, et al. Building Understanding of Smart City Initiatives. In: Scholl HJ, Janssen M, Wimmer MA, Moe CE, Flak LS, editors. Electronic Government: 11th IFIP WG 85 International Conf., EGOV 2012, Kristiansand, Norway, September 3-6, 2012 Proc. Berlin, Heidelberg: Springer Berlin Heidelberg; 2012. p. 40-53.

[9] Bakıcı T, Almirall E, Wareham J. A Smart City Initiative: the Case of Barcelona. *J. of the Knowledge Economy*. 2013;4(2):135-48.
[10] Kitchin R. Making sense of smart cities: addressing present shortcomings. *Cambridge J. of Regions, Economy and Society*. 2015;8:131-6.
[11] Rosati U, Conti S. What is a Smart City Project? An Urban Model or A Corporate Business Plan? Procedia - Social and Behavioral Sciences. 2016;223:968-73.
[12] Lara AP, Costa EMD, Furlani TZ, Yigitcanlar T. Smartness that matters: towards a comprehensive and human-centred characterisation of smart cities. *J. of Open Innovation: Technology, Market, and Complexity*. 2016;2(8).
[13] Lee JH, Hancock MG, Hu M-C. Towards an effective framework for building smart cities: Lessons from Seoul and San Francisco. *Technological Forecasting and Social Change*. 2014;89:80-99.
[14] Niaros V. Introducing a Taxonomy of the “Smart City”: Towards a Commons-Oriented Approach? tripleC. 2016;14(1).
[15] Manville C, Cochrane G, Cave J, Millard J, Pederson JK, Thaarup RK, et al. Mapping Smart Cities in the EU. Brussels: European Parliament, Directorate-General for Internal Policies; 2014.
[16] Cocchia A. Smart and Digital City: A Systematic Literature Review. In: Dameri RP, Rosenthal-Sabroux C, editors. Smart City: How to Create Public and Economic Value with High Technology in Urban Space. New York: Springer; 2014. p. 13-44.
[17] Alizadeh T. A policy analysis of digital strategies: Brisbane vs. Vancouver. *Int. J. of Knowledge-Based Development*. 2015;6(2):85-103.
[18] Townsend AM. Smart cities: big data, civic hackers, and the quest for a new utopia. New York: W.W. Norton; 2013.
[19] Yigitcanlar T, Inkinen T, Makkonen T. Does size matter? Knowledge-based development of second-order cityregions in Finland. *disP-The Planning Review*. 2015;51(3):62–77.
[20] Neirotti P, Marco AD, Cagliano AC, Mangano G, Scorrano F. Current trends in Smart City initiatives: Some stylised facts Cities. 2014;38:25-36.
[21] Kavta K, Yadav PK. Indian Smart Cities and Their Financing: A First Look. In: Seta F, Sen J, Biswas A, Khare A, editors. From Poverty, Inequality to Smart City Singapore: Springer; 2017. p. 123-41.
[22] Watson V. The allure of ‘smart city’ rhetoric: India and Africa. Dialogues in Human Geography. 2015;5(1):36-9.
[23] Sanseverino ER, Sanseverino RR, Vaccaro V, Macaione I, Anello E. Smart Cities: Case Studies. In: Sanseverino lR, Sanseverino RR, Vaccaro V, editors. Smart Cities Atlas. London: Springer; 2016. p. 47-140.
[24] Giffinger R, Fertner C, Kramar H, Kalasek R, Pichler-Milanović N, Meijers E. Smart cities: Ranking of European medium-sized cities. Vienna: Centre of Regional Science; 2007.
[25] Bertot JC, Gorham U, Jaeger PT, Sarin LC, Choi H. Big data, open government and e-government: Issues, policies and recommendations. *Information Polity*. 2014;19(1):5-16.
[26] Kumar TMV. E-Governance for Smart Cities. In: Kumar TMV, editor. E-Governance for Smart Cities. Singapore: Springer Singapore; 2015. p. 1-43.
[27] Bodhani A. Smart transport. Engineering & Technology. 2012;7(6):70-3.
[28] Debnath AK, Chin HC, Haque MM, Yuen B. A methodological framework for benchmarking smart transport cities Cities. 2014;37:47-56.
[29] Lee SW, Sarp S, Jeon DJ, Kim JH. Smart water grid: the future water management platform. Desalination and Water Treatment. 2015;55(2):339-46.
[30] Al-Ani A. Government as a Platform: Services, Participation and Policies. In: Friedrichsen M, Kamalipour Y, editors. Digital Transformation in Journalism and News Media. Berlin, Germany: Springer; 2017. p. 179-96.
[31] Hollands RG. Critical interventions into the corporate smart city. *Cambridge J. of Regions, Economy and Society*. 2015;8(1):61-77.
[32] Datta A. India's smart city craze: big, green and doomed from the start? 2014 [cited 2017 Oct 23]. Available from: https://www.theguardian.com/cities/2014/apr/17/india-smart-city-dholera-flood-farmers-investors.

[33] Carvalho L. Smart Cities From Scratch? A Socio-Technical Perspective. *Cambridge J. of Regions, Economy and Society.* 2015;8(1):43-60.
[34] Datta A. New Urban Utopias of Postcolonial India: Entrepreneurial urbanization in Dholera smart city, Gujarat. Dialogues in Human Geography. 2015;5(1):3-22.
[35] McNeill D. Global firms and smart technologies: IBM and the reduction of cities. Transactions of the Institute of British Geographers. 2015;40(4):562-74.
[36] Alizadeh T, Grubesic T, Helderop E. Urban governance and big corporations in the digital economy: An investigation of socio-spatial implications of Google Fiber in Kansas City. Telematics and Informatics. 2017;https://doi.org/10.1016/j.tele.2017.04.007.
[37] Hollands R. Beyond the corporate smart city: glimps of other possiblities of smartness. In: Marvin S, Luque-Ayala A, McFarlane C, editors. Smart Urbanism: Utopian Vision Or False Dawn? London: Routledge; 2015. p. 168-84.
[38] Söderström O, Paasche T, Klauser F. Smart cities as corporate storytelling. City: analysis of urban trends, culture, theory, policy, action. 2014;18(3):307-20.
[39] McNeill D. IBM and the visual formation of smart cities In: Marvin S, Luque-Ayala A, McFarlane C, editors. Smart Urbanism: Utopian Vision or False Dawn? London: Routledge 2016. p. 34-52.
[40] Harvey D. From Managerialism to Entrepreneurialism: The Transformation in Urban Governance in Late Capitalism. Geografiska Annaler Series B, Human Geography. 1989;71(1):3-17.
[41] Albino V, Berardi U, Dangelico RM. Smart cities: Definitions, dimensions, performance, and initiatives. Journal of Urban Technology. 2015;22(1):3-21.
[42] Gabrys J. Programming environments: environmentality and citizen sensing in the smart city. Environment and Planning D: Society and Space. 2014;32(1):30-48.
[43] Kitchin R. The real-time city? Big data and smart urbanism. *GeoJournal.* 2014;79(1):1-14.
[44] Chatterton P. Towards an agenda for postcarbon cities: lessons from LILAC, the UK's first ecological, affordable, cohousing community. *Int. J. of Urban and Regional Research.* 2013;37(5):1654-74.
[45] Radywyla N, Biggs C. Reclaiming the commons for urban transformation. *J. of Cleaner Production.* 2013;50:159-70.
[46] Kostakis V, Bauwens M, Niaros V. Urban Reconfiguration after the Emergence of Peer-to-Peer Infrastructures: Four Future Scenarios with an Impact on Smart Cities. In: Araya D, editor. Smart Cities as Democratic Ecologies. New York, NY: Palgrave Macmillan; 2015. p. 116-24.
[47] Giffinger R, Gurdum H. Smart cities ranking: an effective instrument for the positioning of cities? ACE: Archit City and Environ. 2010;4(12):7-25.
[48] Anttiroiko A-V. U-cities reshaping our future: reflections on ubiquitous infrastructure as an enabler of smart urban development. Artificial Intelligence and Society. 2013;28(4):491-507.
[49] Howe J. The Rise of Crowdsourcing Wired Magazine. 2006;14(6):1-5.
[50] Zhao Y, Zhu Q. Evaluation on crowdsourcing research: Current status and future direction. Information Systems Frontiers. 2014;16(3):417-34.
[51] Estellés-Arolas E, González-Ladrón-De-Guevara F. Towards an integrated crowdsourcing definition. *J. of Information science.* 2012;38(2):189-200.
[52] Kietzmann JH. Crowdsourcing: A revised definition and introduction to new research. Elsevier; 2017.
[53] Thapa BE, Niehaves B, Seidel C, Plattfaut R. Citizen involvement in public sector innovation: Government and citizen perspectives Information Polity. 2015;20(1):3-17.
[54] Noveck B. Smart citizens, smarter state: The technologies of expertise and the future of governing. Cambridge: Harvard University Press; 2015.
[55] Willems LG, Alizadeh T. Social Media for Public Involvement and Sustainability in International Planning and Development. *Int. J. of E-Planning Research (IJEPR).* 2015;4(4):1-17.
[56] Dowson R, Bynghal S. Getting Results From Crowds: The definitive guide to using crowdsourcing to grow your business. San Francisco: Advanced Human Technologies Inc; 2011.
[57] Charalabidis Y, Loukis EN, Androutsopoulou A, Karkaletsis V, Triantafillou A. Passive crowdsourcing in government using social media. Transforming Government: People, Process and Policy. 2014;8(2):283-308.

[58] Loukis E, Charalabidis Y. Active and passive crowdsourcing in government. Policy Practice and Digital Science: Springer; 2015. p. 261-89.
[59] Charalabidis Y, Triantafillou A, Karkaletsis V, Loukis E. Public Policy Formulation through Non Moderated Crowdsourcing in Social Media. Int. Conf. on Electronic Participation; September 3-5; Kristiansand, Norway2012.
[60] Loukis E, Charalabidis Y, Androutsopoulou A. Promoting open innovation in the public sector through social media monitoring. Government Information Quarterly. 2017;34(1):99-109.
[61] Certomà C, Dyer M, Pocatilu L, Rizzi F. Citizen Empowerment and Innovation in the Data-Rich City. Springer; 2017.
[62] Brabham DC. Using crowdsourcing in government. IBM Center for the Business of Government. 2013:1-42.
[63] See L, Mooney P, Foody G, Bastin L, Comber A, Estima J, et al. Crowdsourcing, Citizen Science or Volunteered Geographic Information? The Current State of Crowdsourced Geographic Information. *Int. J. of Geo Information*. 2016;5(5).
[64] McLennan B, Whittaker J, Handmer J. The changing landscape of disaster volunteering: opportunities, responses and gaps in Australia. Natural Hazards 2016;10.1007/s11069-016-2532-5:1-18.
[65] Horita FEA, Degrossi LC, Assis LFGd, Zipf A, Albuquerque JPd. The use of Volunteered Geographic Information (VGI) and Crowdsourcing in Disaster Management: a Systematic Literature Review. AMCIS2013.
[66] Poblet M, García-Cuesta E, Casanovas P. Crowdsourcing roles, methods and tools for data-intensive disaster management. Information Systems Frontiers. 2017. p. 1-17.
[67] Liu SB. Crisis Crowdsourcing Framework: Designing Strategic Configurations of Crowdsourcing for the Emergency Management Domain. Computer Supported Cooperative Work. 2014;23(4-6):389-443.
[68] Afzalan N, Sanchez TW, Evans-Cowley J. Creating smarter cities: Considerations for selecting online participatory tools. *Cities*. 2017;67:21-30.
[69] Panagiotopoulos P, Bowen F, editors. Conceptualising the digital public in government crowdsourcing: Social media and the imagined audience. Int. Conf. on Electronic Government; 2015; Cham: Springer.
[70] Alizadeh T, Sipe N. Vancouver's Digital Strategy: Disruption, new direction, or business as usual? *Int. J. of E-Planning Research (IJEPR)*. 2016;5(4):1-15.
[71] Brabham D. The four urban governance problem types suitable for crowdsourcing citizen participation. In: Silva CN, editor. Citizen E-Participation in Urban Governance: Crowdsourcing and Collaborative Creativity. New York: IGI Global; 2013.
[72] Harford T. Big data: A big mistake? Significance. 2014;11(5):14-9.
[73] Kleinhans R, Ham MV, Evans-Cowley J. Using Social Media and Mobile Technologies to Foster Engagement and Self-Organization in Participatory Urban Planning and Neighbourhood Governance. Planning Practice & Research. 2015;30(3):237-47.
[74] Schweitzer L. Planning and social media: a case study of public transit and stigma on Twitter. *J. of the American Planning Association*. 2014;80(3):218-38.
[75] Linders D. From e-government to we-government: Defining a typology for citizen coproduction in the age of social media. Government Information Quarterly. 2012;29(4):446-54.
[76] Schmidthuber L, Hilgers D. Unleashing Innovation beyond Organizational Boundaries: Exploring Citizensourcing Projects. *Int. J. of Public Administration*. 2017:1-16.
[77] Castelnovo W. Co-production Makes Cities Smarter: Citizens' Participation in Smart City Initiatives. In: Fugini M, Bracci E, Sicilia M, editors. Co-production in the Public Sector. Milan: Springer; 2016. p. 97-117.
[78] Berst J, Enbysk L, Williams C. Smart Cities Readiness Guide: The planning manual for building tomorrow's cities today. Seattle: Smart Cities Council; 2014.
[79] Norris DF, Reddick CG. Local E-Government in the United States: Transformation or Incremental Change? Public Administration Review. 2013;73(1):165-75.
[80] Pardo T, Nam T, Burke B. E-government interoperability: Interaction of policy, management, and technology dimensions. Social Science Computer Review. 2012;30:7-23.

[81] Arroub A, Zahi B, Sabir E, Sadik M, editors. A literature review on Smart Cities: Paradigms, opportunities and open problems. Wireless Networks and Mobile Communications (WINCOM), 2016 Int. Conf. on; 2016: IEEE.

[82] Certomà C, Rizzi F. Crowdsourcing Processes for Citizen-Driven Governance. In: Certomà C, Dyer M, Pocatilu L, Rizzi F, editors. Citizen Empowerment and Innovation in the Data-Rich City. London: Springer International Publishing; 2017. p. 57-77.

[83] Pick JB, Sarkar A. The Global Digital Divides. London: Springer; 2015.

[84] Mossberger K. Toward digital citizenship. Addressing inequality in the information age. In: Chadwick A, Chadwick A, Howard PN, editors. Routledge Handbook of Internet Politics. New York: Routledge; 2009.

[85] Hossain M, Kauranen I. Crowdsourcing: a comprehensive literature review. *J. of Global Operations and Strategic Sourcing.* 2015;8(1):2-22.

[86] Bozzon A, Houtkamp JM, Kresin F, de Sena NHHA, de Weerdt M. From Needs to Knowledge. A reference framework for smart citizens initiatives. 2016.

[87] Fieseler C, Fleck M. The pursuit of empowerment through social media: structural social capital dynamics in CSR-blogging. *J. of Business Ethics.* 2013;118(4):759-75.

[88] Panagiotopoulos P, Bowen F. Social media for government crowdsourcing: An exploratory study of challenges and opportunities. the 29th Annual BAM Conference 8-10 September Portsmouth, UK2015.

[89] Panagiotopoulos P, Bowen F. Conceptualising the Digital Public in Government Crowdsourcing: Social Media and the Imagined Audience. Int. Conf. on Electronic Government; 30th August - 2nd September; Thessaloniki, Greece 2015.

[90] Warf B. Contemporary digital divides in the United States. Tijdschrift voor Economische en Sociale Geografie. 2013;104(1):1-17.

[91] Alizadeh T. The spatial justice implications of telecommunication infrastructure: the socio-economic status of early national broadband network rollout in Australia. *Int. J. of Critical Infrastructures.* 2015;11(3):278-96.

[92] Grubesic TH. Future Shock: Telecommunications Technology and Infrastructure in Regional Research. In: Jackson R, Shaeffer PV, editors. Regional Research Frontiers. New York: Springer; 2017.

[93] Angelidou M. Smart city policies: A spatial approach. Cities. 2014;41(1):S3-S11.

[94] Schweitzer L. Planning and social media: A case study of public transit and stigma on twitter. *J. of the American Planning Association.* 2014;80(3):218-38.

[95] Kavada A. Creating the collective: social media, the Occupy Movement and its constitution as a collective actor. Information, Communication & Society. 2015;18(8):872-86.

[96] Swabey P. IBM, Cisco and the Business of Smart Cities 2012 [cited 2016 20 Sept]. Available from: http://www.information-age.com/channels/comms-and-networking/company-analysis/2087993/ibm-cisco-and-the-business-of-smartcities.Thtml

[97] Viitanen J, Kingston R. Smart Cities and Green Growth: Outsourcing Democratic and Environmental Resilience to the Global Technology Sector. Environment and Planning A. 2014;46:803-19.

[98] Goodspeed R. Smart cities: Moving beyond urban cybernetics to tackle wicked problems. *Cambridge J. of Regions, Economy and Society.* 2015;8(1):79-92.

[99] Lombardi P, Giordano S, Farouh H, Yousef W. Modelling the smart city performance. *Innovation: The European J. of Social Science Research.* 2012 25(2).

[100] Caragliu A, Bo CD, Nijkamp P. Smart Cities in Europe. *J. of Urban Technology.* 2011;18(2):65-82.

[101] Certoma C, Corsini F, Rizzi F. Crowdsourcing urban sustainability. Data, people and technologies in participatory governance. *Futures.* 2015;74:93-106.

IOP Conf. Series: Earth and Environmental Science **158** (2018) 012047 doi:10.1088/1755-1315/158/1/012047

Delivering smart city system through experimental smart building concept. Design case of Nordhavn Community Centre, Denmark

F Septiandiani[1], W Raharjo[2]
[1]Department of Regional and City Planning, School of Architecture, Planning and Policy and Policy Development, Institut Teknologi Bandung
[2]Department of Architecture, Faculty of Civil Engineering and Planning, Universitas Islam Indonesia

Corresponding author: fadilasepti@gmail.com

Abstract. It is an undisputed fact that the development of a city requires more energy to accommodate the needs of the city's population. Greater energy consumption due to growing cities is a concern for scholars as well as governments all over the world. In the European Union, Denmark's renewable energy policy provides tax exemptions for passive air conditioning and renewable energy sources to foster public participation. To meet its energy provision objectives under this condition, cities need instruments to reduce energy consumption. The building of a community centre in Nordhavn (Denmark) was chosen as such an instrument due to its flexibility and possible exposure to solar radiation as an endless source of energy. An experimental design for the building envelope was developed to test its thermal performance when including a thermal storage wall. Design research was conducted using 3D modelling. Testing was done on a simulation of the building made with the Ecotect software application to provide comparable results for thermal performance supported by qualitative-descriptive methods. It was concluded that including a thermal storage wall in the building model corresponds well with the objectives of the design. Based on the result of the test, in the context of, the thermal storage wall is capable of contributing to passive air conditioning.

1. Introduction

The advance and rapid changes in the modern built environment have made city development into a more demanding job for city planners, local government and related stakeholders. Not only in the practical realm; the smart city discourse is also evolving in theoretical frameworks. Although it is a relatively new concept, many scholars have defined the smart city in different areas and from different points of view. They generally agree that smart cities are created to facilitate urban dwellers's needs in a more efficient manner. Calvillo et al. [1] further explain that the smart city is a medium to optimize urban resource allocation, particularly in energy provision, transportation and waste control. Another perspective can be seen from the use of ICT to connect certain urban services with indicators observed by users. Baron [2] states that the use of ICT has captivated the younger generations in modern society to engage with the smart city system, since almost all city systems are in their grip. These conditions, in fact, did not occur suddenly, according to Darwish [3]. The over-use of natural resources caused by conventional economic development has increased awareness of many scholars in terms of how to tackle this problem without compromising the needs of future generations. From this point of view, it can be seen that there is an ultimate goal in the utilization of smart city systems, starting from managing resources and simplifying monitoring processes to ensuring the needs of future generations.

As an urban area gradually expands, alterations in land-use in nearby territory – rural, hinterland, and water ecosystem – are evident. A variety of public amenities, easy access to vital transportation and proximity of workplaces persuade more people to migrate to locations with a higher standard of living. The socio-economic and political aspects embedded in a city make its development inevitable. Some people, especially low incomers, are forced to make a trade-off between their good quality of life in the hinterland and living and working in a city with the promise of a better subsistence. It is widely known that more than 50 per cent of the world's population now lives in cities [4]. This massive number is projected to rise further in the following decades. Since natural resources are limited, the demands of public needs cannot be ignored, so urban management is certainly becoming a necessity. Some cities already encounter severe urban problems due to overcapacity. For example, the mega-region Jakarta and its satellite cities (together called Jabodetabek) is occupied by 31 million people[1] in a total area of 6283.34 km^2. Jabodetabek faces challenging problems in transportation due to traffic congestion, among others [5]. Such problems do not merely occur in developing countries; the developed world faces similar problems, yet with a different emphasis. Taking Denmark as an example, [6] points out the necessity of maintaining energy resources.

"*The history of green Denmark lies in the oil crisis 1973-1974; Denmark was dependent on imported energy, and the pollution caused by fossil fuels was largely recognized. The need for more environmental policy raised and ever since renewable energy and energy efficiency have been the cornerstones of the Danish energy framework. In 2015, 49% of Denmark's energy production came from crude oil and 23% from renewable energy. E.g. 67% of total production of electricity came from renewable energy sources. Over the next 10 years, the share of electricity production from renewables is expected to increase to around 82%.*" [p. 6]

One of the solutions in Denmark is a energy policy providing tax exemptions for passive air conditioning and renewable energy sources to foster public participation. Responding to the Intergovernmental Panel on Climate Change (IPCC) in 2010, stating that the building sector contributes 32% and 19% to global energy use and GHG emissions respectively, it is no wonder that the Danish Government introduced an energy efficient building policy as a foundation for achieving the ultimate goal of reducing energy use and mitigating GHG emissions.

The aim of this paper is to present an experimental design for the building envelope of a community centre in Nordhavn (Denmark) in view of investigating whether it can used for reducing energy use by passive air conditioning. The design was developed based on analyses of the physical conditions of the location, microclimate conditions, building mass orientation, wind directions, landscaping and a thermal storage wall system. The configuration of the building mass, which reflects Nordhavn's morphology, was attained by optimizing sunray exposure in winter, minimizing sunray exposure in summer, optimizing air circulation. To achieve optimum sunray exposure, normally, the longest side of the building is orientated to face south. However, this may lead to over-radiation in summer. Hence it was decided that the appropriate orientation is having the shortest side of the building face the south (in accordance with the solar path direction). In view of the wind, which is very useful in summer for cooling, the building mass was configured in a U-shape, so that more surface will be exposed to the wind. However, wind exposure should be avoided during the winter, which was done by arranging vegetation in the surroundings of the building envelope. Thermal testing of the design model was done, although only for a sector (the thermal storage wall area).

[1]Obtained from BadanPengelolaTransportasiJabodetabek website (http://bptj.dephub.go.id/?p=1274 , reported on 27 Sept 2017)
[2] Calculated from Central Bureau of Statistics (2017)

IOP Conf. Series: Earth and Environmental Science **158** (2018) 012047 doi:10.1088/1755-1315/158/1/012047

2. Methodology

2.1. Building model
A building model was designed for a location in Nordhavn, Denmark. Before examining the properties of the building envelope, a description of the location and a brief explanation and analysis of its physical conditions are provided.

2.1.1 Physical conditions
Copenhagen is the capital of Denmark, located in the Northern Europe, close to Norway, Sweden and Germany. Copenhagen is part of the region of Oresund (Zealand, Lollland-Falster and Bornholm in Denmark and Scania in Sweden), which has 10 districts and where Nordhavn is one of the cities around Copenhagen.

Figure 1. Existing structured area
Source: Observation, 2014

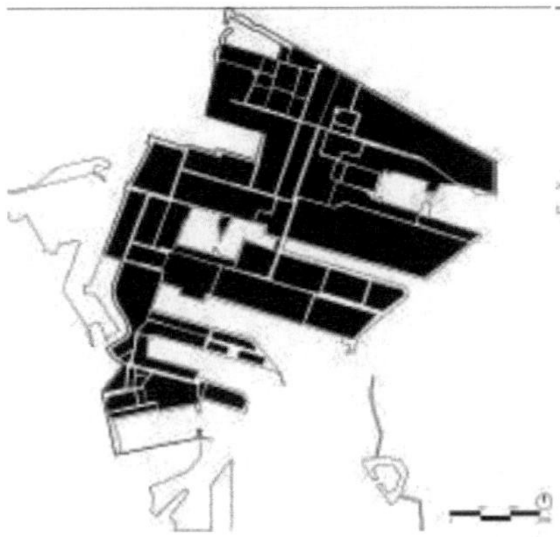

Figure 2. Urban space and city blocks

Figure 3. Existing green area distribution

The images above illustrate the current condition of Nordhavn for several physical aspects. According to figure 1, most of the built-up area is on the west side because the eastern side functions as an industrial area with a port and water activities. Thus, the city structure is varied in size and distribution. The location for the building model is at a fisherman settlement on the north side. Figure 2 shows the condition of the urban space. From this picture it can be seen that the urban structure is geometric with repetitive masses and spaces, which enables dynamic development. The distribution of green areas can be seen in figure 3. Most of these areas are designated as protected territory. On the other side, the pattern of urban space development is dominated by vertical and horizontal lines that cross each other and define the direction of the city's structure as shown in figure 4. As a result of the larger development of the west side, traffic circulation also mainly occurs on that side, with crowded meeting points, both for motor vehicles and other means of transportation. Nodes and traffic circulation can be seen in figure 5. Meanwhile, the types of transportation are depicted in figure 6. Public transportation is quite varied, i.e. there are public buses, fast boats and in the future also water taxies.

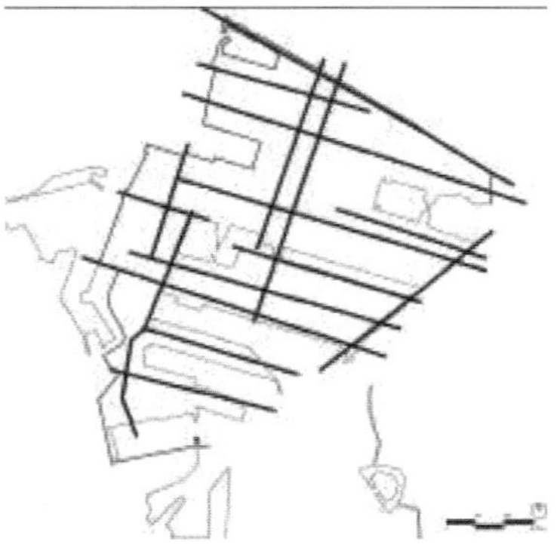

Figure 4. Traffic circulation pattern of area

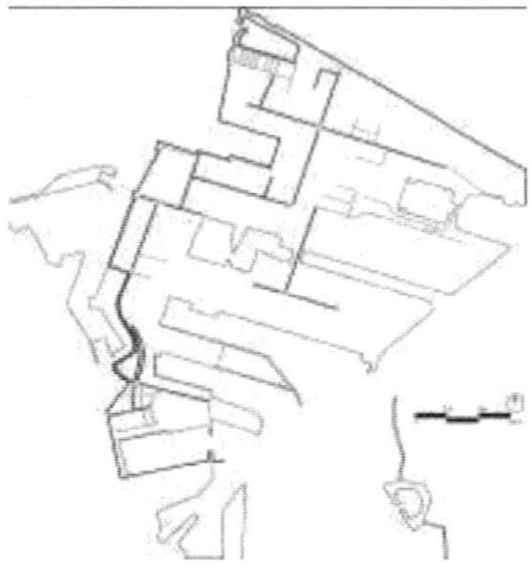

Figure 5. Main traffic circulation distribution and its nodes

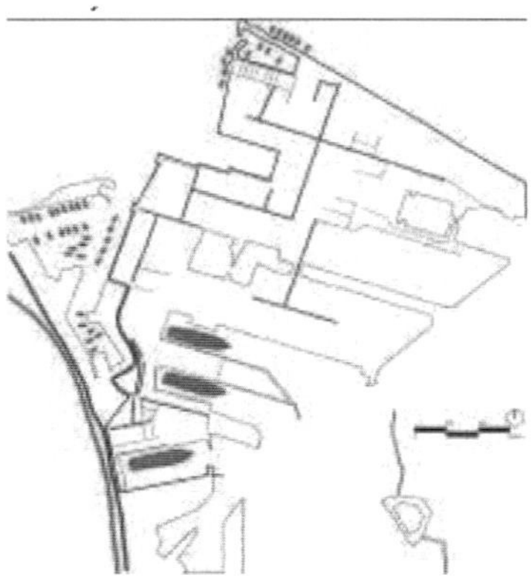

Figure 6. Provision of transport modes

Figure 7. Keyplan

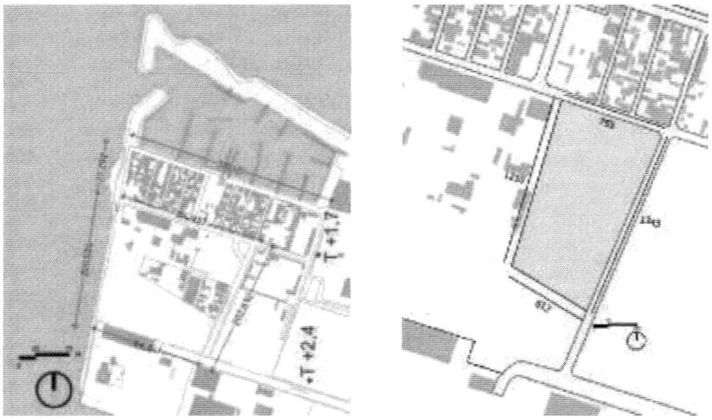

Figure 8, 9. Proposed site and its neighbourhood and detailed dimensions of proposed site.

2.1.2 Location and microclimate

The location of the building model is situated nearby a fisherman village and a new residential area. As for the microclimate aspect, Denmark has a temperate climate with temperatures ranging from –15 to 35 °C with most precipitation in the winter season, in the form of rain, fog and snow (Lohnert et al. [7]). For the region of Nordhavn, the winter season is from October to April with average temperature at 3.8 °C, followed by the spring season from May to June with an average temperature of 14.5° C, while in the summer time the average temperature increases a little to around 15.1 °C. It decreases to 2 °C in the autumn [windfinder.com, 2014]. Apart from considering temperature, wind distribution also has to be taken into account. According to the Norwegian meteorological institute NRK [8], the wind mostly comes from a direction between west and northwest (11%). The second most prevailing wind directions are south and southwest (9% and 8%, respectively). Winds are considerably strong in the southern area of Nordhavn.

2.1.3 Building mass orientation

Because the designated location is situated in the northern hemisphere (latitude 55.71 and longitude 12.6), generally, the sun's position in relation to the site is in the south. The specific sun positions at the location

IOP Conf. Series: Earth and Environmental Science **158** (2018) 012047 doi:10.1088/1755-1315/158/1/012047

were further examined using the Ecotect 2011 software application, considering critical days in summer (June 21) and winter (December 21). The result of the Ecotect analysis was as follows:

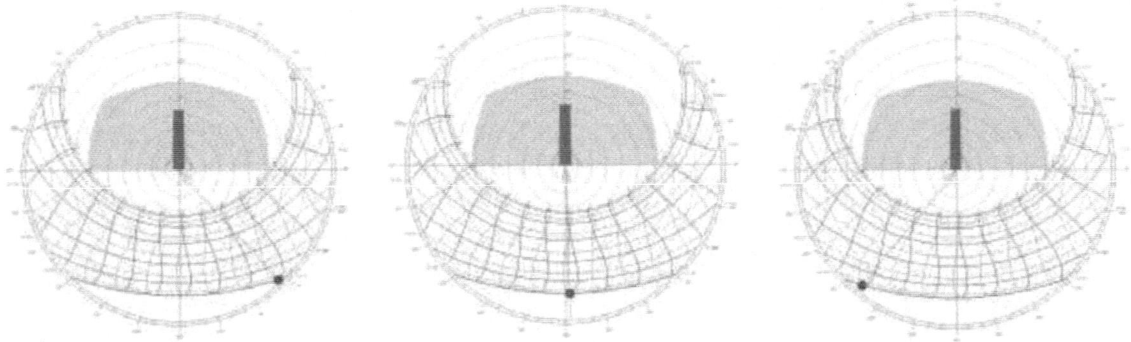

Figure 10a **Figure** 10b **Figure** 10c

Figure 10. Position of the sun on December 21 at 09.00 (figure 10a), 12.00 (figure 10b) and 15.00 (figure 10c).

From the data, it can be **found** that the location of the sun at 09.00 = 138°, at 12.00 = 173° and at 15.00 = 218°. The analysis of **the sun's** position in winter is important **for determining** the orientation as well as the form of **the building** mass in relation to the **use** of **a** thermal storage wall **system. This** system works by capturing sunlight and passing it into the building. **According to Branz [9] the optimal angle for the design of a thermal storage wall on the south is between 30° to the right and 30° to the left. This is also consistent with the result of the azimuth position at 150° and 210° from the analysis of the position of the sun. It shows that at an angle of 30° to the right and to the left represents the time when the sun is at its daily peak (around 12.00).**

2.1.4 Wind

The wind tends to blow from the northwest, south and east, causing the southern part of the proposed site to get more wind (see figure 11). Figure 11 shows the wind rose projected on top of the site and the resulting wind patterns.

2.1.4 Landscape

Vegetation in the surroundings of the proposed site can act as protection against the weather (sun and wind). This depends on plant type, plant size, leaf splits, and growth conditions [7]. Which can be adapted to the design objectives. In summer, evergreen plants are able to provide total protection and produce cool areas around the building to make it more comfortable in summer, while molt type plants may serve as heat triggering agents around the building to conserve heat during winter. A similar idea has also been put forward by Pelsmaker in [10], in which she presents several scenarios for winter and summer and provides an overview of the positioning of plants around a building. Lochnert et al. [7] further classifies different kinds of trees according to their ability to produce shadows in the winter time, as shown in figure 12.

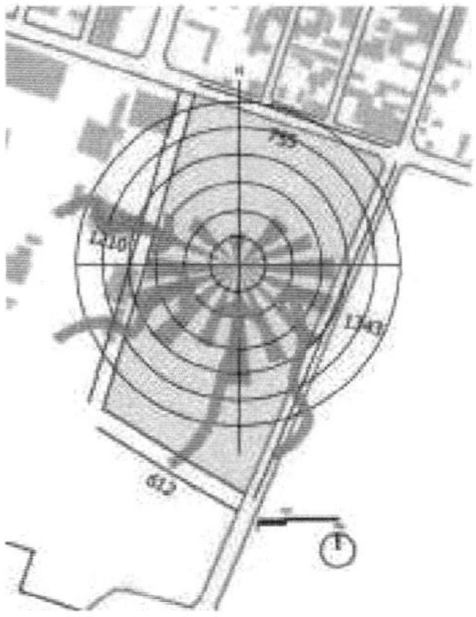

Figure 11a

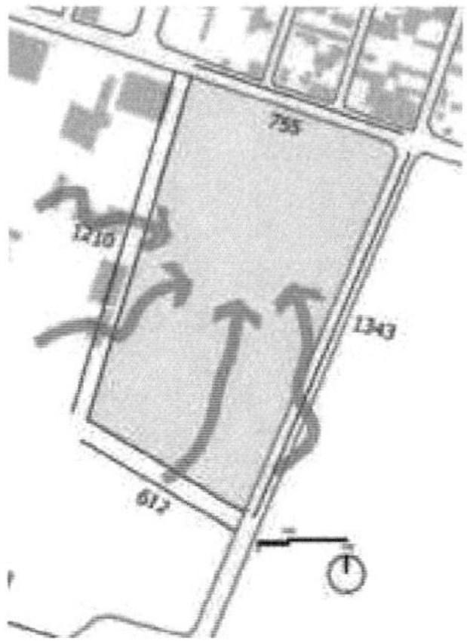

Figure 11b

Figures 11. Wind orientation on the proposed site. Figure 11a demonstrates the wind distribution, while figure 11b shows the main directions of wind.

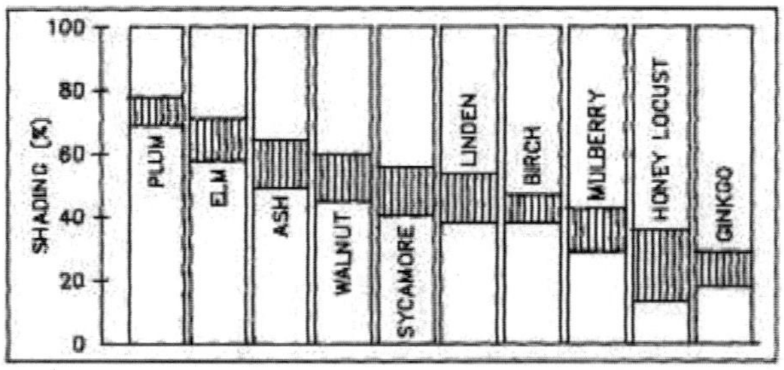

Figure 12. Types of plants/trees with varying ability to produce sunlight in winter [7]

The varying shading percentage (15-80%) indicates the trees' capabilities to produce shadow. The plum tree is the plant that produces the most shadow (80%), while honey locust produces the least (15%). The calculations also depend on the height, extension and thickness of the plant branches.

2.1.6 Thermal Wall Storage

A thermal storage wall is a way of achieving passive air conditioning that utilizes the microclimate (sunshine, wind, and air). Wilson [11] explains that a heat storage wall, better known as thermal storage wall, is an indoor air conditioning technology that consists of glazing, air between glazing, and wall mass, ventilation, and extended roof (overhang). He describes the general workings of a thermal storage wall as follows: the glazing side receives heat from the sun, which is absorbed by dark-coloured masses (dark surfaces are able to absorb heat better); hot air enters into the room through an air inlet; cold air moves out of the room through vented wall mass, while conduction of air movement (direct) occurs on unvented wall mass.

There are two types of thermal storage walls: thermal storage walls with perforation (vented) and without perforation (unvented). Each type has advantages and disadvantages. According to Wilson [11], a vented thermal storage wall is more appropriate for rooms that are used in the morning and afternoon, because the indoor temperature is higher during the daytime and cooler at night. Thus, the resulting temperature is more varied than with an unvented thermal storage wall. An unvented thermal storage wall works based on the conduction properties of the system. Because there is no perforation in the system, the temperature conditions are more stable. Thus, an unvented thermal storage wall is more appropriate for residential buildings.

The effect of a thermal storage wall also differs during the summer and the winter. In summer, Wilson [11] suggests, when the indoor temperature is too high, ventilation vents on the upper side of the wall mass should be closed, while vents on the lower side and in the glazing should remain open. To get cool air into the building, he suggests placing a window on the north side of the building, so there is an indoor breeze and indoor heat build-up can be reduced. Siliski [12] recommends the use of plants on the north side of the building to cool the air in summer. Siliski [12] also adds some benefits of ventilation during the night and on rainy days by adjusting the ventilation to the needs of the users. For the building model a vented thermal wall system was chosen, considering the static building purpose (community centre). It is incorporated in the building model by adding heat storage material in the form of brick/concrete partitions, so the heat can be absorbed during the day and released during the night. The vented thermal storage wall is equipped with ventilation on the north side. However, because of its location in the northern hemisphere, in summer the sun is higher and tends to cause excessive heat through the glazing. On the other hand, the benefits of a thermal storage wall are highly desirable in winter. When the sun warms the glazing on the outside, the air between the glazing and the wall mass expands and flows into the room through the upper and lower ventilation vents. During the night, all the ventilation vents, both inside and outside, should be closed to prevent warm air from going out of the room.

2.2 Synthesis of Building Mass Concept

The building mass concept of the building model was derived from the observations and analyses of its physical conditions in chapter 3. It contains among others: a building mass system that optimizes passive heating (in winter), minimizes passive heating (summer), optimizes wind cooling and air circulation, and reflects the morphology of Nordhavn city.

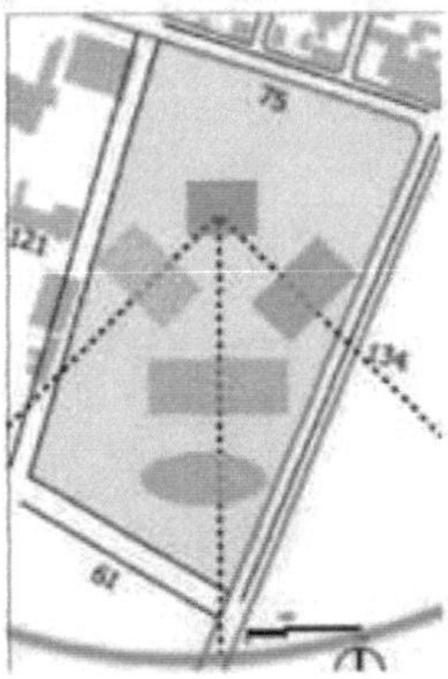

Figure 13a Sun rays in winter

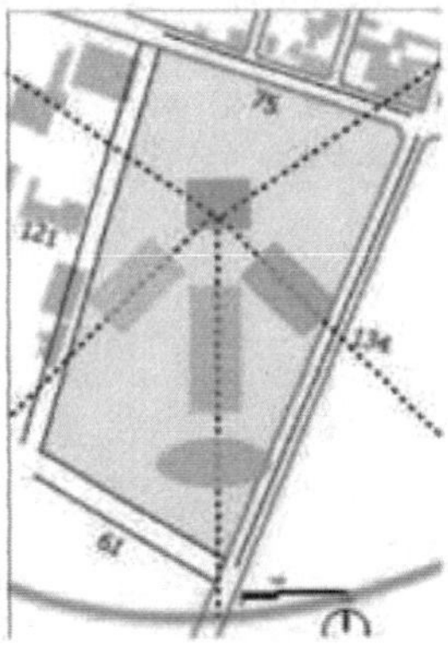

Figure 13b Sun rays in summer

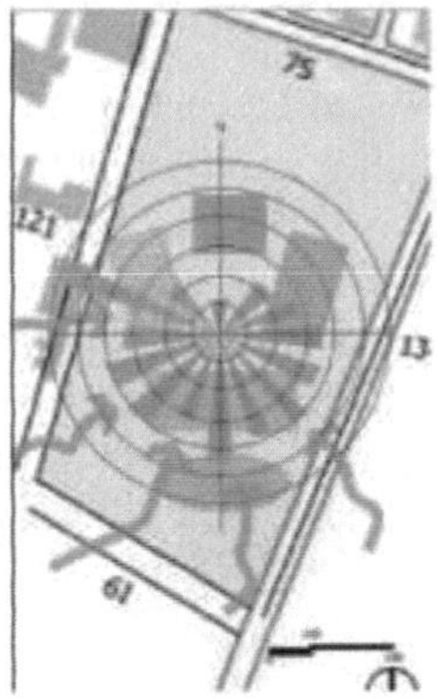

Figure 13c Wind

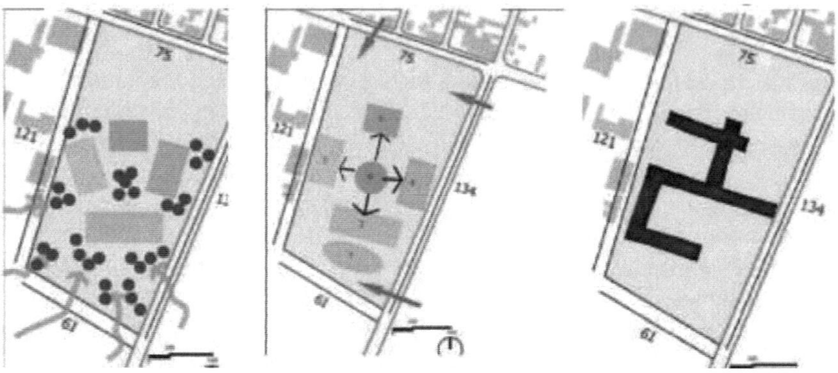

Figure 13d Vegetation and wind **Figure** 13e Circulation **Figure** 13f Morphology

Figure 13. Building mass orientation diagram

The configuration of the building changed during the design process. To maximize the utilization of sunlight in winter, the long side of a building mass should be oriented to the south. Thus, the surface area of the building will be exposed to more sunlight. However, the sun's radiation may become excessive in summer, in which case the shortest sides of the building model face should south. Responding to the prevailing wind directions in view of wind cooling is useful in summer, therefore a U-shaped building mass configuration was considered. However, wind cooling should be avoided in winter due to the thermal discomfort caused by cold winds. A solution can be found in arranging vegetation around the building correspondingly.

In addition, the building mass configuration was also investigated in relation to the air circulation between the building components. The initial building mass configuration was designed with reference to the basic morphological form of Nordhavn city. The basic form was then transformed in response to the microclimate in an effort to obtain thermal comfort by utilizing sun, wind and vegetation. The resulting building mass configuration is shown in figure 14.

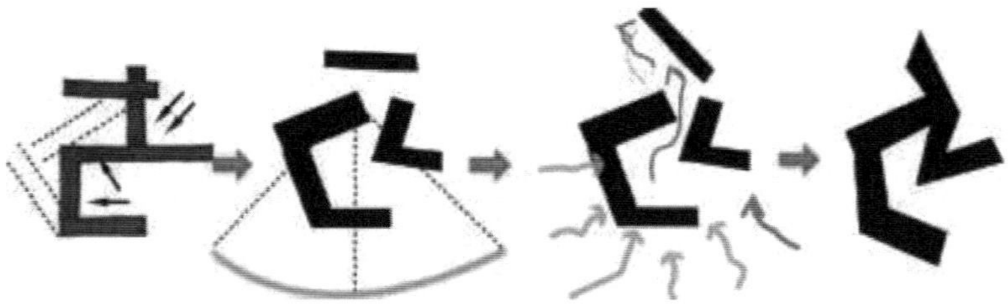

Figures 14. Building mass transformation

Based on consideration of heat radiation exposure in summer and in winter, the sides of the mass were given an angle of 225° (figure 14, second from left). Considering that heat radiation is very useful in winter, the long sides of the building mass were designed to face south. This is in line with the building's orientation to take maximum advantage of thermal storage wall system. Meanwhile, to reduce heat build-up in the building mass in the summer, evergreen vegetation is planted nearby the building envelope. The configuration of the building mass with an opening at the top enables the wind to reach all parts of the building (figure 14, third from left). The wind will also spread to the north side, whereas it potentially dominates the south side. From the above considerations, the initial configuration of the building mass was obtained (figure 14, fourth from the left), which was the basis for the next building mass shape.

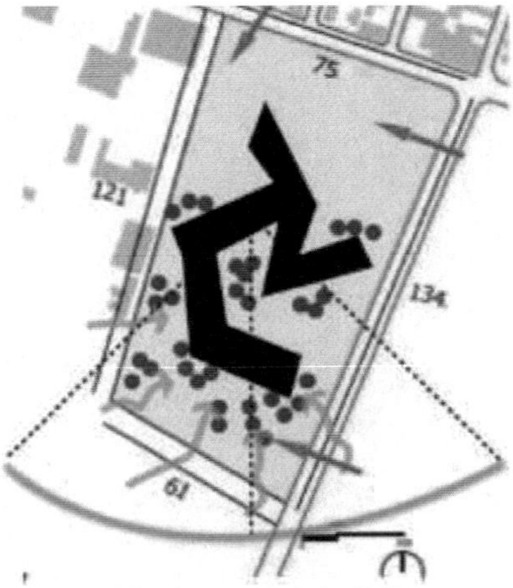

Figures 15. Proposed building mass of the model

3. Result and discussion

In a design test, the effectiveness of the building mass orientation and shape was determined. In this case, the design test focused on the sun's effect on the building in relation to the thermal storage wall system used in the design.

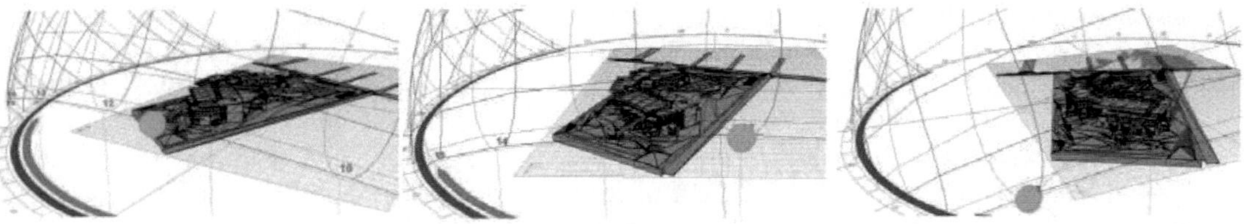

Figure 16. Effects of sunlight on 21 December at 3 different time

Testing was done for the peak of winter (December 21) in order to know the solar path, which can be used in deciding where to locate the thermal storage wall system. The testing procedure was based on the latitude and longitude of the location, which was simulated with the help of the Ecotect software application. From the test it was found that the most significant time of day is 10.30, since at that moment most points of the building are exposed to sunlight. At these points, the thermal storage wall will be located. After ensuring that the building receives the highest possible amount of heat radiation, the test was continued by examining the thermal performance of the building envelope, more specifically the thermal storage wall, which was done by comparing the thermal performance of the building envelope with and without thermal storage wall in an Ecotect simulation.

IOP Conf. Series: Earth and Environmental Science **158** (2018) 012047 doi:10.1088/1755-1315/158/1/012047

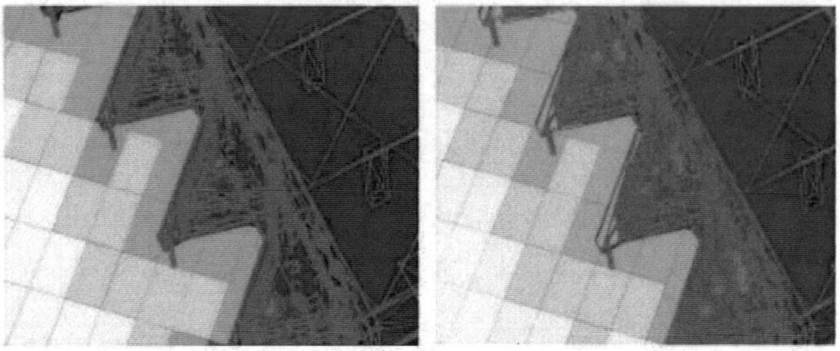

Figure 17. Result of building model in Ecotect software with thermal storage wall (left figure) and without thermal storage wall (right figure).

The test was done to observe the amount of solar heat absorbed by the building envelope. Figure 17 shows the condition of the building envelope both with and without thermal storage wall for comparison. The yellow colour in the result is indicates the highest level of heat absorption, while the blue colour represents the lowest level. The red color represents a fair amount of heat absorption, while the color changing to violet and blue indicates that there is little solar heat penetration. The result shows that if a thermal storage wall is installed, solar heat can still penetrate into the building envelope, as can be seen from the red color. It is suspected that this condition is also influenced by the building envelope containing glass elements.

The test showed that the position and orientation of the thermal storage wall in the model can contribute to the passive heating of the building. This is shown by the decrease of the violet area in the building envelope, which is due to heat radiation being absorbed by the thermal storage wall. With this system, heat that has been captured in the storage wall can be released to the interior.

In this study, the test was not intended to provide specific temperature values but only to examine to which degree the thermal storage wall is efficient in relation to its orientation and position. To provide specific temperature values many influencing factors need to be assessed, such as air tightness, hourly air circulation and heating system, whereas in this simulation these factors could not be included due to the limited information and author's knowledge.

Figure 18 shows the angle of the sun when it is at its highest position (at 12:00) on December 21st. From these measurements it can be seen that the angle between the position of the sun and the horizontal plane is 15°

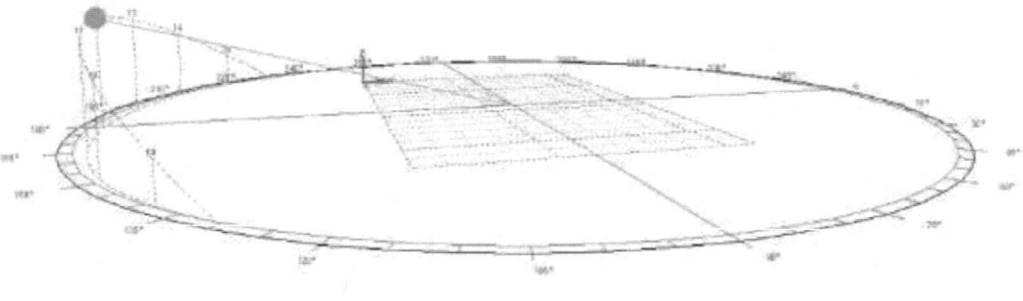

Figures 18. Sun position in winter

Similar to the calculation of the winter altitude, calculation of the summer altitude was also undertaken using Ecotect. The difference between both tests was the testing time. In summer, the critical day is June 21st. Therefore, the summer altitude calculation was done for that day.

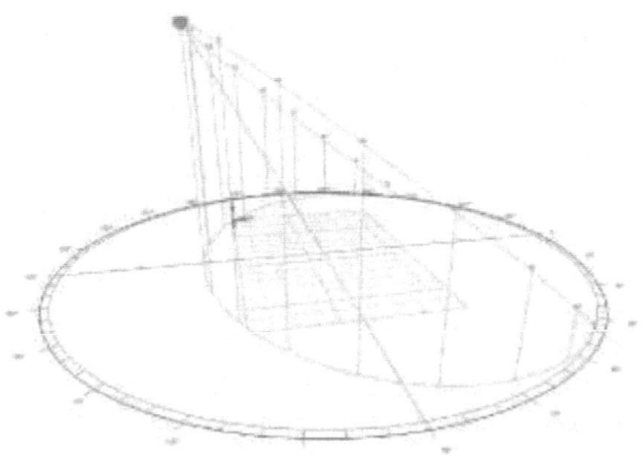

Figures 19. Sun position in summer

According to the result, the angle is 60° at 12.00 a.m. (highest angle). From both calculations it can be clearly seen that the altitude angles change along the year. The data above were then used to examine the result of application of the thermal storage wall in building envelope for the whole model. From figure 20 it was found that in summer, the sides of the building are more exposed to sunlight. In winter, the sides of the building envelope in the south and the west are mostly exposed to sunlight, as marked by the red areas.

In a temperate climate, the effort to increase temperature is more demanding to decrease temperature. Therefore, the thermal storage wall in the building model is able to contribute to the passive air conditioning of the building.

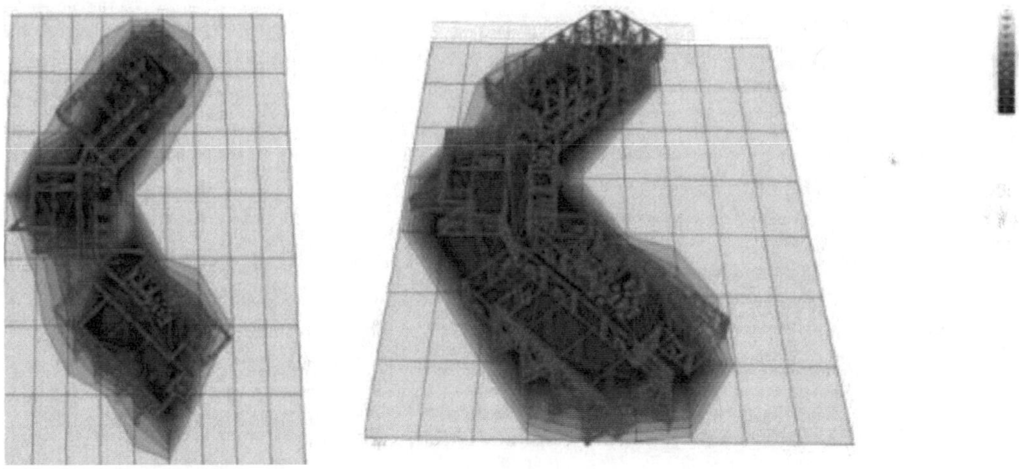

Figures 20. Test result of the model in summer (left) and winter (right).

4. Conclusion

Testing was conducted on a sector (i.e. the thermal storage wall area) of an experimental design for the envelope of a community centre in Nordhavn (Denmark). It was concluded that the location of the thermal storage wall in the building is appropriate for the objective of the design, i.e. to contribute to the passive air conditioning of the building. In summer, the building envelope could be over-exposed to heat radiation. However, the vegetation in the surroundings of the building envelope was not taken into account in the test due to the limitations of the author's hardware. The building envelope is also exposed

IOP Conf. Series: Earth and Environmental Science **158** (2018) 012047 doi:10.1088/1755-1315/158/1/012047

to heat radiation in winter, particularly from south to west. Based on the test result, the thermal storage wall system is capable of contributing to passive air conditioning of the building. In a wider context, the system can also contribute to a smart city system with respect to energy use reduction.

References

[1] Calvilllo C F, Sanchez-Miralles A, and Villar J 2016 Energy management and planning in smart cities *Renewable and Sustainable Energy Reviews* **55** 273-287

[2] Baron M 2012 Do we need smart cities for resilience *J. Economics and Management* **10** 31-46

[3] Darwish A S Sustainable green smart buildings: future energy survivor *J. Science and Technology* **12** 35-42

[4] Dodman D, McGranahan G, and Dalal-Clayton B 2013 *Integrating the environment in urban planning and management* (Nairobi: United Nations Environment Programs (UNEP))

[5] Asri D U and Hidayat B 2005 Current transportation issues in Jakarta and its impacts in environment *Proc. the Eastern Asia Society for Transportation Studies* **5** 1792-1798

[6] Witty City Program of Tekes 2017 *Smart city benchmarking visit to Copenhagen and Hamburg* (Copenhagen, Hamburg: Tekes)

[7] Lohnert G 1989 *Design context passive and hybrid solar low energy buildings* (Paris: International Energy Agency: Solar Heating and Cooling Program)

[8] YR Norwegian Meteorological Institute and the Norwegian Broadcasting Corporation 2014 Weather Forecast for Nordhavnen, Denmark [Online]. Available from: https://symbol.yr.no/place/Denmark/Capital/Nordhavnen/

[9] BRANZ 2018 Sustainability Fact Sheet Trombe Wall Design *Sustainable Construction* **2** 1-5

[10] Pelsmaker S 2012 *The Environmental Design Pocketbook* (London: RIBA Publishing)

[11] Wilson A 1979 *Thermal Storage Wall Design Manual* (Albuquerque: Modern Press)

[12] Siliski A 2007 *Passive Heating* (Greenfield: The North East Sustainable Energy Association)

Planning in the Digital Era: Films and Social Media as Information Sources for Tourism Planning

N F Muslimah[1], S R Keumala

[1]Studio Cilaki Empat Lima

Corresponding author: nfauzia05@gmail.com

Abstract. Films deliver subliminal promotion to audience about the visual appeal of a certain place. If the place looks picturesque enough and gives a memorable impression to the audience, it has a possibility to become the next tourism magnet. Tourists tend to take pictures as keepsakes and upload them on social media. The number of social media posts of a place can be utilized as a tool to measure and map people's interest. This process is a voluntary advertisement of the place done by community caused by the culture of content sharing in the digital era. This paper is going to elaborate the growth of Gereja Ayam (Chicken Church) in Magelang, Indonesia, as the place which were formerly insignificant and publicly unknown, but has been featured in an Indonesian film *Ada Apa dengan Cinta 2.* An analysis of the place's popularity is conducted using Google search trends, looking for related keyword search about places mentioned before. The number of Instagram posts using related hashtags is used to examine how many people have visited the place.

A literature review is also being carried out to find the correlation between the media exposure and the place branding, the analysis of place's popularity through social media, and how this can contribute as tools for urban and tourism planning.

1. Introduction to film-induced-tourism

Films, television series, or documentaries always use certain location as parts of the stories that show the place to public. According to Sellgren, films and television series have been widely recognised as being able to create representations of destinations all over the world. The audience can gaze upon places, people, stories and other depictions of attributes that a film is built up of from the TV sofa or cinema chair. After watching it, people have an urge to visit the same place as portrayed in the film. Sellgren on his research entitled *Film-induced Tourism: The Effect Films Have on Destination Image Formation, Motivation, and Travel Behaviour,* also said that through movies, TV Programs and Documentaries people can gain images of places they did not know anything about before. Still, from the same research above, this condition created film-induced-tourism.

> "Film-induced tourism is a phenomenon that is in the middle of those industries. The majority of films today are not produced with intent to lure people to visit certain destinations. However, in a sense can a destination in a film be seen as a form of product placement that consumers are ready to invest money in see and experience" [1]

People's motivation for traveling, especially for young adult group age, varies from gaining different experience in new places, fulfilling their adventurous side, to escaping the normal daily routine. Some travellers have a tendency to have new experience from a common place. Also, it is suggested that the factor behind seeking something unique and unexplored could be status/prestige [1]. Where can people find unique and unexplored places? We can see many unique and unexplored places from the films' locations. Films and TV dramas can give people the imagination of certain places because those products can show the place in detail, which is enough to give strong images before actual visitation to the places.

The phenomenon of the film-induced tourism has also created a group of people who are committed to the so-called film-induced-tourists. However, the tourists are not simply divided into general tourists, incidental film tourists and film-induced tourists. Sellgren divided the film-induced-tourists into two subgroups.

a) People who visit a specific or exact film location portrayed in a film (could be a restaurant, house, tree, cliff, or other attraction) as the tourists' goal can be referred as *dedicated (specific) film tourists.*
b) People whose the images of destination has mainly been influenced by movies or television programs are simply strong enough to induce a visit to a city, region, or country can be referred as *dedicated (general) film tourists.*

Another research by Macionis which was specifically conducted on film-induced tourist, specifically for divides tourists into 3 sub-groups [2]:

a) Specific Film Tourist: those who actively seek out places that they have seen in film.
b) General Film Tourist: those who are not specifically drawn to a film location but who participate in film tourism activities while at a destination,
c) Serendipitous Film Tourist: those who just happen to be in a destination portrayed in a film.

Increasing interest in film → →

Serendipitous Film Tourist	General Film Tourist	Specific Film Tourist
• those who just happen to be in a destination portrayed in a film	• those who are not specifically drawn to a film location but who participate in film tourism activities while at a destination	• those who actively seek out places that they have seen in film
Motivations include: • social interaction • novelty	Motivations include: • novelty • education • nostalgia	Motivations include: • self-actualisation • pilgrimage • self-identity • fantasy • romance • nostalgia

Increasing self-actualisation motivations → →

Figure 1. Film-induced tourist groups

In the case of film-induced tourism, people don't simply come to the film location after watching the film. The authenticity of the pre-visit images deprived from a certain film is what matters to tourists. As we know, sometimes films give stereotypes into the place (which sometimes are negative) or exaggerate the actual site for sake of good scenery on screen. From example, two French respondents from Sellgren's research explained that most French movies did not show the real Paris [1]. Actual visitation to the location is still needed to understand the actual condition. However, how can people be convinced enough to visit the film location despite the stereotype and exaggerated scenery?

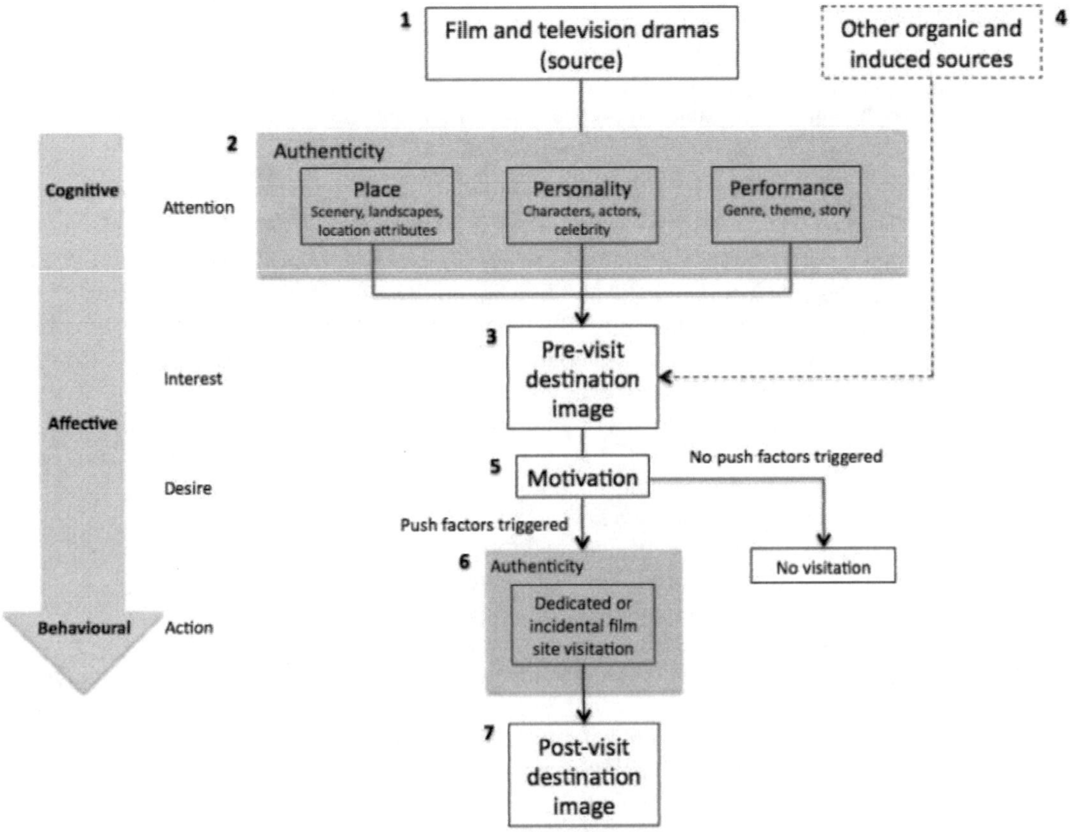

Figure 2. Film-induced Tourism Model [1].

We should be aware of the authentication of the film location because the film maker might exaggerate some points or degrade the actual condition. Also, not all people watch the movie so many tourists are not convinced enough to visit the new attraction. There is a tendency for status/prestige in the tourists' motivation to travel, so they will take some pictures from their destination as a keepsake and show them off to their families and close friends when they get together. Nevertheless, nowadays, tourists tend to upload and show off their traveling stories on social media. Uploading the pictures to social media means widely spreading the news about the places on the internet, whereby everyone can see it.

The number of social media posts of a place can be utilized as a tool to measure and map people's interest. Social networking might give a user the opportunity to exchange and share ideas with many people in real time [3]. Information and images from social media can be used as sources of travelling reference and thus, encourage the mobilization of tourists [4]. This process is a voluntary advertisement of the place done by community caused by the culture of content sharing in the digital era.

1.1. The role of social media as an information source

The development of internet in past decade has helped people create new ways to communicate with each other, and social media were born from that. The presence of ICTs, for example websites and social media, gives a lot of information about tourist destinations and their facilities such as places, properties, restaurants [5]. The social media also help us spread information in fast and easy ways. Thus, social media can be considered as a quite effective tool to spread information, including tourist information.

Planners also use new approaches as along with the growth of ICT. Moreover, the technology has become more reliable, especially related to data. Any kind of data can be conveniently used by planners if analyzed and used correctly; as quoted from an article written below [6],

> "There is just one problem: the world is literally drowning in data. To make use of all the information that people involuntarily produce, planners must improve how data is captured, analysed, and shared between the public and private sectors. If we succeed, some of the biggest obstacles that the world faces – from poverty to climate change – could become a bit more manageable."

Many city governments have now used real-time analytics to manage aspects of how a city functions and is regulated [7]. Those real-time analytics using big data are updated every second as a source of information. Citizens also use real-time information for their reference, such as using twitter trends to know what is happening in their city. Such data and the analysis offers the citizens insights into the city life, aid the everyday living and decision-making, and empower alternative visions for the city development [7]. The real-time data can also be used for searching new places to visit, so they can imagine the places they want to visit before actually coming to the place. Nowadays, data are everywhere and we call them as "Big Data" because the massive size of the available data. Big Data have the following characteristics [8]:

a) huge in volume, consisting of terabytes or petabytes of data;
b) high in velocity, being created in or near real time;
c) diverse in variety, being structured, semi-structured and unstructured in nature;
d) exhaustive in scope, striving to capture entire populations or systems (n = all);
e) fine-grained in resolution and uniquely indexical in identification;
f) relational in nature, containing common fields that enable the conjoining of different data sets, and
g) flexible, holding the traits of extensionality (can add new fields easily) and scale-ability (can expand in size rapidly).

Social Media are also a part of "Big Data". According to the classification of Big Data from UNECE Statistics Wikis page, Social Networks are categorized as a part of Big Data [9].

> "Social Networks (human-sourced information) this information is the record of human experiences, previously recorded in books and works of art, and later in photographs, audio and video. Human-sourced information is now almost entirely digitized and stored everywhere from personal computers to social networks. Data are loosely structured and often ungoverned."

Several Social Networks description mentioned are:

a) Social Networks: Facebook, Twitter, Tumblr etc.
b) Blogs and comments
c) Personal documents
d) Pictures: Instagram, Flickr, Picasa etc.
e) Videos: Youtube etc.
f) Internet searches
g) Mobile data content: text messages
h) User-generated maps
i) E-Mail

With the development of internet, people tend to use fast and easy-to-get data as the source of information when looking up about new places. People who want to visit new places also look for online reviews as references. The characteristic of online reviews written by other users do not only potentially increase or decrease the tourist visits, but also raise consumers' expectations to the tourist destinations [10]. Nowadays, online reviews are not only written on blogs or websites, but also on social media platform, where users can generate contents, for example on Twitter and Instagram.

Twitter is known as an open social networking that can be categorized as a microblogging platform, whereby users can post short descriptions, news or conversation followed by videos, images or audios. One important element of twitter is hashtag (#). Twitter users can use hashtag to mark the topics of their posts so people can easily use them to search similar conversations or certain topics. One of twitter's features is the ability to see the real time trending hashtags. A hashtag can become a trend if many users use the similar tag. Hence, it is easy to looking for tourism information specifically on twitter.

Another popular social media platform is Instagram. Instagram is an internet-based photo album. People can easily share their photos with other users. If Twitter focuses on conversation and microblogging, Instagram focuses on sharing photos and videos. Instagram also uses hashtags to mark the mood, location, etc. Nowadays, Instagram has become a platform for self-promoting as Instagram's value rocketed into marketing place. Instagram is present as a medium that has a lot of personal facilities that allow users to upload what they see in tourism sites which they consider attractive to share (Fatanti and Suyadnya, 2015).

Both Twitter and Instagram can give clues to see how a tourist attraction is growing through recognizing the number of posts about the place. Nevertheless, it has some limitations, particularly from posts that is set private by the uploaders (only selected people can view).

1.2. The correlation between film-induced-tourism and planning

Film-induced tourism might also maintain the image of the places because there might be gaps between the image built by the film makers and the actual image of the places; in other words, the gaps between the authenticity of the location and the tourists' expectation. These gaps may be perceived or actual gaps that need development to be filled [11].

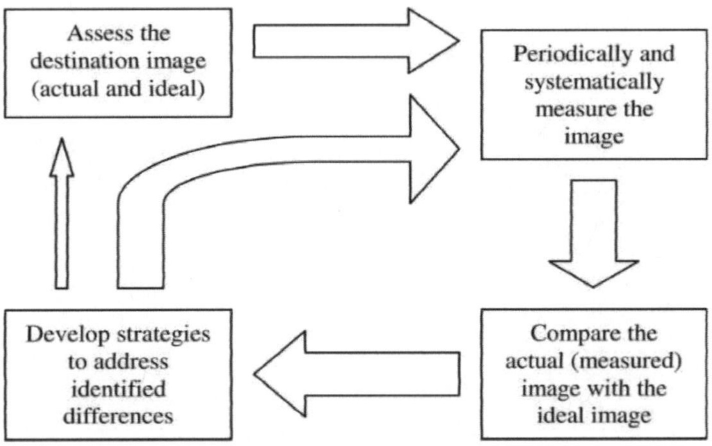

Figure 3. Image management process [11]

According to figure 3, the image management process needs to go through the assessment of the destination's image (between actual image and ideal), and then it is necessary to maintain the image by periodically and systematically measuring the image. The next step is to compare the actual (measured) image with the ideal image, and after getting the result, strategies to address identified difference should be developed. One of the example of the complex large-building's role is the promotion of Tourism in New Zealand and film entitled The Lord of The Rings [11].

2. How 'Ada Apa Dengan Cinta 2' introduces Gereja Ayam as a tourist attraction

Ada Apa dengan Cinta is an Indonesian teenage romance feature film. It was released in 2002 and received a lot of praises. Released on 28 April 2016, the sequel entitled *Ada Apa dengan Cinta 2* told how the two protagonists reunited after being separated for 14 years. This movie was a big hit with 3,626,929 viewers at the end of May 2016 affected by the long wait from the fans for the continuation of the story [12]. *Ada Apa Dengan Cinta 2* had a beautiful cinematography and the places used in this movie were not places that had been commonly visited by people in general. When the audience watch this movie, the protagonists seemed like inviting the audience to explore new places in Yogyakarta with them. The effect of *Ada Apa dengan Cinta 2* was unexpected, this film has made all the places visited by the two protagonists become hot topics on social media.

Unlike the prequel which depicted Jakarta's high school life, the sequel took place in Yogyakarta and Magelang (Central Java). Hence, the film's scenes highlighted a number of tourism sites in both provinces. One of them is Gereja Ayam. Gereja Ayam (translated: Chicken Church) is located in Magelang Regency, Central Java Province, Indonesia. The facade of Gereja Ayam is intended to be a dove. However, it is commonly mistaken as chicken and influences people to name the building ayam (chicken). Although it is a church, it is not regularly used by Christians. The building of Gereja Ayam and Bukit Rhema Hill are located in the vicinity of Punthuk Setumbu Hill, which is also a tourist attraction. People often come to Punthuk Setumbu Hill to witness the scenery of Borobudur Temple during sunset or sunrise from afar.

In the *Ada Apa dengan Cinta 2*, Gereja Ayam has left a meaningful memory because it is the place where the film's two protagonists watched sunrise together. For the audience, the scene in Gereja Ayam is quite emotional, romantic. Moreover, it is also supported by the beautiful scenery and the unique architecture of the building. These emotional and aesthetical sides are being offered by Gereja Ayam as a new tourist attraction, which offers reminiscence of the film's scene for its visitors.

3. Using social media mata as information for planners: the case of Dove Church

This part will explain the example of how planners can be informed about Film-Induced tourism by using data from internet and social media. It will be measured from:

a) the number of keywords used in Google
b) the number of Instagram posts with related hashtags

3.1. Google Search

Google records the number of searches done by using certain keywords and provides the graphic to show trends by time. From it, this study can compare people's interest in the places before and after the film's release date. The keywords compared are 1) gereja ayam; 2) punthuk setumbu, and; 3) bukit rhema.

IOP Conf. Series: Earth and Environmental Science **158** (2018) 012048 doi:10.1088/1755-1315/158/1/012048

Week 17 (24– 30 April 2016)

Week 18 (1– 7 May 2016)

Figure 4. The number of google searches using related keywords in 2016
Source: Google Trends, 2018

On week 18 of year 2016, the number of searches for all keywords boosted into 100. Whilst the rise itself began from week 17, when the film was firstly released, the number on week 18 or one week after the release shows the peak of public enthusiasm and curiosity towards Gereja Ayam, Punthuk Setumbu, and Bukit Rhema.

Before *Ada Apa dengan Cinta* 2 was released on 28 April 2016, the number of searches for all keywords were no higher than 10. This means the film had succeeded to trigger people's curiosity for the places since they had been featured in the film. In the process of searching for webpages using the keyword 'gereja ayam' and finding articles, blog, or news about it, Google users will be introduced to new related keywords, 'punthuk setumbu' and 'bukit rhema' since both of them are also situated in the place where Gereja Ayam is located.

From week 21 afterwards, the number of searches declined drastically. However, the average of searches remains higher compared to prior the film's release date. Even though it is uncertain if the Google users who look for those keywords are curious with Gereja Ayam because they watch *Ada Apa dengan Cinta 2*, Gereja Ayam, Punthuk Setumbu, and Bukit Rhema have already become popular and been established as tourism spots among internet and social media users.

3.2. Instagram Posts

Unlike Google, there is no free feature on Instagram which allow the users to view the timely trends for posts uploaded using certain hashtags. Nevertheless, we can be informed about how many posts are uploaded until the recent time. The table below shows the number of hashtags related to Gereja Ayam, Punthuk Setumbu, and Bukit Rhema with various style. The data retrieved per 14 January 2018, 18.25 PM.

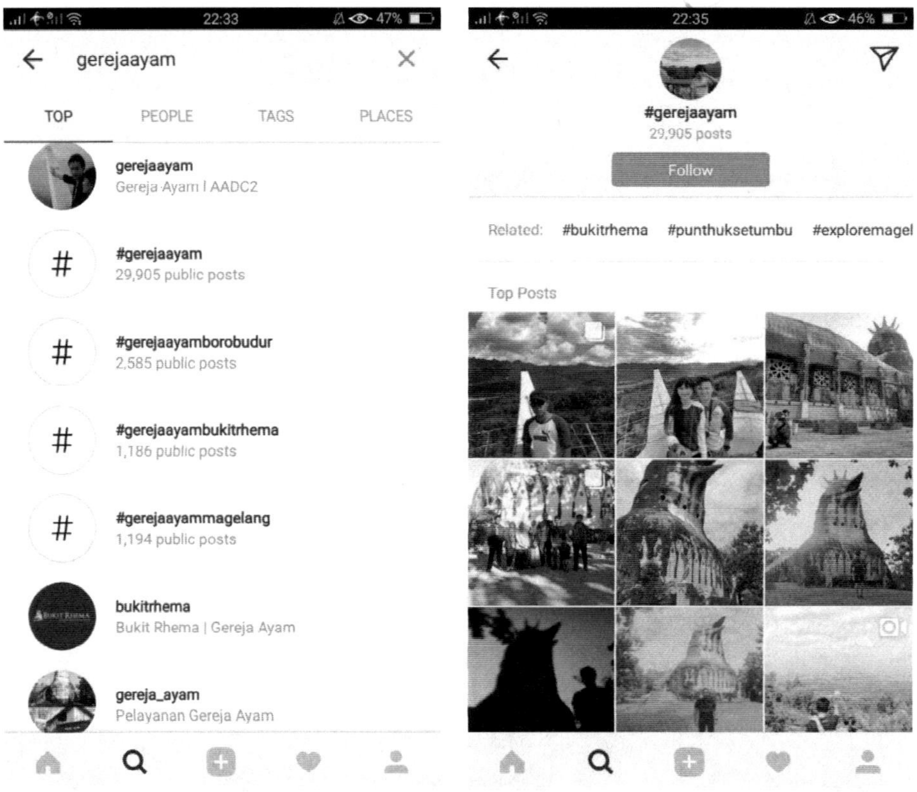

Figure 5. Looking for Instagram hashtags

When we search for the hashtag, all related hashtags and people/ accounts are displayed on the screen. We can also look for related posts to the hashtag, Instagram will display pictures related to what we look for. It is very convenient for people who want to look the real pictures of the place they want to visit and look for particular spots to visit through Instagram posts. We also try to count how much posts under the related hashtag and try several combinations since there is no specific rule and condition regarding how people type the hashtags on the social media.

Table 1. Instagram hashtag counts for various hashtags on 14 January 2018

Gereja Ayam		Punthuk Setumbu		Bukit Rhema	
Keywords	**Number of Posts**	**Keywords**	**Number of Posts**	**Keywords**	**Number of Posts**
gerejaayam	29.905	Punthuksetumbu	6.651	Bukitrhema	17.106
gereja_ayam	131	punthuk_setumbu	25	bukit_rhema	50
gerejaayamaadc	2			Bukitrhemamagelang	2.104
gerejaayamaadc2	8			Bukitrhemagerejaayam	409
gerejaayamborobudur	2.585				
gerejaayambukitrhema	1.186				
Gerejaayammagelang	1.194				

4. How planners can take advantages from the social media data

Social media are within our grasp. Through smartphones, we can be updated about what is happening around us, nationwide and worldwide. This especially applies for news about places and tourism spots. Today, social media are platforms for people not only to share their interests and daily lives, but also to look for detailed information regarding certain places, especially the popular ones. This is due to the fact that people do not only give descriptions about what they post, but they also post series of photos of a place from different angles. Social media can give much information if we use it correctly. Just use the search bar and type name of the place or related hashtags, and we can get much information. We can also use our social media feed to know latest information. For example, when a film induces a tourism activity, people are notified through the news or their social media feed.

Social media and Google Trends actually provide the number of posts or searches which reflects how a tourism site can grow. In the case of Film-induced tourism, planners can monitor the development of tourism and urban planning through quantitative and qualitative data provided freely on those social media services. Although not everyone who visit the place posted on social media, at least planners can predict the big picture about how the tourism is developing in an area.

Assuming that the popularity of a tourism spot is linear to the number of visits, planners should be aware of externalities caused by the increasing number of visitors, such as the growth of commercial area, traffic problems, or environmental issues. Through analyzing the popularity of a place in Google and social media, urban planners can predict possible activities that might happen in a potential tourism attraction, as well as anticipate the possible impacts towards the zoning, traffic, infrastructure development planning, and people activities. Conflict analysis between various stakeholders could also be predicted by looking at the possibilities if the tourism is developing on that site. With proper planning, the popularity of those places would not just be a temporary trend, but it can be maintained as tourism attraction for a long term.

5. Conclusions

Social Media data as parts of big data, can be utilized for planning. Nowadays, big data are used to support decision making processes. Social media data also can be challenging to use since the data are sometimes too big and unrelated. Hence, planners need to filter the information needed before using it for analysis and processing it statistically. We cannot rely to traditional sources of data or small data since currently the information is updated very fast through the internet.

Related to the tourism planning, quantitative and qualitative information from social media and Google trends could be the first step to notify the growth of tourism site. Since the information is accessible from smartphones, planners can attain the awareness of a growing tourism site in new and undiscovered places or old places, which get attention lately because of certain events.

The limitation to this research is not all tools for social media analysis can be accessed freely. In order to have a full access of the data, users need to buy third party service for analytical results (for example showing trend), except for Google Trends which can be accessed freely and easy to use (they give us data and trends).

IOP Conf. Series: Earth and Environmental Science **158** (2018) 012048 doi:10.1088/1755-1315/158/1/012048

It has been a long time since films have allegedly increased the number of tourists when certain places are used as parts of the story, especially if the films that become box office. However, planners should note that not all films can trigger film-induced tourism. We must firstly look at the trend from the post-release of a film to know the influence on tourism. Mostly filming locations that have become film-induced tourism are iconic filming location and have strong emotional connection with the viewers. It is a planner task to make the iconic locations become long-term tourist attractions, instead of just a temporary phenomenon due to the influence of the film.

References

[1] Sellgren E 2010 Film-induced Tourism: The Effect Films Have on Destination Image Formation, Motivation, and Travel Behaviour (Copenhagen: Copenhagen Business School)

[2] Macionis, Niki 2004 Understanding the Film-Induced Tourist *Int. Tourism and Media Conf. Proc.* p 86-97 (Melbourne: Tourism Research Unit, Monash University)

[3] Bizirgianni and Dionysopoulou 2013 The Influence of Tourist Trends of Youth Tourism through Social Media (SM) & Information and Communication Technologies (ICTs) *Procedia - Social and Behavioral Sciences* **73** p 652 – 660 (Elsevier)

[4] Fatanti M N and Suyadnya 2015 Beyond User Gaze: How Instagram Creates Tourism Destination Brand? *Proc. 2nd Global Conference on Business and Social Science, GCBSS-*2015 (Elsevier)

[5] Litvin S. W. and Hoffman L. M. 2012 Responses to consumer generated media in the hospitality marketplace: An empirical study *J. of Vacation Marketing*, **18** (2) p 135–145.

[6] Moya A L. The data-driven city [Internet]. Enjisight. 2018 [cited 22 February 2018]. Available from http://www.ejinsight.com/20180220-the-data-driven-city/

[7] Kitchin, Rob 2013 The Real Time City? Big Data and Smart Urbanism *Geojournal* **79** p 1-14 (Springer Science + Business Media Dordrecht)

[8] Kitchin, Rob 2015 The opportunities, challenges and risks of big data for official statistics *Statistical J.of the IAOS* **31** p 471–481

[9] Vale S. Classification of Types of Big Data [Internet]. UNECE Statistics Wikis. 2013 [cited 22 February 2018]. Available from: https://statswiki.unece.org/display/bigdata/Classification+of+Types+of+Big+Data 22 February 2018

[10] de Bruyn A and Lilien G 2008 A multi-stage model of world-of-mouth influence *Int. J. of Research in Marketing* p 151-163

[11] Croy W G 2010 Planning for Film Tourism: Active Destination Image Management *Tourism and Hospitality Planning & Development* **7** p 21-30 (London: Routledge)

[12] Pengerang A M K. Ditonton 3,6 Juta Orang, "AADC? 2" Merajai Box Office Indonesia [Internet]. Kompas.com. 2016 [cited 22 February 2018]. Available from: https://entertainment.kompas.com/read/2016/06/01/122518410/ditonton.3.6.juta.orang.aadc.2.merajai.box.office.indonesia

IOP Conf. Series: Earth and Environmental Science **158** (2018) 012049 doi:10.1088/1755-1315/158/1/012049

The use of social media and open data in promoting civic co-management: case of Jakarta

Pritta A Widyanarko
Research Coordinator, Peta Bencana Foundation, Jakarta, Indonesia

Corresponding author: pritta@petabencana.id

Abstract. With the high number of population and high use of social media, residents willingly share information in the digital world. While cities are sometimes seen as data-scarce, this digital platform produces informal and scattered, but also valuable data. One way to prepare for co-management during disaster situations is to extend these informal networks to include a channel between residents and government agencies. The platform PetaBencana.id crowd-sources these actual and on-ground observations from residents on social media and instant messaging, integrates the informal and formal disaster-related-data, gives the residents access to the same tool used by the government, and provides an interface that answers to residents and government's needs; thus making the information more useful in co-managing the city during disaster situation. More information-based decisions can be made by both the residents and government through improved situational knowledge, resulting in better disaster response and resilience of the city.

1. Introduction

In 2006, in its report, "The State of the World‘s Cities", the United Nations HABITAT office made a formal pronouncement that, for the first time, more than half of the world's population, nearly 3.3 billion, lived in urban agglomerations rather than in rural areas [1]. To conquer the challenge of expanding settlement and population in the cities, infrastructures are being built to ensure the life of the inhabitants, and finally, it becomes an unconscious part of the user, making urban life possible, according to Gandy (as cited in Furlong) [2]. Gandy [3] furthermore discusses that this condition is becoming the 'new normal' for residents in the city, thus, only in moments of crisis or failure do these complex and vulnerable systems come sharply into public view.

Computational systems, which enable the quantification and analysis necessary for understanding changing environmental conditions [4], are being developed to predict extreme weather events. However, the unpredicted disaster risk from failure of infrastructure is hard to be taken into account; in addition, megacities are often seen as data-scarce, which prohibits city managers and the people to not only respond to the emergency situation but also to do preventive efforts. How can we answer the needs of stakeholders and residents to make more informed decisions to foster resilience together? Through literature review and interviews of community, resident, and disaster managers, this paper will discuss the flood conditions and the power of social media in Jakarta, the applied research project—named PetaBencana.id—that harnesses the social and digital media to disseminate flood information in real-time, and how the open access of flood updates promotes civic co-management, both from residents and government, during disaster situation in the city. Flood report data and website traffic data used are sourced from PetaBencana.id usage during the February 2017 monsoon.

2. Jakarta: The City of Anthropocene

As the capital of the world's fourth most populous country, the megacity of Jakarta is by far the most densely populated city in Indonesia [5]. With over 30 million residents in the metropolitan area (known as "Jabodetabek"), Jakarta is also the second largest contiguous conurbation on earth[6]; meanwhile, the city has one of the highest concentrations of digital and social media usage in the world, generating 2.4% of the world's tweets annually [7]. Jakarta is also home to 13 rivers and a complex system of hydrological infrastructure, and this megacity is also situated on a coastal edge of the world's most disaster prone region [8]. The residents have lived with seasonal flooding since at least the seventeenth century; while the global sea level has risen between ten and twenty five centimeters over the past 100 years [9], Jakarta's land has sunk by 1.5 meters just from 1999 to 2009 [10], with massive groundwater pumping as the main suspect on this land subsidence [11].

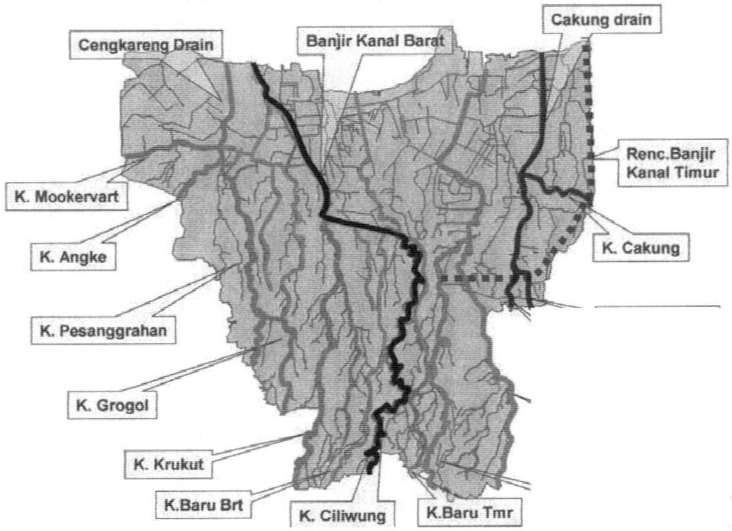

Figure 1. Major Rivers and Canals in Jakarta [12]

2.1. Floods in Jakarta

As a home for millions of residents and a complex system of infrastructure, and being located in the coast, Jakarta experienced flooding created by not only one particular cause. As analyzed by BPBD DKI Jakarta (Jakarta's Disaster Management Agency), there are three immediate manifestations of flood hazards in the city: 1) rain on the upstream area of rivers outside of Jakarta, which causes the overflow of rivers and macro-level hydrological infrastructures; 2) rain on the downstream or in Jakarta itself, which causes the overflow of drainages and micro-level hydrological infrastructures; 3) high tide, which causes a backflow of sea water to the northern part of Jakarta [13]. The upstream area of rivers in Jakarta itself is located outside the city, but in Puncak and Bogor, which are governed by different municipalities. As the river and flooding itself are located across administrative boundaries, different issues have contributed to the flooding, as discussed by Padawangi and Douglass, that not only global climate change affected the flooding, but also the intensity of human settlement manifested in covering land with non-porous materials, the sinking of cities due to excessive groundwater extraction, and changes in land uses in upland areas [14].

As residents in Jakarta are living side-by-side with seasonal floods, traditional ways of knowing the city or what we call "residential epistemologies" have been developed. Particularly with floods, residents have built an informal system to both predict and prepare for the arrival of seasonal flooding. Among the grassroots communities, they developed a communication system to alert people living along Ciliwung River, so that the community located in the upstream in Katulampa will alert the people in the lower area. With this system, they developed a knowledge of predicting and preparing for the flood; if the water has been rising to a certain level or the rain has come for certain hours in Katulampa, it will reach the city in certain hours. Cited from Abdul Kodir and Teja from Ciliwung Institute [15], this created a chain until the information reached people in the downstream area, giving them sufficient information to predict and prepare for the upcoming floods. Furthermore, a resident on the riverside of Ciliwung [15] explained how residents in the downstream area, for example in Bukit Duri and Kampung Pulo, also developed a knowledge system for anticipating the floods. As Kampung Pulo is lower than Bukit Duri and located on the other side of the river, people in Bukit Duri will start to evacuate or move their furniture up if the water reaches the second floor of houses in Kampung Pulo, as it means it will reach their houses soon. This knowledge is developed throughout time and consisted of a combination of experiences and senses in their relation to the river and floods.

However, these residential epistemologies have been disrupted due to unpredictable weather patterns leading to infrastructural changes. Climate change has increased concerns about flooding in Jakarta, as unusual torrential rains and more frequent tidal floods [16] are becoming familiar phenomenon. Competing infrastructural interventions to the already complex systems of hydraulic and hydrological infrastructures compound the difficulty of anticipating the flood. Floods have been the most destructive disaster and growing worse with the increasing population and extensive development along the years; the floods of 1996, 2002, 2007, and 2013 were the greatest and most destructive in the city's history, reported by GFDRR as cited by Padawangi and Douglass [14]. In 2007, 75 percent of the city was flooded and 430,000 people were displaced from their homes with direct estimated economic costs on infrastructure and assets of more than US $900 million [14].

While infrastructure development throughout the city includes the means to anticipate the flood, there is the growing vulnerability of the infrastructure itself. In 2013, the flood was unusual and reached the central landmark of the city, Bunderan HI, due to an unexpected canal breaking within 1 km from the landmark. Padawangi and Douglass [14] explain that while the 2007 flood episode was mostly due to excessive rain, the overall pattern of floods in 2013 approximates the flooding of 2007 and affected the flood-prone settlements in coastal areas and along rivers. This flood episode caused 11 fatalities and at least 20,000 residents were displaced from their homes [17].

Figure 2. Flooding in Jakarta's Central Business District and Bunderan HI in 2013 [18]

The vast project of infrastructure construction has changed the existing residential epistemologies within a short time. Since the river wall construction–popularly known as '*normalisasi*'–started, the relationship between the river and people living alongside it has changed; people could not monitor the changes of water on the river since the wall is built over the eye level. The combination of the diverse infrastructure projects throughout the city–including waterproofing the city with a giant sea wall–and erratic rain patterns due to climate change have debatably changed the way flooding arrives, as now residents sensed it is now more often but far less severe. As a resident quoted during an interview [15], "*...Before, they had big floods, perhaps 3 times a year for a few days, now it is almost every week but only goes up to the ankles*." The construction of infrastructures is also changing the flood-risk perception, as noted by Timothy Forsyth (cited by Padawangi and Douglass) [14]: "*... Once new infrastructure is in place, land areas that are presumably safer from floods become more attractive as sites for projects* *directed toward affluent users and consumers. These engineering projects for flood mitigation, presented as scientifically sound solutions without acknowledging the relations of power that drive them, can be detrimental to marginalized livelihoods and may even increase environmental degradation.*" With the vast changes of the city and behavior of floods, both the government and residents need more real-time information in anticipating the floods.

2.2 Social media usage in Jakarta

In Indonesia, a vast expansion of ICT infrastructure is underway, followed by a social network of information that is realized in its wake. In 2017, Indonesia has the highest growth rate for internet penetration, with 51% year-on-year growth [19]. With approximately 10% in the country's population concentrated in Jakarta, the image of the city, and the nation, is being shaped by digital platforms which act as central nodes for the dissemination and curation of information and trending topics.

Simultaneously, the ubiquity of Internet-connected mobile devices has transformed methods and possibilities of participation in urban life. These smartphones—which actually function as *portable computers*—bring applications to the migrating masses, and more importantly, connect the whole city, rich and poor, to the internet [19]. In 2016, Indonesia was claimed as the third biggest smartphone market in the Asia Pacific region [20]; in 2017, mobile connections are listed at 142% of the country's population [19]. Roughly, 35% of the Indonesian population accessed social media from mobile devices in 2017—that is, about 80 million people used smartphones to connect and exchange with one another.

With data now operationalized as a new form of currency, information is becoming a new kind of power. The expansion of Internet penetration in Indonesia reached 51% of the population in 2017, just slightly higher than the global average [19]. One of the characteristics of Indonesian people is that we really like to share information, whether as an act of service, or as a form of self-actualization, we are enthralled with freedom of expression. Thus, 41% of the population has social media accounts and Indonesia has the most Facebook users: 106 million users, or 6% of the total active users.

3. Civic co-management goes digital

During disaster situations, Jakarta's residents tend to self-organize and inform each other. This habit started in the 19th century, when people used a traditional type of bell and clapper made from wood, called a *kentungan*. Assigned residents would use the *kentungan* to let the others know if the water started rising, especially if this occurred during the night. Today, as discussed before on Section 2, communities along the Ciliwung River build informal SMS chains to inform one another about the water heights and river gauges nearby their houses. That way, downstream residents can know approximately when the water will arrive in their neighborhood, and thus have more time to evacuate families and move valuable goods.

By 2050, studies estimate that over 200 million people globally will be forced to flee due to the sea level rise [21]. Yet, each year tens of thousands of people are moving to Jakarta, the flood-prone coastal area that is already overwhelmed by the pressures of urbanization. As the global discourse about climate change moves from concerns over "sustainability" to issues of "resilience," the co-management of migration could actually help to transform neighborhoods and communities into a more diverse and resilient society.

3.1 The role of social media in disaster management

With the high number of population and high use of social media, residents willingly share information in the digital world. While cities sometimes are seen as data-scarce, this digital platform produces informal and scattered, but also valuable data. One way to prepare for co-management during disaster situations is to extend these informal networks to include a channel between residents and government agencies. With the ubiquity of GPS-embedded smartphones, people can share updates in real time, including their geolocation, via social media and instant messaging. Unlike traditional media, social media relies on user-generated content, which refers to any content that has been created by end users or the general public as opposed to professionals. As the numbers of users has increased, there has been a growing interest in applying social media to address national priorities, not just using them for entertainment or corporate purposes, as discussed by Priolli, Preece, & Shneiderman in Bertot et al. [22].

Government employment of social media offers several key opportunities: democratic participation and engagement to foster participatory dialogue and providing a voice in discussions of policy development and implementation; co-production between the government and the public to jointly develop, design, and deliver public services; seeking innovation through crowdsourcing public knowledge and talent to develop innovative solutions [22]. A study done on local government usage of social media by G. Moss et al. in UK outlined four main organizational purposes that social media and its analytics might serve: communication, public relations, customer services, and public consultation and engagement; all of which involve making 'the public' present to the organization in different ways and with different potential implications and effects [23]. Knox in 2013 discusses that while there is potential to increase legitimacy through involvement, collaboration and dialogue, there are also difficulties with accommodating large flows of direct public feedback into public organizations [24].

As the provider of public services, the government is expected by its public to apply its assets to provide relief and assistance in disaster. Regardless of community size or the nature of disaster, local government leaders are responsible for overseeing all four phases of emergency management: preparedness, response, recovery, and mitigation, while federal and state governments play a supporting role in the immediate aftermath and in providing funding and guidance for long-term recovery and mitigation [25]; however, with their limitation and bounds, government agencies could not work alone in disaster management.

The use of social media in recent disasters (e.g., 2010 Haiti earthquake, 2011 Japan earthquake) around the world has been well documented, and some researches such as Yates and Paquette in 2010 even suggest that "disaster response may be the ideal environment for 'proving the worth' of social media as a serious knowledge management platform" [26]. Neil Dufty further discusses that social media have already demonstrated their use in the emergency management field, but also have potential in the field of disaster risk reduction and community development, as social media can easily form 'communities of practice' before, during, and after the event. In disaster risk reduction field, social media can help people understand the residual disaster risks in their communities and what is being done to manage this risk [26]. In emergency management field, the main goal is to ensure community safety through 'shared responsibility', in the context that communities, individuals and households need to take greater responsibility for their own safety and to act on advice and other cues given to them by state and local government agencies. By the nature, social media can build emergency management communities of interests that share responsibilities, as they rely on peer-to-peer networks that are collaborative, decentralized, and community driven [26]. Social media have made it simpler to interact with others without the limitations of geography and lack of time, therefore, can encourage and sustain learning communities [27].

3.2. PetaBencana.id: real-time flood mapping

Jakarta's love of social media is best exemplified with its Twitter users' posts—2.4% percent of global tweets are coming from Jakarta [7]. With the habit of sharing via social media, people can disseminate time-critical updates and photos of the flood situations. Based on Twitter #DataGrant (2013), there were over 150,000 tweets with geo-locations containing the word *flood* or '*banjir*' in Jakarta and its metropolitan area during 2013-2014, showing that residents are the best sensors of the city. By sharing their updates, they are providing real-time updates and actual on-ground observation of their surroundings, which can be further leveraged for stakeholders in decision-making process.

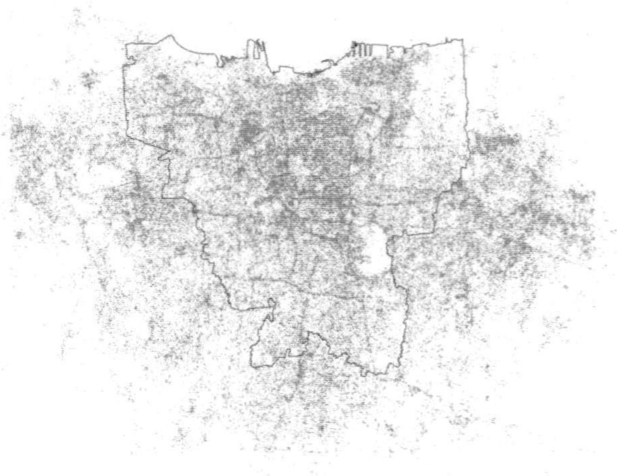

Figure 3. Map of tweets from Twitter #DataGrant containing the word “flood” or “banjir” in Jabodetabek area, each represented by blue dot, during monsoon floods in 2014 [28]

Harnessing the power of social media in Jakarta, PetaBencana.id supports a digital information-sharing ecosystem during flood. Residents can easily describe their situations, include selfies, and as the tweet is geo-referenced, PetaBencana.id uses an open source software named CogniCity OSS to map the tweets and share them through a publicly accessible map. This map functions as a decision-making support for residents to evacuate and navigate through the city during flood, and for government agencies to take an action in response to life-threatening situations. This social media power inspires PetaBencana.id to adopt a “people as sensors” paradigm, where reports are collected directly from the users, or crowd-sourcing, at street level in a manner that removes expensive and time-consuming data processing.

This method of crowd-sourcing answers the limitation of data collection during disaster situation. Limitations of conventional sensors restrain the collection of real time and street level data, which is critical for emergency response. As conventional sensors are mostly needed to be located in static places, it needs regular maintenance to be able to provide real-time data, and the data collection is limited to the number of sensors available in the city while stakeholders need to get an overview of the emergency situation comprehensively in a short time.

3.3 Perspective of residents

By sharing this critical and valuable information, PetaBencana.id aims to democratize decision support and de-credentialize local reporting during emergency situation in floods, as any resident can report and access the flood report map by only connecting to the Internet. When the actual updates of flood situations are shared, not only city stakeholders can use it as a decision support for managing disaster situations, but this will also help residents in making their decisions in organizing themselves during these critical times, thus democratizing real-time and actual information of the floods as a decision support for the public.

Enabling people to report through social media, as well, de-credentializes the local reporting system in disasters, as reports collected from the residents are coming from the street level and help first responders to grasp the current situation in a short time and eliminate the need for time-consuming data collection in disasters. In this manner, interaction among residents and between residents and stakeholders are encouraged, without certain qualifications or eligibility criteria. Most importantly, the main victim of disaster is the people, so the best “sensors” of disaster are the people themselves. Sensors in rivers and canals are still important to measure the water level, but the senses collected from the people are the most important data; by sharing what they see, experience, or need during the flood, people can be a reliable source, and that can be done through the most ubiquitous way: mobile use of social media.

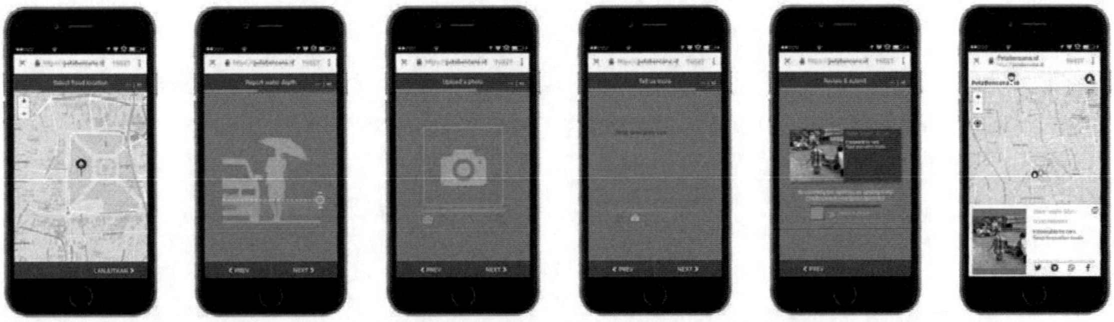

Figure 4. Interactive flood cards designed to guide users to input flood updates via PetaBencana.id. From left to right: location confirmation, input flood height, submit picture, describe the situation, submission confirmation, and users will be redirected to the public map to see their report [29]

CogniCity OSS and PetaBencana.id are adapting to receive more information from apps, making the platform medium-agnostic. Based on the behavior of app usage, as people are deeply engaged to their favorite apps, imagine the situation during disasters; people would look for information more intensely within those apps during their mobile usage and reintroducing them to use new apps with new functions and interface would be an ineffective way [30]. Instead, PetaBencana.id chooses to integrate with other local reporting apps and citizen journalism apps and connect to instant messaging platforms. Reports can be shared to private or group chats in messaging apps, making the information of flooded locations circulate easily.

To report the flood through Twitter and Telegram, residents can call the report link by tweeting #banjir to @petabencana or sending text /banjir to @BencanaBot on Telegram. BencanaBot will reply with a link to the report cards, and users will be guided to submit a report through 3 easy steps: confirm the location on the map, input flood height, submit picture and/or include text description, as shown in Figure 4. These flood cards—acting as a micro-survey tool—collects important information on the disaster, thus the user experience (UX) was designed to correspond with residents' interaction with the flood in the city. The second picture in Figure 4 or the "input flood height" card shows a comparison of an average car height and adult height. Users can input the flood height via the interactive slider while comparing the flood height with car or adult height, as the important message of this information is to tell the others that whether the road is passable by vehicle or how safe it is for the people to stay.

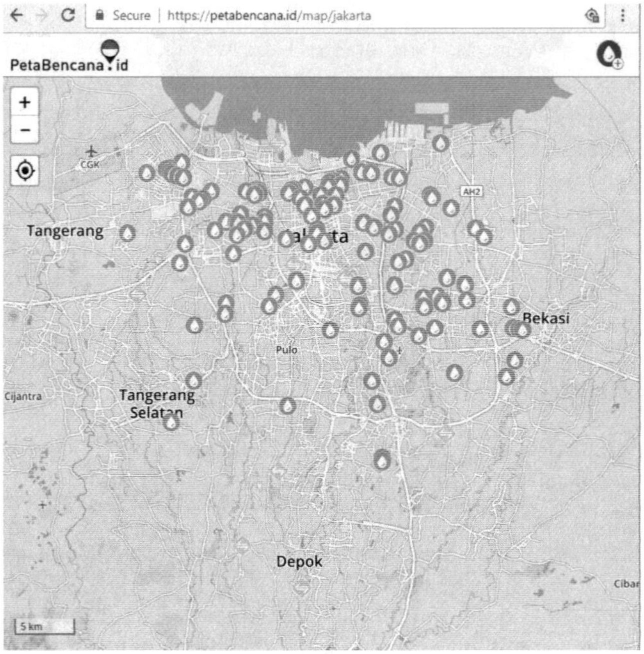

Figure 5. Screenshot of https://petabencana.id/map/jakarta on February 21st, 2017, 8:09 am Jakarta time. Each water drop icon represents a flood report from the user [29]

After the flood report is successfully submitted by the user, user will be directed to the map to see their submitted report, as this sends a message that the user has contributed by sharing valuable information. The freely accessible map of real-time flood reports—available on https://www.petabencana.id/—is designed to be loaded lightly on mobile browsers, thus residents can easily use it on mobile devices during floods to help them make informed decisions on navigation or evacuation. This UI/UX were designed to provide the interface needs to make the city and the happening flood sensible for the residents throughout the city, resembling that users can affect the city and floods and be affected by them [4].

During the last monsoon in 2017, a big flood happened on February 21-22. During that time the residents in Jakarta and the greater metropolitan area actively reported flood in their neighborhood; more than 1,000 reports of flood were received and displayed in real-time on the PetaBencana.id map, which was accessed by more than 250,000 users. These reports were used to help drivers navigate the city during flood - one online taxi company, Uber, promoted the map to its drivers.

Table 1. Statistics of reports and website traffic of PetaBencana.id during the flood in February 21st 00:00 until February 22nd 23:59, 2017. [29]

Reports	
Total Reports Received	1,135
Reports from Qlue	876
Reports from Twitter & Telegram	223
Reports from PasangMata	36
Website Traffic	
Sessions (total times viewed)	312,794
Users	252,449
Pageviews	373,591
Website User Device Type	
Mobile	83.79%
Desktop (including laptop and notebook)	14.69%
Tablet	1.51%

From 223 reports we received through Twitter & Telegram on February 21-22 as shown in Table 1, 171 reports contained pictures of the flood situation. 203 reports containing text descriptions of the flood situation were analyzed using content analysis method.

Table 2. Content analysis of texts in reports of February 21-22, 2017 [29]

Content	**Sample Report (translated)**	**Count of Reports**	**Percentage**
Mentions specific location	*"Flood about 100 cm,* ***at Jl. Penegak Raya****"*	115	56.72%
Includes water depth	*"Already enter the house* ***7 cm****"*	68	33.83%
Describes changes over time	*"Flooded since 05.00,* ***raising until now****"*	31	15.42%
Mentions of flood reason	*"Flood in this road,* ***because the drainage is not that deep****"*	26	12.93%
Mentions time flood started	*"Started to flood* ***at 9 am****"*	21	10.45%
Mentions hydro infrastructure	*"There are* ***2 pumps*** *but only 1 pump works. please repair it soon to help recedes the water"*	18	8.95%
Alerts others to risk	*"Flood in bekasi,* ***be aware of disaster****"*	17	8.45%
Mentions rain	*"If there is* ***rain*** *it is always flooding"*	16	7.96%
Request for help	***"Please help to dry it*** *sooner"*	14	5.97%
Mentions traffic situation	*"Flood in pedongkelan area,* ***vehicles can't pass****"*	10	4.98%
Summons government	*"...* ***Please do something, government!*** *It's very disturbing and I am not living here for free"*	3	1.49%

Based on the text content, users are providing information that is valuable for other residents to avoid the flood. Most of the description (56.72%) contains specific location of street or physical landmark even though users were guided to confirm the location by pinpoint on the map on the first flood card. Other markers such as water depth (33.83%), related hydro infrastructure (8.95%), and traffic situation (4.98%) are also used by users who report to warn other residents, in addition to messages directly alerting others to risk (8.45%). As this platform is also freely accessible for government use and disaster relief agencies, the description is functioning as a way to communicate with these agencies. Users put some more detailed aspects to direct disaster managers to help, such as changes of flood over time (15.42%), the timeframe of flood start (10.45%), request for help (5.97%), even directly summoning the government (1.49%). Some reports also contained description of the cause of the flood: direct mention of rain (7.96%) or other related reasons (12.93%), showing that residents know best about their surroundings and thus can better inform others about their situation.

By crowd-sourcing flood information from the residents, the actual and on-ground updates on flood can be collected in a non-time-consuming manner. By integrating these informal but valuable reports with a detailed map of the city and critical infrastructure data from formal agencies, then sharing it openly via a publicly accessible map, civic co-management can be promoted and residents can make more independent decisions on safety and navigation in response to the flood in real-time. Now that the residents have the updated information on floods despite their limited resources, we move to the next question: how can the government leverage this tool for disaster response?

3.4 Government perspectives

From 2014 to 2016, PetaJakarta.org (the pilot project which preceded PetaBencana.id) was used by the Jakarta Disaster Management Agency (BPBD DKI Jakarta) to monitor the flood situation and improve response time for the first responders. The success of this pilot project was monitored by the National Disaster Management Authority (BNPB), which then promotes the development of this platform to a greater area, as the development of a platform with the basis of social media and digital media enables people to report easily and in real time; a disaster is everybody's business so the people need to be involved in sharing the information of their needs in time of pre-disaster, during disaster, and post-disaster. The spokesperson of BNPB, Dr. Sutopo Nugroho [15] stated that the behavior of the Indonesian people who willingly share information can be used by the government to do their task better: to provide and meet public needs to support residents' preparedness during disaster. It is harder for the higher level of government to wait for reports from local authorities due to bureaucracy; with information from the public, the government will know the real situation and what the public needs.

As the government is one of the main stakeholders in disaster management and has more resource for first responders, PetaBencana.id is designed as a platform to support decision making and dynamic communication between the government and residents. The Command Center of BPBD DKI Jakarta uses the platform to monitor the floods especially during monsoon season and to support the decision-making process for first aid response. Distribution of government staff is beneficial to help verify the flood information, which can be shared to residents in a public map, via a government interface or dashboard, called a Risk Evaluation Matrix (REM).

Before PetaJakarta.org established, the erstwhile official flood updates from government were issued through text messages and the official map is published in a portable document format (PDF) as shown in Figure 6. As these official updates were issued in 6-hour periods, the updates were used mostly for curative efforts, rather than acting as real-time information for disaster response and for communicating with residents to avoid bigger risk. Through PetaBencana.id, these verified flood heights or official updates from the Command Center is being put via REM, by activating a color-coded menu on particular community group boundaries–named Rukun Warga (RW) and extending around 25 ha on average in Jakarta–on the map. The color code marks the water depth ranged from 10-70 cm, 71-150 cm, and >150 cm flood. This official information is shown in real-time on the public map; thus, the flood map acts as a communication medium to give the residents the verified flood status from the government. While the intention of democratizing decision support has not been changed since the usage of the erstwhile PDF flood map, giving residents access to the same tool used by the government and changing the format to an easily readable map that relates to the residents' needs would make the information more useful for the residents.

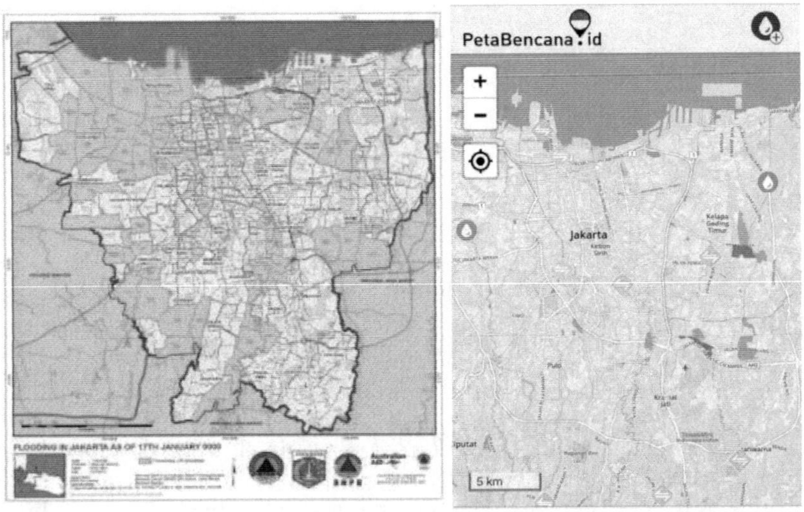

Figure 6. Comparison between flood map PDF used by BPBD DKI Jakarta before PetaJakarta.org (left) and screenshot of official flood updates on PetaBencana.id map, February 21st, 2017 (right) [29]

Table 3. Color codes of flood classification and number of flood polygons (flood affected RWs) activated by BPBD DKI Jakarta during flood in February 21-22 [29]

Flood Classification	Feb 21st 04:00-08:00	Feb 21st 08:00-12:00	Feb 21st 20:00-00:00	Feb 22nd 04:00-08:00	Feb 22nd 08.00-12:00	Feb 22nd 20:00-00:00
Use Cautions	0	3	0	1	2	1
10-70 cm flood	39	151	63	119	90	11
71-150 cm flood	4	8	13	23	16	0
>150 cm flood	0	0	0	7	7	0
Total	43	162	76	150	115	12

Not only the user interface should be concerned, but the server capacity is also important as the population of Jakarta is reaching almost 10 million people. In order to handle the spike in web-traffic on February 21st, PetaBencana.id scaled to up to 8 servers. The capacity to scale at the Elastic Beanstalk layer meant that the underlying database only experienced a small increase in CPU load. Importantly, while the websites of all other government and local agencies responsible for disseminating emergency information crashed under the heavy traffic, PetaBencana.id was the only platform that remained consistently functional; testifying the amplified capacities of multi-stable, multi-scalar, and multi-dimensional computational architectures. More importantly, the map was used as a reference for Governor of Jakarta province to deploy the field officers to help drying the flood and making sure the residents were safe [31].

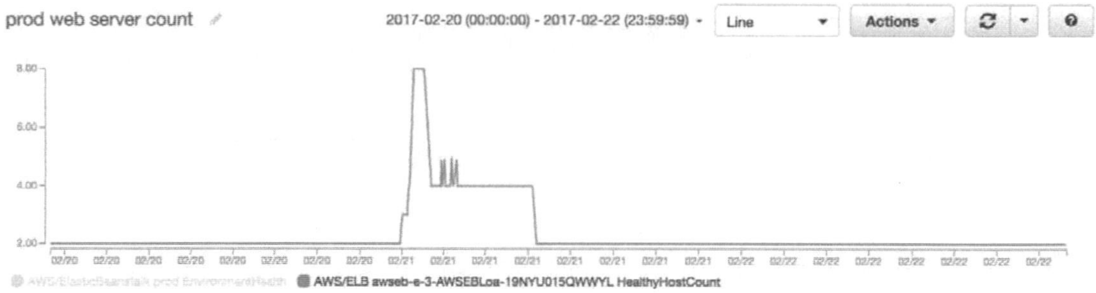

Figure 7. PetaBencana.id web server count graph showing that on February 21st, 2017, the web server scaled up to 8 servers [29]

Having a two-way communication during disaster situation is a new challenge to the government of Jakarta. Jakarta Disaster Management Agency (BPBD), involved in the development of PetaJakarta.org since 2014, stated that PetaJakarta.org was a new breakthrough [32]; it began when the agency needed real-time information as they did not have enough sensors to detect the flood height in the city, aside from the water level sensors in the river. The information from citizens is gathered through PetaJakarta.org, and will be verified through REM, thus acting as a flood map, both used internally within the agency or as information for other agencies and for residents. Moreover, BPBD and PetaJakarta.org team trained the head of villages or *Lurah* in 5 cities in Jakarta to report and use the flood map; therefore, after the command center of BPBD updates the official information of flood through REM, the agency expects the head of villages to respond on the flood.

With the existence of the platform, there is citizen support enabled by this framework. BPBD noticed that residents are being more interactive than before to give information about flood, and at the same time the agency tends to be more open to share information, as the agency always emphasizes data transparency. Upon creating an information-sharing environment on flood with this platform, there is a pressure felt by BPBD created from public expectation; however, it is also felt as an advantage for the agency, as they will know immediately about the flood situation in the city and could act faster [15]. On the other hand, BPBD as disaster managers hope that they can do disaster response together with the residents. When people share information to each other, along with that, their awareness of flood threat will increase. BPBD noticed that since PetaJakarta.org was established, people started to be aware of disaster information; back then, they ignored the warnings. It can be seen when the command center was late to share the weather forecast, the residents would start to comment 'where is today's weather forecast?' [15]

The agency hopes to raise the concern and caring among each other; that is their main purpose, so people can help each other. The government's long-term mission is to start delegating the authority of disaster management, so that it would not only come from the government, but also non-government agencies (private sector, NGO, and citizens), so when disaster happens they are aware and do the first response autonomously. This condition is more sustainable than only waiting for the government to help the people; if there is any severe flooding, they hope people who know about the flood anywhere can also help the others [15]. To support this, PetaJakarta.org and later PetaBencana.id also share and visualize critical infrastructure data collected by the government alongside the reports, to help government agencies and residents anticipate the potential of incoming floods. More information-based decisions can be made by government agencies in Jakarta through improved situational knowledge; therefore, increasing resilience of the city to flooding and its attendant difficulties and fostering equitable climate change adaptation.

Operating in an open-data environment, the reports and verified polygons in PetaBencana.id can be downloaded through an open API for further use of research and development. The data is not only used by other non-governmental organizations, but also BPBD DKI Jakarta to assess the damage and loss with other software and also synchronize the data with other agencies, e.g., Agency of Public Works, Planning Agency, and Spatial Agency for future planning [13].

4. Conclusion

In co-managing the city during a disaster situation, the need of actual and real-time disaster information in megacities like Jakarta can be answered by crowd-sourcing the information from residents via social media and instant messaging. This valuable information can be harnessed by residents and the government to make more informed decisions. It is important to note that achieving disaster resilience is not solely the responsibility of emergency management agencies, but a shared responsibility across the society; therefore, disaster managers and first responders, especially local government units in Jakarta, should work together to improve emergency management and advisory roles. Providing reliable information for residents should not be seen as an incompetence of the government in managing the disaster; moreover, the government should communicate the up-to-date situation to residents, so residents can use the information provided as their base of action for their own safety in the event of disaster and to avoid residual risks. Integrating the informal and formal data source, giving residents access to the same tool used by the government, and designing the interface that relates to residents' and the government's needs are important to make the information more useful in democratizing decision support, to create a multiplying effect of valuable information in disaster situation. Better evidence-based decisions are made through accessible situational knowledge, resulting in better disaster response and resilience in the city.

Acknowledgement

The author would like to thank the team of PetaBencana.id, Dr. Etienne Turpin, Dr. Tomas Holderness, and Dr. Matthew Berryman for their support and assistance in the process of writing this paper. The PetaBencana.id platform is made available through financial support from USAID, and in collaboration with Pacific Disaster Center, Humanitarian OpenStreetMap Team, National Disaster Management Authority (BNPB), and Regional Disaster Management Agency (BPBD).

References

[1] Soja E and Kanai M 2014 *The urbanization of the world* in *Implosions/explosions: Towards a study of planetary urbanization* ed Brenner N (Berlin: jovis Verlag GmbH) pp 142-159
[2] Furlong K 2010 Small technologies, big change: Rethinking infrastructure through STS and geography *Progress in Human Geography* **35** 460-482
[3] Gandy M 2012 *Where does the city end? in Implosions/Explosions: Towards a study of planetary urbanization* ed Brenner N (Berlin: jovis Verlag GmbH) pp 86-89
[4] Bratton B 2015 *The stack | on software and sovereignity* (Cambridge: MIT Press)
[5] Worldometers 2017 Countries in the world by population (2017) *Worldometers* [Online]. Available from: http://www. worldometers.info/world-population/population-by-country/
[6] Cox W 2015 Largest 1,000 cities on earth: World urban areas: 2015 edition *New Geography* [Online]. Available from: http://www.newgeography.com/content/004841-largest-1000-cities-earth-world-urban-areas-2015-edition
[7] Aly W 2013 Welcome to Jakarta, the world's number one Twitter city *RN Drive* [Online]. Available from: http://www. abc.net.au/radionational/programs/drive/social-media-in-indonesia/4720678
[8] Turpin E, Bobbette AW, and Miller M 2013 Introduction: Who does Jakarta think it is? in *Jakarta: Architecture + Adaptation* (Depok: Universitas Indonesia Press) pp 2-58
[9] UNEP 2017 Vital climate graphics: Observed climate trends *GRID-Arendal* [Online]. Available from: http://old.grida.no/ climate/vital/19.htm
[10] Prameshwari P 2009 The tides efforts never end to repel an invading sea *JakartaGlobe* [Online]. Available from: http://jakartaglobe.id/archive/the-tides-effortsnever-end-to-repel-an-invading-sea/

[11] Abdurrahman O 2014 Air tanah Jakarta: Air diambil, angka dicatat *GeoMagz* [Online]. Available from: http://geomagz. geologi.esdm.go.id/air-tanah-jakarta-air-diambil-angka-dicatat/
[12] Dinas Pekerjaan Umum Provinsi DKI Jakarta 2018 [Online]. Available from: http://www.serverjakarta. com/peta_13sungai.aspx
[13] BPBD DKI Jakarta 2017 *Sistem informasi bahaya banjir di Provinsi DKI Jakarta, Presentation by BPBD DKI Jakarta* (Jakarta: BPBD DKI Jakarta)
[14] Padawangi R and Douglass M 2015 Water, water everywhere: toward participatory solutions to chronic urban flooding in Jakarta *Pacific Affairs* **88** 517-550
[15] PetaBencana.id 2017 *The Same River, Twice* (Documentary exhibited in Lisbon: The New Art Fest)
[16] Desyani A 2013 Jokowi pantau lokasi banjir rob Pademangan *Tempo.co* [Online]. Available from: https://metro.tempo.co/ read/488850/jokowi-pantau-lokasi-banjir-rob-pademangan
[17] Leoni B 2013 Indonesia declares state of emergency in wake of heavy floods *UNISDR* [Online]. Available from https:// www.unisdr.org/archive/30593
[18] Munns L 2013 Photo shared by Twitter user @LukeMunns [Online]. Available from: https://twitter.com/lukemunns
[19] Kemp S 2017 Digital in 2017: Global overview *We Are Social* [Online]. Available from: https://wearesocial.com/special-reports/digital-in-2017-global-overview
[20] eMarketer 2016 Indonesia is the 3rd-largest smartphone market in the Asia *Pacific Indonesia-Investments* [Online]. Available from: http://www.indonesiainvestments.com/news/ todays-headlines/indonesia-is-the-3rd-largest-smartphone-market-in-the-asia-pacific/item6777
[21] Kamal B 2017 Climate migrants might reach one billion by 2050 *ReliefWeb* [Online]. Available from: https://reliefweb.int/ report/world/climate-migrants-might-reach-one-billion-2050
[22] Bertot JC, Jaeger PT, Hansen D 2012 The impact of policies on government social media usage: Issues, challenges, and recommendations *Government Information Quarterly* **29** 30-40
[23] Moss G, Kennedy H, Moshonas S, Birchall C 2015 Knowing your publics: The use of social media analytics in local government *Information Polity* **20** 287-298
[24] Djerf-Pierre M, Pierre J 2016 Mediatised local government: Social media activity and media strategies among local government officials 1989-2010 *Policy & Politics* **44** 59-77
[25] Becker C 2009 Disaster recovery: A local government responsibility *Public Management* [Online]. Available from: http://ICMA.org/pm
[26] Dufty N 2012 Using social media to build community disaster resilience *Australian Journal of Emergency Management* **27** 40-45
[27] Tibbitt J 2011 Social media, social capital and learning communities *PASCAL International Observatory*
[28] Data Grant 2015 “Flood” or “banjir” in Jabodetabek area during monsoon floods in 2014 [Online]. Available from: https://PetaJakarta.org
[29] PetaBencana.id 2017 Peta Bencana Indonesia [Online] Available from: https://PetaBencana.id
[30] Perez S 2015 Consumers spend 85% of time on smartphone in apps, but only 5 apps see heavy use *TechCrunch* [Online]. Available from: https://techcrunch.com/2015/06/22/ consumers-spend-85-of-time-on-smartphones-in-apps-but-only-5-apps-see-heavy-use/
[31] Kurniawan T 2017 Ahok pamer bukti banjir cepat surut *MetroTVNews.com* [Online]. Available from: http://news.metrotvnews.com/read/2017/02/23/662355/ahok-pamer-bukti-banjir-cepat-surut
[32] OECD 2017 Case study: PetaBencana.id, Indonesia *Embracing Innovation in Government Global Trends* pp 23-26 [Online]. Available from: https://www.oecd.org/gov/ innovative-government/embracing-innovation-in-government.pdf

IOP Conf. Series: Earth and Environmental Science **158** (2018) 012050 doi:10.1088/1755-1315/158/1/012050

An Examination of Regional Competitiveness: Early Findings from Banten, Indonesia

Y M Holis, I Syabri and H Prabatmojo

Department of Regional and City Planning, Institut Teknologi Bandung, Bandung, West Java, Indonesia

Corresponding author: syabri@gmail.com

Abstract. Over time, the regional development planning paradigm continues to change procedural, substantive, actor and institutional aspects. The increasing globalization and decentralized governance systems will continue to determine how a country or region can compete and even be able to take advantage of the positive value of economic globalization. This certainly provides challenges and opportunities for the renewal of regional development concepts, in particular, the concept of regional competitiveness. This concept still lacks a scientific basis and its interpretation is still in debate. To date, the clear scientific basis is the concepts of corporate and national competitiveness. Whereas, the concept of competitiveness at the meso-level, namely the competitiveness of the region, becomes very important. Regional competitiveness prepares a region and to survive in response to the more open dynamics of economic globalization and the system of decentralization of authority in Indonesia. The objective of this research is to complement the concept of competitiveness which is still at the level of microeconomics (business theory) and macroeconomy (theory of growth) with the concept of competitiveness at the regional level that focuses on (i) finding determinant factors at the regional level in Indonesia; (ii) profiling regional competitiveness in several regions; and (iii) assessing the role of local government (institutional and actor) on determinants of regional competitiveness. This research is carried out by using a mixed-methods approach with a concurrent triangulation strategy model. The research follows the stages of literature study to identify the determinant factors of regional competitiveness relevant to the purpose of research, followed by descriptive analysis of the variables and selected by statistics through Analytical Network Process (ANP) to become the determinant factor of competitiveness. In accordance with the concept and definition of regional competitiveness, this research develops five main indicators determining regional competitiveness and its 30 variables. These indicators are regional economy; manpower and human resources; productive business environment; infrastructure; natural resources and environment; and banking and financial institutions. The data used are primary and secondary data by compiling from the Central Statistics Agency (BPS) and questionnaires to the Regional Development Planning Agency (BAPPEDA) following by in-depth interviews with experts. This research uses the case study of Banten Province. The main reason is that Banten has varied areas, ranging from developed cities to lagging regions. It is expected to represent other areas in Indonesia. This research provides an alternative framework of competitiveness analysis methods at the regional level. The concept of regional competitiveness can be a catalyst and enabler to the development of regional specialization and networking among regions in the era of economic globalization.

IOP Conf. Series: Earth and Environmental Science **158** (2018) 012050 doi:10.1088/1755-1315/158/1/012050

1. Introduction

There are deep concerns about the performance of the regional economy when viewed from the tendency of international competitiveness in recent years. The development of a country's competitiveness certainly cannot be separated from the development of its components, namely regional competitiveness. Therefore, the position of regional competitiveness rating is a reflection of the competitiveness position of the regions. Increased regional competitiveness is believed to be able to improve the competitiveness of the Region at the national level. One of the efforts to improve the competitiveness of regional economy is through increasing its economic competitiveness. The improvement of regional competitiveness is expected not only to improve the competitiveness of the regional economy in a global market but also to accelerate the growth of the real sector which has been considered not yet optimal.

The implementation of regional autonomy and fiscal decentralization since 2001 (Law 32/2004 and Law 33/2004), states that decentralization in the regions is focused at the regency/municipality level. Regions/municipalities are given the widest authority to regulate and carry out development in accordance with the potential and function of each region (except for some sectors that have been agreed to be the authority of the central government). Decentralization has given freedom to regions, especially regencies to develop local economies, thus accelerating regional development and improving the welfare of their communities.

The background mentioned above shows the importance of the ability of the region in synergizing its strengths and potentials in an effort to improve competitiveness. The ability of this region will depend on its ability to identify the factors that affect the competitiveness of each region, as well as its ability in implementing policies to improve its regional competitiveness relative to other regions.

Therefore, a new methodology and broader scope are needed to determine the competitiveness of a regency/municipality. The methodology adapted to the current policy concept will have policy implications far different from those of previous studies. This methodology will systematically identify the determinants of competitiveness and then describe the competitiveness of a region through the determinants of competitiveness above. The framework of this regional competitiveness methodology will provide an overview of the level of competitiveness of a region and what policy interventions are required by the government in enhancing regional competitiveness.

1.1. Goals

Considering the above-mentioned background, in more detail this study aims to:

a. Determine the main factors that affect the competitiveness of a region;
b. Determine the weight of the main factors that affect the competitiveness of the region;
c. Calculate scores and competitiveness between regencies/municipalities for each classification of the regions in Banten Province;
d. Classify the area using the clustering method.

1.2. Benefits

In accordance with the research objectives to be achieved, this research will contribute to:

a. Increasing the role of regional governments to increase economic growth through enhancing regional competitiveness;
b. Improving roles and policies to assist local governments in enhancing regional economic competitiveness;
c. Increasing the effectiveness of fiscal planning and fiscal policy.

2. The concept of regional competitiveness

Understanding of regional competitiveness originates from the concept of competitiveness that has evolved to enterprise and country level. The thinking of Michael Porter [5,6] colored the development and application of the concept of competitiveness at the enterprise level. Furthermore, this concept is developed at the country level as a Global Competitiveness, particularly through the institutions of the

World Economic Forum (Global Competitiveness Report) and the International Institute for Management Development (World Competitiveness Yearbook). The strategic impact of this Global Competitiveness Index measurement on the country's strategic position in addressing its international competitiveness has made this measure a reference to the economic performance and investment climate of an economy. The trend of decentralization has reinforced the need for identification of competing determinants at the local level. This section discusses the development of previous and ongoing research libraries since the concept of state competitiveness is rolled out to the development of concepts and research on regional competitiveness. The discussion includes also the development of the concept of regional competitiveness used in research which is an adaptation of the concept of regional competitiveness that has developed so far.

The notion of regional competitiveness began to develop after [23] defined national competitiveness. He defined it as an outgrowth of a country's ability to innovate in order to achieve or maintain a favorable position compared to other countries in a number of key sectors. This understanding is further developed by Thurow [8], who stated that the key sectors can be categorized as sectors that have a high technological composition with high demand and have a highly competitive base which is then called the "knowledge-based industries". Specialization of these sectors is believed to be a trigger for the improvement of higher living standards for the community.

The notion of competitiveness at the sub-national level was then developed by Storper [7], who defined regional competitiveness as the economic capacity of an area to attract and retain firms with a stable or increasing market share in their operations, while ensuring the improvement or stability of living standards for the communities participating in such activities. Regional competitiveness will be assured only if sustainable development can be achieved at the level of the workforce that can enhance overall living standards.

One of the key factors in achieving such growth is the existence of potential firms indicated by the level of business density that can lead to the birth of entrepreneurs and the new innovations in developing sectors and markets in the area; even more important is to create new jobs [2]. The regional competitiveness of American cases is more influenced by the growth of companies in the region that are characterized by being innovative and able to take advantage of new technological discoveries to create new business opportunities in the market.

Regional competitiveness in European countries is often defined as the ability of an area to achieve economic growth by producing goods and/or services, in which the product is able to compete at both national and international levels, providing a high standard of living to its people as well as employment, and the results of such development can be enjoyed on an ongoing basis. As defined by the UK Department of Trade and Industry (UK-DTI), regional competitiveness is a region's ability to generate high income and employment while remaining open to domestic and international competition. Moreover, the Center of Urban and Regional Studies (CURS) in 1998 defined regional competitiveness as the ability of a business or company in a region in generating high income and wealth that is more equitable for the population.

Bank Indonesia and the Faculty of Economics of UNPAD (2001) defined regional competitiveness by modifying the definition of UK-DTI. The regional competitiveness of the study is defined as the ability of the regional economy to achieve high and sustainable growth while remaining open to domestic and international competition. IMD (2001) analyzed competitiveness for 49 countries based on both quantitative and qualitative data, classified into eight groups. Quantitative data is given weight 1, while the survey data was given a weight 0.64. While WEF (1999) weighted the data in eight classifications: openness, 1/6; government, 1/6; finance, 1/6; infrastructure, 1/9; technology, 1/9; management, 1/18; employment, 1/6; and institutional, 1/18. The weighting is based on regression analysis by looking at the correlation between these factors and per capita economic growth.

A study of regional competitiveness in China in 2006 developed the concept of regional competitiveness using two main components, i.e., the technological progress component of each region (calculated based on the Total Factor Productivity score) as well as the regional investment climate component (consisting of several assessment components such as market share; labor market flexibility; technological characteristics and regional expertise; government bureaucracy efficiency; enforcement and protection of business agreements and copyright; regional access to financial institutions; and quality of life of local communities).

2.1. The concept and definition of regional competitiveness in this research

The concept and definition of regional competitiveness developed in this study are based on two considerations, namely regional economic development in terms of regional economic aspects, and the development of concepts and definitions of regional competitiveness of previous studies. From the regional economics aspect, regional economic development cannot be separated from the development of national economy and economy of other regions. The interaction between regions (both goods, services, and labor) and policies at the national level will have an impact on the economy of a region. The description of the economy of a region [4] can be illustrated as a drum (figure-1), where the regional economy is heavily dependent on the outflows and incomes and commodities of other regions or nationalities, the magnitude of potential products in the region, and the magnitude of potential leakage in the region (such as unemployment rate, and waste in resource use).

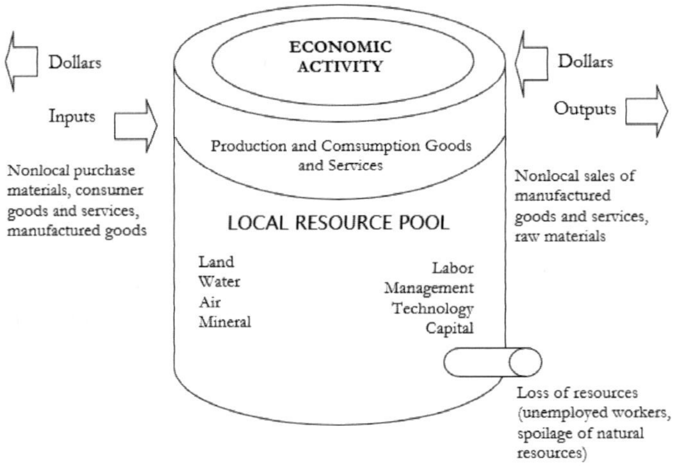

Figure 1. Regional economy [4].

Meanwhile, the development of regional economic progress is determined by four factors [1]. First, the job opportunities in the area (including the understanding of the quality of the workforce so as to provide a good location access for companies that will do business in the area). Secondly, the basis of regional development (in the sense that the development of good economic institutions that can lead to increased business desires for the business community). Third, the location asset is a regional competitive advantage based on environmental quality. Fourth, the knowledge resources, in the sense of knowledge as the basis of the economic drivers (knowledge-based development).

Based on the above understanding, the process of economic development in the region aims to:

- Establish a new "institution" that supports the regional economy;
- Develop alternative industries;
- Increase the capacity of workers to produce better products;
- Stimulate the occurrence of technology transfer; and
- Open investment opportunities for entrepreneurs.

Therefore, the main objective of regional economic development is to increase local initiatives to create employment opportunities for the community (both in the number and type of work) and to stimulate the increase of economic activity in order to increase community incomes.

On the other hand, based on several preceding definitions of regional competitiveness, it appears that regional competitiveness is produced by complex interactions between input, output, and outcome factors in their respective regions. Therefore, the interrelationship between these three aspects will largely determine the success of a region compared with other regions. Local competitiveness is also related to non-economic factors, such as political, social, and cultural parameters of society. The concept of competitiveness, which is the interaction of input, output and outcome components, for example, has been used by Huggins in calculating regional competitiveness index in England [3].

Figure 2 shows the Regional Competitiveness Pyramid, which can explain the differences in regional competitiveness in a region, referring to the definition of regional competitiveness developed in this study. Regional competitiveness is formed by the main factors (inputs) which are both regional endowment or caused by the interaction of community activities such as differences in the productive business environment; the structure and the condition of the regional economy; the difference in the quality and quantity of human resources in each region; infrastructure, natural resources and environment; and the condition of financial and banking institutions. The performance of the main factors of the regional competitiveness will make a difference in the performance of the economy (output) of each region, such as the difference in the level of employment created and the productivity of its workforce. These differences will ultimately result in differences in the sustainability of the quality of life and the welfare of the people of each region.

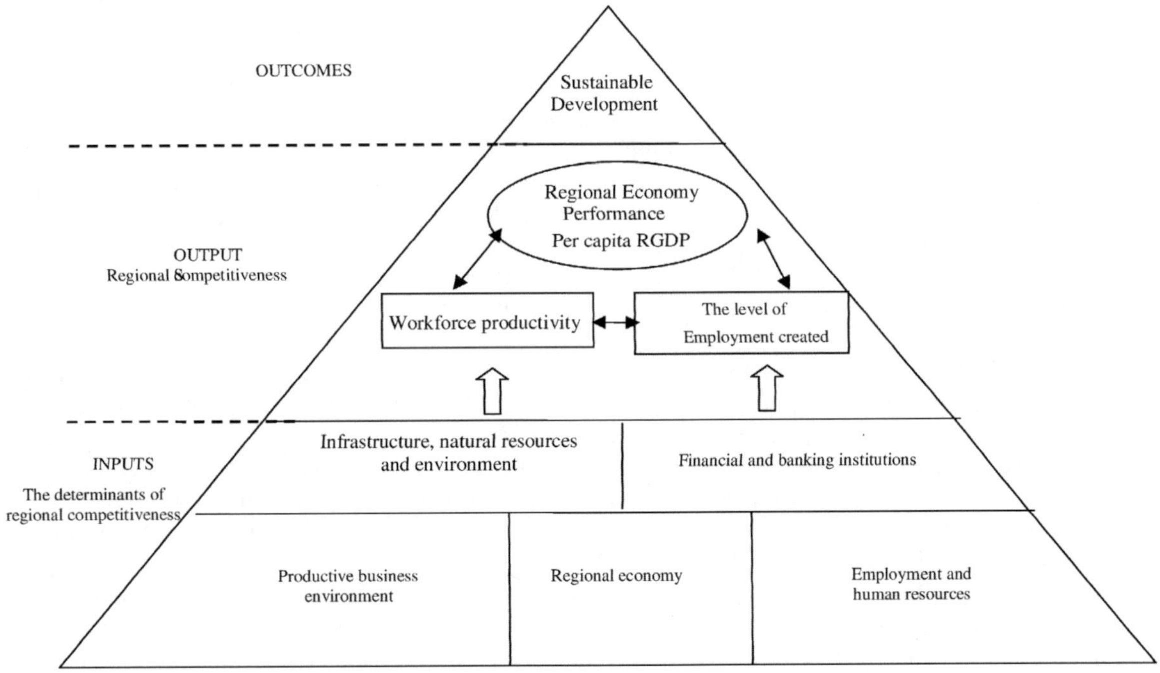

Figure 2. The pyramid of regional competitiveness [14].

Thus, based on the previous literature on the theory of regional development as well as the concept and definition of regional competitiveness, the meanings of regional competitiveness in this study are: the ability of regions to synergize between input, output and outcome that exist in the region on a continuous basis, with attention to technological changes and institutions in the area, in order to compete both nationally and internationally, so as to improve the living standards of the community.

3. Research Methodology

The definition and concept of regional competitiveness of this study are as described in the previous section where the realization will be represented in the competitiveness index consisting of the outputs of input, output, and outcome. Besides a descriptive method and literature study, the main method used in this research is econometrics and expert opinion polling especially by using Delphi method (omitted). This method is a more systematic and consistent implementation of focus groups.

3.1. Determination of competitiveness factors and variables

The factors that shape competitiveness at the regional economic level, particularly at the regency and municipal levels, may be specific to the region itself (internal factors) as well as external factors that are exogenous and cannot be used as policy variables for the regions concerned. Referring to the Huggins study [3], competitiveness indexes are grouped into three main components, namely input, output, and outcome. In Huggins, the input component is equivalent to internal factors [3]. In this study, the input component is further expanded to include the external competitiveness factors that will primarily affect the business environment.

The internal competitiveness factors refer to the concept of 'productivity', i.e., the amount or rate of change in value added per unit of input generated by an entity. This productivity is further elaborated into measurable variables that generally have a dimension of 'ratio'. The external factors can be 'industry-specific', region-specific, 'or' country-specific'.

After establishing the determinants of competitiveness, the next step is to describe it in terms of the variables that represent each of these factors. The method used for this is semi ad-hoc in the sense that the variables used in this study are selected based on the benchmarking method which is then modified as needed for regional and regency levels through literature search and 'simple focus groups'.

3.2. Determination of Weight and Ranking of Competitiveness

Some of the important outputs expected in this study are the rank and balance of competitiveness for each region or group of regions. Before the ranking and balance of competitiveness can be made, scoring and weighting methods are required. The commonly used method for this is 'expert opinion polling', which is to use expert knowledge and opinions to generate a criterion or index of valuing and weighing factors or variables. Two popular methods are Analytic Hierarchy Process (AHP) and Delphi Method. AHP is a method of preparing a functional hierarchy that is done systematically. The main target of AHP is to find out the weight of relevance and importance.

In this study, the Delphi method of analysis is used in determining the weight and rank of competitiveness. This method is chosen because it is believed to be more appropriate to the context under study. Delphi is the name of a set of procedures for obtaining and filtering opinions from a group, usually an expert panel. This method is a way to achieve a consensus of a group of experts after obtaining opinions in defining problems based on the knowledge and intuition of experts. The collective appraisal of these experts, though derived from subjective opinions, remains better than an individual statement and produces a more objective outcome.

The Delphi method has four main characteristics: (1) structured questions, (2) iterations, (3) controlled feedback, and (4) the anonymity of the respondents. The structured questions are manifested through the use of a questionnaire. This questionnaire will keep the problem-solving in focus and allow the moderator to control the process to produce good results. Iteration is the process by which questionnaires are distributed and answered in several rounds so that respondents can rethink the answers they have given. Controlled feedback can be achieved by sharing the results of group responses to all respondents for consideration to rethink the answer. Anonymity is manifested by a freely answerable questionnaire expressing opinions without fear of 'pressure' from other panelists. Decision-making in a group can be influenced by several things, including the dominance of some panelists and a strong panelist personality. Sometimes these things will reduce the ability to generate unbiased decisions or predictions. With the 'anonymity' characteristics doubts about a biased decision can be avoided.

Delphi can be executed as often as three rounds to reach a level of convergence that will generate the weight for each variable or factor, which will then be used to rate the competitiveness between regions. However, before the rating can be made, the different measurement units that occur in the various variables require that the variables need to be adjusted first so that the rating is 'apple-to-apple'. For that, all variables will be converted first into a standard normal score (standard normal score).

3.3. *Clustering*

Comparing competitiveness between one regency/municipality and other regencies/municipalities sometimes does not provide a desirable picture, since the characteristics of the regency/municipality are in different dimensions or classes. For example, comparing a regency/municipality in Java with a regency/municipality in Papua is often considered unwise. Therefore, in addition to comparing total competitiveness, comparisons and competitiveness ratings will be based on groups, both ad-hoc (ex-ante) and ex-post. For ad-hoc, in addition to the total comparison, the grouping will also be performed based on input and output components. The grouping is done based on western and eastern regions in Indonesia. Ex-post, grouping is based on clusters formed by considering the cluster forming factor or variable by applying cluster analysis.

3.4. *Data and data sources*

In accordance with the concept and definition of Regional Competitiveness developed in this study, there are five main indicators for the formation of regional competitiveness and its variables as presented in Table 1 below. These indicators are the Regional Economy; Employment and Human Resources; Productive Business Environment; Infrastructure, Natural Resources, and Environment; and Banking and Financial Institutions. Data sources rely solely on the availability of secondary data from the Central Statistics Agency (BPS) or other agencies.

Table 1. Indicators and Variables determining Regional Competitiveness.

No	Variable	Descriptions	Data Sources
INPUT			
I. Regional Economy			
	Sectoral Productivity		
1	Primary Sectoral Productivity	The GRDP of the primary sector is divided by the number of workers in the primary sector of the past year.	BPS
2	Secondary Sectoral Productivity	The GRDP of the secondary sector is divided by the number of workers in the secondary sector in the past year.	BPS
3	Tertiary Sectoral Productivity	The GRDP of the tertiary sector is divided by the total number of workers in the tertiary sector in the past year.	BPS
	Regional Finance		
4	Local Fiscal Capacity	Local Fiscal Capacity = ((Local Revenue + Profit Sharing + General Allocation Fund + Other Legitimate Regional Income) - Office Officers)/Population.	Ministry of Finance

No	Variable	Descriptions	Data Sources
5	Government size	Total Regional government expenditure divided by GRDP value.	BPS
	Investment and Regional overpriced index		
6	Local Total Export Potential	Total Export Potential Area divided by GRDP value.	BPS
7	Total Investment Per Capita	Total Investment divided by total population.	BPS
8	Regional Overpriced Index	Measuring the level of difficulty and expensiveness of conducting business activities in a regency/municipality. Proxy by using Construction Cost Index (IKK).	BPS
9	Firm Density	The number of companies in the manufacturing sector divided by the area.	BPS
II. HUMAN RESOURCES AND EMPLOYMENT			
	Human Resources		
10	Total Population	Total Population in a regency/municipality.	BPS
11	Dependency Ratio	Comparison between the population 0-14 years old, plus the population 65 years and above compared with the population aged 15-64 years.	BPS
12	Mean years of schooling	The average number of years spent by people aged 15 years and over to go through all types of formal education ever undertaken.	BPS
13	Life Expectancy Index	Estimated life expectancy of the average population with the assumption that there is no change in patterns of mortality by age.	BPS
	Employment		
14	Mean years of labor schooling	The average number of years spent by the workforce for all types of formal education ever undertaken.	BPS
15	Size of the workforce	Population in the working age (15 years and over) either working or looking for a job/unemployment).	BPS
16	Growth rate of the labor force	Average population growth in the working age (15 years and over) both working and looking for a job/	BPS

No	Variable	Descriptions	Data Sources
		unemployment) for the last three years.	
III. PRODUCTIVE BUSINESS ENVIRONMENT			
	Productive Business Environment by the Government		
17	Public service expenditure per capita	The amount of public service expenditure divided by the total population.	BPS and Ministry of Finance
18	The number of problematic Regional Regulations	The number of Regional Regulations canceled by the Central Government.	Ministry of Finance
	Productive Business Environment by Community		
19	The number of base sectors in the regency/municipality	The number of sectors with a Location Quotient value greater than 1.	BPS
20	Percentage of population with university as the highest education	Percentage of population with university as the highest education of the total population.	BPS
21	Poverty Gap Index	The difference in the average income of the poor from the poverty line as a proportion of the poverty line.	BPS
22	Population Density	The ratio of the population size to the total area.	BPS
IV. INFRASTRUCTURE, NATURAL RESOURCES AND ENVIRONMENT			
	Transportation and Communications Conditions		
23	Value-added of the transportation sector per capita	The value added of the transport sector divided by the total population.	BPS
24	Road conditions according to road quality	Population perceptions of road conditions in the regency/municipality.	BPS
25	The number of telephone connections per capita	The number of households in the area that owns a telephone cable.	BPS
	Energy Condition		
26	Electricity consumption of the industry and households per capita	The average electricity consumption of households and industries in a regency/municipality divided by the population.	BPS

No	Variable	Descriptions	Data Sources
27	Electricity production per capita	The value added of the electricity subsector divided by the total population.	BPS
28	Fuel consumption of the industry and households per capita	The average household fuel consumption plus industrial fuel consumption divided by the total population.	BPS
	Natural Resources and Environment		
29	Ratio of productive land to total land area	The area of productive land divided by the total land area.	BPS
30	Water resources per capita	Percentage of households with access to clean water compared to the total amount of households.	BPS
31	Value-added of the mining and quarrying sector per capita	The value added of the mining and quarrying sector divided by the total population.	BPS
V. BANKING AND FINANCIAL INSTITUTIONS			
	Banking and Non-Banking Infrastructure (Cooperatives)		
32	The number of bank offices	The number of bank offices in a regency/municipality	BI
33	Ratio of business volume value to the number of active cooperatives	The ratio of business volume value to the number of active cooperatives in the regency/municipality	Ministry of cooperatives and small and medium enterprises
	Banking and Financial Sector Performance		
34	Total Banking Credits	Total value of bank credit channeled in a regency/municipality	Bank Indonesia
35	Third Party Funds	Funds collected by banks that come from the community.	Bank Indonesia
36	NPL (Non-Performing Loan)	Measuring the level of bank risk in channeling funds in a regency/municipality	Bank Indonesia
37	Value-added Financial Sector per capita	The value added of the financial sector divided by the total population.	BPS
OUTPUT			
1	Labor Productivity	GRDP divided by the amount of labor.	BPS
2	GRDP per capita	GRDP divided by the total population	BPS

No	Variable	Descriptions	Data Sources
3	Level of employment opportunities	Measuring the quality of the labor market in a regency/municipality	BPS

4. Analysis and discussion

4.1. The analytical framework and calculation stages of regional competitiveness in Banten Province
After collecting all the data required for this study, the next step was to identify the data that had a negative effect on the competitiveness of the region and change it in the formula. This is done so all data has a positive correlation with competitiveness. For example, without transformation, the more problematic regional regulations (*Perda*) the smaller the competitiveness of the regency/municipality, but after the transformation its correlation becomes positive. In other words, the 'statement' was changed to the fewer problematic regional Regulations, the higher the competitiveness of the regency/municipality.

The main results of this study are the competitiveness ratings among regions and the Regional Competitiveness Balance for each regency/municipality in Banten. There are two calculations necessary to rank and balance the competitiveness, i.e., variables that can show and indicate the level of competitiveness of regions and the scores that can transform the variables or indicators of competitiveness into a ranking of competitiveness of the region. Determining, screening, and selecting the competitiveness variables was done through literature tracking and discussions. Meanwhile, the weight is generated by applying the analysis network process (ANP) method through the dispersion of competitiveness questionnaires to the relevant service offices.

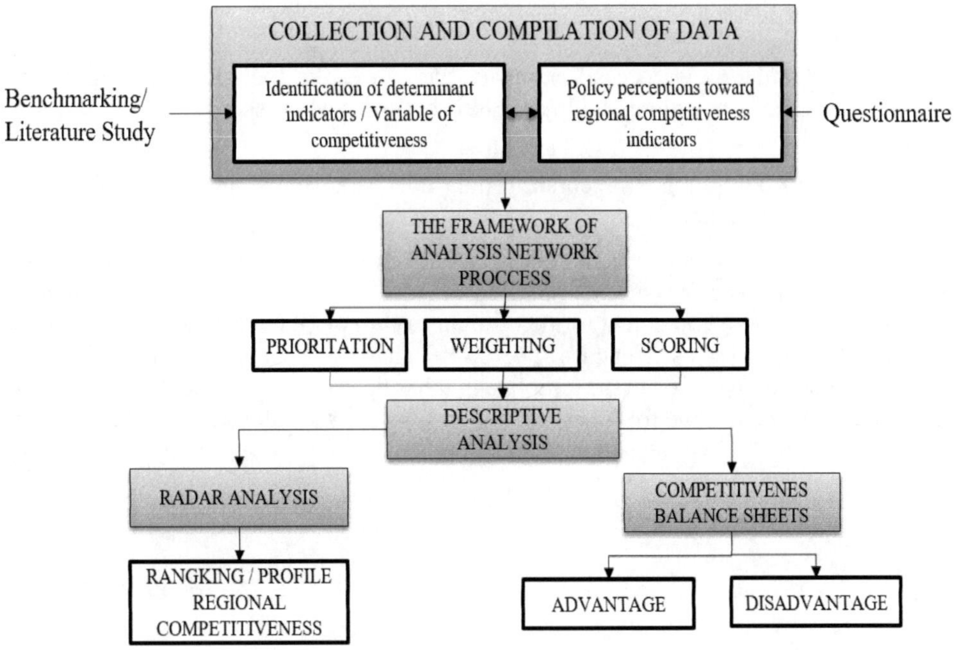

Figure 3. The analytical framework of regional competitiveness.

In this study, the variables that form the competitiveness of the regency/municipality based on the theoretical framework can be classified into groups of input and output variables. The input variables can be defined as a "flow concept" and are a resource to rise to a higher rank in the competitiveness ranking. Meanwhile, the output variable can be interpreted as "stock concept" and reflects the position of competitiveness of a regency/municipality.

The scoring mechanism - commonly referred to as the scoring process - comprises several stages. The respective stages of scoring calculation can be explained as follows.

1) All data used in this study should be transformed to be "comparable".

   The first step in calculating the competitiveness score to transform the data into the normal standard. The method used in this study is to use the normal standard (z-score) as the basis of normalization so that each variable will have an average value of 0 and a standard deviation of 1. The use of normalization methods by using the z-score method is also expected to facilitate readers in understanding the variables, provided that the value of a positive variable (+) means the value is above the average value of all variables that exist throughout Banten.

2) The competitiveness score for a regency/municipality is calculated by counting the weighted sum of all variables, which has been transformed into a standard z-score standard using the weight obtained from the ANP workshop. The weighted numbers of normal high standard indicate a higher level of competitiveness. Increasing the competitiveness among regencies/municipalities is done by comparing the weighted-sum score of this variable for all regencies/municipalities.

3) In addition to competitiveness ratings among regencies/municipalities throughout Banten, this study also discusses the competitiveness ratings for each indicator or sub-indicator of each regency/municipality. The calculation of scores at the indicator and sub-indicators level is, in principle, the same as the calculation of scores at the regency/municipality level. The only difference is that this calculation involves fewer variables. In other words, the score at the indicator level is a 'subset' score at the regency/municipality level and the score at the sub-indicator level is the 'subset' score at the indicator level.

In more detail, the calculation stage of regional competitiveness is as follows:

1) As mentioned in the previous section, this research uses the analytical network process (ANP) method. The data is calculated and processed by using Microsoft Excel and Super Decisions version 2.8.

2) The collect and compiling of secondary data consisting of 37 input variables and 3 output variables as shown in table 1 above.

3) Conducting survey interviews using questionnaires to determine the weight of competitiveness of each variable. The survey was conducted with experts from the representatives of the department who were competent to increase the competitiveness of the regions. There were 13 respondents from various agencies. The scores that were obtained were put together using Microsoft Excel and entered in the template that has been created on the dashed box, as in Figure 4 below. The same is done for other respondents, so we will have 13 sheets 1 through 13.

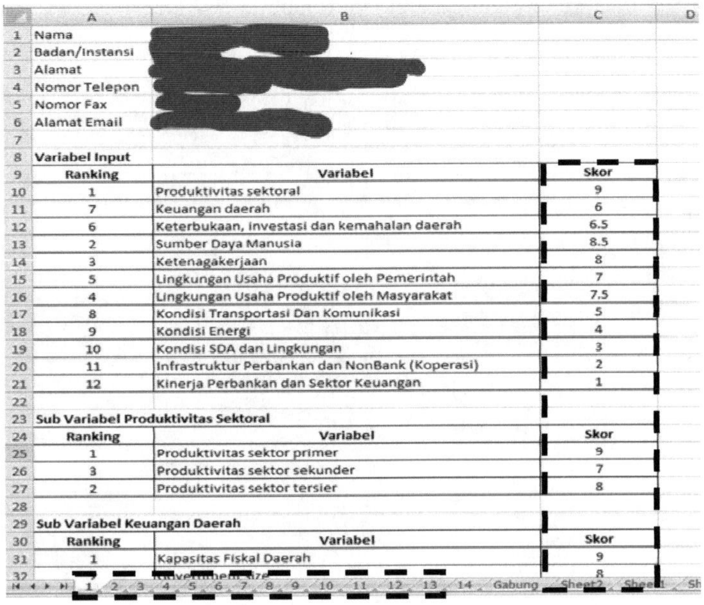

Ranking	Variabel	Skor
Nama	[redacted]	
Badan/Instansi		
Alamat		
Nomor Telepon		
Nomor Fax		
Alamat Email		
Variabel Input		
Ranking	**Variabel**	**Skor**
1	Produktivitas sektoral	9
7	Keuangan daerah	6
6	Keterbukaan, investasi dan kemahalan daerah	6.5
2	Sumber Daya Manusia	8.5
3	Ketenagakerjaan	8
5	Lingkungan Usaha Produktif oleh Pemerintah	7
4	Lingkungan Usaha Produktif oleh Masyarakat	7.5
8	Kondisi Transportasi Dan Komunikasi	5
9	Kondisi Energi	4
10	Kondisi SDA dan Lingkungan	3
11	Infrastruktur Perbankan dan NonBank (Koperasi)	2
12	Kinerja Perbankan dan Sektor Keuangan	1
Sub Variabel Produktivitas Sektoral		
Ranking	**Variabel**	**Skor**
1	Produktivitas sektor primer	9
3	Produktivitas sektor sekunder	7
2	Produktivitas sektor tersier	8
Sub Variabel Keuangan Daerah		
Ranking	**Variabel**	**Skor**
1	Kapasitas Fiskal Daerah	9

Figure 4. The results of questionnaires from several experts for the weighting of variables of regional competitiveness.

4) The next step is to get normalized and idealized values using super decision software. To accommodate the thirteen respondents and to have a super decision file recap for them, a template file was made with the file type 01.sdmod up to 13.sdmod, as can be seen in Figure 5 below.

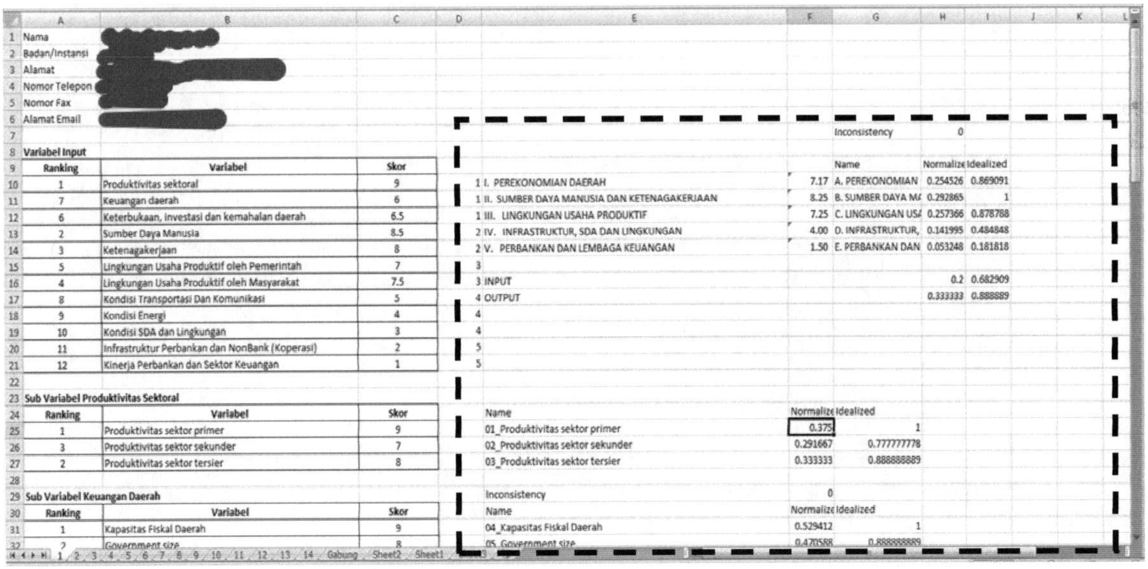

		Inconsistency	0	
		Name	Normalize	Idealized
1 I. PEREKONOMIAN DAERAH	7.17	A. PEREKONOMIAN	0.254526	0.869091
1 II. SUMBER DAYA MANUSIA DAN KETENAGAKERJAAN	8.25	B. SUMBER DAYA MA	0.292865	1
1 III. LINGKUNGAN USAHA PRODUKTIF	7.25	C. LINGKUNGAN USA	0.257366	0.878788
2 IV. INFRASTRUKTUR, SDA DAN LINGKUNGAN	4.00	D. INFRASTRUKTUR,	0.141995	0.484848
2 V. PERBANKAN DAN LEMBAGA KEUANGAN	1.50	E. PERBANKAN DAN	0.053248	0.181818
3				
3 INPUT			0.2	0.682909
4 OUTPUT			0.333333	0.888889
4				
4				
5				
5				

Name	Normalize	Idealized
01_Produktivitas sektor primer	0.375	1
02_Produktivitas sektor sekunder	0.291667	0.777777778
03_Produktivitas sektor tersier	0.333333	0.888888889
Inconsistency	0	
Name	Normalize	Idealized
04_Kapasitas Fiskal Daerah	0.529412	1
05_Government size	0.470588	0.888888889

Figure 5. Normalized and idealized value by using super decision software.

5) Step 5 is to create the framework of the analytical network process (ANP) using predetermined competitiveness variables. Then the file type "s.mod" is entered into the software super decision for all 13 files (respondents). The output of the data processing consists of normalized and idealized values for each variable of regional competitiveness.

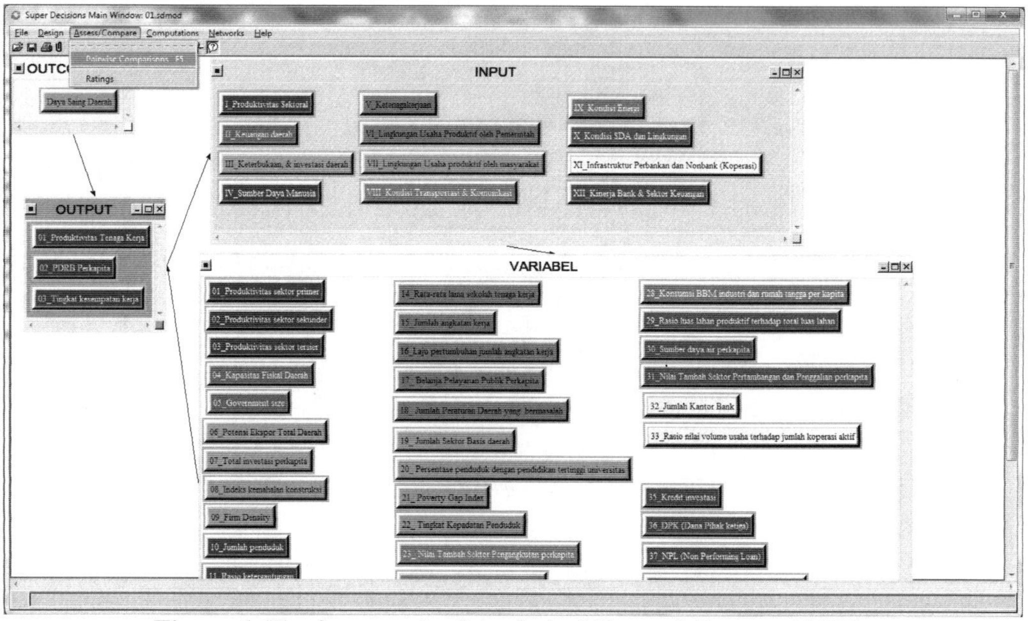

Figure 6. The framework of Analytical Network Process (ANP).

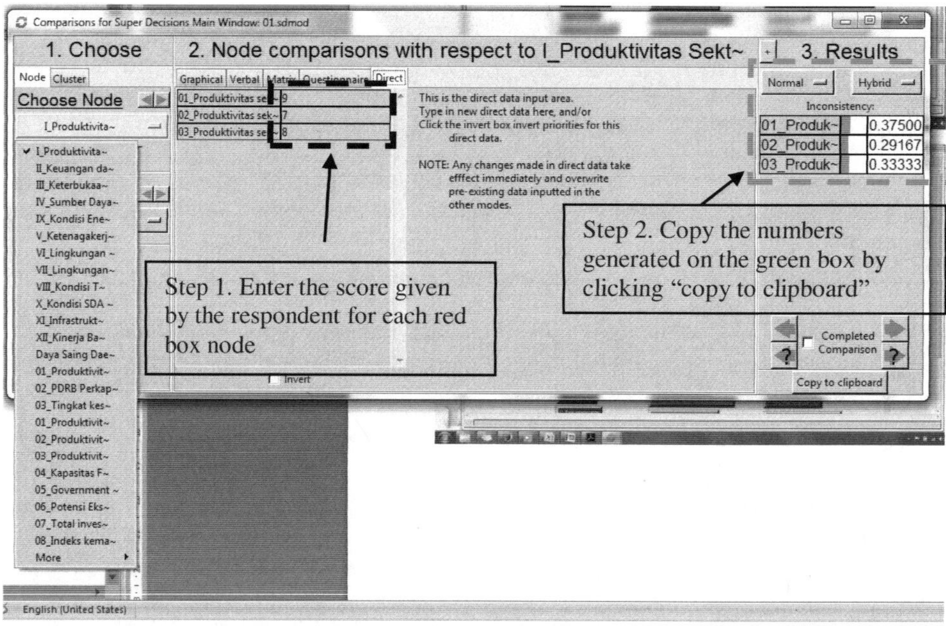

Figure 7. Data processing to determine normalized and idealized data for weighting each variable of regional competitiveness.

6) The recap results can be seen on the joined sheet. The red box shows the average value of all respondents. This value is used as the weight that will be multiplied by the variable value of each region. The processed data is then used to rank the competitiveness of regions and to see the various variables that are advantages and disadvantages of the competitiveness of a region. Both analyses will be discussed in the next section.

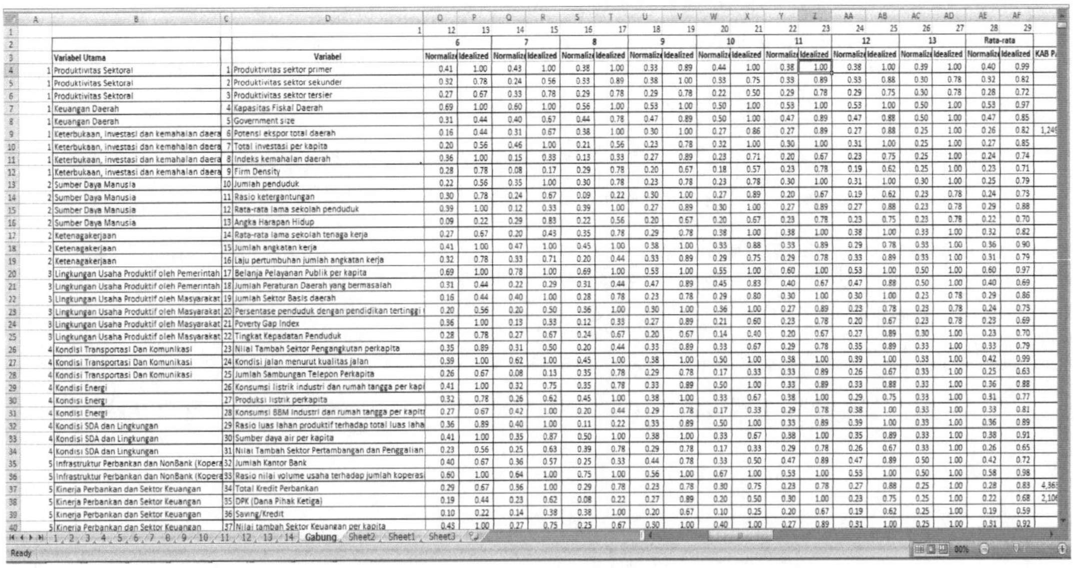

	Variabel Utama		Variabel	6 Normalized	6 Idealized	7 Normalized	7 Idealized	8 Normalized	8 Idealized	9 Normalized	9 Idealized	10 Normalized	10 Idealized	11 Normalized	11 Idealized	12 Normalized	12 Idealized	13 Normalized	13 Idealized	Rata-rata Normalized	Rata-rata Idealized	KAB P…
1	Produktivitas Sektoral	1	Produktivitas sektor primer	0.41	1.00	0.43	1.00	0.38	1.00	0.33	0.89	0.44	1.00	0.38	1.00	0.38	1.00	0.39	1.00	0.40	0.99	
1	Produktivitas Sektoral	2	Produktivitas sektor sekunder	0.32	0.78	0.24	0.56	0.33	0.89	0.38	1.00	0.33	0.75	0.33	0.89	0.33	0.88	0.30	0.78	0.32	0.82	
1	Produktivitas Sektoral	3	Produktivitas sektor tersier	0.27	0.67	0.33	0.78	0.29	0.78	0.29	0.78	0.22	0.50	0.29	0.78	0.29	0.75	0.30	0.78	0.28	0.72	
1	Keuangan Daerah	4	Kapasitas Fiskal Daerah	0.69	1.00	0.60	1.00	0.56	1.00	0.53	1.00	0.50	1.00	0.53	1.00	0.53	1.00	0.50	1.00	0.53	0.97	
1	Keuangan Daerah	5	Government size	0.31	0.44	0.40	0.67	0.44	0.78	0.47	0.89	0.50	1.00	0.47	0.89	0.47	0.88	0.50	1.00	0.47	0.85	
1	Keterbukaan, investasi dan kemahalan daera	6	Potensi ekspor total daerah	0.16	0.44	0.31	0.67	0.38	1.00	0.30	1.00	0.27	0.86	0.27	0.89	0.27	0.88	0.25	1.00	0.26	0.82	1,24
1	Keterbukaan, investasi dan kemahalan daera	7	Total investasi per kapita	0.20	0.56	0.46	1.00	0.21	0.56	0.23	0.78	0.32	1.00	0.30	1.00	0.31	1.00	0.25	1.00	0.27	0.85	
1	Keterbukaan, investasi dan kemahalan daera	8	Indeks kemahalan daerah	0.36	1.00	0.15	0.33	0.13	0.33	0.27	0.89	0.23	0.71	0.20	0.67	0.23	0.75	0.25	1.00	0.24	0.74	
1	Keterbukaan, investasi dan kemahalan daera	9	Firm Density	0.28	0.78	0.08	0.17	0.29	0.78	0.20	0.67	0.18	0.57	0.23	0.78	0.19	0.62	0.25	1.00	0.23	0.71	
2	Sumber Daya Manusia	10	Jumlah penduduk	0.22	0.56	0.35	1.00	0.30	0.78	0.23	0.78	0.23	0.78	0.30	1.00	0.31	1.00	0.30	1.00	0.25	0.79	
2	Sumber Daya Manusia	11	Rasio ketergantungan	0.30	0.78	0.24	0.67	0.09	0.22	0.30	1.00	0.27	0.89	0.20	0.67	0.19	0.62	0.23	0.78	0.24	0.73	
2	Sumber Daya Manusia	12	Rata-rata lama sekolah penduduk	0.39	1.00	0.12	0.33	0.39	1.00	0.27	0.89	0.30	1.00	0.27	0.89	0.27	0.88	0.23	0.78	0.29	0.88	
2	Sumber Daya Manusia	13	Angka Harapan Hidup	0.09	0.22	0.29	0.83	0.22	0.56	0.20	0.67	0.20	0.67	0.23	0.78	0.23	0.75	0.23	0.78	0.22	0.70	
2	Ketenagakerjaan	14	Rata-rata lama sekolah tenaga kerja	0.27	0.67	0.20	0.43	0.35	0.78	0.29	0.78	0.38	1.00	0.38	1.00	0.38	1.00	0.33	1.00	0.32	0.82	
2	Ketenagakerjaan	15	Jumlah angkatan kerja	0.41	1.00	0.47	1.00	0.45	1.00	0.38	1.00	0.33	0.88	0.33	0.89	0.29	0.78	0.33	1.00	0.36	0.90	
2	Ketenagakerjaan	16	Laju pertumbuhan jumlah angkatan kerja	0.32	0.78	0.33	0.71	0.20	0.44	0.33	0.89	0.29	0.75	0.29	0.78	0.33	0.89	0.33	1.00	0.31	0.79	
3	Lingkungan Usaha Produktif oleh Pemerintah	17	Belanja Pelayanan Publik per kapita	0.69	1.00	0.78	1.00	0.69	1.00	0.53	1.00	0.55	1.00	0.60	1.00	0.53	1.00	0.50	1.00	0.60	0.97	
3	Lingkungan Usaha Produktif oleh Pemerintah	18	Jumlah Peraturan Daerah yang bermasalah	0.31	0.44	0.22	0.29	0.31	0.44	0.47	0.89	0.45	0.83	0.40	0.67	0.47	0.88	0.50	1.00	0.40	0.69	
3	Lingkungan Usaha Produktif oleh Masyarakat	19	Jumlah Sektor Basis daerah	0.16	0.44	0.40	1.00	0.28	0.78	0.23	0.78	0.29	0.80	0.30	1.00	0.30	1.00	0.23	0.78	0.29	0.86	
3	Lingkungan Usaha Produktif oleh Masyarakat	20	Persentase penduduk dengan pendidikan tertinggi	0.20	0.56	0.20	0.50	0.36	1.00	0.30	1.00	0.36	1.00	0.27	0.89	0.23	0.78	0.23	0.78	0.24	0.75	
3	Lingkungan Usaha Produktif oleh Masyarakat	21	Poverty Gap Index	0.36	1.00	0.13	0.33	0.12	0.33	0.27	0.89	0.21	0.60	0.23	0.78	0.20	0.67	0.23	0.78	0.23	0.69	
3	Lingkungan Usaha Produktif oleh Masyarakat	22	Tingkat Kepadatan Penduduk	0.28	0.78	0.27	0.67	0.24	0.67	0.20	0.67	0.14	0.40	0.20	0.67	0.27	0.89	0.30	1.00	0.23	0.70	
4	Kondisi Transportasi Dan Komunikasi	23	Nilai Tambah Sektor Pengangkutan perkapita	0.35	0.89	0.31	0.50	0.20	0.44	0.33	0.89	0.33	0.67	0.29	0.78	0.35	0.89	0.33	1.00	0.33	0.79	
4	Kondisi Transportasi Dan Komunikasi	24	Kondisi jalan menurut kualitas jalan	0.39	1.00	0.62	1.00	0.45	1.00	0.38	1.00	0.50	1.00	0.38	1.00	0.39	1.00	0.33	1.00	0.42	0.99	
4	Kondisi Transportasi Dan Komunikasi	25	Jumlah Sambungan Telepon Perkapita	0.26	0.67	0.08	0.13	0.35	0.78	0.29	0.78	0.17	0.33	0.33	0.89	0.26	0.67	0.33	1.00	0.25	0.63	
4	Kondisi Energi	26	Konsumsi listrik industri dan rumah tangga per kapi	0.41	1.00	0.32	0.75	0.35	0.78	0.33	0.89	0.50	1.00	0.33	0.89	0.33	0.88	0.33	1.00	0.36	0.88	
4	Kondisi Energi	27	Produksi listrik perkapita	0.32	0.78	0.26	0.62	0.45	1.00	0.38	1.00	0.33	0.67	0.38	1.00	0.29	0.75	0.33	1.00	0.31	0.77	
4	Kondisi Energi	28	Konsumsi BBM Industri dan rumah tangga per kapit	0.27	0.67	0.42	1.00	0.20	0.44	0.29	0.78	0.17	0.33	0.29	0.78	0.38	1.00	0.33	1.00	0.33	0.81	
4	Kondisi SDA dan Lingkungan	29	Rasio luas lahan produktif terhadap total luas laha	0.36	0.89	0.40	1.00	0.11	0.22	0.33	0.89	0.50	1.00	0.33	0.89	0.39	1.00	0.33	1.00	0.36	0.89	
4	Kondisi SDA dan Lingkungan	30	Sumber daya air per kapita	0.41	1.00	0.35	0.87	0.50	1.00	0.38	1.00	0.33	0.67	0.38	1.00	0.35	0.89	0.33	1.00	0.38	0.91	
4	Kondisi SDA dan Lingkungan	31	Nilai Tambah Sektor Pertambangan dan Penggalian	0.23	0.56	0.25	0.63	0.39	0.78	0.29	0.78	0.17	0.33	0.29	0.78	0.26	0.67	0.33	1.00	0.26	0.65	
5	Infrastruktur Perbankan dan NonBank (Kopera	32	Jumlah Kantor Bank	0.40	0.67	0.36	0.57	0.25	0.33	0.44	0.78	0.33	0.50	0.47	0.89	0.47	0.89	0.50	1.00	0.42	0.72	
5	Infrastruktur Perbankan dan NonBank (Kopera	33	Rasio nilai volume usaha terhadap jumlah koperasi	0.60	1.00	0.64	1.00	0.75	1.00	0.56	1.00	0.67	1.00	0.53	1.00	0.53	1.00	0.50	1.00	0.58	0.98	
5	Kinerja Perbankan dan Sektor Keuangan	34	Total Kredit Perbankan	0.29	0.67	0.36	1.00	0.29	0.78	0.23	0.78	0.30	0.75	0.23	0.78	0.27	0.88	0.25	1.00	0.28	0.83	4,36
5	Kinerja Perbankan dan Sektor Keuangan	35	DPK (Dana Pihak Ketiga)	0.19	0.44	0.23	0.62	0.08	0.22	0.27	0.89	0.20	0.50	0.30	1.00	0.23	0.75	0.25	1.00	0.22	0.68	2,10
5	Kinerja Perbankan dan Sektor Keuangan	36	Saving/Kredit	0.10	0.22	0.14	0.38	0.38	1.00	0.20	0.67	0.10	0.25	0.20	0.67	0.19	0.62	0.25	1.00	0.19	0.59	
5	Kinerja Perbankan dan Sektor Keuangan	37	Nilai tambah Sektor Keuangan per kapita	0.45	1.00	0.27	0.75	0.25	0.67	0.30	1.00	0.40	1.00	0.27	0.89	0.31	1.00	0.25	1.00	0.31	0.92	

Figure 8. The result of data processing using the ANP method.

4.2. Analyzing regional competitiveness based on input and output variables

This study classifies the variables forming the competitiveness of the regency/municipality based on the developed theoretical framework into groups of input and output variables. The input variables can be defined as "flow concept" and are a resource to rise to a higher competitiveness rank. Meanwhile, the output variable can be interpreted as "stock concept" and reflects the position of competitiveness of a regency/municipality.

Both variables are positively correlated. If it happens that a regency/municipality has a high input but low output, this can be interpreted as 'inefficiency'. On the other hand, if a regency/municipality has a low input position but high output, then other external factors have positively increased the competitiveness of the area. In extreme conditions, this can be interpreted as "over-capacity", which in the long term can reduce the competitiveness of the region.

4.2.1. Analyzing regional competitiveness based on input variables. In the context of the input factors (see Figure 9 below), each region has different competitive determinants that accumulatively determine the different levels of competitiveness in every regency/municipality in Banten.

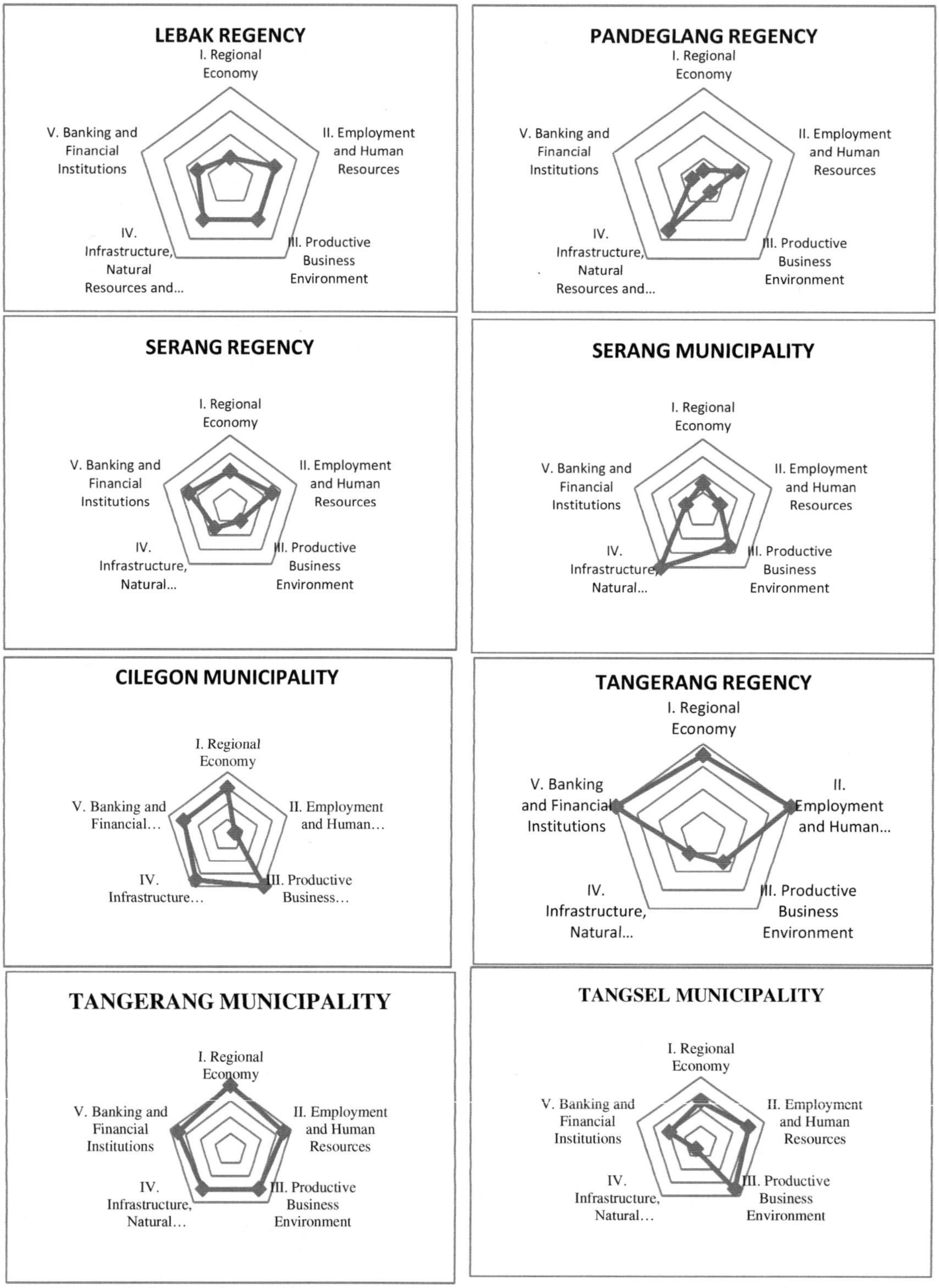

Figure 9. Comparison of input competitiveness factor of regency/municipality in Banten.

Regional economy indicators. Regional economic indicators consisting of several sub-indicators such as sectoral productivities, local fiscal capacity, government size, total investment, and firm density contribute in determining the level of regional competitiveness. This indicator shows that the regencies of Lebak, Pandeglang, Serang and Serang Municipality have relatively low competitiveness compared to other regencies/municipalities. This happens for several reasons. First, regencies/municipalities with low competitiveness in the aspect of the regional economy only rely on primary sectoral productivity that lacks the value added of economic activity, while regencies/municipalities that rely on secondary and tertiary productivities have relatively high competitiveness. Furthermore, areas with high per capita investment and firm density such as Tangerang Regency, Tangerang Municipality, Cilegon Municipality and Tangsel Municipality contribute significantly to the regional economy, which, in turn, can improve the competitiveness of these areas. The investment per capita and firm density also affect local fiscal capacity and government size so these governments have a large budget to develop their regions, inversely with regions of low fiscal capacity and small government size.

Table 2. Sectoral Productivity of the Regions in Banten.

REGIONS	PRIMARY PRODUCTIVITY (RP)	SECONDARY PRODUCTIVITY (Rp)	TERTIARY PRODUCTIVITY (Rp)
PANDEGLANG REGENCY	34.603.219	61.127.299	34.096.593
LEBAK REGENCY	23.209.142	57.747.371	39.438.115
TANGERANG REGENCY	74.976.640	78.347.223	39.735.678
SERANG REGENCY	40.441.011	171.065.531	49.796.226
TANGERANG MUNICIPAL	182.392.378	159.742.629	74.669.295
CILEGON MUNICIPALITY	67.325.194	1.142.768.012	138.391.943
SERANG MUNICIPALITY	46.725.377	77.124.432	69.514.177
TANGSEL MUNICIPALIY	29.937.432	197.369.109	59.122.459
BANTEN PROVINCE	37.409.031	146.084.065	57.390.006

*Human resource and employment indicators.*The human resource and employment indicators consist of several sub-indicators such as dependency ratio, life expectancy index, mean years of schooling, mean years of labor schooling, and growth rate of the labor force. As can be seen in Figure 9 above, this indicator shows that the regencies of Lebak, Pandeglang, Serang as well as Serang Municipality have relatively low competitiveness compared to the other regencies/municipalities. This is caused by a low quality of human resources and labor. Figure 10 explains the mean years of labor schooling in these areas is very low; in Lebak Regency only 5.86 years. Whereas one of the determinants of regional competitiveness is the availability of human resources and labor with a high ability and extensive knowledge, mastering IT, have creativity and innovations so that the resulting product has competitiveness. Furthermore, the growth rate of the labor force in regions with a low competitiveness is low; the labor force growth rate in Serang Regency and Pandeglang Regency is even negative (-2.20 and-4.31). Likewise, the low life expectancy index correlates with other human resource and employment data. Meanwhile, Tangerang Municipality, Tangerang Regency, Tangsel Municipality and Cilegon Municipality have high enough human resource quality and workforce to contribute to improving the competitiveness of its region.

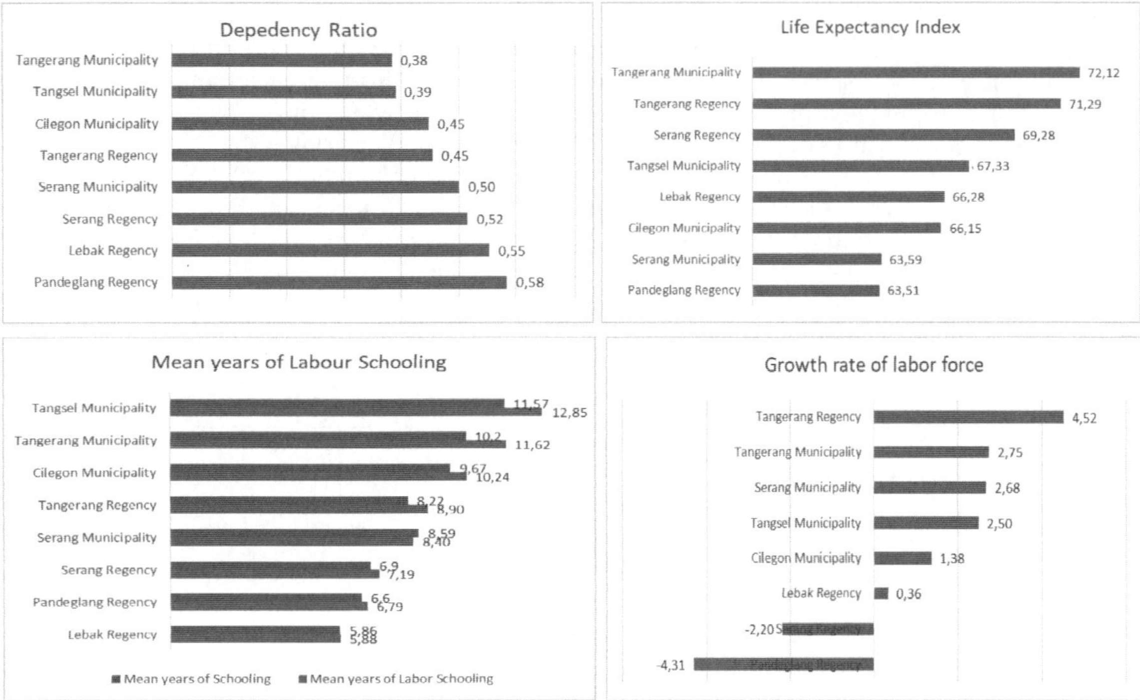

Figure 10. Human resource and employment variables for determining regional competitiveness.

Productive business environment indicator. An important indicator for determining regional competitiveness is a productive business environment. This indicator is divided into two sub-indicators. First, the government's productive business environment indicators consist of Public Service Expenditure Per Capita and the number of problematic Regional Regulations. Second, the productive business environment indicator by the community consists of the number of base sectors in the regency/municipality, the percentage of the population with university as the highest education, the poverty gap index, and population density.

As can be seen in Figure 9, the productive business environment of Pandeglang, Lebak, and Serang is very low. There are several things that cause a low productivity business environment. First, the Public Service Expenditure Per Capita in these three areas is low. This means that spending on public services is still low compared with the population or the needs of the community. As shown in Figure 6, the Public Service Expenditure Per Capita in Pandeglang Regency and Serang Regency only amounts to 600 thousand rupiah. Secondly, Serang Municipality, Lebak Regency, and Pandeglang Regency have a relatively large poverty gap index of about 1.5 times greater than the other regencies/municipalities (see figure 6). Third, the percentage of the population with university as the highest education in these areas is still low. This leads to a lack of research and development in the region as well as low human resources that ultimately hinder regional competitiveness. On the other hand, Tangerang Regency, Tangerang Municipality, Cilegon Municipality and Tangsel Municipality have relatively productive business environment indicators that are conducive to building their regional competitiveness (see Figure 11).

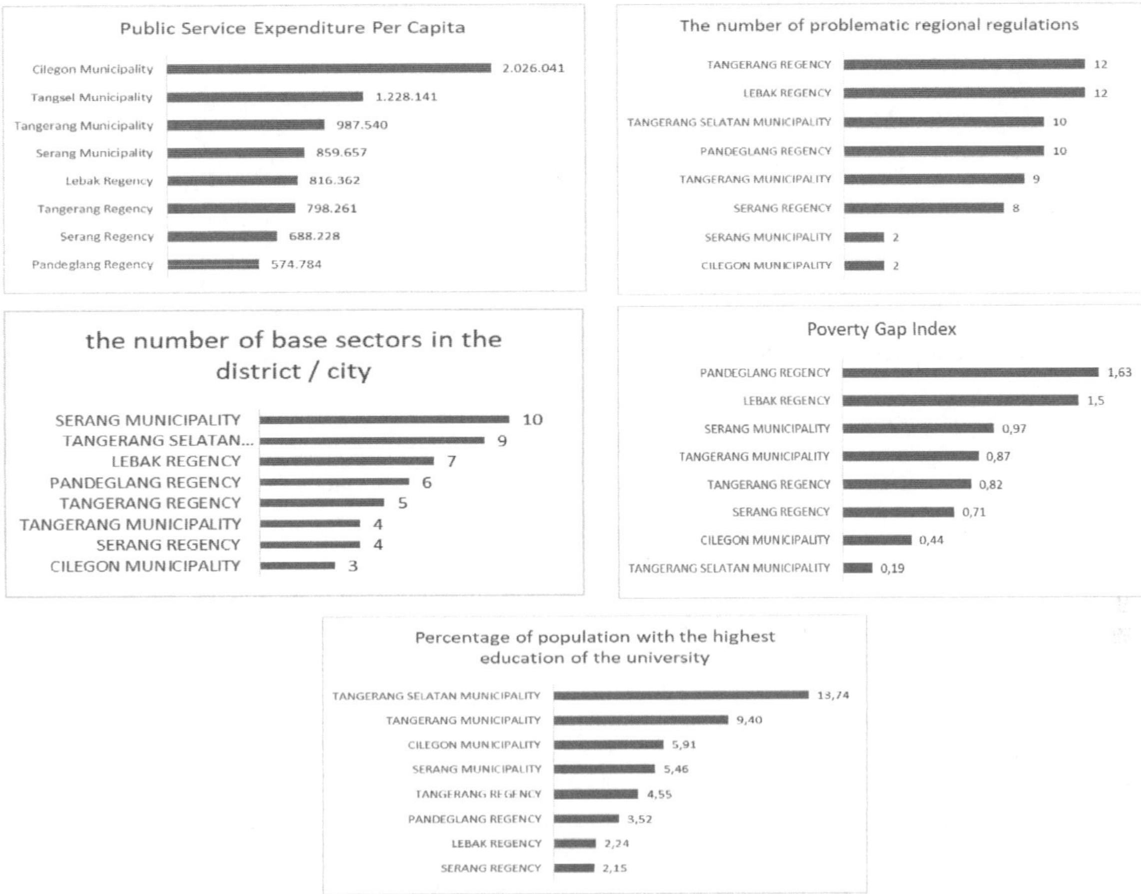

Figure 11. Productive Business Environment variable for determining regional competitiveness.

Infrastructure, natural resources and environment indicator. The factors of availability of infrastructure, natural resources, and environment also determine the level of competitiveness of a region. Some of the infrastructure, natural resources, and environment that can form the competitiveness of the region include the value added of the transportation sector per capita; road conditions according to road quality; electricity production per capita; fuel consumption of industry and household per capita; the ratio of productive land to total land area; and value-added mining and quarrying sector per capita. Based on figure 9 above, cumulatively there are only three regions with relatively good input indicators, namely Cilegon Municipality, Tangerang Municipality and Serang Municipality. This means that the availability of adequate infrastructure, natural resources, and environment can create regional competitiveness. As illustrated in figure 12 below, Cilegon Municipality and Pandeglang Regency have good and adequate road conditions that support regional development. Thus, the focus of development is no longer on road infrastructure provision but can be focused on other sectors. Meanwhile, Lebak Regency and Tangerang Regency have about 30 percent good road conditions. Therefore, it is necessary to focus on the development and improvement of road infrastructure because the road is a growth engine that has high leverage in supporting economic activities.

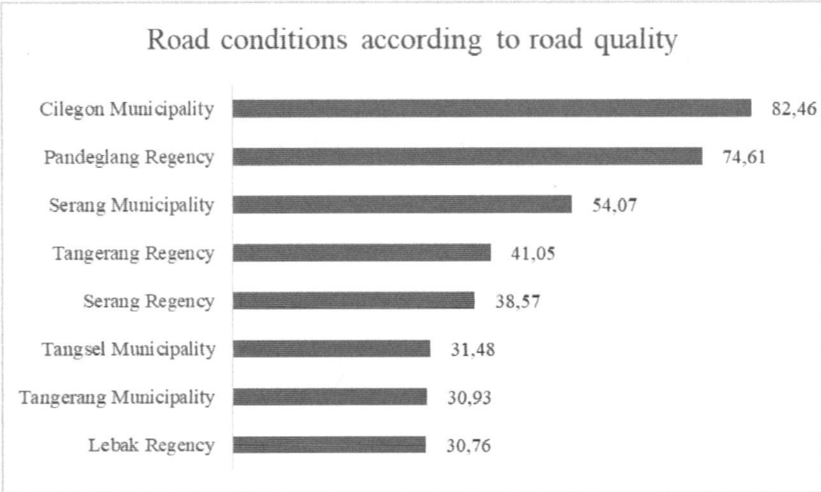

Figure 12. Road conditions according to road quality.

Banking and financial institution indicator. Other important regional competitiveness factors are the performance of banks and other financial institutions. The variables that can represent the competitiveness of the region include the number of bank offices; the ratio of business volume value to the number of active cooperatives; Third Party Funds; NPL (Non-Performing Loans); and value added of the financial sector per capita. Figure 9 illustrates that only three regions have good banking performance, namely Tangerang Municipality, Tangerang Regency and Cilegon Municipality. In fact, these three areas have large industrial estates. Based on banking indicators, Tangerang Regency and Tangerang Municipality have a very large number of banks, with 294 and 426 banks respectively. The three regions also have a value added per capita of the financial sector that is relatively high compared to other regions; Cilegon Municipality about 3.39 million rupiah, Tangerang Municipality about 1.17 million rupiah and Tangerang Regency about 1.04 million rupiah. Furthermore, these three areas are also supported by an abundance of third-party funds; the Municipality of Tangerang about 51.99 trillion rupiah, the Municipality of Tangerang about 34.54 trillion rupiah, and the Municipality of Tangsel 22.81 trillion rupiah. These variables contribute to regional competitiveness to be followed also by areas with low competitiveness.

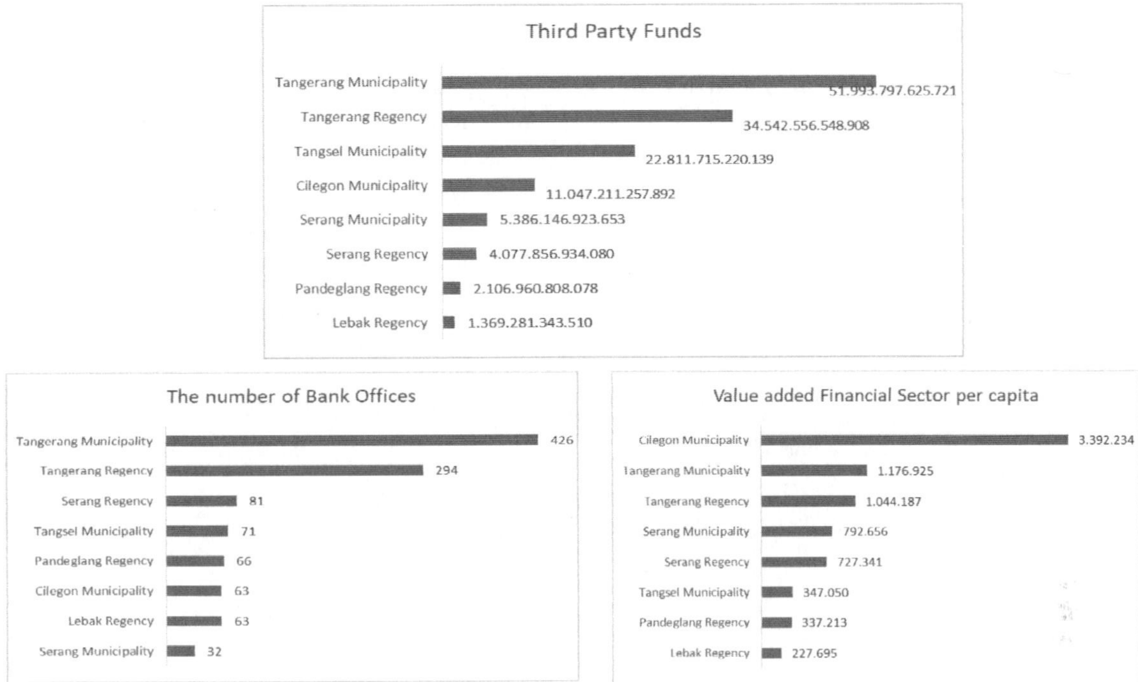

Figure 13. Banking and financial institution variables.

4.2.2. Analyzing Regional Competitiveness based on Output Variables

After analyzing the regional competitiveness based on the input variables, this study then examines the output of the variables as a result of input activity variables. Output variables embody high or low competitiveness of a region. The variables used to measure regional competitiveness are labor productivity, GRDP per capita, and the level of employment opportunities.

Figure 9 shows that Tangerang Municipality and some areas in the northern region of Banten have relatively high output factor. The GRDP per capita Cilegon is the highest in Banten, about 146 million rupiah, followed by Tangerang Municipality with 44 million rupiah, while the lowest is Lebak with 13 million rupiah. Furthermore, Cilegon also has the highest labor productivity in Banten, amounting to 366 million rupiah, while the smallest is Lebak Regency with 33 million rupiah. Furthermore, the highest level of employment is found in Tangsel Municipality with 93.87 percent, while the lowest is Serang Regency at 85.2%. The regions ranked based on the highest output variables to the lowest are; Tangerang Municipality, Tangerang Regency, Tangsel Municipality, Cilegon Municipality, Serang Municipality, Serang Regency Pandeglang Regency, and Lebak Regency. Therefore, Tangerang Municipality is a competitive region among other regions, while Lebak Regency has the region's lowest regional competitiveness.

The northern region of Banten greatly benefited because it has a very strategic location. The northern part of Banten is part of the economic corridor of Sumatra-Java along with some areas of the megapolitan Jabodetabek area such as Tangsel, Tangerang, and Tangerang Municipality. Many large-scale economic activities exist in the northern region ranging from commercial areas to large-scale industrial areas. Meanwhile, the southern part of Banten, e.g., Pandeglang Regency and Lebak Regency is left behind, which causes a very high gap. However, this study did not emphasize the importance of regional ratings. The emphasis of the study is how a region is able to identify the advantages and disadvantages of the determinants of each region's competitiveness. Therefore, the next section examines the advantages and disadvantages of the regions against the regional competitiveness variables.

IOP Conf. Series: Earth and Environmental Science **158** (2018) 012050 doi:10.1088/1755-1315/158/1/012050

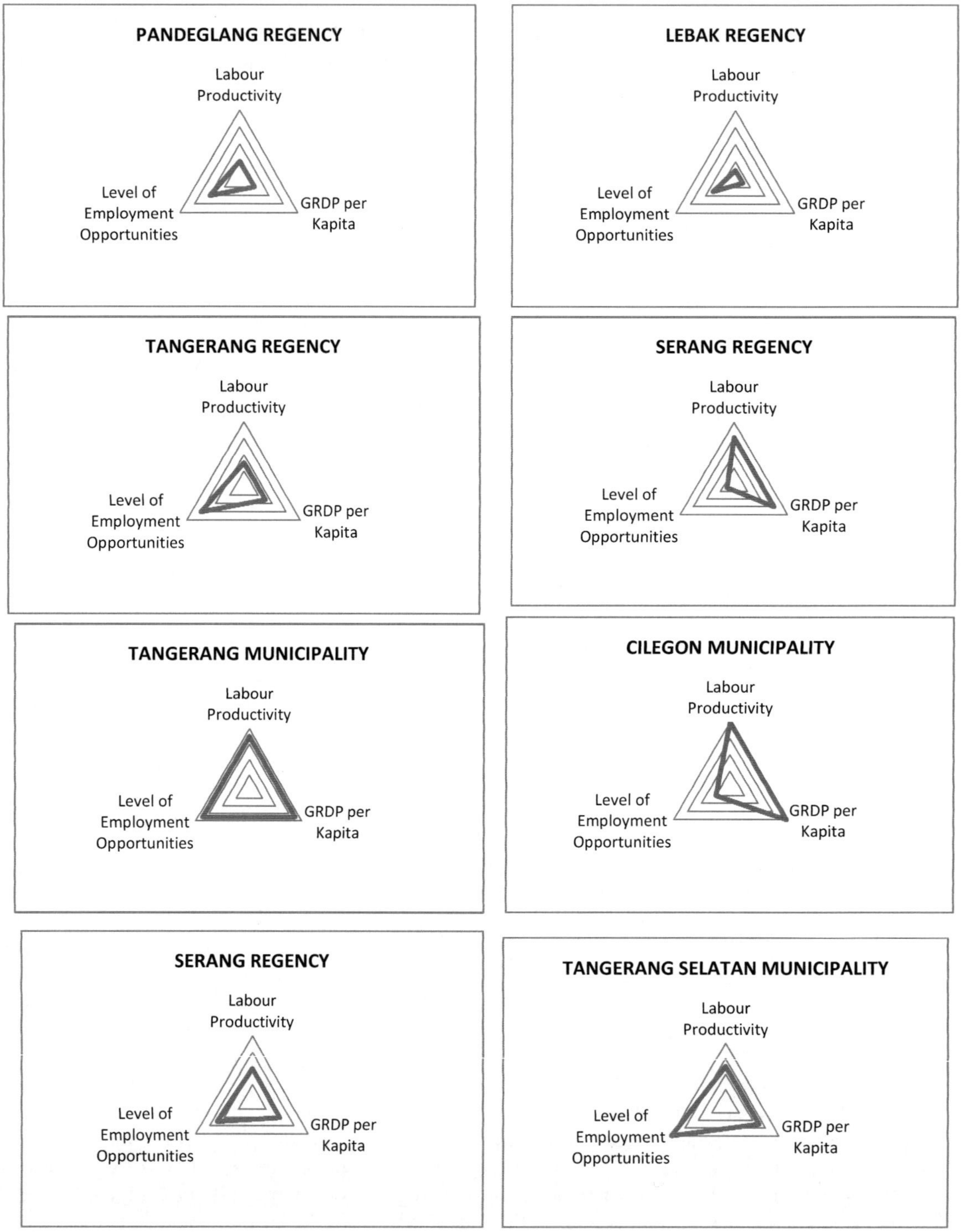

Figure 14. Comparison of regency/municipality output variables in Banten Province.

4.3. Regional competitiveness balance sheet

Making the Regional Competitiveness Balance requires a criterion in the form of numbers to filter the variables that are advantages for a region and the variables that are disadvantages. The basis for calculating the Competitiveness Balance is a normalized variable (semi-normalized) in a regency/municipality. Then the variables will be grouped into two major groups, indicating the advantage of the area if the number of variables is greater than the average - indicated by a positive z-score value. Meanwhile, the other variables will indicate a disadvantage for a regency/municipality if the rank number of those variables is smaller than the normal standard value (z-score); the variables that have a negative sign are deducted 0.5 times the standard deviation.

PANDEGLANG REGENCY	
Overall Rangking	8
Indicator-based Rangking	
INPUT	8
I. REGIONAL ECONOMY	8
II. EMPLOYMENT AND HUMAN RESOURCES	6
III. PRODUCTIVE BUSINESS ENVIRONMENT	8
IV. INFRASTRUCTURE, NATURAL RESOURCES AND ENVIRONMENT	4
V. BANKING AND FINANCIAL INSTITUTIONS	8
OUTPUT	7
I. LABOUR PRODUCTIVITY	7
II. GRDP PER KAPITA	7
III. LEVEL OF EMPLOYMENT OPPORTUNITIES	5

NERACA DAYA SAING DAERAH

ADVANTASE		
Variabel	Peringkat	Z-Score
INPUT		
Government size	2	1,37
Local Fiscal Capacity	1	1,46
The number of problematic regional regulations	3	0,47
Poverty Gap Index	1	1,52
Road conditions according to road quality	2	1,30
Ratio productive land to total land area	5	0,11
Value Added Mining and Quarrying Sector per capita	1	1,91

DISADVANTASE		
Variabel	Peringkat	Z-Score
INPUT		
Primary Sectoral Productivity	6	-0,54
Secondary Sectoral Productivity	7	-0,50
Tertiary Sectoral Productivity	8	-0,86
Local Fiscal Capacity	8	-0,65
Local Total Export Potential	8	-1,17
Total investment per kapita	8	-0,55
Regional overpriced index	7	-0,86
Firm Density	8	-0,56
Total Population	6	-0,33
Mean years of Schooling	7	-0,94
Life expectancy Index	8	-1,21
Mean years of Labor Schooling	7	-0,90
The Number of workforce	6	-0,42
Growth rate of labor force	8	-1,81
Public Service Expenditure Per Capita	8	-0,92
the number of base sectors	4	0,00
Percentage of population with the highest education of the university	6	-0,60
Population Density	7	-0,76
Added Value of Transportation Sector per capita	7	-0,52
The number of Telephone Connections per capita	8	-0,79
Electricity consumption of industry and household per capita	7	-0,64
Electricity production per capita	4	-0,38
Fuel consumption of industry and household per capita	7	-0,63
Water resources per capita	7	-1,31
The number of Bank Offices	5	-0,50
Ratio of business volume value to the number of active cooperatives	6	-0,50
Total Banking Credits	7	-0,71
Third Party Funds	7	-0,79
NPL (Non Performing Loan)	4	-0,47
Value added Financial Sector per capita	7	-0,65
OUTPUT		
Labour Productivity	7	-0,60
GRDP per kapita	7	-0,63
Level of Employment Opportunities	5	-0,07

LEBAK REGENCY

Overall Rangking	6
Indicator-based Rangking	
INPUT	6
I. REGIONAL ECONOMY	7
II. EMPLOYMENT AND HUMAN RESOURCES	5
III. PRODUCTIVE BUSINESS ENVIRONMENT	5
IV. INFRASTRUCTURE, NATURAL RESOURCES AND ENVIRONMENT	5
V. BANKING AND FINANCIAL INSTITUTIONS	6
OUTPUT	8
I. LABOUR PRODUCTIVITY	8
II. GRDP PER KAPITA	8
III. LEVEL OF EMPLOYMENT OPPORTUNITIES	6

NERACA DAYA SAING DAERAH

ADVANTASE

Variabel	Peringkat	Z-Score
INPUT		
Government size	1	1,72
Local Fiscal Capacity	2	1,02
The number of problematic regional regulations	1	0,96
The number of base sectors	3	0,40
Poverty Gap Index	2	1,25
Ratio productive land to total land area	6	0,07
Value Added Mining and Quarrying Sector per capita	2	1,28

DISADVANTASE

Variabel	Peringkat	Z-Score
INPUT		
Primary Sectoral Productivity	8	-0,76
Secondary Sectoral Productivity	8	-0,50
Tertiary Sectoral Productivity	7	-0,70
Local Fiscal Capacity	3	-0,04
Local Total Export Potential	7	-1,13
Total investment per kapita	3	-0,20
Regional overpriced index	5	-0,68
Firm Density	7	-0,56
Total Population	5	-0,25
Mean years of Schooling	8	-1,32
Life expectancy Index	5	-0,36
Mean years of Labor Schooling	8	-1,28
The Number of workforce	5	-0,25
Growth rate of labor force	6	-0,21
Public Service Expenditure Per Capita	5	-0,39
Percentage of population with the highest education of the university	7	-0,92
Population Density	8	-0,78
Added Value of Transportation Sector per capita	6	-0,52
Road conditions according to road quality	8	-0,84
The number of Telephone Connections per capita	4	-0,61
Electricity consumption of industry and household per capita	6	-0,61
Electricity production per capita	8	-0,42
Fuel consumption of industry and household per capita	8	-0,64
Water resources per capita	8	-1,75
The number of Bank Offices	6	-0,52
Ratio of business volume value to the number of active cooperatives	5	-0,44
Total Banking Credits	5	-0,48
Third Party Funds	8	-0,83
NPL (Non Performing Loan)	8	-0,70
Value added Financial Sector per capita	8	-0,76
OUTPUT		-0,83
Labour Productivity	8	-0,63
GRDP per kapita	8	-0,64
Level of Employment Opportunities	6	-0,27

TANGERANG REGENCY

Overall Rangking	1
Indicator-based Rangking	
INPUT	1
I. REGIONAL ECONOMY	2
II. EMPLOYMENT AND HUMAN RESOURCES	1
III. PRODUCTIVE BUSINESS ENVIRONMENT	6
IV. INFRASTRUCTURE, NATURAL RESOURCES AND ENVIRONMENT	7
V. BANKING AND FINANCIAL INSTITUTIONS	1
OUTPUT	2
I. LABOUR PRODUCTIVITY	6
II. GRDP PER KAPITA	6
III. LEVEL OF EMPLOYMENT OPPORTUNITIES	3

NERACA DAYA SAING DAERAH

ADVANTASE

Variabel	Peringkat	Z-Score
INPUT		
Primary Sectoral Productivity	2	0,24
Total investment per kapita	2	1,08
Regional overpriced index	2	1,07
Total Population	1	2,05
Life expectancy Index	3	0,57
The Number of workforce	1	2,01
Growth rate of labor force	1	1,22
The number of problematic regional regulations	1	0,96
The number of Telephone Connections per capita	3	0,57
Ratio productive land to total land area	4	0,31
Water resources per capita	3	0,68
The number of Bank Offices	2	1,10
Total Banking Credits	1	2,17
Third Party Funds	2	0,97
Value added Financial Sector per capita	3	0,04
OUTPUT		
Level of Employment Opportunities	3	0,40

DISADVANTASE

Variabel	Peringkat	Z-Score
INPUT		
Secondary Sectoral Productivity	5	-0,45
Tertiary Sectoral Productivity	6	-0,69
Local Fiscal Capacity	7	-0,61
Government size	5	-0,35
Total investment per kapita	4	-0,34
Firm Density	4	-0,49
Depedency Ratio	5	-0,33
Mean years of Schooling	5	-0,12
Mean years of Labor Schooling	4	-0,04
Public Service Expenditure Per Capita	6	-0,43
The number of base sectors	5	-0,40
Percentage of population with the highest education of the university	5	-0,34
Poverty Gap Index	5	-0,15
Population Density	3	-0,17
Added Value of Transportation Sector per capita	8	-0,54
Road conditions according to road quality	4	-0,34
Electricity consumption of industry and household per capita	4	-0,14
Electricity production per capita	2	-0,13
Fuel consumption of industry and household per capita	6	-0,41
Value Added Mining and Quarrying Sector per capita	5	-0,55
Ratio of business volume value to the number of active cooperatives	4	-0,29
Total Banking Credits	5	-0,57
OUTPUT		
Labour Productivity	6	-0,42
GRDP per kapita	6	-0,41

SERANG REGENCY

Overall Rangking	4
Indicator-based Rangking	
INPUT	4
I. REGIONAL ECONOMY	5
II. EMPLOYMENT AND HUMAN RESOURCES	4
III. PRODUCTIVE BUSINESS ENVIRONMENT	7
IV. INFRASTRUCTURE, NATURAL RESOURCES AND ENVIRONMENT	6
V. BANKING AND FINANCIAL INSTITUTIONS	4
OUTPUT	6
I. LABOUR PRODUCTIVITY	3
II. GRDP PER KAPITA	3
III. LEVEL OF EMPLOYMENT OPPORTUNITIES	8

NERACA DAYA SAING DAERAH

ADVANTASE

Variabel	Peringkat	Z-Score
INPUT		
Total investment per kapita	2	0,11
Dependency Ratio	3	0,51
Ratio productive land to total land area	3	0,49
Ratio of business volume value to the number of active cooperatives	3	0,58
Total Banking Credits	3	0,07

DISADVANTASE

Variabel	Peringkat	Z-Score
INPUT		
Primary Sectoral Productivity	5	-0,43
Secondary Sectoral Productivity	3	-0,20
Tertiary Sectoral Productivity	5	-0,39
Local Fiscal Capacity	5	-0,58
Government size	6	-0,38
Local Total Export Potential	5	-0,12
Regional overpriced index	4	-0,07
Firm Density	5	-0,55
Total Population	4	-0,02
Mean years of Schooling	6	-0,79
Life expectancy Index	7	-1,19
Mean years of Labor Schooling	6	-0,74
The Number of workforce	4	-0,11
Growth rate of labor force	7	-1,08
Public Service Expenditure Per Capita	7	-0,67
The number of problematic regional regulations	6	-0,03
The number of base sectors	6	-0,80
Percentage of population with the highest education of the universit	8	-0,94
Poverty Gap Index	6	-0,37
Population Density	6	-0,68
Added Value of Transportation Sector per capita	4	-0,49
Road conditions according to road quality	5	-0,46
The number of Telephone Connections per capita	7	-0,76
Electricity consumption of industry and household per capita	8	-0,65
Electricity production per capita	3	-0,34
Fuel consumption of industry and household per capita	3	-0,24
Water resources per capita	6	-0,21
Value Added Mining and Quarrying Sector per capita	4	-0,51
The number of Bank Offices	3	-0,39
Third Party Funds	6	-0,68
Saving/Kredit	7	-0,69
Value added Financial Sector per capita	5	-0,27
OUTPUT		
Labour Productivity	3	-0,16
GRDP per kapita	3	-0,24
Level of Employment Opportunities	8	-1,82

TANGERANG MUNICIPALITY

Overall Rangking	2
Indicator-based Rangking	
INPUT	2
I. REGIONAL ECONOMY	1
II. EMPLOYMENT AND HUMAN RESOURCES	2
III. PRODUCTIVE BUSINESS ENVIRONMENT	3
IV. INFRASTRUCTURE, NATURAL RESOURCES AND ENVIRONMENT	3
V. BANKING AND FINANCIAL INSTITUTIONS	2
OUTPUT	1
I. LABOUR PRODUCTIVITY	2
II. GRDP PER KAPITA	2
III. LEVEL OF EMPLOYMENT OPPORTUNITIES	2

NERACA DAYA SAING DAERAH

ADVANTASE

Variabel	Peringkat	Z-Score
INPUT		
Primary Sectoral Productivity	1	2,32
Secondary Sectoral Productivity	2	0,34
Tertiary Sectoral Productivity	1	1,43
Regional overpriced index	3	1,00
Firm Density	1	2,05
Total Population	2	0,60
Mean years of Schooling	2	0,89
Life expectancy Index	2	1,19
Mean years of Labor Schooling	2	1,08
The Number of workforce	2	0,77
Growth rate of labor force	2	0,61
The number of problematic regional regulations	5	0,22
Percentage of population with the highest education of the university	2	0,89
Population Density	1	1,85
Added Value of Transportation Sector per capita	2	0,31
The number of Telephone Connections per capita	1	1,48
Fuel consumption of industry and household per capita	2	0,08
Ratio productive land to total land area	7	0,05
Water resources per capita	1	0,70
The number of Bank Offices	1	2,02
Ratio of business volume value to the number of active cooperatives	2	0,60
Third Party Funds	1	1,92
Saving/Kredit	2	1,03
Value added Financial Sector per capita	2	0,17
OUTPUT		
GRDP per kapita	2	0,08
Level of Employment Opportunities	2	0,78

DISADVANTASE

Variabel	Peringkat	Z-Score
INPUT		
Secondary Sectoral Productivity	4	-0,23
Local Fiscal Capacity	4	-0,06
Government size	7	-0,82
Total investment per kapita	5	-0,45
Dependency Ratio	8	-1,31
Public Service Expenditure Per Capita	3	-0,02
The number of base sectors	6	-0,80
Poverty Gap Index	4	-0,04
Road conditions according to road quality	7	-0,83
Electricity consumption of industry and household per capita	5	-0,20
Electricity production per capita	5	-0,38
Value Added Mining and Quarrying Sector per capita	7	-0,57
Saving/Kredit	4	-0,34
OUTPUT		
Labour Productivity	2	-0,02

IOP Conf. Series: Earth and Environmental Science **158** (2018) 012050 doi:10.1088/1755-1315/158/1/012050

CILEGON MUNICIPALITY

Overall Rangking	3
Indicator-based Rangking	
INPUT	3
I. REGIONAL ECONOMY	3
II. EMPLOYMENT AND HUMAN RESOURCES	8
III. PRODUCTIVE BUSINESS ENVIRONMENT	1
IV. INFRASTRUCTURE, NATURAL RESOURCES AND ENVIRONMENT	2
V. BANKING AND FINANCIAL INSTITUTIONS	3
OUTPUT	4
I. LABOUR PRODUCTIVITY	1
II. GRDP PER KAPITA	1
III. LEVEL OF EMPLOYMENT OPPORTUNITIES	7

NERACA DAYA SAING DAERAH

ADVANTASE			DISADVANTASE		
Variabel	Peringkat	Z-Score	Variabel	Peringkat	Z-Score
INPUT			INPUT		
Primary Sectoral Productivity	3	0,09	Government size	8	-1,06
Secondary Sectoral Productivity	1	2,45	Regional overpriced index	6	-0,78
Tertiary Sectoral Productivity	1	2,23	Firm Density	3	-0,40
Local Fiscal Capacity	1	2,33	Total Population	8	-1,18
Local Total Export Potential	3	0,43	Dependency Ratio	6	-0,43
Total investment per kapita	1	2,42	Life expectancy Index	6	-0,40
Mean years of Schooling	3	0,62	The Number of workforce	8	-1,14
Mean years of Labor Schooling	3	0,52	The number of problematic regional regulations	7	-1,53
Growth rate of labor force	5	0,14	The number of base sectors	8	-1,20
Public Service Expenditure Per Capita	1	2,24	Poverty Gap Index	7	-0,93
Percentage of population with the highest education of the university	3	0,01	Population Density	5	-0,37
Road conditions according to road quality	1	1,68	Added Value of Transportation Sector per capita	3	-0,10
Electricity consumption of industry and household per capita	1	2,39	The number of Telephone Connections per capita	6	-0,68
Electricity production per capita	1	2,46	Ratio productive land to total land area	8	-2,38
Fuel consumption of industry and household per capita	1	2,41	Value Added Mining and Quarrying Sector per capita	3	-0,44
Water resources per capita	4	0,63	The number of Bank Offices	6	-0,52
Saving/Kredit	2	0,69	Ratio of business volume value to the number of active cooperatives	7	-0,96
Value added Financial Sector per capita	1	2,33	Third Party Funds	4	-0,31
OUTPUT			Saving/Kredit	6	-0,64
Labour Productivity	1	2,42	OUTPUT		
GRDP per kapita	1	2,41	Level of Employment Opportunities	7	-0,75

SERANG MUNICIPALITY

Overall Rangking	7
Indicator-based Rangking	
INPUT	7
I. REGIONAL ECONOMY	6
II. EMPLOYMENT AND HUMAN RESOURCES	7
III. PRODUCTIVE BUSINESS ENVIRONMENT	4
IV. INFRASTRUCTURE, NATURAL RESOURCES AND ENVIRONMENT	1
V. BANKING AND FINANCIAL INSTITUTIONS	7
OUTPUT	5
I. LABOUR PRODUCTIVITY	5
II. GRDP PER KAPITA	5
III. LEVEL OF EMPLOYMENT OPPORTUNITIES	4

NERACA DAYA SAING DAERAH

ADVANTASE

Variabel	Peringkat	Z-Score
INPUT		
Tertiary Sectoral Productivity	3	0,19
Dependency Ratio	4	0,30
Mean years of Schooling	4	0,07
Growth rate of labor force	3	0,59
The number of base sectors	1	1,60
Added Value of Transportation Sector per capita	1	2,36
Road conditions according to road quality	3	0,30
Ratio productive land to total land area	1	0,83
Water resources per capita	5	0,55
Saving/Kredit	3	0,02
OUTPUT		
Level of Employment Opportunities	4	0,21

DISADVANTASE

Variabel	Peringkat	Z-Score
INPUT		
Primary Sectoral Productivity	4	-0,30
Secondary Sectoral Productivity	6	-0,45
Local Fiscal Capacity	6	-0,61
Government size	3	-0,21
Potensi ekspor total daerah	6	-0,90
Total investment per kapita	7	-0,50
Regional overpriced index	8	-1,07
Firm Density	6	-0,55
Total Population	7	-0,93
Life expectancy Index	4	-0,04
Mean years of Labor Schooling	5	-0,24
The Number of workforce	7	-0,91
Public Service Expenditure Per Capita	4	-0,30
The number of problematic regional regulations	7	-1,53
Percentage of population with the highest education of the university	4	-0,11
Population Density	4	-0,36
The number of Telephone Connections per capita	5	-0,64
Electricity consumption of industry and household per capita	3	-0,12
Electricity production per capita	6	-0,41
Fuel consumption of industry and household per capita	5	-0,30
Value Added Mining and Quarrying Sector per capita	6	-0,56
The number of Bank Offices	8	-0,74
Ratio of business volume value to the number of active cooperatives	8	-0,96
Saving/Kredit	8	-0,71
Third Party Funds	5	-0,61
Value added Financial Sector per capita	4	-0,21
OUTPUT		
Labour Productivity	5	-0,30
GRDP per kapita	5	-0,30

TANGERANG SELATAN MUNICIPALITY	
Overall Rangking	5
Indicator-based Rangking	
INPUT	5
I. REGIONAL ECONOMY	4
II. EMPLOYMENT AND HUMAN RESOURCES	3
III. PRODUCTIVE BUSINESS ENVIRONMENT	2
IV. INFRASTRUCTURE, NATURAL RESOURCES AND ENVIRONMENT	8
V. BANKING AND FINANCIAL INSTITUTIONS	5
OUTPUT	3
I. LABOUR PRODUCTIVITY	4
II. GRDP PER KAPITA	4
III. LEVEL OF EMPLOYMENT OPPORTUNITIES	1

NERACA DAYA SAING DAERAH

ADVANTASE			DISADVANTASE		
Variabel	Peringkat	Z-Score	Variabel	Peringkat	Z-Score
INPUT			INPUT		
Local Fiscal Capacity	2	0,22	Primary Sectoral Productivity	7	-0,63
Potensi ekspor total daerah	4	0,37	Secondary Sectoral Productivity	2	-0,12
Regional overpriced index	1	1,38	Tertiary Sectoral Productivity	4	-0,12
Firm Density	2	1,07	Government size	4	-0,27
Total Population	3	0,05	Total investment per kapita	6	-0,48
Mean years of Labor Schooling	1	1,59	Dependency Ratio	7	-1,21
Life expectancy Index	1	1,44	Poverty Gap Index	8	-1,44
Mean years of Labor Schooling	1	1,59	Added Value of Transportation Sector per capita	5	-0,51
The Number of workforce	3	0,04	Road conditions according to road quality	6	-0,81
Growth rate of labor force	4	0,53	Electricity consumption of industry and household per capita	2	-0,03
Public Service Expenditure Per Capita	2	0,50	Electricity production per capita	7	-0,41
The number of problematic regional regulations	3	0,47	Fuel consumption of industry and household per capita	4	-0,26
The number of base sectors	2	1,20	Value Added Mining and Quarrying Sector per capita	7	-0,57
Percentage of population with the highest education of the university	1	2,00	The number of Bank Offices	4	-0,46
Population Density	2	1,28	Saving/Kredit	6	-0,69
The number of Telephone Connections per capita	2	1,43	Value added Financial Sector per capita	6	-0,64
Ratio of business volume value to the number of active cooperatives	2	0,51	OUTPUT		
Water resources per capita	2	0,70	Labour Productivity	4	-0,29
Ratio of business volume value to the number of active cooperatives	1	1,98	GRDP per kapita	4	-0,26
Third Party Funds	3	0,33			
Saving/Kredit	1	2,01			
OUTPUT					
Level of Employment Opportunities	1	1,50			

4.4. Conclusions

1. In the era of a massive and unlimited global economy, together with the adoption of decentralization systems especially in Indonesia, the concept of regional competitiveness is essential to complement the clearly defined concepts of competitiveness at the enterprise and national level. There needs to be a concept of regional competitiveness that is not only from a microeconomic and macroeconomic perspective but also a territorial perspective. Five main indicators and 30 sub-indicators determine the level of competitiveness of a region. These indicators are (i) regional economy, (ii) manpower and human resources, (iii) productive business environment, (iv) infrastructure, natural resources, and environment, and (v) banking and financial institutions. The combination of input and output factors represents the competitiveness of a region seen based on GRDP per capita, labor productivity, and the level of employment opportunities.
2. This study did not emphasize the importance of regional ratings. Instead, the most important point of the study is how a region can identify the advantages and disadvantages of the determinants of each region's competitiveness.
3. Regions with low competitiveness generally have an economic base sourced from natural resource wealth.
4. Regencies/municipalities will have high regional competitiveness if this is supported by human resource wealth and industrial-based or service-based economic activities. Those regions are located in the northern part of Banten;

5. Regencies/municipalities with the lowest regional competitiveness position, generally rely on the primary sector (especially agriculture). Those regions are located in southern part of Banten.

4.5. Recommendations

Local governments need to implement some policies to create a highly competitive region:

a. Increase economic growth through optimizing local revenue, increasing local fiscal capacity, and increasing local investment.

b. Improve the quality of human resources and employment through improving the health status of the population and labor; and improving the quality of education as well as competence, technology, and skills.

c. Create a productive business environment that can attract businesses through the simplification and harmonization of various regulations, the implementation of one-stop integrated services to accelerate and facilitate the process of licensing and non-licensing, and ease the process of land acquisition, creating a controlled security.

References

[1] Blakely E J 1989 *Planning Local Economic Development: Theory and Practice* (Newbury Park (CA): Sage)

[2] Gallagher C, Robson G and Kerr J 1993 Regional variations in the population of small and medium-sized enterprises In: *Britain in Small Firms: Recession and Recovery* Eds Chittenden F, Robertson M and Watkins D (Paul Chapman, London) pp 77–93

[3] Huggins R 2013 Regional Competitiveness : Theories and Methodologies for Empirical Analysis. *Centrum* **6**(2), 155–172.

[4] Moomaw R L and Shatter AM 1993 Urbanization as a Factor in Economic Growth. *J. of Economics* **19** 1–6

[5] Porter M E 1990 The competitive advantage of nations *Harvard business review* **68**(2): 73-93

[6] Porter M E 2003 *Clusters and Regional Competitiveness : Recent Learnings International Conference on Technology Clusters 7 November 2003 Montreal Canada* Available from: https://s3.amazonaws.com/academia.edu.documents/30781735/Montreal_Cluster_Conference_2003.11.07_ckrb.pdf?AWSAccessKeyId=AKIAIWOWYYGZ2Y53UL3A&Expires=1521282851&Signature=%2FpO%2B8LvGyjqcUDROzVPhFStpugM%3D&response-content-disposition=inline%3B%20filename%3DClusters_and_Regional_Competitiveness_Re.pdf

[7] Storper M 1997 *The Regional World: Territorial Development in a Global Economy* (New York: Guilford Press)

[8] Thurow L 1999 *Building Wealth: The New Rules for Individuals, Companies, and Nations in the Knowledge-Based Economy* (New York, Harper Collins)

IOP Conf. Series: Earth and Environmental Science **158** (2018) 012051 doi:10.1088/1755-1315/158/1/012051

Assembling the Past and the Future of the City through Designing Coworking Facilities

Y A Lukman, A S Ekomadyo, A S Wibowo

Architecture Department, School of Architecture, Planning, and Policy Development, Institut Teknologi Bandung, Indonesia

Corresponding Author: lukman.yaldi@gmail.com

Abstract. Bandung is known as a creative city in Indonesia, which can be seen from the large number of communities in Bandung that work in the creative industry. A creative city can be further developed by a good understanding of its local identity. One of the characteristic features of Bandung are the numerous old buildings across the city. Unfortunately, these buildings are no longer utilized optimally due to a mismatch between their function and typology. Housing new functions in old buildings to meet present space needs can increase their value. One kind of new function that fits well in old buildings, especially in Bandung, is the coworking space, a new type of workspace that has emerged as a result of the needs of today's society. Mutually beneficial relations can be formed when the old building is well suited to carry out its function as a coworking space for the creative class. The idea is to assemble the past (using an old building as a workplace) and the future (developing the creative economy) through designing coworking facilities. The design simulation conducted in this study used the 'third place' theory by Ray Oldenburg as well as the approach of adaptive building reuse. By changing old buildings in Bandung into new workplaces for the creative class – coworking spaces – Bandung can maintain the city's identity and provide new workplaces or public spaces for communities to develop their creativity and increase city income.

1. Introduction

Bandung is known as a creative city in Indonesia, which actively supports the development of the creative economy. The forms of creativity in Bandung are reflected in some unique characteristics of the city, such as using old buildings and the old part of the city as attractions. Unfortunately, because of rapid developments in economy and technology in recent years, some older parts of the city have started to experience degradation. If this continues, an important characteristic of Bandung will be lost.

One of the developments that certainly cannot be stopped, is the growing importance of the internet. Because of the improvement of internet connectivity, some people no longer need to go to an office to work; they can work anywhere by using only a laptop and an internet connection. These people are referred to as belonging to the creative class [1] (Florida 2002). Trends such as remote working are making work more flexible. On the other hand, this potentially makes people from the creative class more isolated, because they tend to meet less other people. Therefore, there a need has emerged for a new type of workspace that suits the work behaviour of people from the creative class and allows them to work flexibly and still be able to interact with others and not feel lonely/isolated.

One such kind of new workspace is the coworking space. Unfortunately, many so-called coworking spaces – which are growing rapidly in Bandung (quantitavely) – do not posess the essence of a coworking space (qualitatively), which is the formation of a community among the people using it. Therefore, this study focused on how to create a coworking space in Bandung in accordance with this value using Oldenburg's 'third place' theory through a design simulation.

The characteristics of Bandung and the development of coworking spaces in Bandung can have a mutually beneficial relationship, where an old building requires a relatively large amount of capital for rejuvenation and maintenance, while a coworking space needs a strategic location and low initial costs.

2. Method

This study was conducted by using a design method to answer the question of how to assemble the past and the future of Bandung through coworking spaces as a third place. This method reveals the reflections of the author during the design process.

The first stage was a review of the literature on coworking spaces, the third place, and the millennial generation in support of designing a coworking space as a third place for the millennial generation in Bandung. As the design process took place in Bandung, the identity of Bandung connected to its architectural history was identified in the form of the old buildings that are spread throughout the city. Then, a review of the literature on these old buildings was conducted. Several characteristics of the location for the design were identified and this stage was closed with a survey to find a suitable location for the design.

During the second stage, the users of the coworking space were determined based on a literature review and a market trend survey. The users that fit the criteria are mostly in the information and technology sector. After fixing the users and the design location, the next stage was data collection done by expert interviews of information and technology start-up company founders to find out criteria for a coworking space that meets their specific needs. Besides that, primary data on the old buildings in Bandung were collected through a site survey.

The third stage was an analysis of the values of coworking spaces and the criteria for the third place, which generated design criteria that were used to produce a design concept. The design concept was based on the idea of assembling the past and the future of Bandung and furthermore became the basis for the design simulation.

3. Literature Review

3.1. Coworking Space

A coworking space is a workspace for a community, a place where people come together to collaborate with each other. It also is an affordable and accessible workspace – a place to share knowledge from different business backgrounds that can contribute to each other [2].

The rise of the coworking space is caused by the emergence of a knowledge economy that simultaneously develops into a digital economy [3]. In the concurrent development of the knowledge economy and the digital economy, there are professionals that follow this economical transformation, called the creative class. According to [1], this is not limited to people who have artistic abilities, such as architects, artists, or musicians, but also includes consultants, lawyers, and programmers.

According to [4], coworking spaces have five main values that are open to individual interpretation. They can be used by coworking space owners to attract people dedicated towards these same values as the foundation for building a strong coworking space community. These values are:

- Collaboration. A willingness to work together and share knowledge.
- Openness. Freedom to share ideas, information, and people.
- Community. A group that aims to share knowledge through coworking.

- Sustainability. The use of a common space that is ecologically and economically sustainable.
- Accessibility. A coworking space as a place that is accessible both economically and socially by people from various backgrounds.

Moriset states that coworking spaces can be a third place for living and working because the main values of coworking spaces are in line with the third-place characteristics as defined by [5].

3.2. Third Place Approach by Oldenburg

The 'third place' is a term coined by the American psychologist Ray Oldenburg to refer to a place between residence and workplace, where people can informally interact. According to Oldenburg, the first place refers to the residence (home), the second place refers to the workplace (office), and the third place refers to a community space. Some typical places that can be a third place are barbershops, cafes, hotels, lounges, etc.

There are eight criteria for a third place according to Oldenburg, namely:

- Neutral Ground. A place where individuals can come and go freely without being attached to some activity.
- Leveler. A place that is not concerned with an individual's status in society.
- Conversation. A place where conversation is the main activity and the topics of discussion are fun without being heavy.
- Accessibility and accommodation. A place with a high level of convenience in accessibility, which provides accommodation that meets the needs of its users.
- The Regulars. A place with a fixed client-base, which provides a distinctive spatial atmosphere. The regular customers invite new customers and make them feel welcome and comfortable.
- Low Profile. A place that is neutral – not pretentious or expensive-looking – but comfortable to use. It should also be agreeable to various types of individuals.
- A Playful Mood. A place that is not tense, where social interaction happens naturally and in a fun way.
- A Home Away From Home. Although physically a third place is different from a home, psychologically it has the comfortable atmosphere of home, so users feel part of the place.

3.3. Heritage Building Design Proceeding

There are five ways to design infill buildings in a cultural heritage area [6], namely:

- Mimicry
  This method changes traditional forms in an abstract way and turns this symbolic form into a modern urban design so that people still recognize and receive historical information and traces.
- United complementary
  By using this method, the design fully follows the traditional/old layout and provides a certain modern element in the design through materials, etc. It intends to maintain the traditional patterns and values of the building. The basic elements of traditional architecture that can be implanted into contemporary architecture, such as scale, environmental climate, organic morphology and place ownership, for example through materials and construction techniques [7].

- United diversity
  A building that is different from other buildings tends to be more prominent and valuable as exhibition venue. These characteristics can be achieved through adaptation to historical peculiarities in a street pattern and to neighbouring buildings. For example, providing overhangs on the upper floors can make the building stand out, even though its ground floor matches neighbouring buildings [8]. Using this method, the architect commonly maintains the scale of the building while uses colours, materials and textures as the distinguishing elements.
- Refurbishment and conversion
  This method provides an area with a sense of ownership and reflects the past in a clear way. Repairing means anything from the installation of new elevators to total demolition behind a maintained facade. Putting new functions into an old building can help renew an area within the city and provide accommodation in a strategic location [8]. This method prioritizes the authenticity of the building facade in order to create a sense of ownership and recreate the character of the area from the past in a real way and then provide new functions in the building to improve the economy of the local community.
- Contrast
  This method focuses on a free way of thinking and a form of rebellious expression of the past [7]. An infill strategy for urban design with this method is channeled with aspects that have a specific design that highlights modern characteristics and ignore cultural and contextual factors [9].

4. Description and Analysis

4.1. Location

Not all old buildings scattered throughout Bandung are suitable for functioning as a coworking space. Therefore, the selection of appropriate old buildings must consider some aspects that are essential for the function of a coworking space, such as land use in accordance with RTRW Bandung, economic conditions of the land, proximity to public facilities, and existence of other old buildings in the area.

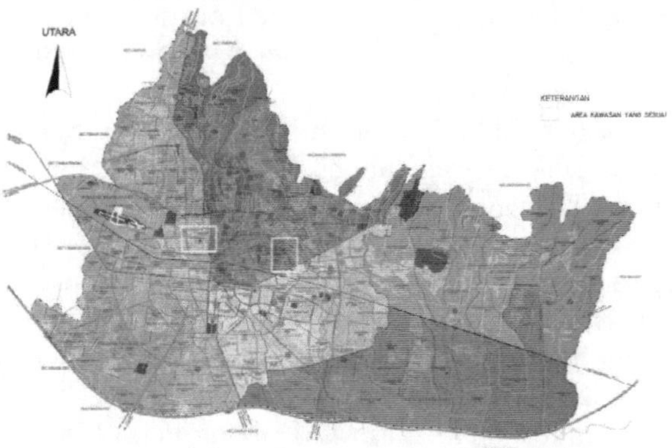

Figure 1. Superimposed Map of Essential Location Aspects.
Source: (processed from https://ppdbkotabandung.wordpress.com/pustaka/peta-kota-bandung/)

Two potential locations in Bandung that match the success criteria for a coworking space are: (1) the area around Pajajaran Street and (2) the area around Baranangsiang Street. Comparing these two areas regarding the target characteristics of a coworking space for users in Bandung, we see that:

- the area around Pajajaran Street is closer to various millenial community gathering places, such as shopping centers, cafes, restaurants, and more.
- the area around Baranangsiang Street is a centre for Sundanese culture and more dominated by generation X communities.
- the area around Barangsiang Street is more dominated by public facilities such as a traditional market. Besides that, public transportation around Baranangsiang Street is relatively less available compared to the area of Pajajaran Street. This indicates that Baranangsiang Street area is less likely to be visited by the coworking space target groups.
- the old buildings in Baranangsiang are more suitable for coworking spaces because their scale is not too large nor too small.
- Based on the above analysis, the area that will be used for the conceptual design is the area of Pajajaran Street. The site is located at Pajajaran Street No. 49. The design is contains two parts: the first part is the old building on the east of the site (red rectangle) and the second part is a new building on the west of the site, which will be redesigned in the context of the heritage area. The total area of the site is 1.476 m2.

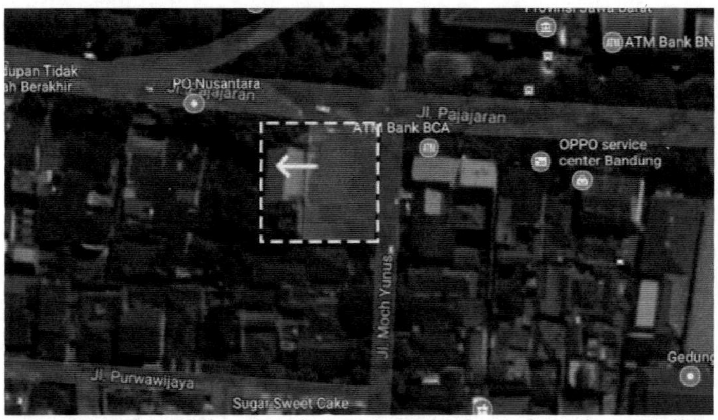

Figure 2. Old Building (Red) Inside Location of Design (Yellow).
Source: (processed from maps.google.com)

In general, the structure of the old building is still in good condition. There are no reinforcements or additions that go against building regulations. The authenticity of the old building has been maintained well by the owner. The new building located on the west side of the building is used by the owner for commercial purposes, that is selling and servicing car tires. The rectangular structure with a steel frame is 4 stories high and the design is completely different from the buildings around it.

The location at Pajajaran Street is close to two public transportation hubs: Bandung railway station (1.4 km) and Husein Sastranegara Airport (2.7 km). This makes it easier for clients or coworkers from outside the city to reach the location and meet with coworkers in Bandung. In addition, the location of the site is close to shopping centres Istana Plaza and Living Plaza (450 meters from the location) and Paskal Hypersquare (2 km from the location). Besides that, the location is also close to restaurants and cafes on Pasirkaliki Street and Pajajaran Street.

Mohammad Yunus Street, located to the east of the site, creates a shortcut from Pajajaran Street to Pasirkaliki Street and Kebon Kaung Street, which gives the site high accessibility. In addition, the location is also near universities such as Binus University and Bandung Law School, students from which could be prospective members of the coworking space.

In economical terms, the design of the old building is maintained in a relatively good condition so the cost of adaptation of the building is not too high. In addition, the architectural style of the old building is different from other historic buildings in Pajajaran Street, so the design of the building with the new function can be more ‘eye-catching’.

IOP Conf. Series: Earth and Environmental Science **158** (2018) 012051 doi:10.1088/1755-1315/158/1/012051

4.2. Design Analysis

4.2.1. User Analysis. Based on the analysis related to changes in the current economic system, which is moving towards a digital economy, and the requirements of new spaces in Bandung, the main target group for the coworking space are workers in the informatics sector and also students and various freelancers. According to Budi Raharjo, the informatics sector in Bandung is particularly growing in the field of programming and animation [10].

In the early stages of a project, informatics workers and informatics companies tend not to collaborate with other parties, among others due to concerns related to plagiarism of work and ideas. Collaboration can occur at the stage of system development with other fields. Up until now, coworking spaces have not been a major alternative workplace for informatics workers due to considerations (beyond economic aspects) concerning security as well as privacy.

4.2.2. Activities Analysis. The main activity in a coworking space is flexible working flexibly. In addition, the determination of other activities that can be done in a coworking space will be determined based on the needs of its users and the basic idea of designingthe coworking space design. In particular, to accommodate the needs of the main userusers of a coworking space which is , in this case informatics workers, so the preparation of activities in the coworking space is also influenced by the system workswork systems of its userusers. According to Budi Raharjo, in the informatics work process of informatics works, there are two types of work system, such as:systems, i.e [10].:

- Workgroup discussions, where thisthe group needneeds a space that can provide the privacy and freedom to work within their styles, own work style and that is comfortable, and safe for private discussiondiscussions;
- Individual work, where an individual needs a open space that can accommodate working alone, such that the activityneeds of each individual in relation to their work alone in open space, so theystyle can meet the need of work style of eachbe met.

4.2.3. Coworking in third-place analysis. Based on the third place theory by Oldenburg and coworking space development as a movement for using workspaces outside the home to avoid isolation. The criteria for a third place and the values of a coworking space are interrelated, so the coworking space has potential to become a third place.

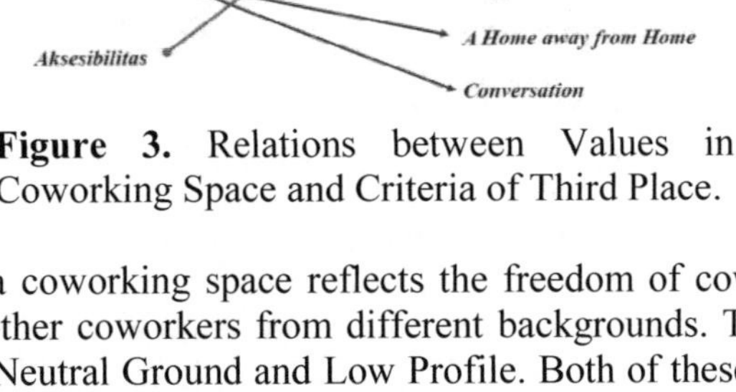

Figure 3. Relations between Values in Coworking Space and Criteria of Third Place.

The value of openness in a coworking space reflects the freedom of coworkers in sharing ideas, information, and welcoming other coworkers from different backgrounds. This value corresponds to the third-place characteristics Neutral Ground and Low Profile. Both of these characteristics require a simple building that is still comfortable to use by people from any part of society without being bound by certain restrictions.

IOP Conf. Series: Earth and Environmental Science **158** (2018) 012051 doi:10.1088/1755-1315/158/1/012051

The value of collaboration in a coworking space reflects the coworker's desire to work together and share knowledge. This value corresponds to the third-place characteristics Leveler and A Playful Mood, where the coworking space does not distinguish its users in interacting and collaborating according to social status or economic status.

The value of a coworking space community is that it is a group of coworkers that has the common goal of sharing knowledge through coworking. This value corresponds to the third-place characteristic The Regulars. A community in a coworking space is a collection of routine members who come and work there so they have the opportunity to have discussions and get to know each other. The intense interaction between these members forms a solid community in the coworking space.

The value of accessibility in a coworking space corresponds to the third-place characteristic Accessibility and Accommodation, where the coworking space must be accessible to everyone, both economically and socially.

Meanwhile, the value of sustainability in a coworking space is not only sustainability in terms of energy and resources but also the sustainability of the community formed within it. In maintaining the sustainability of the coworking space community, this value corresponds to the third-place characteristics Conversation and A Home Away From Home. A community in a coworking space can be sustainable if communication between the members occurs in every activity. Community sustainability in a coworking space makes the coworking space comfortable for its members as they are surrounded by people they know, like in a home environment.

5. Design Concept

The main idea of this study was to create a design that connects a part of Bandung's past with the development of present and future trends. Old buildings in a cultural heritage area as part of Bandung's past can incorporate a new function – as a coworking space – to maintain the old building and increase its value while at the same time providing a solution for the spatial needs of the community. The adaptive reuse building approach was used in adapating the old building to its new function as a coworking space. Meanwhile, to meet the present and future needs of society, the idea was to design the coworking space not only as a working space but also as a place to interact socially, so that the third-place approach by Oldenburg was implemented in the designing process.

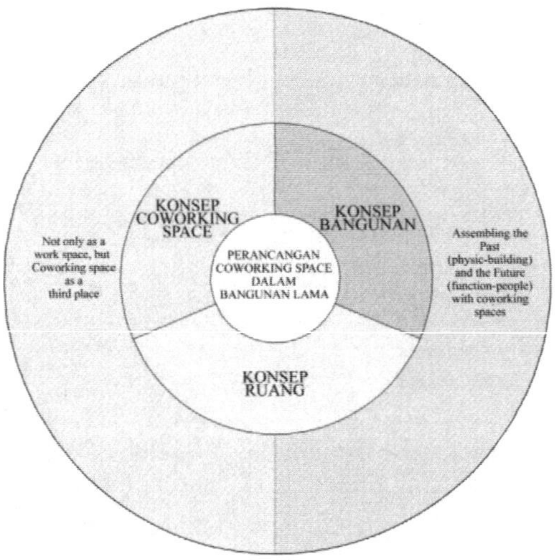

Figure 4. General Concept of Designing Coworking Space.

5.1. New Building Design Proceeding

The old building at the site is not listed as a heritage building. However, it is located in one of the cultural reserve areas in Bandung, which underscores the need to maintain the old building and prevent degradation. Therefore, the design concept for the old building uses the refurbishment and conversion method [6]. Based on this method, the exterior of the building is kept in order to maintain the atmosphere of the cultural heritage area. However, the interior of the old building is changed according to the needs of the new function.

Figure 5. The Exterior of the Old Building is Kept.

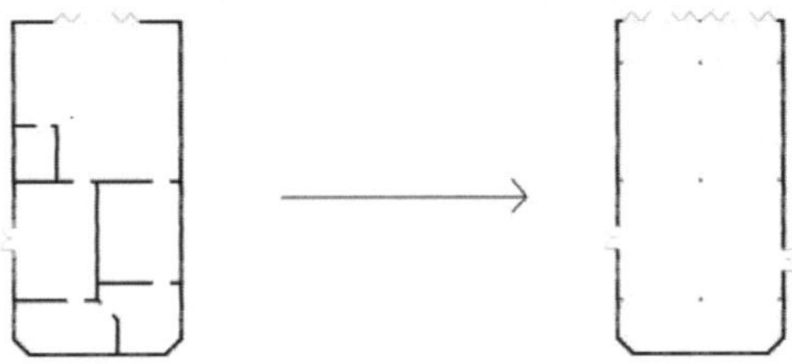

Figure 6. The Interior Design of the Old Building Changes Completely

5.2. New Building Design Proceeding

Meanwhile, the design of the new infill building is based on the united complementary concept [6], where the new building does not reduce the value of existing of old building, which becomes a complementary building that still has its own identity.

Architectural elements of the old building that are adapted to the design of the new building are proportions, scale, and morphology of the building facade. Proportions from the old building are used in the new building, namely the ratio between the width of the front-looking part of the building and the middle height of the building, which is 1:1.04. Meanwhile, the scale element from the old building that is adopted is the distance from floor to ceiling of the old building. The floor-to-ceiling proportion of the new building is almost the same as that of the old building, i.e. 3.75 meters in the old building and 3.74 meters in the new building.

IOP Conf. Series: Earth and Environmental Science **158** (2018) 012051 doi:10.1088/1755-1315/158/1/012051

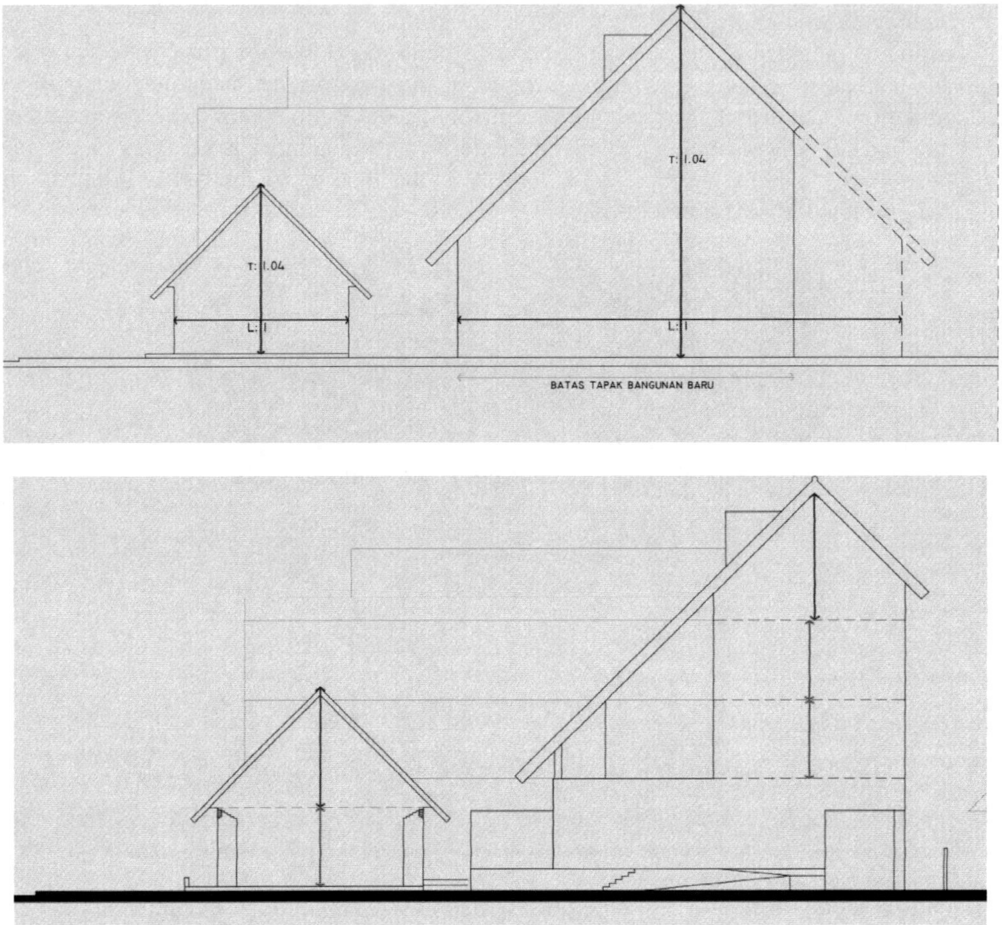

Figure 7. (Above) Comparison of Proportion in Old and New Building | (Below) Comparison of Scale in Old and New Building

Morphological elements from the old building facades that were adapted for the facade design of the new building are the forms of the openings in the old building, the canopy design at the top of the windows and doors, and the slope and shape of the roof of the old building. Among these various aspects, the dominant aspect adopted for the design of new the building is the shape and slope of the roof.

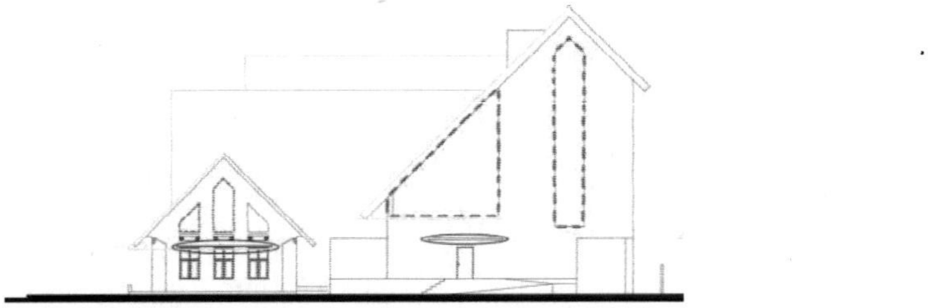

Figure 8. Comparison of Morphological Elements in Old and New Building

In addition to adopting form elements from the old building, modern elements are also incorporated the new building by designing some contrasting architectural elements, so that although the new building is adapted to the old building, the styles of both buildings are different. The contrast in architecture mainly comes from the selection of building materials. The old building is dominated by bricks and white paintwork, while the new building is dominated by unfinished concrete, steel and glass to create a modern industrial impression.

5.3. Coworking Space Design Concept

To meet the design criteria of a coworking space that suits its purpose not only as a workplace but also as a third place, the coworking space was designed with consideration of five aspects, i.e. collaboration, adaptability, accessibility, sustainability, and community.

The key points in creating a third place are social interaction and communication between coworkers. In connection with the coworking space, the social interaction is expected to result in collaboration between coworkers. The layout of the space plays an important role in enabling collaboration, among others by:

- making sure that the total space is dominated by shared workspaces;
- making sure there is sufficient variation in model and size of the workspaces to accommodate the needs for many different activities;
- using a big group of tables in order to make coworkers feel comfortable to work together;
- providing supporting facilities that allow the coworkers to have social interaction, e.g. the kitchen and an open break area that are multifunctional.

Besides being supporting collaborative work, a coworking space is a workplace that has a high rate of change, both in types of coworkers and in activities inside the space. Therefore, the workspaces in a coworking space must be designed to be highly adaptable, so every workspace can be used effectively. This can be achieved by:

- using modular workspaces so they can be merged or separated easily according to the needs of co-workers or activities;
- leaving spaces open in the workspace layout so that the furniture can be changed according to need;
- having more than one function for every zone in the coworking space to maximize the saleable area that can be used by coworkers;
- designing different workspace layouts to meet the preferences of different coworkers so they can choose a coworking area that suits them.

In order to create a harmonious and complementary relationship between both buildings in this study, the new building and the old building must have an equivalent role or functionality so they are inseparable. Therefore the two buildings need to be connected so they have a dependency on each other. This can be achieved by (1) designing a connecting bridge from the old building to the second floor of the new building to connect the coworking space functions of the old building and the new building; (2) the ground floor of the old building and that of the new building are linked to a green space that can be used as a public area and for social interaction activities.

5.4. Concept of Building Zoning

New for the old building are public functions such as café-coworking. This café function is incorporated in the old building in order to attract visitors. There are some considerations regarding this new function, namely:

- Cafés in old buildings have distinctive features and identities that can attract the interest of informatics workers and millennials that work in a mobile and flexible way.
- The entrance fee for the cafe can be used for the maintenance of the old building.
- The cafe in the old building can act as a transition zone between a cozy cafe to 'hang out' in and a place that is comfortable to work in before the user switches a coworking space.

Apart from functioning as a café, the interior of the old building was designed with a mezzanine for use as a coworking space area in order to create equality between both buildings and maximize space use. By adding this function, the historical character of Bandung as well as the uniqueness of the design of the coworking space in this study are maintained.

Meanwhile, the new building functions as the main coworking space. The spatial concept used in relation to the architectural design is the concept of a 'pyramid', where the zoning of the new building is divided vertically according to the proximity/depth of collaboration that occurs between coworkers. The proximity/depth of collaboration on a higher floor is higher to have better quality of collaboration than on the floor below it. This makes the number of users who use the higher floors lower than those using the lower floors, so the rooms are increasingly smaller at the higher levels. In addition, the pyramid concept is also intended to support the concept of a coworking space as a third place. By applying pyramid zoning, the areas that allow interaction and collaboration are at a more accessible level and have a larger area of space. With this division, there is more space available that can become a third place.

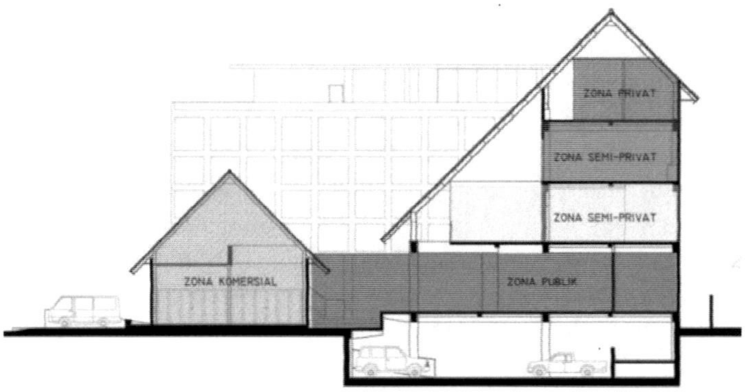

Figure 9. The Concept of Zoning Vertically.

The first-floor area is a public zone aimed at attracting people to the location. 'Public' means that the space on this floor is not only used by coworker members but also by others (people who visit the cafe). Therefore, the ground floor consists of spaces for people to gather and to get to know each other, encouraging the formation of a community.

Figure 10. The Atmosphere of Coworking-Cafe in Old Building

The first-floor area is a semi-public zone used by newcomers to work in and get to know each other by communicating informally at points for communication that are provided, or to work together with a nearby coworker. The second floor has a concept that enables collaboration, so the workspace layout on this floor consists of various kinds of workstations and also has a hot-desk system, thus increasing the chance of people meeting different people every day and getting to know these persons in a more intense interaction in several settings defined by different space layouts.

Figure 11. (Left) Atmosphere of Smoking Zone | (Center) Atmosphere of Coworking Space Zone | (Right) Atmosphere of Break Room in Second Floor

The third-floor area is semi-public zone used by coworkers who collaborate on the second floor and need a quieter place to focus on collaborating in groups for a short period of time (groups created on a project basis). The space on the third floor uses a dedicated desk system with a variety of workspaces that is less diverse, i.e. standard workspaces and a lounge workspace to relax and interact with other coworkers.

Figure 12. (Left) Atmosphere of Pantry | (Right) Atmosphere of Break Room on the Third Floor

The fourth-floor area is a private zone consisting of a cubical space that functions as an office. This zone is used by a coworker who has already built sufficient trust with another coworker and decides to work with that coworker for a longer time in a start-up company. In order to meet the needs specific to this situation, the fourth space provides a highly private office with a standard workplace to focus working in an enterprise. Besides that, this floor also provides a lobby, a pantry, and a brainstorming room for interacting and relaxing with other coworkers.

Figure 13. (Left) Atmosphere of Private Office | (Right) Atmosphere of Rooftop Garden on the Fourth Floor

Based on the vertical zone division, the ground floor is a public area where a coworker community can interact with other communities from the city. The second and the third floor are a semi-public areas where a third place can be created. Zoning on these floors refers to the characteristics of a third place in accordance with the values of a coworking space. Lastly, the fourth floor is a private area where a coworker who has already built sufficient trust with other coworkers can run a start-up company.

6. Conclusion

Based on the analysis and design in this study, it was found that to maintain the values and historical character of Bandung can be done by reusing old buildings in the city and redesigning them to be representative of the past but with a new function that is suitable for the younger generations in Bandung, while at the same time being representative of the future by designing a coworking space to function as a third place. Making a coworking space into a third place can be done by creating a space with four zones that support four stages of interaction, namely (1) a public area as a hangout place for young people, (2) a semi-public area for interaction and getting acquainted, (3) a semi-public area for group collaboration over a short period, (4) a private area for long-term collaboration in a company or a group.

The existence of coworking spaces in old buildings can be an attraction, distinguishing coworking spaces in Bandung from coworking spaces elsewhere. In addition, the utilization of an old building with a new function, such as a coworking space, can contribute to a revitalized image and contribute to the development of the old city area in Bandung.

References

[1] Florida R 2011 *The Rise of the Creative Class-Revisited: Revised and Expanded* (Philadelphia: Basic Books)
[2] Spinuzzi C 2012 Working alone together: Coworking as emergent collaborative activity *J. Business and Tech Com*, 29(4), 399-441
[3] Moriset B 2013 Building new places of the creative economy. The rise of coworking spaces. *Proc off the 2nd Geography of Innovation International Conference 2013*, Utrecht University
[4] Kwiatkowski A and Buczynski 2011 *B Coworking: Building Community as a Space Catalyst* (New York: Cohere LLC)
[5] Oldenburg R 1989 *The great good place: Cafés, coffee shops, bookstores, bars, hair salons and the other hangouts at the heart of a community* (Massachusetts, USA: Cambridge)
[6] Al Ani M Q A G 2014 Creative Infill Projects in Urban Environment and Effect on Livable City Life *Int. J. Eng. Science & research technology*. Iraq
[7] Al Talib TH 1995 The Means and Ways of Dealing with Architectural Heritage *Assoiation of arab univiersities journal* **2**
[8] Tugnutt A and Robertson M 1987 *Making Townscape: A Contextual Approach to Building in an Urban Setting* (London:Mitchell)
[9] Zhu J 2011 Urban Waterfront Landscape Cultural Design Studies Institute of Urban Design Wuhan University Wuhan China p 6564 – 67 IVSL
[10] Raharjo B 2017 *Kumpulan Solusi Pemograman Ruby* (Bandung: Informatika)

IOP Conf. Series: Earth and Environmental Science **158** (2018) 012052 doi:10.1088/1755-1315/158/1/012052

The Development of a Web-Based Urban Soundscape Evaluation System

A S Sudarsono[1] and J Sarwono[1]

[1]Building Physics and Acoustic Laboratory, Institut Teknologi Bandung

Correponding author: anugrah@tf.itb.ac.id, anugrahsabdono@gmail.com

Abstract. Acoustic quality is one of the important aspects of urban design. It is usually evaluated based on how loud the urban environment is. However, this approach does not consider people's perception of the urban acoustic environment. Therefore, a different method has been developed based on the perception of the acoustic environment using the concept of soundscape.

Soundscape is defined as the acoustic environment perceived by people who are part of the environment. This approach considers the relationship between the sound source, the environment, and the people. The analysis of soundscape considers many aspects such as cultural aspects, people's expectations, people's experience of space, and social aspects. Soundscape affects many aspects of human life such as culture, health, and the quality of life.

Urban soundscape management and planning must be integrated with the other aspect of urban design, both in the design and the improvement stages. The soundscape concept seeks to make the acoustic environment as pleasant as possible in a space with or without uncomfortable sound sources. Soundscape planning includes the design of physical features to achieve a positive perceptual outcome. It is vital to gather data regarding the relationship between humans and the components of a soundscape, e.g., sound sources, features of the physical environment, the functions of a space, and the expectation of the sound source. The data can be measured and gathered using several soundscape evaluation methods. Soundscape evaluation is usually conducted using in-situ surveys and laboratory experiments using a multi-speaker system. Although these methods have been validated and are widely used in soundscape analysis, there are some limitations in the application. The in-situ survey needs to be done at one time with many people at the same time because it is hard to replicate the acoustic environment. Conversely, the laboratory experiment does not have a problem with the repetition of the experiment. This method requires a room with a multi-speaker reproduction system. This project used a different method to analyse soundscape developed using headphones via the internet. The internet system for data gathering has been established; a website has enabled to reproduce high-quality audio and it has a system to design online questionnaires. Furthermore, the development of a virtual reality system allows the reproduction of virtual audio-visual stimulus on a website. Although the website has an established system to gather the required data, the problem is the validation of the reproduction system for soundscape analysis, which needs to be done with consideration of several factors: the suitable recording system, the effect of headphone variation, the calibration of the system, and the perception result from internet-based acoustic environment reproduction. This study aims to develop and validate a web-based urban soundscape evaluation method. By using this method, the experiment can be repeated easily and data can be gathered from many respondents. Furthermore, the simplicity of the system allows for the application by the stakeholders in urban design. The data gathered from this system is important for the design of an urban area with consideration of the acoustic aspects.

1. The soundscape concept to evaluate the urban sonic environment

City planners have developed many concepts to create adequate urban areas, which mainly focused on the visual aspect [1]. The acoustic aspect is usually neglected by urban designers although it has an important role in shaping the environmental quality of a city [2]. Furthermore, the analysis of the acoustic aspect only considers the negative aspects of sound as represented by noise [3].

The aspect of sound has become an important issue in city development [4]. ITU-T-FG-SSC has defined the Smart Sustainable City as "A Smart Sustainable City is an innovative city that uses Information and Communication Technologies (ICTs) and other means to improve quality of life; efficiency of urban operation and services; and competitiveness; while ensuring that it meets the needs of present and future generations with respect to economic; social; environmental as well as cultural aspects" [5]. The aspect of the sound environment is important in three elements of the definition of the Smart Sustainable City: social, environmental, and cultural.

The Smart Sustainable City is explained in detail using six dimensions: Smart Economy, Smart People, Smart Governance, Smart Mobility, Smart Environment, and Smart Living [6]. The dimension of the smart environment includes four characteristics: attractivity of natural conditions, pollution, environmental protection, and sustainable resource management. In here, the sound environment aspect is related to the noise pollution of a city and the environmental protection.

The dimension of smart living also highly relates to the acoustic aspect. It has seven characteristics: cultural facilities, health conditions, individual safety, housing quality, education facilities, tourist attractivity, and social cohesion. The study by Berglund shows the effect of soundscape on human health [7]. The soundscape also affects the tourist attractivity of urban parks [8][9] and the other tourist attractions [10].

The urban sonic environment is usually evaluated using the concept of noise measurement. This concept relies on sound level measurement and comparing the result to the available standard. If the sound level exceeds the standard than the noise must be reduced regardless the sound source as shown in Figure 1.

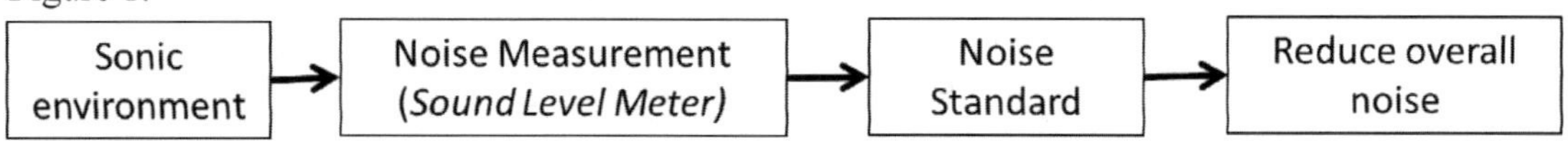

Figure 1. Sonic environment management based on noise measurement

The Indonesian government has developed some policy regarding the level of permissible noise in different areas. The Indonesian Ministry of Environment and Forestry has a policy for noise standards according to KEP-48/MENLH/11/1996 as shown in Table 1.

Table 1. Noise Standards according to the Indonesian Ministry of Environment and Forestry

Area	Noise Level (dBA)
Residential Area	55
Commerce and Services	70
Office and Trading	65
Green Open Space	50
Industry	70
Government and Public Facility	60
Recreation	70
Port	70
Cultural Heritage	60
Hospital	55
Schools and Education Facilities	55

Area	Noise Level (dBA)
Place of Worship	55

The standards seek to establish quiet places, which is considered a better sonic environment. A problem arises from the standard, namely that most areas in Indonesia exceed the noise standards and people still doing their activities there.

The standards and the noise measurement concept only consider the noise level with no consideration of how the people who use the space perceive the sonic environment. The noise control of a noisy space is focused on reducing the overall sound level [11], although it is found that the noise level does not necessarily affect the perception of a space [12] [13]. The perception of a space is affected by the subjective preference of sound sources rather than only the sound level [14].

Soundscape is a different approach to analysing the sonic environment by considering the human perception. Soundscape is defined as the "acoustic environment as perceived or experienced and/or understood by people, in context" and the acoustic environment is defined as "sound from all sound sources as modified by the environment" [15]. The concept tries to integrate three aspects in the analysis: people's perception, sound source, and the environment, as shown in Figure 2.

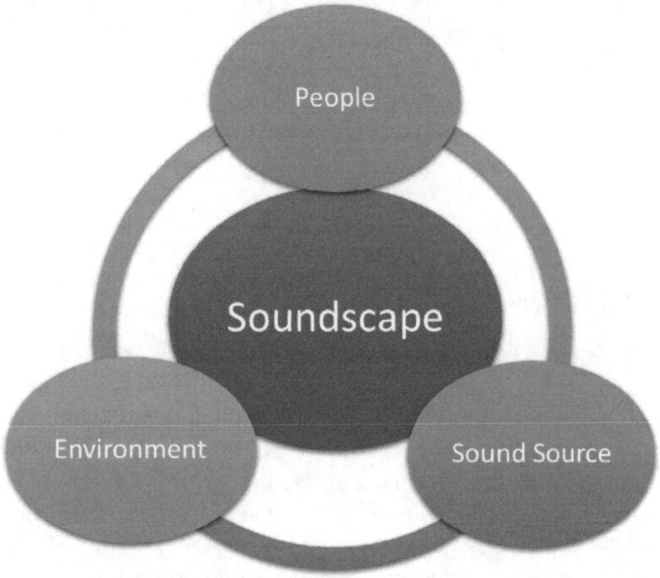

Figure 2. Three Aspects of Soundscapes

The soundscape concept mainly applies to the urban soundscape [16–18] and it must be implemented in the design stage and for urban sonic environment treatment [19]. The concept tries to understand the positive aspects of the sonic environment, not only the negative aspects and noise [20]. The soundscape approach conducts the treatment of the sonic environment in two ways: reducing the negative sound and adding positive sound.

The soundscape approach tries to analyse the sonic environment by focusing on how people perceive the sonic environment, not only based on noise measurement [21]. Table 2 presents the differences between noise management based on noise measurement and the soundscape concept.

Table 2. Sonic environment management based on the noise concept and the soundscape concept

Sonic environment management based on the noise concept	Sonic environment management based on the soundscape concept
The evaluation result is generalized for similar functions of space	The evaluation result cannot be generalized for similar functions of space
Based on noise measurement	Based on perception evaluation

All sound sources contribute to the overall noise	The sound sources are categorised into positive and negative sound sources
Managed by reducing the overall sound level	Managed by reducing the negative sound sources or masking the sonic environment using positive sound sources

2. Soundscape analysis method

According to Aletta et al., the method of analysing soundscape can be divided into based on three approaches as shown in Figure 3: in-situ experiment, laboratory experiment, and recalled in memory.

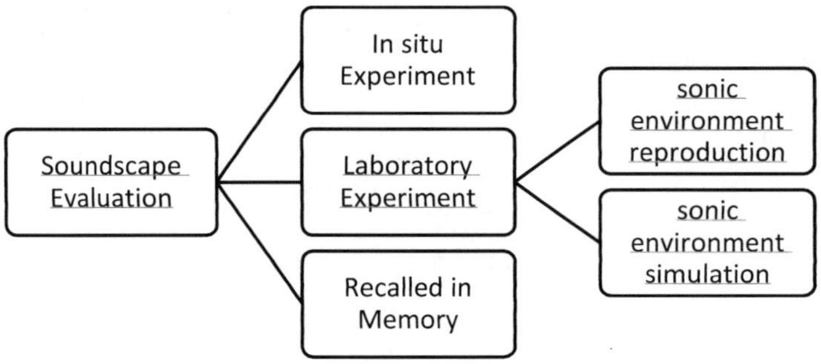

Figure 3 Soundscape evaluation method according to Aletta et al [22]

The in-situ experiment usually comprises three methods: behavioural observation [23], soundwalk [24–27], and narrative interview [21,28]. The soundwalk is an empirical method used to identify a soundscape and its sound objects in-situ [14]. Figure 4 shows a snapshot of the soundwalk. Although the in-situ experiment is one of the most common methods to analyse the urban soundscape, this method has the issue of repeating the experiment. Semidor suggested that a soundwalk should be conducted more than once, in order to minimise the temporal change effect [25]. Time effects are an important aspect of soundwalks since soundscapes are always changing. The in-situ experiment also needs to be done in a group with a limited number of participants. Another problem with the in-situ experiment is the limited variation of the sonic environment, making it difficult to model the interaction between the sound source and the perception.

Figure 4. Soundwalk in an urban area

The laboratory experiment usually involves sonic environment reproduction and sonic environment simulation [29][30]. The sonic environment is commonly reproduced using a multi-speaker system, as seen in Figure 5. This system has been validated using semantic differential analysis [31,32] and semantic categorization [33]. The reproduction by using a headphone has also been implemented but there is limited information about the validity of the system. The sonic environment simulation using a multi-speaker system also has been validated using semantic differential analysis [34].

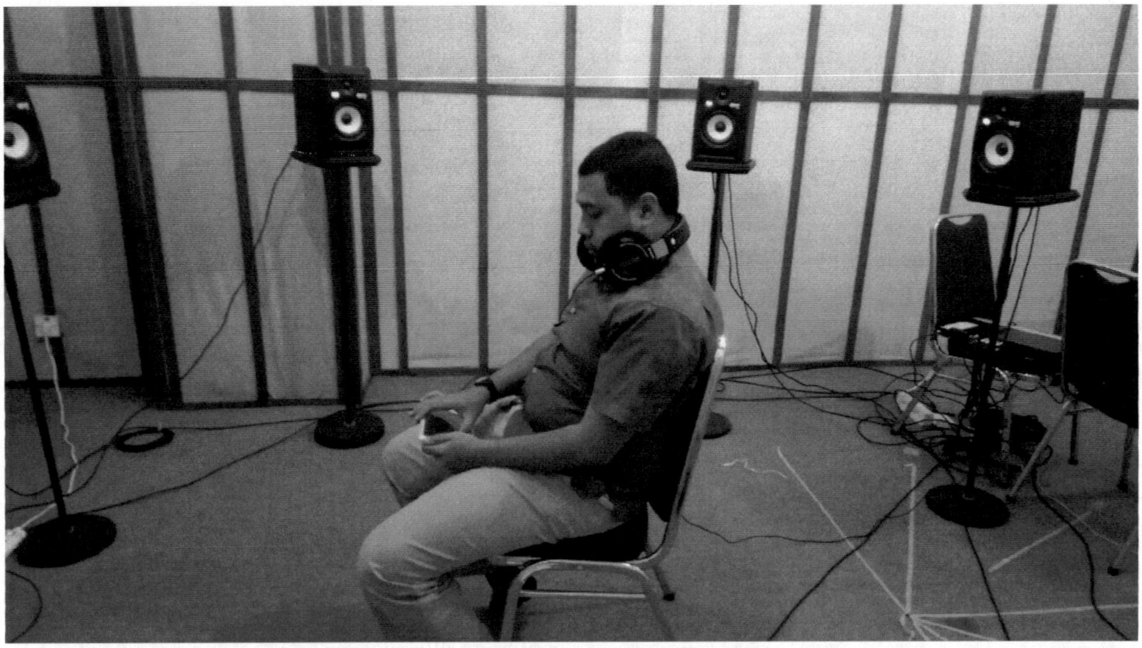

Figure 5. Sonic reproduction system using speakers

The laboratory experiment has the advantage of the repeatability of the experiment; the sonic environment can be repeated as often as possible. However, since the sonic environment needs to be reproduced using a multi-speaker system, the system requires a room with acoustic treatment. The experiment must also be conducted individually in the room where the system is installed. Another problem is that the variation of the actual sonic environment limits the variation of the sonic environment.

The soundscape evaluation method based on memory is usually conducted using interviews or questionnaires. This method aims to identify the sound sources and to understand the expectation of the sonic environment [35]. This method is also commonly used for a soundscape research where the in-situ experiment and the sonic environment recording cannot be done such as in an Intensive Care Unit [36]. The limitation of this method is that the participants must have experienced the actual space since the context of the sonic space cannot be provided by a recording.

Apart from these three methods, a method is developed to analyse soundscape using the soundscape composition concept and internet-based tests as shown in Figure 6. These two methods are developed to deal with the limitations of the common methods.

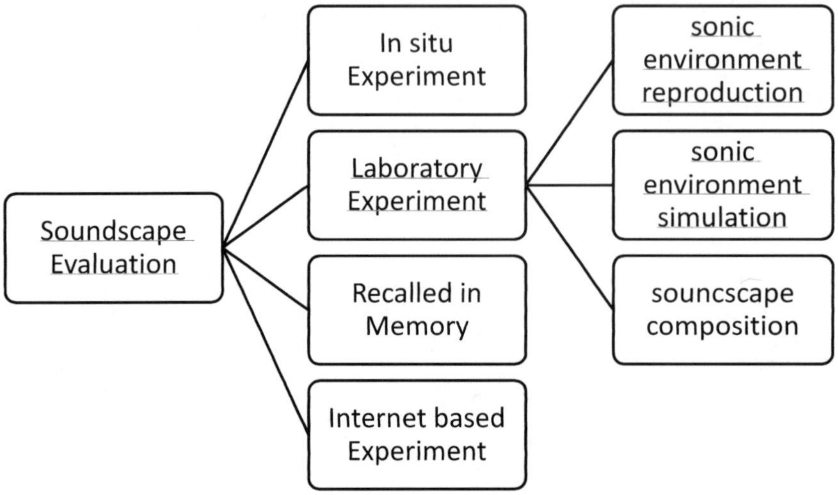

Figure 6. The Development Soundscape evaluation method

The soundscape composition concept is proposed by Bruce et al. [37] by developing a sonic environmental simulator that allows the addition of sound sources in a sonic environment. Sudarsono et al. [34] further developed the simulator, enabling the composition of a sonic environment using only the sound sources. This simulator can also simulate the movement of moving sound sources and the position of the sound sources. By using the simulator, the participant can easily compose the expected soundscape and the model that represents the interaction between sound source and people's perception has been determined [38].

The internet-based experiment is developed to increase the participation in soundscape experiments. The internet-based experiments have become an alternative for various reasons. The first reason is the increasing internet speed, which allows playing high-quality audio and even videos. The system for soundscape evaluation can be developed using internet easily.

The second reason is the development of virtual reality technology [39]. This technology allows us to make immersive virtual environments using three-dimensional audio video. The reproduction system also has been available using Facebook 360 and YouTube 360. The virtual environment can be easily reproduced in a laboratory and even by using the internet. This system has become an alternative for soundscape research [40]. The limitation of this system is that the reproduction needs using specific tools such as mobile phones or virtual reality devices and the sonic environment needs to be evaluated after the participant experiences the sonic environment.

The aim of this study is to develop a soundscape evaluation tool using the internet. The tools will be developed without virtual technology to simplify the experiment. By using this system, the participant can fill in a questionnaire while listening to the sonic environment. In addition, the system is developed using simple recording and a reproduction system allowing everyone to use this system easily. This system will allow the soundscape experiment to be conducted anywhere with more participants compared to the existing methods. The experiment also can be conducted only using headphone and laptop/mobile phone.

3. Development of an internet-based urban soundscape evaluation system

An internet-based urban soundscape evaluation system allows the soundscape experiment to be reproduced anywhere. Using this system, the data can be collected through an online form and can be shared easily. The key element of this system is how to reproduce the sonic environment in a valid way regarding the variation of experimental conditions. In the internet-based experiment, the selection of headphones and the tools to fill in the questionnaire cannot be controlled. Regardless the variation, the data from the questionnaire must be valid. Moreover, the system must be made as simple as possible in making the recording and the questionnaire.

The concept of participation in soundscape analysis has been analysed by Laso et al. [41] who state that urban design must consider the acoustic problem in the society and that participatory analysis has become an important tool in data collection. The smart city concept also includes the aspect of participation in decision-making which is can be done easily using an internet-based soundscape test.

The soundscape evaluation system is based on an online questionnaire form. The online form is selected because of its simplicity and low cost. By using this online form, the questionnaire can be made easily for free. The questionnaire consists of several soundscapes recording and the problem is how to reproduce the soundscape by using headphone, which can give the similar perception to the actual soundscape via the internet.

The focus of this study is to find a recording technique that is simple but provides valid reproduction; find out the effect of different headphones to the sonic environment perception; and establish the calibration of the reproduction system.

3.1 Recording technique

The recording technique has been developed quickly. The recordings in soundscape study are commonly made using a Soundfield microphone, which records in B-Format (4-channel signals). This method is also known by ambisonic recording technique. The B-format signal is then reproduced using speakers in a laboratory. Although the system has valid results [42], the application using a headphone has not been determined. Furthermore, the system is a little bit complex since the signal needs to be recorded simultaneously and decoded before the signal reproduced. The signals can be decoded into various systems such as an ambisonic reproduction system, surround system, or stereo system. Figure 7 shows the ambisonic recording system for stereo reproduction.

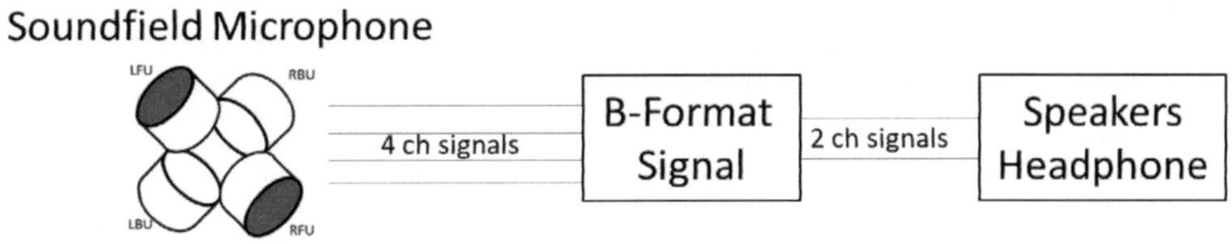

Figure 7. Ambisonic Recording System for Stereo Reproduction

Another study uses binaural recording with an in-ear microphone or dummy head to make an impressive sonic environment reproduction. The system is much simpler than using a Soundfield microphone since the output is already in stereo. Figure 89 shows the recording system. The disadvantages of this system are that the validity of binaural reproduction using a headphone for a soundscape experiment has not been determined. Moreover, the microphone used for this recording is not widely available.

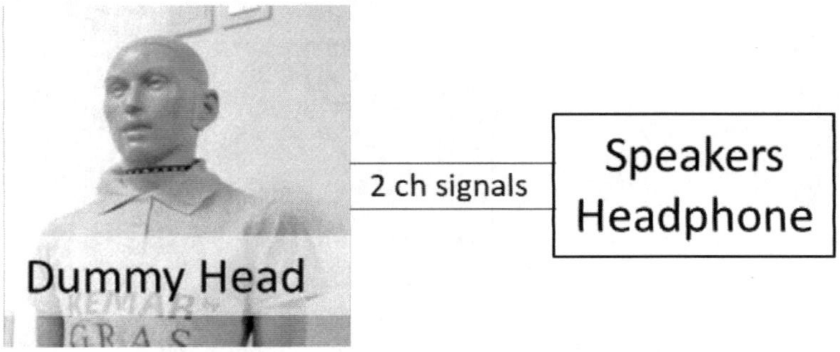

Figure 8. Binaural Recording System for Stereo Reproduction

The focus of this study is to find a simple stereo recording technique that is widely available and easy to use. Our study found out that the Mid Side recording technique is suitable for soundscape recording.

Mid Side recording is a recording technique using two types of microphones (cardioid and figure of eight microphone) as shown in Figure 9. This is one of the most common recording techniques in hand-held recorders beside the XY recording technique. The Mid Side signal must be decoded into left and right channel before the signal can be reproduced but some hand-held recorders include the decoder inside so the output signals are ready to be reproduced.

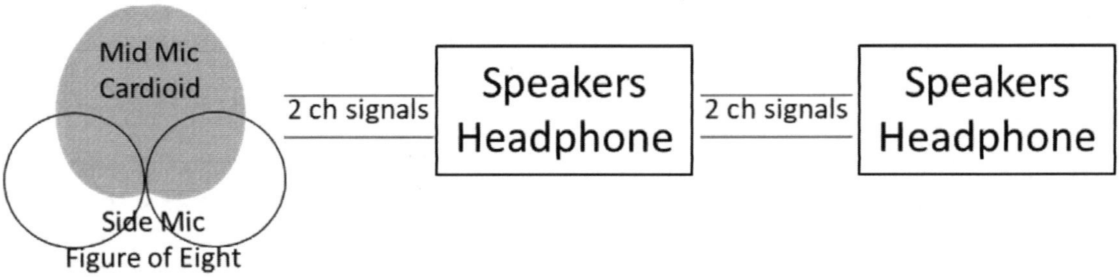

Figure 9. Mid Side Recording System for Stereo Reproduction

The validation in the laboratory using semantic differential analysis shows that the recording with Mid Side technique could give the similar perception to the in-situ experiment and the reproduction using multi-speakers reproduction system [31]. Three soundscape dimensions emerge from the experiment using MS recording: Relaxation, Dynamic, and Communication. The detailed result for this part is presented in another paper about the validation of soundscape reproduction using headphone.

3.2 Headphone variation

The headphone variation is analysed to understand the difference in perception between three different types of headphone, the over-ear, on-ear, and in-ear headphones. Figure 10 shows the different types of headphones used for the experiment. Our study shows that there is no significant difference ($sig<0.05$) between the types of headphones. Furthermore, the same soundscape dimension also emerges from these three types of headphone. A detailed result is presented in another paper about the validation of soundscape reproduction using headphones.

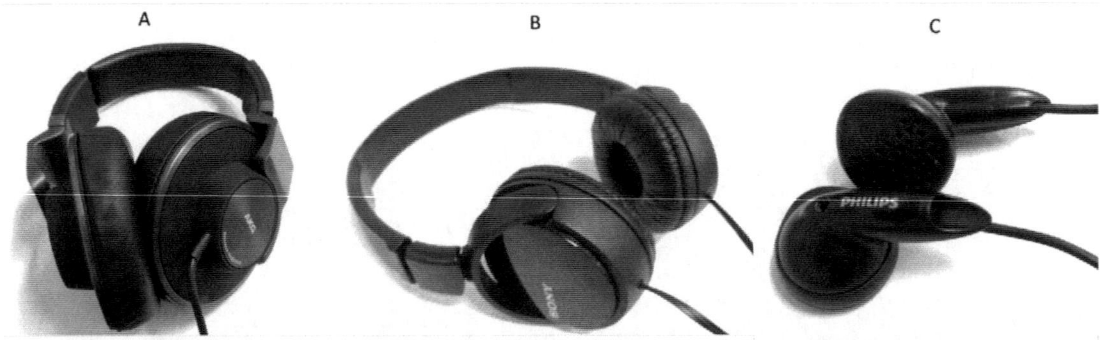

Figure 10. Headphone used in the experiment. A is the over-ear headphone, B is the on-ear headphone, and C is the in-ear headphone

3.3 Calibration system

One of the important aspects of the soundscape reproduction is the calibration of the reproduction. Sudarsono et al. showed that sound reproduction at the wrong level could affect the perception [43].

Furthermore, their study indicates that when the participants of the experiment are given the opportunity to adjust the sound level, which represents the actual sounds, they adjust it to a similar level.

The results of the study indicate that it is possible to develop a calibration system for soundscape reproduction using people's perception. This method could prove to be useful since the calibration usually requires an instrument that is not widely available.

Figure 11 shows the calibration system. Before the experiment, the participants listened to a sonic environment recording and were requested to adjust the sound level of the gadget (laptop, tablet, mobile phone) that represented the actual level. After the calibration, the experiment could be started. Since the other recordings were made with the same microphone and the same sensitivity, the calibration applied to the other recording. The detailed results of this experiment are presented in another paper about the development of web-based soundscape reproduction calibration.

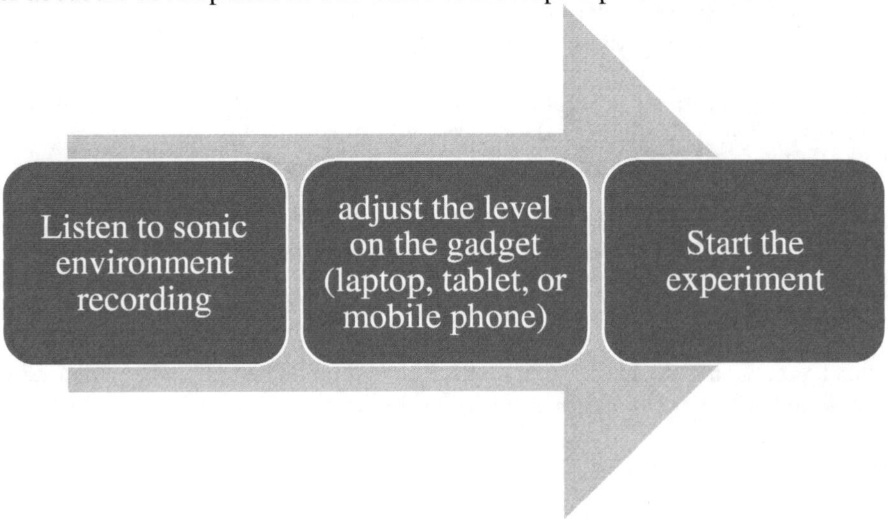

Figure 11. Internet-based sonic reproduction system's calibration based on perception

4. Conclusion

The internet-based sonic environment simulator using a headphone has been developed based on the principles of simplicity, low-cost, and validity. The study demonstrated a suitable recording technique, which is simple and relatively cheap, namely the Mid Side Stereo Recording Technique. This recording technique can provide a perception similar to the actual soundscape. The experiment also demonstrates that the type of headphone does not significantly affect the perception of the soundscape. The calibration system also has been developed based on perception. This alternative calibration method can be used for calibrating the internet-based reproduction system anywhere.

References

[1] Elmqvist T 2013 Designing the Urban Soundscape
[2] Rehan R M 2015 The phonic identity of the city urban soundscape for sustainable spaces HBRC J.
[3] Adams M, Davies B and Bruce N 2009 Soundscapes : an urban planning process map Inter-Noise 2009 (Ottawa: INCE)
[4] Giffinger R, Fertner C, Kramar H, Kalasek R, Pichler-Milanović N and Meijers E 2018 Sustainable urban transformations towards smarter, healthier cities: Theories, agendas and pathways *J. Clean. Prod.* 173 1–10
[5] Study G I-T Focus Group on Smart Sustainable Cities
[6] Science R and Studies M 2007 Smart cities Ranking of European medium-sized cities

[7] Berglund B 2006 From WHO guidelines for community noise to healthy soundscapes *Proc. Inst. Acoust. Inst. Acoust.* **28** 1–9
[8] Ozcevik A and Can Z Y 2012 A field study on the subjective evaluation of soundscape 2121–6
[9] Westerkamp H 2002 Linking soundscape composition and acoustic ecology Organised Sound **7** 51–6
[10] Liu A, Wang X L, Liu F, Yao C, Deng Z, Liu A and Wang X L 2018 Soundscape and its influence on tourist satisfaction 2069
[11] Hall D A, Irwin A, Edmondson-Jones M, Phillips S and Poxon J E W 2013 An exploratory evaluation of perceptual, psychoacoustic and acoustical properties of urban soundscapes *Appl. Acoust.* **74** 248–54
[12] Viollon S, Lavandier C and Drake C 2002 Influence of visual setting on sound ratings in an urban environment *Appl. Acoust.* **63** 493–511
[13] Yang W and Kang J 2005 Soundscape and Sound Preferences in Urban Squares : A Case Study in Sheffield *J. Urban Des.* 10 61–80
[14] Schafer R M 1977 The Soundscape: Our Sonic Environment and The Tuning of the World (New York: Alfred Knopf, Inc)
[15] International Organization for Standarization 2013 Draft BS ISO 12913-1 Acoustics - Soundscape. Part 1: Definition and conceptual framework 44
[16] Yang W and Kang J 2005 Acoustic comfort evaluation in urban open public spaces *Appl. Acoust.* **66** 211–29
[17] Andringa T C, Weber M, Payne S R, Krijnders J D D, Dixon M N, v d Linden R, de Kock E G L and Lanser J J L 2013 Positioning soundscape research and management. *J. Acoust. Soc. Am.* **134** 2739–47
[18] Asdrubali F 2014 New frontiers in environmental noise research 1–2
[19] Kang J, Aletta F, Gjestland T T, Brown L A, Botteldooren D, Schulte-fortkamp B, Lercher P, Kamp I Van, Genuit K, Luis J, Coelho B, Maffei L and Lavia L 2016 Ten questions on the soundscapes of the built environment Build. Environ.
[20] Brown A L, Kang J and Gjestland T 2011 Towards standardization in soundscape preference assessment *Appl. Acoust.* **72** 387–92
[21] Davies, Adams, Bruce, Cain, Carlyle, Cusack, Hume, Jennings and Plack 2007 The Positive Soundscape Project 19th INTERNATIONAL CONGRESS ON ACOUSTICS (Madrid)
[22] Aletta F, Kang J and Axelsson Ö 2016 Landscape and Urban Planning Soundscape descriptors and a conceptual framework for developing predictive soundscape models Landsc. *Urban Plan.* **149** 65–74
[23] Lepore F and Aletta F 2016 A preliminary investigation about the influence of soundscapes on people's behaviour in an open public space
[24] Jeon J Y, Hong J Y and Lee P J 2013 Soundwalk approach to identify urban soundscapes individually. *J. Acoust. Soc. Am.* **134** 803–12
[25] Semidor C 2006 Listening to a city with soundwalk method Acta Acoust. *United with Acoust.* **92** 959–64
[26] Liu, Kang, Behm and Luo 2014 Effects of landscape on soundscape perception: Soundwalks in city parks Landsc. *Urban Plan.* **123** 30–40
[27] Adams, Bruce, Davies, Cain, Jennings, Carlyle, Cusack, Hume and Plack 2008 Soundwalking s Methodology for Understanding Soundscapes Institute of Acoustics Spring Conference 2008 **30** (Reading: the Institute of Acoustics)
[28] Davies W J, Adams M D, Bruce N S, Cain R, Carlyle A, Cusack P, Hall D a., Hume K I, Irwin A, Jennings P, Marselle M, Plack C J and Poxon J 2013 Perception of soundscapes: An interdisciplinary approach *Appl. Acoust.* **74** 224–31
[29] Hong J Y and Jeon J Y 2013 Designing sound and visual components for enhancement of urban soundscapes *J. Acoust. Soc. Am.* **134** 2026–36
[30] Jeon J Y, Lee P J, You J and Kang J 2012 Acoustical characteristics of water sounds for soundscape enhancement in urban open spaces *J. Acoust. Soc. Am.* **131** 2101–9
[31] Sudarsono A S, Lam Y W and Davies W J 2016 The effect of sound level on perception of reproduced soundscapes *Appl. Acoust.* **110** 53–60

[32] Davies W J, Bruce N S and Murphy J E 2014 Soundscape Reproduction and Synthesis Acta Acust *United with Acust.* **100** 285–92

[33] Guastavino C, Katz B F G, Polack J and Levitin D J 2005 Ecological Validity of Soundscape Reproduction Acta Acoust *United with Acoust.* **91** 333–41

[34] Sudarsono A S, Lam Y W and Davies W J 2017 The validation of acoustic environment simulator to determine the relationship between sound objects and soundscape Acta Acust. united with Acust. 103

[35] Bruce N S and Davies W J 2014 The effects of expectation on the perception of soundscapes *Appl. Acoust.* **85** 1–11

[36] Rofiatun I, Azzahra N, Sarwono J, Sudarsono A S and Frans R 1811 Hospital Soundscapes : Perception Analysis of Acoustics Environment in Intensive Care Unit Inter Noise 2017 1811–6

[37] Bruce N S, Davies W J and Adams M D 2009 Development of a soundscape simulator tool Inter-Noise 2009 (Ottawa)

[38] Sudarsono A S, Lam Y W and Davies W J Soundscape Simulator as A Tool to Predict the Perception of Acoustic Environment Inter Noise 2017 1073–8

[39] Ruotolo F, Maffei L, Di Gabriele M, Iachini T, Masullo M, Ruggiero G and Senese V P 2013 Immersive virtual reality and environmental noise assessment: An innovative audio-visual approach Environ. Impact Assess. Rev. 41

[40] Hong J Y, He J, Lam B, Gupta R and Gan W 2017 Spatial Audio for Soundscape Design : Recording and Reproduction

[41] Villazón C, César L and Ignacio A 2017 Design and Validation of a Simulator Tool Useful for Designers and Policy Makers in Urban Sound Planning Acoust. Aust.

[42] Stevens F, Murphy D and Smith S 2017 Ecological Validity of Stereo UHJ Soundscape Reproduction Audio Engineering Society Convention 142

[43] Sudarsono A S, Lam Y W and Davies W J 2016 The effect of sound level on perception of reproduced soundscapes *Appl. Acoust.* **110** 53–60

IOP Conf. Series: Earth and Environmental Science **158** (2018) 012053 doi:10.1088/1755-1315/158/1/012053

Ride-sharing business model for sustainability in developing country: Case Study *Nebengers*, Indonesia

Asirin Asirin[1]; Danang Azhari[2]
[1,2]Urban and Regional Planning, Institut Teknologi Sumatera (ITERA), Lampung Selatan
[1]Resilience Development Initiative (RDI), Kota Bandung

Corresponding author: asirin@pwk.itera.ac.id

Abstract. The growth of population and urban economy increased the need for humans' mobility to support their activities. On the other hand, online Information and Communication Technology (ICT) is growing rapidly and more affordable. Within few years, there is some sharing economy business formed by using online platform. This condition brings through the emergence of ride-sharing business model using an online platform which can be beneficial to sustainability. This research aims to explore one of ridesharing business models which use the online platform and its impact on sustainability. This research used the procedure of case study method with a single case study of *Nebengers*. This research explores the case study with the scope of this research is limited by using several conceptual frameworks, they are sharing economy business model, four elements of a business model for sustainability (BMfS), Social Construction of Technology (SCoT), sustainable mobility and agency theory. *Nebengers* is a sharing economy business using online platform that historically can be explained using Social Construction of Technology (SCoT) Theory. There are conflicts between *nebengers* entrepreneur and the city government. *Nebengers* disrupts traditional and formal public transportation services which are managed by the government. However, *nebengers* also contributes to achieve the city government goal in developing sustainable mobility. The future challenge is how to arrange ride-sharing collaborative governance business model for sustainability in the cities in Indonesia.

1. Introduction

The growth of population and urban economy increased the need for humans' mobility to support their activities. On the other hand, online Information and Communication Technology (ICT) is growing rapidly and more affordable. Within few years, there is some sharing economy business formed by using online platform. This condition brings through the emergence of ride-sharing business model using an online platform which can be beneficial to sustainability. This research aims to explore one of ridesharing business models which use the online platform and its impact on sustainability. This research used the procedure of case study method with a single case study of *Nebengers*. This research explores the case study the scope of this research is limited by using several conceptual frameworks, they are sharing economy business model, four elements of business model for sustainability (BMfS) [1], Social Construction of Technology (SCoT) [2] [3], sharing mobility for sustainable mobility [4] and agency theory [5]. *Nebengers* is a sharing economy business using online platform that historically can be explained using Social Construction of Technology (SCoT) Theory.

Table 1. The elements of sharing economy business model [6] [7]

Aspect	Description
Sharing economy business: platform for collaboration	• Types of platforms used in delivering services • Role and functionality of the platform • Specific use of the platform by the users, customers and company • Types of activities performed by the company and users while interacting • Degree of platform centrality to the company's core

Aspect	Description
	business
Under-utilized resources	• Kind of users' resources which the peers share to enable sharing • Reliance on the excess capacity of users
Peer-to-peer interaction	• Types of peer-to-peer interactions • Types of peer-to-peer transactions • Relevance of interaction and transaction to business model and performance
Collaborative Governance	• Legal structure, equity model, governance structure and mechanism, decision-making processes, consultation, and CSR and engagement strategies
Mission-driven	• Formal mission statement • Relevance of social and environmental value and impacts in relation to economic value and to the business as a whole • Evidence on how the mission is implemented: strategies, practices
Alternative funding	• Type of funding and stage in which the funding is requested/used. • Extent to which the business utilizes or prioritizes alternative funding such as crowdfunding, equity-based crowdfunding or similar
Leverage on technology	• Type of technology used by the business • Relationship between the technology in use and the core business • How the business uses technology to operate

Table 2. Elements of a Business Model for Sustainability (BMfS) [1]

Element	Definition
Value proposition	Provide measurable ecological and/or social value in concert with economic value
Supply chain	Involve suppliers who take responsibility for their own as well as for the focal company's stakeholders
Customer interface:	Motivate customers to take responsibility for their consumption as well as for the focal company's stakeholders
Financial model	Reflect an appropriate distribution of economic costs and benefits among actors involved in the business model and accounts for the company's ecological and social impacts

2. Result and discussion

2.1. Nebengers profile and social construction of technology of Nebengers application

This sub-section describes *Nebengers* profile based on information source from *Nebengers* Company. *Nebengers* is a social movement to give each other a ridesharing which exists nowadays with a lot of popular online transportation. There are about 80,000 members of the community and 150 private vehicles active in *Nebengers* application. There are some conditions when Nebengers can be relied on, such as a daily going home trip, homecoming trip, and weekend activity to commute out of town. Nowadays, *Nebengers* is managed by five young people, including Andreas Aditya Swasti, an IT department employee at a private company. This community was started in 2011 when Andreas was aware that the cause of traffic jam in Jakarta is the higher increase in the number of vehicles than the growth of existing road length. Then, Andreas Aditya Swasti thought how to get to a destination without riding either his private vehicle or public transportation which sometimes is uncomfortable.

The answer is carpooling or ride-sharing with someone's vehicle with the same destination or direction. The idea started with creating a Twitter Account named *Nebengers*. This account would help people to find their carpool partner with the same destination. The carpooling is also based on the agreement among each member of the Nebengers community. Then, *Nebengers* begins to demand, especially by citizens of Jakarta and Bandung. This concept is considered suitable for those who need to commute between the two cities. Currently, *Nebengers* also becomes an alternative transportation in *mudik* season at Ramadhan Idul Fitri. *Nebengers* evolves not only ride other people's vehicles, but also become a community where they can share with others who need it. This community also can make a new friendship because it is not just a ridesharing to the same destination, but also a new meeting which can continue to the next one.

The *Nebengers* app is an app that can help urban citizens in carpooling activity. This app has been designed to answer the needs of carpooling and the performance assessment of *Nebengers* for almost 3 years. In the previous platform, which is Twitter, there are some difficulties in quantifying some activities, such as how many trips occur each day and the most crowded route per days. This new platform is expected to assess its performance, and estimate solutions to develop *Nebengers* as one of the Smart City transportation solutions in several cities. In addition, the previous platform, Twitter is limited for those who have a Twitter account. Therefore, with this new platform, *Nebengers* gives access to everyone who will be able to do carpooling activities.

This platform features "Create *Tebengan*" for those who have an empty seat in their vehicle, and want to share it. They can choose routes, departure schedules and sharing either the cost of gasoline or toll fee per person. Beside, people who want to do ridesharing can use 'Search *Tebengan*' feature, then they will be able to book a seat and negotiate for the payment with 'Chat' feature. Afterwards, the rider who shares the empty seat can either "Approved" or "Reject".

With the Notifications and Chat feature, the communication of rider and the passenger can be simplified. When meeting at the meeting point, the rider is required to press the "Getting Started" feature to indicate that the trip begins. If the trip is over, the rider will press "End the Trip" feature. After the trip, the passenger can review the trip with the rider.

Nebengers is a ride-sharing marketplace which helps to connect between the owners of the empty seat in their vehicle named as "Captain", with "Passengers" or the one who wants to do ridesharing. *Nebengers* makes a community, so users can do a ride-sharing with another users who are registered at nebengers.com to save their travel cost, make new friends, share experiences and support the green environment. To make sure everything works well, fun and rewarding for both *Nebengers* users and managers, *Nebengers* sets terms and conditions agreement. *Nebengers* does the following things:

a. Provide applications in the form of websites, Android and iOS so users can provide a ride and look for a ride
b. Do accuracy calculation on the trip route submitted. Submitted data does not violate ethics such as SARA, pornography and other matters that interfere the user's security and convenience
c. *Nebengers* may provide user-related data, travel transactions that occur in cases of criminal offences and crimes.
d. *Nebengers* invites its users to post the route on the application website, android or iPhone
e. Provide social media information to the users about both rider and passenger in consideration for those who want to go together
f. Provide a verification system for rider/ captain and passenger.

Nebengers does not do the following things:

a. Provide user A's recommendations to come with Captain B.
b. Misusing personal data and harming its users
c. Provide a personal mobile number to other users.

Nebengers users need to do the following things:

a. Obey the ethics and laws under the Indonesian Constitution of 1945

b. Post on the Nebengers app to give a ride or look for a ride
c. Maintain the security and comfort of other users.
d. Make timely appointments.

If people are looking for a ride in *Nebengers* using Twitter, people can use this hashtag *#CariTebengan* and make a new tweet using this format, for example "*#CariTebengan* Jakarta-Bandung | 1 seat | 21 Maret 2016 09:00 | share gasoline and toll". People also can use *Nebengers* Application with the following steps:

1. Search Destination
Enter your destination in "Tujuan" box.

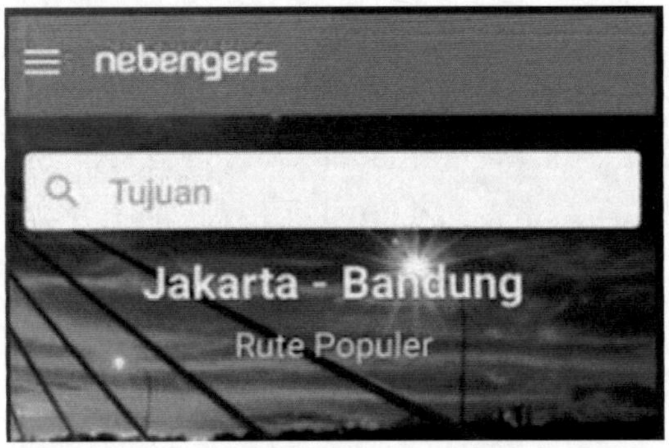

Figure 1. *Nebengers* Application Menu Options: Search Destination

2. Booking a seat
If there are some provided routes, you can choose a route that you want. Then, you will get the carpooling detail. You can either contact the rider by pressing Chat icon in the top right corner or book a seat by pressing 'booked a seat' button.

Figure 2. *Nebengers* Application Menu Options: Available Routes

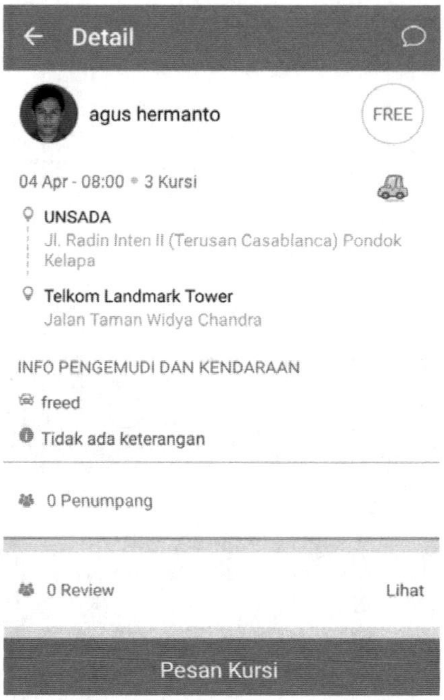

Figure 3. *Nebengers* Application Menu Options: Booked a Seat

3. Meeting and Starting the trip

Soon after the customers press the button "Pesan Kursi", their request has already been logged in to the Rider. If it is approved, then they will get a notification.

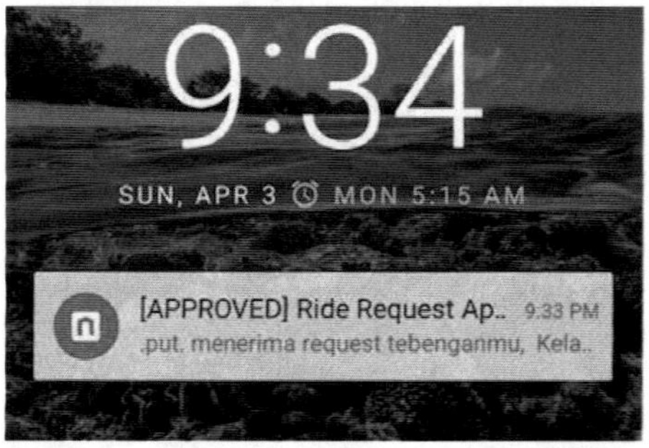

Figure 4. *Nebengers* Application: Approved Ride Request

Nebengers is a community that consists of people who are willing to do ride-sharing. In the first phase, *Nebengers*'s members communicate via social media twitter. The next phase, *Nebengers* develops a mobile app namely *Nebengers*. In the Social Construction of Technology (SCoT) perspective, the online platform *Nebengers* is developed to facilitate ride-sharing behavior.

2.2. The characteristics of the ridesharing business model for sustainability

This section discusses the details description of characteristics of the ride-sharing business model for sustainability.

2.2.1. Platform and technology. Type of platforms used in delivering services is an online platform. The technology used initially is social media and then *Nebengers* application is developed as the online platform. The role and function of online platform are to facilitate the person who wants to involve in ride-sharing activity. The online platform is used for transactional media of ride-sharing information. The platform is used by *Nebengers* company to determine the performance of *Nebengers* application and the activities of *Nebengers* community. Online platform plays an important and decisive role in the succesful Nebengers activities. The online platform facilitates *Nebengers* community interaction quickly and widely. If there is no *Nebengers* application platform, it can still be run by utilizing Twitter.

2.2.2. Mission-driven and value preposition. The *Nebengers* company claims that *Nebengers* application can be a smart city transport solution in several cities in Indonesia. Formally *nebengers* states that the objective values to be achieved are:

a. 3 Cars Being 1. Private vehicles leave many empty seats; therefore, let us get together to overcome the traffic jam.
b. More efficient. By sharing, the transportation is more cost-effective and makes the trip more efficient.
c. More fun. The beauty of sharing empty seats, reducing city congestion and expanding friendship for business as well as spouse.

Nebengers also claims that ridesharing has benefits among others:

a. Transportation cost savings (gasoline and toll fee)
b. Travel expenses (gasoline and toll fee) can be split together so that the trip becomes more efficient and economical.
c. Add companions.
d. Environmentally friendly.
e. Expand business relationships.
f. Minimize stress during traffic jam because there is a friend on the road

In other words, the value of the *Nebengers* can reduce emissions and congestion, and subsidize driver costs.

2.2.3. Underutilized resources and supply chain. The vehicles used are private vehicles owned by people who want to share their vehicle. The offered vehicle has an unused empty chair *(underutilized).*

2.2.4. Peer to peer interaction and customer interface. The transacting party in the ridesharing was originally a colleague of close friends, neighbors, and peer to peer. Then, the party participating in ridesharing could be anyone who can come. Related to this, there is a security issue, but to minimize it, *Nebengers* makes a mechanism that ride-sharing community must register first with their identity. Then, there is a review and rating mechanism to validate the truth or make sure that the parties are safe. Ride-sharing community may mutually reject if it does not match, and it is okay to accept if it fits. In other words, those who can join are close friends, neighbors, and even new people who then become friends. If they are appropriate, they will be involved in continuous ride-sharing when the seat is available. There are also thirteen users out of twenty users who are satisfied by *Nebengers* services. It indicates that *nebengers* does a nice job at interacting with the one who has an empty seat and the passengers.

"Warid, a Student who used Nebengers said that the services of Nebengers is as good as an online taxi in a good way, and it was a great experience. Because the members of Nebengers are nice and friendly. According to him, Nebengers is non-money oriented service like any other online taxi services, that is why the members of Nebengers are nice and friendly."

2.2.5. Type of governance. The governance model is private corporate. There is a partnership with sponsors.

2.2.6. Financial model/alternative funding. At first, the founders of *Nebengers* practised their own ride-sharing culture with their close friends. Then, the founders of *Nebengers* propagated the idea and used Twitter platform. At that time, the founders of *Nebengers* only facilitated it without getting any profit from the transaction. Transaction payments are made by the ride-sharing community self. Then, the *Nebengers* company has created the *Nebengers* application which facilitates the transaction of ride-shraing community. The *Nebengers* company gets the profit from downloading apps activities by ridesharing community on Android and iOS. *Nebengers* company gets revenue from sponsorship from various parties such as Top 1, Nu Green Tea, SmartFren, Indosat Ooredoo, XL, Telkomsel, Ice Cool, Datsun, Garda Oto, and Hard Rock FM Jakarta. In other words, the ride-sharing community payment is based on the agreement. Meanwhile, there is no payment from ride-sharing community to *Nebengers* application company for each ridesharing transaction.

2.3. Nebengers contribution to sustainable mobility

Nebengers community mobility activities is a form of sustainable mobility because it reducec local transportation's environmental footprint through reducing the utilization of car and fossil fuel consumption. The evidence that *Nebengers* contributes to creating sustainable mobility is based on information from one of the *Nebengers* community members that changes his mobility mode from private car utilization to ride-sharing. Based on our survey of twenty users of *Nebengers,* there are four out of twenty users chose this reason about why they do carpooling activity, because they want to reduce air pollution in Indonesia.

Mr. Yudhi is a 36-year-old male entrepreneur who regularly uses the Nebengers application from Bandung (Gasibu) to Jakarta (Cibubur). He uses Nebengers application once or twice a day. His level of education is Bachelor. At first, he used Twitter and then used the Nebengers app. Before he joined Nebengers, he used a private vehicle from Bandung to Jakarta. He spent Rp 75,000 per trip. The motivation of his involvement in Nebengers ride-sharing community is to reduce transportation costs and meet new friends.

In our survey, there are fifteen out of twenty users do carpooling activity. They used to drive their private car. They have changed their mobility mode from private car utilization to ride-sharing. At this point, *Nebengers* community mobility activities is a form of sustainable mobility because it reduces local transportation's environmental footprint through reducing utilization of car and fossil fuel consumption.

2.4. Principal-agent relationship of Nebengers entrepreneur and city government

Online transportation in Indonesia is regulated by Transportation Ministry Regulation Number 108 Year 2017 about organizing non-route public vehicles/transportation. Based on the review of the regulations, it is identified that *Nebengers* is a legal subject that is regulated as a land transportation application company. Meanwhile, the members of *Nebengers* ride-sharing community are not public transport companies. They are not subjected to the law in the regulation. However, based on agency theory, there is a principal-agent relationship between city government in Indonesia and *Nebengers* entrepreneur. The city government is a principal, on the other side, *Nebengers* entrepreneur is an agent. There are conflicts between *nebengers* entrepreneur and city government. *Nebengers* disrupts traditional and formal public transportation services managed by the government. Nevertheless, *Nebengers* also contributes to achieve city government goal in developing sustainable mobility.

3. Conclusion

Nebengers is a community that consists of people who are willing to do ride-sharing. In the first phase, *Nebengers*'s members communicate via social media Twitter. The next phase, *Nebengers* has developed an mobile app namely *Nebengers*. In this case, the online platform *nebengers* was developed to facilitate ridesharing behavior. The characteristics of *nebengers* are using peer to peer (P2P) platform type and underutilized shared resources. According to four elemenets of a Business Model for Sustainability (BMfS), *nebengers*'s values are to reduce emissions and congestion, and subsidize driver costs; *nebengers*'s supply chain is personal vehicle; nebengers's customer interface is colleagues or neighborhood; nebengers's financial model is small fees charged for users. *Nebengers* community mobility activities is a form of sustainable mobility because it reduces utilization of car and fossil fuel consumption. Based on agency theory, there is a principal-agent relationship between city government in Indonesia and nebengers entrepreneur. The city government is a principal, on the other side, *Nebengers* entrepreneur is an agent. There are conflicts between *Nebengers* entrepreneur and city government. *Nebengers* disrupts traditional and formal public transportation services managed by the government. However, *nebengers* also contributes to achieve city government goal in developing sustainable mobility. The future challenge is how to arrange a ride-sharing collaborative governance business model for sustainability in the cities in Indonesia.

Acknowledgements
This research is funded by ITERA 2018 Research Grant. We express our thanks to Mrs Helmia Adita Fitra, B.Sc., M.Sc. and Mr Yudha Rahman B.Sc., M.Sc. that support the funding of participation in the conference.

References
[1] Boons F and Lüdeke-Freund F 2013 Business models for sustainable innovation: State of the art and steps towards a research agenda *J. Cleaner Production* **45** 9-19
[2] Pinch T J and Bijker W E 1984 The social construction of facts and artefacts: Or how the sociology of science and the sociology of technology might benefit each other *Social Studies of Science* **14** 399-441
[3] Winner L 1993 Upon Opening the black box and finding it empty: Social constructivism and the philosophy of technology *Science, Technology, and Human Values* **18** 362-378
[4] Cohen B and Kietzmann J 2014 Ride on! Mobility business models for the sharing economy *Organization & Environment* **27** 279 –296
[5] Eisenhardt K 1989 Agency theory: An assessment and review *Academy of Management Review* **14** 57-74
[6] Munoz P and Cohen B 2017 Mapping out the sharing economy: A configurational approach to sharing business modeling *Technological Forecasting & Social Change* **125** 21-37
[7] Cohen B and Muñoz P 2015 Sharing cities and sustainable consumption and production: Towards an integrated framework *Journal of Cleaner Production* **134** 87-97

IOP Conf. Series: Earth and Environmental Science **158** (2018) 012054 doi:10.1088/1755-1315/158/1/012054

Challenges on Java's small city spatial planning

B Wirawan[1] and J R Tambunan[1]
[1]Urban and Regional Development Institute (URDI), Jakarta, Indonesia

Corresponding author: b.wirawan1996@gmail.com

Abstract. Most Indonesians nowadays live in the urban area, due to urbanization. 60 percent of the inhabitants of Java, the most populous island in Indonesia, will live in the urban area by 2010, a figure that is higher than the national average of 55 percent. Urbanization has brought a large influx of newcomers not only into the metropolitan or large cities, but also into small and medium cities. One of urbanization's most concerning impacts is urban sprawl that harms sustainability. There is a dearth of academic literature studying the phenomenon of urban sprawl in Indonesian small cities. Urban sprawl in small cities is commonly ignored, considered as a contained problem the solution for which is a simple use of local spatial plan. Unfortunately, for Indonesia, this solution is difficult to implement because city spatial plans works only within administrative jurisdictions, whilst the urban sprawls themselves generally occur across the borders of different administrative jurisdictions. This study studies urban sprawls occurring in several small cities in Java. While those cities already have spatial plan, most were the general spatial plan (*Rencana Tata Ruang Wilayah* Kota/Kabupaten) that operates only within the city's administrative jurisdiction. We are unable to find detailed strategies in these cities to deal with urban sprawl. We also found that all spatial plans are unable to come up with urban growth boundary strategy. Finally, we are also unable to find integrated detailed spatial plan among those cities to answer the urban sprawl situation.

1. Introduction about urbanization in Java

By 2010, almost half of Indonesia's population inhabited urban areas. By 2035, the number of Indonesians living in urban areas will have increased into being more than two-thirds of national population, according to BPS[1] projection [1]. For Java[2], the island where sixty percent of Indonesians live, the urbanization is higher than national. More than fifty-eight percent of Java's inhabitants lived in the urbanized area by 2010 and it is will reach around eighty-five percent by 2035. With less than seven percent of Indonesia area, the growth of Java urban population needs to be cautiously watched. Urbanization in Java is caused by its function as central activity of Indonesia such as: economic center whereas almost sixty percent of Indonesia GDP centralized in Java by 2013 [2], prime location of higher education whereas seventy-five percent of the best Indonesia universities located in Java in 2017 [3], healthcare center whereas fifty one percent of general hospital and two-thirds of specialized hospital were located by 2013 [4], also location of armed forces major installations and headquarters.

Rapid urbanization in Java comes with rapid urban growth. Firman elucidated that urbanization in Java seems to be uncontainable now, and its growth resulted on urban expansion at various places in Java [5], which then created lots of small cities or small urban areas agglomeration. This development happened not just for areas near existing metropolitans or large cities, but also in rural areas near existing small and medium cities [6]. With urbanization, come the consequences. Rapid urban growth resulted by urbanization creates several impacts, in which the positives are usually overridden by the negatives specifically due to the uncontrolled and uncoordinated growth of the cities [7] namely as urban sprawl. Urban sprawl has a direct impact on the increase of traffic, reducing resources and cluttered the city with

[1] BPS (Biro Pusat Stastistik) is Indonesia's central bureau of statistics

[2] Furthermore this research will state it as Java. Java consists of 6 provinces namely Jakarta, Banten, West Java, Central Java, Jogjakarta, and East Java. Madura Island which part of East Java Province included.

IOP Conf. Series: Earth and Environmental Science **158** (2018) 012054 doi:10.1088/1755-1315/158/1/012054

built-up areas [6] which leads to uncertainty and muddling through for the future of the city [8]. The intention of tackling urban sprawl has been part of the international consensus stipulated in the New Urban Agenda (NUA) under 2016's Quito Declaration on Habitat III Conference [9]. Understanding the negative impact of urbanization could enhance the sustainability of the cities, especially in Java, the most populated island in the world [10][11].

While there were many references regarding urban expansion about metropolitan or large cities, discourse about urban expansion in small and medium cities is relatively low. Bell and Jayne stated many urban theories based on research from metropolitan and large cities development toward their global or regional context which then neglected the uniqueness and complexity of small cities [12]. Way also expressed that small-medium cities often being neglected despite its better chance to create various productive ways of the quality of life and sustainability [13]. This study tries to understand the challenging situation faced by the spatial planning efforts in small cities in Java small citiess, which deals with increasing and never-ending urbanization. Based on population and urban expansion growth, this study examine whether or not population growth and urban expansion drive urban sprawl in Java small cities. This study also relates the ability of each city spatial plan to overcome this situation.

2. Finding small city in Java

There were several definitions of the small city that often used population number as the main parameter. United Nations defined as urban areas agglomeration with fewer than 300,000 inhabitants [14], Europe Union used functional urban area term with 50,000 to 250,000 inhabitants [15], while Government of Indonesia defined as urban areas with 50,000 to 100,000 inhabitants [16]. There is drawback by only using population size to define a city because city hierarchy may differ among countries and it lacks explanation on economic function and status [12][17]. For the case of Indonesia, there are several considerations that need to be taken into account. First is how the Indonesian administrative jurisdiction rules classify districts into two categories: city (*kota*) for urbanized area and regency *(kabupaten)* for rural areas. Secondly, higher population density is often seen in villages, the smallest administrative jurisdiction (*desa* for rural areas and *kelurahan* for its urban counterparts) of the urban areas than those of rural areas. There are also several thresholds on population density for functional urban areas. European Union uses 1,500 people per km^2 [15] while Indonesia's BPS uses 500 to 8500 people per km^2 threshold, combined with urban facilities and people occupation [16].

In this study, authors define the small city areas with urban function based on several parameters. Firstly, authors used administrative jurisdiction which is a city (*kota*) as the basis of selecting areas of study. By doing this, authors did not study urban areas located in regencies (*kabupaten*) unless that urban area is part of the selected city urban areas expansion. Secondly, this study used the combined definition of small city population threshold from UN which was between 100,000 to 300,000 inhabitants for its city core. Thirdly, we excluded cities that part of metropolitan or large city expansion because we believed that urban sprawled in these cities resulted from the effect of the larger city. By using those three parameters, we found that there were twelve small cities in Java by 2010, namely Sukabumi, Cirebon, Pekalongan, Tegal, Banjar, Salatiga, Magelang, Kediri, Probolinggo, Pasuruan, Madiun, and Blitar. We excluded two small cities due to their proximity to the large city and/or metropolitan area, they were Batu (part of Greater Malang) and Mojokerto (part of Surabaya Metropolitan Area).

IOP Conf. Series: Earth and Environmental Science **158** (2018) 012054 doi:10.1088/1755-1315/158/1/012054

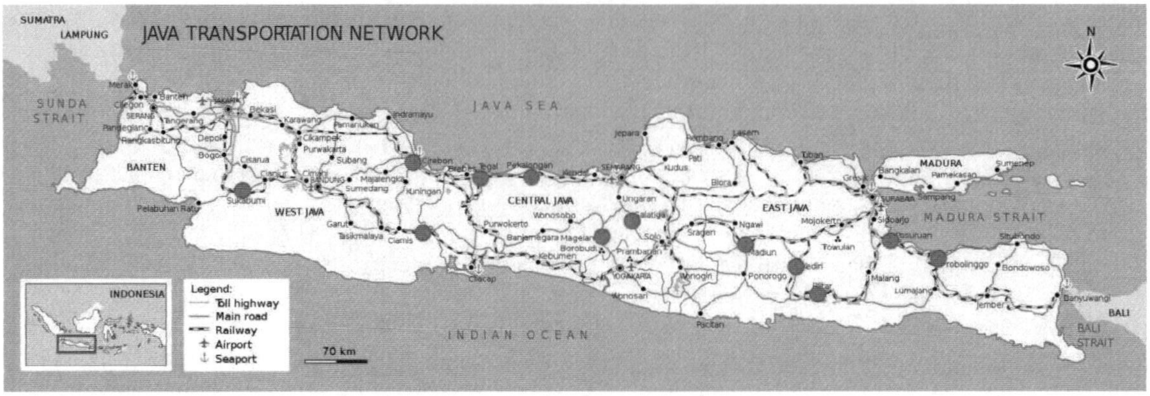

Figure 1. Location of selected small cities [18]

3. Research methodology

This study mainly tries to identify the effect of urbanization on the small cities which are selected as an area of interest. Then authors will elaborate how their spatial plans are able to face the negative effect of urbanization. This process was done through two stages. Firstly, we analyzed urban expansion in two ways, first the growth of urban village and their population, and second by using a change of land cover during the year 2000-2010 to understand the development of built-up areas. For urban village expansion we checked urban-village growth in all twelve cities and its surrounding regencies by using village data (*potensi desa*) released by BPS year 2000 and year 2010. Then, we cross-referenced urban-village growth with a change of land cover to triangulate the range of real urban expansion in all twelve cities and its surrounding regencies. In our effort to associated built-up expansion with population growth, we utilized land cover data from Ministry of Forestry despite its shortfall on its scale. During these research, we were unable to obtained more detailed scale of map which had the same year with population data, therefore we opted to utilize land cover data as our source of land use analysis. These analyses were figured out in a map in order to find the course of urban expansion in all small cities. Secondly, we examined each of city/regency spatial plans (*Rencana Tata Ruang Wilayah*/RTRW) to find whether each urban area has developed their strategies on urban sprawl or not.

4. Urbanization in Java's small cities

Rahayu [17] pointed out that based on population census; by 2010 there were 65% of Java urban dwellers lived in the small and medium cities. By 2010, there were twelve small cities in Java with population between 100,000 to 300,000 that are not adjacent to large city and/or metropolitan city. Three of these cities which are Sukabumi, Cirebon and Banjar located in West Java Province; four which are Tegal, Pekalongan, Salatiga and Magelang located in Central Java, and five which are Madiun, Kediri, Blitar, Pasuruan and Probolinggo located in East Java. Geographically, five of these cities are situated in the coastal areas (Cirebon, Tegal, Pekalongan, Pasuruan, and Probolinggo) while the other seven are situated in the mainland (Sukabumi, Banjar, Magelang, Salatiga, Madiun, Kediri, Blitar). According to Directorate General of Spatial Planning [19] of all cities, only Banjar formally established as *kota* in 2001, meanwhile, other cities has been formally recognised as *kota* since 1950. However, the roots of the cities itself are recognised on several Java old maps such as Chatelain maps that mentioned Tegal, Pasuruan, and Cirebon [20] and Witkamp map that mentioned Pekalongan, Magelang, Madiun, Banjar, Sukabumi, Blitar, Kediri, Probolinggo [21], we believed there were other older maps that preceded Chatelain and Witkamp maps, nevertheless those two maps sufficient enough to pointed out that many small cities in Java already established more than a centuries ago.

The growths of the small cities were supported by their good accessibility, marked by several transportation infrastructures. By 2010, national road network as the main connector had connected all small cities with other areas in Java [22]. All small cities were not connected with toll road network by 2010, however, Government Regulation [23] planned by 2029 a Trans Java toll road networks will connect most of the small cities (except for Madiun, Kediri, and Blitar). All small cities – except for Magelang and Salatiga – also connected with railway network in 2010 and by 2030 all small cities were planned to be connected by railway network [24]. From all small cities, only Cirebon had both airport and seaport facilities while the others relied on closer large cities

Java small cities already had their specific functions as activity centres. Government Regulation [25] designated two cities (Cirebon and Salatiga) as national activity centers (*pusat kegiatan nasional*); nine cities (Sukabumi, Tegal, Pekalongan, Magelang, Madiun, Kediri, Blitar, Pasuruan, and Probolinggo) as regional activity centers (*pusat kegiatan wilayah*); and West Java Provincial Regulation [26] designated Banjar as local activity centers (*pusat kegiatan lokal*).

4.1 Population growth

Based on population data, all Java's small cities experienced positive population growth and increased population density. While the number went up there was a different situation in each small city related to population growth rate. Sukabumi, Banjar, Salatiga, Blitar, Probolinggo, and Pasuruan have high growth rates (more than one percent per year) while others have low growth rates, especially for Magelang and Tegal that barely above zero growth. Compared to Java, only three cities – Sukabumi, Salatiga, and Probolinggo – surpassed Java's urban population growth. By using same growth rate, we predict that in the next ten years Sukabumi, Cirebon, Pekalongan and Kediri will have become medium-sized cities.

In average, population density increased from 4,576 to 4,981 population/km^2 between the years 2000-2010 or at an annual growth rate of 0.98 percent. By 2010, the lowest density happened in Banjar due to its low number of population combined with vast urban areas (with 113.49 km^2, Banjar was the largest among all small cities, Kediri which next to Banjar only had around half of Banjar). Population density in seven cities, namely Sukabumi, Cirebon, Tegal, Pekalongan, Magelang, Madiun and Pasuruan were higher than average. The reason why this situation happened will be explained in the next section of this paper.

Table 1. Population of small cities, 2000-2010 [19][27]

City	Land Area (Km^2)	Population			Density (pop/km^2)		
		2000	2010	Growth (%)	2000	2010	Growth (%)
Sukabumi	48.42	252,420	298,681	18.3	5213	6169	18.33
Cirebon	37.54	272,263	296,389	8.9	7253	7895	8.86
Banjar	113.49	156,555	175,157	11.9	1379	1543	11.88
Tegal	39	235,443	239,599	1.8	6037	6144	1.77
Pekalongan	45.25	261,308	281,434	7.7	5775	6220	7.70
Salatiga	56	144,788	170,332	17.6	2586	3042	17.64
Magelang	18.12	115,271	118,277	2.6	6362	6527	2.61
Madiun	33.29	163,956	170,964	4.3	4925	5136	4.27
Kediri	63.40	244,519	268,507	9.8	3857	4235	9.81
Blitar	35.27	119,372	131,968	10.6	3385	3742	10.55
Probolinggo	56.67	191,522	217,602	13.3	3380	3840	13.62
Pasuruan	35.29	168,323	186,262	10.7	4770	5278	10.66

4.2 Urban village expansion

In this study, we categorized urban village in small cities into two parts which are urban village located in core area and urban village located in extended area. Core area itself defined as the selected small cities based on administrative jurisdiction which has already stated on table 1. Meanwhile extended area is part of the *kabupaten* that is located adjacent or directly influenced by the core area. In addition there were some villages in the city core that not yet became urban village, however this study included them as the urban village.

Our analysis found that all of the cities expanded towards their surrounding area. The number of the extended urban village varied among the cities but mostly they are almost the same number as the urban village in the city core (see table 2). Based on the urban village expansion analysis, we found only Banjar and Salatiga have a low number of urban villages in the extended areas compared to in the core areas. On the contrary, a number of cities have more urban villages in the extended areas than in their core areas. These cities are Cirebon, Tegal, Pekalongan, Kediri and Blitar. The expansion is mostly linear, following the road network (see figure 2).

Table 2. Urban village expansion in Java's small city, 2000-2010 [28][29]

City	2000			2010			Growth	Extended Areas (Regency)
	Core	Extended	Total	Core	Extended	Total		
Sukabumi	33	19	52	33	46	79	27	Sukabumi
Cirebon	22	6	28	22	165	187	159	Cirebon
Banjar	25	1	26	25	4	29	3	Ciamis and Cilacap
Tegal	27	155	182	27	160	187	5	Tegal and Brebes
Pekalongan	47	53	100	47	82	129	29	Pekalongan and Batang
Salatiga	22	4	26	22	8	30	4	Semarang
Magelang	17	24	41	17	29	46	5	Magelang
Madiun	27	17	44	27	21	48	4	Madiun
Kediri	47	48	95	47	76	123	28	Kediri
Blitar	21	34	55	21	47	68	13	Blitar
Probolinggo	29	10	39	29	12	41	2	Probolinggo
Pasuruan	34	20	54	34	22	56	2	Pasuruan

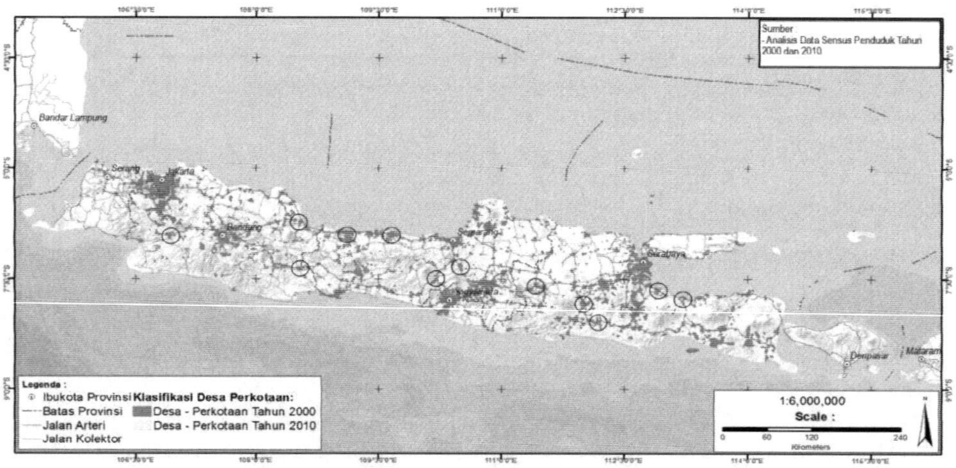

Figure 2. Urban village expansion in Java's small city [28][29]

Using *Potensi Desa* data we found that between years 2000-2010 there are significant increase of urban village number. In a decade, 165 villages morphed into urban village, an increase of 19.2 percent (see table 2). The data revealed that based on number of urban village, small cities in Java mostly are expanded.

Some cities experienced high growth of urban village expansion. Cirebon, for example, increased very large number of urban village that made Cirebon five times bigger, followed by Sukabumi and Pekalongan. Cirebon and Pekalongan, together with Tegal, are geographically located on the northern coastline of Java and are connected in one corridor. These areas are strategically located at the central nexus of Java road network, which positions them as the main transit area for Java transportation system. Nevertheless, for Tegal we found that its urban expansion was not as impressive as Cirebon and Pekalongan because Tegal already expanded in 2000. The expansion of Cirebon was also supported by its function as a location of Indonesia's main port with roots as far as since Dutch colonization era [17][30]. For the case of Sukabumi, Andrea stated that urban area expansion in Sukabumi happened due to its location as the hub between Jakarta Metropolitan Region and its hinterland [31]. Handayani elucidated that cities expansion in Central Java happened pushed by industrialization that happened in those cities [32].

4.3 Population dynamics on small city extended area

Following our analysis on urban village expansion, we also looked at the population dynamics in Java's small cities. Table 1 exhibited number of population on small city based on administrative jurisdiction as core city areas. However when we combined with the number of population in extended areas we discovered several interesting findings (see Table 3), they are:

- In general, the idea of small city on the selected twelve cities dissipated when we analyzed them as extended urban areas by using combined data of population in core and extended area;
- **Sukabumi** is not a small city. In 2000, Sukabumi had already became a large city with more than 400,000 inhabitants and continued to grow into a large city. In 2010, Sukabumi has more than 640,000 inhabitants;
- **Cirebon** is an interesting case in which a small city transformed into metropolitan. In 2000, Cirebon and its urban extended area was still considered a small city with less than 300,000 inhabitants. However, by 2010, Cirebon had become a metropolitan with 1.2 million inhabitants living in its urban extended areas;
- **Banjar** was a small city. With less than 200,000 inhabitants by 2010 we confirmed that Banjar was a small city. Population expansion was relatively low, and growth in core area was almost the same as the extended areas;
- **Tegal** was not a small city. In 2000 Tegal extended urban area had already became a metropolitan with 1.1 million inhabitants. Tegal was still a metropolitan in 2010 with less than 1.2 million inhabitants, however their population growth was low compared with other cities;
- **Pekalongan** was not a small city. Started as small city in 2000 with 227,000 inhabitants, Pekalongan growth tripled into a large city in 2010 with 673,000 inhabitants;
- **Salatiga** was a small city. In 2010, inhabitants of Salatiga urban expanded areas was 201,000 inhabitants, a little increase compared to having 167,000 inhabitants in 2000;
- **Magelang** was a small city geared to medium-sized city. With 209,000 inhabitants lives in its urban extended areas in 2000, Magelang grew into early medium-sized city in 2010 with 301,000 inhabitants;
- **Madiun** was a small city. Likely as Salatiga, population growth in Madiun is relatively low. Therefore, by 2010, its population was around 261,000 inhabitants;
- **Kediri** was not a small city. In 2000, Kediri had already become a medium-sized city. It became a large city in 2010 when it amassed 619,000 inhabitants;
- **Blitar** was another example of a small city morphing into a medium-sized city. With 270,000 inhabitants by the year 2000, Blitar continuously growth into medium-sized city with 338,000 inhabitants;

- **Probolinggo** was a small city. Though Probolinggo almost doubled its population between the year 2000-2010, Probolinggo still categorized as small city with population of 271,000 inhabitants; and
- **Pasuruan** was a small city. Pasuruan was the city with lowest population growth among other twelve cities. On a decade, Pasuruan population 'only' growth with less than 20,000 inhabitants and made Pasuruan a small city with 242,000 inhabitants in 2010.

Table 3. Population expansion in Java's small city, 2000-2010 [28][29]

City	2000			2010			Remarks
	Core	Extension	Total	Core	Extension	Total	
Sukabumi	252,420	151,723	404,143	298,681	324,369	641,050	Large City
Cirebon	272,263	24,453	296,716	296,389	909,590	1,205,979	Metropolitan
Banjar	156,555	5,917	162,472	175,157	23,732	198,899	Small City
Tegal	235,443	874,050	1,110,950	239,599	939,011	1,178,610	Metropolitan
Pekalongan	261,308	56,732	227,675	281,434	392,243	673,677	Large City
Salatiga	144,788	16,007	167,445	170,332	30,856	201,188	Small City
Magelang	115,271	146,064	209,512	118,277	183,028	301,255	Medium City
Madiun	163,840	69,941	233,781	170,964	89,954	261,552	Small City
Kediri	244,519	210,243	454,762	268,507	351,446	619,953	Large City
Blitar	119,372	155,647	270,711	131,968	222,252	338,210	Medium City
Probolinggo	191,522	41,204	158,951	217,602	54,325	271,379	Small City
Pasuruan	168,323	56,732	225,055	186,262	71,477	242,245	Small City

By analyzing population based on urban extended areas, not all of twelve cities classified as small cities in Table 1 are, in actuality, small cities. Based on population expansion depicted in Table 3, we found that only five cities in Java are able to be properly categorized as small city: Salatiga, Madiun, Probolinggo and Pasuruan.

4.4 Built-up Expansion

In this section, we will explain how population development in small cities also influenced space. Based on land cover data on the 2000-2010, we found that there were relatively low built-up development (under 1 percent growth in a decade) in Cirebon, Banjar, Tegal, Pekalongan, Salatiga, Magelang, Madiun, Kediri, and Blitar. We believe that cities with already more than 50 percent built-up areas (Cirebon, Tegal, Pekalongan, Salatiga, Magelang, Madiun, and Kediri) had already met their geographical landscape threshold, therefore unable to expand their built-up areas inside their jurisdiction areas. This situation answers why population growth in core areas of the small cities has been relatively very low: they eventually run out of space. In order to compensate the need for growth, all cities then expand outside of the core area.

Banjar is an exception. With 12.5 percent, it has a relatively low built-up area compared to other cities. Moreover, Banjar experienced a stagnation in built-up area development which occurred because its status as a new city (acquired in 2003 when it separated from Ciamis Regency) combined with its limited availability of urban facilities [32]. Meanwhile, Sukabumi, Probolinggo and Pasuruan has a low built-up area growth rate of below 6% in a decade. Despite their built-up areas growth, these city only use one-thirds of their areas for built-up area.

Table 4. Small cities built-up urban areas development (km^2) 2000-2010 [33][34]

City	Land Area (Km^2)	Built-up Area 2000		Built-up Area 2010		Change (%)
		Area (Km^2)	%	Area (Km^2)	%	
Sukabumi	48.42	15.05	31.08	17.78	36.72	5.64
Cirebon	37.54	25.28	67.34	25.35	67.53	0.19
Banjar	113.49	12.49	11.01	12.49	11.01	0.00
Tegal	39	22.77	58.38	22.77	58.38	0.00
Pekalongan	45.25	23.42	51.76	23.42	51.76	0.00
Salatiga	56	35.95	64.2	35.95	64.20	0.00
Magelang	18.12	13.83	76.32	13.83	76.32	0.00
Madiun	33.29	21.38	64.22	21.39	64.25	0.03
Kediri	63.40	32.30	50.95	32.31	50.96	0.02
Blitar	35.27	11.34	32.15	11.34	32.15	0.00
Probolinggo	56.67	15.01	26.49	17.12	30.21	3.72
Pasuruan	35.29	9.84	27.88	11.18	31.68	3.80

A closer look at the proportion of built-up areas between core and expansion cities reveals an interesting finding. Contrary with population and urban village expansion that usually expands toward extended areas, by 2000, the development of built-up areas had been mostly concentrated in city core. However, a decade later each city has its own dynamics (see Table 5).

Table 5. Small cities expanded built-up urban areas development (km^2) 2000-2010 [33][34]

City	Built-up Areas 2000			Built-up Areas 2010		
	Core Area (km^2)	Expanded (km^2)	% of Core Area	Core Area (km^2)	Expanded (km^2)	% of Core Area
Sukabumi	15.05	19.23	78.26	17.78	30.97	57.41
Cirebon	25.28	27.04	93.49	25.35	95.40	26.57
Banjar	12.49	12.76	97.88	12.49	13.72	91.03
Tegal	22.77	107.09	21.26	22.77	111.74	20.38
Pekalongan	23.42	42.85	54.66	23.42	56.93	41.12
Salatiga	35.95	42.07	85.45	35.95	44.22	81.30
Magelang	13.83	60.84	22.73	13.83	66.81	20.70
Madiun	21.38	28.67	74.57	21.39	38.18	56.02
Kediri	32.30	52.97	60.98	32.31	83.98	38.47
Blitar	11.34	15.31	74.07	11.34	22.90	49.52
Probolinggo	15.01	18.76	80.01	17.12	21.38	80.07
Pasuruan	9.84	12.87	76.46	11.18	14.47	77.26

Tegal and Magelang exhibited cities with small proportion of built-up areas inside its the city core, and built-up development continuously expanded outside the city core due to the limited spaces inside both cities. Meanwhile, built-up development in Banjar, Salatiga, Probolinggo, and Pasuruan were still centralized in their cores with relatively low development outside of its core. Sukabumi, Pekalongan, Blitar, Madiun, and Kediri experienced urban expansion towards its expanded areas which marked with the decreased of built-up area proportion in core areas compared to the expanded urban areas in these cities. Cirebon was an interesting and different situation compare to other eleven cities. Although by 2000, most of built-up development had been concentrated in the core area of Cirebon, a decade later the opposite had been the case. By 2010, Cirebon City core areas only contributed to around 27 percent of its urban expanded built-up areas, which meant that the urban areas of Cirebon City had expanded significantly towards Cirebon Regency.

Small city in Java experienced urban expansion at a different rate. However, growth of built-up areas in extended areas of small city mostly happened near the core areas and/or along main transportation networks that have good connection with core areas (see figure 3). Based on this evidence, local governments need to be careful in dealing with urban development that spills over across the borders of adjacent administrative jurisdictions.

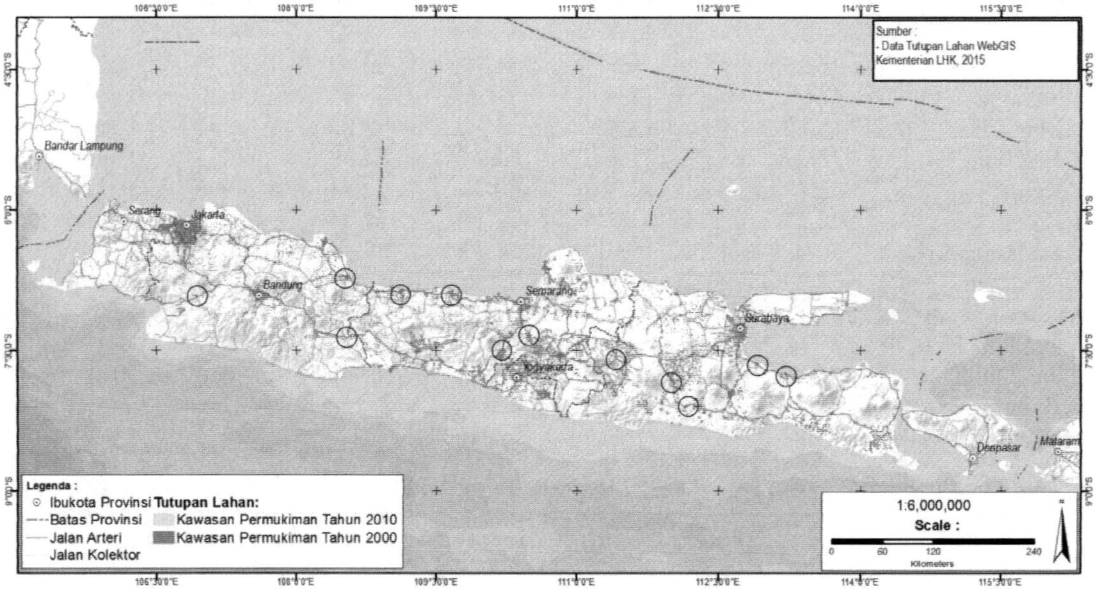

Figure 3. Built-up expansion in Java's small city [33][34]

5. Spatial plan policy

Settlement development was the main reason why urban expansion occurred in the small cities of Java. Between 2000 and 2010, all of the city cores had already got their general spatial plan (*Rencana Umum Tata Ruang*/RUTR) defined by Law 24/1992 about Spatial Planning, formalized between late 90's to early 2000's (see next table). The RUTR is used as the basis of land use permit by small cities. However, due to the nature of Law 24/1992 that uses discretionary planning – especially from central government – in the name of supporting economic development, many RUTR are unable to promote sustainable development in small cities, marked by the expansion of the built-up areas for land-based investment that left not much room for open spaces. The absence of sanctions for spatial plan violations and weak enforcement in creating detailed spatial plan are also key factors as to why the RUTR fails to effectively counteract urban land expansion due to urbanization.

The multidimensional crisis in Indonesia in 1997, followed by a substantial transformation of Indonesia's governance system, rendered Law 24/1992 as not in accordance with newly promulgated regulations [35] as well as the new format of the decentralized government [36]. Therefore, the Government of Indonesia substituted spatial planning regulation with Law 26/2007. Law 26/2007 leaned towards regulatory planning style with designated specific function to a parcel of land, replacing the discretionary style used by Law 24/1992. The new spatial planning act made all cities should replace their RUTR with the new form of the general spatial plan (RTRW). Law 26/2007 mandates that all *kota* and *kabupaten* must legalise their RTRW with district regulation (*peraturan daerah/perda*) in three years after 2007. Among all small cities in Java, only Probolinggo was able to comply this mandate; others needed more time (see table below) [37]. RTRW itself focuses on the development of spatial structure (*struktur ruang*) and only provides general guidelines for the designation of spatial pattern (*pola ruang*). A city needs more detailed spatial plan for spatial permit utilization [38].

Table 6. Small Cities Spatial Plan Policy [37]

City core	RUTR (Law 24/192)	RTRW (Law 26/2007)
Sukabumi	1999	2012
Cirebon	1989	2012
Banjar	2004	2014
Tegal	2004	2012
Pekalongan	2003	2011
Salatiga	1996	2011
Magelang	1999, revised in 2001	2012
Madiun	2004	2011
Kediri	2002	2012
Blitar	No data available	2011
Probolinggo	2000	2010
Pasuruan	2002	2012

Law 26/2007 also mandates all urban areas to create a detailed spatial plan (*Rencana Detail Tata Ruang*/RDTR) equipped with zoning regulation. RDTR will be used as land use planning and main consideration for land use permit, creating certainty for the utilization of space in the city. Nevertheless, the creation of the RDTR itself was not an easy task. Cities need to have RTRW first, and the delayed promulgation of RTRW hampered the RDTR creation on small cities. Between 2007 and 2010, no RDTR were promulgated. Directorate General of Spatial Planning (2017) reporting that small cities in Java being overwhelmed in preparing the RDTR. None from all small cities and their counterpart regencies a part of the urban expansion in this study promulgated their RDTR and zoning regulation documents into district regulation [39]. One exception is Central Java Province, which promulgated Provincial Regulation 6/2015 regulating urban development in the Brebes-Tegal-Slawi-Pemalang urban areas corridor. With the absence of the RDTR and zoning regulation as urban development tools, authors worry that Java's small cities will continue to expand without any regard to their urban sustainability.

6. Concluding remarks

Urbanization has been criticized as being the main culprit behind urban sprawls in Indonesia. This study reveals and emphasizes that urban sprawl is not only a phenomenon in large and metropolitan cities but also in small cities. Expansion of small cities beyond its core city leads to extended urban areas. By looking at the urban expansion data we found that a small city defined by its administrative jurisdiction is not in actuality small. We found that a city is small only as far as its core goes and when combined with its population growth and built-up area development could very easily become metropolitan cities (Cirebon and Tegal), large cities (Sukabumi, Kediri, and Pekalongan) or medium-sized cities (Magelang and Blitar) because its built-up expansion occurs around the core.

We also found that existing city spatial plans are not able to deal with urban expansion. The existence of strict delineation on jurisdiction hampers the possibility to create an integrated spatial plan. There is a lack of political will in the local government to work together in managing their urban growth marked by very few interjurisdiction spatial plan that had been promulgated. As a result of that glacial movement by the government, it was feared that any resulting detailed spatial plan will not be able to resolve the urban sprawls already occurring in Java's small cities.

Acknowledgement
We would like to thank Urban and Regional Development Institute (URDI) for their support, help, and funding towards this research.

IOP Conf. Series: Earth and Environmental Science **158** (2018) 012054 doi:10.1088/1755-1315/158/1/012054

References

[1] BPS (Central Bureau Statistics of Indonesia) 2014 Urban population percentage by province 2010-2035 [Online]. Available from: https://www.bps.go.id/statictable/2014/02/18/1276/persentase-penduduk-daerah-perkotaan-menurut-provinsi-2010-2035.html

[2] BPS (Central Bureau Statistics of Indonesia) 2009 Percentage distribution of gross regional domestic product at current prices by province 2000-2013 [Online]. Available from: https://www.bps.go.id/statictable/2009/06/29/1625/distribusi-persentase-produk-domestik-regional-bruto-atas-dasar-harga-berlaku-menurut-provinsi-2000-2013-persen-.html

[3] Indonesia Ministry of Research and Higher Education (Kemenristekdikti) 2017 List of 100 Non-Polytechnic Universities Ranking in Indonesia in 2017 [Online]. Available from: http://kelembagaan.ristekdikti.go.id/index.php/2017/08/18/daftar-100-peringkat-perguruan-tinggi-non-politeknik-tahun-2017/

[4] BPS (Central Bureau Statistics of Indonesia) 2017 Number of general hospitals, special hospitals, and public health centers [Online]. Available from: https://www.bps.go.id/linkTableDinamis/view/id/933

[5] Firman T 2007 The urbanisation of Java, 2000–2010: towards 'the island of mega-urban regions', *Asian Population Studies* **17** 1-17

[6] Mardiansyah F H 2015 *Perkembangan kota-kota sekunder (secondary cities) di Indonesia* (Jakarta: URDI)

[7] Bhatta B 2010 *Analysis of urban growth and sprawl from remote sensing data* (Heidelberg: Springer)

[8] Polidoro M, de Lollo J A, and Barros M V F 2012 Urban sprawl and the challenges for urban planning *J. Environmental Protection* **3** 1010-1019

[9] United Nation 2017 New Urban Agenda

[10] Prasetyo L B, Kartodihardjo H, Adiwibowo S, Okarda B, and Setiawan Y 2009 Spatial model approach on deforestation of Java Island, Indonesia *J.Integrated Field Science* **6** 37-44

[11] Asian Development Bank 2016 *Indonesia, country water assessment* (Mandaluyong City: Philippines)

[12] Bell D and Jayne M 2009 Small cities? Towards a research agenda *Int. J. Urban and Regional Research* **33** 683-699

[13] Way H 2016 Beyond the big city: the question of size in planning for urban sustainability *Proc. Environmental Sciences* **36** 138-145

[14] United Nations 2014 *World urbanization prospects: the 2014 revision* (New York: Department of Economic and Social Affairs, Population Division)

[15] European Union 2016 *The state of European cities 2016 cities leading the way to a better future (Luxembourg: European Union, United Nations Human Settlements Programme* (UN-Habitat))

[16] BPS (Central Bureau Statistics of Indonesia) 2010 Indonesia rural and urban classification book 2: Java (in Bahasa Indonesia) [Online]. Available from: https://sirusa.bps.go.id/webadmin/doc/MFD_2010_Buku_2.pdf

[17] Rahayu P 2016 *The governance of small cities in decentralizing Indonesia, the case of Cirebon City and its sorrounding regions Dissertation* (Groningen: University of Groningen)

[18] Kartapranata G 2010 Java transportation network [Online]. Available from: https://commons.wikimedia.org/wiki/User:Gunkarta#/media/File:Java_Transportation_Network.s vg

[19] Directorate General Spatial Planning 2011 *Pengembangan kebijakan dan strategi nasional pembangunan perkotaan* (Jakarta: Ministry of Public Works)

[20] Chatelain H 1718 Carte de l'Ile de Java *Atlas Historique* vol **47** 128

[21] de Bussy J H 1893 Kaart van Nederlandisch-Indie naar oorspronkelijke teekening van H. Ph. Th. Witkamp [Online]. Available from: https://legacy.lib.utexas.edu/maps/historical/nederlandsch_indie_1893.jpg

IOP Conf. Series: Earth and Environmental Science **158** (2018) 012054 doi:10.1088/1755-1315/158/1/012054

[22] Roberts B H 2014 *Managing systems of secondaries cities* (Brussels: Cities Alliance/UNOPS)
[23] National spatial plan. Gov. Regist. [regulation on the Internet]. 26/2008. Available from: http://perpustakaan.bappenas.go.id/lontar/file?file=digital/15997-[_Konten_]-KONTEN%204022.pdf
[24] Directorate General Railway 2011 *National railway masterplan* (in Bahasa Indonesia) (Jakarta: Ministry of Transportation)
[25] Revision on national spatial plan. Gov. Regist [regulation on the Internet]. 13/2017. Available from: http://peraturan.go.id/pp/nomor-13-tahun-2017.html.
[26] Spatial plan of West Java Province 2009-2029. West Java Gov. Regist. 22/2010
[27] Jones G and Mulyana W 2015 *Urbanization in Indonesia, UNFPA Indonesia monograph series: No. 4* (Jakarta: UNFPA Indonesia)
[28] BPS (Central Bureau Statistics of Indonesia) 2000 *Potensi desa year 2000* (Jakarta: BPS)
[29] BPS (Central Bureau Statistics of Indonesia) 2010 *Potensi desa year 2010* (Jakarta: BPS)
[30] Fahmi F Z, Hudalah D, Rahayu P, Woltjer J 2014 Extended urbanization in small and medium-sized cities: The case of Cirebon, Indonesia *Habitat International* **42** 1 – 10
[31] Andrea E P 2015 *Study on impact of urbanization and rapid urban expansion in Java and Jabodetabek megacity, Indonesia Dissertation* (Kyoto: Kyoto University)
[32] Handayani W 2013 Rural-urban transition in Central Java: population and economic structural changes based on cluster analysis *Land* **2** 419 – 436
[33] Ministry of Environment and Forestry 2000 *Landsat Data Year 2000* (Jakarta: Ministry of Environment and Forestry)
[34] Ministry of Environment and Forestry 2011 *Landsat Data Year 2011* (Jakarta: Ministry of Environment and Forestry)
[35] Hudalah D and Woltjer J 2007 Spatial planning system in transitional Indonesia *International Planning Studies* **12** 291-303
[36] Rukmana D 2015 The change and transformation of Indonesian spatial planning after Suharto's new order regime: The Case of the Jakarta Metropolitan Area *International Planning Studies* **20** 350-370
[37] Directorate General of Spatial Planning 2017 *Status RTRW provinsi/kabupaten/kota seluruh Indonesia (Status tanggal 29 September 2017)* (Jakarta: Ministry of Agrarian and Spatial Plan)
[38] Spatial planning. Gov. Regist. [regulation on the Internet]. 26/2007. Available from: http://www.hukumonline.com/pusatdata/downloadfile/fl52198/parent/26897
[39] Directorate General Spatial Planning 2017 *Status RRTR dan RDTR seluruh Indonesia (Status tanggal 29 September 2017)* (Jakarta: Ministry of Agrarian and Spatial Plan)

IOP Conf. Series: Earth and Environmental Science **158** (2018) 012055 doi:10.1088/1755-1315/158/1/012055

Rural Industry Clustering Towards Transitional Rural-Urban Interface

P Nugroho

Department of Urban and Regional Planning, Diponegoro University

Correponding author: prihadi.nugroho@live.undip.ac.id

Abstract. Rural industrialization seems to be attractive for policymakers looking for counter-urbanization efforts – and nowadays peri-urbanization forces – in line with growing decentralized autonomy of local Indonesian authorities. To promote better rural development, an extended growth pole strategy has been introduced as well as an agropolitan approach and its derivatives. In fact, there is little evidence for their success; rural autonomy remains elusive instead. However, institutional capacity of rural authorities and organizations still fails to deliver rural development initiatives properly. This research was aimed at examining this issue by looking at rural industry clustering in the Greater Solo Region, Indonesia as a response against extended urbanization in peripheral regions. The study focused on batik industry clustering in the rural periphery of Solo City, which provides a transitional rural-urban interface necessary to drive rural independence. Having inherited the batik tradition underpinned by an agriculture-led peasant society, the rural batik industrialization has reinforced the socio-economic transition from a purely agrarian society to a mixed rural-urban society. This study employed an explanatory sequential mixed-method approach, where a quantitative spatial analysis was used to identify the expansion of urbanized areas in villages, and a qualitative case study analysis to figure out the socio-economic shift in rural livelihoods. The results showed that physical spatial changes in these villages do not conform to the socio-economic change into an urban industrial society in a substantial way. Rather, the local villagers preserve an informal economy to support the existence of a mixed rural-urban livelihood.

1. Introduction

In the theoretical and empirical domains, extended urbanization has raised many controversies regarding its effectivity in accelerating development in peripheral regions. For more than a century the debate on urban-led development has recursively placed more importance on urban regions over their peripheries, positing a predominantly top-down spatial development approach. Neither the opposite nor a more balanced approach seems possible. The growing importance of urban regions for promoting development can be traced back to long-lasting periods of urbanization around the world. During the 1800s backwards, the urban population size was about three percent out of the world population of less than one billion. In the early 1900s, the world population entered the age of rapid urbanization, where the urban population size was multiplied five times at 15 percent, comprising 1.6 billion urban inhabitants. In the decades that followed, the urban size multiplied even faster, at 29 percent in 1950, and 43 percent in 1990. Recently, the urban population size has hit 54 percent in 2014, or equivalent to 3.9 billion urban inhabitants, and has been predicted to reach 66 percent by 2050, and still counting [1,2]. During the long history of worldwide urbanization, urban functions have become more complex. Cities not only function as centers for manufacturing and trade of rural commodities but also for political, social, cultural, and defense activities that benefit the whole nation thanks to the advances of technology, transportation access, and communication networks [1,3,4]. With the fluid heterogeneity of the urban functionalities, the urbanization process has resulted in a new way of life under the notion of urbanism, which has brought the desired standard of living beyond city limits [5,6].

Early urban-centered development dates back to the ancient cities of Egypt, India, Mesopotamia, China, Greece, and many parts of the world that showed a salient state-oriented centralism. Cities were built to represent a predominantly authoritative government regime through the manifestation of an axial-geometric internal structure, surrounded by a fortress-like wall, filled with gigantic-scale buildings, sculptures, monuments, gardens, and public open spaces and facilities. Cities during these times were a symbol of government power to perform multiple functions, from (mainly) military and defense to trade and political activities, which more benefited aristocrats, high-ranked bureaucrats and bourgeois elites rather than the general public. In the meantime, rural and countryside development was ignored and seen as no more than a residual space for nature [7–10]. The rise of the Industrial Revolution in the late 1700s, which coincidentally took place contemporaneously with Adam Smith's laissez-faire capitalism introduced in 1776, shifted the ways cities functioned following the permeation of the market mechanism in decision making. Many large cities turned into manufacturing centers supported by modernized transportation systems, including roads, railways, and riverways. The steam machine invented by James Watt in 1769 broke the city walls down through mechanization of manufacturing products and mobilization of people from the urban fringes. On the other hand, new industrial towns were built in the countryside in response to lower transportation costs of coal exploitation for the sake of manufacturing factories. At the end of 19th century, the growth of industrial cities had become a new fashion in urbanization, marked by capital intensification directing spatial planning and development supported by improved communication networks and new technological inventions [7–1]. Spatial planning related to mining extraction and transportation networks influenced the growth of large cities and smaller towns. High consumption by industrial cities of mining resources contributed to spatially discontinuous patterns between older cities and new industrial towns. The emerging number of slum areas and unhealthy urban conditions soon appeared as the main feature of these industrial cities, where private companies took control over the spatial development process.

After 150 successive years of urban industrial-led spatial planning, city growth received a wide range of criticism over its failure in providing better living space. Urban degradation became a major issue for policymakers due to severe air pollution, traffic congestion, slumification, contagious diseases, and high crime levels. Between the late 1800s and the early 1900s, the city beautification era was a response to the immediate call for urban lifestyle improvement. Ebenezer Howard and Patrick Geddes were the pioneers in this advent of urban rehabilitation. Howard's radical utopian notion of the Garden Cities of Tomorrow (1898) promoted a romantic throwback to preindustrial city development. He proposed new town development in the countryside around large factories. The so-called garden city, built far from city centers, was meant to establish new working-and-living urbanism in the city outskirts by combining urban industrial routines with a natural countryside environment [7–10]. Similarly, but in a slightly different direction, Geddes' notion of Cities in Evolution (1915) underlined the importance of human interaction with the environment, for which human geography supplies the scientific basis of understanding. He pointed out that the tendency of town development in the suburbs should consider a regional sphere of influence of towns interconnected by various transportation networks following the natural landscape. Planning intervention, therefore, should take into account the locational characteristics of the natural region where the proposed urban economic activities will take place [7–10]. During this rapid urbanization era, the second wave of urbanization stressed the importance of the contribution of the countryside to regional urban-centered spatial planning and development. Suburbs and rural regions played a key role in supporting outward extending urbanization. Broader functionalities of the countryside to provide alternative working and living space for urban inhabitants initiated attempts to tackle old city degradation by utilizing better technology and accessibility to support the mobilization of people and resources beyond the city borders.

The continuation of rapid urbanization in the following decades emerged as a result of market capitalization in (mostly) large cities. Many investments in city development projects initiated by state planning institutions made the Geddesian greenbelt city concept flourish across the countryside, particularly in North American and European cities. Increased state-led urban planning interventions during the pre-World War II period were aimed at promoting city-regions to boost the national economy. As a result, the growing number of new metropolises that arose from the intensified urbanization process would serve laissez-faire market capitalism. However, this trend somewhat declined between the Great Depression era of the 1930s and the early period of World War II. Due to the massive economic downturn and increasing political instability and insecurity worldwide, the urbanization slowdown forced former industrial cities and countryside towns to deal with budget cuts. Most government budgets were allocated to finance war, economic recovery, and social security programs. Limited physical development projects were confined to the hands of central government to finance the most prioritized projects only, such as those executed by the Tennessee Valley Authority (TVA). These projects were aimed at supporting national basic needs, including the provision of clean water, electricity, sanitation and affordable housing while the government was struggling with various urban degradation problems [7–10]. Such ivory tower projects thus became a new trend of top-down spatial planning along with the withdrawal of decentralized urbanization and suburbanization processes. Resource accumulation was centralized and managed by central government institutions for ensuring adequate technical capacity and financial assistance to project implementation.

The post-war period of urban expansion from the 1940s to 1970 displayed striking debates on national and spatial development approaches. The contestation of two major development paradigms, i.e. balanced and unbalanced growth theories, was the lasting result of the immediate government reaction to the severe impact of the war vis-à-vis heightened public expectations towards comprehensive recovery programs. Basic needs fulfillment remained the main public concern, aligned with job creation, social security, human shelter, and improved living condition programs. The balanced-growth scholars, such as Paul Rosenstein-Rodan and Ragnar Nurkse, proposed an equal-development approach by promoting simultaneous development projects to supply massive public goods to the entire country. Arguably, this approach could improve the capacity of both public and private institutions in providing basic needs and consumer goods to the market, which in turn would lead to economic recovery as well as social welfare [11–15]. In contrast, the proponents of unbalanced growth suggested intensification of prioritized sectors for ensuring a higher return on investment through limited resource mobilization efficiency. Therefore, strategic projects were more preferred to stimulate growth of downstream industries. Prominent proponents of the unbalanced-growth theory were scholars such as Walt Whitman Rostow, Albert Otto Hirschman, and Gunnar Myrdal [13,15–17], to name a few.

Following these debates, spatial development policies were polarized into two schools: development from below and development from above. The development-from-below school adhered to balanced-growth principles according to which inward-looking rural development strategies were designed to protect rural assets and resources from urban overexploitation. Selective spatial closure, territorialism, integrated rural development and endogenous development were popular phrases circulated by proponents of this school such as Walter Stohr and Franz Tötdling [15,18,19]. They were much inspired by the works of François Perroux and later supported by Jacques Boudeville, who advised the growth pole strategy in 1955, which suggests that economic development spreads from certain leading industries (mostly) concentrated in urban areas. This theory was aligned with Hirschman's trickle-down concept and polarization effects and Myrdal's concept of spread and backwash effects of economic development. It was argued that the urban industrial sector will accumulate resources from peripheral and rural regions in the most efficient way to produce higher added value of goods and increasing returns to scale (polarization or backwash effect), and then spread the induced development benefits to the rest of economy (trickle-down or spread effect) [15,18,20]. Consequently, urban-biased development has been a major concern in national and regional development since the 1950s, associated with the emerging tensions between balanced- and unbalanced-growth scholars [13,15,21,22]. Perroux's growth pole strategy could perhaps be accused of being part of the widespread urban-biased development phenomenon in the modern era, from which the rural-urban dichotomy has emerged as a by-product of urban-based decision-making processes [23–25].

From the 1980s onwards, rural-urban dichotomy issues have emerged in public policies beyond the increasing global trend of megacity development. As metropolises expanded, so did many urban regions to form global urban networks; the parasitic role of urban exploitation over its peripheral and rural regions has occurred at an unprecedented rate that could never have been imagined a century ago

[2,3,26]. The urban sprawl phenomenon, which marked the spatial development landscape between the 1990s and 2000s, has stirred the suburbanization process – followed by the peri-urbanization process a few decades later – conveying multidimensional effects of the uncontrolled growth of megacities towards more integrated spatial development at both national and regional levels. Globalization forces that welcomed the upcoming digital economy era have converged the global megacities into an open world market system, thanks to borderless foreign direct investments and post-Fordist global production networks. Despite the increased competitiveness of megacities and nations, many national governments have been losing power and sovereignty in directing spatial development properly. Many multinational corporations have taken over spatial development control through investments in large-scale projects, privatization, business partnership, and domestic firm acquisition [27–31]. In contrast, the neglected rural development has caused greater dependency on global products and agencies, including the global culture of urbanism, while rural inhabitants are trapped in intergenerational poverty. Meanwhile, the role of small and medium-sized towns in preventing the deteriorating effects of such extended urbanization has disappeared, particularly due to the absence of well-articulated urban systems at the regional level. Therefore, the lasting rural neglect has worsened regional inequality incidence in contrast to the so-called modernity inducement in rural development [14,32].

However, the search for an integrated spatial development to get rid of the rural-urban dichotomy has received much attention from policymakers and practitioners at least since the 1970s. Many national governments were keen to search for alternative spatial development strategies to make the most of the urban growth centers' advantages in the world market system. At the same time, rural development initiatives were intensified to improve their capacity and independence in managing local resources as well as urban network opportunities. These alternatives included urban functions in rural development, and agropolitan and rural-urban linkage strategies. The basic principle of these alternatives was to focus on local resource development by connecting rural centers with existing urban systems and investing in urban facilities and infrastructure to support rural development [14,18,25,32,33]. In fact, counter-urbanization efforts to relieve the rural-urban divide seem difficult to realize since rural-based developments may be promising but cannot be successful without urban assistance. Therefore, this paper aims to examine rural industry clustering as a response against extended urbanization in peripheral regions. Arguably, rural industrialization could be a solution by facilitating transitional rural-urban interfaces in small towns and peri-urban settlements instead of allowing parasitic exploitation of urban forces.

2. Methods

This study employed an explanatory sequential mixed-method approach, where quantitative spatial analysis was used to identify the expansion of urbanized areas in villages, and qualitative case study analysis to figure out the socio-economic shift in rural livelihoods. According to [34], this approach is used to deepen the researcher's understanding of the phenomenon observed by more detailed examination of selected cases, objects, samples or indicators that have served the research interest in the former quantitative phase. The rural-to-urban land use change in the rural periphery of the greater Solo region (GSR) represents the case of interest that guides a limited exploration of the socio-economic shift pattern, particularly in newer urbanized rural areas. The underlying hypothetical assumption behind the use of such a mixed-method approach is that the physical land use changes correlate positively to the socio-economic changes of space from rural to urban sectors. Arguably, urban land use should include non-agriculture activities with more complex and heterogeneous relations between actors.

2.1. *Quantitative data collection and analysis*

In the quantitative research phase, data on land use change were collected by comparing settlement classification changes published in population census documents (*Sensus Penduduk*/SP). These documents are published by the Representative Central Bureau of Statistics (*Badan Pusat Statistik*/BPS) in each local administration of GSR, consisting of one municipality and six regency administrations, i.e. Surakarta municipality, Boyolali regency, Sukoharjo regency, Karanganyar regency, Wonogiri regency, Sragen regency, and Klaten regency. The population census that was undertaken in 2000 (SP-2000) and 2010 (SP-2010) accounted for the primary database to sort out the *desa perkotaan* (urban villages) and *desa perdesaan* (rural villages) classification. For the purposes of more accurate mapping, the database was confronted with the existing rural potentials (*Potensi Desa*/Podes) dataset, spatial plans (*Rencana Tata Ruang Wilayah*/RTRW), and field observations. The collection of all these data took place from 2015 to 2017 with some adjustments made wherever necessary, particularly because of missing data and inaccuracies. With the support of ArcGIS software the collected data were then analyzed by utilizing a spatial urban-rural land use distribution map.

The next data collection focused on batik firm distribution across the region. The initial data sources were obtained from the Local Planning Authority (*Badan Perencanaan Pembangunan Daerah*/BAPPEDA) and BAPPEDA's arm-length quasi-government institution called Forum for Economic Development and Employment Promotion (FEDEP). The data gathered were verified through a series of focus group discussions (FGDs) and in-depth interviews involving batik cluster organizations, prominent batik entrepreneurs, the Local Agency for Industry and Trade Services (*Dinas Perindustrian dan Perdagangang/DISPERINDAG*), the Local Agency for Cooperatives and Micro, Small and Medium-sized Enterprises (*Dinas Koperasi dan Usaha Mikro, Kecil, dan Menengah*/DISKOP-UMKM), village government heads, academics, informal leaders, and non-governmental organizations (NGOs). Collection of the data was undertaken between 2015 and 2017 and double-checked through field observations to ensure that the list of existing batik firms was up to date. With the exception of Surakarta municipality, which has been one of the leading nationwide batik industry centers for decades, the locations of all batik firms were verified directly in the field. After that, the digital mapping of the spatial distribution of batik firms was accomplished, and the result was overlaid with the prior urban-rural land use distribution map (Figure 1).

2.2. Qualitative data collection and analysis

In the qualitative research phase, this study focused on collecting and analyzing data and first-hand information from the FGDs and in-depth interviews. The main objective of this phase was to figure out the socio-economic shift in rural livelihoods, particularly in urban villages across the peripheral regions of Surakarta municipality, which has been performing as the growth center for Solo Raya regionalization. Special attention was paid to batik industry clustering through which rural industrialization is taking place to carry out rural-urban interface patterns in particular. Some key elements of clustering that were considered as influencing the socio-economic shift of rural livelihoods are the shared use of public infrastructure (e.g. roads and communication), facilities (e.g. workshops, training centers, and showrooms), economic linkages (e.g. inputs, joint production, and marketing channels), social capital (e.g. trust and norms, family ties and social networks), and institutional capacity (e.g. community organizations, rules of the game, and collaborative partnership).

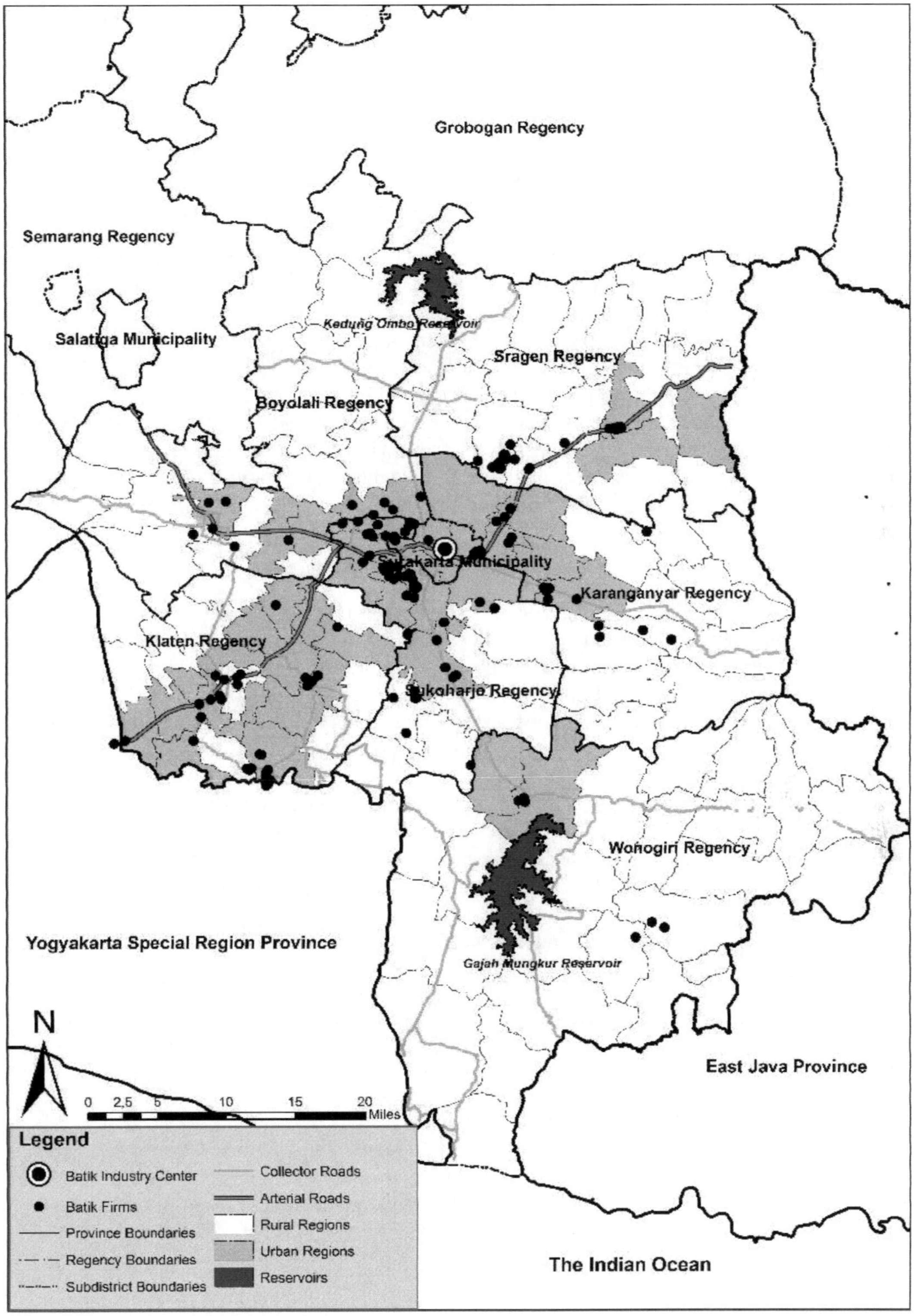

Figure 1. Spatial distribution pattern of batik firms in greater Solo region

3. Results and discussion

In general, the emergence of batik industry clusters in GSR is rooted in the batik tradition of Surakarta, even though many of the existing batik entrepreneurs and traders no longer consistently maintain their business cycle with authentic batik products. Recently, many batik products from other production centers such as Yogyakarta and Pekalongan have overwhelmed the local batik markets, creating a missing stickiness of the local batik identity against a great variety of competitive batik products. Since the 2000s, the local governments have been keen to revitalize the local batik industry by promoting the building of batik industry clusters in GSR's peripheral villages, particularly in response to the central government's campaign for promoting batik products as a national identity through the official declaration of *Hari Batik Nasional* (National Batik Day) on October 2nd, 2009.

Despite the growing interest of local governments and batik industry players to successfully develop batik industry clusters, their emergence, in fact, has raised anxiety regarding the future of rural industrialization in promoting rural independence spatially and economically. This study began with a cautionary awareness that rural industrialization may encourage transitional rural-urban interfaces into which rural communities and settlements could transform from below to become more urbanized, with minimum dependencies on the core growth center. Arguably, the current batik industry clusters can fulfill that promise because of the abundance of resources to support the local batik industry revival. The following sections will further discuss this issue with exclusion of the Boyolali regency case, primarily because the local batik industry has been created from zero only in recent years so that presumably they have neither the so-called inherited batik tradition nor historical connections with the well-established batik industry in GSR.

3.1. Extended urbanization trend

The urbanization pattern in GSR has demonstrated de-urbanization and rurbanization (rural urbanization) phenomena simultaneously since the 1990s. While the urban population in Surakarta is declining, the opposing trend has been occurring in other subregions. Not surprisingly, Surakarta faces a very limited territorial coverage of 44.04 km^2, with more than a half million urban inhabitants, creating out-migration and urban space expansion outwards to its periphery. The immediate effect of such urban expansions is experienced by the adjacent subregions, i.e. Sukoharjo and Karanganyar, along the major road axis Surakarta-Surabaya. Somewhat different to this pattern, the growth of urbanized areas in Klaten and Sragen is not only affected by peri-urbanization (peripheral urbanization) spillover from Surakarta but also by endogenous rurbanization dynamics. From Figure 1, we can observe discontinued peri-urbanization from Surakarta outwards as indicated by the existence of rural regions inbetween. Table 1 shows this pattern by depicting the urban population growth trends.

Table 1. Urban population trends, 1980-2010

		1980	1990	2000	2010
Surakarta		469,532	504,176	490,214	499,337
	%	*43.57*	*30.19*	*18.07*	*16.21*
Boyolali		42,623	128,494	235,307	306,090
	%	*3.96*	*7.69*	*8.67*	*9.94*
Sukoharjo		126,811	324,214	539,024	632,367
	%	*11.77*	*19.41*	*19.87*	*20.53*
Karanganyar		76,178	148,860	336,170	414,969
	%	*7.07*	*8.91*	*12.39*	*13.47*
Wonogiri		36,607	108,852	164,143	190,450
	%	*3.40*	*6.52*	*6.05*	*6.18*

Sragen		56,985	70,621	233,474	277,310
	%	*5.29*	*4.23*	*8.61*	*9.00*
Klaten		268,805	384,896	714,535	759,450
	%	*24.95*	*23.05*	*26.34*	*24.66*
Greater Solo		**1,077,541**	**1,670,113**	**2,712,867**	**3,079,973**
	%	***100.00***	***100.00***	***100.00***	***100.00***

3.2. Infrastructure
The availability of supporting infrastructure is an obligatory condition for any industrialization to continue for a long time. A reliable road network ensures flows of input-output linkages and exchange of knowledge and innovation among actors. The local batik industry clusters in GSR are connected with either the national arterial or the provincial collector road system, except those in Wonogiri regency, so that the local batik firms there are quite isolated. Figure 1 shows that the so-called batik industry cluster in this region is located in the middle of nowhere in Tirtomoyo subdistrict, comprising very few batik firms. Actually, the absence of a reliable road network and transportation system does not obstruct the spread of the batik industry to remote areas. A salient feature of the batik industry is that as a folkloristic home-based industry it enacts a cultural tacit knowledge transfer across generations, which enables the actors to adapt, modify, and develop batik production skills in compliance with local social systems. It is the immaterial possession of batik production creativity and innovation that matters in determining local batik industry growth instead of the physical infrastructure. Therefore, access to knowledge infrastructure is more important for supporting local batik industry growth.

3.3. Facilities
Some specific facilities to support batik industry existence are schools and training centers, workshops and showrooms, exhibition venues, and batik specific markets. All these types of facilities can be found in Surakarta only, while the remaining GSR subregions have no schools, training centers or markets specific to batik industry needs (Table 2). However, informal batik training is available in each local batik cluster occasionally, hosted by prominent batik entrepreneurs, community organizations, government agencies, or external parties. Regarding the absence of batik specific markets, local batik producers and traders prefer to rely on annual batik exhibition events.

Table 2. Specific facilities to support local batik industry

Subregion	Types of specific facilities			
	School and training center	Workshop and showroom	Exhibition venue	Batik specific market
Surakarta	√	√	√	√
Sukoharjo	-	√	√	-
Karanganyar	-	√	√	-
Wonogiri	-	√	√	-
Sragen	-	√	√	-
Klaten	-	√	√	-

To date, there is a division of spatial production chains between Surakarta and its peripheries. As a core industry region, Surakarta more functions as a national center of the batik market and industrial research and development. Newer batik motifs and styles are mostly found in Surakarta as well as the improvement of batik techniques and materials. On the other hand, the GSR subregions focus on batik production and labor regeneration. Many batik workers of the younger generations come from peripheral villages in response to the recent local batik industry growth.

3.4. Input linkages

More than 80 per cent of the raw material supply for the entire GSR batik industry comes from suppliers in Surakarta. Exceptionally, the local batik industry clusters in Klaten and Sragen can find alternative suppliers instead of maintaining high dependency on those in Surakarta. This finding is interesting considering that large textile companies are located in Sukoharjo and Karanganyar. Figure 2 below explains batik industry interdependence in GSR.

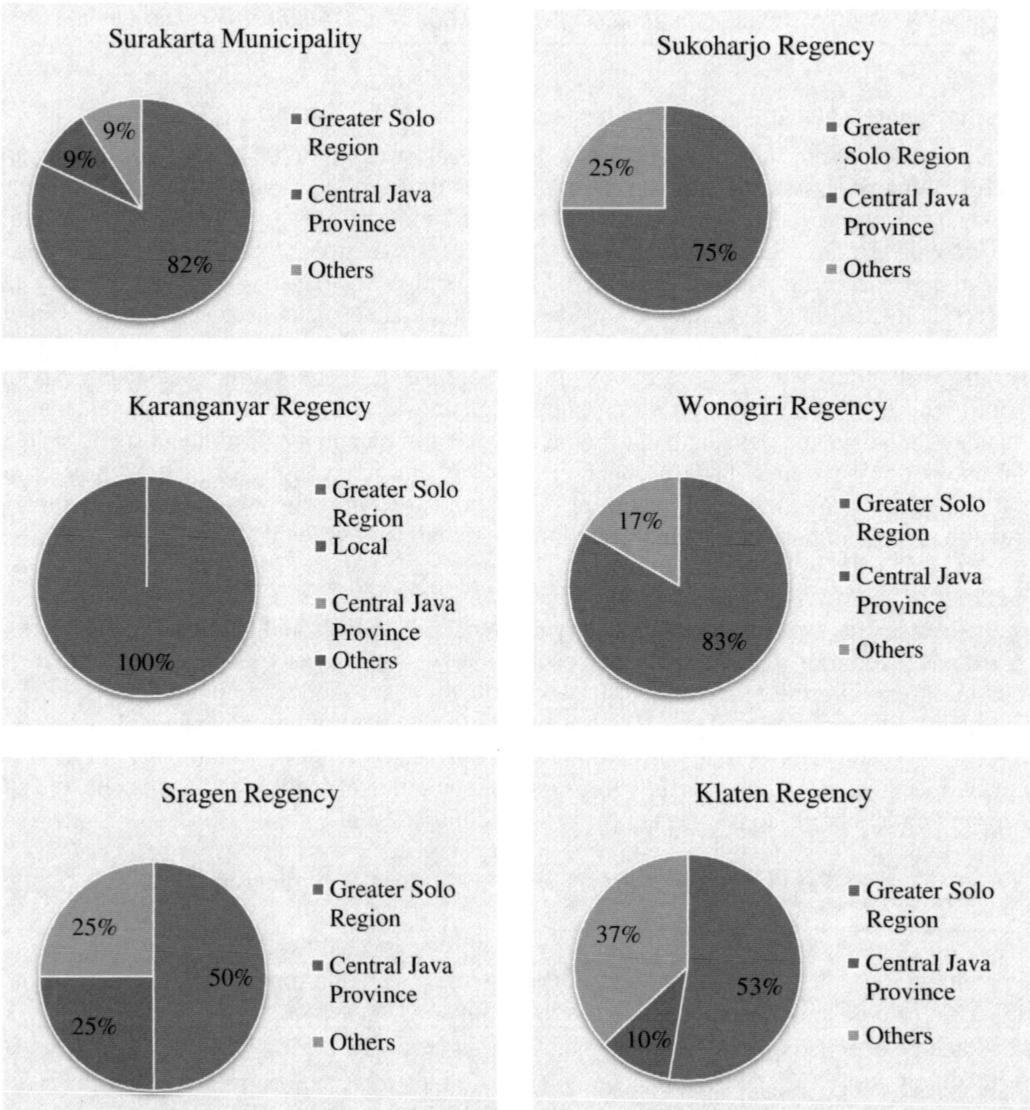

Figure 2. Input linkages pattern

3.5. Production network

Joint production networks cannot be found clearly in each batik industry cluster. The only forms of joint production are subcontracting and coordinated division of labor within the clusters. The former form is very common everywhere when the larger batik producers assign their smaller partners to produce a certain amount of batik works based on contractual agreements. The latter is found limited to Klaten and

Karanganyar, where a local batik business group distributes batik works equally among the group's members.

3.6. Output linkages

Regarding marketing destinations, the local batik industry cluster in GSR has developed many different markets, not only delivering batik products to the local batik market. Even though the local GSR markets remain prospective, many local batik producers have expanded their markets into various destinations. Again, the local batik industry clusters in Klaten and Sragen have managed to find market alternatives for selling their products. Figure 3 presents the output linkage patterns of the local batik industry clusters in GSR.

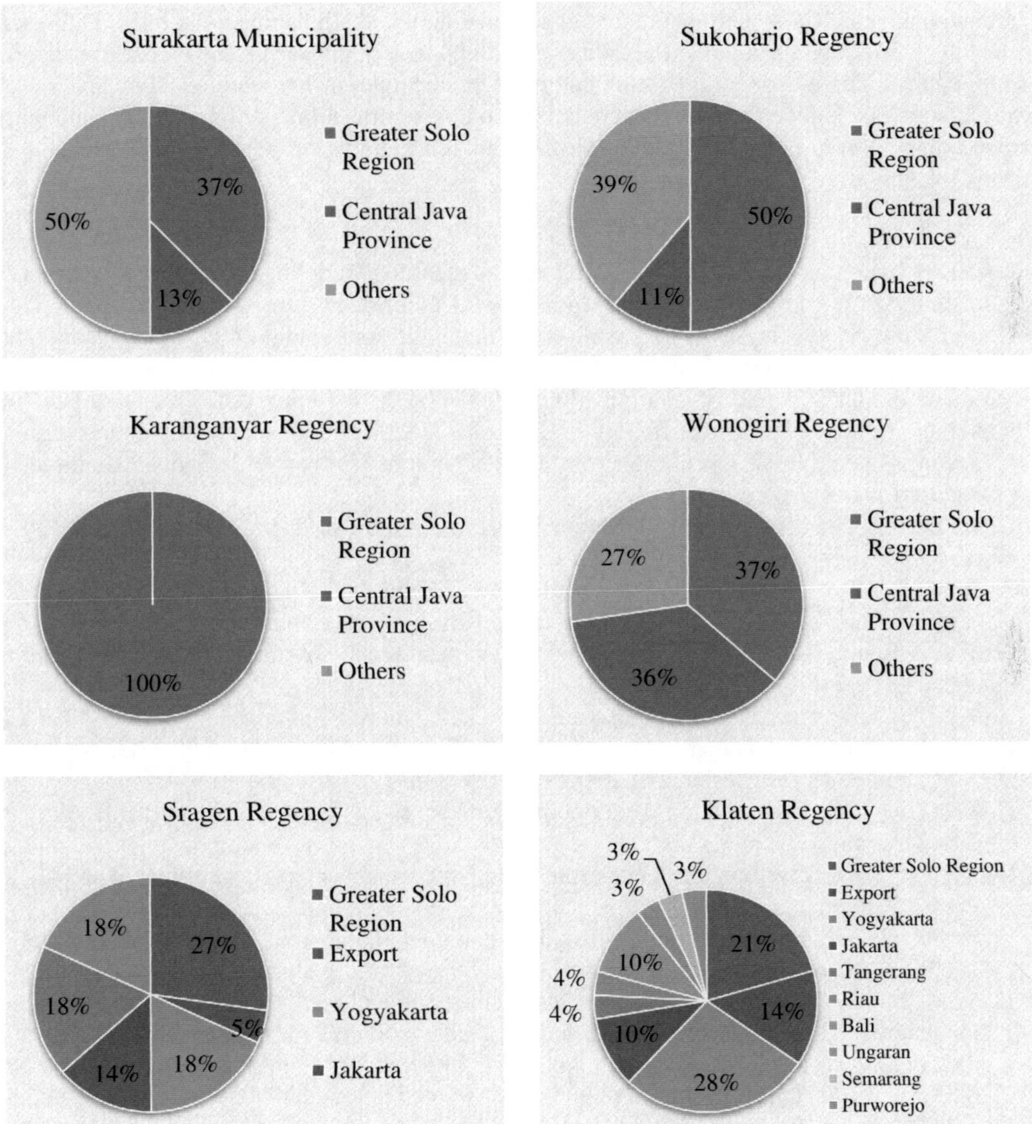

Figure 3. Output linkages pattern

IOP Conf. Series: Earth and Environmental Science **158** (2018) 012055 doi:10.1088/1755-1315/158/1/012055

4. Conclusions

The recent emergence of rural batik industry clustering in the GSR peripheral regions is a good sign for rural-to-urban transformation. Traditionally, batik industry has a strong connection with rural agriculture activities since the old practices of batik industrialization are associated with harvesting period breaks. While waiting for the upcoming cultivation period, the female agriculture workers were used to spending their spare time doing batik works. In the modern era, such practices still exist, even though very limited and in particular places. The greater exposure of the local batik industry players to more competitive markets has somehow changed their views on how to run a batik business properly. More modernized batik industrialization has raised their awareness to seek for self-organizing solutions by establishing community organizations, business networking, and technological adoption to survive in the market.

However, the most important findings of this study are the sustained high dependency of the local batik industry in the GSR peripheral regions towards the batik industry center in Surakarta. Unfortunately, while such rural industrialization growth has not matured yet, the extended urbanization in these regions has failed to encourage the establishment of proper infrastructure and facilities required to achieve business independence. Instead of encouraging the local villagers to become more modernized, they tend to preserve the current informal economy to support the existence of a mixed rural-urban livelihood.

References

[1] Lees L H 2015 World urbanization 1750 to the present In: Mcneill J, Pomeranz K, editors. The Cambridge World Histor p 34–57 Cambridge: Cambridge University Press

[2] United Nations Department of Economic and Social Affairs Population Division. World Urbanization Prospects: The 2014 Revision [Internet]. New York; 2015. Available from: https://esa.un.org/unpd/wup/Publications/Files/WUP2014-Report.pdf

[3] UN-Habitat 2016 World Cities Report 2016 - Urbanization and Development: Emerging Futures. International Journal. Nairobi: United Nations Human Settlements Programme (UN-Habitat) 1-248 p

[4] Davis K 1955 The origin and growth of urbanization in the world *Am J Sociol.* **60** 5 p 429–37.

[5] Wirth L 1938 Urbanism as a way of life *Am. J. Sociol.***44** 1 p 1–24

[6] Anderson N 1959 Urbanism and urbanization *Am. J. Sociol.***65** 1 p 68–73

[7] Catanese A J and Snyder J C 1992 Perencanaan Kota (Jakarta Penerbit Erlangga)

[8] Gallion A B and Eisner S 1992 Pengantar Perancangan Kota: Desain dan Perencanaan Kota (Jakarta: Penerbit Erlangga)

[9] Hall P 2002 Urban and Regional Planning. 4th ed. (London: Routledge)

[10] Hall P 2014 Cities of Tomorrow: An Intellectual History of Urban Planning and Design Since 1880. Fourth **13** Chichester, UK: John Wiley & Sons Ltd

[11] Hoover E M 1948 The Location of Economic Activity (New York: McGraw-Hill Book Company, Inc.)

[12] Krugman P 1998 Development, Geography, and Economic Theory (Cambridge: The MIT Press)

[13] Lipton M. Balanced and unbalanced growth in underdeveloped countries. Econ J [Internet]. 1962;72(287):641–57. Available from: http://www.jstor.org/stable/2228442

[14] Douglass M 1998 A regional network strategy for reciprocal rural-urban linkages: An agenda for policy research with reference to Indonesia. *Third World Plann Rev.* **20** 1 p 1–31

[15] Dawkins C J 2003 Regional development theory: Conceptual foundations, classic works, and recent developments *J Plan Lit.* **18** 2

[16] Pike A, Rodríguez-Pose A and Tomaney J. Shifting horizons in local and regional development. Reg Stud [Internet]. 2016;3404(August):1–12. Available from: http://www.tandfonline.com/doi/abs/10.1080/00343404.2016.1158802

[17] Glaeser E L 2000 The New Economics of Urban and Regional Growth. In: Clark GL, Feldman MP, Gertler MS, editors. The Oxford Handbook of Economic Geography p 83–98 (Oxford: Oxford University Press)

[18] Gore C 1985 Region in Questions: Regions, Space, and Development Theory (New York: Methuen)
[19] Tötdling F 2011 Endogenous approaches to local and regional development policy. In: Pike A, Tomaney J, editors. Handbook of Local and Regional Development p 333–43 (Abingdon: Routledge)
[20] Soja EW 2009 Regional planning and development theories. In: Kitchin R, Thrift N, editors. International Encyclopedia of Human Geography Volume **8** p 259–70. (Amsterdam: Elsevier)
[21] Chapman K 2005 From "growth centre" to "cluster": Restructuring, regional development, and the Teesside chemical industry *Environ Plan A.* **37** 4 p 597–615
[22] Gore C 1985 Urban-industrial growth pole strategies and the diffusion of modernization. In: Regions in Question p 81–117 (New York: Methuen)
[23] Stewart C T 1946 The urban-rural dichotomy: concepts and uses p 152–8
[24] Berdegué J A, Carriazo F, Jara B, Modrego F, and Soloaga I 2015 Cities, territories, and inclusive growth: Unraveling urban-rural linkages in Chile, Colombia, and Mexico. *World Dev.* **73** p 56–71
[25] Lo F-C, Salih K and Douglass M 1981 Rural-urban transformation in Asia. In: Lo F-C, editor. Rural-Urban Relations and Regional Development (Singapore: Maruzen Asia)
[26] Vernon Henderson J 1986 Urbanization in a developing country *J Dev Econ.* **22** 2 p 269–93
[27] Zhang XQ. The trends, promises and challenges of urbanisation in the world. Habitat Int [Internet]. 2016;54(13):241–52. Available from: http://dx.doi.org/10.1016/j.habitatint.2015.11.018
[28] Gibson C and Kong L.Cultural economy: a critical review. Prog Hum Geogr [Internet]. 2005;29(5):541–61. Available from: http://phg.sagepub.com/cgi/doi/10.1191/0309132505ph567oa
[29] Wei YHD 2010 Beyond new regionalism, beyond global production networks: Remaking the Sunan model, China *Environ Plan C Gov Policy* **28** 1 p 72–96
[30] Yeung HW, Coe NM. Toward a Dynamic Theory of Global Production Networks. Econ Geogr [Internet]. 2015;91(1):29–58. Available from: http://www.scopus.com/inward/record.url?eid=2-s2.0-84920999873&partnerID=40&md5=476d8c90b1a7124f27912077aa1dd396
[31] Bathelt H, Li P-F. Global cluster networks--foreign direct investment flows from Canada to China. J Econ Geogr [Internet]. 2013 Mar 23 [cited 2015 Mar 23];14(1):45–71. Available from: http://joeg.oxfordjournals.org/cgi/doi/10.1093/jeg/lbt005
[32] Satterthwaite D and Tacoli C 2003 The Urban Part of Rural Development: The Role of Small and Intermediate Urban Centres in Rural and Regional Development and Poverty Reduction. The Earthscan Reader in Rural-Urban Linkages. London; 2003. Report No.: 9.
[33] Leeuwen ES van. Urban-rural interactions: towns as focus points in rural development (Google eBook) [Internet]. 2010. 177 p. Available from: http://books.google.com/books?id=H6yioZVMlEYC&pgis=1
[34] Creswell JW 2014 Research Design: Qualitative, Quantitative, and Mixed Methods Approaches. 4th ed (Thousand Oaks: Sage Publications, Inc)

IOP Conf. Series: Earth and Environmental Science **158** (2018) 012056 doi:10.1088/1755-1315/158/1/012056

A Rural Transformation Model: The facts of rural development in the Surakarta Metropolitan Region

D P Puspa Sari, I Asyifa, I F Derman, D R Jayanti and F Y Hanatya

Urban and Regional Planning Department, Faculty of Engineering, Diponegoro University, Semarang, Indonesia

Corresponding author: dwiputripuspasari@gmail.com

Abstract. Not only cities are entering the urban age but suburban villages are also feeling the impact of this global phenomenon. In Indonesia, the uncontrolled rural transformation has had some negative impacts because of the unpreparedness of various aspects such as land conversion, the emergence of the informal sector, and crime. This phenomenon is often referred to as developmental externalities that need to be anticipated in planning and controlling the growth of cities and villages. This inevitable rural transformation also occurs in the Surakarta Metropolitan Region. The previous rural transformation studies in the Surakarta Metropolitan Region are based on economic, spatial to socio-ecological perspectives and are still rarely studied from the perspective of urban studies. This article aims to examine the model of rural transformation in the Surakarta Metropolitan Region based on the Rural-Urban Transformation theory by Lo, Shalih & Douglass (1998), especially in the Simo, Sambi, Ngemplak, and Nogosari Sub-districts in Boyolali District. The qualitative methods consisting of interviews, 150 questionnaires, and field observations in 2017 and literature study were used for the discussion in this article. The rural to urban transformation of the Surakarta Metropolitan Region follows the Southeast Asian Model. This research opens a new discussion on how to create a sustainable city system in the Surakarta Metropolitan Region.

1. Introduction

The rapid flow of urbanization which resulted in the increasing number of people living in urban areas make the cities in the world into a new period commonly referred to as the urban age [2]. The assumption is based on the phenomenon of urban population growth rate that is not only influenced by natural population growth but the increasing rate of urbanization. United Nation (2014) noted that the number of urban residents in 2014 reached 54% and is projected to reach 66% by 2050 [27]. One consequence is the total urban area in the world will triple in the next 40 years and will convert land agriculture [25]. Specifically, for developing countries like Indonesia, it is assumed that urban population growth rate is higher than developed countries. It is of recent concern that the phenomenon is not only happening in big cities, but starting to happen in small towns.

Same condition with other developing countries, Indonesia is experiencing a rapid urbanization process; the amount of urban land increased from ± 8,900 km^2 to 10,000 km^2 between 2000 and 2010 and this urbanization continues to increase to date. The increase of urban area in Indonesia is about 1.1 percent per year and the growth rate is ranked second in East Asia after China. This has resulted in a shift to an urban economy. Cities in Indonesia have an average economic growth of 4.1 percent per year, a faster pace than other Asian cities. By 2025, an estimated 68 percent of Indonesians will be urban dwellers [29]. This shows the existence of rural transformation into urban areas. This phenomenon can impact on economic shifts and on the socio-cultural conditions of people living in rural and urban areas. These conditions should be noted for the negative impacts of urbanization such as increasing urban waste generation [18], emergence of urban informal sector [23], and small town dependence on large cities [24] can be anticipated as early as possible to achieve sustainable and harmonious development.

Rural transformation can be defined as a process of comprehensive community change in which rural communities diversify their economies and reduce their dependence on agriculture; they become dependent on distant places to trade and to acquire goods, services, and ideas; moving from scattered villages to small and medium-sized cities; and are culturally more like large urban agglomerations [1]. There are trends and dynamics that encourage and accelerate the process of rural transformation. These factors become causes and effects that illustrate how these dynamics change the rural space and how they are interrelated. Of course, these dynamics are revealed differently in each country and are contextual [12].

Worldwide, the rural transformation process affects the lives of some 4.6 billion people who work 60 percent of the world's fertile land and produce nearly two-thirds of food and non-food products. Rural transformation is a major challenge for the people and rural areas, but also provides great opportunities for sustainable development [12]. Rural transformation can improve the performance of both urban and rural areas when strong rural-urban linkages are developed. However, it is generally understood that strong reciprocal links exist between urban and rural areas, through the flow of human resources, capital, goods, information, and technology [26].

These developments have had an impact on rural areas near metropolitan areas that are experiencing the phase of the urban age. Then, the failure of the development paradigm that initially relies on the urban-based industry in creating equitable distribution has led to new approaches and alternative regional planning strategies. Douglass suggests that there are limitations in the role of selected cities and growth centres to stimulate rural development solely through the promotion of urban growth [4]. So, an effort is made that is a selective investment in rural areas of towns. This further encouraged the rural-urban transformation. Therefore, the rural-urban linkage is a crucial part of regional development.

Lo, Salih, & Douglass [13] in their chapter Rural-Urban Transformation in Asia identified dualism within the framework of macro spatial development, namely (a) north and south dualism; (b) formal and informal sector dualism, and (c) rural-urban dualism. This dualistic structure shows an imbalance in the interaction of the three. These three dualistic relationships are interrelated and integrated into complex systems where differences for each country depend on four important factors, namely natural resources, demographic characteristics, technology and human resources, and development ideology. In the concept of economic growth and development process there are four differentiated patterns of development according to its characteristics, namely: (1) Southeast Asia Model, (2) South Asia Model, (3) China Model, and (4) East Asia Model. Each model gives rise to different rural-urban transformation patterns.

Surakarta is one of the big cities in Indonesia that continues to grow to form a metropolitan area in the central part of Central Java Province. The Surakarta Metropolitan Region is a metropolitan area consisting of Surakarta, Boyolali, Sukoharjo, Karanganyar, Wonogiri, Sragen, and Klaten commonly abbreviated 'Subosukowonosraten'. The development of Surakarta affects the development of other (peri-urban) areas around it of both the spatial and regional characteristics of the economy. The rapid growth of settlements in Surakarta that is not supported by adequate land causes the growth of the city towards the outskirts of Surakarta. This phenomenon has influenced the development of villages around Surakarta and caused the transformation from rural to urban.

Previous research by Giyarsih [10] showed that the spatial pattern in the Yogyakarta-Surakarta corridor shows a regional transformation where a higher accessibility leads to a higher level of transformation. Rahmayana and Handayani [21] found that the rural transformation into urban areas in the Solo Baru Region, Sukoharjo District, is seen from the socio-economic changes of local communities where changes in economic aspects are faster than social ones.

This study aims to determine the model of rural-urban transformation in Metropolitan Surakarta from the perspective of urban studies based on the rural-urban transformation theory in Asia which was put forward by Lo, Salih & Douglass [13] especially in the Simo, Sambi, Ngemplak, and Nogosari Sub-districts in Boyolali Regency. This research explains the transformation occurring in the four subdistricts from the aspects of land use, socio-demography, accessibility, and development policy.

2. Material and methods

2.1. Study area

Simo, Sambi, Ngemplak, and Nogosari Sub-districts, hereinafter referred to as "Mobilano" are located in Boyolali District which is part of the Surakarta Metropolitan Region. The four sub-districts are located southwest of Surakarta City with a respective size of Sambi: 4,649,4935 Ha, Ngemplak: 3,852.70 Ha, Nogosari: 5,508.43 Ha, and Simo: 4,804,0275 Ha.

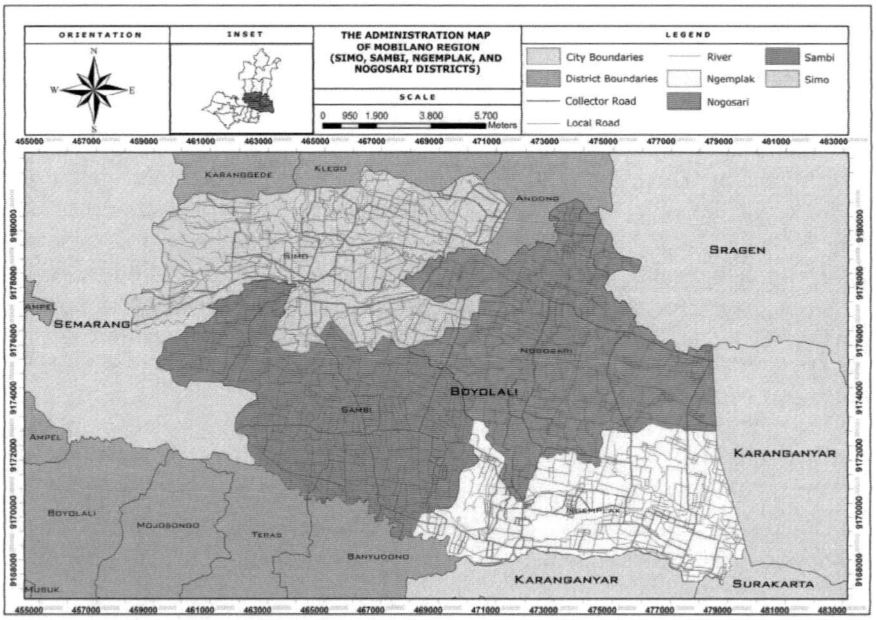

Figure 1. Administrative map of Simo, Sambi, Ngemplak, and Nogosari Sub-districts [15]

The four sub-districts were selected as the case study for the rural-urban transformation in the Surakarta Metropolitan Region because of the following aspects that spur the 'urban age' phenomena, i.e., (1) Ngemplak Sub-district is the administrative location of Adi Soemarmo International Airport or known as Solo Airport, 2) Simo Sub-district is the administrative area of the Hajj Central Java-Special Region of Yogyakarta Embarkation Building that every day attracts people from various regions within their service area, (3) An increased accessibility is indicated from the development of the Trans Java Freeway that intersects Ngemplak Sub-district, dividing it into a northern and southern part with one of the freeway gates located close to the airport, which is also serviced by a commuter railway project, (4) The four sub-districts are Sustainable Food Farm regions for rice crops, (5) 'Pro-investment' policies from the local government of Boyolali District. The high accessibility makes these four areas referred to as the gateway of Metropolitan Surakarta.

2.2. Research approach

This research uses secondary and primary data with a qualitative evaluation process which refers to the development pattern in Asia based on Lo, Salih, & Douglass (1998)[13].

2.2.1. Assessment by theory. The rural-urban transformation refers to the pattern of development in Asia of Douglass et al. [4], which distinguishes the Southeast Asia, South Asia, China, and East Asia models. Each model has different characteristics seen in its economy, natural resources, and social culture [4]. In the theory of regional development with a growth pole, the growth of a region comes from an already developed centre. The development of the regional centre will affect its surrounding areas which forms suburbs. The area has mixed village and city traits. This shows the transformation of rural to urban areas.

The Southeast Asian rural-urban model is characterized by richness in natural endowment, open domestic markets, and low technology. In their transformation, urban villages in Southeast Asia underwent a process of foreign investment in the resource and commercial sectors, exports of resource-based primary products, foreign investment in modern industries, imports of modern industrial products, rural development expenditures, economic income from remittances, and the relationship between landowners and their tenants.

In the South Asian transformation model, urban villages are characterized by low resource availability per capita, distance from urban to rural areas which are not yet supported by transportation, and low rural-urban transformation. The South Asian model exhibits a transformation process that starts from the potential of exporting limited commercial crops, foreign industries, and modern industries with limited manufacturing export capacity, import substitution industries with modern technology and limited manufacturing.

The characteristics of the East Asia model are a closed economy with food deficits and a high population density, a closed economic and political system, and indigenous technology. In the transformation of the rural-urban area, East Asia underwent two phases; the first phase was the redistribution of assets (land, capital, etc.) and the second phase consisted of minimizing external dependency, self-reliant regional development, and rural-based urbanization.

The Chinese model is like the East Asia model, consisting of a centrally planned closed economy with food deficits and high population density, a closed economic and political system, and indigenous technology. China's rural-urban transformation model also consists of two phases; the first phase is the redistribution of assets (land, capital, etc.), a controlled production and distribution system, and controlled rural-urban migration. The second phase consists of minimizing external dependency, self-reliant regional development, and rural-based urbanization.

2.2.2. Evaluation process. The model of rural-urban transformation in the Surakarta Metropolitan Region is determined by comparing facts with theory. The evaluation is based on the rural model of urban transformation model by Mike Douglass [4-8] describing the transformation model from East Asia, Southeast Asia, South Asia, and China. The analysis of the changing conditions of the Surakarta Metropolitan region from 2005 - 2015 based on the data already obtained is used to describe the characteristics of the phenomenon of transformation of the urban villages in the focus area. This is then matched with the four rural-urban transformation models from Douglass to uncover which model the transformation of urban villages in the Surakarta Metropolitan Region follows, especially Simo, Sambi, Ngemplak, and Nogosari Sub-districts. The transformation of the urban villages of East Asia, South Asia, and Southeast Asia has distinctive features and the characteristics of each region are distinguished based on economic, physical, demographic, and socio-cultural conditions.

3. Material and methods

3.1. Study area

This section describes the rural-urban transformation phenomenon in the study area.

3.1.1. Demographic characteristics. The development of the city of Surakarta does not rule out that it will attract people from outside the city. An area with unchanged administrative boundaries will experience high density if it continues to accommodate the incoming community. This affects the Boyolali district, especially the four sub-districts of Simo, Sambi, Ngemplak, and Nogosari (the Mobilano Area) which are near Surakarta City. This phenomenon can be seen from the data by the Central Statistical Agency of Boyolali District. The data from 1990 to 2015 shows the population trends in Boyolali District, in particular of the 'Mobilano' Region.

Table 1. Population in the 'Mobilano' Region Per Five Years from 1990 to 2015 [15].

Sub-district	Population						Percentage of Boyolali District (2015)	Percentage of 'Mobilano' (2015)
	1990	1995	2000	2005	2010	2015		
Sambi	45.354	46.549	47.672	48.303	48.657	49.783	5%	20%
Ngemplak	58.880	62.743	65.975	69.235	71.111	84.717	9%	34%
Nogosari	58.578	60.326	61.078	61.066	60.788	65.580	7%	27%
Simo	41.053	41.400	42.443	43.235	43.667	45.649	5%	19%
Total	203.865	211.018	217.168	221.839	224.223	245.729	25%	
Boyolali District	870.326	896.527	922.852	941.147	953.839	963.690		

Ngemplak Sub-district has the largest population while Simo Sub-district has the smallest. The large number of inhabitants influences the contributing percentage of the larger upper region, where a large population will contribute more and vice versa. Ngemplak Sub-district has the largest contribution towards the population of the 'Mobilano' Region and Boyolali District.

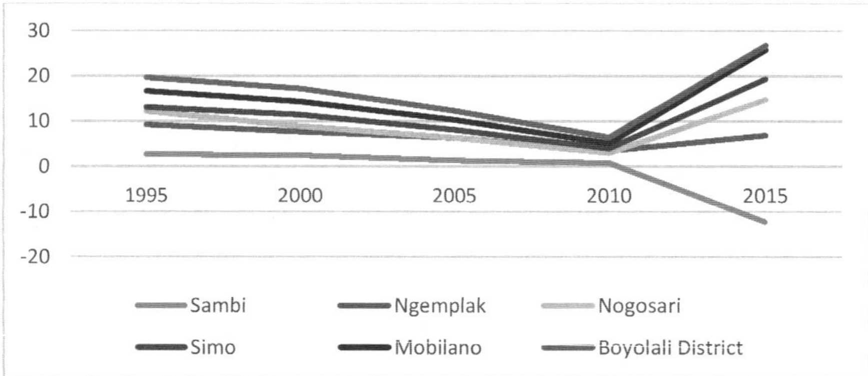

Figure 2. Growth rate per five years in the 'Mobilano' Region 1995 – 2015 [15].

The growth rate of the Mobilano Region exhibits a similar pattern as Boyolali District that decreases in the first four periods and rises in the last period, while Sambi Sub-district experienced a decrease in the last period. The increasing rate of population growth in 2010 - 2015 may be caused by the transformation of Surakarta. The increase in the rate of population growth in the last five years is quite a drastic change from previous years. The population in the 'Mobilano' Region is predicted to increase in the coming year considering the development of Trans Java Freeway that will intersect the area.

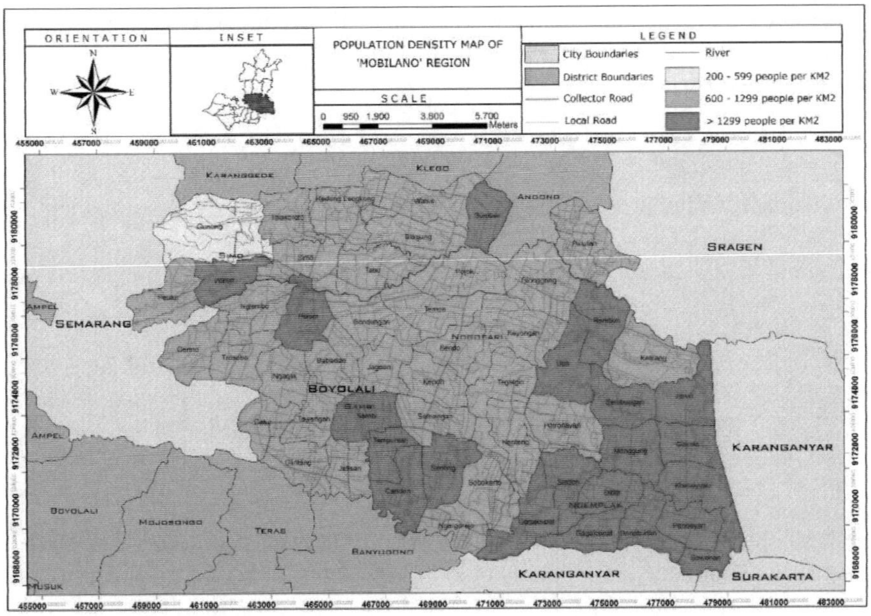

Figure 3. Population density map of 'Mobilano' Region [15].

Rural-urban transformation in Surakarta City impacts on the increasing population numbers in Boyolali District. Ngemplak Sub-district has the highest density in the Mobilano region, which makes its urban development more visible than in the other three sub-districts.

The low population density has the potential to attract migration from other areas in the Greater Solo region, thus supporting urban processes in urban areas. People moving from Greater Solo on average have higher potential in terms of human resources and soft skills. It is expected that this ability can impact the development of urban areas. In addition to affecting the urban development process, it is expected to also affect the indigenous people of the urban area and improve the quality of human resources.

The population of productive age (15-60 years old) of Boyolali District is concentrated in Ngemplak Sub-district, which is a potential driving force of the regional economy. The existence of this potential has not been supported by a good human development program which can be seen from the relatively high level of poverty in the Mobilano region of almost 50 percent; the sub-districts of Sambi and Nogosari have a poverty rate of more than 60 percent. This high level of poverty affects the ability to obtain education and health services, which tends to be low. This will affect the quality of human resources to be low, causing new problems of unemployment. When the unemployment rate is high, then the dependency rate is also high, hindering development and increasing poverty levels.

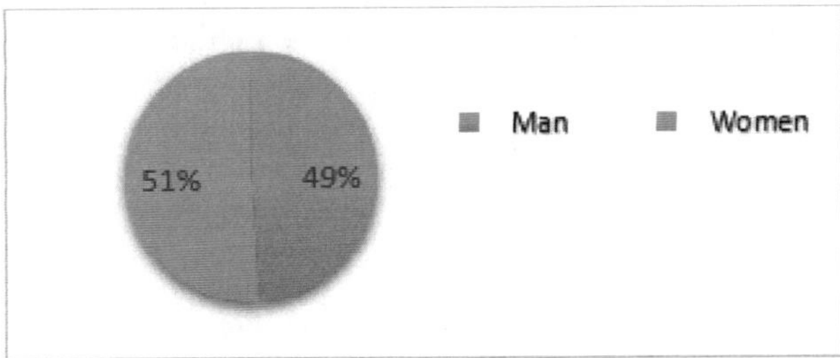

Figure 4. Sex ratio graphic of 'Mobilano' in 2015 [15].

IOP Conf. Series: Earth and Environmental Science **158** (2018) 012056 doi:10.1088/1755-1315/158/1/012056

The female population in Mobilano is higher than the male population. The high female population is a potential that can support activities in the industrial sector, which requires female workers, e.g., the cigarettes and garment industries.

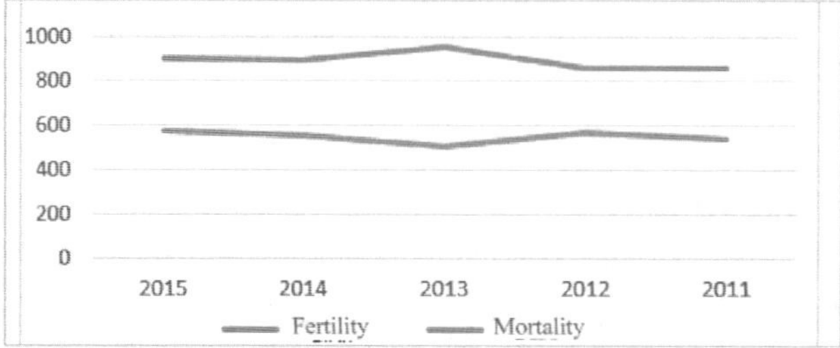

Figure 5. Mortality and birth rate of Boyolali District. in 2011 – 2015 [15].

In population dynamics, birth rates are higher than death rates from year to year. The productive age of the population is around age 34 - 40 years; in the 'Mobilano' area, the productive population is 60 percent and the dependency ratio is 66 percent. Thus, the region has a high labour potential.

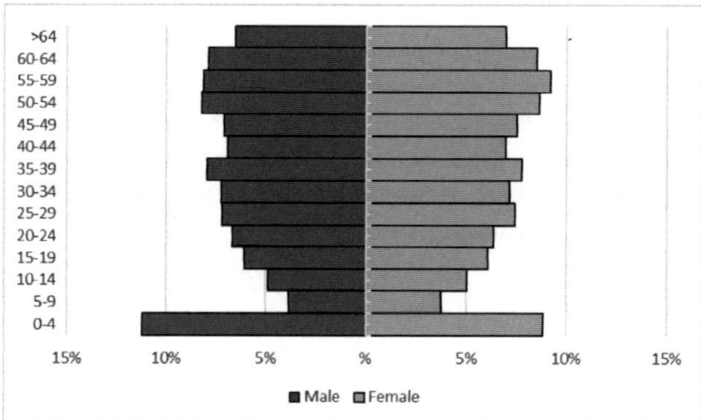

Figure 6. Population pyramid of the 'Mobilano' Region [15]
[a] The female population is shown in red and the male population in blue.

The population pyramid of the 'Mobilano' Region in 2015 is of a constructive type that shows low birth rates and very high mortality rates. The population pyramid shows a large old age group but also an extensive young population. In addition, the low growth rate causes a decline in the population.

3.1.2. Human resources. The quality of human resources in the four sub-districts (Simo, Sambi, Ngemplak, Nogosari) can be seen from the level of education. Figure 7 shows the changes in of human resources over five years.

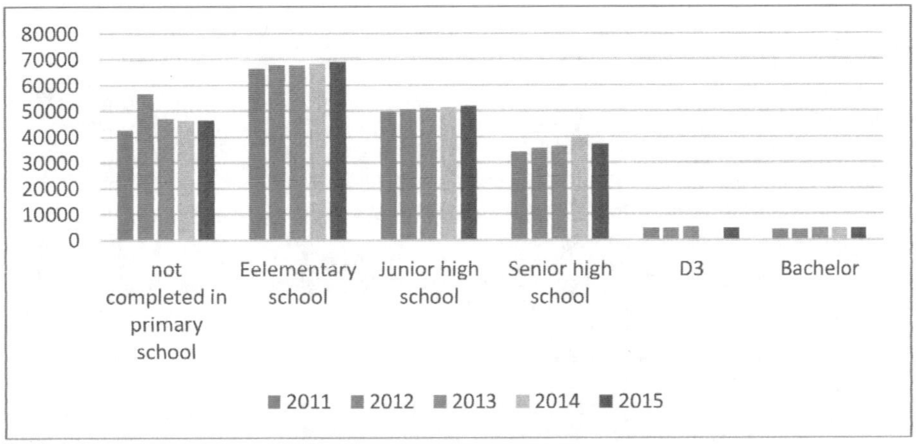

Figure 7. Population based on highest completed education level in 2011-2015 [15].

Figure 7 shows the level of education in the four sub-districts of the study area. The graph shows that the dominant level of education in the study area is elementary school. However, the graph shows that the level of senior high school graduates increased in the last five years and the population that does not complete primary school is quite stagnant although it has increased since last five years. The overall education trends show an increase in human resources in the four sub-districts. Although the population with elementary school graduates increased, the amount of senior high school and university graduates showed a stronger increase than the number of elementary school graduates.

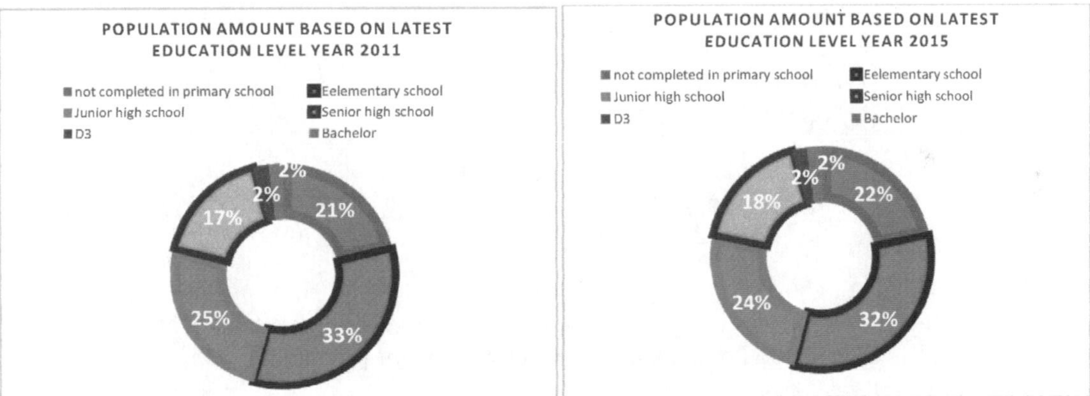

Figure 8. Population based on highest education level obtained in 2011 [15].

The above graph shows the data of all residents in the four sub-districts including the elderly residents and non-resident students. Therefore, the number of elementary and non-primary school graduates is still quite high. However, the current school survey is dominated by senior high school graduates, which could be a result of the recent increase in the number of senior high school educators. Thus, the quality of human resources in the study area has increased during the last five years, which can be seen from the decreased level of elementary school education and the increase in senior high school graduates. This indicates a positive change over the last five years.

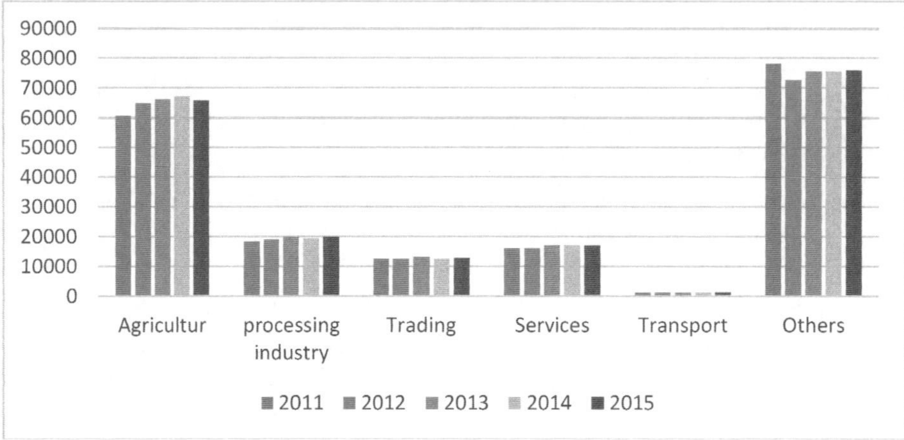

Figure 9. Number of population graphic according to labour in 2011-2015 [15].

Figure 9 shows that the residents of the study area have livelihoods dominated by 'others', comprising non-agricultural labour other than industry, trade, services, and transportation. In the study area, the labour force in the agriculture sector is still quite large, being the second highest group after other labour (non-agriculture other than industry, trade, and service) during 2011 - 2015. They are conventional farmers using simple technology. The number of workers in these six categories tends to increase. Thus, the population of the study area is dominated by non-agricultural labour with a ratio of non-agricultural to agricultural labour of 3:2.

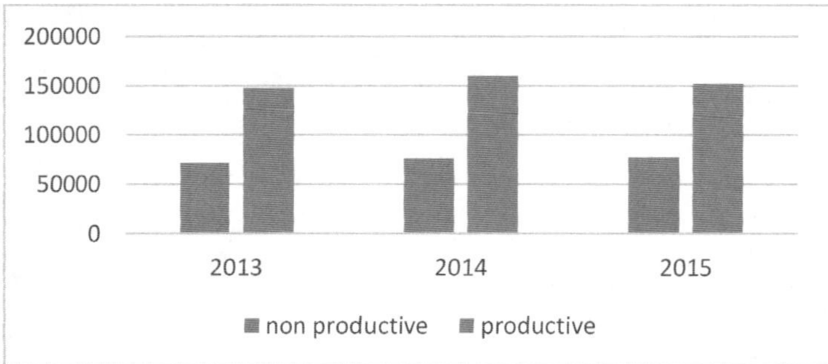

Figure 10. Population of productive and non-productive age in 2013-2015 [15].

Figure 10 shows that the study area is dominated by population in the productive age of the 15 to 65 years old. This high availability of labour offers opportunities for the study area. The number of people in the productive age, as well as the non-productive age group increase, the annual population increase in the study area, could explain this. Looking at the population in the productive age group compared with the population in the workforce it can be concluded that the population in the productive age predominantly work in non-agricultural sectors. However, the population working in the agriculture is still widely found in the study area with 43 percent of the population.

3.1.3. Land use. Transformations of human activity impact on land use change [15]. The conversion of land from natural to human use is the most permanent and often irreversible effect of human interaction with the natural environment [12]. The transformation of land use in the four sub-districts, affected by the development of the Surakarta Metropolitan Region, has converted non-built-up to built-up areas. Most changes in land use that occur are due to urban activities such as industry as well as commercial activities. This land use change is driven by factors of urban development that are intertwined due to the rural-urban relationships between the four districts with the Surakarta Metropolitan Region. In addition, airport and freeway development plans have had a huge impact on the land-use change that accelerates the process of rural transformation.

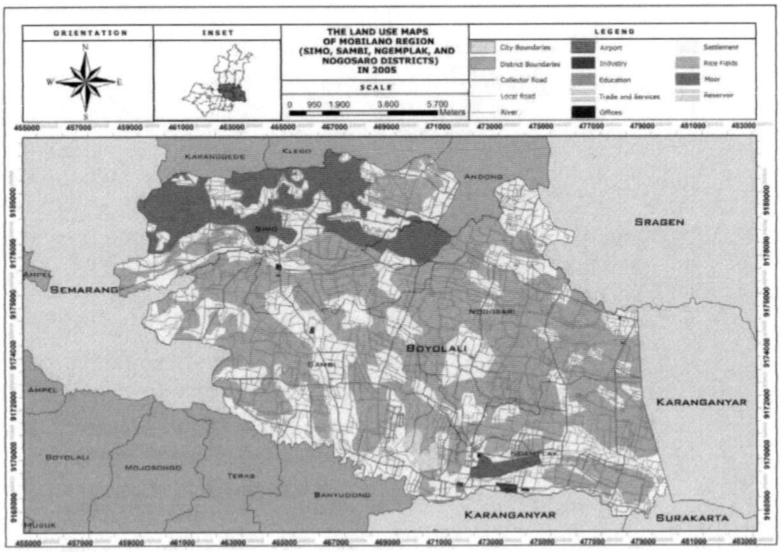

(1)

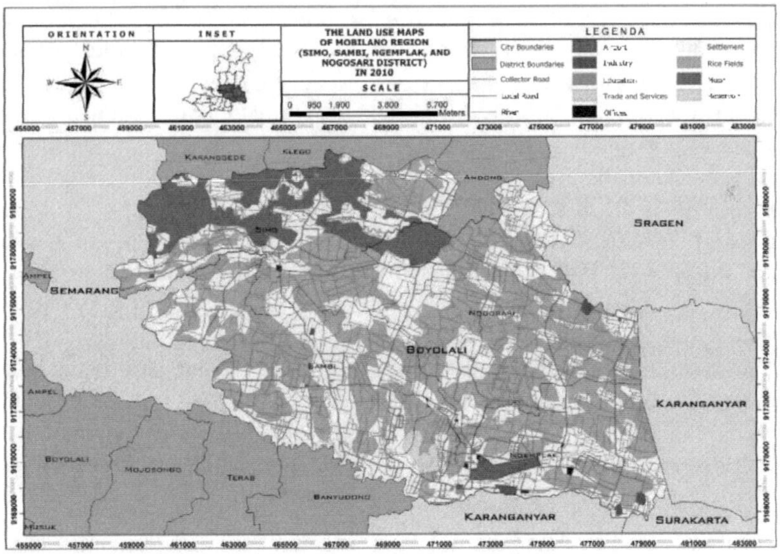

(2)

(3)

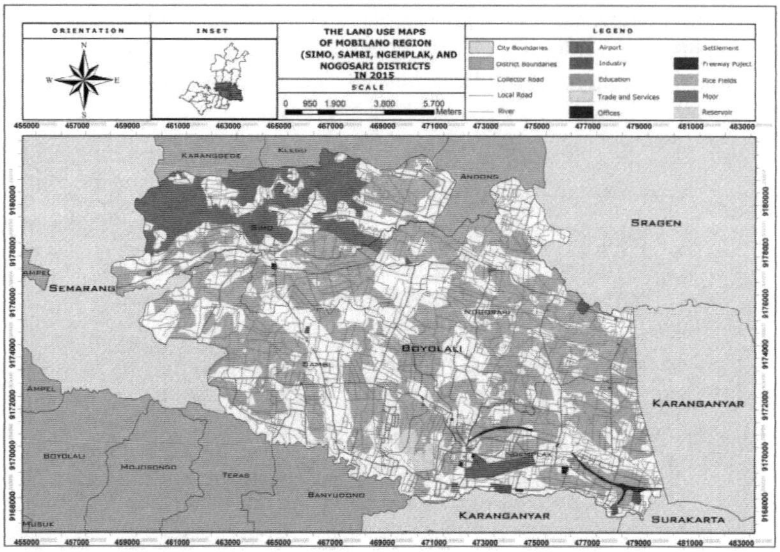

Figure 11. Land use Conversion in Simo, Sambi, Ngemplak, and Nogorasi Sub-districts [15].
(1) in 2005, (2) in 2010, and (3) in 2015

The land in the four sub-districts is designated for agriculture, swamps, education facilities, settlements, trade and services, offices, industries, and airports. In 2005, the land use consisted of more than 50 percent of non-built-up areas, which were dominated by rice fields. This implies that most of the activities in these four sub-districts are related to agriculture. Most land use changes from 2005 to 2015 comprised the conversion of agricultural land into built-up lands such as settlements and industry. When viewed from the population growth from 2005 to 2015, it can be seen an increase in population from migration because of the growth and development of the economic sector due to the construction of new industries. This is the cause of land conversion from year to year.

Ngemplak Sub-district experienced a decrease in agricultural land in the north and south due to an increase in the built-up area. In the southern part, most of the areas around Adi Soemarmo Airport have changed functions into trade and services sectors such as hotels and restaurants. In the north, land conversion occurs due to the development of industrial sectors of large, medium, and small scale. Examples are the ceramic and wood industry. Nogosari is the sub-district with the largest land conversion among the four sub-districts. It experienced a conversion of 25 percent of agricultural land for the development of settlements due to the high industrial growth, in particular, the processing industry (market). Meanwhile, in sub-districts of Simo and Sambi, there is also a change of non-built-up land into settlements for new industrial development. In Simo Sub-district, however, this land conversion is also caused by its many district-level health and education facilities. The trend of land use change in these four sub-districts mostly follows a linear and diffused pattern, which is influenced by the existence of roads and public service facilities.

3.1.4. Development policy in creating a conducive investment climate. Since 2010, Boyolali District has a "Pro Investment" policy to arrange a conducive investment climate to boost the local economy. The 'Pro Investment' policy is intended to simplify all services and investment licensing through the one-stop service implemented by the Board of Investment and Integrated Service Permits Boyolali District, the existence of information centres, and the improvement of supporting infrastructure [18]. This policy has been legalized in District Regulation of Boyolali District Number 3 Year 2014 concerning Investment. Another breakthrough that supports the pro-investment policy is technological innovation in the service community to facilitate licensing services.

IOP Conf. Series: Earth and Environmental Science **158** (2018) 012056 doi:10.1088/1755-1315/158/1/012056

The regional and even national development projects such as Trans Java Freeway, Commuter Line Station, and the development of International Airport Adi Soemarmo increasingly supports the investment climate in Boyolali District. The 'Mobilano' region can be viewed as a transitional area that comprises the development of the Inter-City passenger transport network within the province connecting the cities of Semarang, Surakarta, and Yogyakarta. The construction of a commuter station as an intermodal station will connect land transportation with air transportation in the 'Mobilano' region. The planned air transport network is an international-scale airport development. The transportation system is also strengthened by the Semarang-Surakarta freeway development plan of 75.8 km, Yogyakarta-Surakarta of 40.495 km and, Surakarta-Mantingan of 56.1 km in length which passes the Boyolali District and is an intersection between the freeways. Based on the Spatial Plans of Boyolali District for 2011-2031 the 'Mobilano' region is designated for large industries consisting of machinery, electricity, textile, transportation, food, non-metal, wood, and other similar industries. The designation of a large industrial area in the 'Mobilano' region is earmarked only in Sambi, Ngemplak and Nogosari Sub-districts, while large industries are not allowed in Simo Sub-district because of its function as the focus area of education and health services.

In addition, the four sub-districts of the study area are agricultural areas that are subject to the Sustainable Food Farms policy which stipulates that land use should not be altered either temporarily or temporarily to maintain food availability through incentives-disincentives [3].

Table 2. Food Farms Area and Distribution in Boyolali District in 2015 (Ha) [3].

Area of Agricultural Land (Ha)				
No	Sub-district Name	Sustainable Food Farms	Not Sustainable Food Farms	The Sum
1	Nogosari	2,199.134	935.175	3,134.309
2	Simo	1,775.450	1,310.133	3,085.583
3	Banyudono	1,559.977	1,668.124	3,228.101
4	Andong	1,337.970	1,664.702	3,002.672
5	Teras	1,329.423	213.328	1,542.751
6	Ngemplak	1,283.104	513.118	1,796.222
7	Sawit	1,252.380	39.558	1,291.938
8	Klego	1,212.184	2,869.116	4,081.300
9	Sambi	1,112.030	1,357.046	2,469.076
10	Wonosegoro	959.613	5,352.776	6,312.389
11	Karanggede	868.410	1,207.546	2,075.956
12	Mojosongo	745.634	1,737.147	2,482.781
13	Ampel	505.527	3,284.347	3,789.874
14	Boyolali	265,2	129.628	394.828
15	Juwangi	91.415	921.466	1,012.881
16	Selo	15.25	2,886.021	2,901.271
17	Cepogo	0	3,454.458	3,454.458
18	Kemusu	0	2,873.543	2,873.543
19	Musuk	0	4,799.132	4,799.132
	Total	16,512.701	37,216.364	53,729.065

The creation of a conducive investment climate further encourages investors to enter the Mobilano region. It is expected to increase economic growth, create new jobs, and provide other benefits to raise the welfare of the community. On the other hand, this policy has implications for the conversion of agricultural land and the shift of the economic structure from agrarian to industrial to encourage rural-urban transformation. Therefore, collaboration and cooperation of all stakeholders, especially from government institutions as policymakers, is needed to select what kind of investment is permitted, how and where it should be developed. In addition, the role of society is also needed in the need of supervision of control over the utilization of space related to the implementation of the investment that will be done.

3.1.5. An increase in accessibility that allows changes in the regional economic structure. Adi Soemarmo Airport as a transportation gateway connecting Subosukowonosraten (Surakarta, Boyolali, Sukoharjo, Karanganyar, Wonogiri, Sragen, and Klaten) both nationally and internationally supported by the Interchange Freeway affects the development of the villages in Ngemplak Sub-district and its surrounding areas including the Simo, Sambi, and Nogosari Sub-districts. The existence of the airport has triggered the development of urban activities such as the rapid growth of industrial activity in Ngemplak Sub-district, especially in Gagaksipat Village. Until 2015, there were about 10 large industries spread in Ngemplak, Sambi, and Nogosari Sub-districts, consisting mostly of the garment industry.

The industry sector is changing the economic structure in the region from mostly agriculture into industry. By 2015, the industrial sector contributed 45 percent of the total GRDP in the four sub-districts with a total employment of 17 percent of the total workforce. The change in the economic structure led to a shift from rural to urban characteristics with the growing settlements in the four sub-districts. Thus, in addition to improving the accessibility of the region, the airport and freeways are also able to encourage urban activity on the outskirts of the Surakarta Metropolitan Region.

However, Adi Soemarmo airport has not been optimized due to a lack of transportation integration between the airport, station, and terminal. This is indicated by the poor relationship between existing means of transportation, as there is no transportation service that connects villages. In some sub-districts in the 'Mobilano' area, there are no means of public transportation such as buses that can be used to support people's daily activities such as going to work or school.

In addition, in some areas there is also a lack of terminal facilities; currently, the 'Mobilano' region only has a terminal in the Simo Sub-district. This terminal is not big but it plays an important role as it provides a connection to Klego Sub-district, Karanggede Sub-district and even Karanganyar District and Sragen District. Although the lack of terminal facilities causes problems, there is potential that can be developed, i.e., Ngemplak Sub-district plans to build an intermodal station. This station is expected to facilitate access for people to go to other areas by train. Thus, integrated transportation development programs can solve the problems of accessibility and support activities for the community in the 'Mobilano' region.

Moreover, the region is a gateway for Boyolali District, which offers the possibility to integrate regions, especially with the city of Surakarta. This potential is triggered by the location of Adi Soemarmo Airport and the toll gate that is planned in the Mobilano area. The planned extension of Adi Soemarmo Airport's runway further supports this. This development plan aims to improve the function of the airport, thereby serving the needs of passengers of larger aircrafts. On top of this, the runway extension can encourage the addition of aircraft flight destination as an effort to improve connections with other countries and develop Adi Soemarmo Airport as an international airport.

3.2. Evaluation

The spatial structure of metropolitan cities in Asian countries is characterized by the dualism of the formal and informal sectors. Within these cities, formal sectors such as industry and services exist side by side with the informal sector in the form of street vendors, parking attendants, etc. The phenomenon of industrialization leads to disparities; the industrialization that occurred in the city caused residents from rural areas to flock to the city resulting in a large flow of urbanization. The irony is that urbanization is not accompanied by adequate employment in the city, so the migrant population is forced to work in the informal sector that does not require special skills or requirements. This process eventually led to the formation of illegal settlements in the city's area.

In a response to this phenomenon, four models of development emerged in Asia. The first model explains Asian development which is dependent on an advanced foreign country that has advanced technology. Most Asian countries have abundant but untapped natural resources. Therefore, many foreign parties come and invest in Asian countries, affecting the economic progress in Southeast Asia. However, the presence of these foreign parties exacerbates the dualistic structure in which domestic products tend to lose the competition with foreign products. In addition, the presence of foreign investment is primarily focused on big cities, leading to an increasingly visible economic imbalance.

The second model shows that a large population in a region causes the region to become poorer. One reason for this is a reduction of existing natural resources unaccompanied by technological advances. The lack of natural resources causes the country to be unable to export its commodities and result in low investment in the country. These factors cause minimal relations between rural and urban areas.

The third model is a continuation of the second model which is accompanied by a strong political ideology or so-called Chinese characteristics. In this model, economic development begins with local economic disturbances controlled by the bureaucracy; the government dictates the economic activities and the migration between rural and urban areas. The paradigm embraced by this model aims to minimize dependence on foreign parties.

The fourth model is characterized by high population density and few natural resources but balanced with sophisticated technology. Examples are Japan, Korea, and Taiwan which have modern and state-oriented technology capital in continental Europe and America. The first stage in this model is the industrial sector that changes the economy from import-based to export-based, which can strengthen the domestic market in the region. This model describes the spatial transition from an agriculture-based system to the modern industrial system. These changes can improve living and income standards better than other models. In this model, independence in the industrial sector in urban areas cannot be separated from the influence or existence of rural areas. In addition to supplying raw materials (resources) to industrial cities, villages also provide a lot of manpower. This increased infrastructure in the villages and provided accessibility of the village community to the city. Thus, the imbalance that occurs between the village and the city is not too large.

From this explanation, the spatial transformation from rural to urban starts from a change in pattern or tendency from every aspect of forming a region such as the patterns of manpower, settlements, and the land production in rural areas. These changes can occur because of the intervention of the community or without government interference as a result of urbanization.

One example of the phenomenon of rural-urban transformation occurs in the Ngemplak Sub-district in Boyolali District. Ngemplak is one of the largest rice producers in Boyolali District. However, based on data obtained in 2015, the rice productivity in this sub-district has decreased. By 2015, the base or pre-eminent sector in the region has shifted to the industry, trade, and services sectors. Agricultural land that once dominated this area also gradually disappeared by changing function into residential and industrial land. The migration in this sub-district also increases every year along with the increasing number of large, medium, and small industries.

The process of rural-urban transformation in Ngemplak and its surrounding areas including the Simo, Sambi, and Nogosari Sub-districts tends to be influenced by foreign investment in modern industry such as the printing and garment industries. With the availability of natural resources, the investments in the industry cause increased migration flows or commuters from surrounding areas to enter the region. The industry is automatically able to attract residents or especially labour from other areas to the Ngemplak Sub-district. Therefore, Ngemplak Sub-district which was formerly based on agricultural sector has gradually transformed into a region with a base of processing and commercial industries and services. This spatial transformation can be one of the triggers or the starting point for Ngemplak Sub-district, which was originally rural, to one day become a centre of growth for the surrounding areas or to transform into an urban area.

3.3. Results

The development of urban areas in Indonesia or known as urbanization has occurred very rapidly in the last ten years. This is no exception in the Surakarta Metropolitan Region which affects the areas around it. Areas that are also affected by the development of the Surakarta Metropolitan Region include the Sub-districts of Simo, Sambi, Ngemplak, and Nogosari. Initially, these four sub-districts had rural characteristics and were dominated by agricultural activities; they are the largest rice supplier for Boyolali District. However, the development of the Surakarta Metropolitan area influenced the development of the four sub-districts, changing their rural characteristics in urban characteristics.

Previous research conducted by Douglass, Lo, and Salih showed that the transformation of urban villages follows several models such as Southeast Asia, South Asia, East Asia, and China model. The development of the four sub-districts seen from land use, technology, demographic characteristics, human resources, accessibility, and development policy show the transformation model of the Southeast Asian urban village. So that the four sub-districts underwent extensive transformation and are characterized by richness in natural resources, an open domestic market, and low technology.

The study area exhibits several phenomena of agriculture land conversion into industrial and commercial land since 2005 - 2015. The migration of population leads to the conversion of agricultural land. The most rapid development of land use occurred following the road network. In this condition, the landowner of non-built-up land can sell or rent the land to be built area. This is in accordance with the characteristics of Southeast Asia regarding the relationship between landlords and tenants.

The development of human resources between 2011 to 2015 showed a positive development despite the dominance of agricultural activity. Based on the review of the human resources condition of the four sub-districts, the study area matches the Southeast Asian rural-urban transformation model. This is because of the dualism of agricultural and non-agricultural activities seen from the number of workers who dominate in four districts affected by the development of Surakarta Metropolitan Region. In addition, natural resources are abundant as seen from the number of agricultural labour reaching 34 percent. However, agricultural produce is sold raw and has not undergone processing due to the low availability of technology.

The development of investment into the study area is in the form of the garment industry and other labour-intensive industries. The Boyolali District's "Pro Investment" policy also supports the industrial development in the study area. Foreign investment that began to emerge in the form of modern industry makes urban dualism increasingly visible in the study area. This is because the study area is also the largest rice producer in Boyolali District, while the land for Sustainable Agriculture Food Farms is slowly transformed into urban activities. Meanwhile, developing industries are not in accordance with the available natural wealth. Therefore, the availability of natural resources is not utilized optimally. This results in abundant natural resource outputs in exports plus the absence of emerging technologies to support the process of optimizing natural resources. So, there is a mixed phenomenon between rural and urban characteristics that are not clear.

4. Conclusion

The model of rural-urban transformation in the Surakarta Metropolitan Region was evaluated based on a comparison between field data of Simo, Sambi, Ngemplak, and Nogosari Sub-districts in Boyolali District with model characteristics by Lo, Shalih, and Douglass [12]. This comparison shows that the Southeast Asian Model is in accordance with Indonesian administrative locations in South East Asia. It is in accordance with the characteristics of natural resources are abundant but cannot be utilized because human resources still tend to be low despite the increase and low technology used. In addition, the efforts to create a conducive investment climate illustrate the dependence on the investors from within and outside the region. Therefore, many outsiders, even from abroad, who come and invest in this region will affect the economic progress. The theory found in 1998 by Lo, Salih, and Douglass [12] is still relevant to the current dynamics of life in the new urban agenda.

This study, of course, has shortcomings given the limitations of data obtained such as economic data due to limited availability and a relatively short time for field surveys of approximately two weeks. In addition, the authors are still inexperienced in research considering they are in the 6th semester of their undergraduate study at the Department of Urban and Regional Planning, Faculty of Engineering, Diponegoro University. Research in relation to rural-urban transformation in the Surakarta Metropolitan Region is still insufficient given the enormous challenge in a creating sustainable planet. Thus, this paper opens a new discussion on how to create a system of sustainable development in the Surakarta Metropolitan Region.

References

[1] Berdegué J A, Rosada T and Bebbington A J 2013 Rural Transformation *Preprint* http://hdl.handle.net/10625/51569

[2] Brenner N and Schmid C 2016 The 'Urban Age' in Question *International J. of Urban and Regional Res.* **37** 1-25

[3] Dinas Pertanian, P. dan K. K. B. 2016 *Buku Peta Potensi Lahan Pertanian Kabupaten Boyolali Tahun 2016*

[4] Douglass M 1998 A Regional Network Strategy For Reciprocal Rural-Urban Linkages: An Agenda For Policy Research With Reference To Indonesia *Third World Plan. Rev.* **20** 124-154

[5] Douglass M 2015 Post-Disaster Resilience in Asia's Urban Transition – Neighborhoods, City Regions and Mega-trends in Disaster Governance *National University of Singapore-China Academy of Urban Planning and Design Collaborative Workshop* September p. 18–19

[6] Douglass M 2016 Resilient Urbanism in the Anthropocene, The Rise of Progressive City Regions in Asia's Urban Transition *Workshop on Resilience in Asian Urbanism University of Washington Center for Asian Urbanisms Colledge of Built Environment, Jackson School of International Studies 28-30 March 2016* Available from: https://www.researchgate.net/publication/299818828

[7] Douglass M and Huang L 2007 Globalizing The City in Southeast Asia: utopia On The Urban Edge - The Case Of Phu My Hung, Saigon *Intl. J. of Asia-Pacific Studies* **3** 1–42

[8] Douglass M, Wissink B and Van Kempen, R 2012 Enclave Urbanism In China: Consequences and Interpretations *Urban Geography* **33** 167–182

[9] Garden B 2016 *The Future of Donor Programmes in Rural Development* September p. 1–21

[10] Giyarsih S R 2010 Spatial Pattern of Regional Transformation In Yogyakarta-Surakarta Corridor *Forum Geografi* **24** 28–38

[11] Global Donor Platform for Rural Transformation *Annual Report Promoting Sustaianable Rural Transformation* 2016 36p. Available from: https://www.donorplatform.org/publications-about-us-detail/platform-annual-report.html

[12] Jolly C L and Torrey B B 1993 *Population and Land Use in Developing Countries: Report of a Workshop* (Washington DC: National Academy Press)

[13] Lo Fu-C, Salih K and Douglass M 1981 Rural Urban Transformation in Asia In: *Rural-Urban Urban Relations and Regional Development* Regional Development Series **5** p. 7-43 (Singapore: Maruzen Asia for United Nations Centre for Regional Development, Nagoya, Japan)

[14] Mobilano K S P P 2017 *The Reports Planning Process* (Semarang: Diponegoro University)

[15] Mitsuda Y, Ito S 2011 A Review Of Spatial-Explicit Factors Determining Spatial Distribution Of Land Use/Land-Use Change *Landscape Ecol. Eng.* **7** 117–125

[16] UN News 2014 *More Than Half of World's Population Now Living in Urban Areas, UN Survey Finds* [updated 2014 July; cited 2018] Available from: https://news.un.org/en/story/2014/07/472752-more-half-worlds-population-now-living-urban-areas-un-survey-finds

[17] Permana A, Towolioe S, Aziz N and Ho C S 2015 Sustainable Solid Waste Management Practices And Perceived Cleanliness In A Low Income City *Habitat Int.* **49** 197-205

[18] Penelitian B, Kementerian B P P and Negeri D 2017 Pro Investment In Increasing Regional Investment In Boyolali Regency *Jurnal Teknologi Informasi Dan Komunikasi* **6** 15–28

[19] Pudianti A, Syahbana J A and Suprapti A 2016 Role of Culture In Rural Transformation In Manding Village, Bantul Yogyakarta, Indonesia *Procedia - Soc. and Behavioral Sciences* **227** 458 – 464

[20] Purnamasari L S, Yudana G and Rini E F 2017 Spatial Transformation of Surakarta'S Peripheral Rural Villages Under in-Situ Urbanization Phenomenon : the Case of Gentan Village *Geoplanning: J. of Geomatics and Plan.* **4** 83-96

[21] Rahmayana L and Handayani W 2017 Transformasi Sosial Ekonomi Penduduk Lokal Akibat Adanya Pembangunan di Kawasan Solo Baru Kabupaten Sukoharjo [Social-Economy Transformation of Local Residents as an Impact of Development in Solo Baru Region Kabupaten Sukoharjo] *E-Journal Teknik PWK Undip* **5** 174–185

[22] Sariffuddin S, Wahyono H & Brotosunaryo B 2017 Street Vendors Hypergrowth: Consequence of Uncontrolled Urbanization In Semarang City *Komunitas: International Journal of Indonesian Society and Culture* **9** 81-91

[23] Santi Mardhotillah & Sariffuddin S 2016 Population Mobility Characteristic: Notes From The Urban-Urban Interaction In Semarang Metropolitan Region *DIMENSI Journal of Architecture and Built Environment* **43** 115-122

[24] Swilling Mark 2016 *The Curse of Urban Sprawl: How Cities Grow, and Why This Has to Change* [updated 2016 July; cited 2018] Available from: https://www.theguardian.com/cities/2016/jul/12/urban-sprawl-how-cities-grow-change-sustainability-urban-age.

[25] Tacoli C 1998 Rural–Urban Interactions: A Guide to The Literature *Environ. Urban.* **10** 147–165

[26] United Nations, Department of Economic and Social Affairs, Population Division 2015 *World Urbanization Prospects: The 2014 Revision* (ST/ESA/SER.A/366)

[27] World Bank. 2015. *East Asia's changing urban landscape : measuring a decade of spatial growth (Vol. 2) (English)* Urban Development Series (Washington, DC : World Bank Group) Available from: http://documents.worldbank.org/curated/en/467661468248692758/East-Asias-changing-urban-landscape-measuring-a-decade-of-spatial-growth

Acknowledgements

This paper develops ideas explored in "Rural-Urban Relations and Regional Development", Chapter: Rural-Urban Transformation in Asia by Fu-Chen Lo, Kamal Salih and Mike Douglass (1998). This paper is based on research supported by the team of Mobilano (Simo, Sambi, Ngemplak, and Nogosari) in subjects of "Planning Process Studio" (4th semester) and "Planning Studio" (5th semester) in 2017, Department of Urban and Regional Planning (URP), Faculty of Engineering, Diponegoro University. These Studios serve as a practicum for planning students. The research to which this publication relates was funded by all the members of the Mobilano Team and an anonymous alumnus of URP, Diponegoro University. The authors would like to thank their lecturer, Mr. Sariffuddin for his guidance and motivation so this paper could be finished on time.